T0181031

Geopedology

Joseph Alfred Zinck • Graciela Metternicht
Héctor Francisco del Valle • Marcos Angelini
Editors

Geopedology

An Integration of Geomorphology
and Pedology for Soil and Landscape Studies

Second Edition

 Springer

Editors
Joseph Alfred Zinck (deceased)
Faculty of Geo-Information Science
and Earth Observation (ITC)
University of Twente
Enschede, The Netherlands

Héctor Francisco del Valle
Facultad de Ciencia y Tecnología (FCyT)
Centre. Centro Regional de Geomática
Universidad Autónoma de Entre
Ríos (UADER)
Paraná-Oro Verde, Entre Rios, Argentina

Graciela Metternicht
School of Science
Western Sydney University
Richmond, NSW, Australia

Marcos Angelini
Instituto de Suelos
National Agricultural Technology Institute
Buenos Aires, Argentina

ISBN 978-3-031-20669-6   ISBN 978-3-031-20667-2   (eBook)
https://doi.org/10.1007/978-3-031-20667-2

This Springer imprint is published by the registered company Springer Nature Switzerland AG
The registered company address is: Gewerbestrasse 11, 6330 Cham, Switzerland

In Memoriam

Joseph Alfred Zinck (1938–2021) embodied the true nature of an educator. Education has two different Latin roots: *educare,* which means to train or to mould, and *educere,* meaning to lead out. Alfred mastered both meanings. He was gifted at passing on knowledge (many ITC alumni may remember his lecture notes on "Physiography and Soils" that taught the foundations of the geopedological approach to understanding soil-landscape relationships), and through his mentorship and supervision, he trained a new generation of researchers, teaching them how to question, think critically, and create solutions to problems known and not yet known, bringing the best out of each individual that he supervised or mentored over the many years he devoted to the profession of educator and researcher.

Prof. (em.) Dr. Joseph Alfred Zinck was a soil scientist with a BSc in geography (1960) and an MSc in phytosociology-pedology (1962) from the University of Strasbourg (France) and a doctoral degree in regional planning (1981) from the University of Bordeaux (France). He started and developed his professional career in Venezuela as an expert in technical cooperation of the French Government, working with the Ministry of Public Works (1965–1977) and with the Ministry of Environment and Natural Resources (1977–1981) in soil survey and land use planning projects. During his time in Venezuela, he was also part-time lecturer in postgraduate programs of the Central University of Venezuela (Caracas and Maracay).

From 1986 until his retirement in 2003, he was full Professor of Soil Survey at the International Institute for Geo-Information Science and Earth Observation (ITC) in Enschede (the Netherlands), and for a couple of years he headed the Soil Science Division. He also was a gifted writer (in Spanish, French, and English), publishing 166 papers, 20 soil survey reports, and 10 books.

This second version of the book *Geopedology* began under his leadership, and sadly we lost him halfway through the project. Therefore, this book is dedicated to his memory, and we intend it to be a testimony of his legacy as an outstanding scientist, human being, and educator.

Dean of Science   G. I. Metternicht
Western Sydney University
Richmond, Australia
August 2022

Foreword

Bringing Geopedology to a New Generation

In summer 1997, when I accepted a position as Senior University Lecturer in Professor Alfred Zinck's chair group at the International Institute of Aerospace Survey & Earth Observation ("ITC") in Enschede (NL), now a faculty of the University of Twente, I had only a vague idea of *geopedology*. Of the formative elements, I had a clear idea: "geo" from geography, geology, geomorphology, and "pedology" from my formation at Cornell with Marlin Cline (BSc) and Armand Van Wambeke (PhD), but what did this neologism mean, precisely? I knew that Alfred had developed a systematic approach to the integration of "geo" and "pedology" during his time at the Ministerio de Obras Públicas (MOP), later the Ministerio del Ambiente y de los Recursos Naturales (MARNR) in Venezuela from 1965 to 1981. I worked at the MARNR for 2 years in the early 1990s and was exposed to Alfred's publications, most notably his groundbreaking analysis with Pedro Urriola of the Río Guaraphiche (Zinck and Urriola 1971), and to maps made with the geopedologic method. However, by this time, the concepts had been somewhat diluted and transformed into a set of steps, without the deep understanding of landscape formation implicit in geopedology.

Once at ITC, I was thrown into teaching soil classification, soil survey interpretation, and land evaluation, while Alfred and Abbas Farshad taught geopedology, using the lecture notes "Physiography and Soils," ITC Lecture Note SOL.4.1, from 1988, along with a set of airphoto interpretation exercises. These are still available online.[1] I carefully studied these notes, and was attracted by the systematic hierarchical approach, quite different from the haphazard approach of "look at the landscape and try to see where the soil forming factors have been effective" and then "try to delineate areas with more or less the same set of factors and results" in which I had been trained, well-described by Hudson (1992). Certainly, in the USA, the close relation of geomorphology and pedology had been studied and explained

---

[1] https://webapps.itc.utwente.nl/librarywww/papers_1989/tech/zinck_phy.pdf

by, among others, Ruhe and Daniels (Daniels and Gamble 1978; Daniels et al. 1978, 1999; Ruhe et al. 1967). These were detailed studies aiming at the elucidation of the distribution of mappable soil series over the landscape. By contrast, Alfred's system covered the entire range of scales from continental to local in a systematic hierarchical legend, revealing relations with soil formation at all levels. Another key point of geopedology was the explicit link to the hierarchy soil taxonomy.

Yet there was no easily available reference to which I could refer colleagues. I prodded Alfred about this, since I could see that his system had wide applicability as well as its interesting conceptual basis. Alfred presented the outlines in his inaugural address "Soil survey: Epistemology of a vital discipline," (Zinck 1990) later published in the ITC Journal, but not the details of the system. After his retirement, he found time to codify his previous work and in 2013 presented it in the ITC Special Lecture Notes Series as "Geopedology: Elements of geomorphology for soil and geohazard studies," (Zinck 2013) and in Spanish (2012) (Zinck 2012) as "Geopedología: Elementos de geomorfología para estudios de suelos y de riesgos naturales." Still, this was somewhat difficult to find, and it did not cover the wide range of applications related to the geopedologic approach.

Therefore, the plans for the first edition of the present book with contributions from many of Alfred's former students and colleagues as well as expanded explanations from Alfred. This was published in 2015. Finally, we had a book from a commercial publisher (Springer) with which to explain the concepts of geopedology and its many applications. The subtitle "An Integration of Geomorphology and Pedology for Soil and Landscape Studies" was accurate as far as it goes but does not suggest the deep concepts developed by Alfred to effect this integration.

In the meantime, so-called "digital soil mapping" (DSM) has fairly conquered most of the soil mapping activity; the seminal paper of McBratney et al. (2003) announced this transition, and the increasing computer power, development of machine learning methods, and increasing number of environmental covariates have accelerated the transition. Although most DSM methods rely on geomorphometry, i.e., measurements of the landform, they ignore geomorphology, i.e., the systematic understanding of landforms and their evolution, and of course their tight relation expressed by geopedology. At the same time, traditional soil survey based on landscape analysis and theories of soil formation practically stopped in most areas of the world. That is, the conceptual basis of soil survey is ever less taught and applied.

It is in this intellectual and operational environment that it is vital to re-introduce the conceptual model of soil geography provided by geopedology to the current generation of soil mappers and pedologists. There have been some efforts to integrate traditional soil surveyor's conceptual knowledge into DSM (e.g., Bui et al. 2020), but these are few when compared to the deluge of machine learning empirical models based on large sets of covariates. The recent "10 challenges for the future of pedometrics" (Wadoux et al. 2021) includes "Challenge 6: Can we incorporate mechanistic pedological knowledge in digital soil mapping?" However, this is focused on models of soil formation and landscape evolution, not on geomorphology. What is missing is the systematic description of the landscape on which soils form, and the soil distribution that results. This is exactly the task of geopedology.

I hope that this second edition of Prof. Zinck's final contribution to science will be widely read and will provoke the new generation to think in terms of landscapes. Even better, perhaps it will provoke them to leave the computer and go outside and see for themselves.

Department of Crop & Soil Sciences   D. G. Rossiter
Cornell University
Ithaca, NY, USA

References

Bui EN, Searle RD, Wilson PR, et al (2020) Soil surveyor knowledge in digital soil mapping and assessment in Australia. Geoderma Reg. https://doi.org/10.1016/j.geodrs.2020.e00299

Daniels RB, Gamble EE (1978) Relations between stratigraphy, geomorphology and soils in Coastal Plain areas of Southeastern U.S.A. Geoderma 21:41–65. https://doi.org/10.1016/0016-7061(78)90004-6

Daniels RB, Gamble EE, Wheeler WH (1978) Age of soil landscapes in the Coastal Plain of North Carolina. Soil Sci Soc Am J 42:98–105. https://doi.org/10.2136/sssaj1978.03615995004200010022x

Daniels RB, Buol SW, Kleiss HJ, Ditzler CA (1999) Soil systems in North Carolina. North Carolina State University, Soil Science Dept, Raleigh

Hudson BD (1992) The soil survey as paradigm-based science. Soil Science Society of America Journal 56:836–841. https://doi.org/0.2136/sssaj1992.03615995005600030027x

McBratney AB, Mendonça Santos ML, Minasny B (2003) On digital soil mapping. Geoderma 117:3–52. https://doi.org/10.1016/S0016-7061(03)00223-4

Ruhe RV, Daniels RB, Cady JG (1967) Landscape evolution and soil formation in southwestern Iowa. U.S. Department of Agriculture, Washington, DC.

Wadoux AMJ-C, Heuvelink GBM, Lark RM, et al (2021) Ten challenges for the future of pedometrics. Geoderma 401:115155. https://doi.org/10.1016/j.geoderma.2021.115155

Zinck JA (1990) Soil survey: epistemology of a vital discipline. ITC J 4:335–351

Zinck JA (2012) Geopedología: Elementos de geomorfología para estudios de suelos y de riesgos naturales, p 131 [ITC Special Lecture Notes Series]. Faculty of Geo-information Sciences and Earth Observation, University of Twente. http://www.itc.nl/library/papers_2012/general/zinck_geopedologia_2012.pdf

Zinck JA (2013) Geopedology: elements of geomorphology for soil and geohazard studies, pp 127 [ITC Special Lecture Notes Series]. Faculty of Geo-information Sciences and Earth Observation, University of Twente. https://webapps.itc.utwente.nl/librarywww/papers_2013/general/zinck_geopedology_2013.pdf

Zinck A, Urriola P (1971) Estudio edafológico del valle del Río Guarapiche, Estado Monagas. Ministerio de Obras Públicas, Caracas

Contents

Part VI Synthesis

Part I
Foundations of Geopedology

Chapter 1
Introduction

J. A. Zinck, G. Metternicht, H. F. del Valle, and M. Angelini

Abstract Geopedology aims at integrating geomorphology and pedology to analyze soil-landscape relationships and map soils as they occur on the landscape. This second version of the book on geopedology fills a knowledge gap, presenting a proven approach for reliable mapping of soil-landscape relationships to derive value-added information for policy-making, planning, and management at scales ranging from local to national to continental. This chapter introduces the structure and contents of the book.

Keywords Soil mapping · Geomorphology · Pedology · Soil mapping · Soilscape

J. A. Zinck died before publication of this work was completed.

J. A. Zinck (deceased)
Faculty of Geo-Information Science and Earth Observation (ITC), University of Twente, Enschede, The Netherlands

G. Metternicht (✉)
Earth and Sustainability Science Research Centre (ESSRC), University of New South Wales, Sydney, Australia

School of Science, Western Sydney University, Penrith, Australia
e-mail: G.Metternicht@westernsydney.edu.au

H. F. del Valle
Facultad de Ciencia y Tecnología, Centro Regional de Geomática (CeReGeo), Universidad Autónoma de Entre Ríos (UADER), Oro Verde, Entre Ríos, Argentina
e-mail: delvalle.hector@uader.edu.ar

M. Angelini
Instituto de Suelos, INTA-CIRN, Buenos Aires, Argentina

Global Soil Partnership (GSP), Land and Water Division (NLD), Food and Agriculture Organization of the United Nations (FAO), Rome, Italy

Departamento de Ciencias Básicas, Universidad Nacional de Luján, Luján, Argentina
e-mail: angelini.marcos@inta.gob.ar; marcos.angelini@fao.org; mangelini@unlu.edu.ar

Soil is a vital resource for society at large, and an important determinant of the economic status of nations (Daily et al. 1997). Soils are used for many purposes, ranging from agricultural to engineering to sanitary, and provide a broad range of ecosystem services. However, soils have commanded lesser consideration and attention than other components of the natural capital, such as water and forests. They are increasingly exposed to degradation through erosion, salinization, compaction, and/or pollution. It takes nature centuries, even millennia to form a few centimeters of soil, while billions of tons of arable land are eroded every year.

Whereas the Food and Agriculture Organization of the United Nations estimates that 95% of food comes from soils (FAO 2022) and the United Nations General Assembly has acknowledged the global role of soils in declaring the International Year of Soils 2015, the soil as natural capital remains largely unknown at scales appropriate for practical uses. Traditionally, soil data were collected by systematic soil surveys organized by national government agencies, but the latter have decreased considerably over the last decades because of global economic recession and a tendency of planners to disregard soil information. The multiplication of (pseudo)-natural disasters, including landslides, gullying, flooding, and competing uses of land for food and bio-fuels have contributed to raising awareness about the relevant role of the pedosphere in the natural and anthropogenic environments. Recent papers and global initiatives such as the Global Soil Partnership of the FAO and the GlobalSoilMap.net show a renewed interest in soil research and its applications to improved planning and management of this fragile and finite resource (Hartemink 2008; Sanchez et al. 2009; McBratney et al. 2014).

Traditional soil surveys remain an expensive piece of information, and least developed countries lack the human and financial resources to undertake detailed soil mapping. To make soil surveys cost-effective and more attractive to users, technological and methodological innovations for data gathering and conversion into information have been developed through increased use of information technology in the areas of remote sensing, geographic information systems, spatial modelling, and spatial statistics. These technological advances have facilitated the development of digital soil mapping (DSM). However, digital soil cartography is still mainly limited to terrain/land surface properties, in contrast to the 3D soil body that farmers, engineers, planners, and extension officers manage for decision making. And while the DSM framework has become operational (e.g., Global soil organic carbon map of FAO and ITPS 2018; SoilGrids described in de Sousa et al. 2020), digital soil mapping still faces several challenges ahead (Wadoux et al. 2021). One of them is the capability to incorporate expert knowledge regarding the soil-landscape relationship.

The use of geomorphology, integrated with pedology, has proven to speed up and improve soil inventory. The external geomorphic terrain features (i.e., morphometric and morphographic attributes) help delineate natural soil distribution units, while the internal features of the geomorphic material (i.e., morphogenic and morphochronologic attributes) contribute to understanding soil formation. Geomorphology and pedology are conceptually and practically related. In terms of soil mapping, geomorphic units (i.e., geoforms) provide cartographic frames for soil delineations, while pedologic descriptions supply the soil information of the delineations (e.g., soil properties, use, classification). Geomorphology alone provides information on three of the five soil forming factors (i.e., relief/topography, parent material, and age). Thus, the combination of geomorphology and pedology is a mutually beneficial endeavor and fits nicely with the underlying principles of international standard taxonomic soil classifications such as the World Reference Base for Soil Resources, largely based on soil morphology as an expression of soil formation conditions (Dominati et al. 2010).

Geopedology aims at integrating geomorphology and pedology to analyze soil-landscape relationships and map soils as they occur on the landscape. The geopedologic view is similar to the frequently used expression of "soil geomorphology". A few decades ago, several reference books and seminal papers focused on soil geomorphology; however, the most recent one dates to 2005 (Schaetzl and Anderson). This second version of the book on geopedology fills a knowledge gap, presenting a proven approach for reliable mapping of soil-landscape relationships to derive value-added information for policy-making, planning, and management at scales ranging from local to national, and to continental. The book presents the theoretical and conceptual framework of the geopedologic approach and a bulk of applied research showing its use and benefits for knowledge generation relevant to geohazard assessment and prediction, land use planning and conflict mitigation, and landscape management.

Part I introduces the foundations of geopedology. Basic geopedologic concepts are described, with emphasis on the construction of a hierarchic system organizing geoforms into six categories to serve soil mapping at different levels of detail. The geopedology approach to soil survey combines pedologic and geomorphic criteria to establish soil map units. Geomorphology provides the contours of the map units ("the container"), while pedology provides the soil components of the map units ("the content"). Therefore, the units of the geopedologic map are more than soil units in the conventional sense of the term, since they also contain information about the geomorphic context in which soils have formed and are distributed. In this sense, the geopedologic unit is an approximate equivalent of the soilscape unit, but with the explicit indication that geomorphology is used to define the landscape. This is usually reflected in the map legend, which shows the geoforms as an entry point to the legend and their respective pedotaxa as descriptors.

Chapter 2 introduces the theoretical framework, and a brief review of the relationships between geomorphology and pedology is presented in **Chap. 3**. These relationships including the conceptual aspects and their practical implementation in studies and research have been referred to under different names, the most common expression being *soil geomorphology*. Definitions and approaches are reviewed, distinguishing between academic and applied streams. There is consensus on the basic relationships between geomorphology and pedology: geomorphic processes and resulting landforms contribute to soil formation and distribution while, in return, soil development has an influence on the evolution of the geomorphic landscape. However, there is still no unified body of doctrine, in spite of a clear trend toward greater integration between the two disciplines. There are few references in international journals that provide some formal synthesis on how to carry out integrated pedo-geomorphic mapping.

Chapter 4 outlines the essence of the geopedologic approach in conceptual, methodological, and operational terms. Geopedology is based on the conceptual relationships between geoforms and soils; it is implemented using a variety of methods based on the three-dimensional concept of the geopedologic landscape, and it becomes operational primarily within the framework of soil inventories that can be represented by a hierarchic scheme of activities. The approach focuses on reading the landscape in the field and from remotely-sensed imagery to identify and classify geoforms, as a prelude to their mapping along with the soils they enclose and the interpretation of the genetic relationships between soils and geoforms. There is an explicit emphasis on the geomorphic context as an essential factor of soil formation and distribution. The geopedologic approach is essentially descriptive and qualitative. Geoforms and soils are considered as natural bodies, which can be described by direct observation in the field and by interpretation of aerial photographs, satellite images, topographic maps, and digital terrain models.

The pedologic component of geopedology is described in **Chap. 5**, with special consideration to the organization of the soil material in the pedologic landscape. Soil material is multiscalar with features and properties specific to each scale level. The successive structural levels are embedded in a hierarchic system of nested soil entities or holons known as the holarchy of the soil system. At each hierarchic level of perception and analysis of the soil material, distinct features are observed that are particular to the level considered. The whole of the features describes the soil body in its entirety. Each level is characterized by an element of the soil holarchy, a unit (or range of units) measuring the soil element perceived at that level, and a means of observation or measurement for identifying the features that are diagnostic at the level concerned. The holarchy of the soil system allows highlighting relevant relationships between soil properties and geomorphic responses at different hierarchic levels. These relationships form the core concept of geopedology.

Chapters 6, 7, and 8 refer to the geomorphic component of geopedology, considering the criteria for classifying geoforms, the classification of the geoforms, and the attributes of the geoforms successively. **Chapter 6** describes how the combination of basic taxonomic system criteria with the hierarchic arrangement of the geomorphic environment determines a structure of six nested categorical levels. Geoforms have distinct physiognomic features that make them directly observable through visual and digital perception from remote to proximal sensing. Changing the scale of perception changes not only the degree of detail but most significantly the nature of the object observed. The geolandscape is a hierarchically structured and organized domain. Therefore, a multi-categorial system, based on nested levels of perception to capture the information and taxonomic criteria to organize that information, is an appropriate frame to classify geoforms. Categorial levels are identified by their respective generic concepts, including from upper to lower level: geostructure, morphogenic environment, geomorphic landscape, relief/molding, lithology/facies, and the basic landform or terrain form.

Chapter 7 attempts to organize existing geomorphic knowledge and arrange the geoforms in the hierarchically structured system with six nested levels introduced in **Chap. 6**. Geoforms are grouped thematically, distinguishing between geoforms mainly controlled by the geologic structure and geoforms mainly controlled by the morphogenic agents. It is thought that this multi-categorial geoform classification scheme reflects the structure of the geomorphic landscape *sensu lato*. It helps segment and stratify the landscape continuum into geomorphic units belonging to different levels of abstraction. This geoform classification system has shown to be useful in geopedologic mapping, and it offers great potential for digital soil mapping.

Attributes are needed to describe, identify, and classify geoforms. These are descriptive and functional indicators that make the multi-categorial system of the geoforms operational. Four kinds of attributes are used as outlined in **Chap. 8**: morphographic attributes to describe the geometry of geoforms; morphometric attributes to measure the dimensions of geoforms; morphogenic attributes to determine the origin and evolution of geoforms; and morphochronologic attributes to frame the time span in which geoforms originated and evolved. The morphometric and morphographic attributes apply mainly to the external component of the geoforms, are essentially descriptive, and can be extracted from remotely-sensed imagery or derived from digital elevation models. The morphogenic and morphochronologic attributes apply mostly to the internal component of the geoforms, are characterized by field observations and measurements, and need to be substantiated by laboratory determinations.

Part II presents a variety of approaches to establishing and analyzing relationships between soil and landscape in space and time. Information and knowledge can be obtained from field observations and landscape reading through systematic surveys or transect descriptions or a combination thereof. The chapters have in common the analysis of soil distribution patterns on the landscape (i.e., soilscape) using geopedology from various points of view and for different objectives, such as soilscape ecology, soilscape history, soilscape diversity, and soilscape complexity and heterogeneity.

Mapping of soil bodies, not properties in isolation, is what gives insights into the soil-landscape relationship, and geopedology is important to that end as it integrates field observations with an understanding of the geomorphic conditions under which soils evolve. In **Chap. 9**, Rossiter discusses examples from exhumed paleosol areas, low-relief depositional environments, and recent post-glacial landscapes, where simplistic digital soil mapping would fail but geopedology would succeed in mapping and explaining soil distribution. Attempts at correlating environmental covariates from current terrain features, vegetation density, and surrogates for climate cannot succeed in the presence of unmapped variations in parent material, soil bodies, and landforms inherited from past environments.

Diversity analysis of natural resources attempts to account for the variety of forms and spatial patterns exhibited by natural bodies, biotic and abiotic, at the earth's surface. Recently pedologists drew attention to soil diversity analysis and modelling using statistical tools similar to the ones used by ecologists, reporting insightful relations between spatial patterns of soil and vegetation. Geodiversity studies are primarily concerned with the preservation of the geological heritage, bypassing most of the aspects related to its spatial distribution. In **Chap. 10**, Ibañez and Brevik explore a novel perspective for joining soils, geoforms, climate, and biocenoses in an integrated and comprehensive approach to describe the structure and diversity of earth surface systems.

Chapter 11 focuses quantitative methods of pedology that facilitate analysis of soil distribution and genesis. Beaudette et al. introduce the Algorithms for Quantitative Pedology (AQP) project that intends to bridge the gap between pedometric theory and practice and geopedology. The project provides tools to effectively apply statistical methods for studying the relations between pedons and their complex data structure. The AQP R packages described in the chapter have been applied to projects involving hundreds of thousands of soil profiles and integrated into widely used tools such as SoilWeb.

Soils by virtue of their parent materials can provide key information about past sedimentologic or geologic processes and systems. Dealing with soilscape history, **Chap. 12** analyses the interaction of geological and pedological processes in the genesis of soils with gypsum, in the Northern Patagonia of Argentina. Irisarri et al. identify processes that affect the formation of gypsum soils, as well as the uses of these soils in the region. Their work traces and explain the presence of gypsum soils in Holocene terraces of the Colorado River basin, linking that history to the currently higher relative exposure of gypsum and anhydrite deposits, along the upper Colorado River basin.

Accurately determining the origin of a soil parent material is important because soil properties – tied to the parent material – are frequently part of mapping unit attributes for regional and local scale maps. In **Chap. 13**, Miller et al. focus on examples of studies or situations where careful examination of uniform parent material type and distribution can provide important information about the geomorphic attributes and history of the landscape. The relationship found between soils and their parent materials connects soil survey maps and geological maps. Different information collected for, and represented by, the respective maps – due to differences in purpose, focus, or resources – can assist the investigations of other disciplines. This multiple utility is especially true for studying soil-landform assemblages and soil-landscape evolution.

Part III comprises a set of chapters dealing with different spatial modelling techniques for soil pattern recognition and mapping, and the characterization of soil properties relevant for productivity and environmental risk management. A commonality between the case studies is the use of digital elevation models, remote-sensed imagery, digitally processed data using GIS, and spatial analysis and modelling techniques to transform data into usable information. This section shows complementarity between the geopedologic and the digital soil mapping approaches. A digital soil map is essentially a spatial database of soil properties, based on a statistical sample of landscapes (Sanchez et al. 2009).

In **Chap. 14**, Schulz et al. present an approach to derive digital soil texture maps of Argentina and analyse their relationship with soil-forming factors and processes using a combination of field data and environmental covariates to represent the soil-forming factors (e.g., remote sensing data, climate data or geomorphology maps). Their work shows the impact of limited data on uncertainty, providing new insights into both the distribution of parent materials and the intensity of pedogenetic processes in the region. The geopedologic approach can act as the conceptual framework that 'guides' digital soil mapping. For instance, the segmentation of the landscape into geomorphic units provides spatial frames in which digital terrain models combined with remote-sensed data and geostatistical analyses can be applied to assess in detail spatial variability of soils and geoforms, better framing digital mapping over large territories.

Chapter 15 presents an approach for deriving multidimensional soil and vegetation information from microwave (Sentinel-1) and optical-infrared (Sentinel-2) remote sensing data to assist interpretation of features relevant to mapping and assessing land use and land cover (LULC) changes during the period 2016–2017 and the year 2020. Using an area of the northern part of the Entre Rios province, del Valle et al. present an approach for improving information of existing soil maps at scale 1:100,000 for soil and land use assessment considering vegetation, soil, and water indices based on the integration of radar and optical data and the use of Random Forest analysis techniques. Results show that σ^0_{VV} and σ^0_{VH} backscatter values are 1.0–1.8 dB lower during the 2020 drought compared to values in 2016–2017. Both σ^0_{VV} and σ^0_{VH} polarizations and the Radar Vegetation Index combined with selected optical radiometric indices for soil, vegetation, and moisture from Random Forest analysis are suitable for representing LULC changes in years with changes in moisture availability. The results showed that a significant change in LULC patterns had occurred in the driest year, 2020, in the study area.

Geo-morphometric, soil properties, and climate variables are used in **Chap. 16** to map landslide susceptibility. Correa-Munoz et al. use this information and random forest (RF) analysis and logistic regression (LR) to identify areas susceptible to landslides in southwestern Colombia. The authors report that elevation, soil silt content, slope, a topographic topographic ruggedness index (TRI), landscape unit, soil clay and sand contents, are the more important predictors of landslides in that area.

Chapter 17 outlines a new sedimentological and geopedologic approach that explains more accurately soil development and spatial distribution in the Mesopotamian Pampa, in central Argentina. New studies undertaken in large pits reveal the existence of diapiric structures in the lower sediment, which remained "hidden" up to the present. Therefore, it can now be considered that Vertisols were also the dominant soils here, later buried by the thick loess. In the humid periods, the summit of convex slopes would have been partly eroded, leaving the underlying smectitic material closer to the surface. As a consequence, current Mollisols on top of the landscape developed vertic properties due to the mixing of materials and depending on the greater or lesser proximity to the paleosurface. Therefore, these Pampean Vertisols and vertic Mollisols and Alfisols can be considered polygenic and related by different degrees of the same process. These advances in the understanding of the landscape and soils together with the quantitative analysis of soil profile data, appear highly useful to distinguish vertic soils at the series level and improve surveying and mapping work.

A gilgai is a micro-relief characteristic of clayey areas on vertisol soils, with soft mounds and depressions, which form patterns with some degree of symmetry. **Chapter 18** uses the time series of NDVI of Sentinel 2 images over three years and takes advantage of the methodology for mapping management zones (MZM) in crop fields. That approach evidences the presence of granular structure that can be described by gilgai micro-relief. The granular structure remains undisturbed throughout three consecutive campaigns in the 2019–2021 years, under different crops, which apparently points to the soil variability as the origin of this structure. The chapter discusses how management zone maps can be used to infer gilgai sub-surface micro-relief. The results enable asserting that crop and pasture growth are uneven across a paddock in gilgai areas, providing insights into the spatial distribution of this type of micro-relief.

Part IV is dedicated to applications in land degradation and geohazard studies that use geomorphic and pedologic analysis integrating spatial modelling and earth observation information. Environmental deterioration, land degradation, and geohazard are of increased concern in many regions around the world. In this regard, understanding and quantifying the geopedologic processes that such regions are undergoing is fundamental to promoting efficient solutions.

Chapter 19 of Bocco summarizes how gully erosion research has developed, its major achievements in the conceptual and methodological dimensions, and potential courses of action for further research, with emphasis on the contribution of geopedology. It is claimed that despite the advancements in the development of models and in remote sensing and GIS techniques, gully erosion remains a complex issue difficult to model and predict. In this regard, the author argues that geopedology may play a role in its understanding and management. As with other geomorphic processes, gullies occur in certain terrain, soil, and hydrology conditions, which may be conveniently approached from a geopedologic perspective.

In **Chap. 20** Frugoni et al. apply principles of geopedology to soil erosion assessment and mitigation scenarios in the Northwestern Patagonia of Argentina. Soil erosion and mass movement features identified in a geopedological study are applied to landforms to observe and estimate the spatial dimension of current erosion processes. Scenarios of afforestation are applied to plan for the most efficient ways to reduce and/or reverse current soil erosion, showing that after 15 years a non-erosion hazard scenario could be reached. The modelling also shows that more than 4000 hectares could be recovered via silvopastoral practices without soil erosion hazard preserving the transhumant system that characterizes this geography and reducing soil loss by more than 30%.

The geopedologic approach to soil mapping amplifies the role of geomorphology. It helps understand soil variation in the landscape, which increases mapping efficiency. In **Chap. 21**, Shrestha et al. show the adequacy of soil data resulting from geopedology-based predictive soil mapping for assessing land degradation in two contrasting climatic regions: humid tropics in Thailand and dry and hot arid climates in Iran. The result shows that the geopedologic approach helps in mapping soil distribution in inaccessible mountain areas; and how the geopedologic approach in combination with digital image processing and/or the application of simple decision rules applied in a Geographic Information System (GIS) help in mapping soil salinity trends.

Chapter 22 explores the contribution of geopedology to the implementation of national frameworks for land degradation neutrality. It proposes a conceptual framework for integrating geopedology in the preliminary assessments required to establish the baseline for land degradation neutrality (LDN) at national or sub-national levels and to monitor outcomes of interventions targeting the LDN hierarchy of avoid, reduce, reverse land degradation. Specifically, it proposes using geopedologic mapping units as the main sources to derive the land types and associated land units considered in the conceptual framework for LDN. In such a way, geopedologic units inform decisions around the concepts of 'like-for-like' necessary to plan and implement the counterbalancing mechanism core to the principle of neutrality of land degradation. The novel framework is illustrated via a case study of Cochabamba's Valle Alto, in Bolivia.

Part V is devoted to issues in land use planning, land zoning and policy implementation where geopedology plays a key role, both conceptually and in applied terms. These are important topics and are somehow neglected in the current scientific literature, more prone to purely digital mapping and pixel-based approaches. Semi-quantitative geopedologic studies aiming at the stratification of space for planning and zoning purposes are able to generate valuable scientific and practical information.

In **Chap. 23**, Bhaskar et al propose a geo-pedological approach for land use planning. In India, there are noticeable land use conflicts in semi-arid and sub-humid regions due to uncertain market-driven current and potential land use dynamics. If one considers the current agricultural crisis in the case of soil degradation, climate change and depletion in soil health, systematic and innovative approaches in the land use planning process must be pursued. The pedological information is the first step in developing decisive and sustainable land use alternatives. Three case studies representing semi-arid and sub-humid agro-ecological regions of the Meghalaya Plateau, the hot semi-arid eco-region of cotton growing Yavatmal district, Maharashtra and peanut growing Pulivendula tehsil, the YSR Kadapa district in Andhra Pradesh are used to discuss the usefulness of the geo-pedological approach to land use exercises to strengthen local agriculture.

A conceptual framework is presented in **Chap. 24** to illustrate the role of geopedologic information in a continuous planning process which includes land resources inventory within a particular institutional context, land evaluation, land allocation and tactical operations on specific geopedologic units and their subdivisions. Explanations and examples are given on the components of each phase as well as their interactions. Special attention is given to land use conflicts for strategic land use allocation through a resolution based on consensus between different interest groups, in particular for multi-functional land use conflicts. Rodriguez-Parisca argues that this approach strengthens the sustainability of major land use allocation decisions. Land degradation is highlighted as a central process that must be tackled in order to mitigate its harmful consequences, prevent its occurrence or restore already degraded land by means of allocating and selecting specific land utilization types and associated technologies which can secure land sustainability.

Policy making and soil security require addressing the sustainable management of soil resources at the landscape level. Therefore, Montanarella argues in **Chap. 25** that geopedology can become highly relevant for effective policy-making for achieving long-term soil protection. The necessary precondition is the availability of a solid scientific basis and detailed data on the actual status and trends of soils within relevant landscapes. The recent emergence of high-resolution digital soil mapping techniques offers new possibilities for achieving such a knowledge base. Several examples from the European Union demonstrate that geopedology can be a valuable tool for understanding soil processes at the landscape level and designing effective soil protection policies.

Chapter 26 demonstrates the interlinkages between soil ecosystem services (ES), land cover change and climate change in the state of Maine (USA). Mikhailova et al seek to determine baseline regulating ES values of soil organic carbon (SOC), soil inorganic carbon (SIC), and total soil carbon (TSC) stocks for Maine by soil order and county, based on the concept of the avoided social cost of carbon dioxide (CO_2) emissions. They also assess how recent changes in land use/land cover (LULC) across Maine have potentially affected these regulating ES values. This analysis provides actionable spatial information on how 15 years of land cover change likely impacted soil carbon releases to the atmosphere. These results can help inform Maine's Climate Action Plan (an example of land-based policy where geopedologic information can add value) on the land cover and soil type combinations that should be managed to help reduce carbon emissions.

Chapter 27 integrates digital data and statistical techniques, such as those from the Digital Soil Mapping framework, to improve existing soil maps and land zoning in the area of Campo de Borja (Spain) where viticulture is the main land use. Precision viticulture requires homogeneous terroir units (land units). In this work, an existing Soil Resource Information zoning in the denomination of origin (DO) of Campo de Borja, northeastern Spain, is upgraded using a new soil map derived from a novel technique of disaggregation of multi-component units that reveals potentially homogeneous areas by the unsupervised classification algorithm CLARA. Quantitative metrics of vine quality and occupation on the terroir units are derived for the reference and the new maps. Change analysis using a confusion matrix derived from the intersection of the old and new maps shows a five-fold increase in the number of mapping units, enhancing their delineations and the detection of areas of higher potential within the mapping units of the original map. Overall, the enhancements to the integrated terroir zoning provide an approach for scaling up zoning maps to meet the demands for detailed cartography.

References

Daily G, Matson P, Vitousek P (1997) Ecosystem services supplied by soils. In: Daily G (ed) Nature's services: societal dependence on natural ecosystems. Island Press, Washington, DC

de Sousa LM, Poggio L, Batjes NH, Heuvelink GB, Kempen B, Riberio E, Rossiter D (2020) SoilGrids 2.0: producing quality-assessed soil information for the globe. Soil Discuss 1(10.5194)

Dominati E, Patterson M, Mackay A (2010) A framework for classifying and quantifying the natural capital and ecosystem services of soils. Ecol Econ 69:1858–1868

FAO (2022) Soils for nutrition: state of the art. FAO, Rome. https://doi.org/10.4060/cc0900en

FAO and ITPS (2018) Global soil organic carbon map (GSOCmap) Technical Report. FAO and ITPS, Rome. 162 pp

Hartemink AE (2008) Soils are back on the global agenda. Soil Use Manag 24:327–330

McBratney A, Field DJ, Koch A (2014) The dimensions of soil security. Geoderma 213:203–213

Sanchez P, Ahamed S, Carré F et al (2009) Digital soil map of the world. Science 325:680–681

Schaetzl R, Anderson S (2005) Soils: genesis and geomorphology. Cambridge University Press, New York

Wadoux AMC, Heuvelink GB, Lark RM, Lagacherie P, Bouma J, Mulder VL, Libohova Z, Yang L, McBratney AB (2021) Ten challenges for the future of pedometrics. Geoderma 401:115155

Chapter 2
Theoretical Framework

J. A. Zinck

Abstract This chapter presents the foundations of geopedology. Geopedology, as it is considered in this book, refers to the relations between geomorphology and pedology, with emphasis on the contribution of the former to the latter. More specifically, geopedology is in the first instance a methodological approach to soil inventory, while providing at the same time a framework for geographic analysis of soil distribution patterns. The prefix *geo* in geopedology refers to the earth surface – the geoderma – and as such covers, in addition to geomorphology, concepts of geology and geography.

Keywords Geopedology · Georforms · Geography · Soil · Pedology

Geopedology, as it is considered here, refers to the relations between geomorphology and pedology, with emphasis on the contribution of the former to the latter. More specifically, geopedology is in the first instance a methodological approach to soil inventory, while providing at the same time a framework for geographic analysis of soil distribution patterns. The prefix *geo* in geopedology refers to the earth surface – the *geoderma* – and as such covers, in addition to geomorphology, concepts of geology and geography. Geology intervenes through the influence of tectonics in the geoforms of structural origin, and through the influence of lithology in the production of parent material for soils as a result of rock weathering. Geography relates to the analysis of the spatial distribution of soils according to the soil forming factors. However, in the concept of geopedology, emphasis is on

---

J. A. Zinck died before publication of this work was completed.

J. A. Zinck (deceased)
Faculty of Geo-Information Science and Earth Observation (ITC), University of Twente, Enschede, The Netherlands

geomorphology as a major structuring factor of the pedologic landscape and, in this sense the term geopedology is a convenient contraction of geomorphopedology. Geomorphology covers a wide part of the physical soil forming framework through the relief, the surface morphodynamics, the morphoclimatic context, the unconsolidated or weathered materials that serve as parent materials for soils, and the factor time. Geopedology underpins the argument of Wilding and Lin (2006) that the frontiers of soil science would benefit from moving towards a geoscience.

The relationship between geomorphology and pedology can be considered within the context of landscape ecology. With its integrative approach, landscape ecology tries to bridge the gap between related disciplines, both physical and human, that provide complementary perceptions and visions of the structure and dynamics of natural and/or anthropized landscapes. Landscape ecology, as a discipline of integration, has holistic vocation, but it is often practiced de facto as parts of a whole. For instance, one stream emphasizes the ecosystem concept as the basis of the biotic/ecological landscape (Forman and Godron 1986); while another stream stresses the concept of land as the basis of the cultural landscape (Zonneveld 1979; Naveh and Lieberman 1984); and still another one puts emphasis on the concept of geosystem as the basis of the geographic landscape (Bertrand 1968; Haase and Richter 1983; Rougerie and Beroutchachvili 1991). Geomorphology and pedology participate in this concert, and their respective objects of study, i.e., geoform and soil, constitute an essential, inseparable pair of the landscape.

Geoforms or terrain forms sensu lato are the study object of geomorphology. Soils are the study object of pedology, a branch of soil science. The relations between both objects and between both disciplines are intimate and reciprocal. Geoforms and soils are essential components of the earth's epidermis (Tricart 1972), sharing the interface between lithosphere, hydrosphere, biosphere, and atmosphere, within the framework of the noosphere as soils are resources subject to use decisions by human individuals or communities. It is not a mere static juxtaposition; there are dynamic relationships between the two objects, one influencing the behavior of the other, with feedback loops. Moreover, in nature, it is sometimes difficult to categorically separate the domain of one object from the domain of the other, because the boundaries between the two are fuzzy; geoforms and soils interpenetrate symbiotically. This integration of the geoform and soil objects, that coexist and coevolve on the same land surface, has fostered the study of the relations between the two. As it often happens, the interface between disciplines is a frontier area where new ideas, concepts, and approaches sprout and develop.

The analysis of the relationships and interactions between geoforms and soils and the practical application of these relationships in soil mapping and geohazard studies have received several names such as soil geomorphology, pedogeomorphology, morphopedology, and geopedology, among others, denoting the transdisciplinarity of the approaches. By the position of the terms in the contraction word, some authors want to point out that they put more emphasis on one object than on the other. For instance, Pouquet (1966) who has been among the first ones to use the word geopedology, emphasizes the pedologic component and implements geopedology as an approach to soil survey and to erosion and soil conservation studies.

In contrast, Tricart (1962, 1965, 1994) who has possibly been one of the first authors to use the word pedogeomorphology, puts the accent on the geomorphic component.

Chapter 3 of this book illustrates the variety of modalities implemented to address the relationships between geomorphology and pedology. The applied context in which geopedology was developed is different from other ways of visualizing the relationships between both disciplines; this specificity of geopedology is described in Chap. 4. The geopedologic approach focuses on the inventory of the soil resource. This means logically addressing themes such as soil characterization, formation, classification, mapping, and evaluation. Chapter 5 summarizes relevant aspects of these themes with emphasis on the hierarchic structure of the soil material, which allows highlighting that geomorphology is involved at various levels. The application of geomorphology in soil survey programs at various scales, from detailed to generalized, requires establishing a hierarchic taxonomy of the geoforms, so that the latter can serve as cartographic frames for soil mapping and, additionally, as genetic frames to help interpret soil formation. These aspects are addressed in Chap. 6 (criteria for classifying geoforms), Chap. 7 (geoform classification), and Chap. 8 (geoform attributes).

This text is partially drawn from the lecture notes used in a course on geopedology under the heading of *Physiography and Soils* (Zinck 1988), taught by the author at the International Institute for Geo-Information Science and Earth Observation (ITC, Enschede, The Netherlands) as part of an annual postgraduate course in soil survey in the period 1986–2003. Shorter versions of the course were also taught by the author on several occasions between 1970 and 2003 in various countries of Latin America, especially in Venezuela and Colombia.

References

Bertrand G (1968) Paysage et géographie physique globale. Esquisse méthodologique. Rev Géogr Pyrénées et SO 39(3):249–272

Forman RTT, Godron M (1986) Landscape ecology. Wiley, New York

Haase G, Richter H (1983) Current trends in landscape research. Geo J (Wiesbaden) 7(2):107–120

Naveh Z, Lieberman AS (1984) Landscape ecology. Theory and application. Springer, Munich

Pouquet J (1966) Initiation géopédologique. Les sols et la géographie. SEDES, Paris

Rougerie G, Beroutchachvili N (1991) Géosystèmes et paysages. Bilan et méthodes. Armand Colin, Paris

Tricart J (1962) L'épiderme de la terre. Esquisse d'une géomorphologie appliquée. Masson, Paris

Tricart J (1965) Principes et méthodes de la géomorphologie. Masson, Paris

Tricart J (1972) La terre, planète vivante. Presses Universitaires de France, Paris

Tricart J (1994) Ecogéographie des espaces ruraux. Nathan, Paris

Wilding LP, Lin H (2006) Advancing the frontiers of soil science towards a geoscience. Geoderma 131:257–274

Zinck JA (1988) Physiography and soils, Lecture notes. International Institute for Aerospace Survey and Earth Sciences (ITC), Enschede

Zonneveld JIS (1979) Land evaluation and land(scape) science. International Institute for Aerospace Survey and Earth Sciences (ITC), Enschede

Chapter 3
Relationships Between Geomorphology and Pedology: Brief Review

J. A. Zinck

Abstract The relationships between geomorphology and pedology, including the conceptual aspects that underlie these relationships and their practical implementation in studies and research, have been referred to under different names, the most common expression being *soil geomorphology*. Definitions and approaches are reviewed distinguishing between academic stream and applied stream. There is consensus on the basic relationships between geomorphology and pedology: geomorphic processes and resulting landforms contribute to soil formation and distribution while, in return, soil development has an influence on the evolution of the geomorphic landscape. However, a unified body of doctrine is yet to be developed, in spite of a clear trend toward greater integration between the two disciplines.

Keywords Approaches · Evolution · Integration · Classification · Conceptual

3.1 Introduction

The relationships between geomorphology and pedology, including the conceptual aspects that underlie these relationships and their practical implementation in studies and research, have been referred to under different names. Some of the most common expressions are *soil geomorphology* (Daniels et al. 1971; Conacher and Dalrymple 1977; McFadden and Knuepfer 1990; Daniels and Hammer 1992; Gerrard 1992, 1993; Schaetzl and Anderson 2005; among others), *soils and geomorphology* (Birkeland 1974; Richards et al. 1985; Jungerius 1985a, b; Birkeland 1990, 1999), *pedology and geomorphology* (Tricart 1962, 1965a, b, 1972; Hall 1983), *morphopedology* (Kilian 1974; Tricart and Kilian 1979; Tricart 1994; Legros

---

J. A. Zinck died before publication of this work was completed.

J. A. Zinck (deceased)
Faculty of Geo-Information Science and Earth Observation (ITC), University of Twente, Enschede, The Netherlands

© The Author(s), under exclusive license to Springer Nature Switzerland AG 2023
J. A. Zinck et al. (eds.), *Geopedology*,
https://doi.org/10.1007/978-3-031-20667-2_3

19

1996), *geopedology* (Principi 1953; Pouquet 1966), and *pedogeomorphology* (Conacher and Dalrymple 1977; Elizalde and Jaimes 1989), without mentioning the numerous publications that treat the subject but do not explicitly use one of these terms in their title. Due to this diversity of expressions, it is convenient to first define what the relations between geomorphology and pedology cover, and subsequently analyze the nature of the relationships.

3.2 Definitions and Approaches

Soil geomorphology, sometimes called pedologic geomorphology or pedogeomorphology, is the term most frequently found in English-published literature, with the word geomorphology being a noun and the word soil being an adjective that qualifies the former. According to this definition, the center of interest is geomorphology, with the contribution of pedology. However, under the same title of soil geomorphology, there are research works in which the roles are reversed. Therefore, in practice, the relationship between geomorphology and pedology goes both ways. The emphasis given to one of the two disciplines depends on a number of factors including, among others, the context of the study, the purpose of the research, and the primary discipline of the researcher.

The relations between geomorphology and pedology as scientific disciplines, and between geoform and soil as study objects of these disciplines can be viewed in two ways depending on the focus and weight given to the leading discipline. In one case, emphasis is on the study of the geoforms, while soil information is used to help resolve issues of geomorphic nature, as for example, characterizing the geoforms or estimating the evolution of the landscape. Literally, this approach corresponds to the expression of soil geomorphology or pedogeomorphology. In the other case, focus is on the formation, evolution, distribution, and cartography of the soils, with the contribution of geomorphology. Literally, this approach corresponds to the expression of geomorphopedology, or its contraction as geopedology. In practice, the various expressions have been used interchangeably, showing that the distinction between the two approaches is fuzzy. Based on this apparent dichotomy, two streams, initially separated, have contributed to the development of the relations between geomorphology and pedology: (1) an academic stream, oriented towards the investigation of the processes that take place at the geomorphology-pedology interface, and (2) a more practical stream, applied to soil survey and cartography. The first one flourished more in hillslope landscapes, which offer propitious conditions to conduct toposequence (catena) and chronosequence studies, whereas the second one developed more in depositional, relatively flat landscapes, with conditions suitable for the use of soils for agricultural or engineering purposes.

3.2.1   Academic Stream

The academic stream consists of research conducted mainly at universities for scientific purposes. It is based on detailed site and transect studies to identify features of interdependence between geoforms and soils without preset paradigm. In general, this stream seeks to use geomorphology and pedology for analyzing, in a concomitant way, the processes of formation and evolution of soils and landscapes. This current covers in fact a variety of approaches, as illustrated by the definitions given by various authors with regard to their conceptions of the relationships between geomorphology and pedology and the study domains covering these relationships. Hereafter, some definitions of soil geomorphology are presented in chronological order.

- The analysis of the balance between geomorphogenesis and pedogenesis and the terms of control of the former on the latter in soil formation (Tricart 1965a, b, 1994).
- The use of pedologic research techniques in studies of physical and human geography (Pouquet 1966).
- The study of the landscape and the influence of the processes acting in the landscape on the formation of the soils (Olson 1989).
- The study of the genetic relationships between soils and landscapes (McFadden and Knuepfer 1990).
- The assessment of the genetic relationships between soils and landforms (Gerrard 1992).
- The application of geologic field techniques and ideas to soil investigations (Daniels and Hammer 1992).
- The study of soils and their use in evaluating landform evolution and age, landform stability, surface processes, and past climates (Birkeland 1999).
- The scientific study of the origin, distribution, and evolution of soils, landscapes, and surficial deposits, and of the processes that create and modify them (Wysocki et al. 2000).
- The scientific study of the processes of evolution of the landscape and the influence of these processes on the formation and distribution of the soils on the landscape (Goudie 2004).
- A field-based science that studies the genetic relationships between soils and landforms (Schaetzl and Anderson 2005).
- A subdiscipline of soil science that synthesizes the knowledge and techniques of the two allied disciplines, pedology and geomorphology, and that puts in parallel the genetic relationships between soil materials and landforms and the commensurate relationships between soil processes and land-forming processes (Thwaites 2007).
- The study that informs on the depositional history in a given locality, and also takes into account the postdepositional development processes in the interpretation of the present and past hydrological, chemical, and ecological processes in the same locality (Winter 2007).

This short review, which is far from being exhaustive, shows the diversity of concepts and conceptions encompassed in the expression *soil geomorphology*. From the above definitions, several main approaches may be derived:

- Geologic approach, with geomorphology as a subdiscipline of geology; this reflects the times when soil surveyors' basic training was in geology.
- Geomorphic approach, considering pedology as a discipline that gives support to geomorphology; etymologically, this approach could be called pedogeomorphology.
- Pedologic approach, considering geomorphology as a discipline that gives support to pedology; etymologically, this approach could be called geomorphopedology.
- Integrated approach, based on the reciprocal relations between both disciplines.
- Elevation of soil geomorphology at the level of a science, exhibiting therefore a status higher than that of a simple approach or type of study.

3.2.2 Applied Stream

The applied stream is related to soil survey and consists in using geomorphology for soil cartography. Historically, the analysis of the spatial relationships between geomorphology and pedology and the implementation of the soil-geoform duo were born out of practice. Soil survey has been the field laboratory where the modalities of applying geomorphology to soil cartography were formulated and tested. The structure of the geomorphic landscape served as background to soil mapping, while the dynamics of the geomorphic environment helped explain soil formation, with feedback of the pedologic information to the geomorphic knowledge.

Originally, different modalities of combining geomorphology and pedology were used for cartographic purposes, including the preparation of separate maps, the use of geomorphology to provide thematic support to soil mapping, and various forms of integration. Some authors and schools of thought advocated the procedure of antecedence: first the geomorphic survey (i.e. the framework), then the pedologic survey (i.e. the content), carried out by two different teams (Tricart 1965a; Ruhe 1975). In other cases, there was more integration, with mixed teams making systematic use of the interpretation of aerial photographs (Goosen 1968). Already in the 1930s, the soil survey service of the USA (National Cooperative Soil Survey) had an area of study in soil geomorphology (parallel mode), which was later on formalized with the mission of establishing pedogeomorphic relation models at the regional level to support soil survey (Effland and Effland 1992). The contribution of Ruhe (1956) meant a breakthrough in the use of geomorphology for soil survey in the USA. Ruhe was in favor of completely separating the description of the soils from the study of geomorphology and geology in a work area. Only after completing the disciplinary studies, could the interpretation of the relationships between soil characteristics and landforms be undertaken (Effland and Effland 1992). In the second half of the twentieth century, progress in systematic soil cartography, especially in developing countries, and progress in soil cartography to support agricultural development projects in a variety of countries have led to various forms of integration, with mixed teams of geomorphologists and pedologists. Work performed by French agencies such as ORSTOM (now IRD) and IRAT provide examples of this kind of soil cartography.

The need to boost agricultural production to support fast population growth has led many developing countries in the middle of the last century, especially in the tropics, to initiate comprehensive soil inventory programs. These were carried out mostly by public entities (ministries, soil institutes) and partly by consultancy agencies. In Venezuela, for instance, soil inventory began in the 1950–1960s as local and regional projects to support the planning of irrigation systems in the Llanos plains and, subsequently, as a nationwide systematic soil inventory. These surveys implemented an integrated approach based on the paradigm of the geopedologic landscape, which is closely related to the concepts of pedon, polypedon, and soilscape as entities for describing, sampling, classifying, and mapping soils. The integration between geomorphology and pedology took place all along the survey process, from the initial photo-interpretation up to the elaboration of the final map. The integration was reflected in the structure of the legend with two columns, a column for the geomorphic units that provide the cartographic frames, and a column for the soil units that indicate the soil types. This kind of approach is more appropriate for technical application than for scientific investigation. However, applied research always underlies the survey process, as new soil-geoform situations and relationships might occur and require analysis that goes beyond the strict survey procedure. This is a relatively formalized and systematic approach that can be applied with certain homogeneity by several soil survey teams working at various scales. One of the major requirements to make the implementation of geomorphology more effective in this kind of integrated survey is to apply a system of geoform taxonomy.

A novel way of integration can be found in the morphopedologic maps, based on the concept of the morphogenesis/pedogenesis balance (Tricart 1965b, 1994). Integration takes place not only at conceptual level, but also at the cartographic level (i.e., mapping procedure). The map distinguishes between stable and dynamic elements. The relatively stable geologic substratum, including lithology and structural settings, forms the map background, on which the geomorphic units are superimposed. Each map unit is characterized in the legend by the dominant pedogenic and dominant geomorphogenic processes. This information is used to derive a balance between pedogenesis and geomorphogenesis, which serves as a basis for identifying limitations to soil use.

The implementation of geomorphology in soil survey has strengthened the link between geomorphology and pedology. This practical cooperation has contributed more than academic studies in small areas or at site locations to enhance understanding of their reciprocal relations. These developments were closely related to the golden period of soil inventories during the second half of the twentieth century, particularly in emerging countries that needed soil information at various scales for ambitious agricultural development and irrigation projects. By mid-century, the systematic use of photo-interpretation revolutionized the soil survey procedure and made the contribution and mediation of geomorphology indispensable for identifying and delineating the surficial expression of soil units on the landscape. The rise of the liberal economy and the globalization of the economic relations over the last decade of the past century resulted in letting the market laws decide on the occupation and use of the territory. This meant the suspension of many land-use planning projects and, by the same token, the cancellation of the supporting soil inventory and land evaluation programs (Zinck 1990; Ibáñez et al. 1995). More recently, a growing societal awareness with regard to soil degradation and erosion is calling the attention on the threats affecting the soil resource, while creating new initiatives and opportunities for soil mapping (Hartemink and McBratney 2008; Sanchez et al. 2009).

Contemporaneously, the multiplication of GIS-related databases to store and manage the variety of data and information provided by the inventories of natural resources revealed the need for a unifying criterion able to structure the entries to the databases: geomorphology showed it could provide this structuring frame (Zinck and Valenzuela 1990). Hence the importance of having a classification system of the geoforms, preferably with hierarchic structure, to serve as comprehensive entry to the various information systems on natural resources, their evaluation, distribution, and degradation hazards.

In recent years, emphasis has been drawn to digital soil mapping based on remote-sensed data, together with the use of a variety of spatial statistics and geographical information systems (McBratney et al. 2003; Grunwald 2006; Lagacherie et al. 2007; Boettinger et al. 2010; Finke 2012; among others). The combination of remote sensing techniques and digital elevation models (DEM) allows improving predictive models (Dobos et al. 2000; Hengl 2003), but tends to see the soil as a surface rather than a three-dimensional body. Remote sensors provide data on individual parameters of the terrain surface and the surficial soil layer. There are also techniques and instruments able to detect soil property variations with depth via proximal sensing (e.g., frequency-domain electromagnetic methods FDEM, ground-penetrating radar GPR, among others), but their use is still partly experimental. Digital elevation models allow relating these parameters with relief variations, but the contribution of geomorphology is generally limited to geomorphometric attributes (Pike et al. 2009).

Some authors put emphasis on improving the precision of the boundaries between cartographic units as compared with a conventional soil map (Hengl 2003), or predicting spatial variations of soil properties and features such as for example the thickness of the solum (Dobos and Hengl 2009), or comparing the cartographic accuracy of a conventional soil map with that of a map obtained by expert system (Skidmore et al. 1996). In all these cases, morphometric parameters are mobilized along with pre-existing soil information (soil maps and profiles). The essence of the soil-geomorphology paradigm, in particular the genetic relationships between soils and geoforms and their effect on landscape evolution, is not sufficiently reflected in the current digital approach. It is difficult to find any theoretical or conceptual statement on soil-geoform relationships, except the reference that is usually made to classic models such as the hillslope model of Ruhe (1975) and the soil equation of Jenny (1941, 1980). Technological advances in remote sensing and digital elevation modelling are mainly used to explore and infer soil properties and their distribution in the topographic space. From an operational point of view, digital soil mapping is still mostly limited to the academic environment and essentially consists in mapping attributes of the soil surface layer, not full soil bodies that are actually the units managed by users (e.g., farmers, engineers). In official entities in charge of soil surveys, digital cartography is frequently limited to digitizing existing conventional soil maps (Rossiter 2004). There are few examples of national or regional agencies that have adopted automated methods for the production of operational maps (Hengl and MacMillan 2009).

3.3 Nature of the Relationships and Fields of Convergence

There is a collection of books on soil geomorphology that deal with the topic from different points of view according to the area of expertise of each author (Birkeland 1974; Ruhe 1975; Mahaney 1978; Gerrard 1981; Jungerius 1985a; Catt 1986; Retallack 1990; Daniels and Hammer 1992; Gerrard 1992; Birkeland 1999; Schaetzl and Anderson 2005; among others). These works are frequently quite analytical, recording benchmark case studies and describing exemplary situations that illustrate some kind of relationship between geomorphology and pedology. An epistemological analysis of the existing literature is needed to highlight the variety of points of view and enhance broader trends. Synthesis essays can be found in some scientific journal articles. What follows here is based on a selection of journal papers and book chapters, which provide a synthesis of the matter at a given time and constitute milestones that allow evaluating the evolution of ideas and approaches over time.

3.3.1 Evolution of the Relationships

The purely geologic conception of Davis (1899) on the origin of landforms as a function of structure, process and time, excluded soil and biota in general as factors of formation (Jungerius 1985b). For half a century, the denudation cycle of Davis has influenced the approach of geomorphologists, more inclined to develop theories than observe the cover materials on the landscape and to give preference to the analysis of erosion features rather than depositional systems. By contrast, the paradigm of soil formation, born from the pioneer works of Dokuchaiev and Sibirzew, and subsequently formalized by Jenny (1941, 1980), was based on a number of environmental factors including climate, biota, parent material, relief, and time. These original conceptual differences have led geomorphologists and pedologists to ignore each other's for a long time (Tricart 1965a), although Wooldridge (1949) had already written an early essay on the relationships between geomorphology and pedology. McFadden and Knuepfer (1990) note that soils have historically been neglected by many geomorphologists, who gave preference to the analysis of sedimentological and stratigraphic relations or morphometric studies. The situation changed by mid-twentieth century when it was recognized that the two models could be combined based on interrelated common factors (geologic structure, parent material, relief, time, and stage of evolution) and complementary factors (processes, climate, biota). This has allowed researchers to use the concepts and methods of both disciplines in varying combinations and for various purposes.

Tricart (1965a) has been one of the first researchers to draw the attention on the mutual relations uniting geomorphology and pedology. According to this author, geomorphology provides a framework for soil formation as well as elements of balance for pedogenesis, while pedology provides information about the soil properties involved in morphogenesis. Jungerius (1985b) shows that, although geomorphology and pedology have different approaches, the study objects of these two disciplines, i.e. landforms and soils, share the same factors of formation; the same author also highlights the fact that the relationships are two-way, generating mutual contributions. Since the early works of synthesis, which focused on what one discipline could bring to the other, the field of soil geomorphology has evolved toward greater integration, variable according to the topics, with simultaneous use of geomorphology and pedology and less consideration for the conventional boundaries that separate both disciplinary domains. In some universities there are now departments that house the two disciplines under the same roof (e.g., Department of Geomorphology and Soil Science, Technical University of Munich, Freising, Germany).

3.3.2 Mutual Contributions

Since the relationships between geomorphology and pedology are multiple, the spectrum of the areas and topics of interdisciplinary research is wide and varied, and the preferences depend on the orientation of each researcher. In the absence of a formal body of themes, here is how some authors have synthesized the content of soil geomorphology.

Already half a century ago, Tricart in his treatise on *Principes et Méthodes de la Géomorphologie* (Tricart 1965a) showed the reciprocal relationship of the two disciplines.

- Geomorphology contributes to pedology providing morphogenic balances that reflect the translocation of materials at the earth's surface. The concept of morphogenic balance is well illustrated in the case of the soil toposequences or catenas, where the removal of materials at the slope summit causes soil truncation, while the accumulation of the displaced materials at the footslope causes soil burying. Another example of balance between antagonistic processes that control soil development on slopes is the difference of intensity between the weathering

of the substratum and the ablation of debris on the terrain surface. In active alluvial areas, the morphology of the soil results from the balance between the deposition rate of the sediments and their incorporation in the soil by the pedogenic
processes.

- Geomorphology also provides a natural setting in which soil formation and evolution take place. The geomorphic environment, by way of integrating the factors
of parent material, relief, time, and surface processes, constitutes an essential
part of the spatial and temporal framework in which soils originate, develop, and
evolve. Tricart argued that geomorphic mapping should precede soil mapping
and was not in favor of integrating both activities.

- In return, pedology provides information on soil properties such as texture, structure, aggregate stability, iron content, among others, which play an important
role in the resistance of the surface materials to the morphogenic processes.
Privileging his own discipline, Tricart suggests that pedology ought to be a
branch of geomorphology, for the reason that pedology studies specific features
of the phenomena taking place at the contact between lithosphere and atmosphere, in particular in the stratum where living beings modify a surficial part of
the lithosphere, while geomorphology covers the greater part of the earth's epidermis. This view is shared by other authors such as, for example, Gerrard (1992)
or Daniels and Hammer (1992). Tricart, however, recognizes that the most
important thing is actually to intensify the ties of cooperation between both
disciplines.

The volume on *Pedogenesis and Soil Taxonomy* published in 1983 (Wilding et al.)
has been a reference book in its time, the main purpose of which was to provide a
balance between soil morphology and genesis to help understand and use the comprehensive classification system of Soil Taxonomy (Soil Survey Staff 1975). The
chapter written by Hall (1983) on geomorphology and pedology is an interesting
inclusion in a work specifically oriented towards soil taxonomy. The above author
shows that the soil is more than an object of classification and tries to reconcile soil
and landscape, an aspect largely ignored in Soil Taxonomy. Hall emphasizes that it
is necessary to map soils and geomorphic surfaces independently and establish correlations later, a point of view that coincides with positions previously defended by
Tricart (1965a) and Ruhe (Effland and Effland 1992). He says that it is not allowed,
in a new study area, to predict soils from their location on the landscape or infer the
geomorphic history of the area only on the basis of soil properties. Despite this
somewhat old-fashioned position, Hall acknowledges the lack of clear boundaries
between geomorphic and pedologic processes, and that interdisciplinary studies are
needed to explain the features that both sciences address.

In a supplement of the CATENA journal dedicated to *Soils and Geomorphology*
(Jungerius 1985a), Jungerius (1985b) presents the results of a broad literature
review from the first works of the mid-twentieth century until the publication date
of the supplement, with emphasis on papers published in CATENA. The author
adopts a dichotomous approach, similar to Tricart's approach, to show the mutual

contributions between both disciplines, but with emphasis on the contribution of pedology to geomorphology.

- To illustrate the significance of the landform studies for pedology, it is pointed out that pedologic processes such as additions, losses, translocations, and transformations (Simonson 1959) are under geomorphic control. Subsequently, reference is made to recurring themes in the literature that emphasize the role of relief as a factor of soil formation and geography. Highlighted topics address, for instance, the effect of the terrain physiography on the spatial distribution and cartography of soils, the effect of the topography on the genesis and catenary distribution of soil profiles, and the effect of landscape evolution on soil differentiation.
- The significance of the soil studies for geomorphology is analyzed in more detail. After showing how such studies contribute to prepare geomorphic and soil erosion maps, there is emphasis on two types of study that benefit substantially from the contribution of pedology: the studies of geomorphogenic processes and the paleogeomorphic studies. To investigate the nature of the processes that operate on a slope requires knowing the present soil system, with its spatial and temporal variations. Many of the authors cited by Jungerius (1985b) insist on the importance of the control that the horizon types exert on the geomorphic processes. A key differentiation is made between A horizons and surface crusts and their impact on the patterns of runoff and infiltration, on the one hand, and B horizons and subsurface pans and their impact on the formation of pipes and tunnels, gullies, and mass movements, on the other hand. With respect to the paleogeomorphic studies, these emphasize the importance of the paleosoils as indicators of a landscape stability phase, with the possibility of inferring factors and conditions that prevailed in the same period. The interpretation of the paleosoils helps the geomorphologist reconstruct past climate and vegetation conditions, infer the evolution time of a landscape, detect changes in a landscape configuration, and investigate past geomorphic processes.

3.3.3   Trend Towards Greater Integration

A pioneer work focusing on soils as landscape units is that of Fridland (1974, 1976). Fridland shows that soils are distributed on the landscape according to patterns that shape the structure of the soil mantle. Although the term geomorphology does not appear in his texts, he sets relationships between genetic and geometric soil entities and landforms. Ten years later, Hole and Campbell (1985) adopted Fridland's approach in their analysis of the soil landscape. Contemporaneously to the work of Fridland, Daniels et al. (1971) used the superposition of soil mantles to determine relative ages and sequences of events in the landscape, laying the foundations of pedostratigraphy.

More recent synthesis articles focus on showing how the concepts and methods of the two disciplines have been integrated to investigate interface features, rather than identifying the specific contribution of each discipline individually. Modern studies of soil geomorphology transgress the boundaries between the two sciences of origin and integrate parts of the doctrinal body of both. This new research domain constitutes an interface discipline, or "border country" as it is called by Jungerius (1985b), which gains in autonomy and maturity, with its own methodological approach and topics of interest. This has led Schaetzl and Anderson (2005) to qualify soil geomorphology as a full-fledged science. Hereafter, reference is made to some key articles that attempt to formalize the domain of soil geomorphology.

Olson (1989) considers that a study in soil geomorphology should have three main components, including (1) the recognition of the surface stratigraphy and the parent materials present in an area; (2) the determination of the geomorphic surfaces in space and time; and (3) the correlation between soil properties and landscape features. This approach is in accordance with the definition that Olson (1989) gives of soil geomorphology as the study of the landscape and the influence of landscape processes on soil formation. There is integration of the two disciplines, but geomorphology plays the leading role. In a subsequent publication (Olson 1997), the same author notes that the patterns or models of soil-geomorphology can be applied in a consistent and predictable manner in soil survey and considers that the pedologist should acquire the ability to use the pedogeomorphic patterns to interpolate within a study area or extrapolate to similar geographic areas.

The journal *Geomorphology 3* (1990) published the proceedings of a symposium dedicated to soil geomorphology (Proceedings of the 21st Annual Binghamton Symposium in Geomorphology, edited by Knuepfer and McFadden 1990). In addition to numerous articles analyzing case studies in a variety of sites and conditions, the journal contains two introductory papers that present an overview of the trends in this area in the late 1980s. McFadden and Knuepfer (1990) analyze the link between pedology and surface processes. In a short historical account, they show how pioneer work of some geologists, geomorphologists, and pedologists, concentrating on the study of the genetic relationships between soils and landscapes, resulted in the creation of the soil-geomorphology stream. The authors refer to three topics they consider central to the development of soil geomorphology. First, they point out the significance of the fundamental equation of Jenny (1941) to show the relevance of geomorphology in pedologic research through the factors of climate change, time, and relief. In particular, the study of chronosequences has contributed enormously to understanding geomorphic processes and landscape evolution, especially in river valleys with systems of nested terraces. The theme of fluvial terraces is an outstanding area of convergence, because understanding the genesis of the terraces is important to interpret the soil data. Secondly, the authors take up the issue of modelling and simulation. They contrast the conceptual models, such as those of Jenny (1941) and Johnson et al. (1990), with numerical models designed to simulate the behavior of complex systems, and consider that modelling is still limited by the poor definition of basic concepts such as polygenetic soils, soil-forming intervals, and rates of soil development, among others. Finally, the authors mention some of the problems that the investigation in soil geomorphology faces when dealing with complex landscapes. Hillslopes are a typical example of complex landscape, where the current morphogenic processes sometimes have no or little relationship with the formation of the slope itself, and often there is no clear relationship between slope gradient and degree of soil development. In synthesis, McFadden and Knuepfer (1990) consider that the soil-landform relationship is one of interaction and mutual feedback. The better we understand soils, including the speed at which the formation processes operate, and the variations caused by the position of the soils on the landscape, the deeper will be our understanding of the processes that originate the landforms. Reciprocally, whenever we better understand the evolution of the landscape at variable spatial and temporal scales, we will be able to elucidate complex pedologic problems.

In the same special issue of *Geomorphology 3*, Birkeland (1990) points out that it is difficult to work in one specific field of soil geomorphology without using information from the others. He illustrates this need to integrate information by analyzing various types of chronosequence and chronofunction in arid, temperate, and humid regions. Generalizing, Birkeland considers that, in the majority of cases, the studies of soil geomorphology pursue one of the four following purposes: (1) establishing a soil chronosequence that can be used to estimate the age of the surface formations; (2) using the soils, on the basis of relevant properties of diagnostic horizons, as indicators of landscape stability in the short or long term; (3) determining relationships between soil properties that allow inferring climate changes; and (4) analyzing the interactions between soil development, infiltration and runoff, and erosion on slopes.

Following the same order of ideas, Gerrard (1993) considers that the challenge of soil geomorphology is to integrate elements from the four research areas recognized by Birkeland (1990), to develop a conceptual framework of landscape evolution. The author describes several convergent conceptual models, such as those addressing the relationship between thresholds and changes of the soil landscape, the formation of soils on aggradation surfaces, soil chronosequences, and the relationship between soil development and watershed evolution.

The book of Schaetzl and Anderson (2005) on *Soils, Genesis and Geomorphology*, contains an extensive section devoted to soil geomorphology (pp. 463–655). The authors raise soil geomorphology to the level of a discipline that deals specifically with the two-way relations between geomorphology and pedology. The relationships emerge from the fact that soils are strongly related to the landforms on which they have developed. The authors emphasize that soil geomorphology is a science based primarily in field studies. They take up again, with new examples of more or less integrated studies, the three themes that soil geomorphology has been favoring: soil catena studies, soil chronosequences, and reconstruction of landscape evolution through the study of paleosoils. As a relevant attempt to get closer to a definition of the basic principles of the discipline, Schaetzl and Anderson recognize six main topics that comprise the domain of soil geomorphology: (1) soils as indicators of environmental and climatic changes; (2) soils as indicators of geomorphic stability and landscape stability; (3) studies of soil genesis and development (chronosequences); (4) soil-rainfall-runoff relationships, especially with regard to slope processes; (5) soils as indicators of current and past sedimentological and depositional processes; and (6) soils as indicators of the stratigraphy and parental materials of the Quaternary. This outline is similar, in more detail, to the list of objectives previously proposed by Birkeland (1990). This shows that certain conceptual and methodological coherence has been achieved.

3.4 Conclusion

Several authors have produced books and synthesis articles on soil geomophology, with extensive lists of references that readers are suggested to consult for more information. This has contributed to make soil geomorphology a discipline in its own right. There is consensus on the basic relationship between geomorphology and pedology: geomorphic processes and resulting landforms contribute to soil formation and distribution while, in return, soil development and properties have an influence on the evolution of the geomorphic landscape. The research themes that have received more attention (in the literature) are chronosequence and toposequence (catena) studies. These two kinds of study provide the majority of the examples used to illustrate the relationships between geomorphology and pedology. Some authors favor the chronosequences as integrated study subjects including pedostratigraphy and paleopedology. Many others emphasize the study of soil distribution and evolution within the framework of the catena concept popularized by the hillslope models of Wood (1942), Ruhe (1960, 1975), and Conacher and Dalrymple (1977). Some articles point out general principles, but there is still no unified body of doctrine. There are few references in international journals that provide some formal synthesis on how to carry out integrated pedogeomorphic mapping.

References

Birkeland PW (1974) Pedology, weathering and geomorphological research. Oxford University Press, New York

Birkeland PW (1990) Soil-geomorphic research – a selective overview. Geomorphology 3:207–224

Birkeland PW (1999) Soils and geomorphology, 3rd edn. Oxford University Press, New York

Boettinger JL, Howell DW, Moore AC, Hartemink AE, Kienast-Brown S (eds) (2010) Digital soil mapping: bridging research, environmental application, and operation, Progress in soil science, vol 2. Springer, New York

Catt JA (1986) Soils and quaternary geology. Clarendon Press, Oxford

Conacher AJ, Dalrymple JB (1977) The nine-unit landscape model: an approach to pedogeomorphic research. Geoderma 18:1–154

Daniels RB, Hammer RD (1992) Soil geomorphology. Wiley, New York

Daniels RB, Gamble EE, Cady JG (1971) The relation between geomorphology and soil morphology and genesis. Adv Agron 23:51–88

Davis WM (1899) The geographical cycle. The genetic classification of land-forms. Geogr J Wiley-Blackwell:481–504

Dobos E, Hengl T (2009) Soil mapping applications. In: Hengl T, Reuter HI (eds) Geomorphometry: concepts, sofware, applications, Developments in soil science, vol 33. Elsevier, Amsterdam, pp 461–479

Dobos E, Micheli E, Baumgardner MF, Biehl L, Helt T (2000) Use of combined digital elevation model and satellite radiometric data for regional soil mapping. Geoderma 97(3–4):367–391

Effland ABW, Effland WR (1992) Soil geomorphology studies in the U.S. soil survey program. Agric Hist 66(2):189–212

Elizalde G, Jaimes E (1989) Propuesta de un modelo pedogeomorfológico. Revista Geográfica Venezolana XXX:5–36

Finke PA (2012) On digital soil assessment with models and the pedometrics agenda. In: Geoderma, Entering the digital era: special issue of Pedometrics 2009, vol 171–172, Beijing, pp 3–15

Fridland VM (1974) Structure of the soil mantle. Geoderma 12:35–41

Fridland VM (1976) Pattern of the soil cover. Israel Program for Scientific Translations, Jerusalem

Gerrard AJ (1981) Soils and landforms, an integration of geomorphology and pedology. Allen & Unwin, London

Gerrard AJ (1992) Soil geomorphology: an integration of pedology and geomorphology. Chapman & Hall, New York

Gerrard AJ (1993) Soil geomorphology. Present dilemmas and future challenges Geomorphology 7(1–3):61–84

Goosen D (1968) Interpretación de fotos aéreas y su importancia en levantamiento de suelos. In: Boletín de Suelos 6. FAO, Roma

Goudie AS (ed) (2004) Encyclopedia of geomorphology, vol 2. Routledge, London

Grunwald S (ed) (2006) Environmental soil-landscape modeling: geographic information technologies and pedometrics. CRC/Taylor & Francis, Boca Raton

Hall GF (1983) Pedology and geomorphology. In: Wilding LP, Smeck NE, Hall GF (eds) Pedogenesis and soil taxonomy, I concepts and interactions. Elsevier, Amsterdam, pp p117–p140

Hartemink AE, McBratney A (2008) A soil science renaissance. Geoderma 148:123–129

Hengl T (2003) Pedometric mapping. Bridging the gaps between conventional and pedometric approaches. ITC dissertation 101, Enschede

Hengl T, MacMillan RA (2009) Geomorphometry: a key to landscape mapping and modelling. In: Hengl T, Reuter HI (eds) Geomorphometry: concepts, software, applications, Developments in soil science, vol 33. Elsevier, Amsterdam, pp 433–460

Hole FD, Campbell JB (1985) Soil landscape analysis. Rowman & Allanheld, Totowa

Ibáñez JJ, Zinck JA, Jiménez-Ballesta R (1995) Soil survey: old and new challenges. In: Zinck JA (ed) Soil survey: perspectives and strategies for the 21st century. FAO world soil resources report 80. FAO-ITC, Rome, pp 7–14

Jenny H (1941) Factors of soil formation. McGraw-Hill, New York

Jenny H (1980) The soil resource. Origin and behaviour. Ecological studies 37. Springer, New York

Johnson DL, Keller EA, Rockwell TK (1990) Dynamic pedogenesis: new views on some key soil concepts, and a model for interpreting quaternary soils. Quat Res 33:306–319

Jungerius PD (ed) (1985a) Soils and geomorphology, Catena Supplement, vol 6. Catena Verlag, Cremlingen

Jungerius PD (1985b) Soils and geomorphology. In: Jungerius PD (ed) Soils and geomorphology, Catena Supplement, vol 6. Catena Verlag, Cremlingen, pp 1–18

Kilian J (1974) Etude du milieu physique en vue de son aménagement. Conceptions de travail. Méthodes cartographiques. L'Agronomie Tropicale XXIX(2–3):141–153

Knuepfer PLK, McFadden LD (1990) Soils and landscape evolution. Proceedings of the 21st Binghamton symposium on geomorphology. Geomorphology 3(3–4):197–578

Lagacherie P, McBratney AB, Voltz M (eds) (2007) Digital soil mapping: an introductory perspective, Developments in soil science, vol 31. Elsevier, Amsterdam

Legros JP (1996) Cartographies des sols. Presses Polytechniques et Universitaires Romandes, Lausanne, De l'analyse spatiale à la gestion des territoires

Mahaney WC (ed) (1978) Quaternary soils. Geo Abstracts, Norwich

McBratney AB, Mendonça Santos ML, Minasny B (2003) On digital soil mapping. Geoderma 117(1–2):3–52

McFadden LD, Knuepfer PLK (1990) Soil geomorphology: the linkage of pedology and surficial processes. Geomorphology 3:197–205

Olson CG (1989) Soil geomorphic research and the importance of paleosol stratigraphy to quaternary investigations, midwestern USA, Catena Supplement 16. Catena Verlag, Cremlingen, pp 129–142

Olson CG (1997) Systematic soil-geomorphic investigations: contributions of RV Ruhe to pedologic interpretation. Adv Geoecol 29:415–438

Pike RJ, Evans IS, Hengl T (2009) Geomorphometry: a brief guide. In: Hengl T, Reuter HI (eds) Geomorphometry: concepts, sofware, applications, Developments in soil science, vol 33. Elsevier, Amsterdam, pp 3–30

Pouquet J (1966) Initiation géopédologique. Les sols et la géographie. SEDES, Paris

Principi P (1953) Geopedologia (Geologia Pedologica). Studio dei terreni naturali ed agrari. Ramo Editoriale degli Agricoltori, Roma

Retallack GJ (1990) Soils of the past. Unwin Hyman, Boston

Richards KS, Arnett RR, Ellis S (eds) (1985) Geomorphology and soils. Allen & Unwin, London

Rossiter DG (2004) Digital soil resource inventories: status and prospects. Soil Use & Management 20:296–301

Ruhe RV (1956) Geomorphic surfaces and the nature of soils. Soil Sci 82:441–455

Ruhe RV (1960) Elements of the soil landscape. Trans 7th Intl Congr Soil Sci (Madison) 4:165–170

Ruhe RV (1975) Geomorphology. Geomorphic processes and surficial geology. Houghton Mifflin, Boston

Sanchez PA, Ahamed S, Carré F, Hartemink AE, Hempel J, Huising J, Lagacherie P, McBratney AB, McKenzie NJ, Mendonça-Santos ML, Minasny B, Montanarella L, Okoth P, Palm CA, Sachs JD, Shepherd KD, Vågen TG, Vanlauwe B, Walsh MG, Winowiecki LA, Zhang G (2009) Digital soil map of the world. Science 325:680–681

Schaetzl R, Anderson S (2005) Soils: genesis and geomorphology. Cambridge University Press, New York

Simonson RW (1959) Outline of a generalized theory of soil genesis. Soil Sci Soc Am Proc 23:152–156

Skidmore AK, Watford F, Luckananurug P, Ryan PJ (1996) An operational GIS expert system for mapping forest soils. Photogram Eng & Remote Sensing 62(5):501–511

Soil Survey Staff (1975) Soil taxonomy. A basic system of soil classification for making and interpreting soil surveys. In: USDA agric handbook 436. US Government Print Office, Washington

Thwaites RN (2007) Development of soil geomorphology as a sub-discipline of soil science. Retrieved from https://crops.confex.com/crops/wc2006/techprogram/P17546.HTM. Nov 19, 2007

Tricart J (1962) L'épiderme de la terre. Esquisse d'une géomorphologie appliquée, Masson, Paris

Tricart J (1965a) Principes et méthodes de la géomorphologie. Masson, Paris

Tricart J (1965b) Morphogenèse et pédogenèse. I Approche méthodologique: géomorphologie et pédologie. Science du Sol A:69–85

Tricart J (1972) La terre, planète vivante. Presses Universitaires de France, Paris

Tricart J (1994) Ecogéographie des espaces ruraux. Nathan, Paris

Tricart J, Kilian J (1979) L'éco-géographie et l'aménagement du milieu naturel. Editions Maspéro, Paris

Wilding LP, Smeck NE, Hall GF (eds) (1983) Pedogenesis and soil taxonomy. I Concepts and interactions, Elsevier, Amsterdam

Winter SM (2007) Soil geomorphology of the Copper River Basin, Alaska, USA. Retrieved from https://cwseducation.ucdavis.edu/sites/g/files/dgvnsk9956/files/classes/files/Soil%20 Geomorphology%20of%20the%20Copper%20River%20Basin%2C%20Alaska%2C%20 USA.pdf. Nov 19, 2007

Wood A (1942) The development of hillside slopes. Geol Ass Proc 53:128–138
Wooldridge SW (1949) Geomorphology and soil science. J Soil Sci 1:31–34
Wysocki DA, Schoeneberger PJ, LaGarry HE (2000) Geomorphology of soil landscapes. In: Sumner ME (ed) Handbook of soil science. CRC Press, Boca Raton, pp E5–E39
Zinck JA (1990) Soil survey: epistemology of a vital discipline. ITC Journal 1990(4):335–351
Zinck JA, Valenzuela CR (1990) Soil geographic database: structure and application examples. ITC Journal 1990(3):270–294

Chapter 4
The Geopedologic Approach

J. A. Zinck

Abstract The relationships between geomorphology and pedology can be analyzed from different perspectives: conceptual, methodological, and operational. Geopedology (1) is based on the conceptual relationships between geoform and soil which center on the earth's epidermal interface, (2) is implemented using a variety of methodological modalities based on the three-dimensional concept of the geopedologic landscape, and (3) becomes operational primarily within the framework of soil inventory, which can be represented by a hierarchic scheme of activities. The approach focuses on the reading of the landscape in the field and from remote-sensed imagery to identify and classify geoforms, as a prelude to their mapping along with the soils they enclose and the interpretation of the genetic relationships between soils and geoforms. There is explicit emphasis on the geomorphic context as an essential factor of soil formation and distribution.

Keywords Concept · Geopedologic landscape · Method · Geopedologic integration · Implementation · Contribution to soil survey

4.1 Introduction: Definition, Origin, Development

The first one to use the term *geopedology* was most probably Principi (1953) in his treatise on *Geopedologia (Geologia Pedologica); Studi dei Terreni Naturali ed Agrari.* In spite of the prefix *geo,* the relationships between pedology and geology and/or geomorphology are not specifically addressed, except for the inclusion of three introductory chapters on unconsolidated surface materials, hard rocks, and

J. A. Zinck died before publication of this work was completed.

J. A. Zinck (deceased)
Faculty of Geo-Information Science and Earth Observation (ITC), University of Twente, Enschede, The Netherlands

© The Author(s), under exclusive license to Springer Nature
Switzerland AG 2023
J. A. Zinck et al. (eds.), *Geopedology,*
https://doi.org/10.1007/978-3-031-20667-2_4

rock minerals, respectively, as sources of parent material for soil formation. Principi's *Geopedologia* is in fact a comprehensive textbook on pedology. Following the pioneer work of Principi, the term geopedology continues being used in Italy to designate the university programs dealing with soil science in general.

The geopedologic approach, as formulated hereafter, is based on the fundamental paradigm of soil geomorphology, i.e., the assessment of the genetic relationships between soils and landforms and their parallel development, but with a clearly applied orientation and practical aim. The approach puts emphasis on the *reading of the landscape* in the field and from remote-sensed documents to identify and classify geoforms, as a prelude to their mapping along with the soils they enclose and the interpretation of the genetic relationships between soils and geoforms (geoform as defined below). As such, geopedology is closely related to the concept of pattern and structure of the soil cover developed by Fridland (1974, 1976) and taken up later by Hole and Campbell (1985), but with explicit emphasis on the geomorphic context as an essential factor of soil formation and distribution.

It is common acceptance that there are relationships between soils and landscapes, but often without specifying the nature or type of the landscape in consideration (e.g., topographic, ecological, biogeographic, geomorphic). The use of landscape models has shown that the elements of the landscape are predictable and that the geomorphic component especially controls a large part of the non-random spatial variability of the soil cover (Arnold and Schargel 1978; Wilding and Drees 1983; Hall and Olson 1991). Wilding and Drees (1983), in particular, stress the importance of the geomorphic features (forms and elements) to recognize and explain the systematic variations in soil patterns. Geometrically, the geomorphic landscape and its components, which often have characteristic discrete boundaries, are discernible in the field and from remote-sensed documents. Genetically, geoforms make up three of the soil forming factors recognized in Jenny's eq. (1941), namely the topography (relief), the nature of the parent material, and the relative age of the soil-landscape (morphostratigraphy). Therefore, the geomorphic context is an adequate frame for mapping soils and understanding their formation.

Geopedology aims at supporting soil survey, combining pedologic and geomorphic criteria to establish soil map units and analyze soil distribution on the landscape. Geomorphology provides the contours of the map units (i.e., the container), while pedology provides their taxonomic components (i.e., the content). Therefore, the geopedologic map units are more comprehensive than the conventional soil map units, since they also contain information about the geomorphic context in which soils are found and have developed. In this sense, the geopedologic unit is an approximate equivalent of the soilscape concept (Buol et al. 1997), with the particularity that the landscape is basically of geomorphic nature. This is reflected in the legend of the geopedologic map, which combines geoforms as entries to the legend and pedotaxa as components.

The geopedologic approach, as described below, was developed in Venezuela with the systematic application of geomorphology in the soil inventory programs that this country carried out in the second half of the twentieth century at various scales and different orders of intensity. In a given project, the practical implementation of geomorphology began with the establishment of a preliminary photo-interpretation map prior to fieldwork. This document oriented the distribution of the observation points, the selection of sites for the description of representative pedons, and the final mapping. As a remarkable feature, geoforms provided the headings of the soil map legend. The survey teams included geomorphologists and pedologists, who were trained in soil survey methodology including basic notions of geomorphology. This kind of training program had started in the Ministry of Public Works (MOP), responsible for conducting the basic soil studies for the location and management of irrigation and drainage systems in the alluvial areas of the country. It was subsequently extended and developed in the Commission for the Planning of the Hydraulic Resources (COPLANARH) and the Ministry of the Environment and Renewable Natural Resources (MARNR). From this experience was generated a first synthesis addressing the implementation of geomorphology in alluvial environment, basically the Llanos plains of the Orinoco river where large soil survey projects for the planning of irrigation schemes were being carried out (Zinck 1970). Later, with the extension of soil inventory to other types of environment, the approach was generalized to include landscapes of intermountain valleys, mountains, piedmonts, and plateau (Zinck 1974).

Subsequently, the geopedologic approach was formalized as a reference text under the title of *Physiography and Soils* within the framework of a postgraduate course for training specialists in soil survey at the International Institute for Aerospace Survey and Earth Sciences (ITC), now Faculty of Geo-Information Science and Earth Observation, University of Twente, Enschede, The Netherlands (Zinck 1988). For over 20 years, were formed geopedologist originating from a variety of countries of Latin America, Africa, Middle East, and Southeast Asia, who contributed to disseminate and apply the geopedologic method in their respective countries. In these times, the ITC also participated in soil inventory projects within the framework of international cooperation programs for rural development. This in turn has contributed to spreading the geopedologic model in many parts of the inter-tropical world. In certain countries, this model has received support from official agencies for its implementation in programs of natural resources inventory and ecological zoning of the territory (Bocco et al. 1996).

The geopedologic approach was developed in specific conditions, where the implementation of geomorphology was requested institutionally to support soil survey programs at national, regional, and local levels. Originally, the first demand emanated from the Division of Edaphology, Direction of Hydraulic Works of the Ministry of Public Works in Venezuela. This institutional framework has contributed to determining the application modalities of geomorphology to semi-detailed and detailed soil inventories in new areas for land use planning in irrigation systems and for rainfed agriculture at regional and local levels. The same thing happened later with the small-scale land inventory carried out by COPLANARH as input for the water resources planning at national level. In order to simplify logistics and lower the operation costs, geomorphology was directly integrated into the soil inventory. Hence, *geopedology* turned out to be the term that best expressed the relationship between the two disciplines, with geomorphology at the service of pedology, specifically to support soil mapping. Geomorphology was considered as a tool to improve and accelerate soil survey, especially through geomorphic photo-interpretation.

Geopedology is one of several ways described in Chap. 3 that study the relationships between geomorphology and pedology or use these relationships to analyze and explain features of pedologic and geomorphic landscapes. Compared to other approaches, geopedology has a more practical goal and could be defined as the soil survey discipline, including characterization, classification, distribution, and mapping of soils, with emphasis on the contribution of geomorphology to pedology. Geomorphology especially intervenes to understand soil formation and distribution by means of relational models (for instance, chronosequences and toposequences) and to support mapping. The central concept of geopedology is that of the soil in the geomorphic landscape. The geopedologic landscape is the paradigm.

The application of geomorphology to soil inventory requires hierarchic geoform taxonomy, suitable to be used at various categorial levels according to the degree of detail of the soil inventory and cartography. In Table 4.1, the general structure and main components of such a geoform classification system are presented. In this

Table 4.1 Synopsis of the geoform classification system (Zinck 1988)

Level	Category	Generic concept	Short definition
6	Order	Geostructure	Large continental portion characterized by a given type of geologic macro-structure (e.g., cordillera, geosyncline, shield).
5	Suborder	Morphogenic environment	Broad type of biophysical environment originated and controlled by a style of internal and/or external geodynamics (e.g., structural, depositional, erosional, etc.).
4	Group	Geomorphic landscape	Large portion of land/terrain characterized by given physiographic features: it corresponds to a repetition of similar relief/molding types or an association of dissimilar relief/molding types (e.g., valley, plateau, mountain, etc.).
3	Subgroup	Relief/molding	Relief type originated by a given combination of topography and geologic structure (e.g., cuesta, horst, etc.). Molding type determined by specific morphoclimatic conditions and/or morphogenic processes (e.g., glacis, terrace, delta, etc.).
2	Family	Lithology/facies	Petrographic nature of bedrocks (e.g., gneiss, limestone, etc.) or origin/nature of unconsolidated cover formations (e.g., periglacial, lacustrine, alluvial, etc.).
1	Subfamily	Landform/terrain form	Basic geoform type characterized by a unique combination of geometry, dynamics and history.

context, the word *geoform* refers to all geomorphic units regardless of the taxonomic levels they belong to in the classification system, while *landform/terrain form* is the generic concept that designates the lower level of the system. The *geoform* concept includes at the same time relief features and cover formations. The vocable *landform* may lead to confusion because it is used with different meanings in geomorphology, pedology, landscape ecology, and land evaluation, among others. The expression *terrain form* would be preferable.

The relationships between geomorphology and pedology can be analyzed from various points of view: conceptual, methodological, and operational. Geopedology (1) is based on the conceptual relationships between geoform and soil which center on the earth's epidermal interface, (2) is implemented using a variety of methodological modalities based on the three-dimensional concept of the geopedologic landscape, and (3) becomes operational primarily within the framework of soil inventory, which can be represented by a hierarchic scheme of integrated activities.

4.2   Conceptual Relationships

Geoform and soil are natural objects that occur along the interface between the atmosphere and the surface layer of the terrestrial globe. They are the only objects that occupy integrally this privileged position. Rocks (lithosphere) lie mostly underneath the interface. Living beings (biosphere) can be present inside or below, but essentially occur above. Air (atmosphere) can penetrate into the interface but is mostly over it. Figure 4.1 highlights the central position of the geoform-soil duo in the structure of the physico-geographical environment. The geoform integrates the concepts of relief/molding and cover formation.

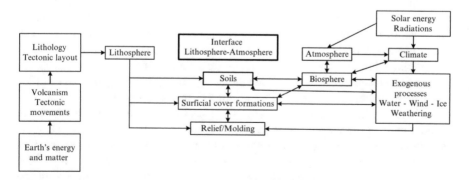

Fig. 4.1 The position of the geoform-soil duo at the interface between atmosphere and lithosphere. (Adapted from Tricart 1972)

4.2.1 Common Forming Factors

Because geoform and soil develop along a common interface in the earth's epidermis, a thin and fragile envelope called earth's critical zone where soils, rocks, air, and water interact, they share forming factors that emanate from two sources of matter and energy, one internal and another external.

- The endogenous source corresponds to the energy and matter of the terrestrial globe. The materials are the rocks that are characterized by three attributes: (1) the lithology or facies that includes texture, structure, and mineralogy; (2) the tectonic arrangement; and (3) the age or stratigraphy. The energy is supplied by the internal geodynamics, which manifests itself in the form of volcanism and tectonic deformations (i.e., folds, faults, fractures).
- The exogenous source is the solar energy that acts through the atmosphere and influences the climate, biosphere, and external geodynamics (i.e., erosion, transportation, and sedimentation of materials).

Geoform and soil are conditioned by forming factors derived from these two sources of matter and energy that act through the lithosphere, atmosphere, hydrosphere, and biosphere. The boundaries between geoform and soil are fuzzy. The geoform has two components: a terrain surface that corresponds to its external configuration (i.e., the epigeal component) and a volume that corresponds to its constituent material (i.e., the hypogeal component). The soil body is found inserted between these two components. It develops from the upper layer of the geomorphic material (i.e., weathering products – regolith, alterite, saprolite – or depositional materials) and is conditioned by the geodynamics that takes place along the surface of the geoform (e.g., aggradation, degradation, removal). Many soils do not form directly from hard rock, but from transported detrital materials or from weathering products of the substratum. These more or less loose materials correspond to the surface formations that develop at the interface lithosphere-atmosphere, with or without genetic relationship with the substratum, but closely associated with the evolution of the relief of which they are the lithological expression (Campy and Macaire 1989). The surficial cover formations constitute, in many cases, the parent materials of the soils. The nature and extent of these surface deposits often determine the conditions and limits of the interaction between processes of soil formation (Arnold and Schargel 1978).

The fact that geoform and soil share the same forming factors generates complex cause-effect relationships and feedbacks. One of the factors, namely the relief that corresponds to the epigeal component of the geoforms, belongs inherently to the geomorphology domain. Another factor, the parent material, is partially geomorphic and partially geologic. Time is a two-way factor: the age of the parent material (e.g., the absolute or relative age of a sediment) or the age of the geoform as a whole (e.g., relative age of a terrace) informs on the likely age of the soil; conversely, the dating of a humiferous horizon or an organic layer informs on the stratigraphic position of the geoform. Therefore, the relationships between these three forming factors are both intricate and reciprocal, the geoform being a factor of soil formation and the soil being a factor of morphogenesis (e.g., erosion-accumulation on a slope). Biota and climate influence both the geoform and the soil, but in a different way. In the case of the biota, the relationship is complex since part of the biota (the hypogeal component) lives within the soil and is considered part of it.

The geoform alone integrates three of the five soil forming factors of the classic model of Jenny (1941), while reflecting the influence of the other two factors. This gives geomorphology a role of guiding factor in the geoform-soil pair. Its importance as a structuring element of the landscape is reflected in the geomorphic entries to the geopedologic map legend. Figures 4.2 and 4.3 provide an example of this kind of integrated approach, showing each soil unit in its corresponding geomorphic landscape unit.

The geomorphic map of Fig. 4.2 represents the graben of Punata-Cliza in the eastern Andes of Bolivia, close to the city of Cochabamba. For some time, this tectonic depression was occupied by a lake that dried up into a lagunary environment. Subsequently, detrital sediments coming from the mountain borders formed fans and glacis in the margins of the depression, leaving uncovered relict lagunary flats in the center of the depression. Photo-interpretation and fieldwork allowed segmenting the alluvial fans in proximal, central, and distal sectors. The geomorphic structure of the depression bottom resulting from this evolution during the Quaternary provides the basic framework for soil formation and spatial distribution. This is reflected in the coupled geomorphic-pedologic legend of Fig. 4.3. The sequential partitioning of the geomorphic environment into landscape, relief, facies, and landform units allowed identifying and mapping geomorphic units with their respective soil taxa, forming thus geopedologic units.

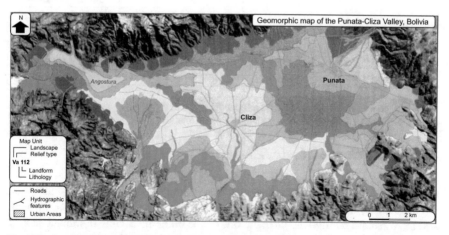

Fig. 4.2 Geomorphic map of the Punata-Cliza tectonic depression, eastern Andes of Bolivia. (Metternicht and Zinck 1997)

GEOPEDOLOGIC LEGEND					
LANDSCAPE	RELIEF TYPE	FACIES	LANDFORM	CODE	SOILS
PIEDMONT	Dissected-depositional glacis	Alluvial	Proximal	Pi 111	*Association:* Typic Calciorthids Typic Camborthids
			Central	Pi 112	*Consociation:* Typic Camborthids (ca)* Ustochreptic Camborthids
			Distal	Pi 113	*Association:* Ustalfic Haplargids Ustochreptic Camborthids
	Depositional glacis	Colluvio-alluvial	Distal	Pi 213	*Consociation:* Ustochreptic Camborthids Typic Camborthids
	Active fans	Alluvial	Active channels	Pi 411	*Miscellaneous land type:* Mixed Alluvial
			Inactive channels	Pi 412	*Consociation:* Typic Torrifluvents Typic Torriorthents
	Recent fans	Colluvio-alluvial		Pi 51	*Association:* Ustic Torriorthents Typic Torrifluvents
	Old dissected fans	Glacio-alluvial	Proximal	Pi 661	*Association:* Typic Camborthids Typic Haplargids
			Central	Pi 612	*Consociation:* Ustochreptic Camborthids (ca)*
			Distal	Pi 613	*Consociation:* Ustochreptic Camborthids
	Hills	Quartzitic sandstones		Pi 71	*Consociation:* Lithic Torriorthents
		Marls sandstones limestones		Pi 72	*Consociation:* Typic Calciorthids Lithic Calciorthids
VALLEY	Lagunary depressions	Alluvio-lagunary	Higher lagunary flats	Va 111	*Association:* Fluventic Camborthids Ustochreptic Camborthids
			Middle lagunary flats	Va 112	*Association:* Ustalfic Haplargids Ustochreptic Camborthids
			Lower lagunary flats	Va 113	*Association:* Ustalfic Haplargids (saso)* Ustochreptic Camborthids (sa)*
		Lagunary	Playas	Va 124	*Association:* Typic Salorthids Natric Camborthids
* Phases: (ca) calcareous (saso) saline-alkaline (sa) saline					

Fig. 4.3 Geopedologic legend of the map shown in Fig. 4.2, referring to the Punata-Cliza tectonic depression, eastern Andes of Bolivia. (Metternicht and Zinck 1997)

4.2.2 The Geopedologic Landscape

Geoform and soil fuse to form the geopedologic landscape, a concept similar to that of soilscape (Buol et al. 1997), to designate the soil on the landscape. Geoform and soil have reciprocal influences, being one or the other alternately dominant according to the circumstances, conditions, and types of landscape. In flat areas, the geopedologic landscapes are mainly constructional, while they are mainly erosional in sloping areas.

4.2.2.1   Flat Areas

In flat constructional areas, the sedimentation processes and the structure of the resulting depositional systems control often intimately the distribution of the soils, their properties, the type of pedogenesis, the degree of soil development and, even, the use potential of the soils. The valley landscape offers good examples to illustrate these relationships. Figure 4.4 represents a transect crossing the lower terrace built by the Guarapiche river in the north-east of Venezuela during the late Pleistocene (Q1). In the wider sectors of the valley, the river activity produced a system that consists of a sequence of depositional units including river levee, overflow mantle, overflow basin, and decantation basin, in this order across the valley from proximal positions close to the paleo-channel of the river, to the distal positions on the fringe of the valley.

The relevant characteristics of the four members of the depositional system are as follows (pedotaxa refer to dominant soils):

- River levee (or river bank): highest position of the system, convex topography, narrow elongated configuration; textures with dominant sandy component (loamy sand, sandy loam, sometimes sandy clay loam); well drained; Typic Haplustepts (or Fluventic); land capability class I.

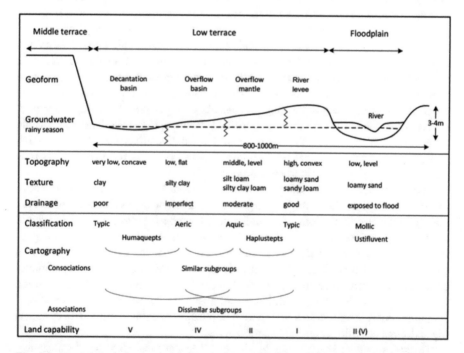

Fig. 4.4 Geopedologic landscape model of a young fluvial terrace. Example of the Guarapiche river valley, northeast of Venezuela; pedotaxa refer to the dominant soil type in each geoform

- Overflow mantle: medium-high position, flat topography, wide configuration; textures with dominant silty component (silt loam, silty clay loam); moderately well drained; Aquic Haplustepts (or Fluvaquentic); land capability class II.
- Overflow basin: low position, flat to slightly concave topography, wide oval configuration; mainly silty clay texture; imperfectly drained; Aeric Humaquepts; land capability class IV.
- Decantation basin: lowest position of the system, concave topography, closed oval configuration; usually very fine clay texture; poorly drained; Typic Humaquepts, sometimes associated with Aquerts; land capability class V.

The transitions between geomorphic positions are very subtle to imperceptible on the terrain surface. External markers such as slight undulations of field border fences and changes in color or compaction of dirt road trails help presume changes of positions. Unit boundaries and kinds were tentatively recognized by photo-interpretation on the basis of tone nuances, but definitively identified by field observations along transects. Parent material must be qualified to identify geoforms. The total relief amplitude between levee and decantation basin is approximately 2 m over a distance of about 600 m (0.3% transversal slope).

The soil classes referred to in this example correspond to the dominant soils in each geomorphic unit. Major soils are generally accompanied by subordinate soils that may have common taxonomic limits with the dominant soils in the classification system (i.e., similar soils) and some inclusions that are usually not contrasting. The geoform, with its morphographic, morphometric, morphogenic and morpho-chronologic features, controls a number of properties of the corresponding soil unit (e.g., topography, texture, drainage) and relates to its taxonomic classification and land use capability. The geoform also guides the composition of the cartographic unit, with the possibility of mapping soil consociations on the basis of similar subgroups (e.g., Aquic Haplustepts and Aeric Humaquepts) or soil associations on the basis of dissimilar subgroups (e.g., Typic Haplustepts and Aeric Humaquepts), according to the soil distribution pattern and the mapping scale. The geomorphic framework, which controls the determination and delineation of the soil map units, makes that these units are relatively homogeneous, allowing for a reasonably reliable soil interpretation for land use purposes.

The Guarapiche valley example is an ideal textbook model, rather infrequent in its full expression. The complete sequence in the right depositional order occurs mainly in the largest sections of the valley that have been sedimentologically stable over some time (see Fig. 4.9 in Sect. 4.3.3.1). In narrow sections, some of the geomorphic positions are usually missing, with for instance the levee running parallel to the basin. In other places, the river axis has been shifting over the depositional area, moving for instance during a heavy flood event from unstable channel between high levees to the low-lying marginal basin position. This results in less organized spatial geomorphic structures and more complex geopedologic units with contrasting sediment stratifications and superpositions.

The soil sequence in a given geopedologic landscape can also vary, for instance, according to the prevailing bioclimatic conditions (e.g., Mollisols sequence in a moister climate) or according to the age of the terrace (e.g., Alfisols sequence on a Q2 terrace and Ultisols sequence on a Q3 terrace). Post-depositional perturbations in flat areas, through fluvial dissection of older terraces or differential eolian sedimentation-deflation, for example, may cause divergent pedogenesis and increase variations in the soil cover that are often not readily detectable. The resulting geopedologic landscapes are often much more complex than the initial constructed ones (Ibáñez 1994; Amiotti et al. 2001; Phillips 2001; among others). McKenzie et al. (2000) mention the case of strongly weathered sesquioxidic soils in Australia that were formed under humid and warm climates during the Late Cretaceous and Tertiary and are now persisting under semiarid conditions, showing the imprints from successive environmental changes.

4.2.2.2   Sloping Areas

In sloping areas and other ablational environments, the relationships between geoform and soil are more complex than in constructed landscapes. The classic soil toposequence is an example of geopedologic landscape in sloping areas. The lateral translocation of soluble substances, colloidal particles, and coarse debris on the terrain surface and within the soil mantle results in the formation of a soil catena, whose differentiation along the slope is mainly due to topography and drainage. Typically, the summit and shoulder of a hillslope lose material, which transits along the backslope and accumulates on the footslope. This relatively simple evolution usually results in the formation of a convex-concave slope profile with shallow soils at the top and deep soils at the base. When the translocation process accelerates, for instance after removal of the vegetation cover, soil truncation occurs on the upper slope facets, while soil fossilization takes place in the lower section because pedogenesis is no longer able to digest all the incoming material via continuous soil aggradation/cumulization. Such an evolution reflects relatively clear relationships between the geomorphic context and the soil cover, which can be approximated using the slope facet models. The segmentation of the landscape into units that are topographically related, such as the facet chain along a hillside, provides a sound basis for conducting research on spatial transfers of soil components (Pennock and Corre 2001). However, this idealized soil toposequence model might not be that frequent in nature.

On many hillsides, soil development, properties, and distribution are less predictable than in the case of the classic toposequence. Sheet erosion controlled by the physical, chemical and biological properties of the topsoil horizons, along with other factors, causes soil truncation of variable depths and at variable locations. Likewise, the nature of the soil material and the sequence of horizons condition the morphogenic processes that operate at the terrain surface and underneath. For instance, the difference in porosity and mechanical resistance between surficial horizons, subsurficial layers and substratum controls the formation of rills, gullies and mass movements on sloping surfaces, as well as the hypodermic development of pipes and tunnels. The geopedologic landscapes resulting from this active geodynamics can be very complex. Their spatial segmentation requires using geoform phases based on terrain parameters (e.g., slope gradient, curvature, drainage, micro-relief, local erosion features, salinity spots, etc.).

Paleogeographic conditions may have played an important role in hillslope evolution and can explain a large part of the present slope cover formations. Slopes are complex registers of the Quaternary climate changes and their effect on vegetation, geomorphic processes, and soil formation. The resulting geopedologic landscapes are polygenic and have often an intricate, sometimes chaotic structure. The superimposition or overlapping of consecutive events causing additions, translocations, and obliterations, with large spatial and temporal variations, makes it often difficult to decipher the paleogeographic terrain history and its effect on the geopedologic relationships.

The following example shows that an apparently normal convex-concave slope can conceal unpredictable variations in the covering soil mantle. The case study is a soil toposequence along a mountain slope between 1100 m asl and 1500 m asl in the northern Coastal Cordillera of Venezuela (Zinck 1986). Soils have developed from schist under dense tropical cloud forest, with 1850 mm average annual rainfall and 19 °C average annual temperature. Slope gradient is 2–5° at slope summit, 40–45° at the shoulder, 30–40° along the backslope, and 10–25° at the footslope. By the time of the study, no significant erosion was observed. However, several features indicate that the current soil mantle is the result of a complex geopedologic evolution, with alternating morphogenic and pedogenic phases, during the Holocene period.

- Except at the slope summit, soils have formed from detrital materials displaced along the slope, and not directly from the weathering in situ of the geologic substratum.
- There is no explicit correlation between slope gradient and soil properties. For instance, shoulder soils are deeper than backslope soils, although at higher slope inclination.
- Many soil properties such as pedon thickness and contents of organic carbon, magnesium and clay show discontinuous longitudinal distribution along the slope (Figs. 4.5 and 4.6). The most relevant interruption occurs in the central stretch of the slope, around 1300 m elevation.
- Soils in the upper part of the slope have two Bt horizons (a sort of bisequum) that reflect the occurrence of two moist periods favoring clay illuviation, separated by a dry phase.

Pollen analysis of sediments from a nearby lowland lake reveals that, by the end of the Pleistocene, the regional climate was semi-arid, vegetation semi-desertic, and soils probably shallow and discontinuous (Salgado-Labouriau 1980). From the beginning of the Holocene when the cloud forest started covering the upper ranges of the Cordillera, deep Ultisols developed. During the Holocene, dry episodes have occurred causing the boundary of the cloud forest to shift upwards and leaving the lower part of the slope, below approximately 1350–1300 m asl, exposed to erosion.

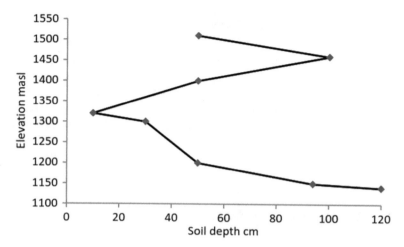

Fig. 4.5 Variation of soil depth with elevation along a mountain slope in the northern Coastal Cordillera of Venezuela. (Zinck 1986)

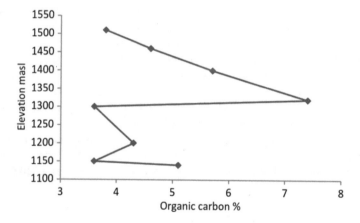

Fig. 4.6 Variation of organic carbon content (0–10 cm) with elevation along a mountain slope in the northern Coastal Cordillera of Venezuela. (Zinck 1986)

The presence, in the nearby piedmont, of thick torrential deposits dated 3500 BP and 1500 BP indicates that mass movements have episodically occurred upslope during the upper Holocene. This would explain why soil features and properties show a clear discontinuity at mid-slope, around 1300 m asl.

The alternance of morphogenic and pedogenic activity along mountain and hill slopes causes geopedologic relationships to be complex in sloping areas, in general more complex than in flat areas. The older the landscape, the more intricate are the relationships between soil and geoform because of the imprints left by successive environmental conditions.

4.3   Methodological Relationships

The methodological relationships refer to the modalities used to analyze the spatial distribution and formation of the geoform-soil complex. Geomorphology contributes to improving the knowledge of soil geography, genesis, and stratigraphy. In return, soil information feeds back to the domain of geomorphology by improving the knowledge on morphogenic processes (e.g., slope dynamics). The above needs the integration of geomorphic and pedologic data in a shared structural model to identify and map geopedologic units.

4.3.1   *Geopedologic Integration: A Structural Model*

Figure 4.7 shows the data structure of the geoform-soil complex in the view of the geopedologic approach (Zinck and Valenzuela 1990). Soil survey data are typically derived from three sources: (1) visual interpretation and digital processing of remote-sensed documents, including aerial photographs, radar and multi-spectral images, and terrain elevation models; (2) field observations and instrumental measurements, including biophysical, social, and economic features; and (3) analytical determinations of mechanical, physical, chemical, and mineralogical properties in the laboratory. The relative importance of these three data sources varies according to the scale and purpose of the soil survey. In general terms, the larger is the scale of the final soil map, the more field observations and laboratory determinations are required to ensure an appropriate level of information.

As soils and geoforms are three-dimensional bodies, external and internal (relative to the terrain surface) features are to be described and measured to establish and delimit soil map units. The combination of data and information provided by sources

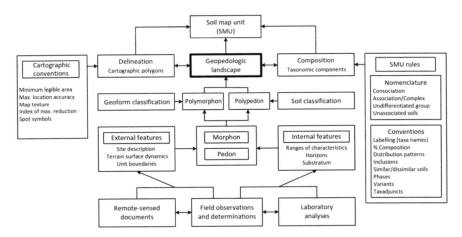

Fig. 4.7 Conceptual-structural model of the geopedologic approach. (Zinck and Valenzuela 1990)

(1) and (2) serves to describe the environmental conditions and areal dynamics (e.g., erosion, flooding, aggradation of sediments, changes in land uses, etc.) and to delineate the map units. At this level, the implementation of geomorphic criteria through interpretation of remote-sensed documents and field prospection plays a relevant role for the identification and characterization of the soil distribution patterns and the understanding of their spatial variability. The interpretation of remote-sensed documents (photo, image, DEM) can benefit from applying a stepwise procedure of features identification using the geoform hierarchy to highlight the nested structure of the landscape (see Table 4.1). The sequence of steps includes photo/image reading, identification of master lines, sketching the structure of the landscape to select representative cross sections, pattern recognition along the cross sections, delimitation of the geomorphic units via interpolation and extrapolation, and establishing a preliminary geomorphic interpretation legend for field verification.

The combination of data and information provided by sources (2) and (3) allows characterizing and quantifying the properties of the pedologic materials, geomorphic cover formations, and geologic substrata. The horizon (or layer) is the basic unit of data collection. Horizon and substratum information is aggregated in observation profiles, modal pedons, and modal morphons. Pedon and polypedon are described and established according to the criteria of Soil Taxonomy (Soil Survey Staff 1999). The morphon is the geomorphic equivalent of the pedon. It is described at the same site as the pedon but without fixed size standards. Conventionally, the areal size of a pedon varies from 1 m^2 for horizontally layerd soils to 10 m^2 for soils having cyclic horizons. The extent of a morphon is obviously larger to capture the variations of the terrain surface. The description of the morphon includes internal and external features. The internal features correspond to the characteristics and properties of the geomaterial in the substratum, thus the parent material of the soil. The external features cover the conditions and dynamics of the terrain area at the site of description and its surroundings. The pedologic material (i.e., the solum) occupies the volume between the substratum and the terrain surface. As in the case of the pedon, the morphon is the description and sampling site. Therefore, pedon and morphon are two fundamentally related entities. This is nothing new since the description of the pedon has always included that of the parent material and surface features. However, the contribution of the geomorphic analysis methods improves the characterization of the geomaterials in the substratum and that of the surface geodynamics. The methodological integration can be achieved by experts skilled in both geomorphology and pedology or by interdisciplinary teams.

The concepts of polypedon and polymorphon are significantly different from each other. The polymorphon corresponds to a whole geoform and is therefore a more comprehensive unit than the polypedon. A polymorphon can include more than one polypedon, and this is actually often the case, especially at the upper levels of the geoform classification system. The foregoing is reflected in the taxonomic composition of the map units: a relatively homogeneous geoform may correspond to a consociation of similar soils, while a less homogeneous geoform may correspond to an association of dissimilar soils. The identification and description of the polymorphon follow the criteria set out in Chaps. 6, 7 and 8, which deal with the taxonomy and attributes of the geoforms. Variations among identification profiles by comparison with a modal profile (pedon or morphon) are expressed in terms of ranges of characteristics for each taxon present in a map unit.

At this stage, the available data consist of: (1) geopedologic point observations, with additional information on the spatial variations of the characteristics, and (2) a framework of spatial units based essentially on external geomorphic criteria (i.e., characteristics of the terrain surface). The combination of the two results in a map of geopedologic units.

For mapping purposes, both objects – soil and geoform – are given identification names (i.e., taxonomic names) that are supplied by their respective classification systems. Assemblies of contiguous similar soils, forming polypedons, are classified by comparison with taxonomic entities established in soil classification systems, such as Soil Taxonomy (Soil Survey Staff 1999), the WRB classification (IUSS 2007), or any national classification. A similar procedure is used for the classification of the geomorphic units, moving from the description and sampling unit (morphon) to the classification entity (polymorphon). A basic geomorphic unit (polymorphon) can contain one or more polypedons. For instance, Entisols (e.g., Mollic Ustifluvents) and Mollisols (e.g., Fluventic Haplustolls) can occur intermixed with contrasting inclusions in a recent river levee position. The combination in the landscape of a polymorphon with the associated polypedons constitutes a geopedologic landscape unit.

Due to the inherent spatial anisotropy of the pedologic material, which is generally more pronounced than the anisotropy of the geomorphic material, soil delineations are usually heterogeneous. This requires that the taxonomic components of a map unit be named and their respective proportions quantified using conventional rules of soil cartography (Soil Survey Staff 1993). The delimitation of polygons follows a number of cartographic conventions that assure a good readability of the soil map. In this way, the geopedologic landscape units, cartographically and taxonomically controlled, as unique combinations of geomorphic polygons and their pedologic contents, result being the soil map units.

This theoretical-methodological model of the geoform-soil complex can be implemented to design the structure of an integrated geopedologic database, such as shown in Zinck and Valenzuela (1990).

4.3.2 Geopedologic Integration: Soil Geography, Genesis, and Stratigraphy

Within the framework of the previously described geopedologic model, themes such as soil geography, genesis, and stratigraphy can benefit substantially from the integration of pedologic and geomorphic methods.

4.3.2.1   Soil Geography

Soil survey generates information on the spatial distribution of soils. The implementation of geomorphic criteria in soil survey improves the identification and delimitation of the soils. At the same time, the rationality of the geopedologic approach contributes to compensate or partially replace what Hudson (1992) called the acquisition of tacit knowledge for the application of the soil-landscape paradigm. The integrated geopedologic analysis facilitates the reading of the landscape, because the geomorphic context controls, in a large proportion, the soil types that are found associated in a given kind of landscape such as, for instance, the sequence of levee-mantle-basin in an alluvial plain or the sequence of summit-shoulder-backslope-footslope along a hillside. These models of geopedologic associations that are genetically related and produce characteristic spatial patterns, are the components (i.e. soil combinations) of what Fridland (1974) calls *the structure of the soil cover* and Schlichting (1970) formulates as *Bodensoziologie* (i.e. pedosociology). Geopedologic spatial patterns depict the landscape and its elements the same way they can be seen in nature, in contrast to the artificial delineations shown on some geostatistically-based soil maps. This is why geopedologic maps are easy to read, even for non-specialists. For instance, on Fig. 4.2 it is easy to recognize the triangular shape of the alluvial fans.

- *Soil identification* is based on the description of the soils in the field, which leads to their characterization and classification. Geomorphology contributes to this activity through the selection of the description sites. The use of geomorphic criteria facilitates the choice of representative sites, regardless of the implemented sampling scheme. In oriented sampling, the observation sites are pre-selected based on geomorphic criteria within units delimited by interpretation of aerial photos or satellite images. Random sampling only makes sense if it is applied within the framework of units previously established with geomorphic criteria. A random sampling scheme is more objective and appropriate for statistical data analysis, but frequently generates a number of little representative profiles and, for this reason, is more expensive.

Grid-based systematic sampling is difficult to apply as an operational technique to an entire soil survey project because it would be too costly. It is useful when applied locally to estimate the spatial variability of the soils within and between selected map units and to establish their degree of purity. Bregt et al. (1987) compare two thematic soil maps, one derived from a conventional soil map and another one obtained by kriging of grid point data. The average purity of the map units, determined on the basis of three criteria including thickness of the A horizon, depth to gravel, and depth to boulder clay, is 77% in both cases, with less dispersion in the first case (72–82%) than in the second (69–85%). The interpretation of geostatistical data is probably more meaningful when geomorphic criteria are used.

- *Soil delimitation* is based on the interpretation of aerial photos and satellite images, the use of digital elevation models, and fieldwork. The features detected by remote sensing are essentially ground surface features, which are often of geomorphic nature. Therefore, what is observed or interpreted in remote-sensed documents are characteristics of the epigeal part of the geoforms and soils. The hypogeal part is still largely inaccessible and some of its features can be detected at distance only with special techniques (e.g., GPR). This is efficient when a three-dimensional representation of the geomorphic landscape is available, which can be obtained by stereoscopic interpretation of aerial photos or satellite images or based on a combination of images and elevation or terrain models.

In this context, geomorphology contributes to the following tasks related to soil delimitation: (1) the selection of sample areas, transects, and traverses; (2) the drawing of the soil map unit boundaries based on the conceptual relations between geoforms and soils (common forming factors; geopedologic landscape); and (3) the identification, temporal monitoring, and explanation of the spatial variability of the soils.

- *Soil variability* is partly controlled by the geomorphic context, especially systematic variability (Wilding and Drees 1983). Landform and soil patterns match often on a one-to-one correspondence (Wilding and Lin 2006). Geomorphology provides criteria for segmenting the soilscape continuum into discrete units that are relatively homogeneous. Such units are suitable frameworks for estimating the spatial variability of soil properties using geostatistical analysis (Saldaña et al. 1998; Kerry and Oliver 2011). They have been used also as reference units to apply spatial analysis metrics, including indices of heterogeneity, diversity, proximity, size and configuration, for quantitatively describing soil distribution patterns at various categorial levels of geoform (i.e. landscape, relief, terrain form) (Saldaña et al. 2011; Toomanian 2013).

The mapping scale and observation density influence the relationship between geoform and soil, as the spatial variability of the geomorphic and pedologic properties are not the same magnitude. In general, at large scales the latter vary more than the former, especially at short distances. Therefore, the geopedologic approach may perform better at smaller than at larger scales. Rossiter (2000) considers that the approach is adequate for semi-detailed studies (scales 1:35,000–1:100,000). Esfandiarpoor Borujeni et al. (2009) analyzed the effect of three observation point intervals (125 m, 250 m, and 500 m) on the results of applying the geopedologic approach to soil mapping and concluded that this approach works satisfactorily in reconnaissance or exploratory surveys. To increase the accuracy of the geopedologic results at large scales, they suggest adding a category of landform phase. The geoform classification system already includes the concept of phase for any practical subdivision of a landform or of any geoform class at other categorial levels (Zinck 1988). Using statistical and geostatistical methods, Esfandiarpoor Borujeni et al. (2010) show that the means of the soil variables in similar landforms within their study area were comparable but not their variances. They conclude that the geopedologic soil mapping approach is not completely satisfactory for detailed mapping scales (1,10,000–1: 25,000) and suggest, as above, the use of landform phases to increase the accuracy of the geopedologic results.

Similarly, the geoform-soil integration facilitates the extrapolation of information obtained in sample areas to unvisited areas or areas of difficult access, using artificial neural networks and decision trees, among other techniques (Moonjun et al. 2010; Farshad et al. 2013). Using a set of terrain parameters extracted from a digital elevation model, Hengl and Rossiter (2003) show that supervised landform classification allowed extrapolating geopedologic information obtained from photo-interpretation of selected sample areas over a large hill and plain region with about 90% reproducibility.

The geomorphic context is far from embracing the full span of soil variability. However, its contribution to soil cartography decreases in general the amplitude of variation of the soil properties within map units enough to make practical interpretations and decisions for land use planning. Systematic soil surveys using the geopedologic approach in large areas have performed satisfactorily when used for general land evaluation. Specific applications such as precision farming or site engineering need to be supported by very detailed soil information.

4.3.2.2 Soil Genesis and Stratigraphy

Geomorphic processes and environments are used, respectively, as factors and spatial frameworks to explain soil formation and evolution. The geomorphic context, through parent material (weathering products or depositional materials), relief (slope, relative elevation, aspect), drainage conditions, and morphogenesis, controls a large part of the soil forming factors and processes. In return, the soil properties influence the geomorphic processes. There is co-evolution between the pedologic and geomorphic domains. At the same time, the geomorphic history controls soil stratigraphy, while soil dating (i.e., chronosequences) helps reconstruct the evolution of the geomorphic landscape. The use of geomorphic research methods and techniques contributes to elucidate issues in soil genesis and stratigraphy.

Figure 4.8 shows a model of geopedologic relationships in a chronosequence of nested alluvial terraces, in the Guarapiche river valley, Venezuela (Zinck 1970). The geoform, here at the categorial level of terrain form (see Table 4.1), controls soil formation in two directions. On the one hand, the relative age of the geomorphic material, i.e., the parent material of the soils, from Holocene (Q0) to lower Pleistocene (Q4), directly influences the *degree* of pedogenic development from the level of Entisol to that of Ultisol. On the other hand, the nature of the geomorphic position closely influences the *type* of pedogenic development, distinguishing between well drained soils with ustic regime in levee position and poorly drained soils with aquic regime in basin position.

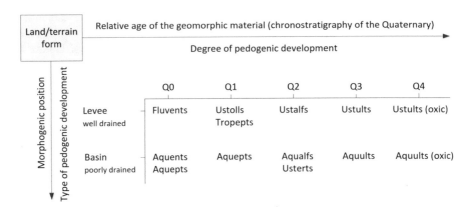

Fig. 4.8 Model of geopedologic relationships in alluvial soils, Guarapiche river valley, Venezuela. (Zinck 1970)

4.3.3 Geopedologic Integration: A Test of Numerical Validation

4.3.3.1 Materials and Method

The contribution of geomorphology to soil knowledge and, in particular, to the spatial distribution of soils can be considered efficient if, among other things, it facilitates and improves the grouping of the soils into relatively homogeneous cartographic units. To substantiate the geopedologic integration and validate quantitatively the relationships between geoform and soil, the technique of numerical classification was implemented, as the latter allows comparing the performance of an object classification system in relation to a reference system (Sokal and Sneath 1963).

A numerical classification test of the geopedologic units supplied by a semi-detailed soil survey (1:25,000) of the Guarapiche river valley, northeast of Venezuela (Zinck and Urriola 1971), was run to estimate the efficiency of both the soil classification and the geoform classification in building consistent groups by comparison with the phenetic groups of the numerical classification (Zinck 1972). The geopedologic units belong to a chronosequence of nested terraces, spanning the Quaternary from the lower Pleistocene (Q4) to the Holocene (Q0). Soils have formed mostly from longitudinal alluvial deposits, coming from the upper catchment area of the river, and secondarily from local colluvial deposits (Fig. 4.9).

The boundaries of the cartographic units are essentially of geomorphic nature, while their contents are of pedologic nature (consociations and associations of soil series, not shown here). Extract of the original soil map at 1:25,000 scale (Zinck and Urriola 1971).

Twenty-six pairs of modal pedons-morphons, representative of the soil series mapped in the survey area, were chosen, and 24 mechanical, physical and chemical properties were selected to characterize the pedologic material (solum) and the geomorphic material (parent material). Soil units classified at subgroup level

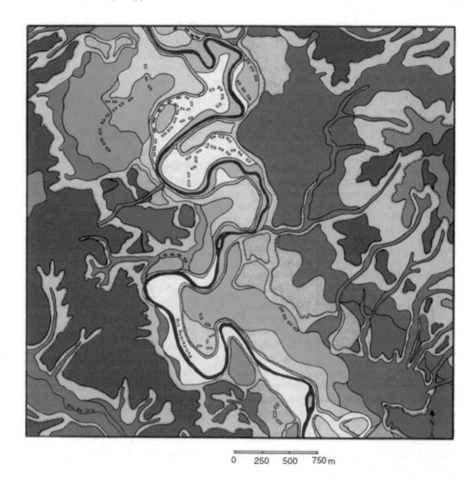

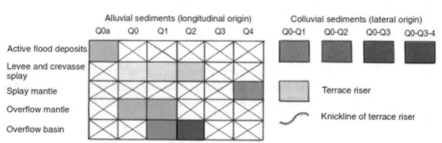

Fig. 4.9 Portion of the Guarapiche river valley, northeast of Venezuela, showing a chronosequence of nested terraces covering the Quaternary period (from Q0 to Q4)

(Soil Survey Staff 1960, 1967) and geomorphic units classified by depositional facies and relative age were compared. Data handling implemented techniques and methods available in the 1960s when the essay was performed: (1) the method of Hole and Hironaka (1960) for estimating the index of similarity between pairs of

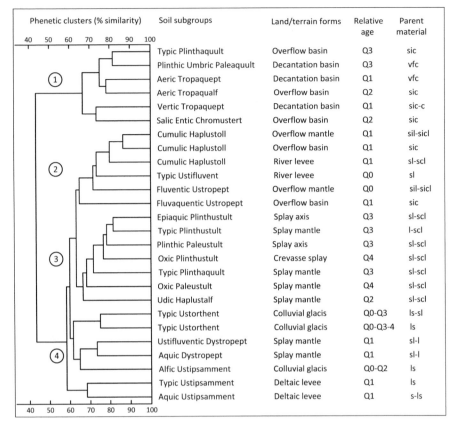

Fig. 4.10 Dendrogram showing four groups of geopedologic units; Guarapiche river valley, Venezuela. (Zinck 1972)
Soil classification according to Soil Survey Staff (1960, 1967)
Relative age of the geomorphic material (i.e., soil parent material) by increasing order from Q0 (Holocene) to Q4 (lower Pleistocene)
Texture of the parent material: s = sand; l = loam; si = silt; c = clay; vf = very fine

units and elaborating the similarity matrix, and (2) the method using unweighted pair-groups with arithmetic mean as described in Sokal and Sneath (1963) to cluster the units, construct the dendrogram represented in Fig. 4.10, and calculate the average similarities.

4.3.3.2 Results

The numerical classification generated four phenetic groups with a variable number of geopedologic units (i.e., soil-geoform combinations). The soils are reported as subgroup classes. Geoforms are identified by their sedimentary position at the terrain form level, their relative age, and the texture of the depositional material (i.e., the parent material of the soils).

- Group 1: six geopedologic units that share the following characteristics: low topographic positions of overflow basin (three) or decantation basin (three), poorly drained (five units with aquic regime), and fine-textured (silty clay or clay), regardless of the chronostratigraphy of the parental materials (relative age varying from Q1 to Q3) and the degree of soil development (one Vertisol, two Inceptisols, one Alfisol, two Ultisols).
- Group 2: six geopedologic units that share the following characteristics: medium to high topographic positions of levee (two), overflow mantle (two), and over-flow basin (two), well drained, textures mostly loamy and silty, soils of incipient to moderate development (one Entisol, two Inceptisols, three Mollisols), all formed from recent to relatively recent materials (Q0 and Q1).
- Group 3: seven geopedologic units that share the following characteristics: medium to high topographic positions of splay axis, splay mantle and crevasse splay, moderately well to well drained, textures sandy loam and sandy clay loam, soils of advanced development (one Alfisol, six Ultisols), all formed from old materials (Q3 and Q4).
- Group 4: seven geopedologic units with predominantly sandy textures (loamy sand and sandy loam) that restrict soil development to an incipient stage (five Entisols including three Psamments, two Inceptisols); the soils occur in a variety of depositional sites (deltaic levee, splay mantle, colluvial glacis) and chro-nostratigraphic units (from Q0 to Q4; the colluvial deposits being of continuous, diachronic formation).

In all cases, the factor that most closely controls the grouping of the geopedologic units is of geomorphic nature, with specific leading factors clustering the soils in each group:

- Group 1: basin depositional facies and low position in the landscape.
- Group 2: relatively recent age of the parental materials (late Pleistocene to Holocene).
- Group 3: advanced age of the parental materials (lower to early middle Pleistocene).
- Group 4: coarse textures of the parent materials.

4.3.3.3   Conclusion

Mean similarities of great soil groups (73%) and terrain forms (75%) are compara-ble to the average similarity of the numerical groups (75%), indicating that the three classification modes are relatively efficient in generating consistent groupings. Groups 2 and 3 are more homogeneous than groups 1 and 4. The factors that most contribute to differentiate the four groups and generate differences within the het-erogeneous groups are attributes of the geoforms, in particular their depositional origin (with their particle size distribution), their position in the landscape, and their relative age. These factors basically correspond to three of the five soil forming factors: i.e., parent material, topography-drainage, and time, which together highlight the contribution of geomorphology to pedology and constitute the foundation of geopedology.

4.4 Operational Relationships

4.4.1 Introduction

The conceptual and methodological relationships between geoform and soil can be implemented basically in two ways: (1) through studies at representative sites, usually of limited extent, to analyze in detail the genetic relationships between geoforms and soils (scientific studies, mostly in the academic domain), and (2) through the inventory of the soils as a resource to establish the soil cartography of a territory (project area, region, entire country) and assess their use potential and limitations (practical studies, in the technical domain).

The operational relationships are examined here in the framework of the soil inventory, from the generation of the geopedologic information through field survey to its interpretation through land evaluation for multi-purpose uses. In this process, geomorphology can play a relevant role. The operational importance of geomorphology refers to the value added to the soil survey information when geomorphology is incorporated into the successive stages of the survey operation.

Soil survey is an information system, which can be represented by a model that describes its structure and functioning using systems analysis, and which allows to estimate the efficiency of the contribution of geomorphology to the soil survey. The opportunity to conduct a trial of this nature was given by a semi-detailed soil survey project to be carried out in the basin of Lake Valencia, Venezuela (Zinck 1977). This is a region of approximately 1000 km^2 of flat land bordered by mountains, traditionally used for intensive irrigated agriculture, but increasingly exposed to land-use conflicts as a result of fast, uncontrolled urban-industrial sprawling. The size of the study area, the level of detail of the survey, the diversity of objectives to meet, and the number of personnel involved, were decisive factors in the design of the study. A reference framework was needed to plan the survey activities, establish the timetable for implementation, and select the variety of soil interpretations required to supply the necessary information for land-use planning and contribute to mitigate the land-use conflicts.

4.4.2   The Structure of the Soil Survey

Proceeding by iteration, a model structure with five categorial levels was obtained, as represented in Fig. 4.11. The three lower levels comprise the domain proper of the soil survey – its internal area – where the information is produced. The two upper levels represent the sphere of influence of the soil survey – its external area – where the information generated is implemented. Each level responds to a generic concept, and, at each level, a series of tasks is performed (Tables 4.2, 4.3, 4.4, 4.5 and 4.6).

The numbers in the boxes refer to the themes labelled in Tables 4.2, 4.3, 4.4, 4.5 and 4.6. The numbers inserted in the arrows indicate the amount of critical pathways through which information circulates from a given level to the following one.

- Level 1: elementary tasks, which consist in the generation of the basic data, including the interpretation of aerial photos, satellite images and DEM, soil description and sampling, laboratory determinations, and gathering of agronomic, social, and economic data.

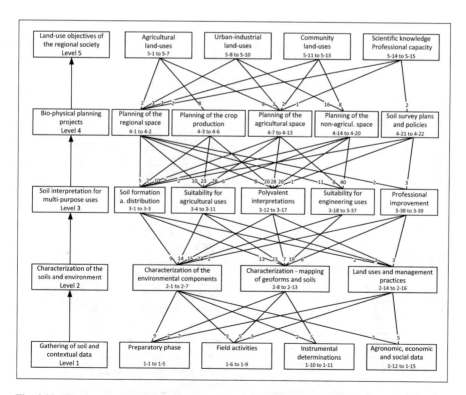

Fig. 4.11 Graph representing the soil survey as an information system, with production, interpretation, and dissemination of data and information, Lake Valencia project. (Zinck 1977)

Table 4.2 Level 1 themes: elementary soil study tasks; information collection. Lake Valencia project (Zinck 1977)

1–1	Collection and analysis of existing no-pedologic information
1–2	Photo-field exploration, analysis of existing soil information, identification soil legend
1–3	Generalized 1: 50,000 photo-interpretation, identification of the physical-natural macro-units
1–4	Selection of the sample areas
1–5	Detailed 1: 25,000 photo-interpretation, identification of the geoforms, location of the sample areas
1–6	Survey of the sample areas
1–7	Control observations, photo-interpretation adjustments
1–8	Composition of the cartographic units, descriptive soil legend
1–9	Description of representative pedons
1–10	Physical field determinations and measurements
1–11	Laboratory determinations
1–12	Survey of crop yields, production costs, and development costs
1–13	Survey of irrigation practices
1–14	Survey of cultivation and conservation practices
1–15	Evaluation of deforestation, levelling, drainage, stone-removal costs.

Table 4.3 Level 2 themes: intermediate soil study tasks; synthesis of the information on soil and environment characterization. Lake Valencia project (Zinck 1977)

2–1	Characterization of the climate.
2–2	Characterization of the surface hydrology and hydrography.
2–3	Characterization of existing hydraulic works.
2–4	Characterization of the water quality.
2–5	Characterization of the topography.
2–6	Characterization of the geology and hydrogeology.
2–7	Characterization of the geomorphology and hidrogeomorphology.
2–8	Geopedologic mapping and soil map preparation.
2–9	Morphologic characterization of the soils.
2–10	Chemical characterization of the soils.
2–11	Mineralogical characterization of the soils.
2–12	Physical characterization of the soils.
2–13	Mechanical characterization of the soils.
2–14	Survey of current land-uses.
2–15	Survey of management practices and levels.
2–16	Evaluation of required improvements and their feasibility.

Table 4.4 Level 3 themes: final soil study tasks; multi-purpose interpretations. Lake Valencia project (Zinck 1977)

3–1	Overall characterization of the natural environment (integrated study).
3–2	Spatial distribution of the soils (soil chorology).
3–3	Genesis and taxonomic classification of the soils.
3–4	Land suitability for rainfed agriculture.
3–5	Land suitability for irrigated agriculture.
3–6	Land suitability for ornamental plants and garden vegetables.
3–7	Agricultural productivity (productivity of the land).
3–8	Development costs for agricultural land-use.
3–9	Current soil fertility.
3–10	Soil salinity.
3–11	Limitations of the land for the use of mechanized farm implements.
3–12	Characterization of the natural drainage.
3–13	Drainability of the land.
3–14	Current morphodynamics (erosion, sedimentation).
3–15	Erodibility of the land.
3–16	Land irrigation requirements.
3–17	Water availability.
3–18	Sources of material for topsoil.
3–19	Sources of sand and gravel.
3–20	Sources of material for road filling.
3–21	Constraints for road network design.
3–22	Limitations for road cuts.
3–23	Limitations for placement of cables and pipes.
3–24	Limitations for foundations of low buildings and houses.
3–25	Limitations for embankment foundations.
3–26	Limitations for residential areas.
3–27	Limitations for streets and parking lots.
3–28	Limitations for excavation of channels.
3–29	Limitations for construction of farm ponds.
3–30	Limitations for construction of dikes.
3–31	limitations for septic filtration areas.
3–32	Limitations for oxidation ponds.
3–33	Limitations for waste disposal areas.
3–34	Limitations for recreation areas (picnic, play grounds).
3–35	Limitations for lawns, golf courses, landscaping.
3–36	Limitations for camping sites.
3–37	Limitations for sports fields.
3–38	Training of the technical personnel.
3–39	Publications, conferences, education.

Table 4.5 Level 4 themes: regional planning and development projects, designed and executed by official and private entities. Lake Valencia project (Zinck 1977)

4–1	Soil correlation.
4–2	Land-use zoning in the regional space (arbitration between competitive uses).
4–3	Ecological zoning of crops.
4–4	Selection of crop and rotation systems.
4–5	Substitution of crops in time and space.
4–6	Increase of land productivity (yields).
4–7	Determination of agricultural plot sizes.
4–8	Irrigation planning and management.
4–9	Improvement of poorly drained soils.
4–10	Improvement of saline soils.
4–11	Management of heavy soils (clay soils).
4–12	Soil conservation techniques.
4–13	Agricultural extension.
4–14	Urban and peri-urban planning (master zoning plan).
4–15	Supply of water and gas.
4–16	Control of soil and water pollution.
4–17	Disposal or recycling of industrial, urban, and agricultural wastes.
4–18	Channeling and excavation of effluents.
4–19	Planning of communication routes.
4–20	Tourism development.
4–21	Professional training and improvement.
4–22	Expanding basic knowledge in geomorphology and pedology.

Table 4.6 Level 5 themes: relevant technical issues faced by the regional (or national) community. Lake Valencia project (Zinck 1977)

5–1	Marginal agriculture.
5–2	Land reform.
5–3	Intensification processes of agriculture.
5–4	Incorporation of new areas to agricultural activities.
5–5	Supply of agricultural products for human consumption.
5–6	Supply of special agricultural products (flowers, out-of-season crops).
5–7	Supply of raw agricultural materials for the industry.
5–8	Creation of industrial zones.
5–9	Urbanization processes (cities, towns, secondary residences).
5–10	Transport of people, products, energy, and information.
5–11	Areas for recreation and tourism (water bodies, areas for outdoor activities and sports).
5–12	Protected areas (parks, reserves, green areas).
5–13	Environmental conservation, protection, and improvement.
5–14	Enlargement of the technical capacity of the regional community.
5–15	Increase in basic scientific knowledge.

- Level 2: intermediate tasks, which consist in the synthesis of the information, including the characterization of the environmental components, characterization and mapping of the geoforms and soils, and description of the land-use types and management practices.
- Level 3: final tasks, which consist in the interpretation of the information for multiple purposes, including the genetic interpretation of the soils and their formation environments, land evaluation for agricultural, engineering, sanitary, recreational and aesthetic purposes, and professional improvement of the geopedologist.
- Level 4: primary external objectives, which correspond to biophysical planning in the local and regional contexts, including territorial zoning, planning of the agricultural and non-agricultural areas, planning of the agricultural production, and formulation of soil survey policies and plans.
- Level 5: final external objectives, which correspond to the concerns, perceptions, and priorities of the regional (or national) society in terms of agricultural land-use, urban-industrial land-use, use of community spaces, and creation of scientific knowledge and improvement of professional skills.

4.4.3 The Functioning of the Soil Survey

The operation of the system refers to the information flows that circulate through the soil survey. To identify the direction of the information flows and evaluate their intensity, several matrices relating the themes of the consecutive layers of the model were built. The matrices were subjected to the judgement of a team of ten experts in soil survey, who identified the relationships between the themes of pairs of levels and assessed the intensity of these relationships through a rating procedure using two score ranges: 0–9 for the internal area and 0–2 for the external area. The individual estimates were averaged to get the direction and intensity of the information flows. This resulted in a complex graph of flows that is shown simplified in Fig. 4.11. The graph indicates the orientation and the amount of flows (critical pathways) that connect each theme with others. The combination of the two criteria of orientation and number of flows allowed establishing a ranking of the soil survey tasks according to their importance in generating or transmitting information.

4.4.4 The Contribution of Geomorphology to Soil Survey

The direct contribution of geomorphology takes place at levels 1 and 2.

- Level 1: geomorphology contributes to the tasks of photo-interpretation, selection of sample areas, identification of representative sites, and delineation of the geopedologic units.

- Level 2: geomorphic synthesis is one of the most prolific themes of the system by the number of flows issued and the number of themes reached at level 3 (30 themes). Based on this performance, the geomorphic synthesis ranked as the most efficient theme of level 2, along with the topography theme.

Thus, the incorporation of geomorphology helps streamline, speed up and improve the soil survey. Unfortunately, nowadays soil inventory is not given priority on political agendas, despite the severe risks of degradation of the soil resource.

4.5 Conclusions

In addition to promoting integration between geomorphology and pedology, geopedology focuses on the contribution of the former to the latter for soil mapping and understanding of soil formation. This contribution is based on the following.

- The geoforms and other geomorphic features, including processes of formation, aggradation and degradation, can be recognized by direct observation in the field and by interpretation of remote-sensed documents (aerial photographs and satellite images) and products derived therefrom (e.g., DEM). Documents that allow stereoscopic vision to have the advantage of providing the third dimension of the geoforms in terms of volume and topographic variations. In this regard, aerial photographs are still the more faithful and explicit documents for the interpretation of the relief at large and medium scales.
- Many geoforms have relatively discrete boundaries, facilitating their delimitation. This is particularly the case of constructed geoforms in depositional systems (e.g., geoforms of alluvial, glacial, and eolian origin) and, to a lesser extent, those built in morphogenic systems controlled by endogenous processes (e.g., geoforms of volcanic and structural origin). By contrast, hillsides frequently show continuous variations, which can be approximated using the slope facet models.
- Geoforms are generally distributed in landscape systems controlled by a dominant forming agent (e.g., water, ice, wind). The foregoing results in families of geoforms associated in characteristic patterns that repeat in the landscape. This allows interpolating/extrapolating information in mapping areas and predicting the occurrence of geopedologic units at unvisited sites.
- Geoforms are relatively homogeneous at a given categorial level and with respect to the properties that are diagnostic at this level. The hypogeal component, corresponding to the morphogenic and morphostratigraphic features of the material, is usually more homogeneous than the epigeal component, corresponding to the morphographic and morphometric features of the terrain surface. The nonrandom, systematic variations of the soil mantle are frequently of geomorphic nature.
- The geomorphic context is an important framework of soil genesis and evolution, covering three of the five classic soil forming factors, namely the features of the

relief-drainage compound, the nature of the parent material, and the age of the geoform. Many soils have not formed directly from the hard bedrock, but rather from the geomorphic cover material (e.g., unconsolidated sediments, slope materials in translation, regolith, weathering layers).

- To sum up the foregoing, geomorphic analysis enables segmenting the continuum of the physiographic landscape into spatial units that are frameworks for (1) interpreting soil formation along with the influence of biota, climate and human activity, (2) composing the soil cartographic units, and (3) analyzing the spatial variations of the soil properties.

The geopedologic approach is essentially descriptive and qualitative. Geoforms and soils are considered as natural bodies, which can be described by direct observation in the field and by interpretation of aerial photos, satellite images, topographic maps, and digital elevation models. The approach relies on a combination of basic knowledge in geomorphology and pedology, incremented by working experience, in particular the experience gained from the practice of field observation and landscape reading. Expert knowledge, the acquisition and development of which constitute an inherent process in human societies in evolution, represents a source of cognitive richness that is nowadays attempted to be formalized before it disappears. Expert knowledge has been considered as a factor of subjectivity (Hudson 1992) and personal bias (McBratney et al. 1992) in the conventional practice of soil survey, in contrast to the pedometric (digital) soil mapping which would be more objective (Hengl 2003). Geopedology is a conventional approach with the particularity and advantage that bias, and subjectivity can be minimized or compensated by the systematic and integrated use of geomorphic criteria. Geoforms provide a comprehensive cartographic framework for soil mapping, which goes beyond the mere morphometric terrain characterization. However, both modalities, the qualitative and the quantitative, can be usefully combined. Geopedologic units are reference units for more detailed geostatistical studies and for the spatial control of the digital data that are used to measure soil and geoform attributes. "The full potential of (digital) terrain analysis in soil survey will be realized only when it is integrated with field programs with a strong emphasis on geomorphic and pedologic processes" (McKenzie et al. 2000).

References

Amiotti N, Blanco MC, Sanchez LF (2001) Complex pedogenesis related to differential aeolian sedimentation in microenvironments of the southern part of the semiarid region of Argentina. Catena 43:137–156

Arnold R, Schargel R (1978) Importance of geographic soil variability at scales of about 1: 25,000. Venezuelan examples. In: Drosdoff M, Daniels RB, Nicholaides JJ III (eds) Diversity of soils in the tropics, vol 34. ASA Special Publication, pp 45–66

Bocco G, Velázquez A, Mendoza ME, Torres MA, Torres A (1996) Informe final, subproyecto regionalización ecológica, proyecto de actualización del ordenamiento ecológico general del territorio del país. INE-SEMARNAP, México

Bregt AK, Bouma J, Jellineck M (1987) Comparison of thematic maps derived from a soil map and from kriging of point data. Geoderma 39:281–291

Buol SW, Hole FD, McCracken RJ, Southard RJ (1997) Soil genesis and classification, 4th edn. Iowa State University Press, Ames IA

Campy M, Macaire JJ (1989) Géologie des formations superficielles. Géodynamique, faciès, utilisation. Masson, Paris

Esfandiarpoor Borujeni I, Salehi MH, Toomanian N, Mohammadi J, Poch RM (2009) The effect of survey density on the results of geopedological approach in soil mapping: a case study in the Borujen region, Central Iran. Catena 79:18–26

Esfandiarpoor Borujeni I, Mohammadi J, Salehi MH, Toomanian N, Poch RM (2010) Assessing geopedological soil mapping approach by statistical and geostatistical methods: a case study in the Borujen region, Central Iran. Catena 82:1–14

Farshad A, Shrestha DP, Moonjun R (2013) Do the emerging methods of digital soil mapping have anything to learn from the geopedologic approach to soil mapping or vice versa? In: Shahid SA, Taha FK, Abdelfattah MA (eds) Developments in soil classification, land use planning and policy implications: innovative thinking of soil inventory for land use planning and management of land resources. Springer, Dordrecht, pp 109–131

Fridland VM (1974) Structure of the soil mantle. Geoderma 12:35–41

Fridland VM (1976) Pattern of the soil cover. Israel Program for Scientific Translations, Jerusalem

Hall GF, Olson CG (1991) Predicting variability of soils from landscape models. In: Mausbach MJ, Wilding LP (eds) Spatial variabilities of soils and landforms, vol 28. SSSA Special Publication, pp 9–24

Hengl T (2003) Pedometric mapping. Bridging the gaps between conventional and pedometric approaches. ITC Dissertation 101, Enschede

Hengl T, Rossiter DG (2003) Supervised landform classification to enhance and replace photo-interpretation in semi-detailed soil survey. Soil Sci Soc Am J 67:1810–1822

Hole FD, Campbell JB (1985) Soil landscape analysis. Rowman & Allanheld, Totowa

Hole FD, Hironaka M (1960) An experiment in ordination of some soil profiles. Soil Sci Soc Am Proc 24(4):309–312

Hudson BD (1992) The soil survey as paradigm-based science. Soil Sci Soc Am J 56:836–841

Ibáñez JJ (1994) Evolution of fluvial dissection landscapes in Mediterranean environments: quantitative estimates and geomorphic, pedologic and phytocenotic repercussions. Z Geomorphologie 38:105–119

IUSS (2007) World reference base for soil resources. World soil resources report 103. IUSS Working Group WRB/FAO, Rome

Jenny H (1941) Factors of soil formation. McGraw-Hill, New York

Kerry R, Oliver MA (2011) Soil geomorphology: identifying relations between the scale of spatial variation and soil processes using the variogram. Geomorphology 130:40–54

McBratney AB, de Gruijter JJ, Brus DJ (1992) Spatial prediction and mapping of continuous soil classes. Geoderma 54:39–64

McKenzie NJ, Gessler PE, Ryan PJ, O'Connell DA (2000) The role of terrain analysis in soil mapping. In: Wilson JP, Gallant JC (eds) Terrain analysis. Principles and applications. John Wiley & Sons, New York, pp 245–265

Metternicht G, Zinck JA (1997) Spatial discrimination of salt- and sodium-affected soil surfaces. Intl J Remote Sensing 18(12):2571–2586

Moonjun R, Farshad A, Shrestha DP, Vaiphasa C (2010) Artificial neural network and decision tree in predictive soil mapping of Hoi Num Rin sub-watershed, Thailand. In: Boettinger JL, Howell DW, Moore AC, Hartemink AE, Kienast-Brown S (eds) Digital soil mapping: bridging research, environmental application, and operation. Springer, New York, pp 151–163

Pennock DJ, Corre MD (2001) Development and application of landform segmentation procedures. Soil & Tillage Res 58:151–162

Phillips JD (2001) Divergent evolution and the spatial structure of soil landscape variability. Catena 43:101–113

Principi P (1953) Geopedologia (Geologia Pedologica). Studio dei terreni naturali ed agrari. Ramo Editoriale degli Agricoltori, Roma

Rossiter DG (2000) Methodology for soil resource inventories. Lecture Notes, 2nd revised version. International Institute for Aerospace Survey and Earth Sciences (ITC), Enschede

Saldaña A, Stein A, Zinck JA (1998) Spatial variability of soil properties at different scales within three terraces of the Henares valley (Spain). Catena 33:139–153

Saldaña A, Ibáñez JJ, Zinck JA (2011) Soilscape analysis at different scales using pattern indices in the Jarama-Henares interfluve and Henares River valley, Central Spain. Geomorphology 135:284–294

Salgado-Labouriau ML (1980) A pollen diagram of the Pleistocene-Holocene boundary of Lake Valencia, Venezuela. Rev Palaeobotany and Palynology 30:297–312

Schlichting E (1970) Bodensystematik und bodensoziologie Z Pflanzenernähr Bodenk 127(1):1–9

Soil Survey Staff (1960) Soil classification: a comprehensive system. 7th Approximation. US Gov Print Of, Washington, DC

Soil Survey Staff (1967) Supplement to soil classification system (7th Approximation). Soil Cons Serv, Washington, DC

Soil Survey Staff (1993) Soil survey manual. In: US Dept Agric Handbook 18. US Government Print Office, Washington, DC

Soil Survey Staff (1999) Soil taxonomy. In: US Dept Agric Handbook 436. US Government Print Office, Washington, DC

Sokal RR, Sneath PHA (1963) Principles of numerical taxonomy. Freeman, San Francisco

Toomanian N (2013) Pedodiversity and landforms. In: Ibáñez JJ, Bockheim J (eds) Pedodiversity. CRC Press, Taylor & Francis Group, Boca Raton, pp 133–152

Tricart J (1972) La terre, planète vivante. Presses Universitaires de France, Paris

Wilding LP, Drees LR (1983) Spatial variability and pedology. In: Wilding LP, Smeck NE, Hall GF (eds) Pedogenesis and soil taxonomy. I Concepts and interactions. Elsevier, Amsterdam, pp 83–116

Wilding LP, Lin H (2006) Advancing the frontiers of soil science towards a geoscience. Geoderma 131:257–274

Zinck JA (1970) Aplicación de la geomorfología al levantamiento de suelos en zonas aluviales. Ministerio de Obras Públicas (MOP), Barcelona

Zinck JA (1972) Ensayo de clasificación numérica de algunos suelos del Valle Guarapiche, Estado Monagas, Venezuela. IV Congreso Latinoamericano de la Ciencia del Suelo (resumen), Maracay

Zinck JA (1974) Definición del ambiente geomorfológico con fines de descripción de suelos. Cagua, Ministerio de Obras Públicas (MOP)

Zinck JA (1977) Ensayo sistémico de organización del levantamiento de suelos. Maracay, Ministerio del Ambiente y de los Recursos Naturales Renovables (MARNR)

Zinck JA (1986) Una toposecuencia de suelos en el área de Rancho Grande. Dinámica actual e implicaciones paleogeográficas. In: Huber O (ed) La selva nublada de Rancho Grande, Parque Nacional "Henri Pittier". El ambiente físico, ecología vegetal y anatomía vegetal. Fondo Editorial Acta Científica Venezolana y Seguros Anauco CA, Caracas, pp 67–90

Zinck JA (1988) Physiography and soils. In: Lecture notes. International Institute for Aerospace Survey and Earth Sciences (ITC), Enschede

Zinck JA, Urriola PL (1971) Estudio edafológico Valle Guarapiche, Estado Monagas. Barcelona, Ministerio de Obras Públicas (MOP)

Zinck JA, Valenzuela CR (1990) Soil geographic database: structure and application examples. ITC Journal 1990(3):270–294

Chapter 5
The Pedologic Landscape: Organization of the Soil Material

J. A. Zinck

Abstract The soil material is organized from structural, geographic, and genetic points of view. Structurally, the soil material is multiscalar with features and properties specific to each scale level. The successive structural levels are embedded in a hierarchic system of nested soil entities or holons known as the holarchy of the soil system. At each hierarchic level of perception and analysis of the soil material, distinct features are observed that are particular to the level considered. The whole of the features describes the soil body in its entirety. Each level is characterized by an element of the soil holarchy, a unit (or range of units) measuring the soil element perceived at this level, and a means of observation or measurement for identifying the features that are diagnostic at the level concerned. The levels are labelled based on a connotation with the proper dimension of the soil element into consideration at every level: nano, micro, meso, macro, and mega. The holarchy of the soil system allows highlighting relevant relationships between soil properties and geomorphic response at different hierarchic levels. These relationships form the conceptual essence of geopedology.

Keywords Hierarchic levels · Soil reactions · Micromorphologic components · Soil horizons · Pedon · Polypedon

5.1 Introduction

The soil material is organized from structural, geographic, and genetic points of view. Structurally, the soil material is multiscalar with features and properties specific to each scale level. The successive structural levels are embedded in a hierarchic system of nested soil entities, or holons, that Haigh (1987) has called the

---

J. A. Zinck died before publication of this work was completed.

J. A. Zinck (deceased)
Faculty of Geo-Information Science and Earth Observation (ITC), University of Twente, Enschede, The Netherlands

© The Author(s), under exclusive license to Springer Nature
Switzerland AG 2023
J. A. Zinck et al. (eds.), *Geopedology*,
https://doi.org/10.1007/978-3-031-20667-2_5

holarchy of the soil system (Fig. 5.1). Geographically, the soil material is not randomly distributed on the landscape; instead, it is organized according to spatial distribution patterns under the control of the soil forming factors (Fridland 1974, 1976; Hole and Campbell 1985). Genetically, the soil material is formed and develops as an open system of exchanges and transformations of matter and energy (Jenny 1941; Simonson 1959).

Hereafter, a model similar to Haigh's holarchy is used to introduce some basic soil notions and analyze their relationships with the geopedologic approach at various scalar levels (Table 5.1). This scheme of nested holons is a condensate of pedology ranging from molecular reactions to the (geo)pedologic landscape. At each hierarchic level of perception and analysis of the soil material, distinct features are observed that are particular to the level considered. The whole of the features describes the soil body in its entirety. At each level correspond an element of the soil holarchy, a unit (or range of units) measuring the soil element perceived at this level, and a means of observation or measurement for identifying the features that are diagnostic at the level concerned. The levels are labelled based on a connotation with the proper dimension of the soil element into consideration at every level: nano, micro, meso, macro, and mega (Table 5.1).

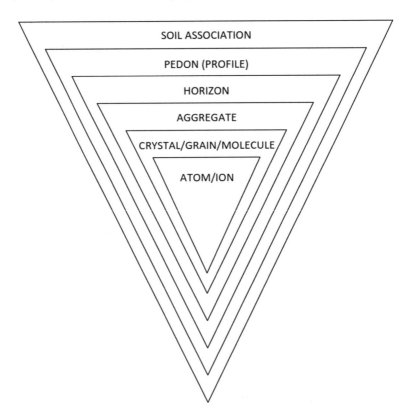

Fig. 5.1 The holarchy of the soil system. (Adapted from Haigh 1987)

Table 5.1 Hierarchic levels of the soil system (Zinck 1988)

Level	Unit	Concept	Soil feature
Nano	nm–μm	Particle	Basic soil reactions
Micro	μm–mm	Aggregate	Micromorphologic structure
Meso	mm–cm–dm	Horizon	Differentiation of the soil material
Macro	m	Pedon	Soil volume for description and sampling
Mega	m–km	Polypedon	Soil classification and mapping – (geo)pedologic landscape

5.2   Nano-level

At the nano-level, the soil material is considered in its elementary form of molecules and combinations of molecules into particles, which can be either identified through chemical reactions, or observed using an electron microscope, or determined by X-ray diffraction. At this level take place the basic reactions of the soil material: chemical, mechanical, and physico-chemical. These reactions control processes and features such as rock weathering and soil formation, but also mass movements and other erosion phenomena that have the particularity of manifesting and taking visual expression at coarser levels of perception.

5.2.1   Chemical Reactions

The chemical reactions, which take place in the soil material as well as in the parent material (hard rock or unconsolidated sediment) to transform the latter into soil material, operate in two modalities: (1) by solubility changes of the chemical compounds in the salts, carbonates, and silicates, and (2) by structural changes in the oxide minerals.

- Solution (salts): $NaCl + H_2O \Leftrightarrow Na^+ + Cl^- + H_2O$
- Carbonation (carbonates): $CO_2 + H_2O \Rightarrow HCO_3^- + H^+$
$$CaCO_3 + (HCO_3^- + H^+) \Rightarrow Ca\,(HCO_3)^2$$
- Hydrolysis (silicates): $KAlSi_3O_8 + HOH \Rightarrow HAlSi_3O_8 + KOH$
- Hydration (oxides): $2Fe_2O_3 + 3H_2O \Rightarrow 2Fe_2O_3 * 3H_2O$
- Oxido-reduction (oxides): $4FeO + O_2 \Leftrightarrow 2Fe_2O_3$

The performance of these reactions depends on the bioclimatic conditions, the nature of the substratum, and the type of relief and associated drainage conditions, among other factors. These are basic processes of rock weathering, alteration of unconsolidated materials, and formation of pedogenic material. Some processes operate only in specific geopedologic environments. For instance, the dissolution, concentration and, eventually, (re)crystallization of salts and the resulting geoforms are typical of halomorphic conditions in coastal and dry inland areas. Likewise, the dissolution of carbonates into bicarbonates and the mobilization of the latter are typical of calcimorphic conditions and responsible, in particular, for the formation of karstic relief. The hydrolysis of potassium feldspar, favored by high humidity and high temperature in tropical environment, results in the formation of acid clay together with potassium hydroxide that is lost by lixiviation. Hydration makes iron oxide more fragile. Oxydo-reduction is a reversible process typical of the inter-tidal zone.

5.2.2 Mechanical Reactions

The mechanical reactions depend on the way particles are arranged and associated. Coarse particles have the tendency to pile up into different kinds of packing, while the behavior of the fine particles depends on the intensity of their agglomeration into various kinds of fabric. In general terms, these mechanical reactions of nano-level determine the susceptibility of the materials to mass movements, the geomorphic expression of which is visible on the landscape at coarser levels of perception (from meso to mega).

5.2.2.1   Types of Packing

Coarse particles including sand and coarse silt grains (2–0.02 mm) cluster in piles, the structure of which varies according to the degree of roundness of the grains. Rounded grains (e.g., sand grains of marine or eolian origin) usually present a cubic arrangement with limited contact surface and high porosity. This allows water to penetrate readily in the pore space, resulting in water pressure in the pores that tends to separate the grains. For this reason, the cubic packing is in general an unstable arrangement, which facilitates the process of moving sands (quicksands). Less rounded grains (e.g., sand grains of alluvial or colluvial origin) generally show a tetrahedral type of packing, with greater contact surface and lower porosity, which is a more stable arrangement. Irregular grains and rock fragments tend to be tightly interlocked, with large friction surface that ensures greater stability of the material.

5.2.2.2   Types of Fabric

The fabric arrangement of the fine particles, including clay and fine silt (<0.02 mm), depends on the mode and intensity of the contacts between particles in the soil solution. Various modes of particle association in clay suspensions are recognized, with four basic types of micro-mechanical fabric, ranging from the total absence of agglomeration (i.e. deflocculated state) to a strongly agglomerated condition (i.e. flocculated state), and a series of combinations of these basic types (Mitchell 1976) (Fig. 5.2). The fabric types are related to the moisture content in the soil, which determines the mechanical state of the material, from liquid to solid, and the consistence limits (i.e., Atterberg limits) between mechanical states. Obviously, the fabric

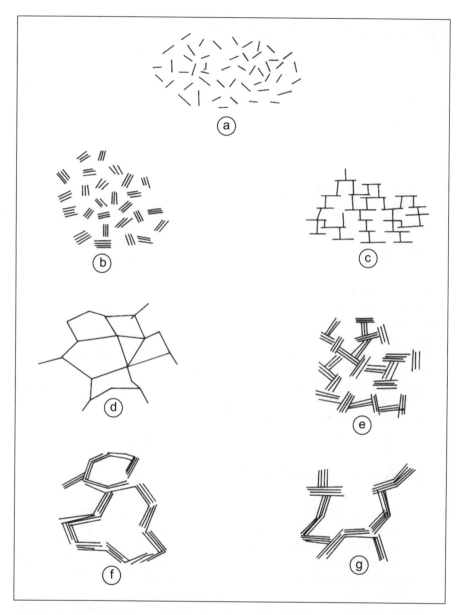

Fig. 5.2 Modes of particle association in clay suspensions (after van Olphen 1963): (**a**) dispersed and deflocculated; (**b**) aggregated but deflocculated; (**c**) edge-to-face flocculated but dispersed; (**d**) edge-to-edge flocculated but dispersed; (**e**) edge-to-face flocculated and aggregated; (**f**) edge-to-edge flocculated and aggregated; (**g**) edge-to-face and edge-to-edge flocculated and aggregated. (Adapted from Mitchell 1976)

Table 5.2 Influence of the fabric type and the consistence of the soil material in the generation of mass movements (the most likely to occur)

Fabric type	State of the material	Mass movement
Deflocculated	Liquid	Mudflow
Dispersed	Plastic	Solifluction
Aggregated	Semi-solid	Landslide
Flocculated	Solid	Metastability
Organization of the soil material	Soil property (consistence, Atterberg limits)	Morphogenic process (geomorphic response)

depends also on other factors such as the type of clay, organic matter content, and the presence of salts, among others.

In geopedologic terms, the fabric of the soil material plays an important role in the generation of mass movements (Table 5.2).

- Deflocculated state: all particles are individually in suspension in the soil solution, without interaction between particles. This fabric condition favors the occurrence of mudflows.
- Dispersed state: there are elementary associations between individual particles, essentially contacts between particle edges and faces. This fabric condition creates a risk of solifluction.
- Aggregated state: there are associations between particle clusters, creating a situation that favors the potential occurrence of landslides.
- Flocculated state: all kinds of contact between faces and between edges and faces take place, generating the most stable arrangement of particles in the soil solution and resulting in high soil strength and stability.

5.2.3 Physico-Chemical Reactions

The physico-chemical reactions are based on the colloidal properties of clay and humus. Both compounds have electronegative charges at the edges of the layers and in the space between layers. The electronegative charges attract cations with decreasing intensity according to the lyotropic sequence of preferential adsorption, which reflects the number of charges and the hydrated size of the cations: $Al^{+++} > Ca^{++} > Mg^{++} > K^+ = NH_4^+ > Na^+$. Divalent cations play an important role in establishing bridges between clay particles, which is a basic process for the formation of aggregates. The physico-chemical reactions that take place at the nano-level control soil fertility, aggregation, structural stability and its influence on soil susceptibility to erosion.

5.2.4   Relationship with Geopedology

The reactions taking place at the nano-level determine the fundamental processes of soil formation, evolution, differentiation, as well as degradation. The production of regolith through rock weathering, the alteration of the unconsolidated cover formations, and the transformation of these loose materials into soil material largely depend on the chemical and physico-chemical reactions that operate in the substratum - inherently the domain of geomorphology. The different mechanical reactions that take place in the soil material and regolith, according to variations in moisture content, control the morphogenesis by mass movements, the impact of which is directly visible in the landscape.

5.3   Micro-level

At the micro-level, the object of interest is the soil aggregate, which can be observed with the use of a petrographic microscope. This is the investigation domain of micromorphology. The observation of an aggregate in thin section under the petrographic microscope allows characterizing the micromorphologic structure of the soil matrix, both in its solid component and porous component, and identifying features derived from the addition of material and transformation of the matrix. Some of these micromorphologic characteristics are shown schematically in Fig. 5.3 and summarized in Table 5.3.

5.3.1   The Micromorphologic Components

At the micro-level, the soil material is divided into two main components: the soil matrix, which corresponds to the soil material in situ, and pedologic features. Each of these two components is subdivided into elements that play important roles in the functioning of the soil, including plasma, pore space, skeleton grains, and pedologic features (Table 5.3).

5.3.1.1   Skeleton Grains

The skeleton grains consist of:

- Mineral grains, essentially sand and silt grains, which constitute the inert soil material, without colloidal properties, that dominates in coarse-grained soils.
- Organic fragments, which are pieces of undecomposed organic material, essentially fragments of leaves, twigs, and branches (folic material), that dominates in the litter.

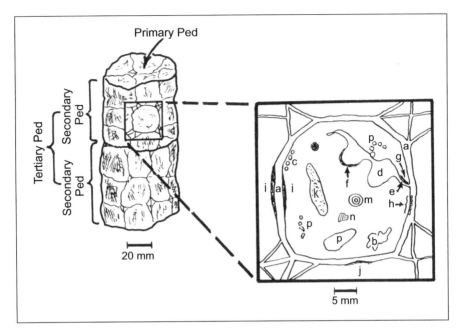

Fig. 5.3 Micropedologic features. Voids: (**a**) packing voids, (**b**) vugh, (**c**) vesicles, (**d**) chamber, (**e**) channel. Cutans: (**f**) chamber cutan, (**g**) channel cutan, (**h**) skeletans, (**i**) argillan or sesquan, (**j**) stress cutan. Other features: (**k**) pedotubule, (**1**) nodule, (**m**) concretions, (**n**) papule. Note that the S-matrix is the mass of plasma, skeleton grains (**p**), and voids. (Adapted from Buol et al. 1997)

Table 5.3 Micromorphologic organization of the soil material

Soil material	Soil matrix (S-matrix) (soil material in situ)	Solids	Skeleton grains (coarse material)
			Plasma (fine material)
		Pore space (voids, pores)	Vesicles
			Chambers, vughs
			Channels
			Planes
	Pedologic features (addition to or transformation of soil material)	Cutans	
		Glaebules	
		Tubules	
		Plasma separations	

5.3.1.2 Plasma

The plasma is the active phase of the solid material, where the chemical and physico-chemical reactions take place, and which controls the mechanical mobility of the fine particles. The plasma is endowed with relevant properties, among others:

- Colloidal property that provides the clay minerals and the humus with electro-negative charges.
- Solubility property that allows salts and carbonates to be converted into ions.
- Chelation property, thanks to which insoluble compounds (e.g., Fe and Al sesquioxides) can migrate in association with organic molecules.

5.3.1.3 Pores

Pores vary in configuration and location within and between aggregates, and for this reason fulfill different functions. Packing voids, vesicles, and chambers are examples of pore differentiation in the soil.

- Packing pores are located around the aggregates and control the permeability, with its influence on drainage, and the adhesion between aggregates.
- Vesicles are closed empty spaces, without active function.
- Chambers are pores open on one extremity, which retain moisture even when the soil appears to be dry; these are places where the microfauna (e.g., bacteria) responsible for the decomposition of the organic matter tends to concentrate, and where the oxido-reduction mechanisms responsible for hydromorphism take place.

5.3.1.4 Pedologic Features

Micromorphologic soil features derive essentially from the addition of new material to the soil and/or the transformation of the soil material in situ.

- The additions can be traced by the coatings (cutans) that form when fine particles move within the soil solution from eluvial horizons and deposit in the pores or on the surface of the aggregates in the underlying illuvial horizons. According to the nature of the constituents, different types of cutan are recognized, including clay cutans (argillans), iron cutans (ferrans), manganese cutans (manganans), etc.
- The transformations can be (1) physical: e.g., pressure faces (stress cutans) on the surface of the aggregates caused by contraction-expansion; (2) chemical: e.g., local concentration of chemical compounds (Fe_2O_3, $CaCO_3$, SiO_2) in the form of nodules and concretions; and (3) biological: e.g., fecal nodules, pedotubules.

5.3.2 Relationship with Geopedology

The micromorphologic characteristics represent an important source of information for the genetic interpretation of the soils and for inferring soil properties and qualities that control geomorphic processes.

- The pedologic features, which refer to the additions and transformations that take place in the soil material, are indicators of soil formation and evolution. The translocation of substances (e.g., clay illuviation) is a particularly good example that reveals a type of pedogenic dynamics. The micromorphologic analysis also allows identifying paleo-environmental influences in polygenic soils (Jungerius 1985) and correlatively in the evolution of the geomorphic landscape.
- The soil matrix has influence on geomorphogenesis. The nature of the plasma conditions the aggregate stability, which plays a relevant role in the processes of soil erosion by water and wind. Porosity controls the movement of water and air in the soil. The microporosity determines the capacity of water retention in the soil, while the macroporosity determines the surface runoff, the infiltration, and the percolation of water through the soil. An imbalance between these different terms of the water dynamics on the surface of and within the soil causes susceptibility to sheet erosion and mass movement.

5.4 Meso-level

At the meso-level, the organization entity of the soil material is the horizon, which usually consists of a mass of aggregates, except when the material is single-grain (sandy soil) or compact (clay soil). Horizons result from the differentiation of the parent material by pedogenic processes. The mode of analysis is direct observation and description in the field.

5.4.1 Horizon Definition and Designation

A horizon is a layer of soil material with a unique combination of properties, different from the properties of the soil in the horizons above and below that horizon (e.g., color, texture, structure). The concept of horizon refers to the pedogenic material and is therefore different from the concept of stratum that refers to the geogenic material (in the C layer). Soil horizons are identified at three successive levels using a designation nomenclature of letters and numbers.

5.4.1.1 Primary Divisions: The Master Horizons

The primary divisions reflect the effect of the basic soil forming processes, resulting in the differentiation of the soil material in master horizons. These are identified by capital letters (O, A, E, B, C, R). At this level, the horizons are distinguished according to the nature of the material and according to their position in the soil profile.

The distinction of the material according to its nature allows separating the organic material from the mineral material. A material is considered to be organic (O horizon) when it complies with the following contents of organic carbon (OC):

- In well drained soils: OC >20%.
- In poorly drained soils: OC $\geq$18%, if clay $\geq$60%; OC $\geq$12%, if clay = 0%; proportional percentages of OC for intermediate clay contents.

The distinction of the material according to the position in the profile leads to separate four kinds of horizon/layer: surficial horizon (topsoil), subsurface horizon, subsoil, and substratum.

- Topsoil horizons: A and E horizons

  - *A horizon*: layer where the incorporation of organic matter occurs and where the biologic activity shows its maximum expression; there may also be some downwashing of constituents.
  - *E horizon*: layer that loses soil material through eluviation according to the degree of solubility of the constituents. A generalized sequence by order of decreasing susceptibility to leaching includes salts, carbonates, bases, clay, OM, Fe and Al sesquioxides. In an extremely leached situation, only SiO_2 remains in situ, giving the horizon a whitish color (albic horizon).

- Subsurface horizons: B horizons
- The nature of the B horizon varies according to the process of formation, which can operate by weathering of the parent material (consolidated or loose), illuviation of chemical compounds (salts, carbonates, clay, OM, sesquioxides, etc.), and neoformation of clay minerals.
- Subsoil: C layer = parent material.
- Substratum: R layer = bedrock.

5.4.1.2   Secondary Divisions: Specific Genetic Features

The secondary divisions inform on specific genetic features of the horizons, using lowercase letters:

- Degree of decomposition of the organic material:

  i = slightly decomposed organic material (Fibrist).
  e = moderately decomposed organic material (Hemist).
  a = strongly decomposed organic material (Saprist).

- Degree of weathering of the mineral material: w (Bw), r (Cr).
- Accumulation: z, y, k, n, t, h, s, q, in order of decreasing mobility of the chemical compounds, referring respectively to salts more soluble than calcium sulphate, gypsum, carbonates, sodic clay, clay, humus, sesquioxides, and silica.

- Concentration: c, o, v, referring respectively to concretions, no-concretionary nodules, and plinthite.
- Transformation: f, g, m, p, x, b, d, referring respectively to frozen soil, gleization, compaction, plowpan, fragipan, buried horizon, and densified horizon.

5.4.1.3 Tertiary Divisions

The tertiary divisions are concerned with a variety of unrelated features, using arabic numerals:

- Subdivision of genetic horizons based on differences in color and/or texture, among other criteria (e.g., Bt1-Bt2) (numerical suffixes).
- Lithologic discontinuity based on textural contrasts indicating several successive depositional phases that result in the superposition of layers or profiles (e.g., Bt-2Bt-2C) (numerical prefixes).
- Bisequum that reflects the superimposition or imprint of a recent soil within a soil formed previously under different bioclimatic conditions, vegetation cover, or land-use. For instance, a Spodosol developing under pine plantation that invades the upper part of an Alfisol previously formed under deciduous forest (e.g., O-A-E-Bs-E′-Bt′-C).

5.4.2   Relationship with Geopedology

The designation symbols are information vectors that summarize the relevant characteristics of a horizon, including properties, mode of formation, and position in the profile. The nomenclature is used to identify genetic horizons based on the qualitative inference of the process(es) responsible for their formation. For instance, a Bw horizon reflects weathering of primary minerals, whereas a Bt horizon reflects clay illuviation. To be diagnostic for taxonomic classification of the soils, genetic horizons must comply with quantitative requirements (e.g., color, depth, thickness, % content, etc.) specified by the taxonomic system that is implemented. For this reason, it can be stated that all argillic horizons are Bt horizons, but not all Bt horizons are argillic horizons.

The soil information describing the nature of the horizons and, especially, their sequence in the profiles is very useful in geomorphic research on the susceptibility of the soils and cover formations to erosion processes. As highlighted by Jungerius (1985), A and B horizons exert different control on the geomorphic processes. The difference in strength between surficial horizons (A) and subsurface horizons (Bt) often determines the depth of soil truncation by sheet erosion. Similarly, differences in physico-mechanical properties between consecutive horizons may cause shear planes that can activate surface mass movements. Suffusion, piping, and tunnelling processes also depend on the sequence of and contrast between horizons.

5.5   Macro-level

5.5.1   *Definition*

At the macro-level, the basic concept is the pedon, which is defined as the minimum soil volume for describing and sampling a soil body (Soil Survey Staff 1975, 1999). Conventionally, the pedon is represented with a hexagonal configuration (Fig. 5.4). It covers a large part of the lateral and vertical variations of a soil body. The normal size of the area is 1 m² in the case of a soil with approximately parallel horizons and isotropic spatial variations. The maximum size of the area is 10 m² when horizons

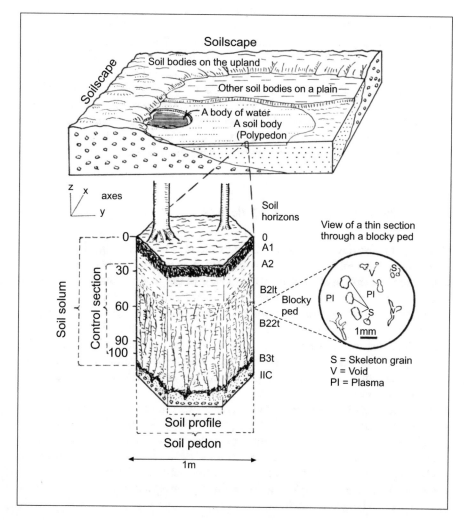

Fig. 5.4   Soil profile, pedon, polypedon, and soilscape. (Adapted from Buol et al. 1997)

show cyclic variations. The theoretical depth is down to the parent material of the soil, but for practical reasons it is usually limited to the upper 2 m.

5.5.2 Related Concepts

Several other concepts that characterize the soil body are related with the pedon concept, such as soil profile, solum, and control section.

- Soil profile is a face of the pedon including the entire sequence of horizons, commonly used to describe and sample. Statistical trials have shown that, when collecting material of a horizon laterally in all faces of the pedon to obtain a composite sample, probable mean errors can be divided approximately by two for most of the physical and chemical parameters (Wilding and Drees 1983).
- Solum includes soil horizons O + A + E + B, the C and R layers being excluded.
- Control section is the specific depth of the pedon within which selected soil characteristics need to occur to be considered diagnostic for taxonomic classification. For instance, for most of the soils, the family of particle-size distribution is determined within the depth of 25–100 cm. Likewise, to be diagnostic, plinthite should be present at <125 cm depth at great group level (e.g., Plinthustult) and at <150 cm depth at subgroup level (e.g., Plinthic Paleustult).

5.5.3 Relationship with Geopedology

Geomorphic literature does not provide any criteria or norms that specify the size of the minimum area for description and sampling. In practice, there is no space limitation for the description of the epigeal component of the geoform since processes and features of the terrain surface are directly observable. However, defining a minimum observation area can be useful for comparison between sites and for generalization of field information. With respect to the hypogeal component of the geoform, thus the proper geomorphic material (i.e., regolith, depositional material) that constitutes the C layer of the soils, it is not directly accessible to observation, description and sampling, except when there are natural or artificial exposures. Therefore, geomorphic research faces an issue of minimum volume for description and sampling similar to the one that has been solved in pedology with the concept of pedon. As the geopedologic survey integrates the description of the geoform and that of the soil in one place, the size criteria of the pedon may also apply to the morphon. The morphon covers the features of both the terrain surface and the subsoil/substratum, while the pedon covers the volume of the intermediate material that corresponds to the solum. In the geopedologic practice, the two are inseparable and their distinction may be regarded as superfluous.

The above comments apply primarily to the lower level of the hierarchic classification of the geoforms i.e., that of landform/terrain form (see Chap. 7). They are less pertinent at the higher categories of the system since the external features of the geoform often allow inferring the nature of the substratum.

5.6   Mega-level

5.6.1   Definition

The polypedon is the basic concept at the mega-level. It is an extended soil body formed by adjacent similar pedons that fit within the range of variation of a single taxonomic unit (e.g., soil series) (Soil Survey Staff 1975, 1999). It is a real physical soil body, limited by "no-soils" (e.g., rock outcrops, water bodies, built areas, etc.) or by pedons that exhibit dissimilar characteristics. The minimum area is 2 m^2 (i.e., two pedons), but there is no specification of maximum area. The concepts of soil body and soil individual are synonymous with polypedon. In similar terms, Boulaine (1975) proposed the concept of *genon* to designate the soil volume of all pedons that have the same structure and characteristics and that result from the same pedogenesis.

5.6.2   Relationship with Geopedology

- The polypedon constitutes the fundamental link between the actual soil volume (i.e., pedon) and the taxonomic unit in the classification system. It is the concept used to taxonomically classify the soil bodies. A polypedon comprises all contiguous pedons of equal classification.
- The polypedon provides the pedologic content of the cartographic unit. A polypedon is a concrete soil individual (i.e., soil body) on the landscape. Polypedon and landscape together form the soilscape. Polypedons can constitute (1) relatively pure map units with one dominant polypedon per unit (consociation), or (2) composite map units comprising more than one dominant polypedon (association, complex).
- The polypedon correlates with the geomorphic unit (polymorphon), especially at the lower taxonomic level (landform/terrain form). In its simplest expression, a polypedon together with the corresponding geomorphic frame forms a geopedologic landscape unit. However, the geopedologic landscape is usually more complex, because a single geoform often comprises more than one polypedon.

5.7   Conclusion

The holarchy of the soil system allows highlighting relevant relationships between soil properties and geomorphic response at different hierarchic levels. These relationships form the conceptual essence of geopedology. A notable phenomenon refers to the cause-effect relationships between reactions that occur in the soil material at micro-scale, thus not directly perceptible, and their geomorphic expression in the landscape at macro-scale. This is especially the case of landscape shaping by mass movements, which are controlled by micro-mechanical reactions in the soil fabric. With respect to soil cartography, the most conspicuous relationship takes place at the mega-level, where polypedon and polymorphon integrate to form a geopedologic landscape unit.

References

Boulaine J (1975) Géographie des sols. Presses Universitaires de France, Paris

Buol SW, Hole FD, McCracken RJ, Southard RJ (1997) Soil genesis and classification, 4th edn. Iowa State University Press, Ames

Fridland VM (1974) Structure of the soil mantle. Geoderma 12:35–41

Fridland VM (1976) Pattern of the soil cover. Israel Program for Scientific Translations, Jerusalem

Haigh MJ (1987) The holon: hierarchy theory and landscape research, Catena Supplement 10:181–192. CATENA Verlag, Cremlingen

Hole FD, Campbell JB (1985) Soil landscape analysis. Rowman & Allanheld, Totowa

Jenny H (1941) Factors of soil formation. McGraw-Hill, New York

Jungerius PD (1985) Soils and geomorphology. In: Jungerius PD (ed) Soils and geomorphology, Catena Supplement 6:1–18. CATENA Verlag, Cremlingen

Mitchell JK (1976) Fundamentals of soil behavior. Wiley, New York

Simonson RW (1959) Outline of a generalized theory of soil genesis. Soil Sci Soc Am Proc 23:152–156

Soil Survey Staff (1975) Soil taxonomy. A basic system of soil classification for making and interpreting soil surveys, USDA Agric Handbook 436. Government Publishing Office, Washington, DC

Soil Survey Staff (1999) Soil taxonomy, USDA Agric Handbook 436. Government Publishing Office, Washington, DC

Wilding LP, Drees LR (1983) Spatial variability and pedology. In: Wilding LP, Smeck NE, Hall GF (eds) Pedogenesis and soil taxonomy. I concepts and interactions. Elsevier, Amsterdam, pp 83–116

Zinck JA (1988) Physiography and soils. In: ITC soil survey lecture notes. International Institute for Aerospace Survey and Earth Sciences, Enschede

Chapter 6
The Geomorphic Landscape: Criteria for Classifying Geoforms

J. A. Zinck

Abstract Combining the basic criteria to build a taxonomic system with the hierarchic arrangement of the geomorphic environment determines a structure of nested categorial levels. Five of these levels are essentially deduced from the epigeal physiographic expression of the geoforms. To substantiate the relationship between geoform and soil, it is necessary to introduce in the system information on the internal hypogeal component of the geoforms, namely the constituent material, which is in turn the parent material of the soils. As a result of the foregoing, an additional level is needed to document the lithology in the case of bedrock substratum or the facies in the case of unconsolidated cover materials. This leads finally to a system with six categorial levels, identified by their respective generic concepts, including from upper to lower level: geostructure, morphogenic environment, geomorphic landscape, relief/molding, lithology/facies, and the basic landform or terrain form. Such a system with six categories complies with *Miller's Law*, which postulates that the capacity of the human mind to process information covers a range of seven plus or minus two elements.

Keywords Geomorphic classifications · Classification system structure · Levels of landscape perception · Geoform taxonomy · Geomorphometry

6.1 Introduction

Unlike other scientific disciplines, geomorphology still lacks a formally structured taxonomic system to classify the forms of the terrestrial relief, hereafter designated as *geoforms*. There is some consensus for grouping the geoforms according to the

J. A. Zinck died before publication of this work was completed.

J. A. Zinck (deceased)
Faculty of Geo-Information Science and Earth Observation (ITC), University of Twente, Enschede, The Netherlands

© The Author(s), under exclusive license to Springer Nature Switzerland AG 2023
J. A. Zinck et al. (eds.), *Geopedology*,
https://doi.org/10.1007/978-3-031-20667-2_6

families of processes that operate on given geologic substrata or in given bioclimatic zones. Examples of the former are the karstic forms generated by the dissolution of calcareous rocks, desert forms shaped by wind, glacial forms resulting from the activity of ice, or alluvial forms controlled by the activity of the rivers. However, these geoforms are not integrated in a structured hierarchic scheme. It is necessary to create a system that allows accommodating and organizing the geoforms according to their characteristics and origin and also considering their hierarchic relationships. This requires a multicategorial framework.

Geoform is the generic concept that designates all types of relief form regardless of their origin, dimension, and level of abstraction, similarly to how the concept of soil is used in pedology or the concept of plant in botany (Zinck 1988; Zinck and Valenzuela 1990). The term of geoform, with generic meaning, has been introduced recently in the Spanish version of the FAO Guidelines for soil description (FAO 2009). Geoforms have an internal (hypogeal) component and an external (epigeal) component in relation to the terrain surface. The internal component is the material of the geoform (the content), the characteristics of which convey genetic and stratigraphic (i.e., chronological) information. The external component of the geoform is its shape, its "form" (the container), which expresses a combination of morphographic and morphometric characteristics. The external component is directly accessible to visual perception, proximal or distal, either human or instrumental. Ideally, the classification of the geoforms should reflect features of both components, i.e., the constituent material and the physiographic expression. The external appearance of the geoforms is very relevant for their direct recognition and cartography. For this reason, a system of geoform classification must necessarily combine perception criteria of the geomorphic reality and taxonomic criteria based on diagnostic attributes.

Seemingly, geoform taxonomy has not fomented the same interest as plant taxonomy and soil taxonomy did. This might be due to the fact that more importance has been given to the analysis of the morphogenic processes than to geomorphic mapping which requires some kind of classification of the geomorphic units. There are few countries that have had, at some time, a systematic program of geomorphic mapping similar to those carried out in several Eastern European countries after the Second World War or in France in the second part of the last century (Tricart 1965; CNRS 1972).

Soil map legends often ignore the geomorphic context that, however, largely controls soil formation and distribution. Usually, the legend of the soil maps shows only the pedotaxa, without mentioning the landscapes where the soils are found, although the concept of "soilscape" is considered to provide the spatial framework for mapping polypedons (Buol et al. 1997). A mixed legend, showing the soil in its geomorphic landscape, facilitates the reading, interpretation, and use of the soil map by nonspecialists working in academic and practitioner environments (see the example in Fig. 4.2, Chap. 4). With the use of GIS, the geomorphic context is emerging as the structuring element of a variety of legends, including legends of taxonomic maps, interpretive maps, and land-use planning maps, among others.

6.2   Examples of Geomorphic Classification

Geomorphologists have always shown some interest in classifying geoforms, but the criteria used for this purpose have changed over the course of time and are still very diverse. After mentioning some geomorphic classification approaches, the structure of a taxonomic system for geoform classification is decribed. This has been developed from geopedologic surveys in Venezuela and later used in the ITC (Enschede, The Netherlands) to train staff from a variety of countries in Latin America, Africa, Middle East, and Southeast Asia (Zinck 1988; Farshad 2010).

6.2.1   Classification by Order of Magnitude

The dimensional criterion has been used by several authors to classify the geomorphic units (Tricart 1965; Goosen 1968; Verstappen and Van Zuidam 1975; among others). These classifications are hierarchic, with emphasis on structural geomorphology in the upper levels of the systems. The classification proposed by Cailleux-Tricart (Tricart 1965) in eight temporo-spatial orders of magnitude is a representative example of this approach (Table 6.1). The spatial dimension and the temporal dimension of the geomorphic units vary concomitantly from global to local and from early to recent. Tricart (1965) considers that the dimension of the geomorphic objects (facts and phenomena) intervenes not only in their classification, but also in the selection of the study methods and in the nature of the relationships between geomorphology and neighboring disciplines.

With a similar but less elaborate approach, Lueder (1959) distributes the geoforms in three orders of magnitude. The first order includes continents and ocean basins. Mountain ridges are an example of second order. The third order includes a variety of forms such as valley, depression, crest, and cliff.

Table 6.1 Taxonomic classification of the geomorphic units by Cailleux-Tricart

Order	Unit types	Unit examples	Extent (km^2)	Time (years)
I	Configuration of the earth's surface	Continent, ocean basin	10^7	10^9
II	Large structural assemblages	Shield, geosyncline	10^6	10^8
III	Large structural units	Mountain chain, sedimentary basin	10^4	10^7
IV	Elementary tectonic units	Serranía, horst	10^2	10^7
V	Tectonic accidents	Anticline, syncline	10	10^6–10^7
VI	Relief forms	Terrace, glacial cirque	10^{-2}	10^4
VII	Microforms	Lapies, solifluction	10^{-6}	10^2
VIII	Microscopic features	Corrosion, disaggregation	10^{-8}	–

Summarized from Tricart (1965)

6.2.2 Genetic and Genetic-Chorologic Classifications

There are variants of genetic classification of the geoforms based on the conventional division of geomorphology as a scientific discipline in specialist areas concerned with different types of geoforms (Table 6.2).

The genetic-chorologic classification of geoforms is based on the concept of morphogenic zone. The latitudinal and altitudinal distribution of the morphogenic zones parallels the division of the earth's surface in large bioclimatic zones, generating a series of morphoclimatic domains, each with a specific association of geoforms: glacial, periglacial, temperate (wet, dry), mediterranean, subtropical, and tropical (wet, dry). The classification combines origin and geographic distribution of the geoforms. It is often used to present and describe the geoforms by chapters in textbooks on geomorphology. This type of classification is based on some kind of hierarchic structure and leads to a typology of the geoforms but does not provide a clear definition of the criteria used in the ranking and typology. There is tendency to emphasize one type of attributes of the geoforms to the detriment of others: for instance, the dimension, or the genesis, or the geographic distribution.

The project of the Geomorphic Map of France (CNRS 1972) establishes a hierarchy of geomorphic information in five levels, called *terms*, as reference frames to gather the data, represent them cartographically, and enter them in the map legend. The five terms are in descending order: the location, the structural context (type of structural region, lithology, tectonics), the morphogenic context (age, morphogenic system), surface formations (origin of the material, particle-size distribution, consolidation, thickness, morphometry), and finally the forms. The last term contains the entire collection of recognized forms, with grouping into classes and subclasses according to the origin of the forms. Each form is given a definition and a symbol for its cartographic representation. Two main groups of forms are distinguished: (1) the endogenous forms (volcanic, tectonic, structural), and (2) the forms originated by external agents (eolian, fluvial, coastal, marine, lacustrine, karstic, glacial, periglacial and nival forms, and slope and interfluve forms).

For the purpose of soil mapping, Wielemaker et al. (2001) proposed a hierarchic terrain objects classification, qualified as morphogenic by the authors, which includes five nested levels, namely region, major landform, landform element, facet, and site. This system was derived from the analysis of a concrete case study located in Southern Spain, using a methodological framework to formalize expert knowledge on soil-landscape relationships and an interactive GIS procedure for sequential disaggregation of the landscape (de Bruin et al. 1999).

Table 6.2 Families of geoforms as per origin

Study fields of geomorphology	Types of geoforms
Structural geomorphology: types of relief	Cuesta, fold, shield reliefs, etc.
Climatic geomorphology: types of molding	Glacial, periglacial, eolian moldings, etc.
Azonal geomorphology: types of form	Alluvial, lacustrine, coastal forms, etc.

A variant of genetic-chorologic classification is the ordering of landscapes and geoforms in the context of a given country (Zinck 1974; Elizalde 2009). This type of classification combines physico-geographic units at the higher levels of the system with taxonomic units at the lower levels. The physico-geographic units belong to a specific regional context and, therefore, cannot be generalized or extrapolated to other regional situations. The division of a country into physiographic provinces and natural regions is an example of this type of nomenclature. Instead, the taxa of the lower categories (e.g., landscape types or relief types) convey sufficient abstraction to be recognizable on the basis of differentiating features in a variety of regional contexts.

6.2.3 Morphometric Classification

First attempts of morphometric relief characterization go back to mid-nineteenth century in the Germanic countries. However, it was only after the Second World War that systematic use of morphometric techniques was made to describe features of the topography, parameters of the hydrographic network, drainage density, and other measurable attributes of the relief (Tricart 1965). In recent decades, the technology of the digital elevation models (DEM) has given a new impulse to morphometry and automated extraction of morphometric information (Pike and Dikau 1995; Hengl and Reuter 2009). Geomorphometry focuses on the quantitative analysis of the terrain surface with two orientations: a specific morphometry that analyzes the discrete features of the terrain surface (e.g., landforms/terrain forms), and a general morphometry that deals with the continuous features. In its present state, geomorphometry pursues essentially the characterization and digital analysis of continuous topographic surfaces (Pike et al. 2009).

The use of DEM has allowed measuring and extracting attributes that describe topographic features of the landscape (Gallant and Wilson 2000; Hutchinson and Gallant 2000; Olaya 2009). The most frequently measured parameters include altitude, slope, exposure, curvature, and roughness of the relief, among others. The spatial distribution of these parameters allows inferring the variability of hydrologic, geomorphic, and biological processes in the landscape. The combination of data derived from DEM and satellite images contributes to improve predictive models (Dobos et al. 2000).

There are attempts to classify landforms and model landscapes using morphometric parameters (Evans et al. 2009; Hengl and MacMillan 2009; Nelson and Reuter 2012). Idealized geometric primitives (Sharif and Zinck 1996) and ideal elementary forms (Minár and Evans 2008) have been used to segment the landscape and approximate the representation of a variety of terrain forms. The implementation of automated algorithms to classify landforms has facilitated the mapping of landform elements and relief classes (Pennock et al. 1987; MacMillan and Pettapiece 1997; Ventura and Irvin 2000; Meybeck et al. 2001; Iwahashi and Pike 2007; MacMillan and Shary 2009). Ventura and Irvin (2000) analyzed different methods of automated landform classification for soil landscape studies, but the experiments were basically restricted to slope situations according to the classic models of Ruhe (1975) and Conacher and Dalrymple (1977).

The use of quantitative parameters allows describing continuous variations of topographic features with the support of fuzzy sets techniques (Irwin et al. 1997; Burrough et al. 2000; MacMillan et al. 2000). However, this approach may be less efficient in identifying differentiating characteristics of geoforms that have discrete boundaries, as is frequent in erosional (e.g., gullies, solifluction features) and depositional areas (e.g., alluvial or eolian systems). The DEM-based analysis leads to a classification of topographic features of the relief and contributes to the morphometric characterization of the terrain forms but does not generate a terrain form classification in the geomorphic sense of the concept. The classification of slope facets by shape and gradient is essentially a descriptive classification which does not convey information on the origin of the relief. However, this kind of classification results in an organization of the relief features that allows formulating hypotheses about their origin (Small 1970). Compared with the multiplication of tests carried out in rugged areas, the possibilities of digital mapping in flat areas, especially areas of depositional origin, have been so far less explored.

In the FAO Guidelines for soil description (2006), landforms are described by their morphology and not by their origin or forming processes. The proposed landform classification in a two-level hierarchy is based mainly on morphometric criteria. At the first level, three classes called, respectively, level land, sloping land, and steep land, are considered. These classes are subdivided according to three morphometric attributes including slope gradient, relief intensity, and potential drainage density. Applying this procedure to the level-land class, for instance, four subclasses are recognized, namely plain, plateau, depression, and valley floor. Sloping-land and steep-land include plain, valley, hill, escarpment zone, and mountain subclasses, differentiated by the above morphometric features.

6.2.4   Ethnogeomorphic Classification

Indigenous people in traditional communities use topographic criteria, before taking the soils into consideration, to identify ecological niches suitable for selected crops and management practices. Their approach to segment a hillside into relief units is similar to the slope facet models of Ruhe (1975) and Conacher and Dalrymple (1977). Likewise in depositional environments, where the topographic variations are often subtle and less perceptible, farmers clearly recognize a variety of landscape positions, as for instance the characteristic *banco-bajio-estero* trio (bank-depression-backswamp) for pasture management in the Orinoco river plains. Trials of participatory mapping, with the collaboration of local land users and technical staff, show that the mental maps of the farmers visualize the relief using a detailed nomenclature, which allows converting them into real maps that are very similar to the geomorphic maps prepared by specialists (Barrera-Bassols et al. 2006,

2009). The two maps in Fig. 6.1 show cartographic as well as taxonomic similarities: main unit delineations coincide, and taxa recognized by scientists and local farmers are comparable (e.g., gently sloping lava flow vs tzacapurhu meaning lava flow of stony land).

Indigenous soil classifications usually include the relief at the top level of the classification system, forming the basis of ethnogeopedology. In their perception of the environment, indigenous farmers use the relief, along with other features of the

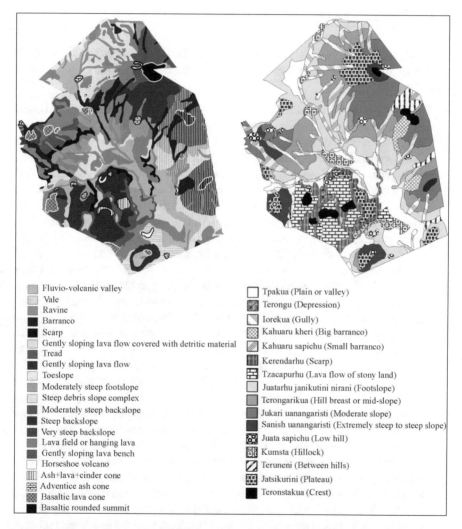

Fig. 6.1 Comparison of a geomorphic map made using technical criteria (left) and a relief map drawn up according to the indigenous Purhépecha nomenclature (right) of the territory of San Francisco Pichátaro, Michoacán, in the volcanic belt of Central Mexico. (Adapted from Barrera-Bassols et al. 2006)

landscape, as a main factor for identifying, locating, and classifying soils. Because of the importance that both disciplines give to the relief factor, ethnopedology and geopedology are strongly related.

6.3 Bases for a Taxonomic Classification System of the Geoforms

6.3.1 Premises and Basic Statements

A set of assumptions is formulated hereafter as a basis for structuring a taxonomic system of the geoforms and improving the traditional approaches to geomorphic classification.

- The object to be classified is a unit of the geolandscape or subdivision thereof that can be recognized by its configuration and composition. The most commonly used term to designate this entity in English-written geomorphic literature is *landform*. The same term is indistinctly used by geomorphologists, geologists, pedologists, agronomists, ecologists, architects, planners, contemplative and active users of the landscape, among others, but there is no standard definition accepted by everybody. In the FAO Guidelines for soil description (2006), the concept of *major landform* is considered to refer to the morphology of the whole landscape. Way (1973) provides a satisfactory definition in the following terms: "*Landforms* are terrain features formed by natural processes, which have a defined composition and a range of physical and visual characteristics that occur wherever the form is found and whatever is the geographic region". This statement poses two basic principles: (1) a landform is identified using internal constituents as well as external attributes, and (2) a landform is recognized by its intrinsic characteristics and not according to the context in which it occurs. In Spanish language, landform literally means *forma de tierra(s)*, a term that has an agricultural or agronomic connotation. *Land* in landscape ecology includes not only the physical features of the landscape, but also the biota and the human activities (Zonneveld 1979, 1989). The term *terrain form* is more appropriate to designate the elementary relief form, while the term *geoform* is the generic concept that encompasses the geomorphic units at all categorial levels. *Terrain form* is etymologically equal to terms with similar geomorphic meaning used in other languages, such as *forma de terreno* in Spanish and *forme de terrain* in French.
- The objects that are classified are the geoforms, or geomorphic units, which are identified on the basis of their own characteristics, rather than by reference to the factors of formation. Local or regional combinations of criteria such as climate, vegetation, soil, and lithology, which are associated with the geoforms and contribute to their formation, can be referred to in the legend of the geomorphic map, but are not intrinsically part of the classification of the geoforms. The climate factor is implicitly present in the geoforms originated by exogenous morphogenic agents (snow, ice, water, wind).

- Classes of geoforms are arranged hierarchically to reflect their level of membership to the geomorphic landscape. For instance, a river levee is a member of a terrace, which in turn is a member of a valley landscape. Therefore, levee, terrace, and valley shall be placed in different categories in a hierarchic system, because they correspond to different levels of abstraction. Similarly, the slope facets (i.e., summit, shoulder, backslope, and footslope) are members of a hill, which is a member of a hilland type of landscape.
- The genesis of the geoforms is taken into consideration preferably at the lower levels of the taxonomic system, since the origin of the geomorphic units can be a matter of debate and the genetic attributes may be not clear or controversial, or their determination may require a number of additional data. At higher levels, the use of more objective, rather descriptive attributes are privileged, in parallel with the criteria of pattern recognition implemented in photo and image interpretation.
- The dimensional characteristics (e.g., length, width, elevation, slope, etc.) are subordinate attributes and are not diagnostic for the identification of the geoforms. A geoform belongs to a particular class regardless of its size, provided it complies with the required attributes of that class. For instance, the extent of a dune or a landslide can vary from a few m² to several km².
- The names of the geoforms are often derived from the common language and some of them may be exposed to controversial interpretation. Priority is given here to those terms that have greater acceptation by their etymology or usage.
- The concepts of physiographic province and natural region, as well as other kinds of chorologic units related to specific geographic contexts, are not taken into account in this taxonomic system, because they depend on the particular conditions of a given country or continental portion, a fact that limits their level of abstraction and geographic repeatability.
- The geographic distribution of the geoforms is not a taxonomic criterion. The chorology of the geoforms is reflected in their cartography and in the structure of the geomorphic map legend.
- Toponymic designations can be used as phases of the taxonomic units (e.g., Cordillera de Mérida, Pantanal Basin).

6.3.2   Prior Information Sources

The development of the geoform classification system uses prior knowledge in terms of concepts, methods, information, and experience.

- Existing geoform typologies, with definitions and descriptive attributes, have been partially taken from the literature. The proposed classification builds on and organizes prior knowledge in a hierarchic taxonomic system. Some of the key documents that were consulted for this purpose are as follows:

- Various classic texbooks of geomorphology: Tricart and Cailleux (1962, 1965, 1967, 1969), Tricart (1965, 1968, 1977), Derruau (1965, 1966), Thornbury (1966), Viers (1967), CNRS (1972), Garner (1974), Ruhe (1975), Huggett (2011), among others.
- Dictionaries and encyclopedias: Visser (1980), Lugo-Hubp (1989), Fairbridge (1997), Goudie (2004), among others.
- Manuals of geomorphic photo-interpretation: Goosen (1968), Way (1973), Verstappen and Van Zuidam (1975), Verstappen (1983), Van Zuidam (1985), among others.

- For the structure of the system, inspiration was taken from the conceptual framework of the USDA Soil Taxonomy (Soil Survey Staff 1975, 1999) with regard to the concepts of category, class, and attribute.
- Development and validation of the system have taken place essentially in Venezuela and Colombia, within the framework of soil survey projects at different scales from detailed to generalized, with the implementation of geomorphology as a tool for soil mapping (applied geomorphology). The system was modified and improved progressively as ongoing field surveys provided new knowledge. Subsequently, the already established system became teaching and training matter in postgraduate courses in soil survey at the ITC (Zinck 1988) for students from different parts of the world, especially Latin America, Africa, Middle East, and Southeast Asia.

6.3.3   Searching for Structure: An Inductive Example

Let's consider the collection of objects included in Fig. 6.2 (Arnold 1968). Squares, triangles, and circles can be recognized. The objects are large or small, green (G) or red (R). Thus, the objects are different by shape, size, and color. Based on these three criteria, the objects may be classified in various ways. One option is to sort the objects first by size, then by color, and finally by shape (Fig. 6.3). They can also be sorted successively by shape, color, and size. Six hierarchization alternatives are possible. This simple experiment shows that artificial or natural objects may be classified in various ways. Any alternative is valid if it meets the objective pursued.

From example in Fig. 6.2, three basic elements of a hierarchic classification system can be induced by effect of generalization: category, class, and attribute.

- The categories are hierarchic levels that give structure to the classification system. Three categories are present, identified by generic criteria (size, color, shape). Several (6) hierarchic arrangements are possible.
- Classes are groups of objects that have one or more differentiating characteristics in common. There are seven differentiating characteristics: large, small, red, green, square, triangular, and circular. The aggregation of characteristics generates an increase of classes from the top to the bottom of the system.
- Attributes are characteristics or properties of the objects, such as red, green, large, small, square, triangular, and circular.

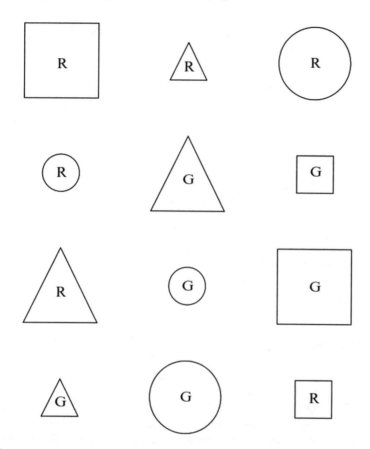

Fig. 6.2 Collection of objects different by shape, size, and color. (Adapted from Arnold 1968)

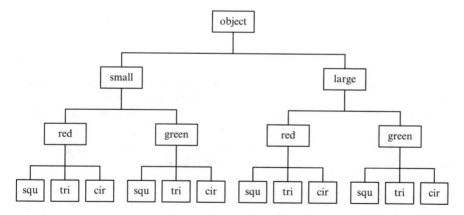

Fig. 6.3 Hierarchic arrangement of the objects displayed in Fig. 6.2 by size (2 classes), color (4 classes), and shape (12 classes) (*squ* square, *tri* triangular, *cir* circular)

6.4   Structure and Elements for Building a Taxonomic System of the Geoforms

A taxonomic system is characterized by its structure (or configuration) and its elements (or components).

6.4.1   Structure

Various configuration models are possible: hierarchic, relational, network, and linear, among others (Burrough 1986). In general, the hierarchic multicategorial model is considered appropriate for taxonomic purposes. Haigh (1987) states that the hierarchic structure is a fundamental property of all natural systems, while Urban et al. (1987) consider that breaking a landscape into elements within a hierarchic framework allows to partially solve the problem of its apparent complexity. Although a hierarchic structure is less efficient than, for instance, a relational system or a network system in terms of automated data handling by computer, it is however particularly suitable for archiving, processing, and retrieving information by the human mind (Miller 1956, 2003).

A system can be compared to a box containing all the individuals belonging to the object that is sought to be classified: for example, all soils, all geoforms. The collection of individuals constitutes the universe that is going to be divided into classes and arranged into categories. The classification results in (1) a segmentation of the universe under consideration (e.g., the soil cover continuum) into populations, groups, and individuals by descending disaggregation, and (2) a clustering of individuals into groups, populations, and universe by ascending aggregation.

6.4.2   Elements

6.4.2.1   Category

A category is a level of abstraction. The higher the level of the category, the higher is the level of abstraction. Each category comprises a set of classes showing a similar level of abstraction. A category is identified by a generic concept that characterizes all classes present in this level (color, size, shape, in Fig. 6.3). For instance, a valley landscape, a fluvial terrace, and a river levee are objects belonging to different levels of abstraction. The levee is a member of the terrace, which in turn is a member of the valley. In a hierarchic system of geoforms, these geomorphic entities shall be placed in three successive categories.

6.4.2.2   Class

A class is a formal subdivision of a population at a given categorial level. A class can be determined using different modalities among which the two following are commonly implemented: (1) the range of variation of a diagnostic attribute or a combination thereof, and (2) a central class concept in relation to which other classes deviate by one or more characteristics.

An example of the first modality is provided by the way the percentage of base saturation is used in soil taxonomy as a threshold parameter to separate Alfisols ($\geq$35%) and Ultisols (<35%). Using a similar procedure, the strata dip in sedimentary rocks allows separating several classes of monoclinal relief, including mesa, cuesta, creston, hogback, and bar (Fig. 6.4). A similar approach can be applied to the classification of the geoforms caused by mass movements through segmentation of the continuum between solid and liquid states using the consistence limits (Fig. 6.5). There are very few references in the geomorphic literature where the segmentation of a continuum is used to differentiate related geoforms.

The central typifying concept is used to position a typical class in relation to intergrades and extragrades, which depart from the central class by deviation of some attributes. This is the case, for instance, of the "Typic" as used at subgroup level in the USDA Soil Taxonomy (Soil Survey Staff 1975, 1999). No examples were found in the geomorphic literature implementing formally the central concept to distinguish modal situations from transitional ones.

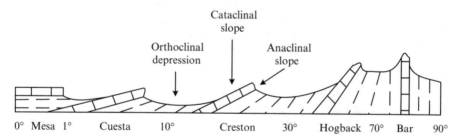

Fig. 6.4 Monoclinal relief classes determined based on strata dip ranges in sedimentary bedrocks (e.g. limestone, sandstone). (Adapted from Viers 1967)

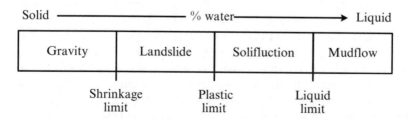

Fig. 6.5 Classes of geoforms originated by different kinds of mass movement

6.4.2.3 Taxon

A taxon (or taxum) is a concrete taxonomic unit as a member of a class established at a given categorial level. Usually, a particular taxon covers only part of the range of variation allowed in the selected attributes that define the class. For instance, the texture of a river bank, above the basal gravel strata, can vary from gravelly to sandy clay loam. A particular bank can be sandy to sandy loam without covering the entire diagnostic textural range.

6.4.2.4 Attribute

An attribute is a characteristic (or variable) used to establish the limits of the classes that make up the system and to implement these limits in the description and classification of individuals. There are several kinds of attribute, as for instance:

- Dichotomous: e.g., presence or absence of iron reduction mottles, concentration of carbonates or other salts.
- Multi-state without ranges: e.g., types of soil structure, types of depositional structure.
- Multi-state with ranges: e.g., size of structural aggregates, plasticity and adhesion classes.
- Continuous variation: e.g., base saturation, bedrock dip.

Implementing these basic taxonomic criteria in geomorphology requires (1) the inventory of the known geoforms and their arrangement in a hierarchic system, and (2) the selection, categorization (diagnostic or not), hierarchization, and measurement of the attributes used to identify and describe the geoforms.

6.5 Levels of Perception: Exploring the Structure of a Geomorphic Space

Geomorphology is primarily a science of observation, aiming at the identification and separation of landscapes from topographic maps, digital elevation or terrain models, and remote-sensed documents allowing stereoscopic vision, but mainly by reading the physiographic features in the field. Geoforms can be perceived by human vision or artificial sensors because they have a physiognomic appearance on the earth's surface (i.e., geolandscape). Physiography describes this external appearance corresponding to the epigeal component of the geoforms. Thanks to their scenic expression, geoforms are the most directly structuring elements of the terrain, more than any other object or natural feature. Even a non-scientific observer can notice that any portion of the earth's crust shows a structure determined by the relief, which allows subdividing it into components. The times that a terrain area

can be subdivided into elements depend on the level of perception used for the segmentation. Although the concept of perception level is subjective when the human
eye is used, it helps hierarchize the structural components of a terrain surface.

Hereafter, an example is developed that illustrates the effect of the perception
scale on the sequential identification of different terrain portions. The example
refers to the contact area between the Caribbean Sea and the northern edge of the
South American continent in Venezuela (Zinck 1980). The use of successive perception levels, increasingly detailed, materialized by observation platforms of decreasing elevation in relation to the earth's surface, allows dividing the selected portion
of continent into classes of geoforms that are distributed over various hierarchic
categories (Fig. 6.6 and Table 6.3). An observer mounted on a spaceship at about
800–1000 km elevation would distinguish two physiographic provinces, namely the
east-west oriented coastal mountain chain of the Cordillera de la Costa to the north
and the basin of the Llanos Plains to the south. These two macro-units of contrasting
relief correspond to two types of geostructure: a folded cordillera-type mountain
chain and a geosincline-type sedimentary basin, respectively. From an airplane flying at about 10 km elevation, one can distinguish the two parallel branches of the
Cordillera de la Costa, namely the Serranía del Litoral range to the north and the

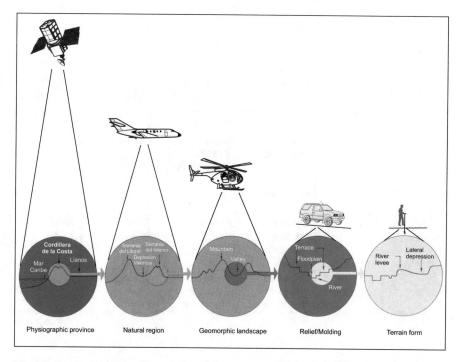

Fig. 6.6 Successive levels of perception of geoforms from different observation elevations. From
left to right: physiographic province (geostructure), natural region (morphogenic environment),
geomorphic landscape, relief/molding, terrain form (Zinck 1980). The features referred to are
explained in Table 6.3

Table 6.3 Sequential identification of geoforms according to increasing levels of perception

Observation platform	Observation area	Observed features	Criteria used Inferred factors	Resulting geoforms	Derived generic categorial concepts
Satellite	Large continental portion	*Cordillera de la Costa* narrow, longitudinal, high relief mass; abrupt limits	Topography Internal geodynamics (orogenic area)	Cordillera (folded mountain chain)	Geostructure
		Llanos del Orinoco Extensive, flat, low relief mass	Topography Internal geodynamics (sinking area)	Geosyncline (sedimentary basin)	
Airplane	Cordillera	*Serranía del Litoral Serranía del Interior* parallel, dissected mountain ranges	Topography Internal/external geodynamics (erosion)	Structural/ erosional environment	Morphogenic environment
		Depresión de Valencia Low-lying, flat terrain areas; concave margins	Topography Internal/external geodynamics (sedimentation)	Depositional environment	
Helicopter	Structural/ erosional environment	Parallel mountain ridges	Topography Tectonics Hydrography	Mountain	Geomorphic landscape
		Narrow longitudinal depressions, parallel or perpendicular to the ridges	Topography Tectonics Hydrography	Valley	
Earth surface	Valley	Topographic step treads separated by risers	Topography	Terrace	Relief/ molding
		Valley bottom, river system, riparian forest	Topography Drainage Vegetation	Floodplain	

(continued)

Table 6.3   (continued)

Observation platform	Observation area	Observed features	Criteria used Inferred factors	Resulting geoforms	Derived generic categorial concepts
Terrain surface and subsurface	Terrace	Longitudinal, narrow, convex bank; well drained, coarse-textured	Topography Drainage Morphogenesis	Levee	Terrain form
		Large, concave depression, poorly drained, fine-textured	Topography Drainage Morphogenesis	Basin	

Based on the features observed in Fig. 6.6. Zinck (1988)

Serranía del Interior range to the south, separated by an alignment of tectonic depressions such as that of Lake Valencia. These units are natural regions that correspond to types of morphogenic environment: the mountain ranges are structural environments undergoing erosion, whereas depressions are depositional environments. When increasing the level of perception as from a helicopter flying at two km elevation, a mountain range can be divided into mountain and valley landscapes. A field transect through a valley allows to cross a series of topographic steps with risers and treads that correspond to fluvial terraces. Detailed field observation of the topography and sediments in a given terrace will reveal a sequence of depositional units from the highest, the river levee (bank), to the lowest, the decantation basin (swamp). The results of this exploratory inductive procedure, leading to a sequential segmentation of a portion of the South American continent, are summarized in Table 6.3. This empirical approach generates a hierarchic scheme of geoforms in five nested categorial levels, each identified by a generic concept from general to detailed (Fig. 6.7).

6.6   Structure of a Taxonomic System of the Geoforms

Combining the basic criteria to build a taxonomic system (Sects. 6.3 and 6.4) with the results of the exploration aimed at detecting guidelines of hierarchic arrangement in the geomorphic environment (Sect. 6.5), a structure of nested categorial levels is obtained. Five of these levels are essentially deduced from the epigeal physiographic expression of the geoforms. The units recognized at the two upper levels are identified by local names, because they belong to a particular national or regional context. These are chorologic units which are formalized as taxonomic units under the generic concept of geostructure and morphogenic environment,

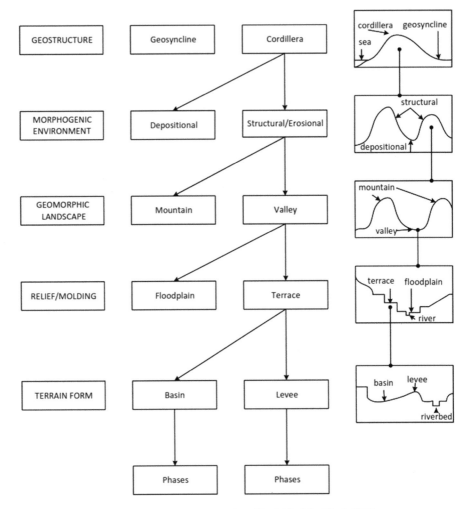

Fig. 6.7 Generalization of the information displayed in Table 6.3. (Zinck 1988)

respectively. To substantiate the relationship between geoform and soil, it is necessary to introduce in the system information on the internal hypogeal component of
the geoforms, namely the constituent material, which is in turn the parent material
of the soils. As a result of the foregoing, an additional level is needed to document
the lithology, in the case of bedrock substratum, or the facies in the case of unconsolidated cover materials. After several iterations, this category was inserted
between the level of relief/molding (level 3) and the level of terrain form (level 1).
Its inclusion in the lower part of the system is justified by the fact that field data are
often needed to supplement or clarify the general information provided by the

Table 6.4 Synopsis of the geoform classification system (Zinck 1988)

Level	Category	Generic concept	Short definition
6	Order	Geostructure	Large continental portion characterized by a type of geologic macro-structure (e.g., cordillera, geosyncline, shield)
5	Suborder	Morphogenic environment	Broad type of biophysical environment originated and controlled by a style of internal and/or external geodynamics (e.g., structural, depositional, erosional, etc.)
4	Group	Geomorphic landscape	Large portion of land/terrain characterized by given physiographic features: it corresponds to a repetition of similar relief/molding types or an association of dissimilar relief/molding types (e.g., valley, plateau, mountain, etc.)
3	Subgroup	Relief/molding	Relief type originated by a given combination of topography and geologic structure (e.g., cuesta, horst, etc.) Molding type determined by specific morphoclimatic conditions and/or morphogenic processes (e.g., glacis, terrace, delta, etc.)
2	Family	Lithology/facies	Petrographic nature of the bedrocks (e.g., gneiss, limestone, etc.) or origin/nature of the unconsolidated cover formations (e.g., periglacial, lacustrine, alluvial, etc.)
1	Subfamily	Landform/terrain form	Basic geoform type characterized by a unique combination of geometry, dynamics, and history

geologic maps (see Fig. 7.3 and Table 7.2 in Chap. 7). This leads finally to a system with six categorial levels (Table 6.4), identified by their respective generic concepts that are explained in Chap. 7. It can be noted that obtaining a system with six categories complies with the rule called *Miller's Law*, which postulates that the capacity of the human mind to process information covers a range of seven plus or minus two elements (Miller 1956, 2003).

6.7   Conclusion

Geoforms are the emerging parts of the earth's crust. Their distinct physiognomic features make them directly observable through visual and artificial perception from remote to proximal sensing. Changing the scale of perception changes not only the degree of detail but most significantly the nature of the object observed. For instance, a levee is a member of a terrace which is a member of a valley, thus three geomorphic objects bearing different levels of abstraction. The geolandscape is a hierarchically structured and organized domain. Therefore, a multicategorial system, based on nested levels of perception to capture the information and taxonomic criteria to organize that information, is an appropriate frame to classify geoforms.

References

Arnold R (1968) Apuntes de agrología (documento inédito). Barquisimeto, Ministerio de Obras Públicas (MOP)

Barrera-Bassols N, Zinck JA, Van Ranst E (2006) Local soil classification and comparison of indigenous and technical soil maps in a Mesoamerican community using spatial analysis. Geoderma 135:140–162

Barrera-Bassols N, Zinck JA, Van Ranst E (2009) Participatory soil survey: experience in working with a Mesoamerican indigenous community. Soil Use Manage 25:43–56

Buol SW, Hole FD, McCracken RJ, Southard RJ (1997) Soil genesis and classification, 4th edn. Iowa State University Press, Ames

Burrough PA (1986) Principles of geographical information systems for land resources assessment. Clarendon Press, Oxford

Burrough PA, van Gaans PFM, MacMillan RA (2000) High-resolution landform classification using fuzzy k-means. Fuzzy Sets Syst 113:37–52

CNRS (1972) Cartographie géomorphologique. Travaux de la RCP77, Mémoires et Documents vol 12. Editions du Centre National de la Recherche Scientifique, Paris

Conacher AJ, Dalrymple JB (1977) The nine-unit landscape model: an approach to pedogeomorphic research. Geoderma 18:1–154

de Bruin S, Wielemaker WG, Molenaar M (1999) Formalisation of soil-landscape knowledge through interactive hierarchical disaggregation. Geoderma 91:151–172

Derruau M (1965) Précis de géomorphologie. Masson, Paris

Derruau M (1966) Geomorfología. Ediciones Ariel, Barcelona

Dobos E, Micheli E, Baumgardner MF, Biehl L, Helt T (2000) Use of combined digital elevation model and satellite radiometric data for regional soil mapping. Geoderma 97(3–4):367–391

Elizalde G (2009) Ensayo de clasificación sistemática de categorías de paisajes. Primera aproximación, edn revisada. Maracay

Evans IS, Hengl T, Gorsevski P (2009) Applications in geomorphology. In: Hengl T, Reuter HI (eds) Geomorphometry: concepts, sofware, applications, Developments in Soil Science, vol 33. Elsevier, Amsterdam, pp 497–525

Fairbridge RW (ed) (1997) Encyclopedia of geomorphology. Springer, New York

FAO (2006) Guidelines for soil description, 4th edn. Food and Agricultural Organization of the United Nations, Rome

FAO (2009) Guía para la descripción de suelos, cuarta edn. Organización de las Naciones Unidas para la Agricultura y la Alimentación, Roma

Farshad A (2010) Geopedology. An introduction to soil survey, with emphasis on profile description (CD-ROM). University of Twente, Faculty of Geo-Information Science and Earth Observation (ITC), Enschede

Gallant JC, Wilson JP (2000) Primary topographic attributes. In: Wilson JP, Gallant JC (eds) Terrain analysis: principles and applications. Wiley, New York, pp 51–85

Garner HF (1974) The origin of landscapes. A synthesis of geomorphology. Oxford University Press, New York

Goosen D (1968) Interpretación de fotos aéreas y su importancia en levantamiento de suelos. In: Boletín de Suelos 6. FAO, Roma

Goudie AS (ed) (2004) Encyclopedia of geomorphology, vol 2. Routledge, London

Haigh MJ (1987) The holon: hierarchy theory and landscape research. Catena Supplement 10. CATENA Verlag, Cremlingen, pp 181–192

Hengl T, MacMillan RA (2009) Geomorphometry: a key to landscape mapping and modelling. In: Hengl T, Reuter HI (eds) Geomorphometry: concepts, software, applications, Developments in soil science, vol 33. Elsevier, Amsterdam, pp 433–460

Hengl T, Reuter HI (eds) (2009) Geomorphometry: concepts, software, applications, Developments in soil science, vol 33. Elsevier, Amsterdam

Huggett RJ (2011) Fundamentals of geomorphology. Routledge, London

Hutchinson MF, Gallant JC (2000) Digital elevation models and representation of terrain shape. In: Wilson JP, Gallant JC (eds) Terrain analysis: principles and applications. John Wiley & Sons, New York, pp 29–50

Irwin BJ, Ventura SJ, Slater BK (1997) Fuzzy and isodata classification of landform elements from digital terrain data in Pleasant Valley, Wisconsin. Geoderma 77:137–154

Iwahashi J, Pike RJ (2007) Automated classifications of topography from DEMs by an unsupervised nested-means algorithm and a three-part geometric signature. Geomorphology 86(3–4):409–440

Lueder DR (1959) Aerial photographic interpretation: principles and applications. McGraw-Hill, New York

Lugo-Hubp J (ed) (1989) Diccionario geomorfológico. Universidad Nacional Autónoma de México, Cd México

MacMillan RA, Pettapiece WW (1997) Soil landscape models: automated landscape characterization and generation of soil-landscape models. Research Report 1E:1997

MacMillan RA, Shary PA (2009) Landforms and landform elements in geomorphometry. In: Hengl T, Reuter HI (eds) Geomorphometry: concepts, software, applications, Developments in Soil Science, vol 33. Elsevier, Amsterdam, pp 227–254

MacMillan RA, Pettapiece WW, Nolan SC, Goddard TW (2000) A generic procedure for automatically segmenting landforms into landform elements using DEMs, heuristic rules and fuzzy logic. Fuzzy Sets Syst 113:81–109

Meybeck M, Green P, Vorosmarty CJ (2001) A new typology for mountains and other relief classes: an application to global continental water resources and population distribution. Mount Res Dev 21:34–45

Miller GA (1956) The magical number seven, plus or minus two: some limits on our capacity for processing information. Psychol Rev 63(2):81–97

Miller GA (2003) The cognitive revolution: a historical perspective. Trends Cogn Sci 7(3):141–144

Minár J, Evans IS (2008) Elementary forms for land surface segmentation: The theoretical basis of terrain analysis and geomorphological mapping. Geomorphology 95:236–259

Nelson A, Reuter H (2012) Soil projects. Landform classification from EU Joint Research Center, Institute for Environment and Sustainability. http://eusoils.jrc.ec.europa.eu/projects/landform

Olaya V (2009) Basic land-surface parameters. In: Hengl T, Reuter HI (eds) Geomorphometry: concepts, software, applications, Developments in soil science, vol 33. Elsevier, Amsterdam, pp 141–169

Pennock DJ, Zebarth BJ, De Jong E (1987) Landform classification and soil distribution in hummocky terrain, Saskatchewan, Canada. Geoderma 40:297–315

Pike RJ, Dikau R (eds) (1995) Advances in geomorphometry. Proceedings of the Walter F. Wood memorial symposium. Zeitschrift für Geomorphologie Supplementband 101

Pike RJ, Evans IS, Hengl T (2009) Geomorphometry: a brief guide. In: Hengl T, Reuter HI (eds) Geomorphometry: concepts, sofware, applications, Developments in soil science, vol 33. Elsevier, Amsterdam, pp 3–30

Ruhe RV (1975) Geomorphology. Geomorphic processes and surficial geology. Houghton Mifflin, Boston

Sharif M, Zinck JA (1996) Terrain morphology modelling. International Archives of Photogrammetry and Remote Sensing XXXI Part B3:792–797

Small RJ (1970) The study of landforms. A textbook of geomorphology. Cambridge University Press, London

Soil Survey Staff (1975) Soil taxonomy. A basic system of soil classification for making and interpreting soil surveys. USDA Agric Handbook 436. US Government Print Office, Washington, DC

Soil Survey Staff (1999) Soil taxonomy. USDA Agric Handbook 436. US Government Print Office, Washington,DC

Thornbury WD (1966) Principios de geomorfología. Editorial Kapelusz, Buenos Aires

Tricart J (1965) Principes et méthodes de la géomorphologie. Masson, Paris

Tricart J (1968) Précis de géomorphologie. T1 Géomorphologie structurale. SEDES, Paris

Tricart J (1977) Précis de géomorphologie. T2 Géomorphologie dynamique générale. SEDES-CDU, Paris

Tricart J, Cailleux A (1962) Le modelé glaciaire et nival. SEDES, Paris

Tricart J, Cailleux A (1965) Le modelé des régions chaudes. Forêts et savanes. SEDES, Paris

Tricart J, Cailleux A (1967) Le modelé des régions périglaciaires. SEDES, Paris

Tricart J, Cailleux A (1969) Le modelé des régions sèches. SEDES, Paris

Urban DL, O'Neill RV, Shugart HH Jr (1987) Landscape ecology. A hierarchical perspective can help scientists understand spatial patterns. Bioscience 37(2):119–127

Van Zuidam RA (1985) Aerial photo-interpretation in terrain analysis and geomorphological mapping. ITC, Enschede

Ventura SJ, Irvin BJ (2000) Automated landform classification methods for soil-landscape stydies. In: Wilson JP, Gallant JC (eds) Terrain analysis: principles and applications. John Wiley & Sons, New York, pp 267–294

Verstappen HT (1983) Applied geomorphology; geomorphological survey for environmental development. Elsevier, Amsterdam

Verstappen HT, Van Zuidam RA (1975) ITC system of geomorphological survey. ITC, Enschede

Viers G (1967) Eléments de géomorphologie. Nathan, Paris

Visser WA (ed) (1980) Geological nomenclature. Royal geological and mining society of the Netherlands. Bohn, Scheltema & Holkema, Utrecht

Way DS (1973) Terrain analysis. A guide to site selection using aerial photographic interpretation. Dowden, Hutchinson & Ross, Stroudsburg, Pennsylvania

Wielemaker WG, de Bruin S, Epema GF, Veldkamp A (2001) Significance and application of the multi-hierarchical landsystem in soil mapping. Catena 43:15–34

Zinck JA (1974) Definición del ambiente geomorfológico con fines de descripción de suelos. Ministerio de Obras Públicas (MOP), Cagua

Zinck JA (1980) Valles de Venezuela. Lagoven, Petróleos de Venezuela, Caracas

Zinck JA (1988) Physiography and soils. ITC soil survey lecture notes. International Institute for Aerospace Survey and Earth Sciences, Enschede

Zinck JA, Valenzuela CR (1990) Soil geographic database: structure and application examples. ITC J 1990(3):270–294

Zonneveld JIS (1979) Land evaluation and land(scape) science. ITC, Enschede

Zonneveld JIS (1989) The land unit – A fundamental concept in landscape ecology, and its applications. Landsc Ecol 3(2):67–86

Chapter 7
The Geomorphic Landscape: Classification of Geoforms

J. A. Zinck

Abstract This chapter attempts to organize existing geomorphic knowledge and arrange the geoforms in the hierarchically structured system with six nested levels introduced in the foregoing Chap. 6. Geoforms are grouped thematically, distinguishing between geoforms mainly controlled by the geologic structure and geoforms mainly controlled by the morphogenic agents. It is thought that this multicategorial geoform classification scheme reflects the structure of the geomorphic landscape sensu lato. It helps segment and stratify the landscape continuum into geomorphic units belonging to different levels of abstraction. This geoform classification system has shown to be useful in geopedologic mapping, and it could also be useful in digital soil mapping.

Keywords Geotaxa · Geostructure · Morphogenic environment · Geomorphic landscape · Relief/molding · Lithology/facies · Terrain form/landform

7.1 Introduction

The terms used hereafter to name the geoforms have been taken from a selection of textbooks, compendia, and other general books of geomorphology, including among others: Tricart and Cailleux (1962, 1965, 1967, 1969), Tricart (1965, 1968, 1977), Derruau (1965, 1966), Thornbury (1966), Viers (1967), CNRS (1972), Garner (1974), Zinck (1974), Ruhe (1975), Verstappen and Van Zuidam (1975), Visser (1980), Verstappen (1983), Van Zuidam (1985), Lugo-Hubp (1989), Fairbridge (1997), Goudie (2004), and Huggett (2011). Readers may not unanimously agree

---

J. A. Zinck died before publication of this work was completed.

J. A. Zinck (deceased)
Faculty of Geo-Information Science and Earth Observation (ITC), University of Twente, Enschede, The Netherlands

© The Author(s), under exclusive license to Springer Nature Switzerland AG 2023
J. A. Zinck et al. (eds.), *Geopedology*,
https://doi.org/10.1007/978-3-031-20667-2_7

with the proposed terminology, as some terms can be subject to controversial interpretation or variability of use among geomorphologists, geomorphology schools, and countries.

The geomorphic vocabulary, especially vocables referring to landforms, is to a large extent of vernacular origin, derived from terms used locally to describe landscape features and transmitted orally from generation to generation (Barrera-Bassols et al. 2006). Many of these terms, initially extracted from indigenous knowledge by explorers and field geomorphologists, subsequently received more precise definitions and were gradually incorporated into the scientific language of geomorphology. A typical example is the term *karst*, which refers to a mound of limestone fragments in Serbian language, and now applies to the dissolution process of calcareous rocks and the resulting geoforms. Many terms are used with different meanings depending on the country. For instance, the term *estero* (i.e., swamp) used in Spain means salt marsh, or tidal flat, or an elongated saltwater lagoon lying between sandbanks in a coastal landscape. In Venezuela, the same term refers to a closed depression, flooded by rainwater most of the time, in an alluvial plain. This kind of semantic alteration of concepts is common in countries colonized by Europeans, who intended to describe unfamiliar landscapes by similarity with their home experience. This resulted in vocabulary confusions and ambiguities that endure today. There is not yet a standardized terminology to label the geoforms, with additional semantic issues when the terms are translated from one language to another. Hereafter, an amalgam of vocables originating from various sources is used to name and describe the classes of geoforms in the six categories of the classification system.

7.2   The Taxonomy: Categories and Main Classes of Geotaxa

The categories in descending order are as follows (see Table 6.4 in Chap. 6):

- Geostructure
- Morphogenic environment
- Geomorphic landscape
- Relief/molding
- Lithology/facies
- Terrain form/landform

7.2.1   Geostructure

The concept of geostructure refers to an extensive continental portion characterized by its geologic structure, including the nature of the rocks (lithology), their age (stratigraphy), and their deformations (tectonics). These macro-units are related to plate tectonics. They include three taxa: cordillera, shield, and geosyncline.

- *Cordillera*: a system of young mountain chains, including also plains and valleys, which have been strongly folded and faulted by relatively recent orogenesis. The component ranges may have various orientations, but the mountain chain as a whole usually has one single general direction.
- *Shield*: a continental block that has been relatively stable for a long period of time and has undergone only slight deformations, in contrast to cordillera belts. It has been exposed to long-lasting downwasting and is composed mainly of Precambrian rocks.
- *Geosyncline* (or *sedimentary basin*): wide basin-like depression, usually elongate, that has been sinking deeply over long periods of time and in which thick sequences of stratified clastic sediments, layers of organic material, and sometimes volcanic deposits have accumulated. Through orogeny and folding, geosynclines are transformed into mountain ranges.

7.2.2 Morphogenic Environment

The morphogenic environment refers to a general type of biophysical setting, originated and controlled by a style of internal and/or external geodynamics. It comprises six taxa.

- *Structural environment*: controlled by internal geodynamics through tectonic movements (tilting, folding, faulting, overthrusting of bedrocks) or volcanism.
- *Depositional environment*: controlled by the deposition of detrital, soluble and/or biogenic materials, under the influence of water, wind, ice, mass removal, or gravity.
- *Erosional environment* (or denudational): controlled by processes of dissection and removal of materials transported by water, wind, ice, mass movement, or gravity.
- *Dissolutional environment*: controlled by processes of rock dissolution generating chemical erosion (karst in calcareous rocks, pseudokarst in non-calcareous rocks).
- *Residual environment*: characterized by the presence of surviving relief features (e.g., inselberg).
- *Mixed environment*: e.g., a structural environment dissected by erosion.

7.2.3 Geomorphic Landscape

7.2.3.1 Definition

Landscape is a complex concept which covers a variety of meanings:

- In common language: scenery of a portion of land or its pictorial representation.
- In media language: political, financial, intellectual, artistic landscape, etc.
- In scientific language: term used differently in landscape ecology, pedology, biogeography, geomorphology, architecture, etc.
- In the geomorphic literature: the expression *geomorphic landscape* is used without taxonomic connotation or mention of the level of generalization; it can thus correspond to any of the six categories of the system described here.
- Adopted definition: large land surface characterized by its physiographic expression. It is formed by a repetition of similar types of relief/molding or an association of dissimilar types of relief/molding. For instance, a large active alluvial plain may consist of a systematic spatial repetition of the same molding type, namely a set of adjoining floodplains constructed by a network of rivers. In contrast, a valley usually shows an association of various molding types, such as floodplain, terrace, fan, and glacis.
- Ambiguity of the concept of landscape: a valley, for instance, can cover three different kinds of spatial frame (Fig. 7.1):

    1. An area of longitudinal transport and deposition of sediments, including the floodplain and terraces of the valley bottom. This space corresponds to the concept of valley sensu stricto.
    2. An area similar to the previous one plus the sectors of lateral deposition forming fans and glacis. This space modeled by side deposits actually corresponds to the concept of piedmont landscape.
    3. An area controlled by human settlements, including the lower parts of the surrounding mountain slopes. This portion of space in fact belongs to the mountain landscape.

There is no consensus on whether restricting the concept of valley to the area covered by longitudinal deposits, or also including one or both of the two other components. An extreme position would be to extend the valley space to the surrounding water divides. In this case, the mountain landscape would vanish.

7.2.3.2 Taxa

The present system of geoform classification recognizes seven taxa at the categorial level of geomorphic landscape: valley, plain, peneplain, plateau, piedmont, hilland, and mountain (Figs. 7.2 and 7.3).

- *Valley*: elongated portion of land, flat, lying between two bordering areas of higher relief (e.g., piedmont, plateau, hilland, or mountain). A valley is usually drained by a single river. Stream confluences are frequent. For recognition, a valley should have a system of terraces which, in its simplest expression, comprises at least a floodplain and a lower terrace. In the absence of terraces, it is merely a fluvial incision, which is expressed on a map by the hydrographic network.

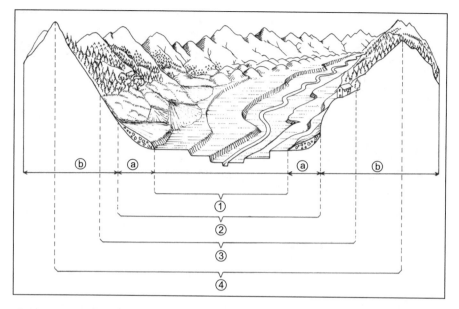

Fig. 7.1 Various definitions of the "valley" concept and their corresponding spatial expressions. (Zinck 1980)

1. Valley as an area where sediments of longitudinal origin, coming from the catchment area of the upper watershed, are deposited in the floodplain and terraces of the valley bottom
2. Valley as an area where longitudinal as well as lateral sediments are deposited, including piedmont glacis and fans
3. Valley as an area directly influenced by human occupation and activities, including the lower reaches of the surrounding mountain slopes
4. Hydrographic basin delineated by the water divides between adjacent watersheds
   (a) Piedmont
   (b) Mountain

- *Plain*: extensive portion of land, flat, unconfined, low-lying, with low relief energy (1–10 m of relative elevation difference), and gentle slopes, usually less than 3%. Several rivers contribute to form a complex fluvial system. Stream difluences are frequent.
- *Peneplain*: slightly undulating portion of land, characterized by a systematic repetition of low hills, rounded or elongated, with summits of similar elevation, separated by a dense hydrographic network of reticulated pattern. The hills and hillocks have formed either by dissection of a plain or plateau, or by downwasting and flattening of an initially rugged terrrain surface. Often, a peneplain consists of an association of three types of relief/molding: namely hills surrounded by a belt of glacis and, further, by peripheral colluvio-alluvial vales.
- *Plateau*: large portion of land, relatively high, flat, commonly limited at least on one side by an escarpment relating to the surrounding lowlands. It is frequently caused by tectonic uplift of a plain, and the elevated land portion is subsequently subdivided by incision of deep gorges and valleys. The summit topography is table-shaped or slightly undulating, because erosion is mostly linear. The plateau

Fig. 7.2 Types of geomorphic landscape. (Zinck 1980)
1 valley; 2 plain; 3 plateau; 4 piedmont; 5 hilland; 6 mountain

Fig. 7.3 Examples of geomorphic landscapes: (**a**) hilland dissected by two parallel valleys guided by tectonics, mountains in the background, Coastal Cordillera, northern Venezuela; (**b**) sandstone plateau dominating a valley, Guayana Shield, southern Venezuela; (**c**) alluvial plain with meandering river, Llanos basin, central Venezuela; (**d**) peneplain with residual hill in the center, surrounded by an annular glacis and a peripheral circular vale with palm trees, Guayana Shield, southern Venezuela

landscape is independent of specific altitude ranges, provided it complies with the diagnostic characteristics of this kind of geoform, such as high position, tabular topography, and escarpments along the edges and the water courses that deeply incise the relief. According to this definition, the table-shaped relief of the Mesa Formation in eastern Venezuela, cut by valleys of variable depth (40–100 m), makes up a plateau landscape at no more than 200–300 masl, while the Bolivian Altiplano is a plateau landscape lying at 3500–4000 masl.

- *Piedmont*: sloping portion of land lying at the foot of higher landscape units (e.g., plateau, mountain). The internal composition is generally heterogeneous and includes: (1) hills and hillocks formed from pre-Quaternary substratum, exposed by exhumation after the Quaternary alluvial cover has been partially removed by erosion; and (2) fans and glacis, often in terrace position (fan-terrace, glacis-terrace), composed of Quaternary detrital material carried by torrents from surrounding higher terrains. Piedmonts located at the foot of recent mountain systems (cordilleras) usually show neotectonic features, as for example faulted and tilted terraces.
- *Hilland*: rugged portion of land, characterized by a repetition of high hills, generally elongated, with variable summit elevations, separated by a moderately dense hydrographic network and many colluvio-alluvial vales.
- *Mountain*: high portion of land, rugged, deeply dissected, characterized by: (1) important relative elevations in relation to external surrounding lowlands (e.g., plains, piedmonts); (2) strong internal dissection, generating important net relief energy between ridge crests and intramountain valleys.

7.2.4 Relief/Molding

7.2.4.1 Definition

The concepts of relief and molding are based on the definition that is commonly given to both terms in the geomorphic French literature (Viers 1967).

- Relief: geoform that results from a particular combination of topography and geologic structure (e.g., cuesta relief); largely controlled by internal geodynamics.
- Molding: geoform determined by specific morphoclimatic conditions or morphogenic processes (e.g., glacis, fan, terrace, delta); largely controlled by external geodynamics.

7.2.4.2 Taxa

Relief and molding include an ample variety of taxa that can be grouped into families according to the dominant forming process: structural, erosional, depositional, dissolutional, and residual (Table 7.1 and Fig. 7.4). In general, the geomorphic

Table 7.1 Relief and molding types (Zinck 1988)

Structural	Erosional	Depositional	Dissolutional	Residual
Depression	Depression	Depression	Depression	Planation surface
Mesa (meseta)	Vale	Swale	Dome	Dome
Cuesta	Canyon (gorge)	Floodplain	Tower	Inselberg
Creston	Glacis	Flat (e.g., tidal	Hill (hum)	Monadnock
Hogback	Mesa (meseta)	flat)	Polje	Tors (boulders field)
Bar	Hill (hillock)	Terrace	Blind vale	…
Flatiron	Crest	Mesa (meseta)	Dry vale	
Escarpment	Rafter	Fan	Canyon	
Graben	(chevron)	Glacis	…	
Horst	Ridge	Bay		
Anticline	Dike	Delta		
Syncline	Trough (glacial)	Estuary		
Excavated anticline	Cirque (glacial)	Marsh		
Hanging syncline	…	Coral reef		
Combe		Atoll		
Ridge		…		
Cone (dome)				
Dike				
…				

literature does not establish a clear differentiation between geoforms of level 4 (relief/molding) and geoforms of level 6 (terrain form/landform). The list of geoforms in Table 7.1 was obtained by iteration, taking into account the possibility to subdivide types of relief and molding into terrain forms/landforms at level 6 of the system. It is an open-ended collection, which can be improved by the incorporation of additional geoforms.

7.2.5 Lithology/Facies

7.2.5.1 Definition

Level 5 provides information on (1) the petrographic nature of the bedrocks that serve as hard substratum to the geoforms, and (2) the facies of the unconsolidated cover formations that often constitute the internal hypogeal component of the geoforms. In both cases, the information concerns the parental material of the soils.

If the taxonomic system were restricted to depositional geoforms, the present categorial level could result redundant and therefore superfluous, as the lithology would be conveniently covered by the facies of the geomorphic material (i.e., the parent material of the soil) at level 6 of the system (i.e., the terrain form level). However, in areas where the soils are formed directly or indirectly from consolidated geologic material, the system should allow entering information about the lithology of the bedrocks.

Fig. 7.4 Examples of morphogenic environments and relief/molding types: (**a**) structural: mesetas developed on horizontal sandstone layers, Guayana Shield, southern Venezuela; (**b**) residual: monadnock inselberg rising above a semiarid peneplain landscape, Bahia State, north-eastern Brazil; (**c**) erosional-depositional: glacial landscape with gelifraction crests and cirques in the background, glacial trough in the center, moraines (lateral, ground, and frontal) in the foreground, Venezuelan Andes; (**d**) dissolutional: doline depression in limestone, southern France

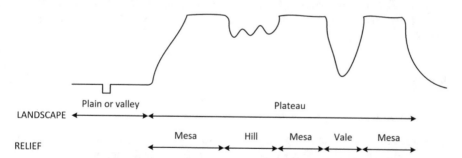

Fig. 7.5 Sequential partition of a plateau landscape into relief patterns to infer the lithology of the substratum (see Table 7.2 for lithology alternatives). (Zinck 1988)

In some geomorphic classification systems, the lithology is referred as high categorial levels. For instance, in the case of the geomorphic map of France, lithology is the second information layer in the structure of the legend, following a first level that deals with the location of the description sites (CNRS 1972).

Analyzing the portion of terrain represented in Fig. 7.5, an observer would recognize successively (hierarchically) the patterns identified in Table 7.2, by reasoning in the field or by photo-interpretation. The example shows that lithology is best

Table 7.2 Inference of the substratum lithology related to the plateau landscape depicted in Fig. 7.5 (Zinck 1988) (*API* aerial photo-interpretation)

Categorial level	Identification features	Geoform or material inferred	Generic concept	Resolution API Field	
High	Flat summit topography High position in relation to the surrounding lowlands Abrupt edges (escarpments) Deep river incision	Plateau	Landscape	+	−
Intermediate	Summit topography divided into: (1) level areas (2) undulating areas	(1) Mesas (2) Hills	Relief/ molding	+/−	+/−
Low	(1) If concordance between slope of the terrain surface and dip of the underlying rock layers, then structural surface supported by horizontally-lying rock strata (2) If no concordance between terrain surface and rock dip, then erosional surface truncating no-horizontally-lying rock strata	(1a) Hard sedimentary rocks (e.g. limestone, sandstone) or (1b) Hard extrusive igneous rocks (e.g. basalt) (2a) Tectonized stratified rocks (sedimentary or volcanic) or (2b) Intrusive igneous rocks	Lithology	−	+

positioned below the categorial levels where the concepts of landscape and relief/molding are located, respectively, taking into account criteria such as the hierarchic subdivision mechanism, the level of perception and the degree of resolution through interpretation of aerial photos (API), and the need for field and laboratory data.

7.2.5.2 Taxa

- Bedrocks (according to conventional rock classification):

  - igneous rocks, including intrusive rocks (e.g., granite, granodiorite, diorite, gabbro) and extrusive rocks (e.g., rhyolite, dacite, andesite, basalt)
  - metamorphic rocks (e.g., slate, schist, gneiss, quartzite, marble)
  - sedimentary rocks (e.g., conglomerate, sandstone, limolite, shale, limestone)

- Facies of unconsolidated materials:

  - nival (snow)
  - glacial (ice, glacier)
  - periglacial (ice, cryoclastism, thermoclastism)
  - alluvial (concentrated water flow = fluvial = river)
  - colluvial (diffuse, laminar water flow)

- diluvial (torrential water flow)
- lacustrine (freshwater lake)
- lagoonal (brackish water lake)
- coastal (fringe between continent and ocean; tidal)
- mass movement (plastic or liquid debris flow; landslide)
- gravity (rock fall)
- volcanic (surface flow or aerial shower of extrusive igneous materials)
- biogenic (coral reef)
- mixed (fluvio-glacial, colluvio-alluvial, fluvio-volcanic)
- anthropic (kitchen midden, sambaqui, tumulus, rubble, urban soil, etc.)

7.2.6   Terrain Form/Landform

7.2.6.1   Definition

In general, geomorphology textbooks do not establish a formal hierarchic differentiation of geoforms below the level of landscape. The terms *terrain form* and *landform* are often used as a general concept that covers any class of geomorphic unit from landscape level down to the lower levels of the system, without distinction between degrees of abstraction or levels of hierarchy. In this sense, both terms are synonyms of the generic term *geoform*.

In the present hierarchic system of geoform classification, terrain form/landform is considered as the generic concept of the lower level of the system. It corresponds to the elementary geomorphic unit whose minor internal and/or external variations are signaled by phases. It is characterized by its geometry, dynamics, and history.

The hierarchic arrangement of the collection of geoforms in Tables 7.3, 7.4, 7.5, 7.6, 7.7, 7.8, 7.9, 7.10, 7.11 is based on expert judgement and field experience (Zinck 1988). Geoforms can be conveniently grouped into: (1) geoforms predominantly controlled by the geologic structure and bedrock substratum (internal geodynamics), and (2) geoforms predominantly controlled by the morphogenic agents and surface formations (external geodynamics). Section 7.3 provides more details.

7.2.6.2   Taxa

- Geoforms predominantly controlled by the geologic structure and bedrock substratum

  - structural (monoclinal, folded, faulted)
  - volcanic
  - karstic

Table 7.3 Structural geoforms

Relief		Terrain form
Primary	Derived	
Monoclinal		
Cuesta (1–10° dip)	Double cuesta	Relief front (front slope)
Creston (10–30°)	Outlier hill	Scarp (overhang)
Hogback (30–70°)	Flatiron	Debris talus
Bar (70–90°)	Orthoclinal (subsequent) depression	Relief backslope
Flatiron	Cataclinal (consequent) depression	Structural surface
	Anaclinal (obsequent) depression	Substructural surface
		Cataclinal gap
Folded (Jurassian)		
Mont (original anticline)	Excavated anticline	Anticlinal hinge zone
Val (original syncline)	Hanging syncline	Synclinal hinge zone
	Rafter (chevron)	Fold flank
	Creston	Scarp
	Combe	Debris talus
	Cluse	
	Ruz	
Folded (Appalachian)		
	Truncated anticline	Scarp
	Bar	Debris talus
	Hanging syncline	
	Cataclinal gap	
Folded (complex)		
Overthrust nappe	Klippe	Scarp
Overthrust fold	Creston of overturned fold	Debris talus
Box fold	Escarpment of faulted fold	
Diapiric fold	Combe	
Faulted/fractured		
Fault scarp	Faultline scarp	Scarp
Horst	Fault escarpment facet	Debris talus
Graben	Cuesta	
Faults en échelon		
Block-faulted area		

- Geoforms predominantly controlled by the morphogenic agents and surface formations

  - nival, glacial, periglacial
  - eolian
  - alluvial and colluvial
  - lacustrine
  - gravity and mass movements
  - coastal

- Banal hillside geoforms

Table 7.4 Volcanic geoforms

Relief	Variety of geoforms
Depression	Crater
	Caldera
	Maar
	Lake
Cone	Ash cone
	Cinder cone
	Lava cone
	Spatter cone
	Stratovolcano
Dome	Cumulo-volcano shield-volcano intrusion dome
	Extrusion dome
	Extrusion cilinder
Flat	Lava flow
	Block lava (aa lava)
	Ropy lava (pahoehoe lava)
	Pillow lava volcanic mudflow (lahar)
	Fluvio-volcanic flow
	Cinder field
	Ash mantle
	Pyroclastic deposit
Mesa	Planèze
Cuesta	Hanging lava flow
	Sill
Bar	Longitudinal dyke
Dyke	Annular dyke (ring-dyke)
Tower	Volcano scarp
Escarpment	Volcanic plug (neck)
	Volcanic chimney (vent)
	Volcanic spine

Table 7.5 Karstic geoforms

Relief	Terrain form
Cockpit karst (dolines) hum karst (hills)	Karren Sima (aven)
Tower karst	Ponor
Cone karst	Doline
Polje (karstic plain)	Uvala
Karrenfeld	
Collapse valley	
Blind valley	
Dry valley	

Table 7.6 Glacial geoforms

Molding	Terrain form
Cirque trough	Trough threshold Cirque threshold
	Trough basin
	Trough shoulder
	Hanging valley (gorge)
	Roches moutonnées
	Ground moraine
	Lateral moraine
	Medial moraine
	Frontal moraine Knob-and-kettle till
	Blocks stream
	Dead-ice depression
Flat	Roches moutonnées field
	Drumlin field
	Ground moraine
	Push moraine
	Kame
	Esker
	Fluvio-glacial outwash fan (sandur)

Table 7.7 Periglacial geoforms

Molding	Terrain form
Crest (gelifraction)	Nunatak (horn)
	Debris talus (scree talus)
	Debris fan (scree fan)
Flat	Polygonal ground
	Mud field
	Stone field (pavement)
	Permafrost
	Tundra hummock
	Peatland (moor, bog)
	Dune field
	Loess mantle
Slope	Gravity scree
	Patterned ground striped ground
	Stone stream
	Mud flow (solifluction)

Table 7.8 Eolian geoforms

Molding	Terrain form
Flat (dune field, erg)	Barchan
	Nebka
	Parabolic dune
	Longitudinal dune
	Transverse dune
	Pyramidical dune (ghourd)
	Reticulate dune
	Blowout dune (eolian levee)
	Loess cover
	Blowout depression
	Reg (deflation pavement)
	Yardang
Meseta	Hamada (rocky deflation surface)

Table 7.9 Alluvial and colluvial geoforms

Depositional facies/erosion	Terrain form
Overload facies	Scroll bar
	Point bar complex
	River levee
	Distributary levee
	Delta channel levee
	Splay axis
	Splay mantle
	Crevasse splay
	Splay fan
	Splay glacis
	Alluvial fan
Overflow facies	Overflow mantle
	Overflow basin
Decantation facies	Decantation basin
	Backswamp (lateral depression)
	Cut-off meander with oxbow lake
	Infilled channel
Colluvial facies	Colluvial fan
	Colluvial glacis
Water erosion features	Sheet erosion
	Rill
	Gully
	Badland

Table 7.10 Gravity and mass movement geoforms

Process (consistence states)	Terrain form
Creep (variable consistence)	Creep mantle
	Pied-de-vache
	Terracette
Flow (plastic/líquid)	Rock flow
	Earth flow
	Debris flow
	Mud flow
	Solifluction sheet
	Solifluction tongue (stripe)
	Solifluction lobe
	Torrential lava
Slide (semi-solid)	Rotational slide (slump)
	Translational slide (slip)
	Rock slide
	Block slide
	Debris slide landslide
	Landslide scar
Fall (solid)	Rock fall
	Scree talus

Table 7.11 Coastal geoforms

Formation mode	Terrain form
Mechanical deposition	Beach
	Beachridge (coastal bar)
	Offshore bar (barrier beach) offshore trough
	Baymouth bar (Restinga)
	Cuspate bar spit
	Tombolo
	Slikke-schorre (tidal mudflat)
	Lagoon
	Dune sand cay
	Beachrock platform
Biogenic formation	Fringing reef
	Barrier reef
	Reef flat
	Reef front
	Lagoon
Erosion	Cliff
	Wave-cut platform/terrace
	Tidal channel
	Grao

7.3   Classification of the Geoforms at the Lower Levels

7.3.1   Introduction

The geotaxa belonging to the upper and middle levels of the system are defined in the previous section. The present section describes the classification of the geoforms at the lower categorial levels of the system: relief/molding and terrain form. The taxa listings are neither exhaustive nor free of ambiguity. It is mainly an attempt to categorize the existing geotaxa according to their respective level of abstraction and place them either at level 4 or level 6 of the classification system. A variety of synonymous terms can be found in the specialized literature, and the same type of geoform may be referred to with different names. With further progress in geomorphic mapping, probably new types of geoform will be identified and new names will appear. The concepts and terms used here are extracted from general texbooks and treatises in geomorphology. In case of multiple terms for a particular geoform, preference is given to the most commonly used one. Terms borrowed from different languages are kept in their original form and spelling, especially when already internationally accepted.

A criterion often used for grouping the geoforms in families is their origin or formation mode. Hereafter, the concept of origin is used in a broad sense, referring indistinctly to a type of environment (e.g., structural), an agent (e.g., wind), a morphogenic system (e.g., periglacial), or a single process (e.g., decantation).

The concept of origin, as a synonym for formation, is implicitly or explicitly present at all levels of the taxonomic system, but its diagnostic weight increases at the lower levels. The origin controlled by the internal geodynamics is more relevant in the upper categories, while the origin controlled by the external geodynamics is more important in the lower categories. It results from the former that there is a differential hierarchization of the diagnostic attributes according to the origin of the geoforms. For instance, in the case of the structural geoforms, genetic features have maximum weight at the level of the relief type, while in the case of the geoforms caused by exogenous agents (e.g., water, wind, ice), the genetic features have maximum weight at the lower levels of the system (i.e., facies and terrain form).

A morphogenic agent can cause erosional as well as depositional features according to the context in which the process takes place. For this reason, a distinction is made between erosional and depositional terrain forms. Likewise, structural geoforms may have been strongly modified by erosion, a fact which leads to distinguish between original (primary) and derived forms.

A geoform is considered erosional when the erosion process, operating either by areal removal of material or by linear dissection, is responsible for creating the dominant configuration of that geoform. Local modifications caused, for instance, by the incision of rills and gullies or surficial deflation by wind are identified as phases of the affected taxonomic unit. Similarly, point features and phenomena of limited extent are not considered as taxonomic units and are represented by cartographic spot symbols on the maps (e.g., geysers, erratic blocks, pingos, etc.).

For the definition of the geoforms whose names are reported in the attached tables, it is recommended to consult the textbooks and dictionaries of geomorphology, namely Derruau (1965), CNRS (1972), Visser (1980), Lugo-Hubp (1989), among others. The multilingual *Geological Nomenclature* (Visser 1980) is particularly useful, in the current context of unstandardized vocabulary, for short definitions of geoforms and multilingual equivalents. Some geoforms may appear named at both levels of relief/molding and terrain form, because their taxonomic position in the classification system is not yet clearly established.

7.3.2 Geoforms Mainly Controlled by the Geologic Structure

Geostructural control acts through tectonics, volcanism and/or lithology. Therefore, the internal geodynamics is determinant in the formation of this kind of geoforms, in combination with external processes of erosion or deposition in varying degrees. The dissection of primary structural reliefs by mechanical erosion, for instance, results in the formation of derived relief forms. Chemical erosion through limestone dissolution or sandstone disintegration causes the formation of karstic and pseudo-karstic reliefs. Deposition of volcanic ash or scoriae can alter the original configuration of a structural relief.

7.3.2.1 Structural Geoforms Proper

Geoforms directly caused by structural geodynamics (folds and faults) cover a large array of relief types (Table 7.3):

- Monoclinal reliefs: rock layers uniformly dipping up to 90° (see Fig. 6.4 in Chap. 6). Strata of hard rocks (e.g., sandstone, quartzite, limestone) overlie softer rocks (e.g., marl, shale, slate). The duo hard rock/soft rock can be recurrent in the landscape causing the same relief type to repeat several times (e.g., double cuesta).
- Jurassian fold reliefs: symmetrical folds in regular sequences of structural highs (anticlines) and structural lows (synclines) in their original or almost original form; related to important volumes of stratified sedimentary rock layers.
- Appalachian fold reliefs: fold reliefs in advanced stage of flattening and dissection.
- Complex fold reliefs: primary or derived fold reliefs controlled by overthrust tectonics and complex folding.
- Fault reliefs: primary or derived reliefs caused by faults or fractures; the faulting style (i.e., normal, reverse, rotational, overthrust, etc.) controls the type of resulting relief.

7.3.2.2   Volcanic Geoforms

Volcanic materials can constitute the whole substratum or an essential part thereof or be limited to cover formations in a variety of landscapes including mountain, plateau, piedmont, plain, and valley. Volcanic geoforms are of variable complexity, and this makes it difficult to strictly separate relief types and terrain forms. An ash cone, for instance, can be a very simple geoform and constitute therefore an elementary terrain form, while a stratovolcano cone is usually a much more complex geoform with various terrain forms (Table 7.4).

7.3.2.3   Karstic Geoforms

Karst formation operates by chemical erosion of soluble rocks and originates sculpted terrain surfaces and underground gallery systems of complex configuration, characterized by residual geoforms of positive or negative relief. The resulting taxa enter the system essentially at the relief/molding level. The karstic geoforms are both endogenous by the influence of the lithology in their constitution and exogenous by the dissolution process which originates them (Table 7.5).

7.3.3   Geoforms Mainly Controlled by the Morphogenic Agents

Water, wind, and ice are morphogenic agents that cause erosion or deposition according to the prevailing environmental conditions. The resulting geoforms are usually more homogeneous than the geoforms controlled by the internal structure. For this reason, many of the geoforms originated by exogenous agents can be classified at the level of terrain form. Hereafter, six main families of geoforms are distinguished according to their origin.

7.3.3.1   Nival, Glacial, and Periglacial Geoforms

The nival, glacial, and periglacial geoforms have in common the fact that they develop in cold environments (high latitudes and high altitudes) by the accumulation of snow (nival geoforms), alternate freezing-thawing causing gelifraction (periglacial geoforms), or accumulation of ice mass (glacial geoforms). Some geoforms result from deposition (e.g., moraines), others from erosion (e.g., glacial cirque) (Fig. 7.6). Some can be recognized and mapped as elementary terrain forms (e.g., a moraine). Others are molding types that consist of more than one kind of terrain form. A glacial trough, for instance, can contain different types of moraine (e.g., ground, lateral, frontal), surfaces with "roches moutonnées", hanging valleys, and lagoons, among others (Tables 7.6 and 7.7). Strictly speaking, the nival forms are

Fig. 7.6 Configuration and components of a glacial valley or glacial trough. (Zinck 1980)

not terrain forms, since they are covered with snow (e.g., nivation cirque, permanent snowpack, and snow avalanche corridor and fan).

7.3.3.2 Eolian Geoforms

Dry environments, from desert to subdesert, are most favorable to forms arising from the action of the wind. Eolian geoforms occur mainly in coastal or continental plains where the effect of the wind is more pronounced (Table 7.8).

7.3.3.3 Alluvial and Colluvial Geoforms

Alluvial geoforms can occur in almost all types of landscape, but mostly in plains and valleys where they form terraces, floodplains, glacis, and fans. The colluvial geoforms are typical features of the piedmont landscape where they form fans and glacis. The elementary terrain forms are grouped according to the type of depositional process that originates them (Table 7.9).

7.3.3.4 Lacustrine Geoforms

The receding of lake shorelines, which is a common process in drying lakes after the last glaciation, leaves exposed lacustrine material in the form of terraces. In arid and semi-arid environments, stratified fluvio-lacustrine deposits occur in playa-type depressions. In areas emerging from proglacial lakes there are stratified varve deposits.

7.3.3.5 Gravity and Mass Movement Geoforms

The mechanical condition of the material, with continuity from solid state to liquid state, controls the mass movement processes, including creep, flow, slide, and fall, that give rise to the geoforms (Table 7.10).

7.3.3.6 Coastal Geoforms

The most typical coastal geoforms are developed in the coastal lowlands, including molding types such as salt marsh, mangrove marsh, estuary, delta, bay, reef, and atoll. Cliff is the most common form in rocky coasts (Table 7.11).

7.3.4 Banal Hillside Geoforms

Geoforms without remarkable physiographic features have been called *banal* (CNRS 1972). Such geoforms are frequent in soft sedimentary rocks, devoid of structural control (e.g., marls and other argillaceous rocks), and in igneous-metamorphic rocks without marked schistosity (e.g., granite, gneiss). Their most common physiographic feature is expressed by convex-concave hillslopes that have inspired the slope facet models. This unoriginal but not uncommon topography does not reflect any specific internal or external geodynamics, as is the case of the two other large families of geoforms. Present hills are in general an inheritance of a long geomorphic evolution. They are increasingly exposed to severe slope erosion processes including sheet erosion, rills, gullies, and mass movements.

7.3.4.1 Main Characteristics

- General topography of hills, ridges, and crests, originated by dissection.
- Little or none structural influence, in particular lack of specific control by fold or fault tectonics in the topography.
- Presence of fractures that favor and control the incision and organization of the hydrographic network.

- The drainage pattern has a relevant influence on the configuration of the resulting dissection topography, especially in peneplain and hilland landscapes.
- Homogeneous rock substratum over wide expanses.
- Material of moderate to weak resistance to physical and/or chemical erosion. Banal geoforms are frequent in shale and marl. In warm and moist tropical environments, chemical erosion of granite or gneiss produces also banal geoforms in peneplain landscape.

7.3.4.2 Classes of Banal Hillside Geoforms

Banal geoforms occur at the levels of relief/molding and terrain form in mountain, hilland, peneplain, and piedmont landscapes.

(a) At the level of relief/molding

- The *backbone* configuration consists of an association between a main longitudinal dorsal and a set of perpendicular hills (chevron, rafter, nose) separated by vales (Fig. 7.7). This type of relief is common in fractured sedimentary rocks. Its further evolution generates elongated horseback-shaped hills.
- The *half-orange* configuration consists of a systematic repetition of rounded hills with similar elevation. This type of relief is typical of the peneplain landscape developed in homogeneous but intensively fractured igneous or metamorphic substratum, with reticulate drainage pattern. It is common in the Precambrian shields of the intertropical zone.

(b) At the level of terrain form

 Slope segmentation into interrelated facets seems to be the most convenient criterion to subdivide any hilly relief. The slope models such as the nine-unit-land-surface model of Conacher and Dalrymple (1977) or the five-hillslope-element model of Ruhe (1975) can be implemented to this effect. Table 7.12

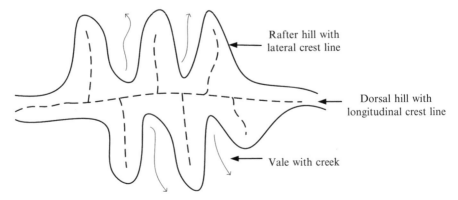

Fig. 7.7 Hilland landscape with backbone configuration comprising a longitudinal dorsal and perpendicular rafters

Table 7.12 Slope facet model

Slope facet	Topographic profile	Dominant morphodinamics
Summit	Level/convex	Ablation/erosion
Shoulder	Convex	Erosion
Backslope	Rectilinear-inclined	Material in transit
Footslope	Concave	Lateral accumulation
Toeslope	Concave/level	Longitudinal accumulation

Adapted from Ruhe (1975)

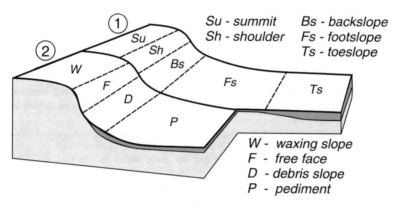

Fig. 7.8 Models of convex-concave "fully developed hillslopes" with lateral deposits. (Adapted from Ruhe 1975)
1. Ruhe's model (note that the toeslope deposits are of longitudinal origin)
2. Model combining elements taken from Wood (1942) and King (1957)

    shows the relationships between slope facet, topographic profile, and dominant morphogenic dynamics according to Ruhe's model (Fig. 7.8). It is worth noting that the toeslope is actually not a slope facet; instead, it is a unit that belongs to the adjoining valley or vale, with slope perpendicular to the hillside and with longitudinal deposits.

Models are suitable generalizations of real situations. However, the general hillside model with convex-concave profile can be disturbed by irregularities. For instance, the cross section of a hill shows often complications that should be considered in the mapping of the geoforms and soils. These complications can be caused by the heterogeneity of the local geologic substratum or the local morphodynamics. A convex-concave slope can be interrupted by treads and scarps that reflect tectonic influence or lithologic changes. Likewise, the general topographic profile can be locally disturbed or modified by water erosion (e.g., rills and gullies) or mass movements (e.g., terracettes, landslides, solifluction scars and tongues).

7.4   Conclusion

This chapter organizes existing geomorphic knowledge and arranges the geoforms in a hierarchically structured system with six nested levels. It is thought that this multicategorial geoform classification scheme reflects the structure of the geomorphic landscape sensu lato. The approach segments and stratifies the landscape continuum into geomorphic units belonging to different levels of abstraction. This geoform classification system has shown to be useful in geopedologic mapping and could be useful also in digital soil mapping.

References

Barrera-Bassols N, Zinck JA, Van Ranst E (2006) Local soil classification and comparison of indigenous and technical soil maps in a Mesoamerican community using spatial analysis. Geoderma 135:140–162

CNRS (1972) Cartographie géomorphologique. Travaux de la RCP77, Mémoires et Documents vol 12. Editions du Centre National de la Recherche Scientifique, Paris

Conacher AJ, Dalrymple JB (1977) The nine-unit landscape model: an approach to pedogeomorphic research. Geoderma 18:1–154

Derruau M (1965) Précis de géomorphologie. Masson, Paris

Derruau M (1966) Geomorfología. Ediciones Ariel, Barcelona

Fairbridge RW (ed) (1997) Encyclopedia of geomorphology. Springer, New York

Garner HF (1974) The origin of landscapes. A synthesis of geomorphology. Oxford University Press, New York

Goudie AS (ed) (2004) Encyclopedia of geomorphology, vol 2. Routledge, London

Huggett RJ (2011) Fundamentals of geomorphology. Routledge, London

King LC (1957) The uniformitarian nature of hillslopes. Trans Edinb Geol Soc 17:81–102

Lugo-Hubp J (ed) (1989) Diccionario geomorfológico. Universidad Nacional Autónoma de México, Cd México

Ruhe RV (1975) Geomorphology. Geomorphic processes and surficial geology. Houghton Mifflin, Boston

Thornbury WD (1966) Principios de geomorfología. Editorial Kapelusz, Buenos Aires

Tricart J (1965) Principes et méthodes de la géomorphologie. Masson, Paris

Tricart J (1968) Précis de géomorphologie. T1 Géomorphologie structurale. SEDES, Paris

Tricart J (1977) Précis de géomorphologie. T2 Géomorphologie dynamique générale. SEDES-CDU, Paris

Tricart J, Cailleux A (1962) Le modelé glaciaire et nival. SEDES, Paris

Tricart J, Cailleux A (1965) Le modelé des régions chaudes. Forêts et savanes. SEDES, Paris

Tricart J, Cailleux A (1967) Le modelé des régions périglaciaires. SEDES, Paris

Tricart J, Cailleux A (1969) Le modelé des régions sèches. SEDES, Paris

Van Zuidam RA (1985) Aerial photo-interpretation in terrain analysis and geomorphological mapping. ITC, Enschede

Verstappen HT (1983) Applied geomorphology; geomorphological survey for environmental development. Elsevier, Amsterdam

Verstappen HT, Van Zuidam RA (1975) ITC system of geomorphological survey. ITC, Enschede

Viers G (1967) Eléments de géomorphologie. Nathan, Paris

Visser WA (ed) (1980) Geological nomenclature. Royal Geological and Mining Society of the Netherlands. Bohn, Scheltema & Holkema, Utrecht

Wood A (1942) The development of hillside slopes. Geol Ass Proc 53:128–138
Zinck JA (1974) Definición del ambiente geomorfológico con fines de descripción de suelos. Ministerio de Obras Públicas (MOP), Cagua
Zinck JA (1980) Valles de Venezuela. Lagoven, Petróleos de Venezuela, Caracas
Zinck JA (1988) Physiography and soils. In: ITC soil survey lecture notes. International Institute for Aerospace Survey and Earth Sciences, Enschede

Chapter 8
The Geomorphic Landscape: The Attributes of Geoforms

J. A. Zinck

Abstract Attributes are characteristics used for the description, identification, and classification of the geoforms. They are descriptive and functional indicators that make the multicategorial system of the geoforms operational. Four kinds of attribute are used: (1) morphographic attributes to describe the geometry of geoforms; (2) morphometric attributes to measure the dimensions of geoforms; (3) morphogenic attributes to determine the origin and evolution of geoforms; and (4) morphochronologic attributes to frame the time span in which geoforms originated. The morphometric and morphographic attributes apply mainly to the external (epigeal) component of the geoforms, are essentially descriptive, and can be extracted from remote-sensed documents or derived from digital elevation models. The morphogenic and morphochronologic attributes apply mostly to the internal (hypogeal) component of the geoforms, are characterized by field observations and measurements, and need to be substantiated by laboratory determinations.

Keywords Morphography · Morphometry · Morphogenesis · Morphochronology · Attribute classes · Attribute weights

8.1 Introduction

Attributes are characteristics used for the description, identification, and classification of the geoforms. They are descriptive and functional indicators that make the multicategorial system of the geoforms operational. This implies two requirements: (1) select descriptive attributes that help identify the geoforms, and (2) select

---

J. A. Zinck died before publication of this work was completed.

J. A. Zinck (deceased)
Faculty of Geo-Information Science and Earth Observation (ITC), University of Twente, Enschede, The Netherlands

differentiating attributes that allow classifying geoforms at the various categorial levels of the taxonomic system.

To determine a geoform, it is necessary to sequentially perform the following operations:

- description and measurement, to characterize the properties and constituents
- identification, to compare the geoforms to be determined with established reference types
- classification, to place the geoforms to be determined in the taxonomic system

For this purpose, four kinds of attribute are used, following Tricart's proposal with respect to the four types of data that a detailed geomorphic map should comprise (Tricart 1965a, b):

- geomorphographic attributes, to describe the geometry of geoforms
- geomorphometric attributes, to measure the dimensions of geoforms
- geomorphogenic attributes, to determine the origin and evolution of geoforms
- geomorphochronologic attributes, to frame the time span in which geoforms originated

In order to simplify the expressions, it is customary to omit the prefix *geo* in the denomination of the attributes.

The morphometric and morphographic attributes apply mainly to the external (epigeal) component of the geoforms, are essentially descriptive, and can be extracted from remote-sensed images or derived from digital elevation models. The morphogenic and morphochronologic attributes apply mostly to the internal (hypogeal) component of the geoforms, are characterized from field observations and measurements, and need to be substantiated by laboratory determinations.

8.2 Morphographic Attributes: The Geometry of Geoforms

The morphographic attributes describe the geometry and shape of the geoforms in topographic and planimetric terms. They are commonly used for automated identification of selected geoform features from DEM (Hengl 2003).

8.2.1 Topography

Topography refers to the cross section of a portion of terrain (Fig. 8.1). It can be viewed in two dimensions from a vertical cut through the terrain generating the topographic profile (Table 8.1), and in three dimensions from a terrain elevation model generating the topographic shape (Table 8.2). The characterization of these features is particularly relevant in sloping areas. The shape and the profile of the topography are related to each other but described at different categorial levels. The

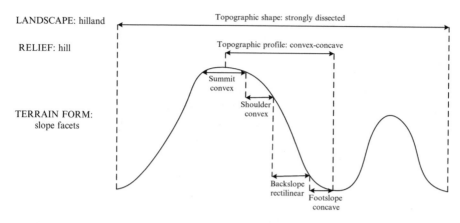

Fig. 8.1 Relationship between topographic attributes and categorial levels of the geoform classification system

Table 8.1 Topographic profile (2D)

Classes	Examples
Level	Mesa, terrace
Concave	Basin, footslope facet
Convex	Levee, summit/shoulder facet
Convex-concave	Slope facet complex
Convex-rectilinear-concave	Slope facet complex
Rectilinear (straight)	Backslope
With intermediate flat step(s)	Slope facet complex
With protruding rock outcrop(s)	Slope facet complex
With rocky scarp(s)	Slope facet complex, cuesta
Asymmetric	Hill, hogback
Irregular	Hillside

Table 8.2 Topographic shape (3D)

Classes	Slope %	Relief amplitude
Flat or almost flat	0–2	Very low
Undulating	2–8	Low
Rolling	8–16	Low
Hilly	16–30	Moderate
Steeply dissected	>30	Moderate
Mountainous	>30	High

topographic shape attributes are used at landscape level, while the topographic profile attributes are used at the levels of relief and terrain form. The third descriptor, the exposure or aspect which indicates the orientation of the relief in the four cardinal directions and their subdivisions, can be used at any level of the system.

8.2.2 *Planimetry*

Planimetry refers to the vertical projection of the geoform boundaries on a horizontal plane. It is a two-dimensional representation of characteristic geoform features that closely control the soil distribution patterns. Fridland (1965, 1974, 1976) and Hole and Campbell (1985) were among the first to recognize configuration models that delimit soil bodies and relate these with the pedogenic context. The configuration of the geoform, the design of its contours, the drainage pattern, and the conditions of the surrounding environment are the main attributes described for this purpose.

8.2.2.1 Configuration of the Geoforms

Many geoforms at the levels of relief/molding and terrain form show typical configurations that can be easily extracted from remote-sensed documents, especially air photos. This enables preliminary identification of geoforms based on the covariance between morphographic and morphogenic attributes. For instance, a river levee is generally narrow and elongated, while a basin is wide and massive. The configuration attributes give an idea of the massiveness or narrowness of a geoform (Table 8.3).

8.2.2.2 Contour Design of the Geoforms

The design of the contours describes the peripheral outline of the geoform at the levels of relief/molding and terrain form (Fig. 8.2 and Table 8.4). It can vary from straight (e.g., recent fault scarp) to wavy (e.g., depositional basin) to indented (e.g., scarp dissected by erosion). These variations from very simple linear outlines up to complex convoluted contours that approximate areal configurations, are reflected in

Table 8.3 Configuration of the geoforms

Classes	Examples
Narrow	Levee
Large	Overflow mantle
Elongate	Dike
Massive	Basin
Annular (ring-shaped)	Volcanic ring-dyke
Oval/elliptic	Doline, sinkhole
Rounded	Hill
Triangular	Fan, delta
Irregular	Dissected escarpment

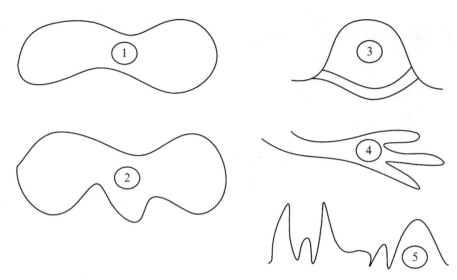

Fig. 8.2 Configuration and contour design of some geoforms (2D)

Table 8.4 Contour design of geoforms

Classes	Examples
Rectilinear	Escarpment
Arched (lunate)	coastal bar
Sinuate (wavy)	River levee
Lobulate	Basin
Denticulate	Dissected escarpment
Digitate	Deltaic channel levee (distal sector)
Irregular	Gully, badland

variations of the fractal dimension (Saldaña et al. 2011). The attribute of contour design can be used also as an indirect morphogenic indicator. For instance, an alluvial decantation basin has usually a massive configuration, but the shape of the boundaries can vary according to the dynamics of the neighboring forms. In general, a depositional basin has a sinuous outline, but when a crevasse splays that forms after opening a gap in a river levee in high water conditions penetrates into the basin, the different fingers of the splay create a lobulated distal contour. Thus, a lobulated basin contour can reflect the proximity of a digitate splay fan, with overlap of a light-colored sandy deposit fossilizing the argillaceous gley material of the basin (Fig. 8.3).

1. Basin with ovate configuration and sinuous contour
2. Basin with ovate configuration and lobulate contour (lower part), reflecting the penetration of a digitate crevasse splay fan (see Fig. 8.3)
3. Bay closed by an arch-shaped offshore bar
4. Deltaic channel levee with digitate distal extremities
5. Dissected scarp with denticulate contour pattern

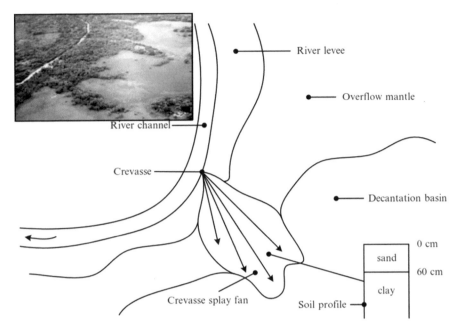

Fig. 8.3 Modification of a basin contour design by the penetration of a crevasse splay fan upon rupture of a levee during high channel water. The intrusion of the fan in the neighboring lateral depression results in the overlaying of sandy cover sediments on top of the clayey basin substratum, creating a lithologic discontinuity at 60 cm depth in this case, with the formation of a buried soil

8.2.2.3   Drainage Pattern

The drainage pattern refers to the network of waterways, which contributes to enhance the configuration and contour outline of the geoforms. It is mainly controlled by the geologic structure (tectonics, lithology, and volcanism) in erosional areas, and by the structure and dynamics of the depositional system in aggradation areas. Representative patterns taken from the Manual of Photographic Interpretation (ASP 1960) are shown in Figs. 8.4 and 8.5: radial pattern of a conic volcano, annular pattern in a set of concentric calderas, dendritic pattern in homogeneous soft sedimentary rocks without structural control, trellis pattern in sedimentary substratum with alternate hard and soft rock layers and with structural control (faults and fractures), parallel pattern in alluvial area, and rectangular pattern in a till plain. The network of waterways creates connectivity between the areas that it crosses and controls the various kinds of flow that traverse the landscape (water, materials, wildlife, vegetation, humans).

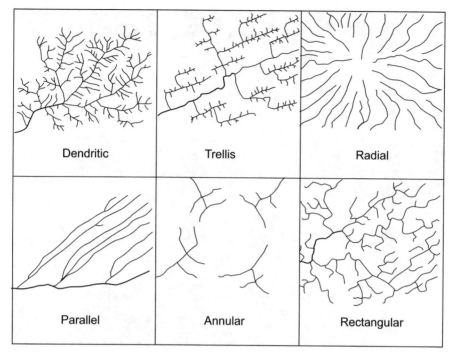

Fig. 8.4 Drainage patterns controlled by features of the geologic and geomorphic structure (see comments in the text). (Adapted from ASP 1960)

8.2.2.4   Neighboring Units and Surrounding Conditions

The geomorphic units lying in the vicinity of a geoform under description shall be mentioned along with the surrounding conditions. This attribute applies at the levels of landscape, relief/molding, and terrain form. According to its position in the landscape, a geoform can topographically dominate another one, be dominated by it, or lie at the same elevation (e.g., a plain dominated by a piedmont). These adjacency conditions suggest the possibility of dynamic relationships between neighboring geoforms and enable to model them. In a piedmont landscape, for instance, can start water flows that cause flooding in the basins of a neighboring alluvial plain, or material flows that cause avulsion in agricultural fields and siltation in water reservoirs. The segmentation of the landscape into functionally distinct geomorphic units provides a frame for analyzing and monitoring transfers of physical, chemical, mineralogical, and biological components within and between landscapes.

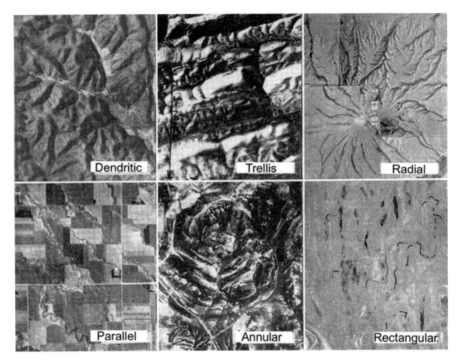

Fig. 8.5 Geologic and geomorphic structure features controlling the drainage patterns (see comments in the text). (Adapted from ASP 1960)

8.2.3 *Morphography and Landscape Ecology*

The morphographic attributes, in particular the configuration and contour design of the geoforms, have close semantic and cartographic relationships with concepts used in landscape ecology, such as mosaic, matrix, corridor, and patch (Forman and Godron 1986). A deltaic plain is a good example that illustrates the relationship between the planimetry of the geoforms and the metrics used in landscape ecology. A deltaic plain that occupies the distal area in a depositional system is a dynamic entity that receives materials and energy from the medial and proximal sectors of the same system. Delta channels are axes which introduce water and material in the system, conduct them through the system, and distribute them to other positions within the system such as overflow mantles and basins. Channels are elongated, sinuous, narrow corridors that feed the deltaic depositional system. In general, the mantles (overflow or splay) are extensive units that form the matrix of the system. The basins are closed depressions, forming scattered patches in the system (Fig. 8.6).

In the center and to the right, a deltaic alluvial system with relative age Q1 (i.e., upper Pleistocene) fossilizes a previous depositional system of relative age Q2 (i.e., late middle Pleistocene) of which the elongated patches of overflow basin are

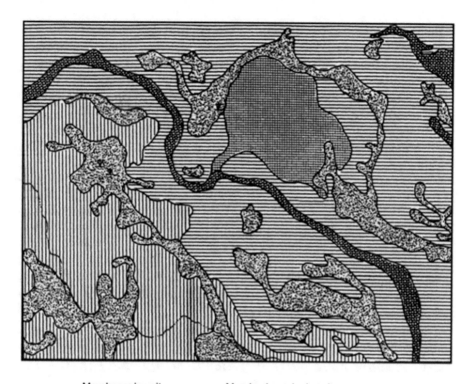

Morphogenic unit	Morphochronologic unit
Delta channel levee	Q1
Crevasse splay fan	Q1
Overflow mantle	Q1
Overflow mantle	Q2
Overflow basin	Q2

0   250   500   750 m

Fig. 8.6 Contact area between two depositional systems differentiated by relative age
Extract from a soil series map of the Santo Domingo river plain, Venezuela; survey scale 1: 25,000.
(Adapted from Pérez-Materán 1967)

remnants. The delta channel is the axial unit of the depositional system and functions as a corridor through which water and sediments transit before being distributed within the system. A unit of triangular configuration is grafted on the delta channel, corresponding to a crevasse splay fan that originated upon the opening of a gap in the levee of the channel. The overflow mantles are the matrices of both depositional systems (Q1 and Q2). The basins and the splay fan correspond to patches.

8.3   Morphometric Attributes: The Dimension of Geoforms

Morphometry covers the dimensional features of the geoforms as derived from a numerical representation of the topography (Pike 1995; Pike and Dikau 1995). Computerized procedures allow the extraction and measurement of a variety of morphometric parameters from a DEM, some being relevant at local scale and others at regional scale, including slope, hypsometry, orientation (aspect), visual exposure, insolation, tangential curvature, profile curvature, catchment characteristics (extent, elevation, slope), and roughness (Gallant and Hutchinson 2008; Olaya 2009). While many of these land-surface parameters are used in topography, hydrography, climatology, architecture, urban planning, and other applied fields, only a few actually contribute to the characterization of terrain forms, in particular the relative elevation, drainage density, and slope gradient. These are subordinate, not diagnostic attributes which can be used at any categorial level with variable weight. Morphometric attributes are interrelated: at a specific range of relative elevation, there is a direct relationship between drainage density and slope gradient; the higher the drainage density, the greater is the slope gradient, and conversely (A and B, respectively, in Fig. 8.7).

8.3.1   Relative Elevation (Relief Amplitude, Internal Relief)

The relative elevation between two geoforms is evaluated as high, medium, or low. Ranges of numerical values (e.g., in meters) can be attributed to these qualitative classes within the context of a given region or project area. Numerical ranges are established on the basis of local or regional conditions and are valid only for these conditions. Relative elevation is a descriptive attribute, and the classes of relative elevation can be differentiating but are not diagnostic. Likewise, the absolute altitude is not a diagnostic criterion, because similar geoforms can be found at various elevations. For instance, the Bolivian Altiplano at 3500–4000 masl, the Gran Sabana area in the Venezuelan Guayana at 800–1100 masl, and the mesetas of eastern Venezuela at 200–400 masl show all three the diagnostic characteristics of the plateau landscape, although at different elevations.

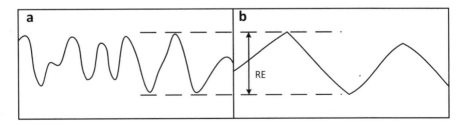

Fig. 8.7   Relationship between drainage density and slope gradient in similar conditions of relative elevation (RE). (Adapted from Meijerink 1988)

8.3.2 Drainage Density

Drainage density measures the degree of dissection or incision of a terrain surface. Density classes are set empirically for a given region or project area. For instance, Meijerink (1988) determines drainage density classes (called valley density VD) based on the relationship $VD = \Sigma L/A$, where ΣL is the cumulative length of drainage lines in km and A is the area in km^2. Not only the conditions of the region studied but also the study scale affects the numerical values of VD (Fig. 8.8). The FAO Guidelines for soil description (2006) define potential drainage density values based on the number of "receiving" pixels within a window of 10×10 pixels.

8.3.3 Relief Slope

The slope gradient is expressed in percentages or degrees. There are geoforms that have characteristic slopes or specific slope ranges. For instance, a coastal cliff or a young fault escarpment is often vertical and has therefore a slope close to 90°. A debris talus has an equilibrium slope of 30–35°, which corresponds to the angle of repose of the loose debris covering the slope. However, the mere knowledge of these numerical values does not contribute directly to identify the corresponding geoform. The slope gradient is essentially a descriptive attribute, at the most covariant with other attributes of higher diagnostic value. Obviously, a hill has a slope greater than a valley floor.

8.3.4 Terrain and Soil Surface Features

Morphometry is not limited to the extraction of topographic parameters from a DEM. Remote sensing also contributes to the characterization of the epigeal component of the geoforms. A variety of terrain and soil surface features can be identified, measured, and delineated from remote-sensed data. An inventory of parameters that can be characterized from optical and microwave sensors includes mineralogy, texture, moisture, organic carbon, iron oxides, salinity, carbonates, terrain roughness, and erosion features (Wulf et al. 2015). Spectral signatures covariate with laboratory determined or field observed property values. Some of these attributes may perform better al local scale to identify patches of specific surface features such as spatial variations in texture, organic matter content, or soil erosion in a given geomorphic unit. Others can contribute to delineate entire geoforms or associations thereof, for instance, in poorly drained, salt-affected or land degraded areas. Landscapes with no or sparse vegetation cover in dry environments offer the best possibilities for remote-sensed morphometry characterization (Metternicht and Zinck 1997, 2003; del Valle et al. 2010; Metternicht et al. 2010). Del Valle et al.

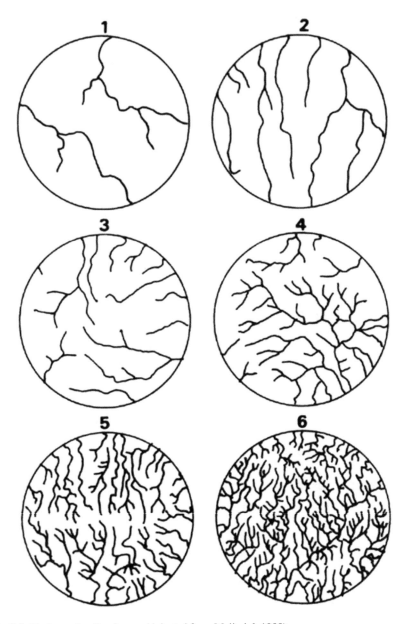

Fig. 8.8 Drainage density classes. (Adapted from Meijerink 1988)

(Chap.19 in this book) show the capability of the PALSAR L-band to penetrate coarse-textured materials several decimeters below the terrain surface to detect buried geological and geomorphic features. So far less widely used than remote sensors, proximal sensors present promising opportunities to further explore the hypogeal component of the geoforms.

8.3.5 *Contribution of Digital Morphometry*

With the development of digital cartography, (geo)morphometry is increasingly used to characterize terrain units based on individual numerical parameters that are extracted from a DEM, such as altitude, relative elevation, slope, exposure, and curvature, among others. Attributes such as slope, and curvature can present continuous variations in space and are therefore suitable for fuzzy mapping. This is in particular the case of banal hillside reliefs with convex-concave slope profiles according to the model of Ruhe (1975). However, many geoforms have relatively discrete boundaries that reflect their configuration and contour design. This is especially the case of constructed geoforms. In brief, the contribution of digital morphometry resides essentially in the automated estimation of dimensional attributes of the geoforms. However, limiting the description of the geoforms to their morphometric characteristics, just because the latter can be extracted automatically from a DEM, carries the risk of replacing field observation and image reading by numerical parameters which do not reflect satisfactorily the structure and formation of the geomorphic landscape. The scope of the morphometric characteristics to interpret the origin and evolution of the relief is limited, because morphometry covers only part of the external features of the geoforms, their epigeal component.

8.4 Morphogenic Attributes: The Dynamics of Geoforms

Selected geoform attributes reflect forming processes and can therefore be used to reconstruct the morphogenic evolution of an area or infer past environmental conditions. In general, the attribute-process relationship is more efficient for identifying geoforms in depositional environment than in erosional environment. Constructed geoforms are usually more conspicuous than erosional geoforms, except for features such as gullies or karstic erosion forms, for instance. Hereafter, some morphogenic attributes are analyzed by way of examples. Particle size distribution, structure, consistence, mineralogical characteristics, and morphoscopic features are good indicators of the origin and evolution of the geoforms.

8.4.1 *Particle Size Distribution*

8.4.1.1 Relevance

Particle size distribution, or its qualitative expression of texture, is the most important property of the geomorphic material, as well as of the soil material, because it controls directly or indirectly a number of other properties. The particle size distribution provides basic information for the following purposes:

- Characterization of the material and assessment of its suitabilities for practical uses (e.g., agricultural, engineering, etc.).
- Inference of other properties of the material that closely depend on the particle size distribution (often in combination with the structure of the material), such as bulk density, specific surface area, cohesion, adhesion, permeability, hydraulic conductivity, infiltration rate, consistence, erodibility, CEC, etc.
- Inference and characterization of geodynamic and pedodynamic features such:
  - transport agents (water, wind, ice, mass movement)
  - depositional processes and environments
  - weathering processes (physical and chemical)
  - soil-forming processes

8.4.1.2 The Information

The particle size distribution of the material is determined in the laboratory using methods such as densitometry or the pipette method to separate the fractions of sand, silt, and clay, and sieves to separate the various sand fractions. The analytical data are used to classify the material according to particle size scales. The most common of these grain size classifications are the USDA classification for agricultural purposes, and the Unified and AASHTO classifications for engineering purposes (USDA 1971). Significant differences between these classification systems concern the following aspects:

- The upper limit of the sand fraction: 2 mm in USDA and AASHTO; 5 mm in Unified.
- The lower limit of the sand fraction: 0.05 mm (50 µm) in USDA; 0.074 mm (74 µm) in Unified and AASHTO (solifluidal threshold).
- The boundary between silt and clay: 0.002 mm (2 µm) in USDA; 0.005 mm (5 µm) in Unified and AASHTO (colloidal threshold).

8.4.1.3 Examples of Inference and Interpretation

Hereafter, some examples are analyzed to show the type of information that can be derived from particle size data to characterize aspects of sedimentology, weathering, and soil formation. The granulometric composition of the material allows inferring and interpreting important features relative to the formation and evolution of the geoforms: for instance, the nature of the agents and processes that mobilize the material, the modalities of deposition of the material and their variations in time and space, the mechanisms of disintegration and alteration of the rocks to form regolith and parent material of the soils, and the differentiation processes of the soil material.

(a) Transport agents
   Wind and ice illustrate two extreme cases of relationship between transport agent and granulometry of the transported material.

- Wind is a highly selective transport agent. The competence of the wind covers a narrow range of particle sizes, which usually includes the fractions of fine sand, very fine sand, and coarse silt (250–20 µm). Coarser particles are too heavy, except for saltation over short distances; smaller particles are often immobilized in aggregates or crusts, a condition that causes mechanical retention in situ. As a result, the material transported by wind is usually homometric.
- Ice is a poorly selective agent. Glacial deposits (e.g., moraines) include a wide range of particles from clay and silt (glacial flour) to large blocks (erratic blocks). This results in heterometric material.

(b) Transport processes

Cumulative grain size curves at semi-logarithmic scale, established from the analytical laboratory data, allow inferring and characterizing processes of transport and deposition, especially in the case of the processes controlled by water or wind. The granulometric facies of a deposit reflects its origin and mode of sedimentation (Rivière 1952). According to Tricart (1965a), granulometric curves are basically of three types, sometimes called canonical curves (Rivière 1952): namely, the sigmoid type, the logarithmic type, and the parabolic type (Fig. 8.9).

Granulometric curves that correspond to three types of sediments deposited by a flood event of the Guil river, in southern France, are displayed in Fig. 8.9 (Tricart 1965a).

- The sigmoid or S-shaped curve shows that a large proportion of the sample (ca 85%) lies in a fairly narrow particle size range (150–40 µm), which corresponds mostly to the fractions of coarse silt and very fine sand. This material results from a very selective depositional process, which is common in areas of calm, no-turbulent, fluvial overflow sedimentation. In such places, the vegetation cover of the soil, especially when it comes to grass, operates an effect of sieving and biotic retention mainly of silt and fine sand particles (overflow process). Eolian deposits of particles that have been transported over long distances, as in the case of loess, generate similar S-shaped curves.
- The logarithmic curve, with a more or less straight slope, reveals that the deposit is distributed in approximately equal proportions over all particle size classes. This reflects a poorly selective depositional mechanism that is characteristic of the splay process. Glacial moraine sediments can also produce logarithmic type curves.
- The parabolic curve shows an abrupt slope inflection in the range of 30–20 µm. All particles are suddenly laid down upon a blockage effect caused by a natural or artificial barrier. For example, a landslide or a lava flow across a valley can obstruct the flow of a river and lead to the formation of a lake where the solid load is retained.

(c) Depositional terrain forms

A transect across an alluvial valley usually shows a typical sequence of positions built by river overflow. A full sequence may include a sandy to coarse

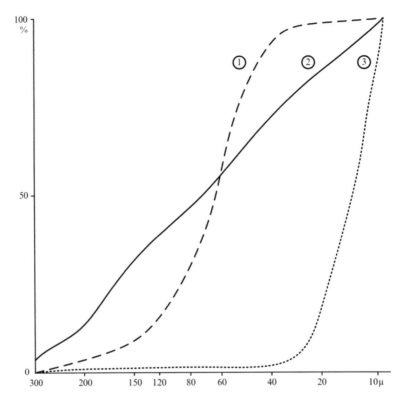

Fig. 8.9 Types of granulometric curves in depositional materials. Sediments of a flood event (June 1957) in the watershed of the Guil river, southern France. (Taken from Tricart 1965a)
1 Sigmoid curve, characteristic of free sediment accumulation
2 Logarithmic curve, characteristic of a torrential lava flow (in this case) or splay deposits
3 Parabolic curve, characteristic of an accumulation forced by an obstacle obstructing the flow

loamy levee, a silty to fine loamy overflow mantle, and a clayey basin, in this order from the highest position, closest to the river channel, to the lowest and farthest position in the depositional system (see Fig. 4.4 in Chap. 4).

(d) Lithologic discontinuity

The soil profile included in Fig. 8.3 shows a contrasting change of texture from sand to clay, which constitutes a lithologic discontinuity at 60 cm depth. This particle size change reveals an event of splay deposition following a basin depositional phase.

(e) Weathering processes

• Physical weathering of rocks produces predominantly coarse fragments. This is particularly common in extreme environmental conditions such as the following:

- Cold environments, where frequent recurrence of freezing and thawing in the cracks and pores causes rock fragmentation. Cryoclastism or gelifraction is common at high latitudes and high altitudes.
- Hot and dry environments, where large thermic amplitudes between day and night favor the repetition of daily cycles of differential expansion-contraction between leucocratic (felsic) minerals and melanocratic (mafic) minerals. Termoclastism is common in desert regions with large daily temperature variations.

- Chemical weathering produces predominantly fine-grained products, especially clay particles that are neoformed upon weathering of the primary minerals of the rocks.

(f) Soil forming processes

A classic example is the comparison of clay content between eluvial and illuvial horizons to infer the process of clay translocation. Soil Taxonomy (Soil Survey Staff 1975, 1999), as well as other soil classification systems, uses ratios of clay content between A and B horizons to recognize argillic Bt horizons. For instance, a B/A clay ratio > 1.2 is required for a Bt horizon to be considered argillic, when the clay content in the A horizon is 15–40%. The B/A clay ratio is also used as an indicator of relative age in chronosequence studies of fluvial terraces.

8.4.2 Structure

8.4.2.1 Geogenic Structure

The geogenic structure refers to the structure of the geologic and geomorphic materials (bedrocks and unconsolidated surface materials, respectively).

(a) Rock structure

The examination of the rock structure allows evaluating the degree of weathering by comparison between the substratum R and the Cr horizon, especially in the case of crystalline rocks (igneous and metamorphic) where the original rock structure can still be recognized in the Cr horizon (saprolite). For instance, gneiss exposed to weathering preserves the banded appearance caused by the alternation of clear stripes (leucocratic felsic minerals) and dark stripes (melanocratic mafic minerals). The weathering of the primary minerals, especially the ferromagnesians, releases constituents, mainly bases, that are lost by washing to the water table. In the Cr horizon, the rock volume remains the same as that of the unweathered rock in the R substratum, but the weight has decreased. For example, the density could decrease from 2.7 Mg m^{-3} in the non-altered rock to 2.2–2.0 Mg m^{-3} in the Cr horizon. This process has received the name of isovolumetric alteration (Millot 1964).

(b) Depositional structures

The sediments show often structural features that reveal the nature of the depositional processes. Rhythmic and lenticular structures are examples of syn-depositional structures, while the structures created by cryoturbation and bio-turbation are generally postdepositional.

- The rhythmic structure reflects successive depositional phases or cycles. It can be recognized by the occurrence of repeated sequences of strata that are granulometrically related, denoting a process of cyclic aggradation. For example, a common sequence in overflow mantles includes layers with texture varying between fine sand and silt. Consecutive sequences can be separated by lithologic discontinuities.
- The lenticular structure is characterized by the presence of lenses of coarse material within a matrix of finer material. Lenses of coarse sand and/or gravel, several decimeters to meters wide and a few centimeters to decimeters thick, are frequent in overflow as well as splay mantles. They correspond to small channels of concentrated runoff, flowing at a given time on the surface of a depositional area, before being fossilized by a new phase of sediment accumulation.
- Cryoturbation marks result from the disruption of an original depositional structure by ice wedges or lenses.
- Bioturbation marks result from the disruption of an original depositional structure by biological activity (burrows, tunnels, pedotubules).

8.4.2.2 Pedogenic Structure

The soil structure type is often a good indicator of how the geomorphic environment influences soil formation. For instance, in a well-drained river levee position, the structure is usually blocky. The structure is massive or prismatic in a basin position free of salts, while it is columnar in a basin position that is saline or saline-alkaline. On the other hand, the grade of structural development may reflect the time span of soil formation.

8.4.3 Consistence

Consistence limits, also called Atterberg limits, are good indicators to describe the mechanical behavior, actual or potential, of the geomorphic and pedologic materials according to different moisture contents. In Fig. 8.10, consistence states, limits, and indices, which are relevant criteria in mass movement geomorphology, are related to each other. These relationships are controlled by the particle size distribution and mineralogy of the materials. In general, clay materials are mostly susceptible to landsliding, while silt and fine sand materials are more prone to solifluction. A low plasticity index makes the material more susceptible to liquefaction, with the risk of

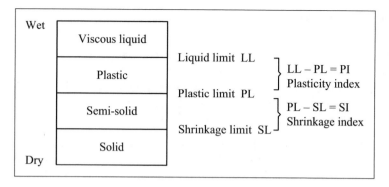

Fig. 8.10 Consistence/consistency parameters

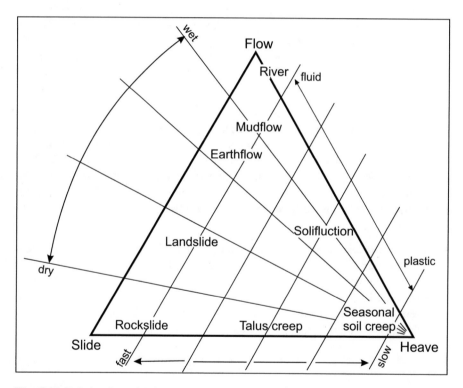

Fig. 8.11 Relational model for classifying mass movements. (Adapted from Carson and Kirkby 1972)

creating mudflows. The graphic model of Carson and Kirkby (1972) shows how continuity solutions in terms of speed, water content, and plasticity that relate the basic mechanisms of swell, slide, and flow, can be segmented for differentiating types of mass movement (Fig. 8.11).

8.4.4 Mineralogy

The mineralogical composition of the sand, silt, and clay fractions in the unconsolidated materials of surface formations is an indicator of the geochemical dynamics of the environment, as related to or controlled by morphogenic processes, and helps follow the pathways of tracer minerals. The associations of minerals present in cover formations allow making inferences about the following features:

– They reflect the dominant lithologies in the sediment production basins.
– They help distinguish between fresh and reworked materials; the latter result from the mixing of particles through the surficial translation of materials over various terrain units.
– They reflect the morphoclimatic conditions of the formation area: for instance, halites in hot and dry environment; kandites in hot and moist environment.
– They reflect the influence of topography on the formation and spatial redistribution of clay minerals along a slope forming a catena of minerals. In humid tropical environment, a catena or toposequence of minerals commonly includes kandites (e.g., kaolinite) at hill summit, micas (e.g., illite) on the backslope, and smectites (e.g., montmorillonite) at the footslope.

Table 8.5 shows an example of determination of minerals in sand and silt fractions to reconstitute the morphogenic processes acting in the contact area between a piedmont and an alluvial valley. The sampling sites are located on the lower terrace of the Santo Domingo river (Barinas, Venezuela) at its exit from the Andean foothills towards the Llanos plain. Sites are distributed along a transect perpendicular to the valley from the base of the piedmont to the floodplain of the river. Site A is close to the piedmont, site C is close to the floodplain, and site B is located in an intermediate position.

• *Site A: colluvial deposit (reworked material).* Rubified colluvium, coming from the truncation of a strongly developed red soil lying on a higher terrace (Q3). The reworking effect can be inferred from the high contents of clean quartz grains, washed during transport by diffuse runoff, and soil aggregates, respectively. The absence of rock fragments and micas indicates that colluviation removed fully pedogenized material from the piedmont.

Table 8.5 Mineralogy of silt and sand fractions (%); eastern piedmont of the Andes, to the west of the city of Barinas, Venezuela

Site	Clean quartz + feldspars	Ferruginous quartz	Soil aggregates	Rock fragments	Micas	Total
A	40	5	55	0	0	100
B	21	14	22	42	1	100
C	22	0	0	0	78	100

Data provided by the Institute of Geography, University of Strasbourg, France. Courtesy J. Tricart

- *Site B: mixed deposit, colluvial and alluvial.* Mixture of red colluvium (presence of aggregates) removed from an older soil mantle on a middle terrace (Q2), and recent alluvium (presence of rock fragments) brought by the Santo Domingo river.
- *Site C: alluvial deposit.* Holocene alluvial sediments exclusively composed of clean quartz and fresh micas. The high proportion of micas results from the retention of silt particles trapped by dense grass cover.

8.4.5 Morphoscopy

Morphoscopy (or exoscopy) consists of examining coarse grains (sand and coarse silt) under a binocular microscope to determine their degree of roundness and detect the presence of surface features.

- The shape of the grains can vary from very irregular to well rounded:
    - Well rounded grains reflect continuous action by (sea)water or wind.
    - Irregular grains indicate torrential or short-distance transport.

- The brightness of the grains and the presence of surface marks, such as striae, polishing, frosting, chattermarks, gouges, among others, indicate special transport modes or special environmental conditions:
    - Shiny grains: seawater action.
    - Frosted grain surface: wind action.
    - Grains with percussion marks: chemical corrosion or collision of grains transported by wind.

8.5 Morphochronologic Attributes: The History of Geoforms

8.5.1 Reference Scheme for the Geochronology of the Quaternary

The Quaternary period (2.6 Ma) is a fundamental time frame in geopedology, because most of the geoforms and soils have been formed or substantially modified during this period. Pre-Quaternary relict soils exist but are of fairly limited extent. The Quaternary has been a period of strong morphogenic activity due to climatic changes, tectonic paroxysms, and volcanic eruptions, which have caused destruction, burial, or modification of the pre-Quaternary and syn-Quaternary geoforms and soils, while at the same time new geoforms and soils have developed.

In temperate and boreal zones, as well as in mountain areas, glacial and interglacial periods have alternated several times. In their classic scheme based on observations made in the Alps, Penck and Brückner (1909) considered a relatively limited number of glacial periods (i.e. Würm, Riss, Mindel, Günz). A similar scheme was established for the chronology of the Quaternary period in North America. Recent research shows that the alternations of glacial-interglacial periods were actually more numerous. In Antarctica, up to eight glacial cycles over the past 740,000 years (740 ka) have been recognized. The average duration of climatic cycles is estimated at 100 ka for the last 500 ka and at 41 ka for the early Quaternary (before 1 Ma), with intermediate values for the period from 1 Ma to 500 ka (EPICA 2004). In addition, shorter climate variations have occurred during each glacial period, similar to the Dansgaard-Oeschger events of the last glaciation. Many regions are now provided with very detailed geochronologic reference systems for the Pleistocene and especially for the Holocene. In the intertropical zone, climate change is expressed more in terms of rainfall variations than in terms of temperature variations. Dry periods have alternated with moist periods, in approximate correlation with the alternation between glacial and interglacial periods at mid- and high latitudes.

Quaternary geochronology is conventionally based on the recurrence of climatic periods, which are assumed of promoting alternately high or low morphogenic activity and high or low pedogenic development. Erhart (1956), in his bio-rhexistasis theory, summarizes this dichotomy by distinguishing between (1) rhexistasic periods with unstable environmental conditions, rather cold and dry, conducive to intense morphogenic activity, and (2) biostasic periods with more stable environmental conditions, rather warm and humid, favorable to soil development. The biostasic periods are assumed of having been longer than the rhexistasic periods (Hubschman 1975). Butler's model of K cycles (1959) is based on the same principle of the alternation of stable phases with soil development and unstable phases with predominance of erosion (soil destruction) or sedimentation (soil fossilization). In the context of soil survey, various rather simple geochronologic schemes have been implemented to record the relative age of geoforms and associated soils, using letters such as K (from kyklos), t (from terrace), and Q (from Quaternary), with increasing numerical subscripts according to increasing age of the geopedologic units, assimilated to chronostratigraphic units (Table 8.6). Although these relative chronology schemes have a spatial resolution limited, for instance, to a region or a country, they also allow coarse stratigraphic correlations over larger territories.

Comments on Table 8.6:

- Q identifiers refer to the inferred relative age of the geomorphic material that serves as parent material, thus not directly to the age of the soil derived from this material. In erosional, structural, and residual relief areas, there is often a large gap between the age of the geologic substratum and the age of the overlying soil mantle. In many cases, the bedrock may even not be the parent material of the soil. This occurs in hill and mountain landscapes, where soils often develop from allochthonous slope formations lying atop the rocks in situ. By contrast, in depositional environments, the initiation of soil formation usually coincides fairly well with the end of the period of material accumulation. However, in sedimentation areas of considerable extent, deposition does not stop abruptly or does not

Table 8.6 Relative geochronology scheme of the Quaternary (Zinck 1988)

		Rhexistasic periods	Biostasic periods
HOLOCENE			Q0
	Upper	Q1	
			Q1-2
	Late middle	Q2	
PLEISTOCENE			Q2-3
	Early middle	Q3	
			Q3-4
	Lower	Q4	
			Q4-5
PLIO-PLEISTOCENE		Q5	

stop in all sectors at the same time. For this reason, Q1 deposition in floodplains, for example, can extend locally into Q0 without notable interruption.

- The numerical indices (Q1, Q2, etc) indicate increasing relative age of the parental materials. Where necessary, the relative scale can be extended (e.g., Q5, etc) to refer to deposits that overlap the end of the Pliocene (Plio-Quaternary formations).
- Each period can be subdivided using alphabetical subscripts to reflect minor age differences (e.g., Q1a more recent than Q1b).
- Some geoforms, such as for example colluvial glacis, may have evolved over the course of several successive periods. A composite symbol can be used to reflect this kind of diachronic formation (e.g., Q1-Q2; Q1-Q1–2).

8.5.2 Dating Techniques

Ideally, age determination of a geoform or a soil requires finding and sampling a kind of geomorphic or pedologic material that allows using any of the absolute or relative dating techniques available, or a combination thereof, including:

- Carbon-14 (organic soils, charcoal, wood; frequently together with analysis of pollen)
- K/Ar (volcanic materials)
- Thermoluminescence (sediments, e.g., beach sands, loess)
- Dendrochronology (tree growth rings)

- Tephrochronology (volcanic ash layers)
- Varves (proglacial lacustrine layers)
- Analysis of historic and prehistoric events (earthquakes, etc.).

These techniques are relatively expensive and their implementation within the framework of a soil survey project is generally limited for budgetary reasons. On average, a determination of carbon-14 costs 300–350 euros. Some techniques are applicable only to specific kinds of material (e.g., ^{14}C only on material containing organic carbon; K/Ar only on volcanic material). Certain techniques cover restricted ranges of time (e.g., ^{14}C for periods shorter than 50–70 ka; thermoluminescence up to 300 ka). Interpretation errors can result from the contamination of the samples or the residence time of the organic matter (in the case of ^{14}C).

The former suggests that the most common materials in the geomorphic and pedologic context likely to be dated in absolute terms are soil horizons and sedimentary strata containing organic matter. In many situations, this limits practically absolute dating to about 60,000 years BP, a time span that covers the Holocene and a small part of the upper Pleistocene corresponding to half of the last glacial period. This underlines the need for indirect dating means such as those provided by pedostratigraphy.

8.5.3 Relative Geochronology: The Contribution of Pedostratigraphy

8.5.3.1 Definition

Relative geochronology is based on establishing relationships of temporal antecedence between the various geoforms or deposits in a study area and building correlations at several spatial scales. This procedure practically consists in extending the stratigraphic system used in pre-Quaternary geology to the Quaternary period. Geologic maps often provide scarce information about the Quaternary (e.g., Qal for alluvial cover formations, Qr for recent deposits), in comparison with the detailed lithologic information concerning the pre-Quaternary. This information is usually to coarse to determine the temporal frame of soil formation. In contrast, the geopedologic information provided by the proper soil survey can contribute to improving the stratigraphy of the Quaternary.

Pedostratigraphy or soil-derived stratigraphy consists in using selected soil and regolith properties to estimate the relative age of the cover formations and the geoforms on which soils have developed. This makes it possible to determine the chronostratigraphic position of a material or a geoform in a geochronologic reference scheme (Zinck and Urriola 1970; Harden 1982; Busacca 1987; NACSN 2005), with the possibility of recognizing successive soil generations.

Etymologically, pedostratigraphy means the use of soils or soil properties as stratigraphic tracers to contribute establishing the relative chronology of geologic, geomorphic, and pedologic events in a territory. However, according to the definitions provided by the North American Stratigraphic Code (NACSN 2005), the concepts of pedostratigraphy and soil stratigraphy are not strictly synonymous. According to this code, the basic pedostratigraphic unit is the geosol, which differs in various ways from the basic unit of soil stratigraphy, the pedoderm. One of the key differences is that the geosol is a buried weathering profile, while the pedoderm may correspond to a buried soil, a surficial relict soil, or an exhumed soil. Disregarding these definition differences, what is in fact relevant is that soils are recognized as stratigraphic units and, in this sense, the term pedostratigraphy has been used in geomorphology and pedology without complying with the strict definition of geosol. Pedostratigraphy is a privileged area of the geopedologic relationships with mutual contribution of geomorphology and pedology. The chronosequences of fluvial terraces provide illustrative examples of this close interrelation. The relative age of the terraces as determined on the basis of their position in the landscape, the lowest being usually the most recent, generally correlates fairly well with the degree of soil development and conversely. Morphostratigraphy and pedostratigraphy complement each other.

8.5.3.2 Indicators

A variety of pedologic and geomorphic indicators has been used to establish relative chronology schemes of the Quaternary in regions with different environmental characteristics (Mediterranean, tropical, etc). These criteria include, among others, the following.

- The degree of activity of the geoforms, distinguishing between active geoforms (e.g., dune in formation), inherited geoforms in survival (e.g., hillside locally affected by solifluction), and stabilized geoforms (e.g., coastal bar colonized by vegetation).
- The degree of weathering of the parent material based on the color of the cover formations and the degree of disintegration of stones and gravels (Fig. 8.12). In humid tropical environment, the fragments of igneous and metamorphic rocks found in detrital formations are usually much more altered than most of the sedimentary rock fragments. Quartzite is most resistant in all kinds of climatic condition and often provides the dominant residual fragments in detrital formations of early Quaternary.
- The degree of soil morphological development, inferred from criteria such as color, pedogenic structure, solum thickness, and leaching indices, among others.

  - Color is a good indicator of the relative age of soils, particularly in humid tropical climate, with gradual increase of the red color (rubification) as the weathering of the ferromagnesian minerals in the parent material proceeds. The possibility of differentiating soil ages by color dims over time in well-developed soils. Red soils can also be recent, when they arise from materials eroded from older rubified soils and redeposited in lower portions of the landscape. Likewise, red soils on limestone can be relatively young.

Fig. 8.12 Quaternary alluvial cover formations differentiated by color resulting from increasing rubefaction through time; materials belonging to (**a**) Holocene to upper Pleistocene (Q0-Q1), western piedmont of Venezuelan Andes; (**b**) late middle Pleistocene (Q2), eastern piedmont of Venezuelan Andes; (**c**) early middle Pleistocene (Q3), eastern piedmont of Venezuelan Andes; (**d**) lower Pleistocene (Q4), mesetas of the eastern Venezuelan Plateau

- The pedogenic structure reflects (1) the conditions of the site and the nature of the parent material which together control the type of structure (e.g., blocky, prismatic, columnar), and (2) the elapsed time that influences the grade of structural development (from weak to strong). The relationship between development grade and time reaches a threshold in well-developed soils, beyond which structure tends to weaken because of the impoverishment in substances that contribute to the cohesion of the soil material (e.g., organic matter, type and amount of clay, divalent cations).
- The thickness of the solum generally increases with the duration of pedogenic development in conditions of geomorphic stability. As in the case of structural

development and rubification, solum thickness reaches a threshold over time beyond which increases become gradually insignificant.
- Leaching indices allow evaluating the intensity of the translocation of soluble or colloidal substances from eluvial horizons to the underlying illuvial horizons. The most commonly implemented are the clay and calcium carbonate ratios. The leaching intensity decreases with time as the eluvial horizons become depleted in mobilizable substances, resulting in stabilization of the translocation rates.

- The status of the adsorption complex. In general terms, the adsorption complex of the soil changes quantitatively and qualitatively with increasing time. Soil reaction (pH), cation exchange capacity, and base saturation are among the most sensitive indicators. With the passage of time, soils lose alkaline and alkaline-earth cations, resulting in a decrease or a change of composition (more H^+ and/or Al^{+++}) of the adsorption complex and an increase in acidity of the soil solution.
- Clay mineralogy changes with soil development as a function of time, among other factors. The associations of clay minerals originally present in the Cr or C horizons will be replaced by other associations with increasing time. In general, the 2:1 type clays (e.g., smectites, micas) are going to be replaced by or transformed into 1:1 type clays (e.g., kandites).

8.5.3.3 Combining Indicators

The simultaneous use of several of the above-mentioned soil properties allows determining pedostratigraphic units. To this effect, Harden (1982) established a quantitative index to estimate degrees of soil development and correlate these with dated soil units. The index was originally developed based on a soil chronosequence in the Merced River valley, central California, combining properties described in the field with soil thickness. Eight properties were integrated to form the index, including the presence of clay skins, texture combined with wet consistence, rubification based on change in hue and chroma, structure, dry consistence, moist consistence, color value, and pH. Other properties described in the field can be added if more soils are studied. The occasional absence of some properties did not significantly affect the index. Quantified individual properties and the integrated index were examined and compared as functions of soil depth and age. The analysis showed that the majority of the properties changed systematically within the 3 Ma that span the chronosequence of the Merced River. The index has been applied to other sites with successive adjustments (Busacca 1987; Harden et al. 1991).

Stepped alluvial terrace systems usually offer the possibility to establish illustrative soil chronosequences. Table 8.7 reports data on selected properties of soils that have developed on five consecutive Quaternary terraces in the Guarapiche river valley, northeast of Venezuela. Melanization with mollic horizon and soil structure formation on terrace Q1 corresponds to the first stage of soil development from raw depositional material of Q0. From period Q2 onwards clay illuviation starts upon

Table 8.7 Pedostratigraphic thresholds; Guarapiche river valley, Venezuela (Zinck and Urriola 1970)

Relative age of parent material	Dominant color	Average solum thickness cm	CaCO$_3$ eq. %	Clay illuviation B/A index	pH 1:1 H$_2$O	Base saturation %	CEC cmol+ kg^{-1} clay	Clay mineral associations	Main taxa
Q0	Grayish-brown	30	>3	–	+/–8	100	80–120	S > K = M	Entisols Inceptisols
Q1	Dark- brown	80	1–3	–	6–7.5	80–100	60–95	S > K > M	Mollisols Inceptisols
Q2	**Reddish-yellow**	**200**	**<1**	**1.2–1.6**	4.5–6	40–60	40–60	S > K > M	Alfisols Vertisols
Q3	Yellowish-red	250	<0.5	2.1–2.7	**4–5**	**20–40**	40–50	V > K > M	Ultisols
Q4	Red	300	0	2.4–2.5	4–5	<20	**20–30**	**K>> > V > M > HIV**	Ultisols oxic subgr.

Based on the soil subgroups included in the phenetic clusters 1-2-3 of the dendrogram in Fig. 4.10 (Chap. 4)

Properties refer mainly to the solum (mean values or ranges of values)

Properties in bold represent pedostratigraphic thresholds reflecting discontinuities in soil evolution during the Quaternary

Q0 Holocene, Q1 Upper Pleistocene, Q2 Late Middle Pleistocene, Q3 Early Middle Pleistocene, Q4 Lower Pleistocene

S smectite, K kaolinite, M mica, V vermiculite, HIV hydroxy-interlayered vermiculite

descarbonation, together with substantial solum deepening. This is followed on level Q3 by important desaturation of the soil complex and soil solution. On the older Q4 terrace kaolinite formation takes place, causing degradation of the adsorption complex. Each terrace is characterized by a different stage of soil development, adding up to further soil evolution. The properties quantifying these consecutive pedogenic stages show value leaps that correspond to pedostratigraphic thresholds. The latter reflect discontinuities in soil formation during the Quaternary. The pedotaxa sequence comprising increasingly developed soils parallels the Quaternary pedogenic evolution, from Entisols to Mollisols to Alfisols to Ultisols to oxic (kanhaplic) Ultisols (Fig. 8.13).

There is no single model describing the relationship between time and soil development. Pedogenic development rates vary according to the considered time segment and the geographic conditions of the studied area. In general, soil development rates decrease when time increases above a given threshold and with increasing aridity (Zinck 1988; Harden 1990).

Fig. 8.13 Well-drained soil profiles belonging to the chronosequence of the Guarapiche river terrace system, eastern Venezuela (see Table 8.7). All soils have similar parent materials (sandy loam C horizons with mixtures of smectite, kaolinite, and mica minerals) originating throughout the Quaternary from the sedimentary rocks (sandstone, lutite, limestone) of the southern slope of the Coastal Cordillera. The sequence shows the factor time effect on soil formation and differentiation: (**a**) Entisol (Mollic Ustifluvent), Fluvisol, in the floodplain (Q0); (**b**) Mollisol (Cumulic Haplustoll), Phaeozem, on the lower terrace (Q1); (**c**) Alfisol (Typic Haplustalf), Luvisol, on the lower middle terrace (Q2); Ultisol ('Kanhaplic' Paleustult), Acrisol, on the upper terrace (Q4) under savanna cover. The Ultisol (Typic Paleustult), Lixisol, of the higher middle terrace (Q3) is not depicted here

8.6   Relative Importance of the Geomorphic Attributes

Not all attributes are equally important to identify and classify geoforms. For instance, the particle size distribution of the material is most important, because it has more differentiating power and therefore more taxonomic weight than the relative elevation of a geoform.

8.6.1   Attribute Classes

Following an approach that Kellogg (1959) applied to distinguish between soil characteristics, the attributes of the geoforms can be grouped into three classes according to their weight for taxonomic purposes: differentiating, accessory, and accidental attributes, respectively.

8.6.1.1   Differentiating Attributes

A differentiating attribute is one that enables to distinguish one type of geoform from another at a particular categorial level. Therefore, a change in an attribute's state, expressed by a range of values, leads to a change in geoform classification. An attribute that has this property is considered diagnostic. Such an attribute, along with other differentiating attributes, contributes to the identification and classification of the geoforms.

A few examples:

- The dip of the geologic layers is a diagnostic criterion for recognizing monoclinal reliefs and the degree of dipping is a differentiating feature for distinguishing classes of monoclinal reliefs (see Fig. 6.4 in Chap. 6).
- A slope facet should be concave to classify as footslope. In this case, the topographic profile is the differentiating attribute and "concave" is the state of the attribute.
- The material of a decantation basin normally comprises more than 60% clay fraction. In this case, the particle size distribution is the differentiating attribute, and the attribute state is expressed by the high clay content.

8.6.1.2   Accessory Attributes

An attribute is accessory if it reinforces the differentiating capability of a diagnostic attribute with which it has some kind of correlation (covariant attribute). For instance, the lenticular type of depositional structure can occur in several alluvial facies but is more common in deposits caused by overload flow accompanied by mechanical friction (river levee, different kinds of splay). By itself, the presence of a lenticular structure is not enough to recognize a type of geoform.

8.6.1.3   Accidental Attributes

An accidental attribute does not contribute to the identification of a particular type of geoform but provides additional information for its description and characterization. This kind of attribute can be used to create phases of taxonomic units for the purpose of mapping and separation of cartographic units (e.g., slope classes or classes of relative elevation).

8.6.2   Attribute Weight

8.6.2.1   Morphographic Attributes

Morphographic attributes are essentially accessory, sometimes differentiating.

- Accessory weight. For instance, a newly formed river levee has a characteristic morphology (elongated, narrow, sinuous, convex shape), which facilitates its identification in aerial images. An older levee, the contours of which have been obliterated with the passing of time, is more difficult to recognize from its external features. In the case of a levee buried underneath a recent sediment cover, it is possible to reconstruct the configuration and design of the contours by means of perforations. In these last two cases, the identification of the geoform rests primarily on the granulometric composition of material, with accessory support of the morphographic features.
- Differentiating power. In hill and mountain landscapes, the morphographic attributes can be differentiating. For instance, in the case of a convex-concave hillside, the characteristic topographic profile of each single slope facet is in itself differentiating.

8.6.2.2   Morphometric Attributes

Morphometric attributes are predominantly accidental. They contribute to the description of the geoforms, but seldom to their identification. For instance, the difference of elevation (i.e., relative elevation) between the summit surface of a plateau and the surrounding lowlands (e.g., valley or plain landscapes) can be as little as 100–150 m (e.g., the mesetas in eastern Venezuela) or as much as 1000–1500 m (e.g., the Bolivian Altiplano). In both cases, however, the geoform meets the diagnostic plateau attributes at the categorial level of landscape. In general, the dimensional features have low taxonomic weight, but are relevant for the practical use of the geomorphic information, for instance, in evaluation of environmental impacts or land-use planning. To this end, phases of relative elevation, drainage density, and slope gradient can be implemented.

8.6.2.3 Morphogenic Attributes

The morphogenic attributes are essentially differentiating, either individually or in group, especially when they are reinforced by accessory attributes. For instance, the consistence is a diagnostic attribute for assessing the susceptibility of a material to mass movement and for interpreting the origin of the resulting geoforms. The depositional geoforms show always specific ranges of granulometric composition, which is a highly diagnostic attribute in this case.

8.6.2.4 Morphochronologic Attributes

Morphochronologic attributes are mostly differentiating, because the relative age of a geoform is an integral part of its identity. The fact that a river levee has formed during the Holocene (Q0) or during the middle Pleistocene (Q2) probably does not have great effect on its configuration, although the contour design may have been obliterated with the passage of time. However, the chronostratigraphic position of the geoform is differentiating because it determines a time frame in which the morphogenic processes take place, and which controls the evolution of the soils and their properties.

8.6.3 Attribute Hierarchization

Not all attributes are used at each categorial level of the geoform classification system. Table 8.8 shows an attempt of differential hierarchization of the geomorphic attributes according to their diagnostic weight. This aspect is of growing importance for the automated treatment of the geomorphic information. Hereafter are mentioned the criteria that have guided the hierarchization in terms of attribute amount, nature, function, and implementation at the upper and lower levels of the system, respectively (Table 8.9).

8.6.3.1 Upper Levels

- Limited number of attributes.
- Preferably descriptive attributes, reflecting external features of the geoforms (i.e., morphographic and morphometric attributes).
- Function of generalizing and aggregating information.
- Information about attributes is mostly obtained by interpretation of aerial photos, satellite images, and digital elevation models.

Table 8.8 Hierarchization of the geomorphic attributes (Zinck 1988)

Attributes	Landscape	Relief	Lithology	Terrain form
Morphometric				
Relative elevation	+	+	−	o
Drainage density	+	+	−	−
Slope	+	+	−	+
Morphographic				
Topographic shape	+	o	−	−
Topographic profile	−	+	−	+
Exposure	−	+	−	+
Configuration	−	+	−	+
Contour design	−	+	−	+
Drainage pattern	+	+	−	−
Surrounding conditions	+	+	+	+
Morphogenic				
Particle size distribution	−	o	+	+
Structure	−	−	+	+
Consistence	−	−	+	+
Mineralogy	−	−	+	+
Morphoscopy	−	−	+	+
Morphochronologic				
Degree of weathering	−	−	+	+
Degree of soil development	−	−	o	+
Leaching indices	−	−	o	+
Adsorption complex status	−	−	o	+
Clay mineralogy	−	−	+	+

+ Very important attribute
o Moderately important attribute
− Less important attribute

Table 8.9 Relations between geomorphic attributes according to the categories of the system

Attributes	Amount	Nature	Function	Implementation
Upper levels	Few	Descriptive external characterization	Generalizing aggregation	Interpretation of photos, images, and DEM
↕	↕	↕	↕	↕
Lower levels	Many	Genetic internal characterization	Detailing disaggregation	Field and laboratory

8.6.3.2   Lower Levels

- Greater number of attributes, resulting from the addition of information.
- Preferably genetic attributes, reflecting internal characteristics of the geoforms (i.e., morphogenic and morphochronologic attributes).
- Function of differentiating and detailing information.
- More field information and laboratory data are required.

8.7   General Conclusion on Geopsedology

Geopedology is an approach to soil survey that combines pedologic and geomorphic criteria to establish soil map units. Geomorphology provides the contours of the map units ("the container"), while pedology provides the soil components of the map units ("the content"). Therefore, the units of the geopedologic map are more than soil units in the conventional sense of the term, since they also contain information about the geomorphic context in which soils have formed and are distributed. In this sense, the geopedologic unit is an approximate equivalent of the soilscape unit, but with the explicit indication that geomorphology is used to define the landscape. This is usually reflected in the map legend, which shows the geoforms as entries to the legend and their respective pedotaxa as descriptors.

In the geopedologic approach, geomorphology and pedology benefit from each other in various ways:

- Geomorphology provides a genetic framework that contributes to the understanding of soil formation, covering three of the five factors of Jenny's equation: nature of the parent material (transported material, weathering material, regolith), age and topography (Jenny 1941, 1980). Biota is indirectly influenced by the geomorphic context.
- Geomorphology provides a cartographic framework for soil mapping, which helps understand soil distribution patterns and geography. The geopedologic map shows the soils in the landscape.
- The use of geomorphic criteria contributes to the rationality of the soil survey, decreasing the personal bias of the surveyor. The need of prior experience to ensure the quality of the soil survey is offset by a solid formation in geomorphology.
- Geomorphology contributes to the construction of the soil map legend as a guiding factor. The hierarchic structure of the legend reflects the structure of the geomorphic landscape together with the pedotaxa that it contains.
- The soil cover or soil mantle provides the pedostratigraphic frame based on the degree of soil development, which enables to corroborate the morphostratigraphy (e.g., terrace system).
- The soil cover through its properties (mechanical, physical, chemical, mineralogical, biological) provides data that contribute to assess the vulnerability of the geopedologic landscape to geohazards and estimate the current morphogenic balance (erosion-sedimentation).
- The geopedologic approach to soil survey and digital soil mapping are complementary and can be advantageously combined. The segmentation of the landscape sensu lato into geomorphic units provides spatial frames in which geostatistical and spectral analyses can be applied to assess detailed spatial variability of soils and geoforms, instead of blanket digital mapping over large territories. Geopedology provides information on the structure of the landscape in hierarchically organized geomorphic units, while digital techniques provide

information extracted from remote-sensed imagery that help characterize the geomorphic units, mainly the morphographic and morphometric terrain surface features.

This first part of the book addresses the basic concepts and ideas underlying geopedology, with emphasis on the identification, characterization, and classification of geoforms to support soil survey and field soil studies at large. The following parts comprise a variety of studies that implement the geopedologic approach here introduced or other modalities based on soil-landscape relationships, using different methods and techniques, for soil pattern recognition, analysis and mapping, soil degradation assessment, and land use planning.

References

ASP (1960) Manual of photographic interpretation. American Society of Photogrammetry, Washington, DC

Busacca AJ (1987) Pedogenesis of a chronosequence in the Sacramento Valley, California, USA. I application of a soil development index. Geoderma 41:123–148

Butler BE (1959) Periodic phenomena in landscapes as a basis for soil studies, Soil Publ 14. CSIRO, Canberra, Australia

Carson MA, Kirkby MJ (1972) Hillslope form and process. Cambridge University Press, Cambridge

del Valle HF, Blanco PD, Metternicht GI, Zinck JA (2010) Radar remote sensing of wind-driven land degradation processes in northeastern Patagonia. J Environ Qual 39:62–75

EPICA (2004) Eight glacial cycles from an Antarctic ice core. Nature 429(6992):623–628

Erhart H (1956) La genèse des sols en tant que phénomène géologique. Masson, Paris

FAO (2006) Guidelines for soil description, 4th edn. Food and Agricultural Organization of the United Nations, Rome

Forman RTT, Godron M (1986) Landscape ecology. Wiley, New York

Fridland VM (1965) Makeup of the soil cover. Sov Soil Sci 4:343–354

Fridland VM (1974) Structure of the soil mantle. Geoderma 12:35–41

Fridland VM (1976) Pattern of the soil cover. Israel Program for Scientific Translations, Jerusalem

Gallant JC, Hutchinson MF (2008) Digital terrain analysis. In: NJ MK, Grundy MJ, Webster R, Ringrose-Voase AJ (eds) Guidelines for surveying soil and land resources, Australian soil and land survey handbook series, vol 2, 2nd edn. CSIRO, Melbourne, pp 75–91

Harden JW (1982) A quantitative index of soil development from field descriptions: examples from a chronosequence in Central California. Geoderma 28(1):1–28

Harden JW (1990) Soil development on stable landforms and implications for landscape studies. Geomorphology 3:391–398

Harden JW, Taylor EM, Hill C (1991) Rates of soil development from four soil chronosequences in the southern Great Basin. Quat Res 35:383–399

Hengl T (2003) Pedometric mapping. Bridging the gaps between conventional and pedometric approaches. ITC dissertation 101, Enschede

Hole FD, Campbell JB (1985) Soil landscape analysis. Rowman & Allanheld, Totowa

Hubschman J (1975) Morphogenèse et pédogenèse quaternaires dans le piémont des Pyrénées garonnaises et ariégoises. Thèse de Doctorat, Université de Toulouse-Le-Mirail, Toulouse

Jenny H (1941) Factors of soil formation. McGraw-Hill, New York

Jenny H (1980) The soil resource. Origin and behaviour. Ecological studies 37. Springer, New York

Kellogg CE (1959) Soil classification and correlation in the soil survey. USDA, Soil Conservation Service, Washington, DC

Meijerink A (1988) Data acquisition and data capture through terrain mapping units. ITC Journal 1988(1):23–44

Metternicht G, Zinck JA (1997) Spatial discrimination of salt- and sodium-affected soil surfaces. Int J Remote Sensing 18(12):2571–2586

Metternicht GI, Zinck JA (2003) Remote sensing of soil salinity: potentials and constraints. Remote Sensing Environ 85:1–20

Metternicht G, Zinck JA, Blanco PD, del Valle HF (2010) Remote sensing of land degradation: experiences from Latin America and the Caribbean. J Environ Qual 39:42–61

Millot G (1964) Géologie des argiles. Altérations, sédimentologie, géochimie. Masson, Paris

NACSN (2005) North american stratigraphic code. North American Commission on Stratigraphic Nomenclature. AAPG Bull 89(11):1547–1591

Olaya V (2009) Basic land-surface parameters. In: Hengl T, Reuter HI (eds) Geomorphometry: concepts, software, applications, Developments in soil science, vol 33. Elsevier, Amsterdam, pp 141–169

Penck A, Brückner E (1909) Die Alpen im Eiszeitalter. Tauchnitz CH, Leipzig

Pérez-Materán J (1967) Informe de levantamiento de suelos, río Santo Domingo. Ministerio de Obras Públicas (MOP), Caracas

Pike RJ (1995) Geomorphometry: progress, pratice, and prospect. Zeitschrift für Geomorphologie Supplementband 101:221–238

Pike RJ, Dikau R (eds) (1995) Advances in geomorphometry. Proceedings of the Walter F. Wood memorial symposium. Zeitschrift für Geomorphologie Supplementband 101

Rivière A (1952) Expression analytique générale de la granulométrie des sédiments meubles. Indices caractéristiques et interprétation géologique. Notion du faciès granulométrique. Bul Soc Géol de France 6è Série(II):156–167

Ruhe RV (1975) Geomorphology. Geomorphic processes and surficial geology. Houghton Mifflin, Boston

Saldaña A, Ibáñez JJ, Zinck JA (2011) Soilscape analysis at different scales using pattern indices in the Jarama-Henares interfluve and Henares River valley, Central Spain. Geomorphology 135:284–294

Soil Survey Staff (1975) Soil taxonomy. A basic system of soil classification for making and interpreting soil surveys. In: USDA Agric handbook 436. US Government Print Office, Washington, DC

Soil Survey Staff (1999) Soil taxonomy. In: USDA Agric Handbook 436. US Government Print Office, Washington

Tricart J (1965a) Principes et méthodes de la géomorphologie. Masson, Paris

Tricart J (1965b) Morphogenèse et pédogenèse. I Approche méthodologique: géomorphologie et pédologie. Science du Sol A:69–85

USDA (1971) Guide for interpreting engineering uses of soils. USDA, Soil Conservation Service, Washington, DC

Wulf H, Mulder T, Schaepman ME, Keller A, Jörg PC (2015) Remote sensing of soils: project report from the Federal Office of the Environment (FOEN/BAFU). University of Zurich, Zurich, Switzerland

Zinck JA (1988) Physiography and soils. ITC soil survey lecture notes. International Institute for Aerospace Survey and Earth Sciences, Enschede

Zinck JA, Urriola PL (1970) Origen y evolución de la Formación Mesa. In: Un enfoque eda-fológico. Ministerio de Obras Públicas (MOP), Barcelona

Part II
Approaches to Soil-Landscape Pattern Analysis

Chapter 9
Knowledge Is Power: Where Digital Soil Mapping Needs Geopedology

D. G. Rossiter

Abstract Much of current digital soil mapping (DSM) practice relies on terrain, climate and remote sensing-derived covariates. These are easy to obtain and can serve as proxies to soil forming factors and from these to soil properties or classes. However, mapping of soil bodies, not properties in isolation, is what gives insight into the soil landscape. A naïve attempt at correlating environmental covariates will not succeed in the presence of unmapped variations in parent material, soil bodies and landforms inherited from past environments. It also takes no account of spatial relations among soil bodies. Geopedology integrates an understanding of the geomorphic conditions under which soils evolve with field observations. Examples where simplistic DSM would fail but geopedology would succeed in mapping and, even better, explaining the soil distribution are shown: exhumed paleosols, low-relief depositional environments, inverted landscapes, and recent post-glacial landscapes.

Keywords Geomorphology · Digital soil mapping · Soil-landscape relations · Pleistocene glaciation

9.1 Introduction

Digital Soil Mapping (DSM) is the term given to producing predictive soil maps by the use of mathematical models applied to field observations of soils and synoptic layers related to soil formation and distribution (McBratney et al. 2003); this was termed "predictive soil mapping" by Scull et al. (2003), which is perhaps not such a suitable term, since all soil mapping is predictive of what the map user will encounter on the landscape. The "digital" in DSM is a byproduct of advances in technology

D. G. Rossiter (✉)
Section of Soil & Crop Sciences, New York State College of Agriculture & Life Sciences, Cornell University, Ithaca, NY, USA
e-mail: d.g.rossiter@cornell.edu

© The Author(s), under exclusive license to Springer Nature Switzerland AG 2023
J. A. Zinck et al. (eds.), *Geopedology*,
https://doi.org/10.1007/978-3-031-20667-2_9

and is meant to replace or extend the inductive reasoning of the expert soil mapper. The idea is to predict soil types, diagnostic features, or properties over a landscape, based on a set of observations and a set of environmental covariates covering the area to be mapped. These covariates are supposed to be either proxies for soil-forming factors, or sometimes simply empirically related to the property, feature or class to be mapped.

We restrict attention here to so-called 'scorpan'-based DSM, that is, where covariates are chosen to represent climate ('c'), organisms ('o'), relief ('r'), parent material ('p'), and time ('a', replacing Jenny's 't'); known soils ('s') are used for calibration and neighborhood ('n') relations (i.e., local spatial correlation) may be used. In practice, 'r' (terrain) and 'o' as represented by vegetation indices, surface reflectance or land use maps are the most widely-used covariates. The attraction of DSM is easy to understand large areas can be covered with reduced field survey, the uncertainty shows the reliability of the map, and the models behind the predictions can be made explicit, often providing insight into soil geography. This is in contrast to previous approaches, which relied on the mapper's mental model of the soil landscape, spatialized by manual interpretation of aerial photographs (Farshad et al. 2013). The geopedological approach of Zinck (2013) is the most theoretically-sound of these methods, because it is based on a systematic hierarchical soil-landscape analysis, not an *ad hoc* partitioning of the landscape based on perceived homogeneity.

Most digitally-produced soil maps are of single properties, notably soil organic C and particle-size distribution, often showing the depth distribution (e.g., Liu et al. 2013) as specified by the GlobalSoilMap.net project (Arrouays et al. 2014). Empirical methods based on point observations and correlation with spatially-complete covariates is well-suited for such mapping, although it provides no insight into soil geomorphology. By contrast, the geopedological approach considers soil as a natural body with its own history, ecology, function and, importantly, spatial relation with other bodies. The actual soils form clusters in the very large potential space formed by each attribute taken separately, and the soil function can only be appreciated as a whole, much greater than the sum of its parts. Maps of these clusters, i.e., soil types, can then be interpreted for multiple uses, and in addition they form a sound basis for stratification in the mapping of single properties. Thus, here we restrict our attention to DSM efforts to map soil types, as in the geopedological approach.

Some DSM approaches start from existing maps, which implicitly contain rich geopedological knowledge, and use digital methods to refine or update them (e.g., Kempen et al. 2009; Yang et al. 2011), and even attempt to disaggregate existing maps to a finer scale, e.g., the DSMART approach (Odgers et al. 2014) used in the POLARIS gridded maps of the continental USA (Chaney et al. 2019). The SoLIM (Soil-Landscape Inference Model) approach (Zhu et al. 2001) reasons by analogy from known soil-landscape relations. This requires either a pre-existing map of soil types or expert knowledge of where each type occurs on the landscape. Here we only consider the case where there is no existing soil-landscape map, only some point observations (usually purposive or opportunistic, not a probability sample) and a set of whole-field covariates, i.e., the common 'scorpan' approach.

DSM is the obvious soil mapping counterpart to similar data-driven approaches to knowledge in this computer age. Most current DSM models rely on terrain, climate and vegetation intensity covariates. These are easy to obtain; see for example Hengl (2013). They can be used as proxies for soil forming factors, thus they are related via pedogenesis to many soil properties, and from the assemblage of these properties to a soil type. An early statement of the hope of the digital soil mapper is from Zhu et al. (1996): "We assume that every soil series occurs under one or more typical environmental configurations or 'niches' and has a typical set of soil properties... and can be characterized by a vector of environmental parameters in an m-dimensional parameter space".

Why is the 'scorpan' approach not always successful? Fundamentally, there is much more to soil formation than the current environment. In particular, the soil forming factor 'time' is only approximately represented by landscape position, and the factor 'parent material' does not always have a close relation to topography. As early as 1935 Milne (1935) recognized that some east African toposequences (his 'catenas') developed on uniform parent rock, others on sequences of outcropping rocks. A direct correlation between soil types and slope positions was thus not possible. Variations in parent material (in the absence of a detailed surficial geology map) and the short time-scale of covariates compared with the time-scale of soil formation result in models that do not fully characterize the soil cover.

Another problem with the empirical 'scorpan' approach is that soils often have inherited much of their current characteristics from previous climates and the associated vegetation, and indeed they may be the result of multiple cycles of soil formation, as evidenced by stone lines, lithologic discontinuities or landscape inversion resulting from cycles of erosion and/or deposition. In younger landscapes, the topography may be relict from recent disruptions such as glaciation or vulcanism.

A final major problem with the 'scorpan' approach is that soil bodies often have a spatial relation, where materials have been transported from one body to form another. Examples are alluvial fans and colluvial deposits. These do have a landscape position and morphology (accounted for with the 'r' factor) but in addition inherent their parent material ('p' factor) from adjacent units.

To date DSM has been almost exclusively empirical: a statistical relation is established between the observations and covariates, and this relation is then applied across the area to be mapped. Soil property DSM has used, among others, geostatistics (e.g., Kriging with External Drift), multiple regression, random forest regression and similar "machine learning" methods, generalized additive models (GAM). Soil feature or class mapping has used multiple logistic regression (e.g., Abbaszadeh Afshar et al. 2018), similarity in feature space (e.g., Zhu et al. 2015), or random forest classification. A good review of the various DSM methods for soil classes is by Heung et al. (2016).

In this chapter I give some examples where 'scorpan'-based DSM of soil classes based on the usual covariates will fail, but where geomorphic analysis results in successful landscape stratification, within which field observations can be placed, will produce a reliable map. We consider four examples: exhumed paleosols, depositional low-relief environments, inverted landscapes, and young post-glacial landscapes. The last example is explained in detail.

A separate issues is the complex and contingent nature of pedogenesis as evolution with continuously-varying environmental conditions (Phillips 2001; Huggett 1998); this suggests that there is a chaotic, non-deterministic element to pedogenesis that cannot be inferred from observations of soils in similar niches. This is outside the scope of this chapter.

9.2 Example 1: Exhumed Paleosols

Exhumed paleosols are soils, now at the surface or covered by a thin mantle of newer material, which developed under a different climate than the present. They were then buried by new deposits, e.g., by a younger glacial till or loess, but then by landscape evolution (dissection, down-wasting) exposed again at the surface. Their soil properties are largely controlled by conditions in the past, although of course now subject to current conditions for further evolution. A classic study is from Ruhe et al. (1967), who identified various glacial till, loess and paleosol layers from four glacial and three interglacial stages in Iowa (USA). A detailed geomorphic investigation reveals, for example, relict fluvial surfaces (floodplain alluvium, slope fan alluvium) from the Sangamon interglacial which are now above the current base level where current fans and flood plains are located; further a relict pediment with stone line developed in Kansan till is mantled by a thin Wisconson loess layer, and on the interfluves a modern soil developed in the loess but overlying a 'gumbotil' layer, i.e., very clayey weathered Kansan till (Kay and Pearce 1920). Some late Wisconsin-Recent slopes have cut back to interfluves, and on these erosional slopes Yarmouth-Sangamon paleosols outcrop, with younger soils above and below. These exhumed paleosols may also be truncated, so the paleo-B horizons are now at the surface. Others have thin caps of loess or slope wash.

How could 'scorpan'-based DSM deal with this area? The exhumed soils occupy a defined elevation range, but since this represents a relict dissected surface, there are several soil classes in this same position. By contrast, the geomorphic analysis of Ruhe explains the soil distribution and provides a key for mapping. In geopedological terms, the surfaces would be separated at the lithology level.

9.3 Example 2: Depositional Low-Relief Environments

Soils in depositional low-relief environments such as fluvial systems with rapidly-changing channels and variable infilling (e.g., the Rhine-Meuse delta of the Netherlands, see Berendsen 2005) cannot be mapped by interpolation, even with intensive boring campaigns, without geomorphic interpretation of the paleo-geography. Another example is the detailed study by of the alluvial and terrace soils associated with the Río Guarapiche in Monagas state, Venezuela (Zinck and Urriola 1971; Zinck 1987). From the geomorphology one can delineate various landscape components such as current and abandoned channels, backswamps, splay fans, and associate these with soil types. The relief is subtle. Vegetation differences can reveal some of the differences, but only in areas where there has been no artificial drainage.

How could 'scorpan'-based DSM deal with this area? The landforms are quite similar, although backswamps may have slightly more concave shape. The elevation differences between terrace levels are quite small, and the absolute elevation decreases downstream, so that a single elevation range cannot be used to identify a terrace. Splay fans have the same elevation as backswamps but quite different soils.

9.4   Example 3: Inverted Landscapes

Pain and Ollier (1995) present a convincing argument that landscape inversion is a common form of landscape evolution. Ferricrete ('laterite', 'plinthite', 'ironstone') and duricrusts often form in lower landscape positions, and lava flows may preferentially follow pre-existing valleys. These materials are more resistent to erosion than their surroundings, and eventually end up as the highest landscape positions. An example is central Uganda (where Milne developed the catena concept), where thick ferricrete mesas are typically the highest landscape positions. During the inversion process continued weathering of saprolite and movement of materials and solutes along the slope have resulted in a complex soil landscape (Brown et al. 2004); see Fig. 9.1.

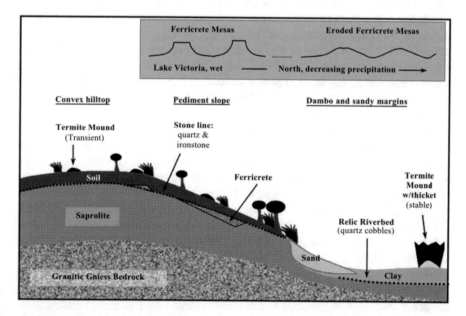

Fig. 9.1 Conceptual diagram of the Buruli catena, central Uganda. (Fig. 2 in Brown et al. (2004), used by permission)

How could 'scorpan'-based DSM deal with this area? If the positions within the catena are regular, the 'r' covariates elevation, slope gradient, curvature and wetness index could separate the soil types. This depends on (1) a limited area with a repeating landscape pattern, (2) training observations in all landscape positions. If this area is mapped as part of a larger area, perhaps including coordinates in a random forest DSM model might be able to "box" this area, within which the relation with indicated covariates would apply. As in the paleosols example, geomorphic analysis clearly explains the soil distribution and provides a key for mapping. In geopedological terms, the components of the catena would be separated at the landform level.

9.5   Example 4: Young Post-Glacial Landscapes

Large areas of northern North America and Europe are covered with soils developed in young post-glacial landscapes; smaller areas are from recent alpine glaciation. In these areas the geomorphology and distribution of parent materials can only be understood by means of the detailed history of glaciation and deglaciation (e.g., proglacial lakes, outwash plains, sandurs) which have only an indirect relation with terrain variables. We illustrate this with an example from Tompkins and Tioga counties, New York State (USA).

Figure 9.2 is a fragment of the USGS 7.5′ 1:24000 topographic map West Danby and Willseyville (NY) sheets. An analyst following the geopedological approach would use stereo-pairs of remotely-sensed images, e.g., airphotos, but even without stereo view the map clearly shows features that are immediately recognizable to a trained analyst familiar with the Pleistocene history of the region (Bloom 2018): (1) a terminal moraine of the Valley Heads stage, behind which are (2) hummocks and kettles from stagnating ice; (3) pro-moraine outwash terraces, breached on the E and NE margin by (4) post-retreat outflow channels which formed (5) outwash terraces transecting the end moraine; (6) truncated spurs and post-glacial incisions; (7) in the NE edge a high-level terrace formed above the moraine when it was blocking outflow; (8) high-level outflows from the main glacial tongue, when it was pressed up against the E margin; (9) post glacial fans from upland erosion; (10) a large kettle, now a shallow lake and swamp, in front of the centre of the moraine, corresponding to a large block of ice separated from the glacier.

Figure 9.3 shows the detailed soil survey of the same area, provided by the NRCS (USA) Web Soil Survey (http://websoilsurvey.sc.egov.usda.gov), here displayed on a Google Earth background by the SoilWeb application (O'Geen et al. 2017; California Soil Resource Lab n.d). Table 9.1 shows a tentative geopedologic legend for this area.

Referring to Fig. 9.3, we identify several situations where a DSM approach using the usual covariates will not work, but where geomorphic knowledge results in an easy landscape interpretation and soil mapping:

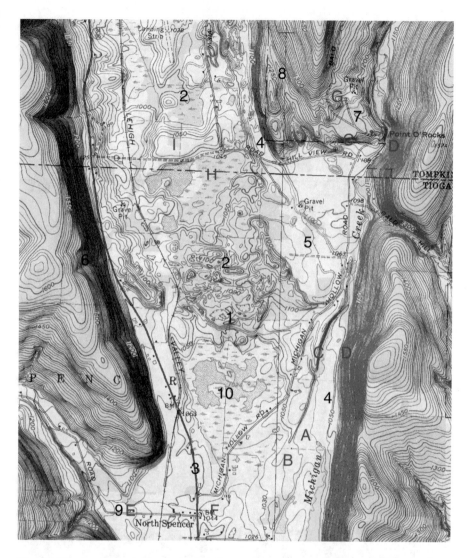

Fig. 9.2 Fragment of the USGS 7.5′ 1:24,000 topographic map West Danby and Willseyville (NY) sheets. Annotations are geomorphic features (black numbers) and sites where soils are discussed (red letters); see text

- Positions A and B have identical slopes (flat), differ in elevation by less than a meter, are the same distance from streams, have almost the same wetness index, both are agricultural fields, yet the soils are quite different. A is mapped as the somewhat poorly-drained Middlebury (coarse-loamy Fluvaquentic Eutrochrepts) and well-drained Tioga series (coarse-loamy Dystric Fluventic Eutrochrepts), aggrading alluvial soils in silty and sandy alluvium from the present-day outlet of Michigan Creek, while B is mapped as the Howard series (loamy-skeletal

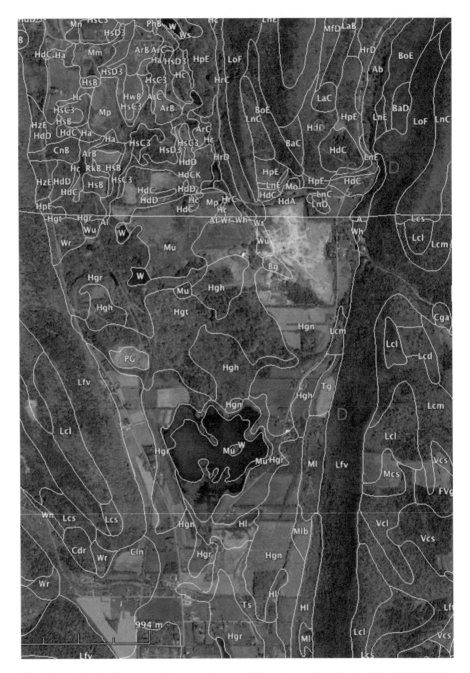

Fig. 9.3 Detailed soil survey of the area shown in Figure 1, provided by the NRCS (USA) Web Soil Survey (http://websoilsurvey.sc.egov.usda.gov), displayed on a Google Earth background by the SoilWeb application (http://www.gelib.com/soilweb.htm). Annotations as in Fig. 1. See SoilWeb for map unit codes and descriptions

Table 9.1 Provisional geopedologic legend for Example 4

Landscape	Relief type	Lithology	Landform	Soil series or family
Dissected plateau	Truncated ridge	Thin till from Devonian shales & mudstones	Convex summit	Lordstown
			Concave backslope	Volusia
		same, plus outcropping bedrock	Straight, very steep front slope	Lordstown, Arnot
		Deep till	Side slopes	Langford
	Side valleys	Recent poorly-sorted alluvium from upland material	Narrow valley, moderate gradient	Chenango, coarser
			Alluvial fan	Chenango, finer
	Terraces	Glacial outwash	Dissected	Howard
			Flat	Howard
	Ice-margin complex	Dissected thin till and outcropping bedrock	Overflow channel (upland)	Valois
		Glacial outwash	Overflow channel (upland margin)	Howard, Valois
Through valley	End moraine complex	Wisconsonian poorly-sorted pushed material	End moraine	Howard, Palmyra
		same, plus recent organic sediment and sorted fine sand	Hummocks and kettles	Howard, Arkport, Saprists, Water
		same, plus recent organic sediment	Post-moraine lake and marsh	Saprists
	Outwash plain	Glacial outwash from end moraine material	Plain	Howard
		Ice-block inclusions	Pro-moraine kettle	Saprists, Water
	Recent overflow channels	Alluvium	Flat-bottomed channel	Tioga, Middlebury

Glossoboric Hapludalfs), a well-drained well-developed (considering the approximately 12 k years since the retreat of the glacier) gravelly loam from pro-glacial outwash, with about 30% rock fragments, mostly rounded cobbles of mixed origin. These soils differ considerably in age and lithology but cannot be separated by terrain covariates.

- Positions C and D (two examples) have identical very steep slopes and slope shapes (straight), both well-vegetated with native hardwoods, yet the soils are radically different. C is again the Howard series but truncated by the modern outlet of Michigan Creek to expose an outcrop of gravelly glacial outwash, while D is mapped as the Lordstown series (coarse-loamy Typic Dystochrepts), a moderately deep to bedrock channery silt loam with about 20% large to medium rock fragments from Devonian shale and mudstone; on the steepest slopes the soils are probably in the shallow to bedrock Arnot series (loamy-skeletal Lithic Dystochrepts).

- Positions E and F are adjacent, with similar terrain parameters, elevation and land use, but are easily recognized as a modern alluvial fan (9, E) and glacial outwash (10, F). Again, F is the Howard series; here E is mapped as Chenango (loamy-skeletal Typic Dystrudepts), a younger soil with periodic flash floods (e.g., due to hurricanes and rapid snowmelt in the contributing watershed) resulting in additions of subrounded poorly-sorted gravels (mudstone and sandstone) from the surrounding uplands.

- Position G is especially interesting. It is at a high elevation, has moderately steep slopes, is in native forest vegetation, yet is also mapped as the Howard series, i.e., it is glacial outwash, not soil in residuum, e.g., the surrounding Lordstown soils with the same topography and vegetation. The geomorphic clue here is outside the figure: Michigan Hollow (seen entering on the NE) is a through valley where the original drainage divide, about 5 km N, was removed by the glacier; subsequently as that tongue melted a large amount of outwash was deposited in what was then a lake behind the terminal moraine (1). Apparently, there were two levels; the higher one (G) was subsequently easily eroded by upland runoff; the lower terrace (between G and C) remains almost flat. The incision at (8) is also explained by a period where the ice filled the valley (NW in the figure) so that meltwater had to follow this channel to produce some of the outwash (5). The W margin of this hill shows the same phenomenon but from when the ice had melted enough to allow water to flow along its margins at the base of the truncated spur.

- Positions H and I differ by only 30 m elevation, are both flat, both with dense vegetation; yet while I is again mapped as Howard (glacial outwash), H is mapped as Typic and Terric Medisaprists, i.e., an organic soil. Geomorphically this is easy to understand: both positions are part of the kettle moraine (2). Some similar positions to I are mapped as Arkport (coarse-loamy Psammentic Hapludalfs); these are further behind the end moraine where meltwater was sandier.

Although 'scorpan'-based DSM would not be able to find these differences, some other approaches might have some success. To do so, they would have to emulate the geopedologic interpretation. For example, it might be possible to identify post-glacial alluvial fans by their relative landscape position: where narrow steep side valleys emerge onto outwash plains. Also, their shape is diagnostic: narrow at the proximal (upstream) end, widening at the distal end. These might be revealed by a segmentation which then considered adjacency and oriented (proximal-distal) shape relations. However, the boundary between the fan and the outwash which it overlays (E vs. F) is quite subtle. Although visible to the geopedologist it seems difficult to delineate automatically. The difference between C and D might be revealed by total slope length and position on the slope.

Another covariate that might have some success is hyperspectral remote sensing, which might allow vegetation communities to be distinguished (e.g., between positions H and I). Differences in amount of weathering (factor 'a') can sometimes be inferred from aerial gamma-ray survey (Moonjun et al. 2017). However, the geopedologic legend and map give a holistic view of the soil landscape.

9.6 Discussion

9.6.1 What Could Be the Contribution of the Geopedologic Approach to DSM?

DSM methods are important additions to the soil mapper's toolkit, especially when large areas need to be mapped, and when estimates of uncertainty are needed. In simple landscapes with close correlation between topographic parameters, land use and soil type, it has shown good success. This is especially true in smaller areas where many soil-forming factors are more or less constant, and only a few covariates are needed to separate the soil types or properties. An example is a landscape with a single lithology ('p') and climate ('c'). Here toposequences ('r') correlate well with soil development ('a'), and the land use ('o') can account for anthropogenic influences. But as Milne's original catena shows, in many toposequences lithology is not constant, for example the ironstone caps of relict plateaus from a landscape inversion. These are easily identifiable and understandable for the geomorphologist.

However, as the above examples show, there are situations where a geomorphic understanding is necessary to identify locations where each soil type is expected. The only geopedological knowledge used in current DSM approaches is the selection of covariates to (presumably) represent soil-forming factors. Recently a set of challenges for the future of pedometrics (Wadoux et al. 2021) recognizes this limitation as one of its ten challenges: "Can we incorporate mechanistic pedological knowledge in digital soil mapping?" Ma et al. (2019) discusses this in detail. This is not exactly an appeal to geopedology, but "mechanistic" could be replaced by "expert", i.e., geopedologic knowledge of soil-landscape relations, built explicitly into the DSM workflow.

One promising approach is so-called "contextual" DSM (Behrens et al. 2018). This uses a multiscale version of a digital elevation model (DEM) to derive a set of terrain derivative (factor 'r') at different scales. The relation to geopedology is that the coarser scales may in some cases correspond to the higher levels in the geopedologic hierarchy, from landscape (most general) through relief type to landform. A similar idea is the use of so-called "deep learning" in the form of convolutional neural networks (CNN) which use contextual information from the environmental covariates, i.e., from a hierarchical set of neighborhoods, not just the covariate values at the observation point (Padarian et al. 2019). These have not yet been applied to soil classes but do have the possibility to account for some adjacency or upstream-downstream relations.

Object-oriented image segmentation applied to stacks of terrain parameters (Dragut et al. 2009) offers a digital approach to discovering landscape units, which can perhaps be interpreted and correlated to soil types. However, several segments may have similar landscape parameters, yet be of contrasting origin, for example, alluvial terraces vs. glacial outwash terraces. Here the relative position in the landscape can be used to differentiate them. This requires concepts of adjacency and flow direction.

This begs the question as to whether geomorphology, as opposed to geomorphometry, can be digitally mapped (Bishop et al. 2012). If so, the digital geomorphic map could be used as a powerful covariate for digital soil mapping; perhaps geopedology would not be necessary. The most promising method is object-oriented analysis, followed by geomorphometric characterization (Hengl and Reuter 2008), leading, it is hoped, to interpretable terrain units. However, Bishop et al. are clear on the limitations: "Although this scale-dependent approach is conceptually pleasing, it is nonetheless fundamentally a cartographic approach to mapping that does not formally address issues of processes, internal and external forcing factors, feedback mechanisms and systems, or spatio-temporal dynamics." In other words, geomorphology, and hence geopedology, is not simply terrain analysis, no matter how sophisticated. Evans (2012) has a similarly pessimistic view of the prospects for automated geomorphic mapping.

9.7   Conclusion

There are situations where neither DSM nor geopedology will be successful, and where intensive systematic field observation is the only way to map important soil differences. An example is given by Toomanian (2013) of a playa in the Zayandeh-rud valley, Iran, where a uniform surface is created by an aeolian mantle; this mantle covers a wide diversity of aeolian, lagoonal, and alluvial layers deposited during the Quaternary and Tertiary. The geomorphometry is uniform, the soil surface reflectance and vegetative cover as well. Although surface salinization can be detected, this is not related to important subsurface differences. There is no solution but to grid sample and interpolate. But for many soil landscape, the integration of geomorphic understanding and its relation to soil genesis allows successful mapping, where simple environmental correlation using 'scorpan' covariates as presumed proxies for soil-forming factors is not successful.

References

Abbaszadeh Afshar F, Ayoubi S, Jafari A (2018) The extrapolation of soil great groups using multinomial logistic regression at regional scale in arid regions of Iran. Geoderma 315:36–48. https://doi.org/10.1016/j.geoderma.2017.11.030

Arrouays D, Grundy MG, Hartemink AE et al (2014) GlobalSoilMap. Adv Agron 125:93–134

Behrens T, Schmidt K, MacMillan RA, Viscarra Rossel RA (2018) Multiscale contextual spatial modelling with the Gaussian scale space. Geoderma 310:128–137. https://doi.org/10.1016/j.geoderma.2017.09.015

Berendsen HJA (2005) Landschappelijk Nederland: de fysisch-geografische regio's, 3rd edn. Uitgeverij Van Gorcum

Bishop MP, James LA, Shroder JF Jr, Walsh SJ (2012) Geospatial technologies and digital geomorphological mapping: concepts, issues and research. Geomorphology 137:5–26. https://doi.org/10.1016/j.geomorph.2011.06.027

Bloom AL (2018) Gorges history: landscapes and geology of the Finger Lakes Region. Paleontological Research Institution, Ithaca

Brown DJ, Clayton MK, McSweeney K (2004) Potential terrain controls on soil color, texture contrast and grain-size deposition for the original catena landscape in Uganda. Geoderma 122:51–72. https://doi.org/10.1016/j.geoderma.2003.12.004

California Soil Resource Lab (n.d.) SoilWeb Apps [WWW Document]. URL https://casoilresource.lawr.ucdavis.edu/soilweb-apps/. Accessed 1 Jan 22

Chaney NW, Minasny B, Herman JD et al (2019) POLARIS soil properties: 30-m probabilistic maps of soil properties over the contiguous United States. Water Resour Res 55:2916–2938. https://doi.org/10.1029/2018WR022797

Dragut L, Schauppenlehner T, Muhar A et al (2009) Optimization of scale and parametrization for terrain segmentation: an application to soil-landscape modeling. Comput Geosci 35:1875–1883

Evans IS (2012) Geomorphometry and landform mapping: what is a landform? Geomorphology 137:94–106. https://doi.org/10.1016/j.geomorph.2010.09.029

Farshad A, Shrestha DP, Moonjun R (2013) Do the emerging methods of digital soil mapping have anything to learn from the geopedologic approach to soil mapping and vice versa? In: Shahid SA, Taha FK, Abdelfattah MA (eds) Developments in soil classification, land use planning and policy implications. Springer, Dordrecht, pp 109–131

Hengl T (2013) WorldGrids. http://worldgrids.org/doku.php. Accessed 28 Nov 2014

Hengl T, Reuter HI (eds) (2008) Geomorphometry: concepts, software, applications. Elsevier Science

Heung B, Ho HC, Zhang J, Knudby A, Bulmer CE, Schmidt MG (2016) An overview and comparison of machine-learning techniques for classification purposes in digital soil mapping. Geoderma 265:62–77. https://doi.org/10.1016/j.geoderma.2015.11.014

Huggett R (1998) Soil chronosequences, soil development, and soil evolution: a critical review. Catena 32(3–4):155–172

Kay GF, Pearce JN (1920) The origin of gumbotil. J Geol 28:89–125. https://doi.org/10.1086/622700

Kempen B, Brus DJ, Heuvelink GBM, Stoorvogel JJ (2009) Updating the 1:50,000 Dutch soil map using legacy soil data: a multinomial logistic regression approach. Geoderma 151:311–326

Liu F, Zhang G-L, Sun Y-J et al (2013) Mapping the three-dimensional distribution of soil organic matter across a subtropical hilly landscape. Soil Sci Soc Am J 77:1241–1253. https://doi.org/10.2136/sssaj2012.0317

Ma YX, Minasny B, Malone BP, McBratney AB (2019) Pedology and digital soil mapping (DSM). Eur J Soil Sci 70:216–235. https://doi.org/10.1111/ejss.12790

McBratney AB, Mendonça Santos ML, Minasny B (2003) On digital soil mapping. Geoderma 117:3–52

Milne G (1935) Some suggested units of classification and mapping, particularly for East African soils. Soil Res Suppl Proc Int Societ Soil Sci 4:183–198

Moonjun R, Shrestha DP, Jetten VG, van Ruitenbeek FJA (2017) Application of airborne gamma-ray imagery to assist soil survey: a case study from Thailand. Geoderma 289:196–212. https://doi.org/10.1016/j.geoderma.2016.10.035

O'Geen A, Walkinshaw M, Beaudette D (2017) SoilWeb: a multifaceted interface to soil survey information. Soil Sci Soc Am J 81:853–862. https://doi.org/10.2136/sssaj2016.11.0386n

Odgers NP, Sun W, McBratney AB et al (2014) Disaggregating and harmonising soil map units through resampled classification trees. Geoderma 214–215:91–100. https://doi.org/10.1016/j.geoderma.2013.09.024

Padarian J, Minasny B, McBratney AB (2019) Using deep learning for digital soil mapping. Soil 5:79–89. https://doi.org/10.5194/soil-5-79-2019

Pain C, Ollier C (1995) Inversion of relief – a component of landscape evolution. Geomorphology 12:151–165. https://doi.org/10.1016/0169-555X(94)00084-5

Phillips JD (2001) Contingency and generalization in pedology, as exemplified by texture-contrast soils. Geoderma 102(3–4):347–370

Ruhe RV, Daniels RB, Cady JG (1967) Landscape evolution and soil formation in southwestern Iowa. In: Technical bulletin (United States. Dept. of Agriculture). U.S. Department of Agriculture, Washington, DC

Scull P, Franklin J, Chadwick O, McArthur D (2003) Predictive soil mapping: a review. Prog Phys Geogr 27:171–197

Toomanian N (2013) Fundamental steps for regional and country level soil surveys. In: Shahid SA, Taha FK, Abdelfattah MA (eds) Developments in soil classification, land use planning and policy implications. Springer, Dordrecht, pp 203–227

Wadoux AMJ-C, Heuvelink GBM, Lark RM et al (2021) Ten challenges for the future of pedometrics. Geoderma 401:115155. https://doi.org/10.1016/j.geoderma.2021.115155

Yang L, Jiao Y, Fahmy S et al (2011) Updating conventional soil maps through digital soil mapping. Soil Sci Soc Am J 75:1044–1053. https://doi.org/10.2136/sssaj2010.0002

Zhu A-X, Band LE, Dutton B, Nimlos T (1996) Automated soil inference under fuzzy logic. Ecol Model 90:123–145

Zhu A-X, Hudson B, Burt J et al (2001) Soil mapping using GIS, expert knowledge, and fuzzy logic. Soil Sci Soc Am J 65:1463–1472. https://doi.org/10.2136/sssaj2001.6551463x

Zhu AX, Liu J, Du F, Zhang SJ, Qin CZ, Burt J, Behrens T, Scholten T (2015) Predictive soil mapping with limited sample data. Eur J Soil Sci 66:535–547. https://doi.org/10.1111/ejss.12244

Zinck JA (1987) Aplicación de la geomorfología al levantamiento de suelos en zonas aluviales y definición del ambiente geomorfológico con fines de descripción de suelos. Instituto Geográfico "Augustín Codazzi, Bogotá

Zinck JA (2013) Geopedology : elements of geomorphology for soil and geohazard studies. ITC, Enschede

Zinck JA, Urriola P (1971) Estudio edafológico del valle del Río Guarapiche. Ministerio de Obras Públicas, Estado Monagas, Caracas

Chapter 10
Geodiversity and Geopedology in a Logarithmic Universe

J. J. Ibáñez and E. C. Brevik

Abstract Pedology and geomorphology are considered independent scientific disciplines, but form in fact a single indivisible system. Diversity analysis of natural resources tries to account for the variety of forms and spatial patterns that are displayed in the natural bodies, biotic and abiotic, appearing at the earth's surface. The application of mathematical tools to diversity analysis requires a classification of the universe concerned. Biodiversity studies have a long tradition in comparison to those in earth sciences. Recently pedologists started paying attention to soil diversity using the same mathematical tools as ecologists and finding interesting relations between the spatial patterns of soil and vegetation. So far geodiversity studies are primarily concerned with the preservation of geological heritage, bypassing most of the aspects related to its spatial distribution and the quantitative estimation of geodiversity. Vegetation scientists have developed a classification that links climate and plant communities, the so-called syntaxonomic system. The purpose of this chapter is to explore a perspective for joining soils, geoforms, climate, and biocenoses in an integrated and comprehensive approach to describe the structure and diversity of earth surface systems. Likewise, a meta-analysis of the findings detected to date was carried out considering the scales of geographical space, a Paretian perspective and the cognitive processing of the data by the human mind.

Keywords Geopedology · Pedodiversity · Landscape diversity · Geodiversity · Logarithmic nature · Human cognition

J. J. Ibáñez
National Museum of Natural History (MNCN), Spanish National Research Council (CSIC), Madrid, Spain
e-mail: jj.ibanez@csic.es

E. C. Brevik (✉)
College of Agricultural, Life, and Physical Sciences, Southern Illinois University Carbondale, Carbondale, IL, USA
e-mail: Eric.Brevik@siu.edu

© The Author(s), under exclusive license to Springer Nature
Switzerland AG 2023
J. A. Zinck et al. (eds.), *Geopedology*,
https://doi.org/10.1007/978-3-031-20667-2_10

10.1 Introduction

Natural resources vary in space and time. Throughout the centuries naturalists have observed that some landscapes are more heterogeneous than others regardless of the nature of the study object (e.g., biological species, rocks, landforms, soils, etc.) (Ibáñez 2017a, b). It is essential practice in science to identify, categorize, and classify the heterogeneity of the objects of study, leading to taxonomy. Using variable criteria several classification systems with different numbers of taxa have been proposed. To achieve a common language between experts in a scientific community, the most desirable outcome would be universally accepted taxonomies. It has also been recognized that it is impossible to develop perfect taxonomies, as these change over time according to perceptions, scientific progress, and societal information demands (Ibáñez and Montanarella 2013). The taxonomy of a given natural resource in a given time provides an inventory of the global diversity that is accepted by the experts that embrace that mental construct. The tree of life has been most extensively studied so that biological taxonomies are considered the most elaborate and sophisticated (Ibáñez and Montanarella 2013). Studies of biological diversity are several decades ahead compared to other natural resources, such as minerals, landforms, or soils, and have used numerous methodologies and mathematical procedures to estimate biodiversity. In contrast, the analysis of pedodiversity started only in the last decades of the twentieth century (Fridland 1976; Ibáñez et al. 1990). Recently Ibáñez et al. (2013) and Ibáñez (2017a, b) have synthesized the knowledge on soil diversity. Biodiversity and pedodiversity studies pursue two different but complementary purposes: (a) the analysis of the structure of ecosystems and soil assemblages, respectively, and (b) the preservation of both natural resources. Geodiversity studies have focused mainly on the preservation of geological heritage, and its quantification remains a huge challenge to be solved (Ibánez and Brevik 2019; Ibáñez et al. 2019).

The purpose of this chapter is to explore a perspective for joining soils, geoforms, parent materials, climate, and biocenoses (associations of organisms that form a closely integrated community) in an integrated and comprehensive approach to describe the structure and diversity of earth surface systems. Likewise, the geopedological approach offers a new frame to quantify the whole of geodiversity in view that their inventory and mapping consider geology *s.l*, geomorphology, rocks, landforms and soils (see Sect. 10.6). Both approaches capture all soil forming factors, permitting a step forward in geodiversity quantification.

10.2 The Concept of Diversity

There is no consensus among experts on the definition of biodiversity and pedodiversity. However, the following proposal is satisfactory from a methodological point of view and does not distinguish between the different objects that can be analyzed.

According to Huston (1994, p. 65), diversity can be conceptually defined as follows: *"The concept of diversity has two primary components, and two unavoidable value judgments. The primary components are statistical properties that are common to any mixture of different objects, whether the objects are balls of different colors, segments of DNA that code for different proteins, species or higher taxonomic levels, or soil types or habitat patches on a landscape. Each of these groups of items has two fundamental properties: 1. the number of different types of objects (e.g., species, soil types) in the mixture or sample; and 2. the relative number or amount of each different type of object. The value judgments are: 1. whether the selected classes are different enough to be considered separate types of objects; and 2. whether the objects in a particular class are similar enough to be considered the same type. On these distinctions hangs the quantification of biological diversity."*

There are essentially three components of diversity: (a) the variety of taxa (richness); (b) the way in which the individuals are distributed among those taxa (evenness or equitability); and (c) diversity indices that attempt to incorporate the aforementioned components into a single value (Magurran 2003; Ibáñez et al. 2013; Ibáñez 2017a, b). In addition, abundance distribution models provide the closest fit to the observed pattern of object abundance (e.g., geometric series, log series, lognormal series, power laws, abundance distribution models). Statistical regression models and other mathematical procedures have also been extensively applied to analyze diversity-area relationships (e.g., Ben-Shahar and Coe 1992; Ibáñez et al. 2005a).

Several other methodologies and useful mathematical tools have been proposed to complement the former such as fractals, multifractals, nested subsets theory, neural networks, and some procedures applied by physicists to the study of non-linear systems (Phillips 2016, 2017). Ibáñez et al. (2021a, b), based on an extensive review of the literature and empirical data, found that most natural and artificial taxonomies and maps follow the same mathematical patterns that result because they are non-linear complex systems. They also found indications that our cognitive apparatus that explore the world around us work intuitively in a logarithmic fashion rather than a linear one. This is the so-termed Gaussian-Bayesian dilemma (Andriani and McKelvey 2009 among many others).

10.3   Biodiversity, Pedodiversity, Landform Diversity, and Lithological Diversity Patterns

Few studies have been carried out to analyze the relationships between pedodiversity and the diversity of other natural resources (Fig. 10.1). The most frequent are those that compare biodiversity and pedodiversity (e.g., Petersen et al. 2010; Williams and Houseman 2013; Ibáñez and Feoli 2013; Ibáñez et al. 2016). Ibáñez et al. (2013) and Ibáñez (2017a, b) showed that biodiversity and pedodiversity are usually positively correlated and have similar spatial patterns. There are several

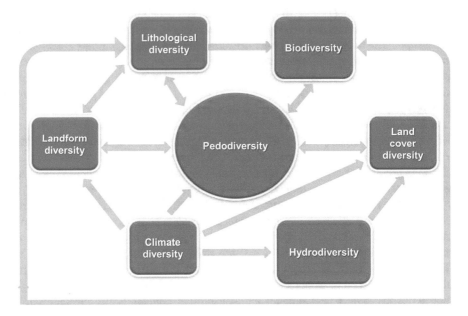

Fig. 10.1 Interrelationships between pedodiversity and the diversity of soil forming factors

studies that analyzed the relations between biodiversity and landforms, lithodiversity and landforms, pedodiversity and landforms and geodiversity and landforms (e.g., Tukiainen et al. 2017, 2019). Ibáñez et al. (2013 and references therein) and Toomanian (2013), among others, have detected positive correlations between pedodiversity and landform diversity, as well as between pedodiversity and lithodiversity, among others.

10.4 Geopedologic and Bioclimatic Approaches

All natural resources are diverse. A simple description of the landscape could be interesting to reduce the divisions between natural resources by making use of more holistic concepts (e.g., pedodiversity + landform diversity = geopedologic diversity and/or plant community diversity + climate diversity = bioclimate diversity).

Geopedology proposes a landscape approach that integrates geoforms and soils (Zinck 2013). Because geomorphology takes into account relief and morphodynamics in a morphoclimatic context and geoforms are in many aspects conditioned by lithology, this approach is more integrative than pedology, geomorphology and geology *s.l.* applied individually. Zinck (2016) established a hierarchic geopedologic approach to analyze soils in the landscape that included geostructure, morphogenic environment, geomorphic landscape, relief, lithology, and landforms. The geopedologic approach has been applied to soil survey (Zinck 2016) as well as in plant-landforms studies (e.g., Stacey and Monger 2012; Michaud et al. 2013). Geopedology is also a method considered in landscape ecology (e.g., Saldaña 2013; Zinck 2013, and references therein). Some geopedologic classifications have been proposed for universal application (e.g., Zinck 1988, 2013), whereas others have been developed to analyze only specific territories (e.g., Michaud et al. 2013).

The European school of geobotany, called phytosociology, is a discipline that focuses on the classification of plant communities (e.g., Westhoff and van der Maarel 1978), with an International Code of Phytosociological Nomenclature (Weber et al. 2000). To inventory, map, and classify plant landscapes according to phytosociology, geobotanists use the syntaxonomic system (e.g., Mirkin 1989). The syntaxonomic approach mainly takes into account plant communities based on the concept of plant natural vegetation (PNV) and climate to develop a classification of plant landscapes in bioclimatic belts (Loidi and Fernández-González 2012). However, the geobotanical school also considers pedologic, geomorphic, and lithological land features to analyze plant-soil relationships at the landscape level (Rivas-Martínez 2005). The landscape is divided into units termed tessela and microtessela that correspond to terrestrial areas where the same PNV are present. Within the syntaxonomic classification, the nomenclature of plant assemblages includes terms such as climatophilous (plant communities that only depend on climatic factors), basophilous (plant associations that grow on pedotaxa rich in nutrients; eutric in pedological terms), siliceous (plant associations that grow on pedotaxa poor in nutrients; dystric in pedological terms), calcicolous (plant assemblages associated with the presence of calcium carbonate in soils; calcic and calcaric in soil classifications), and edaphophylous (plant associations that depend on specific soil features and properties). There are other terms such as gypsiferous or halophytic (reserved for gypsiferous and halophytic vegetation, associated with gypsum- and salt-rich soils, respectively). The edaphophylous units are divided into edaphoxerophilous (plants adapted to xericity that grow in tessela or microtessela on pedotaxa that store very little water) and edaphohygrophilous (plants species associated to pedotaxa with permanent or seasonal waterlogging). Other terms such as permaseries, geopermaseries, and geoseries are indicators of PNV that occur on sites with abiotic constraints that produce permanent environmental stresses that limit the full development of an ecological succession. Rivas-Martínez (2005), among others, explains concepts and nomenclature. Summarizing, the syntaxonomic system classifies PNV units taking into account all the environmental variables that influence the distribution of plant communities by adding these to the formal nomenclature of each syntaxum. Most of these factors are climatic, but also include pedological, geomorphic, and lithological ones, all coming together to influence landscape diversity and through its PNV (Fig. 10.2).

Figure 10.3 shows the richness of PNV in the Almeria province (south-eastern Spain) according to the factors that determine their presence and geographic dispersion in the study area (Ibáñez et al. 2015). The geographic distribution (dominant, abundant, common, frequent, rare, endemic) of PNV was clustered according to the number of bioclimatic belts (BB) (there are a total of seven in the Almeria province) where these plant communities appear. For example, abundant PNV means that they

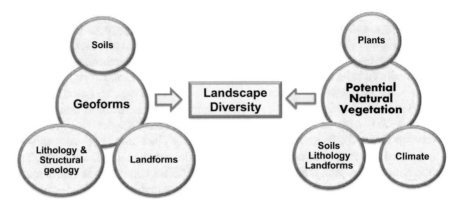

Fig. 10.2 A landscape diversity scheme

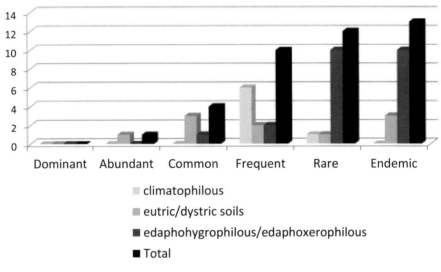

Fig. 10.3 Plant landscape diversity scheme (number of PNV types) and the role of soils in Almeria Province, Spain. (Ibáñez et al. 2015)

appear in most climatic belts, whereas endemic is those that only appear in one of them. It is noticeable that in the Almeria Province there are many more plant communities determined by soil types and properties than those conditioned by climatic factors only. However, many of these soils are in turn associated with specific landforms.

Thus, in principle it would be possible to analyze the landscape, including environmental heterogeneity and diversity, using only two classifications: the geopedologic and the syntaxonomic. These classifications comprise virtually all soil-forming factors, with the exception of humans (Jenny 1941). It is interesting to note that the syntaxonomic system also classifies aquatic vegetation in shallow water bodies, whereas the underlying sediments are also included in the most recent soil taxonomies such as WRB 2006, among others. However, as we stated above, the main concern is that there are no universal classifications widely accepted by experts in lithology and geoforms (Ibáñez et al. 2019), whereas the syntaxonomic approach is only popular in continental Europe.

10.5   Geographical Analysis, Taxonomies, Maps and Soil Surveys: Fractal Structures

Researchers have paid scarce attention to the epistemological and philosophical analysis of diversity in most of the natural sciences. However, the studies that exist have produced very interesting findings.

In Chap. 6 of this book Alfred Zinck argues that: *"The geolandscape is a hierarchically structured and organized domain. Therefore, a multicategorial system, based on nested levels of perception to capture the information and taxonomic criteria to organize that information, is an appropriate frame to classify geoforms"* (....) *"Such a system with six categories complies with Miller's Law, which postulates that the capacity of the human mind to process information covers a range of seven plus or minus two elements....)".*

In the abstract of the book: *"Magic Numbers: A Meta-Analysis for Enlarging the Scope of a Universal Soil Classification System"* Ibáñez and Montanarella (2013) state that: *"Categorization of the world around us in discrete classes is an innate capacity of the human mind to organize the information and carry on languages in all past and present cultures. Likewise, our cognitive apparatus organizes these categories in a hierarchical way. Recently the authors of this monograph demonstrated that the breaking of the continua of biological and pedological entities in order to carry on taxonomies follows the same mathematical rules: an iterative fragmentation according to fractal rules. For this reason, both biological and pedological taxonomies have similar topological structures. However, these mental constructs divert little bit of the expected fractal values by utilitarian, geographic and cognitive bias. It can recognize two types of cognitive bias, termed the prototypic effect and the constraints of humans to process the information to do not exceed our channel memory capacity* (The Miller law). *Therefore, the fractal fragmentation and our channel memory capacity determine the structures of taxonomies to get efficient information systems. On this working hypotheses the authors show in this monograph a set of rules that should be follow could efficient and user friendly information systems. Furthermore, current biological and pedological taxonomies and possibly classification of other disciplines, was carry out by experts conforming to the above-mentioned rules in an intuitive way. In view that there is not a science of the taxonomies the authors offer a set of rules to assist in this task (...)."*

10.5.1 Paretian Thinking in a Logarithmic World

It is intriguing that the regularities previously found in biodiversity studies were later corroborated in the study of other abiotic entities such as pedodiversity, land system units diversity, mineral diversity, and lithodiversity, among many others, when they have not typically been considered idiosyncratic of living beings (e.g., Ibáñez et al. 2021a, b). However, it seems even more interesting that they emerge again in the analysis of structures made by our cognitive apparatus, such as urban maps. Given that both the urban environment and the map of that environment are created by humans rather than found in nature (i.e., are artificial constructs), it is interesting that the patterns found in natural systems are also found in these human systems. Many authors have begun to suggest that Gaussian thinking should be replaced by Paretian thinking (Newman 2005; Andriani and McKelvey 2009; McKelvey and Andriani 2005; Goodchild and Mark 1987). Although it is possible that both modes of thinking are housed in our minds, Paretian thinking seems to predate Gaussian. This is a conjecture proposed by several authors from disparate disciplines, and either corroborating or refuting this would be a major advance forward in our understanding of natural systems and ability to model them. Ibáñez et al. (2021a) synthetized the state of the art in this type of meta-analysis. In Gaussian thinking people analyze datasets using the mathematical tools derived from normal or Gaussian curves, whereas in the Paretian world this analysis is based in power law distributions.

Therefore, it should not be surprising that several authors have had the same question as Varshney and Sun (2015) in recent years, namely, "why do we perceive logarithmically?" Several anthropologists found that at least some aboriginal cultures, such as the Mundurucu indigenous peoples of the Amazon (Pica et al. 2004), calculate logarithmically instead of in a linear way, which is typical of western culture. Likewise, experimental psychologists have provided evidence that childrens' minds begin to calculate logarithmically, changing with time and gradually (probably via cultural training) to do so in a linear way (see Siegler and Booth 2004). Why do we perceive the world around us logarithmically? According to some authors, such a way of calculating is an adaptive advantage in the search for natural resources, in contrast with Gaussian thinking. In the analysis of biological and pedological taxonomies as well as in the inventory of soils, vegetation and other natural resources, power laws have been found everywhere, as in pedodiversity analysis (e.g., Taylor 2019).

10.5.2 Paretian Thinking, Natural Resources Inventories, and Mapping

Andriani and McKelvey (2009, 2011), among many others, contend that the classical geographical analyses and map representations are not appropriate for the analysis of geographical spaces and propose a paradigm shift from our current Gaussian thinking to Paretian-based thinking, as has been suggested for other scientific disciplines (e.g., Goodchild and Mark 1987; McKelvey and Andriani 2005: Newman 2005). Geographical analyses show the ubiquity of power-law (or related) distributions instead of Gaussian bell curves and linear fits. Some classical and well-known examples of power laws include the Pareto distributions, Zipf Law, Gutenberg-Richter law of earthquakes, allometric laws, Horton Laws of drainage basin networks, Korak's law on area number islands relationships, and bio and pedodiversity-richness area-relations, among many others. In all these cases datasets provide better fits to power laws than to normal (Gaussian) ones (e.g., Andriani and McKelvey 2011; Newman 2005; Taylor 2019; Salingaros and West 1999; Jiang and Brandt 2016, among many others). Power laws that reach above three magnitude orders are usually considered the fingerprints of fractal structures (e.g., Schroeder 1991). Likewise, several studies have shown that natural resources are complex or non-linear systems (e.g., Korvin 1992; Taleb 2007; Phillips 2016, 2017), fractal structures being a product of such non-linearity (Turcotte 1997; Ibáñez et al. 2021a among many others). Therefore, it should not be surprising that similarities have been detected in the patterns of biodiversity (e.g., Borda-de-Águas et al. 2002; Brown et al. 2002; Harte et al. 1999, 2001; Pueyo 2006) and pedodiversity (e.g., Caniego et al. 2006, 2007, 2013; Ibáñez and Feoli 2013; Kunin 1998; Ibáñez et al. 2005a, b).

10.5.3 Human Cognition, Logarithmic Thinking and Miller's Law

A universal taxonomy is in fact a universal diversity inventory in a given discipline and changes with time and acquisition of new knowledge. As Alfred Zinck stated in Chap. 6 of this book, the geopedological classification follows Miller's Law (Miller's Rule) that optimizes the flow of information. Miller (1956, 2003) contended that there is a magic number (seven) for cognitive reasons that maximize information exchange. Ibáñez et al. (2006a, b, c) and Ibáñez and Montanarella (2013) detected this magic number when studying pedological and biological classifications. There is evidence suggesting the magic number seven (plus or minus two) is the usual capacity limit of human short-term memory. In general, adults can store between five and nine items in their short-term memory. If our mind tries discerning between more elements it causes confusion or inconsistencies. Power laws, fractal/multifractal structures and the magic number "seven" joined with the Mayr criteria (see below) explain why hierarchical taxonomies (e.g., USDA Soil Taxonomy) conform to laws that dictate and optimize information flow in user friendly retrieval systems according to the Jaynes' Principle (Jaynes 1957 in Ibáñez and Montanarella 2013). Mayr (see Ibáñez and Montanarella 2013) encouraged biological taxonomists to create equal-sized taxonomic units. He argued that a low number of larger taxa size and an excessive number of monotypic taxa reduce the usefulness of taxonomy as an information retrieval system. Thus, if each taxa of any taxonomic category are subdivided into the same number of subtaxa along all the hierarchy, a perfect fractal is obtained (but in general is a little biased by other cognitive and/or practical reasons).

Does the mind work logarithmically? Both biology and biological taxonomies conform to free scaling laws (fractals and / or multifractals) (see Ibáñez and Montanarella 2013 and references therein). Soil and biological classifications were developed independently, with biological taxonomies being natural whereas soil taxonomies are artificial constructs (Ibáñez and Montanarella 2013). In other words, there are obvious, "natural" items to classify in biological taxonomies, while we must create an object of classification (e.g., the pedon, an artificial construct) to create soil taxonomies. Taxonomies are intuitive rational products of numerous groups of experts working independently over decades in different disciplines and without considering mathematical aspects desirable to build efficient taxonomic structures (Ibáñez et al. 2009). The case of the fractal coupling between soil taxonomies and soil survey and mapping procedures will serve as an example that corroborates the above-mentioned thesis. Studies show that USDA Soil Taxonomy has the same mathematical structure as some biological taxonomies that conform to physical laws that dictate and optimize information flow in user friendly retrieval systems according to the principle of maximum entropy (Jaynes 1957 in Ibáñez and Montanarella 2013). These studies show that the probability distribution which best represents the current state of knowledge about a system is the one with largest entropy. Likewise, Ibáñez et al. (2009) demonstrated that the fractal/multifractal nature of USDA Soil Taxonomy is strongly linked with conventional soil survey practices. In fact, most surveys are packed with power law distributions, such as: (i) hierarchic taxonomic levels used according to the map scale; (ii) minimum polygon size that fit the functions to the map scale; and (iii) boundary density-scale map relationships, among others (see also empirical data previously published by Beckett and Bie, 1978 in Ibáñez et al. 2009). Consequently, a cascade of power laws occurs in soil survey products and soil taxonomies (see Ibáñez and Montanarella 2013). Because both activities (soil survey and creating soil taxonomies) are strongly linked it seems the minds of soil surveyors and soil taxonomists create similar fractal structures. This process could explain why maps devoided of legends and other information to aid in their interpretation have a strong resemblance and information content, and with independence of scales, they provide a clear fractal signature.

Regrettably, it remains to be clarified whether Paretian and Gaussian thinking (i) exist simultaneously in our minds; (ii) if Paretian thought is a residual product in our cultural evolution, being replaced by Gaussian in our civilizations; (iii) if we unconsciously project our logarithmic mental maps into nature and structure the world around us in such a way that many scientific findings are more a mental product than an unchanging reality; and (iv) if our minds and the world around us and with which we interact are inescapably the product of a fractal universe (Ibáñez et al. 2021a). Thus, it could be conjectured that scale-invariant information processing is intuitive to human minds and that a more rigorous formalization of survey-taxonomy architectures may help practitioners better understand their activities and constructs and provide a way to improve them. It is intriguing that comparisons of the mathematical structures of soil taxonomies and maps may reveal, in part, how the human mind works (Ibáñez et al. 2021a, b and references therein). The geopedology approach of Alfred Zink seems to be an example.

10.6 Pedodiversity, Geodiversity and the Preservation of Geoforms as Part of Natural Heritage

According to Gray (2004), geodiversity can be defined as "*the natural range (diversity) of geological (rocks, minerals, fossils), geomorphological (land form, processes) and soil features. It includes their assemblages, relationships, properties, interpretations and systems*". So far, experts have not proposed any index or mathematical procedure to quantify the diversity of more than one natural resource at a time (Ibánez et al. 2019). It might therefore be advisable to analyze the diversity of each natural system independently. The main problem for the estimation of lithologic and geomorphic diversity derives from the lack of universal classifications. The absence of this kind of taxonomic construct precludes conducting comparative studies to detect regularities in spatial patterns and laws in different environments (Ibánez and Brevik 2019; Ibáñez et al. 2019). Furthermore, current geoconservation studies and projects have not considered soils, in contrast to other resources (Ibáñez et al. 2019).

The preservation of geodiversity and pedodiversity has generated much interest in recent years (e.g., Ibáñez and Brevik 2019; Ibáñez et al. 2012, 2019; Bockheim and Haus 2013). It is impossible to preserve soils without preserving the geoforms in which they formed. Even though most definitions of geodiversity include soil resources, it is puzzling that in practice, the vast majority of experts in geodiversity and preservation of geological heritage completely disregard soils. This includes in the proposal of geodiversity indices and in proposals for geoparks endorsed by UNESCO. Using bibliometric analysis, Ibáñez et al. (2019) confirmed such incomprehensible approaches. In fact, to date no geodiversity index has been proposed that conforms to the principles that have been applied in studies of biodiversity and pedodiversity. This is based on a lack of interest in pedodiversity, as well as the fact that most of the natural resources that appear in the concept of geodiversity lack universal classifications. How is it possible to compare the geodiversity of different natural areas in the absence of such taxonomic constructs? Rarely do publications on this matter consider pedodiversity, and then mainly collaterally. Very rarely is pedodiversity named and applied in the publications and documents used to propose UNESCO geoparks (Ibáñez et al. 2019). Therefore, at present it is not possible to provide a scientifically sound quantification of geodiversity, and the same occurs with geological heritage. The geopedological approach of Alfred Zinck (1988, 2013 and this book) demands an inventory of the soil forming factors most relevant in geodiversity studies (soil, land forms, and geology), offering the opportunity to quantify geodiversity, especially if the same taxonomies are used.

10.7   Conclusions

All natural resources, biotic and abiotic, are part of natural heritage. Diversity analysis could be applied to all of them as a mathematical tool to understand their respective spatial distribution patterns in the landscape. The only requirement for implementing these formal procedures is the existence of their respective taxonomies. Obviously, universal taxonomies are preferable to national or ad hoc purpose-oriented classifications, as they allow comparing results by extracting regularities from different regions and environments by different researchers. Conventional soil inventories implicitly use soil-physiography relationships, the so-called soil-landscape paradigm. The geopedologic approach makes explicit the implicit traditional knowledge of soil survey activities, formalized in a single taxonomy or classification system. Likewise, the syntaxonomic approach plays the same role concerning the plant community-climate relationships. Using both approaches concomitantly make it feasible to achieve a unifying vision of landscape structure from the perspective of landscape ecology. The geopedologic approach is a step forward in this direction. It is desirable to use the diversity analysis of all natural resources independently but also jointly to reach a scientifically sound geodiversity index. The same is true for the preservation of natural heritage, in view that all natural resources interact with each other, being mutually interdependent. The geopedological approach proposes a classification that promises to meet all the specifications needed to achieve an efficient information system.

References

Andriani P, McKelvey B (2009) From Gaussian to Paretian thinking: causes and implications of power laws in organizations. Org Sci 20:1053–1071. https://doi.org/10.1287/orsc.1090.0481

Andriani P, McKelvey B (2011) From skew distributions to power-law science. In: Allen P, Maguire S, McKelvey B (eds) The sage handbook of complexity and management. SAGE, London. ISBN: 978-144620108-4;978-184787569-3, pp 254–273. https://doi.org/10.4135/9781446201084.n15

Beckett PHT, Bie SW (1978) Use of soil and land-system maps to provide soil information in Australia. Division of Soils Technical Paper vol. 33. Commonwealth Scientific and Industrial Research Organization, Australia

Ben-Shahar R, Coe MJ (1992) The relationships between soil factors, grass nutrients and the foraging behaviour of wildebeest and zebra. Oecologia 90:422–428

Bockheim JG, Haus N (2013) Soil endemism and its importance to taxonomic pedodiversity. In: Ibáñez JJ, Bockheim JG (eds) Pedodiversity. CRC, Boca Raton, pp 195–210

Borda-de-Água L, Hubbell SP, McAllister M (2002) Species-area curves, diversity indices, and species abundances distributions: a multifractal analysis. Am Nat 159:138–155. http://www.jstor.org/stable/10.1086/324787 (free available from http://ctfs.arnarb.harvard.edu/Public/pdfs/Borda_etal_2002_AmNat.pdf)

Brown JH, Gupta WK, Li BL, Milne BT, Restrepo C, West GB (2002) The fractal nature of nature: power laws, ecological complexity and biodiversity. Philos Trans R Soc Lond B Biol Sci 2002 29 357(1421):619–626. https://doi.org/10.1098/rstb.2001.0993

Caniego J, Ibáñez JJ, San José Martínez F (2006) Selfsimilarity of pedotaxa distributions at planetary level: a multifractal approach. Geoderma 134:306–317. https://doi.org/10.1016/j. geoderma.2006.03.007

Caniego FJ, Ibáñez JJ, San José Martínez F (2007) Rényi dimensions and pedodiversity indices of the earth pedotaxa distribution, Nonlin. Proc Geophys 14:547–555

Caniego J, Sán-JoséMartínez F, Ibáñez JJ, Pérez R (2013) Lacunarity of the spatial distributions of soil types in Europe. Vadose Zone J. https://doi.org/10.2136/vzj2012.0210

Fridland VM (1976) The soil cover pattern: problems and methods of investigation. In: Soil combinations and their genesis (Translated from Russian). Keter Publishing House, Jerusalem

Goodchild MF, Mark DM (1987) The fractal nature of geographic phenomena. Ann Assoc Am Geogr 77:265–278. https://doi.org/10.1016/j.geoderma.2006.03.007

Gray M (2004) Geodiversity: valuing and conserving abiotic nature. Wiley, Chichester

Harte J, Kinzig A, Green J (1999) Self-Similarity in the distribution and abundance of species. Science 284:334–336. https://doi.org/10.1126/science.284.5412.334

Harte J, Blackburn T, Ostling A (2001) Self-similarity and relationship between abundance and range size. Am Nat 157:374–386

Huston MAH (1994) Biological diversity. Cambridge University Press, Cambridge

Ibáñez JJ (2017a) Diversity of soils. In: Warf B (ed) Oxford Bibliographies in Geography (article online). Oxford University Press, New York

Ibáñez JJ (2017b) Diversity of soils. Oxford Bibliographies (2nd vervison). Oxford University Press. (article on line). http://www.oxfordbibliographies.com/view/document/obo-9780199874002/obo-9780199874002-0104.xml

Ibáñez JJ, Brevik EC (2019) Divergence in natural diversity studies: the need to standardize methods and goals. Catena 182:104110. 10. https://doi.org/10.1016/j.catena.2019.104110

Ibáñez JJ, Feoli E (2013) Global relationships of pedodiversity and biodiversity. Vadose Zone J. https://doi.org/10.2136/vzj2012.0186

Ibáñez JJ, Montanarella L (2013) Magic numbers: a metha-analysis for enlarging the scope of a universal soil classification system. JRC Technical Reports, European Commission, Brussels. Pdf free available from https://publications.jrc.ec.europa.eu/repository/handle/JRC79481

Ibáñez JJ, Jiménez-Ballesta R, García-Álvarez A (1990) Soil landscapes and drainage basins in Mediterranean mountain areas. Catena 17:573–583. https://doi.org/10.1016/0341-8162(90)90031-8

Ibáñez JJ, Caniego FJ, San-José F, Carrera C (2005a) Pedodiversity-area relationships for Islands. Ecol Model 182:257–269

Ibáñez JJ, Caniego J, García-Álvarez A (2005b) Nested subset analysis and taxa-range size distributions of pedological assemblages: implications for biodiversity studies. Ecol Model 182/3–4:239–256

Ibáñez JJ, Ruiz-Ramos M, Tarquis A (2006a) The mathematical structures of biological and pedological taxonomies. Geoderma 134:360–372

Ibáñez JJ, Ahrens R, Pérez-Gómez R (2006b) The fractal mind of pedologists (soil taxonomist and soil surveyors), 42: 1.4A impact of national soil classification on soil science and society – theater 18th world congress of soil science: frontiers of soil science in the technology and information age. 18th world congress of soil science, July 9–15, 2006 – Philadelphia, Pennsylvania

Ibáñez JJ, Arnold R, Sánchez-Díaz J (2006c) The magical numbers of the USDA soil taxonomy: towards an outline of a theory of natural resource taxonomies. Impact of national soil classification on soil science and society – theater 18th world congress of soil science: frontiers of soil science in the technology and information age. 18th world congress of soil science, July 9–15, 2006, Philadelphia, Pennsylvania

Ibáñez JJ, Arnold RW, Ahrens RJ (2009) The fractal mind of pedologists (soil taxonomists and soil surveyors). Ecol Complex 6:286–293

Ibáñez JJ, Krasilnikov P, Saldaña A (2012) Archive and refugia of soil organisms: applying a pedodiversity framework for the conservation of biological and non-biological heritages. J Appl Ecol 49(6):1267–1277. https://doi.org/10.1111/j.1365-2664.2012.02213.x

Ibáñez JJ, Vargas RJ, Vázquez-Hoehne A (2013) Pedodiversity state of the art and future challenges. In: Ibáñez JJ, Bockheim GJ (eds) Pedodiversity. CRC Press, Boca Raton, pp 1–28

Ibáñez JJ, Pérez-Gómez R, Oyonarte C, Brevik EC (2015) Are there arid land soilscapes in Southwestern Europe? Land Degrad Dev 26:853–862. https://doi.org/10.1002/ldr.2451

Ibáñez JJ, Pérez-Gómez R, Brevik EC, Cerdà A (2016) Islands of biogeodiversity in arid lands on a polygons map study: detecting scale invariance patterns from natural resources maps. Sci Total Environ 573:1638–1647. https://doi.org/10.1016/j.scitotenv.2016.09.172

Ibáñez JJ, Brevik EC, Cerdá A (2019) Geodiversity and geoheritage: detecting scientific and geographic biases and gaps through a bibliometric study. Sci Total Environ 659:1032–1044. https://doi.org/10.1016/j.scitotenv.2018.12.443

Ibáñez JJ, Ramírez-Rosario B, Fernández-Pozo LF, Brevik EC (2021a) Exploring the scaling law of geographical space: Gaussian versus Paretian thinking. Eur J Soil Sci 72:495–509. https://doi.org/10.1111/ejss.13031

Ibáñez JJ, Ramírez-Rosario B, Fernández-Pozo L, Brevik EC (2021b) Land system diversity, scaling laws and polygons map analysis. Eur J Soil Sci 72:656–666. https://doi.org/10.1111/ejss.13035

Jaynes ET (1957) Information theory and statistical mechanics. Phys. Rev 106:620–630

Jenny H (1941) Factors of soil formation. McGraw-Hill, New York

Jiang B, Brandt A (2016) A fractal perspective on scale in geography. ISPRS Int J Geo-Inf 2016 5(6):95. https://doi.org/10.3390/ijgi5060095

Korvin G (1992) Fractal models in the earth sciences. Elsevier, Amsterdam

Kunin W (1998) Extrapolating species abundance across spatial scales. Science 281:513–1515

Loidi J, Fernández-González F (2012) Potential natural vegetation: reburying or reboring? J Veg Sci 23:596–604. https://doi.org/10.1111/j.1654-1103.2012.01387.x

Magurran AE (2003) Measuring biological diversity. Blackwell Publishing, Oxford

McKelvey B, Andriani P (2005) Why Gaussian statistics are mostly wrong for strategic organization. Strateg Organ 3:219–228. https://doi.org/10.1177/1476127005052700

Michaud GA, Monger HC, Anderson DL (2013) Geomorphic-vegetation relationships using a geopedological classification system, northern Chihuahuan Desert, USA. J Arid Environ 90:45–54. https://doi.org/10.1016/j.jaridenv.2012.10.001. Free available from DigitalCommons@University, Nebraska–Lincoln. Publications from USDA-ARS/UNL Faculty, USDA Agricultural Research Service (pdf free available from http://digitalcommons.unl.edu/cgi/viewcontent.cgi?article=2172&context=usdaarsfacpub

Miller GA (1956) The magical number seven, plus or minus two: some limits on our capacity for processing information. Psychol Rev 63(2):81–97

Miller GA (2003) The cognitive revolution: a historical perspective. Trends Cognit Sci 7(3):141–144

Mirkin BM (1989) Plant taxonomy and syntaxonomy: a comparative analysis. Vegetatio 82:35–40. https://doi.org/10.1007/BF00217980

Newman MEJ (2005) Power laws, pareto distributions and Zipf's law. Doi Contem Phys 46:323–351. https://doi.org/10.1080/00107510500052444

Petersen A, Gröngröft A, Miehlich G (2010) Methods to quantify the pedodiversity of 1km2 areas. Results from southern African drylands. Geoderma 155:140–146. https://doi.org/10.1016/j.geoderma.2009.07.009

Phillips JD (2016) Complexity of earth surface system evolutionary pathways. Math Geosci 48:743–765. https://doi.org/10.1007/s11004-016-9642-1

Phillips JD (2017) Soil complexity and pedogenesis. Soil Sci 182:117–127. https://doi.org/10.1097/SS.0000000000000204

Pica P, Lemer C, Izard V, Dehaene S (2004) Exact and approximate arithmetic in an amazonian indigene group. Science 306(5695):499–503. https://science.sciencemag.org/content/306/5695/499

Pueyo S (2006) Self-similarity in species-area relationship and in species abundance distribution. Oikos 112:156–162. https://doi.org/10.1111/j.0030-1299.2006.14184.x

Rivas-Martínez S (2005) Notions on dynamic-catenal phytosociology as a basis of land-scape science. Plant Biosyst 139:135–144. Pdf free available from https://doi.org/10.1080/11263500500193790

Saldaña A (2013) Pedodiversity and landscape ecology. In: Ibáñez JJ, Bockheim JG (eds) Pedodiversity. CRC Press, Boca Raton, pp 105–132

Salingaros N, West BJ (1999) A universal rule for the distribution of sizes. Environ Plan B Plan Design 26:909–923. https://doi.org/10.1068/b260909

Schroeder M (1991) Fractals, chaos, power laws: minutes from an infinite paradise. W. H. Freeman & Company, New York. ISBN 10: 0716721368. ISBN 13: 9780716721369

Siegler RS, Booth JL (2004) Development of numerical estimation in young children. Child Dev 75:428–444. https://doi.org/10.1111/j.1467-8624.2004.00684.x

Stacey LW, Monger HC (2012) Banded vegetation-dune development during the Medieval Warm Period and 20th century, Chihuahuan Desert, New Mexico, USA. Ecosphere 3:art21 https://doi.org/10.1890/ES11-00194.1. Pdf free available from http://www.esajournals.org/doi/abs/10.1890/ES11-00194.1

Taleb NN (2007) The Black Swan: The impact of the highly improbable. Random House LCC, Allen Lane: London. ISBN-10: 9781400063512, ISBN-10: 0141034599

Taylor RAJ (2019) Taylor's power law order and pattern in nature. Academic, Cambridge, MA

Toomanian N (2013) Pedodiversity and landforms. In: Ibáñez JJ, Bockheim JG (eds) Pedodiversity. CRC Press, Boca Raton, pp 133–152

Tukiainen H, Alahuhta J, Field R et al (2017) Spatial relationship between biodiversity and geo-diversity across a gradient of land-use intensity in high-latitude landscapes. Landsc Ecol 32:1049–1063. https://doi.org/10.1007/s10980-017-0508-9

Tukiainen H, Kiuttu M, Kalliola R, Alahuhta J, Hjort J (2019) Landforms contribute to plant bio-diversity at alpha, beta and gamma levels. J Biogeogr 46:1699–1710. https://doi.org/10.1111/jbi.13569

Turcotte DL (1997) Fractals and chaos in geology and geophysics. Cambridge University Press. ISBN 0 521 41270 6. https://doi.org/10.1002/gj.3350280216

Varshney LR, Sun JZ (2015) Why do we perceive logarithmically? In: Pitici M (ed) The best writing on mathematics 2014. Princeton University Press, pp 64–73. ISBN 978-0-691-16417-5

Weber HE, Moravec J, Theurillat JP (2000) International code of phytosociological nomenclature. J Veg Sci 11:739–768. https://doi.org/10.2307/3236580

Westhoff V, van der Maarel E (1978) The Braun-Blanquet approach. In: Whittaker RH (ed) Classification of plant communities, 2nd edn. Dr W Junk, The Hague, pp 287–399

Williams BM, Houseman GR (2013) Experimental evidence that soil heterogeneity enhances plant diversity during community assembly. J Plant Ecol 7:461–469. https://doi.org/10.1093/jpe/rtt056

WRB (2006) Guidelines for constructing small scale map legends using the WRB., 2nd ed, World Soil Resources. FAO, Rome

Zinck JA (1988) Physiography and soils, Soil survey lecture notes. ITC, Enschede

Zinck JA (2013) Geopedology. Elements of geomorphology for soil and geohazard studies, ITC special lecture note series. ITC, Enschede

Zinck JA (2016) The Geopedologic approach. In: Zinck JA, Metternicht G, Bocco G, Del Valle HF (eds) Geopedology. Springer, Cham. https://doi.org/10.1007/978-3-319-19159-1_4

Chapter 11
Algorithms for Quantitative Pedology

D. E. Beaudette, J. Skovlin, A. G. Brown, P. Roudier, and S. M. Roecker

Abstract The Algorithms for Quantitative Pedology (AQP) project consists of a suite of packages for the **R** programming language that simplify quantitative analysis of soil profile data. The "aqp" package provides a vocabulary (functions and data structures) tailored to the complexity of soil profile information. The "soilDB" package provides interfaces to databases and web services, leveraging the "aqp" vocabulary. The "sharpshootR" package provides tools to assist with summary and visualization. Bridging the gap between pedometric theory and practice is central to the purpose of the AQP project. The AQP **R** packages have been extensively tested and documented, applied to projects involving hundreds of thousands of soil profiles, and integrated into widely used tools such as SoilWeb. These packages serve an important role in routine data analysis within the U.S. Department of Agriculture and in other soil survey programs worldwide.

Keywords Geomorphology · Pedology · Soil survey · Soil data analysis · Data visualization

Supplementary information: Annotated **R** code used to create the figures and table within this chapter are posted at https://github.com/ncss-tech/geopedology-chapter. The following package versions were used during the preparation of this content: *aqp* 1.42, *soilDB* 2.6.14, and *sharpshootR* 1.9.1.

D. E. Beaudette (✉) · A. G. Brown
USDA-NRCS Soil and Plant Science Division, Sonora, CA, USA
e-mail: dylan.beaudette@usda.gov

J. Skovlin
USDA-NRCS Soil and Plant Science Division, Missoula, MT, USA

P. Roudier
Manaaki Whenua - Landcare Research, Palmerston North, New Zealand

S. M. Roecker
USDA-NRCS Soil and Plant Science Division, Lincoln, NE, USA

11.1 Introduction

Soil survey data are a rich source of soil parent material information and observations relating to the distribution and extent of surficial deposits and geologic stratigraphy. The overlap of the domains of pedology and geomorphology, referred to as geopedology (Zinck 2016), emphasizes the need for clear relationships between soil parent materials and geomorphic concepts. Soil is the dynamic interface connecting the biosphere and the lithosphere (Wysocki et al. 2005). Soil provides a medium and conduit for water storage and nutrient flow. The development of soil properties is largely determined by the inherent nature of surficial deposits or soil parent material from which they originate. While soil survey has historically been grounded in a soil-landscape paradigm (Hudson 1992), geopedologic concepts aim to further integrate geomorphic concepts with soil survey information for broader application within the earth science community.

Data alone cannot support decisions, generate useful conclusions, or convey embedded relationships without thoughtful analysis and processing. The wide array of data sources, formats, and conventions (even within a single institution) can further complicate efforts to synthesize soil information from soil data sources. Growth in promising tools that link database APIs, spatial data, and enable the analysis of complex soil description data are expanding the potential for soil science data analysis within programming environments like **R** (R Core Team 2022).

The Algorithms for Quantitative Pedology (AQP) project are a suite of packages for the **R** programming language that simplify many facets of soil data analysis. The project began in 2006 as a loosely coordinated collection of **R** scripts used to support the management, analysis, and visualization of digital soil morphology records. By 2010, it became clear that an **R** package (code, manual pages, and example data following strict guidelines) hosted by CRAN would be the best route forward. The first version of the *aqp* package was submitted to CRAN in May of 2010; with a name and core functionality inspired by the concept of "quantitative pedology" (Jenny 1941), analysis by regular depth-intervals (Harradine 1963; Moore et al. 1972), and the characterization of depth-functions (Myers et al. 2011). A companion article by Beaudette et al. (2013b) contained a detailed description and simple demonstrations of the main package features.

As functionality evolved, the *aqp* package was split into three main categories which became **R** packages to increase modularity and divide administrative tasks: *aqp* (soil-specific data structures, profile sketches, color conversion, pedotransfer functions, etc.), *soilDB* (wrapper and convenience functions for accessing APIs and harmonization of results), and *sharpshootR* (specialized tasks and visualizations designed for use with soil database connections provided by *soilDB* and data structures provided by *aqp*). Like most **R** packages, the AQP suite of packages depends on other packages for optimized computation (Dowle and Srinivasan 2021), color conversion (Pedersen et al. 2021; Zeileis et al. 2020), numerical classification (Maechler et al. 2021), and methods for compositional data (e.g., sand, silt, and clay content) (Moeys 2018; van den Boogaart et al. 2021), to name a few. The authors hope that other scientists will find a suitable foundation in *aqp*, *soilDB*, and *sharp-shootR*, upon which more specialized tools can be built, documented, and delivered in the form of new **R** packages.

Since 2011, the AQP suite of **R** packages has been extensively updated and documented by U.S. Department of Agriculture – Natural Resources Conservation Service (USDA-NRCS) Staff to support routine operations within the Soil and Plant Science Division. Some examples include aggregation and synthesis of field data to support initial soil survey (new mapping), graphical comparisons and correlation analysis to support soil survey update projects (refinement of existing mapping), and visual presentation of soil survey data to the public via tools like SoilWeb (O'Geen et al. 2017). The *soilDB* package has become one of the most widely used interfaces to USDA-NRCS data sources, with support for queries that accept (and return) mixtures of spatial and tabular data from the Soil Survey Geographic Database (SSURGO) (Soil Survey Staff 2022b). Spatial formats defined by the *sf* (Pebesma 2018) and *raster* (Hijmans 2021) **R** packages are used extensively by the *soilDB* package to minimize data conversion or pre-processing steps.

11.1.1   Example Data: Clarksville Soil Series (Fig. 11.1)

A curated set of soil morphologic and laboratory characterization data correlated to the Clarksville soil series (Loamy-skeletal, siliceous, semiactive, mesic Typic Paleudults) is used to demonstrate key functionality and visualization possibilities provided by the AQP suite of **R** packages. These data represent a very deep (>150 cm), somewhat excessively drained soil of large extent in the Ozark Highlands of southern Missouri, USA. Clarksville soils are formed in residual and colluvial soil parent materials of cherty dolomite or cherty limestone (Kabrick et al. 2008).

Fig. 11.1 Clarksville series soil profile (left) and associated representative landscape (right). (Soil profile photo: Satchel Gaddie, landscape photo: Jayme LeBrun)

These soils typically occur on ridges and steep side slopes, spanning summit, shoulder, and backslope positions of an idealized 2D hillslope. Mean annual air temperature ranges from 13–15 °C, mean annual precipitation ranges from 1150–1250 mm, with most precipitation falling as rain.

Clarksville soils are generally highly weathered, acidic with low to moderate base saturation, low cation exchange capacity and nutrient limiting for available phosphorus, calcium, and magnesium (Kabrick et al. 2011; Singh et al. 2015). Morphology of Clarksville soils commonly include soil textures high in silt with thick accumulations of translocated clay at depth. Although soils are generally high in rock fragments, silt-rich soil textures dominate surface soil horizons due the influence of wind-blown loess parent material. These landscapes support a mixed forest of black oak (*Q. velutina* Lam.), white oak (*Quercus alba* L.), blackjack oak (*Q. marilandica* Muench.), post oak (*Q. stellata* Wangenh.), shortleaf pine (*Pinus echinata* Mill.), black hickory (*Carya texana* Buckl.), red maple (*A. rubrum* L.), and dogwood (*Cornus florida* L.).

11.2 Representing Collections of Soil Profiles in R

Soil profile data are complex, and typically consists of site description, soil morphologic description, and optionally laboratory data. The SoilProfileCollection (SPC) is a data structure which attempts to capture this complexity and is designed to coordinate linkages between those elements. Functions operating on the SoilProfileCollection include special constraints to ensure linkages are not broken during routine operations such as editing, sub-setting, or combining collections.

The first level of abstraction involves two main tiers: "site" and "horizon" data. "Site" data refers to above-ground or those properties that are specific to a single soil profile description (e.g., surface slope). "Horizon" data refers to below-ground or those properties that are specific to a single genetic soil horizon or layer. An additional level of abstraction is used to store spatial data (coordinates and coordinate reference system) and depth-interval information such as diagnostic horizons. The SoilProfileCollection structure provides a means of storing user-defined metadata such as units of measure, horizon designation column name, data source, and data citation. The SoilProfileCollection object was designed with data analysis in mind; as compared to other (more complex) data structures used for archival purposes, such as those used within the USDA-NRCS National Soil Information System (NASIS).

Of primary importance are the horizon data, or the layers that comprise the profiles in the collection. The SoilProfileCollection is "horizon data forward," in that a user starts with a table of horizon data. Each horizon record must have an upper and lower boundary, a unique ID linking to a single soil profile observation, and any other observed or measured properties. There are no set limits on the number of horizons per profile, or profiles per collection, but available memory will dictate practical limitations. Horizon depths should be specified as integers (typically centimeters) and should not overlap.

A SoilProfileCollection is created through "promotion" of an **R** data.frame with the depths() function. Other data.frame-like objects such as tibble (Müller and Wickham 2021) or data.table (Dowle and Srinivasan 2021) can be used as input. Promoting a data.frame to SoilProfileCollection requires the following parameters: profile_id (the name of a column containing unique profile IDs), top_depth (the name of a column containing horizon upper depths) and bottom_depth (the name of a column containing horizon lower depths) along with any additional horizon data associated with the horizons in the profile. For example, the promotion of a data.frame called x to SoilProfileCollection would follow depths(x) <- profile_id ~ top_depth + bottom_depth.

In the **R** language, the tilde symbol ~ separates the left and right-hand sides of a formula. Commonly ~ is used in formulas to mean "modeled as." In a SoilProfileCollection the geometry and ordering of horizons within each unique profile using the upper and lower depths is "modeled." Performing this operation automatically sorts horizon data first by profile ID and then by horizon top depth.

The site()<- method is used to move site-level data from horizon-level records (necessary when starting with a mixture of horizon and replicated site data in the same table), or to merge a new table of site-level data via common ID and left join (missing records in the new table are filled with NA). In a similar manner, the horizons()<- method is used to merge a new table of horizon-level data into the SoilProfileCollection object via common ID (unique to specific horizons) and left join. Additional site and horizon data can be created or extracted one by one using the $ or [[methods. When creating a new variable, the SoilProfileCollection will check whether the length of the vector matches either the number of "sites" or the number of "horizons." Extracting horizon or site-level data as plain data.frame objects is performed with the horizons() and site() functions. A detailed explanation of the SoilProfileCollection object and associated methods for manipulation of these objects is presented in the "Introduction to SoilProfileCollection Objects" tutorial (Beaudette 2022).

The *soilDB* package for **R** provides a common interface to many of the National Cooperative Soil Survey databases. Several functions from this package return data as a SoilProfileCollection object: fetchKSSL() (laboratory characterization data), fetchOSD() (basic soil morphology from the Official Series Description), fetchSDA() (SSURGO and STATSGO data from Soil Data Access), and fetchNASIS (National Soil Information System).

11.2.1   Subsetting

The "bracket" methods are one of the primary ways that objects in **R** can be subset by rows and columns (e.g., data.frame) or element (e.g., list, vector, etc.). The SoilProfileCollection builds on these patterns to extract specific profiles and/or horizon collections based on numeric or logical indices.

The syntax used by the SoilProfileCollection bracket method is x[i, j, k]; where x is a SoilProfileCollection object, i is a profile index, j is a within-profile horizon index and k represent optional special functions that can operate on the horizon data in the collection to replace the profile-specific j-index (Fig. 11.2). To obtain the first profile in the collection use the syntax x[1,]. For the first horizon in each profile use x[, 1]. To get the last horizon of each profile, use x[, , .LAST] where .LAST is a special "keyword" that can identify the j-index of the deepest horizon in each profile. Subsets based on i, j and k indices of the SoilProfileCollection can be combined, for instance: x[1:2, 1:2] gives the first two horizons of the first two profiles. Also, the k index can be combined with the i index, for instance x[1:2, , .LAST] gives the last horizon of the first two profiles (Fig. 11.2).

The representation of horizon position with the j-index can be extended to develop other "horizon spatial predicates" such as hzAbove(), hzBelow() and hzOffset(). The former two take logical expressions to match horizons and return the part of the collection adjacent to the match (above or below respectively). The hzOffset() function allows arbitrary horizon indices and offsets to be calculated. This type of logic is further helpful for inspecting and fixing horizon geometry for errors or inconsistencies.

Common querying operations with criteria in the form of logical expressions can be used to subset profiles or horizons in a collection that meet specific criteria of interest. The *aqp* functions subset() and subsetHz() can be used with logical expressions in terms of the site or horizon variables to specify the constraints. These expressions make use of site or horizon-level variables in the collection. The subset() function returns whole profiles, if criteria were specified for horizon data, then only some of the horizons of those profiles may meet criteria. More specifically, subsetHz() requires horizon-level expressions and returns only the portion of horizons within profiles that meet criteria.

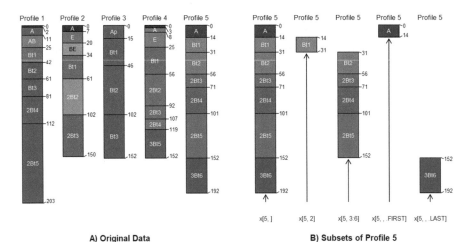

Fig. 11.2 Five soil profiles correlated to the Clarksville soil series (**a**). Examples of bracket methods for subsetting profiles, sequences of horizons and top or bottom horizons (**b**)

Partitioning soil profile collections on logical expressions of site and horizon level properties is powerful, but soil scientists often need to extract data within or overlapping specific depths. Two methods: glom(x, z1, z2, …) and trunc(x, z1, z2, …) facilitate this in *aqp*. The glom() function returns the subset of horizons in a collection that overlap with a specific depth interval [z1, z2]. The depth interval could be a point (only z1 specified) or a range (z1 and z2 specified) (Fig. 11.3).

The interval [z1, z2] can be constant across the collection or unique to each profile. By default, the whole horizon is returned unmodified whether it falls fully within the range or not – creating a "ragged" SoilProfileCollection (Fig. 11.3). The upper and lower boundaries of the resulting horizons will be cleanly cut to the interval specified using glom(truncate=TRUE) or via trunc() and resulting profiles will have consistent upper and lower boundaries assuming there are no missing data in the specified interval.

11.2.2   Data Quality and Repairs

The *aqp* package provides several methods for identifying problematic profile geometry and attempts to "correct" it. Most soil databases and methods for storing soils information do not have front end validations that prevent entry of data with "illogical" content. Some analyses rely on having only one record of data per depth/profile combination such as those involving depth-weighted averages or those that rely on having a "complete" set of records in all profiles over a particular interval.

The *aqp* function checkHzDepthLogic() inspects a SoilProfileCollection object looking for four common errors in horizon depths: bottom depth shallower than top

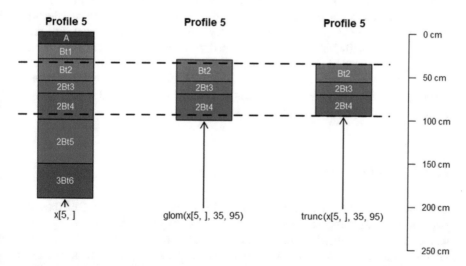

Fig. 11.3 A demonstration of selecting horizons that overlap with a depth interval via glom() and truncation to that interval via trunc()

depth, equal top and bottom depth, missing top or bottom depths, and gap or overlap between adjacent horizons. With byhz = TRUE it is possible to perform the first three of the above logic checks on individual horizons.

Assumptions concerning horizon order based on horizon top depth are tested by the repairMissingHzDepths() function. This can be used to fill in some missing (bottom) horizon depths. This function will set missing bottom depths of a horizon to the next deepest (adjacent) top depth. Also, it adds a constant vertical offset to the top depth of bottom-most horizons missing bottom depth.

The fillHzGaps() function attempts to find "gaps" in the horizon records of a SoilProfileCollection object and fill with placeholder horizons (profile ID, horizon ID, top/bottom depths, all else NA). This function is for filling profiles to a static top and bottom depth. For instance, a morphologic description containing horizons with omitted upper or undetermined lower depth as in the case of undescribed organic horizons or soil profiles that have variable bedrock depth below depth of excavation.

11.3 Soil Morphology

The field description of a soil profile (genetic horizon depths, boundaries, color, soil texture, rock fragment volume, structure, etc.) is typically the foundation upon which additional sampling, laboratory characterization, or soil survey are based. In aggregate, a complete collection of horizons, associated properties, and landscape context (e.g., catenary position or other geomorphic description) represent an atomic unit of pedologic inquiry: the pedon (Soil Science Division Staff 2017). The AQP family of **R** packages and the SoilProfileCollection data structure were designed specifically to elevate the pedon (and collections of pedons) to a convenient abstraction (an object), enabling a simpler interface to what would otherwise be a complex hierarchy of above and below-ground records. In *aqp*, the more generic term "profile," is used instead of pedon to accommodate incomplete data (missing above-ground information) or otherwise truncated horizon observations. Central to this approach is the specification of profile IDs and horizon depths, above-ground ("site") vs. below-ground ("horizon") attributes, and ideally horizon designation with associated attributes such as soil color.

11.3.1 Soil Color

The color of soil material observed during field investigations is one of the most striking and useful properties recorded as part of a soil profile description. Typically recorded in the Munsell system (Munsell 1947; Simonson 1993; Soil Science Division Staff 2017) in the form of "hue, value/chroma," the three components of this notation provide interpretive suggestions about iron oxides and oxidation state (hue and chroma) (Schwertmann 1993; Scheinost and Schwertmann 1999), soil carbon (value) (Wills et al. 2007; Liles et al. 2013), as well as hints about the relative importance of catenary relationships (Brown et al. 2004). Several color-based metrics of soil development (Buntley and Westin 1965; Harden 1982), rubification (Barron and Torrent 1986; Hurst 1977), and melanization (Harden 1982; Thompson and Bell 1996) are implemented in the *aqp* package.

11.3.1.1   Color Conversion

The *aqp* package provides several interfaces for conversion between Munsell notation and sRGB or CIELAB color spaces, largely based on the 1943 Munsell renotation table (Centore 2012). Forward conversion from standard Munsell notation (e.g., 10YR 3/4) is performed via look-up table, derived from the renotation table, and interpolated to include odd chroma and 2.5 value. The function munsell2rgb() performs a direct transformation to sRGB-encoded colors in hexadecimal (#5C4222), sRGB coordinates scaled to the interval of 0–1 ([0.36187, 0.25989, 0.13375]), or CIELAB coordinates ([30.273, 7.2731, 23.753]) (Beaudette et al. 2013a, b). Inverse transformation from sRGB coordinates is performed by the rgb2munsell() function, approximated by nearest-neighbor search of the Munsell-sRGB look-up table using the CIE2000 color contrast metric (Pedersen et al. 2021). All color space coordinates are referenced to the CIE standard illuminant D65, which is a close approximation to average midday sunlight in the northern hemisphere (Marcus 1998). sRGB and CIELAB color spaces were selected to address two common applications: sRGB, for digital representation of color on computer screens or reproduction on printed media, and CIELAB for the convenient alignment of axes and common pigments in the soil environment (Viscarra Rossel et al. 2006; Liles et al. 2013).

Non-standard notation of Munsell colors (e.g., 10.6YR 3.3/5.5), as collected by digital colorimeter, can be converted to approximate sRGB coordinates using the getClosestMunsellChip() function. However, this approach uses rounding of value and chroma and snapping to the nearest standard hue (10YR). Exact conversion of non-standard Munsell notation can be performed using the *munsellinterpol* **R** package (Gama et al. 2021).

11.3.1.2   Color Contrast

Color contrast (perceptual difference between two colors) within a soil sample is an important component of field-described redoximorphic features, concentrations, and mottles (Schoeneberger et al. 2012). Soil Survey products and wetland delineation protocols adopted by the National Cooperative Soil Survey (NCSS) currently use *contrast classes* (faint, distinct, and prominent) to describe color contrast, based on differences in Munsell hue, value, and chroma (Soil Survey Staff 2022c). The colorContrast() function in *aqp* computes differences in Munsell (hue, value/chroma), soil color contrast class, and the CIE2000 color contrast metric (Sharma

Table 11.1 Output from the colorContrast() function includes: change in hue (dH), change in value (dV), change in chroma (dC), CIE2000 color contrast (dE00), and NCSS soil color contrast class (CC)

Color 1	Color 2	dH	dV	dC	dE00	CC
10YR 3/3	10YR 3/4	0	0	1	3.13	Faint
7.5YR 6/6	5YR 4/6	1	2	0	21.1	Distinct
2.5Y 2/2	5G 4/8	9	2	6	30.5	Prominent

et al. 2005) for pairs of colors specified in Munsell notation. The function is fully vectorized meaning that multiple comparisons can be generated without explicit looping (Table 11.1).

Tabular color contrast output can be convenient when used as an intermediate step in a more complex workflow but can be difficult for non-specialists to interpret. A graphical representation of these data is created by the colorContrastPlot() function provided by *aqp*. For example, the differences between typical dry and moist soil colors for the Musick soil series (Fine-loamy, mixed, semiactive, mesic Ultic Haploxeralfs) are demonstrated in Fig. 11.4. While exact replication of Munsell colors is not possible on un-calibrated displays or printers, the sRGB approximation is sufficient to demonstrate relative differences in hue, value, and chroma.

To further aid with the calculation and interpretation of color contrast, "color contrast charts" can be created with the contrastChart() function provided by *aqp*. These charts are based on a source color in Munsell notation (e.g., 7.5YR 4/3) and select pages of Munsell hue. Pair-wise metrics of color contrast are evaluated between all color "chips" and the source color (outlined in red). Hue is split across panels in a familiar format with Munsell chroma on the x-axis and value on the y-axis. Soil color contrast class and CIE2000 values are printed below each color "chip" (Fig. 11.5).

11.3.2 Soil Profile Sketches

Conceptual sketches of soil profiles that illustrate variation in morphology (e.g., horizon depths, horizon designations, color, texture, etc.) in relation to transect or catenary position are a pedologic staple. Either hand-drawn in field notes or carefully produced as part of a final soil survey manuscript, these sketches represent an important vehicle for communicating observation and context to technical and non-technical audiences. A data-driven approach to creating soil profile sketches was one of the original motivations for the *aqp* **R** package (Beaudette et al. 2013a, b). Since 2010, the profile sketch authoring tools in *aqp* have progressed from basic layout of filled rectangles (profiles and horizons) to thematic coloring of horizons based on properties or classes, encoding of horizon boundary information, and handling of label collision.

Fig. 11.4 Color contrast plot, comparing the moist and dry soil colors of the Musick soil series. CIE2000 color contrast values are printed below soil color contrast classes. Smaller values describe smaller perceptual differences between colors

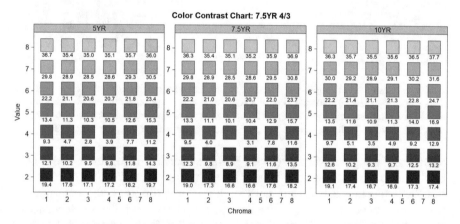

Fig. 11.5 Color contrast chart for 7.5YR 4/3 (chip outlined in red), including reference 5YR, 7.5YR, and 10YR hue pages. Numbers below soil color chips represent CIE2000 color contrast values, as compared with the target color 7.5YR 4/3. Soil color contrast classes have been omitted for clarity

The plotSPC() function in the *aqp* package is the primary tool for creating soil profile sketches from SoilProfileCollection objects, using **R**'s "base graphics" system. Figure 11.6 demonstrates several possible data sources, processing steps, and output generated from plotSPC(). Soil components (retrieved from the detailed Soil Survey via fetchSDA() as a SoilProfileCollection) within map unit "2vxq8" occur on summit and shoulder hillslope positions while components in map unit "2vxq9" occur on backslope, footslope and toeslope positions. USDA soil texture classes (<2 mm fraction) of each horizon are symbolized with color to show the variation of textures within the catena. Labeling of horizon depths (vs. common depth axis), leader lines, and collision detection (common with thin horizons) are optional enhancements to the standard output, specified via function arguments (Soil Survey Staff 2022a). Narrower profiles to the left of each component sketch represent data from the Official Series Descriptions via fetchOSD(). These data represent the typical morphology (horizon depths, designations, colors, etc.) for all soil series used in the US Soil Survey. Munsell colors (moist conditions) have been converted to sRGB

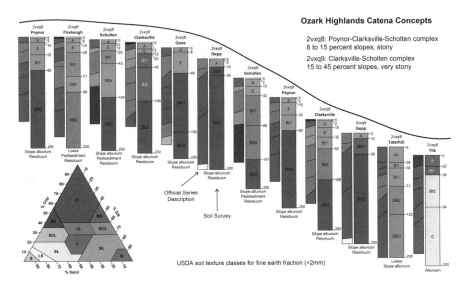

Fig. 11.6 Illustration of an idealized hillslope catena for soil components from two adjacent map units within the Ozark Highlands. SSURGO map unit soil components are placed within a geomorphic hillslope sequence to convey soil property to soil parent material relationships. Munsell soil colors from each official series description are displayed in tandem, companion profiles

coordinates using munsell2rgb(), and horizon boundary distinctness codes have been converted into vertical offsets using hzDistinctnessCodeToOffset(). The plotSPC() function can encode horizon distinctness offsets as diagonal horizon boundaries, where increasingly steeper angles represent the following sequence of boundary distinctness: "very abrupt," "abrupt," "clear," "gradual," "diffuse" (Schoeneberger et al. 2012). A visual explanation of the many arguments to plotSPC() is provided via explainPlotSPC() which shows the usage of ordering vectors, graphical offsets and scaling factors within the graphical space. Detailed examples of plotSPC() usage are available in the function documentation (Soil Survey Staff 2022a) and associated tutorials (Beaudette 2022). Future developments to plotSPC() will include conversion to the more advanced "grid" graphics system, pattern fills (e.g., geologic and stratigraphic symbols), and tighter integration with other plotting libraries such as *lattice* and *ggplot2*.

11.3.3 *Functional Horizon Aggregation*

Soil scientists use a common language of horizon designation nomenclature to describe and articulate the observed differences in soil horizons within a soil profile. These basic notations and the act of "naming" genetic horizons distills important information in the form of master horizons, characteristic subscripts, horizon and pedogenic sequences, and parent material discontinuities. Horizon designations convey a qualitative description of soil properties and process while allowing flexibility in how horizon designations are applied. Experienced soil scientists will generally apply horizon nomenclature consistently from site to site and through time due to the rigid guidelines and definitions of their application. However, among a group of soil scientists there will be variability in the exact designations used—each having their own unique training, field experiences, and tendencies towards "lumping" (describing fewer and thicker horizons) or "splitting" (describing more and thinner horizons).

Building on the interpretation of horizon designations, the Generalized Horizon Label (GHL) concept seeks to unify functionally similar horizon designations for the purpose of aggregation, analysis, and summary operations (Beaudette et al. 2016; Roecker et al. 2016). The conceptual approach of functional aggregations of horizons within collections of soil profiles has been attempted as a framework for developing pedotransfer functions (Wagenet et al. 1991). The process of applying GHL to a collection of soil profiles involves a series of micro-correlations made by a soil scientist to determine which horizons have similar soil morphology and properties to be grouped together for aggregation across horizons within a SoilProfileCollection.

The application of GHL to a SoilProfileCollection is driven by Regular Expression (REGEX) pattern matching. REGEX provides a rich syntax for string-matching of horizon designation sets into unified horizon GHL groups. A user developed set of REGEX rules are matched to an identified vector of horizonation by the generalize.hz() function in the *aqp* **R** package.

Features unique to the Clarksville soil series are argillic diagnostic horizons and parent material discontinuities as horizons transition along boundaries between cherty dolomite slope alluvium and colluvium over dolomite residuum. GHL can be used to identify these common features in a set of soils once applied (Fig. 11.7).

The depthOf() family of functions also utilize REGEX pattern matching of horizon designations to determine the depth to horizons with matching designation. For example, the pattern "Bt" would match all horizon designations containing "Bt" (case sensitive). Within a SoilProfileCollection, the minDepthOf() and maxDepthOf() operations provide additional utility to find either the top (shallowest) or bottom (deepest) depth to a matching horizon pattern. These functions provide convenience handling for missing values or when target patterns are not found within a profile. Results are returned as a numeric vector for single profiles or a data.frame of results with profile ID, horizon ID, top or bottom depths, horizon designation and pattern provided. Profile sketches in Fig. 11.7 have been sorted by values returned by the depthOf() function, first according to depth of "3Bt4" GHL and second according to depth of "2Bt3" GHL.

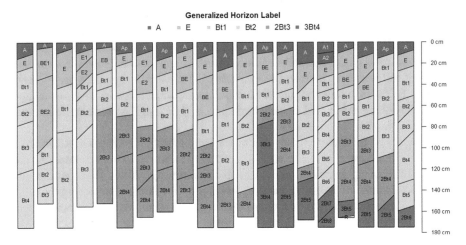

Fig. 11.7 A series of generalized horizon labels (GHL) applied to a collection of soil profiles of the Clarksville soil series from the Ozark Highlands, Missouri. Horizon labels within each soil profile show the original horizon designations while colors indicate assigned GHL

11.3.4 Change of Depth Support

Soil data are typically collected either by genetic horizon with varying horizon depths, at regular depth intervals (every 10 cm), or from composite samples representing specified depth intervals (0–10 cm, 10–25 cm, etc.). The structure of these depth intervals will typically vary from one profile to the next. To facilitate analysis throughout the profile collection, profile horizon depths may need to be modified and/or harmonized comprising a change of depth support. A simple down-scaling of horizons (without interpolation) into a regular sequence of thinner depth slices, referred to as "slicing," is implemented in the dice() function (Fig. 11.8a). The segment() function offers another approach to restructuring horizon depths, using horizon-thickness weighted mean values for conversion to fixed depth intervals (e.g., 0–25 cm). This is a common step in the thematic mapping of soil property data. A more complex change of horizon depths can be achieved using constrained interpolation. This method, popular for applications such as digital soil mapping (or other tasks requiring harmonized horizon depths) uses mass-preserving splines (Bishop et al. 1999). In *aqp*, this type of down-scaling is performed with the spc2mpspline() function which provides a convenient interface to the *mpspline2* package (O'Brien 2022) suitable for SoilProfileCollection objects.

A change of support operation performed over all profiles in a collection result in a statistical summary of that collection. The slab() function performs this kind of operation by "slicing" horizon data (continuous or categorical) into 1 cm-thick depth intervals (Fig. 11.8a), then aggregating "across" those depth intervals (Harradine 1963; Beaudette et al. 2013a, b). Continuous values are reduced to select percentiles (or any user-defined function) per slice, and categorical values are

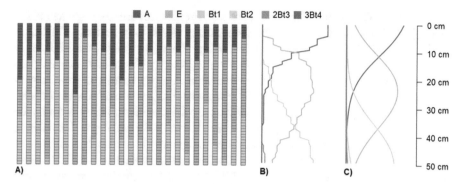

Fig. 11.8 Change of support demonstration using 25 pedons correlated to the Clarksville soil series. Generalized horizon labels (GHL) have been resampled to 1cm slices (**a**) via dice(), aggregated across slices to GHL proportions (**b**) via slab(), and modeled via proportional-odds logistic regression (**c**)

reduced to proportions (Fig. 11.8b) (Beaudette et al. 2013a, b). Alternatively, sliced categorical data such as GHL can be aggregated using statistical models for ordinal data, such as the proportional-odds logistic regression model (Fig. 11.8c) (Beaudette et al. 2016).

11.4   Numerical Classification of Soils

Since the 1960s (likely corresponding with increased availability of computing hardware) there has been considerable interest in the development of numerical alternatives to traditional soil classification systems such as Soil Taxonomy (Soil Survey Staff 1999) and World Reference Base (Chesworth et al. 2008). A "numerical taxonomy" (Sneath and Sokal 1973) of soil horizons or collections of horizons (i.e., soil profiles or aggregation thereof) relies on a deliberate selection of characteristics (soil properties), distance metric (e.g., Euclidean) and criteria used to identify clusters (e.g., hierarchical vs. partitioning methods) (Arkley 1976). Selection of characteristics is complex because a limited set of soil properties cannot universally describe differences between individuals, and the use of all measurable properties is unfeasible – requiring a selection to be made prior to analysis (Sarkar et al. 1966; Arkley 1971). Furthermore, a generalized approach to the numerical classification of soil profiles is complicated by the hierarchical nature of linked site and horizon-level properties, sampling style (depth-intervals vs. genetic horizons) and subtle differences in horizon designation (through time, regionally, etc.). Despite the many challenges, there have been many successful applications of numerical taxonomy to soil science and soil classification (Hole and Hironaka 1960; Rayner 1966; Moore et al. 1972; Dale et al. 1989; Carré and Jacobson 2009).

The NCSP (Numerical Comparison of Soil Profiles) algorithm, implemented in the *aqp* package for **R**, attempts to address many of the long-standing difficulties with a numerical classification of entire soil profiles (Beaudette et al. 2013a, b; Maynard et al. 2020). Building on methods suggested by Moore et al. (1972), pairwise distances (between soil profiles) are evaluated along regular depth-slices by Gower's distance metric (Gower 1971), using any combination of continuous, categorical, or boolean attributes (Fig. 11.9). Total pair-wise dissimilarity is computed by taking the sum of slice-wise dissimilarities, to a user-defined depth. Variation in profile depth is accounted for by assigning maximum slice-wise dissimilarity to comparisons between soil and non-soil. Further customization of the NCSP algorithm is described in Beaudette et al. (2013b). The resulting dissimilarity matrix can be used to assist with topics ranging from initial mapping ("similar/dissimilar" soils), comparisons below family-level classification in Soil Taxonomy, soil series correlation, map unit harmonization, and correlation between different taxonomic systems.

Applied to the same set of soil profiles highlighted in Fig. 11.7, the NCSP algorithm was used to generate a distance matrix using only the GHL classes (ordinal values) to a depth of 175cm (Fig. 11.10). A dendrogram was created from the distance matrix via divisive hierarchical clustering (Kaufman and Rousseeuw 2005) and combined with profile sketches with the plotProfileDendrogram() function from the *sharpshootR* package. Profiles with similar GHL assignments, occurring at similar depths, are allocated to clusters defined by branching near the bottom of the dendrogram. When a combination of site (e.g., slope, drainage class, geoform, etc.) and horizon-level properties are requested, the final distance matrix is developed from a weighted average of the site and horizon-level distance matrices. Pair-wise distances between soil profiles can be difficult to interpret when a large number of properties are included in the calculation and may require a different approach to thematic coloring of profile sketches such as principal component analysis of the property matrix or principal coordinate analysis of the distance matrix scores. An alternative presentation of the data is possible by arranging profile sketches (with plotSPC()) according to the new axes created by 2-dimensional ordination, typically via non-Metric Multidimensional Scaling (nMDS) of the distance matrix.

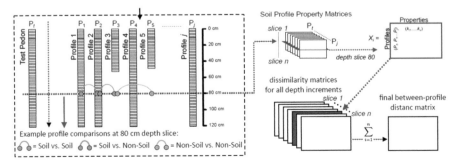

Fig. 11.9 Graphical outline of the Numerical Classification of Soil Profiles (NCSP) algorithm. (Figure c/o Jon Maynard, adapted from Maynard et al. (2020))

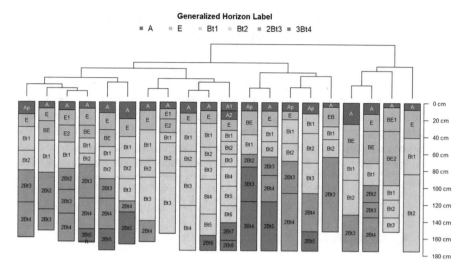

Fig. 11.10 Soil profile sketches from Fig. 11.7, arranged according to divisive hierarchical clustering of the distance matrix generated by the Numerical Classification of Soil Profiles (NCSP) algorithm

11.5   Water Balance

The soil forming state factor of climate and the interactions and timing of moisture and temperature are pivotal in understanding soil formation and describing site dynamics at local and regional scales. Water balance models accounting for inputs of precipitation and losses to evapotranspiration (Thornthwaite 1948) have evolved as a valuable tool for exploring the nuances of climate at a given point on the landscape. Water balance metrics relate to soil storage and the downward and upward flux of water through soil and associated soil property development as the soil acts as a sponge responding to atmospheric supply and demand (Arkley and Ulrich 1962). Correlation of water balance metrics to vegetation growth at sites has become an important tool for the study of site conditions related to existing vegetation distribution (Stephenson 1998; Lutz et al. 2010) and forecasting site climate trajectories. High quality, widely accessible gridded climate data has increased the use of soil water balance models.

The *sharpshootR* package provides methods (via dependencies on the *elevatr*, *daymetr*, *Evapotranspiration*, and *hyrdomad* packages) for calculating water balance variables of precipitation (PPT), potential evapotranspiration (PET), actual evapotranspiration (AET), deficit (D), soil moisture storage (S), surplus (U), volumetric water content (VWC) on monthly and daily time steps (monthlyWB() and dailyWB() functions). The prepareDailyClimateData() function assembles the available water-holding capacity (AWC) values derived for major components in soil map units of the US Soil Survey for specific point coordinates. Gridded DAYMET climate data (Thornton et al. 2020) is then downloaded for the specified location and daily water balance metrics are estimated via dailyWB_SSURGO().

The *soilDB* package provides query access to the station data in the USDA-NRCS Soil Climate Analysis Network (SCAN). These stations provide above and below ground climate sensor networks that measure soil temperature, soil moisture, air temperature and precipitation. Many stations have associated soil characterization data in the Kellogg Soil Survey Laboratory (KSSL) records. The data in Fig. 11.11 were assembled through separate calls to fetchSCAN() for the SCAN climate station data and dailyWB_SSURGO() which derives AWC for the Scholten soil component sampled at SCAN station 2194, assembles DAYMET data for the SCAN station location and runs a daily water balance model. Comparisons of modeled and measured values allow for evaluation of water balance model utility and function.

11.6   Conclusions

Pedology and geomorphology are inherently visual, field-based sciences that share a common paradigm and fundamental units of description. Geomorphic description of landforms or geoforms and a merging of these disciplines functionally defined as geopedology is a progression that elevates the geoform as a primary landscape concept that can guide the operational inventory and integrated study of soil geomorphic relationships. Understanding the relevance and importance of contextual

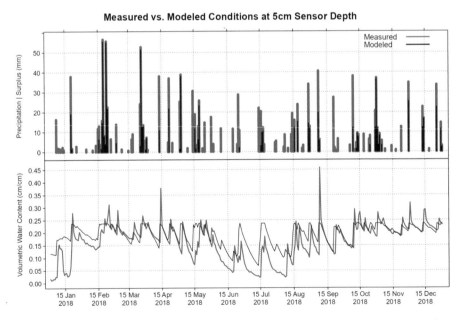

Fig. 11.11 Comparison of annual water balance metrics (volumetric water content and precipitation) for 2018 at Soil Climate Analysis Network (SCAN) station 2194 in the Ozark Highlands

linkages to geomorphology in soil survey products integrates soil information with other environmental data and is critical to informing the public and the wider scientific community. Embracing a visual, quantitative analytical approach to collections of soil profiles and varying formats of soil survey data allows for creative abstraction of geopedologic concepts.

Complex data structures and increasing volumes of available data demand progress in methods that allow iterative aggregation, summary, and graphical expression of soil data. The AQP suite of packages works to provide examples and routines that meet these challenges with an emphasis on generalized methods that can be applied to common data structures. This collection of tools leverages the open-source flexibility and extensibility of the **R** programming environment and the progress that can be gained from collaborative effort to solve common challenges in working with soil data. Providing visual alternatives for viewing soil morphologic data and improving the quality and accessibility of soil survey data will foster greater use and application of soil survey data to inform all users.

References

Arkley RJ (1971) Factor analysis and numerical taxonomy of soils. Soil Sci Soc Am J 35(2):312–315. https://doi.org/10.2136/sssaj1971.03615995003500020038x

Arkley RJ (1976) Statistical methods in soil classification research. Adv Agron 28:37–70. https://doi.org/10.1016/S0065-2113(08)60552-0

Arkley RJ, Ulrich R (1962) The use of calculated actual and potential evapotranspiration for estimating potential plant growth. Hilgardia 32(10):443–462

Barron V, Torrent J (1986) Use of the Kubelka-Munk theory to study the influence of iron oxides on soil colour. J Soil Sci 37(4):499–510. https://doi.org/10.1111/j.1365-2389.1986.tb00382.x

Beaudette DE (2022) Introduction to soil profile collection objects. http://ncss-tech.github.io/AQP/aqp/aqp-intro.html

Beaudette DE, Dahlgren RA, O'Geen AT (2013a) Terrain-shape indices for modeling soil moisture dynamics. Soil Sci Soc Am J 77:1696–1710. https://doi.org/10.2136/sssaj2013.02.0048

Beaudette DE, Roudier P, O'Geen AT (2013b) Algorithms for quantitative pedology: a toolkit for soil scientists. Comput Geosci 52:258–268

Beaudette DE, Roudier P, Skovlin J (2016) Probabilistic representation of genetic soil horizons. In: Hartemink AE, Minasny B (eds) Digital soil morphometrics. Springer, pp 281–293. https://doi.org/10.1007/978-3-319-28295-4_18

Bishop TFA, McBratney AB, Laslett GM (1999) Modelling soil attribute depth functions with equal-area quadratic smoothing splines. Geoderma 91(1-2):27–45

Brown DJ, Clayton MK, McSweeney K (2004) Potential Terrain controls on soil color, texture contrast and grain-size deposition for the original Catena landscape in Uganda. Geoderma 122(1):51–72. https://doi.org/10.1016/j.geoderma.2003.12.004

Buntley GJ, Westin FC (1965) A comparative study of developmental color in a Chestnut-Chernozem-Brunizem Soil Climosequence. Soil Sci Soc Am J 29(5):579–582. https://doi.org/10.2136/sssaj1965.03615995002900050029x

Carré F, Jacobson M (2009) Numerical classification of soil profile data using distance metrics. Geoderma 148:336–345. https://doi.org/10.1016/j.geoderma.2008.11.008

Centore P (2012) An open-source inversion algorithm for the munsell renotation. Color Res Appl 37:455–464

Chesworth W, Arbestain MC, Macías F, Spaargaren O, Otto S, Mualem Y, Morel-Seytoux HJ et al (2008) Classification of soils: world reference base (WRB) for soil resources. In: Chesworth W (ed) Encyclopedia of soil science. Springer, Dordrecht, pp 120–122. https://doi. org/10.1007/978-1-4020-3995-9_104

Dale MB, McBratney AB, Russell JS (1989) On the role of expert systems and numerical taxonomy in soil classification. J Soil Sci 40(2):223–234. https://doi.org/10.1111/j.1365-2389.1989. tb01268.x

Dowle M, Srinivasan A (2021) data.table: extension of 'data.frame'. https://CRAN.R-project.org/ package=data.table

Gama J, Centore P, Davis G (2021) Munsellinterpol: Interpolate Munsell Renotation Data from Hue/Chroma to CIE/RGB. https://CRAN.R-project.org/package=munsellinterpol

Gower JC (1971) A general coefficient of similarity and some of its properties. Biometrics 27(4):857–871. http://www.jstor.org/stable/2528823

Harden JW (1982) A quantitative index of soil development from field descriptions: examples from a Chronosequence in Central California. Geoderma 28(1):1–28. https://doi. org/10.1016/0016-7061(82)90037-4

Harradine F (1963) Morphology and genesis of noncalcic brown soils in California. Soil Sci 96:277–287

Hijmans RJ (2021) Raster: geographic data analysis and modeling. https://CRAN.R-project.org/ package=raster

Hole FD, Hironaka M (1960) An experiment in ordination of some soil profiles. Proc Soil Sci Soc Am 24:309–312

Hudson BD (1992) The soil survey as paradigm-based science. Soil Sci Soc Am J 56:836–841

Hurst VJ (1977) Visual estimation of iron in saprolite. GSA Bull 88(2):174–176. https://doi.org/1 0.1130/0016-7606(1977)88<174:VEOIIS>2.0.CO;2

Jenny H (1941) Factors of soil formation: a system of quantitative pedology. McGraw-Hill, New York

Kabrick JM, Dey DC, Jensen RG, Wallendorf M (2008) The role of environmental factors in Oak decline and mortality in the Ozark Highlands. For Ecol Manag 255:1409–1417

Kabrick JM, Dey DC, Jensen RG, Wallendorf M (2011) Landscape determinants of exchangeable calcium and magnesium in Ozark Highland Forest Soils. Soil Sci Soc Am J 75:164–180

Kaufman L, Rousseeuw PJ (eds) (2005) Finding groups in data, Wiley series in probability and statistics. Wiley, Hoboken. https://doi.org/10.1002/9780470316801

Liles GC, Beaudette DE, O'Geen AT, Horwath WR (2013) Developing predictive soil C models for soils using quantitative color measurements. Soil Sci Soc Am J 77:2173–2181. https://doi. org/10.2136/sssaj2013.02.0057

Lutz JA, Franklin JF, Van Wagtendonk JW (2010) Climatic water deficit, tree species ranges, and climate change in Yosemite National Park. J Biogeogr 37(5):936–950

Maechler M, Rousseeuw P, Struyf A, Hubert M, Hornik K (2021) Cluster: cluster analysis basics and extensions. https://CRAN.R-project.org/package=cluster

Marcus RT (1998) The measurement of color. In: Nassau K (ed) Color for science, art and technology. Elsivier, pp 31–96

Maynard JJ, Salley SW, Beaudette DE, Herrick JE (2020) Numerical soil classification supports soil identification by citizen scientists using limited, simple soil observations. Soil Sci Soc Am J 84(5):1675–1692. https://doi.org/10.1002/saj2.20119

Moeys J (2018) Soiltexture: functions for soil texture plot, classification and transformation. https://CRAN.R-project.org/package=soiltexture

Moore AW, Russell JS, Ward WT (1972) Numerical analysis of soils: a comparison of three soil profile models with field classification. J Soil Sci 23:194–209

Müller K, Wickham H (2021) Tibble: simple data frames. https://CRAN.R-project.org/ package=tibble

Munsell AH (1947) A color notation, 10th edn. Munsell Color Company, Inc

Myers DB, Kitchen NR, Sudduth KA, Miles RJ, Sadler EJ, Grunwald S (2011) Peak functions for modeling high resolution soil profile data. Geoderma 166(1):74–83. https://doi.org/10.1016/j. geoderma.2011.07.014

O'Brien L (2022) Mpspline2: mass-preserving spline functions for soil data. https://CRAN.R--project.org/package=mpspline2

O'Geen A, Walkinshaw M, Beaudette DE (2017) SoilWeb: a multifaceted interface to soil survey information. Soil Sci Soc Am J 81. https://doi.org/10.2136/sssaj2016.11.0386n

Pebesma E (2018) Simple features for R: standardized support for spatial vector data. R J 10(1):439–446. https://doi.org/10.32614/RJ-2018-009

Pedersen TL, Nicolae B, François R (2021) Farver: high performance colour space manipulation. https://CRAN.R-project.org/package=farver

R Core Team (2022) R: a language and environment for statistical computing. R Foundation for Statistical Computing, Vienna. http://www.R-project.org/

Rayner JH (1966) Classification of soils by numerical methods. J Soil Sci 17:79–92

Roecker S, Skovlin J, Beaudette D, Wills S (2016) Digital summaries of pedon descriptions. In: Hartemink AE, Minasny B (eds) Digital soil morphometrics. Springer, pp 267–279. https://doi.org/10.1007/978-3-319-28295-4_17

Sarkar PK, Bidwell OW, Marcus LF (1966) Selection of characteristics for numerical classification of soils. Soil Sci Soc Am Proc 30:269–272

Scheinost AC, Schwertmann U (1999) Color identification of iron oxides and hydroxysulfates: use and limitations. Soil Sci Soc Am J 63(5):1463–1471. http://soil.scijournals.org/cgi/content/abstract/soilsci;63/5/1463

Schoeneberger PJ, Wysocki DA, Benham EC, Soil Science Division Staff (2012) Field book for describing and sampling soils, Version 3.0. Natural Resources Conservation Service, National Soil Survey Center

Schwertmann U (1993) Relations between iron oxides, soil color, and soil formation. In: Bigham JM, Ciolkosz EJ (eds) Soil color, SSSA special publication number 31. Soil Science Society of America, Inc., Madison, pp 51–69

Sharma G, Wencheng W, Dalal EN (2005) The CIE DE2000 color-difference formula: implementation notes, supplementary test data, and mathematical observations. Color Res Appl 30(1):21–30. https://doi.org/10.1002/col.20070

Simonson RW (1993) Soil color standards and terms for field use—history of their development. In: Bigham JM, Ciolkosz EJ (eds) Soil color, SSSA special publication number 31. Soil Science Society of America, Inc., Madison, pp 1–20

Singh G, Goyne KW, Kabrick JM (2015) Determinants of total and available phosphorus in forested alfisols and ultisols of the Ozark Highlands, USA. Geoderma Reg 5:117–126

Sneath PHA, Sokal RR (1973) Numerical taxonomy. W.H. Freeman Company

Soil Science Division Staff (2017) Soil survey manual. Ditzler C, Scheffe K, Monger HC (eds). USDA Handbook 18. Government Printing Office

Soil Survey Staff (1999) Soil Taxonomy: a basic system of soil classification for making and interpreting soil surveys, U.S. Department of Agriculture Handbook 436, 2nd edn. Natural Resources Conservation Service, Washington, DC

Soil Survey Staff (2022a) Online Manual for the AQP Package. http://ncss-tech.github.io/aqp/

Soil Survey Staff (2022b) Soil Survey Geographic (SSURGO) Database. Edited by Natural Resources Conservation Service, United States Department of Agriculture. https://sdmdataaccess.sc.egov.usda.gov/

Soil Survey Staff (2022c) Soil Survey Technical Note 2. Edited by Natural Resources Conservation Service, United States Department of Agriculture. https://www.nrcs.usda.gov/wps/portal/nrcs/detail/soils/ref/?cid=nrcs142p2_053569

Stephenson NL (1998) Actual evapotranspiration and deficit: biologically meaningful correlates of vegetation distribution across spatial scales. J Biogeogr 25(5):855–870

Thompson JA, Bell JC (1996) Color index for identifying hydric conditions for seasonally saturated Mollisols in Minnesota. Soil Sci Soc Am J 60(6):1979–1988. https://doi.org/10.2136/sss
aj1996.03615995006000060051x

Thornthwaite CW (1948) An approach toward a rational classification of climate. Geogr Rev
38(1):55–94. http://www.jstor.org/stable/210739

Thornton MM, Shrestha R, Wei Y, Thornton PE, Kao S, Wilson BE (2020) DAYMET: daily surface weather data on a 1-Km grid for North America, Version 4. ORNL Distributed Active
Archive Center. https://doi.org/10.3334/ORNLDAAC/1840

van den Boogaart KG, Tolosana-Delgado R, Bren M (2021) Compositions: compositional data
analysis. https://CRAN.R-project.org/package=compositions

Viscarra Rossel RA, Minasny B, Roudier P, McBratney AB (2006) Colour space models for soil
science. Geoderma 133(3):320–337. https://doi.org/10.1016/j.geoderma.2005.07.017

Wagenet RJ, Bouma J, Grossman RB (1991) Minimum data sets for use of soil survey information
in soil interpretive models. In: Mausbach MJ, Wilding LP (eds) Spatial variabilities of soils and
landforms, SSSA Spec. Publ. 28. SSSA, Madison, pp 161–182

Wills SA, Lee Burras C, Sandor JA (2007) Prediction of soil organic carbon content using field
and laboratory measurements of soil color. Soil Sci Soc Am J 71(2):380–388. https://doi.
org/10.2136/sssaj2005.0384

Wysocki AD, Schoeneberger PJ, LaGarry HE (2005) Soil surveys: a window to the subsurface.
Geoderma 126(1):167–180

Zeileis A, Fisher JC, Hornik K, Ihaka R, McWhite CD, Murrel, l P., Stauffer, R., and Wilke,
C.O. (2020) Colorspace: a toolbox for manipulating and assessing colors and palettes. J Stat
Softw 96:1–49

Zinck JA (2016) The geopedologic approach. In: Metternicht G, Zinck JA, Del Valle HF (eds)
Geopedology. Springer, pp 27–59. https://doi.org/10.1007/978-3-319-19159-1_4

Chapter 12
Interaction of Geological and Pedological Processes in the Genesis of Soils with Gypsum, Northern Patagonia, Argentina

J. A. Irisarri, A. C. Dufilho, and G. S. de la Puente

Abstract The origin and classification of gypsum soils, which have been usually treated secondarily with calcareous soils, are discussed. Different types of gypsum soils present in Northern Patagonia associated with gypsum and anhydrite outcrops in the upper part of the related hydrographic basins, and their redistribution in alluvial plains and fans, and Pleistocene and Holocene fluvial terraces, are analyzed. The water chemistry in the Colorado and Neuquén rivers is also considered. Gypsum soils formed on Pleistocene terraces have a higher grade of differentiation, and a good internal drainage. The presence of gypsum soils in Holocene terraces of the Colorado River basin, and their lack in the Neuquén River basin, is linked to the currently higher relative exposure of gypsum and anhydrite deposits, along the upper Colorado River basin. Processes that affect the gypsum soils, as well as the uses of these soils in the region, are also discussed.

Keywords Pedogenesis · Gypsum soils · Taxonomy · Neuquén River · Colorado River

J. A. Irisarri (✉)
Facultad de Ciencias Agrarias, Universidad Nacional del Comahue,
Buenos Aires, Neuquén, Argentina

A. C. Dufilho
CITAAC CONICET, CIGPat Dpto. de Geología y Petróleo, Facultad de Ingeniería – Facultad de Ciencias del Ambiente, Universidad Nacional del Comahue,
Buenos Aires, Neuquén, Argentina

G. S. de la Puente
CITAAC CONICET, CIGPat Dpto. de Geología y Petróleo, Facultad de Ingeniería,
Universidad Nacional del Comahue, Buenos Aires, Neuquén, Argentina
e-mail: susana.delapuente@comahue-conicet.gob.ar

223

12.1 Introduction

A soil is considered as a gypsum soil when it contains sufficient amounts of calcium sulfate to interfere with the normal growth of plants (FAO 1990). There is no general consensus between specialists about to their formal denomination. In the soil literature, gypsum soils are usually analyzed together with calcareous soils as an extension of them. The calcium sulfate molecule has no relation with the calcium carbonate molecule, beside their low water solubility, and common deposition in arid climates. Thus, the origin and properties of gypsum and calcareous soils are not largely shared. In the classification of the United States Department of Agriculture (Soil Survey Staff 1999–2014), gypsum soils are included in the Entisol (soils with gypsum of geological genesis) and Aridisol Orders (soils with gypsum of pedological genesis). They are not formally classified in Spanish. Epistemologically, they have not been well defined either in a systematic position. Thus, the gypsum soils are recognized as "zonal" soils according to their climatic relation. Besides, they can also be assigned as "lithomorphic intrazonal" soils considering the required presence of the calcium sulfate molecule from the parent material, similarly to the "rendzinas". However, the origin of the gypsum molecule is not authigenic.

Known aspects about the genesis of the gypsum soils have been briefly described and, in general, it is recognized a limited knowledge on their origin. It has been observed that it is not essential the presence of a gypsum bedrock to form gypsum-rich genetic horizons. Under arid climate (aridic soil moisture regime), the dissolved gypsum that accumulate in the upper section of the soil profile, in a preexisting source of surface water or groundwater with a stable piezometric level (at least of the Pleistocene), it is sufficient to form a gypsum soil. The specialists have not defined yet which is the developed, mature or climax state of gypsum soils.

In relation to the soil taxonomy, as Casby-Horton et al. (2015) have pointed out, in the Keys to Soil Taxonomy, Tenth Edition (Soil Survey Staff 2006), the gypsum horizon is defined as an illuvial horizon in which the secondary gypsum has been significantly accumulated; and the petrogypsic horizon as an illuvial horizon at least 10 cm thick, in which secondary gypsum has accumulated to such an extent that the horizon is cemented or indurated.

In the more recent Keys to Soil Taxonomy, Twelfth Edition (Soil Survey Staff 2014), the gypsum horizon is instead defined as "a horizon in which gypsum has been significantly accumulated or transformed"; and the petrogypsic horizon as "a horizon in which visible secondary gypsum has accumulated or has been transformed. The horizon is cemented (i.e., extremely weakly through indurated cementation classes), and the cementation is both laterally continuous and root limiting, even when the soil is moist." From the Tenth (Soil Survey Staff 2006) to the Twelfth (Soil Survey Staff 2014) Editions, it is remarkable that the "leaching" term disappears. This suggests that there are uncertainties, between North American pedologists, about the origin of gypsic and petrogypsic horizons. In the North American pedology, the "leaching" term means movement of soluble and/or pseudo soluble substances vertically downward in the soil profile.

In Northern Patagonia, gypsum soils are associated to the ancient fluvial terraces of the current Neuquén and Colorado rivers. In the Neuquén River hydrographic basin, these soils are only developed on Pleistocene terraces, while in the Colorado River hydrographic basin they are developed on Holocene terraces as well. Moreover, the morphological, physical and chemical properties of the soils that are developed on terraces of different ages, i.e., Pleistocene and Holocene, show distinctive characteristics. Additionally, the stability of the soils and the land use show distinctive features in each case. The aim of this work is to characterize the different types of gypsum soils present in Northern Patagonia associated with gypsum and anhydrite outcrops in the upper part of the related hydrographic basins and their redistribution in alluvial plains and fans, and Pleistocene and Holocene fluvial terraces. Thus, it is considered their relationship with the parent material and the hydrology.

12.2 Distribution of Gypsum Soils and Its Relationship with the Calcium Sulfate Source

According to their origin, gypsum soils related to the hydrographic basins of the Neuquén and Colorado rivers are different. On the one hand, the gypsum soil of geological origin, which is directly associated to gypsum and anhydrite rocks as parent materials. On the other hand, the gypsum soil of pedogenic origin generated by sulfate accumulation in the profile from the sulfate-rich surface water or groundwater, coming from dissolution of the gypsum outcrops in the basin. Moreover, in Northern Patagonia, gypsum soils developed on Pleistocene terraces show a higher grade of differentiation than those developed on Holocene ones.

12.2.1 Gypsum Soils of Geological Origin

Gypsum soils of geological origin belong to the Entisol Order. They are directly related to stratigraphic units of the Neuquén Basin (Fig. 12.1). The Neuquén Basin (Bracaccini 1970) is one the main Mesozoic and Cenozoic sedimentary basins in Argentine according to both, its peculiar geological record and its economic importance (Yrigoyen 1991; Howell et al. 2005; Ramos et al. 2011). The origin of this basin is closely linked to the interaction of tectonic plates, oceanic and continental, in constant collision on the west from, approximately, 180 Ma (Middle Jurassic) ago. The most relevant stratigraphic units that contain gypsum and/or anhydrite were deposited during the Mesozoic; these units are the Tábanos Formation (Stipanicic 1966; Dellapé et al. 1979); the Auquilco Formation (Schiller 1912; Weaver 1931), the Huitrín Formation (Groeber 1946; Uliana et al. 1975a, b; Leanza 2003) and, in a less extent, the Rayoso Formation (Groeber 1946, 1953; Herrero

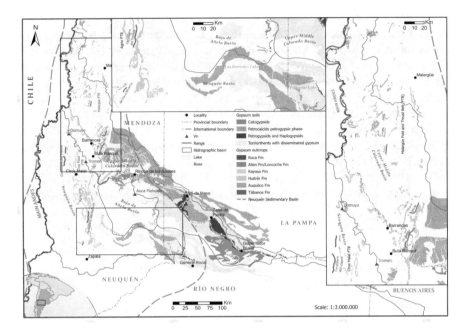

Fig. 12.1 Gypsum soils (Based on Ferrer et al. 2006) and outcrops of gypsum stratigraphic unit (Based on Narciso et al. 2001; Leanza et al. 2001; Leanza 2003; Nullo et al. 2005; Rodríguez et al. 2007) in the Bajo de Añelo, Colorado and Neuquén hydrographic basins

Ducloux 1946) (Fig. 12.1). Other units that include gypsum levels are the Allen Formation (Roll 1939; Uliana and Dellapé 1981) and the Roca Formation (Ihering 1903; Weaver 1927) (Fig. 12.1). The record of the later unit includes the Mesozoic-Cenozoic boundary.

The Tábanos Formation (Fig. 12.1; Table 12.1) is a 10–50 m thick deposit of anhydrite with carbonate and shale (Riccardi et al. 1991; Narciso et al. 2001; Leanza et al. 2001; Nullo et al. 2005). The unit contains thick calcareous gypsy packages and a layer of gypsum nodules in a calcareous matrix with irregular stratification (Leanza et al. 2001). This depocenter developed in the western-central part of the basin, extending longitudinally from the south of Mendoza (34°30′) to the center of Neuquén, in the Vaca Muerta Range. In the central part, the evaporitic deposit is covered by fluvial and eolian facies. The endorheic environment was the result of the limited connection of the basin with the ocean to the west (Pacific Ocean) (Legarreta 1991; Legarreta and Uliana 1999). This unit crops out in the area of the upper basins of the Colorado and Neuquén rivers, in the outer zone of the Chos Malal fold and thrust belt (FTB), to the east of the Cordillera del Viento, and in the Agrio FTB, in the Vaca Muerta Range area.

The Auquilco Formation (Figs. 12.1 and 12.2; Table 12.1), originally described as the "Main Gypsum" ("Yeso Principal", Schiller 1912), is precisely a distinctive deposit because of its important outcrops in the upper basin of the Colorado River, and in a much lesser extent, in the upper basin of the Neuquén River. It constitutes

Table 12.1 Approximated outcrop area of gypsum stratigraphic units in the analyzed hydrographic basins

Time Scale (Ma)		Stratigraphic unit	Gypsum area (Km²) / Gypsum thickness (m) / Gypsum volume (Km³)			Total gypsum area (Km²)	Total gypsum thickness (m²)	Total gypsum volume (Km³)
			Hydrographic basin					
			Bajo de Añelo	Colorado	Neuquén			
Uppermost K-lowermost Paleogene	~66	Roca Fm	-	4.43 / <1/ 0.004	16.28 / 7 / 0.11	20.71	<8	0.12
Late Upper K	~72	Allen-Loncoche Fms	79.06 / 3 / 0.24	7.99 / <1 / 0.01	36.60 / >5 / 0.18	123.65	9	0.43
Late Lower K	~115	Rayoso Fm	-	-	178.03 / >7 / 1.25	178.03	>7	1.25
	~125	Huitrín Fm	-	434.81 / 20 / 8.70	661.90 / 20 / 13.24	1096.70	40	21.93
Early Upper J	~160	Auquilco Fm	-	381.03 / 200 / 76.21	42.23 / 60 / 2.53	423.27	260	78.74
Late Middle J	~165	Tábanos Fm	-	9.52 / 50 / 0.48	12.74 / 35 / 0.45	22.26	85	0.92
Total			79.06 / 3 / 0.24	837.78 / 272 / 85.39	974.29 / 134 / 17.76	1864.63	409	103.39

Data of the maximum thickness per unit and estimated volume in the outcrops

an evaporite sequence up to 400 m thick, which is composed mainly by gypsum and anhydrite with thin limestone and fine-grain siliciclastic interbedded layers (Legarreta and Uliana 1999; Brodtkorb et al. 1999). In the subsurface, the thickness of this unit exceeds 500 m (Arregui et al. 2011). The gypsum has a nodular structure and laterally passes into gypsiferous limestone and calcareous gypsiferous sandstone (Leanza et al. 2001). Anhydrite facies form stratified packages with sulfate and carbonate interbedded layers, stratified-nodular gypsum deposits and zones with mosaic-nodular texture (Legarreta et al. 1993; Narciso et al. 2001). The cumulated gypsum forms, known as "gypsum blisters" and common in old, more stable soil landscape of high-gypsum parent materials resulting from dissolution and reprecipitation of the gypsum exposure, are frequent in the surface, near the Buta Ranquil locality (Fig. 12.2b, c).

The gypsum presence in alluvial soils (Torrifluvents with disseminated gypsum) is common as well. This formation is exposed in the Chos Malal FTB, from the eastern side of the Cordillera del Viento to the Las Yeseras anticline (to the east of the Tromen Volcano), and in the Agrio FTB (Vaca Muerta Range). In addition, it is widely exposed in the Malargüe FTB, in the Reyes Range area, where it can reach 200 m thick, and in the Cara Cura and Azul ranges, with more reduced thickness (Narciso et al. 2001). The unit was deposited in a closed hypersaline basin, with a

Fig. 12.2 (**a**) Gypsum outcrops of the Auquilco Formation, Las Yeseras anticline (to the east of the Tromen Vn.), near Buta Ranquil locality (Neuquén). Gypsum deposits are those of white color, without vegetal cover. View toward the NW. (**b**) "Gypsum blister" on the Auquilco Formation, as a result of gypsum dissolution and reprecipitation. (**c**) Crust of the gypsum blister

cyclic and continues supply from the shallow sea water, implying desiccation of the basin (Leanza et al. 2001). The rhythmical alternation in the sulfate and carbonate precipitation was due to the seasonal variations, which were determined by the fresh or sea water. The whole sedimentary sequence, from the Early Jurassic, indicates a progressive regression from the south (40°S) to the north (Zöllner and Amos 1973). This depocenter had a wider developing toward the north and east than the Tábanos Formation one. It is remarkable to mention that has been estimated, from northern Chile to Neuquén, in a narrow belt of 2000 km length, the deposition of approximately 40,000 km^3 of gypsum in around 7 Ma (Legarreta and Uliana 1996; Brodtkorb et al. 1999).

These gypsum levels are preserved, by their relative resistance to the erosion compare with the fine siliciclastic deposits, in the core of the synclines of the Chos Malal and Agrio fold and thrust belts. In the upper basin of the Colorado River, these levels contain thin gypsum and limestone layers in a fine siliciclastic sequence, which vertically pass to nodular, at the base, and laminated, toward the top, anhydrite facies (Narciso et al. 2001). In outcrops, the gypsum is "sugar" type ("saccharoid") by hydration of anhydrite (Leanza 2003; Gómez Figueroa et al. 2011). In the upper basin of the Neuquén River, the gypsum occurs in lenses in certain levels, and it is replaced vertically and laterally by carbonates and siliciclastics (Zöllner and Amos 1973). The unit ends (Salina Member) with predominant fine siliciclastic levels, and gypsum lenses of minor lateral and vertical extension exceptionally reaching 1 m thick. The gypsum occurs in pure noduls, and as sugar type, in a clastic and gypsiferous matrix (Leanza 2003). The formation is widely exposed in the middle basin of the Neuquén River, the Agrio FTB, to the west and east of the Cordillera del Salado. This formation also records a terminal regressive stage of a new marine sedimentary cycle, associated with a restricted sea connection and, therefore, the occurrence of a hypersaline body in an environment with a negative hydrologic balance (Upper Troncoso Member).

Deposits of the Rayoso Formation (Fig. 12.1; Table 12.1) are composed by very thick to very fine siliciclastics, and gypsum. They are exposed in the Chos Malal and Agrio fold and thrust belts, and along the Neuquén River (Leanza et al. 2001, 2005; Leanza 2003). The uppermost part of its lower section (Quili Malal Member) is characterized by a thin siliciclastic sequence and, at least, seven levels of massive gypsum. Gypsum levels stand out in the outcrops by their relative higher resistance; they are laterally well extended and exceed 1 m thick. Towards the top of the unit (Cañadón de la Zorra Member), the gypsum occurs in very thin banks interbedded with carbonaceous deposits in a new fine siliciclastic sequence (Leanza 2003). These deposits accumulated in the western-central part of the basin. They also constitute the infilling of a siliciclastic cycle, which ends with evaporite in a predominantly continental environment indicating the sea disconnection of the basin to the west. Some authors propose that the origin of the gypsiferous levels of the upper part of the Huitrín Formation and those of the Rayoso Formation would have been deposited in a shallow, perennial and of variable salinity lacustrine body, in which humid and extremely arid periods alternated. Evaporites would be deposited during the contractive system tract of the lake associated with carbonates and salts, alternating with siliciclastics. This lake would have received the phreatic supply of saline-rich water resulting from dissolution of previously deposited levels, corresponding to the lower part of the Huitrín Formation (Upper Troncoso Member) (Zavala and Ponce 2011).

Subsequently to the deposition, the region is affected by successive phases of compressive deformation (Andean Cycle) as a result of transient shallowing of the subduction zone, which definitively disconnects the basin from the marine influence in the west. This changes the relative location of the depocenters and the sediment source area, which was placed to the east of the region. The volcanic arc migrates to the east from the Late Cretaceous to the Paleogene. Positive inversion of previous direct faults, in some cases, and folding and thrusting, in others, are developed constituting the Malargüe, Chos Malal and Agrio fold and thrust belts. In general, these fold and thrust belts are composed of an inner zone, where the deepest levels are involved during the deformation, in two compressive stages that occur in the Late Cretaceous-Eocene and the middle-late Miocene respectively; and an outer zone where, in its eastern part, deep levels are also involved and the exhumation ages correspond to the late Miocene (Sánchez et al. 2018, 2020; Turienzo et al. 2020).

The evaporitic levels (Auquilco and Huitrín formations) had an important role during the deformation, being, among others, main detachment levels, which in some cases duplicate the thickness of the sedimentary units. Thus, the Lower

Cretaceous evaporitic deposits (Huitrín Formation) began to be exhumed between the Late Cretaceous and the Paleocene, between 72 and 51 Ma, in the western sector of the Chos Malal FTB (Cordillera del Viento); from the early to middle Miocene, between 15 and 10 Ma, in its central zone (between the Cordillera del Viento and Tromen Volcano), and from the late Miocene, approximately from 9 Ma, in the eastern part of the belt (to the east of the Tromen Volcano) (Sánchez et al. 2018, 2020). As the Mesozoic units are exposed to the west, the depocenters of the basin progressively migrate towards the east of the deformation front, in a foredeep context of a retroarc foreland basin (Franzese et al. 2003), receiving from here the main supply of the source areas from the west. The subsequent units were developed in these depocenters.

The Allen Formation, and equivalent units (Loncoche Formation) (Fig. 12.1; Table 12.1), contains in its upper part gypsum layers interbedded with limestones, pyroclastics and siliciclastics. The gypsum is massive, reaching 60 cm thick, and nodular and laminated upwards in the section. In some sectors, there is abundant fibrous gypsum and gypsum levels that exceed 5 m thick. In others, the gypsum forms acicular aggregates in 10 m thick clastic layers. The unit crops out in the eastern area of the Bajo de Añelo hydrographic basin, and on the banks of the Neuquén and in Río Negro rivers, below terrace levels, and in the Pellegrini Reservoir and surrounding areas. In the lower Neuquén River basin, some outcrops are beveled by the first level of pediments. In the Colorado River valley, it underlies the youngest terrace levels. Around the Casa de Piedra Reservoir, the outcrops are also truncated by pedimentation surfaces, and the unit ends with a centimetric gypsum level (Rodríguez et al. 2007). This formation was deposited in a sabkha type supratidal, continental to marginal marine environment, recording the first marine ingression of the basin to the east (Atlantic Ocean) (Barrio 1990).

The Roca Formation (Fig. 12.1; Table 12.1) records supralittoral evaporitic accumulations (Legarreta et al. 1989) of sabkha type (Rodríguez 2011). The unit contains gypsum banks, and it is covered by gypsiferous siliciclastic and carbonatic material of varying thickness. Its upper section is not always present, and it is composed of 25–30 cm thick layer with large gypsum crystals in a sugar gypsum matrix, locally interbedded with fine siliciclastic layers. In some sectors, the gypsum reaches 7 m thick (Fig. 12.3). The outcrops that contain gypsum are concentrated in the middle and lower Neuquén River basin (Rodríguez et al. 2007).

The deformation in the inner zone of the Chos Malal FTB, in the Cordillera del Viento area, continues to the middle-late Miocene, beginning the exhumation of the Jurassic evaporitic deposits (Tábanos Formation) in this sector, from 10.7 Ma approximately. These deposits are more exposed to the south, in the Agrio FTB, in the Cordillera del Salado area (Lebinson et al. 2015, 2018). In the Malargüe FTB, in the Cara Cura Range, the Jurassic and Lower Cretaceous deposits have been exhumed during the Late Cretaceous (Gómez et al. 2019; Fennell et al. 2020). In the Agrio FTB, the main detachment level is developed in the Auquilco Formation, and the deformation occurred during the Late Cretaceous, between 100 and 73 Ma, and the Miocene, between 11 and 6 Ma, uplifting the Cordillera del Salado and the Los Chihuidos Range, in its outer part (Zamora Valcarce et al. 2011; Lebinson et al.

Fig. 12.3 Roca Formation. (**a**) Gypsum bank (near General Roca locality). (**b**) Paleochannel infilled with gypsum

Fig. 12.4 Colorado River terraces. (**a**) Haplogypsids with gypsic horizon and marked alluvial character. (**b**) Petrogypsids with petrogypsic pieces (black arrow) that were extracted during the soil pit construction; they also have a marked alluvial character

2020). From the Pliocene to the Quaternary, as a result of the steepening of the sub-ducted oceanic plate, the volcanic arc migrates towards the west and the retroarc volcanism, represented in the area by the Domuyo (from the Plio-Pleistocene), Tromen (from the Pleistocene) and Auca Mahuida (from the late Pliocene) volcanic fields, is developed (Mas et al. 2011; Folguera et al. 2011). The development of these volcanic fields, which constitute an intraplate volcanic belt oblique to the Andean calc-alkaline orogenic belt (Llambías et al. 2010), shows a remarkable influence on the hydrographic evolution of the area (Fig. 12.1).

From the analyses of the geodynamic evolution of the area and the distribution of gypsum outcrops (Fig. 12.1; Table 12.1), it is notable the role of the Auquilco Formation in the Colorado River hydrographic basin, which also contains the thicker exposed gypsum sequence.

In recent deposits, in alluvial plains and modern terraces of the Colorado River, there are soils with disseminated gypsum (Torrifluvents with gypsum) and gypsum accumulation (Haplogypsids and Petrogypsids), both in similar proportions and without layers of calcium carbonate accumulation (Fig. 12.4a, b).

Fig. 12.5 Recent deposits coming from the Roca Formation. (**a**) Sinkholes in Entisols with disseminated gypsum, without irrigation. (**b**) Profile with tunnels and sinking

The soils developed on the Holocene alluvial-colluvial fans and plains, that border the Pleistocene or older local terraces and pediments, contain disseminated gypsum, with dissolution phenomena causing subsidence sinkholes, tunnels, and collapse or sinking episodes (Fig. 12.5a, b). The horizons of gypsum are absent due to the lack of phreatic water at less than 5 m deep.

12.2.2 Gypsum Soils of Pedogenic Origin

Kovda (1968) affirmed that gypsum accumulation occurs in two ways: by evaporation of the mineralized groundwater, and by precipitation within the groundwater itself. Gypsum soils of pedogenic origin, overlying elevated terrace levels, are first recognized and mapped in Syria in 1953. These soils occur in regions with aridic, and in some cases xeric, moisture regimes (Nettleton et al. 1982). Buringh (1960) stated that the gypsum soils in Iraq are mainly associated with bedrock, which contains gypsum and anhydrite interbedded layers in Pleistocene terraces linked to such deposits.

Barzanji (1973) pointed out that the Iraq Basin was partially infilled of clastic sediments during the early Miocene and that the inland seas formed, by evaporation, the gypsum and limestone, which characterize the middle Miocene formations. This last author concluded that these stratigraphic units constitute the origin of most of gypsum soils in Iraq. Segalen and Brion (1981) sustained that the gypsum is originated from eroded and redeposited material. Van Alphen and de los Ríos Romero (1971) mentioned the occurrence of solid or interbedded gypsum deposits from the upper Miocene and Oligocene-Eocene of the Ebro River valley in Spain. In a similar line but under more current methods, Herrero Isern (1991) carried out a detailed study of gypsum soils from the Ebro River basin. A revision about the knowledge state to 2015 of the gypsum soils, with emphasis on micromorphological descriptions of petrogysic horizons, has been elaborated (Casby-Horton et al. 2015).

Accumulation of gypsum and soluble salts can occur in a very short period (Bower et al. 1952). Gypsum and soluble salts precipitate simultaneously if they are in solution, movement is given over a short distance, and evaporation is rapid. Most of the gypsum soils are present in arid and some semiarid areas. The presence of gypsum indicates the humidity characteristics during the deposition, or it is related to brackish phreatic water. The bottom of rivers and lakes are proper sites to form secondary gypsum deposits.

The accumulation of sodium-rich soluble salts inhibits the gypsum precipitation. In the hydrographic basins of the Colorado and Neuquén rivers, topographic depressions (bolsons) and playa lakes have soils with salic horizons and no gypsum accumulation horizons. The absence of gypsum accumulation can be explained due to the effect of the increase in solubility of gypsum in sodium bicarbonate-rich solutions (common ion effect) (FAO 1990; Casby-Horton et al. 2015).

Soils with accumulation of indurated calcium carbonate (petrocalcic) overlying petrogypsic horizon are the most common on the Pleistocene terraces in the hydrographic basin of the Colorado and Neuquén rivers. These soils have not been recognized in the Soil Taxonomy (Soil Survey Staff 2014), where Gypsids gypsum soils are defined as other "Aridisols that have a gypsic or petrogypsic horizon within 100 cm from the soil surface and lack of a petrocalcic horizon overlying any of those horizons". On the other hand, soils with a petrocalcic horizon are defined as Petrocalcids. In the subgroups, soils with petrogypsic horizon are not recognized and, therefore, they are classified at the phase level. Soils with calcic and gypic horizons are classified as Calcigypsids, at the great group level (Fig. 12.6).

The fallowing observations have been made in the hidrographic basins of the Colorado and Neuquén rivers:

1. As the gypsum content increases through the profile soil, the calcium carbonate content tends to decrease and vice versa. There are no soil horizons with abundant gypsum and calcium carbonate.
2. Saline water composition of the Colorado River and of the water table stops the sodic soil formation.
3. Gypsum precipitates and accumulates if the electric conductivity is less than 600 mS cm^{-1}.
4. No gypsum soils are observed in saline basin areas (with salic horizon) affected in the drainage.

Fig. 12.6 (**a**) Typic Petrocalcids, petrogypsic phase. (**b**) Calcic Petrogypsids. (**c**) Calcigypsids

5. The age of the geomorphic surface besides the chemical quality of the surface and subsurface runoff water contribute to the genesis of calcic and gypsic accumulation horizons.
6. The percolating water quality in soils influences in the management of soils with gypsum accumulation horizons. Gypsum dissolution effects are observed in soils irrigated with calcium sulfate-rich water (Colorado River) when overlying calcium carbonate accumulation horizons are absent. This effect is explained by the common ion effect in the percolation solution of soils.

Gypsum accumulation in soils is also related to the age of the terrace deposits. On Holocene terraces of the Colorado River basin, it is recognized Gypsids, soils with gypsum accumulation; Haplogypsids, if the horizon is gypsic; and Petrogypsids, if the horizon is petrogypsic (Fig. 12.7a). On Pleistocene, and older, terraces of the Colorado River basin, and partially of the Neuquén River basin, it is recognized Petrogypsids, well drained soils with indurated gypsum accumulation; Petrocalcids, petrogypsic phase, with soft calcium carbonate and a petrocalcic horizon, overlying a petrogypsic horizon and a salic horizon even lower, which is the soil with the greater differentiation and corresponds to the highest terrace levels (Fig. 12.7b); and Calcigypsids, with soft gypsum accumulation underlying a calcic horizon (Fig. 12.7c).

Irisarri and Ferrer (2012) concluded that gypsum accumulations in soils developed on Holocene terraces of the Colorado River basin, occur in ascendant processes from gypsum-rich solution of the water table. They are present in soft gypsum-rich horizons or in indurated layers at variable depths, depending on textural variations and the history of the water table (Fig. 12.7a). This origin agrees

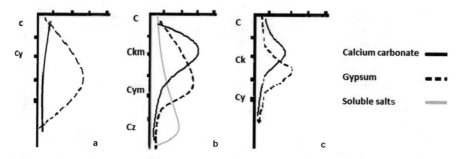

Fig. 12.7 Gypsum accumulation in soils. On Holocene terraces: (**a**) Gypsids: Haplogypsids (with gypsic horizon); Petrogypsids (with petrogypsic horizon). On Pleistocene terraces: (**b**) Petrocalcids (with petrocalcic horizon overlying a petrogypsic horizon and a lower salic horizon). (**c**) Calcigypsids (with petrocalcic horizon overlying a gypsic horizon)

with Boyadgiev (1974), who pointed out that the main reason for gypsum accumulation in soils is its precipitation from groundwater and prolonged runoff, and the result of intense evaporation. On Holocene surfaces, the soils contain soft and/or indurated gypsum accumulations, and the layers observed are thin (<1 m) without the presence of neither soft nor indurated calcium carbonate accumulation horizons.

On Pleistocene terraces, the most common gypsum accumulation forms a very hard, thick (>1 m) layer (petrogypsic horizon), underlying a cemented calcium carbonate layer (petrocalcic horizon) (Fig. 12.7b, c). In other cases, a horizon of carbonate accumulation (calcic horizon) is present, overlying a petrogypsic horizon. Distribution of gypsum in the profile of these ancient terraces results from leaching processes, downward vertical movement of gypsum, which is evidenced by the marks observed on coarse fragments (alluvial gravel). They are placed underlying petrocalcic horizons, and overlying layers of soluble salt accumulations if they are present.

The higher content of sulfate and chloride than bicarbonate in the water from the Colorado River and the water table (Fig. 12.8), avoid the sodium bicarbonate formation and, therefore, the soil sodification is reduced. In Buta Ranquil locality, the medium annual throughflow was 148 m³/s during 2000–2015, the average electrical conductivity was 982 µS/cm, and the total dissolved solid concentration was 615 mg/l, being the conductivity/total dissolved solid ratio 0.63. Among the cations, calcium (48%) and sodium (43%) dominated in relation to magnesium (8%) and potassium (1%). Among the anions, the chloride (44%) and sulfate dominated (40%) and the bicarbonate was the less abundant (16%) (COIRCO-UNS 2017).

In Paso de los Indios locality (upstream of Los Barreales Reservoir), the Neuquén River has medium annual throughflow of 310 m³/s, and an average electrical conductivity of 270 µS/cm, which has a sodium bicarbonate composition that enables the formation of sodic soils and avoid the presence of sulfate soils. These absence of gypsum-rich horizons occurs due to this ion is a minority in relation to chlorides and carbonates-bicarbonates along the course of the river (Fig. 12.8). Although outcrops of the Huitrín and Rayoso formations are important in the Neuquén River

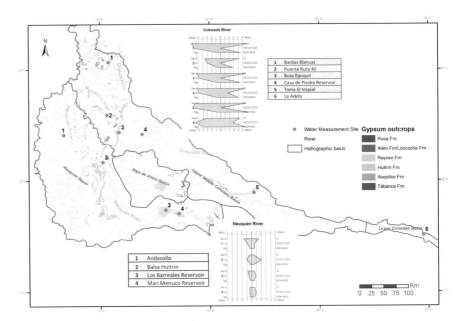

Fig. 12.8 Water chemical composition of the Neuquén and Colorado rivers. (Based on INTA 1990)

basin, the gypsum availability is lower than in the Colorado River, considering the thickness of gypsum layers (Table 12.1). The latter is important in the Auquilco Formation, which is well exposed in the Colorado River basin. Therefore, the washing in the Neuquén River basin by surface runoff produces a lower gypsum concentration than in the Colorado River basin, where the outcrops of Auquilco Formation present thick banks of purer gypsum.

In soils with Holocene gypsum accumulations, the horizons affected by soluble salts are placed above the gypsic horizon.

12.3   Processes Affecting the Soil. Soil Uses

All gypsum soils in zones under irrigation, in the studied area, have low and very low organic matter and total nitrogen content, probably linked to sulfate content of the irrigation water, tilling for weed control, passive defense to frost and to the long irrigation intervals that have been observed. Among the crops, the best adapted species to the gypsum presence in soils and in the irrigation water from the Colorado River, under temperate climate, are onion (*Allium cepa*), garlic (*Allium sativa*), alfalfa (*Medicago sativa*), and grapevine (*Vitis vinifera*). These crops allowed the agricultural development in the Colorado River valley, while in Neuquén and Río Negro river valleys fruit growing is, in addition, developed as the main productive activity.

Fig. 12.9 (a) Sinking of fruit plant due to dissolution of the petrogypsic horizon for irrigation. (b) Sinkhole in alluvial soils on Holocene terraces with petrogypsic horizon due to furrow irrigation method

In restricted sectors of irrigated areas, in the Colorado River basin, suffusion processes, sinkholes, tunnels and collapse or sinking phenomena (Fig. 12.9a) due to dissolution of the superficial gypsum layers, which have affected the soil surface and furrow and/or gutter parts (Fig. 12.9b), are identified. Vertical collapses and fruit tree falls are also recorded. Sinking phenomena of the land are related to unevenness due to gypsum dissolution that forces the excessive use of irrigation water, feeding back the problem. According to the Riverside Standards, the samples correspond to the C3 S1 category, classifying as usable water for irrigation with caution.

Under natural conditions, this phenomenon is also observed in fine soils forming undermining, sinking and gullies as the result of the water erosion by surface and subsurface runoff (Fig. 12.5a, b). This erosive dynamic interferes with productive activities, mainly grazing cattle and hydrocarbon industry, in the middle Colorado River basin.

12.4   Conclusions

In relation to the genesis of gypsum soils, it is concluded that the gypsum soils generally develop in arid and semiarid climate zones, with less than 400 mm of annual precipitation (aridic and xeric moisture regimes), mean temperatures (mesic and thermic temperature regimes), and where gypsum source (gypsum-rich bedrock) is abundant. The time factor has an important role in the differentiation of soil horizons (soils with petrocalcic horizon overlying a petrogypsic horizon, on Pleistocene or older surfaces). Future studies on gypsum accumulation horizons in soils should consider the age of the surface, the quality and direction of the percolating and capillary water movement, including the common ion effect in calcium carbonate and calcium sulfate solubilities. This knowledge will contribute to resolve frequent management problems in gypsum soils under irrigation.

Gypsum outcrop analysis in the studied hydrographic basin allows a better understanding of the distribution of the gypsum availability. Gypsum stratigraphic units, which constitute the gypsum source for the region, have been exposed by the development of dynamic fold and thrust belts and their related depocenters, from the Late Cretaceous and Miocene Andean deformation phases. Gypsum levels are better preserved in the core of the synclines of the Chos Malal and Agrio fold and thrust belts due to their relative higher resistance to the erosion compare with fine siliciclastic deposits. Currently, the main stratigraphic unit regarding to its thickness exposure and purer gypsum content, is the Auquilco Formation, which has been exposed in the western part of the region from the Cretaceous time (100–72 Ma ago), and it is broadly exposed in the upper Colorado River basin.

Subsequent units, as Huitrín and Rayoso formations, have been previously exposed, within the same deformation stages, according to their timing of deposition. Although they are widely exposed, mainly in the Neuquén River basin, their gypsum content is much lower. The development of the Domuyo, Tromen and Auca Mahuida volcanic fields from the Pliocene, in an aligned placement, shows a remarkable influence on the hydrographic evolution of the area.

The Colorado River presents the highest dissolved salt content with a predominance of sulfates from the upper basin, provided by affluents due to the runoff of areas with gypsum-rich outcrops. This source of salts, active since the initial phase of the basin, would allow distribute sulfates in the Pleistocene alluvial sediments present in remain ancient terraces. This dynamic currently continues, defining water chemistry of the river and associated free aquifers, and the soil evolution of the Colorado River valley. In the Neuquén River basin, the relative lower gypsum availability in the outcrops, results in lower salinity and sulfate content in the water, producing deposits restricted to the Pleistocene surfaces. This sulfate absence allows the existence of saline sodic soils in the current alluvial plain of the river.

The maximum pedogenetic gypsum accumulation observed in the Pleistocene and older terraces (Irisarri and Ferrer 2012), it is confirmed. The vertical distribution of the profile of the most developed, well-drained soil is composed by an upper petrocalcic horizon, an underlying petrogypsic horizon, and a soluble salts-rich horizon in deep.

In the alluvial plain of the Colorado River, a water table less than 5 m deep with mineralized water containing sulfate and chloride produces gypsum accumulation in the soils. In Holocene and Pleistocene terraces, the process of the initial accumulation could have been similar. Gypsum soils have limitations related to the total annual applicable irrigation sheet due to their collapse risk.

The technology application of the water and the control or monitoring of the moisture front in the profile is essential for the sustainable soil productive management. These soils are only irrigable with high-frequency and whole cover irrigation technology, due to the low water-holding capacity of gypsum layers and the need to control the application sheet to avoid dissolution and soil sinking, as it is evidenced by the regional experience.

References

Arregui CD, Carbone OC, Sattler F (2011) El Grupo Lotena (Jurásico Medio-Tardío) en la Cuenca Neuquina. In: Leanza HA, Arregui C, Carbone O, Danieli J, Vallés J (eds) Geología y Recursos Naturales de la Provincia del Neuquén, 18° Congreso Geológico Argentino. Neuquén, Relatorio, pp 91–98

Barrio CA (1990) Paleogeographic control of Upper Cretaceous tidal deposits, Neuquén Basin, Argentina. J S Am Earth Sci 3(1):31–49

Barzanji AF (1973) Gypsiferous soils of Iraq. Ph.D. Thesis, University of Ghent, Belgium

Bower CA, Reitemeier RF, Fireman M (1952) Exchangeable cation analysis of saline and alkali soils. Soil Sci 73:251–261

Boyadgiev TG (1974) Contribution to the knowledge of gypsiferous soils. AGON/SYR/67/522. FAO, Rome

Bracaccini O (1970) Rasgos tectónicos de las acumulaciones mesozoicas en las provincias de Mendoza y Neuquén, República Argentina. Rev Asoc Geol Argent 25(2):275–284

Brodtkorb MK, de Barrio RE, del Blanco M, Etcheverry RO (1999) Geología de los depósitos de baritina, celestina, yeso y halita de la Cuenca Neuquina, Neuquén y Mendoza. In: Zappettini EO (ed) Recursos Minerales de la República Argentina, vol 35. Instituto de Geología y Recursos Minerales SEGEMAR, Anales, pp 1041–1046

Buringh P (1960) Soils and soil conditions in Iraq. Ministry of Agriculture, Iraq

Casby-Horton S, Herrero J, Rolong NA (2015) Gypsum soils their morphology, classification, function, and landscapes. In: Sparks DL (ed) Advances in agronomy, vol 130, pp 231–290

COIRCO-UNS (2017) Análisis Estadístico de los Parámetros Fisicoquímicos Metales y Metaloides Estación piloto "Buta Ranquil". COIRCO – Universidad Nacional del Sur

Dellapé DA, Mombrú C, Pando GA, Riccardi AC, Uliana MA, Westermann GE (1979) Edad y correlación de la Formación Tábanos en Chacay Melehue y otras localidades de Neuquén y Mendoza, con consideraciones sobre la distribución y significado de las sedimentitas Lotenianas. Obra Centenario Museo La Plata 5:81–105

Fennell L, Borghi P, Martos F, Rosselot EA, Naipauer M, Folguera A (2020) The late cretaceous Orogenic system: early inversion of the Neuquén Basin and associated Synorogenic deposits (35°–38° S). In: Kietzmann D, Folguera A (eds) Opening and closure of the Neuquén Basin in the southern Andes. Springer Earth System Sciences, pp 303–322

Ferrer J, Irisarri J, Mendia JM (2006) Suelos de la Provincia del Neuquén. Ediciones INTA CFI

Folguera A, Spagnuolo M, Rojas Vera E, Litvak V, Orts D, Ramos VA (2011) Magmatismo neógeno y cuaternario. In: Leanza HA, Arregui C, Carbone O, Danieli J, Vallés J (eds) Geología y Recursos Naturales de la Provincia del Neuquén, 18° Congreso Geológico Argentino. Relatorio, Neuquén, pp 275–286

Food and Agriculture Organization (FAO) of the United Nations (1990) Management of Gypsiferous Soils. FAO Soils, Rome, Italy. Bulletin 62

Franzese JR, Spalletti LA, Gómez Pérez I, Macdonald D (2003) Tectonic and paleoenvironmental evolution of Mesozoic sedimentary basins along the Andes foothills of Argentina (32°–54° S). J S Am Earth Sci 16:81–90

Gómez Figueroa J, Monardez C, Balod M (2011) El Miembro Troncoso Superior de la Formación Huitrín (Cretácico Temprano). In: Leanza HA, Arregui C, Carbone O, Danieli J, Vallés J (eds) Geología y Recursos Naturales de la Provincia del Neuquén, 18° Congreso Geológico Argentino. Relatorio, Neuquén, pp 189–198

Gómez R, Lothari L, Tunik M, Casadío S (2019) Onset of foreland basin deposition in the Neuquén Basin (34°-35°S): new data from sedimentary petrology and U–Pb dating of detrital zircons from the Upper Cretaceous non-marine deposits. J S Am Earth Sci 95:102257

Groeber P (1946) Observaciones geológicas a lo largo del meridiano 70° 1. Hoja Chos Malal Revista de la Sociedad Geológica Argentina 1(3):177–208

Groeber P (1953) Andico. In: Groeber P (ed) Mesozoico. Geografía de la República Argentina. Sociedad Argentina de Estudios Geográficos (GAEA), 2(1):349–536

Herrero Ducloux A (1946) Contribución al conocimiento geológico del Neuquén extrandino. Boletín de Informaciones Petroleras 23(226):1–39

Herrero Isern J (1991) Morfología y génesis de suelos sobre yeso. Servicio de investigación Agraria Diputación General de Aragón Apdo. 727.50080, Zaragoza

Howell J, Schwarz E, Spalletti L, Veiga G (2005) The Neuquén Basin: an overview. In: Veiga G, Spalletti L, Howell J, Schwarz E (eds) The Neuquén Basin: a case study in sequence stratigraphy and basin dynamics, vol 252. Geological Society of London, Special Publication, pp 1–14

Ihering HV (1903) Les mollusques des terrains crétaciques superieurs de l'Argentine Orientale. Anales del Museo Nacional de Buenos Aires. Serie 3(2):193–228

INTA (1990) Atlas digital de la República Argentina

Irisarri J, Ferrer J (2012) Suelos con acumulaciones de yeso en la cuenca del río Colorado, Argentina. In: INTA (ed) Suelos con acumulaciones calcáreas y yesíferas de Argentina, vol 7, pp 183–196

Kovda VA (1968) Geography and classification of soils of Asia; Transl. from Russian by A. Gourevitch. Israel Program for Scientific Translations: 226 pp.

Leanza HA (2003) Las sedimentitas huitrinianas y rayosianas (Cretácico Inferior) en el ámbito central y meridional de la Cuenca Neuquina, vol 2. SEGEMAR, Serie Contribuciones Técnicas-Geología, Argentina, pp 1–31

Leanza HA, Hugo CA, Repol D (2001) Hoja Geológica 3969-I, Zapala. Provincia del Neuquén, vol 275. Instituto de Geología y Recursos Minerales, Servicio Geológico Minero, Argentino, pp 1–128

Leanza HA, Llambías EJ, Carbone O (2005) Unidades limitadas por discordancias en los depocentros de la Cordillera del Viento y la Sierra de Chacaicó durante los inicios de la Cuenca Neuquina. 6° Congreso de Exploración y Desarrollo de Hidrocarburos, CD-ROM, Mar del Plata

Lebinson F, Turienzo M, Sánchez NP, Araujo V, Dimieri L (2015) Geometría y cinemática de las estructuras tectónicas en el extremo septentrional de la faja corrida y plegada del Agrio, Cuenca Neuquina. Rev Asoc Geol Argent 72:299–313

Lebinson F, Turienzo M, Sánchez NP, Araujo V, D'Annunzio MC, Dimieri L (2018) The structure of the northern Agrio fold and thrust belt (37°30'S), Neuquén Basin, Argentina. Andean Geol 45(2):249–273

Lebinson F, Turienzo M, Sánchez NP, Cristallini E, Araujo V, Dimieri L (2020) Kinematics of a backthrust system in the Agrio fold and thrust belt, Argentina: insights from structural analysis and analogue models. J S Am Earth Sci 100:102594

Legarreta L (1991) Evolution of a Callovian-Oxfordian carbonate margin in the Neuquén basin of west-Central Argentina: facies, architecture, depositional sequences and global sea level changes. Sediment Geol 70:209–240

Legarreta L, Uliana MA (1996) The Jurassic succession in west-Central Argentina: stratal patterns, sequences and paleogeographic evolution. Palaeogeogr Palaeoclimatol Palaeoecol 120:303–330

Legarreta L, Uliana MA (1999) El Jurásico y Cretácico de la Cordillera Principal y la Cuenca Neuquina. Geología Argentina, vol 29. Instituto de Geología y Recursos Minerales. SEGEMAR, Anales, pp 399–432

Legarreta L, Kokogián DA, Bogetti DA (1989) Depositional sequences of the Malargüe group (upper cretaceous – lower tertiary), Neuquén Basin, Argentina. Cretac Res 10:337–356

Legarreta L, Gulisano CA, Uliana MA (1993) Las secuencias sedimentarias jurásico-cretácicas. In: Ramos V A (ed), Geología y Recursos Naturales de Mendoza. 12° Congreso Geológico Argentino. Relatorio 1(9):87–114

Llambías EJ, Bertotto GW, Risso C, Hernando I (2010) El volcanismo cuaternario en el retroarco de Payenia: una revisión. Rev Asoc Geol Argent 67(2):278–300

Mas LC, Mas GR, Bengochea L, López N (2011) Actividad eruptiva en la región del Volcán Domuyo. In: Leanza HA, Arregui C, Carbone O, Danieli J, Vallés J (eds) Geología y Recursos Naturales de la Provincia del Neuquén, 18° Congreso Geológico Argentino. Relatorio, Neuquén, pp 609–612

Narciso V, Santamaría G, Zanettini J (2001) Hoja Geológica 3769-I, Barrancas. Provincias de Mendoza y Neuquén, vol 253. Instituto de Geología y Recursos Minerales, Servicio Geológico Minero, Argentino, pp 1–49

Nettleton WD, Nelson RE, Brasher BR, Derr PS (1982) Gypsiferous soils in the Western United States. In: Kittrick JA, Fanning DS, Hossner LR (eds) Acid sulfate weathering. Soil Science Society of America, Madison., Special Publication 10:147–168

Nullo FE, Stephens G, Combina A, Dimieri L, Baldauf P, Bouza P, Zanettini JCM (2005) Hoja Geológica 3569-III/3572IV, Malargüe, provincia de Mendoza. Instituto de Geología y Recursos Minerales, Servicio Geológico Minero Argentino 346:1–85

Ramos VA, Folguera A, García Morabito E (2011) Las provincias geológicas del Neuquén. In: Leanza HA, Arregui C, Carbone O, Danieli J, Vallés J (eds) Geología y Recursos Naturales de la Provincia del Neuquén, 18° Congreso Geológico Argentino. Relatorio, Neuquén, pp 317–326

Riccardi AC, Westermann GEG, Elmi S (1991) Biostratigraphy of the upper Bajocian-middle Callovian (Middle Jurassic), South America. J S Am Earth Sci 4(3):149–157

Rodríguez MF (2011) El Grupo Malargüe (Cretácico Tardío-Paleógeno temprano) en la Cuenca Neuquina. In: Leanza HA, Arregui C, Carbone O, Danieli J, Vallés J (eds) Geología y Recursos Naturales de la Provincia del Neuquén, 18° Congreso Geológico Argentino. Relatorio, Neuquén, pp 245–264

Rodríguez MF, Leanza HA, Salvarredy Aranguren M (2007) Hoja Geológica 3969-II, Neuquén, provincias del Neuquén, Río Negro y La Pampa. Instituto de Geología y Recursos Minerales. SEGEMAR 370:1–172

Roll A (1939) La Cuenca de los Estratos con Dinosaurios al sur del río Neuquén. Yacimientos Petrolíferos Fiscales (unpub, Buenos Aires

Sánchez NP, Coutand I, Turienzo M, Lebinson F, Araujo V, Dimieri L (2018) Tectonic evolution of the Chos Malal fold–and–thrust belt (Neuquén Basin, Argentina) from (U–Th)/He and fission–track thermochronometry. Tectonics 37:1907–1929

Sánchez NP, Turienzo M, Coutand I, Lebinson F, Araujo V, Dimieri L (2020) Structural and thermochronological constraints on the exhumation of the Chos Malal fold and thrust belt (~37° S). In: Kietzmann D, Folguera A (eds) Opening and closure of the Neuquén Basin in the Southern Andes. Springer Earth System Sciences, pp 323–340

Schiller W (1912) La alta cordillera de San Juan y Mendoza y parte. Anales Ministerio de Agricultura Sección Geología y Mineralogía 7(5):1–68

Segalen P, Brion JC (1981) Pédochimie: quatrième partie. T.1.ORSTOM, Paris

Soil Survey Staff (1999) Soil taxonomy: a basic system of soil classification for making and interpreting soil surveys. In: Agriculture handbook 436, 2nd edn. USDA-Natural Resources Conservation Service, Washington, DC

Soil Survey Staff (2006) Keys to soil taxonomy, 10th edn. USDA-Natural Resources Conservation Service, Washington, DC

Soil Survey Staff (2014) Keys to soil taxonomy, 12th edn. USDA-Natural Resources Conservation Service, Washington, DC

Stipanicic PN (1966) El Jurásico en Vega de La Veranada (Neuquén), el Oxfordense y el diastrofismo Divesiano (Agassiz-Yaila) en Argentina. Revista Asociación Geológica Argentina 20(4):403–478

Turienzo MM, Sánchez N, Lebinson F, Peralta F, Araujo VS, Irastorza A, Dimieri LV (2020) Basement-cover interaction in the mountain front of the northern Neuquén fold and thrust belt (37°10′–37°40′ S), Argentina. J S Am Earth Sci 100:102560

Uliana MA, Dellapé DA, Pando GA (1975a) Distribución y génesis de las sedimentitas rayosianas. 2° Congreso Iberoamericano de Geología Económica. Actas 1:151–176

Uliana MA, Dellapé DA, Pando GA (1975b) Estratigrafía de las sedimentitas rayosianas. 2° Congreso Iberoamericano de Geología Económica. Actas 1:77–196

Uliana MA, Dellapé DA (1981) Estratigrafía y evolución paleoambiental de la sucesión maestrichtiano – Eoterciaria del Engolfamiento Neuquino (Patagonia septentrional). 7° Cong Geol Arg 3:673–711

Van Alphen JG, de los Rios Romero F (1971) Gypsiferous soils – Notes on their characteristics and management. International Institute for Land Reclamation and Improvement, Wageningen. Bulletin 12

Weaver C (1927) The Roca formation in Argentina. Am J Sci 13(5):417–434

Weaver Ch (1931) Paleontology of the Jurassic and Cretaceous of West Central Argentina. University of Washington. Memoir 1:1–469

Yrigoyen MR (1991) Hydrocarbon resources from Argentina. Petrotecnia, Special Issue 13:38–54

Zamora Valcarce G, Zapata T, Ramos VA (2011) La faja plegada y corrida del Agrio. In: Leanza HA, Arregui C, Carbone O, Danieli J, Vallés J (eds) Geología y Recursos Naturales de la Provincia del Neuquén, 18° Congreso Geológico Argentino. Neuquén, Relatorio, pp 367–374

Zavala C, Ponce JJ (2011) La Formación Rayoso (Cretácico Temprano) en la Cuenca Neuquina. In: Leanza HA, Arregui C, Carbone O, Danieli J, Vallés J (eds) Geología y Recursos Naturales de la Provincia del Neuquén, 18° Congreso Geológico Argentino. Neuquén, Relatorio, pp 205–222

Zöllner W, Amos AJ (1973) Descripción geológica de la Hoja 32b, Chos Malal, provincia del Neuquén. Servicio Nacional Minero Geológico 143:1–91

Chapter 13
Use of Soil Maps to Interpret Soil-Landform Assemblages and Soil-Landscape Evolution

B. A. Miller, C. J. Baish, and R. J. Schaetzl

Abstract Soils are a key link to the surficial sedimentologic system(s) that originally deposited the unconsolidated parent material, or to the weathering system that formed the residual parent material. In young soils, i.e., those formed since the end of the Pleistocene, parent materials can often be identified as to type, enabling accurate links to their past depositional system(s). On older landscapes, determining the origin(s) of the soil parent material is more challenging. Accurately determining the origin of a soil parent material is also important because soil properties – tied to the parent material – are frequently part of mapping unit attributes for regional and local scale maps. Examining soil maps in a geographic information system (GIS) helps to broadly associate soil types with parent materials, enabling the user to create maps of surficial geology. We suggest that maps of this kind have a wide variety of applications in the Earth Sciences, and to that end, provide six examples from temperate climate soil-landscapes.

Keywords GIS · Soil maps · Soil parent materials · Soil geomorphology · Soil landscapes · Lithologic discontinuities

13.1 Introduction

Soils form from (and in) unconsolidated parent materials. Parent material is one of the five main soil-forming factors (Dokuchaev 1883/1967), and thus, pre-conditions soil development and the pedogenic system from the inception of pedogenesis. That

B. A. Miller (✉)
Department of Agronomy, Iowa State University, Ames, IA, USA
e-mail: millerba@iastate.edu

C. J. Baish · R. J. Schaetzl
Department of Geography, Michigan State University, East Lansing, MI, USA

243

is, the inherent physical and chemical characteristics of the parent material govern the trajectory and outcomes of soil formation. For example, soils forming in dune sand will never be clayey and are likely to always be highly permeable. Similarly, soils forming in lacustrine clays will never be sandy. Glacial till parent materials are lacking in areas that have never been glaciated, and marine clays do not exist in interior, continental locations. By extension, proper interpretation of soils, as they exist today, can provide key links between them, the soil landscape, and the geologic or geomorphic processes that emplaced the soil parent material (Ehrlich et al. 1955; Gile 1975; Schaetzl 1998). That is, soils can provide key information about past sedimentologic or geologic processes and systems, by virtue of their parent materials, e.g., Schaetzl et al. (2000).

Some parent materials overlie a previously formed soil, i.e., a buried paleosol (Follmer 1982; Schaetzl and Sorenson 1987; Krasilnikov and Calderon 2006). In some instances, the remnant soil may be buried with its original profile intact, whereas in others it may have been truncated or removed by erosional processes (Nettleton et al. 2000). If pedogenesis extends through the interface of the two parent materials, "welding" of the surface soil can occur to either the paleosol below (Ruhe and Olson 1980; Olson and Nettleton 1998) or to the lower parent material. Instances of vertically "stacked" parent materials like this can complicate both parent material interpretations as well as pedogenesis in the surface soil (Wilson et al. 2010). But they also provide important indications of changes in past depositional and erosional systems. We provide these examples only to note that, in this chapter, we will focus on the more common and straightforward situations, in which soils form in one, fresh, parent material. These kinds of soils provide good opportunities for establishing the linkages between soil type and parent material type, as well as the processes that emplaced it.

Such examples abound. Many landscapes, especially those that have recently undergone glaciation, are rich in parent materials that were relatively unaltered and "fresh" at the time that pedogenesis began. Examples of such parent materials include dune sand, glacial till, volcanic ash, and flood deposits. In most cases, this material is easily identified by excavating deeply, i.e., below the solum and into the C horizon. All of the materials mentioned above are unconsolidated, porous, and permeable. Hence, pedogenesis, largely driven by percolating water, can operate freely in such materials, and can begin immediately after time zero. Thus, a clear and often indisputable link can be made between the soil and some form of past geologic/sedimentologic process.

Although much can also be gained from the careful interpretation of soil parent materials on old, stable sites in the continental interior (Brown et al. 2003; Eze and Meadows 2014), most applications involving soil parent materials are found in younger landscapes. Young soils, e.g., Entisols and Inceptisols, resemble their parent materials fairly closely, because pedogenesis has had little time or energy to alter these materials. In these and other soils that are minimally weathered, soil parent materials can often be readily identified as to type. In older soils, however, especially highly weathered Oxisols and Ultisols, determining the type of parent material present at $time_{zero}$ can be more difficult, mainly because many of the primary minerals in such soils have been altered or destroyed by weathering. Also, erosion may have removed some of the material or brought in other materials from upslope or from upwind. Textures may have been changed by pedogenesis. Coarse fragments may have weathered away.

With this introduction in mind, we observe that the study of soil parent materials in the context of geopedology has much to offer the pedology, geoscience, geomorphology, and even the landscape ecology communities. Our focus will be on providing examples of studies or situations where careful examination of parent material type and distribution can provide important information about the geomorphic attributes and history of the landscape. Utilizing parent material information aligns with the family level of geoform classification described in Table 6.4 of Chap. 6.

We also provide one important caveat: many soils have formed in "stacked" parent materials, in which a thin layer of one parent material lies immediately atop a distinct but different parent material. The two parent materials are, by definition, separated by a lithologic discontinuity (Schaetzl 1998). Although this situation sometimes makes parent material interpretations more difficult, it also often provides even greater opportunities for paleoenvironmental interpretation, because such soils can enlighten us about a depositional process or system (the lower material) that then changed to another type of system (the upper material), i.e., twice the amount of information is potentially available. Examples follow in the text below.

13.2 Methodological Approach

The approach we present can be operationalized with a soil map and a digital elevation model (DEM). Both must be in digital form, so they can be manipulated in a geographic information system (GIS). Soil maps focus on surficial materials and are usually more spatially detailed than available geologic maps, due to investments in agricultural development and land valuation. Normally, we overlay the soil information on a hillshaded DEM product, so that the soil information can also be matched to topography. For sites in the USA, digital soils data produced by the National Cooperative Soil Survey (NCSS) are hosted on the Natural Resources Conservation Service (NRCS) web site via the Geospatial Data Gateway (http://datagateway.nrcs.usda.gov). Furthermore, DEM products from the United States Geological Survey's (USGS) National Geospatial Program are available via The National Map (TNM) download client (http://apps.nationalmap.gov). Downloadable files from these sites can be added to a GIS.

A key additional step is incorporating supplementary information into the GIS file, such as parent material types, thicknesses, and if the soil profile includes multiple, stacked parent materials Because soils are spatially heterogeneous and thus differ in the size and shape of their areas on the landscape, soil surveys often delineate map unit boundaries according to groupings of the dominant soil component (e.g., soil series). It is important to note that our approach centers on incorporating supplementary information from the representative (dominant) soil series occurring within each map unit. We normally code as many of the dominant soil series as possible to parent material by using a two-step process. First, for each soil series we look up its official description on the NRCS web site (https://soilseries.sc.egov. usda.gov/osdname.aspx) if we do not already know it. From the official series description, we note the parent material and code it into the GIS as one of several parent material classes, e.g., till, outwash, glaciofluvial sediment, loess, lacustrine sediment, dune sand, residuum, and a few other, minor categories (Miller et al. 2008). For soils with a mantle of thin loess and an underlying sediment – both listed as the parent material – the loess thickness and the type of underlying sediment can also be noted in separate fields. Especially for till, the parent material can frequently be associated with a geologic formation and that information included as well.

NCSS soil maps in the USA are very detailed, often having been produced at a scale of 1:15,840, resulting in maps that regularly subdivide parent material areas by changes in other soil forming factors. Therefore, interpreting these detailed maps for parent material generally results in aggregation of soil map units. Although the relationship between soils and their parent materials is ubiquitous, the scale and purpose of the soil map could potentially deemphasize the parent material-related information available in the map.

The approach described here enables the user to display maps of parent material (and possibly loess thickness) in a GIS, with the data matched nicely to topography. We have also added additional fields to the GIS attribute table, centering around soil texture, e.g., texture of the surface mineral (usually A) horizon, as well as its parent material (lowest horizon) texture. We also note when the texture modifier on the lowest horizon contained the words "gravelly," "cobbly," or "stony," allowing us to compile a data layer for soils that contain significant amounts of coarse fragments in their parent materials. The result is a digital map of surficial geology attributes with greater spatial detail and coverage than is typically available from other sources.

13.3   Example 1: A Detailed Surficial Geology Map of Iowa, USA

This example illustrates how the methodological approach described above can convert soil survey information into a format customized for investigations of soil-landform assemblages and soil-landscape evolution for large areas.

In Iowa, surficial geology maps with a high level of detail are only available for a fraction of the state. In contrast, detailed soil maps are available statewide. Although the relationship between those producing the respective maps is strong and information is freely shared between the two groups (geologists and soil scientists), differences in disciplinary practices have left a gap in available map products. Notably, geologists here often use NCSS soil maps as base maps, but verify and enhance the information with consideration of deeper bore holes and interpretation for more specifics, e.g., age, about the respective geologic formation, stratigraphy, etc. These investigations require additional time and resources, which helps to explain the limited coverage of the surficial geologic maps produced in this way. Benefiting from the investment in land use and management information over the past century, detailed soil maps fully cover the state. However, they focus on the top two meters, and only include a brief attribution of the parent material to the geology, as understood at the time of map production.

Using the methodological approach we described, Miller and Burras (2015) constructed a relational database for each of the soil series mapped in Iowa. Although the NCSS soil database does contain a parent material attribute field, the labels are usually generic and frequently missing. More useful information can be found in the geomorphic setting of the official soil series descriptions. However, even the official soil series descriptions often do not directly link the soil series to the recognized geologic formation and geomorphic landform; some interpretations are required, based usually on local knowledge. For example, the Clarion soil is described as having formed in calcareous till parent material and occurring on convex slopes of gently undulating to rolling Late Wisconsin till plain. Its parent material has loam to clay loam textures. These characteristics, combined with the geographic extent of the soil series, clearly match what geologists would recognize as the Pleistocene Dows Formation. Additional geomorphic information can also be gained from soils mapped in the Clarion-Nicollet-Webster soil association. The Webster soil is generally mapped in swales below Clarion delineations and is described as being formed in glacial till or local alluvium derived from till. Thus, the spatial juxtaposition of these two and similar soils indicates the pattern of hillslope erosion and basin fill processes along with landform structure (Fig. 13.1b).

Creating this translation between information recorded in the soil survey to terms useful for geomorphic purposes requires knowledge of the local geology and sometimes careful consideration of context. Nonetheless, by evaluating 863 soil series across Iowa, Miller and Burras (2015) leveraged the soil maps to create a detailed surficial geology map efficiently and accurately for Iowa, spanning 145,700 km² (Fig. 13.1a). Although the resulting map does not contain as much attribute information as the maps produced by geologists, 67%–99% of the pixels in it agree, and the map provides considerably more spatial information and detail than existing geology maps. This level of information is often vital to environmental and geomorphic research.

13.4 Example 2: The Loess-Covered Landscapes of Western Wisconsin, USA

This example illustrates how detailed soil surveys can help determine loess thicknesses across a landscape. Loess covers most upland sites in western Wisconsin (Hole 1976). Most of this loess originated from the Mississippi River, which was a major conduit for silt-rich glacial meltwater, and which forms the western boundary

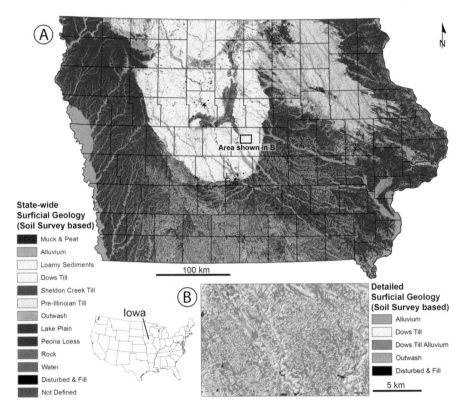

Fig. 13.1 Surficial geology maps for Iowa, USA, based on digital Soil Survey maps and interpretation of official soil series descriptions. (After Miller and Burras 2015). (**a**) Although the same soil series in different counties are technically different soil map units, they are still constrained by definitions set in the official series description. This allows for several county-scale maps to be efficiently translated to desired attribute classes. (**b**). The attribute scale can be customized by the user to include as much, or as little detail as needed for the map's purpose. At this larger cartographic scale, it is useful to distinguish soils formed in till of the Dows Formation versus soils formed in the slopewash alluvium derived from that till. Patterns of parent material at this scale are complimented by the elevation hillshade and make landscape structure more visible

of the state (Scull and Schaetzl 2011). In most cases, the loess overlies bedrock residuum, as this part of the state has never been glaciated.

Here, map units in county-scale soil maps are described with a typical loess thickness. Thus, the soil maps can provide detailed information about loess thickness and distribution (Fig. 13.2; Schaetzl et al. 2018). Some soil series here are formed in loess that is thicker than the typical 150 cm (60 inches) profile description, and in these cases, loess thicknesses provided by the soil maps represent only a minimum value. Most soils, however, are formed in <150 cm (60 inches) of loess over another parent material, e.g., residuum, bedrock, colluvium or alluvium. For example, the official description for the Dubuque series states that it, "consists of moderately deep, well drained soils formed in 46–91 cm (18–36 inches) of loess and a thin layer of residuum from limestone bedrock or reddish paleosol…" Another

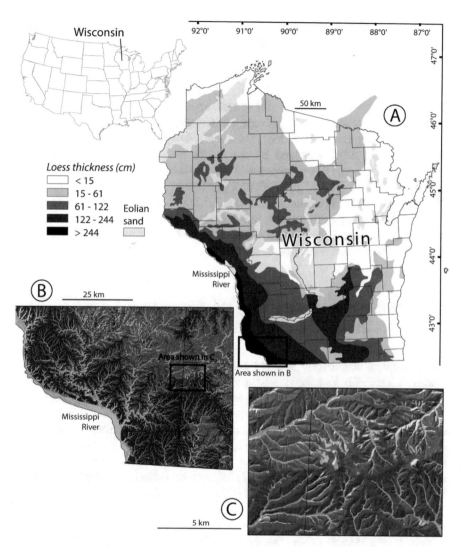

Fig. 13.2 Distribution and thickness of loess and eolian sand across Wisconsin, USA. The loess thickness color legend is similar for all three maps. (**a**) Regional loess thickness. (After Hole 1950/1968; Thorp and Smith 1952). (**b**) Loess thicknesses across southwestern Wisconsin, as determined in a GIS by using soil series descriptions, and coding each to a category of loess thickness. (**c**) A more detailed map of loess thickness created using similar methods but shown at a larger cartographic scale

common soil in the area, Norden, is "formed in loess and in the underlying loamy residuum weathered from glauconitic sandstone." Note that the parent material description for Norden soils does not include loess thickness. In this case, one must examine the official profile description to determine the typical loess thickness. Norden soils have the following typical horizonation: Ap 0–20 cm, Bt1 20–28 cm, Bt2 28–51 cm, 2Bt3 51–64 cm, 2Bt4 64–84 cm, 2Bt5 84–94 cm, and 2Cr 94–150 cm. All horizons above the lithologic discontinuity (at 51 cm) are silt loam in texture, as is typical for loess. Thus, where mapped, Norden soils can be assumed to have formed in approximately 51 cm of loess.

This type of procedure can be adopted for all soils in the region, and after the loess thicknesses have been entered into the GIS attribute table by map unit, detailed maps of loess thickness can be readily created. Although map units contain information about component soil series, for spatial representation each map unit can have only one value per attribute. In this case, all polygons associated with the same map unit will be assigned the same loess thickness. However, for regional analyses the level of spatial detail provided by the NCSS maps are more than sufficient. Figure 13.2 illustrates this approach at a variety of scales. This approach has been successful in a number of loess studies performed in the upper Midwest, USA (Jacobs et al. 2011; Luehmann et al. 2013; Schaetzl and Attig 2013; Schaetzl et al. 2014, 2018). Such data are extremely valuable for determining the source areas for loess, which is usually thickest near its source. These types of maps are also useful for guiding land management decisions.

13.5   Example 3: The Recently Deglaciated Landscape of Northeastern Lower Michigan, USA

This example illustrates how detailed soil surveys can help interpret the geomorphic history of a recently glaciated landscape.

Northeastern Lower Michigan was deglaciated roughly 12,300 calibrated years ago (Larson et al. 1994). At that time, ice associated with the Greatlakean advance of the Laurentide ice sheet had moved rapidly into the region from the northwest, out of the Lake Michigan basin. The ice then stagnated and is assumed to have melted in place (Schaetzl 2001). Associated with the Greatlakean advance and the stagnant ice margin were a number of shallow, short-lived, proglacial lakes, or at least this has frequently been assumed (Eschman and Karrow 1985; Schaetzl et al. 2000). The Greatlakean advance left no conspicuous end moraine, and thus, the exact location of the outer limit of the ice advance is not known and has been the subject of considerable debate (Melhorn 1954; Burgis 1977; Schaetzl 2001). Thus, it is conceivable that soil data (maps) may be able to help resolve the extent of this ice advance, as it has been shown to do elsewhere (Millar 2004).

Fortunately, detailed (1:15,840) soil maps (Knapp 1993) and DEMs exist for this area. These maps can be used to help interpret the most recent sedimentary systems that were operational during deglaciation, because post-glacial modifications to these materials have been minimal. Topographic data are not particularly insightful for determining the limit of the Greatlakean ice in this area, given the glacier left no end moraine. However, because water presumably ponded in front of the ice, the northernmost limits of clayey glaciolacustrine sediment can suggest a likely glacial margin (Fig. 13.3). Indeed, Schaetzl (2001) used this type of data (as well as some others), gleaned from soil parent material descriptions, to infer an ice margin just to the north of large areas of glaciolacustrine sediment. Similar sediment behind (north

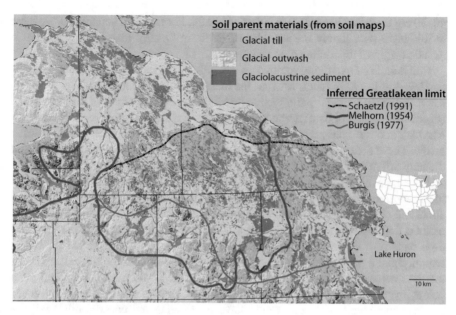

Fig. 13.3 Soil parent materials in northeastern Lower Michigan, as determined from soil maps and the official soil series descriptions, in a GIS. Also shown are the inferred limits of the Greatlakean ice advance, ca. 12,300 cal. years ago

of) this inferred margin is associated with a later, high-level paleolake and is thus clearly not associated with Greatlakean ice (Fig. 13.3).

13.6 Example 4: An Enigmatic Soil Parent Material on the Outwash Plains of Southwestern Michigan, USA

This example illustrates how field and lab data can help determine the parent materials for soil series that have only been described "generically", and how soils with a lithologic discontinuity can potentially provide excellent information about changes in depositional systems in the past.

Many soils on the low relief outwash plains of southwestern Michigan have loamy upper profiles, despite (as expected) being underlain by sandy and gravelly outwash. The origin of this upper material has long been an enigma to soil scientists and geologists alike. It was too thin to be a separate layer of glacial till, and too fine-textured to be glacial outwash. Evidence of any kind of ponded water was also absent from these landscapes.

The main soils that occur on these outwash surfaces are in the Kalamazoo and Schoolcraft series. Kalamazoo soils are described as having formed in, "loamy outwash overlying sand, loamy sand, or sand and gravel outwash on outwash plains", whereas Schoolcraft soils have "formed in loamy material over sand or gravelly sand on outwash plains." Typically, this generically described "loamy material" is

40–90 cm thick and has a diffuse lower boundary. For lack of a better term, we refer to this layer as a loamy mantle. Use of the term "loamy mantle" implies that a genetic origin of this sediment is lacking.

Soil data from two representative pedons (Fig. 13.4) illustrate that the outwash at depth is dominated by sand, whereas the loamy mantle is either silty (Fig. 13.4a) or has a distinctly bimodal particle size distribution – with both sand and silt peaks (Fig. 13.4b). Textural data for the loamy mantle (not shown here) are almost always bimodal, and the sand peak aligns with the same particle size peak in the outwash below. These data suggest that the loamy mantle is a mixed sediment – sand from the outwash mixed with a silty sediment above, but of unknown origin.

Luehmann et al. (2016) sampled and determined the textural distributions of 167 locations across the outwash plains of southwestern Michigan. The loamy mantle in almost all of these soils have a bimodal particle size distribution. Using a "filtering" method first reported in Luehmann et al. (2013), they were able to isolate the textural pattern of the original, silty sediment, and map its characteristics across the region. Spatial patterns for the loamy mantle were easily interpretable, suggesting that it is actually silt-rich loess that has been subsequently mixed with sand from below, by pedoturbation. In the initial stages of loess deposition onto the sandy outwash plain, saltating sand would have been deposited concurrently with the loess, further promoting a bimodal grain-size distribution for the loamy mantle. The mantle is thickest near a large meltwater valley that existed during deglaciation (Fig. 13.5), suggesting that it was the main loess source. Textural data of various sorts (not shown here) also confirmed that the loess that comprises the loamy mantle gets finer-textured and better sorted to the east, away from this channel. This type of spatial pattern is typical for loess.

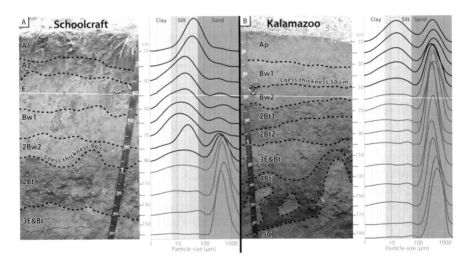

Fig. 13.4 Photos, with horizon boundaries marked, and textural curves for a Schoolcraft and a Kalamazoo soil profile. For the textural curves, black lines are used for loess; brown lines for outwash. (Note the distinct area of mixing between the two parent materials. After Luehmann et al. 2016)

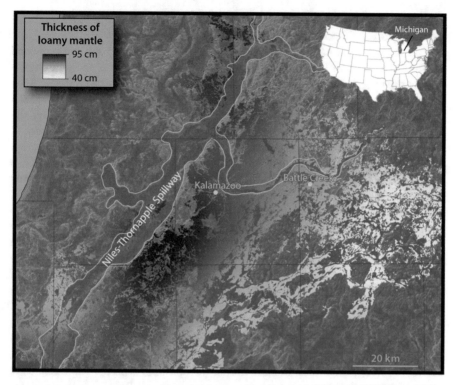

Fig. 13.5 Interpolated map, using ordinary kriging, of the thickness of the upper sediment, which is interpreted as loess, on the outwash plains of southwestern Michigan. Interpolated data are shown only in areas where outwash soils with a loamy mantle are mapped. (After Luehmann et al. 2016)

This work showed that the heretofore enigmatic mantle on the outwash plains of southwestern Michigan is silt-rich loess that was derived from the Niles-Thornapple Spillway and its major tributary channels. The spillway was active for approximately 500 years, between ca. 17,300 and 16,800 cal. years ago, carrying silt-rich meltwater. This study highlights the fact that not all soil parent materials are "obvious" or stated in their official series descriptions, but with some work, the genetic origin of the sediment can often be determined.

13.7 Example 5: A Watershed with a Complex Geology in the Western Grand-Duchy of Luxembourg

This example illustrates how the use of detailed soil surveys for interpreting soil-landform assemblages can also be applied to non-glaciated landscapes, and thus can provide key information for other scientific inquires. In particular, relationships between bedrock parent materials, soil morphology, and indicative vegetation patterns can provide important information for hydrological modelling.

The available geologic map (1:25,000) for the Huewelerbach experimental catchment in western Luxembourg shows the locations of several geologic formations in the watershed, including units of sandstone, limestone, and claystone. Some of these formations have alternating layers of marl. The catchment also contains a colluvial-alluvial complex at the base of many hillslopes. Complicating the spatial distribution of these formations is a fault that is believed to run mostly northwest of, and parallel to, the main trunk stream. Because of this fault, the hydrologic characteristics of the opposing hillslopes are not identical. Parent materials yielding soils with B and/or C horizons consisting of heavy clay create an environment dominated by overland flow. In contrast, parent materials yielding thicker soils with sandy to silty-sandy textures facilitate better infiltration and deeper percolation, and hence more lateral subsurface flow and less surface runoff.

Juilleret et al. (2012) conducted a soil survey of the catchment, classifying six soil map units with 70 hand auger drillings to a depth of 110 cm. They subsequently verified the relationships between the properties of the soil profile with the parent material, using a mechanized coring machine to sample a maximum depth of 400 cm at 12 locations along two transects. Using the World Reference Base (IUSS 2006) to describe the soils, they found Calcisols corresponded with geologic formations containing units of marl, Podzols with a sandstone formation that lacked marl layers, and Colluvisols within the colluvial-alluvial complex at footslope locations. Podzols corresponded with the occurrence of conifers, whereas the other soils occurred under deciduous vegetation. Under grasslands, Pelosols and Brunisols were identified. In the Bw and C horizons of these soils, a distinctive sequence of a red, clayey layer and a grey sandy layer helped reveal the presence of an additional formation recognized in the area, but not previously depicted on the existing geologic map. For this catchment of soils formed in a variety of sedimentary rocks, standard soil survey methods were able to improve upon the information available from the standard geologic map. This information was valuable for improving the mapping of geologic formations and for providing key information for modelling hillslope hydrology.

13.8 Example 6: Improving the Precision of Delineations for Large Extent Maps in the Glaciated Central Lowlands, USA

Map scale is often an obstacle to geomorphic analysis of regional landscapes, due to the challenge of making precise delineations when viewing large extents, i.e., from small-scale maps. Coarse delineations of landform regions introduce noise when comparing characteristics between regions, which can obscure patterns for areas that are naturally high in spatial variability. To delineate more precise boundaries, one must increase the level of detail in the map and/or reduce the extent of the map. This practical limitation was a problem for paper maps because it precludes analyzing large areas with precise delineations. The advent of GIS, however, enables the map scale to be changed while working with a given map. This feature enables large regions to be delineated with more detail.

The flexibility of manipulating map scale in a GIS makes detailed boundaries for regional delineations possible, but base maps with finer resolution are still needed to support any improvements in regional boundary placement. Digital elevation models are becoming increasingly available at finer resolutions and provide essential information about changes in topography. But while topographic data can be used to infer differences in geologic material (see examples provided in Table 6.3 of Chap. 6), they do not directly include that information. As with the previous examples, soil maps that have been produced at larger map scales provide information about soil parent material with fine spatial detail, helping to alleviate this issue.

Surficial geology maps in the USA have long utilized soil survey maps (produced at larger map scales) to refine their boundary delineations. As states have been updating their surficial geology maps, they have been doing so using GIS platforms, allowing them to take advantage of the spatial detail in the soil maps. For example, the recent surficial geology map of Minnesota (Minnesota Geological Survey 2021) makes several improvements over the 1982 version (Hobbs and Goebel 1982). Among those improvements are increased spatial detail in many areas. The dependency on soil survey maps is evident in delineation changes found at county boundaries, where disjuncts between soil maps are often quite evident (Fig. 13.6). These differences coincide with counties where soil survey maps were not yet available or incongruities between county-level soil survey maps.

Another opportunity that GIS presents is the quantitative analysis of broader, full landscape extents. These analyses have the most value when multiple landscapes can be compared and each one precisely delineated. McDanel et al. (2022) created a landform regions map for the glaciated portion of the Central Lowlands in the USA by synthesizing surficial geology maps from the multiple states (Fig. 13.7). This region is so large that it encompasses the areas discussed in the first four examples of this chapter. Because of its immense size, the region is a mosaic of landscapes of different ages, but usually with a glacial history. Since deglaciation, processes of landscape evolution have interacted with the topography and sediments to modify these different landform regions. To summarize physical characteristics for each of the regions (e.g., sediment texture, topography), descriptive statistics of those characteristics can be calculated with a zonal statistics function in a GIS (e.g., Lehmkuhl et al. 2021). However, to reduce noise in those statistics and assist in detecting differences between regions, the region delineations need to be precise. In the process of identifying accurate and detailed boundaries, the new map also needed to harmonize information from different sources as it crossed many political boundaries. The area of interest for this project is covered by a patchwork of surficial geology maps, compiled by different investigators at different times in the past, and with varying cartographic styles. Most of the maps were also made using traditional methods, where the larger the extent of the map, the coarser the delineations.

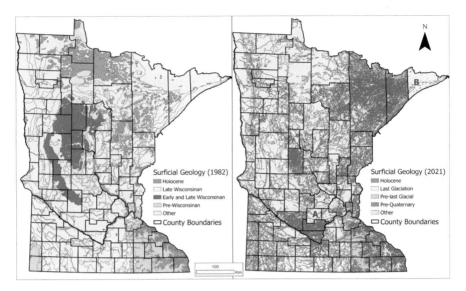

Fig. 13.6 Evolution of the surficial geology map for the state of Minnesota, USA. Although the 1982 map was digitized and made available as a shapefile, the 2021 map includes more detail because of the ability to add finer delineations while 'zoomed' into smaller extents. However, adding that detail requires the support of additional sources of spatial information. The reliance on soil survey maps is evident where detailed delineations stop at county boundaries, as found on (A) the south and west borders of McLeod County and (B) on the western border of Cook County. For McLeod County, the soil mapper did not recognize the glaciolacustrine deposits that soil mappers in neighboring counties did. In the case of Cook County, soil maps were not yet available, which prevented the continuation of the same level of detail used in the county to the west

Therefore, to produce a landform region map to support spatial analysis across the entirety of this large area requires synthesizing multiple map sources, including maps produced at large cartographic scales for the finer detail in delineations.

For the purpose of landscape comparisons, a map of new landform regions needs to balance generalization and precision (detail). McDanel et al. (2022) integrated information from multiple sources, including variables derived from a digital elevation model (see Chap. 8), and soil parent material data interpreted from NCSS maps. The process of creating a continuous soil parent material map – covering the whole area – from soil survey maps was the same as described earlier in this chapter for Iowa (Example #1). As shown in examples 2 through 5, information from soil mapping provided evidence for resolving disparities between maps.

The resulting landform region map divided the area into areas with a common deposit type and topography. These units were then associated with the dates compiled in the MOCA (Meltwater routing and Ocean-Cryosphere-Atmosphere response project) by Dalton et al. (2020). Surface exposure ages were able to be assigned to all but 17 of the regions. The ages of those remaining regions were

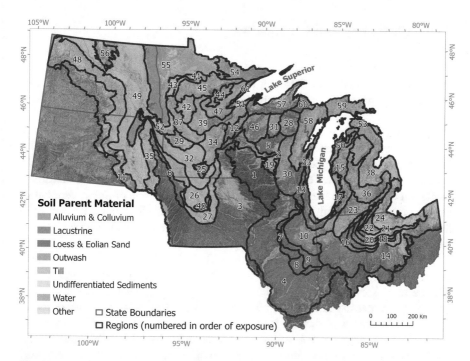

Fig. 13.7 Landform regions delineated for the glaciated portion of the Central Lowlands physiographic region of the USA (McDanel et al. 2022). Using the same methods demonstrated earlier, interpretations of soil parent material from soil survey maps were compiled for all intersecting states. The combined soil parent material map was used to support the reconciling of differences in state-level surficial geology maps and to delineate higher precision boundaries. The landform regions are numbered in sequence from the oldest to youngest dates of surface exposure. Landform regions that were indistinguishable in age are labeled with the same number. However, the dates associated with some of these regions do not follow a logical, spatial juxtaposition. For example, the Algona till plain (32) should pre-date the glacial lake plains (25 and 29) formed on it. Analyses like these highlight the utility of landform region maps to evaluate the broader spatial context for constraining the age of landscapes. The full extent of Pre-Illinoian glaciation was not included because those boundaries are poorly defined in many locations

inferred from the available literature (e.g., Bettis et al. 1996; Mickelson and Attig 2017; Curry et al. 2018). Uniqueness of physical characteristics were also evaluated. The median slope gradient per region tended to increase for regions associated with glacial lake plain, till plain, outwash plain, and glacial moraine landform types, respectively. Mean sand content was generally lower in glacial lake and till plain regions, as compared to outwash and morainic regions. The uniqueness of the landform regions in this map – with minimized noise from adjacent regions – presents a new opportunity to quantitatively compare the landscapes for geomorphic analysis. For example, these regions can be used to test questions about factors controlling rates of stream network development.

13.9   Summary and Conclusions

The relationship between soils and the material in which they form connect soil maps and geological maps. Different information collected for, and represented by, the respective maps – due to differences in purpose, focus, or resources – can assist multiple other kinds of investigations. This multiple utility is especially true for studying soil-landform assemblages and soil-landscape evolution.

Although the pedogenic pathway of a soil is constrained by the parent material, interpretation of soil properties to infer parent material origins needs to carefully consider the potential for confounding factors. For example, other factors of soil formation can alter the material, especially over long periods of time. Also, buried paleosols within the modern soil profile can result in new horizons with properties that are influenced by the interaction of modern pedogenesis with the properties of the older, remnant horizons.

Because of the interconnection between soils and geology, one should beware of the potential for circular reinforcement of information. The reason soil maps often provide more spatial information than available geologic maps are because of the greater spatial density of field sampling and greater availability of easily-observed covariates for spatial prediction. However, soil mappers also use geologic maps as one of the base maps for their soil maps (Miller and Schaetzl 2014). Therefore, the potential exists for an error on one type of map to become circularly reinforced. Only field investigation is capable of catching an error that is being perpetrated from one source to the next, which in doing can help improve many maps.

In many cases, soil maps provide information that is not available from any other source, particularly with regard to spatial detail and characteristics of the top meter of unconsolidated material. Therefore, these maps often represent an untapped potential for improving our geomorphic understanding of landscapes (Brevik and Miller 2015).

References

Bettis EA, Quade DJ, Kemmis TJ (eds) (1996) Hogs, Bogs, & Logs: quaternary deposits and environmental geology of the Des Moines Lobe. Geological Survey Bureau: guidebook series no. 18. 171 p

Brevik EC, Miller BA (2015) The use of soil surveys to aid in geologic mapping. Soil Horiz 56(4):1–9

Brown DJ, Helmke PA, Clayton MK (2003) Robust geochemical indices for redox and weathering on a granitic landscape in central Uganda. Geochim Cosmochim Acta 67:2711–2723

Burgis WA (1977) Late-Wisconsinan history of northeastern lower Michigan. PhD Dissertation, University of Michigan. 396 pp

Curry BB, Lowell TV, Wang H, Anderson AC, Kehew AE (2018) Revised time-distance diagram for the Lake Michigan Lobe, Michigan subepisode, Wisconsin episode, Illinois, USA. In: Kehew AE, Curry BB (eds) Quaternary glaciation of the Great Lakes Region: process, landforms, sediments, and chronology. Geological Society of America, Boulder, pp 69–101

Dalton AS and 70 co-authors (2020) An updated radiocarbon-based ice margin chronology for the last deglaciation of the North American Ice Sheet Complex. Quat Sci Rev 234:106223

Dokuchaev VV (1967) The Russian chernozem (trans: Kaner N). Israel program for scientific translations, Jerusalem. (Original work published in 1883)

Ehrlich WA, Rice HM, Ellis JH (1955) Influence of the composition of parent materials on soil formation in Manitoba. Can J Agric Sci 35:407–421

Eschman DF, Karrow PF (1985) Huron basin glacial lakes: a review. In: Karrow PF, Calkin PE (eds) Quaternary evolution of the Great Lakes, Geological society of Canada special paper, vol 30, pp 79–93

Eze PN, Meadows ME (2014) Texture contrast profile with stone layer in the Cape Peninsula, South Africa: autochthony and polygenesis. Catena 118:103–114

Follmer LR (1982) The geomorphology of the Sangamon surface: its spatial and temporal attributes. In: Thorn C (ed) Space and time in geomorphology. Allen and Unwin, Boston, pp 117–146

Gile LH (1975) Causes of soil boundaries in an arid region: I. Age and parent materials. Soil Sci Soc Am Proc 39:316–323

Hobbs HC, Goebel JE (1982) S-01 Geologic map of Minnesota, Quaternary geology. https://gis-data.mn.gov/dataset/geos-quaternary-geology-mn. Accessed 20 Nov 2021

Hole FD (1950/1968) Aeolian sand and silt deposits of Wisconsin. Wisc Geol Nat Hist Survey map, Madison, WI

Hole FD (1976) Soils of Wisconsin. University of Wisconsin Press, Madison. 223 pp

IUSS (2006) World reference base for soil resources, IUSS Working Group, World soil resources report 103, 2nd edn. FAO, Rome. ISBN 92-5-1-5511-4

Jacobs PM, Mason JA, Hanson PR (2011) Mississippi Valley regional source of loess on the southern Green Bay Lobe land surface, Wisconsin. Quat Res 75:574–583

Juilleret J, Iffly JF, Hoffmann L, Hissler C (2012) The potential of soil survey as a tool for surface geological mapping: a case study in a hydrological experimental catchment (Huewelerbach, Grand-Duchy of Luxembourg). Geol Belg 15(1-2):36–41

Knapp BD (1993) Soil survey of Presque Isle County, Michigan. USDA Soil Conservation Service, US Government. Printing Office, Washington, DC

Krasilnikov P, Calderon NG (2006) A WRB-based buried paleosol classification. Quat Int 156:176–188

Larson GJ, Lowell TV, Ostrom NE (1994) Evidence for the Two Creeks interstade in the Lake Huron basin. Can J Earth Sci 31:793–797

Lehmkuhl F and 19 co-authors (2021) Loess landscapes of Europe- mapping, geomorphology, and zonal differentiation. Earth Sci Rev 215:103496

Luehmann MD, Schaetzl RJ, Miller BA, Bigsby M (2013) Thin, pedoturbated and locally sourced loess in the western Upper Peninsula of Michigan. Aeolian Res 8:85–100

Luehmann MD, Peter B, Connallon CB, Schaetzl RJ, Smidt SJ, Liu W, Kincare K, Walkowiak TA, Thorlund E, Holler MS (2016) Loamy, two-storied soils on the outwash plains of southwestern Lower Michigan: pedoturbation of loess into the underlying sand. Annals Assoc Am Geogs 106(3):551–572

McDanel J, Meghani NA, Miller BA, Moore PL (2022) A harmonized map of landform regions in the glaciated Central Lowlands, USA. J Maps. https://doi.org/10.1080/17445647.2022.2090866

Melhorn WN (1954) Valders glaciation of the southern peninsula of Michigan. PhD Dissertation, University of Michigan. 177 pp

Mickelson DM, Attig JW (2017) Laurentide ice sheet: ice-margin positions in Wisconsin, Wisconsin geological and natural history survey education series, vol 56

Millar SWS (2004) Identification of mapped ice-margin positions in western New York from digital terrain-analysis and soil databases. Phys Geogr 25:347–359

Miller BA, Burras CL (2015) Comparison of surficial geology maps based on soil survey and in depth geological survey. Soil Horiz 56(1):1–12

Miller BA, Schaetzl RJ (2014) The historical role of base maps in soil geography. Geoderma 230-231:329–339

Miller BA, Burras CL, Crumpton WG (2008) Using soil surveys to map Quaternary parent materials and landforms across the Des Moines lobe of Iowa and Minnesota. Soil Survey Hor 49:91–95

Minnesota Geological Survey (2021) D-1 surficial geology. https://mngs-umn.opendata.arcgis.com. Accessed 20 Nov 2021

Nettleton WD, Olson CG, Wysocki DA (2000) Paleosol classification: problems and solutions. Catena 41:61–92

Olson CG, Nettleton WD (1998) Paleosols and the effects of alteration. Quat Int 51:185–194

Ruhe RV, Olson CG (1980) Soil welding. Soil Sci 130:132–139

Schaetzl RJ (1998) Lithologic discontinuities in some soils on drumlins: theory, detection, and application. Soil Sci 163:570–590

Schaetzl RJ (2001) Late Pleistocene ice flow directions and the age of glacial landscapes in northern lower Michigan. Phys Geogr 22:28–41

Schaetzl RJ, Attig JW (2013) The loess cover of northeastern Wisconsin. Quat Res 79:199–214

Schaetzl RJ, Sorenson CJ (1987) The concept of "buried" vs "isolated" paleosols: examples from northeastern Kansas. Soil Sci 143:426–435

Schaetzl RJ, Krist F, Rindfleisch P, Liebens J, Williams T (2000) Postglacial landscape evolution of northeastern lower Michigan, interpreted from soils and sediments. Annals Assoc Am Geog 90:443–466

Schaetzl RJ, Forman SL, Attig JW (2014) Optical ages on loess derived from outwash surfaces constrain the advance of the Laurentide ice from the Lake Superior Basin, Wisconsin. USA Quat Res 81:318–329

Schaetzl RJ, Larson PH, Faulkner DJ, Running GL, Jol HM, Rittenour TM (2018) Eolian sand and loess deposits indicate west-northwest paleowinds during the Late Pleistocene in Western Wisconsin, USA. Quat Res 89:769–785

Scull P, Schaetzl RJ (2011) Using PCA to characterize and differentiate the character of loess deposits in Wisconsin and Upper Michigan, USA. Geomorphology 127:143–155

Thorp J, Smith HTU (1952) Pleistocene eolian deposits of the United States, Alaska, and parts of Canada. Map. 1:2,500,000. Geological Society of America, New York

Wilson MA, Indorante SJ, Lee BD, Follmer L, Williams DR, Fitch BC, McCauley WM, Bathgate JD, Grimley DA, Kleinschmidt K (2010) Location and expression of fragic soil properties in a loess-covered landscape, Southern Illinois, USA. Geoderma 154:529–543

Part III
Methods and Techniques Applied to Pattern Recognition and Mapping

Chapter 14
Digital Soil Texture Maps of Argentina and Their Relationship to Soil-Forming Factors and Processes

G. A. Schulz, D. M. Rodríguez, M. Angelini, L. M. Moretti, G. F. Olmedo, L. M. Tenti Vuegen, J. C. Colazo, and M. Guevara

Abstract Soil texture is determined by the parent material of the soil and the resulting pedogenetic processes. Because the spatial distribution of soil texture governs several soil physical and chemical processes, our goals were to map clay, silt, and sand at different depths using the SISINTA soil profile database; generate textural soil class maps for the same soil depths; and describe the distribution of soil texture in relation to different landscape units. We used 4663 soil texture observations and 64 environmental covariates to represent the soil-forming factors (e.g., remote sensing data, climate data or geomorphology maps). We modelled clay, silt and sand at 0–15, 15–30, 30–60 and 60–100 cm, establishing an empirical relationship with environmental covariates using Random Forest to predict their spatial distribution.

G. A. Schulz (✉) · D. M. Rodríguez
Instituto de Suelos (CIRN), Instituto Nacional de Tecnología Agropecuaria,
Hurlingham, Buenos Aires, Argentina
e-mail: schulz.guillermo@inta.gob.ar

M. Angelini
Instituto de Suelos (CIRN), Instituto Nacional de Tecnología Agropecuaria,
Hurlingham, Buenos Aires, Argentina

Food and Agriculture Organization of the UN, Global Soil Partnership, Roma, Italia

Departamento de Ciencias Básicas, Universidad Nacional de Luján, Luján, Argentina

L. M. Moretti
EEA Cerro Azul, Instituto Nacional de Tecnología Agropecuaria,
Cerro Azul, Misiones, Argentina

G. F. Olmedo
Food and Agriculture Organization of the UN, Global Soil Partnership, Roma, Italia

EEA Mendoza, Instituto Nacional de Tecnología Agropecuaria,
Mendoza, Mendoza, Argentina

Finally, we performed an analysis of uncertainty through repeated cross-validation. We observed model efficiency coefficient (MEC) values between 0.452 and 0.557, with an RMSE between 8.77% and 11.21% for the clay fraction. The MEC for the silt fraction ranged from 0.561 to 0.638 with an RMSE of 10.50% to 12.01%. For the sand fraction the MEC ranged from 0.587 to 0.640 with RMSE values between 16.19% and 16.76%. The general patterns of uncertainty are consistent with areas of limited data. Our results increased the quality, quantity and accessibility of information on soil texture in Argentina by providing new insights into both the distribution of parent materials and the intensity of pedogenetic processes in each region.

Keywords Digital soil map · Clay · Silt · Sand · Soil texture · Soil layers

14.1 Introduction

The National Institute of Agricultural Technology (INTA) began systematic soil surveys with the implementation of the Soil Map Plan in 1964. Some of the main outcomes of this plan were the large number of people trained on soil description and mapping following the standards of the Soil Taxonomy, the numerous maps at semi-detailed scale in the Pamean region (e.g., SAGyP-INTA 1989), and the successively national map (SAGyP-INTA 1990). The Soil Taxonomy was, therefore, widely adopted by universities and public organisations to describe and classify the soils of the country (Rodríguez et al. 2019). In Argentina, almost all of the soil maps published to date were made following the standards of the system proposed by the USDA. In these maps, the surface corresponding to the mapped sector is divided into polygons, each of which has a limited number of soil types in its composition.

Digital soil mapping (DSM, McBratney et al. 2003) has gained momentum in the last 10 years. Unlike the conventional soil map, DSM has been developed to describe, classify and study the spatial distribution patterns of soil properties and

L. M. Tenti Vuegen
Instituto de Suelos (CIRN), Instituto Nacional de Tecnología Agropecuaria,
Hurlingham, Buenos Aires, Argentina

Departamento de Tecnología, Universidad Nacional de Luján, Luján, Argentina

J. C. Colazo
EEA San Luis, Instituto Nacional de Tecnología Agropecuaria,
Villa Mercedes, San Luis, Argentina

Departamento de Ciencias Agropecuarias, Universidad Nacional de San Luis,
San Luis, Argentina

M. Guevara
Centro de Geociencias – Universidad Nacional Autónoma de México, Juriquilla, México

soil types in a more objective way. The great diffusion that the DSM has had in recent years is due in part to the greater availability of spatial data (digital elevation models and remote sensing data) as well as the growing computer capacity to process data, the development of GIS tools and the availability of new statistical software's (Minasny and McBratney 2015). Nevertheless, several challenges related to map interpretation remain unsolved in DSM (Wadoux et al. 2021). In this regard, geopedology can provide a framework to link the spatial patterns of soil properties to the soil-landscape relations.

In Argentina, the application of DSM at national scale was facilitated by the creation of SISINTA (Olmedo et al. 2017), a freely available national soil profile repository, and the participation of INTA in the Global Soil Partnership (GSP) of the Food and Agriculture Organization of the United Nations (FAO). The Soil Institute of INTA participated in all GSP initiatives for the creation of several maps, such as the soil organic carbon stock map (Olmedo et al. 2018), soils affected by salts (Rodríguez et al. 2020), potential for soil organic carbon sequestration (Frolla et al. 2021), and black soil map (Tenti Vuegen et al. 2021). However, soil texture, one of the most important soil information to support decision-making, is still missing.

Soil texture depends on the nature of the parent material and the type and intensity of the soil forming processes. From the physical point of view, the silt and clay content play a very important role, contributing to the development of a structure resistant to erosion processes, both by wind and by water (Colazo and Buschiazzo 2010). Likewise, the capacity for water storage and the processes that determine its movement within the soil profile also depend on the texture (Saxton and Rawls 2006). The dynamics of organic carbon accumulation is strongly controlled by the particle size distribution, since soils with finer textures have a greater capacity to retain useful water, which translates into greater productivity, and therefore a greater entry of C into the system. In turn, the higher silt and clay content favors greater accumulation due to chemical stabilization and protection due to the formation of organo-mineral complexes (Six et al. 2002; Hevia et al. 2003). These latter processes are highly relevant due to the importance of carbon sequestration in the framework of achieving neutrality in land degradation. The texture, due to the adsorption processes generated by the different particle sizes, determines the availability of nutrients for plants (Geering and So 2006; Galantini et al. 2004). In addition, due to the high temporal stability of this variable in the face of management, its analysis allows us to use data from surveys carried out many years ago (Pierce and Larson 1996).

Although many efforts were made to generate soil texture maps in Argentina, these have been partial contributions on certain regions. Therefore, the objectives of this work were (1) to map clay, silt, and sand at 0–15, 15–30, 30–60, and 60–100 cm depth based on the SISINTA soil profile database, (2) generate textural soil class maps for the same soil depths, (3) describe the distribution of soil texture in relation with different landscape units.

14.2 Materials and Methods

14.2.1 Study Area

Argentina is located in the southern sector of South America, and covers approximately 2.8 million square km (IGN 2011). The temperate climate is predominant in the country, but due to its great longitude extension, there are large sectors with warm and subtropical climates in the north, as well as cold climates in the extreme south. The orography is characterized by the presence of Andes range to the west, while extensive plains and giant alluvial fans dominate the east of the country. This remarkable heterogeneity of climates and landscapes determines the existence of a great variety of soil types, with the country registering the presence of the twelve soil orders of the Soil Taxonomy (Rodríguez et al. 2019).

14.2.2 Soil Profile Data

Soil data were accessed from SISINTA (Olmedo et al. 2017, 2018; Angelini et al. 2018) and from MARAS monitoring project (Oliva et al. 2006). The first of these sources is an INTA's repository of over 7100 georeferenced soil profiles with more than 100 soil attributes, while the second one contained 222 topsoil samples (0–10 cm). Sand, silt, and clay were chosen from these datasets, which included a total of 4663 sites (Fig. 14.1). Note that several soil profiles were removed due to a lack of data or location information.

14.2.3 Environmental Covariates

Environmental covariates were selected considering the soil-forming factors related to texture. We use 64 environmental covariates that are proxies for soil-forming factors (McBratney et al. 2003). All covariates were resampled at 1 km spatial resolution. Table 14.1 describes the environmental covariates used in this study.

A total of 43 covariates were used to represent the relief factor, including the global landform layer (Hengl 2018c), MERIT DEM terrain model (Yamazaki et al. 2017) and the geomorphometric layers derived from it (Hengl 2018a; Amatulli et al. 2018a, b), along with other terrain parameters such as surface water maps (Pekel et al. 2016). Other environmental covariates included annual precipitation, mean annual temperature, evapotranspiration, and the Martonne aridity index of the Digital Climate Atlas of the Argentine Republic (Bianchi and Cravero 2010; Cravero et al. 2017). The organism factor was represented by the annual FAPAR layer (Hengl 2018b), and 14 indices derived from MODIS product MOD09A1. The parent materials were incorporated from the global layer of Hengl 2018c and the layer of soil orders (SAGyP-INTA 1990).

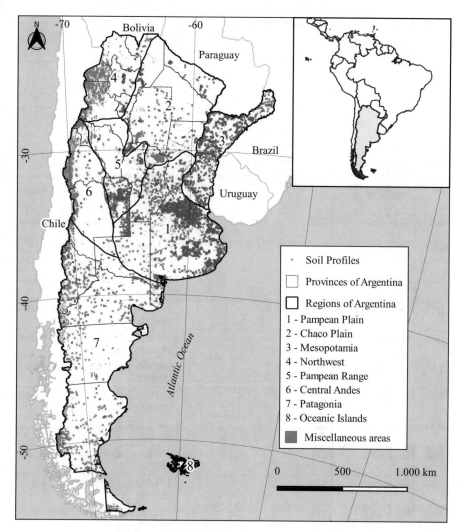

Fig. 14.1 Soil sample distribution and geographics regions. (Modified from Siragusa 1958)

14.2.4 Processing

First, we generated standard depth layers by applying an Equal Area Spline function (Bishop et al. 1999) to estimate the textural components (sand, silt, and clay) at 0–15, 15–30, 30–60, and 60–100 cm. Subsequently, maps of the different textural components (sand, silt and clay) were generated, and the textural classification was carried out for the aforementioned depths using the "Soil texture" package in R (Moeys 2014).

Table 14.1 Description of environmental covariates

Name	Soil forming factor	Description	Source
B15	Soil	Global Lithology USGS	Hengl (2018c)
B43	Soil	Soil Order according to Soil Taxonomy	SAGyP-INTA (1990)
B01	Climate	Annual Mean Evapotranspiration	Bianchi and Cravero (2010)
B02	Climate	Annual Mean Precipitation	Bianchi and Cravero (2010)
B03	Climate	Annual Mean Temperature	Bianchi and Cravero (2010)
B04	Climate	De Martonne Aridity Index	Bianchi and Cravero (2010)
B29	Organism	BI	Soil Salinity Indices obtained from MODIS
B30	Organism	CRSI	Soil Salinity Indices obtained from MODIS
B31	Organism	NDSI	Soil Salinity Indices obtained from MODIS
B32	Organism	NDVI	Soil Salinity Indices obtained from MODIS
B33	Organism	NSI	Soil Salinity Indices obtained from MODIS
B34	Organism	SAVI	Soil Salinity Indices obtained from MODIS
B35	Organism	SI1	Soil Salinity Indices obtained from MODIS
B36	Organism	SI2	Soil Salinity Indices obtained from MODIS
B37	Organism	SI3	Soil Salinity Indices obtained from MODIS
B38	Organism	SI4	Soil Salinity Indices obtained from MODIS
B39	Organism	SI5	Soil Salinity Indices obtained from MODIS
B40	Organism	SI6	Soil Salinity Indices obtained from MODIS
B41	Organism	SR	Soil Salinity Indices obtained from MODIS
B42	Organism	VSSI	Soil Salinity Indices obtained from MODIS
B64	Organism	Vegetation FAPAR	Hengl (2018b)
B05	Topography	Curvature Merit Dem	Hengl (2018a)
B06	Topography	Dev-Magnitude Merit Dem	Amatulli et al. (2018a)
B07	Topography	Dev-Scale Merit Dem	Amatulli et al. (2018a)
B08	Topography	Downlslope Curvature Merit Dem	Hengl (2018a)
B09	Topography	Dvm Merit Dem	Hengl (2018a)
B10	Topography	Dvm2 Merit Dem	Hengl (2018a)

(continued)

Table 14.1 (continued)

Name	Soil forming factor	Description	Source
B11	Topography	Dxx Merit Dem	Amatulli et al. (2018b)
B12	Topography	Dxy Merit Dem	Amatulli et al. (2018b)
B13	Topography	Dyy Merit Dem	Amatulli et al. (2018b)
B14	Topography	Elevation Merit Dem	Hengl (2018a)
B16	Topography	Mrn Merit Dem	Hengl (2018a)
B17	Topography	Neg-Openess Merit Dem	Hengl (2018a)
B18	Topography	Pos-Openess Merit Dem	Hengl (2018a)
B19	Topography	Rough-Magnitude Merit Dem	Amatulli et al. (2018a)
B20	Topography	Roughness Merit Dem	Amatulli et al. (2018a)
B21	Topography	Slope Merit Dem	Amatulli et al. (2018a)
B22	Topography	Tpi Merit Dem	Hengl (2018a)
B23	Topography	Tri Merit Dem	Amatulli et al. (2018a)
B24	Topography	Twi Merit Dem	Hengl (2018a)
B25	Topography	Upslope-Curvature Merit Dem	Hengl (2018a)
B26	Topography	Vbf Merit Dem	Hengl (2018a)
B27	Topography	Vrm Merit Dem	Amatulli et al. (2018a)
B28	Topography	Landform USGS	Hengl (2018c)
B44	Topography	Surface Water-Change	Pekel et al.(2016)
B45	Topography	Surface Water-Extent	Pekel et al. (2016)
B46	Topography	Surface Water-Ocurrence	Pekel et al. (2016)
B47	Topography	Surface Water-Recurrence	Pekel et al. (2016)
B48	Topography	Surface Water-Seasonality	Pekel et al. (2016)
B49	Topography	Surface Water-Transitions	Pekel et al. (2016)
B50	Topography	Analytical hillshading	DEM MERIT derivatives with Terrain Analysis
B51	Topography	Aspect	DEM MERIT derivatives with Terrain Analysis
B52	Topography	Channel network base level	DEM MERIT derivatives with Terrain Analysis
B53	Topography	Closed depressions	DEM MERIT derivatives with Terrain Analysis
B54	Topography	Convergence index	DEM MERIT derivatives with Terrain Analysis
B55	Topography	Cross sectional curvature	DEM MERIT derivatives with Terrain Analysis
B56	Topography	Flow accumulation	DEM MERIT derivatives with Terrain Analysis
B57	Topography	Longitudinal curvature	DEM MERIT derivatives with Terrain Analysis
B58	Topography	LS factor	DEM MERIT derivatives with Terrain Analysis
B59	Topography	Relative slope position	DEM MERIT derivatives with Terrain Analysis

(continued)

Table 14.1 (continued)

Name	Soil forming factor	Description	Source
B60	Topography	Slope	DEM MERIT derivatives with Terrain Analysis
B61	Topography	Topographic wetness index	DEM MERIT derivatives with Terrain Analysis
B62	Topography	Valley depth	DEM MERIT derivatives with Terrain Analysis
B63	Topography	Vertical distance to channel network	DEM MERIT derivatives with Terrain Analysis

14.2.5 Modelling and Accuracy Assessment

Quantile regression forest (QRF; Meinshausen 2006) is a generalization of random forests that can be used to predict conditional quantiles. It has been applied to predict several soil properties at different depths by Lagacherie et al. (2020) and Vaysse and Lagacherie (2017). The model has been implemented using the quantregForest package (Meinshausen 2006) in the R software (R Core Team 2022). We first applied a recursive feature elimination (RFE) using caret package (Kuhn et al. 2017) to select the most significant predictors of each soil property and layer. Using the selected covariates, we trained twelve QRF models. The main parameters of a Random Forest model are the number of covariates used in each regression tree (mtry) and the number of regression trees in the set (ntree). Both of them were set as default by the caret package.

The model was calibrated by 10-times repeated 10-fold cross validation. The overall accuracy was measured by the root mean square error (RMSE), modelling efficiency coefficient (MEC) (Janssen and Heuberger 1995) and mean absolute error (MAE). Spatial explicit accuracy was estimated by the predicted conditional standard deviation, while conditional mean was used as prediction of the most likely values of the twelve layers.

14.2.6 Generation of Textural Class Maps

Once the most probable value maps were obtained, the textural classes were calculated according to the USDA (Soil Survey Staff 2009). After discussing the results in statistical space, this we described the new knowledge by interpreting the digital maps of soil textures generated and their main implications in the detailed information they provide about the spatial variability of the parent materials of the soils, as well as the intensity of the pedogenetic processes specific to each region of the country. The regionalization that we used in the present work is the one proposed by Siragusa (1958) with some modifications, in which the national territory is divided into Pampean Plain, Chaco Plain, Mesopotamia, Northwest, Pampean Range, Central Andes, Patagonia, Oceanic Islands and Argentine Sea, although the latter region was excluded from the analysis.

14.3   Results and Discussion

14.3.1   *Predicted Maps and Accuracy Assessment*

Figure 14.2 shows the maps, at a depth of 0–15 cm of the content of clay (a), silt (b) and sand (c), while Fig. 14.3 shows the associated uncertainty of the maps. All models used the 12 most important predictors in the calibration process, with the exception of silt 0–15 cm, which used the 64 variables. Maps of clay, silt and sand content, with their respective error maps, for depths of 15–30, 30–60 and 60–100 cm deep can be accessed on the Zenodo repository (Schulz et al. 2022).

Table 14.2 presents the MEC, RMSE and MAE for the different depths and texture components. We observe MEC values between 0.452 and 0.557, with an RMSE between 8.77% and 11.21% for the clay fraction. The MEC for the silt fraction ranged from 0.561 to 0.638 with an RMSE of 10.50% to 12.01%. For the sand fraction the MEC ranged from 0.587 to 0.640 with RMSE values between 16.19% and 16.76%. In all cases, the most important predictors of the spatial variation of the

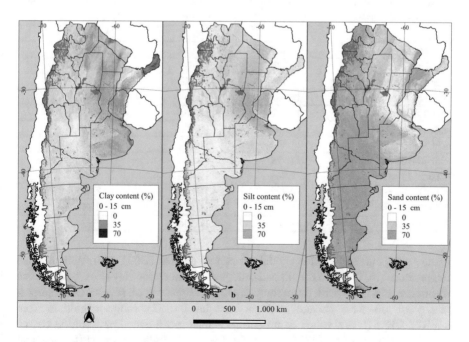

Fig. 14.2 Clay content (**a**), Silt content (**b**) and Sand content (**c**), at 0–15 cm depth

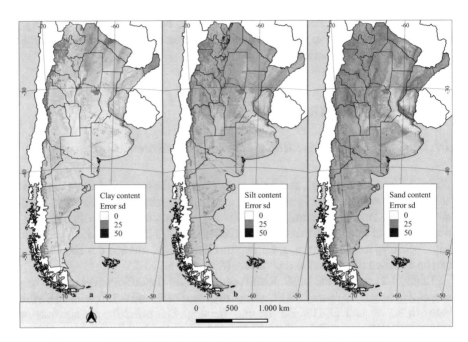

Fig. 14.3 Standard deviations for clay (**a**), silt (**b**) and sand (**c**) at 0–15 cm depth

Table 14.2 Root mean square error (RMSE), modeling efficiency coefficient (MEC) and mean absolute error (MAE) for different soil properties and depths

sp	RMSE	MEC	MAE
Sand 0–15 cm	16.189	0.640	11.069
Sand 15–30 cm	16.320	0.629	11.213
Sand 30–60 cm	16.676	0.618	11.364
Sand 60–100 cm	16.762	0.587	11.472
Silt 0–15 cm	12.011	0.638	8.352
Silt 15–30 cm	11.807	0.608	8.388
Silt 30–60 cm	11.504	0.561	8.168
Silt 60–100 cm	11.728	0.583	8.263
Clay 0–15 cm	8.766	0.475	5.721
Clay 15–30 cm	10.723	0.452	7.432
Clay 30–60 cm	11.211	0.557	7.842
Clay 60–100 cm	11.005	0.536	7.734

texture components were the annual mean precipitation (B02), while annual mean temperature (B03) did so for silt and sand, but not much for clay (Fig. 14.4). These results are consistent with those found by Castro Franco et al. (2018). Other important predictors were channel network base level (B52), rough-magnitude from MERIT DEM (B19), valley depth (B62), annual mean evapotranspiration (B01), global lithology from, USGS (B15), De Martonne aridity index (B04) and soil

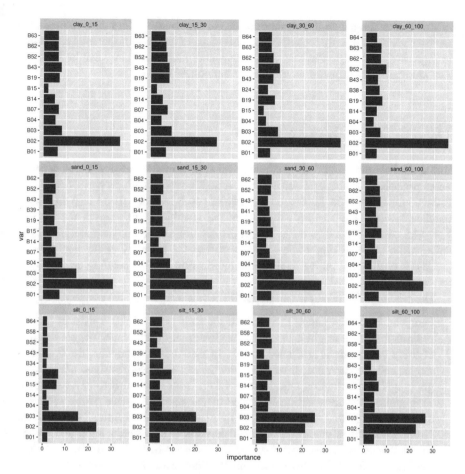

Fig. 14.4 Relative importance (X axis) of the 12 most important predictors expressed in percentage regarding the total contribution for each model, and the predictors (Y axis)

orders according to USDA Soil Taxonomy (B43). These variables contributed to all models in most depths and were responsible for the vast majority of the variance explained. The exception occurs in the model silt 0–15 cm in which all variables were used to train the model (Fig. 14.4). Ließ et al. (2012) found that the best predictors of textural components at topsoil level were terrain attributes and that the same did not occur in depth. This contradicts our findings where terrain attributes had a secondary contribution. It should be noted that covariates derived from salinity indices and those corresponding to vegetation indices were not relevant for most models.

The highest error values occurred, in general, in regions with low data density (Southern Mesopotamia Sector -middle and upper delta of the Paraná River-, Chaco Plain -Chaco province-, Central Andes, Northwest and Patagonia) (Fig. 14.3). Also, we observed the highest uncertainty in sand in large regions, and in silt in few areas related to the Yungas, in the northern centre of the country. Note that we predicted the four soil depths, only topsoil (0–15 cm) maps of texture components are shown in Figs. 14.2 and 14.3. The rest of the maps and their uncertainty can be found in Zenodo repository https://doi.org/10.5281/zenodo.6312654.

14.3.2 Maps of Textural Classes

The spatial distribution of soil texture is related to both morphogenetic and pedogenetic processes. In general terms, the former governs soil properties in depth (60–100 cm), while the latter are more evident at the surface (0–15 cm) and intermediate levels (15–30 and 30–60 cm) of the profiles. In this sense, the map of Fig. 14.5 can be related to the type and distribution of the parent materials of the soils, while those of Fig. 14.5a–c, would reflect the intensity of the pedogenetic processes (argilluviation, vertisolization, laterization, etc.) depending on the forming factors (Jenny 1941).

The results obtained show particularities for each region of the country:

1. Pampean Plain: the parent materials of the soils (Fig. 14.5d) consist of loessic sediments, whose origin is linked to the Quaternary glaciations. The source area of these sediments is attributed mainly to the Andes Mountains, the Pampean ranges and the Paraná river basin (Zárate 2003). The textures show a pattern of grain spatial distribution that decreases from west-southwest to east-northeast. This granulometric selection is related to the source distance of the sediment area, so the proximal sector, known as the "sand sea of the Pampas" (Iriondo and Kröhling 1995) is where the thickest deposits with a sandy loam texture are found. The intermediate sector or "peripheral loess belt" (Iriondo and Kröhling 1995) is rich in medium textures, while the finer textures are found in the eastern sector of this region due both to the distance to the western source area (Andes Mountains and Pampean ranges), as well as to the contribution of clays from the Paraná river basin (Morrás and Moretti 2016). At soil depth between 15 and 60 cm (Fig. 14.5b, c) there are finer textural classes due to the clay illuviation process, which is more intense towards the eastern area because of an increase in precipitation. Finally, on the surface (Fig. 14.5a) the textures are coarser than the previous ones, in this case due to the eluviation process of the fine fractions, as well as to more recent wind deposits. For instance, it can be seen the textural variation in depth in the central sector of the Buenos Aires province, being loam on the surface, clay loam between 15 and 60 cm, and again loam at the base of the profiles.
2. Chaco Plain: the parent materials (Fig. 14.5d) consist of sediments contributed by the great fluvial systems of the Pilcomayo, Bermejo and Salado rivers that cross the region from west to east. They vary from loam to silt loam or clay loam in the same direction. Aeolian sediments with a loamy texture are present in the central sector of this region, scattered among fluvial materials ("*Loess*

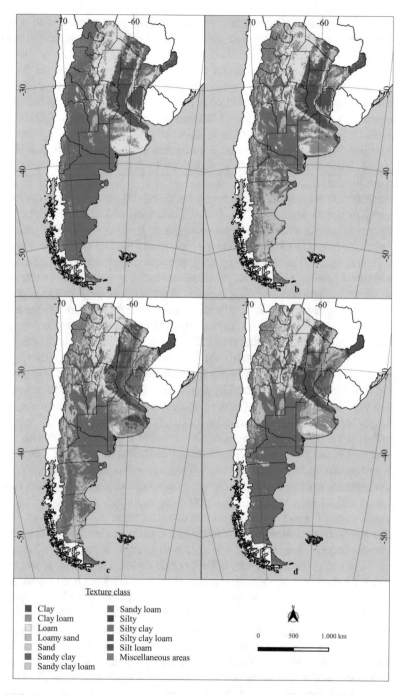

Fig. 14.5 Textural classes at depths 0–15 cm (**a**), 15–30 cm (**b**), 30–60 cm (**c**), and 60–100 cm (**d**)

Chaqueño"; Iriondo 1993). A narrow strip of clay loam texture with a north-south direction is noted on the eastern margin of the region, which is attributed to sediments contributed by the Paraná River. A geopedological model has been proposed for the southeastern sector of the Chaco region (north of the province of Santa Fe), indicating that the concurrence of neotectonics processes and Quaternary climatic changes resulted in parental materials of different origin and composition (Morrás 2017). As in the previous region, the textural differences between topsoil and the subsurface in the western sector are primarily caused by the argilluviation process, whereas in the central and eastern sectors, this process is much less intense, and textural variations in depth are minimal due to the semi-arid climate.

3. Mesopotamia: this region shows marked textural differences due to the hetero-geneity of parent materials (Fig. 14.5d). Sediments of lacustrine origin with a silty clay texture can be found in the south (Entre Ríos province). The sandy clay loam classes associated with sediments contributed by the Paraná river dominate the central sector (Corrientes province), whereas clayey textures are present in the extreme northeast (Misiones province) provoked by an intense weathering of basaltic rocks. The argilluviation process causes textural variations in depth, pri-marily in the center of this region, with an increase in fine fractions in the middle levels of the profiles (Fig. 14.5b, c). Meanwhile, the parent materials in the southern area are of smectite composition, which, combined with a seasonal climate, favor the process of vertisolization, resulting in predominantly silty clay textures in the entire profile. In the north, instead, the dominant soil-forming process is the laterization caused by a subtropical climate without a dry season. Deep profiles with a clayey texture from the surface are found, except in the younger fluvial valleys where the dominant texture is silty clay (Fig. 14.5a).

4. Northwest: this region is characterized by orogenic systems (Puna, Andes Range and Sub-Andean Hills) with rocky outcrops (miscellaneous areas), correspond-ing to igneous (intrusive and extrusive), metamorphic and sedimentary rocks. The dominant soils in this region have a sandy loam texture in the mountainous systems and a loamy texture in the foothill environments to the east. Clay loam textures formed from finer materials can be found in fluvial valleys (Fig. 14.5d). In the humid intermontane valleys it is possible to find an increase of clay texture in depths between 0 and 60 cm (Fig. 14.5a–c) caused by the argilluviation pro-cess. Note that given the scarcity of data of this region some non-soil areas, such as recent sediment deposits and outcrops areas may not have been masked.

5. Pampean Range: this comprises an orogenic system with a north-south orienta-tion located in the center-west sector of Argentina. It is composed mainly of Paleozoic granitic rocks, whose current relief is linked to Cenozoic tectonic pro-cesses. Plains or valleys extend between the mountain ranges. The distribution of rainfall varies considerably, decreasing towards the south and west of the region, while in the southeastern sector there is a strip of temperate climate. The texture maps in the rocky outcrops of the mountain ranges are presented as miscella-neous areas. The presence of colluvial and regolith materials in mountainous areas produced heterogeneous textures in this region, tending to be sandy loam

in the hills, and loamy in fluvial and aeolian sediments in the valleys and plains (Fig. 14.5d). On the other hand, since it is a region where the soils are poorly developed, no significant textural changes are observed within the profile (Fig. 14.5a–c).

6. Central Andes: this region comprises a series of north-south mountain ranges with the highest altitudes on a continental scale. From west to east, this unit is constituted by three orogenic systems: *Cordillera Principal*, *Cordillera Frontal*, and *Precordillera*. Due to the relief, the climate shows great spatial variability, with heavy snowy precipitation in the western mountainous part, becoming semi-arid in the plains and slopes of the eastern part. With regard to texture, the region is characterized by rocky outcrops in the Cordilleras sector (masked as miscellaneous zones), and sandy loam textures from aeolian sediments in foothills (Fig 14.5d). Due to the region's semi-arid climate, no significant changes in texture are observed within the soil profile, with the exception of a few minor humid fluvial valleys (Fig. 14.5a–c).

7. Patagonia: the east of the Andes Range consists of plains formed by sands and gravels of glaciofluvial origin, as well as fluvial and aeolian silts and sands. The dominant texture of the soils in this region is sandy loam. However, despite the arid climate, an increment of clay between 15 and 60 cm depth can be seen in Fig. 14.5b, c, related to a more humid past climate in the region. The dominant soil parent material in the Andes Range is tephra, usually mixed with coarse colluvial deposits or till. Organic parent material is also present in the southern area (Tierra del Fuego province), where finer texture soils are observed on the surface (Fig. 14.5a).

8. Oceanic Islands: in terms of area, the Malvinas Islands are the most significant. They present a landscape characterized by highly eroded low hills with a large number of flooded depressions. The climate is humid, cold to very cold, without water deficit. The parent materials of soils are coarse sand sediments derived from till, cryo-genesis and alluvial-colluvial sources. Due to the bioclimatic conditions and the landscape, in this region organic soils (Histosols) are developed, in addition to other poorly developed soils. The soil textures are loam to sandy clay loam in the subsurface (Fig. 14.5b, c, d), while the loam textures are dominant in the surface (Fig. 14.5a).

14.4 Conclusions

Using the DSM methodology, national maps of different soil texture fractions (sand, silt and clay) and textural classes at different depths were obtained, as well as the error associated with them. The models were run with more than 4600 soil profiles of national repositories and with 64 covariables, much of them locally generated, which resulted in a better adjustment with the observations in relation to the factors and processes of soil forming in the different regions of the country. The values of texture obtained at four depth intervals revealed an uneven spatial distribution in the different regions of the country, mainly related to the nature of the soil parent materials, the distance to the sources of sediments and the type and intensity of pedogenetic processes.

The methodology implemented in this work allowed us to develop the first national maps of soil texture classes and fractions, as well as the spatial explicit uncertainty of the soil fractions at the four soil depths. The accuracy assessment was similar to other publication results at local level and presented levels of uncertainty commonly found in national DSM products. We believe that more accurate maps can be produced by increasing the number of soil samples, especially in regions with low sample density, as well as increasing the quality of the most important predictors.

We have explained the main soil-forming factors and processes that control the distribution of the soil fractions, including expert knowledge and supported by interpretation of the most important predictors of the different maps. In this regard, we found that the climatic environmental covariates were among the most important predictors of the soil fraction spatial distribution.

With this work we expect to advance in the knowledge of the processes that control the spatial distribution of the soil texture in Argentina, while we provide spatial explicit data that can serve as input of other models, such as maps of soil organic carbon potential sequestration, estimation of soil water availability, and cropping models, that so far have used coarser maps, or simply extracting data from global datasets produced with scarce field observations.

Acknowledgments This work was carried out within the framework of INTA research projects (PNSUELO 1134032 and RIST I051) and through training in DSM by Pillar 4 for Latin America and the Caribbean of FAO's Global Soil Partnership.

References

Amatulli G, McInerney D, Sethi T, Strobl P, Domisch S (2018a) Geomorpho90m, empirical evaluation and accuracy assessment of global high-resolution geomorphometric layers. https://doi.org/10.5281/zenodo.1807119

Amatulli G, McInerney D, Sethi T, Strobl P, Domisch S (2018b) Geomorpho90m, empirical evaluation and accuracy assessment of global high-resolution geomorphometric layers (V1.0) [Data set]. Zenodo. https://doi.org/10.5281/zenodo.1807125

Angelini ME, Rodríguez DM, Olmedo GF, Pasquier ML, Schulz GA, Aleksa AS, Angelini HP, Babelis GC, Barrios RA, Bustos MV, Carboni G, Casabella MP, Colazo JC, de Bustos ME, de la Fuente JC, Díaz RC, Di Fede BE, Escobar D, Escobar LE, Faule L, Garay JM, Godagnone RE, Hurtado P, Irigoin J, Kurtz DB, Liotta MA, Medina Herrera D, Miraglia HN, Morales Poclava MC, Navarro MF, Rigo S, Rossi JP, Sánchez JM, Valdettaro RA, Vicondo ME, Vizgarra LA (2018) Sistema de información de suelos del INTA (SISINTA): Presente y futuro. Congreso Argentino de la Ciencia del Suelo, Tucumán

Bianchi AR, Cravero SAC (2010) Atlas Climático digital de la República Argentina. Ediciones INTA

Bishop TFA, McBratney AB, Laslett GM (1999) Modelling soil attribute depth functions with equal-area quadratic smoothing splines. Geoderma 91(1):27–45. https://doi.org/10.1016/S0016-7061(99)00003-8

Castro Franco M, Domenech MB, Borda MR, Costa JL (2018) A spatial dataset of topsoil texture for the southern Argentine Pampas. Geoderma Reg 12:18–27. https://doi.org/10.1016/j.geodrs.2017.11.003. Editorial: Elsevier. ISSN: 2352-0094

Colazo JC, Buschiazzo DE (2010) Soil dry aggregate stability and wind erodible fraction in a semiarid environment of Argentina. Geoderma 159(1–2):228–236. https://doi.org/10.1016/j.geoderma.2010.07.016

Cravero SAC, Bianchi CL, Elena HJ, Bianchi AR (2017) Clima de la Argentina: Mapas digitales mensuales de precipitaciones y precipitación menos evapotranspiración potencial. Adenda del Atlas Climático digital de la República Argentina. Ediciones INTA. http://inta.gob.ar/documentos/clima-de-argentina-adenda-del-atlas-climatico-digital-de-la-republica-argentina [Acceso: 21_08_2021]

Frolla FD, Angelini ME, Beltrán MJ, Di Paolo LE, Peralta GE, Rodríguez DM, Schulz GA (2021) Argentina: soil organic carbon sequestration potential national map. National report. Versión 1.0. In: Global soil organic carbon sequestration potential map – GSOCseq. FAO. http://www.fao.org/fileadmin/user_upload/GSP/GSOCseq/Argentina_SOC_SequestrationPotentialNationalMap.pdf [Acceso: 22_06_2021]

Galantini JA, Senesi N, Brunetti G, Rosell R (2004) Influence of texture on the nitrogen and sulphur status and organic matter quality and distribution in semiarid Pampean grassland soils of Argentina. Geoderma 123(1–2):143–152. https://doi.org/10.1016/j.geoderma.2004.02.008

Geering HR, So HB (2006) Texture. In: Lal R (ed) Encyclopedia of Soil Science, 2nd edn. Taylor & Francis, pp 1759–1763

Hengl T (2018a) Global DEM derivatives at 250 m, 1 km and 2 km based on the MERIT DEM (Version 1.0) [Data set]. Zenodo. https://doi.org/10.5281/zenodo.1447210

Hengl T (2018b) Fraction of Absorbed Photosynthetically Active Radiation (FAPAR) at 250 m monthly for period 2014-2019 based on COPERNICUS land products (Version 1.0-1) [Data set]. Zenodo. https://doi.org/10.5281/zenodo.3459830

Hengl T (2018c) Global landform and lithology class at 250 m based on the USGS global ecosystem map (Version 1.0) [Data set]. Zenodo. https://doi.org/10.5281/zenodo.1464846

Hevia GG, Buschiazzo DE, Hepper EN, Urioste AM, Antón EL (2003) Organic matter in size fractions of soils of the semiarid Argentina. Effects of climate, soil texture and management. Geoderma 116(3–4):265–277. https://doi.org/10.1016/S0016-7061(03)00104-6

IGN (2011) Límites, superficies y puntos extremos. https://www.ign.gob.ar/NuestrasActividades/Geografia/DatosArgentina/LimitesSuperficiesyPuntosExtremos [Acceso: 20_12_2021]

Iriondo M (1993) Geomorphology and late Quaternary of the Chaco (South America). Geomorphology 7:289–303

Iriondo M, Kröhling DM (1995) El sistema eólico pampeano. Museo Provincial de Ciencias Naturales "Florentino Ameghino"

Janssen PHM, Heuberger PSC (1995) Calibration of process-oriented models. Ecol Model 83:55–66

Jenny H (1941) Factors of soil formation. McGraw-Hill, New York

Kuhn M, Wing J, Weston S, Williams A, Keefer C, Engelhardt A, Cooper T, Mayer Z, Kenkel B, R Core Team, Benesty M, Lescarbeau R, Ziem A, Scrucca L, Tang Y, Candan C, Hunt T (2017) caret: classification and regression training. URL https://CRAN.R-project.org/package=caret. R package version 6.0-78.

Lagacherie P, Arrouays D, Bourennane H, Gomez C, Nkuba-Kasanda L (2020) Analysing the impact of soil spatial sampling on the performances of Digital Soil Mapping models and their evaluation: A numerical experiment on Quantile Random Forest using clay contents obtained from Vis-NIR-SWIR hyperspectral imagery. Geoderma 375, art. 114503 [12 p.]. ISSN 0016-7061. https://doi.org/10.1016/j.geoderma.2020.114503

Ließ M, Glaser B, Huwe B (2012) Uncertainty in the spatial prediction of soil texture: comparison of regression tree and random forest models. Geoderma 170:70–79. https://doi.org/10.1016/j.geoderma.2011.10.010

McBratney AB, Santos MM, Minasny B (2003) On digital soil mapping. Geoderma 117(1-2):3–52. https://doi.org/10.1016/S0016-7061(03)00223-4

Meinshausen N (2006) Quantile regression forests. J Mach Learn Res 7:983–999

Minasny B, McBratney AB (2015) Digital soil mapping: a brief history and some lessons. Geoderma 264(b):301–311. https://doi.org/10.1016/j.geoderma.2015.07.017

Moeys J (2014) The soil texture wizard: R functions for plotting, classifying, transforming and exploring soil texture data. http://cran.r-project.org/web/packages/soiltexture/vignettes/soil-texture_vignette.pdf

Morrás H (2017) Una interpretación geopedológica sobre los sedimentos superficiales y suelos actuales de la Cuña Boscosa, Chaco Austral, Provincia de Santa Fe. XX Congreso Geológico Argentino Actas Sesión Técnica 3:38–43. Tucumán

Morrás H, Moretti LM (2016) A new geopedologic approach on the genesis and distribution of Typic and Vertic Argiudolls in the Rolling Pampa of Argentina. In: Zinck A, Metternicht G, Bocco G, del Valle H (eds) Geopedology Book. Springer, 556 p, pp 193–209

Oliva G, Escobar J, Siffredi G, Salomone J, Buono G (2006) Monitoring patagonian rangelands: the maras system. In: Aguirre-Bravo C, Pellicane PJ, Burns DP, Draggan S (eds) Monitoring science and technology symposium: unifying knowledge for sustainability in the Western Hemisphere proceedings RMRS-P-42CD, vol 42. US Department of Agriculture, Forest Service, Rocky Mountain Research Station, Fort Collins, pp 188–193

Olmedo GF, Rodríguez DM, Angelini ME (2017) Advances in digital soil mapping and soil information systems in Argentina. In: Arrouays D, Savin I, Leenaars J, McBratney AB (eds) GlobalSoilMap -digitalsoilmappingfromcountrytoglobe. CRC Press, Boca Raton, pp 13–16

Olmedo GF, Angelini MA, Schulz GA, Rodríguez DM, Taboada MA, Pascale C, Escobar D, Guevara M, Colazo JC, Aleksa AS, Babelis GC, Gaitán JJ, Peralta AR, Peralta G, Rojas JM, Sainz Rosas HR, Vizgarra LA (2018) Prediction stock of soil organic carbon in Argentina (1.5) [Data set]. Zenodo. https://doi.org/10.5281/zenodo.6323695

Pekel JF, Cottam A, Gorelick N, Belward AS (2016) High-resolution mapping of global surface water and its long-term changes. Nature 540:418–422. https://doi.org/10.1038/nature20584

Pierce F, Larson W (1996) Quantifying indicators for soil quality. In: Berger A, Iams W (eds) Geoindicators. Assessing rapid environmental changes in earth systems. Balkema, Rotterdam, pp 323–335

R Core Team (2022) R: a language and environment for statistical computing. R Foundation for Statistical Computing, Vienna. https://www.R-project.org/

Rodríguez DM, Schulz GA, Aleksa AS, Tenti Vuegen LM (2019) Distribution and classification of soils. In: Rubio G, Lavado R, Pereyra F (eds) The soils of Argentina, World Soils Book Series. Springer, Cham. ISBN: 978-3-319-76851-9, pp 63–79. https://doi.org/10.1007/978-3-319-76853-3_5

Rodríguez DM, Schulz GA, Tenti Vuegen LM, Angelini ME, Olmedo GF, Lavado RS (2020) Salt-affected soils in Argentina (1.0) [Data set]. Zenodo. https://doi.org/10.5281/zenodo.6323102

SAGyP-INTA (1989) Mapa de suelos de la provincia de Buenos Aires (Escala 1:500.000). Proyecto PNUD ARG/85/019, Buenos Aires. 544 pp

SAGyP-INTA (1990) Atlas de Suelos de la República Argentina (Escala 1: 500.000 y 1: 1.000.000). Proyecto PNUD ARG/85/019, Buenos Aires. Tomo I: 731 pp, Tomo II: 677 pp

Saxton KE, Rawls WJ (2006) Soil water characteristic estimates by texture and organic matter for hydrologic solutions. Soil Sci Soc Am J 70:1569–1578. https://doi.org/10.2136/sssaj2005.0117

Schulz GA, Rodríguez DM, Angelini ME, Moretti LM, Olmedo GF, Tenti Vuegen LM, Colazo JC, Guevara M (2022) Digital soil texture maps of Argentina (2.0) [Data set]. Zenodo. https://doi.org/10.5281/zenodo.6312654

Siragusa A (1958) República Argentina: Regiones geográficas. (mimeo)

Six J, Conant RT, Paul EA, Paustian K (2002) Stabilization mechanisms of soil organic matter: implications for C-saturation of soils. Plant Soil 241:155–176. https://doi.org/10.1023/A:1016125726789

Soil Survey Staff (2009) Soil survey field and laboratory methods manual. Soil Survey Investigations Report No. 51, version 1.0. Burt R (ed) USDA Natural Resources Conservation Service

Tenti Vuegen LM, Rodríguez DM, Moretti, LM, De la Fuente JC, Schulz GA, Angelini ME (2021) Black soils in Argentina (1.0) [Data set]. Zenodo. https://doi.org/10.5281/zenodo.6323558

Vaysse K, Lagacherie P (2017) Using quantile regression forest to estimate uncertainty of digital soil mapping products. Geoderma 291:55–64

Wadoux AMC, Heuvelink GB, Lark RM, Lagacherie P, Bouma J, Mulder VL, Libohova Z, Yang L, McBratney AB (2021) Ten challenges for the future of pedometrics. Geoderma 401:115155. https://doi.org/10.1016/j.geoderma.2021.115155

Yamazaki D, Ikeshima D, Tawatari R, Yamaguchi T, O'Loughlin F, Neal JC, Sampson CC, Kanae S, Bates PD (2017) A high-accuracy map of global terrain elevations. Geophys Res Lett 44:5844–5853. https://doi.org/10.1002/2017GL072874

Zárate MA (2003) Loess of Southern South America. Quat Sci Rev 22(18–19):1987–2006. https://doi.org/10.1016/S0277-3791(03)00165-3

Chapter 15
Synergistic Use of Radar and Optical Image Data for Improved Land Use and Land Cover Assessment: A Case Study in the North of Entre Rios Province (Argentina)

H. F. del Valle, G. I. Metternicht, F. Tentor, W. F. Sione, P. Zamboni, F. Viva Mayer, and P. G. Aceñolaza

Abstract The synergistic use of optical and radar data is already a well-known alternative in the literature for land cover characterization. The objective of this chapter is to quantify the added value of combining radar imagery from Sentinel-1

H. F. del Valle (✉) · F. Tentor · P. Zamboni · F. V. Mayer
Universidad Autónoma de Entre Ríos (UADER), Facultad de Ciencia y Tecnología, Centro Regional de Geomática (CeReGeo), Oro Verde, Entre Ríos, Argentina
e-mail: delvalle.hector@uader.edu.ar; ftentor@uader.edu.ar; zamboni.pamela@uader.edu.ar; vivamayer.francisco@uader.edu.ar

G. I. Metternicht
Earth and Sustainability Science Research Centre (ESSRC), University of New South Wales, Sydney, Australia

School of Science, Western Sydney University, Richmond, Australia
e-mail: G.Metternicht@westernsydney.edu.au

W. F. Sione
Universidad Autónoma de Entre Ríos (UADER), Facultad de Ciencia y Tecnología, Centro Regional de Geomática (CeReGeo), Oro Verde, Entre Ríos, Argentina

Departamento de Ciencias Básicas, Universidad Nacional de Luján (UNLu), Buenos Aires, Argentina
e-mail: sione.walter@uader.edu.ar

P. G. Aceñolaza
Universidad Autónoma de Entre Ríos (UADER), Facultad de Ciencia y Tecnología, Centro Regional de Geomática (CeReGeo), Oro Verde, Entre Ríos, Argentina

Centro de Investigaciones Científicas y Transferencia de Tecnología a la Producción (CICyTTP-CONICET), Diamante, Entre Ríos, Argentina

Facultad de Ciencias Agropecuarias, Universidad Nacional de Entre Ríos (FCA-UNER), Oro Verde, Entre Ríos, Argentina
e-mail: pablo.acenolaza@uner.edu.ar

J. A. Zinck et al. (eds.), *Geopedology*,
https://doi.org/10.1007/978-3-031-20667-2_15

283

and multispectral imagery from Sentinel-2 (both at 10 m resolution) to provide information on land use and land cover change (LULC) in 2016–2017 and 2020. The Sentinel-1 image data included the Global Backscatter Model (S1GBM) for the 2016–2017 wet period, and for the driest year 2020, which were sourced from the Google Earth Engine (GEE) platform. The Sentinel-2 Global Mosaic (S2GM) service provided surface reflectance mosaic products for the same years. Sentinel-2 data were compared to derived radiometric indices and combined with Sentinel-1 imagery. The LULC classes considered for this study are three classes of Espinal ecotone (closed gallery forest, mid to open gallery forest, open low forest, and shrubland) and four classes of agricultural land defined by soil degradation processes (slight soil water erosion, moderate saline and slight soil water erosion, slight to moderate soil water erosion, and moderate to severe soil water erosion). Results show that σ^0_{VV} and σ^0_{VH} backscatter values are 1.0 to 1.8 dB lower during the 2020 drought compared to values in 2016–2017. Both σ^0_{VV} and σ^0_{VH} polarizations and the Radar Vegetation Index combined with selected optical radiometric indices for soil, vegetation, and moisture from Random Forest analysis are suitable for representing LULC changes in years with changes in moisture availability. The results showed that a significant change in LULC patterns had occurred in the driest year, 2020, in the study area.

Keywords C-band SAR · Radar vegetation index · Optical radiometric indices · Moisture availability · Roughness · Change detection · Random Forest

15.1 Introduction

A fundamental problem in soil mapping is obtaining and maintaining current and accurate information. Soil surveys need to be regularly updated or improved to meet emerging needs of soil information that provide important insights for climate regulation and ecosystem services (Dwivedi 2017).

Accurate profiling of the diversity of soil and landforms (i.e., geodiversity) is important to that end, and remote sensing data supports a pathway to cost-effective, comprehensive, and repeatable monitoring of geodiversity at multiple scales, in contrast to traditional in situ methods (Poppiel 2019; Lausch et al. 2019, 2020; Wadoux et al. 2021). Often, new or additional remote sensing features such as spatial, spectral, temporal, and angular (directional) resolutions must be considered. These features are then used to update an existing soil map at multiple scales. Hengl et al. (2018) suggest that improvements in Earth Observation (satellite imagery and products), digital elevation models (DEMs), and digital soil mapping (DSM), combined with data mining and cloud-based computing, could help address the problem of insufficient soil data.

Synthetic Aperture Radar (SAR) and optical imagery operate in different portions of the electromagnetic spectrum, providing complementary spectral information on the earth's surface objects. Therefore, multi-sensor data integration opens many possibilities for a better understanding of how landform, drainage, soil, and vegetation interact with each other (Mulder et al. 2011; Poggio and Gimona 2017; Bousbih et al. 2019; Eibedingil et al. 2021). Such insights are important in supporting Earth Observation-assisted updating of existing soil maps, derived through conventional soil mapping techniques that rely heavily on field observations and laboratory analyses of soil properties. More to the point, remote sensing provides low-cost alternatives for mapping and detecting spatial soil and crop patterns, given the current advantages of multiple satellite platforms.

In this context, the main objective of the present study is to quantify the added value of combining radar imagery from Sentinel-1 (S1) and multispectral imagery from Sentinel-2 (S2) to add information on land use and land cover (LULC) changes during the period 2016–2017 and 2020. The pilot area of the Espinal, in the northern part of Entre Ríos province (Argentina), is used to investigate the correlation between S1 and S2 radiometric indices derived for soils of this forest-agriculture mosaic, and to ascertain whether additional variables might influence the soil-landscape relationship. Monitoring forest-agriculture mosaics is critical for understanding landscape heterogeneity and biodiversity management. Several radiometric indices from S2 data, sensitive to different soil and plant parameters, are used. Their relationship to S1 backscatter, log-ratio scaling, and radar vegetation index is examined. No soil survey of this area has yet combined data from radar and optical remote sensing data to obtain Digital Soil Mapping.

15.2 Materials and Methods

15.2.1 Subset Selection

A variety of conventional soil maps (CSMs) exist for the Espinal ecotone on the northeastern edge of the Pampas ecoregion, a transitional region between the fertile central plains and the Mesopotamian region of Argentina, based primarily on physiographic analysis with aerial photo interpretation (Bedendo et al. 2014; Bedendo 2019). The study area selected covers 14,669 km^2 between 30° 09' 32.8" S, 59° 59' 45.7" W, and 31° 30' 25.6" S, 57° 48' 06" W and it includes the departments of La Paz, Feliciano, Federal, and Federación. The department is the smallest administrative cadastral unit in Argentina. This chapter presents a subset of 450 km^2 (test study site) between 31° 12' 44.1" S, 59° 47' 21.0" W, and 31° 24' 14.9" S, 59° 33' 56.4" W in the department of La Paz (Fig. 15.1), as an example of the studies conducted in a larger area.

The study area corresponds to a subtropical climate. The average temperature of the warmest month is 26 °C, and the annual mean temperature is 20 °C. Precipitation is abundant and exceeds 1400 mm per year, with the highest rainfall occurring

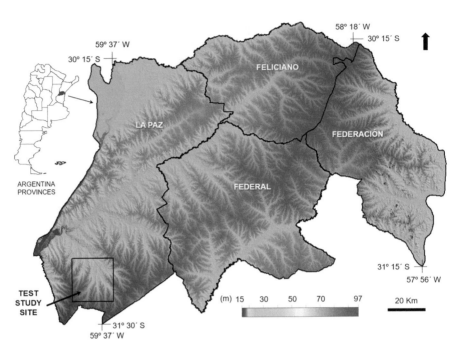

Fig. 15.1 Location of test study site as an example of the studies conducted in a larger area. Copernicus GLO-30 Digital Terrain Model. (del Valle et al. 2022)

between October and May. Potential evapotranspiration averages 1250 mm per year (De Ruyver and Di Bella 2019).

The relief is mostly flat to gently rolling and characterized by two landscape types: low-relief denudation surfaces (peneplains) and valleys cut beneath the peneplains. Vegetation is dominated by crops of soybean, maize, wheat, and sorghum, in order of importance (https://www.bolsacer.org.ar/Fuentes/siber.php). The typical crop rotation is fall-winter cropping (wheat) and spring-summer cropping (soybeans, maize, and sorghum). There are also scattered pastures for livestock grazing (mainly cattle). The natural vegetation (Espinal ecotone) consists of low forests dominated by species of the genus *Prosopis*, ranging from closed to open canopy. It also includes gallery riparian forest and patches of wooded savannas and grasslands (Matteucci et al. 2019).

This region has a history of forest clearance for the cultivation of cereals and oilseeds, initially limited to landscapes dominated by Mollisols, and later to Vertisols and other soils with vertic properties. The expansion of the agricultural frontier has initiated soil erosion processes due to continuous cultivation (Maldonado et al. 2012). In addition, tillage and agricultural activities have altered the physical and chemical properties of the topsoil. On the other hand, there are physical contrasts in the soil landscape between slopes with and without gilgai-related subsurface properties (Bedendo et al. 2016). Human activities have included forest clearing, selective logging, alteration of the natural fire regime, and have resulted in soil water erosion and the introduction of exotic species (Tasi and Bedendo 2008; Matteucci et al. 2019). This area is thought to be particularly prone to severe soil shrinkage events, development of vertic horizons, and formation of subsurface features associated with the Gilgai microrelief. This microrelief was certainly less pronounced before intensive forest clearing and agriculture/grazing. The introduction of cropping systems and grazing has resulted in cyclical changes, particularly in soil surface conditions and land use over time (Fig. 15.2a–f). Therefore, a framework for soil and crop management at a more detailed scale than the one provided by conventional soil surveys at a scale of 1:100,000 is needed to guide management decisions at farm and field levels (Bedendo et al. 2016).

The study considers digital databases of interannual rainfall variability, such as the years 2016/2017 (wet) and 2020 (dry), and high-resolution digital images (Kirches 2020; Bauer-Marschallinger et al. 2021) and land cover products (Karra et al. 2021; Zanaga et al. 2021) published in these years, as input for machine learning algorithms (ML). Rainfall data from the Center for Hydrometeorology and Remote Sensing (CHRS) portal (https://chrsdata.eng.uci.edu/) was accessed via the dataset PERSIANN-CDR (Precipitation Estimation from Remotely Sensed Information using Artificial Neural Networks-Climate Data Record) (Nguyen et al. 2019). Table 15.1 shows the average precipitation in the study area for the period 1983–2021 and average annual records for the years 2016 to 2020.

15.2.2 Land Use and Land Cover Classes

The SIBER (Integrated System for monitoring and estimating cereals and oilseeds production of Entre Rios province) was used to collect data (land use history) on tillage and cultivation activities for 2016–2017, 2019–2020, and 2020–2021 periods (https://www.bolsacer.org.ar/Fuentes/siber.php). This information system works with the processing of images from the S2, Landsat 8 OLI, and LISS 3 satellites, as well as ground checks performed by SIBER specialists. The vector layer of crops from this information system provided the agricultural land use information and sampling design for this study.

The land cover information was sourced from the ESA S2 imagery at 10 m resolution (Karra et al. 2021; Zanaga et al. 2021). It is a compilation of land use and land cover (LULC) predictions to provide a representative snapshot for 2020. Modifications were introduced for the area of the Espinal ecotone that has a history of land clearing (Maldonado et al. 2012); with a clear distinction between remnant forest, cropland, and grassland. Forests ranging from closed to open canopy are mapped accurately; some cultivated pastures (permanent pasture and crop-pasture rotations) are confused with natural grassland.

Seven land use and land cover (LULC) classes were determined (Table 15.2) using the information on soil distribution (Bedendo et al. 2014), ground checks (2016–2019), and land use history. The LULC classes defined are closed gallery forest, mid to open gallery forest, open low forest and shrubland, and agricultural

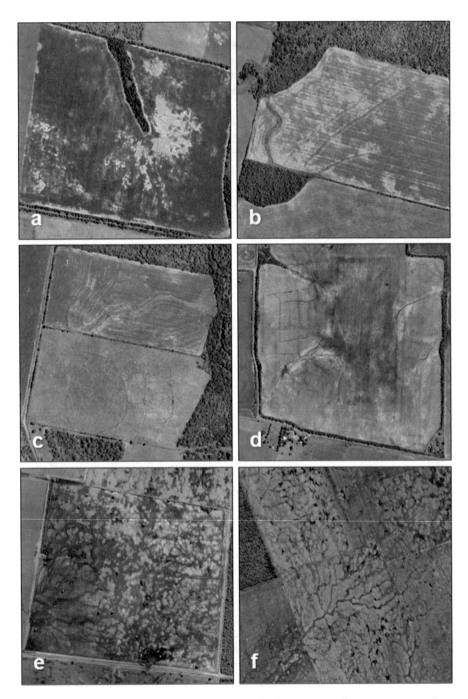

Fig. 15.2 (**a–f**) Subsets from Google Earth images showing soil degradation processes. 15.2a-d. Salt-affected soils (white to whitish-gray color patches with fine to medium texture). Moderately saline and alkaline. 15.2e-f Soil water erosion, topsoil loss and terrain deformation (gilgai microrelief)

Table 15.1 Annual precipitation for the 1983–2021 period and per year from 2016–2020

| Statistics | mm | | | | | |
	1983–2021	2016	2017	2018	2019	2020
Mean	1364	1647	1516	1376	1617	990
Sigma	292	81	84	71	84	65
Maximum	2027[a]	1809	1675	1467	1789	1128
Minimum	808[b]	1443	1358	1246	1452	905

Source: https://chrsdata.eng.uci.edu
[a]1990
[b]2008

Table 15.2 Land use and land cover (LULC) classes

Classes[a]	Landform	Land Use and Land Cover	Description
1	Narrow valleys	Closed gallery forest	Medium forest, a closed canopy, following watercourses and creeks in narrow valleys undercutting the peneplain. Grazing, livestock shade
2	Peneplain, plain to gently rolling plains	Mid to open gallery forest	Mosaic of mid to low open forest and shrubs (>50%)/grass cover (<40%) on convex hills. Grazing low suitability scattered crops
3		Open low forest and shrubland with scattered grassland	Mosaic of low forest and shrubs cover (>50%)/grass cover (<40%) and cropland (<10%) on smooth convex hills. Also, secondary low forests. Grazing low suitability scattered crops
4		Agricultural lands, slight soil water erosion	Continuous cropping rotations. Some agricultural fields present a significant reduction of soil organic carbon content, losses of structure stability, and soil porosity (effects of soil-use intensity)[b]
5		Agricultural lands, moderately saline, and slight soil water erosion	Crop-pasture rotations. Salt-affected soils. Soil water erosion: Loss of topsoil and terrain deformation slight
6		Agricultural lands, slight to moderate soil water erosion	Continuous cropping and crop-pasture rotations. Soil water erosion: Loss of topsoil slight, and terrain deformation moderate
7		Agricultural lands, moderate to severe soil water erosion	Crop-pasture rotations. Changes in moisture content. Soil water erosion: Loss of topsoil moderate, and terrain deformation strong

[a]Other classes (no considered in this study): Built area (0.33%), bare ground (0.16%), flooded vegetation (0.01%), and water (0.02%)
[b]Wilson and Paz-Ferrero (2012)

lands (slight soil water erosion, moderate saline, slight to moderate soil water erosion, and moderate to severe soil water erosion).

Furthermore, triangulation of legacy data with field observations and ancillary information on changes in land use evidence that, as a consequence of the expansion and intensification of agriculture, information related to soil water erosion and its susceptibility has changed and needs updating (Wilson and Paz-Ferrero 2012).

For each LULC class, training samples were selected from field observations, PlanetScope imagery, and Google Earth imagery (see Sect. 15.2.4), delineating mostly polygons around representative sites. Selected and random points as training and testing samples were considered.

15.2.3 Synthetic Aperture Radar Data

The Sentinel-1 (S1) image data included the Global Backscatter Model (S1GBM) for the 2016–2017 wet period, based on the mean C-band radar cross-section in σ^0_{VV} and σ^0_{VH} polarizations, normalized to 38° incidence angle, at 10 m sampling (Bauer-Marschallinger et al. 2021). S1GBM data provide spatial resolution comparable to optical imagery and can be used to classify land cover and determine vegetation and soil conditions.

For the driest year 2020, the "COPERNICUS /S1_GRD" S1 Image Collection, A and B (https://developers.google.com/earth-engine/sentinel1) described in Gorelick et al. (2017) were sourced from the Google Earth Engine (GEE) platform. The imagery consists of Level-1 High-Resolution Ground Range Detected (GRDH) scenes processed for backscatter coefficient (sigma-naught σ°) in decibels (dB). As a result, a dataset of one hundred and seventy-four S1 scenes with a pixel resolution of 10 m and an incidence angle of 38°, with descending right-looking and dual-polarization σ^0_{VV} and σ^0_{VH} was performed for the study area.

15.2.4 Optical Data

The S2 Global Mosaic (S2GM) service (sentinel-hub.com) provides surface reflectance mosaic products derived from the Sentinel-2 (S2) A and B platforms, Level 2A of the Copernicus ground segment, i.e., ESA's S2 core products[1] (Kirches 2020). Annual mosaics for 2016–2017 and 2020 with a spatial resolution of 10 m and a high probability of cloud-free skies (0–2%) were derived from this service. In total, more than three hundred S2 A/B scenes per year are sourced. The bands included were B2 (blue), B3 (green), B4 (red), B5 (red edge 1), B6 (red edge 2), B7 (red edge

---

[1] Due to reprocessing of the S2 archive, the Global Mosaic service Phase 2 is currently available on the website https://s2gm.acri-st.fr/mosaic-hub (Rivet 2022)

3), B8 (near-infrared, NIR 1), B8A (near-infrared, NIR 2), B11 and B12 (covering the shortwave infrared).

Furthermore, we used PlanetScope data (https://www.planet.com/) from 2020 at 3 m spatial resolution with 4 spectral bands: B1 (blue), B2 (green), B3 (red), and B4 (near-infrared) for support training and validation phases. These images are atmospherically corrected for bottom-of-atmosphere reflectance (BOA) (Planet Team 2020). Historical natural color composites (2016–2017) from Google Earth supplemented PlanetScope data.

15.2.5 Methodology

The methodological framework was designed to assess the potential of S1 and S2 data used alone and in combination for supporting the assessment of diagnostic features that can help agronomists and soil surveyors to enhance the mapping of LULC changes (Fig. 15.3a, b). We focused on the selection of relevant radar and optical features and environmental conditions (wet and dry periods). The Sentinel Application Platform (SNAP) v9.0 (SNAP-ESA 2022) was used for digital image processing along with the GEE platform and the ENVI 5.3.1 software (Exelis Visual Information Solutions, Boulder, Colorado).

Co-registration of the multisensor data (Fig. 15.3a, b) was the most important pre-processing step. S1 orthorectification was performed in the SNAP and GEE platforms with some residual errors and shifts occurring when overlaying different images. The S2GM tiles had the greatest geometric distortion. As a result, the ENVI's automated image registration workflow (Jin 2017) that geometrically aligns two images with different viewing geometry and/or different terrain distortions into the same coordinate system was used. The matching score was calculated based on the normalized mutual information between the base image and the warped image. Mutual information measures the interdependence of the two random variables (Suri and Reinartz 2010). The Global Backscatter Model (S1GBM) was used as the base image (Bauer-Marschallinger et al. 2021).

The method applied for combining S1 and S2 data is similar to the incremental classification developed by Inglada et al. (2016). In the feature selection (Fig. 15.3b), a mask was made for natural vegetation (the Espinal ecotone) and croplands.

15.2.5.1 SAR Data Processing

GEE uses the following preprocessing steps (as implemented by the SNAP S1 toolbox) to derive the backscatter coefficient in each pixel: apply orbit file, GRD border noise removal, thermal noise removal, and radiometric calibration (σ^0). Terrain correction (orthorectification) was performed on the SNAP platform using the Copernicus DEM Global 30-m resolution (GLO-30). This digital surface model was converted to a digital terrain model (DTM) for this study (see del Valle et al. 2022).

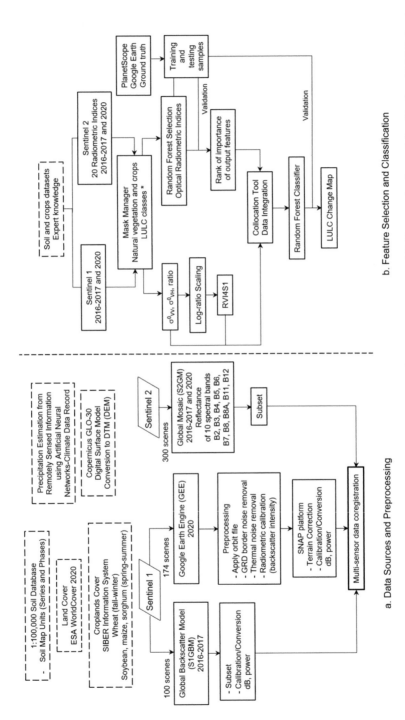

Fig. 15.3 (a, b) Flowchart of the image processing procedure. 15.3a Data sources and preprocessing. 15.3b Feature selection and classification. * LULC: Land Use and Land Cover classes: Espinal ecotone (closed gallery forest, mid to open gallery forest, open low forest and shrubland) and agricultural land (slight soil water erosion, moderate saline and slight soil water erosion, slight to moderate soil water erosion, and moderate to severe soil water erosion)

15.2.5.2 Statistical Analysis

Statistical analysis was performed using Orange Data Mining v3.32.0 software (Demsar et al. 2013). Before performing the correlation analysis, basic statistics were calculated only for the exploration of the SAR data. Pearson's correlation coefficient (r) was used to compare the spatial similarity between the different data-sets, measuring the strength and direction of the linear relationship between the variables (σ^0_{VV} and σ^0_{VH}, wet and dry, respectively). They help judge how useful one variable would be in improving the prediction of the second variable given that information from all the other variables has already been considered. A significant correlation coefficient (r) was considered at the $P < 0.05$ level.

15.2.5.3 Change Detection: Log-Ratio Scaling

We used the algorithm to detect changes in the software SNAP (SAR applications). The goal is to identify year-to-year changes in logging extent. The proportion of corresponding pixels with changes in each band of two images from different times (2016/17 versus 2020) was calculated. Scaling in a logarithmic ratio is an effective means of suppressing image background and enhancing change signatures in a SAR image (Bovolo et al. 2013; Meyer 2019).

15.2.5.4 Radar Vegetation Index

The radar vegetation index (RVI) is a measure of the randomness of the target concerning the scattering mechanism. The RVI is near zero for a smooth, bare surface and increases as vegetation increases. This index is less sensitive to the geometry and topography of the radar measurement and is insensitive to absolute calibration errors.

Considering the characteristics of the interaction between the radar pulse and the vegetation, the instrument can detect the change in plant structure and the amount of water in the plant, aspects that can be associated with the change in biomass. This index is also susceptible to the roughness of the soil, and on recently harvested soils, the stubble on it can be seen brighter.

The formulation of RVI for S1 data according to Nasirzadehdizaji et al. (2019) is:

$$RVI = \frac{4 * \sigma_{VH^{(i)}}}{\sigma_{VV^{(i)}} + \sigma_{VH^{(i)}}} \tag{15.1}$$

The script used, named RVI4S1, was developed by Dipankar Mandal (https://custom-scripts.sentinel-hub.com/custom-scripts/sentinel-1/radar_vegetation_index/script.js).

First, an equivalent to the degree of polarization (DOP) is calculated as $\sigma_{VV^{(i)}}$ / SPAN. SPAN is the total power received at both polarizations and can be treated as $\sigma_{VV^{(i)}} + \sigma_{VH^{(i)}}$. The DOP is utilized to obtain the depolarized fraction as m = 1 − DOP. The m-factor ranges from 0 to 1. For pure or elementary targets, the m-value is close to zero, whereas, for a fully random canopy (at high vegetative growth), it reaches close to 1. This m-factor is multiplied with the vegetation depolarization power fraction, Eq. [15.1]. The m factor is modulated with a square root scale for a better dynamic range of the RVI4S1 index.

15.2.5.5 S2 Derived Indices

Twenty radiometric indices (RIs) for soil, vegetation, and water applications (Table 15.3) were derived using S2 surface reflectance imagery, mostly from the online Index Database (IDB) (www.indexdatabase.de). The IDB is a tool that was created to give an overview of satellite radiometric indices that can be obtained from a specific sensor for a particular application (Henrich et al. 2009).

RIs for Espinal ecotone and crops were analyzed using the Random Forest algorithm (RF). RF can not only classify remote sensing images but also plays a

Table 15.3 Sentinel-2 remote sensing indices

Indice	Name	Reference
ATSAVI	Adjusted Transformed Soil Adjusted Vegetation Index	Baret and Guyot (1991)
BI2	Second brightness index	Escadafal et al. (1994)
GNDVI	Green normalized difference Vegetation index	Gitelson et al. (1996)
MSBI	Misra soil brightness index	Misra et al. (1977)
MSI	Moisture stress index	Welikhe et al. (2017)
MSRREn	Modified simple ratio red edge normalized	Wu et al. (2008)
NITI	Narrow near-infrared tillage index	Najafi et al. (2019)
NDSI	Normalized Difference Salinity Index	Al-Khaier (2003)
NDTI	Normalized difference tillage index	Van Deventer et al.(1997)
NDVI	Normalized difference Vegetation index	Rouse et al. (1974)
NDVI45	Normalized difference Vegetation index 4 and 5-bands	Delegido et al. (2011)
REOSAVI	Red edge optimized soil adjusted Vegetation index	Wu et al. (2008)
RERVI	Red edge ratio Vegetation index	Cao et al. (2013)
SARVI	Soil and Atmospherically Resistant Vegetation Index	Huete and Liu (1994)
SRNIR narrowGreen	Simple ratio NIR narrow and green	Le Maire et al. (2004)
SRNIR narrowRed	Simple ratio NIR narrow and red	Blackburn (1998)
TSAVI	Transformed Soil Adjusted Vegetation Index	Baret et al. (1989)
VRETI	Vegetation red edge tillage index	Najafi et al. (2019)
VRESTI	Vegetation red edge shortwave tillage index	
VSSI	Vegetation soil Salinity index	Dehni and Lounis (2012)

significant role in feature selection and dimension reduction (Belgiu and Dragut 2016; Gislason et al. 2004). The algorithm RF requires setting two main parameters: (1) the number of trees (K) and (2) the number of randomly selected variables to partition a node (mtry). In this study, the classical parameter values were used: K = 100 and mtry = 20.

We ensure that the generation of training and testing samples for the RF algorithm is the same for both periods studied. More than 25 training samples from agricultural lands ranging in size from 10 to 15 ha were evaluated using different field soil observations (2016–2019) and cropping backgrounds. The non-agricultural areas (Espinal ecotone) were sampled using 30 random points and 20 polygons ranging in size from 5 to 10 ha. We validated the LULC map using the reserved testing samples: points (45) and polygons (30), which were not used to train the classification algorithm. In addition to the confusion matrix, producer accuracy, user accuracy, overall accuracy, and the kappa coefficient were used to summarize accuracy statistics.

For each LULC class and environmental moisture conditions, the 3 most influential RIs were selected.

15.2.5.6 Radar and Optical Integration

S1 and S2 imagery were integrated using the collocation tool in SNAP (Braun 2021). In this algorithm, the pixel values of slave products (S2 data) are combined into the geographic grid of the master product (S1 data). In this way, a new product was created containing a copy of all the components of the master and slave products, including backscatter (σ^0_{VV}, σ^0_{VH}), radar vegetation index (RVI4S1), and the three selected optical radiometric indices. The combined images, composed of 6-bands for each moisture condition, were then used for image analysis and classification.

15.2.5.7 Feature Extraction and Classification

We chose the RF algorithm to classify (before explained in Sect. 15.2.5.5). In this last step, the number of trees was set also to K = 100 to reduce calculation time without a major loss in accuracy. The training vectors were the seven LULC classes (see Table 15.2).

Before visualizing the classified outputs, the valid pixel expression had to be changed in the SNAP v9.0 software. By default, only the pixels with a confidence threshold above 0.5 are displayed. To change this parameter, we delete the expression "confidence >= 0.5".

Finally, cross-validation was performed by visual interpretation of classes (Planet's and Google Earth high-resolution imageries) from 300 random points to calculate overall accuracy and kappa coefficient.

15.3 Results and Discussion

15.3.1 Backscattering Behavior

SARs respond to two fundamental characteristics of an agriculture target: structure and moisture. However, backscattering from a vegetated area is not limited to scattering by the vegetation itself. It also includes backscatter from the underlying surface that is attenuated by vegetation, as well as the interaction between soil and vegetation (Ulaby and Long 2014).

The texture and structure, backscatter-derived tone, and bright tone outliers are all represented in the RGB color composite of the image SAR (Simms 2019). Figure 15.4a, b shows the S1 color composite of dual-polarimetric data with σ^0_{VV} (red filter), σ^0_{VH} (green filter), and ratio ($\sigma^0_{VV}/\sigma^0_{VH}$, blue filter) for wet and dry years, respectively. When it rains, soil moisture increases and so does the amount of surface scattering in the soil. Figure 15.4b shows the effects of vegetation and soil moisture on signal brightness in the σ^0_{VH} polarization (green filter) in contrast to the driest year, 2020 (Fig. 15.4b). Note that these differences in greenness are related to moisture changes and structural features that strongly influence backscatter. The differences are notable, particularly because of the appearance of some darker greens in crops expected to have high sensitivity to volume scattering (de Jeu et al. 2008). Green/yellow tones, which correspond to natural Espinal ecotone, present brighter colors during the dryer period (Fig. 15.4b) compared to the wetter one (Fig. 15.4a, b).

The ratio between σ^0_{VV} and σ^0_{VH} is derived based on the premise that the σ^0_{VV} signal accounts for changes in soil scattering, which is mostly governed by soil

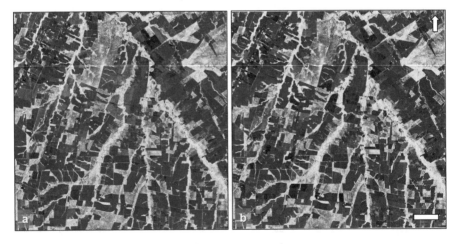

Fig. 15.4 (**a, b**) S1 calibrated products of σ^0_{VV} (red filter), σ^0_{VH} (green filter), and ratio (blue filter) for wet and dry years, 15.4a and 15.4b, respectively. Note the differences in shades of green color (cross-polarized) between the images. The length of the white bar is 2.6 km

moisture. If the signal does not contain any soil signal, the ratio may be less accurate and σ^0_{VH} may more accurately reflect the dynamics of the vegetation. Due to the dry soil, the signal penetrates deeper into it, and the mechanism of scattering switches from surface scattering to volume scattering. An increase in volume scattering leads to an increase in the ratio, which is less affected by sub-surface scattering (soil) and whose C-band product uses slightly different frequencies, i.e., has a different penetration depth (Ulaby and Long 2014).

In the early growth stages of the crops, the image is dominated by low σ^0_{VV} backscatter, giving the composite RGB image (σ^0_{VV}, σ^0_{VH}, and ratio) a purple appearance. The C-band signal shows higher sensitivity when the young plants are too small. This can be seen in Fig. 15.4a, b, where the wide range of distinct colors also represents the differences between the growth stages of the agricultural lands (Bousbih et al. 2017). The agricultural landscape can also be identified by geometric patterns and texture features in the image, such as fields, roads, and streams. Forests and isolated forest stands appear green due to higher σ^0_{VH} backscatter. The natural, semi-open canopy with some gaps between trees allows limited C-band penetration and detection of scrubs and shrubs (reddish color) due to dominant direct co-polarization scattering and grasses (yellow-green color) associated with a weak response under σ^0_{VV} and a strong response under σ^0_{VH} polarization (Ulaby and Long 2014). Non-vegetated fields with smooth surfaces can be identified by their weak response to both σ^0_{VV} and σ^0_{VH} and their blue color. The high moisture content of the soil in this area during the wet period of 2016–2017 contributes to the low backscatter of bare soil surfaces. After harvest (the bare phase), some agricultural fields exhibit a bright magenta color. This color is the result of a tonal change in the RGB image, specifically white, rough (σ^0_{VV}) and black, smooth (σ^0_{VH} and ratio) (Henderson and Lewis 1998; CEOS 2018).

SAR image tones are determined by the nature of the backscatter. These can be diffuse (bright, rough) or specular (dark, smooth), and a continuing beam pathway may be subject to corner or double-bounce backscattering (Simms 2019). Then a relative suite of colors was established for visualizing the image tones. The descriptions used in this chapter with the spectrum palette of the software SNAP are very dark (black), dark (indigo-purple), intermediate (blue, cyan, green), bright (yellow, orange), and very bright (red). Figure 15.5a–d shows the type of backscatter that can be distinguished in increasing order of roughness, from very dark (black) to very bright (red). Backscatter will increase as soil roughness increases. For soils, this means random roughness caused by tillage (and other farm operations) modified by soil erosion and weathering effects, and periodic row structures caused by tillage and planting. The dry year 2020 is characterized by lower backscatter values than the wet period 2016–2017, for both polarizations.

These results indicated that S1 data could potentially represent the interannual behavior of native vegetation and agricultural lands. The temporal profiles of σ^0_{VV}, σ^0_{VH}, and ratio ($\sigma^0_{VV}/\sigma^0_{VH}$) were influenced by contributions from the canopy and ground, such as soil moisture content and vertical plant structure.

Figure 15.6 shows the wet and dry backscatter coefficient values (σ^0) of LULC classes (see Table 15.2). The distribution of the collected data is shown by a

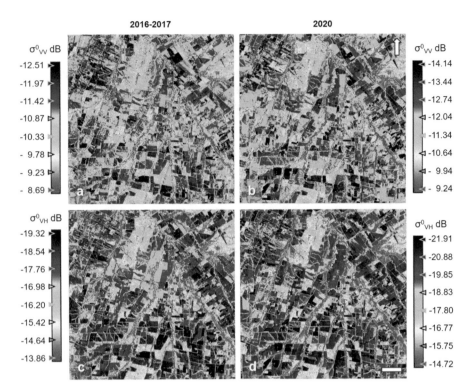

Fig. 15.5 (**a–d**) Average backscatter values (dB). The hues of the S1 images are determined by the type of backscatter, which can be distinguished in ascending order of roughness from very dark (black) to very bright (red). The dry year 2020 is characterized by lower backscatter values than the wet period 2016–2017, for both polarizations. The length of the white bar is 2.6 km

box-and-whisker plot based on five representative quantities: minimum, first quartile, median, third quartile, and maximum. Outliers are also singled out. In general, backscatter coefficients (wet and dry) show adequate separation between classes: Espinal ecotone (Closed gallery forest, Mid to open gallery forest, and Open low forest and shrubland) and agricultural lands (slight soil water erosion, moderately saline and slight soil water erosion, slight to moderate soil water erosion, and moderate to severe soil water erosion).

Table 15.4 shows partial correlation coefficients (Pearson product-moment correlation) between each pair of variables. The magnitude of the correlations ranges from −1 to +1. A value below 0 indicates that there is a negative correlation, i.e., the two variables are inversely related. The closer the value is to −1, the stronger the inverse relationship (if the value of one variable is very high, the value of the other variable is very low). If the value is exactly −1, it means that the two variables have a perfect negative correlation.

There is enough consensus when interpreting Pearson's correlation coefficient values using the following criteria (and considering absolute values): non-existent

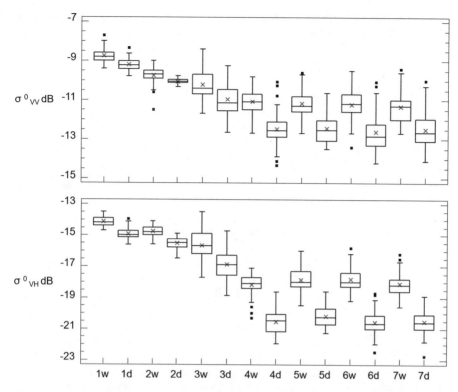

Fig. 15.6 Box-and-whisker plots showing mean, median, quartiles, maximum, and minimum for wet (w) and dry (d) backscatter coefficient (σ^0) values of the LULC classes. The box is drawn to extend from the lower quartile of the sample to the upper quartile. This is the interval covered by the middle 50% of the data values when sorted from smallest to largest. A vertical line is drawn at the median (the middle value). The x (red) is placed at the location of the sample mean. Outside points, which are points more than 1.5 times the interquartile range (box width) above or below the box, are indicated by black point symbols. LULC classes: (1) Closed gallery forest, (2) Mid to open gallery forest, (3) Open low forest and shrubland, (4) Agricultural lands, slight soil water erosion, (5) Agricultural lands, moderately saline and slight soil water erosion, (6) Agricultural lands, slight to moderate soil water erosion, and (7) Agricultural lands, moderate to severe soil water erosion

correlation (0–0.10), weak correlation (0.10–0.29), moderate correlation (0.30–0.50), and strong correlation (0.51–1.00).

For σ^0_{VV} the following pairs of variables have P-values below 0.05: 1w-1d (closed gallery forest), and 3w-3d (open low forest and shrubland). Consequent to this, the moisture condition information was more relevant than the LULC class information. On the other hand, for σ^0_{VH} the pairs of variables corresponding to 1w-1d (closed gallery forest), 1d-2w (dry closed gallery forest and wet mid to open gallery forest), 3w-3d (open low forest and shrubland), 4w-6d (wet agricultural lands, slight soil water erosion and dry agricultural lands, slight to moderate water erosion), and 5w-7d (wet agricultural lands, moderate saline, and slight soil water erosion and dry

Table 15.4 Pearson correlation coefficients between LULC classes and the variables σ^0_{VV} and σ^0_{VH}, wet (w) and dry (d), respectively

LULC classes	σ^0_{VV}													
	1w	1d	2w	2d	3w	3d	4w	4d	5w	5d	6w	6d	7w	7d
1w	■	*0.69	0.14	0.30				-0.27	-0.21			0.19		
1d	*0.69	■	0.13	0.29	0.23	0.23	0.10	-0.21		0.15	-0.11			
2w	0.14	0.13	■		0.27	0.27	-0.33		0.17		-0.38			
2d	0.30	0.29		■	-0.32	-0.29	0.30	-0.27	-0.19	-0.20	-0.20	0.14		0.14
3w		0.23	0.27	-0.32	■	*0.92	-0.15	0.19			-0.32	-0.44		
3d		0.23	0.27	-0.29	*0.92	■	-0.19	0.10			-0.33	-0.35		
4w		0.10	-0.33	0.30	-0.15	-0.19	■			0.12		-0.24		0.12
4d	-0.27	-0.21		-0.27	0.19	0.10		■	0.10		-0.11	-0.21		
5w	-0.21		0.17	-0.19				0.10	■	0.17			-0.10	-0.23
5d		0.15		-0.20			0.12		0.17	■	0.21			0.10
6w		-0.11	-0.38	-0.20	-0.32	-0.33		-0.11		0.21	■	0.34		-0.28
6d	0.19			0.14	-0.44	-0.35	-0.24	-0.21			0.34	■	-0.18	-0.11
7w									-0.10			-0.18	■	
7d				0.14			0.12		-0.23	0.10	-0.28	-0.11		■

LULC classes	σ^0_{VH}													
	1w	1d	2w	2d	3w	3d	4w	4d	5w	5d	6w	6d	7w	7d
1w	■	*0.55		0.11	-0.20	-0.13	0.15		-0.14				0.15	0.11
1d	*0.55	■	*0.35					0.14		0.17	-0.15	0.15	0.34	
2w		*0.35	■	0.24	0.15	0.26	-0.10	0.19	0.18	0.16				-0.11
2d	0.11		0.24	■	-0.17	-0.17		-0.18		0.20	-0.20			
3w	-0.20		0.15	-0.17	■	*0.86		-0.15	-0.20	0.12	0.10	-0.13		
3d	-0.13		0.26	-0.17	*0.86	■	-0.11		0.17					
4w	0.15		-0.10			-0.11	■					*-0.41		0.17
4d		0.14	0.19	-0.18	-0.15			■		0.15		0.24		-0.18
5w	-0.14		0.18		-0.20	0.20	0.17		■	-0.17			-0.33	*-0.26
5d		0.17	0.16	0.20	0.12		0.15	-0.17		■	0.29	0.16	0.20	
6w		-0.15		-0.20	0.10					0.29	■	0.31		0.11
6d		-0.15			-0.13		*-0.41	-0.24			0.31	■	-0.13	
7w	0.15	0.34							-0.33	0.20		-0.13	■	0.14
7d	0.11		0.11				-0.17	-0.18	*-0.26		0.11		0.14	■

LULC classes (see Table 15.2): Espinal ecotone (1, 2, and 3). Agricultural lands (4, 5, 6, and 7). Weak correlation: 0.10–0.29, Moderate correlation: 0.30–0.50, and Strong correlation: 0.51–1.00. *P-values below 0.05 indicate statistically significant non-zero correlations at the 95.0% confidence level

agricultural lands, moderate to severe soil water erosion). For most σ^0_{VH} pairs, both the moisture conditions and LULC information were relevant.

Some significant ($P < 0.05$) relationships between vegetation and SAR backscatters and interannual moisture conditions were observed. The r values related to SAR backscatters presented considerably large variations during wet and dry years, indicating the unstable relationships between vegetation, soil, and SAR backscatters. Most of their differences can be explained by fluctuations in SAR backscatters. For example, the dynamics of soil moisture content driven by drought may cause the decrease of σ^0_{VV} and σ^0_{VH} backscatters (as shown in Fig. 15.5a–d).

15.3.2 SAR Log-Ratio Operator

The main goal of image-based change detection techniques is to suppress background information in the image while preserving the key change signatures of interest (Zanetti and Bruzzone 2018). The logarithmic scaling method shown in

Fig. 15.7 (**a, b**) S1 Change Detection. 15.7a Log-ratio scaling. $C_{VV2016-2017}$ versus C_{VV2020}. 15.7b $C_{VH2016-2017}$ versus C_{VH2020}. Note the differences in polarization in both figures. In the log-ratio image, unchanged features have medium shades of gray (gray value around zero), while changed features are either bright white or dark black. Black features indicate areas where radar brightness has decreased, while white areas have increased brightness. The length of the white bar is 2.6 km

Fig. 15.7a, b is effective in suppressing background and enhancing change features. It minimizes the effects of interannual variations and undesirable changes in surface scattering on change detection.

The differences between images are largely due to changes in moisture availability between acquisition dates and/or differences in agricultural activity. Unchanged features have medium shades of gray (gray values around zero), while changed features are either bright white or dark black. Black features indicate areas where radar brightness has decreased, while white areas have increased brightness. The annual signal is visually more stable in the case of natural vegetation, such as grasslands, shrublands, and forests (Espinal ecotone) than for crops areas (soy, corn, sorghum, wheat, implanted pastures). Considering that the average backscatter difference in dB for σ^0_{VV} is 1.04, and for σ^0_{VH} is 1.78, this could be hypothesized as a result of the combined effects of crop growth and changing soil moisture conditions (Harfenmeister et al. 2019). The rainfall in 2020 was much lower (990 mm) than in 2016 (1647 mm) and 2017 (1516 mm).

15.3.3   Radar Vegetation Index

Radar vegetation index (RVI) is related to the vegetation water content (VWC) (Kim et al. 2012) which is directly related to the wetness of vegetation. From the interannual images (Fig. 15.8a, b), it is evident that the RVI4S1 visually correlates changes in vegetation water content and phenological stages very well. Shang et al.

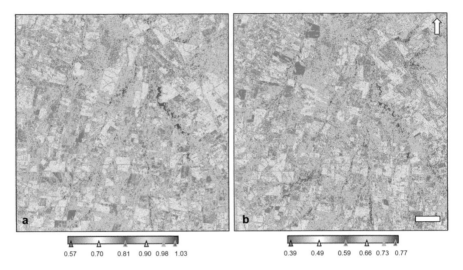

0.57   0.70   0.81   0.90 0.98 1.03   0.39   0.49   0.59 0.66 0.73 0.77

Fig. 15.8 (**a**, **b**) Radar Vegetation Index (RVI4S1). 15.8a 2016–2017 period. 15.8b 2020. The length of the white bar is 2.6 km

(2020) observed sensitivity of the backscatter intensities of σ^0_{VV} and σ^0_{VH} to the presence of water stagnation and soil moisture due to precipitation.

In 2020, RVI4S1 levels were low (mean 0.69 ± 0.15), resulting in a dark appearance on the images. In 2016–2017, the dense, complex distribution of crop canopies produced high randomness in SAR backscatter, resulting in brighter pixels on the agricultural lands (mean 1.00 ± 0.22). Thus, the complex geometry of the crops increases the randomness of the SAR backscatter. According to Bousbih et al. (2017), we found a relationship between ground surface roughness with σ^0_{VV} and σ^0_{VH} backscatter, as well as moisture content with σ^0_{VV} backscatter.

Similarly, radar and optical vegetation indices are used as input parameters in the assessment of other hazards, such as soil erosion, effects of drought, etc. However, in many cases, those indices are used separately, with no synergistic application, restricting the full potential of current Earth Observation Systems.

15.3.4 Soil and Vegetation Radiometric Indices

As a result of the analysis of RF, Table 15.5 shows the selected indices for soil: Soil and Atmospherically Resistant Vegetation Index (SARVI), Red Edge Optimized Soil-Adjusted Vegetation Index (REOSAVI), and Misra Soil Brightness Index (MSBI); for vegetation: Normalized Difference Vegetation Index 4 and 5-bands (NDVI45), for tillage (residue cover): Normalized Difference Tillage Index (NDTI), and Vegetation Red Edge Tillage Index (VRETI); and for moisture: Moisture Stress Index (MSI).

Table 15.5 Selection of optical radiometric indices by Random Forest (RF) analysis for wet (numerator) and dry (denominator) years

LULC	2016–2017 (wet) 2020 (dry)							
	Accuracy	Precision	Correlation	Error Rate	Correct Predictions %	RMSE	Bias	¹ Rank in order of importance
Espinal ecotone (38%)	0.9178	0.8051	0.8054	0.0822	84.74	0.7056	−0.0458	NDVI45-SARVI-REOSAVI
	0.9578	0.8933	0.8943	0.0422	85.64	0.6909	−0.0162	NDVI45-MSI-MSBI
Agricultural lands (62%)	0.8326	0.8224	0.7211	0.1674	83.26	0.4092	0.0158	NDVI45-NDTI-VRETI
	0.8978	0.8877	0.8164	0.1022	89.78	0.3197	0.0130	MSI-NDVI45-NDTI

ª First predictors that were highly significant for RF. RMSE: Root-mean-square deviation. Bias: Objective property of an estimator. Soil and Atmospherically Resistant Vegetation Index (SARVI), Red Edge Optimized Soil-Adjusted Vegetation Index (REOSAVI), Misra Soil Brightness Index (MSBI). For vegetation: Normalized Difference Vegetation Index 4 and 5-bands (NDVI45), for tillage (residue cover): Normalized Difference Tillage Index (NDTI), and Vegetation Red Edge Tillage Index (VRETI); and for moisture: Moisture Stress Index (MSI)

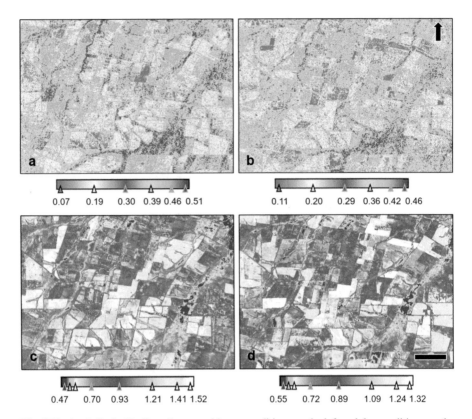

Fig. 15.9 (**a–d**) Optical indices. Images with wet conditions on the left and dry conditions on the right. 15.8a-b Normalized Difference Vegetation Index 4 and 5-bands (NDVI45). 15.9c-d Moisture Stress Index (MSI). The length of the white bar is 1.5 km

The results show the sensitivity of S2 detecting changes caused by moisture variations in different years. When considering soil and vegetation, slight changes in soil color and changes in structure and spatial distribution of vegetation may be indicators of changes in soil degradation in agricultural and natural landscapes, as observed at the test site.

NDVI, which is recognized as a standard for vegetation, is not found in the first ranking of the analysis of RF. One explanation for the relationship between NDVI and vegetation characteristics could be related to confounding influences on the index, as well as atmosphere and ground brightness. NDVI45 was the most significant vegetation index in the land covers and under both moisture conditions (Fig. 15.9a, b) This index is more linear and has lower saturation than NDVI at higher values. Delegido et al. (2011) specifically investigated the optimal bands for use in the NDVI formula with synthesized S2 data. They found that bands 4 and 5 are the optimal combination.

MSI shows the spatial extent of lower moisture due to higher evapotranspiration in 2020 compared with the 2016–2017 period (Fig. 15.9c,d). The effects of drought are related to a prolonged period of unusually warm weather and drought in southern Brazil, Paraguay, and northern Argentina (Grimm et al. 2020).

NDTI has been successfully used to discriminate some broad tillage classes (Sullivan et al. 2008), as shown for both conditions (wet and dry). However, under dry conditions, the ranking is different. In general, water in crop residues and soils attenuates the reflectance signal at all wavelengths and reduces the contrast between soils and crop residues (Quemada and Daughtry 2016). The water content of soils and crop residues often vary spatially and temporally across fields, even with minor changes in topographic relief (residues cover <30% are common).

VRETI is important for the wet period. Because crop residues decompose after harvest, they can be either brighter or darker than soils depending on soil type, crop type, soil water content, and crop residues (including the degree of decomposition), which makes them difficult to distinguish (Wang et al. 2013). In recent years, conservation tillage methods have been recommended as a suitable alternative to intensive tillage methods because of water and soil erosion problems caused by agricultural activities (Zheng et al. 2014; Najafi et al. 2019). Since soybean cultivation is the predominant crop in the study area, it leaves less residue on the soil surface, unlike maize and sorghum.

Reflectance values for SARVI in the Espinal ecotone for the wet period are associated with negative values indicating temporary accumulation of surface water. High positive values indicate closed to open vegetated areas (natural and cultivated).

REOSAVI is strongly correlated with plant physiological parameters, crop yields, and biomass (Verhulst et al. 2009).

MSBI is present both in natural vegetation (Espinal) and in croplands (bare phase). In the Espinal, this index interacts with the influence of the vegetation cover, and it is based on soil properties and exposure (Jinru and Baofeng 2017).

15.3.5 Integration of S1 and S2 Bands

To reduce the confusion between classes, S1 data were synergistically combined with S2 indices to improve the soil and land use assessment. Figure 15.10a–d shows for both wet and dry conditions, a true-color image compared to a false-color composite. Such combinations reveal patterns that are not visible by optical data only (Braun 2021).

Table 15.6 shows the results of the classification with the RF algorithm. The three SAR covariates had the strongest relationship with the optical indices. The number of predictions was very high, above 80%.

The Espinal ecotone presents for the wet period, in order of importance, the following variables: σ^0_{VV}, RVI4S1, NDVI45, σ^0_{VH}, REOSAVI, and SARVI. Unlike in the driest period, where the MSI optical index predominates in the first place, followed by σ^0_{VH}, NDVI45, RVI4S1, σ^0_{VV}, and MSBI. On the other hand, agricultural lands show the covariates radar σ^0_{VV} and σ^0_{VH} in the first places, for the wet and dry

Fig. 15.10 (**a–d**) Images with wet conditions on the left and dry conditions on the right. 15.10a-b True-color images with red = S2-B4, green = S2-B3, and blue = S2-B2. 15.10c-d Merged stack of S2 and S1 bands, false-color composite with red = S2-B4, green = S2-B3, and blue = S1-C_{VV}. The length of the white bar is 1.5 km

period, respectively. The co-polarization in the wet period is mostly dominated by canopy and ground contributions and is also influenced by soil moisture (Bousbih et al. 2019). In dry conditions, the role of ground-vegetation interaction in cross-polarization may be less than the contribution of volume scattering as a result of the increase in σ^0_{VH} backscatter (Ulaby and Long 2014). Both the RVI4S1 and optical NDVI45 are sensitive to vegetation biomass, although most studies in the literature (Gonenc et al. 2019) refer to the radar vegetation index with the optical classical index NDVI.

MSI served as a deterministic drought index; it was affected by fluctuating water content (Espinal ecotone and agricultural lands), with higher values indicating greater water stress and lower water content (Welikhe et al. 2017).

Figures 15.11 and 15.12 show the LULC change for 2016–2017 and 2020, respectively. For this last year, the classes as a percent of the area were stable for mid to open gallery forest (2), agricultural lands, slight soil water erosion (4), slight to moderate soil water erosion (6), and moderate to severe soil water erosion (7); decrease closed gallery forest (1) and increase in open low forest and shrubland (3) and agricultural lands, moderately saline and slight soil water erosion (5).

The overall accuracy and Kappa coefficient of the classification results reached 90%-0.78 (wet) and 92%-0.79 (dry), respectively.

Table 15.6 Selection of radar and optical radiometric indices by Random Forest (RF) analysis for wet (numerator) and dry (denominator) years

LULC classes[a]		2016–2017 (wet) 2020 (dry)							Rank in order of importance
		Accuracy	Precision	Correlation	Error rate	Correct predictions %	RMSE	Bias	
Espinal Ecotone	1	0.9166 0.9610	0.8014 0.9137	0.7581 0.8817	0.0834 0.0390	97.08 94.30	0.2227 0.2379	−0.0071 0.0094	σ0VV-RVI4S1-NDVI45-σ0VH-REOSAVI-SARVI MSI-σ^0_{VH}-NDVI45-RVI4S1-σ^0_{VV}-MSBI
	2	0.9348 0.9184	0.9185 0.8194	0.8313 0.8037	0.0652 0.0816	80.54 83.38	1.0473 1.0222	−0.0102 −0.0634	
	3	0.9428 0.9324	0.8387 0.8567	0.8349 0.7978	0.0572 0.0676	89.44 86.05	0.3087 0.2789	0.0134 0.0112	
Agricultural Lands	4	0.8942 0.9512	0.7910 0.9467	0.7470 0.8723	0.1058 0.0488	81.70 90.04	0.4170 0.2556	0.0390 0.0123	σ0VV -NDVI45-σ0VH -RVI4S1-NDTI-VRETI σ^0_{VH}-MSI-NDVI45-RVI4S1-σ^0_{VV}-NDTI
	5	0.8860 0.9354	0.6903 0.8613	0.6991 0.8067	0.1140 0.0646	81.30 84.72	0.3334 0.2567	0.0230 0.0100	
	6	0.8930 0.9018	0.7183 0.8019	0.7086 0.7643	0.1070 0.0982	80.46 83.70	0.3091 0.2765	0.0301 0.0422	
	7	0.8348 0.9184	0.8185 0.8194	0.8313 0.8037	0.0652 0.0816	80.54 83.38	1.0473 1.0222	0.0502 0.0634	

[a]LULC classes: (1) Closed gallery forest, (2) Mid to open gallery forest, (3) Open low forest and shrubland, (4) slight soil water erosion, (5) moderately saline and slight soil water erosion, (6) slight to moderate soil water erosion, and (7) moderate to severe soil water erosion

RMSE Root-mean-square deviation. Bias: Objective property of an estimator

%

1	10.3
2	12.5
3	15.2
4	20.9
5	11.8
6	21.8
7	7.5

Fig. 15.11 LULC map for 2016–2017. (1) Closed gallery forest, (2) Mid to open gallery forest, (3) Open low forest and shrubland, (4) Agricultural lands, slight soil water erosion, (5) Agricultural lands, moderately saline and slight soil water erosion, (6) Agricultural lands, slight to moderate soil water erosion, and (7) Agricultural lands, moderate to severe soil water erosion. The length of the white bar is 2.6 km

15.4 Concluding Remark

There is limited local literature on the function and potential of satellite data covering the microwave, optical and infrared regions of the spectrum for mapping LULC that can assist in enhancing existing medium scale (i.e., 1:100,000) soil maps. This chapter presents a first evaluation of the potential of S1 and S2 imagery, used individually and in combination, in wet and dry years in the northern part of the Entre Rios province of Argentina. The results show that both sensors can detect LULC changes in response to drought (2020). Image radar and optical mosaics in both the wetter and drier years, such as was used in this study, are recommended.

The interannual variability in the SAR backscatter values indicates that SAR signals are sensitive to drought stress for vegetation and underlying soils. The mean σ^0_{VV} and σ^0_{VH} backscatter values are 1.0 to 1.8 dB lower during the 2020 drought compared to values in 2016–2017.

Both σ^0_{VV} and σ^0_{VH} polarizations and a radar vegetation index (RVI4S1) combined with selected optical radiometric indices from RF analysis are suitable for representing LULC changes in years with a variability of moisture. The produced

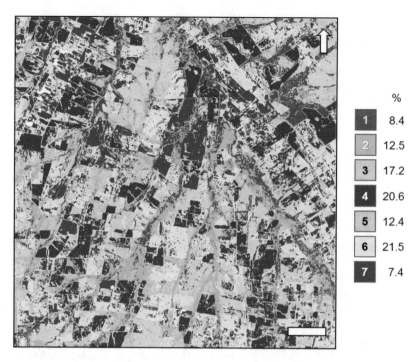

	%
1	8.4
2	12.5
3	17.2
4	20.6
5	12.4
6	21.5
7	7.4

Fig. 15.12 LULC map for 2020. Classes as a percent of the area were stable for mid to open gallery forest (2), agricultural lands, slight soil water erosion (4), slight to moderate soil water erosion (6), and moderate to severe soil water erosion (7); decrease closed gallery forest (1); and increase in open low forest and shrubland (3) and agricultural lands, moderately saline/slight soil water erosion (5). The length of the white bar is 2.6 km

RF rankings of importance on features not only help the user to choose the features to train the classification algorithm with, but also provide invaluable information about the usefulness of S1 and S2, and their spectral features. These results highlight the need for multisensor classification and monitoring of LULC changes to fully exploit the rich potential of existing and future complementary satellite sensors. Likewise, the results provide promising answers to the question of when to update soil productivity based on soil surveys, as the use of satellite imagery provides predictable and reproducible information. It is well known that the condition of vegetation in an agricultural area is related to crop development and directly correlated with potential soil productivity.

RVI4S1 has shown some promise in the vegetation fields, but its relationship with NDVI45 is not known in the context of LULC. These indices should be treated with caution because no direct correlation could be established. However, processed products reveal similar patterns between the optical and radar data. This allows us to fill in gaps when optical data is not available. The combination of RVI4S1-NDVI45 can reinforce the presence of vegetated areas with lower saturation than NDVI classic at higher values and can be integrated back into the σ^0_{VV} and σ^0_{VH}

polarizations. The results of these findings can be further used to support vegetation extraction using integrated radar and optical datasets.

The accuracy of the estimation of the combined data (S1 and S2) in the present study may be related to the environmental conditions of the region. Therefore, the results of our study are site-specific; more in-depth research is needed to obtain transferable results to other regions with similar characteristics. It is also necessary to compare the results that analyze the interannual and seasonal changes of LULC.

In the future, our research will focus on the development of fused radar and optical data for mapping the complexities of LULC to detect variability between and within agricultural lands, as well as systematic methods for evaluating the benefits of fusion at larger spatial scales. Future research on polarimetry and interferometry SAR should be included, as well as work on using higher resolution imagery from optical sensors to study changes closer to field scales, which will improve the interpretation of SAR signatures.

The knowledge about heterogeneity on agricultural lands is essential for sustainable and effective field management. Such research could increase our understanding of the value that multi-sensor remote sensing can add to a more effective mapping of soil and landscape properties that helps soil surveyors and agronomists identify soil and landscape qualities and limitations for productivity.

Acknowledgments We thank all open-source software developers, as well as the European Space Agency (ESA) and the Google Earth Engine (GEE) platform, for providing free satellite imagery and processing data that form the basis for sustainable land management, which requires broader knowledge and the collection of expertise at different levels. We are also grateful for the financial support provided by the Universidad Autónoma de Entre Rios (UADER), and PUE 56 project CONICET, Argentina.

References

Al-Khaier F (2003) Soil Salinity Detection Using Satellite Remotes Sensing. Master Thesis, International Institute for Geo-Information Science and Earth Observation, Enschede, p. 61

Baret F, Guyot G (1991) Potentials and limits of vegetation indices for LAI and APAR assessment. Remote Sens Environ 35(2–3):161–173. https://doi.org/10.1016/0034-4257(91)90009-U

Baret F, Guyot G, Major DJ (1989) TSAVI: a vegetation index which minimizes soil brightness effects on LAI and APAR estimation. In IGARSS 1989, Proc. IEEE International Geoscience and Remote Sensing Symposium and 12th Canadian Symposium on Remote Sensing, Vancouver, B.C. Canada, Vol. 3, pp. 1355–1358, IEEE, Piscataway, New Jersey (10–14 July 1989)

Bauer-Marschallinger B, Cao S, Navacchi C, Freeman V, Reuß F, Geudtner D, Rommen B, Vega FC, Snoeij P, Attema E, Reimer C, Wagner W (2021) The Sentinel-1 Global Backscatter Model (S1GBM) – mapping Earth's Land Surface with C-Band Microwaves (1.0) [S1GBM_VV_VH_mean_mosaic_v1_EQUI7_SA010M]. TU Wien. https://doi.org/10.48436/n2d1v-gqb91

Bedendo DJ (2019) Soils of Entre Rios. In: Rubio G, Lavado RS, Pereyra FX (eds) The soils of Argentina, World soils book series. Springer Int. Publ, pp 165–173

Bedendo DJ, Schulz GA, Olmedo GF, Rodríguez DM, Angelini ME (2016) Updating a physiography-based soil map using digital soil mapping techniques. In: Zinck JA, Metternicht GI, Bocco G, del Valle HF (eds) Geopedology. Springer International Publishing Switzerland, pp 305–319. https://doi.org/10.1007/978-3-319-19159-1_18

Bedendo DJ, Schulz GA, Pausich GM, Tentor F (2014) Cartas de Suelos de Entre Ríos. Proyecto INTA-PFIP ER07–08, INTA EEA Paraná, Centro Regional Entre Ríos. Publicación realizada por el servicio de mapeo web (WMS) en la red GeoINTA: http://geointa.inta.gov.ar/web/index.php/cartas-de-suelos-de-entre-rios/

Belgiu M, Dragut L (2016) Random forest in remote sensing: a review of applications and future directions. ISPRS J Photogramm Remote Sens 114:24–31

Blackburn GA (1998) Quantifying chlorophylls and carotenoids at leaf and canopy scales. Remote Sens Environ 66:273–285. https://doi.org/10.1016/S0034-4257(98)00059-5

Bousbih S, Zribi M, Chabaane Z, Baghdadi N, El Hajj M, Gao Q, Mougenot B (2017) Potential of Sentinel-1 radar data for the assessment of soil and cereal cover parameters. Sensors 17(11):2617. https://doi.org/10.3390/s17112617

Bousbih S, Zribi M, Pelletier C, Gorrab A, Chabaane Z, Baghdadi N, Aissa N, Mougenot B (2019) Soil texture estimation using radar and optical data from Sentinel-1 and Sentinel-2. Remote Sens. https://doi.org/10.3390/rs11131520

Bovolo F, Marin C, Bruzzone L (2013) A hierarchical approach to change detection in very high-resolution SAR images for surveillance applications. IEEE Transactions on Geoscience & Remote Sensing 51(4):2042–2054. https://doi.org/10.1109/TGRS.2012.2223219

Braun A (2021) Synergetic use of radar and optical data. Combination of Sentinel-1 and Sentinel-2 and application of analysis. Sentinel-1 Toolbox. SkyWatch Space Applications Inc. (https://skywatch.co), and ESA http://step.esa.int

Cao Q, Miao Y, Wang H, Huang S, Cheng S, Khosla R, et al. (2013) Non-destructive estimation of rice plant nitrogen status with Crop Circle multispectral active canopy sensor. Field Crops Research, 154:133–144

CEOS (2018) A layman's interpretation guide to L-band and C-band Synthetic Aperture Radar data. Version 2.0. 15 November 2018. Global Forest Observation Initiative (GFOI). Committee on Earth Observation Satellite (CEOS), Systems Engineering Office (SEO), pp 31

Dehni A, Lounis M (2012) Remote sensing techniques for salt-affected soil mapping: application to the Oran region of Algeria. Procedia Eng 33:188–198

de Jeu RAM, Wagner W, Holmes TRH, Dolman AJ, van de Giesen NC, Friesen J (2008) Global soil moisture patterns observed by space borne microwave radiometers and Scatterometers. Surv Geophys 29:399–420. https://doi.org/10.1007/s10712-008-9044-0

Delegido J, Verrelst J, Alonso L, Moreno J (2011) Evaluation of Sentinel-2 red-edge bands for empirical estimation of green LAI and chlorophyll content. Sensors 11:7063–7081. https://doi.org/10.3390/s110707063

del Valle HF, Tentor F, Sione WF, Zamboni P, Aceñolaza PG, Metternicht GI (2022) Vertical accuracy assessment of freely available digital elevation models: implications for low-relief landscapes. IEEE Geoscience and Remote Sensing Society. Kuala Lumpur, International Geoscience and Remote Sensing Symposium (IGARSS)

Demsar J, Curk T, Erjavec A, Gorup C, Hocevar T, Milutinovic M, Mozina M, Polajnar M, Toplak M, Staric A, Stajdohar M, Umek L, Zagar L, Zbontar J, Zitnik M, Zupan B (2013) Orange: data mining toolbox in python. J Mach Learn Res 14(Aug):2349–2353

De Ruyver R, Di Bella C (2019) Climate. In: Rubio G, Lavado RS, Pereyra FX (eds) The soils of Argentina, World soils book series. Springer Int. Publ, pp 26–47

Dwivedi RS (2017) Soil Resources Mapping. In: Remote Sensing of Soils. Springer, Berlin, Heidelberg. https://doi.org/10.1007/978-3-662-53740-4_7

Eibedingil IG, Gill TE, Van Pelt RS, Tong DQ (2021) Combining optical and radar satellite imagery to investigate the surface properties and evolution of the Lordsburg playa, New Mexico, USA. Remote Sens 13(17):3402. https://doi.org/10.3390/rs13173402

Escadafal R, Belghit A, Ben-Moussa A (1994) Indices spectraux pour la télédétection de la dégradation des milieux naturels en Tunisie aride. In: Guyot G (ed) Actes du 6eme Symposium international sur les mesures physiques et signatures en télédétection, vol 17–24. Janvier, Val d'Isère, pp 253–259

Gislason PO, Benediktsson JA, Sveinsson JR (2004) Random forest classification of multisource remote sensing and geographic data. In: Processing IEEE International Geoscience and Remote Sensing Symposium, vol 2. IEEE, pp 1049–1052

Gitelson A, Kaufman YJ, Merzlyak MN (1996) Use of a green channel in remote sensing of global vegetation from EOS-MODIS. Remote Sens Environ 58(3):289–298. https://doi.org/10.1016/s0034-4257(96)00072-7

Gonenc A, Ozerdem MS, Acar E (2019) Comparison of NDVI and RVI Vegetation Indices Using Satellite Images. 8th International Conference on Agro-Geoinformatics, pp. 1–4. https://doi.org/10.1109/Agro-Geoinformatics.2019.8820225

Gorelick N, Hancher M, Dixon M, Ilyushchenko S, Tau D, Moore R (2017) Google earth engine: planetary-scale geospatial analysis for everyone. Remote Sens Environ 202:18–27. https://doi.org/10.1016/j.rse.2017.06.031

Grimm AM, Almeida Scortegagna A, Beneti CAA, Alvim Leite E (2020) The combined effect of climate oscillations in producing extremes: the 2020 drought in southern Brazil. RBRH 25 e48. Epub November 16, 2020. https://doi.org/10.1590/2318-0331.252020200116

Harfenmeister K, Spengler D, Weltzien C (2019) Analyzing temporal and spatial characteristics of crop parameters using Sentinel-1 backscatter data. Remote Sens 11:1569. https://doi.org/10.3390/rs11131569

Henderson F, Lewis A (eds) (1998) Principles and applications of imaging radar. Manual of remote sensing, vol 2, 3rd edn. Wiley

Hengl T, Nussbaum M, Wright MN, Heuvelink GBM, Gräler B (2018) Random forest as a generic framework for predictive modeling of spatial and spatio-temporal variables. Peer J 6:e5518

Henrich V, Götze C, Jung A, Sandow C, Thürkow D, Glaesser C (2009) Development of an online indices database: Motivation, concept and implementation. Conference: 6th EARSel Imaging Spectroscopy SIG Workshop Innovative Tool for Scientific and Commercial Environment Applications. Tel Aviv, Israel

Huete AR, Liu HQ (1994) An error and sensitivity analysis of the atmosphere- and soil-correcting variants of the NDVI for the MODIS-EOS. IEEE Trans Geosci Remote Sens 32(4):897–905

Inglada J, Vincent A, Arias M, Marais-Sicre C (2016) Improved early crop type identification by joint use of high temporal resolution SAR and optical image time series. Remote Sens 8:362

Jin X (2017) ENVI automated image registration solutions. L3 Harris Geospatial Solutions, Inc. whitepaper (2017). Available online at http://www.l3harrisgeospatial.com/Portals/pdfs/ENVI_Image_Registration_Whitepaper

Jinru X, Baofeng S (2017) Significant remote sensing vegetation indices: a review of developments and applications. J Sens 2017, 1353691, 17 pages. https://doi.org/10.1155/2017/1353691

Karra K, Kontgis C, Statman-Weil Z, Mazzariello JC, Mathis M, Brumby SP (2021) Global land use/land cover with Sentinel 2 and deep learning. In: 2021 IEEE International Geoscience and Remote Sensing Symposium IGARSS 2021 Jul 11 pp 4704–4707

Kim Y, Jackson T, Bindlish R, Lee H, Hong S (2012) Radar vegetation index for estimating the vegetation water content of Rice and soybean. IEEE Geosci Remote Sens Lett 9:564–568

Kirches G (2020) Algorithm Theoretical Basis Document Sentinel 2 Global Mosaics Copernicus Sentinel-2 Global Mosaic (S2GM) within the Global Land Component of the Copernicus Land Service. JRC Document, Ref.: S2GM-ATBD-BC Framework

Lausch A, Baade J, Bannehr L, Borg E, Bumberger J, Chabrillat S, Dietrich P, Gerighausen H, Cornelia G, Hacker J, Haase D, Jagdhuber T, Jany S, Jung A, Karnieli A, Krämer R, Makki M, Mielke C, Möller M, Schaepman M (2019) Linking remote sensing and geodiversity and their traits relevant to Biodiversit (part I: soil characteristics). Remote Sens 2356. https://doi.org/10.3390/rs11202356

Lausch A, Schaepman ME, Skidmore AK, Truckenbrodt SC, Hacker JM, Baade J, Bannehr L, Borg E, Bumberger J, Dietrich P, Gläßer C, Haase D, Heurich M, Jagdhuber T, Jany S, Krönert R, Möller M, Mollenhauer H, Montzka C, Pause M, Rogass C, Salepci N, Schmullius C, Schrodt F, Schütze C, Schweitzer C, Selsam P, Spengler D, Vohland M, Volk M, Weber U, Wellmann T, Werban U, Zacharias S, Thiel C (2020) Linking the remote sensing of geodiversity and traits relevant to biodiversity (part II: geomorphology, terrain and surfaces). Remote Sen 12(22):3690. https://doi.org/10.3390/rs12223690

Le Maire G, François C, Dufrêne E (2004) Towards universal broad leaf chlorophyll indices using PROSPECT simulated database and hyperspectral reflectance measurements. Remote Sens Environ 89:1–28. https://doi.org/10.1016/j.rse.2003.09.004

Maldonado FD, Sione WF, Aceñolaza PG (2012) Mapeo de desmontes en áreas de bosque nativo de la Provincia de Entre Ríos. Ambiência 8:523–532. Guarapuava (PR). ISSN 1808–0251

Matteucci SD, Rodríguez AF, Silva ME (2019) Vegetation. In: Rubio G, Lavado RS, Pereyra FX (eds) Pp 49–62, the soils of Argentina. Springer Int. Publ, World Soils Book Series

Meyer F (2019) Spaceborne synthetic aperture radar: principles, data access, and basic processing techniques. Chapter 2, appendix a, pp 21–64. In: Flores-Anderson AI, Herndon KE, Thapa RB, Cherrington E (eds) The synthetic aperture radar (SAR) handbook: comprehensive methodologies for Forest monitoring and biomass estimation. SERVIR Global Science Coordination Office National Space Science and Technology Center. http://www.servirglobal.net/

Misra PN, Wheeler SG, Oliver RE (1977) Kauth-Thomas brightness and greenness axes. (IBM personal communication). Contract NAS-9-14350,. RES, pp 23–46

Mulder VL, de Bruin S, Schaepman M, Mayr TR (2011) The use of remote sensing in soil and terrain mapping – a review. Geoderma 162:1–19

Najafi P, Navid H, Feizizadeh B, Eskandari I, Blaschke T (2019) Fuzzy object-based image analysis methods using sentinel-2A and Landsat-8 data to map and characterize soil surface residue. Remote Sens 11:2583

Nasirzadehdizaji R, Balik Sanli F, Abdikan S, Çakir Z, Sekertekin A, Üstüner M (2019) Sensitivity analysis of multi-temporal Sentinel-1 SAR parameters to crop height and canopy coverage. Appl Sci 9:655. https://doi.org/10.3390/app9040655

Nguyen P, Shearer EJ, Tran H, Ombadi M, Hayatbini N, Palacios T, Huynh P, Updegraff G, Hsu K, Kuligowski B, Logan WS, Sorooshian S (2019) The CHRS data portal, an easily accessible public repository for PERSIANN global satellite precipitation data. Nature Sci Data 6:80296. https://doi.org/10.1038/sdata.2018.296

Planet Team (2020) Planet Surface Reflectance Product v2. Planet Labs Inc Accessed 18(08):2020

Poggio L, Gimona A (2017) Assimilation of optical and radar remote sensing data in 3D mapping of soil properties over large areas. Sci Total Environ 579:1094–1110. https://doi.org/10.1016/j.scitotenv.2016.11.078

Poppiel R (2019) Pedometric mapping of key topsoil and subsoil attributes using proximal and remote sensing in Midwest Brazil. Faculdade de Agronomia e Medicina Veterinária, Universidade de Brasília, Tese de Doutorado em Agronomia. p. 105

Quemada M, Daughtry CS (2016) Spectral indices to improve crop residue cover estimation under varying moisture conditions. Remote Sens 8:660. https://doi.org/10.3390/rs8080660

Rivet JM (2022) Sentinel-2 Global Mosaic. Product User Manual. First version of S2GM phase 2. Copernicus. Land Monitoring Service. Global Mosaic. S2GM2-PUM-001-ACR. Version: 1.0. Framework Contract ref.: 942551 led by ACRI-ST

Rouse JW, Haas RH, Schell JA, Deering DW (1974) Monitoring vegetation systems in the great plains with ERTS. In: Proceedings of the Third Earth Resources Technology Satellite-1 Symposium; NASA SP-351. pp. 309–317

Simms EL (2019) SAR image interpretation for various land covers: a practical guide, 1st edn. CRC Press. https://doi.org/10.1201/9780429264771

Shang J, Liu J, Poncos V, Geng X, Qian B, Chen Q, Dong T, Macdonald D, Martin T, Kovacs J, Walters D (2020) Detection of crop seeding and harvest through analysis of time-series Sentinel-1 interferometric SAR data. In: Remote Sensing 12(10):1551. https://doi.org/10.3390/rs12101551

SNAP-ESA (2022) Sentinel Application Platform v9.0, http://step.esa.int

Sullivan DG, Strickland TC, Masters MH (2008) Satellite mapping of conservation tillage adoption in the Little River experimental watershed. Georgia J Soil Water Conserv 63:112–119

Suri S, Reinartz P (2010) Mutual information-based registration of TerraSAR-X and IKONOS imagery in urban area. IEEE Trans Geosci Remote Sens 48(2):939–949

Tasi H, Bedendo D (2008) Aptitud agrícola de las tierras de la provincia de Entre Ríos. INTA Paraná – Serie Extensión 19 (2nd ed.) ISSN 0325–8874

Ulaby FT, Long DG (2014) Microwave radar and radiometric remote sensing. University of Michigan Press, Ann Arbor

Van Deventer AP, Ward AD, Gowda PH, Lyon JG (1997) Using thematic mapper data to identify contrasting soil plains and tillage practices. Photogramm Eng Remote Sens 63:87–93

Verhulst N, Govaerts B, Sayre KD, Deckers J, François IM, Dendooven L (2009) Using NDVI and soil quality analysis to assess influence of agronomic management on within-plot spatial variability and factors limiting production. Plant Soil 317(1):41–59. https://doi.org/10.1007/s11104-008-9787-x

Wadoux AMC, Heuvelink GB, Lark RM, Lagacherie P, Bouma J, Mulder VL, Libohova Z, Yang L, McBratney AB (2021) Ten challenges for the future of pedometrics. Geoderma 401:115155

Wang CK, Pan XZ, Liu Y, Li YL, Shi R, Zhou R, Xie XL (2013) Alleviating moisture effects on remote sensing estimation of crop residue cover. Agron J 105:967–976

Welikhe P, Essamuah-Quansah J, Fall S, McElhenney W (2017) Estimation of soil moisture percentage using LANDSAT-based moisture stress index. Journal of Remote Sensing & GIS 6:2. https://doi.org/10.4172/2469-4134.1000200

Wilson M, Paz-Ferrero J (2012) Effects of soil-use intensity on selected properties of Mollisols in Entre Ríos, Argentina. Commun Soil Sci Plant Anal 43(1–2):71–80

Wu C, Niu Z, Tang Q, Huang W (2008) Estimating chlorophyll content from hyperspectral vegetation indices: Modeling and validation. Agric For Meteorol 148(8–9):1230–1241. https://doi.org/10.1016/j.agrformet.2008.03.005

Zanaga D, Van De Kerchove R, De Keersmaecker W, Souverijns N, Brockmann C, Quast R, Wevers J, Grosu A, Paccini A, Vergnaud S, Cartus O, Santoro M, Fritz S, Georgieva I, Lesiv M, Carter S, Herold M, Li Linlin, Tsendbazar NE, Ramoino F, Arino O (2021) ESA WorldCover 10 m 2020 v100. https://doi.org/10.5281/zenodo.5571936

Zanetti M, Bruzzone L (2018) A theoretical framework for change detection based on a compound multiclass statistical model of the difference image. IEEE Trans Geosci Remote Sens 56:1129–1143

Zheng B, Campbell JB, Serbin G, Galbraith JM (2014) Remote sensing of crop residue and tillage practices: present capabilities and future prospects. Soil Till Res 138:26–34

Chapter 16
Landslide Susceptibility Mapping Using Supervised Learning Methods – Case Study: Southwestern Colombia

N. A. Correa-Muñoz, L. J. Martinez-Martinez, and C. A. Murillo-Feo

Abstract Landslides are among the more severe geological hazards that threaten and influence infrastructure stability in populated areas. Landslide susceptibility is defined as the spatial distribution of favourable conditions for future landslide events. This research aimed to integrate explanatory variables, such as geomorphometric, soil properties, and climate, into a random forest (RF) analysis and logistic regression (LR) to identify areas susceptible to landslides. Landslide inventory data were used to develop prediction models and validate them. The prediction of landslide events with the LR had an area under the curve (AUC) of 0.91, and the more important predicting factors were road distance, soil silt, soil sand and soil clay contents, elevation, soil drainage, TRI, landscape, soil depth, and slope. The landslide type prediction based on the RF analysis had an overall accuracy of 72%, with elevation, soil silt content, slope, TRI, landscape unit, soil clay and sand contents, and roads distance as the more important predictors.

Keywords Landslide · Terrain analysis · DEM · Soil survey · Random forest · Logistic regression

N. A. Correa-Muñoz
Civil Engineering Faculty, Universidad del Cauca, Popayán, Colombia
e-mail: nico@unicauca.edu.co

L. J. Martinez-Martinez (✉)
Agricultural Sciences Faculty, Universidad Nacional de Colombia, Bogotá, Colombia
e-mail: ljmartinezm@unal.edu.co

C. A. Murillo-Feo
Engineering Faculty, Universidad Nacional de Colombia, Bogotá, Colombia
e-mail: camurillof@unal.edu.co

© The Author(s), under exclusive license to Springer Nature Switzerland AG 2023
J. A. Zinck et al. (eds.), *Geopedology*,
https://doi.org/10.1007/978-3-031-20667-2_16

16.1 Introduction

Landslide hazard maps represent susceptibility, which is the likelihood of a potentially damaging landslide occurring within a given area (Department of Regional Development and Environment 1991). The purpose of susceptibility studies is to identify areas where landslides can initiate and propagate (Guzzetti et al. 2005), based on a hazard and risk evaluation. Landslides can occur because of the interaction of natural and anthropic factors and cause economic and environmental damage and human losses. The Servicio Geológico Colombiano (2017) published a hazard map for landslides, 1:100,000 scale, and concluded that approximately 4% of the country has an extremely high hazard, 20% has a high hazard, 22% has a medium hazard, and 50% has low susceptibility. According to Suárez (1998), tropical mountainous areas are very susceptible to mass movements because of the interaction of slope, seismicity, rock type, and heavy rain, which are crucial factors for landslide events.

Different landslide-conditioning factors have been identified and can be used to establish landslide susceptibility assessments, including rainfall, drainage and soil properties (Metz and Bear-Crozier 2014), landcover (Shu et al. 2019) lithology, lineaments, geomorphology, soil type, and depth, slope angle, slope aspect, curvature, altitude, properties of the lithological material, land use patterns, and drainage networks (Youssef and Pourghasemi 2021), human activities (Achour and Pourghasemi 2020), and soil moisture (Ray et al. 2010). According to Gruber et al. (2009), mass movements are strongly controlled by land-surface form. A landslide process can be triggered by heavy and prolonged rainfall, cutting into slopes for the construction of roads, mining excavation without adequate preventive management, volcanoes, land-use change, and deforestation (Guzzetti et al. 2005).

The assessment of landslide susceptibility at different scales can be accomplished with statistical methods, heuristic approaches, physical models, and spatial models. For mapping landslide susceptibility, various methods have been developed that involve the identification of causative factors and the spatial analysis of interactions, usually supported by remote sensing data. For detailed landslide vulnerability mapping, Yuvaraj and Dolui (2021) used frequency ratio and binary logistic regression, Yu et al. (2021) studied the influence of rock and soil factors on landslide susceptibility mapping with LR modelling, an artificial neural network and support vector machine, Guo et al. (2021) presented a machine learning approach based on the C5.0 decision tree model and the K-means cluster algorithm to produce a regional landslide susceptibility map, Zhou et al. (2021) developed a hybrid model to optimize the factors and enhance the predictive ability of landslide susceptibility modelling, and (S. Lee et al. 2003) carried out a landslide susceptibility analysis using an artificial neural network, weights of evidence (Q. Wang et al. 2019), and LR (Dahoua et al. 2017).

The landslide inventory for Colombia indicates a high occurrence, in mountainous area, and, although there are various statistical, physical, and heuristic models to determine susceptibility, not all of them apply to every area since several factors that affect landslides have a local behaviour that must be analyzed. On the other hand, it is important to have methods that use information that is more commonly available in various countries, such as soil studies, digital elevation models, and climate data. This research aimed to evaluate the use of supervised learning methods for mapping susceptibility to landslides in mountainous areas to generate information for decisions in risk and hazard assessments, planning, infrastructure development, and promotion of economic activities.

16.2   Materials and Method

16.2.1   Study Area

The study area was in southwestern Colombia, specifically in Cauca (Fig. 16.1), within the coordinates 0°57'27.07" N, 77°19'48.75" W, and 2°15'57.60" N, 76°04'33.53" W, in eleven municipalities covering 8488 km². The climate is tropical, with an average annual rainfall of 2382 mm per year. The study area included the Cauca Boot, an area with important ecological and geological meaning since it includes the Santa Rosa link, which connects the Central and East mountainous ranges in Colombia, generating geographic knotting (Hubach 1982), and is a natural access to the Colombian Amazon region.

16.2.2   Data Collection

A SRTM DEM with a spatial resolution of 1 arcsec was compared in terms of accuracy to a self-produced model using an interpolated 1:25,000 topographic digital map. The fill sinks algorithm (L. Wang and Liu 2006) was applied to identify and fill surface depressions in the DEM to prepare the data for the analysis. The accuracy assessment of the DEMs was performed using the DEMANAL software developed by Leibniz University (Jacobsen 2019).

The following parameters were obtained from the DEM with the software SAGA (Conrad et al. 2012): elevation as the primary data given by the DEM; slope defined as the tangent of a plane relative to the surface topography; the aspect, which refers to the direction of slope (Olaya 2009); curvature calculated based on second derivatives for a topographic attribute that describes the convexity or concavity of a terrain surface (Romstad and Etzelmüller 2012); flow accumulation determined by accumulating the weight for all cells that flow into each downslope cell (O'Callaghan and Mark 1984) and was derived with the Top-Down method in SAGA software, as described by (Szypuła 2017), topographic wetness index (TWI), calculated as a second-order derivative of the DEM and used as an indicator of water accumulation in an area of the landscape where water is likely to concentrate through runoff

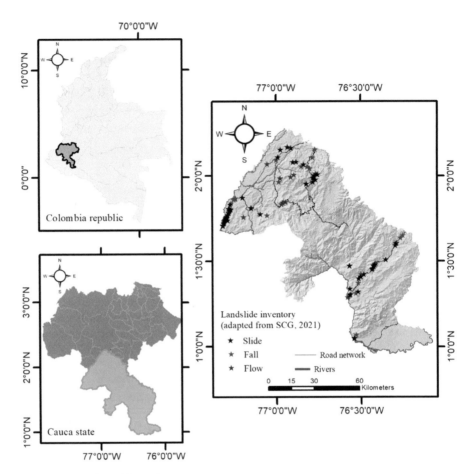

Fig. 16.1 Location of the study area

(Quinn et al. 1991), or, as described by (Vijith 2019), is a parameter that describes the tendency of a cell to accumulate water, topographic ruggedness index (TRI) expresses the elevation difference between adjacent cells of a DEM (Shawn Riley et al. 1999), stream power index (SPI) measures the erosive power of flowing water based on slope and specific catchment area (Moore et al. 1991), LS factor slope length (LS) factor as used by the Universal Soil Loss Equation (USLE) (Böhner and Selige 2006), topographic position index (TPI) indicates the altitude of each data point evaluated against its neighbours, (Guisan et al. 1999), and a description of these landform elements found in (Pike et al. 2009).

The landscape units and soil properties sand, silt and clay contents, soil depth, drainage class, and soil moisture regime were obtained from the General Soil Survey of the Cauca Department, scale 1:100,000 (Instituto Geográfico Agustin Codazzi 2009). Mean annual rainfall data were obtained from the Worldclim dataset (Fick and Hijmans 2017), the landslide map of the study area was created with the national landslide inventory (SIMMA) (Servicio Geológico Colombiano 2021).

16.2.3 Method

A flowchart about the overall steps followed by the landslide susceptibility mapping is in Fig. 16.2.

All the variables were resampled at a cell size of 100 m and combined in a multidimensional raster in QGIS. The statistics of the multidimensional image were re-built using ArcCatalog of ArcGIS 10.8. Then, the RF method was run using the script adapted from the NASA-ARSET webinar in 2019 for SAR applications. Next, the accuracy metrics, such as OOB estimate of error rate, confusion matrix, mean decrease accuracy, overall accuracy, kappa, users, and producers' accuracy, were obtained.

For mapping susceptibility to landslides, a RF classification was applied using version 2021.09.1 of the R statistical software (RStudio, Inc.). The RF tree was built by training each decision tree (ntree) with a random subset of the predictor-variable (mtry) from the training dataset. The algorithm was applied with the training dataset of landslides to obtain the supervised classifier algorithm and validation dataset of landslides to assess the accuracy of the produced landslide classification map. The prediction model of the RF classifier only required the number of classification trees

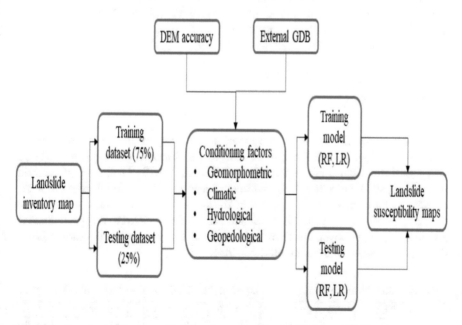

Fig. 16.2 Flowchart of the supervised learning method for landslide susceptibility mapping

(1000) and the number of prediction variables (18). The proportion used in this study was 75:25, as in the study by (Pham et al. 2018).

To map the probability of landslides occurrence, a binomial LR was applied. This statistical method has been well documented in geomorphological studies and is one of the most widespread methods for developing prediction models in geomorphology when system properties are represented by a binary variable (Schoch et al. 2018). The analysis was performed in the R software, and a training model was built using the 'glm' function with the binomial family. This model was assessed with the Chi-squared test, generated with 75% of the landslides inventory data and assessed with the remaining 25% of landslide inventory data using the area under the curve (AUC) as a validation metric of the prediction model (Huang and Zhao 2018).

16.3 Results and Discussion

16.3.1 SRTM DEM Accuracy Assessment

The standard deviation of the height was 11.58 m (Table 16.1), the bias was −2.3 m, and the standard deviation of the height without bias was 11.35 m. The normalized median absolute deviation (NMAD) related to bias-corrected height differences was 10.4 m. The SZ was greater than the NMAD because of a higher percentage of more significant discrepancies. This result agrees with the findings of Mukul et al.(2015), who compared the IGS and SRTM heights with the SRTM-DEM data in forest areas. The accuracy assessment of the SRTM DEM indicated an appropriate data quality for a landslide analysis since the results were equivalent to a scale about of 1:25 K, and the landslide analysis was done at a 1:100,000 scale.

16.3.2 Landslide Inventory

Table 16.2 shows the results of the landslide inventory of the study area, and its location is in Fig. 16.1. Following the Varnes classification (Hungr et al. 2014), it was found that 52.8% corresponded to slides that are displacements of material downslope, and 26.4% fit to falls that involve a collapse of material from the steepest area and accumulation in the base of the slope. 15.2% were classified as flows that are movements of materials down a hill as a fluid, 4.8% were creeps, defined as a slow downslope movement of material, and 0.8% were topples, the forward rotation and movement of material out of a slope.

Table 16.1 Results of the comparative of the 30 m SRTM DEM against 1:25 K topo-map DEM

Reference DEM	1:25K topo-DEM	RMSZ (m)	Bias (m)	SZ without bias	NMAD
DEM for analysis	30 m SRTM-DEM	11.583	-2.302	11.353	10.435

Table 16.2 Classification of landslide inventory. (Adapted from Servicio Geológico Colombiano 2021)

Type	Frequency Absolute	Relative	Subtype	Frequency Absolute	Relative
Falls	33	26.4%	Rockfall	24	19.2%
			Debris fall	6	4.8%
			Earthfall	3	2.4%
Slides	66	52.8%	Translational	46	36.8%
			Rotational	9	7.2%
			Planar	7	5.6%
			Wedge	2	1.6%
			Block slide	2	1.6%
Flows	19	15.2%	Debris flow	7	5.6%
			Mudflow	10	8.0%
			By flow	1	0.8%
			Earth flow	1	0.8%
Creeps	6	4.8%	Soil creep	6	4.8%
Topples	1	0.8%	Rock topples	1	0.8%
Total	125	100%		125	100%

Although the landslide distribution showed two geographically separated groups, the tendency of the mass movement distribution was preserved in each group. The dominant subtypes were translational debris (36.8%), rockfalls (19,2%), debris flows (5.6%) in the south-eastern zone, and mudflow (8%) in the north-western site. The landslide susceptibility analysis was developed using the type of movement for the RF method and the presence or absence of landslides as a binary dependent variable in the LR model.

16.3.3   Landslide Conditioning Factors

In this research, 18 landslide conditioning factors were selected based on literature review and the results of Colombian landslide inventory. A statistical summary of the distribution of each analyzed variable is in Table 16.3.

Elevations varied between 224 m.a.s.l. and 4158 m.a.s.l. (Fig. 16.3a) 26% of the area was below 1000 m.a.s.l, 33% was between 1000 and 2000 m.a.s.l., 29% was between 2000 and 3000, and 12% was over 3000 m.a.s.l.

The landscape units and its main characteristics are shown in Fig. 16.3f and in Table 16.4. A mountain landscape occupies 78% of the extension, hills represent 11%, and an alluvial valley contains 9%, and plateau 2%. The mountain is characterized by slopes greater than 30%, modelled by different geological phenomena associated with volcanic, structural, erosional, and depositional activity, which determines the current landscape characteristics. Most of the mountainous area was

Table 16.3 Statistics summary of the landslides conditioning factors

Layer	Factor	Unit	Analysis	Min	Mean	SD	Max	CV (%)
1	Elevation	m	Morpho	224.0	1775.95	934.35	4157.85	53
2	Slope	Degrees	Morpho	0	16.72	10.51	75.74	63
3	Aspect	Degrees	Morpho	0.0004	187.77	101.09	360	54
4	Curvature	Ordinal	Morpho					
5	Topographic position index	Ordinal	Morpho					
6	Terrain ruggedness index	None	Morpho	0	22.52	14.05	406.30	62
7	Flow accumulation	Km²	Hydro	0.001	7.91	102.5	5148.68	1296
8	LS factor	None	Hydro	0	26.80	46.23	3973.30	173
9	Stream power index	None	Hydro	0	5381.34	106425.39	20651448	1978
10	Topographic wetness index	None	Hydro	2.99	7.94	2.78	23.79	35
11	Landscapes	Category	Geomorpho					
12	Soil sand content	Percentage	Pedology	24.5	52.83	7.35	77	14
13	Soil clay content	Percentage	Pedology	5	20.4	10.61	47	52
14	Soil silt content	Percentage	Pedology	12.5	26.18	6.18	37	24
15	Soil drainage	Ordinal	Pedology					
16	Soil depth	Ordinal	Pedology					
17	Soil moisture regime	Ordinal	Pedology					
18	Rainfall	mm per year	Hydro	1320	2381.52	733	4705	31

Min minimum, *SD* standard deviation, *Max* maximum, *CV* coefficient of variation

developed on Cretaceous and Cenozoic sedimentary or on volcano-sedimentary and plutonic igneous rocks and is covered by volcanic ash. The hilly landscape is made up of areas with heights of less than 300 m with a slope between 7 and 12% although they can reach 50% locally, developed on Tertiary sedimentary rocks. The alluvial valley corresponds to flat areas formed by sediments transported by rivers and plateau, which are flat areas located at the base of the hills. The dominant soils in the area are well-drained, deep to moderately deep, with loam, clay loam, sandy clay loam or sandy clay texture (Fig. 16.3d, e), and udic moisture regime. To a lesser extent, there are superficial or poorly drained soils or with an ustic or aquic moisture regime.

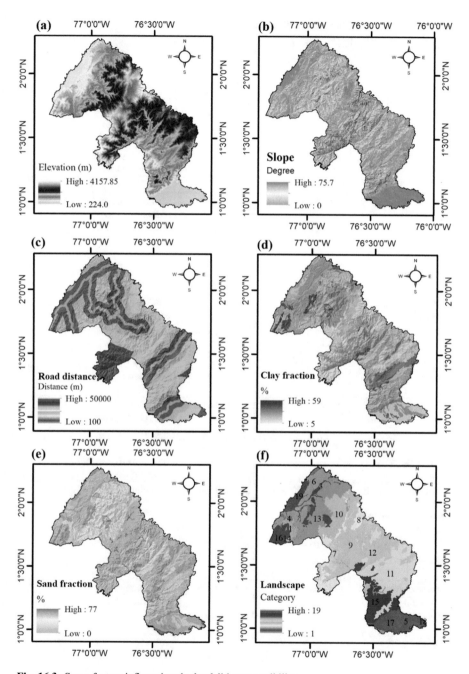

Fig. 16.3 Some factors influencing the landslide susceptibility

Table 16.4 Main characteristics of the landscape units

Landscape	Mean annual temperature (°C), mean annual rainfall (mm)	Relief type	Lithology	Number of mapping unit
Volcanic erosional plateau	18-24; 1000–2000	Plateau	Igneous rocks alternating with sedimentary materials	1
Erosional hill	18-24; 2000–4000	Hills	Volcanic ash layers over igneous rocks	3
	18-24; 1000–2000	Hills	Sedimentary rocks	4
	>24; 2000–4000	Hills	Metamorphic and sedimentary rocks	5
	>24; 1000–2000	Small valleys and hills	Colluvial-alluvial deposits, igneous rocks	6
Glacio-volcanic mountain	<12; 2000–4000	Summits	Igneous rocks	7
Fluvio-volcanic mountain	<12; 2000–4000	Summit, depressions	Volcanic ashes, organic materials	8
Volcanic-structural-erosional mountain	<12; 2000–4000	Summit, hills	Volcanic ashes	9
Volcanic mountain	12-18; 2000–4000	Summit, glacis	Volcanic ashes	10
Erosional-structural Mountain	18-24; 4000–8000	Summit, hills	Metamorphic rocks, volcanic ashes	11
Fluvio-volcanic mountain	18-24; 4000–8000	Summit, little valleys	Igneous rocks, colluvial deposits, volcanic ash	12
Erosional-structural Mountain, gravity flow	18-24; 2000–4000	Summit, hills, glacis	Sedimentary rocks, volcanic ash, igneous rocks	13
Structural erosional mountain and fluvial-gravitational mountain	18-24; 1000–2000	Summit, glacis	Volcanic ashes, metamorphic rocks	14
Erosional-structural Mountain	>24; 4000–8000	Summit, hills	Igneous rocks, volcanic and metamorphic rocks	15
Erosional-structural Mountain and small valley	>24; 1000–2000	Summit, fan, small valleys, flood plain	Sedimentary rocks, colluvial-alluvial deposits, igneous rocks	16
Alluvial valley	>24; 2000–4000	Terrace	Alluvial deposits	17
	>24; 1000–2000	Flood plain	Alluvial deposits	18
	>24; 1000–2000	Flood plain, terrace	Alluvial deposits	19

Adapted from IGAC (2009)

The slope varied between 0° and 76.4° (Fig. 16.3b). A classification of the landscape by its slope indicated a flat area in 13.6% of the study area, sloping areas in 68.3% of the extension, and steep areas with slopes greater than 30° in 18.1% of the zone. The slope aspect, indicating the flow-line direction, was distributed at 14.2% north-eastern, 31.4% south-eastern, 32.6% southwestern, and 21.8% north-western. The curvature plan indicated a concave surface in 48.7% of the area and a convex surface in 51.3% of the study area. The curvature profile indicated a concave form in 52.0% of the area and a convex form in 48.0% of the zone. The tangential curvature was concave in 48.2% of the cases and convex in 51.8% of the area.

The terrain ruggedness index varied between 0 and 275 and classified 53.6% of the area as smooth terrain, 42.0% as rough terrain, and 4.4% as irregular. The flow accumulation indicated that 92.5% of the drainage proportion was less than 0.47 km2; this accumulation reached 2163 km2. The road distance (Fig. 16.3c) shows areas contiguous to the roads and others located up to 50 km.

Slope length (LS) is a topographic parameter used in soil erosion. Its mean value was 38 with a positively skewed distribution, which means that 88% of the distribution was less than 42.6. Stream power index describes potential flow erosion; its distribution was highly positively skewed. The topographic wetness index had a light, positively skewed distribution where about 88% of the distribution had an index less than 8.5 or a moderate wetness index. Finally, rainfall ranged from 1322 mm to 4705 mm per year, with a favourable bias distribution of 80.7% of the study area, less than 2584 mm per year.

16.3.4   LR Model for Landslide Probability Occurrence

The LR method was used to predict the probability of landslide occurrence based on the presence or absence of landslide events as a binary dependent variable based on 18 landslide conditioning factors as explanatory variables.

16.3.4.1   Training Model

Table 16.5 shows the results of the generalized linear model developed with a R script software using the glm-function and the logit-family of the binomial method. As with RF, the model was developed with 75% of the landslide data and validated with the remaining 25%. The results showed that road distance, highly significant relationship with the occurrence of landslides and in the limit, at a significance level of 0.01 are elevation and slope. The above indicated that the probability to obtain the coefficient of the model with respect to the hypothesis the true coefficient is zero was low. The coefficients of the other conditioning factors were not significant different to 0 they had no effect on the probability of landslide occurrence.

The chi square test (Table 16.6) indicated that the variables road distance, soil sand content and slope had a highly statistically significant association between the

Table 16.5 Results of the logit function in the landslide susceptibility analysis

Coefficients	Estimated coefficient	Code	Coefficients	Estimated coefficient	Code
Intercept	-5.291e+00	ns			
Clay content	-6.159e-02	**ns**	Silt content	2.380e-02	ns
Sand content	-6.326e-02	ns	Landscape	-5.514e-02	ns
Roads Distance	-5.039e-04	***	Soil depth	1.246e-01	ns
Drainage	-6.352e-02	ns	Slope	2.084e-01	*
Elevation	-5.660e-04	*	TRI	4.178e-02	ns

Significance: *** (p-value =0), *(p-value <0.01), ns: no significance

Table 16.6 Results of the Chi-square test of the prediction model of landslide susceptibility

Coefficients	Deviance	Code	Coefficients	Deviance	Code
Null					
Clay content	4.741	*	Silt content	0.000	
Sand content	54.469	***	Landscape	3.898	*
Roads Distance	187.748	***	Soil depth	0.381	ns
Drainage	0.033	ns	Slope	17.477	***
Elevation	1.316	ns	TRI	0.193	ns

Significance: *** (p-value =0), ** (p-value < 0.001), *(p-value <0.01), ns: no significance

observed and estimated values, soil clay content, and landscape unit had statistically effect on the landslide prediction.

The probability of landslide occurrence map was prepared with the LR equation using map algebra in ESRI's ArcMap v10.8 and reclassified with four classes with the quantile method (Fig. 16.4). The highest probability of occurrence was found near roads, the remaining area presents medium probability.

Additionally, the information library of the R software was applied to the training geodatabase to compute the weight of evidence and information value metrics. Distance to roads was the most important variable to explain the occurrence of landslides (Fig. 16.5), followed by TRI, soil silt, sand and clay content, elevation, soil drainage, TRI, and landscape type. The greater probability of occurrence of landslides in certain areas is related to natural factors such as edaphic, geomorphometric and climatic that facilitate the occurrence of events and with the anthropic activities, in this case roads construction, which are the trigger that activates the landslide phenomena.

Table 16.7 shows the most significant importance (values > +1) of the bivariate method of the weight of evidence. A positive weight indicates a positive correlation between the presence of the predictable variable and landslides (Jaafari et al., 2015). The conditioning factors within this category were road distance between 0 and 1000 m, soil sand content between 14.5% and 37%, soil drainage moderately well-drained, elevation in the range 607 m to 850 m, soil silt content in the range 36% to

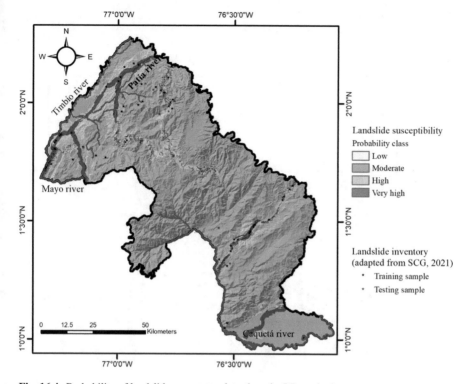

Fig. 16.4 Probability of landslide occurrence based on the LR method

46%, and soil clay content between 30.5% to 32, landscape units structural erosional mountain, mainly in ridges and back-slopes other factor showed also positive values.

16.3.4.2 Performance of the LR Model

The ROC curve and the AUC are standard measures for binary classifier performance. The ROC plot (Fig. 16.6) was obtained by plotting the valid positive rate (TPR) against the false positive rate (FPR), while AUC is the area under the ROC curve. As a rule of thumb, a model with good predictive ability should have an AUC greater than 0.5. The AUC of the landslide susceptibility model with regression analysis was 0.91. An analysis of scaling land-surface variables for landslide detection obtained AUC values between 0.73 and 0.80 (Sîrbu et al. 2019). AUC values between 0.7 and 0.9 indicate a reasonable agreement between the predicted landslides and test landslides (Lee et al. 2018).

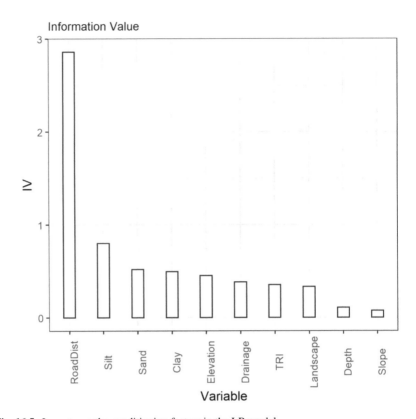

Fig. 16.5 Importance the conditioning factors in the LR model

Table 16.7 Weight of evidence of the variables and range of the category

Landslide conditioning factor	Unit	Area	WOE	IV	Category of the variable
Road distance	m	9.4%	2.11	1.43	0 to 1000
Soil sand content	%	9.6%	1.26	0.30	14.5 to 37
Soil drainage	Category	10.5%	1.25	0.32	2 to 3 (moderately well-drained)
Elevation	m a.s.l.	10.0%	1.18	0.28	606.8 to 850
Landscape	Category	10.3%	1.02	0.33	16 to 19
Soil silt content	%	12.8$	0.88	0.80	36 to 46
Soil clay content	%	9.3%	0.84	0.48	30.5 to 32

16.3.5 Landslide Susceptibility Zoning Based on RF Analysis

16.3.5.1 Training Model

To develop the predictive model of the landslide susceptibility the slides and falls types were selected since these were the more frequent events, and the 16 factors selected as the most relevant for the study area. From the inventory of landslides,

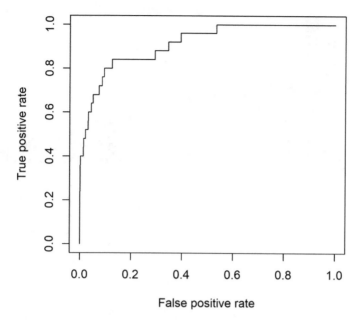

Fig. 16.6 ROC curve of the LR-landslide susceptibility model

Table 16.8 Results of the accuracy RF classification of the landslides

Type of movement	Class accuracy	
	User	Producer
Falls	56.0%	66.7%
Slides	85.7%	79.2%
Overall accuracy	75.7%	

75% were selected to generate the model, and the remaining 25% were used to validate it. To guarantee the stability of the model, 1000 trees were used, as recommended by Lagomarsino et al. (2017).

The overall classification accuracy of developed model (Table 16.8) was 75.7%, slides were the mass movements that were better classified with a user accuracy of 85.7%, which means the percentage of landslides that were correctly classified as compared to the landslide inventory. The producer accuracy refers to the commission error, which was 79.2%. The falls and flows of mass movements had lower user and producer accuracy. The general accuracy depended on the frequency of landslide events, the more frequent the occurrence of a landslide type, the greater accuracy obtained in the prediction, consistent with other researches (Tansey et al. 2004).

Although the general accuracy was low, when the classification for each landslide type was analyzed, good prediction accuracy was found for the landslides that are more frequent in the study area. The analysis was based on existing data, which

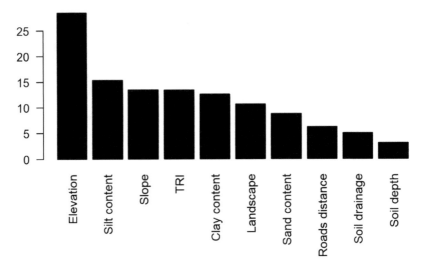

Fig. 16.7 Mean Decrease Accuracy in the study area

is one of the main limitations since some, such as climate and geology, were very general for the scale of the study.

16.3.5.2 Mean Decrease Accuracy (MDA)

The MDA was one of the outcomes of the RF analysis and indicated the degree of importance of each of the variables in the prediction. Figure 16.7 displays the MDA results, ranking the variables by importance. The more important variables in the RF prediction were elevation, soil silt content, slope, TRI, soil clay, landscape unit, soil sand content and roads distance. It was found that there was a relationship between some edaphic and geomorphometric characteristics with the presence of the main types of landslides. Most landslides were found in the structural erosional mountain, in ridges and back-slopes, in loamy and sandy loam soils with a humid climate, and at an altitude between 300 and 1200 m asl.

16.3.5.3 Accuracy of the Landslide Classification by the RF Method

The model assessment helped to evaluate the classifier performance for other data. Table 16.9 summarizes the accuracy metrics derived from the confusion matrixes, which compares the reference values with the predicted values.

The overall accuracy was 72% and indicated the percentage of landslide type correctly classified. When the accuracy of each class was evaluated, the slides had a better performance with 75% and 88.2% of user's and producer's accuracy, respectively, while falls had low accuracy. The error of commission of slide classification

Table 16.9 Evaluation of the model classification performance

Type of movement	Class accuracy	
	User	Producer
Falls	60.0%	37.5%
Slides	75.0%	88.2%
Overall accuracy	72.0%	

was 25%, and the omission error was 11.8% for the classification model, differentiating slides and non-slides for the study area.

According to Korup and Stolle (2014), predictive methods based on machine learning analysis achieve an overall success rate of 75–95%, these authors proposed doing more research on the selection of models, the model overfitting, and the effect of slope failure at a regional scale to improve predictions. In our case, another factor that influences the success of the predictions was the inventory of landslides, the relationship with the distance to the roads is sometimes due to that landslides were much more commonly recorded near roads (Stanley et al. 2020). The objective of our study was to evaluate predictive methods with data from the soil survey, geomorphometric parameters calculated from DEM and climate data available on the internet, however geology data were not included, and it could have effect on predictions. On the other hand, the rainfall data used had a spatial resolution of 900 m, which is exceptionally low compared to other data and therefore had no significant effect on the RF prediction.

The landslide classification map based on the results of the RF analysis (Fig. 16.8) shows the areas that meet the conditions required for the occurrence of the main landslide types and, therefore, are more likely to present this phenomenon. The classification showed that most of the area is susceptible to slides, and in less proportion the area can be affected by falls.

16.4   Conclusions

The probability of landslide events occurrence, estimated with LR, had an AUC of 0.91, and the more important predicting factors were road distance, soil silt, sand and clay content, elevation, soil drainage, TRI, landscape, soil depth and TWI. Landslide occurrence is favour by natural factors, while anthropic activities like the construction of roads is the trigger that initiates the process of occurrence of landslides. The susceptibility of the study area to the occurrence of landslides type based on RF analysis had an overall accuracy of 72% with elevation, soil silt content, slope, TRI, landscape unit, soil clay and sand content, and road distance were the more important predictors.

The integration of the DEM as a data source with the results of the soil surveys using LR and RF made it possible to generate information with acceptable reliability and level of detail for the susceptibility of mountain areas to landslides as a first

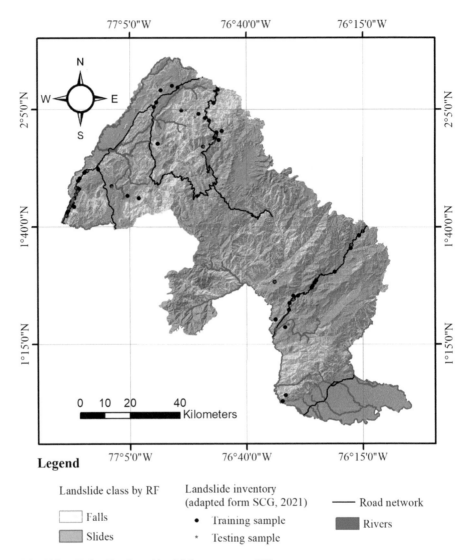

Fig. 16.8 RF classification of landslide type susceptibility areas

approximation for subsequent risks and hazard analyses. This is important considering that the required data is available for all of Colombia for applying more complex predictive models, where data availability and quality are limiting.

The distance to the roads was the factor that had the greatest incidence in the presence of landslides and in its distribution pattern. Consequently, it is also a factor that determines the probability of occurrence of landslides. Most of the study area has medium probability of occurrence but if roads are built it can change to high or very high probability.

Acknowledgments The authors want to thank the National University of Colombia and the Cauca University for the academic support for this research.

References

Achour Y, Pourghasemi HR (2020) How do machine learning techniques help in increasing accuracy of landslide susceptibility maps? Geosci Front 11(3):871–883. https://doi.org/10.1016/j.gsf.2019.10.001

Böhner J, Selige T (2006) Spatial prediction of soil attributes using terrain analysis and climate regionalisation

Dahoua L, Yakovitch SV, Hadji R, Farid Z (2017) Landslide susceptibility mapping using analytic hierarchy process method in BBA-Bouira Region, case study of East-West Highway, NE Algeria. Euro-Mediterranean Conference for Environmental Integration, 1837–1840

Department of Regional Development and Environment (1991) Primer on natural hazard management in integrated regional development planning. In Organization of American States, Department of Regional Development and Environment Executive Secretariat for Economic and Social Affairs Organization of American. https://www.oas.org/dsd/publications/Unit/oea66e/begin.htm

Fick SE, Hijmans RJ (2017) WorldClim 2: new 1-km spatial resolution climate surfaces for global land areas. Int J Climatol 37(12):4302–4315. https://doi.org/10.1002/joc.5086

Gruber S, Huggel C, Pike R (2009) Modelling mass movements and landslide susceptibility. Dev Soil Sci 33(C):527–550. https://doi.org/10.1016/S0166-2481(08)00023-8

Guisan A, Weiss SB, Weiss AD (1999) GLM versus CCA spatial modeling of plant species distribution. Plant Ecol 143(1):107–122

Guo Z, Shi Y, Huang F, Fan X, Huang J (2021) Landslide susceptibility zonation method based on C5.0 decision tree and K-means cluster algorithms to improve the efficiency of risk management. Geosci Front 12(6):101249. https://doi.org/10.1016/j.gsf.2021.101249

Guzzetti F, Reichenbach P, Cardinali M, Galli M, Ardizzone F (2005) Probabilistic landslide hazard assessment at the basin scale. Geomorphology 72(1–4):272–299. https://doi.org/10.1016/j.geomorph.2005.06.002

Huang Y, Zhao L (2018) Review on landslide susceptibility mapping using support vector machines. Catena 165:520–529. https://doi.org/10.1016/j.catena.2018.03.003

Hubach E (1982) El Cauca. Las unidades geográficas y geológicas del Departamento y los recursos del suelo y del subsuelo. In Publicaciones Geológicas Especiales Del Ingeominas, pp. 23–37

Hungr O, Leroueil S, Picarelli L (2014) The Varnes classification of landslide types, an update. In Landslides 11(2):167–194. https://doi.org/10.1007/s10346-013-0436-y

Instituto Geográfico Agustin Codazzi (2009) Estudio General de Suelos y Zonificación de Tierras del departamento del Cauca. Instituto Geográfico Agustin Codazzi

Jaafari A, Najafi A, Rezaeian J, Sattarian A, Ghajar I (2015) Planning road networks in landslide-prone areas: a case study from the northern forests of Iran. Land Use Policy 47:198–208. https://doi.org/10.1016/j.landusepol.2015.04.010

Jacobsen K (2019) DEMANAL Program System BLUH. In Institute of Photogrammetry and Geoinformation, Leibniz University, Hannover

Korup O, Stolle A (2014) Landslide prediction from machine learning. Geol Today 30(1):26–33

Lagomarsino D, Tofani V, Segoni S, Catani F, Casagli N (2017) A tool for classification and regression using random forest methodology: applications to landslide susceptibility mapping and soil thickness modeling. Environ Model Assess 22(3):201–214

Lee S, Ryu JH, Min K, Won JS (2003) Landslide susceptibility analysis using GIS and artificial neural network. Earth Surf Process Landf 28(12):1361–1376. https://doi.org/10.1002/esp.593

Lee JH, Sameen MI, Pradhan B, Park HJ (2018) Modeling landslide susceptibility in data-scarce environments using optimized data mining and statistical methods. Geomorphology 303:284–298. https://doi.org/10.1016/j.geomorph.2017.12.007

Metz L, Bear-Crozier AN (2014) Landslide susceptibility mapping: A remote sensing-based approach using QGIS 2. 2 (Valmiera) (Vol. 2)

Moore ID, Grayson RB, Ladson AR (1991) Digital terrain modelling: a review of hydrological, geomorphological, and biological applications. Hydrol Process 5(1):3–30

Mukul M, Srivastava V, Makul M (2015) Analysis of the accuracy of shuttle radar topography Mission (SRTM) height models using international global navigation satellite system service (IGS) network. J Earth Syst Sci 124(6):1343–1357. https://doi.org/10.1007/s12040-015-0597-2

O'Callaghan JF, Mark DM (1984) The extraction of drainage networks from digital elevation data. Comput Vision Graph Image Process 28(3):323–344

Olaya V (2009) Basic land-surface parameters. Dev Soil Sci 33(C):141–169. https://doi.org/10.1016/S0166-2481(08)00006-8

Pham BT, Prakash I, Tien Bui D (2018) Spatial prediction of landslides using a hybrid machine learning approach based on random subspace and classification and regression trees. Geomorphology 303:256–270. https://doi.org/10.1016/j.geomorph.2017.12.008

Pike RJ, Evans IS, Hengl T (2009) Geomorphometry: concepts, software, applications. In: Hengl T, Reuter HI (eds) Developments in soil science, vol 33. Elsevier, pp 1–28

Quinn PFBJ, Beven K, Chevallier P, Planchon O (1991) The prediction of hillslope flow paths for distributed hydrological modelling using digital terrain models. Hydrol Process 5(1):59–79

Ray RL, Jacobs JM, Cosh MH (2010) Landslide susceptibility mapping using downscaled AMSR-E soil moisture: a case study from Cleveland Corral, California. US Remote Sens Environ 114(11):2624–2636. https://doi.org/10.1016/j.rse.2010.05.033

Romstad B, Etzelmüller B (2012) Mean-curvature watersheds: a simple method for segmentation of a digital elevation model into terrain units. Geomorphology 139:293–302

Schoch A, Blöthe JH, Hoffmann T, Schrott L (2018) Multivariate geostatistical modeling of the spatial sediment distribution in a large-scale drainage basin, Upper Rhone, Switzerland. Elsevier Geomorphology 303:375–392. https://doi.org/10.1016/j.geomorph.2017.11.026

Servicio Geológico Colombiano (2017) Las Amenazas por movimientos en masa de Colombia. Colección Guías y Manuales. SGC. https://srvags.sgc.gov.co/Archivos_Geoportal/Manuales/Libro_MNMM.pdf

Servicio Geológico Colombiano (2021) Inventario nacional de movimientos en masa. https://www2.sgc.gov.co/ProgramasDeInvestigacion/geoamenazas/Paginas/Inventario-nacional-de-movimientos-en-masa-y-SIMMA.aspx

Shawn R, Stephen D, DeGloria ER (1999) A terrain ruggedness index that quantifies topographic heterogeneity. Intermountain J Sci 5:1–4

Shu H, Hürlimann M, Molowny-Horas R, González M, Pinyol J, Abancó C, Ma J (2019) Relation between land cover and landslide susceptibility in Val d'Aran, Pyrenees (Spain): historical aspects, present situation, and forward prediction. Sci Total Environ 693:1–14. https://doi.org/10.1016/j.scitotenv.2019.07.363

Sîrbu F, Drăguţ L, Oguchi T, Hayakawa Y, Micu M (2019) Scaling land-surface variables for landslide detection. Prog Earth Planet Sci 6(1):1–13

Stanley TA et al (2020) Building a landslide hazard indicator with machine learning and land surface models. Environ Model Softw 129:1–15

Suárez J (1998) Deslizamientos y estabilidad de taludes en Zonas Tropicales. Instituto de Investigaciones sobre Erosión y Deslizamientos, pp 1–10

Szypuła B (2017) Digital elevation models in geomorphology. Hydro-Geomorphology-Models Trends InTechOpen 2017b:81–112

Tansey KJ, Luckman AJ, Skinner L, Balzter H, Strozzi T, Wagner W (2004) Classification of forest volume resources using ERS tandem coherence and JERS backscatter data. Int J Remote Sens 25(4):751–768. https://doi.org/10.1080/0143116031000149970

Vijith H (2019) Modelling terrain erosion susceptibility of logged and regenerated forested region in northern Borneo through the Analytical Hierarchy Process (AHP) and GIS techniques. Geoenviron Disasters 6(8):18

Wang L, Liu H (2006) An efficient method for identifying and filling surface depressions in digital elevation models for hydrologic analysis and modelling. Int J Geogr Inf Sci 20(2):193–213

Wang Q, Guo Y, Li W, He J, Wu Z (2019) Predictive modeling of landslide hazards in Wen County, northwestern China based on information value, weights-of-evidence, and certainty factor. Geomat Nat Haz Risk 10(1):820–835. https://doi.org/10.1080/19475705.2018.1549111

Youssef AM, Pourghasemi HR (2021) Landslide susceptibility mapping using machine learning algorithms and comparison of their performance at Abha Basin, Asir region. Saudi Arab Geosci Frontiers 12(2):639–655. https://doi.org/10.1016/j.gsf.2020.05.010

Yu X, Zhang K, Song Y, Jiang W, Zhou J (2021) Study on landslide susceptibility mapping based on rock–soil characteristic factors. Sci Rep 11(1):1–27. https://doi.org/10.1038/s41598-021-94936-5

Yuvaraj RM, Dolui B (2021) Statistical and machine intelligence-based model for landslide susceptibility mapping of Nilgiri district in India. Environ Chall 5(May):100211. https://doi.org/10.1016/j.envc.2021.100211

Zhou X, Wen H, Zhang Y, Xu J, Zhang W (2021) Landslide susceptibility mapping using hybrid random forest with GeoDetector and RFE for factor optimization. Geosci Front 12(5):101211. https://doi.org/10.1016/j.gsf.2021.101211

Chapter 17
Polygenic Vertisols and "Hidden" Vertisols of the Paraná River Basin, Argentina

Héctor José María Morrás, Emiliano Miguel Bressan,
Marcos Esteban Angelini, Leonardo Mauricio Tenti Vuegen,
Lucas Martín Moretti, Darío Martín Rodríguez,
and Guillermo Andrés Schulz

Abstract Vertisols and soils with vertic properties occupy considerable areas in the east of the Pampa Region. Many of these Vertisols have particular features that differentiate them from most of the Vertisols in the world. In the Mesopotamian Pampa, Vertisols have developed from a calcareous and gypsiferous loam, later covered by a thin loess mantle. Here it is considered that Vertisols would have been the dominant soils in the different segments of a hilly landscape. Later, under humid climate, and due to the buffering effect of the non-expansive silty loessic surface, clay illuviation would have been a generalized process across the landscape. At the same time erosion processes took place. The conservation of the thin layer of loessic sediments on the higher slopes resulted in the formation of Vertic Alfisols and few Mollisols, while its erosion in the backslopes caused the exhumation of buried Vertisols. On the other hand, in the High Undulating Pampa, there are no Vertisols but there are Vertic Argiudolls. Two superficial sedimentary levels have been distinguished here. The lower one is a smectitic loessic sediment, covered by a relatively thick illitic loessic deposit. New studies undertaken in large pits have revealed the existence of diapiric structures in the lower sediment, which remained "hidden" up to the present. Therefore, it can now be considered that Vertisols were also the dominant soils here, later buried by the thick loess. In the humid periods, the summit of convex slopes would have been partly eroded, leaving the underlying smectitic material closer to the surface. Consequently, current Mollisols on top of the landscape developed vertic properties due to the mixing of materials and depending on the greater or lesser proximity to the paleosurface. Therefore, these pampean

H. J. M. Morrás (✉) · E. M. Bressan · M. E. Angelini · L. M. Tenti Vuegen ·
D. M. Rodríguez · G. A. Schulz
Instituto de Suelos, INTA-CIRN, Hurlingham, Buenos Aires, Argentina
e-mail: morras.hector@inta.gob.ar

L. M. Moretti
EEA Cerro Azul, Instituto Nacional de Tecnología Agropecuaria, Cerro Azul, Misiones, Argentina

Vertisols and vertic Mollisols and Alfisols can be considered polygenic and related by different degrees of a same process. These advances in the understanding of the landscape and soils together with the quantitative analysis of soil profiles data, appear highly useful to distinguish vertic soils at the series level and improve surveying and mapping work.

Keywords Vertisols · Vertic soils · Pedogenesis · Parent material · Landscape

17.1 Introduction

Vertisols are soils characterized by specific structural features resulting from a marked shrink-swell activity, due to a high proportion of expanding clays and under the influence of contrasting soil moisture patterns. Alternate swelling and shrinking result in deep cracks during the dry season, and formation of sclikensides and wedge-shaped structural elements in the subsurface soil. In most of Vertisols, a "mulch" with a granular or crumb structure is formed at the surface. Structure formation is due to differential stress governed by moisture content and volume change. The stress varies from point to point both vertically and horizontally. This process of expansion-contraction causes soils to mix, which is the central tenet of "argilliturbation". The shearing forces result in "chimney" or "diapir" structures within the soil and are reflected on a surface micro- topography of depressions rimmed by ridges termed "gilgai", although not all Vertisols show this surface phenomena. (Dudal and Eswaran 1988; Wilding and Tessier 1988; Schaetlz and Thompson 2015).

Vertisols are spread over a wide range of temperature and moisture regimes, from semi-arid to humid climates. Most of these soils occur in the semi-arid tropics. However, a common climatic feature is an alternation of wet and dry seasons. Vertisols are generally formed on sedimentary plains, with sediments that have a high content of smectitic clays or that produce such clays upon post-depositional weathering, and on extensive basalt plateaus. Therefore, typical geopedologic landscapes for Vertisols are level to undulating lands and depressions. A frequent physiographic position for Vertisol is flat alluvial plains, often in lower landscape positions and sometimes subjected to a fluctuating water table. Also, small areas of Vertisols are on hill slopes (Dudal and Eswaran 1988, Pal et al. 2012; IUSS 2015).

The Pampa region of Argentina is a vast and continuous plain, where total flat areas alternate with slightly undulating plains and rolling landscapes. Rocky hills occur only in minor areas in the south of Buenos Aires Province (Tandilia and Ventania systems). Based on various environmental factors, Durán et al. (2011) have recognized thirteen sub-regions in the Pampa, four of which are part of the Paraná river basin (Fig. 17.1). In the Pampean region, mean annual temperature ranges from 14 °C in its southern part, to 19 °C in the north. Mean annual rainfall ranges from 500 mm in the southwest to 1100 mm in the northeastern part of the Pampa. According to Soil Taxonomy (NCSR-USDA 2014) soils have a thermic temperature regime, whereas the most extensive moisture regime is udic. However,

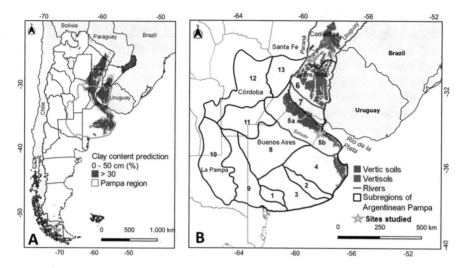

Fig. 17.1 (**a, b**) A-Soils in Argentina with a clay content >30% in 0–50 cm depth, predicted by random forest model under a digital soil mapping framework. B- Sub-regions of the Pampa region with vertic soils (Vertisols and vertic subgroups in other soil Orders) and location of the two sites studied. 4: Flooding (or Depressed) Pampa. 5a: High Undulating Pampa. 5b: Low Undulating Pampa. 6: Mesopotamian Pampa

in normal years the high temperatures of the hottest months lead to a moderate deficiency of water in the soil, also in the most humid areas (Durán et al. 2011; Rubio et al. 2019). From a geological point of view, The Pampa is an extensive and deep sedimentary basin where the crystalline basement is covered by hundreds to thousands meters of sediments of Tertiary and Quaternary age of continental and marine origin. The most recent and widespread Quaternary deposits lying on the surface, and therefore the parent material of present soils, are aeolian sands to the west and loessoid sediments to the east. This granulometric zonation is the result of a combined fluvial-aeolian transport of sediments from the Andes cordillera and Pampean hills in the west, as well as alluvial and aeolian contributions of fine sediments transported by the Paraná and Uruguay rivers on their way to the Rio de la Plata estuary and then to the ocean, to the east (Morrás and Moretti 2016; Morrás 2020a). As a result of the confluence of fine sediments from different sources, the finest soil parent materials are found widely in eastern Argentina, and particularly along the Paraná river basin. The dominant soils in this area are Mollisols, as in most of eastern Argentina.

Despite the fact that the clay content of the solum of most of the soils of the Paraná river basin is high (Fig. 17.1a), and that the predominant clays are of the 2:1 type and with a substantial proportion of expandable components (Iñiguez and Scoppa 1970; Stephan et al. 1977; Morrás et al. 1982; Morrás and Moretti 2016), and that other environmental conditions such as climate and physiography are suitable for the development of Vertisols, the soils classified and mapped as such constitute only a portion of this area (Fig. 17.1b). On the other hand, a significant proportion of the soils in this region present vertic characteristics such as slicken-sides and high expandability values (Castiglioni et al. 2005, 2007; Bressán et al. 2014), but due to the fact that the percentage of clay in the epipedon is less than 30% according to the criteria required by both the Soil Taxonomy (NCSR-USDA 2014) and the WRB (IUSS 2015) they are classified as vertic intergrades, generally with Mollisols (mostly Vertic Argiudolls) but also with Alfisols (e.g., vertic Albacualfs and Natracualfs). Thus, in the Pampean region, Vertisols associated with vertic sub-groups of other Orders, are particularly relevant in three sub-regions (SAGyP-INTA 1989; Imbellone et al. 2010): (1) In the *Mesopotamian Pampa*, along a large part of Entre Ríos Province. In this sub-region the parent material of Vertisols is made up of lacustrine calcareous loams having silty-clay or clay-loam textures. (2) In a 60 km wide strip of the Rolling Pampa, in the eastern part of Buenos Aires and Santa Fe provinces, bordering with the Paraná and de la Plata rivers. The dominant soils in this strip are Vertic Argiudolls, together with some Vertisols in its southern part, developed in a continental environment with clayey loessic deposits. (3) In the south-eastern coast of the Río de la Plata and in the littoral area of the *Flooding Pampa* (central-eastern Buenos Aires Province), with estuarine and marine deposits (Fig. 17.1b).

Along with this general information produced by systematic soil surveys, several questions have arisen, however, regarding some specific features of these vertic soils. This also led initially to some local modifications in the Soil Taxonomy sub-groups nomenclature (Plan Mapa de Suelos 1980; Bedendo 2019). In this sense, several recent studies with a geo-pedological approach and greater detail on sedimentology, mineralogy, macro and micro-morphology and the physical properties of soils, are generating answers to several of these questions. At the same time, recent studies not only make it possible to reveal the existence of Vertisols in remote and less known areas of the country (e.g., Moretti et al. 2012, 2020), but also allow them to be discovered in areas that have been much more studied, as in the Pampa region. These developments clarify the causes of the distribution of some of these soils in the Pampean landscape (Morrás and Moretti 2016), they generate new questions regarding the genesis, variability and concept of Vertisol, and suggest a greater extension of these soils in the country than has been recognized to date. As an example of these developments and challenges, two case studies of vertic soils in the area of the Paraná river basin in Argentina, are presented.

17.2 Polygenic Vertisols of the Mesopotamian Pampa

17.2.1 Setting the Subject

For a long time, it has been considered that a distinguishing feature of Vertisols is a high degree of uniformity of most properties throughout the profiles, and particularly a near absence of textural differentiation due to a process of haploidisation by argillipedoturbation. This mixing process would inhibit horizonation, resulting in cyclic pedons having A-C profiles (Murthy et al. 1982; Blokhuis 1982). Although some earlier studies reported a clay increase with depth (Dudal 1965), these variations in clay content were considered not to be due to clay migration but to inheritance of the parent material. (Ahmad 1983). Later the presence of B horizons was recognized, and now most of Vertisols described in the world display A-AC-C, A-Bw-C or A-Bss-C horizons sequences.

However, in the case of Vertisols in the Mesopotamian Pampa, from the first stages of the systematic survey in the early 1970s, specific characteristics of these soils were evidenced, in particular the presence of horizons described as Bt considering the notable increase in the amount of clay determined by analytical methods and by the presence of clay skins. Also, some Vertisols present an A horizon thicker than 20 cm with high content of organic matter (3.5–6%, even up 7% in some cases) which are classified as mollic epipedons. Since the classification system used (seventh Approximation and Soil Taxonomy of USDA) has not considered the presence of Bt argilic horizons and neither of mollic epipedons, the local soils surveyors created the following specific subgroups within the Pelludert great group: chromic, argillic, chromic-argillic, argiudollic, and argiaquollic (Plan Mapa de Suelos 1980).

Thus, while some authors begun to recognize gradual increases of clay content with depth in Vertisols of different latitudes on the basis of morphology and conventional laboratory analyses (Dudal and Eswaran 1988; Satyavathi et al. 2005), micromorphological studies on thin sections confirmed illuviation process. It seems that the first micromorphological study that identified illuviation features in Vertisols was precisely the work by Jongerius and Bonfils (1964) in a Vertisol of southern Entre Ríos. Then, in the 70s, void argillans have been observed mainly in the lower horizons of various Vertisols from different countries (Nettleton and Sleeman 1985). Later, Morrás et al. (1993) and De Petre and Stephan (1998) also mentioned the presence of clay cutans and papules in different soil series from Entre Ríos. However, Dudal and Eswaran (1988) and Mermut et al. (1996) indicated that the amount of clay cutans found in Vertisols is not enough to consider the process as significant. More recently, several authors have found these features in the Bss horizons of different Vertisols (Nordt et al. 2004; Pal et al. 2012). Consequently, a first issue that still needs to be investigated in more detail is the expression and the origin of the illuviation process in the Vertisols of the Mesopotamian Pampa.

A factor that influences the development of vertic properties, and eventually the clay illuviation, is the composition of soil parent material. The parent material of Vertisols in the Mesopotamian Pampa is made up of lacustrine calcareous loams having silty-clay or clay-loam textures, with a predominance of smectitic minerals in its clay fraction, and with a considerable proportion of calcium carbonate, some gypsum and manganese and iron oxides segregations. The color of the sediment is varied, with a predominance of gray and greenish colors. These sediments make up a stratigraphic unit named *Hernandarias Formation,* deposited by the Uruguay River in lacustrine and palustrine environments during a very dry period of the lower-middle Pleistocene. In the western part of Entre Ríos and overlying the *Hernandarias Formation* there is a deposit of about 2 m thick of loessic sediments of late Pleistocene age named *Tezanos Pinto Formation*, which constitute the parent material of Mollisols. On the other hand, the existence of a layer of aeolian silt only 20–30 cm thick named *San Guillermo Formation* has also been proposed, which during a period of semi-arid climate in the Upper Holocene would have homogeneously covered the entire landscape of the sub-region (Iriondo 1991, 1994; Iriondo and Kröhling 1995). Thus, although it is clear that the Vertisols of Entre Ríos have developed from the *Hernandarias Formation,* it remains to confirm if they have received loessic contributions and if these have eventually influenced their development.

17.2.2   The Study Cases

17.2.2.1   The Soils and Their Environment

A large part of the northwest of the province of Entre Ríos, presents a landscape of undulating and gently undulating plains, with vegetation of xerophytic open woodlands dominated by *Prosopis affinis* (a legume species commonly called "ñandubay"). The characteristic soils are Alfisols on the flat to very gently undulating uplands, without a defined drainage network and with poor drainage, Vertisols on the slopes with a gradient of 1–3% and with moderate runoff, and Mollisols at the foot slopes, developed on retransported loessic sediments (INTA 1990; Bedendo 2019).

Three soil profiles were studied along a transect 150 m long, in a gently undulating landscape near the town of San Gustavo, La Paz Department, Entre Ríos (Fig. 17.2a, b). A 1:20.000 scale map of this area made by means of physiographic analysis, together with the determination of the soil productivity index by DEM derivatives and geo-statistical techniques, was carried out by Bedendo et al. (2016). Some preliminary studies on these same profiles were previously presented (Morrás, et al. 1993, 1998; Morrás 2018). Profile 1, located on a flat surface at the summit of the slope, poorly drained, without runoff and with surface ponding during the humid periods, and developed under natural xerophytic forest vegetation, corresponds to a Vertic Ochracualf (Colonia Trece series) (INTA 1990). Profile 2 is also found on the hill, in a slightly lower position close to the shoulder of the slope, with slightly better drainage, and corresponds to a Vertic Argiudoll. This soil is not mentioned in the existing semi-detailed maps (cartographic unit-ct; Fig. 17.2b), so its extension must be limited and therefore its identification is of interest. Profile 3 is at the upper part of the backslope, with a 2% gradient, moderately well drained and with a slight degree of erosion. This soil is a Vertisol, originally classified according to the local adaptation of the Soil Taxonomy of 1975 as a "Argiudollic" Pelludert, and according to the current version of the Soil Taxonomy (2014) as a Typic Hapludert (Ramblones series).

The morphology of the profiles is presented in Fig. 17.3. The thickness of the solum decreases in the sequence P2 > P1 > P3. The Alfisol (P1) is characterized by

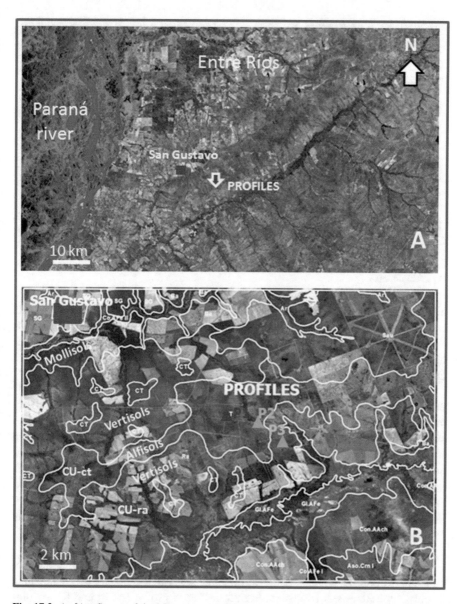

Fig. 17.2 (**a, b**) **a**-Sector of the Mesopotamian Pampa in north-western Entre Ríos Province, and location of studied soils. **b**- Cartographic and taxonomic units, from the soil map in scale 1:50.000. Cu-ct: cartographic unit Colonia Trece. Cu-r: cartographic unit Ramblones. Profiles studied: P1: Alfisol. P2: Mollisol. P3: Vertisol. Note: in the field the soil profiles are much closer together than depicted in the figure

Fig. 17.3 Soils studied in the Mesopotamian Pampa. 1: Vertic Ochraqualfs. 2: Vertic Argiudoll. 3: Typic Hapludert

a thin E horizon followed by a Bt horizon about 55 cm thick. The Mollisol (P2) has an epipedon 23 cm thick (A + BA horizons), a Bt horizon about 60 cm thick, followed by a BC horizon 30 cm thick. For its part, the Vertisol (P3) presents a transitional BA horizon, which is followed by a horizon identified as Bt only about 25 cm thick and then by a dark BC horizon.[1]

In the Alfisol and in the Mollisol humic-clay skins are observed in the upper part of the Bt horizon and clay skins up to the base of the profile, these being clearly more abundant in the first soil. Both profiles show slickensides and wedge-like aggregates in their lower horizons, as well as cracks when dry. Slickensides are scarce in the BC horizons but a little more frequent in the Bt horizons. On the other hand, in the Vertisol there are slickensides in moderate quantity from the BA horizon to the base of the profile and clay skins in moderate quantity only in the deep BC and C horizons. Calcium carbonate nodules as well as infillings and coatings of lenticular gypsum are abundant in the lower horizons of all three profiles. The Mollisol is also characterized by the highest content of $CaCO_3$ nodules in the middle and lower part of the Bt horizon.

For its part, the analytical data provide interesting information regarding the specific and differential characteristics of the three soils. In a synthetic way, some data of only three main and equivalent horizons of each profile (A, Bt and C) are presented in Fig. 17.4a–d.

In the first place, it stands out that the organic carbon content in the surface horizon is lower in the Alfisol and maximum in the Vertisol (Fig. 17.4a). This fact, together with the morphological characteristics of the epipedon in terms of structure

---

[1] Several horizons of the three soils carry the suffixes ss, k, or y, which are not shown. To simplify the text and figures, since illuviation is specifically discussed in this article, only the suffix t is shown.

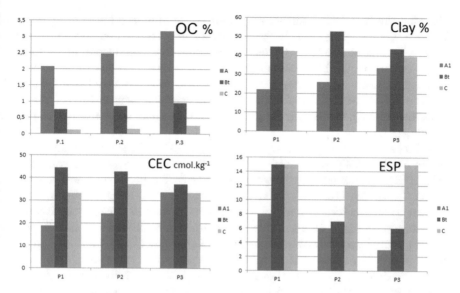

Fig. 17.4 (**a–d**) Selected analytical data of three main horizons (A, Bt and C horizons) of the three studied soils (P1: Alfisol; P2: Mollisol; P3: Vertisol) in the Mesopotamian Pampa. (**a**). OC: organic carbon (%). (**b**). Clay: clay content (%). (**c**). CEC: cation exchange capacity (cmol.kg^{-1}). (**d**). ESP: exchangeable sodium (%)

and thickness, leads to the latter being considered mollic. The granulometric analysis shows similar variations with depth between the three soils, especially highlighting the increase in clay content in the Bt horizons, even in the Bt of Vertisol (43%) (Fig. 17.4b). It is worth mentioning that in this soil the maximum percentage of clay is found in the BC horizon (47%) and then decreases in the C horizon (39%). For its part, the total CEC presents maximum values in the Bt horizon of the three soils, combining there the contributions of clay and organic matter (Fig. 17.4c). These granulometric and chemical properties are also reflected in water retention parameters at different tensions, which also show maximum values in the Bt and BC horizons of the three profiles (Morrás 2018). Finally, a parameter to consider is the exchangeable sodium percentage of these soils. As can be seen in Fig. 17.4d, in the Alfisol both the Bt and C horizons are at the limit of 15%, which could lead to qualifying this soil as natric. In the Mollisol and Vertisol solum the values are lower but not negligible, while in their C horizons the ESP is close to or in the limit of 15%.

17.2.2.2 Clay Mineralogy

The mineralogical analysis by XR diffractometry of the clay fraction of the three profiles has revealed compositional variations as a function of depth (Morrás 2018). The diagrams of C horizons of the three profiles are similar to each other (data not shown), indicating an association of random interstratified minerals of the illite-smectite type with a smaller proportion of smectite, and traces of illite and kaolinite. On the other hand, in the A horizons the clay fraction appears almost exclusively constituted by I-S interstratified minerals. The abundance of these mixed-layer minerals differs from the composition observed in other Vertisols in the province, in which only the presence of montmorillonite is mentioned as an expandable component (Mazza 1975; Scoppa 1976; De Petre and Stephan 1998).

On the contrary, the Bt horizons present clear differences between the three profiles (Fig. 17.5).

In P3 (Vertisol) the diagram is quite similar to that of the C horizon, characterized by a tail of continuous reflections towards the small angles, although with some reflections of greater intensity between 14 and 16 Å. The glycol treatment expands the clay generating a trace of continuous reflections towards the small angles, while the heated clay contracts towards 10 Å. This behavior indicates that the clay is made up of an association of illite-smectite interstratified minerals and a smaller proportion of smectite, probably of low crystallinity. In the case of clay from the Bt horizon of profile P1 (Alfisol), the glycol treatment produces a strong expansion and a clear definition of intense reflections towards 18 Å, as well as the appearance of a clearer peak towards 10 Å. Consequently, in this sample the proportion of interstratified illite-smectite decreases with respect to Vertisol and the proportion of smectite and illite minerals increases. Unlike the two previous cases, in the clay of the Bt horizon of P2 (Mollisol) a peak of great development is manifested towards 15.5 Å, which clearly moves towards 17.8 Å in the glycolated sample and contracts with a marked peak at 10 Å in the heated sample. This result, in coincidence with the results obtained by other determinations (Morrás 2018), indicates that in this horizon the clay is basically made up of smectite, with smaller proportions of illite-smectite interstratifications and illite.

The increase in the smectite component in the B horizon of the Alfisol and particularly in the Mollisol, could be the result of two converging processes: on the one hand, the preferential illuviation of the finer fractions of the clay, which are

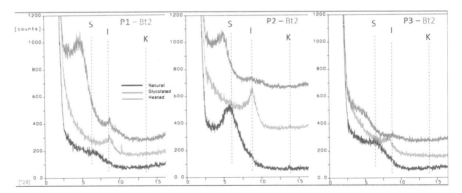

Fig. 17.5 X-ray diagrams of the clay fraction from Bt2 horizons of the three soils studied. Treatments: natural, glycolated and heated. Dotted lines indicate the position of main reflections of different clay minerals S: smectite, I: illite, K: kaolinite

normally enriched in smectites. On the other hand, by *in situ* neformation, which would be facilitated by the physical and chemical conditions of that horizon. In this last regard, in the Bt horizons of both soils, in addition to a slightly alkaline pH, there is a richness of bivalent cations, a fact that can also be deduced from the high concentration of calcareous nodules shown in Fig. 17.3.

17.2.2.3   Mineralogy of the Sand and Silt Fractions

Mineralogical analyzes of the very fine sand fraction (50–100 μm) of the three profiles were carried out by polarizing optical microscopy (Morrás et al. 1993). The heavy fraction is scarce, made up mostly of opaque minerals, among which magnetite predominates. Among the transparent minerals, hornblende is the most abundant, along with a variety of other minerals in very small proportions (zircon, rutile, epidote, etc.). In the light fraction, the most abundant components are quartz and feldspars and a smaller proportion of unidentifiable grains covered with an iron oxide patina (alterites). Quartz is found in two varieties: on the one hand, in monocrystalline, equidimensional grains, rounded and sub-rounded, with a limpid surface and generally with normal extinction, and on the other hand, in polycrystalline grains, with a saccharoidal appearance, generally sub-rounded and limpid.

In the three profiles, the proportions of both components vary in a similar way with depth, observing a dominance of polycrystalline quartz (between 30–45% of the fraction) in the surface horizons, including the top of the Bt horizon, to decrease abruptly from there on (between 5–13%). On the contrary, the monocrystalline quartz content in the upper horizons ranges between 20–30% and increases slightly in the deeper horizons (31–44%). In P1 and P2 profiles, and parallel to these variations of the quartz varieties, the alterites are present in a small proportion in the surface horizons (3–8%) and increase markedly in the deep horizons (10–35%). On the contrary, in P3 the proportion of alterites is irregular along the profile.

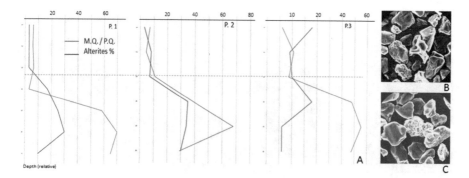

Fig. 17.6 (a–c) (**a**): Selected mineralogical data of the sand fraction of the profiles studied. Alterites (%) and M.Q./A.Q.: relationship between monocrystalline quartz and polycrystalline quartz. (**b**): Scanning electron microscopy (SEM) of the silt fraction of the E horizon of Profile 1 (Alfisol). (**c**): SEM of the silt fraction of the C horizon of Profile 1

Figure 17.6a shows the ratio of monocrystalline quartz/polycrystalline quartz as well as the proportion of alterites as a function of depth. Although inconspicuous, feldspars are slightly more abundant in deeper horizons. Finally, acid-type volcanic glass particles are also observed within the light fraction, without alteration and in a very small quantity (1–3%), although in slightly higher proportion in the superficial horizons. These results would be indicating the existence of a mineralogical discontinuity in the three profiles, located approximately 40 cm deep.

On the other hand, the existence of this discontinuity is confirmed by the mineralogical and morphoscopic analysis by SEM of the silt fraction (2–50 μm) and of the sand of the E and C horizons of Profile 1 (Morrás 2018). In the silt of the C horizon, the morphologies suggest that the quartz grains come from strongly weathered materials, which have then been transported undergoing surface rounding and amorphization (Fig. 17.6c). On the contrary, in the silt fraction of the E horizon, most of the grains present much more irregular surfaces together with sub-rounded quartz grains with fresher shock marks and unpolished surfaces. Fresh-looking volcanic glass particles are also seen here (Fig. 17.6b). These characteristics suggest a mixture of materials from different sources in the coarse fractions of this horizon, and perhaps *in situ* weathering processes.

17.2.2.4   Micromorphology

Micromorphological analyzes in thin section have evidenced the presence of illuvial features in the three profiles, although in different proportions depending on the horizons (Morrás et al. 1993). On the one hand, the three soils present clay coatings (argilans and organo-argilans) on the pore walls and fragmented coatings (papules) incorporated into the groundmass, both in the top of the Bt horizon and in the BC and C horizons (Fig. 17.7a, b).

These textural features are almost non-existent in the middle part of the Bt horizons. On the other hand, although illuviation features are evident, it occurs with

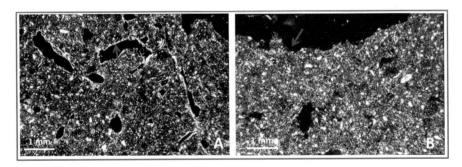

Fig. 17.7 (**a,b**) (**a**): Micrographs of thin sections observed under crossed polarizers. (**a**): profile 1 (Alfisol), C horizon. Red arrow is pointing to a clay coating. Note most of voids surfaces coated with oriented illuvial clay. (**b**): profile 2 (Vertisol), C horizon. Note clay coatings on a void surface and a high number of fragmented thin clay coatings included in the groundmass

different magnitude depending on the soil. Thus, the Alfisol is characterized by the abundance of coatings and fragmented coatings in the Bt1 horizon, which suggests that the overlying E horizon results from a greater intensity of the argilluviation process, probably related to the surface hydromorphism. On the contrary, the textural coatings are scarcer in the Vertisol, which may be due both to a lower intensity of illuviation and to a faster and more complete integration of the clays translocated to the groundmass. For its part, in the Mollisol the amount of textural features is intermediate between the other two soils.

On the other hand, soil porosity and several matrix constituents of the Vertisol were analyzed through micromorphometric methods (Morrás et al. 1998). Results obtained show marked differences in porosity, groundmass and pedological features among different horizons. In this analysis no discrimination was made between the fine fraction (plasma) in the groundmass and in the illuviation features. However, a clear progressive increase in the plasma / skeleton ratio up to a maximum in the BC horizon is observed, which would be related to the clay increases in the same direction found by other methods.

17.2.2.5 Discussion and Conclusions

In the first place, the mineralogical analyzes of the sand and silt fractions lead to establish the existence of a lithological discontinuity in the upper part of the Bt horizon of the three soil profiles. From a depth of about 40 cm downwards, the parent material of horizons B and C is constituted by clayey sediments of the Hernandarias Formation. The data obtained indicate that this sediment is characterized by an association of interstratified illite-smectite and of smectites in the clay fraction with rounded monocrystalline quartz in the coarse fraction. On the contrary, the coarse fraction of the epipedon of the three soils is characterized by grains of irregular morphology, a high proportion of polycrystalline quartz, a lower proportion of alterites and a slightly higher proportion of acid volcanic glass than in the underlying horizons. The polycrystalline quartz grains could correspond to materials of volcanic origin (Kröhling and Orfeo 2002). These results suggest that the material of the superficial horizons of these soils would be in part constituted by loessic contributions from sources of the Andean mountain range. Since their age is not established, these sediments could correspond to the Tezanos Pinto Formation (Upper Pleistocene) or to the San Guillermo Formation (Holocene) as suggested by Iriondo (1991).

Consequently, it can be interpreted that the landscape would have been covered with a thin mantle of aeolian sediments, which was partially mixed with the clayey materials of pre-existing Vertisols developed from the Hernandarias Formation, a process that would have been facilitated by the expansion and contraction typical of this material. Based on the granulometric and mineralogical data, it appears that the loessic sediment is more individualized in the flat summits of the landscape, where the Alfisol and Mollisol are found. It can be speculated that the divergent development of these two soils would be related to their relative position in a paleosurface microtopography, to the thickness of the loessic layer deposited at each site, and to their subsequent combined influence on hydrological behavior and on pedogenetic processes. On the contrary, on the slope where the current Vertisol occurs, the loessic layer would have been mixed as well as eroded more quickly than on the summit of the slope (Fig. 17.8).

Secondly, the granulometric data and other chemical and physical data indicate higher clay contents in the B and BC horizons of the three vertic soils. In turn, micromorphological analyzes show a relatively high proportion of illuviation features in the lower portion of the B horizon and in the BC horizon of the three soils, particularly in the Alfisol but also evident in the Vertisol. These results indicate that clay illuviation is an active process in these vertic soils of Entre Ríos, whose origin could be explained by their polygenic character.

In this interpretation, in a first stage of pedogenesis in the smectitic material of the Hernandarias Formation, the soils would have developed more intensely vertic characteristics, under the influence of more contrasting situations of wetting and drying. At a later stage, and as a consequence of capping with loessic sediments, the differential wetting and the shrink-swell process of the subsoil became limited, due to the buffering effect of the non-expansive silty surface (Hartley et al. 2014). Consequently, in a second stage of pedogenesis, the clay translocation process would have increased. Also, the relatively high percentage of exchange sodium in these soils would facilitate the dispersion and migration of clays. According to what was observed, illuviation features are scarce in the middle part of the B horizons, which would be the result of a greater intensity and frequency of wetting and drying

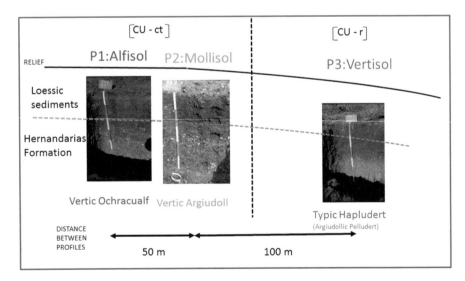

Fig. 17.8 Sketch of the position of the soil profiles in the relief and the distance between them. The Fig. also shows the succession of sediments in which the soils have developed, according to the results obtained in this study. In Alfisol and Mollisol, the superficial loessic sediment is more individualized, while in Vertisol it has a higher degree of mixing with the underlying loam

and a more intense argilliturbation at that level, producing the rupture and the incorporation of coatings into the groundmass. On the contrary, clay films are preserved in the lower horizons, where peds and voids surfaces are more stable.

Finally, the results obtained would justify the designation of the B horizon of the Vertisol here studied as an argillic Bt horizon, as well as its epipedon as mollic, and would also justify the criteria and adaptations of the classification system that were developed by the soil surveyors of Entre Ríos, in the Pampa region.

17.3 "Hidden" Vertisols of the Undulating Pampa

17.3.1 Setting the Subject

The Undulating Pampa (frequently also named as Rolling Pampa) is a subregion of the vast Pampa plains bordering with the Paraná and Río de la Plata rivers to the north-east and the Salado river to the south-west (Fig. 17.1b). It is a sedimentary plain characterized by a gently undulating landscape generated by fluvial erosion in response to neotectonic processes, with a slope generally less than 2%, although in some sectors it can reach 5%. The surface of the subregion is covered by a mantle of loessic sediments 6–7 m thick deposited during the Late Pleistocene and known as the Upper Pampeano or Buenos Aires Formation, stratigraphically and sedimentologically equivalent to the Tezanos Pinto Formation in Entre Ríos. During the Holocene, eolian sediments named the Post Pampean or La Postrera Formation were deposited in some areas of the plain (Iriondo and Kröhling 1996; Zárate 2005). The main source of Pampean sediments are the andesitic and basaltic rocks in northern Patagonia and in the Andes Cordillera, and their aeolian transport promoted a granulometric sorting across the Pampa. The loessial sediments in the Undulating Pampa also present a well expressed granulometric zonation, showing three main parallel strips from the Salado river to the NE: the first composed of sandy loess, the second composed of typical loess, and the third, the closest to the Paraná-Río de la Plata fluvial axis, characterized by the intercalation of typical and clayey loess (Morrás and Cruzate 2000).

Following the initial criteria about a main Andean source, it has long been considered that the surface sediments of the Undulating Pampa, and therefore the parent materials of the current soils, were mineralogically homogeneous (Teruggi 1957; Scoppa 1976; Teruggi and Imbellone 1983). However, a considerable number of works on the sand fraction of sediments and soils from different areas of the Pampa region have shown the existence of different sources and input processes in the formation of the sedimentary mass (e.g., González Bonorino 1965, Etchichury and Tófalo 2004). Moreover, a detailed analysis of sand mineralogy data from sediments and soils in the Undulating Pampa and its proximal sectors have shown great vertical and spatial heterogeneity, allowing the identification of different "mineralogical models" as well as the delimitation of several "mineralogical zones" (Morrás

2003, 2020a, b). This mineralogical heterogeneity would be related to volcanic-pyroclastic contributions coming from different sources in the Andean mountain range, as well as significant contributions of igneous and metamorphic rocks from the Córdoba and San Luis pampean ranges. Moreover, the sediments in the Undulating Pampa are characterized and differentiated from sediments in other sub-regions by containing varied proportions of minerals transported by the river systems of the Mesopotamian region, which would have been moved by the wind from the floodplains during periods of low water levels and perhaps from the continental platforms in times of marine regressions (Morrás and Moretti 2016; Morrás 2020a). Thus, the Undulating Pampa appears as a mineralogical transition area between the northern Pampa and the southern Pampa.

On the other hand, several geopedological units have been recognized within the Undulating Pampa due to specific combinations of geomorphic landscape, lithology and soils, although the main subdivision usually considered is between the High Undulating Pampa and the Low Undulating Pampa (Fig. 17.1b). The Low Undulating Pampa is clearly a transitional area between the High Undulating Pampa to the north and the Depressed or Flooding Pampa to the south. It is characterized by a landscape of very flat hills, with a poorly defined drainage network, and with a significant proportion of close depressions and small lagoons in some sectors. The soils are developed in loessic Upper Pleistocene sediments of the Pampean Formation, with a silty-clay loam texture and a moderate proportion of smectitic clay. Vertic Argiudolls characterize the flat broad summit areas in the central part. Towards the east of the region the Hapluderts appear, while in the southern sector the Natracuerts dominate. Aquic and natric Mollisols occur in waterlogged depressions and runoff paths (Muro et al. 2004; Hurtado et al. 2005; Imbellone et al. 2010). The Vertisols in the Low Undulating Pampa, and similarly to those in the Mesopotamian region, also present illuvial Btss horizons (Imbellone and Giménez, 1990). In this region, diapiric structures and the "gilgai" micro-relief are less evident than in Mesopotamian Vertisols (Scoppa 1976). Also, unlike the soils of Entre Ríos developed in the Hernandarias Formation, these Pampean Vertisols have lower CaCO3 content in the Bt and do not present gypsum at the base of the profiles. Besides the Vertisols in the loessic plain, it is to be mentioned that Aquerts characterize the estuarine coastal plain of the Río de la Plata, developed on a long strip of clayey sediments deposited during a middle Holocene marine ingression.

Towards the north, in the High Undulating Pampa the undulations of the terrain are clearly marked, related to a well-dendritic hydrographic network built by several tributaries of the Paraná and La Plata rivers. Unlike the Low Undulating Pampa, Vertisols have not been described in this sector, but Vertic Argiudolls appear in the strip with clayey loess that borders the Paraná-de la Plata fluvial axis. Numerous detailed studies of sediments and soils have been carried out in a representative area of this sector in recent years, giving rise to a new sedimentological and geopedologic approach explaining the development and spatial distribution of Vertic and Typic Argiudolls in the strip (Morrás and Moretti, 2016; Morrás 2020a, b). According to this, smectitic sediments coming from different sources in the Paraná basin were deposited in the Undulating Pampa, and later covered by illitic loess sediments from Andean and Pampean ranges sources. In a subsequent humid period in the Holocene, while Typic Argiudolls develop on the last sediments blanketing the landscape, the highest crests and slopes would be eroded and the smectitic sediments exposed. Thus, Vertic Argiudolls would develop later in higher positions of the landscape on smectitic sediments of older age.

Although this model of landscape and soil evolution is strongly supported by analytical data, work carried out in recent years has provided new and interesting evidence indicating a greater complexity of pedogenetic processes, which leads to refine the interpretive model and that would generate important implications for the survey and cartography of these soils.

17.3.2   The Study Cases

Numerous pedological studies have been carried out in the fields of the National Institute of Agricultural Technology (INTA) and its surroundings, in the metropolitan area of Buenos Aires, at the boundary between High and Low Undulating Pampa. A synthesis of results obtained have been presented by Morrás and Moretti (2016). Most of the profiles studied in the area over the years were described and sampled in excavations of the usual size for soil survey (about 1 x 2 x 1.5 m). However, in recent years several studies have been undertaken in large excavations, firstly taking advantage of excavations for the construction of buildings or infrastructure works, and then in trenches excavated specifically for the purpose of pedological studies. Thus, and as a result of this way of working, morphologies compatible with diapiric structures began to be observed in the soils located in the summit of the slopes. Consequently, new studies have been undertaken and a new hypothesis has been generated about the genesis of the vertic soils of the Undulating Pampa. We will present here some of the results of ongoing studies.

17.3.2.1   Morphology of a Soil Profile in the INTA Fields

An L-shaped excavation, several times larger than usual for soil description, was made on top of a slightly convex hillslope (34°36′55"S, 58°39′39"W). This excavation was carried out a few meters from where in previous years a soil classified as Vertic Argiudoll was studied in detail. Unlike the previously studied profile, the greater depth of the excavation made it possible to observe the existence of an underlying sediment differentiated from the surface by its color and texture. On the other hand, the greater lateral dimension of the pit faces exposed by the recent excavation, allowed to verify marked undulations of the horizons and the existence of sectors with "bowl" and "chimney" type morphologies, typical of Vertisols. In these sectors, the projection towards the surface of the deeper sediment is evident by the differences in color and structure between B and C horizons, and by the high concentration of $CaCO_3$ nodules in the upper portion of the last (Fig. 17.9).

Fig. 17.9 Photograph of the soil profile in the INTA fields, in the Undulating Pampa. Note the "bowl" and "chimney" morphology and the sediment overlay revealed by digging a large pit. Each section of the tape measures 10 cm

Gilgai-like structures are not observed on the soil surface despite the internal undulations of the horizons. The planimetry of the limits of the soil horizons in the six faces exposed in the excavation, allowed a reconstruction of the internal surfaces of the three most contrasting levels: an upper level constituted by the solum (A + B), an intermediate level constituted by the upper part of the C horizon and characterized by a high concentration of calcium carbonate, and a deeper level with a lower proportion of carbonate and slightly greenish color of the sediment matrix (Fig. 17.10).

17.3.2.2 Chemical and Physical Parameters

Several analytical variables were studied separately in the two sectors in the profile, related to what would be the "bowl" and "chimney" of a Vertisol, The depth functions of some data obtained are presented here.

In the first place, the sand content reveals the granulometric differences between the two sedimentary bodies that make up the profile. In this sense, the superficial loessic sediment where the solum has developed has around 15% sand, while the subsurface sediment contains around 20% sand. Laterally, and coinciding with the differences in the sequence of horizons between the "bowl" and "chimney" sectors,

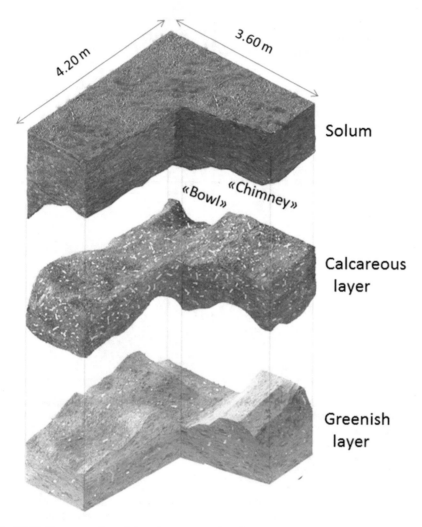

Fig. 17.10 Reconstruction of the micro-topography of the three main pedo-sedimentary levels that constitute the profile

the highest sand content appears in the first at 120 cm depth, while in the second the highest sand content appears closer to the surface, from 80 cm deep (Fig. 17.11a).

As for clay, the content of this fraction in the A horizon ranges around 30%. The increase in clay in Bt occurs in a similar way between the "bowl" and "chimney", with a maximum of around 48% at 70 cm depth. Coinciding with the difference in the sequence of horizons between the two sectors of the profile, the decrease in clay content is more abrupt in the "chimney" and more progressive in the "bowl". In the deep BC and C horizons the clay content fluctuates around 35%.

Other analytical variables that clearly reflect the morphological differences between the "chimney" and the "bowl" are the content of $CaCO_3$ in the fine earth

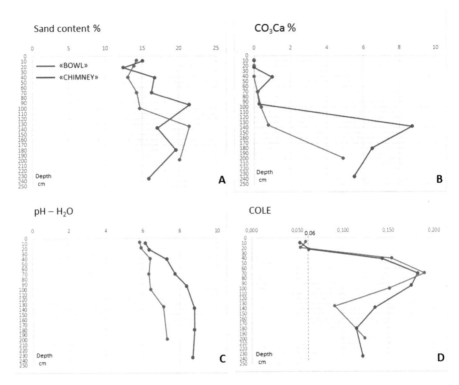

Fig. 17.11 (**a–d**) Some analytical data showing the contrast of physical and chemical properties between the two main sedimentary layers and the "bowl" and the "chimney" sectors in the soil. (**a**): total sand (50–2000 μm) (%). (**b**): CO$_3$Ca content (%) in the fine earth (<2 mm). (**c**): pH in H$_2$O (1:2.5 soil: water relationship). (**d**): Coefficient of linear extensibility (COLE) (dimensionless). The dotted vertical line indicates the value 0.06, conventionally used as the limit between moderate and high expandability of soil materials

(<2 mm) and the pH-H2O. On the one hand, and in coincidence with the observed distribution of calcareous nodules, CaCO$_3$ appears at more superficial levels and in greater concentration in the "chimney" sector (Fig. 17.11b). For its part, the pH ranges around 6–7 throughout the depth of the soil up to about 2 m deep in the "bowl", while in the "chimney" the pH reaches values close to 8 at 70 cm deep, and from there the pH oscillates around 8.5 to the base of the profile (Fig. 17.1c).

Regarding the physical parameters, and given its diagnostic character in the Soil Taxonomy, the COLE is worth highlighting. According to the established categories (Grossman et al. 1968; USDA-NRCS 2014), it is verified that in the epipedon up to about 25 cm deep, the value oscillates around 0.06, that is, in the upper limit of moderate expansibility. From there, there is an abrupt increase with values above 0.09 that place the samples in the category of very high expansibility, reaching a maximum of 0.19 in the Bt horizon, between 60–80 cm deep (Fig. 17.11d). Consequently, the linear extensibility of the soil in both "bowl" and "chimney" easily exceeds the 6 cm limit used as diagnostic by Soil Taxonomy to define vertic subgroups in different Orders.

17.3.2.3 Magnetic Susceptibility

The iron minerals existing in sediments and soils, although minor in their mass, have magnetic properties that can be measured by various techniques. Among these is the determination of the magnetic susceptibility (MS), The data of the MS are dependent on the concentration, grain size and mineralogy of the components of the magnetic fraction; the values obtained in the soils integrate the signal provided by the minerals inherited from the parent material and those newly formed as a result of pedogenetic processes. Studies carried out in various soils of the Undulating Pampa and specifically also in Argiudolls of the INTA fields, have shown that most of the SM value is found in the coarse fraction, that is, it comes from the parent material (Liu et al. 2010; Morrás 2020b).

The analysis of the MS in the soil studied here (Fig. 17.12) shows important oscillations with depth as well as some differences between the "bowl" and "chimney" sectors. In the upper section of the soil, and up to approximately 180 cm deep in the "bowl" and 130 cm deep in the "chimney", the minimum and maximum values are of similar intensity. However, those extreme values are manifested at different depths, in direct relation to the differences and oscillations of the total sand content in each of the two sectors of the profile (compare with Fig. 17.11a).

Below these depths, the MS decreases progressively towards the base of the profile, even though the sand content is higher than in the upper section, thus suggesting a compositional change. On the other hand, the values are much lower in the "chimney" than in the "bowl". In the upper part of this lower level, approximately up to 220 cm deep, the significant decrease in MS in the "chimney" would be partially related to the high content of $CaCO_3$, which is diamagnetic. In the deepest sector and given that the $CaCO_3$ content is somewhat lower there, the low intensity of the MS signal would be related to a compositional difference in the magnetic

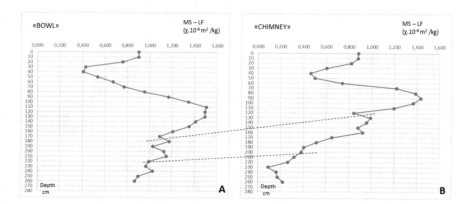

Fig. 17.12 Soil mass magnetic susceptibility (χ) measured at low frequency, in the "bowl" and "chimney" sectors in the profile. The dotted lines indicate the depths at which the limits between the main sedimentary layers appear. Red dotted line: limit between the upper loess and the lower calcareous loam. Green dotted line: limit between the calcareous loam and the greenish loam

fraction of the sediment with respect to the superficial loess, and possibly also with secondary processes of reduction of iron minerals as suggested by the greenish color of the sediment.

17.3.2.4   Clay Mineralogy

The mineralogical analysis of the clay fraction by XRD shows compositional variations with depth that allow three levels to be differentiated. In the epipedon, up to about 25 cm depth, the clay fraction is mainly constituted by illite, together with smaller proportions of interstratified illite-smectite and kaolinite (Fig. 17.13a). At a

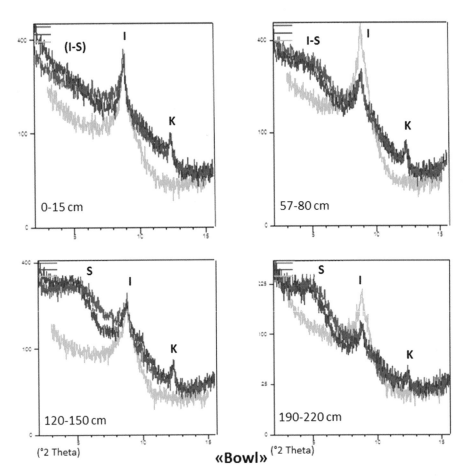

Fig. 17.13 (**a–d**) X-ray diffractograms of the clay fraction (Mg saturated) in soil samples at different depths of the "bowl" sector. (**a**). 0–15 cm. (**b**). 57–80 cm. (**c**). 120–150 cm. (**d**). 190–220 cm. Red line: natural clay. Blue line: clay treated with ethylene glycol. Green line: clay heated to 520 °C. I-S: random interstratified illite-smectite. S: smectite. I: illite. K: kaolinite

second level corresponding to the Bt horizon and up to about 80 cm deep, the proportion of random interstratified I-S increases, which is evidenced by a greater slope and a shoulder of reflections at small angles, by the moderate expansion in the test with glycol and by the collapse to 10 Å upon heating (Fig. 17.13b). Finally, in the lower section up to the base of the profile, the expandable minerals are dominant, in this case with smectite expressed by a more defined development of the reflections towards 15 A, together with the interstratified I-S. In parallel, illite and kaolinite decrease in proportion (Fig. 17.13c, d).

17.3.2.5 Application of the Pedogenetic Model to the Classification and Mapping of Vertic Argiudolls

Nine soil profiles from the INTA soil database (SISINTA; Olmedo et al. 2017), located in the High Undulating Pampa and classified as Solis Series (3 profiles) and Portela Series (6 samples), were analyzed. The hypothesis is that each of these soil profiles correspond to one of the hemicycles ("bowl" or "chimney") of the "hidden Vertisol"pedon found in the INTA field, and not to different soil series. To test this hypothesis, the in-depth variation of $CaCO_3$ content and pH of selected profiles was compared using the AQP R package (Beaudette et al. 2022). In addition, a pairwise dissimilarity analysis was performed to assess the Euclidean distances between the soil profiles (Chap. 11, Sect. 11.4). The Euclidean distance was evaluated considering $CaCO_3$ content (%), clay content (%), pH (soil: H_2O, 1:2.5) and CIC (cmolc/100 g) up to 200 cm depth.

Figure 17.14a is a graphical representation of the soil profiles, where the colors represent the CEC level and the top dendrogram indicates the differences between

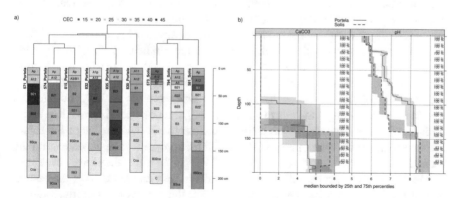

Fig. 17.14 (a, b) (a) Graphical representation of the soil profiles, where the vertical axis represents soil depth in cm and the color of the horizons represent the level of CEC (cmolc/100 g). On top of figure **a**, the tree structure is the dendrogram that represents the Euclidean differences among soil profiles. Figure **b** shows the variation of $CaCO_3$ and pH (X axis) as a function of the soil depth in cm (Y axis). Blue line is the median values of the group of soil profiles classified as Portela, while the red one represents the median values for Solís profiles. Colored shadows zones are 25th and 75th percentiles limits

profiles. The profiles associated with the Portela Series are grouped into a slightly heterogeneous group on the left, characterized by a higher CEC level in their B horizons (between 35 and 45 cmolc/100 g), while the profiles related to Solis Series and with a lower CEC value (between 30 and 35 cmolc/100 g) are grouped to the right. The analysis of the data did not show any evident difference in the content of clay or organic matter between the profiles considered here, which suggests that the differences in the CEC between both groups could be due to differences in the mineralogy of the fine fraction, but more research would be needed to confirm this. The differences between the Portela and Solís profiles in terms of $CaCO_3$ and pH can be seen in Fig. 17.14b, which shows the depth variation of both variables. It is clearly observed that the $CaCO_3$ appears at a lower depth in the Portela profiles than in the Solis profiles. In parallel, the pH is higher in the solum of the Portela profiles than in those of Solis. The differences in these two variables are similar to those observed between the two hemicycles of the "hidden Vertisol" in Castelar. These results suggest that the profiles of these two series could have been described and sampled in each of the two parts of a buried Vertisol: the Portela Series could correspond to the "chimney" and the profiles of the Solis series could correspond to the "bowl".

17.3.2.6 Discussion and Conclusions

The soils located in the summit of slopes in the east of the Undulating Pampa are classified and mapped as Vertic Argiudolls. However, large excavations reveal unexpected soil morphologies. Although in the solum the horizons present horizontal limits, in depth undulating limits and diapiric structures typical of Vertisols appear. In the soil studied here, the sand content, the magnetic susceptibility as well as various chemical parameters confirm the existence of two contrasting sedimentary levels. The clay fraction in the Bt horizon presents interstratified I-S minerals, while the underlying material is characterized by the appearance of smectite.

These results suggest that in this sector of the Undulating Pampa, the most recently deposited loessic sediments would have produced the burial of pre-existing Vertisols. It can be assumed that these Vertisols would have been dominant in different positions of the landscape. Later erosive processes would have thinned the loessic cover in the high and convex positions of the landscape. As a result, a juxtaposition of Vertic Argiudolls with different characteristics and depths of the solum is found today in the upper part of the relief. Their properties are dependent on the sector ("bowl" or "chimney") of the "hidden" Vertisol in which they have developed.

Moreover, several series of Vertic Argiudolls have been identified in the eastern part of the Undulating Pampa. However, neither the criteria for its differentiation nor its intrinsic variability were established. This resulted in series that tend to be very similar, while other series grouped together uneven soil profiles. Therefore, and based on a pedogenetic hypothesis such as the one presented here, it is considered that the application of quantitative analysis is the pathway to establish the origin, the taxonomic differences and the spatial distribution of these vertic soils, as well as other soils in the region.

17.4   General Conclusions and Prospect

The vertic soils of the Mesopotamian Pampa and the Undulating Pampa in the lower basin of the Paraná River have developed from smectite sediments differentiated in their origin and composition. However, a fundamental factor that would have determined its evolution and its differentiation on both sides of the Paraná River, as well as its differentiation from most Vertisols and other vertic soils in the world, is the thickness of the loessic sediments that would have covered them in the recent Quaternary.

Vertisols are the characteristic zonal soils of the Mesopotamian Pampa, developed from a calcareous and gypsiferous loam (Hernandarias Formation), with materials coming from the Brazilian craton. This is a sediment with a significant depth and great extension, and it can be considered that for a long period of time Vertisols would have been the dominant soils in the different segments of a hilly landscape. Later, in the Upper Pleistocene, loessic sediments from western sources (Andes and Pampean Hills) covered the entire landscape, progressively mixing with the preexisting Vertisols and finally burying them under relatively thin layers of aeolian material. More recently, during the humid periods of the Holocene, both pedogenesis and erosion processes took place, with different prevalence depending on topography. In a first stage and due to the buffering effect of the non-expansive silty surface, clay illuviation would have been a generalized process in the soil cover across the landscape. Thus, the conservation of the thin layer of loessic sediments on the higher slopes or their accumulation at the foot slopes resulted in the formation of Vertic Mollisols and Alfisols. On the contrary, the erosion of the loessic layer in the backslopes caused the exhumation of the buried soils. Depending on the proportion of loessic material remaining in their epipedon, some of these Vertisols with illuviation features also developed mollic characteristics. Consequently, these Vertisols from the Mesopotamian Pampas can be considered polygenic.

On the right margin of the Paraná River, in the Undulating High Pampa, there are no Vertisols but there are vertic Argiudolls in a strip about 60 km wide along the fluvial axis. These occur in the upper part of the hills, while in the backslopes and footslopes the dominant soils are Typic Argiudolls. Two superficial sedimentary levels have been distinguished in this strip, deposited in the Pleistocene and assigned to the Buenos Aires Formation. The lower one is a smectitic loessic sediment, which would be deflated from the neighboring alluvial plains of the Paraná river, among other sources to the east. The new results obtained in ongoing research and that have been presented here, have revealed the existence of diapiric structures developed in this sediment. Consequently, it can now be considered that, as in the Mesopotamian Pampa, Vertisols were also the dominant soils here for a prolonged period in the Pleistocene. Later, during the Last Glacial, the soilscape was blanketed by a relatively thick illitic loess, originated from the Andes cordillera and Pampean ranges to the west. Finally, along the humid periods in the Holocene, the higher crests and slopes were partly eroded and the underlying smectitic soils and sediments approached the current surface. Current soils consequently developed vertic properties due to the mixing of materials by physical and biological processes, and in relation to the greater or lesser proximity with the paleosurface. Unlike what happened in the Mesopotamian Pampa, due to the greater thickness of the most recent loessic sediment, these Vertisols of the Undulating Pampa remained as buried paleosols. On the other hand, because their presence could only be revealed from the observation of large excavations, they also deserve to be called "hidden" Vertisols. These advances on the understanding of the evolution of the landscape and the genesis of the vertic soils of the Undulating Pampa, together with the quantitative analysis of soil profiles data, appear highly useful to distinguish these soils at the series level and improve surveying and mapping work.

Acknowledgments The study included here on soils from Entre Ríos was promoted by our valued colleague and friend Carlos J. Vesco (†), from his position as Head of the Soil Mapping Group of the INTA Regional Agricultural Experimental Station in Paraná, Entre Ríos. We hereby express our gratitude to him, which we extend to the other members of the Soil Mapping team for their assistance in the field work.

References

Ahmad N (1983) Vertisols. In: Wilding L, Smeck N, Hall G (eds) Pedogenesis and soil taxonomy. Elsevier, New York, pp 91–123

Beaudette D, Roudier P, Brown A (2022) aqp: Algorithms for Quantitative Pedology. R package version 1.41. https://CRAN.R-project.org/package=aqp

Bedendo D (2019) Soils of Entre Ríos. In: Rubio G, Lavado R, Pereyra F (eds) The soils of Argentina, World soils books series, Springer International Publishing, pp 165–173

Bedendo D, Schulz G, Olmedo D, Rodríguez D, Angelini M (2016) Updating a phyisiography-based soil map using digital soil mapping techniques. In: Zinck A, Metternich G, Bocco G, del Valle H (eds) Geopedology: an integration of geomorphology and Pedology for soil and landscape studies. Chapter 18. Springer, Heidelberg, pp 305–319

Blokhuis W (1982) Morphology and genesis of Vertisols. In: Vertisols and Rice Soils in the Tropics. Transactions 12th International Congress of Soil Science, New Delhi, Vol. 3, pp 23–47

Bressan E, Castiglioni M, Morrás H (2014) Análisis de algunos criterios para la caracterización de Argiudoles vérticos de la Pampa Ondulada. Actas XXIV Congreso Argentino de la Ciencia del Suelo, Bahía Blanca, mayo 2014 edited in CD

Castiglioni M, Morrás H, Santanatoglia O, Altinier M (2005) Contracción de agregados de Argiudoles de la Pampa Ondulada diferenciados en su mineralogía de arcillas. Ciencia del Suelo 23(1):13–22

Castiglioni M, Morrás H, Santanatoglia O, Altinier V, Tessier D (2007) Movimiento del agua en Argiudoles de la Pampa Ondulada con diferente mineralogía de arcillas. Ciencia del Suelo 25(2):109–122

De Petre A, Stephan S (1998) Características pedológicas y agronómicas de los Vertisoles de Entre Ríos. Facultad de Ciencias Agropecuarias, Universidad Nacional de Entre Ríos, Argentina, p 65

Dudal R (1965) Dark clay soils of tropical and subtropical regions. In: FAO Agricultural Development. Paper No. 83 (FAO, Rome, 1965)

Dudal R, Eswaran H (1988) Distribution, properties and classification of Vertisols. In: Wilding L, Puentes R (eds) Vertisols: their distribution, properties, classification and management. Technical Monograph n 18, Texas A & M University Printing Center, pp 1–22

Durán A, Morrás H, Studdert G, Liu X (2011) Distribution, properties, land use and management of Mollisols in South America. Chin Geogr Sci 21(5):511–530

Etchichury M, Tófalo O (2004) Mineralogía de arenas y limos de suelos, sedimentos fluviales y eólicos actuales del sector austral de la cuenca Chacoparanense. Regionalización y áreas de aporte. Revde la Asoc Geol Argent 59(2):317–329

González Bonorino F (1965) Mineralogía de las fracciones arcilla y limo del pampeano en el área de la ciudad de Buenos Aires y su significado estratigráfico y sedimentológico. Rev de la Asoc Geol Arg XX 1:67–148

Grossman R, Brasher B, Franzmeier D, Walker J (1968) Linear extensibility as calculated from natural-clod bulk density measurements. Soil Sci Soc Amer Proc 32:570–573

Hartley P, Presley D, Ransom M, Hettiarachchi G, West L (2014) Vertisols and vertic poperties of soils of the Cherokee Prairies of Kansas. Soil Sci Am J 78:556–566

Hurtado M, Moscatelli G, Godagnone R (2005) Los suelos de la Provincia de Buenos Aires. In: de Barrio R, Etcheverry M, Caballé M, Llambías E (eds.). Geología y Recursos Minerales de la Provincia de Buenos Aires. Relatorio XVI Congreso Geológico Argentino, La Plata, Cap. XII, pp 201–218

Imbellone P, Giménez J (1990) Propiedades físicas, mineralógicas, y micromorfológicas de suelos vérticos del Partido de La Plata. Ciencia del Suelo 8(2):231–237

Imbellone P, Giménez J, Panigatti J (2010) Suelos de la Región Pampeana. In: Procesos de Formación. Ediciones INTA, Buenos Aires, pp 288

Iñiguez A, Scoppa C (1970) Los minerales de arcilla de los suelos zonales ubicados entre los ríos Paraná y Salado (Prov. De Buenos Aires). Revista de Investigaciones Agropecuarias, Serie 3 VII(1):1–41

INTA. 1990. Carta de suelos, Departamento La Paz, Provincia de Entre Ríos. Convenio INTA-Gobierno de Entre Ríos. INTA-EERA Paraná, Serie Relevamiento de Recursos Naturales n° 7, Tomo I (146 p.) y Tomo II (175 p)

Iriondo M (1991) El Holoceno en el Litoral. Com. Mus. In: Prov. Cs. Naturales Florentino Ameghino, Santa Fe, vol. 3, n°1, p. 40

Iriondo M (1994) Los climas cuaternarios de la Región Pampeana. In: Com. Mus. Prov. Cs. Nat. Florentino Ameghino, Santa Fe, vol. 4, N° 2, p. 48

Iriondo M, Kröhling D (1995) El sistema eólico Pampeano. In: Com. Mus. Prov. Cs. Naturales Florentino Ameghino, Santa Fe. vol. 5, n°1, p. 68

Iriondo M, Kröhling D (1996) Los sedimentos eólicos del noreste de la llanura pampeana (Cuaternario Superior). XIII Congreso Geológico Argentino, Buenos Aires. Actas IV: pp 27–48

IUSS Working Group WRB (2015) World Reference Base for Soil Resources 2014, update 2015. International Soil classification system for naming and creatin legends for soil maps. In: Worl Soil Resources Report n° 106, FAO, Rome, p. 192

Jongerius A, Bonfils C (1964) Micromorfología de un suelo negro grumosólico de la provincia de Entre Ríos. Revista de Investigaciones Agropecuarias, Serie 3, Clima y Suelo I(2):33–53

Kröhling D, Orfeo O (2002) Sedimentología de unidades loéssicas (Pleistoceno tardío – Holoceno) del centro-sur de Santa Fe. AAS Revista 9(2):135–154

Liu Q, Torrent J, Morrás H, Hong A, Jiang Z, Su Y (2010) Superparamagnetism of two modern soils from the northeastern Pampean region, Argentina and its paleoclimatic indications. Geophys J Int 183(2):695–705

Mazza C (1975) Mineralogía y probable génesis de la fracción arcillosa de un Vertisol de la provincia de Entre Ríos (Argentina). Rev. de la Fac. de Agron. La Plata, LI(1):1–16

Mermut A, Dasog G, Dowuona G (1996) Soil morphology. In: Vertisols and Technologies for their Management. In: Developments in soil science 24, Ahmad N, Mermut A (eds) Elsevier, Amsterdam, pp 89–114

Moretti L, Morrás H, Angelini M (2012) Génesis de suelos en el pedemonte aluvial del chaco salteño". Ciencia del Suelo 30(2):161–171

Moretti L, Vizgarra L, Morrás H, Rodríguez D, Schulz G, Paladino I, Bressan E, Laghi, J (2020) Origen de los materiales parentales y génesis de suelos en el extremo noroccidental de la cuenca de los Bajos Submeridionales, Santiago del Estero, Argentina. Latin American Journal of Sedimentology and Basin Analysis, Vol. 27, n° 1, 29–53

Morrás H (2003) Distribución y origen de los sedimentos superficiales de la Pampa Norte en base a la mineralogía de arenas. Resultados preliminares AAS Revista 10(1):53–64

Morrás H (2018) Material parental y pedogénesis de suelos verticos del Departamento La Paz, Entre Ríos. Actas XXVI Congreso Argentino de la Ciencia del Suelo, S.M. de Tucumán, mayo 2018, Comisión 5, pp 43–48 (digital ed)

Morrás H (2020a) Modelos composicionales y áreas de distribución de los aportes volcánicos en los suelos de la Pampa Norte (Argentina) en base a la mineralogía de arenas. In: Imbellone P, Barbosa O (eds.). Suelos y Vulcanismo. Asociación Argentina de la Ciencia del Suelo, Buenos Aires, Capítulo 5, pp 127–167

Morrás H (2020b) El material parental de los suelos de la región Pampeana en base a la mineralogía de arenas. Aplicaciones a la interpretación de procesos pedológicos. In: Imbellone P, Barbosa O (eds.). Suelos y Vulcanismo. Asociación Argentina de la Ciencia del Suelo, Buenos Aires, Capítulo 6, pp 168–184

Morrás H, Cruzate G (2000) Clasificación textural y distribución espacial del material originario de los suelos de la Pampa Norte. XVII Congreso Argentino de la Ciencia del Suelo, Mar del Plata, Actas (edited in CD)

Morrás H, Moretti L (2016) A new soil-landscape approach to the genesis and distribution of Typic and Vertic Argiudolls in the rolling Pampa of Argentina. In: Zinck A, Metternich G, Bocco G, del Valle H (eds) Geopedology: an integration of geomorphology and Pedology for soil and landscape studies. Springer, Heidelberg, pp 193–209

Morrás H, Robert M, Bocquier G (1982) Caracterisation minéralogique de quelques solonetz et planosols du Chaco Deprimido (Argentine). Cah.ORSTOM, ser. Pédologie XIX(2):151–169

Morrás H, Bayarski A, Benayas J, Vesco C (1993) Algunas características genéticas y litológicas de una toposecuencia de suelos vérticos de la provincia de Entre Ríos (Argentina). In: Gallardo J (ed) El estudio del suelo y de su degradación. Actas del XII Congr. Latinoam. de la Cien. del Suelo. Sociedad Española de la Ciencia del Suelo, vol II, Salamanca, pp 1054–1061

Morrás H, Benayas J, Ateiro R, Cruzate G (1998) Micromorfometría comparativa de un suelo Vertisol de la Provincia de Entre Rios. XVI Congreso Argentino de la Ciencia del Suelo, Carlos Paz, pp 303–304

Muro E, Sánchez J, Carboni G., Díaz R, de la Fuente J (2004) Actualización de la Taxonomía de suelos Vertisoles en la Provincia de Buenos Aires. Actas XIX Congr. Argentino de la Ciencia del Suelo, Paraná, Actas (edited in CD)

Murthy R, Bhattacharjee J, Landey R, Pofali R (1982) Distribution, characteristics and classification of Vertisols. In: Vertisols and Rice Soils of the Tropics. Symposia Paper In, 12th international Congress of Soil Science, Vol. 3, Indian Society of Soil Science, New Delhi, pp 3–22

Nettleton W, Sleeman J (1985) Micromorphology of Vertisols. In: Douglas L, Thompson M (eds) Soil micromorphology and soil classification, SSSA special publications, vol 15. SSSA, Madison, pp 165–196

Nordt L, Wilding L, Lynn W, Crawford C (2004) Vertisol genesis in a humid climate of the coastal plain of Texas, U.S.A. Geoderma 122:83–102

Olmedo G, Rodriguez D, Angelini M (2017) Advances in digital soil mapping and soil information systems in Argentina. In: Global Soil Map, digital soil mapping from country to globe, pp 13–16

Pal D, Wani S, Sahrawat K (2012) Vertisols of tropical Indian environments: Pedology and edaphology. Geoderma 189–190:28–49

Plan Mapa de Suelos (1980) Suelos y erosión de la provincia de Entre Ríos. INTA-EERA Paraná. Serie Relevamiento de Recursos Naturales n°1. Tomo n°1, p 109

Rubio G, Pereyra F, Taboada M (2019) Soils of the Pampean region. In: Rubio G, Lavado R, Pereyra F (eds) The soils of Argentina, World soils books series. Springer International Publishing, pp 81–100

SAGyP-INTA (1989) Mapa de Suelos de la Provincia de Buenos Aires, Escala 1:500.000. CIRN-INTA, Proyecto PNUD Arg 85/019, Buenos Aires, p 533

Satyavathi P, Ray S, Chandran P, Bhattacharyya T, Durge S, Raja P, Maurya U, Pal D (2005) Clay illuviation in calcareous Vertisols of peninsular India. Clay Research 24(2):145–157

Schaetlz R, Thopmson M (2015) Soils. In: Genesis and Geomorphology, 2nd edition. Cambridge University Press, New York, p 778

Scoppa C (1976) La mineralogía de los suelos de la llanura pampeana en la interpretación de su génesis y distribución. Actas 7ª Reun. Arg. de la Cien. del Suelo, IDIA, Suplemento n° 33, pp 659–673

Stephan S, De Petre A, de Orellana J, Priano L (1977) Brunizems soils of the central part of the province of Santa Fe. Argentina Pédologie XXVII 3:225–253

Teruggi M (1957) The nature and origin of Argentine loess. J Sediment Petrol 27:322–332

Teruggi M, Imbellone P (1983) Perfiles de estabilidad mineral en suelos desarrollados sobre loess de la región pampeana septentrional. Argentina Ciencia del Suelo 1(1):65–74

USDA-NRCS (2014) Keys to soil taxonomy, 12th edn. Soil Survey Staff, p 360

Wilding L, Tessier D (1988) Genesis of Vertisols: shrink-swell phenomena. In: Wilding L, Puentes R (eds) Vertisols: their distribution, properties, classification and management, Technical monograph n° 18. Texas A & M University Printing Center, pp 55–81

Zárate M (2005) El Cenozoico Tardío continental de la Provincia de Buenos Aires. In: de Barrio R, Etcheverry M, Caballé M, Llambías E (eds.). Geología y Recursos Minerales de la Provincia de Buenos Aires. Relatorio XVI Congreso Geológico Argentino, La Plata, Cap. IX, pp 139–158

Chapter 18
Mapping Gilgai Micro-relief and Its Impact on Dryland Agricultural Landscapes Using Time Series of NDVI Derived from Sentinel-2 Imagery

A. Calera, J. Villodre, J. Campoy, M. Calera, A. Osann, K. Finger, B. L. Teece, and G. I. Metternicht

Abstract Remote sensing is a natural tool for mapping soil spatial differences, and it is widely used for digital soil mapping. The new generation of sensors, e.g., those carried by the twin Sentinel2 satellites, can describe crop canopy development with unprecedented high spatial and temporal resolution. These are excellent capabilities to describe the vegetation and have been used in agronomy for mapping within-field differential soil fertility in agricultural areas. These maps, known as Management Zones Maps, are key elements used in precision agriculture to improve the supply of scarce resources such as water, nutrients, seeds, among others. In this chapter, we present a case study of mapping gilgai micro-relief through its impact on crop canopy structure in dryland agricultural paddocks, in the Mallee region, Victoria, Australia. Our methodology uses the time series of NDVI of Sentinel 2 images over 3 years and takes advantage of the methodology for mapping management zones in crop fields. In this way, MZM from Sentinel 2 images exhibit granular structures at

A. Calera (✉)
Remote Sensing and GIS Lab. Universidad de Castilla La Mancha, Albacete, Spain
e-mail: Alfonso.Calera@uclm.es

J. Villodre · J. Campoy · M. Calera · A. Osann
AgriSat Iberia SL, Albacete, Spain; https://www.agrisat.es

K. Finger
BCG. Birchip Cropping Group, Birchip, Australia

B. L. Teece
School of Biological, Earth and Environmental Sciences, University of New South Wales, Sydney, Australia

G. Metternicht
School of Biological, Earth and Environmental Sciences, University of New South Wales, Sydney, Australia

School of Science, Western Sydney University, Richmond, Australia
e-mail: G.Metternicht@westernsydney.edu.au

367

high spatial frequency in some paddocks. The granular structure mimics a bi-dimensional lattice very similar to that described by gilgai micro-relief. The granular structure remains undisturbed throughout three consecutive campaigns in the 2019–2021 years, under different crops, which apparently points to the soil variability as the origin of this structure. Therefore, we discuss MZM describes the gilgai subsurface micro-relief. The results enable us to assert that crop and pasture growth is likely to be uneven across a paddock in gilgai areas. This chapter demonstrates how to map this differential behaviour using a time series of Sentinel2 images, providing insights about the gilgai soil structure. These results can reinforce the approach about using vegetation as a proxy of the soil structure, providing remote sensing-based tools able to measure spatial vegetation structure characteristics.

Keywords Remote sensing · Sentinel-2 · Gilgai micro-relief · Management zones map

18.1 Introduction

The composition, characteristics, and three-dimensional structure of the soil layer in which the plants have their roots have always deserved great multidisciplinary attention (e.g., from geologists, agronomists, botanists, soil scientists). However, traditional soil survey techniques drill a point to take samples and perform their subsequent analysis in the laboratory, so in many cases, the soil architecture is disturbed; but the greatest difficulty may be obtaining maps that show the three-dimensional structure of the soil from a set of point data.

Remote sensing is a natural tool for mapping spatial differences in the soil. In this way, surface images taken of the Earth contribute to digital soil mapping (DSM), a successful sub-discipline of soil science with an active research output. The success of DSM is a confluence of several factors: beginning in the year 2000 with increased availability of spatial data (digital elevation model, satellite imagery), the availability of computing power for processing data, the development of data-mining tools and GIS, and numerous applications beyond geostatistics (Minasny and McBratney 2016). Mapping the structure and spatial characteristics of the soil using optical remote sensing has usually relied on the information provided by the soil reflectivity in the solar spectrum and by the soil emissivity in the thermal channels. Soil texture usually relates to different soil colorations, favouring photo-interpretation processes and automated digital analysis. In this case, information retrieves from the topmost soil layer, within a few millimetres. The use of micro-waves enables the analysis of the first few centimetres depending on the wavelength used, since it penetrates these layers; most work in this zone aims to understand soil moisture in these first layers. For details, see "Assessment of topsoil properties using radar and optical data to provide spatial representation of key soil attributes," Chap. 15, in this book.

Another approach to infer the spatial soil characteristics is the observation of the vegetation that grows on that soil (Coops et al. 2012). This is an indirect approach, because plants grow in tight interaction with the soil, as well as with the atmosphere and other environmental biotic and abiotic factors. The soil-plant interrelationship has been widely used in agriculture for mapping the differences in within-field soil fertility under homogeneous cultivation and management; in this case, within-field fertility differences can be approached by mapping yield differences, excluding other causes other than the soil.

Imagery time series are a unique tool for mapping crop development. However, literature linking the differential development of vegetation to the differential properties of the root soil layer is scarce (Mulder et al. 2011). In some cases, estimates of the evapotranspiration of the canopy together with models of water balance in the root soil layer are used to estimate soil water properties, as root soil layer water storage capacity and temporal variation of soil water content (Campos et al. 2018).

This chapter discusses the use of Sentinel 2 imagery for inferring soil subsurface layer spatial structures by mapping crop canopy growth. We present a study case on the identification of gilgai micro-relief through their impact in dryland agricultural paddocks in the Mallee region, in the state of Victoria, Australia.

The time series of images captured by the new generation of sensors, like those carried out by the twin Sentinel2 satellites, can describe crop canopy development at high spatial and temporal resolution. This technology allows mapping within-field relative differences in crop yield with unprecedented spatial resolution for free imagery. The map of the relative differences in crop yield is a proxy of the within-field relative soil fertility when the crop, atmosphere and management are homogeneous, what usually happens at paddock scale. In the framework of precision agriculture these differential soil fertility maps are usually dubbed Management Zone Maps (MZM). These MZMs have a widespread and increasing use, because they make applying variable rate technology in agriculture feasible. Variable rate applications aim to match water, nutrients and other inputs to the crop demands in space and time for better and more sustainable crop management.

This chapter shows how some MZMs generated in specific paddocks in the Mallee region exhibit a striking granular structure, while this does not happen in other paddock MZMs. This granular structure, mimics a bi-dimensional near-lattice, remains undisturbed throughout the analysed three consecutive campaigns, 2019–2021 years, and under different growing crops. It seems that the subsurface soil layer (the soil layer where the roots grow), is the origin of the MZM granular structure.

The discussion about these granular structures depicted by the crop canopy development and picked by MZMs, leads us to recognize the gilgai soil subsurface micro-relief structure. In this way, the remote sensing-based methodology here described could open a complementary path for mapping gilgai in agricultural areas. Time series of images acquired by the new generation of Earth Observation sensors, as Sentinel2 and others, meet the high spatial resolution and frequent revisit time required for these applications.

18.2   Materials and Methods

18.2.1   Study Area. The COALA Project

The "Copernicus for sustainable agriculture in Australia" (COALA) H2020 project is a working team involving European and Australian researchers and farmers. Within this project we analysed the use of this time series for agricultural purposes in dryland paddocks of several farms nearby to the towns of Horsham and Birchip, in the region of the Wimmera and Mallee, Victoria, Australia. The COALA H2020 project aims to contribute to a better agriculture by implementing tools capable of determining the water and nutrient requirements of crops in space and time, and thus being able to adjust the adequate supply of these resources to optimize yield.

Figure 18.1 shows the location of the farms under analysis. The area is characterized by a semi-arid climate with a gradient from a warm summer, cold winter, from the South of the study area, near to Horshan town, to a hot dry summer, cold winter in the Birchip area, according Australian Bureau of Meteorology, BOM, with an average rainfall around 300–480 mm, dryer towards the North. Soils exhibits grey heavy clays light clays, sandy loams areas. In the Birchip area, cited gilgai soil structure, as shown in Fig. 18.2. Detailed soil description is provided in Victoria resources online (2022).

18.2.2   Sentinel 2 Images and Its Processing Chain

A main input in this work is the time series of images acquired by the Sentinel2a and Sentinel 2b twin satellites in the study area. In orbit since 2017, the European Space Agency, ESA, operates these twin satellites constellation in the framework of

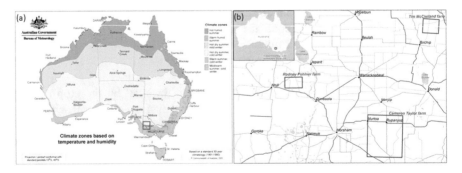

Fig. 18.1 (**b**) Location of the farms in the region of Wimmera and Mallee, Victoria, Australia, where MZM analysis has been performed in the framework of the COALA project. The granular structure detected by MZM and attributed to gilgai soil structure occurs in the farm close to the Birchip city, in the upper right corner. (**a**) shows the Australia climate zones emphasizing the study area exhibits a climatic gradient dryer and hotter from South to the North

Fig. 18.2 Landscape in the Birchip region showing gilgai microrelief in bare soil. (Photo: David S 1996. Adapted from http://vro.agriculture.vic.gov.au/dpi/vro/malregn.nsf/pages/mallee_soil_birchip)

the COPERNICUS program of the European Commission. The images from Sentinel2 constellation offer excellent characteristics for describing crop development on both temporal resolution (3-5 days), and spatial resolution (a pixel of 10 m pixel-side). The characteristics of Sentinel2 images and its free availability do not have precedents in the applications of Earth Observation for agriculture.

For the purposes we describe here, the derived inputs from Sentinel2 images are:

(a) The time series of the most utilized vegetation index, it is the Normalized Differences Vegetation Index (NDVI), an algebraic combination of the reflectance in the Near Infrared band (NIR), and the reflectance in the Red band (R), as is defined by the Eq. 18.1. The NDVI is a measure of the relative photosynthetic size of the canopy, and it is related with the ground green cover.

$$NDVI = \frac{NIR - R}{NIR + R} \qquad (18.1)$$

(b) The time series of the composition colour RGB (SWIR1, NIR, R bands). This RGB images provides a view like a photography of the Earth surface, as Fig. 18.5a shows.

In addition to the Sentinel2 images, we also utilized images from Landsat8, operated by NASA, USA. Landsat8 provides a 30 m pixel side spatial resolution and a revisiting time of 16 days, as a complementary dataset to that of the Sentinel 2, to check the effect of the lower spatial resolution and the higher revisit time on the detection of the crop growth spatial pattern.

The use of sensor constellation Sentinel 2 (Sentinel 2a and Sentinel 2b) ensures the frequent availability of cloud-free images, essential to describe the crop growing cycle accurately. Temporal evolution of NDVI was interpolated linearly on a daily scale using the time adjacent satellite images. The effect of the atmospheric distortion and the possible differences between the sensors used in the study were compensated for using a procedure of absolute normalization of the NDVI values (Chen et al. 2005). Thus, the maximum and minimum NDVI values obtained for each image for pseudo-invariant surfaces (dense vegetation and agricultural bare soil) were linearly rescaled to the corresponding values of NDVI for these surfaces based on the common values in the study area (0.15 for bare soil and 0.91 for dense-green vegetated, respectively). Once the basic processing is done, the images upload in a webGIS platform, dubbed SpiderwebGIS®, built by the Remote Sensing Lab of the University of Castilla La Mancha, through which can be displayed graphically and numerically.

Figure 18.6 displays the NDVI time series for selected pixels during the period 2019–2021 analysed in this study. In this picture, each point represents one selected image after discarding cloudy images.

18.2.3 Methodology for Mapping Management Zones Map, MZM, in Australian Paddocks on the Study Area

The purpose of the MZM is to capture the spatial distribution of within-field crop growth, relative to the average field value. MZM map aims to capture the differences in within-field soil fertility that explain the growth differences between different parts of the paddock when other factors such as management and meteorology are uniform. The MZM, among other agricultural uses, allows modulating the amount of nutrients to apply to the crop in each subzone of the plot, according to its differential growth, what is in the basis of variable rate technology.

Differences in relative biomass accumulation during the crop growing cycle is the best indicator for describing differences in crop-soil interaction when other factors such as management, diseases and meteorology across the field are uniform, what usually is given in a field with a homogeneous crop and management. Biomass accumulation is a particularly good proxy of yield for grain crops. Within-field differential values of accumulated biomass will pick up differential soil fertility, related directly with the properties of the subsurface soil layer where roots grow, usually around the first meter of soil depth.

A good proxy of biomass accumulation utilized here is the value obtained by numerical integration of NDVI time trajectory for each pixel during the crop growing cycle (Campos et al. 2018). Then by mapping for each pixel j the ratio of the value of NDVI time integrated, $NDVI\_TI,j$, defined by Eq.18.2 and the NDVI-TI average value for the entire field, $NDVI\_TI,average,$ defined by Eq.18.3, we can obtain a simplified MZM map.

$$NDVI\_TI, j = \sum_{i=1} \left(NDVI - NDVIbare\ soil\right)i, j \tag{18.2}$$

$$NDVI\_TI, average = \frac{\sum_{j=1}^{j=n} NDVI_{TI}, j}{n} \tag{18.3}$$

where:

i, the number of days, starting when the value of NDVI $\geq$ 0,3 at the beginning of the growing cycle; ending when NDVI decreases below 0,4 at the end of the growing cycle.

NDVIbare soil is a defined typical value when no green cover appears, usually we assume a typical value of NDVIbare soil = 0.2.

The computations for mapping MZM have been performed by using GIS tools and a specialized own software, TONI, TOol for Numerical Integration, for computing numerical integration during the crop growing cycle. An intermediate step obtains NDVI daily data by applying linear interpolation between consecutive images.

18.2.4   Gilgai Soil Structure

A gilgai soil structure is a micro-relief characteristic of clayey areas on vertisol soils, with soft mounds and depressions, which forms patterns with some degree of symmetry (Paton 1974; Knight 1980). Gilgai is a word from the Kamilaroi, Wiradhuri and related Australian Aboriginal languages in which it means "small water hole" and refers to the seasonal accumulation of water in the lower part of these features. Early European settlers in Australia used many terms for these microrelief features, the most common being crabhole, melonhole and gilgai. Figure 18.3 shows a typical gilgai structure (Paton 1974).

A gilgai micro-relief occurs when the clay soil layers shrink and swell during alternate drying and wettingcycles. This process gradually forces 'blocks' of subsoil material upwards to form mounds. Likewise, contraction and expansion processes alter the properties of the subsoil, and subsurface soil differences are associated with the areas occupied by the mounds and the depressions of the micro-relief. Gilgai structures commonly form in black and grey vertosols and, to a much lesser extent, brown and red vertosols. The most common soil parent materials of vertosols are alluvial clayey sediments. The term gilgai is firmly established in soil literature and is a common feature of many clay soils in sub-tropical and tropical areas in the world.

An important feature accompanying gilgai micro-reliefs is the development of vertical ground-surface cracks in the dry season. These cracks have opening widths up to several centimetres, depths of up to several tens of centimetres and a crack

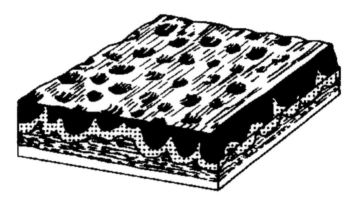

Fig. 18.3 Typical gilgai soil structure, surface and subsurface, showing soft mounds and depressions, which form patterns with some degree of symmetry. (Adapted from Paton 1974)

segment up to three metres long. Cracks also occur horizontally below the surface. Both horizontal and vertical cracks seal during wet periods but do not always close completely. These cracks play an important role in forming these mounds and depressions. During a wet season, horizontal cracks swell at the surface due to rainwater interacting with the clay. This forces the cracks to close, which traps air in the soil profile. As the clay continues to swell, the width of the air-filled cracks is forced to enlarge upwards, leading to mound formation during wet periods. As clay soil is more rigid in its dry state than wet, the mounds remain when they dry out (The State of Queensland 2011 (updated 2013)).

18.3 Results

18.3.1 Management Zones Map

In several paddocks of Tim's farm, close to the Birchip town, see Fig. 18.1, the obtained MZMs in the 2019 campaign drew a striking granular structure. However, in the cases of Cameron and Rodney farm paddocks, see Fig. 18.1 for their location, the pattern was the expected one, distinguishing in its management subzones without the granular structure, according to the previous experience in the elaboration of the MZMs.

Figure 18.4a shows the MZM for the paddock 05-Rogers of Tim's farm exhibiting this striking granular structure, in comparison with that of usual structure exhibited in Fig. 18.4b. The spatial pattern of variability of the crop canopy showed in Fig. 18.4a is similar to that of other paddocks in the Tim's farm. This granular structure implies that when carrying out a linear transect, it is possible to pass through zones that alternate between higher and lower production in a few tens of meters, which may allow a characteristic length in a cycle to be assigned. Moreover,

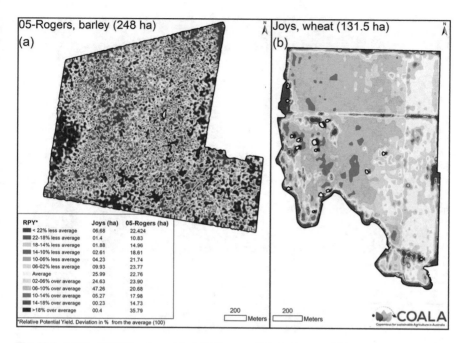

Fig. 18.4 (**a**) The MZM for the paddock 05-Rogers of Tim's farm exhibiting a granular structure in the 2019 campaign. The picture suggests a two-dimensional structure, without a clear preferential direction. In turn, this high-frequency spatial pattern overlaps another variability on a larger spatial scale, which marks differences between paddock subzones. (**b**) The MZM for the paddock Joys belonging to the Cameron farm, near Horsham city, also for 2019 campaign, exhibits the usual spatial structure about crop cover variability relative to the field average value. No granular structure appears here

the picture suggests a two-dimensional structure without a clear preferential direction. In turn, this high-frequency spatial pattern overlaps another variability on a larger spatial scale, which marks differences between paddock larger subzones.

The usual picture of MZM indicates differences between the subzones of a plot on appropriate spatial scales for agronomic management. This also happens in the other farms in the study area. In these farms, the MZM of paddocks exhibits the usual within-field spatial scales without granular structure, as Fig. 18.4b shows corresponding to a selected typical plot of Cam's farm, near Horsham city.

18.3.1.1   How MZM Picks Up the Crop Growth Through Sentinel2 Time Series

Figure 18.5 is a portion of Fig. 18.4b showing the variability pattern in detail. The lower part of the Fig. 18.5 graphs the different time trajectories of NDVI values during the growing cycle for two small areas, flagged in the upper part of the figure. As can be seen in the graph, the NDVI time trajectory provides indication about higher

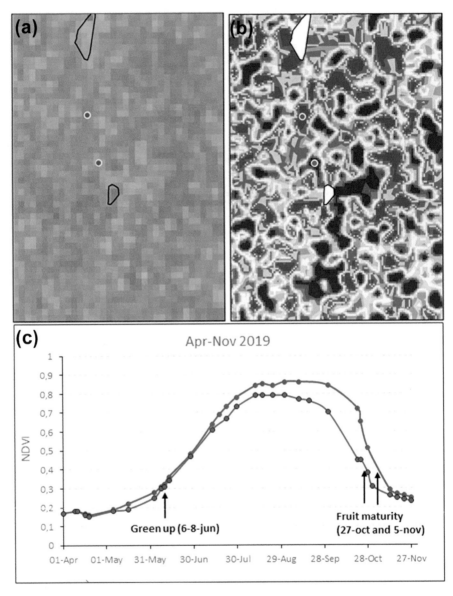

Fig. 18.5 Portion of Fig. 18.4a showing in detail the granular structure; (**a**) RGB colour composition picture corresponding to the date 1st October 2019 (**b**) Detail of MZM for the 2019 campaign, for the same area than that of RGB picture. (**c**) The lower part graphs the different time trajectory of NDVI values during the growing cycle for an expected high-yield area, and for an expected low-yield area, corresponding the points indicated in the upper part of the Figure. This NDVI time trajectory enables to calculate NDVI time integrated value for building the MZM, represented in the (**b**) part

NDVI values in one of these areas than the other, indicating higher fertility value, and hence better soil surface layer properties about water and nutrients, because atmosphere and management are the same across the paddock. This NDVI time trajectory enables the calculation of time integrated NDVI (*NDVI_TI*), for each pixel within the field, elaborating the MZM, as Eq. 18.2 and 18.3 states. The start and the end of the crop growing season is determined by the thresholds NDVI values indicated previously. The MZM captures differential crop development as a good proxy of accumulated biomass when it operates this way.

18.3.2   Detection of Granular Structure in Paddocks by Using RGB from Sentinel2

Since generating the MZM requires knowledge of the borders of a paddock to perform the calculations indicated in Eqs. 18.2 and 18.3, a simpler and more direct way to visualize the spatial structure was visually inspecting the series of RGB colour composition images.

In the RGB colour composition corresponding to the images acquired by the Sentinel2 constellation, the granular structure can be directly appreciated in the RGB image of a proper date, Fig. 18.5a. In this case, RGB looks like a mosaic of bright and dark areas with a near-regular spatial distribution. The difference in hue among bright and dark areas of the RGB image usually accentuates towards the end of the growth cycle, when the differences become larger, as shown in Fig. 18.5c, around early October. The temporal pattern of NDVI trajectory for bright and dark RGB colour seems to point to the impact of what affect to the crop canopy development is permanent during the growing cycle, what MZM is able to pick up, because MZM proceeds to perform the numerical integration during the growing cycle.

The use of RGB images in the aforementioned windows of opportunity opens up the possibility of determining the area affected by the characteristic two-dimensional granular structure, easily recognizable by the human eye. However, the granular spatial pattern does not appear as clear and clean as it does in 2019 in the rest of the years analysed, depending on the climatic conditions that modulate the development of the crop,

For this reason, it seems necessary to display the time series of images over several crop growth cycles and to combine the visualization of the time series of the RGB color composition together with those of the NDVI. These tasks require using GIS tools that allow precise geo-positioning to analyse the consistency over time of the spatial pattern shown by the vegetation canopy beyond a specific campaign. This analysis should be sufficient to indicate the permanent nature of the spatial vegetation cover structure, and therefore support the assumption that it is the variability in the differential subsurface soil properties responsible for the spatial variability of the cover, beyond other factors such as diseases and management.

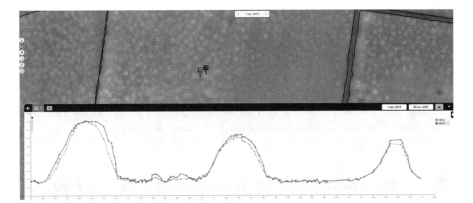

Fig. 18.6 It shows the graph of NDVI time series since 2019 to 2021 for two selected points indicated by flags in the picture over the RGB colour composition in the upper part of the fig. NDVI time trajectory points exhibit the same relative pattern during the growing season for the 3 years despite different crops growing in the analysed period. Numbers in the upper right corner provide the coordinated longitude and latitude for the right flag. RGB and NDVI time series displayed from the SPIDERwebGIS® software

Figure 18.6 shows a continuous and stable difference in vegetation cover between two points during three consecutive years, as it is described by its NDVI time trajectory, in despite a different crop grown for each year. It provides evidence to affirm that permanent differences on the crop cover over time, during several years, take aim to the differential fertility of the subsurface soil as the cause of that.

18.3.3 Detection of Granular Structure by Using Other Imagery: Landsat, Google Earth

The sequence of available Landsat8 images for the 3 years of study have been analysed, without being able to identify the granular pattern described previously. The reasons lie on the low temporal frequency of Landsat8 images has not allowed the observation in the temporal window, around October, in which the granular structure appears more evident. In addition, the spatial resolution of these images is lower than in the case of Sentinel2. In the study area, a typical length between crest and valley for *NDVI-TI* values is around 30–40 m. The intrinsic geolocation uncertainty for remote sensing images degrades the spatial resolution for temporal tracking to a typical value of 3×3 pixel. According to this, the Sentinel2 images could provide an effective spatial resolution for temporal tracking of around 3×3 pixel, it is 30 m × 30 m, while in the case of Landsat8 the area size for temporal tracking around 3×3 pixel is 90 mx 90 m. For Landsat8, it could mean that variability occurring at spatial resolution greater than 90×90 m could not be easily captured, as seems to be the case in the study area.

Very high spatial resolution images, such as those provided by the Google Earth platform and those provided by the Bing platform, have also been accessed and analysed. The result of this inspection is that different grey tones are observed on some areas that spatially coincide with the differences recorded through the previously described MZM methodology, when the soil surface is not masked by vegetation. All these images can complement the analysis of the granular structure of the vegetation cover over time and contribute to evaluating the impact of the soil on the conformation of said crop canopy structure.

18.4 Discussion and Conclusions

The question therefore arises, 'what makes the spatial pattern of variability of Tim's farm clearly different from that of Cameron and Rodney farms?' as the MZM highlights. By sharing and discussing these results with Australian colleagues we learnt what the high-frequency spatial pattern that crop cover development shows in Tim's paddocks. The explanation that emerges is that we are in a gilgai area, and therefore, the gilgai soil structure pattern may reflect on the crop cover pattern during the growing cycle.

The concept of soil structural pattern means a continuous three-dimensional spatial arrangement of planar and linear features. Such patterns are capable of geometrical description as is the case of gilgai. Ground surface (microrelief) and layers within the soil profile as well as crack populations at various scales of observation (Knight 1980) make suitable gilgai soil subsurface structure as the source of high-frequency spatial crop variability as picked by MZM.

The hypothesis on which the MZMs rely is that the variability pattern of the vegetation canopy reflects that of the properties of the soil subsurface layer on which the plant grows. Differential crop growth is a valuable probe for testing differential soil fertility, and MZM derived through time series of Sentinel 2 images pick up the spatial distribution of how the crop growth happens. If so, in the paddock shown in Fig. 18.3 and in other similar ones, the properties of the soil that define its fertility should respond to the same spatial pattern of two-dimensional granular structure observed in the vegetation cover.

It is also necessary to consider the possible erasure effect that years of agricultural tasks could have had on the pristine structure of the soil, at least in the ploughed surface. First, the impact on crop development came from subsurface soil, slightly affected by ploughing. Besides, it, as pointed in The State of Queensland, (2011), despite the continuous cultivation of gilgai paddocks results in the smoothing out of the mound and depression formations. However, these features do not always last for long, because even when gilgai-forming soil is smoothed, the forces that create the mounds and depressions are still operating within the soil. Over time, gilgai reform and require re-levelling if cultivation is to continue. If the land is left undisturbed for several wetting and drying seasons the gilgai will reform.

Two effects overlap in the soil structure of the gilgai that can affect crop canopy growing above. On the one hand, micro-relief and cracks act effectively to prevent surface runoff in the event of abundant rainfall at the beginning of the rainy season, allowing efficient water management when it is scarce. Furthermore, Webster (1977) studied spatial subsurface soil properties variation on a transect across typical gilgai terrain of the Bland Plain of New South Wales. He states that "gilgai evidently recurs sufficiently regularly on the transect for spectral analysis to reveal a periodicity that exhibits a wavelength…," what indicates that subsurface soil exhibits the periodicity that translates into crop canopy development.

Both surface and subsurface gilgai soil characteristics exhibit some degree of spatial symmetry, what it is compatible with the granular structure and spatial scale exhibited by the canopy crop development. As the most plausible explanation, the differential crop growth pattern as MZM picks up and shown in Fig. 18.4, comes from gilgai surface and subsurface soil structure. So far, no other alternative explanation to that of gilgai soil structure has been found credible to explain the high-frequency spatial pattern exhibited by the crop canopy cover.

The results enable us to assert that crop and pasture growth is likely to be uneven across a paddock in areas with gilgai micro-relief. And this chapter demonstrates that this differential behaviour can be mapped from a time series of Sentinel2 images with the methodology described above, providing insights about the gilgai soil structure.

Although further research and ground truth will be needed, in which electromagnetic induction can provide some insights (Watson et al. 2017), the encouraging results seem to support that the Sentinel2 time series and the MZM derived from them could be a valuable tool for mapping gilgai subsurface soil structure in agricultural areas. In this way, the remote sensing-based methodology here described could open a complementary path for mapping gilgai in cultivated areas. The time series of images acquired by the new generation of Earth Observation sensors, such as Sentinel2 and others, meet the high spatial resolution and frequent revisit time required for this purpose.

Acknowledgements This research was developed in the framework of the project COALA H2020 (Copernicus Applications and services for Low impact agriculture in Australia) funded by the European Union's Horizon 2020 research and innovation programme (Grant Agreement ID 870518). The authors would like to thank the active collaboration of farmers Tim McClelland, Rodney Pohlner and Cameron Taylor, belonging Birchip Cropping Group Inc. (BCG), who provided valuable data. BCG is a not-for-profit agricultural research and extension organisation led by farmers from the Wimmera and Mallee regions of Victoria.

References

Campos I, González-Gómez L, Villodre J, Calera M, Campoy J, Jiménez N, Plaza C, Sánchez-Prieto S, Calera A (2018) Mapping within-field variability in wheat yield and biomass using remote sensing vegetation indices. Precis Agric 20(2):214–236. https://doi.org/10.1007/s11119-018-9596-z

Chen X, Vierling L, Deering D (2005) A simple and effective radiometric correction method to improve landscape change detection across sensors and across time. Remote Sens Environ 98(1):63–79. https://doi.org/10.1016/j.rse.2005.05.021

Coops NC, Waring RH, Hilker (2012) Prediction of soil properties using a process-based forest growth model to match satellite-derived estimates of leaf area index. Remote Sens Environ 126:160–173. https://doi.org/10.1016/j.rse.2012.08.024

Knight MJ (1980) Structural analysis and mechanical origins of gilgai at Boorook, Victoria, Australia. Geoderma 23(4):245–283. https://doi.org/10.1016/0016-7061(80)90067-1

Minasny B, McBratney AB (2016) Digital soil mapping: a brief history and some lessons. Geoderma 264:301–311. https://doi.org/10.1016/j.geoderma.2015.07.017

Mulder VL, de Bruin S, Schaepman ME, Mayr TR (2011) The use of remote sensing in soil and terrain mapping – a review. Geoderma 162(1–2):1–19. https://doi.org/10.1016/j.geoderma.2010.12.018

Paton TR (1974) Origin and terminology for gilgai in Australia. Geoderma 11(3):221–242. https://doi.org/10.1016/0016-7061(74)90019-6

The State of Queensland, 2011 (updated 2013) "Gilgai wetlands" Conceptual Model Case Study Series. [WWW Document]. URL. https://wetlandinfo.des.qld.gov.au/resources/static/pdf/resources/tools/conceptual-model-case-studies/cs-gilgai-12-04-13.pdf. Accessed 24 Jan 2022

Victoria Resources Online, Soil Glossary. [WWW Document]. URL http://vro.agriculture.vic.gov.au/dpi/vro/wimregn.nsf/pages/natres_soil_eastwimm_agric. http://vro.agriculture.vic.gov.au/dpi/vro/malregn.nsf/pages/mallee_soil_birchip. Accessed 24 Jan 2022

Watson HD, Neely HL, Morgan CL, McInnes KJ, Molling CC (2017) Identifying subsoil variation associated with gilgai using electromagnetic induction. Geoderma 295:34–40

Webster R (1977) Spectral analysis of gilgai soil. Soil Res 15(3):191. https://doi.org/10.1071/SR9770191

Part IV
Applications in Land Degradation and Geohazard Studies

Chapter 19
Gully Erosion Analysis. Why Geopedology Matters?

G. Bocco

Abstract Gully erosion dynamics is a complex phenomenon, usually human-induced, which cannot be described nor predicted using conventional soil erosion models such as USLE or similar. In the early 1980s it was stated that gully initiation and growth could be studied from a purely empirical perspective, because no deductive approach could serve the purpose. In this chapter, this premise is used to briefly summarize how gully erosion research has developed, what were the major achievements in the conceptual and methodological dimensions, and which may be potential courses of action for further research, with emphasis on the contribution of geopedology. Despite the advancements in the development of models and in RS and GIS techniques, gully erosion still remains a complex issue difficult to model and predict. In this sense, geopedology may play a role in its understanding and management. As other geomorphic processes, gullies occur in certain terrain, soil, and hydrology conditions which may be conveniently approached from a geopedologic perspective.

Keywords Gully erosion analysis · Erosion monitoring · Erosion modeling · GIS · Remote sensing

19.1 Introduction

Gully erosion initiation and development are complex phenomena which originate complex landforms (Bocco 1991). Gullies are usually human-induced and/or triggered by extreme rainfall events; they occur in different environments worldwide. Gully erosion cannot be described nor predicted using conventional soil erosion

G. Bocco (✉)
Universidad Nacional Autónoma de México (UNAM), Centro de Investigaciones en Geografía Ambiental (CIGA), Morelia, Michoacán, Mexico
e-mail: gbocco@ciga.unam.mx

© The Author(s), under exclusive license to Springer Nature
Switzerland AG 2023
J. A. Zinck et al. (eds.), *Geopedology*,
https://doi.org/10.1007/978-3-031-20667-2_19

models such as the Universal Soil Loss Equation (USLE) (Wischmeier and Smith 1978) and similar tools. The initiation and development of gullies are the result of the activity of a family of processes including those triggering inter-rill and rill erosion but also piping and shallow mass movement, involving soils and other surficial materials (Bocco 1993). Imeson and Kwaad (1980) stated that such phenomena as gully erosion could be studied from a purely empirical perspective, because no deductive approach was available. Has the situation changed after 43 years?

The purpose of this chapter is to analyze the potential contribution of geopedology to gully erosion research. After describing gully erosion in this introduction, research approaches to gully erosion studies are reviewed, with some emphasis on monitoring and modeling. Further the way how geopedology could serve gully erosion research is put forward. The idea is tested that the situation indicated by Imeson and Kwaad (1980) has probably not changed so far, because gullies are landforms originated by a variety of hydrologically-driven geomorphic and soil processes, and modeling cannot be based on deductive approaches. Gully erosion, in many instances, is a black box process.

Research on gully erosion is less developed than that on inter-rill and rill erosion. The hydrology of gully erosion is complex because it involves both surficial and subsurficial flows, which means that the hortonian-type of overland flow is only one of drivers of this type of soil erosion. In addition, there is no critical distance to water divide where incision and gullying would start, be it in large or small catchment. Subsurficial flows such as piping may be more important depending on the setting. As landforms, gullies are composed of head, slopes, bottom, channel, and sometimes fan. Incision usually takes place at the gully bottom, and slumps and slides affect mostly gully head and gully sides. Fan formation depends on the sediment delivery capacity of the gully. The upslope area contributing to gully head retreat, described as zero order, may be affected by inter-rill and rill erosion; but the sequence from laminar to gully erosion may not be present, and gullies may start independently of these processes. In addition, they may be triggered by micro-mass movements or because of human action (dirt roads, ill-defined culverts, boundaries between agricultural parcels, and others).

Gullies occur as individual features or as systems composed of several channels and thus of multiple heads and slopes. In the landscape, they are present as valley-side or valley-bottom types; they can be continuous or discontinuous. Usually, gullies develop on accumulative slopes, such as lower portions of footslopes or even plains. As a genetically erosional landform, this is a peculiar fact because gullying may upset the denudation chronology in a region.

Because of their initiation, development, and landscape position, gullies or gully systems were first analyzed from the standpoint of the so-called davisian cycle of erosion or "geographic cycle". In other words, the theory developed at the onset of the twentieth century by Davis (1905) for landscape development and denudation chronology was applied. Thus, gullies were described following the conventional stages of youth, maturity, and decline put forward by Davis to explain landscape evolution. Most of research in the US Department of Agriculture focused on developing typologies in this framework. The approach was criticized because the model assumed tectonic stability, temperate climate, hortonian-type of overland flow, a critical distance to the water divide to start incision, and a natural tendency of gully activity to become extinct. Empirical work showed, nonetheless, that gullies could start anywhere on a given slope, that the hydrologic flow could be complicated by the interference with subsurficial flows, and that gullies could be self-perpetuating systems. From the 1980s onwards, inductive empirical perspectives were established. The davisian approach was abandoned, and deductive models were no longer attempted.

Under these circumstances, empirical work in many contrasting regions showed that gullies could occur in a variety of climates and rainfall regimes, rock and soil types, slope facets and gradients, and land-cover and land-use types. Research progressed from establishing simple relationships between gully growth and time (i.e., gully erosion rates using sequential aerial photographs and photogrammetric means), to analyzing the contributing role of rainfall, soils, slopes, cover and practices, usually applying statistical approaches on the basis of field-verified remote-sensed data. In addition, comparisons between the severity of gully erosion and other types of soil erosion followed, with some emphasis on their respective sediment production and eventual delivery to streams or reservoirs. Gully erosion monitoring and modeling became increasingly popular in the scientific literature as well as in the grey literature produced by technical agencies (governmental and social) at various levels.

19.2   Research on Gully Erosion: Detection, Monitoring, and Modeling

Once the complexity involved in gully erosion was understood and the davisian model rejected, research moved with the beginning of this century towards several key interrelated topics to better understand the processes, their dynamics (monitoring) and simulation/prediction (modeling).

A thorough review of research on gully erosion is beyond the scope of this chapter. However, some lessons learned from the literature are convenient in particular to summarize limitations concerning the trends in gully erosion research and delineate some guidelines about the advantages of using geopedology as a tool in gully erosion.

19.2.1   Gully Erosion Processes and Modeling

Besides an early paper by the author of this chapter (Bocco 1991), one of the first published reviews was that of Bull and Kirkby (1997) who suggested that long-term rates of gully development were not well understood, and that theoretical modeling could provide the way forward for more holistic investigations. Poesen (Poesen and Hooke 1997; Poesen et al. 2003) also called for the need of gully erosion modeling, in particular as related to environmental change. In both reviews, Poesen and collaborators indicate that most field measurements and modeling efforts had concentrated on water erosion processes operating at the runoff plot scale and not at smaller geographic scales (large areas). The implications of this limitation are not dealt with in the reviews. The authors conclude by suggesting research needs and priorities, including the quest for predictive models at various geographic scales and the study of the impact of gully development on hydrology, sediment yield, and landscape evolution. No specific practical reference is made as how to relate gully erosion development to landscape studies. Despite the calls for theoretical modeling, Desmet et al. (1999) highlighted the importance of slope gradient and contributing area for optimal prediction of the initiation and trajectory of ephemeral gullies at plot scale in the Belgian loess region. Moreover, Nachtergaele et al. (2001) when testing the physically-based Ephemeral Gully Erosion Model (EGEM) in the same region, where a robust collection of data depicting input parameters exists, determined a value for simple topographic and morphologic indices in the prediction of ephemeral gully erosion. They stressed the problematic nature of physically-based models, since they often require input parameters that are not available or can hardly be obtained. Capra et al. (2005) also used the EGEM in Sicily but concluded that it seemed simpler using empirical relations between eroded volume and gully length in different environments, until more precise physically-based models were developed. This conclusion together with the data issue questions this type of modeling and its usefulness in potential practical applications in conservation. Likewise, Brazier (2004) stated that even for the UK, although provided with robust data, there is not enough information to validate soil erosion models (at large), especially when the goal is assessing the spatial heterogeneity of soil loss.

The trend from monitoring to modeling seems to be hampered by the lack of good quality data even in places where solid databases are available. As a consequence, the value of the models is challenged by the absence of validation through appropriate data. Similarly, this highlights the importance of the empirical data obtained from monitoring to face gully erosion analysis, even at large geographic scale in data-rich small areas. Brazier (2004) concluded that the paradox between data collection to improve models and erosion modelling to replace data collection must be addressed within the discipline if full use of datasets and improvement of models were to be made.

From a different perspective, Chaplot et al. (2005) quantified linear erosion (LE) at the catchment level and found that some of the LE controlling factors could also be used for prediction over larger areas since topography and land-use data, closely correlated with LE, were easily accessible. Valentin et al. (2005) also referred to the relation between gully erosion and global topics and the need to develop research in areas larger than the cultivated plots, but without providing clear indication on the methods, techniques, and approaches to be used in this type of research in large areas, nor on how to extrapolate findings from the plot to the landscape level. However, the paper addresses soil and gully erosion triggered by agricultural land use in the Chinese loess plateau and suggests that land-use and land-use change are of a paramount importance, even more than climate change.

In a similar line of thinking, Boardman (2006) emphasized the limited value of existing soil erosion models, including those of gullies, in the real world as opposed to the academic sphere. He suggested that approaches should include socioeconomic variables, land-use concerns, and be less "data-rich and people-poor". DeVente et al. (2007) pointed at the absence of reliable spatially distributed process-based models for the prediction of sediment transport at the drainage basin scale and claimed that spatially distributed information on land use, climate, lithology, topography, and dominant erosion processes was required for modeling purposes, including that of gully erosion. Many papers have addressed those controlling variables at the landscape level and have detected relationships particularly between gullying, soil properties, and slope characteristics (Table 19.1). However, there is no reference to a given spatial frame such as terrain, landscape or other type of environmental units, which happen to be defined by the very same controlling variables mentioned above.

Nazari Samani et al. (2011) discuss the limitations of gully erosion models and emphasize, for land managers, the importance of gullies as sediment sources particularly as compared to inter-rill and rill erosion. By contrast, Porto et al. (2014) did not find significant differences in the contribution to soil loss between inter-rill, rill and gully erosion in a small, cultivated plot in Sicily, but reported large inter-annual variability of this contribution. Capra (2013) also emphasized the need to study gully erosion in large areas. However, neither of the two publications provided clear methods and techniques to tackle research at such scale.

Overall, modeling of gully erosion seems not to have moved beyond empirical approaches. No deductive model has been formalized. The attempts to move from

Table 19.1 Relationships between gully erosion and controlling variables in different environments

Author	Study area	Controlling factors
Gábris et al. (2003)	Hungary, temperate humid	Land-use (long term, decades)
Shrestha et al. (2004)	Nepal, temperate humid	Land-use, slope aspect
Chaplot et al. (2005)	Laos, tropical humid	Catchment surface area and perimeter, mean slope gradient
Zucca et al. (2006)	Sardinia, temperate Mediterranean	Rock type, slope gradient, land-use
Schmitt et al. (2006)	Poland, temperate humid	Land-use (long term, decades)
Lesschen et al. (2007)	SE Spain, sub-humid, semi-arid	Abandoned cropland
Moges and Holden (2008)	Ethiopia, sub-humid, semi-arid	Land cover change
Nazari Samani et al. (2009)	SE Iran, arid, semi-arid	Slope gradient, land-use
Van Zijl et al. (2013)	Lesotho, sub-humid, semi-arid	Duplex soils
Shrestha et al. (2014)	Thailand, tropical humid	Land-use, slope gradient

plot to basin and from monitoring to modeling have not yielded the results expected by scholars. An additional difficulty arises from the limited scientific results applied to hazard prevention and mitigation of gully erosion processes. Furthermore, the application of models and knowledge in practical management strategies is still a large issue. The call for "people-rich" approaches by Boardman (2006) is important. The question remains on to how to involve people, an issue that requires research on the social perception of erosion processes and land and landscape management. This would open a different avenue towards social and cultural research, particularly on rural land uses and local conservation knowledge.

19.2.2   Remote Sensing, GIS and Gully Mapping

Remote sensing and geographic information systems have been used to map gullies and controlling variables. The expectation is that research using these technical tools would contribute with a clear spatially-explicit framework to study gully erosion, but research objectives are in fact geared to a variety of topics. Sidorchuk et al. (2003) dealt with the identification of gully erosion forms and processes in the Mbuluzi River catchment (Swaziland) by using the Erosion Response Units (ERU) concept. Input data were obtained from remote sensing (API method) and GIS analyses.

Marzolff and Ries (2007) monitored head cut retreat in 12 gullies using detailed aerial photography taken from remote-controlled platforms to identify runoff patterns and infiltration behavior in the gully head cut surroundings. They emphasized the benefits of high-resolution aerial photography for monitoring and understanding gully erosion processes. Daba et al. (2003) and Ndomba et al. (2009) used sequential aerial photos to study the development and dynamics of gully erosion systems. Evans and Lindsay (2010) extracted gully maps from high-resolution digital elevation models (2 m LiDAR DEMs). Wang et al. (2014) proposed the use of object-oriented analysis (OOA) to quantitatively study small gullies. Shruthi et al. (2014) also used OOA and very high-resolution imagery to digitally detect gully systems. Peter et al. (2014) combined rainfall simulation, gully mapping, and volume quantification at local scale using unmanned aerial vehicle (UAV) remote sensing data. Gómez-Gutiérrez et al. (2014) used 3D photo-reconstruction methods (based on Structure from Motion (SfM) and MultiView-Stereo (MVS) techniques) to estimate gully head cut erosion. Results of this simulation, not surprisingly, pointed out to a clear decrease in the accuracy of the model when the photos were not acquired sequentially around the head cut.

Dube et al. (2014) went back to an empirical model based on seven environmental factors (land cover, soil type, distance from river, distance from road, sediment transport index (STI), stream power index (SPI), and wetness index (WI)) using a GIS-based weight of evidence modeling (WEM). The predictive capability of the weight of evidence model in this study suggests that land-cover, soil type, distance from river, STI, and SPI are useful but not sufficient to produce a valid map of gully erosion hazard. Conoscenti et al. (2014) also insisted in the use of GIS and multi-variate statistical analysis in a small catchment. In contrast with the previous case study, they found "acceptable to excellent accuracies of the models" and good pre-dictive skill.

Despite the advances in the use of new remote sensing techniques, sensors, and approaches (such as the OOA), and GIS models, the goals have not changed sub-stantially with time. In addition, the question concerning how to conduct gully ero-sion research in areas larger than local plots, usually basins, is not addressed. How to stratify large areas for further sampling and extrapolation? Is topography as vari-able and DEM as tool enough to achieve this goal? Past and current research has focused more on technical issues than on essential topics such as testing spatial classification schemes of terrain, soils, and cover. Conceptual approaches based for instance on soil-landscape relationships can help support gully erosion research.

19.3 Using Geopedology in Gully Erosion Research

19.3.1 Why Is It Important?

Review of research on gullies suggests that (1) theoretical modeling is complex, requires usually unavailable data, and so far has not led to successful results; (2) empirical modeling, usually statistically-based and using remote-sensed data, seems to be the most common approach; (3) small catchments or runoff plots seem to be the most common type of study area; (4) no indications concerning methods and techniques to be applied in larger areas or catchments are provided. This includes the absence of proposals for the use of geographically-explicit models.

The first three points basically subscribe an idea which can be simply summa-rized as follows: landforms are difficult if not impossible to model; models may be conceptual but not operational. The last point suggests a lack of understanding of another simple fact: gullies occur in terrains, some of which are more susceptible than others to trigger this type of erosion. This calls for the need to stratify terrains to understand gully erosion processes and develop gully hazard models. Geopedology offers such a geographic approach. The basic assumption behind is that gullies are not randomly distributed but they develop in response to a combination of environ-mental factors (Vázquez and Zinck 1994). This assumption closely matches the need to stratify the landscape when working in relatively large areas. One would expect the occurrence of a combination of gully-prone factors per map unit and predict that some units might be more susceptible than others to gully initiation and development.

These relatively simple relations have very seldom been referred to in the litera-ture (see e.g., Bocco et al. 1990). One possible explanation is the development of purely quantitative approaches, which assume that a semi-quantitative approach is not scientifically sound. Something similar occurs with remote sensing, where digital interpretations are assumed to be more accurate than visual ones. Standalone quantifications seem to have become more important than the understanding of natural processes, in many instances strongly influenced by human action. One example is the overuse of geomorphometric digital terrain modeling and the under-use of visual landform surveying and mapping as research tools in gully erosion analyses. Geopedology provides an alternative basis for gully erosion research.

19.3.2 Why Geopedology Matters?

The geopedologic approach allows differentiating spatial units at variable geographic scales, which are relatively homogeneous in terms of terrain and soil properties. These properties can be assessed as to their suitability for land use and crop production purposes, or as to their susceptibility to different types of land degradation processes, among others, gully erosion. Vázquez and Zinck (1994) used such an approach to model gully development in Central Mexico. They developed a stepwise procedure to explore the spatial relationships between gullies and six selected environmental factors assumed to control and explain gully formation and distribution (i.e., landscape type, relief type, slope gradient, slope shape, soil types and properties, and land use). They first analyzed the cartographic coincidence between factors and gullies, and derived threshold values signaling most favorable conditions for gully formation. Further they built rule-based models in a GIS where class boundaries were the selected threshold values. Then they tested and validated the models by evaluating their efficiency in reproducing existing gullied areas. Finally, they applied the validated models to predict areas potentially favorable to gully formation, and derived recommendations for selecting priority areas for soil conservation.

The above research developed a semi-quantiative method which in practice offered a solution to many of the problems revealed in the literature review on gully erosion modelling previously discussed in this chapter. This approach can be used in fairly large areas at landscape level. It analyzes gully erosion factors considering soil and terrain attributes and provides a practical appraisal for soil conservation. The basis are data derived from a geopedologic survey, using both qualitative and quantiative methods. The survey does not involve complex data collection, instead requires elevation data, aerial photography or very high resolution satellite imagery, and a GIS platform. But it does require expert knowledge in terrain and soil survey, as well as in the hydrological processes that trigger gully erosion. This knowledge and field expertise cannot be replaced by algorithmic approaches or sophisticated data manipulations. It is a robust albeit simple approach to gully erosion or other conspicuous erosion phenomena, which can be extrapolated to any area where the above described knowledge and data are available or can be collected.

19.4 Conclusions

Gully erosion analysis, monitoring, and modeling are complex issues because gullies are complex landforms, usually polygenic, highly dynamic, and human-induced. Numerous variables have been tested using a variety of approaches. Terrain, soil, and cover harbor most of the properties from which variables are derived. These factors also control the effect of the hydrologic processes that trigger gully erosion. Relatively poor comprehension of the relationship between terrain, soil and cover, and strong emphasis on quantitative analyses have probably contributed to disregard conceptual models able to provide guidelines for stratifying land, soil, and cover, and allow sensitive sampling procedures to understand, monitor, and model gully erosion processes. Geopedology or analogous approaches have made substantial though less popular contributions to this end. A matching between quantitative analysis and a thorough spatial and conceptual framework to gully erosion seems to be a path to be further tested.

Acknowledgements María Lira helped in the preparation of the manuscript.

References

Boardman J (2006) Soil erosion science: reflections on the limitations of current approaches. Catena 68(2–3):73–86. In: Helming K, Rubio JL, Boardman J (eds) Soil Erosion Research in Europe 68(2-3):71-202

Bocco G (1991) Gully erosion: processes and models. Prog Phys Geogr 15(4):392–406

Bocco G (1993) Gully initiation in Quaternary volcanic environments under temperate sub-humid seasonal climates. Catena 20(5):495–513

Bocco G, Palacio JL, Valenzuela C (1990) Gully erosion modelling using GIS and geomorphic knowledge. ITC J 3:253–261

Brazier R (2004) Quantifying soil erosion by water in the UK: a review of monitoring and modelling approaches. Prog Phys Geogr 28(3):340–365

Bull LJ, Kirkby MJ (1997) Gully processes and modeling. Prog Phys Geogr 21(3):354–374

Capra A, Mazzara LM, Scicolone B (2005) Application of the EGEM model to predict ephemeral gully erosion in Sicily, Italy. Catena 59(2):133–146

Capra A (2013) Ephemeral gully and gully erosion in cultivated land: a review. In: Lannon EC (ed) Drainage basins and catchment management: classification, modelling and environmental assessment. Nova Science Publishers, New York, pp 109–141

Chaplot V, Coadou le Brozec E, Silvera N, Valentin C (2005) Spatial and temporal assessment of linear erosion in catchments under sloping lands of northern Laos. Catena 63(2):167–184

Conoscenti C, Angileri S, Cappadonia C, Rotigliano E, Agnesi V, Märker M (2014) Gully erosion susceptibility assessment by means of GIS-based logistic regression: a case of Sicily (Italy). Geomorphology 204:399–411

Daba S, Rieger W, Strauss P (2003) Assessment of gully erosion in eastern Ethiopia using photogrammetric techniques. Catena 50(2):273–291

Davis WM (1905) The geographical cycle in an arid climate. J Geol 13(5):381–407. http://www.jstor.org/stable/30067951

Desmet PJJ, Poesen J, Govers G, Vandaele K (1999) Importance of slope gradient and contributing area for optimal prediction of the initiation and trajectory of ephemeral gullies. Catena 37(3):377–392

deVente J, Poesen J, Arabkhedri M, Verstraeten G (2007) The sediment delivery problem revisited. Prog Phys Geogr 31(2):155–178

Dube F, Nhapi I, Murwira A, Gumindoga W, Goldin J, Mashauri DA (2014) Potential of weight of evidence modelling for gully erosion hazard assessment in Mbire District – Zimbabwe. Phys Chem Earth 67-69:145–152

Evans M, Lindsay J (2010) High resolution quantification of gully erosion in upland peatlands at the landscape scale. Earth Surf Process Landf 35(8):876–886

Gábris G, Kertész Á, Zámbó L (2003) Land use change and gully formation over the last 200 years in a hilly catchment. Catena 50(2):151–164

Gómez-Gutiérrez Á, Schnabel S, Berenguer-Sempere F, Lavado-Contador F, Rubio-Delgado J (2014) Using 3D photo-reconstruction methods to estimate gully headcut erosion. Catena 120:91–101

Imeson AC, Kwaad FJ (1980) Gully types and gully prediction. KNAG Geografisch Tijdschrift XIV(5):430–441

Lesschen JP, Kok K, Verburg PH, Cammeraat LH (2007) Identification of vulnerable areas for gully erosion under different scenarios of land abandonment in Southeast Spain. Catena 71(1):110–121

Marzolff I, Ries JB (2007) Gully erosion monitoring in semi-arid landscapes. Z Geomorphol 51(4):405–425

Moges A, Holden NM (2008) Estimating the rate and consequences of gully development, a case study of Umbulo catchment in southern Ethiopia. Land Degrad Dev 19(5):574–586

Nachtergaele J, Poesen J, Steegen A, Takken I, Beuselinck L, Vandekerckhove L, Govers G (2001) The value of a physically based model versus an empirical approach in the prediction of ephemeral gully erosion for loess-derived soils. Geomorphology 40(3):237–252

Nazari Samani A, Ahmadi H, Jafari M, Boggs G, Ghoddousi J, Malekian A (2009) Geomorphic threshold conditions for gully erosion in Southwestern Iran (Boushehr-Samal watershed). J Asian Earth Sci 35(2):180–189

Nazari Samani A, Wasson RJ, Malekian A (2011) Application of multiple sediment fingerprinting techniques to determine the sediment source contribution of gully erosion: review and case study from Boushehr province, southwestern Iran. Prog Phys Geogr 35(3):375–391

Ndomba PM, Mtalo F, Killingtveit A (2009) Estimating gully erosion contribution to large catchment sediment yield rate in Tanzania. Physics and Chemistry of the Earth, Parts A/B/C 34(13–16):741–748

Peter KD, d'Oleire-Oltmanns S, Ries JB, Marzolff I, AitHssaine A (2014) Soil erosion in gully catchments affected by land-levelling measures in the Souss Basin, Morocco, analysed by rainfall simulation and UAV remote sensing data. Catena 113:24–40

Poesen JWA, Hooke JM (1997) Erosion, flooding and channel management in Mediterranean environments of southern Europe. Prog Phys Geogr 21:57–199

Poesen J, Nachtergaele J, Verstraeten G, Valentin C (2003) Gully erosion and environmental change: importance and research needs. Catena 50(2):91-133. In: Poesen J, Valentin C (eds) Gully erosion and global change. Catena 50(2–4):87–564

Porto P, Walling DE, Capra A (2014) Using 137Cs and 210 Pbex measurements and conventional surveys to investigate the relative contributions of interrill/rill and gully erosion to soil loss from a small, cultivated catchment in Sicily. Soil Tillage Res 135:18–27

Schmitt A, Rodzik J, Zgłobicki W, Russok C, Dotterweich M, Bork HR (2006) Time and scale of gully erosion in the Jedliczny Dol gully system, south-east Poland. Catena 68(2–3):124–132

Shrestha DP, Zinck JA, Van Ranst E (2004) Modelling land degradation in the Nepalese Himalaya. Catena 57(2):135–156

Shrestha DP, Suriyaprasit M, Prachansri S (2014) Assessing soil erosion in inaccessible mountain-ous areas in the tropics: the use of land cover and topographic parameters in a case study in Thailand. Catena 121:40–52

Shruthi RBV, Kerle N, Jetten V, Abdellah L, Machmach I (2014) Quantifying temporal changes in gully erosion areas with object oriented analysis. Catena. https://doi.org/10.1016/j.catena.2014.01.010

Sidorchuk A, Märker M, Moretti S, Rodolfi G (2003) Gully erosion modelling and landscape response in the Mbuluzi River catchment of Swaziland. Catena 50(2–4):507–525

Valentin C, Poesen J, Li Y (2005) Gully erosion: impacts, factors and control 63(2–3):132–153. In: Valentin C, Poesen J, Li Y (eds) Gully erosion: a global issue. Catena 63(2–3):129–330

Van Zijl GM, Ellis F, Rozanov DA (2013) Emphasising the soil factor in geomorphological studies of gully erosion: a case study in Maphutseng, Lesotho. S Afr Geogr J 95(2):205–216

Vázquez L, Zinck A (1994) Modelling gully distribution on volcanic terrains in the Huasca area, Central Mexico. ITC J 3:238–251

Wang T, He F, Zhang A, Gu L, Wen Y, Jiang W, Shao H (2014) A quantitative study of gully ero-sion based on object-oriented analysis techniques: a case study in Beiyanzikou catchment of Qixia, Shandong, China. Sci World J. https://doi.org/10.1155/2014/417325

Wischmeier WH, Smith D (1978) Predicting rainfall erosion losses – a guide to conservation plan-ning. USDA Agriculture Handbook 537

Zucca C, Canu A, Della Peruta R (2006) Effects of land use and landscape on spatial distribu-tion and morphological features of gullies in an agropastoral area in Sardinia (Italy). Catena 68(2–3):87–95

Chapter 20
Soil Erosion Assessment and Mitigation Scenarios Based on Geopedology in Northwestern Patagonia, Argentina

M. C. Frugoni, R. F. González Musso, and G. Falbo

Abstract Transhumant systems related to livestock activity constitute a particular productive land use type that occurs in northwestern Patagonia. Buta Mallín catchment corresponds to summer grasslands, with a significant land use pressure. Soil erosion and mass movement features identified in a geopedological study were applied to landforms to observe the spatial dimension of current processes. About 90% of the area experiences moderate and severe active soil erosion while the remaining 10% remains stable. Areas where gully erosion and soil creep occur correspond to soils derived from volcanic tuff, while sheet and rill erosion processes were evident in soils derived from Holocene volcanic ash. The USLE/RUSLE soil erosion equation was applied to create a map for each parameter of the equation. 85% of the area showed different degrees of potential soil erosion, 76% had moderate and severe hazard and 15% showed no erosion hazard. Mean, maximum and minimum soil loss was 84.2, 299 and 0.5 Tons/ha/year, respectively. Afforestation with conifers and crown closure were simulated and applied in moderately suitable areas, which showed that after 15 years a non-erosion hazard scenario could be reached. The modelling also showed that more than 4000 hectares could be gained for implementing silvopastoral practices without soil erosion hazard preserving transhumant system and reducing more than 30% soil loss. This land use system could allow to obtain timber products as well as make sheepherding lands more productive, reversing land degradation processes.

Keywords Soil-landscape relationships · Overgrazing · Silvopastoral system · GIS · USLE/RUSLE

M. C. Frugoni (✉) · R. F. González Musso · G. Falbo
Asentamiento Universitario San Martín de los Andes, Universidad Nacional del Comahue, San Martín de los Andes, Neuquén, Argentina

© The Author(s), under exclusive license to Springer Nature Switzerland AG 2023
J. A. Zinck et al. (eds.), *Geopedology*,
https://doi.org/10.1007/978-3-031-20667-2_20

397

20.1 Introduction

Erosion is one of the main causes of soil degradation in western South America. In Argentina, soils in most of its provinces are affected by water erosion (Gardi et al. 2014). There has been a progressive increment of 790,000 ha per year of water-eroded soils from 1956 to 2015 due to changes in cropping systems, deforestation, and overgrazing (Rubio et al. 2019). Currently, there are 64.6 Mha of soils eroded by water in Argentina, including 20.8 Mha that are severely eroded (Casas 2015). Overgrazing and forest fires are the main drivers of soil erosion by water in northwestern Patagonia (Morales et al. 2013; Palacio et al. 2014) which can cause annual erosion rates exceeding 100 t/ha (Gardi et al. 2014).

Transhumance related to livestock activity is a land production type of cold Mediterranean climate that occurs in various Argentine mountainous areas, but particularly most developed in Neuquén Province, located in Northwestern Patagonia (Bendini and Alemany 2005). The main land use type in Buta Mallín in Neuquén province is transhumant cattle raising, with livestock movements from areas where they remain in autumn-winter (lowlands) to the summer ones to where they move in spring-summer (uplands). Rodeos are made up of *majadas* of sheep and *piños* of goats, with some equines and cattle.

The main objective of these productive activities is the shearing and sale of sheep wool and mohair (goat hair) and lambs and *chivitos* (meat). The study area corresponds to summer grasslands or *veranadas*. Combination of steep slopes and loss of vegetation cover has contributed to the development of soil degradation processes due to water erosion. Designing an afforestation combined with other land uses is a feasible scenario which can contribute to reduce risk of soil loss (Bonilla et al. 2010).

Silvopastoral system is a land use type in which perennial woody species are combined in the same management unit with herbaceous plants and animals and in which there are both ecological and economic interactions between the different components (Young 1987). This land use system allows to run a business of a short and medium term which is livestock activity combined with a long-term investment. In this way, mixing both characteristics achieves income in the short, medium and long term, diversify production, better land use, generation of more and better jobs and greater environmental sustainability (Luccerini et al. 2013). Five components interact in this system: trees, livestock, forage, soil and climate. Within this framework, soil erosion assessment and hazard estimation were carried out to evaluate to what extent this land degradation process was occurring. Current active processes were considered, and the Universal Soil Loss Equation model was applied. Besides, silvopastoral scenarios were used to analyze to what extent the soil surface was protected through this land use system and soil erosion by water was mitigated.

20.2   Materials and Method

20.2.1   Study Area

Buta Mallín is located in the northwest of Neuquén province, Argentine Patagonia. The study area includes the Lileo and Buta Mallín rivers catchment which covers an area of 21.061 ha, between coordinates 37° 6″ 37.8″ and 37° 14″ 0.4″ S, 70° 55″ 52.9″ and 71° 07″ 2.2″ W (Fig. 20.1). The lower elevations are at 1480 m ASL while the higher ones reach 2800 m ASL. This catchment is part of the upper basin of Neuquén River, which covers two-thirds of the provincial territory.

Geodiversity includes forms derived from glacial morphogenesis, mainly land-scapes with predominance of glacial erosion and proglacial plains, geoforms origi-nated in mass movements and fluvial ones, with apparent structural control (González Díaz and Ferrer 1991). Because of this, alluvial parent materials, tuffs, basalts and andesites have been described, the latter dominating on the high peaks with rock outcrops. In addition, this area has received important contributions of pyroclastic material provided by volcanoes present in the Andes and deposited by eolic action, during the Holocene Epoch. Hence, soil parent material corresponds to modern volcanic ash in part of the catchment. In other sectors parent materials of volcanic tuffs, tuffites and alluvial are dominant, some contaminated with volcanic ash. Climate is Mediterranean, with wet and cold winters and dry, warm summers.

The study area has a precipitation gradient in the west-east direction. In the extreme west the average annual rainfall reaches 1150 mm while in the extreme east it barely exceeds 700 mm. Dominant vegetation is herbaceous or herbaceous-shrub steppe. In the wetter portions of the basin, *Nothofagus antarctica* (ñire) shrub develops.

20.2.2   Survey Approach

The method for soil mapping is based on the geopedologic approach (see Chap. 4) which is based on the fundamental paradigm of soil geomorphology that considers the genetic relationships between soils and landforms and their parallel develop-ment, but with a clearly applied orientation and practical aim. Geopedology expresses the relationship between the two disciplines, with geomorphology at the service of pedology, specifically to support soil mapping (Zinck 2016).

Photointerpretation of the study area was performed using vertical photographs with an approximate scale of 1:50,000. Subsequently, delineations were digitalized in a GIS environment, using a georeferenced and ortho-rectified ASTER satellite image (NASA 2015). A total of 13 transects were determined as part of the field survey design, with 65 sampling points (Fig. 20.2). This is one of the methods used in geopedology, considering that similar geoforms (continent) have similar soil pro-files (content), which makes a more efficient field survey (Zinck 2016, Yemefack and Siderius 2016).

Fig. 20.1 Location of Buta Mallín catchment in Neuquén province (gray), Argentina

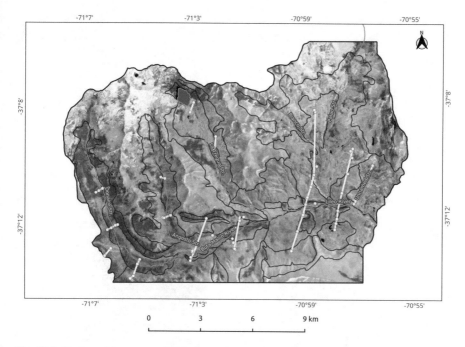

Fig. 20.2 Design of transects and sampling points for the fieldwork

Transects were defined to survey from two to ten times each map unit, depending on the extension of the landform. In each of the sampling points, a soil pit was opened in which site and soil description was made (Schoeneberger et al. 1998) as well as preliminary classification (Soil Survey Staff 2014). In modal profiles, samples were taken for horizon analysis. At each point, intensity of the accelerated water erosion processes was described and classified: absent, mild, moderate and severe, by observing features such as pedestals, grass tussocks, stoniness, accumulation of sediments slope above the plants and presence of rills and gullies (Aguiló Alonso et al. 2014). Besides, evidence of soil creeping, generally combined with erosion processes, was recorded.

20.2.3 Soil Erosion Mapping

Recorded soil erosion and mass movement features were applied to each map unit to observe the spatial dimension of different degrees of current processes of soil degradation. The Universal Soil Loss Equation (USLE/RUSLE) was also applied creating parametric maps based on geopedological (erodibility K factor) (Wishmeier and Smith 1978), climatic (erosivity R factor) (Gaitan et al. 2017), terrain data (length and slope LS factor) (Desmet and Govers 1996), and vegetation cover (C factor) (De Jong et al. 1998).

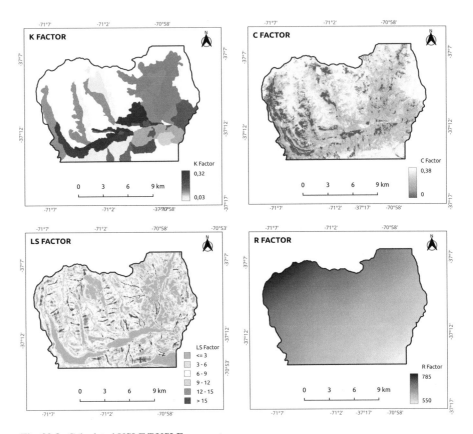

Fig. 20.3 Calculated USLE/RUSLE parameters

As no conservation practices are carried out in the study area, the P factor was assumed to be 1 throughout the study area (Fig. 20.3). Maps were put into a GIS and spatial operators were applied to obtain an estimation of potential soil loss. Spatial analysis and modelling were used to simulate the progress in the crown closure in a hypothetical establishment of a conifer afforestation for a silvopastoral system (density of 500 trees per ha) and the consequent reduction of the C factor was performed.

Criterion applied was that of an established afforestation of 15, 20 and 25 years with litter cover (US Conservation Service 1977, Wischmeier and Smith 1978). This simulation was carried out using a model that relates growth of trees in diameter at breast height (DBH) with crown diameter, which allows simulating the tree cover growth over time and according to land suitability (Frugoni et al. 2001). This analysis was performed in the moderately suitable areas for rainfed forestry defined in the catchment in previous studies (Frugoni et al. 2016) and it was observed to what extent potential for soil loss in those areas decreased.

20.3 Results and Discussion

20.3.1 Soil Types

Three soil orders were identified in the geopedologic study: Andisols, Mollisols and Histosols (Soil Survey Staff 2014). The first two orders were the most widespread in the study area (Table 20.1, Fig. 20.4).

Volcanic ash soils Andisols identified in this study were found in areas where soil climate is less xeric, in the western sector of the basin. Identified subgroups were Humic Vitrixerands and Humic Haploxerands, dominating on slopes modeled by glacial erosion. It should be noted that these two subgroups, have also been detected on slopes affected by mass movement processes, where soil climate favored the evolution of volcanic ash towards materials with andic properties.

The lithic subgroup was found in the steepest portions of these units and in the vicinity of the steep peaks and flanks. Typic Endoacuands were identified in the hanging valleys present on slopes and in the intermontane valley. Volcanic ash soils that did not meet the requirements for andic soil properties (due to a more xeric soil climate) were classified as Mollisols (Frugoni et al. 2020). The identified Subgroup was Vitrandic Haploxerolls. These soils are associated with map units located further east than the previous ones (both on the slopes modeled by glacial action, and those affected by mass movement processes) and where the ash cover is already discontinuous, but landform have favored accumulation of these materials.

Soils derived from volcanic tuff and tuffites The identified subgroups were Vertic and Typic Argixerolls. They were described in areas where volcanic ash deposits were absent. They occupied the easternmost sectors of the glacial modelled slopes and on slopes affected by mass movement processes. In sectors where drainage was impeded, Typic Argiacuolls have been described.

Organic soils Histosols were detected in the valley. The subgroup identified was Fluvacuentic Medifibrists, in somewhat small areas and close to the watercourses.

20.3.2 Current Erosion and Active Processes

About 90% of the area -excluding rock outcrops and debris cover- experiences moderate and severe soil erosion while the remaining 10% appears to be stable areas. Dominant processes are that of sheet erosion, although rill and gully erosion as well as soil creep were recorded and observed in the field (Table 20.2). Figure 20.5 shows the spatial distribution of intensity and processes. It can be noted that the area where gully erosion and soil creep occur corresponds to Argixerolls derived from volcanic tuff and tuffites, probably due to the presence of a clayey subsoil. In the area covered by volcanic ash soils, sheet and rill erosion processes were evident. On slopes facing south and west, moderate and severe sheet and rill erosion occur; while in east and north facing slopes, moderate sheet erosion was observed.

Table 20.1 Legend of the geopedologic map of Buta Mallín catchment

Landscape	Relief type/molding	Lithology/facies	Landform	Map unit	Area (Ha)	Soils (Dominant soils in bold)
Mountains **Mo**	Slopes shaped by glacial erosion **Mo1**	Lapilli and volcanic ash on	Strongly steep peaks and flanks (includes water divides)	**Mo111**	7422	Rock outcrops and debris cover
		Basalts, andesites, breccias and volcanic agglomerates **Mo11**	Steep mountainflank Overall east slope aspect (middle and lower third)	**Mo112**	1915	**Humic Vitrixerands** Humic Haploxerands
			Steep mountainflank West slope aspect (middle and lower third)	**Mo113**	1175	**Humic Vitrixerands** Lithic Vitrixerands
			Steep mountainflank North slope aspect (middle and lower third)	**Mo114**	1035	**Humic Vitrixerands** Lithic Vitrixerands Rock outcrops Humic Haploxerands
			Steep mountainflank South slope aspect (middle and lower third)	**Mo115**	1141	**Humic** Vitrixerands Humic Haploxerands
			Sloping mountain base	**Mo116**	480	**Humic Vitrixerands** Vitrandic Haploxerolls Alluvial-colluvial debris cover
			Hanging Valley	**Mo117**	326	**Typic Endoaquands**
		Discontinuous cover of lapilli and volcanic ash on tuff and basalt **Mo12**	Complex slope (upper third)	**Mo121**	251	**Vitrandic Haploxerolls** Debris cover and rock outcrops
			Gently sloping slope (upper third)	**Mo122**	304	**Humic Vitrixerands**
			Steep slope (middle and lower third)	**Mo123**	886	**Vertic Argixerolls** Vitrandic Haploxerolls Rock outcrops
			Hillside meadow	**Mo124**	88	**Typic Argiacuolls**

Landscape / Relief	Material	Landform	Code	Area	Soils
Slopes affected by mass movement processes **Mo2**	Discontinuous coverage of lapilli and volcanic ash on Tuff **Mo21**	Moderately steep, dissected denudational slopes	**Mo211**	1170	**Vitrandic Haploxerolls** / Vertic Argixeroles / Humic Haploxerands
		Moderately steep, heavily eroded denudational slopes	**Mo212**	2379	**Vertic Argixerolls** / Vitrandic Haploxerolls / Humic Vitrixerands
	Discontinuous coverage of lapilli and volcanic ash on tuffite **Mo22**	Moderately steep, heavily eroded denudational slopes	**Mo221**	631	**Typic Argixerols** / Lithic Haploxerols
Plateau **Pl** — Mesa **Pl1**	Basalts, olivinic andesites, volcanic breccias and agglomerates **Pl11**	Interfluve	**Pl111**	583	Debris cover and rock outcrops
Escarpment **Pl2**	Basalts, olivinic andesites, volcanic breccias and agglomerates **Pl21**	Front of the escarpment	**Pl211**	307	Rock outcrops
Postglacial Intermontane Valley **Va** — Alluvial aggradation surface **Va1**	Alluvial **Va11**	Tread	**Va111**	766	**Typic Endoaquands** / Typic Endoaquolls / Fluvacuentic Medifibrists
		Delta	**Va112**	95	Boulders
Piedmont alluvial forms **Va2**	Alluvium-coluvium **Va21**	Alluvial fan	**Va211**	112	**Stony phase of Vitrandic Haploxerolls**

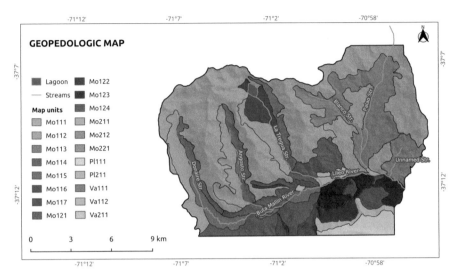

Fig. 20.4 Geopedologic map of Buta Mallín catchment

Table 20.2 Intensity and processes of current soil erosion. (Excluding rock outcrops and debris cover)

INTENSITY AND PROCESSES	AREA (ha)	%
Stable area	1092	9
Moderate sheet erosion	3821	30
Moderate sheet and rill erosion, soil creep	1170	9
Moderate and severe sheet and rill erosion	2307	18
Severe sheet erosion	1248	10
Severe sheet, rill and gully erosion, soil creep	3010	24
TOTAL	12,647	100

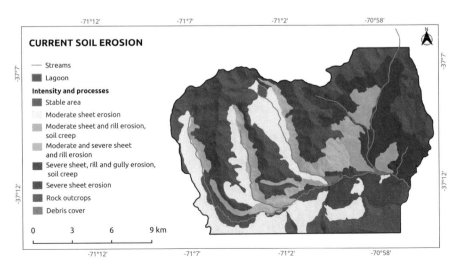

Fig. 20.5 Spatial distribution of intensities and processes of current soil erosion

20.3.3   Soil Erosion Hazard

According to modelling applied using the USLE equation, 85% of the area shows different degrees of potential soil erosion, and 76% experience moderate and severe soil erosion hazard. The remaining 15% of the catchment shows no erosion hazard (Table 20.3). Calculated mean, maximum and minimum soil loss were 84.2, 299 and 0.5 Tons/ha/year, respectively. Areas showing slight hazard are those located at the valley bottoms with low values of the topographic (LS) and vegetation cover (C) factors and in certain slopes with high effect of vegetation cover. Although LS factor was not high in most of the eastern portion of the catchment, the USLE model shows severe and very severe soil erosion hazard, most likely due to higher K (erodibility) and higher C (low vegetation cover) factors (Fig. 20.6).

Table 20.3 Areas occupied by different levels of soil erosion hazard. (Excluding rock outcrops and debris cover)

Soil erosion hazard (A)	Area (ha)	%
No erosion hazard	1948	15
Slight hazard	1003	8
Moderate	922	7
High	942	7
Severe	6041	48
Very severe	1792	14
Total	12,647	100

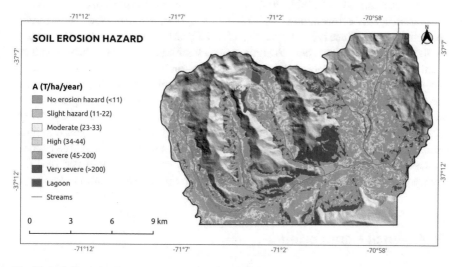

Fig. 20.6 Soil erosion hazard map with current vegetation cover

Table 20.4 Values of a calculated C factor in a simulated afforestation in the moderately suitable lands

Age of afforestation (years)	Trees/ha	Crown density (%)	C factor
15	500	23	0.009
20	500	42	0.004
25	500	67	0.002

Table 20.5 Areas occupied by different levels of soil erosion hazard with a simulated afforestation in the moderately suitable lands. (Excluding rock outcrops and debris cover)

Soil erosion hazard with forest grazing system (A)	area (ha)	%
No erosion hazard	6143	49
Slight hazard	583	5
Moderate	412	3
High	411	3
Severe	3580	28
Very severe	1515	12
Total	12,647	100

20.3.4   Soil Erosion Hazard with a Simulated Silvopastoral System

Based on data of crown closure in the moderately suitable areas for rainfed forestry which covered an area of 4500 ha (Frugoni et al. 2016), an estimation of a simulated C factor was carried out. (Table 20.4). Besides, soil erosion hazard maps considering 3 different ages of afforestation (15, 20 and 25 years) were constructed. The area with this simulated land use type showed that after 15 years of afforestation with a density of 500 trees per hectare, crown closure (with its litter layer, which reaches about 2 cm depth at this age) was enough to reach a non-erosion hazard situation in the moderately suitable lands (Table 20.5, Fig. 20.7). Comparison between the two situations (erosion hazard without and with silvopastoral system) showed that implementation of this proposed land use incorporates more than 4000 hectares into the *without hazard* category of soil erosion. In addition, it reduced by half the area exposed to intermediate soil erosion hazard, reducing by almost 2500 hectares the land area exposed to severe erosion hazard. Comparison of means in tons per hectare per year (A), showed a reduction of more than 30% soil loss in the proposed scenario (Table 20.6).

20.4   Final Considerations

Current soil erosion in Buta Mallín, located in the north-west of Neuquén province, was identified through a soil survey that uses the geopedologic approach introduced in Part I of this book. The principles of the geopedologic approach were applied in

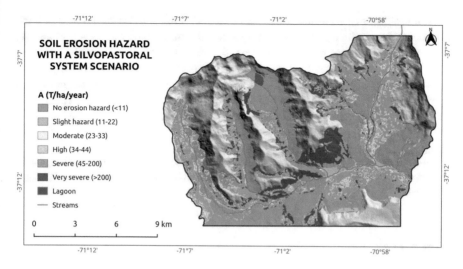

Fig. 20.7 Soil erosion hazard map with a simulated 15 years afforestation in the moderately suitable lands

Table 20.6 Comparison of the different levels of soil erosion hazard with and without implementation of a silvopastoral system. (Excluding rock outcrops and debris cover)

Soil erosion hazard (A)	Area (ha)		Difference
	Without forest grazing system	With forest grazing system after 15 years	
Mean (tons /ha/year)	84	58	−26
No erosion hazard	1948	6143	4195.8
Slight hazard	1003	584	−418.6
Moderate	922	413	−508.9
High	942	412	−530.6
Severe	6041	3580	−2460.5
Very severe	1792	1515	−277.2
Total	12,647	12,647	

the design of field soil survey and in the GIS-based spatial analysis and modelling. This enabled visualizing the spatial dimension of the different soil erosion intensities -and associated processes- that occur in the study area. Most important, some of the most recent soil erosion processes (e.g., gully and soil creep), which are not considered in prior erosion hazard assessments were identified through this approach.

The higher percentage of hazard-free areas on the potential erosion map may be due to the digital mapping methodology applied, which relies on the interpretation of satellite imagery. Because of the pixel-by-pixel analysis, this technique is more sensitive to discriminate areas than field-based geopedology, given the necessary generalization of mapping units.

The scenarios of silvopastoral practices show that a significant number of hectares could be productive land with no hazard of soil erosion due to a low-density afforestation (500 trees per hectare) after 15 years, preserving the ancestral transhumant system. Besides, this is a feasible land use type to improve the grazing system in North Patagonia and it could contribute to the diversification of the economy and increase land productivity, allowing to obtain timber products while reversing the process of land degradation that characterizes the study area.

Acknowledgments This study was funded through an agreement between the Universidad Nacional del Comahue (UNCo) and the Ministry of Production of Neuquén Province, Argentina. Assistance of surveyors of the Ministry and peasants who were our guides in fieldwork, are gratefully acknowledged.

References

Aguiló Alonso M, Albaladejo Montoso J, Aramburu Maqua MP, Carrasco González RM, Castillo Sánchez V, Ceñal González MA, Cifuentes Morales M, Cifuentes Vega O, Cristóbal López MA, Duque JFM, Escribano Bombín R, Glaría Galcerán G, González Alonso S, González Barberá G, Iglesias Gómez J, Iglesias del Pozo E, López de Diego LA, Llorente FM, Martínez Mena M, Milara Vilches R, Pedraza Gilsanz J, Rastrollo Gonzalo A, Rubio Maroto R, Sanz Sa JM, Sanz Santos MA, Valero Huete F (2014) Guía para la elaboración de estudios del medio físico. 4ta. ed. España. Fundación Conde del Valle de Salazar. Ministerio de Agricultura, Alimentación y Medio Ambiente. ISBN: 978-94-96442-55-9

Bendini M, Alemany C (eds) (2005) Crianceros y chacareros en la Patagonia. Cuaderno GESA 5 – INTA – NCRCRD. Editorial La Colmena, Buenos Aires. Páginas 23–40

Bonilla CA, Reyes JL, Magri A (2010, January–March) Water erosion prediction using the revised universal soil loss equation (RUSLE) in a GIS framework, Central Chile. Chil J Agric Res 70(1):159–169

Casas R (2015) La erosión del suelo en Argentina. In: Casas R, Albarracín G (eds) El deterioro del suelo y del ambiente en la Argentina. Prosa, Buenos Aires, pp 433–452

De Jong SM, Brouwer LC, Riezebos HT (1998) Erosion hazard assessment in the Peyne catchment, France. Working paper DeMon-2 Project. Dept. Physical Geography, Utrecht University

Desmet PJJ, Govers G (1996) A GIS procedure for automatically calculating the USLE LS factor on topographically complex landscape units. J Soil Water Conserv 51(5):427–433

Frugoni MC, Peña O, De Jong G, Bertani L, Ambrosio M (2001, November) Aptitud de las tierras para el crecimiento de plantaciones de *Pinus* en la cuenca del río Neuquén, Argentina. Paper presented at the XV Congreso Latinoamericano de la Ciencia del Suelo, La Habana, Cuba, pp. 11–16

Frugoni MC, González Musso R, Falbo G, Zapiola D (2016) La geopedología como base para zonificar la aptitud forestal en una cuenca del noroeste de la Patagonia Argentina. Boletín geográfico. Año XXXVII. N°38–2016, pp. 29–48 Departamento Geografía. Universidad Nacional del Comahue. Neuquén. ISSN 0326-1735; e-ISSN 2313-903X bibliocentral.uncoma. edu.ar/revele/index.php/geografia/index

Frugoni MC, González Musso R, Falbo G (2020) Capítulo 11: Los suelos derivados de cenizas volcánicas de la provincia del Neuquén, Argentina. In: Suelos y vulcanismo: Argentina / Adriana Mehl ... [et al.]; Editors: Perla Imbellone and Osvaldo Andrés Barbosa (1a ed) Ciudad Autónoma de Buenos Aires: Asociación Argentina de la Ciencia del Suelo -AACS, 2020. E book PDF ISBN 978–987–46870-2-9

Gaitan J, Navarro MF, Tenti Vuegen L, Pizarro MJ, Carfagno P, Rigo S (2017) Estimación de la pérdida de suelo por erosión hídrica en la República Argentina. 1ra. ed. Buenos Aires: Ediciones INTA. ISBN 978–987–521-857-4

Gardi C, Angelini M, Barceló S, Comerma J, Cruz Gaistardo C, Encina Rojas A, Jones A, Krasilnikov P, Mendoza Santos Brefin ML, Montanarella L, Muñiz Ugarte O, Schad P, Vara Rodríguez MI, Vargas R (eds) (2014) Atlas de Suelos de América Latina y el Caribe. Comisión Europea. Oficina de Publicaciones de la Unión Europea, L-2995, Luxembourg, p 176

González Díaz E, Ferrer JA (1991) Geomorfología. En: Estudio Regional de Suelos de la Provincia del Neuquén. Plano N° 6. C.F.I. – COPADE

Luccerini SA, Subovsky ED, Bodorosky E (2013) Sistemas silvopastoriles: una altenativa productiva para nuestro país. Apuntes Agroeconómicos. Facultad de Agronomía UBA. Febrero 2013. Año 7. Número 8. ISSN 1667-3212

Morales D, Rostagno CM, La Manna L (2013) Runoff and erosion from volcanic soils affected by fire: the case of *Austrocedrus chilensis* forest in Patagonia. Argentina Plant Soil 370(1):367–380

NASA LP DAAC (2015) ASTER Level 1 Precision Terrain Corrected Registered At-Sensor Radiance V003 [Data set]. NASA EOSDIS Land Processes DAAC. https://doi.org/10.5067/ASTER/AST_L1T.003. Accessed 29 Feb 2016

Palacio RG, Bisigato AJ, Bouza P (2014) Soil erosion in three grazed plant communities of northeastern Patagonia. Land Degrad Dev 25:594–603

Rubio G, Lavado RS, Pereyra FX (eds) (2019) The soils of Argentina, World Soils Books Series. Springer, p 263

Schoeneberger PJ, Wysocky DA, Benham EC, Broderson WD (1998) Libro de Campaña para la descripción y muestreo de suelos. Versión 1.1. Centro Nacional de Relevamiento de Suelos. Servicio de Conservación de Recursos Naturales. Departamento de Agricultura de los Estados Unidos. Lincoln, Nebraska. Traducción al español por investigadores de AICET, Instituto de Suelos y Evaluación de Tierras. INTA, 2000

Soil Survey Staff (2014) Keys to soil taxonomy, 12th edn. U.S. Department of Agriculture (USDA), Natural Resources Conservation Service (NRCS)

US Soil Conservation Service (1977, September) Procedure for computing sheet and reel erosion on project areas. Technical Release 51 (Rev 2)

Wischmeier WH, Smith DD (1978) Predicting rainfall erosion losses – a guide to conservation planning. In: Agriculture handbook N°537. U.S. Department of Agriculture, Washington, DC, p 59

Yemefack M, Siderius W (2016) Chapter 16: applying a geopedologic approach for mapping tropical forest soils and related soil fertility in Northern Thailand. 265–284. In: Zinck JA et al (eds) Geopedology. An integration of geomorphology and Pedology for soil and landscape studies. Springer, p 556

Young A (1987) Soil productivity, soil conservation and land evaluation. Agrofor Syst 5:277–291

Zinck JA (2016) Chapter 4: the geopedologic approach 2–60. In: Zinck JA et al (eds) Geopedology. An integration of geomorphology and Pedology for soil and landscape studies. Springer, p 556

Chapter 21
Adequacy of Soil Information Resulting from Geopedology-Based Predictive Soil Mapping for Assessing Land Degradation: Case Studies in Thailand and Iran

D. P. Shrestha, R. Moonjun, A. Farshad, S. Udomsri, and A. Pishkar

Abstract Soil is a natural body that delivers important ecosystem services apart from being a medium for plant growth. Soil mapping can be time-consuming and expensive. During the 1960s and 1970s, the introduction of air photointerpretation in soil survey through element analysis, physiognomic and physiographic analysis, helped increase mapping efficiency. In the late 1980s, the geopedologic approach to soil mapping amplified the role of geomorphology. It helps understand soil variation in the landscape which increases mapping efficiency. In the present study, the adequacy of soil data resulting from geopedology-based predictive soil mapping for assessing land degradation is applied in two contrasting climatic regions: in humid tropics in Thailand and in dry and hot arid climate in Iran. The result shows that the geopedologic approach helps map soil in inaccessible mountain areas. However, for application in land degradation studies all the required soil properties may not be available in a soil map. The effect of land cover and land use management practices on soil properties, such as porosity and compaction affecting hydraulic conductivity, a parameter used in modelling rainfall-runoff-soil erosion, is usually not reported in soil surveys. These data have to be collected separately. For mapping areas susceptible to frequent floods, the geomorphic understanding of the river valley and soil characterization (Fluventic and Aquic) helps identify susceptible areas. Similarly, the study shows how the geopedologic approach in combination with

D. P. Shrestha (✉) · A. Farshad
Faculty of Geoinformaion Science and Earth Observation (ITC), University of Twente, Enschede, The Netherlands
e-mail: d.b.p.shrestha@utwente.nl

R. Moonjun · S. Udomsri
Land Development Department, Ministry of Agriculture and Cooperatives, Bangkok, Thailand

A. Pishkar
Provincial Agricultural Affairs Organization, Shiraz, Iran

digital image processing and/or the application of simple decision rules applied in a Geographic Information System (GIS) help in mapping soil salinity trends.

Keywords Surface runoff · Erosion modelling · Flood-prone area · Soil salinity trends · Decision tree · Remote sensing

21.1 Introduction

Soil is a natural body that delivers important ecosystem services apart from being a medium for plant growth. It can be considered the "skin of the earth" with interfaces between lithosphere, hydrosphere, atmosphere, and biosphere (Chesworth 2008). Soil mapping in general is time demanding and costly. The introduction of air photo interpretation in soil survey in the 1950s helped increase mapping efficiency. During the 1960s and 1970s, several methods such as "elements analysis" (Buringh 1960), "pattern analysis" (Frost 1960), "physiognomic analysis" and "physiographic analysis" (Bennema and Gelens 1969; Goosen 1967) were developed. Gradually, it was noticed that understanding the relationship between landform and soil variation is crucial in drawing interpretation lines. Finally, in the late 1980s, the "geopedologic approach" to soil survey (Zinck 1989) presented in Part I of this book, amplified the role of geomorphology in understanding and mapping soil variations. Through a systematic and strict application of the geopedologic rules soil is mapped more efficiently (Farshad et al. 2013), although Esfandiarpoor Borujeni et al. (2010) reported that a geopedologic map does not fully represent all the variability of soils and suggested that more research was needed on various landscape stratification techniques to better understand soil formation processes. Similarly, Nazari et al. (2016) reported the increase of soil diversity from the level of soil order to soil family after their study on soil diversity in a hierarchical sequence in categories of the USDA soil taxonomy.

Recently, advances in digital soil mapping using an array of techniques including GIS, digital elevation models, multivariate statistics, geostatistics, neural network, fuzzy logic, among others, claim to increase mapping efficiency (McBratney et al. 2000; Behrens et al. 2005; Lagacherie et al. 2007; Grimm et al. 2008; Lagacherie 2008). These techniques are useful for mapping individual soil properties but mapping soilscape (the whole soil body) remains a challenge. A soil body incorporates not only solids, liquids, and air that cover the land, but it also has horizons and thus extends in depth. Mapping individual soil properties do not equate to mapping the whole soil body. In this respect, the geopedologic approach to soil survey can be considered very useful and efficient. However, the objective is not only to produce a soil map but to evaluate the value of such a map for various applications too. In this chapter, the adequacy of soil data resulting from geopedology-based predictive soil mapping for assessing land degradation is applied in two contrasting climatic regions: in humid tropics in Thailand and in dry and hot arid climate in Iran.

Three case studies of predictive soil mapping using the geopedologic approach are presented to assess the adequacy of soil data for applications in land degradation issues, namely soil erosion, flash flood, and soil salinity hazard studies in Thailand. In Iran, the geopedologic approach in combination with a simple model (decision rule-based) was applied to assess soil salinity risk.

21.2   Materials and Methods

Geopedologic image interpretation is a stepwise procedure that starts with delineating master lines across major landscape units such as mountain, hilland, valley, plain, among others, and drawing cross-sections (Zinck 1989; Farshad et al. 2016). Along the master lines follow the identification of sub-units (relief types) within major landscape units. Subsequently, main lithology types are identified for each unit. Lithological units can be derived from geological maps or inferred using expert knowledge, such as alluvial, colluvial or aeolian origin. Lastly, landform units are identified. This information is used to construct an interpretation legend where symbols are attributed to the landforms. After that, photo-delineation follows using a mirror stereoscope. Effective areas are determined within aerial photographs; these are marked perpendicular to flight lines, using the two transferred principal points, as explained in Paine and Kiser (2012).

Recently, the traditional way of interpreting aerial photos under mirror stereoscope has been replaced by on-screen digitizing of digital stereo pairs. The digital stereo pairs can be generated using either (a) scanning two overlapping air photos or orthophotos and creating a stereo pair, or (b) using a georeferenced image (i.e., orthophoto or satellite image) and the digital elevation data of the corresponding area. Many GIS software packages offer this facility. The Open Source ILWIS software package helps generate stereo pairs which can be viewed using Red-Green or Red-Blue glasses in case of an anaglyph image or using a stereoscope in case of a stereo pair. For viewing a stereo image, a special stereoscope is needed which can be mounted on a computer screen. Interpretation of the stereo image can be done by directly digitizing on the screen.

The advantage of using computer-assisted onscreen digitizing is that the interpretation lines are georeferenced with proper map projection parameters making the final map layout easier. Once interpretation work is completed, sample areas are selected to facilitate fieldwork. A general rule for selecting a sample area is that it should include all the landform units. The area should also be easily accessible. Mini pits are used for soil description and sampling. Soils are classified directly in the field following a classification system (e.g., FAO, USDA). Detailed soil study in the sample areas helps determine soil-landscape relationships and understand soil variability and patterns. This information is used to support extrapolation and mapping outside the sample areas. Laboratory determinations of soil samples are used to adjust soil classification. In this way, the photo-interpretation map is converted into a soil map. By applying the above-mentioned method, we demonstrate that soilscape knowledge enables analysing cause-effect relationships between soil types, their distribution, and land degradation hazards/risks as shown in the hereafter described case studies.

21.3 Soilscape Mapping for Assessing Soil Erosion in Thailand

In mountainous areas, especially in the tropics, soil data are scarce mainly due to limited terrain accessibility. Sloping areas are often mapped as slope complexes. Soil surveys and mapping have so far been carried out in valleys, floodplains, and other low-lying areas in the proximity of human settlements. Although mountain areas are usually considered low priority, they are important because of the ecosystem services they provide through rainwater harvesting and storage, regulating weather conditions, supporting a diversity of flora and fauna, offering scenic and panoramic views, among others. Inadequate management of watersheds can result in soil degradation in upland areas, which in turn affects the streamflow discharge causing offsite effects in low-lying areas (e.g., stream avulsion, flooding). Conservation of watersheds and making effective management plans usually require data modelling and generating scenarios with detailed soil data.

In mountain areas, excess surface runoff as a consequence of torrential rainfall can be the main causal factor of land degradation processes such as rill and gully erosion resulting in soil losses. Runoff generation is a function of rainfall volume and intensity during a rain event, an interception by the vegetation cover, slope gradient, soil moisture storage capacity, and infiltration into the soil, which depend on soil porosity, saturated hydraulic conductivity, and soil depth. Because of the unavailability of detailed soil data, many soil properties, with exception of soil depth, are derived from soil particle size distribution using pedo-transfer functions. Interpolation techniques are commonly applied for mapping spatial variation of soil properties, disregarding in many instances the effect caused by topographic variation. The case study described hereafter attempts to map soils in an inaccessible mountain area of Thailand and assess the adequacy of the survey information for erosion estimation.

21.3.1 Study Area

The Nam Chun watershed (67 km2) is located in Petchabun province, about 400 km north of Bangkok, between $16°40'–16°50'N$ and $101°02'–101°15'E$ (Fig. 21.1a). The area has a tropical climate with distinct dry and wet seasons. The average annual precipitation is 1095 mm (Lomsak station) most of which falls in the wet season (May–September). The average annual temperature is 28 °C, the hottest

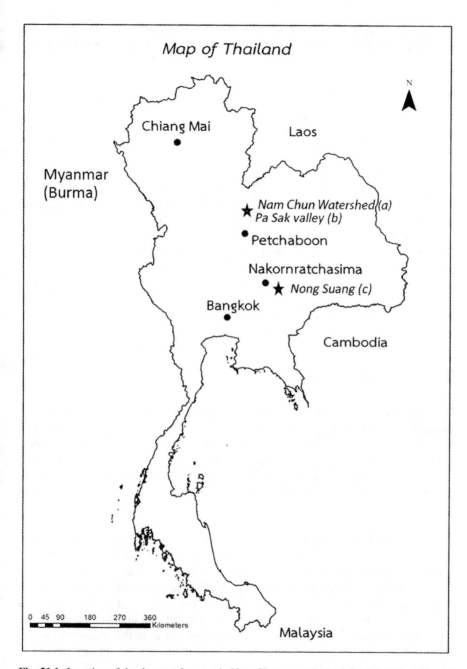

Fig. 21.1 Location of the three study areas in Nam Chun watershed (**a**), Pa Sak valley (**b**), and Nong Suang (**c**)

month being April (38 °C) and the coldest month being December (17 °C). Topography is rugged with mountain ridges of different heights separated by narrow valleys. Elevation varies from 185 to 1490 m asl. General accessibility is limited apart from the main road connecting Lomsak and Phitsanulok.

21.3.2 Soil Studies

The rugged topography and lack of road access precluded stratified random sampling over the entire catchment area. Instead, observations were made along roads and nearby areas which could be reached on foot. A total of 219 soil samples was collected for laboratory analysis of particle size distribution, pH, organic matter content, bulk density, porosity, field capacity, and saturated hydraulic conductivity. In addition to the conventional soil survey work, infiltration tests and shear-strength measurements of the topsoil in major land cover types were carried out to cope with the influence of human activities on soil compaction and cohesion. Such properties are necessary for assessing surface runoff and soil loss. Soils were classified at the subgroup level according to USDA soil taxonomy (Soil Survey Staff 1999).

21.3.3 Results

Through visual interpretation of aerial photographs, four main landscape units were identified as follows: (1) a high plateau (elevation 1200 m) which borders the northwest of the watershed, (2) high mountain areas including very steep ridges (elevation 900 m) and lower dissected slope complexes, (3) low mountain areas (maximum elevation 600 m), and (4) a narrow valley cutting across the watershed. The cartography of these landscape units is presented in Fig. 21.2, and Table 21.1 contains information associated with the map legend.

The plateau landscape accounts for about 16% of the watershed area. The main soils in this area are Typic Haplustalfs and the soil texture varies from loam to clay loam. The mountain landscape covers about 80% of the watershed area. The landforms are narrow summits, mid-slopes including backslope, and footslope. Soils are Lithic and Ultic Haplustalfs. In the low mountains, Lithic Haplustolls occur on the narrow summits, while Lithic Haplustalfs and Typic Paleustalfs are common on mid-slopes. Typic Dystropepts and Typic Haplumbrepts are found on the footslope. Soil texture varies from clay loam to silty clay loam. Soils in the area are characterized by high clay content and were classified in the clay loam textural class. The area delineated as a valley was very narrow and accounted for only 2% of the total area. It consisted of the Hua Nam Chun river and a narrow floodplain that could not be differentiated as a separate unit.

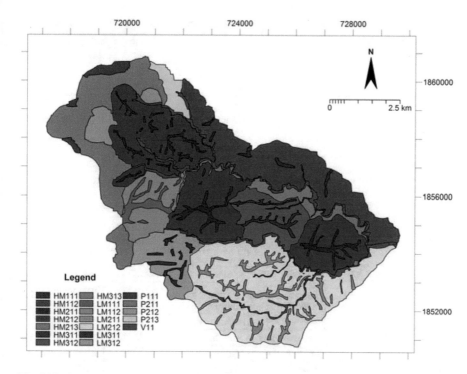

Fig. 21.2 Geopedologic map of Nam Chun watershed. (Adapted from Solomon 2005)

Soils are Fluvents and Haplumbrepts and texture vary from sandy loam to clay loam. Some of the soil properties especially in the topsoil, such as porosity, bulk density, and hydraulic conductivity, are very much related to land cover and land-use practices. The highest soil porosity (53%) and lowest bulk density (1.19 Mg m⁻³) were found under forest cover (Table 21.2). The highest rate of saturated hydraulic conductivity was found in grassland followed by forest areas. Agricultural land had compacted soil (average bulk density of 1.30 Mg m⁻³) and reduced porosity. This contributes to increase surface runoff. Infiltration in grassland and forest land is higher than in other areas.

Erosion assessment was carried out by applying the revised MMF model (Morgan 2001) which requires soil moisture content at field capacity, bulk density, cohesion, and erodibility. The result shows the highest soil loss rates through sheet erosion in agricultural land due to compaction and reduced soil porosity. The use of topography derived contributing area for runoff estimation and slope gradient relationship helped in assessing the types of erosion processes into the sheet, rills, and gully erosion (Shrestha et al. 2014). For this, remote sensing and digital elevation data are required for deriving land cover and topographic parameters such as slope gradient and surface runoff contributing area.

Table 21.1 Legend of the geopedologic map of Nam Chun watershed, Petchabun province

Landscape	Relief	Lithology	Landform	Map unit	Soil types (USDA classification)
Plateau	Cuesta	Sandstone	Undifferentiated	Pl111	
	Escarpment	Sandstone	Scarp	Pl211	
			Talus	Pl212	
			Undulating slope complex	Pl213	Typic Haplustalfs
High mountain	Ridge	Andesite	Summit	HM111	
			Slope complex	HM112	Ultic Haplustalfs
	Ridge	Andesitic tuff	Summit	HM211	Lihic Haplustolls
			Middle slope	HM212	Ultic Haplustalfs
			Foot slope	HM213	Ultic Haplustalfs
	Erosional glacis	Andesitic and rhioliic tuff	Summit	HM311	Lihic Haplustolls
			Middle slope	HM312	Typic Haplustalfs
			Foot slope	HM313	Lihic Haplustolls
Low mountain	High ridges	Andesitic tuff	Summit	LM111	Typic Haplustalfs
			Middle slope	LM112	Ultic Haplustalfs
	Mod. high ridges	Andesitic and rhyolitic tuff	Summit	LM211	Typic Haplustalfs
			Middle slope	LM212	Ultic Haplustalfs
	Low ridges	Andesitic and rhyolitic tuff	Summit	LM311	Typic Dystrustepts
			Middle slope	LM312	Ultic Haplustalfs
Valley		Alluvial/ colluvial	Side slope/bottom complex	V111	Fluvents and Haplumbrepts

Table 21.2 Soil properties in different land cover types (Shrestha et al. 2014)

Land cover types	Hydraulic conductivity mm/h			Bulk density Mg.m^{-3}			Porosity %			Organic matter %		
	Mean	n	Std. dev.	Mean	n	Std. dev.	Mean	n	Std. dev.	Mean	n	Std. dev.
Forest	13.88	9	14.67	1.19	11	0.13	52.57	11	5.36	4.08	12	1.10
Degraded forest	7.06	8	9.19	1.28	10	0.12	49.04	10	4.95	3.15	8	1.38
Cornfield	4.36	9	2.79	1.30	11	0.10	48.12	11	4.08	2.24	19	0.84
Orchard	3.76	10	3.99	1.31	12	0.09	47.90	12	3.67	3.55	14	1.04
Grassland	15.43	11	15.70	1.26	11	0.11	49.61	11	4.50	2.99	13	1.12

21.4 Soilscape Mapping for Flood Hazard Assessment in Central Thailand

Lowlands in Thailand are usually easy to access because of the good road network in flat areas. Abundant land for rice cultivation increases settlements and interconnecting roads. But the area is subjected to frequent flooding during the rainy season. The study hereafter examines the adequacy of soil data for assessing flood hazards.

21.4.1   Study Area

The study area is located in the Pa Sak river valley (graben), Petchabun province, about 400 km north of Bangkok in central Thailand. It served in the past for training ITC students in soil survey (ITC for International Institute of Geo-Information Science and Earth Observation, Enschede, The Netherlands) (Hansakdi 1998). The area is bounded in the east and west by mountain ranges (horst) that delineate the valley (Fig. 21.1b). Elevation varies from 120 m asl in the south to 170 m asl in the north from where the Pa Sak river flows. Floodplains are used for rice cultivation. In the surrounding foothills, tamarind plantations are common. Other fruit crops are lychees and bananas. The main settlements are Lomsak and Lomkao.

21.4.2   Soil Studies in Sample Areas and Transects

Following the visual interpretation of aerial photographs at the scale of 1:50,000, sample areas were selected to include all the landform units and located based on the local road network. The sample areas were surveyed in more detail using aerial photographs at a scale of 1:15,000. Soils were described in (mini-)pits and auger holes following the FAO soil description manual.

21.4.3   Results

The geopedologic map (Fig. 21.3, Table 21.3) shows that the soils in the lowlands belong to Entisols, Inceptisols, and Alfisols (Soil Survey Staff 1999). Inceptisols are the most common soil, occurring in various landforms. Ustropepts, Eutropepts, and Tropaqepts are dominant. Entisols are found on the sides and bottoms of narrow trench valleys. Alfisols include Haplustalfs and Tropaqualfs in low positions and Paleustalfs in higher positions. The occurrence of Aquic, Fluventic, and Fluvaquentic subgroups is typical in the central valley, while Aeric and Ultic subgroups are dominant in the lateral valley (Table 21.3).

In an aquic moisture regime, the soil lacks dissolved oxygen for being saturated by groundwater. Fluventic characteristics are typical of soils formed from alluvial sediments, stratified, and showing frequent variations in texture and organic matter content. Fluvaquentic soils have both fluventic and aquic characteristics. Soil texture in the valley landscape varies from clay loam to silty clay loam. Although soil porosity is high, saturated hydraulic conductivity is very slow in the valley soils (less than 2 mm/h), meaning the inability to percolate stagnated water fast to the groundwater. The soil characteristics and the landscape configuration make the area prone to frequent flooding. The soils in the lateral and trench valleys, higher in the landscape, are more developed, with the argillic horizon, than the valley bottom soils, and they are not exposed to flooding. The result shows how geomorphic

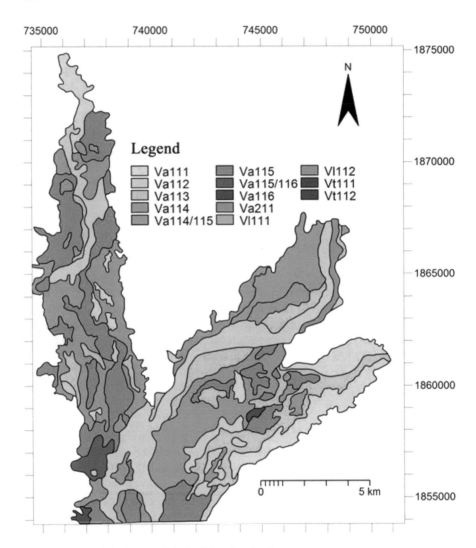

Fig. 21.3 Geopedologic map of the Pa Sak valley flood-prone area

understanding of the river valley and characterization of soils (Fluventic and Aquic) help identify areas susceptible to frequent flooding.

21.5 Soilscape Mapping for Soil Salinity Trend Assessment

Within the past few decades, the concurrent development and application of the tools for mapping and monitoring soil salinity such as GPS, GIS, geophysical techniques involving proximal and remote sensors, and a greater understanding of

Table 21.3 Geopedologic legend of Pa Sak valley, Petchabun province

Landscape	Relief type	Lithology	Landform	Map unit	Soil map unit	Main soil types USDA classification
Valley	Terrace	Alluvium	Tread-riser complex	Va111	Association	Aeric Tropaquepts (40%), Fluventic Ustropepts (30%)
			Levee	Va112	Consociation	Fluventic Ustropepts (60%), Typic Ustropepts (40%)
			Levee/overflow mantle complex	Va113	Association	Aeric Tropaquepts (40%), Fluventic Ustropepts (30%), Typic Ustropepts (20%)
			Overflow mantle	Va114		Fluvaquentic Eutropepts (30%), Typic Eutropepts (30%), Fluventic Ustropepts (30%)
			Overflow basin	Va115		Fluvaquentic Eutropepts (30%), Aquic Eutropepts (30%), Typic Tropaquepts (20%)
			Overflow mantle/basin	Va114/115		Fluvaquentic Eutropepts, Typic Ustropep, Fluventic Ustropepts
			Decantation basin	Va116		Aquic Eutropepts (40%), Aeric Tropaquepts (30%), Fluventic Eutropepts (20%)
			Overflow/decantation basin	Va115/116		Fluvaquentic Eutropepts, Aquic Ustropepts, Typic Tropaquepts, Aquic Eutropepts
	Flood plain	Alluvium	Levee/basin complex	Va211		Fluventic Ustropepts (30%), Aquic Ustropepts (30%), Fluvaquentic Eutropepts (30%)
Lateral valley	Terrace complex	Colluvio-alluvium	Bottom /side complex	VI111		Aeric Tropaqualfs (40%), Typic Haplustalfs (40%)
	Depression			VI112	Consociation	Aeric Tropaquepts (60%), Aeric Tropaqualfs (40%)
Trench valley	Terrace complex	Alluvium/ residual		Vt111	Association	Typic Ustifluvents (50%), Typic Haplustalfs (30%), Ultic Paleustalfs (20%),
	High terrace			Vt112	Consociation	Ultic Paleustalfs (50%), Ultic Haploustalfs (50%)

apparent soil electrical conductivity (EC_a) and multi- and hyperspectral imagery have made it possible to map soil salinity across multiple scales (Corwin and Scudiero 2019). To this list, geopedologic soil mapping can be considered an additional tool for assessing soil salinity. Soil salinity hazard and risk are sometimes synonymously used but the risk has the additional implication of the chance of a particular hazard (Smith and Petley 2009). Thus, the risk is the actual exposure of something of human value to a hazard and is often regarded as the product of probability and loss. Thus, we define a hazard as a potential threat and risk as to the probability of that hazard. The following section describes the application of the geopedologic approach to mapping salinity-prone areas.

21.5.1   Salinity Hazard Assessment in *North-Eastern Thailand*

Soil salinity is a regional issue in the northeast of Thailand. The study area is in the Nong Suang region, Nakhon Ratchasima province, between 101°45′–102°E and 15°–15°15′N, with an elevation ranging from 160 to 175 m asl (Fig. 21.1c). The area is part of the Northeast Korat plateau landscape and is underlain by salt-bearing rocks at about 80 m depth (Imaizumi et al. 2002). The average annual rainfall is 1035 mm (1971–2000), coming in from May to October. The average annual potential evapotranspiration is 1817 mm, thus higher than the mean annual precipitation (1035 mm), indicating that climate is a potential driver of salinity in the area (Shrestha and Farshad 2008). The groundwater is affected by underlying salt-bearing rocks. The capillary rise of the groundwater with high salt contents (conductivity of more than 15 dS m^{-1}) is the main cause of soil salinity in the area.

21.5.1.1   Soil Sampling and Data Analysis

Geopedologic interpretation was carried out to delineate the geomorphic units occurring in the lower parts of the landscape (i.e., floodplains, terraces, and vales), which have a high potential for salinity development. A total of 126 samples was collected in the field from three depths: 0–30 cm, 30–60 cm, and 60–90 cm to study soil salinity (Soliman 2004). Since salt-affected areas give high reflectance values during the dry season in the visible to near-infrared bands due to the concentration of salts on the terrain surface and the formation of salt crusts, a Landsat TM image of the 2003 dry season was obtained. The digital image was processed, and the rotation of near-infrared and red spectral bands was performed to derive a soil line that enhances soil reflectance from saline areas (Shrestha et al. 2005). Level slicing of the resulting band (soil line) was carried out to generate salinity intensity classes within the geomorphic units of the floodplain, terrace, and vale.

21.5.1.2 Results

At the landscape level, the area was separated into peneplain and valley that were further divided into corresponding relief types, lithology, and landform levels (Fig. 21.4, Table 21.4).

Five soil order classes were distinguished (Soil Survey Staff 1999). Ultisols occur on ridges, while Alfisols (Ustalfs and Aqualfs) are in sloping areas adjacent to the ridges. Vertisols occur in the northern part of the area, along rivers and channels, where vertic features form due to the presence of swelling clays. Two suborders, namely Aquerts and Usterts, were distinguished based on the soil moisture regime. Inceptisols are common in the lower part of the lateral valleys that dissect the peneplain lobes. Inceptisols are Aquepts due to poor drainage conditions that lead to the

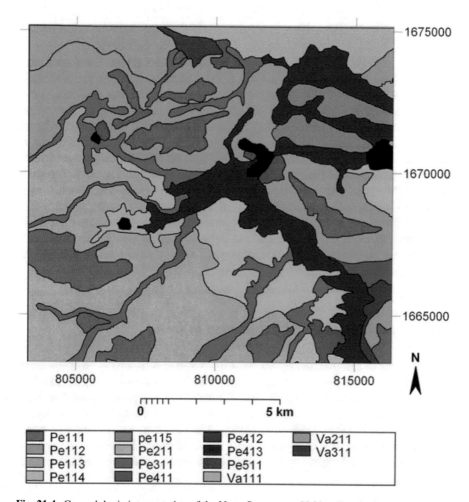

Pe111	pe115	Pe412	Va211	
Pe112	Pe211	Pe413	Va311	
Pe113	Pe311	Pe511		
Pe114	Pe411	Va111		

Fig. 21.4 Geopedologic interpretation of the Nong Suang area, Nakhon Ratchasima province

Table 21.4 Geopedologic interpretation legend of the Nong Suang area, Nakhon Rachasima

Landscape	Relief type	Lithology	Landform	Map unit
Peneplain	Ridge	Sedimentary rocks	Top complex	Pe111
		Korat Group	Side complex	Pe112
			Slope facet complex	Pe113
			Summit	Pe114
			Tread riser complex	Pe115
	Glacis	Sedimentary rocks	Slope complex	Pe211
	Vale	Sedimentary rocks		Pe311
	Lateral vale	Sedimentary rocks	Side complex	Pe411
		Korat Group	Bottom-side complex	Pe412
			Bottom complex	Pe413
Valley	Floodplain	Alluvial deposit	Levee-overflow complex	Va111
	Old terrace	Alluvial deposit	Overflow-basin complex	Va211
	New terrace	Alluvial deposits	Overflow-basin complex	Va311

development of gleyic color, with no abrupt textural change. Wet Psamments, classified as Gleysols according to the FAO World Reference Base for Soil Resources (FAO 1998), occur in a few sloping spots on residual material derived from sandstone. Geomorphic units such as floodplains, terraces, and vales have a high potential for salinity development since they are located in the lower parts of the landscape and are thus most likely close to saline groundwater. They were masked out using digital elevation data in a GIS map overlay procedure to improve classification accuracy (Shrestha and Farshad 2008). The result is shown in Fig. 21.5. The study also showed a good correlation between soil texture and salinity occurrence. Clayey soils are strongly saline because of higher capillary rise from the groundwater. Salinity is lower in coarse-textured soils. Since salt-affected areas usually present higher reflectance in all visible and near-infrared spectral bands when the soil is bare and dry, salinity variations can be mapped using remote sensing data and the enhancement technique.

21.5.2 Soil Salinity Risk Assessment in Iran

Several types of salt-affected soils are distinguished in Iran, totally covering about 15% of the 165 million hectares, the country's total surface area (Naseri 1998). Due to the variations in hydro geopedological settings in different landscape/landform units (soilscapes) the following modes of salinization (/alkalinization) are recognized (Farshad 2015; Farshad et al. 2018): (i) In the extensive depressions, where upwards movement of saline groundwater takes place due to an extremely low (to nil) precipitation and extremely high air temperature. Here the so-called 'sierozemic

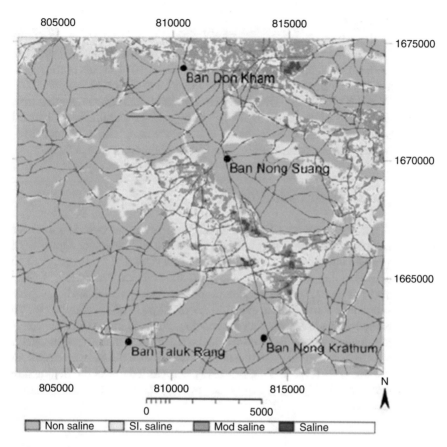

Fig. 21.5 Soil salinity hazard assessment. (Shrestha and Farshad 2008)

mode' takes place for example in the 'Kavir-e- Lut' (30°36′18″N 59°04′04″E). (ii) Salinization also occurs at the foot of the mountains (piedmonts), which includes alluvial fans and glacis, where the secondary salt formation is common e.g., the Shazand alluvial fan (33° 55′ 59.99″ N, 49° 23′ 59.99″ E). (iii) In the flood plain// terraces, for instance of the Sharra river, close to Komijan (34° 43′ 13″ N, 49° 19′ 40″ E) soil salinization is caused due to over-irrigation. (iv) Seep-water-based mode, an example of which is visible in the Qom geological formation, along the road Qom to Kashan (34° 38′ 23.9964″ N, 50° 52′ 35.0004″ E). (v) Other forms of salinization are due to the ingression of seawater through waves, wind, and/pr dew, which is common in the coastal plains of the Caspian sea and the Persian Gulf. The case study described below deals with the soil salinization trend in the piedmont landscape.

21.5.2.1   Study Area

The study area is a part of the piedmont landscape (series of alluvial fans) surrounding the Goorband village, near the city of Minab, about 100 km east of the provincial capital (Bandar Abbas), between 27° 17′ 43″ and 27° 21′ 32″ N, and 56° 56′ 01″ and 56° 56′ 58″ E (Fig. 21.6). The average annual rainfall is 220 mm, with a maximum monthly rain of about 58 mm in February, and a minimum of 2 mm in September. The total annual evaporation is over 980 mm, which is more than 4 times the precipitation (Pishkar 2003). This clearly shows moisture deficiency.

21.5.2.2   Soilscape and Hydropedology

The study area is in a watershed that occurs in the Makran geological formation of the Cenozoic age. The largest part of the area consists of quaternary deposits. The study was carried out in one of the alluvial fans in the piedmont landscape. First, the geopedologic mapping was carried out (Fig. 21.6). Soils are classified in Entisols and Aridisols (Table 21.5), with varying properties corresponding to the geopedomorphic subdivisions as from the apical to the distal parts of the alluvial fan. The study is focused on the use of relevant geopedologic-derived parameters that control the moisture content within the intermediate zone, which is that part of the vadose zone that occurs between the belt of capillary rise layer on the ground surface and the groundwater table (Fig. 21.7). Grid survey was carried out in the field and the spatial distribution of EC, ESP, capillary rise, and the intermediate zone was mapped using geostatistics (Pishkar 2003).

21.5.2.3   Results

Field observations revealed that disturbing moisture balance by human activities is the main cause of soil salinization. Misusing the water of the Minab dam's channel in the absence of a drainage system, triggers waterlogging, and progressive salinization (Fig. 21.8).

Soil salinity risk assessment (Fig. 21.9) was carried out using three soilscape related properties, namely "Intermediate layer" (I), "electrical conductivity of soil (E)," and "terrain slope (S)." By constructing a simple model using a decision tree (Fig. 21.10 and Table 21.6), the model was implemented using the "IF THEN, ELSE" statement to track down salinity dynamics and to generate the soil salinity risk. The hypothesis is that when the groundwater table rises, upward movement of saline water takes place leading to the formation of salt-affected soil and deposition of salts, visible in whitish and greasy colour on the surface. The study proves once again the role of the geopedologic (soilscape) mapping in landscape studies.

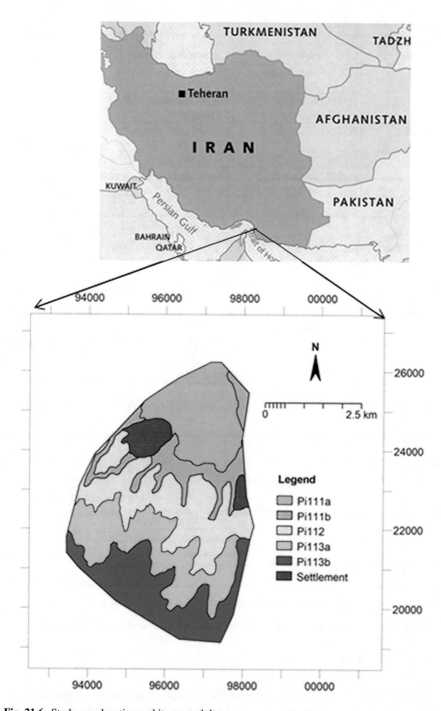

Fig. 21.6 Study area location and its geopedology

Table 21.5 Description of the geopedologic units

Alluvial fan subunits	Unit	Description	Taxonomic description	Slope
Apical	Pi111a	Severely eroded, stony	Sandy skeletal and sandy	>3%
	Pi111b	Moderately eroded	Mixed, hyperthermic, Typic Torrifluvents	
Medial	Pi112a	Cultivated	Coarse loamy, mixed, hyperthermic, Typic Torrifluvents	≥1%– 3%≤
Distal	Pi113a	Moderately saline	Fine loamy, mixed, and coarse loamy	0–1%
	Pi113b	Saline and alkaline	Coarse loamy mixed, hyperthermic, Typic Aquisalids	

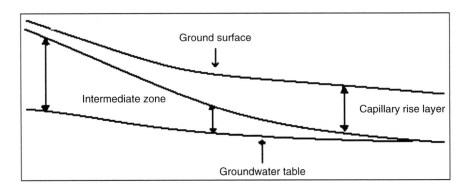

Fig. 21.7 Schematic views of the capillary zone, intermediate zone, and water table depth

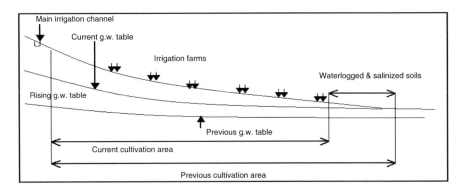

Fig. 21.8 Schematic view of secondary salinization in the Goorband area. (Pishkar 2003)

21.6 Discussion

The case studies show that the geopedologic approach to soil survey promotes soils-cape mapping, knowledge of which enables better interpretations in various fields, such as land degradation, and for making better land-use plans. Soilscape

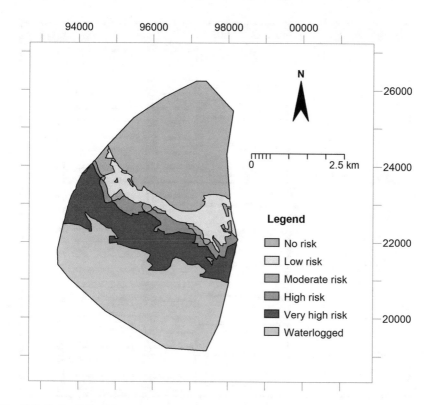

Fig. 21.9 Soil salinity risk assessment

knowledge enables analyzing cause-effect relationships between soil types, their distribution, and land degradation hazards as shown in the case studies with different land degradation problems. Fieldwork remains an essential component of soil mapping. Variability at the subgroup level, for instance between Typic and Ultic soil types, or at some of the intergrades (Alfisols-Ultisols) can only be discovered during fieldwork, by well-trained surveyors. Similarly (Bagheri Bodaghabadi et al. 2016), who worked on digital soil mapping using artificial neural networks (ANN), revealed improved training and interpolation (test area) accuracy, but a low level of extrapolation accuracy. These researchers reported that extrapolating soil maps based solely on topographic features is insufficient. In other words, the geopedologic maps including other soil-forming elements, such as landforms and lithology could be used as an input to ANN to improve prediction.

For running rainfall-runoff or process-based erosion modelling soil data derived from geopedologic map alone may not be sufficient. For hydrological modelling soil properties such as cohesion, porosity and saturated hydraulic conductivity become very important. While cohesion can be measured easily in the field soil physical and hydraulic properties can be derived from laboratory analysis of undisturbed soil samples. In our study for erosion assessment, it shows that soil

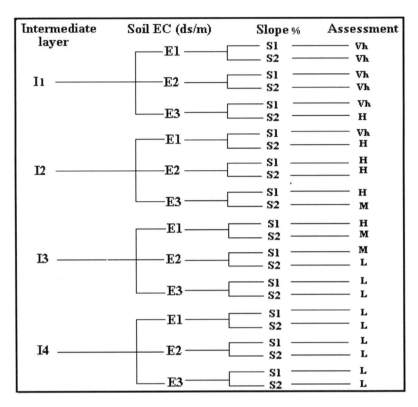

Fig. 21.10 A simple model using decision tree for creating the risk map

Table 21.6 The three layers properties, Increase in the score values is the decrease in the effect on salinization risk

	Intermediate layer (m)				EC ds/m			Slope (%)	
Value	0–3	3–6	6–9	9–12	>8	4–8	<4	<1	>1
Score	I1	I2	I3	I4	E1	E2	E3	S1	S2

properties influenced by human activities e.g., organic matter content and compaction become very important in assessing soil aggregate stability and infiltration of rainwater, which are not readily available in the geopedologic maps. A land cover map became necessary for this, which was derived from remotely sensed imagery (Shrestha et al. 2014). Also, the study shows that topographic data becomes necessary to assess erosion processes.

For assessing flood-prone areas some soil properties such as imperfect drainage conditions of the soils can be inferred from the legend at the lower level of geopedologic unit. Also, soil characterization (Fluventic and Aquic) is helpful. The use of DEM can highlight these characteristics which might be hidden in the map and provide additional information on the terrain for mapping areas susceptible to frequent floods. But if we need to estimate the volume of surface runoff from the upland areas to assess lowland flood hazards soil data from geopedologic map will not be sufficient. For this, soil properties such as saturated hydraulic conductivity and porosity will be required.

Our study has shown that geopedological approach to soil survey can be very useful for mapping salinity-affected areas in the arid/semi-arid climatic regions. In the tropics, geopedologic mapping alone has proved to be insufficient. Remote sensing data can be of help to track down salinity but cannot be directly useful in mapping soil salinity because of vegetation cover and/or due to the presence of soil moisture which prevents the formation of salt crusts. Additional tools such as digital elevation data, specific remote sensing techniques, geophysical techniques, and/or modelling (e.g., application of decision rules) might be required.

21.7 Conclusion

Through the case studies analysed in this chapter, we argue that the geopedologic approach to soil survey helps in mapping soils very efficiently. But for achieving the desired adequacy for applications in different land degradation studies, we may have to use additional tools and techniques to supplement the required soil data. The type of soil data depends on the selection of models. For erosion assessment using the empirical model e.g., Universal Soil Loss Equation or its revised model (RUSLE), geopedological soil mapping can provide the necessary data. If soil loss assessment has to be carried out from a storm event additional soil data (soil cohesion, saturated hydraulic conductivity, soil porosity) will be required. For this, additional undisturbed soil sampling and laboratory analysis will be necessary. The required data can be also derived using pedo-transfer functions (Saxton and Rawls 2006) but field data will be required to validate the results. The same holds true for running a daily erosion model (Shrestha and Jetten 2018). For assessing flood-prone areas, DEM-derived terrain parameters e.g., low lying areas and slope, in addition to soil characteristics (imperfect drainage condition, Fluvents/Aquepts) can be useful in delineating the susceptible areas. For soil salinity hazard assessment soil parameters derived from geopedological mapping may not be sufficient. Additional tools and remote sensing imagery including gamma-ray, multi/hyperspectral data, geophysical techniques as well as modelling may be necessary depending on the climatic conditions and the availability of the data. In general, the capillary rise of groundwater is the major cause of soil salinity. Whether a certain area will be saline depends on the quality of groundwater and if environmental conditions are favourable for the capillary rise. For future research, it will be interesting to assess the concentration of salt in the groundwater and model the magnitude of capillary rise so that necessary preventive measures can be taken in time.

Acknowledgments Materials used in the case studies from Thailand were derived from a joint research project of ITC, Enschede, The Netherlands, with the Land Development Department (LDD), Ministry of Agriculture and Cooperatives, Bangkok, Thailand. The contribution of ITC course participants, especially Anukul Suchinai, Ekanit Hansakdi, Harssema Solomon, and Aiman Soliman, is duly acknowledged. The case study from Iran is based on the MSc. thesis work of Rahim Pishkar.

References

Behrens T, Förster H, Scholten T, Steinrücken U, Spies ED, Goldschmitt M (2005) Digital soil mapping using artificial neural networks. J Plant Nutr Soil Sci 168(1):21–33

Bennema J, Gelens HF (1969) Aerial photointerpretation for soil surveys, ITC lecture notes. ITC, Enschede

Buringh P (1960) The application of aerial photographs in soil surveys. In: Colwell RN (ed) Manual of photographic interpretation. American Society of Photogrammetry, Washington, DC, pp 633–666

Bagheri Bodaghabadi M, Martínez-Casasnovas JA, Esfandiarpour Borujeni I, Salehi MH, Mohammadi J, Toomanian N (2016) Database extension for digital soil mapping using artificial neural networks. Arab J Geosci 9. https://doi-org.ezproxy2.utwente.nl/10.1007/s12517-016-2732-z

Chesworth W (ed) (2008) Encyclopedia of soil science. Springer, Dordrecht

Corwin, DL., Scudiero, E., 2019. Chapter one - review of soil salinity assessment for agriculture across multiple scales using proximal and/or remote sensors, in: Sparks, D.L. (Ed.), Advances in agronomy. Academic, pp. 1–130. https://doi-org.ezproxy2.utwente.nl/10.1016/bs.agron.2019.07.001

Esfandiarpoor Borujeni I, Mohammadi J, Salehi MH, Toomanian N, Poch RM (2010) Assessing geopedological soil mapping approach by statistical and geostatistical methods: a case study in the Borujen region, Central Iran. Catena 82:1–14. https://doi-org.ezproxy2.utwente.nl/10.1016/j.catena.2010.03.006

FAO (1998) World reference base for soil resources. World soil resources report 84 Rome

Farshad A (2015) Soilscape speaks out; an account of the past, presen, and future. ECOPERSIA 3(1):881–899

Farshad A, Pazira E, Noroozi AA (2018) Human-induced land degradation (chapter 12). In: Roozitalab MH, Siadat H, Farshad A (eds) The soils of Iran. World soils book series. Springer

Farshad A, Shrestha DP, Moonjun R (2013) Do the emerging methods of digital soil mapping have anything to learn from the geopedologic approach to soil mapping and vice versa? In: Shahid SA, Taha FK, Abdelfattah MA (eds) Developments in soil classification, land use planning and policy implications. Springer, Dordrecht, pp 109–131

Farshad A, Zinck JA, Shrestha DP (2016) Geopedology promotes prediction and efficiency in soil mapping. Photo-interpretation application in the Henares River Valley, Spain. In: Geopedology. Springer, Cham, pp 347–360

Frost RE (1960) Photo interpretation of soils. In: Colwell RN (ed) Manual of photographic interpretation. American Society of Photogrammetry, Washington, DC, pp 343–402

Goosen D (1967) Aerial photo interpretation in soil survey. In: Food and Agriculture Organization of the United Nations, soils bulletin 6. FAO, Rome

Grimm R, Behrens T, Marker M, Elsenbeer H (2008) Soil organic carbon concentrations and stocks on Barro Colorado Island: digital soil mapping using random forests analysis. Geoderma 146:102–113

Hansakdi E (1998) Soil pattern analysis and the effect of soil variability on land use in the upper Pa Sak area, Petchabun, Thailand. Unpublished MSc thesis, ITC, Enschede

Imaizumi K, Sukchan S, Wichaidit P, Srisuk K, Kaneko F (2002) Hydrological and geochemical behavior of saline groundwater in Phra Yun. Khon Kaen, Thailand

Lagacherie P (2008) Digital soil mapping: a state of the art. In: Hartemink AE, McBratney A, Mendonca-Santos ML (eds) A state of the art: digital soil mapping with limited data. Springer, Dordrecht, pp 3–14

Lagacherie P, McBratney AB, Voltz M (eds) (2007) Digital soil mapping: an introductory perspective. Elsevier, Amsterdam

McBratney AB, Odeh IOA, Bishop TFA, Dunbar MS, Shatar TM (2000) An overview of pedometric techniques for use in soil survey. Geoderma 97(3–4):293–327

Morgan RPC (2001) A simple approach to soil loss prediction: a revised Morgan-Morgan-Finney model. Catena 44:305–322

Naseri MY (1998) Characterization of salt-affected soils for modelling sustainable land management in semi-arid environment: A case study in the Gorgan region, Northeast Iran. PhD thesis, State University Of Ghent, Belgium

Nazari N, Mahmoodi S, Masihabadi MH (2016) Employing diversity and similarity indices to evaluate geopedological soil mapping in Miyaneh, East Azerbaijan Province. Iran Open J Geol 06:1221–1239. https://doi-org.ezproxy.utwente.nl/10.4236/ojg.2016.610090

Paine D, Kiser J (2012) Aerial photography and image interpretation, 3rd edn. Wiley, Hoboken

Pishkar A (2003) Analysis of the relationship between soil salinity dynamics and geopedologic properties: A case study of the Goorband area, Iran. Unpublished MSc thesis. ITC, Enschede, The Netherlands

Saxton KE, Rawls WJ (2006) Soil water characteristic estimates by texture and organic matter for hydrologic solutions. Soil Sci Soc Am J 70:1569–1578

Shrestha DP, Farshad A (2008) Mapping salinity hazard: an integrated application of remote sensing and modeling-based techniques. In: Metternicht G, Zinck JA (eds) Remote sensing of soil salinization: impact on land management. CRC Press, Boca Raton, pp 257–272

Shrestha DP, Jetten VG (2018) Modelling erosion on a daily basis, an adaptation of the MMF approach. Int J Appl Earth Obs Geoinf 64:117–131. https://doi-org.ezproxy2.utwente.nl/10.1016/j.jag.2017.09.003

Shrestha DP, Soliman AS, Farshad A, Yadav RD (2005) Salinity mapping using geopedologic and soil line approach. Asian Conference on Remote Sensing, Hanoi, Vietnam

Shrestha DP, Suriyaprasit M, Prachansri S (2014) Assessing soil erosion in inaccessible mountainous areas in the tropics: the use of land cover and topographic parameters in a case study in Thailand. Catena 121:40–52. https://doi-org.ezproxy2.utwente.nl/10.1016/j.catena.2014.04.016

Soil Survey Staff (1999) Soil taxonomy: a basic system of soil classifi cation for mapping and interpreting soil surveys. US Department of Agriculture, Washington, DC

Soliman AS (2004) Detecting salinity in early stages using electromagnetic survey and multivariate geostatistical techniques: a case study of Nong Sung district, Nakhon Ratchasima province, Thailand. Unpublished MSc thesis, ITC, Enschede

Solomon H (2005) GIS-based surface runoff modeling and analysis of contributing factors: a case study of the Nam Chun Watershed, Thailand. Unpublished MSc thesis, ITC, Enschede

Smith K, Petley DN (2009) Environmental hazards: assessing risk a reducing disaster. Routledge, New York, p 383

Zinck JA (1989) Physiography and soils, ITC lecture notes. ITC, Enschede

Chapter 22
Exploring the Contribution of Geopedology to the Implementation of National Frameworks for Land Degradation Neutrality

G. Metternicht, H. F. del Valle, and M. Angelini

Abstract This chapter proposes a conceptual framework for integrating geopedology in the preliminary assessments required to establish the baseline for land degradation neutrality (LDN) at national or sub-national levels and to monitor outcomes of interventions targeting the LDN hierarchy of avoid, reduce, reverse land degradation. Specifically, it proposes using geopedologic mapping units as main sources to derive the land types and associated land units considered in the conceptual framework for LDN. In such a way, geopedologic units inform decisions around the concepts of 'like-for-like' necessary to plan and implement the counterbalancing mechanism core to the principle of neutrality of land degradation. We illustrate the application of the framework via a case study of Cochabamba's Valle Alto, in Bolivia.

Keywords Geopedology · Land degradation neutrality · Soilscape · Geoforms · Decision-making

G. Metternicht (✉)
Earth and Sustainability Science Research Centre (ESSRC), University of New South Wales, Sydney, Australia

School of Science, Western Sydney University, Richmond, Australia
e-mail: G.Metternicht@westernsydney.edu.au

H. F. del Valle
Facultad de Ciencia y Tecnología, Centro Regional de Geomática (CeReGeo), Universidad Autónoma de Entre Ríos (UADER), Oro Verde, Entre Ríos, Argentina

M. Angelini
Instituto de Suelos (CIRN). Instituto Nacional de Tecnología Agropecuaria, Buenos Aires, Argentina

Food and Agriculture Organization of the UN. Global Soil Partnership, Roma, Italy

Departamento de Ciencias Básicas, Universidad Nacional de Luján, Luján, Argentina
e-mail: angelini.marcos@inta.gob.ar; marcos.angelini@fao.org; mangelini@unlu.edu.ar

437

22.1 Land Degradation Neutrality and Geopedology: Foundations

The concept of land degradation neutrality (LDN) was adopted by the 12th Conference of the Parties of the UN Convention to Combat Desertification, and it is a target of the UN Sustainable Development Goal 15 (i.e., target 15.3). It aims to maintain or enhance the land resources necessary to support ecosystem functions and services by implementing well-planned land use and management interventions underpinned by a response hierarchy of avoiding or reducing new land degradation through sustainable land management or reversing past degradation via land restoration and/or rehabilitation.

Implementation of the framework requires conducting a preliminary assessment to provide decision-makers with information necessary to identify and prioritise appropriate options for sustainable land management, land restoration and/or rehabilitation. These assessments should inform on the potential of the land, its current condition, use, and resilience. Land potential assessment and land stratification are undertaken using the concept of land type, that is, 'clustering' a landscape according to its long-term potential. Land types are a function of climate, topography, and relatively static soils properties (e.g., soil texture, slope, clay types, stoniness, soil structure), and vegetation cover. All these factors are also common to geopedology which studies the formation and distribution of soils in a landscape.

Geopedology is an approach to soil survey that combines pedologic and geomorphic criteria to establish soil map units. Geomorphology helps determine the boundaries of the map units (the container) while pedology provides information on the soil components of the map units (the content, such as soil types) (see Chap. 4). Therefore, the units presented in a geopedologic map are more than 'soil units' in the conventional usage of the term: soil-landscape relations and soil functions remain homogeneous, and the units provide relevant information about the geomorphic 'context' that influences soil formation, distribution, and condition.

A geopedologic approach to landscape analysis considers the intrinsic relationship between soil forming factors, geoforms and their distribution in the landscape. The concept of the geopedologic landscape highlights relevant relationships between soil properties and geomorphic features (e.g., valley, crest, alluvial terraces) at different scales (national, regional, local) and hierarchical levels (e.g., physiographic province, natural region). The geopedologic unit approximates to the concept of soilscapes, but with the premise that geomorphology guides the definition of 'landscape' units (see Chap. 6, Table 6.4).

This chapter proposes a novel framework for the integration of geopedology in baseline studies of LDN, to assess land potential and its condition, and for monitoring advances toward LDN. In the context of the conceptual framework for land degradation neutrality (Orr et al. 2017), geopedologic units can be thought of as land types. In planning for LDN, land types are used in the preparatory activities (Fig. 22.1), particularly for setting the baseline that will inform where and what interventions are needed for achieving LDN in a tract of land (e.g., province, nation).

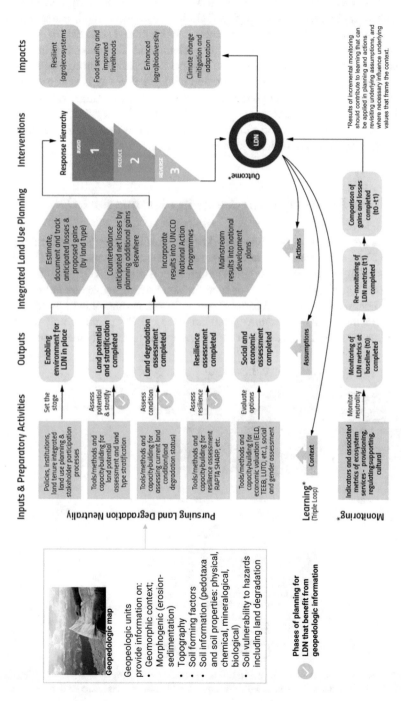

Fig. 22.1 Contributions of the geopedologic approach to support planning for land degradation neutrality

Different land types vary in their potential to deliver different ecosystem services (e.g., nutrient and water cycling, regional climate regulation, food, water) and in different proportions (which in turn depends on soil type and soil health).

Land stratification (Fig. 22.1) divides the land into units for accounting and planning land use, including interventions to reverse degradation through restoration and/or rehabilitation. In this regard, integrated land use planning is key to ensuring the 'like for like' principle is applied when assessing and managing counterbalancing between anticipated land losses and gains so that policies, regulations, and management practices relate in a coordinated way to each class in the administrative or biophysical domain within which decisions are made.

22.2 Entry Points of Geopedology in Planning for LDN

Figure 22.2 presents the conceptual model of a geopedologic approach for designing context-based interventions for achieving land degradation neutrality. Defining land types using a geopedologic approach enables assessing soil types and their potential (or capability/suitability) for specific land uses or management practices. Geoforms become the land type cartographic units (ie. geomorphic units equate to the land units that are the minimum spatial unit used in LDN planning and monitoring). As shown in Fig. 22.2, geoforms can be the spatial units (polygons) used in the

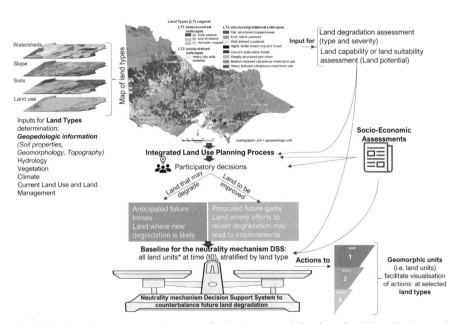

Fig. 22.2 The LDN mechanism for neutrality is the counterbalancing of anticipated gains and losses in land-based natural capital within unique land types via land use and management decisions

production of maps of land condition, land degradation, and also be input in decision support systems (DSS) that seek feasible alternatives to deal with anticipated ongoing (e.g., salinization) or unavoidable (conversion of forest to agricultural areas) land degradation.

The definition of land type (see above) encompasses all the elements (soil forming factors and factors that affect soil health) that are used to define a geopedologic unit (see Chap. 6 of this book). A geopedologic unit integrates information on geoforms (the first step is defining geomorphic units of a landscape) to then characterise the main soil types (e.g., Mollisols, Alfisols, etc) and their distribution within the specific soil units (e.g., association, consociation, complex).

Geomorphic units (or geoforms) are identified based on their own characteristics (e.g., type of geo-structure, morphogenetic environment, geomorphic landscape, relief type, lithology/facies, terrain form). Table 6.4 (Chap. 6) presents the synopsis of the geoform classification system. Local or regional combinations of criteria such as climate, vegetation, soil, and lithology —associated with geoforms and contributing to their formation—, can be referred to in the legend of the geomorphic map.

A difficult aspect of soil-landscape maps produced using conventional methods of soil survey is their inability to seamlessly transition between scales (Arnold 2015). In this regard, the classification of geoforms adopted in the geopedologic approach is scale-able (from regional, national to local scales), it follows a hierarchical categorisation, from order (geostructure) to sub-family (a landform). This enables generalising information collection for specific levels, and to aggregate information from a lower category (e.g., terrain form) to a higher category (e.g., geomorphic landscape). For example, a river levee (one type of 'terrain form' or landform) is a component of an alluvial terrace (a type of relief) which in turn can belong to a river valley (type of geomorphic landscape), that is part of a catchment or sub-catchment area. That hierarchical organisation of the landscape fits well with applications related to national integrated land use planning for land degradation neutrality, where the land planning unit can be a catchment area or administrative unit, which is profiled according to different land types (Fig. 22.2).

When planning for neutrality (Fig. 22.2), land use decisions need to be site-specific and provide spatially explicit information (e.g., in the form of maps). Nowadays, LDN maps are produced largely with inputs from Earth Observation (see for example Akinyemi et al. 2021, Gonzalez-Roglich et al. 2019). Satellite images are used, for instance, to classify landcover at a pixel level, evaluating the trend of this LDN indicator over time. Although it may be accurate in terms of spatial resolution (e.g., 10 m, 20 m, or better depending on the remotely sensed data used), this technique may be prone to uncertainties from the 'noise' (e.g., misclassification errors) from sensing data. Using the geopedologic landscape as a framework, trends of LDN indicators (e.g., landcover change) derived from remote sensing data, could be spatially averaged using geopedological units, decreasing the error of the trend estimations. This idea aligns with prior research (Vaysse et al. 2017) that has demonstrated the reduction of uncertainty in the prediction of soil properties when raster maps are aggregated into larger spatial units.

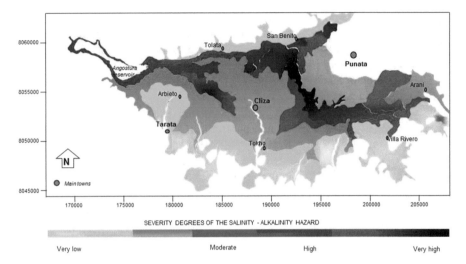

SEVERITY DEGREES OF THE SALINITY - ALKALINITY HAZARD

Very low   Moderate   High   Very high

Fig. 22.3 Severity of saline and alkaline (sodic) degradation process in the geoforms of Cochabamba's Alto Valle. (Source: Metternicht, 1996)

The neutrality mechanism considers, for each land unit, the direction of potential change anticipated at that site at the time land use and management decisions are made. It tracks those decisions that are likely to lead to losses in land-based natural capital and those likely to lead to gains (Orr et al. 2017). Ideally, the neutrality mechanism would consider the direction of change and the magnitude of change (Cowie et al. 2018). Hence, maps of 'direction' or 'magnitude' and type of land degradation (e.g., by soil salinization) as proposed in Metternicht and Zinck (2016) and reproduced in Fig. 22.3 can be used here to help decisions on the extent and type of land rehabilitation and/or land reclamation needed. That information input in a Decision Support System makes the basis of the neutrality mechanism that should be implemented following the LDN principle of 'like for like' as described in Orr et al. (2017) and Cowie et al. (2018).

The mechanism for neutrality takes the form of a DSS (Fig. 22.2) that accommodates geopedologic information and other relevant biophysical variables, plus socio-economic assessments needed to design voluntary measures, regulatory instruments, and/or market-based incentives and associated strategies and action plans. The DSS needs to encompass all types of land degradation identified (e.g., salinisation, overgrazing, deforestation, etc), be coherent with national circumstances (e.g., national development plans) and dynamics of land degradation occurring in the region (see for example Jobbágy et al. 2021 for a description on the impact of land use on salt accumulation and redistribution in the dry plains of Chaco-Pampas).

The mechanism is designed to assist land use decision makers to maintain or do better than 'no net loss' as a minimum standard. To be effective, it should be integrated into existing national and/or subnational LUP processes and systems (Orr et al. 2017). The LDN mechanism for neutrality comprises counterbalancing of anticipated losses (see Fig. 22.2) in land-based natural capital with planned gains (informed by the LDN response hierarchy) within unique land types.

A landscape analysis through the lens of geopedology also facilitates monitoring progress towards LDN, which requires evaluating, per land type, significant changes (positive and negative) in the three global indicators/metrics that serve as proxies for ecosystem services (see Fig. 22.1): land cover/land cover change, land productivity/NPP, and carbon stocks/SOC (Cowie et al. 2018).

Estimating anticipated losses should include the impacts of active decisions on land and effects of passive decisions (continuation of agricultural practices that deplete soil organic matter, exacerbate soil salinization) and natural drivers (e.g., impacts of drought, geological origin of the area). Changes associated with natural drivers (e.g., drought exacerbating further soil salinization) may be difficult to predict. Nevertheless, they will affect the stock and flow of ecosystem services and the indicators for LDN, and their likely impacts on interventions for achieving LDN need to be counterbalanced (Orr et al. 2017).

22.3 Worked Example Using the Cochabamba Valleys (Bolivia)

This section unpacks the proposal of using the information and cartography of the geopedologic units to help inform decisions around the concepts of 'like-for-like' necessary to plan and implement the counterbalancing mechanism, core to the principle of land degradation neutrality. We illustrate the application of the framework through a case study of the Cochabamba valleys in Bolivia.

Chap. 4 (Figs. 4.2 and 4.3) presents a map of the geomorphic units of the semi-arid Cochabamba's Valle Alto (with main towns of Cliza and Punata) in the eastern Andes of Bolivia, and Table 22.1 presents the geopedologic units with their associated soil types. The geography, land use and landcover, and geologic origins of the valley, as well as its main drivers of land degradation have been described extensively in Metternicht (1996), Metternicht and Zinck (1996), Metternicht and Zinck (1997), and Metternicht and Zinck (2016).

Aridisols and Entisols dominate the soilscape, with Entisols developing in piedmont areas that correspond with recent fans of alluvial or colluvial-alluvial origin. Aridisols contain subsurface horizons in which clays, calcium carbonate, silica, salts and/or gypsum have accumulated Calciorthis, and Haplargids dominate on glacis reliefs of the piedmont areas. Calciorthids have developed in the proximal part of depositional glacis and saline-sodic phases of Camborthids and Haplargids occur in the low-lying flats of lagunary origin and playa landforms (Table 22.1).

The lands of the high flats of Cliza (Fig. 4.2) are highly dependent on agriculture; 76% of the land use corresponds with intensive agriculture of maize, barley, potato and alfalfa and a small percentage (mostly home gardens) of land is used for growing vegetables, quinoa, and fruits for subsistence. Like in the nearby valleys of

Table 22.1 Geopedologic units of the Cochabamba's Valle Alto, with identification of main land degradation processes and hypothetical examples of sustainable practices that could be used in the design of interventions to avoid, reduce and/or reverse land degradation

Landscape	Relief type	Facies	Landform/ code**	Soil type and phases*	Land capability class***	Land degradation (type and severity)	Examples of practices to address land degradation (restoration, rehabilitation, reclamation)
Piedmont	Dissected glacis	Alluvial	Proximal (Pi111)	Association: Typic Calciorthids *[Typic Haplocalcids]* Typic Camborthids *[Typic Haplocambids]*	1	Very low soil salinity, localised gully soil erosion	Gully control, mechanical and biological measures to control run-off and promote infiltration
			Central (Pi112)	Consociation: Typic Camborthids (ca) *[Typic Haplocambids]* Ustochreptic Camborthids *[Ustic Haplocambids]*	2	Calcareous. Calcic horizon	Controlled grazing, deforestation, and misuse of the biomass. Promote regenerative agricultural practices, conservation agriculture, agroforestry, afforestation, or reforestation.
			Distal (Pi113)	Association: Ustalfic haplargids Ustochreptic Camborthids *[Ustic Haplocambids]*	3	Moderate saline	Phyto-remediation, agroforestry and silvopasture (e.g., introduction of shrubs and trees for fuelwood and fodder production, to supply timber for rural needs); establish greenbelts around settlements

Depositional glacis	Colluvio-alluvial	Distal (Pi213)	Consociation: Ustochreptic Camborthids [Ustic Haplocambids] Typic Camborthids [Typic Haplocambids]	2	Some stoniness No degradation	Water conservation and harvesting; de-stoning where economically viable; terracing; contour banks. Afforestation / reforestation with native species.
Active fans	Alluvial	Active channels (Pi411)	Miscellaneous land type: Mixed alluvial	5	Stoniness	
		Inactive channels (Pi412)	Consociation: Typic Torrifluvents Typic Torriorthents	5	Stoniness / no degradation	
Recent fans	Colluvio-alluvial	Pi51	Association Ustic Torriorthents Typic Torrifluvents	1	No degradation	Water conservation and harvesting vegetation manipulation to increase water yield; efficient irrigation with devices to promote groundwater recharge and storage; sustainable land management; regenerative farming practices; agroforestry, etc. Revegetation: Introduction of shrubs and trees for fuelwood and fodder production
Old dissected fans	Glacio-alluvial	Proximal (Pi611)	Association: Typic Camborthids [Typic Haplocambids] Typic Haplargids	2	No degradation	

(continued)

Table 22.1 (continued)

Landscape	Relief type	Facies	Landform/code**	Soil type and phases*	Land capability class***	Land degradation (type and severity)	Examples of practices to address land degradation (restoration, rehabilitation, reclamation)
			Central (Pi612)	Consociation: Ustochreptic Camborthids (ca) [Ustic Haplocambids]	2	Calcic horizon	Maintaining a minimum of 70% ground cover; keeping a stubble cover of 30% on cultivated areas, other mechanical and biological measures to control run-off and promote infiltration, control of sediment from bad lands and severely eroded areas.
			Distal (Pi613)	Consociation: Ustochreptic Camborthids [Ustic Haplocambids]	1	No degradation	Sustainable agricultural and pastoral practices to avoid soil erosion (e.g., terracing, contour banks, mulching, etc).
	Hills	Quartzitic sandstones	Pi71	Consociation: Lithic Torriorthents	1	No degradation	Gully control, mechanical and biological measures to control run-off and promote infiltration, control of sediment from bad lands and severely eroded areas,
		Marls sandstones limestones	Pi72	Consociation: Typic Calciorthids [Typic Haplocalcids] Lithic Calciorthids [lithic Haplocalcids]	2,3		Active interventions to prevent and control gully erosion, mechanical and biological measures to control run-off and promote infiltration, control of sediment from bad lands and severely eroded areas

Valley	Lagunary depressions	Alluvio-lagunary				
			Higher lagunary flats (Va111)	Association: Fluventic Camborthids [Fluventic Haplocambids] Ustochreptic Camborthids [Ustic Haplocambids]	4	Non to slightly saline-alkaline	Leaching (removal of excess salts from arable and subsurface soil horizons by flushing water)
			Middle lagunary flats (Va112)	Association: Ustalfic Haplargids [Ustertic Haplargids] Ustochreptic Camborthids [Ustertic Haplocambids]	4	Moderate to strong saline alkaline	Leaching and chemical soil remediation (e.g., adding gypsum). Promote adequate soil drainage. Phyto-remediation using suitable plant species (e.g., halophytes and hyperaccumulators) salt-tolerant forage crops (e.g., alfalfa).
			Lower lagunary flats (Va113)	Association: Ustalfic Haplargids (sa-so) [Ustertic Haplargids] Ustochreptic camborthids (sa)	4	Strongly saline-alkaline	Agroforestry and silvopasture practices with suitable native salt and sodium tolerant species.

(continued)

Table 22.1 (continued)

Landscape	Relief type	Facies	Landform/code**	Soil type and phases*	Land capability class***	Land degradation (type and severity)	Examples of practices to address land degradation (restoration, rehabilitation, reclamation)
		Lagunary	Playas (Va124)	Association: Typic Salorthids [Typic Haplosalids] Natric Camborthids Sodic Haplocambids	7	Strongly saline-alkaline	

Noteworthy is that the dimensions of conservation and restoration actions in arid and semi-arid lands are very varied and are influenced by local socio-ecological and economic conditions. Designation of land capability class (following the USDA Land capability classification is hypothetical, based on expert knowledge of the area

a Soil types follow the USDA Soil Taxonomy (1992), in italics the soil name according to the Keys to Soil Taxonomy, Twelfth Edition (Soil Survey Staff. 2014)

b A common geopedologic legend was prepared for the three valleys of Cochabamba (Alto, Central and Sacaba), see Metternicht (1996). Some landforms and relief types are absent in the Valle Alto (e.g., Pi 3 – swales)

c USDA Land capability classes (Klingebiel and Montgomery 1961)

Fig. 22.4 Field photographs illustrating impacts of salinity and sodicity processes in Cochabamba's Valle Alto working lands. Salt deposits on topsoils in the dry season and when practices like irrigation by *'lameado'* occur. Low lying areas accumulate water that once evaporated leaves saline crusts. Lower right: Puffy soil crusts develop in low lying areas of lagunary flats and playas, where salts crystals of sodium sulphates dominate. Author: Metternicht

Sacaba and Central, irrigation (many times via traditional ways of digging small parallel channels between the plants or using the *'lameado'* practice of inundation of a parcel, see Fig. 22.4) and heavy pesticide use are necessary for most of the crops, particularly for the modern crops (Jokinen 2018; Metternicht 1996), exacerbating land degradation due to processes of soil salinization (Metternicht, 1996).

Most land degradation due to salinization (Table 22.1) occurs in depressions of lagunary origin (playas and lagunary flats). Chlorides and sulphates of sodium and calcium are the predominant anions, with the main salt types being thenardite, mirabilite, bloedite, halite, and sodium carbonate (hence most soils classify as Natric Camborthids and Salorthids) and present weak, puffy soil crusts that once broken by grazing herds leave exposed structureless soil materials that are easily blown away by wind action (ie. combined land degradation by wind erosion and salinization processes) (see Fig. 22.4).

The above description, Figs. 4.2, and 4.3 (Chap. 4 of this book) and Table 22.1 indicate that interventions to reduce and reverse current land degradation trends could be designed and targeted to specific geopedologic units. Within the LDN conceptual framework, this corresponds to the design phase of the counterbalancing mechanism (within a Decision Support System, guided by the principles of land degradation neutrality and integrated land use planning). In this way, land use planning and decisions on interventions for land restoration or land rehabilitation could be underpinned by the main characteristics of the soils (e.g., depth, diagnostic horizons, degree, and type of land degradation). Targets to achieve LDN in the area of the Valle Alto could be informed by geopedology, a socio-economic analysis and other context-relevant information (e.g., access to markets, finance, technology) as shown in Fig. 22.2.

Spatial analysis of geopedologic units within a GIS can help determine the location and percentage of specific soil types and land degradation extent, thus emulating the concept of land types. Table 22.1 shows that strongly saline and sodic land degradation is ongoing in the land type 'playas.' Therefore, land use and land management activities could be planned to follow the LDN hierarchy (see Fig. 22.2). The LDN response hierarchy encourages adoption of localised actions to avoid, reduce and /or reverse land degradation across each land type assessed. In this example, for reducing the rate of land degradation that leads to reduced productivity of agricultural areas, and reversing land degradation processes, well-planned restoration or rehabilitation of severely degraded productive land is needed (i.e., strongly saline and/or sodic areas, see Figs. 4.2 and 4.3 of this book).

Using a synthesis of research on restoration measures for saline and sodic soils (Stavi et al. 2021), Table 22.1 presents hypothetical options of interventions that could be actioned to reduce the rate of land degradation in the Cochabamba's Valle Alto. Noteworthy is that the dimensions of conservation and restoration actions in arid and semi-arid lands are very varied, influenced by local contexts. Reducing and reversing ongoing degrading processes through management practices of phytoremediation, agroforestry and silvopasture with native species, experimenting with more salt-tolerant crops and avoiding irrigation with brackish water. Soil remediation by adding gypsum (in areas affected by sodicity) and afforestation with native grasses, shrubs, and trees to lower the groundwater table (in areas with saline groundwater tables) are options that could be co-designed and planned through participatory processes involving soil experts and landowners. More substantial interventions could be planned and costed to rehabilitate or restore more severely degraded areas using, for instance, high rates of organic amendment, mulching, soil amendments (e.g., gypsum) or more expensive processes of leaching salts from the plant root zone in the case of highly saline soils.

Depending on the political and economic conditions of the socio-ecological system of the Valle Alto, longer-term LDN targets could be established that avoid further land degradation by, for instance, promoting low-impact agriculture and forestry practices that conserve the naturally low soil fertility of this semi-arid valley; minimise disturbance and erosion and avoid the perpetuation of management practices such as irrigation by *'lameado'* or inundation (see Fig. 22.4) that foster accumulation of salts on the soil surface. Some of these interventions require education on alternative land management practices and crops (e.g., replacing current low salt-tolerant crops with salt-tolerant grasses, legumes, woody shrubs, and trees; or intercropping crops by sowing into existing lucerne or alfalfa stands), and the availability of land/field advisors (agronomists). Others may require changes to current land-use regulations and strengthening of institutions such as the *Ministerio de Desarrollo Rural y Tierras* (Ministry of Land and Rural Development) or the *Ministerio de Medio Ambiente y Aguas* (Ministry of Environment and Water) through the establishment of systems for mapping and monitoring land degradation and to inform *Planes Territoriales de Desarrollo Integral* (Regional integrated land-duse plans).

22.4   Final Remarks

This chapter presents a novel approach to landscape analysis through the lens of geopedology to design context-based, reliable, and attainable interventions to achieve land degradation neutrality. A worked example using geopedologic information of the Cochabamba's Valle Alto shows how geopedologic units can inform the biophysical assessment of land condition and potential (e.g., land capability or land suitability), including information on land degradation types and severity. That information can feed into the counterbalancing mechanism (DSS) and be integrated with socio-economic data to design interventions towards LDN that are focused on managing salinity for sustainable agricultural production in salt-affected soils of drylands the Cochabamba's Valleys.

In addition to considering biophysical factors (e.g., eco-hydrological factors, land use history, geology, soils, geomorphology), interventions for advancing toward land degradation neutrality can be co-designed with local communities, be gender-responsive (Collantes et al. 2018), account for tenure regimes to anticipate potential barriers for the adoption of planned interventions. The maps of geopedologic units can also be the foundations of monitoring systems to track the success of planned interventions using the core LDN indicators of land cover change, carbon stocks (soil organic carbon) and land productivity, complemented with regional indicators relevant to the land degradation processes of the area (e.g., soil salinity and soil erosion).

References

Akinyemi FO, Ghazaryan G, Dubovyk O (2021) Assessing UN indicators of land degradation neutrality and proportion of degraded land for Botswana using remote sensing based national level metrics. Land Degrad Dev 32(1):158–172

Arnold R (2015) My dilemmas with soil-landscape relationships. In: Zinck JA, Metternicht G, Bocco G, del Valle E (eds) Geopedology: an integration of geomorphology and pedology for soil and landscape studies. Springer

Collantes V, Kloos K, Henry P, Mboya A, Mor T, Metternicht G (2018) Moving towards a twin-agenda: gender equality and land degradation neutrality. Environ Sci Pol 89:247–253

Cowie AL, Orr BJ, Sanchez VMC, Chasek P, Crossman ND, Erlewein A, Louwagie G, Maron M, Metternicht GI, Minelli S, Tengberg AE (2018) Land in balance: the scientific conceptual framework for land degradation neutrality. Environ Sci Pol 79:25–35

Gonzalez-Roglich M, Zvoleff A, Noon M, Liniger H, Fleiner R, Harari N, Garcia C (2019) Synergizing global tools to monitor progress towards land degradation neutrality: trends. Earth and the world overview of conservation approaches and technologies sustainable land management database. Environ Sci Pol 93:34–42

Jobbágy EG, Giménez R, Marchesini V, Diaz Y, Jayawickreme DH & Nosetto MD (2021) Salt Accumulation and Redistribution in the Dry Plains of Southern South America: Lessons from Land Use Changes. In: Saline and Alkaline Soils in Latin America, Springer, Cham, pp. 51–70

Jokinen JC (2018) Migration-related land use dynamics in increasingly hybrid peri-urban space: insights from two agricultural communities in Bolivia. Popul Environ 40(2):136–157

Klingebiel AA, Montgomery PH (1961) Land-capability classification (No. 210). Soil Conservation Service, US Department of Agriculture

Metternicht G, Zinck JA (1997) Spatial discrimination of salt-and sodium-affected soil surfaces. Int J Remote Sens 18(12):2571–2586

Metternicht G, Zinck JA (2016) Geomorphic landscape approach to mapping soil degradation and hazard prediction in semi-arid environments: salinization in the Cochabamba valleys, Bolivia. In: Geopedology, Springer, Cham, pp. 425–439

Metternicht GI, Zinck JA (1996) Modelling salinity-alkalinity classes for mapping salt-affected topsoils in the semiarid valleys of Cochabamba (Bolivia). ITC journal 2:125–135

Metternicht GI (1996) Detecting and monitoring land degradation features and processes in the Cochabamba valleys, Bolivia: a synergistic approach. PhD thesis, p. 399

Orr BJ, Cowie AL, Castillo Sanchez VM, Chasek P, Crossman ND, Erlewein A, Louwagie G, Maron M, Metternicht GI, Minelli S, Tengberg AE (2017) Scientific conceptual framework for land degradation neutrality. In: A report of the science-policy interface. United Nations Convention to Combat Desertification (UNCCD), Bonn, Germany, pp. 1–98

Soil Survey Staff (2014) Keys to soil taxonomy, 12th edn. USDA-Natural Resources Conservation Service, Washington, DC

Soil Survey Staff (1992) Keys to soil taxonomy, 5th edn. USDA-Natural Resources Conservation Service, Washington, DC

Stavi I, Thevs N, Priori S (2021) Soil salinity and sodicity in drylands: A review of causes, effects, monitoring, and restoration measures. Front Environ Sci 9:330

Vaysse K, Heuvelink GBM, Lagacherie P (2017) Spatial aggregation of soil property predictions in support of local land management. Soil Use Manag 33:299–310

Part V
Applications in Land Use Planning and Policy

Chapter 23
Geo-Pedological Approach for Land Use Planning-Case Studies from India

B. P. Bhaskar, V. Ramamurthy, and B. B. Mishra

Abstract Soil is the nexus tool, wherein soil formation needs synergy among water, air, parent material including plant while net ecosystem productivity needs soil itself together with water, energy inputs and biodiversity (living organisms). In India, there are noticeable land use conflicts in semi-arid and sub-humid regions due to uncertain market-driven current and potential land use dynamics. If one considers the current agricultural crisis in the case of soil degradation, climate change and depletion in soil health, systematic and innovative approaches in the land use planning process must be pursued. The pedological information is the first step in developing decisive and sustainable land use alternatives. Three case studies representing semi-arid and sub-humid agro-ecological regions were conducted in warm per-humid eco-region with high altitude soils in the Meghalaya Plateau, in the hot semi-arid eco-region of cotton growing Yavatmal district, Maharashtra and peanut growing Pulivendula tehsil, YSR Kadapa district in Andhra Pradesh and discussed the usefulness of the geo-pedological approach to land use exercises to strengthen local agriculture. The warm, humid eco-region of the Meghalaya Plateau has forest cover on granite hills, rice fields in valleys and potatoes on steep slopes. The cultivation history is sparse but shows low K reserves in 29% of the examined area. The hot semi-arid agro-eco-region of the Yavatmal district in Central India has typical basalt landscapes with hills and ridges (12.6% of the total area), plateaus (29.3%) in connection with isolated hills, mesas and buttes and steep slopes (17.7%), Pedi plains (28.8%) and plains (8.1%) and pedocomplex of vertisols and vertical intergrade. The suitability analysis showed that 25% of the area is suitable for *Kharif* cotton. The Pulivendula in southern India is characterized by rugged hills

B. P. Bhaskar · V. Ramamurthy (✉)
ICAR-National Bureau of Soil Survey and Land Use Planning, Regional Centre, Bangalore, Karnataka, India

B. B. Mishra
Commission 1.4 (Soil Classification), International Union of Soil Sciences; 2018-22, Vienna, Austria

with valleys consisting of granites, granite gneiss, cherty dolomites, quartzites and shales. 43% of the tehsil area is moderately suitable for peanuts, consisting of 13 soil associations in interhill basins and 10 soil associations in colluvio-alluvial sectors. Three case studies will certainly provide conceptual soil-landscape models to support regional land use issues and management decisions.

Keywords Agro-ecological regions · Soil geographic data base · Soil landscape units · Soil survey · Land use planning

23.1 Introduction

Soil is responsible in general for 99% of the global food production and in this perception, the importance of soil in food, energy and water nexus seems to be apparent. Such nexus interactions are about how one can use and manage resource systems, describing interdependencies, constraints and synergies. The geopedological method is a hierarchical structure wherein geomorphology and soils relationships are studied to generate high purity soil maps (Zinck 1989; Gholizadeh et al. 2001). In this approach, both remote sensing techniques (RS) and Geographic Information Systems (GIS) were used and derived prediction methods for making spatial planning and decision making of natural resource management for Agri - production (Rossiter 2000). The pedologic map has not only soil information but also associated with soil-landscape attributes that can be converted into various thematic maps in a GIS environment (Shit et al. 2016).

Thereby recommended further divisions of phases of landforms and phases of soil series in the Borujen region, Central Iran (Borujeni et al. 2010). It was reported in inaccessible mountainous areas of Doi Ang Khang, Thailand, the decision tree model is proved to be a very useful tool. The results of this study over conventionally derived soil map showed that 95% match at the soil order level, 84% match at suborder level, and 43% at subgroup level (Farshad et al. 2005). The emphasis should be given to applying decision trees, statistical validation, and an Artificial Neural Network (ANN) in soil mapping exercises (Farshad et al. 2013).

The geopedological approach was used for soil mapping from Qazvin province, Iran, and concluded that landscape plays a significant contribution in designing soil map units (Sarmadian et al. 2014). The use of Geomorphology is essential to understand the relationship and variability of soil properties and landforms (Zinck et al. 2016). The geopedological surveys were made using diversity and similarity indices in the Miayneh region of East Azerbaijan Province, Iran. The results showed an increase of Shannon's diversity index from soil order to soil family level with significant differences at the confidence level of 95%. It is further stated that the purity of geopedological units even at the detailed scale, can satisfy the managerial needs (Nazari et al. 2016).

The geo-pedological approach was employed and evaluated soil units for irrigated wheat using stories index and square root method in Qazvin plain, Iran. The results showed that the soil limitations for wheat are soil salinity, gypsum, coarse fragment, soil depth, soil organic carbon, soil texture, lime content, and climate (Mousavi et al. 2017).

It is possible to design more homogeneous soil units but not be able to define and represent soil variability. Later on, Toomanian and Boroujeni (2017) reviewed soil surveys in Iran and stated that the geo-pedologic approach is suitable for pedometrics to study the basic and applied aspects of pedology and proposed to use this method up to semi-detailed scales. Bhaskar et al. (2016) conducted land resource inventory of the tribal populated Seoni district of Madhya Pradesh using a geo-pedological approach and derived fourteen landform units. Out of which upper and lower denudational plateaus covered 37% of the total area. Thirty soil mapping units as series associations were used for land use and fertility capability interpretations. The soil-site suitability analysis showed that eleven land use zones were delineated and estimated that 19% of the area is suitable for sorghum - cotton systems with deficiency of phosphorus, potassium, and zinc while loamy to sandy soils in the southern zone have low water holding capacity, rapid infiltration and low available potassium and zinc status. The soil survey data sets are useful for the exchange of soil management information among closely related soils for agrotechnology transfer by soil analogy (Ramamurthy et al. 2020).

The challenges of dryland farming in three agro-ecoregions of India are to develop strategies to sustain and improve living conditions of debt-driven cotton farmers in Yavatmal district, tribal farming in Meghalaya, and intensive peanut systems of the Pulivendula tehsil. These agro-ecoregions showed the occurrence of prolonged dry spells with a mean of 38–41.36% in 36–44th week during critical cotton growth stages (branching and flowering stages) often coincides with a reduction in lint yield, as in the Yavatmal district of Maharashtra (Bhaskar et al. 2018), in Meghalaya plateau with tops and side hill slopes under pine forest vegetation have an erratic wet and dry cycles with a mean rainfall of 11,418 mm near Cherrapunji and 2014 mm near Shillong (Bhaskar et al. 2009) and similar rainfall variations under semiarid Pulivendula tehsil(mean of 564 mm). Realizing the importance of sustainable development goals of United Nations (United Nations General Assembly 2015), several scientific approaches in the appraisal of land resource evaluations for agricultural purposes were made and reported.

Meyfroidt et al. (2021) reviewed human knowledge about land use and sustainability and created 10 facts to focus on altering agroecosystem scenarios on a global scale. Furthermore, they found that adaptive governance improves land-use systems by facilitating communities' adaptation to shifting aims and unpredicted changes in the environment. The spatial analysis using GIS (Geographical information system) is a common practice in GIS spatial land suitability analysis and mapping (Feizizadeh and Blaschke 2013).

Several studies were made in India using GIS tools for deriving suitability maps for various crops in Seoni district Madhya Pradesh (Bhaskar et al. 2015) and for rice-based cropping systems in Majuli Island, Assam (Bhaskar 2019). The geopedological approach in land-use studies of three geographically different agroecoregions is evaluated with the objective of identifying biophysical limitations in deriving optimal land-use systems at a regional scale in the event of shifting seasonal weather patterns and their effect on dry land agriculture. In this paper, we analyze three agro-ecological regions where geopedological surveys were made in India and assess the potential of these soil geographic data sets for specific cropping systems of regional importance. An attempt was made how an integrated approach was applied to combine both land evaluation and land resource database to make substantial improvements in land use studies.

23.2 Materials and Methods

23.2.1 Details of Study Area

Three case studies in geographically different regions of India are presented in Fig. 23.1, where dry land farming is very critical and threatening to small farm holders experiencing severe loss of crop yields due to consecutive seasonal droughts. The agro-geographic characteristics of three regions are given as under:

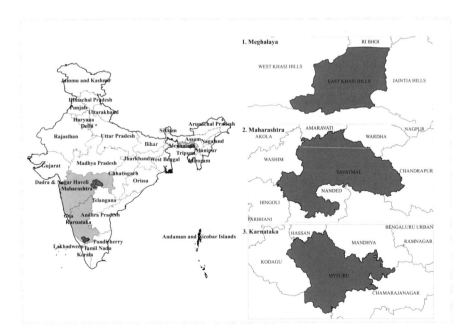

Fig. 23.1 Location map of case studies in India

23.2.1.1   Case Study 1: Meghalaya Plateau

The elevational gradient from Myllieum (around 25°29'N and 91°48′E) to Pynursala (around 25°19' N and 91°53′E) is distinct in the south of Shillong city. The study site had sandstone capping over granite with significant variations in vegetation, rainfall, and temperature. On shoulder slopes (Talus slopes), granite boulders are observed. The Myllieum site on the eastern aspect had an altitude of 1660–1760 m above mean sea level with pine vegetation (*Pinus insularis*) in association with oak, rhododendron, Mongolia, and other temperate forest trees. This area receives a mean annual rainfall of 2026 mm in 128 days and experiences a maximum temperature of 24.1 °C in July and a minimum temperature of 3.6 °C in January. The landforms are summits with pine vegetation (slopes of <8%), side-slopes under potato-radish cultivation (8–15%), and narrow valleys under rice–pea system (<3% slopes). The Pynursala had two land units viz., summits and convex plateau tops covered with grasses but have an elevation of 1280 m to 1631 m and receive a mean annual rainfall of 11,000 mm. The maximum temperature is 22.9 °C in June and the minimum is 7.6 °C in January. Myllieum is cooler by 4 °C than Pynursala in the winter. The soil moisture regime is udic in the hill slopes and in the valleys with thermic soil temperature regime.

23.2.1.2   Case Study 2: Yavatmal District, Maharashtra State, India

Yavatmal district in the eastern Vidarbha region of Maharashtra state (19⁰26' to 20042'N latitude and 77⁰18' to 79⁰98′E longitude) district comes under Deccan Plateau, Hot Semi-Arid Eco-Region (6) of Western Maharashtra plateau, and hot moist semi-arid eco-sub region (Mandal et al. 1999). The mean annual rainfall is 1125 mm in Wani valley to 889 mm in Darwah and 1099.5 mm in Yavatmal with an increasing trend from West to East. Yavatmal district has per capita ecologically productive land (ha) of about 0.16 ha with adequate soil moisture from June to August. The review of land use statistics in Yavatmal shows that this district has 4.41% of the total state area (13,582 sq. km) with 43% of rural families under below poverty line. The district has 34.4% of the area under food crops and 52.19% under fiber crops. The land holding of 2–5 ha constitutes 40.12% of the entire district followed by 28.26% of 5–10 ha holding. The total cultivated land is 8.84 lakh ha with a double-cropped area of 9475 ha and a cropping intensity of 101%. The low cropping intensity indicates that rainfed farming is prevalent in the entire region. This region is a hotspot for critical analysis of land use activity where the economic dependence of farmers is solely on cotton wherein, more than 50% of the total net sown area is under a single crop over years.

23.2.1.3 Case Study 3: Pulivendala

Pulivendula lies between14°16'to 14°44' N and 77°56' to 78°31'E covering 127,463 ha. The agro climate is characterized as hot, semiarid with mean annual rainfall of 564 mm and 43 rainy days. The LGP varied from 90–105 days for Pulivendula and Vemula, 105–120 days for Lingala and Tondur and 120–135 days for Simhadripuram and Vempalli mandals. Physiographically the area is characterized by rugged hills with valleys, pediments and the geology being granites, granite gneisses, cherty dolomites, quartzites and shales (Nagaraja Rao et al. 1987). The study area has the Papaghni and Chitravati group of rocks of Cuddapah Super Groups. Papaghni group includes a) Gulcheru formation comprising quartzite, arkose and conglomerate; b) Vempalli formation comprising dolomites, chert, mudstone, quartzite, basic flows and intrusive. The Chitravathi group includes (a) Pulivendula formation comprising quartzite with conglomerate, (b) Tadipatri formation consisting of shales, dolomite and quartzite; (c) Gandikota formation comprising quartzite and shale.

23.2.2 Geo-Pedological Survey

The 'geo-pedologic approach to soil survey' (Zinck 1989) was used for visual interpretations of Indian remote sensing satellite imageries of three case studies as given under.

- IRS-P6 (Indian Remote-Sensing Satellite-P6)–LISS-data in conjunction with Survey of India toposheets on 1:50000 scale for a study area Yavatmal district Maharashtra.
- IRS-LISS-III data for Meghalaya plateau.
- IRS-P6-LISS-IV data on 1:25000 scale for Pulivendula of YSR Kadapa district.

The above said imageries were used to delineate landform polygons for further studying soil profiles and understanding soil-landform relations. This activity was performed to build strong integration of geomorphology and pedology following the four of six hierarchical levels (Zinck 1989) as given: (1) landscape as the first entry of legend, (2) relief/molding, (3) lithology/facies, and (4) landform. It was a stepwise approach of applying geo-pedological principles (Farshad and Rossiter 2008) as given under:

(i) Interpretation of satellite imagery, (ii) delineation of major geoforms (morphology plus geology), (iii) sketching of cross-sections to verify geoforms, (iv) recognition of patterns of units along each cross-section, (v) delineation of recognized units with filled in legend columns, (vi) construction of coherent legend (composition of legend) and made final legend by matching, (vii) photo interpretation of entire area under study using constructed legend (knowledge of geomorphology is

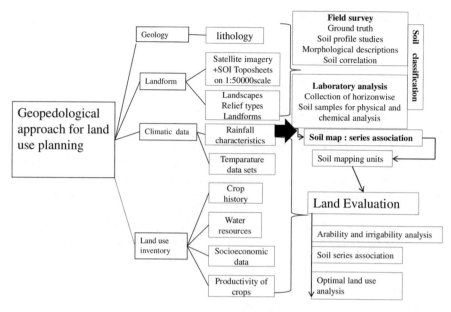

Fig. 23.2 Flow chart of methodology used in the study. (Source: Bhaskar et al. 2014a, b)

needed), (viii) soil mapping unit type and components as per USDA nomenclature (Soil Survey Division Manual 2017) and finally but optional in adding specific soil information viz., depth and parent material.

The case studies in different agro-ecological regions of India were used to demonstrate how the geo-pedologic approach in soil survey forms the backbone of appraisal of land resources for land use planning (Fig. 23.2).

23.2.3   Land Evaluation

The appraisal of land resource information for land suitability of locally grown crops was made. It is recognized as a range of interpretative classifications: qualitative, quantitative, and of current and potential suitability. The structure of the suitability classification is subdivided into categories (FAO 1983). Further, these land units were evaluated for their suitability for potato in Meghalaya (Kamau et al. 2015), for cotton in Yavatmal (Bhaskar et al. 2014a, b) and Groundnut in Pulivendula tehsil (Rajendra Hedge et al. 2019).

23.3   Results and Discussion

23.3.1   Case Study-1: Meghalaya Plateau

23.3.1.1   Landforms and Soils

High-altitude rock types are exposed in the Shillong plateau between the Bangladesh border and Mawphlong, including Archean gneiss and schists in the center and Precambrian quartzites and phyllites in the east, later intruded by young granites and ultrabasic suites. The ancient peneplain surface of the plateau is still preserved with varying degrees of denudational cycles in the central and northern parts. The drainage pattern is a spectacular feature with magnificent gorges scooped out by rivers in the southern Khasi hills. The Indian Remote sensing satellite imagery of March 2001 was used to delineate landforms with visual interpretation. The rolling topography with isolated mounds has a pinkish tone interspersed with a bluish tinge where potato cultivation is dominant near Mylliem. The whitish-blue tinge with dark red tone is visually seen in the hillside slopes covered with exposed rock outcrops and scrubs lands. The whitish chocolate brown tone indicates a barren rocky surface with moderate pine vegetation.

The major landforms of the study area are summits, side slopes, tabletops, and mounds. It is estimated that summits cover 24% of the study area followed by tabletops (51%), side slopes (12.1%), and mounds (7.8%). Four major landforms are delineated after field verification but made 24 photographic units in *jhum* cultivated highlands of the Meghalaya plateau (Table 23.1, Fig. 23.3). A soil map was generated with nine units and 12 series identified (Fig. 23.3). There are four landforms on the study site, namely hills /ridges (39.85% of the study area), tabletops (29.72%), mounds (23.96%), and side slopes (6.43%).

There are three soil mapping units in the summits, of which Mullieum-Umjilang-Sonidan covers 15.43% of the total area. There are rock outcrops associated with the Pynursala series and Umthlew series (17.69% of the area). About 6.43% of the area is covered by the side slopes, which have very shallow and extremely stony soil associated with Laitlyngkot-Mylliem-Umthlew. A great deal of aesthetic value can be found in these landscapes because natural vegetation has a great impact on soil. Due to the dramatic shift in natural pine vegetation on hill slopes and the presence of *jhum* cultivation on tabletops, pedogenesis and vegetation pathways are difficult to determine in the study area (Bhaskar et al. 2004a, b).

Spring and summer tend to be dry in the northern parts of the study site, as the seasonal rainfall decreases from south to north. During January and February, the minimum temperature across the region is 4 °C, with the northern parts receiving five times as much precipitation (11,000 mm) than the southern parts (2026 mm, Nair and Chamuah 1988). In hilly terrains of the southern part of the state, the soils are deeper with clay content ranging from 20 to 38%, while in the northern part, the soils are shallow with clay content ranging from 15 to 20% (Bhaskar et al. 2004a, b).

Table 23.1 Description and extent of soil mapping units in relation to land forms in high lands of Meghalaya

Landform/Soil mapping unit	Description of soil mapping unit	Area		Suitability class for potato
		ha	(%)	
I. Hill summits (5295.43 ha, 39.85%)				
6. Myllieum-Umjliang-Sonidan	Deep, loam, moderately steeply sloping, severely eroded and moderately stony Inceptic Hapludults is associated with very shallow, sandy loam, gently sloping and slightly eroded Typic Udifluvents and moderately deep, loamy sand, very gently sloping, slightly eroded valleys with Typic Udorthents.	2050.6	15.43	N
7. Umthlew-Dansi	Moderately deep, sandy loam, steeply sloping, slightly eroded and moderately stony Ruptic-ultic Dystrudepts is associated with shallow, loam, moderately sloping, severely eroded and slightly stony lithic Udorthents	1612.78	12.14	S3
9. Cherrapunji –Pynursala-Weioli	Shallow, sandy loam, gently sloping, severely eroded, strongly stony lithic Dystrudepts associated with deep, sandy loam, moderately sloping, severely eroded, moderately stony, plateau top soils of Ruptic-Ultic Dystrudepts and deep, sandy loam, moderately steeply sloping, moderately eroded and slightly stony Typic Dystrudepts	1632.05	12.28	S3
II. Side slopes (854.77 ha, 6.43%)				
3. Laitlyngkot-Myllieum-Umthlew	Very shallow, loam, moderately steeply sloping, slightly eroded and moderately stony Typic Udorthents are associated with deep, loam, moderately steeply sloping, severely eroded and moderately stony Inceptic Hapludults and moderately deep, sandy loam, steeply sloping, slightly eroded and moderately stony Ruptic-Ultic Dystrudepts	854.77	6.43	N
III. Table tops (3947.61 ha, 29.72%)				
5. Pynursala-Urskew-rock outcrops	Deep, sandy loam, moderately sloping,severely eroded, moderately stony, plateau top soils of Ruptic-Ultic Dystrudepts is associated with deep, loam, moderately steeply sloping, slightly eroded and moderately stony humic Pachic Dystrudepts and rock outcrops	1597.97	12.03	N
4. Pynursala-Umthlew-rock outcrops	Deep, sandy loam, moderately sloping,severely eroded, moderately stony, plateau top soils of Ruptic-Ultic Dystrudepts moderately deep, sandy loam, steeply sloping, slightly eroded and moderately stony Ruptic-Ultic Dystrudepts and rock outcrops	2349.64	17.69	N

(continued)

Table 23.1 (continued)

Landform/Soil mapping unit	Description of soil mapping unit	Area		Suitability class for potato
		ha	(%)	
IV.Mounds (3184.35 ha, 23.96%)				
8.Umthlew-Thambali-rockoucrops	Moderately deep, sandy loam, steeply sloping, slightly eroded and moderately stony soils of Ruptic-ultic Dystrudepts is associated with deep, loamy, steeply sloping, severely eroded and moderately stony Ruptic Ultic Dystrudepts and rock outcrops (>25%)	658.51	4.95	S3
2.Umjliang-Sonidan - Laitlyngkot	Very shallow, sandy loam, gently sloping slightly eroded soils of Typic Udifluvents is associated with moderately deep, loamy sand, very gently sloping slightly eroded in valleys of Typic Udorthents and very shallow, loam, moderately steeply sloping, slightly eroded and moderately stony Typic Udorthents	1224.91	9.22	S2
1.Mawphlong-Myllieum-Umjliang	Deep, loam, steeply sloping, severely eroded, moderately stony Ruptic Ultic Dystrudepts associated with deep, loam, moderately steeply sloping,severely eroded and moderately stony, Inceptic Hapludults and very shallow, sandy loam, gently sloping slightly eroded Typic Udifluvents	1300.93	9.79	S2
Total		13281.9	100	

The development of redness in sub soils is attributed to weathering of mafic minerals in parent rock and its subsequent alteration to hematite. The morphology of soils suggests that summit soils have minimal illuviation of clay (Inceptic Hapludults) with a pedoturbation that disrupts the formation of cutans. The occurrence of cambic horizons with an insufficient increase in clay to form the argillic horizons. These soils on mounds and tabletops have dark-colored and thicker umbric horizons (Soil Survey Staff 2014).

The soils on side slopes are strongly related to the geomorphic position and often have colluvial and residual soils with lithological discontinuities. The application of Graham and Boul (1990), pedogenic pathways for soils in residuum (summits) and in colluvium (convex plateau tops) are useful in identifying, differentiating, and understanding the transition of cambic to argillic or kandic horizon as needed to place them in the sub-group of Inceptisols and Ultisols. Even though the argillic layer is 40 cm deep, these soils were classified as Inceptic Hapludults at the subgroup level. For soils on the eroded side-slopes of the Shillong plateau, any diagnostic horizons were noted. As a result, this soil is categorized as Typic Udorthents. Typic Udifluvents soils are found in the valleys. Ruptic-Ultic Dystrudepts and Typic Kanhapludults are often seen on convex plateau summits in the research area. Humic Pachic Dystrudepts are summit soils with umbric epipedon and a base saturation of less than 35%.

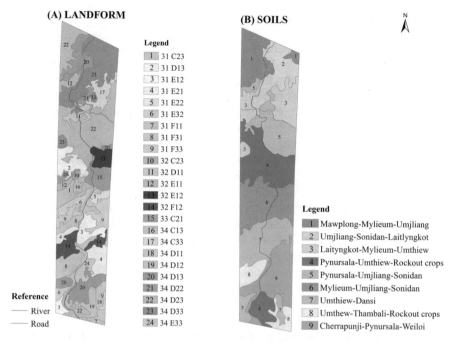

Fig. 23.3 a. Landforms and **b**. Soils in highlands of Meghalaya plateau. (Source: Bhaskar et al. 2004a, b)

After field correlation, a soil map with 9 soil units was created as a series association from twelve soil series reported at the study area (Table 23.1, Fig. 23.3 A and B.). In the high-altitude areas of Meghalaya, the hill summits constitute 5295.43 ha (or 39.85% of the total measured area of 13281.9 ha) (Table 23.1). This land unit is made up of three soil mapping units: 6. Myllieum-Umjliang-Sonidan (2050.6 ha, 15.43%), 7. Umthlew-dansi (1612.78 ha, 12.14%), and 9. Cherrapunji-Pynursala-Weioli (2050.6 ha, 15.43%). (1632.05 ha, 12.28%). The tabletops cover 3947.61 ha (29.72%) with exposed rock outcrops and soil series association of Pynursala with Urskew (5) and Umthlew (4). The mounds encompass 3184.35 hectares and are predominantly found in the benchmark site's southern areas (23.96%). This unit has three soil mapping units: 8. Umthlew-Thambali-rock outcrops (658.51 ha, 4.95%), 2. Umjliang-Sonidan-Laitlyngkot (1224.91 ha, 9.22%), and 1. Mawphlong-Myllieum-Umjliang (1224.91 ha, 9.22%) (1300.93 ha, 9.79%). Only 6.43% of the measured area is covered by side slopes, which have a very shallow to moderately deep soil association of 3. Laitlyngkot-Myllieum-Umthlew.

23.3.1.2 Soil Site Suitability for Potato

The nine soil mapping units of high altitude areas of Meghalaya are evaluated for their suitability to potato using the criteria of Kamau et al. (2015) and Trigso et al. (2020). In the Meghalaya plateau, Potato is grown in 17,685 ha with a productivity of 9.1 t/ha. Three varieties of potato viz., lah saw, lah lupon, and lah taret are grown during February/March to June/July and Autumn season of August to November/December. As the late blight is common in the region, the blight-resistant Kufri Giriraj is recommended for this region (Uma et al. 2006). The study area is part of North-Eastern Hills (Purvachal), a warm to the hot per humid ecosystem (17.1, Sehgal 1995), and has the ideal climate for potato cultivation. This area receives a mean annual rainfall of 8445 ± 7558 mm with a coefficient of variation of 89% over a period of 2004–2016 (13 years, Sharma and Sinha 2016). This area experiences peak rainfall of more than 20,000 mm during 2015 and 2016 wherein June and August recorded more than 8000 mm in 2015 but July in 2016. In the study area, the mean rainfall during the potato growing period is 5939 mm during summer (February to July) and 10,416 mm during autumn (August to December) with a mean air temperature of 16.98 °C to 17.10 °C (summer) and 16.78 °C to 18.78 °C (autumn).

The climate is assessed as highly suitable for climate as per Kamau et al. (2015) and He et al. (2017). The land suitability evaluation for potatoes in the Peruvian region of Amazonas indicated that the contribution of climatological is 30.14% while edaphological is 29. 16% with 25.72% of topographical influence (Trigoso et al. 2020). Among climatological factors, precipitation followed by temperature is critical in high-altitude areas (Kamau et al. 2015 and FAO 2002). The seasonal precipitation could be very essential in high altitude regions in which potato is grown in strongly sloping hill slopes in-furrow –ridge system with low water holding capacity and exchangeable bases of moderately to strongly acid soils prefer extreme water erosion with a high degree of dissected landscapes as appeared in the Cherrapunji area of Meghalaya plateau (Bhaskar et al. 2004a, b). The results confirmed that the area not suitable for potato is 6852.98 ha while moderately suitable covers 2525. 84 ha (19.01%) and a marginally suitable area of 3903.34 ha (29.39%). Among soil limitations, these soils are rich in organic carbon however acidic in soil pH and gravelly subsoils.

23.3.2 Case Study-2-Yavatmal District

23.3.2.1 Soil mapping Units in Relation Landforms

The area under study comprises 80% basaltic landforms and the remainder sandstone/limestone/shale along the coal seams in the eastern part of the Wani valley. A total of three soil mapping units (SMU) that account for more than 5% of the total area are recorded: Lakhi-rock outcrop-Moho (3.12%) on mid-plateaus,

Dhanki-Lakhi-Dhanora-Pandhurna (11.5%) on isolated hills, and Lakhi-Borgaon-Arni-Dhanki (17.28%) on lower plains. Table 3.1 contains detailed descriptions of each SMU. Hill and ridge cover 12.6% of the western Pus valley in Darwah and Yavatmal basaltic plateaus with dominant soil associations of Met (very shallow), Gahuli (very shallow), Lakhi (very shallow), Waghari (moderately deep), except Sindola on limestone (moderately shallow). Upper plateaus (4.77% of total area) have four soil mapping units: Kalamb-Lakhi-Hirdi (45.14%), Hirdi-Met (45.1%), Gahuli-Borgaon-Dhanki (22.6%), Apti-Waghari-Wani (35.44%).

Three soil mapping units cover 17.35% of the total area, namely Lakhi-rockoutcrop-Moho (3, 12.3%), Dhanora-Lakhi-Selodi-Arni (13, 4.37%), and Borgaon-Ralegaon-Lakhi (43, 0.63%). The lower plateaus are mostly found below 320 m above mean sea level and are mainly found in the southeast and west parts of the district. The unit is made up of four SMUs (22, 1, 25 & 39) wherein the very deep Dhanki series is associated with deep Penganga, moderately shallow Borgaon, and very deep Pandhurna (SMU-22). The isolated hills cover 6.81% of the area and are dotted with 3 SMUs: Dhanki-Lakhi-Dhanora-Pandhurna. Buttes and Mesas are broad tabletops that have cliff-like sides (mesas), while buttes have flat tops with vertical sides. The middle plateaus cover 17.35% of the total area with 3 soil mapping units viz., Lakhi-roc-Moho (3, 12.3%), Dhanora-Lakhi-Selodi -Arni (13, 4.37%), and Borgaon-Ralegaon-Lakhi (43, 0.63%). Lower plateaus are mostly confined to elevations below 320 m above mean sea level and concentrated in south eastern and western parts of the district. This unit covers 7.19% with 4 SMU's (22, 1, 25 & 39) wherein very deep Dhanki series associated with deep Penganga, moderately shallow Borgaon and very deep Pandhurna (SMU-22) covers 4.42%. The isolated hills cover 6.81% of the area with 3 SMU's viz., Dhanki-Lakhi-Dhanora-Pandhurna (11, 5.01%). Both Butte and Mesas are broad, flat surfaces with cliff-like sides (mesas), and small, flat surfaces with vertical sides (butte). Dhanki-Waghari-Gahuli (8, 3.97%) and Penganga-Waghari-Saykheda-Lakhi (32, 2.35%) constitute 6.32% of this unit. 4.63% of the area is covered by escarpments, with soil associations including Chikalgaon-Wani-Apti (36, 3.03%), Pandharkawada-Chanda, and Maregaon-Wanjari (38, 1.60%). It is severely eroded and covered with teak and rock exposures on steep slopes and a high percentage of stone at the surface.

There are 8 SMUs in the upper pediplains (Fig. 23.4), namely 12, 34, 48, 14, 47, 26, 33, and 10. The Lakhi-Waghari-Arni (10) series association covers 4.54% of the area while the Wanoda-Kharbhi and Arni (48) series association covers 2.54%. Pediplains in Pusad and Umarkhed tehsils are mostly double/triple cropped and irrigated by canals. This unit covers 15.82% of the area with five SMUs (17, 29, 31, 15, and 27). It is assessed that 7.28% of the area is under shallow Lakhi - moderately shallow Borgaon - moderately deep Arni - very deep Dhanki series and 0.67% of the area under very deep Wanodi-shallow Borgaon-deep Penganga-moderately deep Kolambi series. The other three SMUs cover 2.5% of the area.

The pediplains have 3.38% with 2 SMU's viz., 31 and 37 with few rock outcrops happening near Penganga in southwestern parts of the locale. This unit is generally utilized for cotton and chilies with a couple of patches of paddy cultivation. This unit inclusion is around 1.16% with 3 SMUs having a series relationship of

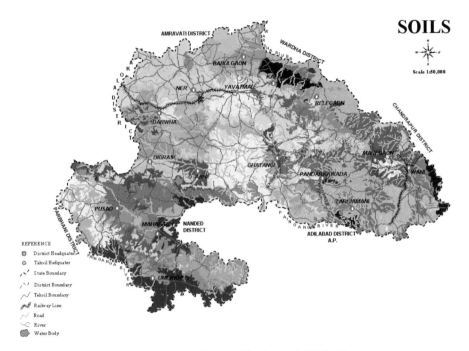

Fig. 23.4 Soil map of Yavatmal district. (Source: Bhaskar et al. 2014a, b)

moderately shallow Borgaon-moderately deep Kolambi and Waghari (16, 0.2%), reasonably shallow Sindola in relationship with very shallow Pandharkawada (41, 0.45%), in relationship with moderately shallow Sindola (44, 0.51). This unit is for the most part utilized for cotton and chilies in Wani valley (Table 23.2).

23.3.2.2 Soil Site Suitability Analysis

The key stage in this process was to match land attributes to crop requirements, resulting in a compatibility grade for each land feature (Fig. 23.4). The initial analysis was completed for each characteristic, giving the information required to estimate land suitability. Cotton suitability is assessed on 47.14% of total arable land, which covers 29 soil mapping units. Twelve soil mapping units (9, 30, 40, 42, 33, 34, 12, 16, 14, 35, 36, and 37) are rated as moderately suitable for a total area of 294,947 ha (19.8%), while 17 soil mapping units (1, 4, 10, 11, 29, 32, 45, 47, 48, 28, 39, 13, 15, 17, 21, 27, and 43) are rated as marginally suitable for a total area of 407,563 ha (Table 23.2). (27.33%). The findings revealed that 16.24% of arable soil mapping units (1, 4, 9, 10, 11, 28, 33, 34, 39, 13, 15, 35, 36, 37, and 22,42,278 ha) are appropriate for sugar cane, 13.23% for redgram (197,289 ha), and 33% for groundnut. The area suitable for cotton, sorghum, and wheat systems is estimated to be 256,373 ha (17.22%) and includes 8 soil mapping units (12, 13,15, 16, 17, 21, 27 and 43).

Table 23.2 Area and extent of soil mapping units in relation to landform and their suitability to crops in Yavatmal district (Source: Bhaskar et al. 2014a, b)

S. No.	Soil mapping unit	Cotton	Maize	Sorghum	Wheat	Soybean	Sugarcane	Area ha	%
1	Lk-Ka-Ar -Dk	S3		S2	S2	S3	S3	10,426	0.69
2	Mt-Mo			S3	S3			33,926	2.35
3	Lk-roc-Mo			S2	S2			177,994	13.55
4	Dk - Ar -Lk-AV	S3		S2	S2	S3	S3	18,802	1.28
5	Gh-Lk							33,557	2.3
6	Dh-Ka-Gh							12,283	0.84
7	Mt-Lk-Ka-Dk			S3	S3	S3		17,121	0.97
8	Dk-Wg-Gh							55,141	3.73
9	Lo-Ar-Av	S2		S2	S2	S3	S2	12,279	0.87
10	Lk-Wg-Ar	S3		S3	S3	S3	S3	61,198	4.17
11	Dk-LK-dh-Pd	S3		S2	S2	S3	S3	68,055	4.62
12	Dh-Ar-lo	S2		S2	S2		S3	10,049	0.71
13	Dh- LK-Sd -Ar	S3		S2	S2		S3	59,792	4.08
14	Sd-Wg-dh	S2	S3	S2	S2		S3	10,450	0.71
15	Dh-Wg-Lk	S3		S2	S2		S3	34,086	2.32
16	Bo-Ka-Wg	S2		S1	S1		S2	1908	0.13
17	Lk-Bo-Ar-Dk	S3		S2	S2		S3	99,983	5.28
18	Wg-Mt-dh			S3	S3			13,989	0.95
19	Sy-Mt-Jm-Wg			S3	S3	S3		13,034	0.89
20	Lk-Bo-Gh-Ar	S3						35,150	2.39
21	Wg-Lk-Ka-Pd	S3		S2	S2		S3	51,195	3.46
22	Dk-Pg-Bo-Pd			S1	S1		S2	60,738	4.13
23	Gh-Bo-Dk							31,204	2.12
24	Bo-Pd-Mo			S2	S2			6664	0.45
25	Gh-Wg-kw							21,272	1.45
26	Wg-kw-Ka-Gh							16,618	1.13
27	Wd-Bo-Pg-Ka	S3		S1	S1		S2	9183	0.62
28	Pg-Ka-Wg	S3	S3	S1	S1	S3	S2	18,546	1.26
29	Wg-Lk-Sy -Pd	S3		S2	S2	S3	S3	33,994	2.29
30	Sy -lo-Ar	S2		S2	S2	S3	S2	9802	0.65
31	Ch-Pk-ma			S3	S3			40,120	2.73
32	Pg-Wg- Sy-Lk	S3		S2	S2	S3	S3	32,581	2.21
33	Sy -Wg	S1	S3	S1	S1	S3	S1	13,719	0.92
34	Wn-Ch	S2	S3	S2	S2	S3	S2	14,915	1.02
35	Ai-Wg - Wn	S2	S3	S2	S2		S2	6092	0.41
36	Ci- Wn -Ai	S2	S3	S2	S2		S2	41,505	2.82
37	Wn-lo -ma	S2	S3	S2	S2		S2	4426	0.3

(continued)

Table 23.2 (continued)

S. No.	Soil mapping unit	Cotton	Maize	Sorghum	Wheat	Soybean	Sugarcane	Area ha	%
38	Pk -Ch-ma-Wj			S3	S3			20,232	1.52
39	Ko-kb-granite	S3	S3	S2	S2	S3	S2	6283	0.42
40	Sn-Ch-ROC	S2		S2	S2	S3	S2	1878	0.13
41	Sn-PK			S3	S3	S3		4375	0.29
42	Ng-Wg	S2		S2	S2	S3	S2	8842	0.61
43	Bo-Ra-Lk	S3		S2	S2		S3	9162	0.62
44	Pk-Sn			S3	S3	S3		4924	0.33
45	Km-Lk-hi	S3		S3	S3	S3	S3	19,393	1.32
46	Hi-Mt			S3	S3	S3		9103	0.62
47	Ra-kb-Lk	S3		S2	S2	S3	S3	18,705	1.27
48	Wd-kb-Ar	S3		S2	S2	S3	S2	28,668	1.95

In central Yavatmal, the soil mapping units 5 (Wg-Lk-Ka-Pd) and 16 (Bo-Ka-Wg) account for 9.36%. Thirteen soil mapping units (1, 4, 9, 10, 11, 29, 30, 32, 40, 42, 45, 47, and 48) encompass 382071.08 hectares and are appropriate for cotton + sorghum + wheat + soybean. In the Wani valley, four soil mapping units (14, 35, 36, 37) are suited for cotton, maize, sorghum, and wheat, covering 63,325 ha (4.24%), whereas four soil mapping units (28, 33, 34, and 39) are suitable for cotton, maize, sorghum, wheat, and soybean, covering 53,993 ha (3.62%, Fig. 23.5). The sorghum + wheat system has seven suitable soil mapping units: 2, 3, 18, 22, 24, 31, and 38 (Table 23.2), which encompass 382,071 ha (25.04% of total arable land). The five soil mapping units 7, 19, 41, 44, and 46, which cover 46,322 hectares, are deemed suitable for sorghum + wheat + soybean systems (3.16%).

Soil depth, soil salinity, calcium carbonate, drainage, and texture are the main limiting factors of shrink-swell soils (Bhaskar 2015). He believes that 40% of the area under very deep to moderately deep vertic subgroups and vertisols requires contour bunding with vegetative barriers, while the remaining 60% of shallow soil associations (Lithic subgroups) on steep lands can be used for low-input grain production (mostly Jowar) during the delayed monsoon.

23.3.3   Case Study-1: Pulivendula

23.3.3.1   Landforms and Soils

In the study region, the common rock types composed Pulivendula quartzites, Vempalle limestones, Gulucheru quartzite/Arkose with conglomerate. These landforms are distinctly different from one another with respect to their relief amplitude, drainage morphometry, network evolution, and position in toposequence. The different landforms identified in Pulivendula are; Elongated ridges (Cuesta forms),

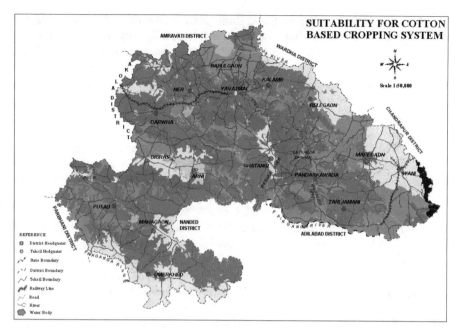

Fig. 23.5 Soil-site suitability for cotton based cropping systems in Yavatmal district. (Source: Bhaskar et al. 2004a, b)

Dissected Hills/summits (550–260 m), Highly dissected plateau remnants (550–240 m), Isolated hills/mounds (430–260 m), Inter-hill basins (360–320 m), Undulating Upper sectors (360–300 m), Gently sloping middle sectors (300–240 m), Colluvial lower-sectors (240–200 m above) and Narrow valley floors (220–180 m above). Elongated ridges have a distinct protuberance and spread all along the southern portion of Pulivendula, bordering Ananthapuramu district. This ridge is in fact a Cuesta landform, with dip and strike slopes, formed and slightly tilted due to the deformation of sedimentary strata during the regional metamorphism. This land unit covers 8442.90 ha (6.62% of total area) and is mostly kept as barren land.

Rugged hills/peaks are located in the southern part of Pulivendula Mandal. The dominant geomorphic process is therefore the encirclement of several sandstone beds with several branches of rivulets at the fingertip. This land area covers 2.32% of the total area (32962.15 ha). These hills are only slightly covered with forest with xerophytic vegetation. More rugged remains of the plateau occur in the northern part of Pulivendula, which mainly covers the mandals of Simhadripuram and Tondur. Many upland structures have Gandikota quartzite and Tadpatri dolomite/limestone, and there are many rugged and exposed hills with "thresholds" reaching the ends of the table with short slopes. This land unit covers 16.18% of the total area (20263.65 ha) and is used mainly for jowar, cotton, and groundnuts.

Isolated hills/mounds are scattered throughout Pulivendula. This unit occupies 6.87% of the total area (8756.61 ha). This unit is usually barren with a thin forest cover. The interhill basins represent an elongated narrow strip of lowland lying between high hills and ridges. Dominantly distributed to the southwest and southeast of Pulivendula Mandal. This unit is often used for planting bananas under drip irrigation. This unit occupies 5815.34 hectares (4.56% of the total area). The undulating upper sectors represent the center of the upper part of the relief and the lower part of the valley represented by the tributary of the river. This unit occupies 10,205.83 ha (8.01%) but is used for banana-based systems. The gently sloping middle sector occurs in the central and northwestern parts of the Mandals of Pulivendula, Lingala, and Simhadripuram. This unit occupies 35535.58 ha (27.88%). This unit is predominant and used for rainfed cotton/jowar.

The colluvial plains are found predominantly in the central and eastern parts of the region and cover Vemula, Vempalle, Tondur, and some parts of the Pulivendula Mandal. This unit covers 27457.79 hectares (21.54%). This unit ranks second after gently sloping soils under rainfed farming. The narrow valley floors have numerous narrow drainage networks of the tributaries, which originate in the upper southern and northern reaches and join the mainstream which flows from northwest to southeast in the north-central part of the Mandal. This same watercourse is a tributary of the main Papagni River, which flows touching the southeast corner of the taluk. This unit covers only 3.48% of 4440.84 hectares. This unit is mainly used for rice cultivation under canal irrigation.

Twenty five series of soils were identified, and soil maps were prepared with 43 mapping units (Fig. 23.6). In general, the hilly and ridge quartzite soils have eight mapping units associated mainly with rocky outcrops, very shallow, somewhat excessively drained, moderately alkaline, sandy to clayey soils, and cover 54,812 ha (42.6% of the area). (Table 23.3).

Most of the area belongs to shale formations covers 73,797 hectares (57.4% of the total area). Under the morphologies of the shales, in the inter-hill basin, there are seven mapping units with six soil associations and one soil association (4.79% of the total area). These soils are shallow and well-drained with subsoil layers from strongly alkaline gravel clay to clayey gravel subsoil layers. The gently sloping lands cover 39,092 hectares (30.4% of the area) with 12 soil map units.

The soils of Vemula (20–1.667 ha; 1.2%) are moderately shallow red limestone soils, well-drained with strongly alkaline clayey loam and strongly alkaline gravelly clayey subsoil with the clayey horizon. The associated map units are Velpula Soils (21–1326 ha, 1.0%), Parnapalle in Lingala Mandal (22–446 ha, 0.3%), Velpula-Vemula association in Tondur Mandal (28–712 ha, 0, 5%). This mapping unit is associated with very deep, black, moderately well-drained, calcareous, strongly to moderately alkaline soils with the potential for swelling and shrinkage. The soils of the colluvial and alluvial plains cover 28,542 ha (22.19% of the total area) with an association of the Tondur-Pernapadu series (30), Pernapadu-Gondipalle association (33), Goturu-Gondipalle (36) and Agadir- Gondipalle Pernapadu Association (41).

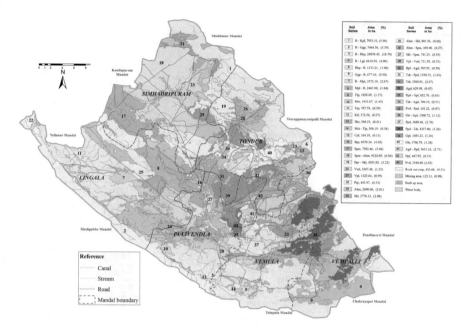

Fig. 23.6 Soil map of Pulivendula tehsil, Kadapa district. (Source: Bhaskar et al. 2021)

Table 23.3 Area and extent of soil mapping units in relation to landforms and soil loss. (Source: Bhaskar et al. 2021)

Land form	Soil mapping unit	Area ha (hectares)	(%) Percent
Hills and ridges	1.Rockoutcrops (R)-Kanampalli(Kpl)	7953	6.18
	2. Rockoutcrops®- -Ganganapalle(Ggp)	7464	5.80
	3. Rockoutcrops®-Rachanakuntapalle(Rkp)	24,939	19.39
	4. Rockoutcrops ®--Lingala(Lgl)	6410	4.98
	5. Rachanakuntapalle(Rkp) – Rockoutcrops®	1333	1.04
	6.Ganganapalle(Ggp)-Rockoutcrops®	677	0.53
	7.Rockoutcrops®-Mupendranpalle(Mpl)	3572	2.78
	8. Mupendranpalle(Mpl)- Rockoutcrops®	2464	1.92

(continued)

Table 23.3 (continued)

Land form	Soil mapping unit	Area	
		ha (hectares)	(%) Percent
Interhill basin	9.Tallalapalle(Tlp)	1829	1.42
	10.Murarichintla(Mct)	1934	1.50
	11. Tatireddipalle(Trp)	788	0.61
	12. Kottalu(Ktl)	372	0.29
	13. Santhakovur(Skv)	548	0.43
	14. .Murarichintala(Mct)- Tallapalle(TlP)	508	0.39
	15. Cherlapalle(Cpl)	184	0.14
	16. Balapanur(Bpr)	6559	5.10
	17. Simhadripuram(Spm)	7583	5.90
	18. Simhadripuram(Spm)- Agraharam(Ahm)	9125	7.10
	19. Balapanur(Bpr)-Sunkesula(Skl)	4294	3.34
	20. Vemula(Vml)	1667	1.30
	21. Velpula(Vpl)	1326	1.03
	22. Parnapalle(Prp)	446	0.35
	23. Agraharam(Ahm)	2690	2.09
	24. Sunkesula(Skl)	2778	2.16
	25. Agraharam(Ahm)- Sunkesula(Skl)	802	0.62
	26. Agraharam(Ahm)- Simhadripuram(Spm)	369	0.29
	27. Sunkesula(Skl)- Simhadripuram(Spm)	741	0.58
	28.. Velpula(Vpl)-. Vemula(Vml)	712	0.55
Colluvial-alluvial pediplains	29. Bhadrampalle(Bpl)- Agadur(Agd)	788	0.61
	30.Tondut(Tdr)-Pernapadu(Ppd)	1351	1.05
	31.Tondur(Tdr)	3568	2.77
	32. Agadur(Agd)	633	0.49
	33.Pernapadu(Ppd)-Gondipalle(Gpl)	853	0.66
	34. Tondur(Tdr)- Agadur(Agd)	709	0.55
	35.Pulivendula(Pvd)-Pernapadu(Ppd)	101	0.08
	36.Goturu(Gtr)-Gondipalle(Gpl)	1501	1.17
	37. Pernapadu(Ppd)	3689	2.87
	38. Pernapadu(Ppd)- Tondur(Tdr)	4358	3.39
	39. Gondipalle(Gpl)	1683	1.31
	40. Goturu(Gtr)	1707	1.33
	41. Agadur(Agd)- Pernapadu(Ppd)	3613	2.81
	42. Bhadrampalle(Bpl)-	448	0.35
	43. Pulivendula(Pvd)	3540	2.75
	Total	128,609	100

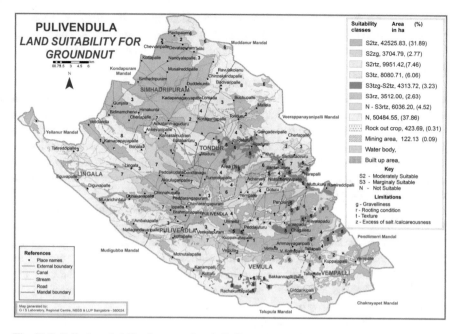

Fig. 23.7 Soil-site suitability for groundnut in Pulivendula tehsil (Source: Bhaskar et al. 2021)

23.3.3.2   Soil Site Suitability Analysis

The suitability evaluation for groundnut shows that only 23 soil mapping units are moderately suitable (Fig. 23.7) with the limitations of rooting depth (r), topography (t), and salt content (z). The moderately suitable soil mapping units cover 56,224 ha (43% of total area) consisting of 13 soil consociations (31,501 ha, 24.49% of total area) and 10 soil associations (24,723 ha, 19.22% of total area). Even though, the suitability findings show 43% of the total area has good potential for groundnut and is extensively cultivated in Vempalle (6894 ha, 27.39% of cultivated area) and Vemula (3613 ha, 17.29% of cultivated area). The arability-irrigability analysis shows that there is a scope to expand the area of 42% (56,092 ha) under groundnut under rainfed but 23% under irrigation.

23.4   Conclusions and Recommendations

Three case studies from potato-growing Meghalaya plateau, cotton-growing Yavatmal region, and peanut-growing Pullivendula tehsil of Kadapa district were used to explore comprehensive and well-structured geo-pedological surveys. To demarcate landforms and generate soil data for developing soil-landform correlations at study locations, the soil resource information is generated step by step.

These data sets were utilized to discover biophysical restrictions that limit locally grown/preferred land use options that satisfy agricultural stakeholders. The systematic surveys produce a scientifically sound soil resource database that may be used to make crop-land allocation decisions.

The soils of the Meghalaya plateau are moderately to severely acidic, with low base saturation and exchangeable Al concentrations, as well as coarse fragments in the subsoil, according to geopedological study. Working with machines at high elevations is hampered by the need to manage steep slopes in order to grow potatoes. Cotton productivity is restricted in the Yavatmal district due to shrink-swell soils with high clay, strong alkaline with ample calcium carbonate, and little organic carbon. Furthermore, seasonal droughts at critical times of cotton production are a determining factor of regional yields.

Droughts have resulted in a significant decline in the area under groundnut in Pulivendula, Kadapa district, in recent years (2018–2020). These areas feature both red and black soils developed under a variety of geological formations with salinity, alkalinity, coarse fragments, and steep hill slopes with exposed rock outcrops as biophysical limits. The study reveals biophysical restrictions for achieving the attainable yield at the local level and gives a first step assessment of soil-site suitability for locally grown crops.

The suitability analysis revealed that only 19% of the total surveyed area in the Meghalaya plateau is moderately suitable for potatoes, while 19.8% of the area in the Yavatmal district is moderately suitable for cotton, and 42% of the area in Pulivendula tehsil is moderately suitable for peanut under rainfed conditions.

Three case examples presented here bolster the two of the ten hard truths of land use that point to sustainability (Meyfroidt et al. 2021). The following are the facts 2: Land use dynamics are complicated, with sudden but difficult-to-predict changes. Example: The government officially approved the production of Bt cotton in 2002. There has been a significant shift in the area, as well as increased agricultural expenses. After considering the cost of production and protection, net returns in MECH-162 (Bt cotton) were 216.4$/ha, compared to 165.8$/ha in non-Bt MECH-162 and 140.1$/ha in traditional cotton (Bambawale et al. 2004). There are 8.84 lakh hectares of cultivated land, including 9475 ha of double cropped land and a cropping intensity of 101%. Due to the rapidly rising expense of Bt cotton farming, the region's cotton growers' economic difficulties have recently forced farmers to commit suicide (Bhaskar et al. 2014a, b). And, in Pulivendula tehsil, fact 3 refers to permanent alterations (for example, salt-affected soils in valley bottoms of the Chitravthi basin, Bhaskar et al. 2021) and route dependency of traditional peanut agriculture.

Management strategies like bench terracing on steep slopes with vegetative barriers, primarily broom grass in case of potato growing Meghalaya region, contour bunding in the case of cotton growing in Yavatmal district, and application of gypsum/zinc in addition to contour bunding/levelling of land in peanut growing Pulivendula tehsil, Kadapa district enhances the productivity of crops.

The study showed that understanding soil suitability limitations is crucial before transferring agro-technologies aimed at improving agricultural yields and restoring ground-level strategies to combat climate change and extreme weather events. The data sets also aid in determining the heterogeneity of soil units in order to allocate crop-specific resources and interventions appropriately. The last element of cropland management plans differed depending on regional landscapes for effective and efficient agro technology in high-drought-risk areas with low-cost interventions.

References

Bambawale OM, Singh A, Sharma OP, Bhosle BB, Lavekar R, Dhandapani A, Kanwar V, Tanwar RK, Rathod KS, Patange NR, Pawar VM (2004) Performance of Bt cotton (MECH-162) under integrated pest management in farmers' participatory field trail in Nanded district, Central India. Curr Sci 86:1628–1623

Bhaskar BP (2015) Landscape planning for Agri development at regional scale: an example from cotton growing Yavatmal district, Maharashtra. India J Agric Environ Int Dev 109(2):235–269

Bhaskar BP (2019) Evaluating potentials of Riverine flood plain soils for management of Rice (Oryza sativa L) in Majuli River Island, Assam. India Int J ecol environ sci 45(2):145–156

Bhaskar BP, Baruah U, Vadivelu S (2004a) Refectance libraries for the development of soil sensors for periodic assessment of Soil Resources.NATP-MM-III-2. Technical Report No.810. National Bureau of Soil Survey and Land Use Planning, Nagpur

Bhaskar BP, Mishra JP, Baruah D, Vadivelu S, Sen TK, Butte PS, Dutta DP (2004b) Soils on Jhum cultivated hill slopes of Narang-Kongripara watershed in Meghalaya. J Indian Soc Soil Sci 52:125–133

Bhaskar BP, Saxena RK, Vadivelu S, Baruah U, Dipak S, Raja P, Butte PS (2009) Intricacy in classification of pine-growing soils in Shillong Plateau, Meghalaya, India. Soil Survey Horizons 50(1):11–16

Bhaskar BP, Dipak Sarkar, Mandal C, Bobade SV, Gaikwad MS, Gaikwad SS (2014a) Reconnaissance soil survey of Yavatmal district, Maharashtra, India. NBSS Publication. No.1059, National Bureau of Soil Survey and Land Use Planning, Nagpur. pp.208

Bhaskar BP, Dipak S, Bobade SV, Gaikwad SS, Anantwar SG (2014b) Land evaluation for irrigation in cotton growing Yavatmal District, Maharashtra. Int J Res Agric 1(2):128–136

Bhaskar BP, Bobade SV, Gaikwad SS, Dipak S, Anantwar SG, Tapas B (2015) Soil informatics for agricultural land suitability assessment in Seoni district, Madhya Pradesh, India. Indian J Agric Res 49(4):315–320

Bhaskar BP, Anantwar SG, Gaikwad SS, Bobade SV (2016) Land resource assessment for agricultural development in Seoni district (Madhya Pradesh). J Appl Nat Sci 8(2):750–759

Bhaskar BP, Satyavati PLA, Singh SK, Anantwar SG (2018) The analysis of standardized precipitation index (SPI) in cotton growing Yavatmal district, Maharashtra. Adv Plants Agric Res 8(6):505–510

Bhaskar BP, Rajendra Hegde, Srinivas S, Sunil Maske, Rameshkumar SC, Ramamurthy V (2021) Visual signs of biophysical indicators for assessing the status of degradation in drylands of Pulivendula tehsil, Kadapa district, Andhra Pradesh, Technical report No.1151. NBSS&LUP, Nagpur

Borujeni E, Mohammad J, Salehi MH, Toomanian N, Poch RM (2010) Assessing geopedological soil mapping approach by statistical and geostatistical methods: a case study in the Borujen region. Central Iran Catena 82(1):1–14

FAO (1983) Guidelines: Land evaluation for rainfed agriculture. FAO Soils Bulletin. No. 52, Rome

FAO (2002) Protected cultivation in a Mediterranean climate. FAO, Roma, ISBN 9253027193

Farshad A, Rossiter DG (2008) Computer-assisted geopedology for predictive soil mapping. In: 3rd Global workshop on digital soil mapping, Utah State University, Logan

Farshad A, Shrestha DP, Moonjun R, Suchinai A (2005) An attempt to apply digital terrain modeling to soil mapping, with special attention to sloping areas; advances and limitations. Wrap up seminar of the LDD-ITC research project, Hua Hin

Farshad A, Shrestha DP, Moonjun R (2013) Do the emerging methods of digital soil mapping have anything to learn from the Geopedologic approach to soil mapping and vice versa. In: Shahid SA, Abdelfattah M, Faisal T (eds) Developments in soil classification, land use planning and policy implications. Innovative Thinking of Soil Inventory for Land Use Planning and Management of Land Resources, pp 109–131

Feizizadeh B, Blaschke T (2013) Land suitability analysis for Tabriz County, Iran: a multi-criteria evaluation approach using GIS. J Environ Plan Manag 56(1):1–23

Gholizadeh A, Moemeni A, Bahrami H, Banaei H (2001) The efficiency of geopedologic approach and the soil survey method adopted in Iran in increasing the purity of soil map units and in decreasing survey's costs. Iran J Soil Water Res 15(Special Issue):13–28

Graham RC, Boul SW (1990) Soil geomorphic relations on the blue ridge front. II. Soil characteristics and pedogenesis. Soil Sci Soc Am J 54:1367–1377

He Y, Zhou Y, Cai W, Wang Z, Duan D, Luo S, Chen J (2017) Using a process-oriented methodology to precisely evaluate temperature suitability for potato growth in China using GIS. J Integr Agric 16:1520–1529

Hedge R, Bhaskar BP, Niranjana KV, Rameshkumar SC, Ramamurthy V, Srinivas S, Singh SK (2019) Land evaluation for groundnut(Arachis hypogaea L.) production in Pulivendula tehsil, Kadapa district, Andhra Pradesh, India. Legume Res Int J 42(3):326–333

Kamau SW, Kuria D, Gachari MK (2015) Crop-land suitability analysis using GIS and remote sensing in Nyandarua County, Kenya. J Environ Earth Sci 5:121–132

Mandal C, Mandal DK, Srinivas CV, Sehgal J, Velayutham M (1999) Soil climatic database for crop planning in India. National Bureau of Soil Survey and Land Use Planning (ICAR), p 994

Meyfroidt P,de Bremond A, Ryan CM, Archer E, Aspinall R, et al.2021 Ten facts about land use systems for sustainability.PNAS.119(7):1–2/e2109217118

Mousavi SR, Sarmadian F, Alijani Z, Taati A (2017) Land suitability evaluation for irrigating wheat by geopedological approach and geographic information system: a case study of Qazvin plain, Iran. Eurasian J Soil Sci 6(3):275–284

Nagaraja Rao BK, Ramalingaswamy G, Rajurkar ST, Ravindra Babu B (1987) Stratigraphy, structure and evolution of the Cuddapah basin. In: Purana basins of Peninsular India. Geological Society of India, Memoir 6:33–86

Nair KM, Chamuah GS (1988) Characteristics and classification of some pine forest soils of Meghalaya. J Indian Soc Soil Sci 36:142–145

Nazari N, Mah-moodi S, Masihabadi MH (2016) Employing diversity and similarity indices to evaluate Geopedological soil mapping in Miyaneh, East Azerbaijan Province, Iran. Geology 6:1221–1239

Ramamurthy V, Challa O, Naidu LGK, Anil Kumar KS, Singh SK, Mamatha D, Ranjitha K, Bipin B, Mishra (2020) Land evaluation and land use planning. In: Mishra BB (ed) The soils of India. Springer publication, pp 191–214

Rossiter DG (2000) Lecture notes and reference methodology for soil resource inventories 2nd revised version. Institute for Areospace Survey and Earth Sciences (ITC), Enschede, p 132

Sarmadian F, Mousavi SR, Iqbal M, Keshavarzi A, Sadeghnejad M (2014) Investigation the variation of soil mapping units using geopedological approach. Acta Adv Agric Sci 2(5):1–9

Sehgal J (1995) Soil resource mapping and agro-ecological zoning for land use planning. India perspective Indian farming, January 1995, pp.15–21

Sharma VC, Sinha SK (2016) State at a glance: Meghalaya. ENVIS Centre Himalayan Ecol 1(9):1–138

Shit PK, Bhunia GS, Maiti R (2016) Spatial analysis of soil properties using GIS based geostatistics models. Model. Earth Syst Environ 2:107. https://doi.org/10.1007/s40808-016-0160-4

Soil Survey Division Staff (2017).Soil Survey Manual.Agri.Handb.No.18.USDA. pp.1–573

Soil Survey Staff (2014) Keys to soil taxonomy, 12th edn. USDA-NRCS, Washington, DC

Toomanian N, Boroujeni E (2017) Outcomes of applying a geopedologic approach to soil survey in Iran. Desert 22(2):239–247

Trigoso DI, Salas LR, Rojas BNB, Silva LJO, Fernandez DG, Manuel O, Huatangari LQ, Terrones MRE, Castillo EB, Barrena GAB (2020) Land suitability analysis for potato crop in the Jucusbamba and Tincas microwatersheds (Amazonas, NW Peru): AHP and RS–GIS approach. Agronomy 10:1898

Uma S, Shantanu K, Pandey NK, Pandey SK (2006) Traditional Potato cultivation practices in Meghalaya. Central Potato Rresearch Institute, Simla. Technical bulletin No.72. pp 1–35

United Nations General Assembly (2015) Transforming our world: the 2030 Agenda for Sustainable Development,70(1). Resolution adopted by the General Assembly on 25 September 2015. Seventieth session Agenda items 15 and 116. A/RES/70/1, pp.1–35

Zinck JA (1989) Physiographyvand Soils. Lecture-notes for soil students. Soil Science Division. Soil Survey Courses Subject matter: K6 ITC, Enschede, the Netherlands

Zinck JA, Metternicht G, Valle HFD, Bocco G (2016) Synthesis and Conclusions. In: Geopedology: An Integration of Geomorphology and Pedology for Soil and Landscape Studies. https://doi.org/10.1007/978-3-319-19159-1

Chapter 24
Geopedologic Information, Foundation for Soil Conservation: Land Evaluation, Land Use Allocation and Associated Conservation Practices

O. S. Rodríguez Parisca

Abstract A conceptual framework is presented to illustrate the role of geopedologic information in a continuous planning process which include land resources inventory within a particular institutional context, land evaluation, land allocation and tactical operations on specific geopedologic units and their subdivisions. Explanations and examples are given on the components of each phase as well as their interactions. Special attention is given to land use conflicts for strategic land use allocation through their resolution based on consensus between different interest groups, in particular for functional conflicts. This approach strengthens the sustainability of major land use allocation decisions. Land degradation is highlighted as a central process that must be tackled in order to mitigate its harmful consequences, preventing its occurrence or restoring already degraded land by means of allocating and selecting specific land utilization types and associated technologies which can secure land sustainability. Land degradation processes that differ from undesirable land use conversions (functional conflicts) are related with intensity of land use conflicts. Other sustainable land management components must be considered as biodiversity, water supply and quality and climate change, among others. In the final phase, tactical management decisions are supported by the capture of high resolution remote and ground data, to optimize land use on specific geopedologic units and their subdivisions. This phase can be associated with precision agriculture, but it has to be also linked to market conditions and food supply chain. It is concluded that geopedologic units characteristics are a very strong determinant on land degradation processes, land evaluation and land use allocation and conservation technologies associated, as they have an almost permanent condition on the landscape and represent a set of information to be used for long periods of time,

O. S. R. Parisca (✉)
Profesor Titular (retirado) Facultad de Agronomía, Cátedra Conservación de Suelos y Agua,
Universidad Central de Venezuela, Maracay, Venezuela
e-mail: osrp1958@gmail.com

© The Author(s), under exclusive license to Springer Nature
Switzerland AG 2023
J. A. Zinck et al. (eds.), *Geopedology*,
https://doi.org/10.1007/978-3-031-20667-2_24

being obtained from the non-random systematic variations of the soil mantle. On the contrary, many other management decisions are subject to dynamic processes that vary in a very small space and time arrangement and must be monitored in a high frequency cycle or even real time or derive from soil variations distributed in short distances in the landscape and needs intensive data acquisition.

Keywords Land use conflicts · Land evaluation · Soil conservation · Land degradation · Sustainable land management · Sustainable land use planning · Land management · Land use planning

24.1   Introduction

Land use planning is a continuous process that requires large amounts of information in different formats at different levels and involve many actors. As such, is a complex and demanding activity that looks to adjust or prevent an unsatisfactory condition and transform it in a more gratifying and positive one. Land use planning looks to avoid or mitigate land use conflicts and optimize resource uses, through the diagnosis of land use problems, delivering viable options and monitoring the implementation of proven alternatives (Dent 1988).

The concept has evolved and land use planning according to FAO-UNEP (2000) is a systematic and iterative procedure carried out in order to create an enabling environment for sustainable development of land resources which meets people's needs and demands. It assesses the physical, socio-economic, institutional and legal potentials and constraints with respect to an optimal and sustainable use of land resources and empowers people to make decisions about how to allocate those resources. The role of land users is enhanced within this new vision.

Baker (2014) mention that traditional definitions tend to emphasize planning as a systematic process with the aim of regulating the future use and development of land. He describes land-use planning as one of a number of terms (e.g., town planning, town and country planning, urban planning, city planning, environmental planning) commonly used to cover public policy intervention related to the ordering and regulation of land use in an efficient, sustainable, and ethical way.

FAO (2022) defines land-use planning as "the systematic assessment of land and water potential, alternatives for land use and economic and social conditions in order to select and adopt the best land use options" within its inclusive and sustainable territories and landscapes platform.

Land use conflicts have their origin in human decisions about land use. When they occur, restrict future options about land use allocation and their outcomes are generally negative, leading to irreversible land use changes or land degradation processes, that need to be prevented by voluntary actions of society trough planning and policy making. Three main land use conflicts can be identified (Rodríguez 1995; Rodríguez and Zinck 2016):

(a) Functional conflicts: A multiple use of a piece of land is rarely attainable. When different land uses are demanded on the same land unit which have high aptitude for many uses or functions conflicts arise between different interest groups (stakeholders) demanding different uses.

(b) Intensity conflicts: When there is an overutilization of land resources the current use exceeds the carrying capacity of the land, land degradation being the direct outcome of the conflict. Conversely, if the land is suitable for intensive uses but used below its optimum capacity, underutilization takes place.

(c) Generational conflicts: Generational conflicts can be of functional or intensity nature and compromise the needs satisfaction of land resources for future generations. Present decisions will compromise the possibility of future adjustments, considering the physical nature and finite supply of the land resources and the irreversibility of some kind of land use conversions or land degradation processes.

Land degradation refers to the many processes that drive the decline or loss in biodiversity, ecosystem functions or services, and includes the degradation of all terrestrial ecosystems including associated aquatic ecosystems that are impacted by land degradation (FAO 2011; IPBES 2018). To avoid, mitigate or recover from a land degradation situation, sustainable land management should be practiced that according to IPBES (2018) it considers the use of land resources, including soils, water, animals and plants for the production of goods to meet changing human needs while ensuring the long-term productive potential of these resources and the maintenance of their environmental functions. Rodríguez (2018), remarks that soil and water conservation technologies and approaches must include, as a basic principle, their contribution to develop sustainable land use systems.

In order to analyze the huge amount of information in a land use planning process, a multipurpose land evaluation procedure must be performed, being the FAO framework for land evaluation (FAO 1976) a flexible and simple tool to do it, without diminishing other approaches. In parallel, the identification of land use conflicts according to their nature (intensity or functional) is accomplished. From the beginning and throughout all the land use planning process a Geographic Information System integrates the data capture, storage, analysis and display of results (Bold 2019).

Many land uses can be physically suitable to specific land units but did not match the environmental requirements or socio-economic conditions that determine their selection within a land use planning process. In the case of environmental requirements, preselected land utilization types can be updated incorporating soil and water conservation technologies to avoid land degradation processes and contribute to a sustainable land management (Rodríguez 2018). Other criteria must be considered to achieve a sustainable land management like carbon neutrality, reduction of water footprint, waste management, promotion of biodiversity, social equity, among others. If this is the case, an optimized land utilization type is obtained. In an open market system, many decisions will be driven by economic forces, but it remains on society to regulate and optimize resource uses in a sustainable way.

The International Society of Precision Agriculture (https://www.ispag.org/) adopted the following definition of precision agriculture in 2019: 'Precision agriculture is a management strategy that gathers, processes and analyzes temporal, spatial and individual data and combines it with other information to support management decisions according to estimated variability for improved resource use efficiency, productivity, quality, profitability and sustainability of agricultural production.' This means, on a high resolution data acquisition management system, land parcels can be subdivided in specific subdivisions an inputs and operations can be applied or performed according to spatial and temporal variability optimizing resource uses.

The geopedologic approach proposed by Zinck (2016a), emphasizes the genetic relationship between soils and geoforms, being the geomorphic context an essential factor of soil formation and distribution, and as Zinck remarks, the geomorphic context is emerging as the structuring element of a variety of legends, including legends of taxonomic maps, interpretive maps and land-use planning maps, among others.

The characteristics of geopedologic units are a very strong determinant on land degradation processes as they influence some of the specific degradation types according to the morphogenetic environment (depositional, erosional, structural) and are associated with land use and management activities that can accelerate natural processes. Geopedologic units also determine land evaluation and land use allocation and related conservation technologies, as they have an almost permanent condition on the landscape and represent a set of information to be used for long periods of time, being obtained from the non-random systematic variations of the soil mantle. As it can be seen, geopedologic information is a common thread along all the land use planning process. The geolandscape is a hierarchically structured and organized domain (Zinck 2016b).

On the contrary, many other management decisions are subject to dynamic processes that vary in a very small space and time arrangement and must be monitored in a high frequency cycle or even real time or derive from soil variations distributed in short distances in the landscape and needs intensive data acquisition.

The purpose of this research is to present an integrated sustainable land use planning and management system illustrated by a conceptual framework, using geopedologic information together with other environmental and socio-economic data as an input for land evaluation, land use conflict analysis, strategic land allocation, and the selection and optimization of land utilization types considering soil and water conservation technologies and other sustainable land management components, controlled and enabled by tactical land management decisions on specific subdivisions of particular geopedologic units supported by high resolution data capture.

24.2 Methodology

In Fig. 24.1, a conceptual framework is presented to illustrate the role of geopedologic information in a continuous land use planning and management process which includes land resources inventory within a particular institutional context, land evaluation and conflict analysis, land allocation and tactical operations on specific geopedologic units and their subdivisions. It is inspired and based on previous schemes (Rodríguez 1995, 2018; Rodríguez and Zinck 1998, 2016).

The first step or phase corresponds to land resources inventory, and it comprises several data streams corresponding to three blocks: institutional context, land use and land cover and geopedologic units; being time and space the main sources of variability in each data set. Many of the data are geographically related and georeferenced and are managed within a GIS system. They are described as follow:

- Institutional context, which represent the social, cultural, economic, technological, legal and political institutions and have a control role upon land use decisions.
- Land use and land cover represent the main function land is performing (production, protection, place or play, Ruswurm 1975). This is the biophysical information subject to more changes in space and time, so it is monitored more frequently. Di Gregorio and Jansen, referred by (FAO-UNEP 2000) describe Land Use as characterized by the arrangements, activities and inputs by people to produce, change or maintain a certain land cover type. They highlight that land use defined in this way establishes a direct link between land cover and the actions of people in their environment. Land Cover is the observed (bio)physical cover on the earth's surface.
- Geopedologic units, although highly variable in space represent the most stable data through time, so other information about natural and cultural features can be georeferenced to them carrying out the role of containers.

Following the inventory stage, integration of the data provided by the three sources mentioned above is needed for land use conflict analysis and land evaluation, and also to determine the occurrence of land degradation processes, usually derived from intensity land use conflicts, where land was or is undergoing an overutilization of its resources and current land use is exceeding its carrying capacity. Land degradation can be typified by defining the type, factors, processes, intensity, extent and externalities.

Land use policies are established, and a policy weighting process is adopted for building alternative scenarios and for land allocation discussion plans. After consultation with the concerned user groups, a final land use allocation plan is proposed for implementation. Within the framework presented in Fig. 24.1, selected land utilization types (SLUTs) for specific land units are chosen according to the strategic major land uses proposed in the land use allocation plan, preventing future land use conflicts and avoiding the loss of agricultural land and nature conservation space as a consequence of irreversible land use changes.

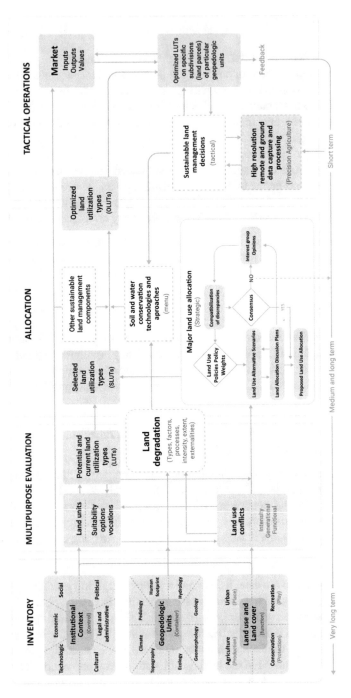

Fig. 24.1 Conceptual framework for an integrated sustainable land use planning and management system

The previously selected land utilization types (SLUTs) have to be submitted to an improvement process in order to obtain optimized land utilization types (OLUTs) considering soil and water conservation technologies and other sustainable land management components. Optimization can be assimilated with the new term "*Land Degradation Neutrality*" defined by the Parties to the Convention as: A state whereby the amount and quality of land resources, necessary to support ecosystem functions and services and enhance food security, remains stable or increases within specified temporal and spatial scales and ecosystems (Orr et al. 2017). If previous or potential land degradation processes are detected, soil and water conservation practices must be associated to the current or proposed land use, in order to restore or avoid productivity losses or eliminate externalities consequences. This means, for example, that beyond cultural management practices usually performed in a normal agricultural operation like fertilization or irrigation, additional supporting technologies are needed to control land degradation. As mentioned before, other sustainable land management criteria must be considered to optimize land utilization types. Many soils and water conservation practices also contribute to those criteria and there is a spectrum of available technologies to be selected, usually in a combined way (FAO 2014; Rodríguez 2018; WOCAT 2007).

Two study cases are given to illustrate the implementation of the framework presented, the first one on the strategic process to allocate major land uses which group selected land utilization types (SLUTs) and give more attention to solve functional land use conflicts. The second focuses more on the tactical side, on resolving intensity conflicts, and determining optimized land utilization types (OLUTs). Both study cases are located in the same study area.

Finally, a brief discussion on precision agriculture based on high resolution data capture is given to describe how sustainable land management decisions are performed on specific subdivisions of particular geopedologic units.

Even though the second study case emphasizes agricultural land uses, other studies can follow a similar approach adapted to other land uses like nature conservation, recreation, etc. The framework presented can be used as a whole integrated land use planning system, or it can be used focusing on particular targets, either strategic or tactical. The advantage of using geopedologic units as original containers for mapping delineations is that information on land use systems (land utilization types + land units) can be correlated and used at different hierarchical levels for technology transfer, research and innovation, market studies, land restoration projects, land use conflicts resolution, production planning, databases configuration and regional or local land use planning, among others.

24.3   Study Area Location and Characteristics

The study area was selected by the author to carry out previous research and is described in more detail in Rodríguez (1995). The following description is extracted from Rodríguez and Zinck (2016). The western part of Caracas, capital of Venezuela, presents a special case of land use conflicts in periurban areas. Intensive market-oriented agricultural activities (vegetables, fruit trees and flowers) on steep land, the protection of a cloud forest environment with a high biodiversity and important in water catchment. Urban expansion through residential subdivisions and intensive recreation activities competes for the same locations because these are equally attractive for all kinds of land use. Also legally protected areas such as national parks and periurban buffer zones are threatened by the uncontrolled expansion of urban activities.

The study area was selected on the basis of its location in the highly dynamic periurban zone of Caracas. It comprises the upper and middle catchment of the Petaquire river and smaller parts of other catchments and cover a large variety of conflictive land uses and protected areas. The latter benefit from a special administrative regime, which theoretically regulates the permitted uses. Protected areas include the Macarao National Park, the Pico Codazzi natural monument and the Caracas protected zone. Tracks of land located within the Caracas metropolitan area and other areas not specifically regulated were also included to show, in conjunction with protected areas, a broad spectrum of land uses and land use conflicts. Steep sloping areas specializing in horticultural crops requiring a temperate climate and coffee generate important farm production and complementary activities within the region, but their protection has not been sufficiently secured.

The expansion of these particular agricultural systems to new areas is limited by the scarcity of soil and water resources and by the legal status of nature preservation imposed in large areas in these unique climatic conditions. Conflicts between the main types of land use (e.g., agriculture and nature conservation) and within specific land uses (e.g., horticulture and intensive recreation resorts and second homes) derive from the competition for the same tracts of land. The location of the study area is shown in Fig. 24.2. The study area covers approximately 17,000 ha.

Other considerations when selecting the study area were: (i) its ecological importance because of the presence of a cloud forest with a high biological diversity and its role in supporting and regulating natural cycles, e.g., the water cycle; (ii) its agricultural importance because of the production of unusual crops, adapted to highland conditions relatively scarce elsewhere in the region; and (iii) its touristic and recreational attractiveness, being in the proximity of La Colonia Tovar, a major tourist and specialized agricultural center. Most of the basic information on land resources was produced by Rodríguez (1995) during his doctoral research program at ITC-The Netherlands.

24.4 Study Case 1

The first study case illustrates how to implement part of the conceptual framework shown in Fig. 24.1, emphasizing the resolution of functional land use conflicts between major land uses. It represents a strategic approach to prevent conflicts between competitive land uses and reduce undesirable and irreversible land use

Fig. 24.2 Study area location

conversions. More details on this approach are discussed in Rodríguez (1995) and Rodríguez and Zinck (2016). Geopedologic units are used to delineate land units integrating geoform and soil information which represent the more stable dataset through time, so that other natural and cultural information subject to temporal changes can be georeferenced to them. For this study case all the study area previously described was covered.

As many of the data are geographically georeferenced, a geographic information system is a central tool for storing and manipulating spatial and other attribute data. Three data bases were implemented: (i) A spatial data base storing different maps within ILWIS system (ITC 1993), (ii) a soil observation database using a relational database management system (Arenas[1] 1993) and (iii) a tabular database to manage farming systems data.

Available software was integrated such as ALES system (Rossiter and Van Wambeke 1993), designed to implement the FAO framework for land evaluation (FAO 1976); and the LUPIS system, an integrated package for land use allocation (IVE 1992). Terrain land units were sorted and classified according to Zinck's geoform classification system (Zinck 1988). Geopedologic units and their individual delineations (polygons) were used as the basic map units throughout the land use planning process. Land cover and land use were introduced as subdivisions of the geopedologic units when necessary for the detection and analysis of land use conflicts. In general, the biophysical and socio-economic data were strongly correlated with the geopedologic context. The geopedologic map units were obtained by a sequential combination of thematic maps including life zones, relief, landform, slope, and current erosion. Physical and economic suitabilities of the map units were determined for different land utilization types including agricultural, engineering, recreation and nature conservation. Land use scenario building in this research used LUPIS as the main tool to generate alternative plans. This is accomplished by varying the policy weights considered for land allocation for a given option.

The core of LUPIS is a matrix-manipulation package, tailored to handle the tedious arithmetic required to repeatedly calculate suitability score, summed across all policies for each land planning unit. A single LUPIS run, using a given set of policy-importance votes, determines the preferred land use option on a planning unit. This outcome is reached by ranking land uses in descending numerical order on the basis of suitability scores defined. Thomson's competitive solution (Thomson 1973) is used for the computation and aggregation of suitabilities, according to the following mathematical expression:

$$Maximize\, S_{ij} = E_{ij} \cdot \Sigma R_{ijk} \cdot V_k$$

Where:

$i = 1, 2, ...n$:   Planning unit
$j = 1, 2, ...m$:   Feasible land useLand use options
$k = 1, 2, ...p$:   Preference policies whose satisfaction is to be maximized
$E\ (E = 0\ o\ 1)$:   Exclusionary policies proscribing given uses for some parcels
$R\ (0 < R < 1)$:   Policy satisfaction rating or degree to which a given use on a given site satisfies a particular policy

---

[1] Simón Arenas. Fondo Nacional de Investigaciones Agropecuarias (FONAIAP). Personal communication.

V: Policy weight or "vote" for a policy by a given interest group
S: Aggregate policy satisfaction for a given use on a given parcel of land

Policies are defined as planning objectives concerned with the site requirements and environmental effects of particular land uses. Ratings are either performed from a primary data item using Qbasic code within the system, input directly from the keyboard, or calculated within a spreadsheet. The latter can also be used to interface the system with other software packages.

A basic set of planning conditions was established before running LUPIS to obtain particular land use scenarios. The conditions for this analysis include the following variables:

- Land use options: 4 (urban, rural residential/intensive recreation, agriculture, conservation)
- Planning parcels: 699 (corresponding or associated to 193 geopedologic units)
- Data items: 23 (e.g., biozone, slope, landform, land suitability, land market values)
- Land use policies

    - Commitment policies: 5
    - Exclusion policies: 5
    - Preference/avoidance policies: 36

Commitment policies allocate a land use to specific planning units, excluding all other uses. Exclusion policies exclude a land use which is not allowed or not feasible within the current planning conditions. Preference/avoidance policies partly contribute to the allocation of land uses and are subject to weighting votes. In other words, policy ratings are affected by users through the votes they give to each individual preference policies. The formulation of policies for this case study is described in the original research (Rodríguez 1995).

The land use scenarios were developed considering users' opinions through weighting votes attached to each preference/avoidance policy by three selected interest groups: farmers, developers and conservationists. Areas not suitable for, or committed to particular land uses, were respectively excluded or selected in advance.

When the scenario maps are overlayed, areas of land use agreement and disagreement between the interest groups can be identified. The land use options selected by farmers, developers and conservationists may coincide and then a planning consensus is reached. In contrast, opinion discrepancies occur, and planning conflicts arise with respect to units showing multiple suitabilities. Such units must be submitted to a compatibilization process, leading to land use compromises. A discussion plan, which is a cartographic document representing the areas of land use agreement and disagreement is supported by the individual scenario maps. Discussions must focus on areas exposed to disagreements between the user groups. The process involves trade-offs between interest groups and a transfer of rights and power. The equity of the process is expected to ensure a more sustainable plan.

A final land use plan is obtained after agreements are achieved among users. Consensus rules are developed to reach a compromise between different users' opinions and produce a proposed land allocation plan. The proposed plan needs supporting planning strategies to be implemented. These involve the use of a set of tools and techniques to make the proposed plan feasible. A legal framework, economic incentives, rights-in-land controls and compensations, farming systems adjustments and institutional support are among some of the strategies that can be used to protect farmland and natural space.

In Fig. 24.3 can be seen the flow of geographic information from inventory to land use allocation in a very synoptic way with a variety of thematic maps.

Conclusions from study case 1

- Land use conflicts are common consequences of the global process of urbanization. They are more acute in urban fringes because of insufficient open space, loss of suitable land for agriculture and nature conservation, and the high costs of services and waste disposal facilities. Three different kinds of conflicts were identified: intensity, functional and generational conflicts.
- A combination of conventional and modern procedures of land resource inventory, land evaluation, land use assignment and land allocation were used for data collection, processing and analysis. Resource inventories, especially of soils, and land use conflict analysis generate basic input data for policy formulation and scenario building. The geoform classification system played a fundamental role in the delineation of basic map units serving as containers for the mapping of the other natural resources. A set of data items with different levels of aggregation were selected from the resource inventory and land evaluation and used to rate the planning units.
- Available software was combined and interfaced with a geographical information system for data handling. The incorporation of multiparty opinions as weighting votes for preference and avoidance land use policies is an innovative approach that encourages community participation in land use decisions, even though it is difficult to fully implement, participation generates support for the proposed land use plan.
- Scenario building is a practical approach for identifying land use conflicts areas, and modeling potential land use plans. It enables the search for land use alternatives and the analysis of their possible negative and positive outcomes. It is also an efficient way of anticipating future land use conflicts and modeling the impact of land use policies in alleviating current and potential land use conflicts.

24.5   Study Case 2

Unlike study case one, where the aim was to strategically allocate major land uses and prevent mainly functional conflicts, the second study case is oriented to identify sustainable land uses and management systems within a major land use previously

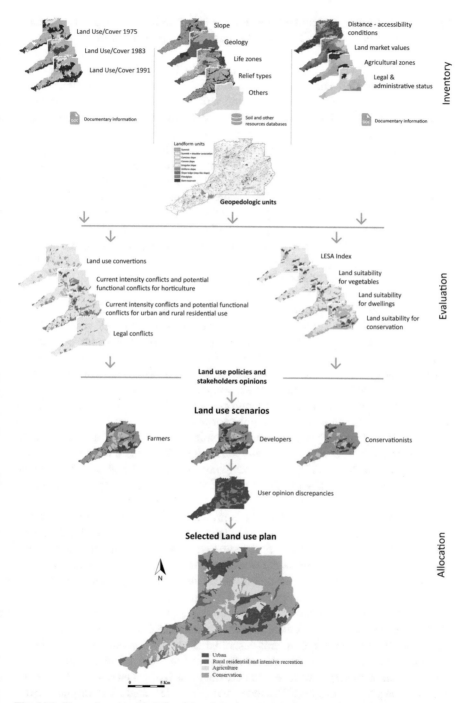

Fig. 24.3 Flow of geographic information from inventory to land use allocation using a logical sequence of thematic maps analysis and processing operations within a GIS system (ILWIS)

selected. In this case it is agriculture, and the aim is to prevent land from degradation processes or restore it. The output is derived as a product of intensity land use conflicts.

When a particular selected land utilization type-SLUT matches the agroecological conditions, mainly climate and soil conditions, and meets environmental requirements, management operations are circumscribed to those applied in a common agricultural schedule like most cultural practices including tillage, fertilization, irrigation, crop handling. If the environmental requirements are not fulfilled by the SLUT it has to be improved through soil and water conservation technologies and other sustainable land management components to avoid land degradation processes that decrease land productivity but also to contribute with a healthy environment that secure water quantity and quality, reduce CO_2 emissions, promote biodiversity and consider social aspects like equity, employment and gender. If this is the case, in order to accomplish so many goals, a diversity of optimized land utilization types-OLUTs must be given to the land user as a technical guide, being the decision a tactical one but highly influenced by the market as well as the rules and trends on environmental goals set by society (e.g., Sustainable development goals-SDG, UN 2015; UNEP 2021).

A specific geopedologic unit was selected within the study area described previously. Further information was extracted from prior research (Rodríguez 1995, 2018). The study area can be described as a convex slope (landform) within and association of primary and secondary crests (relief type) as part of a mountain landscape. Geology corresponds to "Las Brisas" formation with a lithology dominated by quartz, micaceous schist, quartzite and amphibolite. The slope goes from 15% to 20 % and belongs to a convex slope landform type. Climate corresponds to a transition between lower montane dry to moist forest bioclimatic zone, with rain period from 7 to 9 months per year, an annual precipitation of 860 mm, an average temperature around 16 °C with an erosivity index $EI_{30} = 2613$ MJ mm/ha h. The soil is described as an Aquic Palehudult, fine loamy, mixed isothermic with moderate erodibility, acid, low base saturation and restricted internal drainage The soil profile was described by Abreu y Ojeda, cited by Rodríguez (1995).

The agricultural land use in the area is primary horticulture with annual vegetables and permanent fruits that get advantage of the high altitudinal level making possible to produce vegetables from temperate zones. Conventional horticulture vegetable production systems in the area expose the terrain surface to water erosion, which is exacerbated under the conditions of the geopedologic unit selected and previously described, because of a steep slope of convex shape, and a soil profile with restricted internal water drainage conditions.

As an example, a carrot-lettuce sequence was selected as a land utilization type for the particular geopedologic unit described. The common agricultural practices that include raised beds or furrows, are not enough to avoid a land degradation process to occur, in this case, surface water erosion, which has been reported to be more than 40 t/ha year (Rodríguez, Fernández and Fernández cited by Rodríguez 2018). Sometimes, raised beds or furrows are oriented with the slope direction increasing the erosion process even more. Due to this undesirable outcome, the SLUT (carrot-lettuce planting sequence) should be optimized with the support of soil and water conservation technologies and considering other sustainable management criteria. A set of OLUTs is presented in Fig. 24.4.

As discussed before, the land user or manager decides the best option to be chosen. In the context of this study case, usually, agricultural parcels are of small size, a condition that promotes growing high-value crops like vegetables or strawberries,

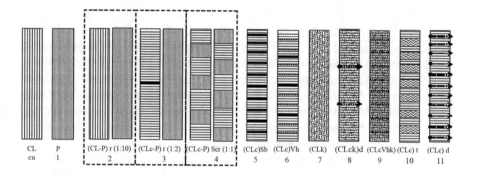

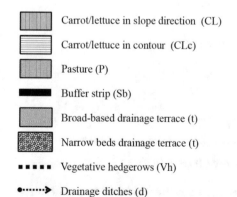

LEGEND

cu. Carrot/lettuce sequence (CL)
1. Pasture for cutting (P)
2. Rotation carrot/lettuce - pasture (CL-P) r (1:10)
3. Rotation carrot/lettuce with contour furrows or raised beds - pasture (CLc-P) r (1:2)
4. Pasture strips rotation with carrot/lettuce with contour furrows or raised beds (CLc-P) Scr (1:1)
5. Carrot/lettuce with contour furrows or raised beds with pasture bands (CLc) Sb
6. Carrot/lettuce with contour furrows or raised beds and vegetative hedgerows (CLc) Vh
7. Carrot/lettuce with contour furrows or raised beds + mulch 100% (CLk)
8. Carrot/lettuce with contour furrows or raised beds + mulch 50% and drainage ditches (CLck) d
9. Carrot/lettuce with contour furrows or raised beds, vegetative hedgerows + mulch 50% (CLcVhk)
10. Carrot/lettuce with contour furrows or raised beds and broad bed terraces (CLc) t
11. Carrot/lettuce with contour furrows or raised beds and drainage ditches (CLc) d

	Carrot/lettuce in slope direction (CL)
	Carrot/lettuce in contour (CLc)
	Pasture (P)
	Buffer strip (Sb)
	Broad-based drainage terrace (t)
	Narrow beds drainage terrace (t)
▪ ▪ ▪ ▪ ▪	Vegetative hedgerows (Vh)
●·······▶	Drainage ditches (d)

Fig. 24.4 Current use (cu) and conservationist use alternatives 1–11. (Adapted from Páez 1992)

with a high demand of capital inputs and labor. This is in detriment of less intensive uses like pastures or forests that enhance water quantity and quality, promote biodiversity and soil health, a very desirable outcome in a mountainous region with a high natural value like the cloud forest and as part of the catchment zone for human water supply.

Decision-making based on economic criteria can be facilitated using net economic impact curves over time, which represent the economic behavior of land units with or without the application of conservation technologies. Indifference curves show how conservation goals can be achieved through different levels of intensity of conservation and land use practices. A schematic example of different alternatives of use and management for the control of the erosion is presented in Fig. 24.4 and the associated indifference and income curves in Figs. 24.5 and 24.6. Other factors influencing managing decisions are land tenure, the length of the planning period and the interest rates in the capital market, but also the farmer´s decisions will be influenced by cultural and social factors. In many situations the best sustainable land management option is only available through regulations, subsidies, incentives or other compensation measures and land use policies.

The options to optimize SLUTs to OLUTs need to be presented with as much detail and with as many alternatives as possible to facilitate the decision-making process and implement the more sustainable choice. Graphics and drawings may help to illustrate each option as presented in Figs. 24.4, 24.5 and 24.6.

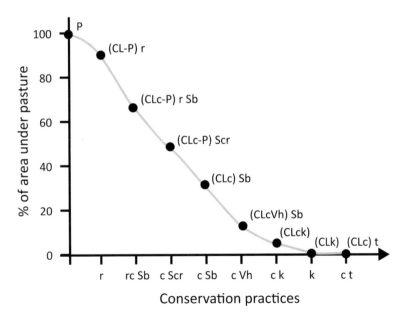

Fig. 24.5 Indifference curve for an erosion tolerance level and different combinations of conservation technologies and land use intensities (land percentage under pasture). Conservationists land use options are associated with Fig. 24.4. (Adapted from Páez 1992)

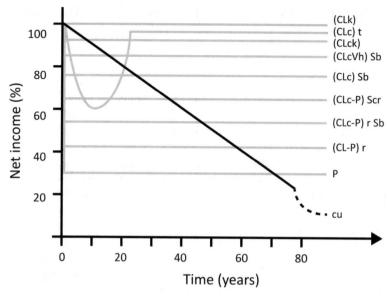

Fig. 24.6 Net income curves trough time associated to conservationists land use options from Fig. 24.4. (Adapted from Páez 1992)

Conclusions study case 2

- This study case presented just a very simple example to support the idea of what it means to optimize a land utilization type using ecological, economic and social criteria in order to achieve a sustainable land management. Possible cases are never ending, but the principles are based on the need for different actions and measures, in a range of combinations and strengths to avoid land degradation processes derived from intensity conflicts.

- The diversity of possible situations according to local conditions and specific land parcels demonstrate the usefulness of an integrated land use planning framework which allows flexibility to explore and process basic data from diagnoses of land use conflicts and multipurpose land evaluation analysis, to select and optimize land utilization types based on geopedologic information complemented by other biophysical and socio-economic data sources. Geopedologic units represent the most stable geographic item through time. Other information of more dynamic nature may be attached to them. This approach enables recycling and reusing the land information and planning system many times, reducing costs and saving time.

24.6 New Trends in Precision Sustainable Land Management

Many sustainable land management decision tasks can be carried out with new —
and not so new— information technology tools to perform tactical operations within
particular geopedologic units and optimize sustainable land management decisions
based on high resolution remote and ground data capture, in response to more randomly distributed variables as compared with the information of the soil mantle
features of geomorphic nature and corresponding to non-random systematic
variations.

According to the EIP-AGRI (2021) precision farming is a management approach
that focuses on (near real-time) observation, measurement, and responses to variability in crops, fields and animals. It can help increase crop yields and animal performance, reduce costs, including labor costs, and optimize process inputs. All of
these can help increase profitability. At the same time, precision farming can
increase work safety and reduce the environmental impacts of agriculture and farming practices, thus contributing to the sustainability of agricultural production.

It is not the purpose of this paper to describe all the new trends in digital technology but rather to highlight its role in tactical operations on specific geopedologic
units and their subdivisions, making practicable sustainable land management
underpinned by high spatial and temporal resolution data sets. This is possible
thanks to the evolution and accessibility of ground and remote sensors of all kinds
(e.g., climate, soil water content, temperature and electric conductivity, crop performance and health, drone and satellite multispectral imagery, among others), real-
time or highly frequency data acquisition and the geolocation of the acquired data.
All these managed within digital networks and controlled by specific software. On
the other hand, the availability of machinery and equipment capable of splitting land
management prescriptions into usually small specific subdivisions (land parcels) of
particular geopedologic units. Most frequent land management practices of this
nature include seeding density, nutrient and irrigation dosage and phytosanitary
control, but it can be extended to other tasks like erosion control, carbon sequestration, biodiversity protection and so on. Many references illustrate this subject as
Balafoutis et al. (2017), Caron et al. (2018), Chartuni and Magdalena (2014), Cheng
et al. (2018), García (2021), Martínez J. (2021), Sassenrath et al. (2016).

Particular decisions will be influenced directly by inputs and outputs prices controlled by market conditions. Tactical decisions on sustainable land management
can reduce operational costs optimizing resources use (water and nutrient savings)
or increment revenues by matching harvest period with demand peaks or improving
product quality. Ecosystem services and other positive externalities must be considered by society to promote those sustainable land use systems that better perform in
particular land units.

24.7   General Conclusion

There is a need for a sustainable land use planning and management system, to minimize irreversible land use conversions and degradation processes while securing productivity and ecosystem services and allowing democratic participation of society in decision making. A conceptual framework is presented where geopedologic information and its central role delimitating the landscape and gathering other biophysical and socio-economic data is highlighted, considering the time stability nature of geopedologic units and their influence on land use decisions. First phases of the framework are controlled by physical factors, but social, economic and technological factors become more relevant in the middle and final phases, similarly, in the first stages decision making is more strategic and become more tactical towards the middle and final stages. Two study cases were presented to analyze and resolve land use conflicts of functional or intensity nature, but with the potential to become generational conflicts, because their negative effects can compromise the resource availability and land use options for future generations. Land degradation is considered an important process in its diverse forms, and soil and water conservation technologies a set of tools to counteract undesirable outcomes together with other sustainable land management components. Subdivisions of land units at a high spatial and temporal resolution, using ground and remote sensors, allows tactical operational management decisions, that improve resource uses and optimize productivity with sustainability criteria, as precision agriculture techniques associated with a huge emerging number of digital technologies arise.

Acknowledgement I want to honor Dr. Alfred Zinck who inspired and supported me to develop the ideas exposed in this research and remember him as a very good friend and tutor. He not only influenced my professional career but left on me a deep mark as a human being.

References

Baker M (2014) Land-Use Planning. In: Michalos AC (ed) Encyclopedia of quality of life and well-being research. Springer, Dordrecht. https://doi.org/10.1007/978-94-007-0753-5_1600

Balafoutis A, Beck B, Fountas S, Vangeyte J, Wal T, Soto I, Gómez-Barbero M, Barnes A, Eory V (2017) Precision agriculture technologies positively contributing to GHG emissions mitigation, farm productivity and economics. Sustainability 9, 1339. Available on internet URL: https://doi.org/10.3390/su9081339

Bold P (2019) GIS Fundamentals. A first text on geographic information systems, 6th edn. Xan Edu, Ann Arbor

Caron J, Anderson L, Sauvageau G, Gendron L (2018) Real time Precision irrigation with variable setpoint for strawberry to generate water savings. In: Proceedings of the 14th international conference on precision agriculture June 24–June 27

Chartuni Mantovani E, Magdalena C (2014) Manual de Agricultura de Precisión. In: PROCISUR-Programa Cooperativo para el Desarrollo Tecnológico Agroalimentario y Agroindustrial del Cono Sur, Montevideo IICA

Cheng Z, Meng J, Shang J, Liu, J., Qian B, Jing Q (2018) Developing an integrated approach for estimation of soil available nutrient content using the modified WOFOST model and time-series multispectral UAV observations. Proceedings of the 14th international conference on precision agriculture June 24–27

Dent D (1988) Guidelines for land use planning in developing countries. Soil Surv Land Eval 8(1988):67–76

EIP-AGRI (2021) Agricultural European innovation partnership. Available on internet URL: https://ec.europa.eu/eip/agriculture/en/digitising-agriculture/developing-digital-technologies/precision-farming-0

FAO (1976) A framework for land evaluation, Soils Bulletin 32. FAO, Roma

FAO (2011) The state of the world's land and water resources for food and agriculture (SOLAW) – Managing systems at risk. Food and Agriculture Organization of the United Nations, Rome and Earthscan, London

FAO (2022) Inclusive and sustainable territories and landscapes platform. Available on internet URL: https://www.fao.org/in-action/territorios-inteligentes/componentes/ordenamiento-territorial/contexto-general/en/

FAO, Organización de las Naciones Unidas para la Alimentación y la Agricultura (2014) Sistematización de prácticas de conservación de suelos y aguas con enfoque de adaptación al cambio climático. Metodología basada en WOCAT para América Latina y el Caribe, Santiago

FAO-UNEP (2000) El Futuro de Nuestra Tierra. In: Enfrentando el desafío. Organización de las Naciones Unidas para la Agricultura y la Alimentación (FAO) en colaboración con el del Programa de las Naciones Unidas para el Medio Ambiente (PNUMA), Roma

García Ramos J (2021) Agricultura de precisión, el inicio de la digitalización del campo. Alianza Agroalimentaria Aragonesa. Available on internet URL: https://alimentandolaciencia.esciencia.es/2021/09/26/agricultura-de-precision-el-inicio-de-la-digitalizacion-del-campo/

IPBES (2018) The IPBES assessment report on land degradation and restoration. In: Montanarella L, Scholes R, Brainich A (eds) Secretariat of the intergovernmental science-policy platform on biodiversity and ecosystem services, Bonn

ITC (1993) ILWIS 1.4 user´s manual. International Institute for Aerospace Survey and Earth Sciences, Enschede

Ive JR (1992) LUPIS: computer assistance for land use allocation. Resource Technology 92. Information Technology for environmental management, Taipei

Martínez Casanova JA (2021) Digitalización y trabajo agrícola automatizado en la próxima década. In Horticultura. Available on internet URL: https://www.interempresas.net/Horticola/Articulos/358524-Digitalizacion-y-trabajo-agricola-automatizado-en-la-proxima-decada.html

Orr BJ, Cowie AL, Castillo Sanchez VM, Chasek P, Crossman ND, Erlewein A, Louwagie G, Maron M, Metternicht GI, Minelli S, Tengberg AE, Walter S, Welton S (2017) Scientific conceptual framework for land degradation neutrality. A report of the science-policy interface. United Nations Convention to Combat Desertification (UNCCD), Bonn

Páez ML (1992) Diseño de prácticas de conservación con la ecuación universal de pérdidas de suelo. CIDIAT. Serie Suelos y Clima Sc-64 Mérida

Rodríguez OS (1995) Land use conflicts and planning strategies in urban fringes. A case study of Western Caracas, Venezuela. Doctoral thesis, ITC publication 27 international institute for aerospace survey and earth science

Rodríguez OS (2018) Conservación de Suelos y Agua. Una premisa del desarrollo sustentable. Universidad Central de Venezuela, Consejo de Desarrollo Científico y Humanístico. Colección Estudios-Agronomía, Fondo Editorial Digital CDCH-UCV Caracas. 2nd edition (Digital). Available on Internet. URL: http://saber.ucv.ve/omp/index.php/editorialucv/catalog/book/11

Rodríguez OS, Zinck JA (1998) El ensamblaje de escenarios para la toma de decisiones ambientales sobre el uso de la tierra. (Scenario building for environmental decision-making support on land use planning) In: Carrillo RJ (Compilador) Memorias del IV Congreso Interamericano sobre el Medio Ambiente, Caracas, Venezuela, 1997. Colección Simposia, Volumen I, Editorial Equinoccio, Ediciones de la Universidad Simón Bolívar, pp 337–342

Rodríguez OS, Zinck JA (2016) Contribution of geopedology to land use conflict analysis and land use planning. In: Zinck JA, Metternicht G, Bocco G, del Valle HF (eds) Geopedology: an integration of geomorphology and pedology for soil and landscapes studies. Springer International Publishing, Switzerland, pp 521–536

Rossitter DG, van Wambeke AR (1993) ALES version 4 User's manual. In: Automated land evaluation system, SCAS Teaching Series No. T93-2. Cornell University/Department of Soil/Crop & Atmospheric Sciences, Ithaca

Russwurm LH (1975) Land policies across Canada: thoughts and viewpoints. In: Battle for land conference report. Community Planning Association of Canada, Ottawa

Sassenrath GF, Mueller T, Shoup D, Alarcón VJ, Kulesza S (2016) In-field variability of terrain and soils in southeast Kansas: Challenges for effective conservation. In: Proceedings of the 13th international conference on precision agriculture July 31 – August 4, St Louis

Thomson W (1973) Middey randstatd part 1: final report. Colin Buchanan, London

UN (2015) Transforming our world: The 2030 agenda for sustainable development. Available on internet URL: https://www.isustainabledevelopment.un.org A/RES/70/1 41p

UNEP-United Nations Environment Programme (2021) Making peace with nature: a scientific blueprint to tackle the climate, biodiversity and pollution emergencies. Nairobi. Available on internet. URL: https://www.unep.org/resources/making-peace-nature

WOCAT (2007) Where the land is greener – case studies and analysis of soil and water conservation initiatives worldwide. In: Liniger H, Critchley W (eds) WOCAT-world overview of conservation approaches and technologies, Berna

Zinck JA (1988) Physiography and soils, ITC soil survey course lecture notes. International Institute for Aerospace Survey and Earth Sciences, Enschede

Zinck JA (2016a) The Geopedologic Approach. In: Zinck JA, Metternicht G, Bocco G, del Valle HF (eds) Geopedology: an integration of geomorphology and pedology for soil and landscapes studies. Springer International Publishing, Switzerland, pp 27–59

Zinck JA (2016b) The Geomorphic Landscape. In: Zinck JA, Metternicht G, Bocco G, del Valle HF (eds) Geopedology: an integration of geomorphology and pedology for soil and landscapes studies. Springer International Publishing, Switzerland, pp 77–99

Chapter 25
The Relevance of Geopedology for Policy Making and Soil Security

L. Montanarella

Abstract Policy making and soil security require to address the sustainable management of soil resources at the landscape level. Therefore, geopedology can become highly relevant for effective policy making for achieving long term soil protection. The necessary pre-condition is the availability of a solid scientific basis and detailed data on the actual status and trends of soils within relevant landscapes. The recent emergence of high-resolution digital soil mapping techniques offers new possibilities for achieving such a knowledge base. Several examples from the European Union demonstrate that geopedology can be a valuable tool for understanding soil processes at the landscape level and design effective soil protection policies.

Keywords Soil functions · Sustainable development goals · Soil quality · European soil strategy

25.1 Introduction

Policy making related to soils is a very difficult and challenging area for regulators at all levels. Soils are mostly in private property rights, do not move (unless in case of landslides and erosion) and are invisible to the general public. To understand soils, you need to dig a soil profile, you need to look below the surface. Most of the public awareness is about land and landscapes, what is visible above ground and can be directly related to human needs, preoccupations and interests. Therefore, most soil related policies are in fact addressing what is happening above ground, at the landscape level.

L. Montanarella (✉)
Joint Research Centre, European Commission, Ispra, Italy
e-mail: luca.montanarella@ec.europa.eu

© The Author(s), under exclusive license to Springer Nature
Switzerland AG 2023
J. A. Zinck et al. (eds.), *Geopedology*,
https://doi.org/10.1007/978-3-031-20667-2_25

Sustainable management practices are applied above ground but have a substantial impact below ground. The same applies for the threats to soil functions, which become relevant to the general public, and the policy once their consequences become visible at the landscape level. One of the best examples is the growing concern in Europe for soil sealing by infrastructure and housing. Covering the soils with impermeable materials due to expanding urbanization, especially on fertile farmland, is increasingly perceived as a major threat, while less visible processes, like soil contamination, get attention only once their effect become tangible on human health, especially through the food chain.

All this for introducing that dealing with soils from a policy perspective is necessarily connected with dealing with land and larger landscapes. Geopedology can therefore be of crucial importance for the policy maker if it delivers the complex interactions between below ground processes and above ground landscapes.

Traditionally, geopedology has been related to soil mapping. Placing soil mapping units in a coherent manner in the landscape by integrating geomorphological data with point observations on selected soil profiles has been the core of the geopedological approach. Probably the best example of that approach has been the global program SOTER by FAO and ISRIC, aiming at a full global coverage with a Soil and Terrain (SOTER) database (https://www.isric.org/explore/soter).

More recently, with the development of highly advanced digital soil mapping techniques, the traditional SOTER approach has been largely abandoned. High resolution grids of soil properties are now freely available at global scale with resolutions being constantly improved (250 m is the most recent release). These gridded soil property data are derived from point observations and a large amount of ancillary data derived from remote sensing and other sources, including high resolution digital terrain models (DTM) available at global scale.

This new generation of high-resolution digital soil data and information has strongly supported an improved understanding of the processes occurring at global, national and local scales affecting soil properties and functions, and are therefore considered the basis for the current policy making processes related to soils.

Probably the best example is the increasing relevance of soil organic carbon (SOC) in the current policy debate, especially in relation to climate change. Different land uses and different soil management practices can strongly influence the amount of organic carbon accumulated in the soil. A geopedological approach to this would integrate landscape features with below-ground SOC accumulation rates. Advanced geostatistical approaches take largely into account geomorphological features, mostly with the help of high-resolution digital terrain models (DTM). The most recent spatially explicit SOC estimates using such an approach are available at the Global Soil Partnership's Global Soil Information System (GLOSIS).

25.2   Geopedology and Soil Security

Soil security has emerged during the recent years as a new paradigm for addressing sustainable soil management. In 2013–2014, soil security was first presented in the literature (Koch et al. 2013; McBratney et al. 2014). Both publications defined soil security as the maintenance and improvement of the world's soil resources so that they can continue to provide food, fiber and fresh water, make major contributions to energy and climate sustainability, and help maintain biodiversity and the overall protection of ecosystem goods and services. This was followed by some regional studies addressing soil security in Australia (Bennett et al. 2019) and Tasmania (Kidd et al. 2018). Meanwhile, soil security starts to gain the momentum and is linked to crop production and global climate (Beerling et al. 2018), soil contamination and human health (Carre et al. 2017; Brevik et al. 2017), farming and ecosystem services (Dazzi et al. 2019) and with the Sustainable Development Goals (Baouma 2020).

Therefore, soil security is developed as a concept in analogy with the other six existential global environmental challenges. Soil security is described by five dimensions known as 5Cs: Soil capability, Condition, Capital, Connectivity & Codification (Koch et al. 2013; McBratney et al. 2014). Here, we want to discuss the implications of soil security for geopedology in the European Union and the possible implementation of this concept in the future EU sustainable soil management strategies.

25.3   Capability

Soil capability is related to soil functions and the capacity of soils to deliver specific ecosystem services. In the EU soil functions have been placed at the core of the EU Soil Thematic Strategy (European Commission 2006; Panagos and Montanarella 2018) and are those functions that deserve to be protected for future generations and the overall wellbeing of the EU citizens. The European Commission in its proposal for an EU Soil Thematic Strategy has identified the following 7 soil functions to be protected in the EU:

1. *Biomass production, including in agriculture and forestry.*
2. *Storing, filtering and transforming nutrients, substances and water.*
3. *Biodiversity pool, such as habitats, species and genes.*
4. *Physical and cultural environment for humans and human activities.*
5. *Source of raw materials.*
6. *Acting as carbon pool.*
7. *Archive of geological and archaeological heritage.*

The capability of soils to deliver all, or some, of those functions is defining the inherent quality of those soils, which can be assessed using the geopedological framework (Chap. 4). A good soil is a soil that delivers these soil functions (van Leeuwen et al. 2017). Certainly, some of the functions are mutually exclusive, and therefore choices have to be made towards the most appropriate use of a given soil resource. Since soils (land) are a definite quantity and cannot be expanded nor easily renewed, sustainable soil management requires to apply management practices that assure the delivery of those seven functions also for future generations (Stavi et al. 2016).

Sustainable soil management has been well defined by the FAO (Baritz et al. 2018; FAO 2017) and has to be fully implemented also in the European Union (EU). A large proportion of EU soils is currently subject to unsustainable management practices, jeopardizing the soil security of the EU. The mission board on soil health and food of the European Commission has recently released the report "Caring for soil is caring for life"(Soil Mission Health 2020) where it highlighted that 60–70% of EU soils are currently in an unhealthy condition, meaning that these soils are not anymore able to fully deliver the functions that are of fundamental importance for the EU citizens. The ambition of the proposed mission on soil health and food is to have 75% of EU soils in a healthy condition by 2030 (Soil Mission Health 2020). This important document has eight important objectives to reach by 2030: (a) Reduce land degradation (b) conserve and increase soil organic carbon stocks (c) No net soil sealing d) reduce soil pollution and enhance soil restoration (e) prevent erosion (f) improve soil structure, (g) reduce the EU global footprint of soils and (h) Increase soil literacy in society across Member States.

25.4 Condition

The soil condition can be referred as the current health of soils. In the EU, a healthy soil is capable to deliver the seven functions listed previously. Determining the soil condition requires appropriate indicators and data to be derived from operational monitoring systems and/or modelling frameworks. Currently the EU derives its data about the condition of its soils from the regular monitoring within the Land Use/ Cover Area frame statistical Survey (LUCAS). This survey collects more than 22,000 soil samples from all EU and derives properties based on chemical, physical and biological measurements (Orgiazzi et al. 2018). Those basic parameters allow to derive indicators that determine the capability of those soils to deliver the required soil functions (soil health). The current estimate, based on those data and indicators, is that 60–70% of soils in the EU are in an unhealthy soil condition.

Soil loss by water erosion is a major threat in EU soils as 24% of land has unsustainable soil water erosion rates (>2 t ha^{-1} year^{-1}) with a mean erosion rate at 2.45 t ha^{-1} year^{-1} (Panagos et al. 2020). In addition, the wind erosion shows mean rate of 0.53 t ha^{-1} year^{-1} in arable lands (Borrelli et al. 2017). As the sediment fluxes to riverine system is major environmental problem, it was estimated that around

15.3% of soil loss is routed to river basins (Borrelli et al. 2018). The mean soil organic carbon content in EU soils is about 48 g Kg^{-1} (de Brogniez et al. 2015) while agricultural soils are relatively poor in soil organic carbon having 2.4 times than permanent grasslands (Hiederer 2018). In addition, it is important to address the nutrient and carbon losses due to erosion. The average phosphorus losses due to soil erosion in European croplands is about 1.2 $Kg\ ha^{-1}\ year^{-1}$ with Eastern European countries experiencing a negative P balance (Alewell et al. 2020).

The soil sealing is a threat for EU soils as the land take rate is about 539 Km^2 per year (2012–2018). The loss of high value agricultural land possesses an important issue in future food security as the land take can be translated into potential crop losses (Gardi et al. 2015). In terms of soil compaction, 23% of EU land had critically high densities (Stolte et al. 2016). As for local contamination, in a larger area which includes EU countries plus 12 neighbouring ones (EEA39), JRC reported c.a 2.8 million potentially contaminated sites with high uncertainties (Payá Pérez and Rodríguez 2018). The diffuse pollution is recently addressed with copper (Cu) being higher than the threshold of 100 mg Kg^{-1} in 1.1% of the lands with particular concerns in vineyards (14.6% over the threshold) and olive groves (4.8 higher than threshold) (Ballabio et al. 2018). In addition, Mercury (Hg) is of particular concern close to old Hg mining areas and coal combustion industries with 0.8% of the sampled land having values higher than the risk threshold (0.5 mg Kg^{-1}) (Ballabio et al. 2021). Acidification is also an issue in EU soils with low pH (<5.5) covering 29.9% of the land (Ballabio et al. 2019).

25.5   Capital

The question of the monetary value of the functions delivered by soils in the EU is central to the discussion on the need for policy intervention for protecting those resources. The extended impact assessment (STS 2006) to the proposal of the European Commission for an EU Soil Thematic Strategy has attempted a full evaluation of the monetary value of those functions and the derived costs to the EU of the on-going soil degradation processes. The total costs of degradation that could be assessed in 2006 for erosion, organic matter decline, salinization, landslides and contamination, on the basis of available data, would be up to €38 billion annually for EU25 with the following breakdown by major threats:

- erosion: €0.7–14.0 billion
- organic matter decline: €3.4–5.6 billion
- compaction: no estimate possible
- salinization: €158–321 million
- landslides: up to €1.2 billion per event,
- contamination: €2.4–17.3 billion
- sealing: no estimate possible
- biodiversity decline: no estimate possible

The major difficulty still remains in the highly uncertain data and estimates currently available. More recently various analyses have been made for a more robust assessment of the actual cost of soil degradation processes. For example, according to a recent study (Panagos et al. 2018), soil erosion costs European countries €1.25 billion in annual agricultural productivity loss and €155 million in the gross domestic product (GDP) loss. Similar estimates have been also completed at global scale the agricultural productivity loss due to soil erosion is at 8 Billion US dollars (Sartori et al. 2019). Even the range of soil contamination costs is quite large (€2.4–17.3 billion), recent assessments have calculated In terms of budget, the management of contaminated sites to cost around 6 billion € annually (Panagos et al. 2013).

In general, there is a large agreement that the soil capital is an important asset of the EU, but a more precise quantification in monetary terms is still lacking. Another approach is to develop a soil natural capital accounting in Europe based on biophysical asset accounts (Robinson et al. 2017).

25.6 Connectivity

Connectivity should be relating soils with the social dimension. It is essentially about connecting people with soils. Within the EU Soil Thematic Strategy this is equivalent to the pillar of action on awareness raising.

There has been an early recognition for the need to raise awareness in the EU for the relevance of soils for sustainable development and the wellbeing of EU citizens. Over the past centuries, the European citizens have gradually lost their connection to soils following the early industrialization process and the subsequent large urbanization trend in Europe. The European population has gradually shifted towards a majority of urban population with currently only 4–5% of total employment in agriculture in the EU. In 2018, 39.3% of the population lived in the cities, 31.6% lived in towns and suburbs, and 29.1% lived in rural areas (source: EUROSTAT). The vast majority of EU citizens today is totally disconnected from soils and not aware of the relevance of soils for their wellbeing.

The new emerging connection with soils for the EU citizens is through their role as consumers. The raising awareness that healthy soils produce healthy food (Rojas et al. 2016) and therefore are of relevance for our health is a major new driver for increased connectivity with soils in the EU. It is also important to explore innovative ways to present soils to citizens and raise soil awareness (Baouma et al. 2012). The large uptake of organic farming (Casagrande et al. 2016) and organic food consumption in the EU is demonstrating that more sustainable and healthy food production systems are appealing for the EU citizen and can be one of the major avenues for raising the awareness for the need of sustainable soil management and soil protection. The European Green Deal proposes ambitious goals with a target to reach 25% of the EU's agricultural land as organically farmed by 2030 and considerable decrease of pesticides and fertilizers (Montanarella and Panagos 2021).

25.7   Codification

Achieving soil protection for the EU needs an appropriate legal framework, which corresponds to the codification dimension of soil security. Establishing such a legal framework specific for soils has been an initiative of the European Commission starting already in the late 90's with the development of a comprehensive soil thematic strategy that after a very large a detailed stakeholder consultation was finally presented by the European Commission to the European parliament and the EU Council in 2006 (Ronchi et al. 2019).

The EU Soil Thematic Strategy builds on the recognition of the seven soil functions already described within the capability dimension of soil security and outlines four pillars of activity that would ensure an adequate legal framework for their protection. Important to note that the proposal of the European Commission is explicitly addressing the protection of soil functions delivering public goods and services, not soil protection as such. A clear distinction needs to be made between soil protection, which is mostly in the hands of private landowners, and the need to protect soil functions that deliver services far beyond the single ownership of a specific area of land.

The four pillars of action address the four major areas of activity of the European Commission on soils:

1. A proposal for a legally binding instrument (soil framework directive)
2. Integration of soil protection in related EU legislation
3. Research and innovation on soil related topics
4. Awareness raising actions to promote soil protection in the EU

Since its presentation by the European Commission in 2006 the strategy has been making important progress in its implementation, especially for pillars two to four. Where the strategy has failed was in the adoption of a legally binding legislation concerning soil protection in the EU. A blocking minority of EU Member States has prevented the final adoption of the proposed soil framework directive, which was ultimately withdrawn by the European Commission in 2014 (Glæsner et al. 2014).

Codification of soil in the EU has proven difficult due to many reasons, mostly related to the strong relevance of soils to the EU Member States sovereignty. Soils are considered a crucial National asset and turn out to be a highly sensitive topic for inclusion in binding EU legislative frameworks. The full adoption of the subsidiarity principle in the EU also requires a strong argumentation for justifying that soils are a topic of transboundary relevance. The argument that "soils don't move" and therefore are strictly of local or National competence prevents the adoption of legal frameworks at EU level or even at global scale. This is clear from the geopedologic perspective, since geopedologic units are not independent bodies that deliver their intrinsic functions but interconnected through biophysical processes that affect one to each other. Clear example of its application are the administrative units based on watersheds implemented in the Netherlands,[1] Australia[2] which considers the interdependence of the land beyond the political administrative units. The Mekong River Commission, an intergovernmental organization for regional dialogue and cooperation in the Lower Mekong River Basin[3] between Cambodia, Lao PDR, Thailand and Vietnam, is another successful example how policies must consider regional processes. Expanding the geopedological concepts to the European Union could help to integrate sustainable soil management considering the processes that work at regional scale.

25.8   Conclusions

Geopedology can play an important role in achieving soil security in the EU and world-wide. In Europe, soil security can only be achieved by considering the full implementation of the EU Soil Thematic Strategy. While good progress has been made on the first four dimensions of soil security: capability, condition, capital and connectivity, a lot still remains to be done in the area of codification. A binding legal framework is a necessary condition for assuring soil security for the EU and protecting this natural resource from further degradation processes.

A revised approach to soil protection in the EU has been open for public consultation (STS 2021) and will be finally presented by the European Commission possibly by end of 2021. The aim of this new EU Soil Strategy will be to address soil- and land-related issues in a comprehensive way and to help achieve land degradation neutrality by 2030, one of the key targets of the Sustainable Development Goals (SDGs).

The recently established EU Soil Observatory will further contribute to soil security in EU with (a) an integration of national soil monitoring systems with LUCAS, (b) a stronger European Soil Data Centre, (c) an enhanced EU Soil Indicator Dashboard, (d) Research and Innovation programmers and (e) an open Forum for increasing awareness.

A multidisciplinary approach to soil related policy making and soil security is absolutely necessary, and certainly geopedology can further contribute to a more holistic approach to sustainable soil management in Europe and in the world.

---

[1] https://en.wikipedia.org/wiki/Water_board_(Netherlands)

[2] https://en.wikipedia.org/wiki/Murray%E2%80%93Darling_basin

[3] https://www.mrcmekong.org/

References

Alewell C, Ringeval B, Ballabio C, Robinson DA, Panagos P, Borrelli P (2020) Global phosphorus shortage will be aggravated by soil erosion. Nat Commun, 11(1):1–12

Ballabio C, Panagos P, Lugato E, Huang JH, Orgiazzi A, Jones A, Fernández-Ugalde O, Borrelli P, Montanarella L (2018) Copper distribution in European topsoils: an assessment based on LUCAS soil survey. Sci Total Environ 636:282–298

Ballabio C, Lugato E, Fernández-Ugalde O, Orgiazzi A, Jones A, Borrelli P, Montanarella L, Panagos P (2019) Mapping LUCAS topsoil chemical properties at European scale using Gaussian process regression. Geoderma 355:113912

Ballabio C, Jiskra M, Osterwalder S, Borrelli P, Montanarella L, Panagos P (2021) A spatial assessment of mercury content in the European Union topsoil. Sci Total Environ 769:144755

Baritz R, Wiese L, Verbeke I, Vargas R (2018) Voluntary guidelines for sustainable soil management: global action for healthy soils. In: International yearbook of soil law and policy 2017, pp 17–36. Springer, Cham

Beerling DJ, Leake JR, Long SP, Scholes JD, Ton J, Nelson PN, Bird M, Kantzas E, Taylor LL, Sarkar B, Kelland M (2018) Farming with crops and rocks to address global climate, food and soil security. Nature Plants 4(3):138–147

Bennett JM, McBratney A, Field D, Kidd D, Stockmann U, Liddicoat C, Grover S (2019) Soil security for Australia. Sustainability 11(12):3416

Bouma J, Broll G, Crane TA, Dewitte O, Gardi C, Schulte RP, Towers W (2012) Soil information in support of policy making and awareness raising. Curr Opin Environ Sustain 4(5):552–558

Bouma J (2020) Soil security as a roadmap focusing soil contributions on sustainable development agendas. Soil Secur 1:100001

Borrelli P, Lugato E, Montanarella L, Panagos P (2017) A new assessment of soil loss due to wind erosion in European agricultural soils using a quantitative spatially distributed modelling approach. Land Degrad Dev 28(1):335–344

Borrelli P, Van Oost K, Meusburger K, Alewell C, Lugato E, Panagos P (2018) A step towards a holistic assessment of soil degradation in Europe: coupling on-site erosion with sediment transfer and carbon fluxes. Environ Res 161:291–298

Brevik EC, Steffan JJ, Burgess LC, Cerdà A (2017) Links between soil security and the influence of soil on human health. In: Global soil security, pp 261–274. Springer, Cham

Carré F, Caudeville J, Bonnard R, Bert V, Boucard P, Ramel M (2017) Soil contamination and human health: a major challenge for global soil security. In: Global soil security, Springer, Cham, pp. 275–295

Casagrande M, Peigné J, Payet V, Mäder P, Sans FX, Blanco-Moreno JM, Antichi D, Bàrberi P, Beeckman A, Bigongiali F, Cooper J (2016) Organic farmers' motivations and challenges for adopting conservation agriculture in Europe. Org Agric 6(4):281–295

Dazzi C, Galati A, Crescimanno M, Papa GL (2019) Pedotechnique applications in large-scale farming: economic value, soil ecosystems services and soil security. Catena 181:104072

de Brogniez D, Ballabio C, Stevens A, Jones RJA, Montanarella L, van Wesemael B (2015) A map of the topsoil orga carbon content of Europe generated by a generalized additive model. Eur J Soil Sci 66(1):121–134

European Commission (2006) Communication from the Commission to the Council, the European Parliament, the European Economic and Social Committee and the Committee of the Regions – Thematic strategy for soil protection (COM(2006) 231 final)

FAO (2017) Voluntary Guidelines for Sustainab Soil Management Food and Agriculture Organization of the United Nations Rome, Italy. http://www.fao.org/3/a-bl813e.pdf

Hiederer R (2018) Data evaluation of LUCAS soil component laboratory data for soil organic carbon, JRC technical report. No. JRC1 12711, Publications Office of the European Union, Luxembourg. https://esdac.jrc.ec.europa.eu/public_path/shared_folder/JRC112711_lucas_oc_data_evaluation_final.pdf

Gardi C, Panagos P, Van Liedekerke M, Bosco C, De Brogniez D (2015) Land take and food secu-
    rity: assessment of land take on the agricultural production in Europe. J Environ Plan Manag
    58(5):898–912
Glæsner N, Helming K, De Vries W (2014) Do current European policies prevent soil threats and
    support soil functions? Sustainability 6(12):9538–9563
Kidd D, Field D, McBratney A, Webb M (2018) A preliminary spatial quantification of the soil
    security dimensions for Tasmania. Geoderma 322:184–200
Koch A, McBratney A, Adams M, Field D, Hill R, Crawford J, Minasny B, Lal R, Abbott L,
    O'Donnell A, Angers D (2013) Soil security: solving the global soil crisis. Global Pol
    4(4):434–441
Montanarella L, Panagos P (2021) The relevance of sustainable soil management within the
    European green deal. Land Use Policy 100:104950
McBratney A, Field DJ, Koch A (2014) The dimensions of soil security. Geoderma 213:203–213
Orgiazzi A, Ballabio C, Panagos P, Jones A, Fernández-Ugalde O (2018) LUCAS Soil, the larg-
    est expandable soil dataset for Europe: A review. European Journal of Soil Science 69(1):
    140–153. https://doi.org/10.1111/ejss.12499
Panagos P, Van Liedekerke M, Yigini Y, Montanarella L (2013) Contaminated sites in Europe:
    review of the current situation based on data collected through a European network. J Environ
    Public Health 2013:158764
Panagos P, Montanarella L (2018) Soil thematic strategy: an important contribution to pol-
    icy support, research, data development and raising the awareness. Curr Opin Environ Sci
    Health 5:38–41
Panagos P, Standardi G, Borrelli P, Lugato E, Montanarella L, Bosello F (2018) Cost of agricul-
    tural productivity loss due to soil erosion in the European Union: from direct cost evaluation
    approaches to the use of macroeconomic models. Land Degrad Dev 29:471–484
Panagos P, Ballabio C, Poesen J, Lugato E, Scarpa S, Montanarella L, Borrelli P (2020) A soil ero-
    sion indicator for supporting agricultural, environmental and climate policies in the European
    Union. Remote Sens 12(9):1365
Payá Pérez A, Rodríguez EN (2018) Status of local soil contamination in Europe: revision of
    the indicator Progress in the management contaminated sites in Europe. EUR 29124 EN,
    Publications Office of the European Union, Luxembourg. https://doi.org/10.2760/093804
Robinson DA, Panagos P, Borrelli P, Jones A, Montanarella L, Tye A, Obst CG (2017) Soil natural
    capital in Europe; a framework for state and change assessment. Sci Rep 7(1):1–14
Rojas RV, Achouri M, Maroulis J, Caon L (2016) Healthy soils: a prerequisite for sustainable food
    security. Environ Earth Sci 75(3):180
Ronchi S, Salata S, Arcidiacono A, Piroli E, Montanarella L (2019) Policy instruments for soil
    protection among the EU member states: a comparative analysis. Land Use Policy 82:763–780
Sartori M, Philippidis G, Ferrari E, Borrelli P, Lugato E, Montanarella L, Panagos P (2019) A
    linkage between the biophysical and the economic: assessing the global market impacts of soil
    erosion. Land Use Policy 86:299–312
Stavi I, Bel G, Zaady E (2016) Soil functions and ecosystem services in conventional, conserva-
    tion, and integrated agricultural systems: a review. Agron Sustain Dev 36(2):32
Soil Mission Health (2020) Caring for soil is caring for life. https://ec.europa.eu/info/publications/
    caring-soil-caring-life_en. Accessed 14 Apr 2021
Stolte J, Tesfai M, Oygarden L, Kvaerno S, Keizer J, Verheijen F, Panagos P, Ballabio C, Hessel
    R 2016 Soil threats in Europe: status, methods, drivers and effects on ecosystem services. Soil
    threats in Europe; EUR 27607 EN; https://doi.org/10.2788/488054
STS (2006) Soil thematic strategy impact assessment. COM(2006)231 final – SEC(2006)1165.
    https://ec.europa.eu/environment/archives/soil/pdf/-SEC_2006_620.pdf
STS (2021) EU soil strategy consultation. https://ec.europa.eu/environment/news/commission-
    consults-new-eu-soil-strategy-2021-02-02_en
van Leeuwen JP, Saby NPA, Jones A, Louwagie G, Micheli E, Rutgers M, Schulte RPO, Spiegel H,
    Toth G, Creamer RE (2017) Gap assessment in current soil monitoring networks across Europe
    for measuring soil functions. Environ Res Lett 12(12):124007

Chapter 26
Significance of Land Cover Change for Soil Regulating Ecosystem Services Using Maine's Climate Action Plan as a Case Study

E. A. Mikhailova, L. Lin, Z. Hao, H. A. Zurqani, C. J. Post, M. A. Schlautman, and G. C. Post

Abstract Land cover change can result in soil-based greenhouse gas (GHG) emissions. The Climate Action Plan for the State of Maine in the United States of America (U.S.A.) can benefit from the analysis of land cover change and the value of soil carbon (C) regulating ecosystem services (ES) and disservices (ED). The objectives of this study were to (1) determine baseline regulating ES values of soil organic carbon (SOC), soil inorganic carbon (SIC), and total soil carbon (TSC)

E. A. Mikhailova (✉) · C. J. Post
Department of Forestry and Environmental Conservation, Clemson University, Clemson, SC, USA
e-mail: eleanam@clemson.edu; cpost@clemson.edu

L. Lin
Department of Biological Science and Biotechnology, Minnan Normal University, Zhangzhou, China
e-mail: lll2639@mnnu.edu.cn

Z. Hao
University Key Lab for Geomatics Technology and Optimized Resources Utilization in Fujian Province, Fuzhou, China
e-mail: zhenbanghao@fafu.edu.cn

H. A. Zurqani
University of Arkansas Agricultural Experiment Station, Arkansas Forest Resources Center, University of Arkansas at Monticello, Monticello, AR, USA
e-mail: Zurqani@uamont.edu

M. A. Schlautman
Department of Environmental Engineering and Earth Sciences, Clemson University, Anderson, SC, USA
e-mail: mschlau@clemson.edu

G. C. Post
Geography Department, Portland State University, Portland, OR, USA
e-mail: grpost@pdx.edu

J. A. Zinck et al. (eds.), *Geopedology*,
https://doi.org/10.1007/978-3-031-20667-2_26

513

stocks for Maine by soil order and county, based on the concept of the avoided social cost of carbon dioxide (CO_2) emissions, and (2) assess how recent changes in land use/land cover (LULC) across Maine have potentially affected these regulating ES values. The total estimated monetary midpoint value for TSC stocks in Maine was \$295.9B USD (i.e., \$295.9 billion, where B = billion = 10^9), which is comprised of SOC stocks (\$270.3B) and SIC stocks (\$25.6B). Histosols (\$136.5B), Spodosols (\$101.9B), and Inceptisols (\$30.4B) had the highest SOC midpoint value. Inceptisols (\$17.4B), Spodosols (\$4.9B), and Histosols (\$2.4B) had the highest SIC midpoint value. Histosols (\$138.8B), Spodosols (\$106.9B), and Inceptisols (\$47.9B) had the highest TSC midpoint value. Aroostook (\$53.7B), Piscataquis (\$50.0B), and Somerset (\$35.5B) counties had the highest SOC midpoint values. Aroostook (\$6.0B), Somerset (\$3.2B), and Penobscot (\$2.9B) counties had the highest SIC midpoint values. Aroostook (\$59.7B), Piscataquis (\$52.3B), and Somerset (\$38.7B) counties had the highest TSC midpoint values. Changes in LULC for Maine between 2001 and 2016 resulted in the maximum potential loss of regulating ES value of \$194.4 M (\$194.4 M, where M = million = 10^6) statewide based on TSC. Changes in LULC across Maine varied by county, soil order, and pre-existing land cover. Changes in LULC showed a large percentage increase in developed medium intensity (11.2%) and the developed high intensity (12.2%) land cover classes, which accounted for more than 50% of the associated potential total loss of regulating ES value. Cumberland, York, and Penobscot counties had the largest increase in development. Most soil orders have experienced losses in "low disturbance" land covers and gains in "high disturbance" land covers with Spodosols contributing over 50% of the total potential loss of regulating ES value in Maine while covering 64% of the total land area. This analysis provides actionable spatial information on how 15 years of land cover change likely impacted soil carbon releases to the atmosphere. These results can help inform Maine's Climate Action Plan what land cover and soil type combinations should be managed to help reduce carbon emissions.

Keywords Ecosystem · Carbon · Emissions, CO_2 · Climate · Cost · Inorganic · Organic · Social · Soil

26.1   Introduction

The State of Maine has established a goal to decrease greenhouse gas (GHG) emissions by 80% by 2050 (An Act 2019), which contributes to the United Nations (UN) Sustainable Development Goals (SDGs), especially concerning combating "climate change and its impacts on future climate" (Wood et al. 2017). The ecosystem services (ES) framework can be useful in achieving these goals because it places a value on the benefits (ES) and/or ecosystem disservices (ED) people obtain from nature (provisioning, regulating/maintenance, and cultural services) (Wood et al. 2017). For example, GHG emissions can be reduced by considering soil carbon regulating ES, which are associated with sequestered stocks of soil organic carbon (SOC) (derived from living matter), soil inorganic carbon (SIC) (different types of carbonates), and total soil carbon (TSC = SOC + SIC), which can vary with geographic location and soil type in Maine. Mikhailova et al. (2021a) referred to sequestered soil carbon in the forms of SOC and SIC as an ES, which results in "avoided" emissions and social costs associated with carbon dioxide (CO_2) entering the atmosphere. "Realized" social costs are associated with the release of CO_2 to the atmosphere from losses of SOC and SIC, which is an ED (Mikhailova et al. 2021a; Zamanian et al. 2021).

Soil resources are a part of "natural" capital, but they are often valued solely for their provisioning ES (e.g., food production) with little consideration of regulating ES (e.g., carbon (C) sequestration), but increased concerns over GHG emissions from soils (ED) require assessment of soil carbon stocks and associated social costs of soil C in various soil types (soil pedodiversity) (Adhikari and Hartemink 2016; Baveye et al. 2016; Ibáñez and Gómez 2016). From a business point of view, soil pedodiversity is a soil "portfolio" with its SOC, SIC, and TSC stocks in a geographic area under various land covers (Mikhailova et al. 2021a). For example, the state of Maine has five soil orders (Entisols, Inceptisols, Histosols, Mollisols, and Spodosols) with soil-specific characteristics and constraints related to soil ES/ED, which are all part of the intricate "quilt" of land use/land covers (LULC) within the landscape (Amundson and Biardeau 2018) (Fig.26.1).

Soils of Maine have undergone three varying degrees of weathering (physical and chemical breakdown of particles (Plaster 2003)): slightly weathered (Entisols, Inceptisols, Histosols), moderately weathered (Mollisols), and strongly developed (Spodosols) (Table 26.1). Entisols (1% of the total area) and Inceptisols (27%) contain low soil C contents with limited capacity to sequester C because of their slight degree of weathering and soil development (Plaster 2003). Spodosols are common soils in Maine (64% of the total area) and often exhibit C translocation from surface organic horizons to the subsoil mineral horizons because of their strong degree of soil development (Plaster 2003). Maine selected Spodosols to be the State Soil (Soil series name: Chesuncook) for its wide distribution in the state forests (USDA/NRCS 2021). Fernandez (2008) conducted research on soil C and nutrient storage in forest soils of Maine and reported that each differs by forest type in the state. In addition, Fernandez (2008) reported significant differences in the vertical distribution of SOC in both organic (O) and mineral horizons (e.g., B) in forest soils. Subaqueous soils are also found in the state (Osher and Flannagan 2007).

Mollisols are nutrient-rich soils high in C, but they are almost negligible in Maine. Histosols (8% of the total area) are organic carbon-rich soils commonly found in different types of wetlands and can be a large source of greenhouse gases emissions resulting from changes in LULC (e.g., drainage, development, etc.) (Wright and Inglett 2009). Using ES as "an operational framework" to place an economic value on soil C is becoming more common, but because of "the difficulty in relating soil properties to ES, soil ES are still not fully considered in the territorial planning decision process" (Fossey et al. 2020). Past research on the economic value of SOC, SIC, and TSC stocks in the USA based on avoided social SC-CO_2

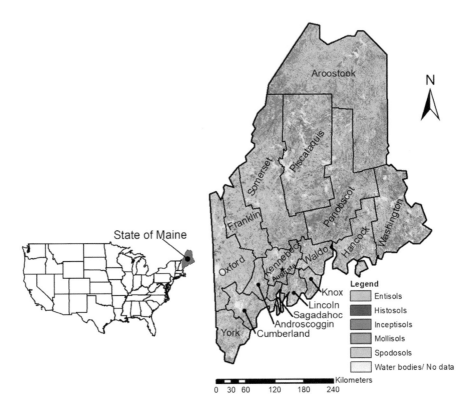

Fig. 26.1 Soil map of Maine (U.S.A.) (Latitude: 42° 58′ N to 47° 28′ N; Longitude: 66° 57′ W to 71° 5′ W). (Adapted from Soil Survey Staff 2021a, b)

emissions has been conducted using both biophysical (e.g., soil orders) and administrative accounts (e.g., states, regions, counties, farm, etc.) (Mikhailova et al. 2019a, b, 2020; Groshans et al. 2019) and has showed the need for soil- and carbon-specific management strategies at the state level. The hypothesis of this study is that pedodiversity (e.g., taxonomic categories) overlaid with administrative units (Fig. 26.1) and LULC change can be used to locate spatial patterns of soil C hotspots for sustainable C management in the state of Maine.

The specific objectives of this study were to (1) determine baseline regulating ES values of SOC, SIC, and TSC stocks for Maine by soil order and county, based on the concept of the avoided social cost of CO_2 emissions (SC–CO_2), and (2) assess how recent changes in LULC across Maine have potentially affected these regulating ES values. Sequestered soil C is one example of avoided SC–CO_2 and has been assigned a value of $46/metric ton of CO_2 (valid until 2025) by the U.S. Environmental Protection Agency (EPA) using 2007 U.S.D. assuming a 3% average discount rate (EPA 2016). Our calculations provide estimates for the monetary values of SOC, SIC, and TSC across the state, by soil order, by LULC and various spatial aggregation levels (i.e., county) using the State Soil Geographic (STATSGO) database and information previously determined by Guo et al. (2006).

Table 26.1 Soil diversity (pedodiversity) is expressed as taxonomic diversity at the level of soil order and ecosystem service types in the state of Maine (U.S.A.)

	Stocks	Ecosystem services		
Soil order	General characteristics and constraints	Provisioning	Regulation/maintenance	Cultural
	Slight weathering			
Entisols	Embryonic soils with ochric epipedon	x	x	x
Inceptisols	Young soils with ochric or umbric epipedon	x	x	x
Histosols	Organic soils with ≥ 20% of organic carbon	x	x	x
	Moderate weathering			
Mollisols	Carbon-enriched soils with B.S. ≥ 50%	x	x	x
	Strong weathering			
Spodosols	Coarse-textured soils with albic and spodic horizons	x	x	x

Adapted from Mikhailova et al. (2021a)
Note: B.S. base saturation

26.2 Materials and Methods

26.2.1 The Ecosystem Services Accounting Framework

This study used monetary valuation based on both biophysical (science-based, Fig. 26.1) and administrative (boundary-based, Fig. 26.1) accounts to calculate the social cost of C (SC-CO_2) for SOC, SIC, and TSC (Tables 26.2 and 26.3).

The present study estimates monetary values associated with stocks of SOC, SIC, and TSC in Maine based on reported contents (in kg m^{-2}) from Guo et al. (2006). The avoided social cost of carbon (SC-CO_2) was estimated assuming a cost of $46/metric ton of CO_2 (valid until 2025), which was provided by the U.S. Environmental Protection Agency (EPA) using 2007 U.S.D. assuming a 3% average discount rate (EPA 2016). The U.S. EPA reports that the SC-CO_2 value is intended to estimate damages from climate change, but it should be noted that this value excludes many potential climate change impacts and may underestimate actual damages from CO_2 releases (EPA 2016). Area-normalized monetary values ($ m^{-2}) were calculated using Eq. (26.1), and total economic values were summed over the appropriate area(s) (Note: 1 metric ton = 1 megagram (Mg) = 1000 kilograms (kg), C = carbon):

$$\frac{\$}{m^2} = \left(SOC/SIC/TSC\,Content, \frac{kg\,C}{m^2} \right) \times \frac{1\,Mg}{10^3\,kg} \times \frac{44\,Mg\,CO_2}{12\,Mg\,C} \times \frac{\$46}{Mg\,CO_2} \quad (26.1)$$

Table 26.2 The ecosystem services accounting framework used in this study

Stocks		Flows		Value
Biophysical accounts (science-based)	**Administrative accounts (boundary-based)**	**Monetary account(s)**	**Benefit(s)/ damage(s)**	**Value (total)**
Soil extent:	Administrative extent:	Ecosystem good(s) and service(s):	Sector:	Value (types):
Composite (total) stock: Total soil carbon (TSC) = Soil organic carbon (SOC) + Soil inorganic carbon (SIC)				
Soil orders (5): Entisols, Inceptisols, Histosols, Mollisols, Spodosols	State (Maine) County (16 counties)	Regulating (e.g., carbon sequestration)	Environment: Sequestered carbon	The social cost of C (SC-CO$_2$) and avoided emissions: $46/metric ton of CO$_2$ (2007 U.S.D. with a 3% average discount rate (EPA 2016))

Adapted from Groshans et al. (2019)

Table 26.4 presents area-normalized contents (kg m^{-2}) and monetary values ($ m^{-2}) of soil carbon, which were used to estimate stocks of SOC, SIC, and TSC and their corresponding values by multiplying the contents/values by the area of a particular soil order within a county (Table 26.3). For example, for the soil order Inceptisols, Guo et al. (2006) reported a midpoint SOC content of 8.9 kg m^{-2} for the upper 2 m (Table 26.4). Using this SOC content in eq. (1) results in an area-normalized SOC value of $1.50 m^{-2}. Multiplying the SOC content and its corresponding area-normalized value each by the total area of Inceptisols present in Maine (20,281 km^2, Table 26.3) results in an estimated SOC stock of 1.80×10^{11} kg with an estimated monetary value of $30.4B (Table 26.5).

Land use/land cover change in Maine between 2001 and 2016 was analyzed using classified land cover data from the Multi-Resolution Land Characteristics Consortium (MRLC) (2021) to identify areas of land cover change that imply land-use change. Changes in land cover, with their associated soil types, were calculated in ArcGIS Pro 2.6 (ESRI 2021) by comparing the 2001 and 2016 data, converting the land cover to vector format, and unioning the data with the soils layer in the Soil Survey Geographic (SSURGO) Database (Soil Survey Staff 2021a, b). It is important to note that use of remote sensing to detect land cover changes cannot account for potential soil carbon changes that might result from different land management strategies within a single land cover class. In other words, if there is no detectable change in land cover by MRLC, the assumption is made that there has been no change in soil carbon contents and stocks.

Table 26.3 Soil diversity (pedodiversity) by soil order (taxonomic pedodiversity), and county in Maine (U.S.A.)

County	Area by county (km²) (% from total)	Degree of weathering and soil development				
		Slight			Moderate	Strong
		Entisols	Inceptisols	Histosols	Mollisols	Spodosols
		2016 Area by soil order (km²) (% from county total)				
Androscoggin	1170 (2)	24 (2)	252 (22)	43 (4)	0	851 (73)
Aroostook	17013 (22)	156 (1)	5119 (30)	997 (6)	0	10742 (63)
Cumberland	2139 (3)	329 (15)	618 (29)	58 (3)	0.00013 (0)	1134 (53)
Franklin	4314 (6)	15 (0)	1469 (34)	305 (7)	0	2524 (59)
Hancock	4020 (5)	199 (5)	667 (17)	264 (7)	0	2890 (72)
Kennebec	2217 (3)	2 (0)	842 (38)	66 (3)	0.00028 (0)	1307 (59)
Knox	931 (1)	1 (0)	550 (59)	6 (1)	0	373 (40)
Lincoln	1165 (2)	0.4519 (0)	720 (62)	13 (1)	0	431 (37)
Oxford	5063 (7)	48 (0)	717 (62)	202 (1)	6 (0)	4090 (37)
Penobscot	8635 (11)	114 (1)	2506 (29)	318 (4)	0	5698 (66)
Piscataquis	9178 (12)	31 (0)	1395 (15)	1476 (16)	0	6276 (68)
Sagadahoc	630 (1)	35 (5)	144 (23)	12 (2)	0	439 (70)
Somerset	9047 (12)	79 (1)	2568 (28)	849 (9)	0.00145 (0)	5551 (61)
Waldo	1766 (2)	9 (1)	433 (25)	132 (7)	8 (0)	1183 (67)
Washington	6559 (9)	0.00240 (0)	1521 (23)	856 (13)	0	4183 (64)
York	2529 (3)	17 (1)	759 (30)	182 (7)	3 (0)	1568 (62)
Totals (%)	**76375 (100%)**	**1058 (1%)**	**20281 (27%)**	**5778 (8%)**	**18 (<0.02%)**	**49240 (64%)**

Based on Soil Survey Geographic (SSURGO) Database (2021a, b)

26.3 Results

Based on avoided SC–CO$_2$, the total estimated monetary mid-point value for TSC in Maine was \$295.9B (i.e., 295.9 billion U.S. dollars, where B = billion = 10^9), \$270.3B for SOC (91% of the total value), and \$25.6B for SIC (9% of the total value). Previously, we have reported that among the 48 conterminous states of the U.S., Maine ranked 32nd for TSC (Mikhailova et al. 2019a), 27th for SOC (Mikhailova et al. 2019b), and 41st for SIC (Groshans et al. 2019).

26.3.1 Value of SOC by Soil Order and County for Maine

Soil orders with the highest midpoint content and monetary value for SOC were: Histosols (\$136.5B), Spodosols (\$101.9B), and Inceptisols (\$30.4B) (Table 26.5). The counties with the highest midpoint SOC values were Aroostook (\$53.7B), Piscataquis (\$50.0B), and Somerset (\$35.5B) (Table 26.5). Although Histosols occupy only 8% (Table 26.3) of the state total area, they contribute 50% to the total SOC value making them a SOC "hotspot" (Table 25.5). Histosols commonly form in wetland landscapes and can be subject to draining, which can result in GHG emissions (Köchy et al. 2015).

Table 26.4 Area-normalized content (kg m^{-2}) and monetary values ($ m^{-2}) of soil organic carbon (SOC), soil inorganic carbon (SIC), total soil carbon (TSC) by soil order based on data from Guo et al. (2006) for the upper 2 m of soil and an avoided social cost of carbon (SC-CO$_2$) of $46/metric ton of CO$_2$ (2007 U.S.D. with a 3% average discount rate (EPA 2016))

Soil order	SOC content	SIC content	TSC content	SOC value	SIC value	TSC value
	Minimum—Midpoint—Maximum values			Midpoint values		
	(kg m^{-2})	(kg m^{-2})	(kg m^{-2})	($ m^{-2})	($ m^{-2})	($ m^{-2})
Slightly weathered						
Entisols	1.8—8.0—15.8	1.9—4.8—8.4	3.7—12.8—24.2	1.35	0.82	2.17
Inceptisols	2.8—8.9—17.4	2.5—5.1—8.4	5.3—14.0—25.8	1.50	0.86	2.36
Histosols	63.9—140.1—243.9	0.6—2.4—5.0	64.5—142.5—248.9	23.62	0.41	24.03
Moderately weathered						
Mollisols	5.9—13.5—22.8	4.9—11.5—19.7	10.8—25.0—42.5	2.28	1.93	4.21
Strongly weathered						
Spodosols	2.9—12.3—25.5	0.2—0.6—1.1	3.1—12.9—26.6	2.07	0.10	2.17

Note: TSC = SOC + SIC

Table 26.5 The monetary value of soil organic carbon (SOC) by soil order and county for Maine (USA), based on the areas shown in Table 26.3 and the area-normalized midpoint monetary values shown in Table 26.4

County	Total value ($)	Degree of weathering and soil development				
		Slight			Moderate	Strong
		Entisols	Inceptisols	Histosols	Mollisols	Spodosols
		Value ($)				
Androscoggin	3.19×10^9	3.23×10^7	3.78×10^8	1.01×10^9	0	1.76×10^9
Aroostook	5.37×10^{10}	2.10×10^8	7.68×10^9	2.35×10^{10}	0	2.22×10^{10}
Cumberland	5.08×10^9	4.44×10^8	9.28×10^8	1.37×10^9	3.00×10^2	2.35×10^9
Franklin	1.47×10^{10}	2.01×10^7	2.20×10^9	7.21×10^9	0	5.23×10^9
Hancock	1.35×10^{10}	2.68×10^8	1.00×10^9	6.23×10^9	0	5.98×10^9
Kennebec	5.53×10^9	2.69×10^6	1.26×10^9	1.55×10^9	6.38×10^2	2.71×10^9
Knox	1.75×10^9	1.59×10^6	8.25×10^8	1.50×10^8	0	7.73×10^8
Lincoln	2.29×10^9	6.10×10^5	1.08×10^9	3.14×10^8	0	8.92×10^8
Oxford	1.44×10^{10}	6.49×10^7	1.08×10^9	4.78×10^9	1.38×10^7	8.47×10^9
Penobscot	2.32×10^{10}	1.54×10^8	3.76×10^9	7.51×10^9	0	1.18×10^{10}
Piscataquis	5.00×10^{10}	4.22×10^7	2.09×10^9	3.49×10^{10}	0	1.30×10^{10}
Sagadahoc	1.45×10^9	4.67×10^7	2.16×10^8	2.73×10^8	0	9.10×10^8
Somerset	3.55×10^{10}	1.07×10^8	3.85×10^9	2.00×10^{10}	3.32×10^3	1.15×10^{10}
Waldo	6.26×10^9	1.23×10^7	6.49×10^8	3.13×10^9	1.94×10^7	2.45×10^9
Washington	3.11×10^{10}	3.24×10^3	2.28×10^9	2.02×10^{10}	0	8.66×10^9
York	8.70×10^9	2.26×10^7	1.14×10^9	4.29×10^9	7.09×10^6	3.25×10^9
Totals	$\mathbf{2.70 \times 10^{11}}$	$\mathbf{1.43 \times 10^9}$	$\mathbf{3.04 \times 10^{10}}$	$\mathbf{1.36 \times 10^{11}}$	$\mathbf{4.03 \times 10^7}$	$\mathbf{1.02 \times 10^{11}}$

26.3.2 Value of SIC by Soil Order and County for Maine

Soil orders with the highest midpoint monetary value for SIC were: Inceptisols ($17.4B), Spodosols ($4.9B), and Histosols ($2.4B) (Table 26.6). The counties with the highest midpoint SIC values were Aroostook ($6.0B), Somerset ($3.2B), and Penobscot ($2.9B) (Table 26.6). Soil inorganic carbon is less prominent in Maine because of the humid continental climate and commonly found glacial parent material, which tends to be acidic and not conducive for SIC formation.

26.3.3 Value of TSC (SOC + SIC) by Soil Order and County for Maine

Soil orders with the highest midpoint monetary value for TSC were: Histosols ($138.8B), Spodosols ($106.9B), and Inceptisols ($47.9B) (Table 26.7). The counties with the highest midpoint TSC values were Aroostook ($59.7B), Piscataquis ($52.3B), and Somerset ($38.7B) (Table 26.7). These monetary values can be used by planners to identify potential "hotspots" of TSC in the landscape to avoid or

Table 26.6 The monetary value of soil inorganic carbon (SIC) by soil order and county for the state of Maine (USA), based on the areas shown in Table 26.3 and the area-normalized midpoint monetary values shown in Table 26.4

		Degree of weathering and soil development				
		Slight			Moderate	Strong
		Entisols	Inceptisols	Histosols	Mollisols	Spodosols
County	Total value ($)	Value ($)				
Androscoggin	3.39×10^8	1.96×10^7	2.17×10^8	1.76×10^7	0	8.51×10^7
Aroostook	6.01×10^9	1.28×10^8	4.40×10^9	4.09×10^8	0	1.07×10^9
Cumberland	9.39×10^8	2.70×10^8	5.32×10^8	2.37×10^7	2.54×10^2	1.13×10^8
Franklin	1.65×10^9	1.22×10^7	1.26×10^9	1.25×10^8	0	2.52×10^8
Hancock	1.13×10^9	1.63×10^8	5.73×10^8	1.08×10^8	0	2.89×10^8
Kennebec	8.83×10^8	1.63×10^6	7.24×10^8	2.70×10^7	5.40×10^2	1.31×10^8
Knox	5.14×10^8	9.66×10^5	4.73×10^8	2.60×10^6	0	3.73×10^7
Lincoln	6.69×10^8	3.71×10^5	6.20×10^8	5.45×10^6	0	4.31×10^7
Oxford	1.16×10^9	3.94×10^7	6.17×10^8	8.30×10^7	1.17×10^7	4.09×10^8
Penobscot	2.95×10^9	9.33×10^7	2.16×10^9	1.30×10^8	0	5.70×10^8
Piscataquis	2.46×10^9	2.56×10^7	1.20×10^9	6.05×10^8	0	6.28×10^8
Sagadahoc	2.01×10^8	2.84×10^7	1.24×10^8	4.75×10^6	0	4.39×10^7
Somerset	3.18×10^9	6.48×10^7	2.21×10^9	3.48×10^8	2.81×10^3	5.55×10^8
Waldo	5.69×10^8	7.48×10^6	3.72×10^8	5.43×10^7	1.64×10^7	1.18×10^8
Washington	2.08×10^9	1.97×10^3	1.31×10^9	3.51×10^8	0	4.18×10^8
York	9.04×10^8	1.37×10^7	6.53×10^8	7.45×10^7	6.00×10^6	1.57×10^8
Totals	$\mathbf{2.56 \times 10^{10}}$	$\mathbf{8.68 \times 10^8}$	$\mathbf{1.74 \times 10^{10}}$	$\mathbf{2.37 \times 10^9}$	$\mathbf{3.41 \times 10^7}$	$\mathbf{4.92 \times 10^9}$

Table 26.7 The monetary value of total soil carbon (TSC) by soil order and county for the state of Maine (USA), based on the areas shown in Table 26.3 and the area-normalized midpoint monetary values shown in Table 26.4

| County | Total value ($) | Slight | | | Moderate | Strong |
| | | Entisols | Inceptisols | Histosols | Mollisols | Spodosols |
		Value ($)				
Androscoggin	3.52×10^9	5.20×10^7	5.95×10^8	1.03×10^9	0	1.85×10^9
Aroostook	5.97×10^{10}	3.38×10^8	1.21×10^{10}	2.40×10^{10}	0	2.33×10^{10}
Cumberland	6.02×10^9	7.13×10^8	1.46×10^9	1.39×10^9	5.53×10^2	2.46×10^9
Franklin	1.63×10^{10}	3.23×10^7	3.47×10^9	7.34×10^9	0	5.48×10^9
Hancock	1.46×10^{10}	4.32×10^8	1.57×10^9	6.34×10^9	0	6.27×10^9
Kennebec	6.41×10^9	4.32×10^6	1.99×10^9	1.58×10^9	1.18×10^3	2.84×10^9
Knox	2.26×10^9	2.56×10^6	1.30×10^9	1.52×10^8	0	8.10×10^8
Lincoln	2.96×10^9	9.81×10^5	1.70×10^9	3.19×10^8	0	9.35×10^8
Oxford	1.56×10^{10}	1.04×10^8	1.69×10^9	4.86×10^9	2.56×10^7	8.87×10^9
Penobscot	2.62×10^{10}	2.47×10^8	5.91×10^9	7.64×10^9	0	1.24×10^{10}
Piscataquis	5.24×10^{10}	6.78×10^7	3.29×10^9	3.55×10^{10}	0	1.36×10^{10}
Sagadahoc	1.65×10^9	7.51×10^7	3.41×10^8	2.78×10^8	0	9.54×10^8
Somerset	3.87×10^{10}	1.71×10^8	6.06×10^9	2.04×10^{10}	6.12×10^3	1.20×10^{10}
Waldo	6.83×10^9	1.98×10^7	1.02×10^9	3.18×10^9	3.57×10^7	2.57×10^9
Washington	3.32×10^{10}	5.21×10^3	3.59×10^9	2.06×10^{10}	0	9.08×10^9
York	9.61×10^9	3.63×10^7	1.79×10^9	4.36×10^9	1.31×10^7	3.40×10^9
Totals	$\mathbf{2.96 \times 10^{11}}$	$\mathbf{2.30 \times 10^9}$	$\mathbf{4.79 \times 10^{10}}$	$\mathbf{1.39 \times 10^{11}}$	$\mathbf{7.44 \times 10^7}$	$\mathbf{1.07 \times 10^{11}}$

minimize the GHG emissions upon soil disturbance. It is important to note that TSC is composed of different types of C: SOC derived from plant and animal matter, and SIC (e.g., calcium carbonate: $CaCO_3$) formed because of mineral weathering or soil reaction with atmospheric CO_2.

26.3.4 Land Use/Land Cover Change by Soil Order in Maine from 2001 to 2016

Maine experienced changes in land use/land cover (LULC) over the 15-year period from 2001 to 2016 (Table 26.8, Fig. 26.2). Most soil orders experienced area losses in "low disturbance" LULC classes while gaining in the areas of "developed" LULC classes. The most dramatic increases in developed land areas occurred in Cumberland, York, and Penobscot counties. Cumberland and York counties are included in the Portland-South Portland-Biddeford metropolitan statistical area, the most populated region in Maine. Penobscot county comprises the Bangor, metropolitan statistical area with the city of Bangor, home to the University of Maine.

Table 26.8 Land use/land cover (LULC) change by soil order in Maine (USA) from 2001 to 2016

NLCD land cover classes (LULC)	2016 Total area by LULC (km²) change (%)	Degree of weathering and soil development				
		Slight			Moderate	Strong
		Entisols	Inceptisols	Histosols	Mollisols	Spodosols
		2016 Area by soil order, km² (Change in area, 2001–2016, %)				
Barren land	207 (−1.2)	10.4 (−1.2)	39.9 (−2.2)	9.6 (0.3)	0.1 (0.0)	147.5 (−1.0)
Woody wetlands	9615 (−1.2)	282.2 (0.0)	4632.4 (−0.9)	1779.3 (−1.9)	6.5 (−0.5)	2914.7 (−1.4)
Shrub/scrub	3730 (89.9)	16.5 (29.9)	621.3 (73.0)	224.6 (60.2)	0.1 (23.9)	2867.9 (97.5)
Mixed forest	24,932 (−2.4)	194.8 (−1.0)	5913.9 (−1.4)	1266.3 (−2.1)	3.9 (−0.4)	17552.9 (−2.9)
Deciduous forest	11,957 (−10.5)	70.0 (−4.1)	2284.7 (−8.0)	568.4 (−12.82)	1.8 (0.1)	9032.2 (−11.1)
Herbaceous	1983 (11.0)	14.0 (25.5)	406.8 (7.3)	118.5 (1.0)	0.1 (48.2)	1444.0 (12.8)
Evergreen forest	16,714 (0.8)	218.1 (1.7)	3976.6 (−0.1)	1398.7 (2.0)	1.7 (−1.2)	11118.7 (0.9)
Emergent herbaceous wetlands	935 (12.5)	73.4 (0.6)	408.9 (9.0)	291.9 (12.5)	0.2 (19.3)	160.4 (30.2)
Hay/pasture	1982 (−5.8)	38.4 (−8.2)	761.8 (−3.9)	14.9 (−13.8)	2.3 (−1.6)	1164.8 (−6.9)
Cultivated crops	917 (3.4)	21.3 (8.1)	185.5 (4.4)	10.7 (13.4)	0.5 (0.0)	698.9 (2.8)
Developed, open space	2080 (0.6)	50.6 (2.1)	583.2 (0.2)	73.4 (0.5)	0.3 (3.3)	1372.5 (0.7)
Developed, medium intensity	305 (11.2)	19.2 (9.8)	114.1 (11.6)	3.6 (13.3)	0.0 (8.3)	168.2 (11.0)
Developed, low intensity	914 (2.0)	41.4 (3.1)	308.6 (1.6)	17.8 (1.8)	0.2 (1.8)	546.3 (2.1)
Developed, high intensity	103 (12.2)	8.1 (7.9)	43.2 (12.0)	0.5 (33.3)	0.0 (0.0)	51.2 (13.0)

Note: the sum of the total area by LULC in this column agrees with the sum of the area reported by the county in Table 26.3

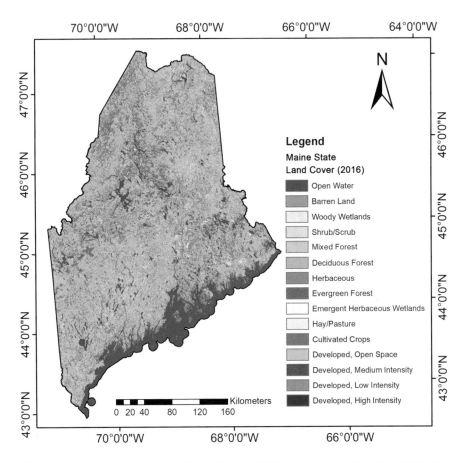

Fig. 26.2 Map of land cover for Maine (U.S.A.): 2016 (Latitude: 42° 58′ N to 47° 28′ N; Longitude: 66° 57′ W to 71° 5′ W) (Multi-Resolution Land Characteristics Consortium 2021)

26.4   Discussion

26.4.1   The Role of Pedodiversity (Soil Diversity) in Carbon Regulating ES/ED in Maine

Achieving reductions in GHG emissions in the state of Maine requires data on soil C regulating ES/ED integrated with the concepts of pedodiversity (soil diversity) at the state and county levels. The Maine soil "portfolio" (Mikhailova et al. 2021a) is composed of five soil orders: Entisols (1% of the total soil area), Inceptisols (27%), Histosols (8%), Mollisols (<0.02%), and Spodosols (64%) (Fig. 26.1, Table 26.3, Fig. 26.2). Highly weathered Spodosols account for the most significant area in the state, but it is not the most important contributor to soil C regulating ES. Instead, because of its high SOC content, Histosols are a carbon "hotspot" that contributes over 50% of the total monetary value for SOC in the state while covering only 8% of the state's area. The relative contribution of SIC to soil C regulating ES is small in the state. Soil "portfolios" differ within each county in Maine.

The social costs of soil C can be presented using the concepts of "avoided" and "realized" social costs, which demonstrate different interpretations of the regulating ES/ED associated with soil C. The "avoided" social cost encompasses the benefits of sequestered soil C because it is not emitted to the atmosphere as CO_2, while "realized" soil cost refers to damages resulting from CO_2 emissions. Data reported in Table 26.8 depict somewhat extreme scenarios with the "realized" social costs as the maximum potential costs that would occur if all stocks of sequestered soil C were released to the atmosphere as CO_2. For example, in Cumberland County, the soil order Spodosols makes the most considerable contribution to the SC–CO_2 because of its dominant area. In Lincoln County, the most significant area is occupied by Inceptisols, but its relatively small area overall and low soil carbon stocks translate into relatively small monetary values for the SC–CO_2. In Piscataquis County, Histosols occupy a relatively small area compared to Spodosols but make the largest contribution to the SC–CO_2. In Maine, climate change can increase decomposition rates of the high C content Histosols under various LULC scenarios, making them vulnerable to C loss. Therefore, Histosols may experience higher decomposition rates due to increases in temperature and precipitation. All soils in the state of Maine have low recarbonization potential because of various reasons (e.g., high economic cost of soil C sequestration, climate change, etc.) (Table 26.9).

26.4.2 The Role of Land Cover Change in Soil Carbon Regulating ES/ED in Maine

Remote sensing can help the state of Maine to track land cover conversions from low disturbance LULC to high disturbance LULC (development) by county and soil type. For example, Table 26.10 demonstrates areas or carbon "hotspots" where development between 2001 and 2016 likely resulted in high potential "realized" social costs based on the conservative assumption that land developed over the time period of interest had no soil carbon stocks remaining after development. This conservative assumption is reasonably consistent with IPCC guidelines, which specify that soil carbon stocks in the original land cover/use decrease to zero after 20 years for Tier 1 evaluations of changes in LULC (IPCC 2006; Sallustio et al. 2015). Monetary values of maximum potential "realized" social cost depend on the area of disturbance and the soil type with its corresponding TSC content (Table 26.10).

For example, Spodosols and Inceptisols are "hotspots" of "realized" social costs because of their dominance in the landscape and significant area change due to development (Table 26.10). Histosols are also a "hotspot" of sequestered C but are vulnerable to development, which has resulted in a maximum potential realized social cost of over \$29.9M from 2001 to 2016 (Table 26.10). Integration of

Table 26.9 Distribution of soil carbon and regulating ecosystem services in Maine (USA) by soil order

Soil regulating ecosystem services in the state of Maine				
Degree of weathering and soil development				
Slight 36%			**Moderate** < 0.02%	**Strong** 64%
Entisols 1%	**Inceptisols** 27%	**Histosols** 8%	**Mollisols** < 0.02%	**Spodosols** 64%
Social cost of soil organic carbon (SOC): $270.3B				
$1.4B	$30.4B	$136.5B	$40.3M	$101.9B
1%	11%	50%	< 0.01%	38%
Social cost of soil inorganic carbon (SIC): $25.6B				
$867.9M	$17.4B	$2.4B	$34.1M	$4.9B
3%	68%	9%	< 0.1%	20%
Social cost of total soil carbon (TSC): $295.9B				
$2.3B	$47.9B	$138.8B	$74.4M	$106.9B
1%	16%	47%	< 0.02%	36%
Sensitivity to climate change				
Low	Low	High	High	Low
SOC and SIC sequestration (recarbonization) potential				
Low	Low	Low	Low	Low

Photos courtesy of USDA/NRCS (Soil Survey Staff 2021a, b). Values are taken/derived from Tables 26.3, 26.5, 26.6, and 26.7
Note: Entisols, Inceptisols, Mollisols, and Spodosols are mineral soils. Histosols are mostly organic soils. M = million = 10^6; B = billion = 10^9

pedodiversity concepts with LULC classes and administrative units (e.g., counties) show that land development in Cumberland County from 2001 to 2016 resulted in the highest SC-CO$_2$ ($59.7M), followed by York ($25.7M) and Penobscot ($21.5M) counties (Tables 26.11 and 26.12, Fig. 26.3).

Remote sensing techniques offer a range of other applications to provide even more detailed analysis of documenting past and potential soil carbon loss. For example, remote sensing can identify both historical and current LULC classes (Table 26.8), which can be integrated with pedodiversity concepts, landscape units, geomorphic units, and administrative units. It can be used to estimate the rate of change of LULC conversion, which can be translated into approximate

Table 26.10 Increases in developed land and maximum potential for realized social costs of carbon due to complete loss of total soil carbon of developed land by soil order in Maine (USA) from 2001 to 2016

NLCD land cover classes (LULC)	Degree of weathering and soil development				
	Slight			Moderate	Strong
	Entisols	Inceptisols	Histosols	Mollisols	Spodosols
	Area change, km^2 (social cost of CO_2, SC-CO_2, $)				
Developed, open space	1.0 ($2.2M)	1.1 ($2.5M)	0.4 ($8.9M)	0.01 ($45,468)	10.2 ($22.0M)
Developed, medium intensity	1.7 ($3.7M)	11.8 ($28.0M)	0.4 ($10.3M)	–	16.6 ($36.1M)
Developed, low intensity	1.3 ($2.7M)	4.9 ($11.5M)	0.3 ($7.6M)	–	11.2 ($24.3M)
Developed, high intensity	0.6 ($1.3M)	4.6 ($10.9M)	0.1 ($3.1M)	–	5.9 ($12.8M)
Totals	4.6 ($9.9M)	22.4 ($52.9M)	1.2 ($29.9M)	0.01 ($45,468)	43.9 ($95.2M)

Values are derived from Tables 26.4 and 26.8
Note: Entisols, Inceptisols, Mollisols, and Spodosols are mineral soils. Histosols are mostly organic soils. M = million = 10^6

Table 26.11 Land development increase (between 2001 and 2016) in Maine (USA) by county

County	Total area change (km^2)	Degree of weathering and soil development				
		Slight			Moderate	Strong
		Entisols	Inceptisols	Histosols	Mollisols	Spodosols
		Developed area increase between 2001 and 2016 (km^2)				
Androscoggin	5.31	0.03	1.44	0.09	0	3.75
Aroostook	4.2	0.28	1.52	0.09	0	2.31
Cumberland	19.04	3.92	6.31	0.47	0	8.34
Franklin	3.61	0.10	0.26	0.12	0	3.13
Hancock	2.47	0.07	0.69	0.03	0	1.68
Kennebec	6.19	0	2.96	0.04	0	3.19
Knox	1.58	0	1.10	0	0	0.48
Lincoln	0.47	0	0.31	0	0	0.16
Oxford	2.4	0	0.27	0.01	0	2.12
Penobscot	9.06	0.11	4.17	0.05	0	4.73
Piscataquis	0.43	0	0.13	0	0	0.30
Sagadahoc	1.93	0.05	0.50	0	0	1.38
Somerset	2.24	0.04	0.66	0.04	0	1.50
Waldo	1.03	0	0.36	0.02	0	0.65
Washington	4.27	0	1.16	0.19	0	2.92
York	10.52	0.06	2.56	0.11	0.01	7.78
Totals	74.75	4.66	24.40	1.26	0.01	44.42

Table 26.12 Impacts of land development increase on the maximum potential realized social costs of carbon dioxide (SC-CO$_2$) from total soil carbon (TSC) in Maine (USA) from 2001 to 2016 by county

County	Total SC-CO$_2$ ($)	Degree of weathering and soil development				
		Slight			Moderate	Strong
		Entisols	Inceptisols	Histosols	Mollisols	Spodosols
		SC-CO$_2$ ($)				
Androscoggin	1.38×10^7	6.51×10^4	3.40×10^6	2.16×10^6	0	8.14×10^6
Aroostook	1.14×10^7	6.08×10^5	3.59×10^6	2.16×10^6	0	5.01×10^6
Cumberland	5.28×10^7	8.51×10^6	1.49×10^7	1.13×10^7	0	1.81×10^7
Franklin	1.05×10^7	2.17×10^5	6.14×10^5	2.88×10^6	0	6.79×10^6
Hancock	6.15×10^6	1.52×10^5	1.63×10^6	7.21×10^5	0	3.65×10^6
Kennebec	1.49×10^7	0	6.99×10^6	9.61×10^5	0	6.92×10^6
Knox	3.64×10^6	0	2.60×10^6	0	0	1.04×10^6
Lincoln	1.08×10^6	0	7.32×10^5	0	0	3.47×10^5
Oxford	5.48×10^6	0	6.37×10^5	2.40×10^5	0	4.60×10^6
Penobscot	2.15×10^7	2.39×10^5	9.84×10^6	1.20×10^6	0	1.03×10^7
Piscataquis	9.58×10^5	0	3.07×10^5	0	0	6.51×10^5
Sagadahoc	4.28×10^6	1.09×10^5	1.18×10^6	0	0	2.99×10^6
Somerset	5.86×10^6	8.68×10^4	1.56×10^6	9.61×10^5	0	3.26×10^6
Waldo	2.74×10^6	0	8.50×10^5	4.81×10^5	0	1.41×10^6
Washington	1.36×10^7	0	2.74×10^6	4.57×10^6	0	6.34×10^6
York	2.57×10^7	1.30×10^5	6.04×10^6	2.64×10^6	4.21×10^4	1.69×10^7
Totals	$\mathbf{1.94 \times 10^8}$	$\mathbf{1.01 \times 10^7}$	$\mathbf{5.76 \times 10^7}$	$\mathbf{3.03 \times 10^7}$	$\mathbf{4.21 \times 10^4}$	$\mathbf{9.64 \times 10^7}$

monetary value associated with the creation of high-disturbance areas. Yearly analysis can show if conversion rates are steady over decades or have been recently accelerating, which may indicate nearby regions where imminent future conversions may occur. Furthermore, areas adjacent to locations that have been subject to development may be more vulnerable to future "contagious" development (Robalino and Pfaff 2012). New efforts to develop higher resolution (10-m) land cover maps on a global scale and yearly basis will allow future, more detailed tracking of land cover conversion on an annual basis (Business Wire 2021). Despite the enormous potential of remote sensing techniques (Grinand et al. 2017; Angelopoulou et al. 2019; Rasel et al. 2017; Mandal et al. 2022), it also is important to recognize one of their major limitations. As noted earlier, potential changes to soil carbon that do not result from a corresponding change in land cover (e.g., soil carbon change resulting from a changing land management strategy (Post et al. 2004; Lal 2010) instead of a land disturbance) will not be detected by remote sensing identification of land cover.

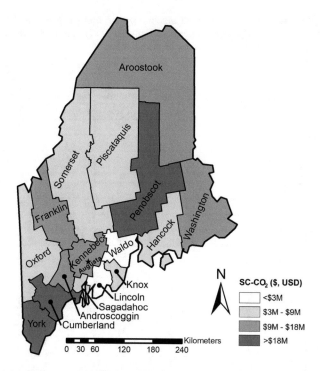

Fig. 26.3 Total dollar value of mid-point total soil carbon (TSC) for newly "developed" land covers (open space, low, medium, and high intensity) from 2001 to 2016 in Maine (U.S.A.) based on a social cost of carbon (SC-CO$_2$) of $46/metric ton of CO$_2$ (2007 U.S.D. with an average discount rate of 3% (EPA 2016))

26.4.3   The Role of Land Cover and Pedodiversity (Soil Diversity) Concepts in Sustainable Management of Soil C Regulating ES/ED

Given the high cost and difficulty of carbon mitigation, the combined use of remote sensing and soil spatial data can be used to prevent the loss of soil carbon by identifying "hotspots" which should be protected to prevent LULC cover change and the associated soil carbon loss. Table 26.13 shows the proportion of social cost of carbon likely lost because of development both between the counties and by soil order within each county. This allows the identification of counties with high overall loss (e.g., Cumberland) as well as counties that have significant development that has disturbed a higher proportion of carbon-rich Histosols. For example, in Washington County, gains in "high disturbance" land covers in the soil order of Histosols contributed 33% of the total county SC-CO$_2$ (Table 26.13).

Table 26.13 Impacts of land development increase on the maximum potential realized social costs of carbon dioxide (SC-CO$_2$) from total soil carbon (TSC) in Maine (USA) from 2001 to 2016 by county expressed as a contribution (%) by each soil order

County	SC-CO$_2$ (% from state total)	Degree of weathering and soil development				
		Slight			Moderate	Strong
		Entisols	Inceptisols	Histosols	Mollisols	Spodosols
		SC-CO$_2$ (% from county total)				
Androscoggin	7	0	25	16	0	59
Aroostook	6	5	32	19	0	44
Cumberland	27	16	28	21	0	34
Franklin	5	2	6	27	0	65
Hancock	3	2	26	12	0	59
Kennebec	8	0	47	6	0	47
Knox	2	0	71	0	0	29
Lincoln	1	0	68	0	0	32
Oxford	3	0	12	4	0	84
Penobscot	11	1	46	6	0	48
Piscataquis	0	0	32	0	0	68
Sagadahoc	2	3	28	0	0	70
Somerset	3	1	27	16	0	56
Waldo	1	0	31	18	0	51
Washington	7	0	20	33	0	46
York	13	1	23	10	0	66
Overall	**100**	**5**	**30**	**16**	**0**	**50**

The state of Maine is somewhat similar to the state of New Hampshire in terms of soil "portfolio" and also contains the same five soil orders (Entisols, Inceptisols, Histosols, Mollisols, and Spodosols) with relatively similar area proportions of each soil order (Mikhailova et al. 2021b), but different in terms of hotspots of realized SC-CO$_2$ due to development. The state of New Hampshire experienced an increase in SC-CO$_2$ associated primarily with the soil order of Histosols (Mikhailova et al. 2021b) compared to Maine's hotspots associated primarily with the soil order of Spodosols.

Remote sensing techniques, combined with soil spatial information, can be utilized to understand the potential for soil carbon losses from past and/or future LULC conversion that results in high disturbance LULC classes at various administrative levels (e.g., state, county, etc.). Planning these offsets to limit the net loss of carbon is complex given the time which may be required to sequester carbon (e.g., forest planting, growth, and accumulation of carbon in the soil) compared to the rapid loss of soil carbon through land development. This planning can be done at various administrative levels (state, county). Table 26.14 presents LULC and pedodiversity data for the state of Maine, which can be used to develop sustainable management of soil C regulating ES/ED. This table reveals that sequestered soil C is limited by both soil type and LULC with different variations:

Table 26.14 Land use/land cover (LULC) by soil order in Maine (USA) in 2016

NLCD land scover classes (LULC)	2016 total area by LULC (%)	Degree of weathering and soil development				
		Slight			Moderate	Strong
		Entisols	Inceptisols	Histosols	Mollisols	Spodosols
		2016 area by soil order, % from total area in each LULC				
Barren land	0.3	5.0	19.2	4.6	0.0	71.1
Woody wetlands	12.6	2.9	48.2	18.5	0.1	30.3
Shrub/Scrub	4.9	0.4	16.7	6.0	0.0	76.9
Mixed forest	32.6	0.8	23.7	5.1	0.0	70.4
Deciduous forest	15.7	0.6	19.1	4.8	0.0	75.5
Herbaceous	2.6	0.7	20.5	6.0	0.0	72.8
Evergreen forest	21.9	1.3	23.8	8.4	0.0	66.5
Emergent herbaceous wetlands	1.2	7.9	43.7	31.2	0.0	17.2
Hay/Pasture	2.6	1.9	38.4	0.8	0.1	58.8
Cultivated crops	1.2	2.3	20.2	1.2	0.0	76.2
Developed, open space	2.7	2.4	28.0	3.5	0.0	66.0
Developed, medium intensity	0.4	6.3	37.4	1.2	0.0	55.1
Developed, low intensity	1.2	4.5	33.8	1.9	0.0	59.7
Developed, high intensity	0.1	7.9	41.9	0.5	0.0	49.7
Totals (%)	**100**	**1.4**	**26.6**	**7.6**	**0.0**	**64.5**

- **Soil limitations to sequester carbon.** For example, in the state of Maine, the soil orders of Spodosols and Inceptisols dominate the state area, and both soils have limited potential to sequester additional soil carbon (Table 26.3).
- **Landcover and land use limitations to sequester carbon.** For example, the cultivated LULC class serves critical provisioning (food) ES and is less likely to be converted to a LULC class to sequester additional carbon.
- **The intersection of soil/LULC class limitations to sequester carbon.** Integration of remote sensing with spatial soil type data also can identify areas where future soil carbon sequestration is unrealistic, given the intersection between both limitations based on soil properties and LULC classes. For example, the soil order of Histosols is often associated with wetland LULC classes which are often protected by state and/or federal regulations because of various reasons, including high soil organic C content.

Lal (2004) proposed research and development priorities around C sequestration that include the recommendation to direct efforts towards "human-dimensions issues" around soil carbon. This should consist of the analysis of LULC changes that likely impact soil carbon pool. Lal (2004) calls for imposing "penalties on land managers who convert to soil degradative land uses," which could be done by linking land-use change to property ownership.

26.4.4 Significance of Results for Maine's Climate Action Plan

The results of this study contribute important scientific findings for Maine's Climate Action Plan (Maine Climate Council. Maine Won't Wait 2020), which includes four goals:

- **Reduce Maine's Greenhouse Gas Emissions**. This study quantified and valued the soil C regulating ES for SOC, SIC, and TSC for the state of Maine by soil type and LULC. According to Maine's Climate Action Plan, approximately 10,000 acres of land is developed every year (Maine Climate Council. Maine Won't Wait 2020). This study identified potential realized $SC\text{-}CO_2$ due to development that occurred during a 15-year period by county, soil type combination, and LULC. This provides actionable information on the location of $SC\text{-}CO_2$ hotspots to help mitigate greenhouse emissions in the future.
- **Avoid the Impacts and Costs of Inaction**. This study quantifies $SC\text{-}CO_2$ as "avoided" and "realized" cost, which can be used for cost/benefit analysis when evaluating the most cost-efficient ways to limit greenhouse gas emissions and ensure greater resilience for the state.
- **Foster Economic Opportunity and Prosperity**. Spatial information about the $SC\text{-}CO_2$ hotspots can be used to optimize economic opportunity while minimizing environmental impact.
- **Advance Equity through Maine's Climate Response**. The spatial information in this study can be combined with data on vulnerable and diverse communities to assure equitable decisions with regards to Maine's Action Plan goals. For example, Fig. 26.4 shows projected impact of future sea rise due to climate change in Maine, which can be used to address climate change response in the coastal areas.

26.5 Summary and Conclusions

This study applied soil diversity taxonomic concepts (U.S. Department of Agriculture (USDA) Soil Taxonomy) to monetarily value soil C regulating ES/ED in the state of Maine (USA), and its counties for sustainable soil C management. Taxonomic pedodiversity in Maine exhibits high soil diversity (five soil orders: Entisols, Inceptisols, Histosols, Mollisols, and Spodosols), which is not evenly distributed within the state and counties. Spodosols occupy the highest proportion of the state area (64%) but ranked only second (after Histosols) in terms of its SOC storage and related social costs of carbon ($101.9B). Despite a relatively small area

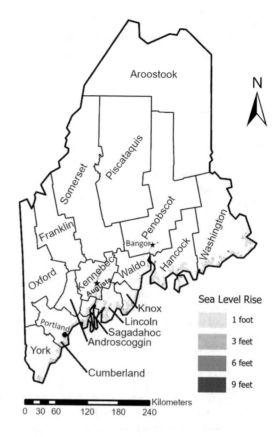

Fig. 26.4 Projected impact of future sea rise due to climate change in Maine, USA

(8% of the total soil area), Histosols contribute $136.5B (50%) to the social cost of SOC and $138.8B (47%) to the social cost of TSC. The contribution of SIC to associated social costs of carbon is small ($25.6B) at the state level and primarily associated with Inceptisols ($17.4B), Spodosols ($4.9B), and Histosols ($2.4B). In the state of Maine, Histosols are particularly sensitive to climate change because of relatively high soil C content, which is most likely to experience higher rates of decomposition due to global warming with increases in temperature and precipitation. All soils in the state of Maine have low recarbonization potential.

Maine is the least densely populated state in the U.S. east of the Mississippi River. Over 80% of its total land cover is classified as forested or unclaimed, which is the most forest cover of any U.S. state (Nowak and Greenfield 2012). Nevertheless, Maine experienced land cover changes between 2001 and 2016, which varied by soil order and land cover, with most soil orders losing in "low disturbance" land covers and gaining in "high disturbance" land covers with most maximum potential "realized" social cost of C associated with all soil orders ($194.4M), but Spodosols ($96.4M) in particular. Cumberland County generated the highest SC-CO$_2$ ($59.7M), followed by York ($25.7M) and Penobscot ($21.5M) counties. This analysis provides actionable spatial information on how 15 years of LULC change likely impacted soil carbon releases to the atmosphere. These results can help inform what land cover and soil type combinations should be managed to help reduce carbon emissions in support of Maine's Climate Action Plan.

References

Adhikari K, Hartemink AE (2016) Linking soils to ecosystem services—a global review. Geoderma 262:101–111

Amundson R, Biardeau L (2018) Soil carbon sequestration is an elusive climate mitigation tool. PNAS 115(46):11652–11656

An Act to establish the Maine Climate Change Council to assist Maine to mitigate, prepare for and adapt to climate change, Pub. L. No. LD 1679 (2019). https://legislature.maine.gov/legis/bills/bills_129th/billtexts/SP055001.asp

Angelopoulou T, Tziolas N, Balafoutis A, Zalidis G, Bochtis D (2019) Remote sensing techniques for soil organic carbon estimation: a review. Remote Sens 11(6):676

Baveye P, Baveye J, Gowdy J (2016) Soil "ecosystem" services and natural capital: critical appraisal of research on uncertain ground. Front. Environ Sci 4, Article 41

Business Wire. ESRI releases new 2020 global land cover map (2021). https://www.businesswire.com/news/home/20210624005176/en/Esri-Releases-New-2020-Global-Land-Cover-Map. Accessed on 28 June 2021

EPA. The Social Cost of Carbon. EPA Fact Sheet. 2016. Available online: https://19january2017snapshot.epa.gov/climatechange/social-cost-carbon_.html Accessed on 15 Mar 2021

ESRI (2021). Arc GIS Pro 2.6. Available online. Accessed on 1 Mar 2021. https://pro.arcgis.com/en/pro-app/2.6/get-started/whats-new-in-arcgis-pro.htm

Fernandez IJ (2008) Carbon and Nutrients in Maine Forest Soils. Maine Agricultural and Forest Experiment Station, The University of Maine, Technical Bulletin 200. ISSN 1070–1524. https://digitalcommons.library.umaine.edu/aes_techbulletin/5/. Accessed on 10 June 2021

Fossey M, Angers D, Bustany C, Cudennec C, Durand P, Gascuel-Odoux C, Jaffrezic A, Pérès G, Besse C, Walter C (2020) A framework to consider soil ecosystem services in territorial planning. Front Environ Sci 8:28

Groshans GR, Mikhailova EA, Post CJ, Schlautman MA, Zhang L (2019) Determining the value of soil inorganic carbon stocks in the contiguous United States based on the avoided social cost of carbon emissions. Resources 8:119

Grinand C, Le Maire G, Vieilledent G, Razakamanarivo H, Razafimbelo T, Bernoux M (2017) Estimating temporal changes in soil carbon stocks at ecoregional scale in Madagascar using remote-sensing. Int J Appl Earth Obs Geoinf 54:1–14

Guo Y, Amundson R, Gong P, Yu Q (2006) Quantity and spatial variability of soil carbon in the conterminous United States. Soil Sci Soc Am J 70:590–600

Ibáñez JJ, Gómez RP (2016) Diversity of soil-landscape relationships: state of the art and future challenges. In: Zinck JA, Metternicht G, Bocco G, Del Valle HF (eds) Geopedology, 1st edn. Springer International Publishing, Cham, pp 183–191

IPCC (International Panel on Climate Change) (2006) Guidelines for National Greenhouse gas Inventories. AFOLU. Agriculture, Forestry and Other Land Use. Retrieved from http://www.ipcc-nggip.iges.or.jp/public/2006gl/vol4.html

Köchy M, Hiederer R, Freibauer A (2015) Global distribution of soil organic carbon – part 1: masses and frequency distributions of SOC stocks for the tropics, permafrost regions, wetlands, and the world. Soil 1:351–365

Lal R (2004) Soil carbon sequestration to mitigate climate change. Geoderma 123:1–22

Lal R (2010) Managing soils and ecosystems for mitigating anthropogenic carbon emissions and advancing global food security. Bioscience 60:708–721

Maine Climate Council. Maine Won't Wait. A Four-year Plan for Climate Action. December 2020. https://www.maine.gov/future/sites/maine.gov.future/files/inline-files/MaineWontWait_December2020.pdf

Mandal A, Majumder A, Dhaliwal SS, Toor AS, Mani PK, Naresh RK, Gupta RK, Mitran T (2022) Impact of agricultural management practices on soil carbon sequestration and its monitoring through simulation models and remote sensing techniques: a review. Crit Rev Environ Sci Technol 52(1):1–49

Mikhailova EA, Groshans GR, Post CJ, Schlautman MA, Post CJ (2019a) Valuation of total soil carbon stocks in the contiguous United States based on the avoided social cost of carbon emissions. Resources 8:157

Mikhailova EA, Groshans GR, Post CJ, Schlautman MA, Post GC (2019b) Valuation of soil organic carbon stocks in the contiguous United States based on the avoided social cost of carbon emissions. Resources 8:153

Mikhailova EA, Post CJ, Schlautman MA, Post CJ, Zurqani HA (2020) Determining farm-scale site-specific monetary values of "soil carbon hotspots" based on avoided social costs of CO_2 emissions. Cogent Environ Sci 6:1–1817289

Mikhailova EA, Zurqani HA, Post CJ, Schlautman MA, Post CJ (2021a) Soil diversity (pedodiversity) and ecosystem services. Land 10(3)

Mikhailova EA, Lin L, Hao Z, Zurqani HA, Post CJ, Schlautman MA, Post GC (2021b) Vulnerability of soil carbon regulating ecosystem services due to land cover change in the state of New Hampshire, USA. Earth 2:208–224

Multi-Resolution Land Characteristics Consortium (MRLC) (2021) Available online. Accessed on 1 Mar 2021. https://www.mrlc.gov/

Nowak DJ, Greenfield EJ (2012) Tree and impervious cover in the United States. Landsc Urban Plan 107:21–30

Osher LJ, Flannagan CT (2007) Soil/landscape relationships in a mesotidal Maine estuary. Soil Sci Soc 71(4):1323–1334

Plaster EJ (2003) Soil science and management. 4th Edition. Delmar Learning, a division of Thomson Learning, Inc., Clifton Park, NY. ISBN: 0766839362

Post WM, Izaurralde CR, Jastrow JD, McCarl BA, Amonette JE, Bailey VL, Jardine PM, West TO, Zhou J (2004) Enhancement of carbon sequestration in US soils. Bioscience 5(10):895–908

Rasel SM, Groen TA, Hussin YA, Diti IJ (2017) Proxies for soil organic carbon derived from remote sensing. Int J Appl Earth Obs Geoinf 59:157–166

Robalino JA, Pfaff A (2012) Contagious development: neighbor interactions in deforestation. J Dev Econ 97:427–436

Sallustio L, Quatrini V, Geneletti D, Corona P, Marchetti M (2015) Assessing land take by urban development and its impact on carbon storage: findings from two case studies in Italy. Environ Impact Assess Rev 54:80–90

Soil Survey Staff, Natural Resources Conservation Service, United States Department of Agriculture 2021a Soil survey geographic (SSURGO) database. Available online: https://nrcs.app.box.com/v/soils. Accessed on 10 Mar 2021

Soil Survey Staff, Natural Resources Conservation Service, United States Department of Agriculture (2021b) Photos of soil orders. Available online: https://www.nrcs.usda.gov/wps/portal/nrcs/detail/soils/edu/?cid=nrcs142p2_053588. Accessed on 20 Feb 2021

USDA/NRCS (2021) State soils. Available online at https://www.nrcs.usda.gov/wps/portal/nrcs/detail/soils/edu/?cid=stelprdb1236841. Accessed on 1 Apr 2021

Wood SL, Jones SK, Johnson JA, Brauman KA, Chaplin-Kramer R, Fremier A, Girvetz E, Gordon LJ, Kappel CV, Mandle L et al (2017) Distilling the role of ecosystem services in the sustainable development goals. Ecosyt Serv 29:701–782

Wright AL, Inglett PW (2009) Soil organic carbon and nitrogen and distribution of carbon-13 and nitrogen-15 in aggregates of Everglades Histosols. Soil Sci Soc Am J 73(2):427–433

Zamanian K, Zhou J, Kuzyakov Y (2021) Soil carbonates: the unaccounted, irrecoverable carbon source. Geoderma 384:114817

Chapter 27
Upscaling the Integrated Terroir Zoning Through Digital Soil Mapping Disaggregation: A Case Study in the Designation of Origin Campo de Borja

A. Lázaro-López, M. L. González-SanJosé, and V. Gómez-Miguel

Abstract The terroir zoning method aims to identify and characterise environmental factors influencing vineyard development, such as soil, climate, or lithology. More detailed cartography and characterisation enables a more comprehensive agro-system management, which is crucial in a context of growing demand for detailed cartographic information on precision viticulture. The integration of new methodologies using digital data and statistical techniques, such as those from the Digital Soil Mapping framework, may improve the baseline information on factors resulting in more detailed homogeneous terroir units. In this work, the Soil Resource Information of an existing zoning in the DO Campo de Borja (NE Spain) is upgraded using a new soil map derived from a novel technique of disaggregation of multi-component units that reveals potentially homogeneous areas by the unsupervised classification algorithm CLARA. Quantitative metrics of vine quality and occupation on the terroir units are then derived for the reference and the new maps. Change analysis using a confusion matrix derived from the intersection of the old and new maps shows a five-fold increase in the number of mapping units, enhancing their delineations and the detection of areas of higher potential within the mapping units of the original map. Overall, the enhancements to the integrated terroir zoning provide an approach for scaling up zoning maps to meet the demands for detailed cartography.

A. Lázaro-López (✉) · M. L. González-SanJosé
Department of Biotechnology and Food Science, Faculty of Science, University of Burgos, Burgos, Spain
e-mail: alazarolop@posteo.me; marglez@ubu.es

V. Gómez-Miguel
Departamento de Producción Agraria, Universidad Politécnica de Madrid (UPM), Madrid, Spain
e-mail: vicente.gomez@upm.es

© The Author(s), under exclusive license to Springer Nature Switzerland AG 2023
J. A. Zinck et al. (eds.), *Geopedology*,
https://doi.org/10.1007/978-3-031-20667-2_27

537

Keywords Digital soil mapping · Spatial disaggregation · Zoning · Terroir ·
Viticultural quality index

27.1 Introduction

The "viticultural terroir" refers to a " geographical area" suitable for the production
of quality wines, and it has a historical background (Gómez-Miguel 2011) that is
embodied within modern legal systems in the form of the Denominations of Origin
(Vaudour et al. 2015). Because of a general lack of social and oenological georeferenced data, this notion is frequently restricted to environmental and agroecosystem features, i.e., soil, geology, geomorphology, climate, and vine-growing
(Gómez-Miguel 2011). Hereafter, generic mentions will refer to the latter. Soil
plays a central role within the terroir as an intermediary between the other environmental factors and strongly influences the grapevine and its by-products, especially
through the soil properties. It also presents the largest source of spatial variability
(Gómez-Miguel 2011).

Zoning is the recommended approach to characterise and delimit terroirs (OIV
2012). It comprises region analysis to divide it into relatively homogeneous zones:
those showing similar strengths and limitations for grapevine development in terms
of their environmental attributes and geographical location (Fregoni et al. 2003).

Zoning methodologies have been closely related to the development of techniques for environmental characterisation. They were initiated via descriptions and
historical justifications, and have traditionally taken the form of land use and suitability maps or landscape models (Gómez-Miguel 2011). Advances in such techniques and new technologies have led the "International Organisation of Vine and
Wine" (OIV) to promote new complex methods based on the combination of the
existing ones (OIV 2012). The so-called Integrated Terroir Zoning (ITZ) adopts this
approach (Gómez-Miguel and Sotés 1992, 2002, 2015).

ITZ combines three partial zonings of environmental factors: soil, climate, and
landscape (combining geomorphology and lithology), whose integration gives rise
to the delimitation and characterisation of "Homogeneous Terroir Units" (HTU).
Then, the viticultural capability of every unit can be expressed quantitatively
through the Quality Index (QI) (Gómez-Miguel and Sotés 2015).

Each partial zoning depends on particular parameters and methodologies from
their respective disciplines. Specifically, the soil zoning corresponds to a Soil
Resource Inventory (SRI) map, traditionally generated by conventional methods.
The adoption of the Digital Soil Mapping (DSM) framework (McBratney et al.
2003) and associated cartography featuring mostly single component Soil Map
Units (SMU), has the potential to improve this partial zoning and, thus, the outcome
of the ITZ.

Existing terroir zonings derived from conventional SRIs can be improved by
disaggregating map units within the DSM framework. In such way, soil information
already available in combination with new data sources can be exploited to increase

the cartographic detail of the SRI, while using fewer human and financial resources. The Designation of Origin (DO) Campo de Borja terroir zoning (Gómez-Miguel and Sotés 2015) meets these conditions and offers a significant potential for improvement.

This chapter presents a new approach for terroir zoning following the ITZ methodology for the DO Campo de Borja that outperforms the original one by improving its SRI through enhancing the existing SMU. The outputs are evaluated to ascertain improvements in the characterisation of SMU achieved by disaggregating the polytaxic or multi-component SMUs of the conventional soil map into mostly monotaxic or single component SMUs. The result is a new map of new HTUs, that are more detailed and better assessed in quantitative terms than the original ones. In addition, those HTUs with the highest occupation rates and those with the highest potential quality within the DO are characterised by taking advantage of the definitions provided by the new zoning. All in all, this series of actions establishes an approach to satisfy part of the demand for detailed cartography required by the Precision Viticulture field.

27.2 Material and Methods

27.2.1 Study Area Characterisation

The study area of DO Campo de Borja (Fig. 27.1), is located in north-eastern Spain within the Autonomous Community of Aragón (41° 55′ 28.9″ to 41° 37′ 15″ N and 1° 44′ 34.5″ to 1° 19′ 1.6″ W). It is a historic wine-growing region covering 653.37 km² where 6416.14 ha of vineyards are grown (2019) and it extends from the piedmont of the Moncayo Range (2316 m asl) to the Ebro river terraces, ranging

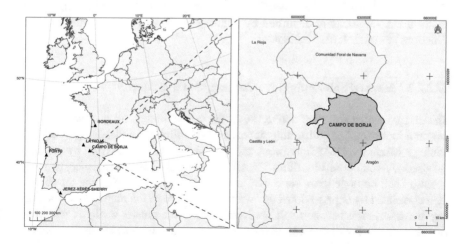

Fig. 27.1 Location of the Designation of Origin Campo de Borja in NE

from 250 to 900 m asl. Its geology is dominated by both structural units, composed of materials from different ages, mainly Mesozoic formations with an alternation of limestone, marl, and sandstone. In addition, its relief is shaped by the river network, particularly in the westernmost part of the region. The climate is continental with an Atlantic influence in winter, especially conditioned by the regional north-westerly wind known as Cierzo (cold and dry). The diurnal and seasonal thermal contrasts are also very marked, with extreme temperatures, whilst rainfall is scarce, with an average of 350 mm in lower and 450 mm in higher areas.

27.2.2 Data Sources

27.2.2.1 Original Soil Zoning: A Conventional Soil Map with Phases

The conventional soil map originates from the ITZ project of the region (Gómez-Miguel and Sotés 2015) (Fig. 27.2). It was produced at a 1:25,000 scale following the Soil Survey Manual (Soil Survey Division Staff 1993), and its delineations were based on Aerial Photo Interpretation and fieldwork with an observation density of 3.8 observations/km^2. The map gathers SMUs of associations and complexes (46) with phases (343), most of them integrating between 4 and 6 components (average of 5.6), both main Soil Taxonomy Units (STU) and inclusions (Lázaro-López et al. 2022a).

STUs are classified according to Soil Taxonomy (Soil Survey Division Staff 2014) and comprise soil series (20) and, when these could not be defined, families assigned to their specific lithology (122). They belong to four Orders (Alfisol, Inceptisol, Entisol and Mollisol), to eight Suborders (Xeralf, Aqualf, Aquept, Xerept, Aquent, Fluvent and Orthent) and to twelve Great Groups (Haploxeroll, Palexeralf, Haploxeralf, Rhodoxeralf, Epiaqualf, Epiaquept, Haploxerept, Calcixerept, Fluvaquent, Epiaquent, Xerofluvent, Xerorthent). In addition, the map shows areas significantly disturbed by anthropogenic action (structures and constructions) as miscellaneous areas.

27.2.2.2 Revised Soil Zoning: A Disaggregated Soil Map

The disaggregated soil map derived from the conventional soil map serves as the new and more detailed soil zoning. It was created following a disaggregation methodology within the DSM framework and it is based on SMU division by the unsupervised classification algorithm CLARA with Mahalanobis distance to reveal potential soil homogeneous zones (Lázaro-López 2022). The resulting clusters are then correlated by expert-knowledge with the STUs from the map legend considering the association between soil observations and the cluster medoids within the covariate space. Finally, cluster delineations are refined recursively up to a scale of approximately 1:10,000.

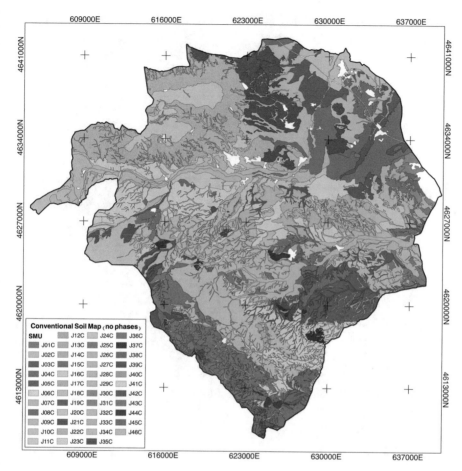

Fig. 27.2 Conventional soil map of the Designation of Origin Campo de Borja composed of 46 Soil Map Units excluding phases. (Lázaro-López et al. 2022a)

Altogether, 143 SMUs were generated having an overall STU content identical to the conventional soil map (Fig. 27.3). Most of them are monotaxic SMUs (52%) and the rest polytaxic, with an average of 1.6 STUs including inclusions (Lázaro-López et al. 2022b).

27.2.2.3 Legacy Soil Observations

Five hundred and nineteen soil point observations are taken from the entire set of observations made during the conventional SRI process, corresponding to soil profiles described, sampled, and analysed in all their horizons (Fig. 27.4). They are also classified according to the USDA Soil Taxonomy (Soil Survey Division Staff 2014)

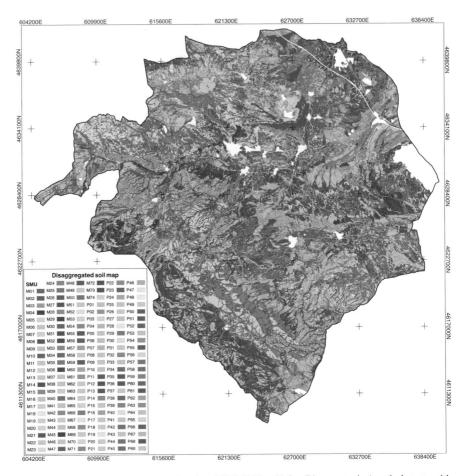

Fig. 27.3 Disaggregated soil map showing 143 Soil Map Units, 74 monotaxic (symbol starts with M) and 69 polytaxic (symbols with P). (Lázaro-López et al. 2022b)

and are associated with one of the STUs listed in the map legend (Lázaro-López et al. 2022b).

27.2.2.4 Climate Zoning

The climate zoning map is sourced from the original ITZ study (Gómez-Miguel and Sotés 2015). It shows 9 units within the project region (Fig. 27.5) with specific characteristics, mainly defined by climatic parameters and bioclimatic indexes (Table 27.1).

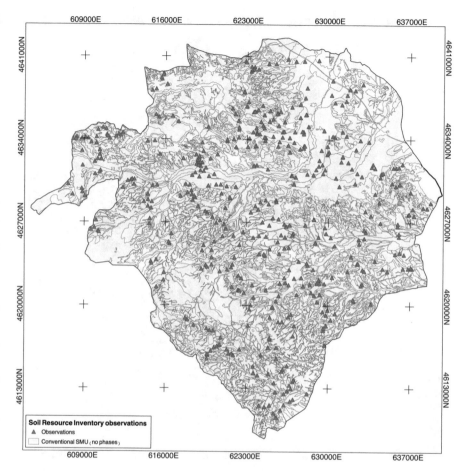

Fig. 27.4 Locations of the soil observations (profiles, excluding augers) of the conventional Soil Resource Inventory, based on guided sampling in the Designation of Origin Campo de Borja

27.2.2.5   Landscape Zoning

The environmental zoning was sourced from prior work of Gómez-Miguel and Sotés (2015), produced by visual interpretation and stratification of aerial photographs and the geological map developed for the SRI of the region (Gómez-Miguel and Sotés 2015). The geological map (Fig. 27.6) has 50 lithostratigraphic units (Table 27.2). At the same time, the aerial photo interpretation is used to distinguish areas of homogeneous aspect, slope, elevation, and the presence of distinctive conditions such as hydromorphism, gypsum and soluble salts, gravels and rock fragments, and terraces.

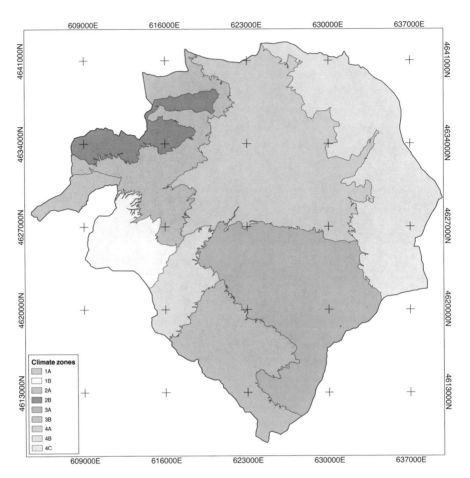

Fig. 27.5 Climate zoning of the Designation of Origin Campo de Borja (zone characteristics 1A–4C are listed in Table 27.1). (Information source: Gómez-Miguel and Sotés 2015)

27.2.2.6 Vineyard Register and Geographical Information System for Agricultural Plots

The 'Vineyard Register' (REVI) is an inventory of farm data by parcel, that gathers agronomic aspects, viticultural features, and a unique identifier for each one.

Likewise, the 'Geographical Information System for Agricultural Plots' (SIGPAC) is also a register focused on the delimitation and identification of crops declared by farmers and stockbreeders in any public aid programme involving cultivated land or land used by livestock.

The joining of both sources allows building a dataset where the vineyard characteristics are associated to a geometry. Thus, it is possible, for example, to learn the distribution of vineyard plots or those of a particular grapevine variety. The present study restricts this information to vineyards within the DO Campo de Borja (2019).

Table 27.1 Climate parameters and bioclimatic indexes on units of the climate zoning

Variables	Climate zones						
	1A	1B; 3B; 4B	2A	2B	3A	4A	4C
TMaxA (°C)	39	38	40	39.3	41	41	43
TMaxm (°C)	24.7	24.8	26.4	26.7	27.5	27.4	26.8
TMax (°C)	17.4	17.6	19.6	19.4	20.8	20.3	20.1
Tm (°C)	12.1	12.2	13.8	14.1	14.9	13.5	13.2
Tmin (°C)	6.7	6.8	8.0	8.6	8.9	9.5	9.8
Tmimm (°C)	1.0	1.3	2.2	3.4	3.6	4.3	4.7
TminA (°C)	−15	−12	−10	−5.7	−10	−10	−10
PRE (mm)	429.1	515.7	362.2	396.8	376.3	398	324.2
ETPTho (mm)	691	690.1	748.7	750.4	794.8	798.6	729
ETPHar (mm)	1036.0	1044.4	1129.6	1098.2	1187.9	1138.4	1115.9
ITE (GGD)	1295.8	1368.7	1775.2	1787.3	2047.1	2075.6	1663.8
IH	1826.5	1908.7	2302.4	2262.6	2573.4	2549.4	2347.4
PB	2.38	3.8	5.6	5.8	6.8	6.9	5.0
IC	10.1	5.8	8.8	8.0	9.7	9.4	9.8
IB	14.9	7.4	15.6	14.6	18.2	17.4	15.5
PA (days)	233	204	236	241	247	247	224
PL (days)	202	169	193	214	203	217	220
SAL (days)	24	27	32	21	33	25	6
RES (days)	76	93	59	83	75	86	44
DAC (mm)	371.5	183.0	386.5	353.6	418.5	400.6	404.8

Information source: Gómez-Miguel and Sotés (2015)

DAC accumulated water deficit up to September, *TMaxA* absolute maximum temperature, *TMaxm* mean of the daily absolute maximum temperatures, *TMax* mean of the daily mean maximum temperatures, *Tm* mean of the daily mean temperatures, *Tmin* mean of the daily mean minimum temperatures, *Tmimm* mean of the daily absolute minimum temperatures, *TminA* absolute minimum temperature, *PRE* annual precipitation, *ETPTho* Thornthwaite evapotranspiration, *ETPHar* Hargreaves evapotranspiration, *ITE* Winkler Index or growing degree-days, *IH* Huglin Index, *PB* Branas hydrothermic index, *IC* Constantinescu Index, *IB* Hidalgo index, *PA* mean active period of the vine, *PL* Emberger frost-free period, *SAL* difference between the beginning of the PA and the PL, *RES* difference between the beginning of the PA and the date on which the water reserve in the soil is consumed

Altogether, it comprises 10,046 geometries of plots with a total area of 6252.2 ha, most of which are dedicated to Garnacha and Tempranillo varieties (Fig. 27.7).

27.2.2.7 Data Management, GIS, and Statistical Software

The project data has been managed from a database developed for this purpose that implements a specific modelling for SRI in Spain and the viticultural terroir (Lázaro-López et al. 2018) and is powered by PostgreSQL (13) and its extension PostGIS (3.1).

The workflow was developed within the R software (R Core Team 2020). In this context, specialized packages were used to subsequently manage data: 'sf' (Pebesma

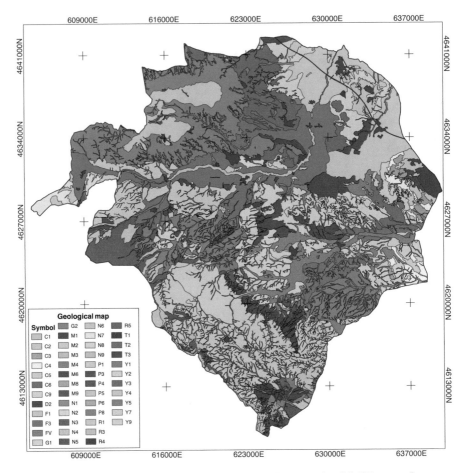

Fig. 27.6 Geological map with lithostratigraphic units in the conventional Soil Resource Inventory of the Designation of Origin Campo de Borja (unit characteristics listed in Table 27.2). (Information source: Gómez-Miguel and Sotés 2015)

2018), and 'Tidyverse' (Wickham et al. 2019). Finally, statistical analyses have relied principally on 'caret' (Bivand et al. 2013) and the 'QGIS' (QGIS Development Team 2019) desktop software.

27.2.3   Procedure

The ITZ methodology addresses the delimitation and characterisation of terroirs based on the integration of partial zonings associated with environmental factors that shape vineyard development, specifically those discussed in the previous sections: soil, climate, and landscape (lithology and geomorphology). This entails a

Table 27.2 Lithostratigraphic units in the conventional Soil Resource Inventory of the Designation of Origin Campo de Borja

Symbol	Description	Chronostratigraphic unit
C1	Marls and dolomite-mud with gypsum and gypsum intercalations	Aquitanian-Chattenian
C2	Red shales with tabular and nodular gypsum intercalations	Aquitanian (Agenian)
C3	Tabular and nodular gypsum	Lower Aragonian
C4	Red mudstones and grey carbonate mudstones and grey and micaceous sandstones laminations	Lower Aragonian
C5	Tabular gypsum with dolomite-mud lamellae	Lower Aragonian
C6	Conglomerates, sandstones and shales. Heterolithic conglomerates	Aragonian
C9	Gypsum with silex, clays and red gypsiferous siltstones (Monteagudo Fm.)	Pontian/Vindobonian
D2	Lagoon deposits, endorheic and navas areas	Holocene
F	Alluvial deposits of main tributaries	Pleistocene-Holocene
F1	Modelling of the river Ebro	Pleistocene-Holocene
F2	Modelling of main tributaries (3rd order)	Pleistocene-Holocene
F3	Modelling of main tributaries (Huecha and Huechaseca)	Pleistocene-Holocene
FV	Alluvial of minor valley bottoms	Holocene
G1	Colluvions	Pleistocene-Holocene
G2	Colluvions	Pleistocene-Holocene
M1	Quartzite conglomerates	Bundsandstein
M2	Conglomerates and sandstones (palaeochannels)	Bundsandstein
M3	Alternation of sandstones and red mudstones	Bundsandstein
M4	Red mudstones with sandstone intercalations	Bundsandstein
M6	Dolomites, dolomitic limestones and marls (Upper Carbonate Ud.)	Muschelkalk
M8	Ophites	Keuper
M9	Variolated clays and gypsum with sandy marl intercalations	Keuper
N1	Well stratified grey dolomites (Imón Fm.)	Rethiense
N2	Sedimentary and collapse breccias, dolomites and carniolas (Cortes de Tajuña Fm.)	Hettangian
N3	Microcrystalline limestones, locally oolitic, well stratified.	Sinemurian
N4	Marls with intercalations of clayey limestones	Toarcian/Pliensbachian
N5	Limestones and marls	Callovian/Bajocian
N6	Sandy loams and limestones with sponges at the base	Oxfordian
N7	Microconglomerate sandstones and clays	Porlandian/Kimmeridgian
N8	Dolomicrites and black micrites with laminations	Lower and Middle Sinemurian
N9	Calcic dolomitic breccias and vacuolar dolomites and carniolas	Hettangian/Rethian
P1	Quartzites, sandstones and shales	Tremadocian

(continued)

Table 27.2 (continued)

Symbol	Description	Chronostratigraphic unit
P3	Sandstones, shales (Bolloncillos Fm.)	Famenian
P4	Coarse sandstones, medium sandstones and microconglomerates (Filluelo Member)	Famenian
P5	Medium, coarse sandstones, bluish-grey siltstones (Valdeinglés Member)	Famennian
P6	Blue-grey siltstones and fine sandstones (Coscojar Member)	Famenian
P8	Highly cemented breccias and conglomerates, coarse sandstones, shales and sandstones	Upper Bundsandstein
R1	Variolithic clays, gypsum and clayey limestones	Keuper
R3	Grey loams with laminations of sandstones	Muschelkalk
R4	Shales, sandstones, siltstones and gypsum stones	Keuper
R5	Reddish-wine-coloured mudstones and sandstones	Bundsandstein
T1	Dejection cones	Pliocene-Pleistocene
T2	Alluvial fans	Pliocene-Pleistocene
T3	Glaciers	Pliocene-Pleistocene
WM	Miscellaneous areas	–
Y1	Gypsum with silex, clays and red gypsiferous silts (Monteagudo Fm.)	Pontian/Vindobonian
Y2	Reddish and grey calcareous clays and silts with frequent limestone intercalations.	Vindobonian (Burdigalian-Aquitanian)
Y3	Calcareous clays and silts of reddish-brown and grey tones (Tudela Fm.)	Pontian/Vindobonian
Y4	Compact limestones, sometimes clayey and siliceous (C. de la Muela)	Pontian
Y5	Detrital materials with carbonate levels	Pontian
Y7	Conglomerates, sandstones and reddish siltstones, limestones intercalations	Pontiense-Burdigaliense
Y9	Tabular gypsum with micrite layers, dolomicrites with gypsum layers and red shales	Lower Aragonian (Burdigalian)

Information source: Gómez-Miguel and Sotés (2015)

spatial intersection of the corresponding zoning maps to give rise to geographical units of unique combinations of all the attributes, the HTUs (Fig. 27.8). In this study, this methodology is implemented with the original climate and environmental zonings in combination with the disaggregated soil map to generate a new zoning and HTUs.

Subsequently, the viticultural quality of the HTUs is quantitatively assessed using the QI, as a numerical value projected over a range from 0 to 100. Similarly, the occupancy rate of growing areas per HTU is estimated in the so-called Occupation Index (OI), as the percentage of the total surface.

For every index and according to its distribution within the project region, a non-linear categorisation by hierarchical clustering into 5 classes is established to

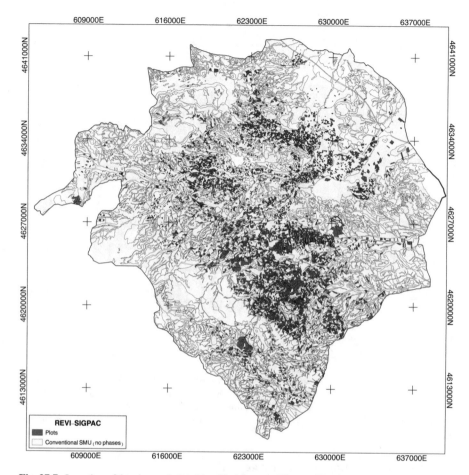

Fig. 27.7 Location of the vineyard plots identified from the Vineyard Register and the Geographical Information System of Agricultural Plots of the Designation of Origin Campo de Borja (2019)

facilitate its analysis. This evaluation system is applied individually to both the original terroir zoning and the new one generated from the disaggregated soil map. In addition, an OI model is built from the QI that seeks to understand vineyard distribution based on the quantification of environmental parameters provided by partial QIs.

After these results, firstly, the memberships of QI classes for both maps are compared using a confusion matrix. The analysis of the reorganisation flows within the zonings, i.e., changes in the quality ratings in the same areas among the maps, attempts to estimate the effects induced by the enhancement of the new soil zoning when the other two partial zonings are kept constant. On the other hand, the OI classes in the new zoning are analysed considering the model elaborated in order to study the characteristics of those most heavily used HTUs.

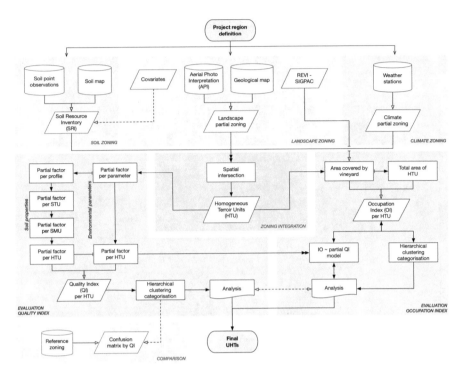

Fig. 27.8 Flowchart of the Integrated Terroir Zoning methodology in solid line. The actions are organised into 5 sections: soil, landscape and climate partial zonings, spatial zoning integration, and evaluation. The evaluation is then divided into two complementary quantitative assessments: Quality and Occupation Indexes. Additional information and comparison routes between zonings in dotted line

27.2.3.1 New Homogeneous Terroir Units

The HTUs reflect homogeneous zones in terms of environmental factors that influence vineyard development and are obtained by spatially intersecting the partial zonings of soil, landscape, and climate factors. The characterisation of these areas thus responds to variables associated to these factors and considered during the partial zonings. In this way, constant values of such variables are expected within them.

Among the various terroir factors, soil occupies a central role due to its interaction with the others. Given its high spatial variability, the soil map also shows the largest granularity. In this sense, HTU delineations generally respond to the limits set by delineations of SMUs, which already consider stratifications of lithology, geomorphology and, indirectly through them, climate. The result is a map of HTUs.

27.2.3.2 Quality Index

The QI is a parametric index to evaluate HTUs that was initially developed from the terroir zoning of the DO Ribera del Duero (Gómez-Miguel and Sotés 1992) and has been fine-tuned in successive projects (Gómez-Miguel and Sotés 2002, 2004, 2006, 2010, 2011) until its current adaptation to the DO Campo de Borja.

It considers a group of soil properties and another group of environmental parameters that individually contribute a factor to the QI value of the HTU (Table 27.3). Each property or parameter is categorised, either in ranges when they are continuous variables or by the classes themselves in the categorical ones. Then, each category has assigned a degree over the maximum factor value in the most favourable assumption. Therefore, the property value is recorded in a category with an associated coefficient score, which is, in the end, the contributing factor of that property to the QI.

Table 27.3 Properties and parameters considered in the Quality Index, the partial zoning from which the data originate and their level of assessment

Partial zoning	Properties and parameters	Calculation level
Soil	Textural classes within the profile	Soil properties (Profile > Soil Taxonomy Unit > Soil Map Unit > Homogeneous Terroir Unit)
	Active limestone within the profile	
	Soil organic matter in epipedon	
	Solum depth	
	Cation exchange capacity of the endopedon	
	Exchange acidity of epipedon	
	Aluminium exchange rate in the epipedon	
	Concentration and percentage of exchangeable potassium in the epipedon	
	Concentration and percentage of exchangeable magnesium in epipedon	
	Potassium/magnesium ratio in epipedon	
	Calcium/magnesium ratio in the epipedon	
	Highest electrical conductivity measured in saturated paste extract within the profile	
	Highest percentage of exchangeable sodium within the profile	
	Water balance of the profile	
Landscape	Lithology	Environmental parameters (Homogeneous Terroir Unit)
	Slope	
	Gypsum and more soluble salts	
	Gravels and rock fragments	
	Hydro-morphology	
	Elevation	
	Terraces	
Climate	Climate zone	

Soil properties are assessed profile-wise according to the available field and laboratory data of the horizons, and a multiplicative parametric system is then run to calculate the QI of the profile. Subsequently, it is possible to estimate a partial QI for each STU by aggregating the profile QIs according to the STU they belong to. Should SMUs be mono-taxic, the partial QI corresponds to the one of the STU they contain. In case of poly-taxic, the partial QI of each STU is weighted by its share in the SMU. All in all, a partial QI of soil properties per HTU is obtained.

The environmental parameters at the HTU level are assessed simultaneously and a separate partial QI is calculated for each HTU. The aggregation of both partial indices results in the final QI per HTU. Lastly, the distribution of the QI within the region is considered and then projected in a range from 0 to 100, respectively the minimum and maximum values, in order to ease its contrast.

A decision rule algorithm, as an extension of the existing database, was written for the parametric rules and actions supporting the calculation of the QI. Ultimately, this quantitative index will help to understand the transition from the original zoning to a new highly detailed zoning.

27.2.3.3 Occupation Index

The OI relates the area covered by vineyards over the total area of the HTU. Vineyard references come from the dataset generated using the REVI, with the agronomic and vineyard properties data, and the geometry from the SIGPAC. Occupation rate is obtained by intersecting the plot geometries with the new zoning map and, in turn, is set as a quantitative index.

Vineyard allocation, given the ongoing decrease in vineyards areas within traditional wine-growing regions, responds to a long-term selection of those plots most favourable for higher-quality grape production based on the experience of winegrowers (Gómez Sanchéz 1995; Gómez-Miguel 2011; Vaudour et al. 2015). Therefore, the analysis of those HTUs where vineyards are located allows a better understanding of the characteristics that drive the quality of the current production.

27.2.3.4 Clustering: HTU Classes by Index

QI and the OI indices are sub-divided into five classes using a hierarchical clustering technique based on the Euclidean distance and the centroid method regarding their distributions within the project region. This approach enables creating consistent groups in non-linear distributions, i.e., that do not fit into regular intervals or sizes, and aims to facilitate comparisons.

The intersection of the zoning maps having their HTUs categorised in terms of their QI allows a comparison via a confusion matrix. Columns and rows of this matrix represent the classes in the original map versus the new zoning respectively, and the cells contain the total area under the cross-over condition. In such way, columns show how areas of a certain class in the original zoning have been sorted into the classes of the new one. Conversely, rows show where the areas for a class of the new zoning map come from in terms of the classes of the original one. The analysis of both perspectives provides insight into the meaning of the changes brought by the new zoning as a result of the usage of the disaggregated soil map.

Moreover, the analysis of HTUs with the most significant ratio of vineyards in terms of OI helps to identify the main characteristics that potentially determine current production as function of quality.

27.2.3.5   Modelling: OI vs. QI

The OI modelling takes the partial QI factors corresponding to the terroir attributes evaluated, thus making it easier to consider all of them whether they were categorical or continuous variables in the first place. From here, an optimised linear regression model is built with stepwise variable selection by addition regarding the Akaike Information Criterion (Kuhn and Johnson 2013) and 75% cross-validation with 50 runs.

Such a model is intended to reveal those factors of major influence on vineyard distribution for the project region in a given temporal period.

27.3   Results and Discussion

27.3.1   New HTUs from the Disaggregated Soil Map

The integration of the three partial zonings by intersection resulted in the new zoning units, with higher cartographical detail and smaller average size (Table 27.4). Altogether, 4460 unique combinations of soil, climate and environmental properties and parameters were identified as the new HTUs, compared to 957 in the original one. In turn, the number of delineations belonging to these units in the new map increased from 3142 to 38,445.

Table 27.4 Cartographic features of the reference zoning and the new zoning maps

Map	Homogeneous Terroir Units (n°)	Delineations (n°)	Average-size area (ha)	Optimal scale of publication[a]
Reference zoning (from conventional soil map)	957	3124	20.39	1:35,700
New zoning (from disaggregated soil map)	4460 (+466%)	38,445 (+1207%)	1.59 (−93%)	1:9993

[a]Cornell University

27.3.2 Quality Index

For each of the new HTUs a QI value was calculated. The distribution of this parameter within the project region is skewed towards very low values, with a few HTUs accounting for the highest values. Dividing it into 5 classes by hierarchical classification resulted in clusters with very clear spacings, from the highest mean in class 1 which decreases markedly in class 2 and to a lesser extent between the successive classes 3, 4 and 5 (Fig. 27.9).

More specifically, class 1 gathers 59 new HTUs with the highest QI values. These HTUs are defined over 14 different SMUs of the disaggregated map which contain 17 distinct STUs that mostly correspond to subcategories of the subgroups Calcic and Typic Haploxeralf, Typic Calcixerepts and Calcic Palexeralf. Notably, most HTUs are spread over two monotaxic units, M50 and M01 (Lázaro-López et al. 2022b), which comprise Calcic Haploxeralf over different lithologies, thus forming different series. The set of HTUs is located on 15 distinct lithostratigraphic units, although most of them are found on calcareous clays and silts of reddish and grey tones with frequent intercalations of limestone (Y2), and sandstones and micro-conglomerates (N7) (see Table 27.2). In climatic terms, zones 3A and 4A (see Fig. 27.5) predominate, characterised by higher temperatures and bioclimatic indexes combined with a longer frost-free period, and at the same time by larger water stress due to lower rainfalls and a shorter time to drain the water reserve from the soil during the growing season.

Areas corresponding to each of the 5 classes were uneven. Out of the 61,423-ha belonging to the project region (excluding miscellaneous areas), approximately 2% was covered by class 1 and 6% by class 2, whereas the remaining 92% was similarly distributed among the other classes (Fig. 27.9). In contrast to the original zoning map, there was a substantial increase in class 1 (approximately 600%) as well as a 30% decrease in class 2 and slight variations in the remaining classes (Table 27.5).

The confusion matrix (Table 27.5) from the intersection between the HTUs categorised by the QI of both zoning maps shows a high degree of agreement, based on the 63% Overall Accuracy or the 0.48 value of the Kappa Index.

From the perspective of how the new zoning was shaped from the baseline, there is a very high agreement between classes 5 of both zonings, i.e., those with the lowest QI values. This fact and a minimal variation in the category area reflect the stability of delimitation and quantification of those potentially less favourable terroirs. The agreement is also strong between classes 2, 3 and 4, although somewhat lower.

However, such agreement is remarkably weak in class 1. Nearly half of the size of this class in the original zoning map was kept, yet the total area in the new zoning increased significantly. This was caused by a shift from lower categories, most noticeably from class 2. A similar flow involving a relatively large amount of area within a lower category gaining in valuation was repeated in classes 2 and 3 (Fig. 27.10). In zoning terms, this may be an indication of an upgrading process, whereby delineations were further fine-tuned, and higher QI zones were identified within them. In this sense, this is a positive result of the new zoning. Between

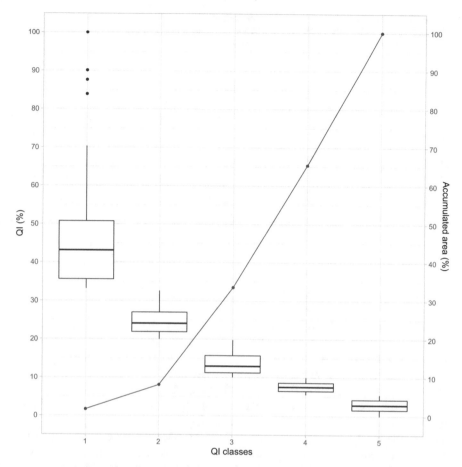

Fig. 27.9 Quality Index distribution of the new Homogeneous Terroir Units split into the 5 defined classes (main axis) and the area covered by each class (secondary axis)

classes 4 and 5 there were exchanges with a low net value, which could similarly denote the restructuring of the delineations.

27.3.3 Occupation Index

Like QI, an OI value was calculated for every HTU based on data of vineyard plots within the project region recorded in REVI and SIGPAC. Approximately 43% of the 4460 new HTUs were covered to some extent by vineyards. Nevertheless, this distribution was not uniform and was skewed towards low values with a few HTUs showing high occupation ratios (Fig. 27.11).

Table 27.5 Confusion matrix from the intersection between the Homogeneous Terroir Units categorised by the Quality Index of the reference zoning (columns) and the new zoning (rows), showing total areas (ha) and relative accuracy metrics

		Area of classes from the Reference zoning (ha)					Total area (ha)	User's accuracy (%)
		1	2	3	4	5		
Area of classes from the New zoning (ha)	1	**75.49**	879.14	57.74	5.00	2.60	**1019.97**	0.074
	2	60.78	**2341.36**	1351.62	160.84	21.33	**3935.93**	0.595
	3	2.27	1700.35	**8462.34**	5052.71	416.49	**15,634.16**	0.541
	4	0.42	183.43	3781.98	**11,895.04**	3711.68	**19,572.55**	0.608
	5	20.52	28.69	1181.90	4141.52	**15,877.08**	**21,249.71**	0.747
Total area (ha)		**159.48**	**5132.97**	**14,835.58**	**21,255.11**	**20,029.18**	**61,412.32**	
Producer's accuracy (%)		0.473	0.456	0.570	0.560	0.793	**Overall accuracy (%)**	**0.629**

Such a distribution may indicate that vineyard locations might not have been random, but rather conditioned by specific criteria of the winegrowers. Part of these criteria would be related to those environmental factors that influence the grapevine development and that are indeed embedded within the terroir notion. In this regard, there is no direct correlation between OI and QI (0.16). However, modelling the OI from partial QI factors of the parameters is significant (p-value: $<2.2e^{-16}$) with a slight variance explanation (R^2: 0.142) (Table 27.6). In other words, parameters considered in the QI give a limited but relevant explanation of how vineyard distribution is characterised. Specifically, the following factors present a significant positive correlation (p value <0.05): slope, climate, SMUs, active limestone in the profile and lithostratigraphic units; whereas negative associations are observed with soil electrical conductivity measured in the saturated paste extract within the profile, the ratio between concentration and percentage of exchangeable potassium in the profile, and the textural class within the profile (Table 27.6).

The hierarchical classification of the distribution into 5 classes gives rise to clusters with highly distinct ranges, where class 1 gathers the highest values and class 5 the lowest. The area of classes is not homogeneous either; approximately 50% of the vineyards are located within class 1 and class 2 (Fig. 27.11).

These two classes contain 332 HTUs distributed over 81 SMUs, among which M02, M12, M17 and M73 stand out, all of them monotaxic. In turn they comprise 102 different STUs, dominated by soil series (J03, J04, J07 and J08; see Lázaro-López et al. 2022b) and in particular J03, also found in the upper class made by QI. Most of them are classified as Typic Calcixerept, Calcic Haploxeralf, Typic Xerorthent and Xerofluvent and match to a large extent the soil classes identified as most favourable by QI. In addition, Xerofluvent, related to bottom valley areas, are identified.

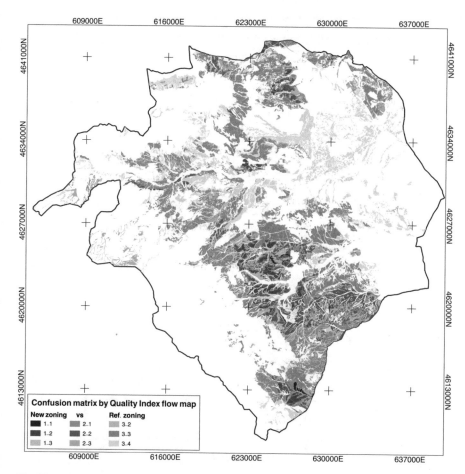

Fig. 27.10 Map of the confusion matrix flows between classes by Quality Index. For each class of the new zoning (hue), the original category in the original map is symbolised (intensity). Those areas retaining the class are represented in stronger intensities and those that changed it in lighter shades

The most relevant soil properties in the aforementioned series are medium and high values of active limestone within the profile (between 6% and 15%); and also, medium values of concentration and percentage of exchangeable potassium in the epipedon, between 0.15 and 0.6 meq/100 g and between 2% and 12% respectively. The highest soil electrical conductivity measured in the saturated paste extract within the profile reaches average values, between 3 and 5 dS/m. Lastly, they mainly exhibit textural classes within the profile of clay loam over loam, silty loam over silty loam and silty clay over clay, according to the United States Department of Agriculture (USDA).

These 332 HTUs are distributed over 16 lithostratigraphic units, though most of them occur on alluvial of minor bottom valley and tributaries, which extend over

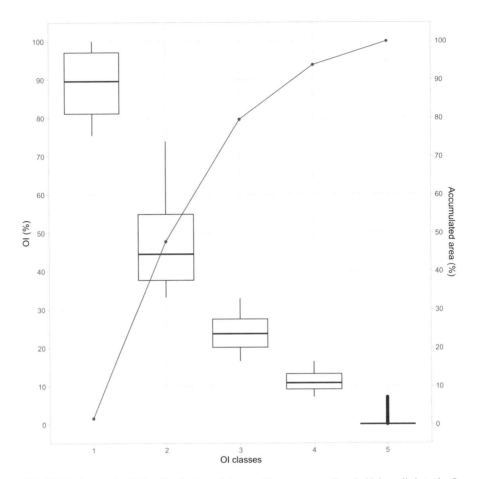

Fig. 27.11 Occupation Index distribution of the new Homogeneous Terroir Units split into the 5 defined classes (main axis) and the area covered by each class (secondary axis)

22% of the project region. This is consistent with some of the most prevalent soil classes mentioned above. Finally, HTUs are mainly located in climate zones 3A and 4A, the same as class 1 of the QI.

27.4 Conclusions

The application of a disaggregated soil map as Soil Resource Inventory within the Integrated Terroir Zoning methodology for enhancing the zoning of the Designation of Origin Campo de Borja has led to the following conclusions:

Table 27.6 Occupancy Index model from partial Quality Indexes of parameters considered in the terroir zoning by linear regression with stepwise variable selection by addition regarding the Akaike Information Criterion and with cross-validation

Variables	Coefficients	Standard deviation	t value	Pr(>|t|)	
(Intercept)	−47.832	11.172	−4.282	1.95E-05	***
Slope	40.647	5.627	7.224	7.30E-13	***
Climate zone	37.132	2.922	12.709	<2e-16	***
SMU	20.309	8.253	2.461	0.01395	*
Active limestone within the profile	12.85	5.499	2.337	0.01956	*
Lithology	9.608	2.587	3.714	0.00021	***
Highest ECe	−7.396	3.865	−1.913	0.05584	.
Ratio K/PKI	−18.498	7.476	−2.474	0.01343	*
Textural classes within the profile	−20.35	6.392	−3.184	0.00148	**

R^2: 0.1421; F: 39.21 on 8 and 1894 degree of freedom; p-value: <2.2e-16

ECe highest soil electrical conductivity measured in the saturated paste extract within the profile, *K* concentration of exchangeable potassium in the epipedon, *PKI* percentage of exchangeable potassium in the epipedon
*** < 0.001
** < 0.01
* < 0.05
. < 0.1

(i) There is a larger number of Homogeneous Terroir Units (HTU) within the project region, as compared to the original soil map, with a five-fold increase in number.

(ii) A better differentiation takes place, as mostly single component Soil Map Units are used to characterise the HTUs.

(iii) The increase in the detection of areas with higher potential by Quality Index (QI) within originally undervalued uniform areas is noticeable, even though the two maps are fairly consistent. This can be noticed from the inter-class flows in the confusion matrix of the intersection between the categorised HTUs of the two maps. The analysis is grounded on the QI, having parameters identified as significant to understand the vineyard distribution in the project region and the degree of occupation of the HTUs.

These insights were among the goals set for the new zoning and are directly derived from the improvements of the disaggregated soil map in terms of cartographic and categorical detail via digital soil mapping techniques. The proposed methodology for disaggregating conventional soil maps is validated, showing the benefits of its implementation for the identification of terroirs within the integrated zoning for viticultural applications.

References

Bivand RS, Pebesma E, Gomez-Rubio V (2013) Applied spatial data analysis with R, 2nd edn. Springer, New York

Fregoni M, Schuster D, Paoletti A (2003) Terroir, zonazione, viticoltura: trattato internazionale. Phytoline, Verona

Gómez Sanchéz PJ (1995) Desarrollo de una metodología edafoclimática para Zonificación Vitícola: aplicación a la D.O. Universidad Politécnica de Madrid, Ribera del Duero

Gómez-Miguel V (2011) Terroir: Parte III. In: Atlas das Castas da Península Ibérica: História, Terroir, Ampelografia. Dinalivro, Lisbon, p 56

Gómez-Miguel V, Sotés V (1992) Metodología y primeros resultados para la zonificación vitícola de la denominación de origen Ribera del Duero, p 2:20

Gómez-Miguel V, Sotés V (2002) Delimitation of terroir in the AOC Rueda and Toro (Castilla y León, Spain). IV Symposium International Zonage Vitivinicole, Avignon, France

Gómez-Miguel V, Sotés V (2004) Zonificación del Terroir: Estudio de Suelos y Ordenación del Cultivo de la Vid en la DO Somontano. Universidad Politécnica de Madrid (UPM), Madrid

Gómez-Miguel V, Sotés V (2006) Zonificación del Terroir: Estudio de Suelos y Ordenación del Cultivo de la Vid en la DO Cigales. Universidad Politécnica de Madrid (UPM), Madrid

Gómez-Miguel V, Sotés V (2010) Zonificación del Terroir: Estudio de Suelos y Ordenación del Cultivo de la Vid en la DO Arribes. Universidad Politécnica de Madrid (UPM), Madrid

Gómez-Miguel V, Sotés V (2011) Zonificación del Terroir: Estudio de Suelos y Ordenación del Cultivo de la Vid en la RD Douro. Universidad Politécnica de Madrid (UPM), Madrid

Gómez-Miguel V, Sotés V (2015) Zonificación del Terroir: Estudio de Suelos y Ordenación del Cultivo de la Vid en la DO Campo de Borja (Zaragoza). Universidad Politécnica de Madrid (UPM), Madrid

Kuhn M, Johnson K (2013) Applied predictive modeling. Springer, New York

Lázaro-López A (2022) Methodology for the disaggregation of polytaxic cartographic units in intensive conventional soil maps. Application towards the improvement of the Integrated Terroir Zoning of the Designation of Origin Campo de Borja. Universidad de Burgos, Burgos, Spain

Lázaro-López A, González-SanJosé ML, Gómez-Miguel V (2018) A spatial database powering a geographic information system for the terroir management: a new consistent and interactive approach. E3S Web Conf 50:6. https://doi.org/10.1051/e3sconf/20185002008

Lázaro-López A, González-SanJosé ML, Gómez-Miguel V (2022a) Map legend of the conventional soil map of the Designation of Origin Campo de Borja [Data set]. Zenodo. https://doi.org/10.5281/zenodo.6948345

Lázaro-López A, González-SanJosé ML, Gómez-Miguel V (2022b) Disaggregated soil map of the Designation of Origin Campo de Borja [Data set]. Zenodo. https://doi.org/10.5281/zenodo.6948303

McBratney AB, Mendonça Santos ML, Minasny B (2003) On digital soil mapping. Geoderma 117:3–52. https://doi.org/10.1016/S0016-7061(03)00223-4

OIV (2012) Resolución 423/2012: OIV guidelines for vitiviniculture zoning methodologies on a soil and climate level

Pebesma E (2018) Simple features for r: standardized support for spatial vector data. R J 10:439–446. https://doi.org/10.32614/RJ-2018-009

QGIS Development Team (2019) QGIS geographic information system

R Core Team (2020) R: a language and environment for statistical computing

Soil Survey Division Staff (1993) Soil survey manual. U.S. Government Printing Office, Washington, DC

Soil Survey Division Staff (2014) Keys to soil taxonomy, 12th edn. USDA-Natural Resources Conservation Service, Washington, DC

Vaudour E, Costantini E, Jones GV, Mocali S (2015) An overview of the recent approaches to terroir functional modelling, footprinting and zoning. Soil 1:287–312. https://doi.org/10.5194/soil-1-287-2015

Wickham H, Averick M, Bryan J et al (2019) Welcome to the Tidyverse. J Open Source Softw 4:1686. https://doi.org/10.21105/joss.01686

Part VI
Synthesis

Chapter 28
Concluding Remarks and Outlook

G. Metternicht, J. A. Zinck, H. F. del Valle, and M. Angelini

Abstract This second version of Geopedology contains 27 chapters covering many subjects, including the basics of geopedology, implementation methods and techniques, and applications in land degradation and land use planning. Issues addressed by the contributing authors are diverse and complementary. This shows that geopedology is a far-reaching discipline that supports the inventory, scientific study, and practical management of natural resources. Geopedology aims at integrating soils and geoforms, two basic components of the earth's geosphere. Use cases that apply different modalities or variants of geopedology are presented, from an open soilscape approach for scientific research, to a more structured survey approach for mapping purposes.

J. A. Zinck died before publication of this work was completed.

G. Metternicht (✉)
School of Science, Western Sydney University, Richmond, Australia

Earth and Sustainability Science Research Centre, University of New South Wales, Sydney, Australia
e-mail: G.Metternicht@westernsydney.edu.au

J. A. Zinck
Faculty of Geo-Information Science and Earth Observation (ITC), University of Twente, Enschede, the Netherlands

H. F. del Valle
Facultad de Ciencia y Tecnología, Centro Regional de Geomática (CeReGeo), Universidad Autónoma de Entre Ríos (UADER), Oro Verde, Entre Ríos, Argentina
e-mail: delvalle.hector@uader.edu.ar

M. Angelini
Instituto de Suelos, INTA-CIRN, Hurlingham, Buenos Aires, Argentina

Global Soil Partnership (GSP), Land and Water Division (NLD), Food and Agriculture Organization of the United Nations (FAO), Rome, Italy

Departamento de Ciencias Básicas, Universidad Nacional de Luján, Luján, Argentina
e-mail: angelini.marcos@inta.gob.ar; marcos.angelini@fao.org; mangelini@unlu.edu.ar

© The Author(s), under exclusive license to Springer Nature 565
Switzerland AG 2023
J. A. Zinck et al. (eds.), *Geopedology*,
https://doi.org/10.1007/978-3-031-20667-2_28

Keywords Geopedology · Mapping · Geomorphic processes · Geomorphic landscape · Geomorphic environment

28.1 Introduction

Geopedology proposes a multi- and inter-disciplinary approach to soil survey and other kinds of studies where soil information is the centerpiece. It combines pedologic and geomorphic criteria to determine and characterize soil map units in the practical-applied realm or analyze the relationships between soils and landscape evolution in the scientific domain. The geopedologic approach as described in Chap. 4 has been used primarily for soil mapping. In this context, geomorphology provides the boundaries of the map units ("the container"), while pedology provides the soil components of the map units ("the content"). Therefore, the units of the geopedologic map are more than soil units in the conventional sense of the term, since they also contain information about the geomorphic context in which soils have formed and are distributed. In this sense, the geopedologic unit is an approximate equivalent of the soilscape unit (as defined in Roudier et al. 2022), explicitly indicating that geomorphology is used to define the landscape. This is usually reflected in the map legend (see for example Fig. 4.3), which shows the geoforms as entries to the legend and their respective pedotaxa as descriptors.

Geopedology is a conceptual framework (Fig. 28.1), not a mapping technique per se. It can be implemented with digital and conventional survey techniques, as a standalone approach or in combination with others (e.g., digital soil mapping approaches, geodiversity and pedometric concepts), and using different survey norms and survey orders as shown in the various chapters of the book. As a conceptual framework geopedology can 'guide' digital soil mapping being a complementary, rather than mutually exclusive, approach to soil survey. For instance, the segmentation of the landscape into geomorphic units provides spatial frames in which digital terrain models combined with remote-sensed data and geostatistical analyses can be applied to assess detailed spatial variability of soils and geoforms, better framing digital mapping over large territories (e.g., Cremon et al. 2021; Coelho et al. 2021). Moreover, geopedology provides information on the structure of the landscape in hierarchically organized geomorphic units, while digital techniques provide information extracted from remotely sensed imagery (e.g., air-satellite-borne, unmanned aerial vehicles) that help characterize the geomorphic units, mainly their morphographic and morphometric terrain surface features.

This second edition of the book offers examples that apply the phases of geopedologic conceptual framework illustrated in Fig. 28.1, from phase 1 focused in organizing preparatory mapping activities (e.g., Chaps. 15 and 27), to data analysis and interpretation using a variety of statistical and geospatial techniques (phase 2). This phase encompasses synthesis of information, characterization of geoforms and soils, description of current landuse and land cover and management practices, all necessary to the production of the geopedologic map (e.g., Chaps. 11, 14, and 15).

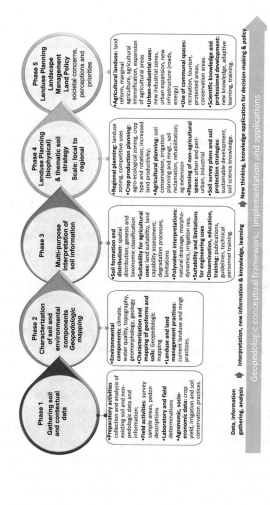

Fig. 28.1 The five phases of the geopedologic conceptual framework, from implementation to applications of information and knowledge it produces

It also shows the applicability of the conceptual framework from open soilscape approach for scientific research (e.g., Chap. 17) to a more structured survey approach for mapping purposes (e.g., Chaps. 16 and 24).

The different phases of the conceptual framework (Fig. 28.1) can accommodate new protocols (e.g. digital soil mapping), new ways of knowledge sharing (e.g. communities of practice for operationalizing digital soil mapping) and new technological developments that contribute to, for instance data collection and analysis for the characterization of the soilscape and geoforms (e.g. soil spectrometry, advanced remote sensing, random forest, pedometrics, big data, artificial intelligence, deep learning). Likewise, other novel conceptual frameworks such as the land degradation neutrality (Chap. 22), integrated spatial land use planning, terroir zoning (see Chap. 27), geodiversity (Chap. 10), regulation of soil ecosystem services (Chap. 26) can be coupled with the geopedologic approach to deliver new knowledge, information and insights to guide soil and land-based decision making and policy at scales ranging from local to national, to continental.

The conceptual framework of geopedology helps resolve one of Arnold's (2015) dilemmas: how to present our understanding of the soil-landscape models as we perceive them. It helps organizing existing data and information, identify data needs and strategically design the acquisition of new data and information (e.g., via field sampling, soil spectroscopy, laboratory determinations, etc.). It guides data analysis, synthesis and interpretation of information to produce new knowledge in the form of the geopedologic map. In short, the conceptual framework becomes a tool for orderly communicate data, information and knowledge about the soil-landscape characteristics and relationships that can inform aspects of landuse planning, landscape management and policy.

Lastly, but not of lesser importance, is the aspect of capacity building and education that implementation of the geopedologic approach opens to technical experts, extension officers, academics, and practitioners from disciplines like geography, planning, geology, geomorphology, engineering, agronomy (see Chap. 4 and Fig. 28.1) for learning and/or enhancing understanding of the soil-landscape relationship. These disciplines and experts need soil information and knowledge, and correct 'reading' of soil-landscape composition and history for a better understanding of the value of the soil resource (i.e., soils as natural capital whose stock and flow of ecosystem services need to be maintained or enhanced).

The geopedologic framework underpins a broad range of applications (Phases 3, 4 and 5) where pedology and geomorphology, and understanding of the soilscape evolution, are important for land use planning, management and policy. Such applications range from genetic interpretation of soils and their formation (e.g., Chap. 12), limitations for land use (e.g., susceptibility to erosion, landslides discussed in Chaps. 16, 19, and 20), to land use planning, land suitability and land evaluation (e.g., Chap. 24) for agricultural, urban, industrial, recreational and environmental conservation/protection applications. Agroecological zoning, integrated land use planning and land allocation in multi-functional landscapes (Chaps. 23 and 24), the design of new soil protection strategies (Chap. 25) are processes that benefit from geopedologic information.

28.2 What We Know and the Challenges Ahead

What have we learned [from soil-landscape relationships]? ask Richard Arnold (former Director of the Natural Resources Conservation Service of the United States Department of Agriculture, USDA-NCRS) in the preface of the first edition of Geopedology: *"…that we need to build on the strengths from the past, stand on the shoulders of giants, and share data, concepts, models and applications"*. In this regard, the geopedologic conceptual framework can be seen as a 'collaborative space' to foster more co-operation and joint work within and across disciplines such as geomatics, pedology, geology, pedometrics, geography, ecology, statistics, and artificial intelligence, to develop further concepts and data sets that advance our knowledge of soilscapes and inform better ways to maintain this natural capital for future generations.

In the preface of the first edition of Geopeology, Arnold (2015) also reflects on dilemmas that as a soil-landscape expert, he pondered throughout his professional life. These dilemmas are knowledge gaps that persist these days and that interdisciplinary research could address in the future.

1. **Classifying and classifications**: *".a major dilemma in classifying is agreeing on what are the objects (entities) that we want to recognize as the individuals of a larger population"*. In Arnold's words, taxonomies are multi-categorical systems that organize classes we recognize at each categorical level according to a set of requirements. He points out that the dilemma in classifying and having taxonomies is directly related to perceptions of scale.
2. **Scales of observation and presenting information**: *"…do we think about pedons or is something larger and more inclusive relevant to what is to be displayed and conveyed to others"* in the form of a map. *'Presentations explaining or hypothesizing soil-landscape relationships are never quite satisfying because the applications of space and time scales are complex'*.
3. **Properties and their interpretations**: *"soil surveys open doors for people to understand the complexity of soils in landscapes better. Conflicts, different viewpoints, and other dilemmas are common, normal and part of our learning process"*.
4. **Sampling soil-landscape relationships**: *'there have been many schemes proposed and used to estimate map unit compositions* (e.g., transects, statistical techniques)'. Some knowledge gaps have been addressed, data relevant for pedogenesis and for geomorphic evolution of landscapes have been collected and could be better used for interpreting soil-landscape functions and their dynamics over time and space;
5. **Building mental models and applying those models**: *"…soil taxonomies and geomorphic taxonomies are mental abstractions of what we think we know at a given point in time. A useful tool is to propose several hypotheses to explain the data sets used to develop our explanations and opinions"*.
6. **Evaluating relationships**: it is important that reasons for an evaluation are stated and agreed as that determines the approach to evaluation. *"…are you look-*

ing for the central concepts and confidence limits of the composition of a group of the same named map units delineated on maps?"

7. **Presenting our understanding** [of soils and their distribution on the landscape]. The description of soil map units depends on the choices made to show our understanding of variability existing in different landscapes. Although standards and guidelines for the description of mapping units and geoforms exist, the actual description and delineation of landforms in a landscape are determined by surveyors as soilscape interpreters who produce the maps.

28.3 Outlook

Although most food that we consume come from soils, soil health and sustainable soil management hardly ever are considered to address malnutrition and food insecurity. Soil and land degradation damage the provision of environmental services of terrestrial ecosystems and the provision of nutrients to food (FAO 2022). In this regard, Arnold's (2015) reflection in the preface of the first edition of Geopedology is on point. He states that "perhaps the most significant obstacle we have in understanding and utilizing soil-landscape interactions is the need to discover practical solutions to use and conserve soils for future generations. Our history is rich with unsatisfactory decisions about the provision of accurate information that determines national policy for the conservation of natural resources" (Arnold 2015).

This second edition offers insights into resolving some of Arnold's dilemmas and others as topics for ongoing research. The geopedological framework has been applied within the realm of conventional soil mapping, which has been criticized for being laborious and subjective, as they depend on the mental model of the soil surveyors. Mental models are not well documented; thus, they are difficult to be reproduced in the same way. On the other hand, digital soil mapping is based on statistical models, which can be considered much more objective. They have been demonstrated to be optimal for mapping soil properties in a different way. However, one of the main challenges in digital soil mapping is to include expert knowledge about soil functioning (Wadoux et al. 2021; Ma et al. 2019).

Increasing opportunities exist for soil scientists to interact with ecologists, biologists, engineers, computer scientists, land use planners, and social scientists to enhance the importance of soil science (geopedology, digital soil mapping, etc.) beyond its traditional applications (see Fig. 28.1), to emphasize on the centrality of soil information for global food security, land degradation neutrality, and climate change adaptation. Interdisciplinary research could address several knowledge gaps that are identified throughout this book, and/or operationalize some of the conceptual propositions (e.g., geodiversity and geopedology). Real or perceived silos of knowledge and thinking need to be overcome. For example, a way forward to combine both schools of thought about soil mapping could be by using object-based image analysis (OBIA) techniques (Cremon et al. 2021; Casagli et al. 2016) to produce land units at different scales using landscape attributes and digital soil maps. The size of the segments could be controlled accordingly to the desired scale and based on the variability of the landscape attributes. These land units could be classified on the basis of soil classes as well as landscape properties. In this way, land units based on OBIA technique could be less subjective than classical soil maps and could be reproduced to the desired scale according to the level of details of the soil data and landscape attributes. They could also be used to aggregate pixel-based soil property maps, reducing their uncertainty as well.

In concluding, mapping of soil bodies, not properties in isolation, is what gives insight into the soil landscape relationships (Chap. 9). While techniques such as OBIA could be a way forward for addressing mapping soil-landscape features, aspects of the geopedologic conceptual framework still need to be studied under this technique, such as the soil sampling method. Catena-sampling has been used for conventional soil mapping, while different methods of model-based or design-based were developed for digital soil mapping (Brus 2022). A different sampling strategy may be required for an OBIA approach. Also, aggregation scales, uncertainty estimations and map reporting—the Arnold's dilemmas presented in this chapter—may require new methodological developments. This is where the geopedological framework can guide design and development for a consistent, insight-driven soil mapping methodology.

References

Arnold R (2015) My dilemmas with soil-landscape relationships. Preface. In: Zinck et al (ed.), Geopedology: an integration of geomorphology and pedology for soil and landscape studies. Springer

Brus DJ (2022) Spatial sampling with R, 1st edn. Chapman and Hall/CRC, Milton. https://doi.org/10.1201/9781003258940

Casagli N, Cigna F, Bianchini S, Hölbling D, Füreder P, Righini G, Del Conte S, Friedl B, Schneiderbauer S, Iasio C, Vlcko J (2016) Landslide mapping and monitoring by using radar and optical remote sensing: examples from the EC-FP7 project SAFER. Remote Sens Appl Soc Environ 4:92–108

Coelho FF, Giasson E, Campos AR, Costa JJF (2021) Geographic object-based image analysis and artificial neural networks for digital soil mapping. Catena 206:105568

Cremon EH, Pereira AC, de Paula LD, Nunes ED (2021) Geological and terrain attributes for predicting soil classes using pixel-and geographic object-based image analysis in the Brazilian Cerrado. Geoderma 401:115315

FAO (2022) Soils for nutrition: state of the art. FAO, Rome. https://doi.org/10.4060/cc0900en

Ma Y, Minasny B, Malone BP, Mcbratney AB (2019) Pedology and digital soil mapping (DSM). Eur J Soil Sci 70(2):216–235

Roudier P, Odgers N, Carrick S, Eger A, Hainsworth S, Beaudette D (2022) Soilscapes of New Zealand: pedologic diversity as organized along environmental gradients. Geoderma 409:115637

Wadoux AMC, Heuvelink GB, Lark RM, Lagacherie P, Bouma J, Mulder VL, Libohova Z, Yang L, McBratney AB (2021) Ten challenges for the future of pedometrics. Geoderma 401:115155

Index

Printed in the United States
by Baker & Taylor Publisher Services

Solid Mechanics and Its Applications

Founding Editor

G. M. L. Gladwell

Volume 172

The fundamental questions arising in mechanics are: Why?, How?, and How much? The aim of this series is to provide lucid accounts written by authoritative researchers giving vision and insight in answering these questions on the subject of mechanics as it relates to solids. The scope of the series covers the entire spectrum of solid mechanics. Thus it includes the foundation of mechanics; variational formulations; computational mechanics; statics, kinematics and dynamics of rigid and elastic bodies; vibrations of solids and structures; dynamical systems and chaos; the theories of elasticity, plasticity and viscoelasticity; composite materials; rods, beams, shells and membranes; structural control and stability; soils, rocks and geomechanics; fracture; tribology; experimental mechanics; biomechanics and machine design. The median level of presentation is the first year graduate student. Some texts are monographs defining the current state of the field; others are accessible to final year undergraduates; but essentially the emphasis is on readability and clarity.

Springer and Professors Barber and Klarbring welcome book ideas from authors. Potential authors who wish to submit a book proposal should contact Dr. Mayra Castro, Senior Editor, Springer Heidelberg, Germany, email: mayra.castro@springer.com

Indexed by SCOPUS, Ei Compendex, EBSCO Discovery Service, OCLC, ProQuest Summon, Google Scholar and SpringerLink.

J. R. Barber

Elasticity

Fourth Edition

 Springer

J. R. Barber
Department of Mechanical Engineering
University of Michigan
Ann Arbor, MI, USA

ISSN 0925-0042   ISSN 2214-7764 (electronic)
Solid Mechanics and Its Applications
ISBN 978-3-031-15216-0   ISBN 978-3-031-15214-6 (eBook)
https://doi.org/10.1007/978-3-031-15214-6

This Springer imprint is published by the registered company Springer Nature Switzerland AG
The registered company address is: Gewerbestrasse 11, 6330 Cham, Switzerland

Preface

The subject of Elasticity can be approached from several points of view, depending on whether the practitioner is principally interested in the mathematical structure of the subject or in its use in engineering applications and, in the latter case, whether essentially numerical or analytical methods are envisaged as the solution method. My first introduction to the subject was in response to a need for information about a specific problem in Tribology. As a practising Engineer with a background only in elementary Mechanics of Materials, I approached that problem initially using the concepts of concentrated forces and superposition. Today, with a rather more extensive knowledge of analytical techniques in Elasticity, I still find it helpful to go back to these roots in the elementary theory and think through a problem physically as well as mathematically, whenever some new and unexpected feature presents difficulties in research. This way of thinking will be found to permeate this book. My engineering background will also reveal itself in a tendency to work examples through to final expressions for stresses and displacements, rather than leave the derivation at a point where the remaining manipulations would be mathematically routine.

The first edition of this book, published in 1992, was based on a one semester graduate course on Linear Elasticity that I have taught at the University of Michigan since 1983. In three subsequent revisions, the amount of material has more than doubled and the character of the book has necessarily changed, but I remain committed to my original objective of writing for those who wish to find the solution of specific practical engineering problems. With this in mind, I have endeavoured to keep to a minimum any dependence on previous knowledge of Solid Mechanics, Continuum Mechanics or Mathematics. Most of the text should be readily intelligible to a reader with an undergraduate background of one or two courses in elementary Mechanics of Materials and a rudimentary knowledge of partial differentiation. Cartesian tensor notation and the summation convention are used in a few places to shorten the derivation of some general results, but these sections are carefully explained, so as to be self-explanatory.

Modern practitioners of Elasticity are necessarily influenced by developments in numerical methods, which promise to solve all problems with no more information about the subject than is needed to formulate the description of a representative element of material in a relatively simple state of stress. As a researcher in Solid Mechanics, with a primary interest in the physical behaviour of the systems I am investigating, rather than in the mathematical structure of the solutions, I have frequently had recourse to numerical methods of all types and have tended to adopt the pragmatic criterion that the best method is that which gives the most convincing and accurate result in the shortest time. In this context, 'convincing' means that the solution should be capable of being checked against reliable closed-form solutions in suitable limiting cases and that it is demonstrably stable and in some sense convergent. Measured against these criteria, the 'best' solution to many practical problems is often not a direct numerical method, such as the finite element method, but rather one involving some significant analytical steps before the final numerical evaluation. This is particularly true in three-dimensional problems, where direct numerical methods are extremely computer-intensive if any reasonably accuracy is required, and in problems involving infinite or semi-infinite domains, discontinuities, bonded or contacting material interfaces or theoretically singular stress fields. By contrast, I would immediately opt for a finite element solution of any two-dimensional problem involving finite bodies with relatively smooth contours, unless it happened to fall into the (surprisingly wide) class of problems to which the solution can be written down in closed form. The reader will therefore find my choice of topics significantly biassed towards those fields identified above where analytical methods are most useful.

As in the earlier editions, I encourage the reader to become familiar with the use of symbolic mathematical languages such as Maple^{TM} and Mathematica^{TM}, since these tools open up the possibility of solving considerably more complex and hence interesting and realistic elasticity problems. They also enable the student to focus on the formulation of the problem (e.g. the appropriate governing equations and boundary conditions) rather than on the algebraic manipulations, with a consequent improvement in insight into the subject and in motivation. Finally, they each posess post-processing graphics facilities that enable the user to explore important features of the resulting stress state. The reader can access numerous files in the https://link.springer.com/book/10.1007/978-90-481-3809-8 Supplementary Material or at my University of Michigan homepage http://www-personal.umich.edu/jbarber/elasticity/book4.html, including the solution of sample problems, electronic versions of the tables in Chapters 22 and 23 algorithms for the generation of spherical harmonic potentials. Some hints about the use of this material are contained in Appendix A, and more detailed tips about programming are included at the above websites. Those who have never used these methods will find that it takes only a few hours of trial and error to learn how to write programs to solve boundary-value problems in elasticity.

This new edition contains four additional chapters, including one on the Eshelby inclusion problem, and a brief introduction to anisotropic elasticity. Also discussed are Love's theory of 'moderately thick plates' and the three-dimensional Hertz contact problem.

The new edition contains numerous additional end-of-chapter problems. As with previous editions, a full set of solutions to these problems is available to *bona fide* instructors on request to the author. Some of these problems are quite challenging, indeed several were the subject of substantial technical papers within the not too distant past, but they can all be solved in a few hours using Maple or Mathematica. Many texts on Elasticity contain problems which offer a candidate stress function and invite the student to 'verify' that it defines the solution to a given problem. Students invariably raise the question 'How would we know to choose that form if we were not given it in advance?' I have tried wherever possible to avoid this by expressing the problems in the form they would arise in Engineering—i.e. as a body of a given geometry subjected to prescribed loading. This in turn has required me to write the text in such a way that the student can approach problems deductively. I have also generally opted for explaining difficulties that might arise in an 'obvious' approach to the problem, rather than steering the reader around them in the interests of brevity or elegance of solution.

I have taken this opportunity to correct the numerous typographical errors in the third edition, but no doubt despite my best efforts, the new material will contain more. Please communicate any errors to me.

As in previous editions, I would like to thank my graduate students and more generally scientific correspondents worldwide whose questions continue to force me to re-examine my knowledge of the subject. I am also grateful to Professor John Dundurs for permission to use Table 9.1 and to the Royal Society of London for permission to reproduce Figures 13.2, 13.3.

Ann Arbor, USA   J. R. Barber
2022

Contents

Part II   Two-dimensional Problems

Part I
General Considerations

Chapter 1
Introduction

The subject of Elasticity is concerned with the determination of the stresses and displacements in a body as a result of applied mechanical or thermal loads, for those cases in which the body reverts to its original state on the removal of the loads. In this book, we shall further restrict attention to the case of linear infinitesimal elasticity, in which the stresses and displacements are linearly proportional to the applied loads, and the displacement gradients are small compared with unity. These restrictions ensure that linear superposition can be used and enable us to employ a wide range of series and transform techniques which are not available for non-linear problems.

Most engineers first encounter problems of this kind in the context of the subject known as Mechanics of Materials, which is an important constituent of most undergraduate engineering curricula. Mechanics of Materials differs from Elasticity in that various plausible but unsubstantiated assumptions are made about the deformation process in the course of the analysis. A typical example is the assumption that plane sections remain plane in the bending of a slender beam. Elasticity makes no such assumptions, but attempts to develop the solution directly and rigorously from its first principles, which are Newton's laws of motion, Euclidian geometry and Hooke's law. Approximations are often introduced towards the end of the solution, but these are mathematical approximations used to obtain solutions of the governing equations rather than physical approximations that impose artificial and strictly unjustifiable constraints on the permissible deformation field.

However, it would be a mistake to draw too firm a distinction between the two approaches, since practitioners of each have much to learn from the other. Mechanics of Materials, with its emphasis on physical reasoning and a full exploration of the practical consequences of the results, is often able to provide insights into the problem that are less easily obtained from a purely mathematical perspective. Indeed, we shall

Supplementary Information The online version contains supplementary material available at https://doi.org/10.1007/978-3-031-15214-6_1.

J. R. Barber, *Elasticity*, Solid Mechanics and Its Applications 172, https://doi.org/10.1007/978-3-031-15214-6_1

make extensive use of physical parallels in this book and pursue many problems to conclusions relevant to practical applications, with the hope of deepening the reader's understanding of the underlying structure of the subject. Conversely, the mathematical rigour of Elasticity gives us greater confidence in the results, since, even when we have to resort to an approximate solution, we can usually estimate its accuracy with some confidence — something that is very difficult to do with the physical approximations used in Mechanics of Materials[1]. Also, there is little to be said for using an *ad hoc* approach when, as is often the case, a more rigorous treatment presents no serious difficulty.

1.1 Notation for Stress and Displacement

It is assumed that the reader is more or less familiar with the concept of stress and strain from elementary courses on Mechanics of Materials. This section is intended to introduce the notation used, to refresh the reader's memory about some important ideas, and to record some elementary but useful results.

1.1.1 Stress

Components of stress will all be denoted by the symbol σ with appropriate suffices. The second suffix denotes the direction of the stress component and the first the direction of the *outward* normal to the surface upon which it acts. This notation is illustrated in Figure 1.1 for the Cartesian coördinate system x, y, z.

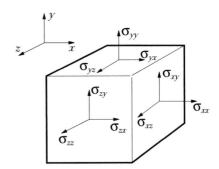

Fig. 1.1 Notation for stress components

Notice that one consequence of this notation is that normal (i.e. tensile and compressive) stresses have both suffices the same (e.g. σ_{xx}, σ_{yy}, σ_{zz} in Figure 1.1) and

---

[1] In fact, the only practical way to examine the effect of these approximations is to relax them, by considering the same problem, or maybe a simpler problem with similar features, in the context of the theory of Elasticity.

are positive when tensile. The remaining six stress components in Figure 1.1 (i.e. $\sigma_{xy}, \sigma_{yx}, \sigma_{yz}, \sigma_{zy}, \sigma_{zx}, \sigma_{xz}$) have two *different* suffices and are shear stresses.

Books on Mechanics of Materials often use the symbol τ for shear stress, whilst retaining σ for normal stress. However, there is no need for a different symbol, since the suffices enable us to distinguish normal from shear stress components. Also, we shall find that the use of a single symbol with appropriate suffices permits matrix methods to be used in many derivations and introduces considerable economies in the notation for general results.

The equilibrium of moments acting on the block in Figure 1.1 requires that

$$\sigma_{xy} = \sigma_{yx} \ ; \ \ \sigma_{yz} = \sigma_{zy} \ \text{ and } \sigma_{zx} = \sigma_{xz} \ . \tag{1.1}$$

This has the incidental advantage of rendering mistakes about the order of suffices harmless! (In fact, a few books use the opposite convention.) Readers who have not encountered three-dimensional problems before should note that there are *two* shear stress components on each surface and one normal stress component. There are some circumstances in which it is convenient to combine the two shear stresses on a given plane into a two-dimensional vector in the plane — i.e. to refer to the *resultant* shear stress on the plane. An elementary situation where this is helpful is in the Mechanics of Materials problem of determining the distribution of shear stress on the cross section of a beam due to a transverse shear force[2]. For example, we note that in this case, the resultant shear stress on the plane must be tangential to the edge at the boundary of the cross section, since the shear stress complementary to the component normal to the edge acts on the traction-free surface of the beam and must therefore be zero. This of course is why the shear stress in a thin-walled section tends to follow the direction of the wall.

We shall refer to a plane normal to the x-direction as an 'x-plane' etc. The *only* stress components which act on an x-plane are those which have an x as the first suffix (This is an immediate consequence of the definition).

Notice also that any x-plane can be defined by an equation of the form $x = c$, where c is a constant. More precisely, we can define a 'positive x-plane' as a plane for which the positive x-direction is the outward normal and a 'negative x-plane' as one for which it is the inward normal. This distinction can be expressed mathematically in terms of inequalities. Thus, if part of the boundary of a solid is expressible as $x = c$, the solid must locally occupy one side or the other of this plane. If the domain of the solid is locally described by $x < c$, the bounding surface is a positive x-plane, whereas if it is described by $x > c$, the bounding surface is a negative x-plane.

The discussion in this section suggests a useful formalism for correctly defining the boundary conditions in problems where the boundaries are parallel to the coördinate axes. We first identify the equations which define the boundaries of the solid and then write down the three traction components which act on each boundary. For example, suppose we have a rectangular solid defined by the inequalities $0 < x < a$,

---

[2] For the elasticity solution of this problem, see Chapter 18.

$0 < y < b$, $0 < z < c$. It is clear that the surface $y = b$ is a positive y-plane and we deduce immediately that the corresponding traction boundary conditions will involve the stress components $\sigma_{yx}, \sigma_{yy}, \sigma_{yz}$ — i.e. the three components that have y as the first suffix. This procedure insures against the common student mistake of assuming (for example) that the component σ_{xx} must be zero if the surface $y = b$ is to be traction free. (*Note* : Don't assume that this mistake is too obvious for you to fall into it. When the problem is geometrically or algebraically very complicated, it is only too easy to get distracted.)

Stress components can be defined in the same way for other systems of orthogonal coördinates. For example, components for the system of cylindrical polar coördinates (r, θ, z) are shown in Figure 1.2.

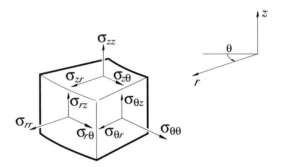

Fig. 1.2 Stress components in polar coördinates.

(This is a case where the definition of the 'θ-plane' through an equation, $\theta = \alpha$, is easier to comprehend than 'the plane normal to the θ-direction'. However, note that the θ-direction is the direction in which a particle would move if θ were increased with r, z constant.)

1.1.2 Index and vector notation and the summation convention

In expressing or deriving general results, it is often convenient to make use of vector notation. We shall use bold face symbols to represent vectors and single suffices to define their components in a given coördinate direction. Thus, the vector $\mathbf{V}$ can be written

$$\mathbf{V} = \mathbf{i} V_x + \mathbf{j} V_y + \mathbf{k} V_z = \left\{ V_x, V_y, V_z \right\}^T , \tag{1.2}$$

where $\mathbf{i}, \mathbf{j}, \mathbf{k}$ are unit vectors in directions x, y, z respectively.

In the last expression in (1.2), we have used the convenient linear algebra notation for a vector, but to take full advantage of this, we also need to replace V_x, V_y, V_z by V_1, V_2, V_3 or more generally by V_i, where the index i takes the values 1, 2, 3. Further compression is then achieved by using the *Einstein summation convention* according to which terms containing a repeated latin index are summed over the three values 1, 2, 3. For example, the expression σ_{ii} is interpreted as

$$\sigma_{ii} \equiv \sum_{i=1}^{3} \sigma_{ii} = \sigma_{11} + \sigma_{22} + \sigma_{33} \, .$$

Occasionally, we may want to use a repeated index without implying a summation. In such cases, we shall add the explicit instruction 'no sum'.

Any expression with one free index has a value associated with each coördinate direction and hence represents a vector. A more formal connection between the two notations can be established by writing

$$V = e_1 V_1 + e_2 V_2 + e_3 V_3 = e_i V_i \, , \tag{1.3}$$

where $e_1 = i, e_2 = j, e_3 = k$ and then dropping the implied unit vector e_i in the right-hand side of (1.3).

The position of a point in space is identified by the position vector,

$$r = ix + jy + kz \, .$$

In index notation, we replace x, y, z by x_1, x_2, x_3 respectively, so that

$$r = \{x_1, x_2, x_3\}^T \, ,$$

which is simply written as x_i, whilst the Cartesian stress components

$$\sigma = \begin{bmatrix} \sigma_{11} & \sigma_{12} & \sigma_{13} \\ \sigma_{21} & \sigma_{22} & \sigma_{23} \\ \sigma_{31} & \sigma_{32} & \sigma_{33} \end{bmatrix} \tag{1.4}$$

are written as σ_{ij}.

Many authors use a subscript comma followed by one or more indices to denote differentiation with respect to Cartesian coördinates. Thus

$$u_{j,i} = \frac{\partial u_j}{\partial x_i} \quad \text{and} \quad \phi_{,ii} = \frac{\partial^2 \phi}{\partial x_i \partial x_i} = \frac{\partial^2 \phi}{\partial x_1{}^2} + \frac{\partial^2 \phi}{\partial x_2{}^2} + \frac{\partial^2 \phi}{\partial x_3{}^2} \, .$$

This clearly leads to a more compact notation which has advantages when writing general formulæ, but we shall not use it in this book because the explicit derivative is easier to interpret, particularly for those meeting the index notation for the first time.

1.1.3   Vector operators in index notation

All the well-known vector operations can be performed using index notation and the summation convention. For example, if two vectors are represented by P_i, Q_i respectively, their scalar (dot) product can be written concisely as

$$\boldsymbol{P} \cdot \boldsymbol{Q} = P_1 Q_1 + P_2 Q_2 + P_3 Q_3 \equiv P_i Q_i \,, \tag{1.5}$$

because the repeated index i in the last expression implies the summation over all three product terms. The vector (cross) product

$$\boldsymbol{P} \times \boldsymbol{Q} = \begin{vmatrix} i & j & k \\ P_1 & P_2 & P_3 \\ Q_1 & Q_2 & Q_3 \end{vmatrix}$$

can be written in index notation as

$$\boldsymbol{P} \times \boldsymbol{Q} = \epsilon_{ijk} P_i Q_j \,, \tag{1.6}$$

where ϵ_{ijk} is the *alternating tensor*, which is defined to be 1 if the indices are in cyclic order (e.g. 2,3,1), −1 if they are in *reverse* cyclic order (e.g. 2,1,3) and zero if any two indices are the same[3]. Notice that the only free index in the right-hand side of (1.6) is k, so (for example) the x_1-component of $\boldsymbol{P} \times \boldsymbol{Q}$ is recovered by setting $k=1$.

The gradient of a scalar function ϕ can be written

$$\nabla \phi = i \frac{\partial \phi}{\partial x} + j \frac{\partial \phi}{\partial y} + k \frac{\partial \phi}{\partial z} = \frac{\partial \phi}{\partial x_i} \,, \tag{1.7}$$

the divergence of a vector $\boldsymbol{V}$ is

$$\text{div } \boldsymbol{V} \equiv \nabla \cdot \boldsymbol{V} = \frac{\partial V_1}{\partial x_1} + \frac{\partial V_2}{\partial x_2} + \frac{\partial V_3}{\partial x_3} = \frac{\partial V_i}{\partial x_i}$$

using (1.5, 1.7), and the curl of a vector $\boldsymbol{V}$ is

$$\text{curl } \boldsymbol{V} \equiv \nabla \times \boldsymbol{V} = \epsilon_{ijk} \frac{\partial V_j}{\partial x_i} \,,$$

using (1.6, 1.7). We also note that the Laplacian of a scalar function can be written in any of the forms

$$\nabla^2 \phi = \frac{\partial^2 \phi}{\partial x^2} + \frac{\partial^2 \phi}{\partial y^2} + \frac{\partial^2 \phi}{\partial z^2} = \text{div } \nabla \phi = \frac{\partial^2 \phi}{\partial x_i \partial x_i} \,.$$

---

[3] In two dimensional problems, the same rules can be used to define ϵ_{ij}, with $i, j = 1, 2$.

1.1.4 *Vectors, tensors and transformation rules*

Vectors can be conceived in a mathematical sense as ordered sets of numbers, or in a physical sense as mathematical representations of quantities characterized by magnitude and direction. A link between these concepts is provided by the *transformation rules*. Suppose we know the components (V_x, V_y) of the vector V in a given two-dimensional Cartesian coördinate system (x, y), and we wish to determine the components (V'_x, V'_y) in a new system (x', y') which is inclined to (x, y) at an angle θ in the anticlockwise direction as shown in Figure 1.3.

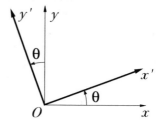

Fig. 1.3 The coördinate
systems x, y and x', y'.

The required components are

$$V'_x = V_x \cos \theta + V_y \sin \theta ; \qquad V'_y = V_y \cos \theta - V_x \sin \theta . \qquad (1.8)$$

We could *define* a vector as an entity, described by its components in a specified Cartesian coördinate system, which transforms into other coördinate systems according to rules like equations (1.8) — i.e. as an ordered set of numbers which obey the transformation rules (1.8). The idea of magnitude and direction could then be introduced by noting that we can always choose θ such that (i) $V'_y = 0$ and (ii) $V'_x > 0$. The corresponding direction x' is then the direction of the resultant vector and the component V'_x is its magnitude.

Now stresses have two suffices and components associated with all possible combinations of two coördinate directions, though we note that equation (1.1) shows that the order of the suffices is immaterial. (Another way of stating this is that the matrix of stress components (1.4) is always symmetric). The stress components satisfy a more complicated set of transformation rules which can be determined by considering the equilibrium of an infinitesimal wedge-shaped piece of material. The resulting equations in the two-dimensional case are those associated with Mohr's circle — i.e.

$$\sigma'_{xx} = \sigma_{xx} \cos^2 \theta + \sigma_{yy} \sin^2 \theta + 2\sigma_{xy} \sin \theta \cos \theta \qquad (1.9)$$
$$\sigma'_{xy} = \sigma_{xy} \left(\cos^2 \theta - \sin^2 \theta\right) + \left(\sigma_{yy} - \sigma_{xx}\right) \sin \theta \cos \theta \qquad (1.10)$$
$$\sigma'_{yy} = \sigma_{yy} \cos^2 \theta + \sigma_{xx} \sin^2 \theta - 2\sigma_{xy} \sin \theta \cos \theta , \qquad (1.11)$$

where we use the notation σ'_{xx} rather than $\sigma_{x'x'}$ in the interests of clarity.

As in the case of vectors we can define a mathematical entity which has a matrix of components in any given Cartesian coördinate system and which transforms into other such coördinate systems according to rules like (1.9–1.11). Such quantities are called *second order Cartesian tensors*. Notice incidentally that the alternating tensor introduced in equation (1.6) is *not* a second order Cartesian tensor.

We know from Mohr's circle that we can always choose θ such that $\sigma'_{xy}=0$, in which case the directions x', y' are referred to as principal directions and the components σ'_{xx}, σ'_{yy} as principal stresses. Thus another way to characterize a second order Cartesian tensor is as a quantity defined by a set of orthogonal principal directions and a corresponding set of principal values.

As with vectors, a pragmatic motivation for abstracting the mathematical properties from the physical quantities which exhibit them is that many different physical quantities are naturally represented as second order Cartesian tensors. Apart from stress and strain, some commonly occurring examples are the second moments of area of a beam cross section (I_x, I_{xy}, I_y), the second partial derivatives of a scalar function ($\partial^2 f/\partial x^2$; $\partial^2 f/\partial x \partial y$; $\partial^2 f/\partial y^2$) and the influence coefficient (compliance) matrix C_{ij} defining the displacement u due to a force F for a linear elastic system, i.e.

$$u_i = C_{ij}F_j , \tag{1.12}$$

where the summation convention is implied.

It is a fairly straightforward matter to prove that each of these quantities obeys transformation rules like (1.9–1.11). It follows immediately (for example) that every beam cross section has two orthogonal *principal axes* of bending about which the two principal second moments are respectively the maximum and minimum for the cross section.

A special tensor of some interest is that for which the Mohr's circle degenerates to a point. In the case of stresses, this corresponds to a state of *hydrostatic stress*, so-called because a fluid at rest cannot support shear stress (the constitutive law for a fluid relates *velocity gradient* to shear stress) and hence $\sigma'_{xy}=0$ for all θ. The only Mohr's circle which satisfies this condition is one of zero radius, from which we deduce immediately that all directions are principal directions and that the principal values are all equal. In the case of the fluid, we obtain the well-known result that the pressure in a fluid at rest is equal in all directions.

It is instructive to consider this result in the context of other systems involving tensors. For example, consider the second moments of area for the square cross section shown in Figure 1.4. By symmetry, we know that Ox, Oy are principal directions and that the two principal second moments are both equal to $a^4/12$. It follows that the Mohr's circle of second moments has zero radius and hence (i) that the second moment about any other axis must also be $a^4/12$ and (ii) that the product

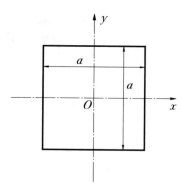

Fig. 1.4 A beam of square cross section.

inertia I'_{xy} must be zero about all axes, showing that the axis of curvature is always aligned with the applied bending moment. These results are not at all obvious merely from an examination of the section.

As a second example, Figure 1.5 shows an elastic system consisting of three identical but arbitrary structures connecting a point, P, to a rigid support, the structures being inclined to each other at angles of 120°.

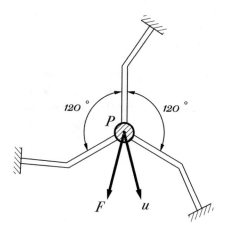

Fig. 1.5 Support structure with three similar but unsymmetrical components.

The structures each have elastic properties expressible in the form of an influence function matrix as in equation (1.12) and are generally such that the displacement u is not colinear with the force F. However, the *overall* influence function matrix for the system has the same properties in three different coördinate systems inclined to each other at 120°, since a rotation of the figure through 120° leaves the system unchanged. The only Mohr's circle which gives equal components after a rotation of 120° is that of zero radius. We therefore conclude that the support system of Figure 1.5 is such that (i) the displacement of P always has the same direction as the force F and (ii) the stiffness or compliance of the system is the same in all directions[4]. These two examples illustrate that there is sometimes an advantage to be gained from considering a disparate physical problem that shares a common mathematical structure with that under investigation.

Three-dimensional transformation rules

In three dimensions, the vector transformation rules (1.8) are most conveniently written in terms of *direction cosines* l_{ij}, defined as the cosine of the angle between unit vectors in the directions x_i' and x_j respectively. We then obtain

$$V_i' = l_{ij} V_j \ . \tag{1.13}$$

For the two-dimensional case of Figure 1.3 and equations (1.8), the matrix[5] formed by the components l_{ij} takes the form

$$l_{ij} = \begin{bmatrix} \cos\theta & \sin\theta \\ -\sin\theta & \cos\theta \end{bmatrix} \ . \tag{1.14}$$

We already remarked in §1.1.1 that only those stress components with an x-suffix act on the surface $x=c$ and hence that if this surface is traction free (for example if it is an unloaded boundary of a body), then $\sigma_{xx}=\sigma_{xy}=\sigma_{xz}=0$, but no restrictions are placed on the components $\sigma_{yy}, \sigma_{yz}, \sigma_{zz}$. More generally, the *traction* t on a specified plane is a vector with three independent components t_i, in contrast to the stress σ_{ij} which is a second order tensor with six independent components. Mathematically, we can define the orientation of a specified inclined plane by a unit vector n in the direction of the outward normal. By considering the equilibrium of an infinitesimal tetrahedron whose surfaces are perpendicular to the directions x_1, x_2, x_3 and n respectively, we then obtain

$$t_i = \sigma_{ij} n_j \ , \tag{1.15}$$

for the traction vector on the inclined surface. The components t_i defined by (1.15) are aligned with the Cartesian axes x_i, but we can use the vector transformation rule (1.13) to resolve t into a new coördinate system in which one axis (x_1' say) is aligned with n. We then have

$$n_j = l_{1j} \ ; \quad t_i = \sigma_{ij} l_{1j} \quad \text{and} \quad t_k' = l_{ki} t_i = l_{ki} l_{1j} \sigma_{ij} \ .$$

---

[4] A similar argument can be used to show that if a laminated fibre-reinforced composite is laid up with equal numbers of identical, but not necessarily symmetrical, laminæ in each of 3 or more equispaced orientations, it must be elastically isotropic within the plane. This proof depends on the properties of the *fourth order* Cartesian tensor c_{ijkl} describing the stress-strain relation of the laminæ (see equations (1.40 1.43) below).

[5] Notice that the matrix of direction cosines is not symmetric.

If we perform this operation for each of the three surfaces orthogonal to x_1', x_2', x_3' respectively, we shall recover the complete stress tensor in the rotated coördinate system, which is therefore given by

$$\sigma_{ij}' = l_{ip} l_{jq} \sigma_{pq} \quad \text{or} \quad \sigma' = \mathbf{\Omega} \, \sigma \, \mathbf{\Omega}^T \, , \tag{1.16}$$

where $\mathbf{\Omega}$ is an unsymmetric matrix whose components $\Omega_{ij} = l_{ij}$. Comparison of (1.16) with (1.9–1.11) provides a good illustration of the efficiency of the index notation in condensing general expressions in Cartesian coördinates.

1.1.5 Principal stresses and von Mises stress

One of the principal reasons for performing elasticity calculations is to determine when and where an engineering component will fail. Theories of material failure are beyond the scope of this book, but the most widely used criteria are the von Mises distortion energy criterion for ductile materials and the maximum tensile stress criterion for brittle materials[6].

Brittle materials typically fracture when the maximum tensile stress reaches a critical value and hence we need to be able to calculate the maximum principal stress σ_1. For two dimensional problems, the principal stresses can be found by using the condition $\sigma_{xy}'=0$ in equation (1.10) to determine the inclination θ of the principal directions. Substituting into (1.9, 1.11) then yields the well known results

$$\sigma_1, \sigma_2 = \frac{\sigma_{xx} + \sigma_{yy}}{2} \pm \sqrt{\left(\frac{\sigma_{xx} - \sigma_{yy}}{2}\right)^2 + \sigma_{xy}^2} \, .$$

In most problems, the maximum stresses occur at the boundaries where shear tractions are usually zero. Thus, even in three-dimensional problems, the determination of the maximum tensile stress often involves only a two-dimensional stress transformation. However, the principal stresses are easily obtained in the fully three-dimensional case from the results

$$\sigma_1 = \frac{I_1}{3} + \frac{2}{3}\left(\sqrt{I_1^2 - 3I_2}\right)\cos\phi \tag{1.17}$$

$$\sigma_2 = \frac{I_1}{3} + \frac{2}{3}\left(\sqrt{I_1^2 - 3I_2}\right)\cos\left(\phi + \frac{2\pi}{3}\right) \tag{1.18}$$

$$\sigma_3 = \frac{I_1}{3} + \frac{2}{3}\left(\sqrt{I_1^2 - 3I_2}\right)\cos\left(\phi + \frac{4\pi}{3}\right) \, , \tag{1.19}$$

where

---

[6] J. R. Barber, *Intermediate Mechanics of Materials*, Springer, Dordrecht, 2nd. edn. (2011), §2.2.

$$\phi = \frac{1}{3} \arccos \left(\frac{2I_1^3 - 9I_1 I_2 + 27 I_3}{2(I_1^2 - 3I_2)^{3/2}} \right)$$

and

$$I_1 = \sigma_{xx} + \sigma_{yy} + \sigma_{zz} \tag{1.20}$$

$$I_2 = \sigma_{xx}\sigma_{yy} + \sigma_{yy}\sigma_{zz} + \sigma_{zz}\sigma_{xx} - \sigma_{xy}^2 - \sigma_{yz}^2 - \sigma_{zx}^2 \tag{1.21}$$

$$I_3 = \sigma_{xx}\sigma_{yy}\sigma_{zz} - \sigma_{xx}\sigma_{yz}^2 - \sigma_{yy}\sigma_{zx}^2 - \sigma_{zz}\sigma_{xy}^2 + 2\sigma_{xy}\sigma_{yz}\sigma_{zx} \ . \tag{1.22}$$

The quantities I_1, I_2, I_3 are known as *stress invariants* because for a given stress state they are the same in all coördinate systems. If the principal value of $\theta = \arccos(x)$ is defined such that $0 \le \theta < \pi$, the principal stresses defined by equations (1.17–1.19) will always satisfy the inequality $\sigma_1 \ge \sigma_3 \ge \sigma_2$.

Von Mises theory states that a ductile material will yield when the strain energy of distortion per unit volume reaches a certain critical value. This enables us to define an *equivalent tensile stress* or *von Mises stress* σ_E as

$$\sigma_E \equiv \sqrt{I_1^2 - 3I_2} = \sqrt{\frac{(3\sigma_{ij}\sigma_{ji} - \sigma_{ii}\sigma_{jj})}{2}}$$

$$= \sqrt{\sigma_{xx}^2 + \sigma_{yy}^2 + \sigma_{zz}^2 - \sigma_{xx}\sigma_{yy} - \sigma_{yy}\sigma_{zz} - \sigma_{zz}\sigma_{xx} + 3\sigma_{xy}^2 + 3\sigma_{yz}^2 + 3\sigma_{zx}^2} \ . \tag{1.23}$$

The von Mises yield criterion then states that yield will occur when σ_E reaches the yield stress in uniaxial tension, S_Y. The Maple and Mathematica files 'principal-stresses' use equations (1.17–1.23) to calculate the principal stresses and the von Mises stress from a given set of stress components.

1.1.6 Displacement

The displacement of a particle P is a vector

$$\mathbf{u} = \mathbf{i}u_x + \mathbf{j}u_y + \mathbf{k}u_z$$

representing the difference between the final and the initial position of P — i.e. it is the distance that P moves during the deformation. In index notation, the displacement is represented as u_i.

The deformation of a body is completely defined if we know the displacement of its every particle. Notice however that there is a class of displacements which do not involve deformation — the so-called 'rigid-body displacements'. A typical case is where all the particles of the body have the same displacement. The name arises, of course, because rigid-body displacement is the *only* class of displacement that can be experienced by a rigid body.

1.2 Strains and their Relation to Displacements

Components of strain will be denoted by the symbol e, with appropriate suffices (e.g. e_{xx}, e_{xy}). As in the case of stress, no special symbol is required for shear strain, though we shall see below that the quantity defined in most elementary texts (and usually denoted by γ) differs from that used in the mathematical theory of Elasticity by a factor of 2. A major advantage of this definition is that it makes the strain e a second order Cartesian Tensor (see §1.1.4 above). We shall demonstrate this by establishing transformation rules for strain similar to equations (1.9–1.11) in §1.2.4 below.

1.2.1 Tensile strain

Students usually first encounter the concept of strain in elementary Mechanics of Materials as the ratio of extension to original length and are sometimes confused by the apparently totally different definition used in more mathematical treatments of solid mechanics. We shall discuss here the connection between the two definitions — partly for completeness, and partly because the physical insight that can be developed in the simple problems of Mechanics of Materials is very useful if it can be carried over into more difficult problems.

Figure 1.6 shows a bar of original length L and density ρ hanging from the ceiling. Suppose we are asked to find how much it increases in length under the loading of its own weight.

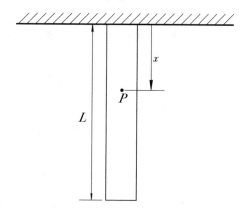

Fig. 1.6 The bar suspended from the ceiling.

It is easily shown that the tensile stress σ_{xx} at the point P, distance x from the ceiling, is

$$\sigma_{xx} = \rho g(L - x) ,$$

where g is the acceleration due to gravity, and hence from Hooke's law,

$$e_{xx} = \frac{\rho g(L - x)}{E} . \tag{1.24}$$

However, the strain varies continuously over the length of the bar and hence we can only apply the Mechanics of Materials definition if we examine an infinitesimal piece of the bar over which the strain can be regarded as sensibly constant.

We describe the deformation in terms of the downward displacement u_x which depends upon x and consider that part of the bar between x and $x + \delta x$, denoted by PQ in Figure 1.7.

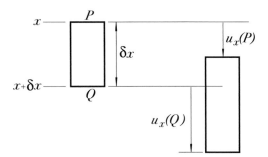

Fig. 1.7 Infinitesimal section of the bar.

After the deformation, PQ must have extended by $u_x(Q) - u_x(P)$ and hence the local value of 'Mechanics of Materials' tensile strain is

$$e_{xx} = \frac{u_x(Q) - u_x(P)}{\delta x} = \frac{u_x(x + \delta x) - u_x(x)}{\delta x} .$$

Taking the limit as $\delta x \to 0$, we obtain the definition

$$e_{xx} = \frac{\partial u_x}{\partial x} . \tag{1.25}$$

Corresponding definitions can be developed in three-dimensional problems for the other normal strain components, i.e.

$$e_{yy} = \frac{\partial u_y}{\partial y} \; ; \; e_{zz} = \frac{\partial u_z}{\partial z} . \tag{1.26}$$

Notice how easy the problem of Figure 1.6 becomes when we use these definitions. We get

$$\frac{\partial u_x}{\partial x} = \frac{\rho g(L - x)}{E} ,$$

from (1.24, 1.25) and hence

$$u_x = \frac{\rho g (2Lx - x^2)}{2E} + A \,, \tag{1.27}$$

where A is an arbitrary constant of integration which expresses the fact that our knowledge of the stresses and hence the strains in the body is not sufficient to determine its position in space. In fact, A represents an arbitrary rigid-body displacement. In this case we need to use the fact that the top of the bar is joined to a supposedly rigid ceiling. Thus, $u_x(0) = 0$ and hence $A = 0$ from (1.27). The increase of length of the bar is represented by the displacement at $x = L$ and is

$$u_x(L) = \frac{\rho g L^2}{2E} \,.$$

1.2.2 Rotation and shear strain

Noting that the two x's in e_{xx} correspond to those in its definition $\partial u_x / \partial x$, it is natural to seek a connection between the shear strain e_{xy} and one or both of the derivatives $\partial u_x / \partial y$, $\partial u_y / \partial x$. As a first step, we shall discuss the geometrical interpretation of these derivatives.

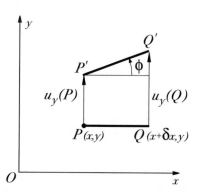

Fig. 1.8 Rotation of a line segment.

Figure 1.8 shows a line segment PQ of length δx, aligned with the x-axis, the two ends of which are displaced in the y-direction. Clearly if $u_y(Q) \neq u_y(P)$, the line PQ will be rotated by these displacements and if the angle of rotation is small it can be written

$$\phi = \frac{u_y(x + \delta x) - u_y(x)}{\delta x} \,,$$

(anticlockwise positive).
   Proceeding to the limit as $\delta x \to 0$, we have

$$\phi = \frac{\partial u_y}{\partial x} \, .$$

Thus, $\partial u_y/\partial x$ is the angle through which a line originally in the x-direction rotates towards the y-direction during the deformation[7].

Now, if PQ is a line drawn on the surface of an elastic solid, the occurrence of a rotation ϕ does not necessarily indicate that the solid is deformed — we could rotate the line simply by rotating the solid as a rigid body. To investigate this matter further, we imagine drawing a series of lines at different angles through the point P as shown in Figure 1.9(a).

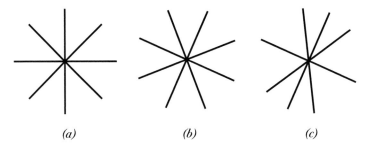

           (a)   *(b)*   *(c)*

Fig. 1.9 Rotation of lines at point P; (a) Original state, (b) Rigid-body rotation; (c) Rotation and deformation.

If the vicinity of the point P suffers merely a local rigid-body rotation, all the lines will rotate through the same angle and retain the same relative inclinations as shown in 1.9(b). However, if different lines rotate through *different* angles, as in 1.9(c), the body must have been deformed. We shall show in the next section that the rotations of the lines in Figure 1.9(c) are not independent and a consideration of their interdependency leads naturally to a definition of shear strain.

1.2.3 Transformation of coördinates

Suppose we knew the displacement components u_x, u_y throughout the body and wished to find the rotation, ϕ, of the line PQ in Figure 1.10, which is inclined at an angle θ to the x-axis.

We construct a new axis system $Ox'y'$ with Ox' parallel to PQ as shown, in which case we can argue as above that PQ rotates anticlockwise through the angle

---

[7] Students of Mechanics of Materials will have already used a similar result when they express the slope of a beam as du/dx, where x is the distance along the axis of the beam and u is the transverse displacement.

$$\phi(PQ) = \frac{\partial u'_y}{\partial x'} \, . \tag{1.28}$$

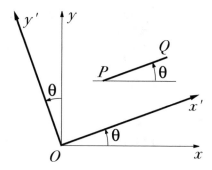

Fig. 1.10 Rotation of a line inclined at angle θ.

Furthermore, we have

$$\frac{\partial}{\partial x'} = \nabla \cdot i' = \left(i \frac{\partial}{\partial x} + j \frac{\partial}{\partial y}\right) \cdot i' = i \cdot i' \frac{\partial}{\partial x} + j \cdot i' \frac{\partial}{\partial y}$$

$$= \cos\theta \frac{\partial}{\partial x} + \sin\theta \frac{\partial}{\partial y} \, , \tag{1.29}$$

and by similar arguments,

$$u'_x = u_x \cos\theta + u_y \sin\theta \, ; \qquad u'_y = u_y \cos\theta - u_x \sin\theta \, . \tag{1.30}$$

In fact equations (1.29, 1.30) are simply restatements of the vector transformation rules (1.8), since both u and the gradient operator ∇ are vectors.

Substituting these results into equation (1.28), we find

$$\phi(PQ) = \left(\cos\theta \frac{\partial}{\partial x} + \sin\theta \frac{\partial}{\partial y}\right)(u_y \cos\theta - u_x \sin\theta)$$

$$= \frac{\partial u_y}{\partial x} \cos^2\theta - \frac{\partial u_x}{\partial y} \sin^2\theta + \left(\frac{\partial u_y}{\partial y} - \frac{\partial u_x}{\partial x}\right) \sin\theta \cos\theta$$

$$= \frac{1}{2}\left(\frac{\partial u_y}{\partial x} - \frac{\partial u_x}{\partial y}\right) + \frac{1}{2}\left(\frac{\partial u_y}{\partial x} + \frac{\partial u_x}{\partial y}\right) \cos(2\theta)$$

$$+ \frac{1}{2}\left(\frac{\partial u_y}{\partial y} - \frac{\partial u_x}{\partial x}\right) \sin(2\theta) \, . \tag{1.31}$$

In the final expression (1.31), the first term is independent of the inclination θ of the line PQ and hence represents a rigid-body rotation as in Figure 1.9(b). We denote this rotation by the symbol ω, which with the convention illustrated in Figure 1.8 is anticlockwise positive.

Notice that as this term is independent of θ in equation (1.31), it is the same for any right-handed set of axes — i.e.

$$\omega = \frac{1}{2}\left(\frac{\partial u_y}{\partial x} - \frac{\partial u_x}{\partial y}\right) = \frac{1}{2}\left(\frac{\partial u'_y}{\partial x'} - \frac{\partial u'_x}{\partial y'}\right) , \tag{1.32}$$

for any x', y'.

In three dimensions, ω represents a small clockwise rotation about the positive z-axis and is therefore more properly denoted by ω_z to distinguish it from the corresponding rotations about the x and y axes — i.e.

$$\omega_x = \frac{1}{2}\left(\frac{\partial u_z}{\partial y} - \frac{\partial u_y}{\partial z}\right) \ ; \ \omega_y = \frac{1}{2}\left(\frac{\partial u_x}{\partial z} - \frac{\partial u_z}{\partial x}\right) \ ; \ \omega_z = \frac{1}{2}\left(\frac{\partial u_y}{\partial x} - \frac{\partial u_x}{\partial y}\right) . \tag{1.33}$$

The rotation ω is therefore a vector in three-dimensional problems and is more compactly defined in vector or index notation as

$$\omega = \frac{1}{2}\text{curl } u \equiv \frac{1}{2}\nabla \times u \quad \text{or} \quad \omega_k = \frac{1}{2}\left(\epsilon_{ijk}\frac{\partial u_j}{\partial x_i}\right) , \tag{1.34}$$

where ϵ_{ijk} is defined in §1.1.3. In two dimensions ω behaves as a scalar since two of its components degenerate to zero[8].

1.2.4   Definition of shear strain

We are now in a position to *define* the shear strain e_{xy} as the difference between the rotation of a line drawn in the x-direction and the corresponding rigid-body rotation, ω_z, i.e.

$$e_{xy} = \frac{\partial u_y}{\partial x} - \omega_z = \frac{1}{2}\left(\frac{\partial u_y}{\partial x} + \frac{\partial u_x}{\partial y}\right) \tag{1.35}$$

and similarly

$$e_{yz} = \frac{1}{2}\left(\frac{\partial u_z}{\partial y} + \frac{\partial u_y}{\partial z}\right) \ ; \ e_{zx} = \frac{1}{2}\left(\frac{\partial u_x}{\partial z} + \frac{\partial u_z}{\partial x}\right) . \tag{1.36}$$

Note that e_{xy} so defined is one half of the quantity γ_{xy} used in Mechanics of Materials and in many older books on Elasticity.

The strain-displacement relations (1.25, 1.26, 1.35, 1.36) can be written in the concise form

$$e_{ij} = \frac{1}{2}\left(\frac{\partial u_i}{\partial x_j} + \frac{\partial u_j}{\partial x_i}\right) . \tag{1.37}$$

---

[8] In some books, $\omega_x, \omega_y, \omega_z$ are denoted by $\omega_{yz}, \omega_{zx}, \omega_{xy}$ respectively. This notation is not used here because it gives the erroneous impression that ω is a second order tensor rather than a vector.

With the notation of equations (1.25, 1.26, 1.33, 1.35), we can now write (1.31) in the form

$$\phi(PQ) = \omega_z + e_{xy}(\cos^2\theta - \sin^2\theta) + (e_{yy} - e_{xx})\sin\theta\cos\theta$$

and hence

$$e'_{xy} = \phi(PQ) - \omega_z \qquad \text{(by definition)}$$
$$= e_{xy}(\cos^2\theta - \sin^2\theta) + (e_{yy} - e_{xx})\sin\theta\cos\theta . \qquad (1.38)$$

This is one of the coördinate transformation relations for strain, the other one

$$e'_{xx} = e_{xx}\cos^2\theta + e_{yy}\sin^2\theta + 2e_{xy}\sin\theta\cos\theta \qquad (1.39)$$

being obtainable from equations (1.25, 1.29, 1.30) in the same way. A comparison of equations (1.38, 1.39) and (1.9, 1.10) confirms that, with these definitions, the strain e_{ij} is a second order Cartesian tensor. The corresponding three-dimensional strain-transformation equations have the same form as (1.16) and can be obtained using the strain-displacement relation (1.37) and the vector transformation rule (1.13). This derivation is left as an exercise for the reader (Problem 1.9).

1.3 Stress-strain Relations

The fundamental assumption of linear elasticity is that the material obeys Hooke's law, implying that there is a linear relation between stress and strain. The most general such relation can be written in index notation as

$$\sigma_{ij} = c_{ijkl}e_{kl} ; \quad e_{ij} = s_{ijkl}\sigma_{kl} , \qquad (1.40)$$

where c_{ijkl}, s_{ijkl} are the *elasticity tensor* and the *compliance tensor* respectively. Both the elasticity tensor and the compliance tensor must satisfy the symmetry conditions

$$c_{ijkl} = c_{jikl} = c_{klij} = c_{ijlk} , \qquad (1.41)$$

which follow from (i) the symmetry of the stress and strain tensors (e.g. $\sigma_{ij} = \sigma_{ji}$) and (ii) the reciprocal theorem, which we shall discuss in Chapter 38. Substituting the strain-displacement relations (1.37) into the first of (1.40) and using (1.41) to combine appropriate terms, we obtain

$$\sigma_{ij} = c_{ijkl}\frac{\partial u_k}{\partial x_l} . \qquad (1.42)$$

Equation (1.16) and the corresponding strain-transformation equation imply that the fourth-order tensors c_{ijkl}, s_{ijkl} take the form

$$c'_{ijkl} = l_{ip}l_{jq}l_{kr}l_{ls}c_{pqrs} \; ; \quad s'_{ijkl} = l_{ip}l_{jq}l_{kr}l_{ls}s_{pqrs} \; , \tag{1.43}$$

in a coördinate system defined by the direction cosines l_{ij}.

In this book, we shall mostly restrict attention to the case where the material is *isotropic*, meaning that the tensors c_{ijkl}, s_{ijkl} are invariant under all such coördinate transformations, and (e.g.) $c'_{ijkl} = c_{ijkl}$. In other words, a test specimen of the material would show identical behaviour regardless of its orientation relative to the original piece of material from which it was manufactured. However, the more general anisotropic case will be discussed in Chapter 36.

1.3.1 Isotropic constitutive law

We shall develop the various forms of Hooke's law for the isotropic medium starting from the results of the uniaxial tensile test, in which the only non-zero stress component is σ_{xx}. Symmetry considerations then show that the shear strains e_{xy}, e_{yz}, e_{zx} must be zero and that the transverse strains e_{yy}, e_{zz} must be equal. Thus, the most general relation can be written in the form

$$e_{xx} = \frac{\sigma_{xx}}{E} \; ; \quad e_{yy} = e_{zz} = -\frac{\nu\sigma_{xx}}{E} \; ,$$

where *Young's modulus E* and *Poisson's ratio ν* are experimentally determined constants. The normal strain components due to a more general triaxial state of stress can then be written down by linear superposition as

$$e_{xx} = \frac{\sigma_{xx}}{E} - \frac{\nu\sigma_{yy}}{E} - \frac{\nu\sigma_{zz}}{E} \tag{1.44}$$

$$e_{yy} = -\frac{\nu\sigma_{xx}}{E} + \frac{\sigma_{yy}}{E} - \frac{\nu\sigma_{zz}}{E} \tag{1.45}$$

$$e_{zz} = -\frac{\nu\sigma_{xx}}{E} - \frac{\nu\sigma_{yy}}{E} + \frac{\sigma_{zz}}{E} \; . \tag{1.46}$$

The relation between e_{xy} and σ_{xy} can then be obtained by using the coördinate transformation relations. We know that there are three principal directions such that if we align them with x, y, z, we have

$$\sigma_{xy} = \sigma_{yz} = \sigma_{zx} = 0 \tag{1.47}$$

and hence by symmetry

$$e_{xy} = e_{yz} = e_{zx} = 0 \; . \tag{1.48}$$

Using a system aligned with the principal directions, we write

$$e'_{xy} = (e_{yy} - e_{xx}) \sin \theta \cos \theta$$

from equations (1.38, 1.48), and hence

$$
\begin{aligned}
e'_{xy} &= \frac{(\sigma_{yy} - \sigma_{xx})(1 + \nu) \sin \theta \cos \theta}{E} \\
&= \frac{(1 + \nu)\sigma'_{xy}}{E} \, ,
\end{aligned}
\tag{1.49}
$$

from (1.44, 1.45, 1.10, 1.47).

We define the *shear modulus* or the *modulus of rigidity*

$$\mu = \frac{E}{2(1 + \nu)} \, ,$$

so that equation (1.49) takes the form

$$e'_{xy} = \frac{\sigma'_{xy}}{2\mu} \, .
\tag{1.50}$$

1.3.2   Lamé's constants

It is often desirable to solve equations (1.44–1.46) to express σ_{xx} in terms of e_{xx} etc. The solution is routine and leads to the equation

$$\sigma_{xx} = \frac{E\nu(e_{xx} + e_{yy} + e_{zz})}{(1 + \nu)(1 - 2\nu)} + \frac{E e_{xx}}{(1 + \nu)}$$

and similar equations, which are more concisely written in the form

$$\sigma_{xx} = \lambda e + 2\mu e_{xx}
\tag{1.51}$$

etc., where

$$\lambda = \frac{E\nu}{(1 + \nu)(1 - 2\nu)} = \frac{2\mu\nu}{(1 - 2\nu)}
\tag{1.52}$$

and

$$e \equiv e_{xx} + e_{yy} + e_{zz} \equiv e_{ii} \equiv \text{div } \boldsymbol{u}
\tag{1.53}$$

is known as the *dilatation*.

The stress-strain relations (1.50, 1.51) can be written more concisely in the index notation in the form

$$\sigma_{ij} = \lambda \delta_{ij} e_{mm} + 2\mu e_{ij} \, ,
\tag{1.54}$$

where δ_{ij} is the *Kronecker delta*, defined as 1 if $i=j$ and 0 if $i \neq j$. Equivalently, we can use equation (1.40) with

$$c_{ijkl} = \lambda \delta_{ij}\delta_{kl} + \mu \left(\delta_{ik}\delta_{jl} + \delta_{jk}\delta_{il}\right) . \tag{1.55}$$

The constants λ, μ are known as Lamé's constants. Young's modulus and Poisson's ratio can be written in terms of Lamé's constants through the equations

$$E = \frac{\mu(3\lambda + 2\mu)}{(\lambda + \mu)} ; \quad \nu = \frac{\lambda}{2(\lambda + \mu)} .$$

1.3.3 Dilatation and bulk modulus

The dilatation, e, is easily shown to be invariant as to coördinate transformation and is therefore a scalar quantity. In physical terms it is the local volumetric strain, since a unit cube increases under strain to a block of dimensions $(1+e_{xx})$, $(1+e_{yy})$, $(1+e_{zz})$ and hence the volume change is

$$\Delta V = (1 + e_{xx})(1 + e_{yy})(1 + e_{zz}) - 1 = e_{xx} + e_{yy} + e_{zz} + O(e_{xx}e_{yy}) .$$

It can be shown[9] that the dilatation e and the rotation vector ω are harmonic — i.e. $\nabla^2 e = \nabla^2 \omega = 0$. For this reason, many early solutions of elasticity problems were formulated in terms of these variables, so as to make use of the wealth of mathematical knowledge about harmonic functions. We now have other more convenient ways of expressing elasticity problems in terms of harmonic functions, which will be discussed in Chapter 21 *et seq.*.

From equations (1.44–1.46), we have

$$e = e_{ii} = \frac{(\sigma_{xx} + \sigma_{yy} + \sigma_{zz})(1 - 2\nu)}{E} = \frac{\overline{\sigma}}{K_b} ,$$

where

$$\overline{\sigma} = \frac{\sigma_{xx} + \sigma_{yy} + \sigma_{zz}}{3} = \frac{1}{3}\sigma_{ii} \tag{1.56}$$

is the mean normal stress, also known as the *bulk stress* or the *hydrostatic stress*, and

$$K_b = \frac{E}{3(1 - 2\nu)} = \frac{2\mu(1 + \nu)}{3(1 - 2\nu)} = \lambda + \frac{2\mu}{3} \tag{1.57}$$

is the *bulk modulus*.

We note that $K_b \to \infty$ if $\nu \to 0.5$, in which case the material is described as *incompressible*.

---

[9] See Problems 2.1, 2.2.

1.3.4 Deviatoric stress

It is sometimes convenient to represent a general state of stress σ_{ij} as the sum of a hydrostatic stress $\bar{\sigma}$ and a *deviatoric stress* σ'_{ij} defined as

$$\sigma'_{ij} = \sigma_{ij} - \bar{\sigma}\delta_{ij} . \qquad (1.58)$$

We also define the *deviatoric strain* e'_{ij} as

$$e'_{ij} = e_{ij} - \frac{e}{3}\delta_{ij} , \qquad (1.59)$$

in which case it is easily shown that

$$\sigma'_{ij} = 2\mu e'_{ij} .$$

In particular, the von Mises stress of equation (1.23) depends only on the deviatoric stress. In other words, if equation (1.58) is used to eliminate σ_{ij} in (1.23), the terms involving $\bar{\sigma}$ cancel.

Problems

1.1. Show that

$$\text{(i)} \quad \frac{\partial x_i}{\partial x_j} = \delta_{ij} \quad \text{and} \quad \text{(ii)} \quad R = \sqrt{x_i x_i} ,$$

where $R = |\mathbf{r}|$ is the distance from the origin. Hence find $\partial R/\partial x_j$ in index notation. Confirm your result by finding $\partial R/\partial x$ in x, y, z notation.

1.2. Prove that the partial derivatives $\partial^2 f/\partial x^2$, $\partial^2 f/\partial x \partial y$, $\partial^2 f/\partial y^2$ of the scalar function $f(x, y)$ transform into the rotated coördinate system x', y' by rules similar to equations (1.9–1.11).

1.3. Show that the direction cosines defined in (1.13) satisfy the identity

$$l_{ij}l_{ik} = \delta_{jk} .$$

Hence or otherwise, show that the product $\sigma_{ij}\sigma_{ij}$ is invariant under coördinate transformation.

1.4. By restricting the indices i, j etc. to the values 1,2 only, show that the two-dimensional stress-transformation relations (1.9–1.11) can be obtained from (1.16) using the two-dimensional direction cosines (1.14).

1.5. Use the index notation to develop concise expressions for the three stress invariants I_1, I_2, I_3.

1.6. Choosing a local coördinate system x_1, x_2, x_3 aligned with the three principal axes, determine the tractions on the *octahedral plane* defined by the unit vector

$$n = \left\{ \frac{1}{\sqrt{3}}, \frac{1}{\sqrt{3}}, \frac{1}{\sqrt{3}} \right\}^T$$

which makes equal angles with all three principal axes, if the principal stresses are $\sigma_1, \sigma_2, \sigma_3$. Hence show that the magnitude of the resultant shear stress on this plane is $\sqrt{2}\sigma_E/3$, where σ_E is given by equation (1.23).

1.7. A rigid body is subjected to a small rotation $\omega_z = \Omega \ll 1$ about the z-axis. If the displacement of the origin is zero, find expressions for the three displacement components u_x, u_y, u_z as functions of x, y, z.

1.8. Use the index notation to develop a general expression for the derivative

$$\frac{\partial u_i}{\partial x_j}$$

in terms of strains and rotations.

1.9. Use the three-dimensional vector transformation rule (1.13) and the index notation to prove that the strain components (1.37) transform according to the equation

$$e'_{ij} = l_{ip}l_{jq}e_{pq} .$$

Hence show that the dilatation e_{ii} is invariant under coördinate transformation.

1.10. Find an index notation expression for the compliance tensor s_{ijkl} of equation (1.40) for the isotropic elastic material in terms of the elastic constants E, ν.

1.11. Show that equations (1.44–1.46, 1.49) can be written in the concise form

$$e_{ij} = \frac{(1+\nu)\sigma_{ij}}{E} - \frac{\nu\delta_{ij}\sigma_{mm}}{E} . \tag{1.60}$$

Chapter 2
Equilibrium and Compatibility

We can think of an elastic solid as a highly redundant framework — each particle is built-in to its neighbours. For such a framework, we expect to get some equations from considerations of *equilibrium*, but not as many as there are unknowns. The deficit is made up by *compatibility* conditions — statements that the deformed components must fit together. These latter conditions will relate the dimensions and hence the strains of the deformed components and in order to express them in terms of the same unknowns as the stresses (forces) we need to make use of the stress-strain relations as applied to each component separately.

If we were to approximate the continuous elastic body by a system of interconnected elastic bars, this would be an exact description of the solution procedure. The only difference in treating the continuous medium is that the system of algebraic equations is replaced by partial differential equations describing the same physical or geometrical principles.

2.1 Equilibrium Equations

We consider a small rectangular block of material — side δx, δy, δz — as shown in Figure 2.1. We suppose that there is a body force[1], p per unit volume, and that the stresses vary with position so that the stress components on opposite faces of the block differ by the differential quantities $\delta\sigma_{xx}$, $\delta\sigma_{xy}$ etc. Only those stress components which act in the x-direction are shown in Figure 2.1 for clarity.

---

[1] A body force is one that acts directly on every particle of the body, rather than being applied by tractions at its boundaries and transmitted to the various particles by means of internal stresses. This is an important distinction which will be discussed further in Chapter 7, below. The commonest example of a body force is that due to gravity.

© The Author(s), under exclusive license to Springer Nature Switzerland AG 2022
J. R. Barber, *Elasticity*, Solid Mechanics and Its Applications 172,
https://doi.org/10.1007/978-3-031-15214-6_2

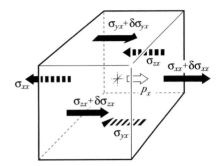

Fig. 2.1 Forces in the
x-direction on an elemental
block.

Resolving forces in the x-direction, we find

$$(\sigma_{xx} + \delta\sigma_{xx} - \sigma_{xx})\delta y\delta z + (\sigma_{xy} + \delta\sigma_{xy} - \sigma_{xy})\delta z\delta x$$
$$+(\sigma_{xz} + \delta\sigma_{xz} - \sigma_{xz})\delta x\delta y + p_x\delta x\delta y\delta z = 0 .$$

Hence, dividing through by $(\delta x\delta y\delta z)$ and proceeding to the limit as these infinitesimals tend to zero, we obtain

$$\frac{\partial\sigma_{xx}}{\partial x} + \frac{\partial\sigma_{xy}}{\partial y} + \frac{\partial\sigma_{xz}}{\partial z} + p_x = 0 . \tag{2.1}$$

Similarly, we have

$$\frac{\partial\sigma_{yx}}{\partial x} + \frac{\partial\sigma_{yy}}{\partial y} + \frac{\partial\sigma_{yz}}{\partial z} + p_y = 0 \tag{2.2}$$

$$\frac{\partial\sigma_{zx}}{\partial x} + \frac{\partial\sigma_{zy}}{\partial y} + \frac{\partial\sigma_{zz}}{\partial z} + p_z = 0 , \tag{2.3}$$

or in index notation

$$\frac{\partial\sigma_{ij}}{\partial x_j} + p_i = 0 . \tag{2.4}$$

These are the *differential equations of equilibrium.*

2.2 Compatibility Equations

The easiest way to satisfy the equations of compatibility — as in framework problems — is to express all the strains in terms of the displacements. In a framework, this ensures that the components fit together by identifying the displacement of points in two links which are pinned together by the same symbol. If the framework is redundant, the number of pin displacements thereby introduced is less than the number of component lengths (and hence extensions) determined by them — and the number of unknowns is therefore reduced.

For the continuum, the process is essentially similar, but much more straight-forward. We define the six components of strain in terms of displacements through equations (1.37). These six equations introduce only three unknowns (u_x, u_y, u_z) and hence the latter can be eliminated to give equations constraining the strain components.

For example, from (1.25, 1.26) we find

$$\frac{\partial^2 e_{xx}}{\partial y^2} = \frac{\partial^3 u_x}{\partial x \partial y^2} \; ; \; \frac{\partial^2 e_{yy}}{\partial x^2} = \frac{\partial^3 u_y}{\partial y \partial x^2}$$

and hence

$$\frac{\partial^2 e_{xx}}{\partial y^2} + \frac{\partial^2 e_{yy}}{\partial x^2} = \frac{\partial^2}{\partial x \partial y}\left(\frac{\partial u_x}{\partial y} + \frac{\partial u_y}{\partial x}\right) = 2\frac{\partial^2 e_{xy}}{\partial x \partial y} \;,$$

from (1.35) — i.e.

$$\frac{\partial^2 e_{xx}}{\partial y^2} - 2\frac{\partial^2 e_{xy}}{\partial x \partial y} + \frac{\partial^2 e_{yy}}{\partial x^2} = 0 \;. \tag{2.5}$$

Two more equations of the same form may be obtained by permuting suffices[2]. It is tempting to pursue an analogy with algebraic equations and argue that, since the six strain components are defined in terms of three independent displacement components, we must be able to develop three (i.e. $6 - 3$) independent compatibility equations. However, it is easily verified that, in addition to the three equations similar to (2.5), the strains must satisfy three more equations of the form

$$\frac{\partial^2 e_{zz}}{\partial x \partial y} = \frac{\partial}{\partial z}\left(\frac{\partial e_{yz}}{\partial x} + \frac{\partial e_{zx}}{\partial y} - \frac{\partial e_{xy}}{\partial z}\right) \;. \tag{2.6}$$

The resulting six equations are independent in the sense that no one of them can be derived from the other five, which all goes to show that arguments for algebraic equations do not always carry over to partial differential equations.

A concise statement of the six compatibility equations can be written in the index notation in the form

$$\epsilon_{ijk}\epsilon_{pqr}\frac{\partial^2 e_{jq}}{\partial x_k \partial x_r} = 0 \;. \tag{2.7}$$

The full set of six equations makes the problem very complicated. In practice, therefore, most three-dimensional problems are treated in terms of displacements instead of strains, which satisfies the requirement of compatibility identically. However, in two dimensions, all except one of the compatibility equations degenerate to identities, so that a formulation in terms of stresses or strains is more practical.

---

[2] i.e. $x \to y, y \to z, z \to x$.

2.2.1   The significance of the compatibility equations

The physical meaning of equilibrium is fairly straightforward, but people often get mixed up about just what is being guaranteed by the compatibility equations.

Single-valued displacements

Mathematically, we might say that the strains are compatible *when they are definable in terms of a single-valued, continuously differentiable displacement.*

We could imagine reversing this process — i.e. integrating the strains (displacement gradients) to find the relative displacement $(u_B - u_A)$ of two points A, B in the solid (see Figure 2.2).

Fig. 2.2   Path of the integral in equation (2.8).

Formally we can write

$$u_B - u_A = \int_A^B \frac{\partial u}{\partial s} ds \, , \tag{2.8}$$

where s is a coördinate representing distance along a path S between A and B. If the displacements are to be single-valued, it mustn't make any difference if we change the path S provided it remains within the solid. In other words, the integral should be *path-independent.*

The compatibility equations are sufficient to guarantee this only if the body is simply-connected.

Suppose tentatively that the contour integral

$$\oint \frac{\partial u}{\partial s} ds = 0 \tag{2.9}$$

around any *infinitesimal* closed loop[3]. We could then make infinitesimal changes in our line from A to B by taking in such small loops until the whole line was sensibly

---

[3] Earlier editions of this book made the erroneous claim that the compatibility equations are sufficient to guarantee this. The author is grateful to Professor Arash Yavari for drawing his attention to this error.

changed, thus satisfying the requirement that the integral (2.8) is path-independent. However, the fact that the line is changed infinitesimally (continuously) prevents us from taking a qualitatively (topologically) different route through a multiply-connected body. For example, in Figure 2.3, it is impossible to move S_1 to S_2 by continuous changes without passing outside the body[4]. The equivalence of these two paths must therefore be enforced explicitly by requiring that

$$\int_{S_1} \frac{\partial \mathbf{u}}{\partial s} ds = \int_{S_2} \frac{\partial \mathbf{u}}{\partial s} ds \ ,$$

or equivalently

$$\oint_{S_0} \frac{\partial \mathbf{u}}{\partial s} ds = 0 \ , \qquad (2.10)$$

where $S_0 = S_1 - S_2$ (i.e. out along S_1 and back along S_2) is a closed path that encircles the hole in Figure 2.3.

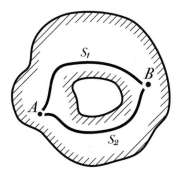

Fig. 2.3 Topologically different integration paths S_1, S_2 between points A and B in a multiply-connected body.

However, the compatibility equations are not quite sufficient even to guarantee (2.9). Using equations (1.34, 1.37), we can write the displacement gradient

$$\frac{\partial u_i}{\partial x_j} = e_{ij} - \epsilon_{ijk}\omega_k \ , \qquad (2.11)$$

from which we conclude that the rotation $\boldsymbol{\omega}$ must also be single-valued if the derivatives in (2.8, 2.9) are to be well-defined. The rotation gradients can be expressed in terms of strain gradients through the relations

---

[4] For a more rigorous discussion of this question, see A. E. H. Love, *A Treatise on the Mathematical Theory of Elasticity*, 4th edn., Dover, 1944, §156A, B. A. Boley and J. H. Weiner, *Theory of Thermal Stresses*, John Wiley, New York, 1960, §§3.6–3.8, or A. Yavari (2013), Compatibility equations of nonlinear elasticity for non-simply-connected bodies, *Archive for Rational Mechanics and Analysis*, Vol. 209, pp. 237–253.

$$\frac{\partial \omega_k}{\partial x_i} = \epsilon_{jkl}\frac{\partial e_{ij}}{\partial x_l} \,, \tag{2.12}$$

but these will integrate to a single-valued expression for ω in the body of Figure 2.3 only if

$$\oint_{S_0}\frac{\partial \omega}{\partial s}ds \equiv \oint_{S_0}\frac{\partial \omega_k}{\partial x_i}dx_i = \oint_{S_0}\epsilon_{jkl}\frac{\partial e_{ij}}{\partial x_l}dx_i = 0 \,. \tag{2.13}$$

If this condition is satisfied, equation (2.10) can also be expressed in terms of strains. Substituting (2.11) into (2.8) and integrating the second term by parts, we obtain

$$\int\frac{\partial u_i}{\partial x_j}dx_j = \int e_{ij}dx_j - \int \epsilon_{ijk}\omega_k dx_j$$

$$= \int e_{ij}dx_j - \epsilon_{ipk}x_p\omega_k + \int \epsilon_{ipk}x_p\frac{\partial \omega_k}{\partial x_j}dx_j \,.$$

If the integral is performed around a closed path S_0, and if ω is single-valued (i.e. (2.13) is satisfied), the second term will take the same value at each end of the path and will therefore sum to zero. Substituting for the rotation gradients from (2.12) and using the identity $\epsilon_{ijk}\epsilon_{imn} = \delta_{jm}\delta_{kn} - \delta_{jn}\delta_{km}$, the remaining terms can then be written

$$\oint_{S_0}\frac{\partial u_i}{\partial x_j}dx_j = \oint_{S_0}\left[e_{ij} + x_p\left(\frac{\partial e_{pj}}{\partial x_i} - \frac{\partial e_{ij}}{\partial x_p}\right)\right]dx_j \,.$$

A similar argument can be used to show that

$$u_i(B) = u_i(A) - \epsilon_{ipk}\omega_k(A)\left[x_p(B) - x_p(A)\right]$$

$$+ \int_A^B\left[e_{ij} + [x_p - x_p(B)]\left(\frac{\partial e_{pj}}{\partial x_i} - \frac{\partial e_{ij}}{\partial x_p}\right)\right]dx_j \,,$$

again subject to the restriction that ω be single-valued[5]. These results were obtained by V. Volterra and E. Cesàro and are known as *Volterra-Cesàro integrals*.

In practice, if a solid is multiply connected it is usually easier to work in terms of displacements and by-pass this problem. Otherwise the equivalence of topologically different paths in integrals like (2.8, 2.13) must be explicitly enforced.

Compatibility of deformed shapes

A more 'physical' way of thinking of compatibility is to state that *the separate particles of the body must deform under load in such a way that they fit together*

---

[5] Notice that the first two terms in this equation correspond to rigid-body translation and rotation, so they can be set to zero if the intention is to find a particular solution for the displacements corresponding to a given strain field.

after deformation. This interpretation is conveniently explored by way of a 'jig-saw' analogy.

We consider a multiply-connected two-dimensional body cut up as a jig-saw puzzle (Figure 2.4), of which the pieces are deformable. Figure 2.4(*a*) shows the original puzzle and 2.4(*b*) the puzzle after deformation by some external loads $\boldsymbol{F}_i$. (The pieces are shown as initially rectangular to aid visualization.)

The deformation of the puzzle must satisfy the following conditions:-

(i) The forces on any given piece (including external forces if any) must be in equilibrium.
(ii) The deformed pieces must be the right *shape* to fit together to make the deformed puzzle (Figure 2.4(*b*)).

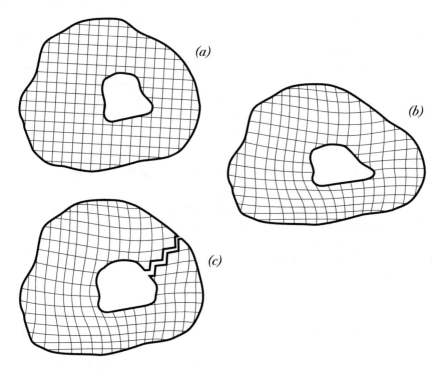

Fig. 2.4 A multiply-connected, deformable jig-saw puzzle: (*a*) Before deformation; (*b*) After deformation; (*c*) Result of attempting to assemble the deformed puzzle if equation (2.9) is not satisfied for any closed path encircling the hole.

For a simply-connected body, the compatibility conditions (2.7) guarantee that (ii) is satisfied, since the deformed shape is defined by the strain components, and each piece can be given whatever rigid-body rotation is required to ensure that it fits. However, if the puzzle is multiply connected — e.g. if it has a central hole —

condition (ii) is *not* sufficient to ensure that the deformed pieces can be assembled into a coherent body.

Suppose we imagine assembling the 'deformed' puzzle working from one piece outwards. The partially completed puzzle is initially simply connected and the shape condition is sufficient to ensure the success of our assembly until we reach a piece which would convert the partially completed puzzle to a multiply-connected body. This critical piece will be the right *shape* for both sides, but it may be the wrong *size* for the separation. Also, the rigid-body rotation needed for it to fit on one side may differ from that required on the other side. If so, it will be possible to leave the completed puzzle in a state with a discontinuity as shown in Figure 2.4(*c*) at any arbitrarily chosen position[6], but there will be no way in which Figure 2.4(*b*) can be constructed.

The lines defining the two sides of the discontinuity in Figure 2.4(*c*) are the same shape and hence, in the most general three-dimensional case, the discontinuity can be defined by six arbitrary constants corresponding to the three rigid-body translations and three rotations needed to move one side to coincide with the other. We therefore get six additional algebraic conditions for each hole in a multiply-connected body. These conditions are of course equivalent to the two vector equations (2.10, 2.13). In the two-dimensional case, the rotation degenerates to a scalar ω and there are only three integral conditions: one from equation (2.13) and two from (2.10)

2.3 Equilibrium Equations in terms of Displacements

Recalling that it is often easier to work in terms of displacements rather than stresses to avoid complications with the compatibility conditions, it is convenient to express the equilibrium equations in terms of displacements. This is most easily done in index notation. Substituting for the stresses from (1.42) into the equilibrium equation (2.4), we obtain

$$c_{ijkl}\frac{\partial^2 u_k}{\partial x_j \partial x_l} + p_i = 0 . \tag{2.14}$$

For the special case of isotropy, we can substitute for c_{ijkl} from (1.55), obtaining

$$\left[\lambda\delta_{ij}\delta_{kl} + \mu\left(\delta_{ik}\delta_{jl} + \delta_{jk}\delta_{il}\right)\right]\frac{\partial^2 u_k}{\partial x_j \partial x_l} + p_i = 0$$

or

$$\lambda\frac{\partial^2 u_l}{\partial x_i \partial x_l} + \mu\frac{\partial^2 u_i}{\partial x_l \partial x_l} + \mu\frac{\partial^2 u_k}{\partial x_k \partial x_i} + p_i = 0 .$$

[6] since we could move one or more pieces from one side of the discontinuity to the other.

The first and third terms in this equation have the same form, since k, l are dummy indices. We can therefore combine them and write

$$(\lambda + \mu)\frac{\partial^2 u_k}{\partial x_i \partial x_k} + \mu\frac{\partial^2 u_i}{\partial x_k \partial x_k} + p_i = 0 \ .$$

Also, noting that

$$\lambda + \mu = \mu\left(1 + \frac{2\nu}{1 - 2\nu}\right) = \frac{\mu}{(1 - 2\nu)}$$

from (1.52), we have

$$\frac{\partial^2 u_k}{\partial x_i \partial x_k} + (1 - 2\nu)\frac{\partial^2 u_i}{\partial x_k \partial x_k} + \frac{(1 - 2\nu)p_i}{\mu} = 0 \ , \tag{2.15}$$

or in vector notation

$$\nabla\,\mathrm{div}\,\boldsymbol{u} + (1 - 2\nu)\nabla^2\boldsymbol{u} + \frac{(1 - 2\nu)\boldsymbol{p}}{\mu} = \boldsymbol{0} \ . \tag{2.16}$$

Problems

2.1. Show that, if there are no body forces, the dilatation e must satisfy the condition

$$\nabla^2 e = 0 \ .$$

2.2. Show that, if there are no body forces, the rotation $\boldsymbol{\omega}$ must satisfy the condition

$$\nabla^2\boldsymbol{\omega} = \boldsymbol{0} \ .$$

2.3. One way of satisfying the compatibility equations in the absence of rotation is to define the components of displacement in terms of a potential function ψ through the relations

$$u_x = \frac{\partial\psi}{\partial x} \ ; \quad u_y = \frac{\partial\psi}{\partial y} \ ; \quad u_z = \frac{\partial\psi}{\partial z} \ .$$

Use the stress-strain relations to derive expressions for the stress components in terms of ψ.

Hence show that the stresses will satisfy the equilibrium equations in the absence of body forces if and only if

$$\nabla^2\psi = \text{constant} \ .$$

2.4. Plastic deformation during a manufacturing process generates a state of residual stress in the large body $z > 0$. If the residual stresses are functions of z only and the surface $z = 0$ is not loaded, show that the stress components $\sigma_{yz}, \sigma_{zx}, \sigma_{zz}$ must be zero everywhere.

2.5. By considering the equilibrium of a small element of material similar to that shown in Figure 1.2, derive the three equations of equilibrium in cylindrical polar coördinates r, θ, z.

2.6. In cylindrical polar coördinates, the strain-displacement relations for the 'in-plane' strains are

$$e_{rr} = \frac{\partial u_r}{\partial r} \; ; \; e_{r\theta} = \frac{1}{2}\left(\frac{1}{r}\frac{\partial u_r}{\partial \theta} + \frac{\partial u_\theta}{\partial r} - \frac{u_\theta}{r}\right) \; ; \; e_{\theta\theta} = \frac{u_r}{r} + \frac{1}{r}\frac{\partial u_\theta}{\partial \theta} \; .$$

Use these relations to obtain a compatibility equation that must be satisfied by the three strains.

2.7. If no stresses occur in a body, an increase in temperature T causes unrestrained thermal expansion defined by the strains

$$e_{xx} = e_{yy} = e_{zz} = \alpha T \; ; \; e_{xy} = e_{yz} = e_{zx} = 0 \; .$$

Show that this is possible only if T is a linear function of x, y, z and that otherwise stresses must be induced in the body, regardless of the boundary conditions.

2.8. If there are no body forces, show that the equations of equilibrium and compatibility imply that

$$(1 + \nu)\frac{\partial^2 \sigma_{ij}}{\partial x_k \partial x_k} + \frac{\partial^2 \sigma_{kk}}{\partial x_i \partial x_j} = 0 \; .$$

2.9. Using the strain-displacement relation (1.37), show that an alternative statement of the compatibility condition is that the tensor

$$C_{ijkl} \equiv \frac{\partial^2 e_{ij}}{\partial x_k \partial x_l} + \frac{\partial^2 e_{kl}}{\partial x_i \partial x_j} - \frac{\partial^2 e_{il}}{\partial x_j \partial x_k} - \frac{\partial^2 e_{jk}}{\partial x_i \partial x_l} = 0 \; .$$

2.10. Show that the derivatives $\partial \omega_i / \partial x_j$ needed for the evaluation of the integral (2.13) can be expressed as derivatives of the strain components e_{kl} and find these expressions.

2.11. By differentiating equation (1.32), show that

$$\frac{\partial \omega}{\partial x} = \frac{\partial e_{xy}}{\partial x} - \frac{\partial e_{xx}}{\partial y} \; ; \; \frac{\partial \omega}{\partial y} = \frac{\partial e_{yy}}{\partial x} - \frac{\partial e_{xy}}{\partial y} \; .$$

A particular two-dimensional strain distribution is defined by

$$e_{xx} = \frac{Cy^2}{(x^2 + y^2)} \; ; \quad e_{xy} = -\frac{Cxy}{(x^2 + y^2)} \; ; \quad e_{yy} = \frac{Cx^2}{(x^2 + y^2)} \; .$$

Verify that these components satisfy the compatibility equation (2.5) except possibly at the origin. Then use the above results to evaluate the contour integral (2.13), using a square path S_0 with corners at the four points $(\pm 1, \pm 1)$.

Comment on your results.

2.12. Show that if there are no body forces, the maximum magnitude $|\omega|$ for the rotation vector must occur at a point on the boundary.

2.13. A semi-infinite prismatic body of uniform density ρ is defined by the inequalities $f(x, y) < 0$, $z < 0$, where the z-axis is vertically upwards and f is any given function of x, y only. If the surfaces of the body are all traction free, show that the state of stress is defined by $\sigma_{zz} = \rho g z$, all the other stress components being zero. Show that this result remains true if the body is multiply connected.

2.14. The constitutive law (1.51) is ill-defined in the incompressible limit $\nu = 0.5$, since λ is then unbounded and the dilatation e must be zero. An alternative formulation for this case is to express the stress components in the form

$$\sigma_{ij} = \sigma'_{ij} + \bar{\sigma}\delta_{ij} \; ,$$

where σ'_{ij}, $\bar{\sigma}$ are respectively the deviatoric stress and the hydrostatic stress, defined in §§1.3.3, 1.3.4.

Use this formulation to develop an alternative to equation (2.16) for an incompressible material, and hence show that $\nabla^2 \bar{\sigma} = -\text{div } \boldsymbol{p}$.

Part II
Two-dimensional Problems

Chapter 3
Plane Strain and Plane Stress

A problem is two-dimensional if the field quantities such as stress and displacement depend on only two coördinates (x, y) and the boundary conditions are imposed on a line $f(x, y) = 0$ in the xy-plane.

In this sense, there are strictly no two-dimensional problems in elasticity. There *are* circumstances in which the stresses are independent of the z-coördinate, but all real bodies must have some bounding surfaces which are not represented by a line in the xy-plane. The two-dimensionality of the resulting fields depends upon the boundary conditions on such surfaces being of an appropriate form.

3.1 Plane Strain

It might be argued that a closed line in the xy-plane *does* define a solid body — namely an infinite cylinder of the appropriate cross section whose axis is parallel to the z-direction. However, making a body infinite does not really dispose of the question of boundary conditions, since there are usually some implied boundary conditions 'at infinity'. For example, the infinite cylinder could be in a state of uniaxial tension, $\sigma_{zz} = C$, where C is an arbitrary constant. However, a unique two-dimensional infinite cylinder problem can be defined by demanding that u_x, u_y be independent of z and that $u_z = 0$ for all x, y, z, in which case it follows that

$$e_{zx} = e_{zy} = e_{zz} = 0 . \tag{3.1}$$

This is the two-dimensional state known as *plane strain*.

In view of the stress-strain relations, an equivalent statement to equation (3.1) is

$$\sigma_{zx} = \sigma_{zy} = 0 \; ; \quad u_z = 0 ,$$

© The Author(s), under exclusive license to Springer Nature Switzerland AG 2022
J. R. Barber, *Elasticity*, Solid Mechanics and Its Applications 172,
https://doi.org/10.1007/978-3-031-15214-6_3

and hence a condition of plane strain will exist in a *finite* cylinder provided that (i) any tractions or displacements imposed on the sides of the cylinder are independent of z and have no component in the z-direction and (ii) the cylinder has plane ends (e.g. $z = \pm c$) which are in frictionless contact with two plane rigid walls.

From the condition $e_{zz} = 0$ (3.1) and the stress-strain relations, we can deduce

$$0 = \frac{\sigma_{zz}}{E} - \frac{\nu(\sigma_{xx} + \sigma_{yy})}{E}$$

— i.e.

$$\sigma_{zz} = \nu(\sigma_{xx} + \sigma_{yy}) \tag{3.2}$$

and hence

$$\begin{aligned}
e_{xx} &= \frac{\sigma_{xx}}{E} - \frac{\nu\sigma_{yy}}{E} - \frac{\nu^2(\sigma_{xx} + \sigma_{yy})}{E} \\
&= \frac{(1 - \nu^2)\sigma_{xx}}{E} - \frac{\nu(1 + \nu)\sigma_{yy}}{E} .
\end{aligned} \tag{3.3}$$

Of course, there are comparatively few practical applications in which a cylinder with plane ends is constrained between frictionless rigid walls, but fortunately the plane strain solution can be used in an approximate sense for a cylinder with any end conditions, provided that the length of the cylinder is large compared with its cross-sectional dimensions.

We shall illustrate this with reference to the long cylinder of Figure 3.1, for which the ends, $z = \pm c$ are traction free and the sides are loaded by tractions which are independent of z.

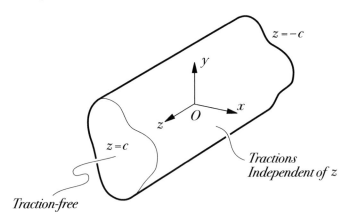

Fig. 3.1 The long cylinder with traction-free ends.

We first solve the problem under the plane strain assumption, obtaining an exact solution in which all the stresses are independent of z and in which there exists a normal stress σ_{zz} on all z-planes, which we can calculate from equation (3.2).

The plane strain solution satisfies all the boundary conditions of the problem *except* that σ_{zz} also acts on the end faces, $z = \pm c$, where it appears as an unwanted normal traction. We therefore seek a *corrective* solution which, when superposed on the plane strain solution, removes the unwanted normal tractions on the surfaces $z = \pm c$, without changing the boundary conditions on the sides of the cylinder.

3.1.1 The corrective solution

The corrective solution must have zero tractions on the sides of the cylinder and a prescribed normal traction (equal and opposite to that obtained in the plane strain solution) on each of the end faces, $\mathcal{A}$.

This is a fully three-dimensional problem which generally has no closed-form solution. However, if the prescribed tractions have the linear form

$$\sigma_{zz} = B + Cx + Dy , \tag{3.4}$$

the solution can be obtained in the context of Mechanics of Materials by treating the long cylinder as a beam subjected to an axial force

$$F = \iint_{\mathcal{A}} \sigma_{zz}\, dx dy \tag{3.5}$$

and bending moments

$$M_x = \iint_{\mathcal{A}} \sigma_{zz}\, y\, dx dy \; ; \;\; M_y = -\iint_{\mathcal{A}} \sigma_{zz}\, x\, dx dy \tag{3.6}$$

about the axes Ox, Oy respectively, where the origin O is chosen to coincide with the centroid of the cross section[1]. Notice that this correction will generally cause the cylinder to be curved, even though the bending moments in the final state are everywhere zero.

If the original plane strain solution does not give a distribution of σ_{zz} of this convenient linear form, we can still use equations (3.5, 3.6) to define the force and moments for an *approximate* Mechanics of Materials corrective solution. The error involved in using this approximate solution will be that associated with yet another corrective solution corresponding to the problem in which the end faces of the cylinder are loaded by tractions equal to the *difference* between those in the plane strain

---

[1] The Mechanics of Materials solution for axial force and pure bending is in fact *exact* in the sense of the Theory of Elasticity, since the stresses clearly satisfy the equilibrium equations (2.4) and the resulting strains are linear functions of x, y and hence satisfy the compatibility equations (2.7) identically.

solution and the Mechanics of Materials linear form (3.4) associated with the force resultants (3.5, 3.6).

In this final corrective solution, the ends of the cylinder are loaded by *self-equilibrated tractions*, since the Mechanics of Materials approximation is carefully chosen to have the same force and moment resultants as the required exact solution and in such cases, we anticipate that significant stresses will only be generated in the immediate vicinity of the ends — or more precisely, in regions whose distance from the ends is comparable with the cross-sectional dimensions of the cylinder. If the cylinder is many times longer than its cross-sectional dimensions, there will be a substantial portion near the centre where the final corrective solution gives negligible stresses and hence where the sum of the original plane strain solution and the Mechanics of Materials correction is a good approximation to the actual three-dimensional stress field.

3.1.2 Saint-Venant's principle

The thesis that a self-equilibrating system of loads produces only local effects is known as *Saint-Venant's principle*. It seems intuitively reasonable, but has not been proved rigorously except for certain special cases — some of which we shall encounter later in this book (see for example Chapter 6). It can be seen as a consequence of the rule that alternate load paths through a structure share the load in proportion with their stiffnesses. If a region of the boundary is loaded by a self-equilibrating system of tractions, the stiffest paths are the shortest — i.e. those which do not penetrate far from the loaded region. Hence, the longer paths — which are those which contribute to stresses distant from the loaded region — carry relatively little load.

Note that if the local tractions are *not* self-equilibrating, some of the load paths must go to other distant parts of the boundary and hence there will be significant stresses in intermediate regions. For this reason, it is important to superpose the Mechanics of Materials approximate corrective solution when solving plane strain problems for long cylinders with traction-free ends. By contrast, the final stage of solving the 'Saint-Venant problem' to calculate the correct stresses near the ends is seldom of much importance in practical problems, since the ends, being traction free, are not generally points of such high stress as the interior.

As a point of terminology, we shall refer to problems in which the boundary conditions on the ends have been corrected only in the sense of force and moment resultants as being solved in the *weak* sense with respect to these boundaries. Boundary conditions are satisfied in the *strong* sense when the tractions are specified in a pointwise rather than a force resultant sense.

3.2 Plane Stress

Plane stress is an approximate solution, in contrast to plane strain, which is exact. In other words, plane strain is a special solution of the complete three-dimensional equations of elasticity, whereas plane stress is only approached in the limit as the thickness of the loaded body tends to zero.

It is argued that if the two bounding z-planes of a thin plate are sufficiently close in comparison with the other dimensions, and if they are also free of tractions, the stresses on all parallel z-planes will be sufficiently small to be neglected, in which case we write

$$\sigma_{zx} = \sigma_{zy} = \sigma_{zz} = 0 \,,$$

for all x, y, z.

It then follows that

$$e_{zx} = e_{zy} = 0 \,,$$

but $e_{zz} \neq 0$, being given in fact by

$$e_{zz} = -\frac{\nu}{E}(\sigma_{xx} + \sigma_{yy}) \,. \tag{3.7}$$

The two-dimensional stress-strain relations are then

$$e_{xx} = \frac{\sigma_{xx}}{E} - \frac{\nu\sigma_{yy}}{E} \tag{3.8}$$

$$e_{yy} = \frac{\sigma_{yy}}{E} - \frac{\nu\sigma_{xx}}{E} \,. \tag{3.9}$$

The fact that plane stress is not an exact solution can best be explained by considering the compatibility equation

$$\frac{\partial^2 e_{yy}}{\partial z^2} - 2\frac{\partial^2 e_{yz}}{\partial y \partial z} + \frac{\partial^2 e_{zz}}{\partial y^2} = 0 \,.$$

Since *ex-hypothesi* none of the stresses vary with z, the first two terms in this equation are identically zero and hence the equation will be satisfied if and only if

$$\frac{\partial^2 e_{zz}}{\partial y^2} = 0 \,.$$

This in turn requires

$$\frac{\partial^2(\sigma_{xx} + \sigma_{yy})}{\partial y^2} = 0 \,,$$

from equation (3.7), unless $\nu = 0$, in which case the plane stress and plane strain solutions are identical.

Applying similar arguments to the other compatibility equations, we conclude that for $\nu \neq 0$, the plane stress assumption is exact if and only if $(\sigma_{xx} + \sigma_{yy})$ is a linear function of x, y. i.e. if

$$\sigma_{xx} + \sigma_{yy} = B + Cx + Dy .$$

The attentive reader will notice that this condition is exactly equivalent to (3.2, 3.4). In other words, the plane stress solution is exact if and only if the 'weak form' solution of the corresponding plane strain solution is exact. Of course this is not a coincidence. In this special case, the process of off-loading the ends of the long cylinder in the plane strain solution has the effect of making σ_{zz} zero throughout the cylinder and hence the resulting solution also satisfies the plane stress assumption, whilst remaining exact.

3.2.1   Generalized plane stress

The approximate nature of the plane stress formulation is distasteful to elasticians of a more mathematical temperament, who prefer to preserve the rigour of an exact theory. This can be done by the contrivance of defining the *average* stresses across the thickness of the plate — e.g.

$$\overline{\sigma}_{xx} \equiv \frac{1}{2c} \int_{-c}^{c} \sigma_{xx} \, dx .$$

It can then be shown that the average stresses so defined satisfy the plane stress equations *exactly*. This is referred to as the *generalized plane stress* formulation.

In practice, of course, the gain in rigour is illusory unless we can also establish that the stress variation across the section is small, so that the local values are reasonably close to the average. A fully three-dimensional theory of thin plates under in-plane loading shows that the plane stress assumption is a good approximation except in regions whose distance from the boundary is comparable with the plate thickness[2].

3.2.2   Relationship between plane stress and plane strain

The solution of a problem under either the plane strain or plane stress assumptions involves finding a two-dimensional stress field, defined in terms of the components $\sigma_{xx}, \sigma_{xy}, \sigma_{yy}$, which satisfies the equilibrium equations (2.4), and for which the corresponding strains, e_{xx}, e_{xy}, e_{yy}, satisfy the only non-trivial compatibility equation (2.5). The equilibrium and compatibility equations are the same in both formulations,

---

[2] See Chapter 16.

the only difference being in the relation between the stress and strain components, which for normal stresses are given by (3.3) for plane strain and (3.8, 3.9) for plane stress. The relation between the shear stress σ_{xy} and the shear strain e_{xy} is the same for both formulations and is given by equation (1.50). Thus, from a mathematical perspective, the plane strain solution simply looks like the plane stress solution for a material with different elastic constants and *vice versa*. In fact, it is easily verified that equation (3.3) can be obtained from (3.8) by making the substitutions

$$E = \frac{E'}{(1-\nu'^2)} \quad ; \quad \nu = \frac{\nu'}{(1-\nu')} \,, \tag{3.10}$$

and then dropping the primes. This substitution also leaves the shear stress-shear strain relation unchanged as required.

An alternative approach is to write the two-dimensional stress-strain relations in the form

$$e_{xx} = \left(\frac{\kappa+1}{8\mu}\right)\sigma_{xx} - \left(\frac{3-\kappa}{8\mu}\right)\sigma_{yy} \; ; \quad e_{yy} = \left(\frac{\kappa+1}{8\mu}\right)\sigma_{yy} - \left(\frac{3-\kappa}{8\mu}\right)\sigma_{xx}$$

$$e_{xy} = \frac{\sigma_{xy}}{2\mu} \,,$$

where κ is *Kolosov's constant*, defined as

$$\kappa = (3-4\nu) \quad \text{for plane strain}$$

$$= \left(\frac{3-\nu}{1+\nu}\right) \quad \text{for plane stress.} \tag{3.11}$$

In the following chapters, we shall generally treat two-dimensional problems with the plane stress assumptions, noting that results for plane strain can be recovered by the substitution (3.10) when required.

Problems

3.1. The plane strain solution for the stresses in the rectangular block $0 < x < a$, $-b < y < b$, $-c < z < c$ with a given loading is

$$\sigma_{xx} = \frac{3Fxy}{2b^3} \; ; \quad \sigma_{xy} = \frac{3F(b^2 - y^2)}{4b^3} \; ; \quad \sigma_{yy} = 0 \; ; \quad \sigma_{zz} = \frac{3\nu Fxy}{2b^3} \,.$$

Find the tractions on the surfaces of the block and illustrate the results on a sketch of the block.

We wish to use this solution to solve the corresponding problem in which the surfaces $z = \pm c$ are traction free. Determine an approximate corrective solution for this problem by offloading the unwanted force and moment resultants using the elementary bending theory. Find the maximum error in the stress σ_{zz} in the corrected solution and compare it with the maximum tensile stress in the plane strain solution.

3.2. For a solid in a state of plane stress, show that if there are body forces p_x, p_y per unit volume in the direction of the axes x, y respectively, the compatibility equation can be expressed in the form

$$\nabla^2(\sigma_{xx} + \sigma_{yy}) = -(1 + \nu)\left(\frac{\partial p_x}{\partial x} + \frac{\partial p_y}{\partial y}\right).$$

Hence deduce that the stress distribution for any particular case is independent of the material constants and the body forces, provided the latter are constant.

3.3. (i) Show that the compatibility equation (2.5) is satisfied by unrestrained thermal expansion ($e_{xx} = e_{yy} = \alpha T$, $e_{xy} = 0$), provided that the temperature, T, is a two-dimensional harmonic function — i.e.

$$\frac{\partial^2 T}{\partial x^2} + \frac{\partial^2 T}{\partial y^2} = 0.$$

(ii) Hence deduce that, subject to certain restrictions which you should explicitly list, no thermal stresses will be induced in a thin body with a steady-state, two-dimensional temperature distribution and no boundary tractions.

(iii) Show that an initially straight line on such a body will be distorted by the heat flow in such a way that its curvature is proportional to the local heat flux across it.

3.4. Find the inverse relations to equations (3.10) — i.e. the substitutions that should be made for the elastic constants E, ν in a plane strain solution if we want to recover the solution of the corresponding plane stress problem.

3.5. Show that in a state of plane stress without body forces, the in-plane displacements must satisfy the equations

$$\nabla^2 u_x + \left(\frac{1+\nu}{1-\nu}\right)\frac{\partial}{\partial x}\left(\frac{\partial u_x}{\partial x} + \frac{\partial u_y}{\partial y}\right) = 0 \ ; \quad \nabla^2 u_y + \left(\frac{1+\nu}{1-\nu}\right)\frac{\partial}{\partial y}\left(\frac{\partial u_x}{\partial x} + \frac{\partial u_y}{\partial y}\right) = 0 .$$

3.6. Show that in a state of plane strain without body forces,

$$\frac{\partial e}{\partial x} = \left(\frac{1-2\nu}{1-\nu}\right)\frac{\partial \omega_z}{\partial y} \ ; \quad \frac{\partial e}{\partial y} = -\left(\frac{1-2\nu}{1-\nu}\right)\frac{\partial \omega_z}{\partial x} .$$

3.7. If a material is incompressible ($\nu = 0.5$), a state of hydrostatic stress $\sigma_{xx} = \sigma_{yy} = \sigma_{zz}$ produces no strain. One way to write the corresponding stress-strain relations is

$$\sigma_{ij} = 2\mu e_{ij} - q\delta_{ij} \ ,$$

where q is an unknown hydrostatic pressure which will generally vary with position. Also, the condition of incompressibility requires that the dilatation

$$e \equiv e_{kk} = 0 \ .$$

Show that under plain strain conditions, the stress components and the hydrostatic pressure q must satisfy the equations

$$\nabla^2 q = \text{div } \boldsymbol{p} \quad \text{and} \quad \sigma_{xx} + \sigma_{yy} = -2q \ ,$$

where $\boldsymbol{p}$ is the body force.

Chapter 4
Stress Function Formulation

Newton's law of gravitation states that two heavy bodies attract each other with a force proportional to the inverse square of their distance — thus it is essentially a vector theory, being concerned with forces. However, the idea of a scalar gravitational potential can be introduced by defining the work done in moving a unit mass from infinity to a given point in the field. The principle of conservation of energy requires that this be a unique function of position and it is easy to show that the gravitational force at any point is then proportional to the gradient of this scalar potential. Thus, the original vector problem is reduced to a problem about a scalar potential and its derivatives.

In general, scalars are much easier to deal with than vectors. In particular, they lend themselves very easily to coördinate transformations, whereas vectors (and to an even greater extent tensors) require a set of special transformation rules (e.g. Mohr's circle).

In certain field theories, the scalar potential has an obvious physical significance. For example, in the conduction of heat, the temperature is a scalar potential in terms of which the vector heat flux can be defined. However, it is not necessary to the method that such a physical interpretation can be given. The gravitational potential can be given a physical interpretation as discussed above, but this interpretation may never feature in the solution of a particular problem, which is simply an exercise in the solution of a certain partial differential equation with appropriate boundary conditions. In the theory of elasticity, we make use of scalar potentials called *stress functions* or *displacement functions* which have no obvious physical meaning other than their use in defining stress or displacement components in terms of derivatives.

4.1 Choice of a Suitable Form

In the choice of a suitable form for a stress or displacement function, there is only one absolute rule — that the operators which define the relationship between

J. R. Barber, *Elasticity*, Solid Mechanics and Its Applications 172,
https://doi.org/10.1007/978-3-031-15214-6_4

the scalar and vector (or tensor) quantities should indeed define a vector (or tensor).

For example, it is appropriate to define the displacement in terms of the first derivatives (the gradient) of a scalar or to define the stress components in terms of the second derivatives of a scalar, since the second derivatives of a scalar form the components of a Cartesian tensor.

In effect, what we are doing in requiring this similarity of form between the definitions and the defined quantity is ensuring that the relationship is preserved in coördinate transformations. It would be quite possible to work out an elasticity problem in terms of the displacement components u_x, u_y, u_z, treating these as essentially scalar quantities which vary with position — indeed this was a technique which was used in early theories. However, we would then get into trouble as soon as we tried to make any statements about quantities in other coördinate directions. By contrast, if we define (for example) $u_x = \partial \psi / \partial x$; $u_y = \partial \psi / \partial y$; $u_z = \partial \psi / \partial z$, (i.e. $\boldsymbol{u} = \boldsymbol{\nabla} \psi$), it immediately follows that $u'_x = \partial \psi / \partial x'$ for any x'.

4.2 The Airy Stress Function

If there are no body forces and the non-zero stress components $\sigma_{xx}, \sigma_{xy}, \sigma_{yy}$ are independent of z, they must satisfy the two equilibrium equations

$$\frac{\partial \sigma_{xx}}{\partial x} + \frac{\partial \sigma_{xy}}{\partial y} = 0 ; \quad \frac{\partial \sigma_{yx}}{\partial x} + \frac{\partial \sigma_{yy}}{\partial y} = 0 . \tag{4.1}$$

The two terms in each of these equations are therefore equal and opposite. Suppose we take the first term in the first equation and integrate it once with respect to x and once with respect to y to define a new function $\psi(x, y)$. We can then write

$$\frac{\partial \sigma_{xx}}{\partial x} = -\frac{\partial \sigma_{xy}}{\partial y} \equiv \frac{\partial^2 \psi}{\partial x \partial y} , \tag{4.2}$$

and a particular solution is

$$\sigma_{xx} = \frac{\partial \psi}{\partial y} ; \quad \sigma_{xy} = -\frac{\partial \psi}{\partial x} . \tag{4.3}$$

Substituting the second of these results in the second equilibrium equation and using the same technique to define another new function $\phi(x, y)$, we obtain

$$\frac{\partial \sigma_{yy}}{\partial y} = -\frac{\partial \sigma_{yx}}{\partial x} = \frac{\partial^2 \psi}{\partial x^2} \equiv \frac{\partial^3 \phi}{\partial x^2 \partial y} , \tag{4.4}$$

so a particular solution is

$$\sigma_{yy} = \frac{\partial^2 \phi}{\partial x^2} \; ; \quad \psi = \frac{\partial \phi}{\partial y} \; . \tag{4.5}$$

Using (4.3, 4.5) we can then express all three stress components in terms of ϕ as

$$\sigma_{xx} = \frac{\partial^2 \phi}{\partial y^2} \; ; \quad \sigma_{yy} = \frac{\partial^2 \phi}{\partial x^2} \; ; \quad \sigma_{xy} = -\frac{\partial^2 \phi}{\partial x \partial y} \; . \tag{4.6}$$

This representation was introduced by G. B. Airy[1] in 1862 and ϕ is therefore generally referred to as the *Airy stress function*. It is clear from the derivation that the stress components (4.6) will satisfy the two-dimensional equilibrium equations (4.1) for all functions ϕ. Also, the mathematical operations involved in obtaining (4.6) can be performed for any two-dimensional stress distribution satisfying (4.1) and hence the representation (4.6) is *complete*, meaning that it is capable of representing the most general such stress distribution. It is worth noting that the derivation makes no reference to Hooke's law and hence the Airy stress function also provides a complete representation for two-dimensional problems involving inelastic constitutive laws — for example plasticity theory — where its satisfaction of the equilibrium equations remains an advantage.

The technique introduced in this section can be used in other situations to define a representation of a field satisfying one or more partial differential equations. The trick is to rewrite the equation so that terms involving different unknowns are on opposite sides of the equals sign and then equate each side of the equation to a derivative that contains the lowest common denominator of the derivatives on the two sides, as in equations (4.2, 4.4). For other applications of this method, see Problems 4.3 and 4.4.

4.2.1   Transformation of coördinates

It is easily verified that equation (4.6) transforms as a Cartesian tensor as required. For example, using (1.29) we can write

$$\frac{\partial^2 \phi}{\partial x'^2} = \left(\cos\theta \frac{\partial}{\partial x} + \sin\theta \frac{\partial}{\partial y} \right)^2 \phi$$

$$= \frac{\partial^2 \phi}{\partial x^2} \cos^2\theta + \frac{\partial^2 \phi}{\partial y^2} \sin^2\theta + 2 \frac{\partial^2 \phi}{\partial x \partial y} \sin\theta \cos\theta \; ,$$

from which using (1.11) and (4.6) we deduce that $\sigma'_{yy} = \partial^2 \phi / \partial x'^2$ as required.

---

[1] For a good historical survey of the development of potential function methods in Elasticity, see H. M. Westergaard, *Theory of Elasticity and Plasticity*, Dover, New York, 1964, Chapter 2.

If we were simply to seek a representation of a two-dimensional Cartesian tensor in terms of the derivatives of a scalar function without regard to the equilibrium equations, the Airy stress function is not the most obvious form. It would seem more natural to write $\sigma_{xx} = \partial^2\psi/\partial x^2$; $\sigma_{yy} = \partial^2\psi/\partial y^2$; $\sigma_{xy} = \partial^2\psi/\partial x\partial y$ and indeed this also leads to a representation which is widely used as part of the general three-dimensional solution (see Chapter 22 below). In some books, it is stated or implied that the Airy function is used because it satisfies the equilibrium equation, but this is rather misleading, since the 'more obvious form' — although it requires some constraints on ψ in order to satisfy equilibrium — can be shown to define displacements which automatically satisfy the compatibility condition which is surely equally useful[2]. The real reason for preferring the Airy function is that it is complete, whilst the alternative is more restrictive, as we shall see in §21.1.

4.2.2   Non-zero body forces

If the body force p is not zero, but is of a restricted form such that it can be written $p = -\nabla V$, where V is a two-dimensional scalar potential, equations (4.6) can be generalized by including an extra term $+V$ in each of the normal stress components whilst leaving the shear stress definition unchanged. It is easily verified that, with this modification, the equilibrium equations are again satisfied.

Problems involving body forces will be discussed in more detail in Chapter 7 below. For the moment, we restrict attention to the the case where $p = 0$ and hence where the representation (4.6) is appropriate.

4.3   The Governing Equation

As discussed in Chapter 2, the stress or displacement field has to satisfy the equations of equilibrium and compatibility if they are to describe permissible states of an elastic body. In stress function representations, this generally imposes certain constraints on the choice of stress function, which can be expressed by requiring it to be a solution of a certain partial differential equation.

We have already seen that the equilibrium condition is satisfied identically by the use of the Airy stress function ϕ, so it remains to determine the governing equation for ϕ by substituting the representation (4.6) into the compatibility equation in two dimensions.

---

[2] See Problem 2.3.

4.3.1 The compatibility condition

We first express the compatibility condition in terms of stresses using the stress-strain relations, obtaining

$$\frac{\partial^2 \sigma_{xx}}{\partial y^2} - \nu\frac{\partial^2 \sigma_{yy}}{\partial y^2} - 2(1+\nu)\frac{\partial^2 \sigma_{xy}}{\partial x \partial y} + \frac{\partial^2 \sigma_{yy}}{\partial x^2} - \nu\frac{\partial^2 \sigma_{xx}}{\partial x^2} = 0 \ , \qquad (4.7)$$

where we have cancelled a common factor of $1/E$.

We then substitute for the stress components from (4.6) obtaining

$$\frac{\partial^4 \phi}{\partial y^4} - \nu\frac{\partial^4 \phi}{\partial x^2 \partial y^2} + 2(1+\nu)\frac{\partial^4 \phi}{\partial x^2 \partial y^2} + \frac{\partial^4 \phi}{\partial x^4} - \nu\frac{\partial^4 \phi}{\partial x^2 \partial y^2} = 0 \ , \qquad (4.8)$$

i.e.

$$\frac{\partial^4 \phi}{\partial x^4} + 2\frac{\partial^4 \phi}{\partial x^2 \partial y^2} + \frac{\partial^4 \phi}{\partial y^4} = \left(\frac{\partial^2}{\partial x^2} + \frac{\partial^2}{\partial y^2}\right)^2 \phi = 0 \ .$$

This equation is known as the *biharmonic equation* and is usually written in the concise form

$$\nabla^4 \phi = 0 \ . \qquad (4.9)$$

The biharmonic equation is the governing equation for the Airy stress function in elasticity problems. Thus, by using the Airy stress function representation, the problem of determining the stresses in an elastic body is reduced to that of finding a solution of equation (4.9) (i.e. a *biharmonic function*) whose derivatives satisfy certain boundary conditions on the surfaces.

4.3.2 Method of solution

Historically, the boundary-value problem for the Airy stress function has been approached in a semi-inverse way — i.e. by using the variation of tractions along the boundaries to give a clue to the kind of function required, but then exploring the stress fields developed from a wide range of such functions and selecting a combination which can be made to satisfy the required conditions. The disadvantage with this method is that it requires a wide experience of particular solutions and even then is not guaranteed to be successful.

A more modern method which has the advantage of always developing an appropriate stress function if the boundary conditions are unmixed (i.e. all specified in terms of stresses or all in terms of displacements) is based on representing the stress function in terms of analytic functions of the complex variable. This method will be discussed in Chapter 20. However, although it is powerful and extremely elegant, it requires a certain familarity with the properties of functions of the complex variable and generally involves the evaluation of a contour integral, which may not be as convenient a numerical technique as a direct series or finite difference attack on the original problem using real stress functions. There are some exceptions — notably for bodies which are susceptible of a simple conformal transformation into the unit circle.

4.3.3 Reduced dependence on elastic constants

It is clear from dimensional considerations that, when the compatibility equation is expressed in terms of ϕ, Young's modulus must appear in every term and can therefore be cancelled as in equation (4.7), but it is an unexpected bonus that the Poisson's ratio terms also cancel in (4.8), leaving an equation that is independent of elastic constants. It follows that the stress field in a simply-connected elastic body[3] in a state of plane strain or plane stress is independent of the material properties if the boundary conditions are expressed in terms of tractions and, in particular, that the plane stress and plane strain fields are identical.

Dundurs[4] has shown that a similar reduced dependence on elastic constants occurs in plane problems involving interfaces between two dissimilar elastic materials. In such cases, three independent dimensionless parameters (e.g. $\mu_1/\mu_2, \nu_1, \nu_2$) can be formed from the elastic constants $\mu_1, \nu_1, \mu_2, \nu_2$ of the materials 1, 2 respectively, but Dundurs proved that the stress field can be written in terms of only two parameters which he defined as

$$\alpha = \left(\frac{\kappa_1+1}{\mu_1} - \frac{\kappa_2+1}{\mu_2}\right) \Big/ \left(\frac{\kappa_1+1}{\mu_1} + \frac{\kappa_2+1}{\mu_2}\right) \tag{4.10}$$

$$\beta = \left(\frac{\kappa_1-1}{\mu_1} - \frac{\kappa_2-1}{\mu_2}\right) \Big/ \left(\frac{\kappa_1+1}{\mu_1} + \frac{\kappa_2+1}{\mu_2}\right), \tag{4.11}$$

where κ is Kolosov's constant [see equation (3.11)]. It can be shown that Dundurs' parameters must lie in the range $-1 \le \alpha \le 1,\ -0.5 \le \beta \le 0.5$ if $0 \le \nu_1, \nu_2 \le 0.5$.

Problems

4.1. Newton's law of gravitation states that two heavy particles of mass m_1, m_2 respectively will experience a mutual attractive force

---

[3] The restriction to simply-connected bodies is necessary, since in a multiply-connected body there is an implied displacement condition, as explained in §2.2 above.

[4] J. Dundurs (1969), Discussion on 'Edge bonded dissimilar orthogonal elastic wedges under normal and shear loading', *ASME Journal of Applied Mechanics,* Vol. 36, pp. 650–652.

$$F = \frac{\gamma m_1 m_2}{R^2} ,$$

where R is the distance between the particles and γ is the gravitational constant. Use an energy argument and superposition to show that the force acting on a particle of mass m_0 can be written

$$F = -\gamma m_0 \nabla V ,$$

where

$$V(x, y, z) = -\iiint_\Omega \frac{\rho(\xi, \eta, \zeta) d\xi d\eta d\zeta}{\sqrt{(x-\xi)^2 + (y-\eta)^2 + (z-\zeta)^2}} ,$$

Ω represents the volume of the universe and ρ is the density of material in the universe, which will generally be a function of position (ξ, η, ζ).

Could a similar method have been used if Newton's law had been of the more general form

$$F = \frac{\gamma m_1 m_2}{R^\lambda} .$$

If so, what would have been the corresponding expression for V? If not, why not?

4.2. An ionized liquid in an electric field experiences a body force p. Show that the liquid can be in equilibrium only if p is a conservative vector field.
Hint: Remember that a stationary liquid must be everywhere in a state of hydrostatic stress.

4.3. An *antiplane* state of stress is one for which the only non-zero stress components are σ_{zx}, σ_{zy} and these are independent of z. Show that two of the three equilibrium equations are then satisfied identically if there is no body force. Use a technique similar to that of §4.2 to develop a representation of the non-zero stress components in terms of a scalar function, such that the remaining equilibrium equation is satisfied identically.

4.4. If a body of fairly general axisymmetric shape is loaded in torsion, the only non-zero stress components in cylindrical polar coördinates are $\sigma_{\theta r}, \sigma_{\theta z}$ and these are required to satisfy the equilibrium equation

$$\frac{\partial \sigma_{\theta r}}{\partial r} + \frac{2\sigma_{\theta r}}{r} + \frac{\partial \sigma_{\theta z}}{\partial z} = 0 .$$

Use a technique similar to that of §4.2 to develop a representation of these stress components in terms of a scalar function, such that the equilibrium equation is satisfied identically.

4.5. (i) Show that the function

$$\phi = y\omega + \psi$$

satisfies the biharmonic equation if w, ψ are both *harmonic* (i.e. $\nabla^2 w = 0$ and $\nabla^2 \psi = 0$).

(ii) Develop expressions for the stress components in terms of w, ψ, based on the use of ϕ as an Airy stress function.

(iii) Show that a solution suitable for the half-plane $y > 0$ subject to normal surface tractions only (i.e. $\sigma_{xy} = 0$ on $y = 0$) can be obtained by writing

$$w = -\frac{\partial \psi}{\partial y}$$

and hence that under these conditions the normal stress σ_{xx} near the surface $y = 0$ is equal to the applied traction σ_{yy}.

(iv) Do you think this is a rigorous proof? Can you think of any exceptions? If so, at what point in your proof of section (iii) can you find a lack of generality?

4.6. The constitutive law for an orthotropic elastic material in plane stress can be written

$$e_{xx} = s_{11}\sigma_{xx} + s_{12}\sigma_{yy} \;;\; e_{yy} = s_{12}\sigma_{xx} + s_{22}\sigma_{yy} \;;\; e_{xy} = s_{44}\sigma_{xy} \,,$$

where $s_{11}, s_{12}, s_{22}, s_{44}$ are elastic constants.

Using the Airy stress function ϕ to represent the stress components, find the equation that must be satisfied by ϕ.

4.7. Show that if the two-dimensional function $w(x, y)$ is harmonic ($\nabla^2 w = 0$), the function

$$\phi = (x^2 + y^2) w$$

will be biharmonic.

4.8. The constitutive law for an isotropic incompressible elastic material can be written

$$\sigma_{ij} = \overline{\sigma}\delta_{ij} + 2\mu\, e_{ij} \,,$$

where

$$\overline{\sigma} = \frac{\sigma_{kk}}{3}$$

represents an arbitrary hydrostatic stress field. Some soils can be approximated as an incompressible material whose modulus varies linearly with depth, so that

$$\mu = Mz$$

for the half-space $z > 0$.

Use the displacement function representation

$$\boldsymbol{u} = \nabla\phi$$

to develop a potential function solution for the stresses in such a body. Show that the functions $\phi, \bar{\sigma}$ must satisfy the relations

$$\nabla^2 \phi = 0 \; ; \; \bar{\sigma} = -2M\frac{\partial \phi}{\partial z} \; ,$$

and hence obtain expressions for the stress components in terms of the single harmonic function ϕ.

If the half-space is loaded by a normal pressure

$$\sigma_{zz}(x, y, 0) = -p(x, y) \; ; \quad \sigma_{zx}(x, y, 0) = \sigma_{zy}(x, y, 0) = 0 \; ,$$

show that the corresponding normal surface displacement $u_z(x, y, 0)$ is linearly proportional to the local pressure $p(x, y)$ and find the constant of proportionality[5].

4.9. Show that Dundurs' constant $\beta \rightarrow 0$ for plane strain in the limit where $\nu_1 = 0.5$ and $\mu_1/\mu_2 \rightarrow 0$ — i.e. material 1 is incompressible and has a much lower shear modulus[6] than material 2. What is the value of α in this limit?

4.10. Solve Problem 3.7 for the case where there is no body force, using the Airy stress function ϕ to represent the stress components. Hence show that the governing equation is $\nabla^4 \phi = 0$, as in the case of compressible materials.

---

[5] For an alternative proof of this result see C. R. Calladine and J. A. Greenwood (1978), Line and point loads on a non-homogeneous incompressible elastic half-space, *Quarterly Journal of Mechanics and Applied Mathematics*, Vol. 31, pp. 507–529.

[6] This is a reasonable approximation for the important case of rubber (material 1) bonded to steel (material 2).

Chapter 5
Problems in Rectangular Coördinates

The Cartesian coördinate system (x, y) is clearly particularly suited to the problem of determining the stresses in a rectangular body whose boundaries are defined by equations of the form $x = a$, $y = b$. A wide range of such problems can be treated using stress functions which are polynomials in x, y. In particular, polynomial solutions can be obtained for 'Mechanics of Materials' type beam problems in which a rectangular bar is bent by an end load or by a distributed load on one or both faces.

5.1 Biharmonic Polynomial Functions

In rectangular coördinates, the biharmonic equation takes the form

$$\frac{\partial^4 \phi}{\partial x^4} + 2\frac{\partial^4 \phi}{\partial x^2 \partial y^2} + \frac{\partial^4 \phi}{\partial y^4} = 0 \tag{5.1}$$

and it follows that any polynomial in x, y of degree less than four will be biharmonic and is therefore appropriate as a stress function. However, for higher order polynomial terms, equation (5.1) is not identically satisfied. Suppose, for example, that we consider just those terms in a general polynomial whose combined degree (the sum of the powers of x and y) is N. We can write these terms in the form

$$P_N(x, y) = A_0 x^N + A_1 x^{N-1} y + A_2 x^{N-2} y^2 + \ldots + A_N y^N$$

$$= \sum_{i=0}^{N} A_i x^{N-i} y^i , \tag{5.2}$$

Supplementary Information The online version contains supplementary material available at https://doi.org/10.1007/978-3-031-15214-6_5.

where we note that there are $(N+1)$ independent coefficients, $A_i (i=0, N)$. If we now substitute $P_N(x, y)$ into equation (5.1), we shall obtain a new polynomial of degree $(N-4)$, since each term is differentiated four times. We can denote this new polynomial by $Q_{N-4}(x, y)$ where

$$Q_{N-4}(x, y) = \nabla^4 P_N(x, y) \tag{5.3}$$

$$= \sum_{i=0}^{N-4} B_i x^{(N-4-i)} y^i . \tag{5.4}$$

The $(N-3)$ coefficients $B_0, \ldots, B_{N-4}$ are easily obtained by expanding the right-hand side of equation (5.3) and equating coefficients. For example,

$$B_0 = N(N-1)(N-2)(N-3)A_0 + 4(N-2)(N-3)A_2 + 24A_4 . \tag{5.5}$$

Now the original function $P_N(x, y)$ will be biharmonic if and only if $Q_{N-4}(x, y)$ is zero for all x, y and this in turn is only possible if every term in the series (5.4) is identically zero, since the polynomial terms are all linearly independent of each other. In other words

$$B_i = 0 \; ; \; i = 0 \text{ to } (N-4) .$$

These conditions can be converted into a corresponding set of $(N-3)$ equations for the coefficients A_i. For example, the equation $B_0=0$ gives

$$N(N-1)(N-2)(N-3)A_0 + 4(N-2)(N-3)A_2 + 24A_4 = 0 , \tag{5.6}$$

from (5.5). We shall refer to the $(N-3)$ equations of this form as *constraints* on the coefficients A_i, since the coefficients are constrained to satisfy them if the original polynomial is to be biharmonic.

One approach would be to use the constraint equations to eliminate $(N-3)$ of the unknown coefficients in the original polynomial — for example, we could treat the first four coefficients, A_0, A_1, A_2, A_3, as unknown constants and use the constraint equations to define all the remaining coefficients in terms of these unknowns. Equation (5.6) would then be treated as an equation for A_4 and the subsequent constraint equations would each define one new constant in the series. It may help to consider a particular example at this stage. Suppose we consider the fifth degree polynomial

$$P_5(x, y) = A_0 x^5 + A_1 x^4 y + A_2 x^3 y^2 + A_3 x^2 y^3 + A_4 x y^4 + A_5 y^5 ,$$

which has six independent coefficients. Substituting into equation (5.3), we obtain the first degree polynomial

$$Q_1(x, y) = (120A_0 + 24A_2 + 24A_4)x + (24A_1 + 24A_3 + 120A_5)y .$$

The coefficients of x and y in Q_1 must both be zero if P_5 is to be biharmonic and we can write the resulting two constraint equations in the form

$$A_4 = -5A_0 - A_2 \; ; \quad A_5 = -\frac{A_1}{5} - \frac{A_3}{5} \; . \tag{5.7}$$

Finally, we use (5.7) to eliminate A_4, A_5 in the original definition of P_5, obtaining the definition of the most general biharmonic fifth degree polynomial

$$P_5(x, y) = A_0 \left(x^5 - 5xy^4\right) + A_1 \left(x^4 y - \frac{y^5}{5}\right)$$

$$+A_2 \left(x^3 y^2 - xy^4\right) + A_3 \left(x^2 y^3 - \frac{y^5}{5}\right) . \tag{5.8}$$

This function will be biharmonic for any values of the four independent constants A_0, A_1, A_2, A_3. We can express this by stating that the biharmonic polynomial P_5 has four degrees of freedom.

In general, the polynomial Q is of degree 4 less than P because the biharmonic equation is of degree 4. It follows that there are always four fewer constraint equations than there are coefficients in the original polynomial P and hence that they can be satisfied leaving a polynomial with 4 degrees of freedom. However, the process degenerates if $N < 3$.

In view of the above discussion, it might seem appropriate to write an expression for the general polynomial of degree N in the form of equation (5.8) as a preliminary to the solution of polynomial problems in rectangular coördinates. However, as can be seen from equation (5.8), the resulting expressions are algebraically messy and this approach becomes unmanageable for problems of any complexity. Instead, it turns out to be more straightforward algebraically to define problems in terms of the simpler unconstrained polynomials like equation (5.2) and to impose the constraint equations at a later stage in the solution.

5.1.1 Second and third degree polynomials

We recall that the stress components are defined in terms of the stress function ϕ through the relations

$$\sigma_{xx} = \frac{\partial^2 \phi}{\partial y^2} \; ; \quad \sigma_{yy} = \frac{\partial^2 \phi}{\partial x^2} \; ; \quad \sigma_{xy} = -\frac{\partial^2 \phi}{\partial x \partial y} \; . \tag{5.9}$$

It follows that when the stress function is a polynomial of degree N in x, y, the stress components will be polynomials of degree $(N-2)$. In particular, constant and linear terms in ϕ correspond to null stress fields (zero stress everywhere) and can be disregarded.

The second degree polynomial

$$\phi = A_0 x^2 + A_1 xy + A_2 y^2$$

yields the stress components

$$\sigma_{xx} = 2A_2 \ ; \quad \sigma_{xy} = -A_1 \ ; \quad \sigma_{yy} = 2A_0$$

and hence corresponds to the most general state of biaxial uniform stress.

The third degree polynomial

$$\phi = A_0 x^3 + A_1 x^2 y + A_2 x y^2 + A_3 y^3 \qquad (5.10)$$

yields the stress components

$$\sigma_{xx} = 2A_2 x + 6A_3 y \ ; \quad \sigma_{xy} = -2A_1 x - 2A_2 y \ ; \quad \sigma_{yy} = 6A_0 x + 2A_1 y \ .$$

If we arbitrarily set A_0, A_1, $A_2 = 0$, the only remaining non-zero stress will be

$$\sigma_{xx} = 6A_3 y \ ,$$

which corresponds to a state of pure bending, when applied to the rectangular beam $-a < x < a$, $-b < y < b$, as shown in Figure 5.1.

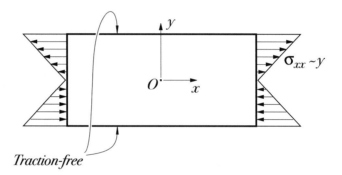

Fig. 5.1 The rectangular beam in pure bending.

The other terms in equation (5.10) correspond to a more general state of bending. For example, the constant A_0 describes bending of the beam by tractions σ_{yy} applied to the boundaries $y = \pm b$, whilst the terms involving shear stresses σ_{xy} could be obtained by describing a general state of biaxial bending with reference to a Cartesian coördinate system which is not aligned with the axes of the beam.

The above solutions are of course very elementary, but we should remember that, in contrast to the Mechanics of Materials solutions for simple bending, they are obtained without making any simplifying assumptions about the stress fields. For example, we have not assumed that plane sections remain plane, nor have we demanded that the beam be long in comparison with its depth. Thus, the present section could be taken as verifying the *exactness* of the Mechanics of Materials solutions for uniform stress and simple bending, as applied to a rectangular beam.

5.2 Rectangular Beam Problems

5.2.1 Bending of a beam by an end load

Figure 5.2 shows a rectangular beam, $0 < x < a$, $-b < y < b$, subjected to a transverse force, F at the end $x = 0$, and built-in at the end $x = a$, the horizontal boundaries $y = \pm b$ being traction free. The boundary conditions for this problem are most naturally written in the form

$$\sigma_{xy} = 0 \; ; \quad y = \pm b \tag{5.11}$$
$$\sigma_{yy} = 0 \; ; \quad y = \pm b \tag{5.12}$$
$$\sigma_{xx} = 0 \; ; \quad x = 0 \tag{5.13}$$
$$\int_{-b}^{b} \sigma_{xy} dy = F \; ; \quad x = 0 \, . \tag{5.14}$$

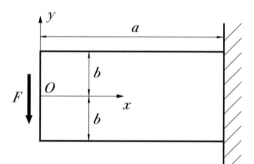

Fig. 5.2 Cantilever with an end load.

The boundary condition (5.14) is imposed in the *weak* form, which means that the value of the traction is not specified at each point on the boundary — only the force resultant is specified. In general, we shall find that problems for the rectangular beam have finite polynomial solutions when the boundary conditions on the ends are stated in the weak form, but that the *strong* (i.e. pointwise) boundary condition can only be satisfied on all the boundaries by an infinite series or transform solution. This issue is further discussed in Chapter 6.

Mechanics of Materials considerations suggest that the bending moment in this problem will vary linearly with x and hence that the stress component σ_{xx} will have a leading term proportional to xy. This in turn suggests a fourth degree polynomial term xy^3 in the stress function ϕ. Our procedure is therefore to start with the trial stress function

$$\phi = C_1 xy^3 \, , \tag{5.15}$$

examine the corresponding tractions on the boundaries and then seek a *corrective* solution which, when superposed on equation (5.15), yields the solution to the problem. Substituting (5.15) into (5.9), we obtain

$$\sigma_{xx} = 6C_1xy \tag{5.16}$$

$$\sigma_{xy} = -3C_1y^2 \tag{5.17}$$

$$\sigma_{yy} = 0 , \tag{5.18}$$

from which we note that the boundary conditions (5.12, 5.13) are satisfied identically, but that (5.11) is not satisfied, since (5.17) implies the existence of an unwanted uniform shear traction $-3C_1b^2$ on both of the edges $y = \pm b$. This unwanted traction can be removed by superposing an appropriate uniform shear stress, through the additional stress function term C_2xy. Thus, if we define

$$\phi = C_1xy^3 + C_2xy , \tag{5.19}$$

equations (5.16, 5.18) remain unchanged, whilst (5.17) is modified to

$$\sigma_{xy} = -3C_1y^2 - C_2 .$$

The boundary condition (5.11) can now be satisfied if we choose C_2 to satisfy the equation

$$C_2 = -3C_1b^2 ,$$

so that

$$\sigma_{xy} = 3C_1(b^2 - y^2) . \tag{5.20}$$

The constant C_1 can then be determined by substituting (5.20) into the remaining boundary condition (5.14), with the result

$$C_1 = \frac{F}{4b^3} .$$

The final stress field is therefore defined through the stress function

$$\phi = \frac{F(xy^3 - 3b^2xy)}{4b^3} , \tag{5.21}$$

the corresponding stress components being

$$\sigma_{xx} = \frac{3Fxy}{2b^3} ; \quad \sigma_{xy} = \frac{3F(b^2 - y^2)}{4b^3} ; \quad \sigma_{yy} = 0 . \tag{5.22}$$

The solution of this problem is given in the Mathematica and Maple files 'S521'.

Boundary conditions at $x = a$

We note that no boundary conditions have been specified on the built-in end, $x = a$. In the weak form, these would be

$$\int_{-b}^{b} \sigma_{xx} dy = 0 \; ; \quad x = a \tag{5.23}$$

$$\int_{-b}^{b} \sigma_{xy} dy = F \; ; \quad x = a \tag{5.24}$$

$$\int_{-b}^{b} \sigma_{xx} y\, dy = Fa \; ; \quad x = a \,. \tag{5.25}$$

However, if conditions (5.11–5.14) are satisfied, (5.23–5.25) are merely equivalent to the condition that the whole beam be in equilibrium. Now the Airy stress function is so defined that whatever stress function is used, the corresponding stress field will satisfy equilibrium in the local sense of equations (2.4). Furthermore, if every particle of a body is separately in equilibrium, it follows that the whole body will also be in equilibrium. It is therefore not necessary to enforce equations (5.23–5.25), since if we were to check them, we should necessarily find that they are satisfied identically.

5.2.2 Higher order polynomials — a general strategy

In the previous section, we developed the solution by trial and error, starting from the leading term whose form was dictated by equilibrium considerations. A more general technique is to identify the highest order polynomial term from equilibrium considerations and then write down the most general polynomial of that degree and below. The constant multipliers on the various terms are then obtained by imposing boundary conditions and biharmonic constraint equations.

The only objection to this procedure is that it involves a lot of algebra. For example, in the problem of §5.2.1, we would have to write down the most general polynomial of degree 4 and below, which involves 12 separate terms even when we exclude the linear and constant terms as being null. However, this is not a serious difficulty if we are using Maple or Mathematica, so we shall first develop the steps needed for this general strategy. Shortcuts which would reduce the complexity of the algebra in a manual calculation will then be discussed in §5.2.3.

Order of the polynomial

Suppose we have a normal traction on the surface $y = b$ varying with x^n. In Mechanics of Materials terms, this corresponds to a distributed load proportional to x^n and elementary equilibrium considerations show that the bending moment can then be expected to contain a term proportional to x^{n+2}. This in turn implies a bending stress σ_{xx} proportional to $x^{n+2} y$ and a term in the stress function proportional to $x^{n+2} y^3$ — i.e. a term of polynomial order $(n+5)$. A corresponding argument for shear tractions proportional to x^m shows that we require a polynomial order of $(m+4)$.

We shall show in Chapter 30 that these arguments from equilibrium and elementary bending theory define the highest order of polynomial required to satisfy any polynomial boundary conditions on the lateral surfaces of a beam, even in three-dimensional problems. A sufficient polynomial order can therefore be selected by the following procedure:-

(i) Identify the highest order polynomial term n in the normal tractions σ_{yy} on the surfaces $y = \pm b$.
(ii) Identify the highest order polynomial term m in the shear tractions σ_{yx} on the surfaces $y = \pm b$.
(iii) Use a polynomial for ϕ including all polynomial terms of order $\max(m + 4, n + 5)$ and below, but excluding constant and linear terms.

In the special case where both surfaces $y = \pm b$ are traction free, it is sufficient to use a polynomial of 4th degree and below (as in §5.2.1).

Solution procedure

Once an appropriate polynomial has been identified for ϕ, we proceed as follows:-

(i) Substitute ϕ into the biharmonic equation (5.1), leading to a set of constraint equations, as in §5.1.
(ii) Substitute ϕ into equations (5.9), to obtain the stress components as functions of x, y.
(iii) Substitute the equations defining the boundaries (e.g. $x = 0$, $y = b$, $y = -b$ in the problem of §5.2.1) into appropriate[1] stress components, to obtain the tractions on each boundary.
(iv) For the longer boundaries (where strong boundary conditions will be enforced), sort the resulting expressions into powers of x or y and equate coefficients with the corresponding expression for the prescribed tractions.

---

[1] Recall from §1.1.1 that the only stress components that act on (e.g.) $y = b$ are those which contain y as one of the suffices.

(v) For the shorter boundaries, substitute the tractions into the appropriate weak
   boundary conditions, obtaining three further independent algebraic equations.

The equations so obtained will generally not all be linearly independent, but they
will be sufficient to determine all the coefficients uniquely. The solvers in Maple and
Mathematica can handle this redundancy.

Example

We illustrate this procedure with the example of Figure 5.3, in which a rectangular
beam $-a < x < a$, $-b < y < b$ is loaded by a uniform compressive normal traction p
on $y = b$ and simply supported at the ends.

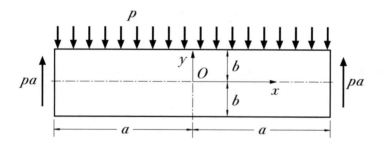

Fig. 5.3 Simply supported beam with a uniform load.

The boundary conditions on the surfaces $y = \pm b$ can be written

$$\sigma_{yx} = 0 \; ; \; y = \pm b \tag{5.26}$$
$$\sigma_{yy} = -p \; ; \; y = b \tag{5.27}$$
$$\sigma_{yy} = 0 \; ; \; y = -b . \tag{5.28}$$

These boundary conditions are to be satisfied in the strong sense. To complete the
problem definition, we shall require three linearly independent weak boundary conditions on one or both of the ends $x = \pm a$. We might use symmetry and equilibrium
to argue that the load will be equally shared between the supports, leading to the
conditions[2]

---

[2] It is not necessary to use symmetry arguments to obtain three linearly independent weak conditions.
Since the beam is simply supported, we know that

$$M(a) = \int_{-b}^{b} \sigma_{xx}(a, y) y \, dy = 0 \; ; \quad M(-a) = \int_{-b}^{b} \sigma_{xx}(-a, y) y \, dy = 0$$

$$F_x(a) = \int_{-b}^{b} \sigma_{xx}(a, y) \, dy = 0 .$$

$$F_x(a) = \int_{-b}^{b} \sigma_{xx}(a, y)dy = 0 \tag{5.29}$$

$$F_y(a) = \int_{-b}^{b} \sigma_{xy}(a, y)dy = pa \tag{5.30}$$

$$M(a) = \int_{-b}^{b} \sigma_{xx}(a, y)ydy = 0 \tag{5.31}$$

on the end $x=a$. As explained in §5.2.1, we do not need to enforce the additional three weak conditions on $x=-a$.

The normal traction is uniform — i.e. it varies with x^0 ($n=0$), so the above criterion demands a polynomial of order $(n+5)=5$. We therefore write

$$\begin{aligned}
\phi = {} & C_1 x^2 + C_2 xy + C_3 y^2 + C_4 x^3 + C_5 x^2 y + C_6 xy^2 + C_7 y^3 + C_8 x^4 \\
& + C_9 x^3 y + C_{10} x^2 y^2 + C_{11} xy^3 + C_{12} y^4 + C_{13} x^5 + C_{14} x^4 y \\
& + C_{15} x^3 y^2 + C_{16} x^2 y^3 + C_{17} xy^4 + C_{18} y^5 .
\end{aligned} \tag{5.32}$$

This is a long expression, but remember we only have to type it in once to the computer file. We can cut and paste the expression in the solution of subsequent problems (and the reader can indeed cut and paste from the web file 'polynomial'). Substituting (5.32) into the biharmonic equation (5.1), we obtain

$$\begin{aligned}
(120C_{13} + 24C_{15} + 24C_{17})x + (24C_{14} + 24C_{16} + 120C_{18})y \\
+ (24C_8 + 8C_{10} + 24C_{12}) = 0
\end{aligned}$$

and this must be zero for all x, y leading to the three constraint equations

$$120C_{13} + 24C_{15} + 24C_{17} = 0 \tag{5.33}$$

$$24C_{14} + 24C_{16} + 120C_{18} = 0 \tag{5.34}$$

$$24C_8 + 8C_{10} + 24C_{12} = 0 . \tag{5.35}$$

The stresses are obtained by substituting (5.32) into (5.9) with the result

$$\begin{aligned}
\sigma_{xx} = {} & 2C_3 + 2C_6 x + 6C_7 y + 2C_{10} x^2 + 6C_{11} xy + 12C_{12} y^2 + 2C_{15} x^3 \\
& + 6C_{16} x^2 y + 12C_{17} xy^2 + 20C_{18} y^3 \\
\sigma_{xy} = {} & -C_2 - 2C_5 x - 2C_6 y - 3C_9 x^2 - 4C_{10} xy - 3C_{11} y^2 - 4C_{14} x^3 \\
& - 6C_{15} x^2 y - 6C_{16} xy^2 - 4C_{17} y^3 \\
\sigma_{yy} = {} & 2C_1 + 6C_4 x + 2C_5 y + 12C_8 x^2 + 6C_9 xy + 2C_{10} y^2 + 20C_{13} x^3 \\
& + 12C_{14} x^2 y + 6C_{15} xy^2 + 2C_{16} y^3 .
\end{aligned}$$

The tractions on $y=b$ are therefore

It is easy to verify that these conditions lead to the same solution as (5.29–5.31).

$$\sigma_{yx} = -4C_{14}x^3 - (3C_9 + 6C_{15}b)x^2 - (2C_5 + 4C_{10}b + 6C_{16}b^2)x$$
$$-(C_2 + 2C_6b + 3C_{11}b^2 + 4C_{17}b^3)$$
$$\sigma_{yy} = 20C_{13}x^3 + (12C_8 + 12C_{14}b)x^2 + (6C_4 + 6C_9b + 6C_{15}b^2)x$$
$$+(2C_1 + 2C_5b + +2C_{10}b^2 + 2C_{16}b^3)$$

and these must satisfy equations (5.26, 5.27) for all x, giving

$$4C_{14} = 0 \tag{5.36}$$
$$3C_9 + 6C_{15}b = 0 \tag{5.37}$$
$$2C_5 + 4C_{10}b + 6C_{16}b^2 = 0 \tag{5.38}$$
$$C_2 + 2C_6b + 3C_{11}b^2 + 4C_{17}b^3 = 0 \tag{5.39}$$
$$20C_{13} = 0 \tag{5.40}$$
$$12C_8 + 12C_{14}b = 0 \tag{5.41}$$
$$6C_4 + 6C_9b + 6C_{15}b^2 = 0 \tag{5.42}$$
$$2C_1 + 2C_5b + 2C_{10}b^2 + 2C_{16}b^3 = -p . \tag{5.43}$$

A similar procedure for the edge $y = -b$ yields the additional equations

$$3C_9 - 6C_{15}b = 0 \tag{5.44}$$
$$2C_5 - 4C_{10}b + 6C_{16}b^2 = 0 \tag{5.45}$$
$$C_2 - 2C_6b + 3C_{11}b^2 - 4C_{17}b^3 = 0 \tag{5.46}$$
$$12C_8 - 12C_{14}b = 0 \tag{5.47}$$
$$6C_4 - 6C_9b + 6C_{15}b^2 = 0 \tag{5.48}$$
$$2C_1 - 2C_5b + 2C_{10}b^2 - 2C_{16}b^3 = 0 . \tag{5.49}$$

On $x = a$, we have

$$\sigma_{xx} = 2C_3 + 2C_6a + 6C_7y + 2C_{10}a^2 + 6C_{11}ay + 12C_{12}y^2 + 2C_{15}a^3$$
$$+6C_{16}a^2y + 12C_{17}ay^2 + 20C_{18}y^3$$
$$\sigma_{xy} = -C_2 - 2C_5a - 2C_6y - 3C_9a^2 - 4C_{10}ay - 3C_{11}y^2 - 4C_{14}a^3$$
$$-6C_{15}a^2y - 6C_{16}ay^2 - 4C_{17}y^3 .$$

Substituting into the weak conditions (5.29–5.31) and evaluating the integrals, we obtain the three additional equations

$$4C_3b + 4C_6ab + 4C_{10}a^2b + 8C_{12}b^3 + 4C_{15}a^3b + 8C_{17}ab^3 = 0 \tag{5.50}$$
$$-2C_2b - 4C_5ab - 6C_9a^2b - 2C_{11}b^3 - 8C_{14}a^3b - 4C_{16}ab^3 = pa \tag{5.51}$$
$$4C_7b^3 + 4C_{11}ab^3 + 4C_{16}a^2b^3 + 8C_{18}b^5 = 0 . \tag{5.52}$$

Finally, we solve equations (5.33–5.35, 5.36–5.49, 5.50–5.52) for the unknown constants $C_1, ..., C_{18}$ and substitute back into (5.32), obtaining

$$\phi = \frac{p}{40b^3}\left(5x^2y^3 - y^5 - 15b^2x^2y - 5a^2y^3 + 2b^2y^3 - 10b^3x^2\right) . \tag{5.53}$$

The corresponding stress field is

$$\sigma_{xx} = \frac{p}{20b^3}\left(15x^2y - 10y^3 - 15a^2y + 6b^2y\right) \tag{5.54}$$

$$\sigma_{xy} = \frac{3px}{4b^3}\left(b^2 - y^2\right) \tag{5.55}$$

$$\sigma_{yy} = \frac{p}{4b^3}\left(y^3 - 3b^2y - 2b^3\right) . \tag{5.56}$$

The reader is encouraged to run the Maple or Mathematica files 'S522', which contain the above solution procedure. Notice that most of the algebraic operations are generated by quite simple and repetitive commands. These will be essentially similar for any polynomial problem involving rectangular coördinates, so it is a simple matter to modify the program to cover other cases.

5.2.3 Manual solutions — symmetry considerations

If the solution is to be obtained manually, the complexity of the algebra makes the process time consuming and increases the likelihood of errors. Fortunately, the complexity can be reduced by utilizing the natural symmetry of the rectangular beam. In many problems, the loading has some symmetry which can be exploited in limiting the number of independent polynomial terms and even when this is not the case, some saving of complexity can be achieved by representing the loading as the sum of symmetric and antisymmetric parts. We shall illustrate this procedure by repeating the solution of the problem of Figure 5.3.

The problem is symmetrical about the mid-point of the beam and hence, taking the origin there, we deduce that the resulting stress function will contain only *even* powers of x. This immediately reduces the number of terms in the general stress function to 10.

The beam is also symmetrical about the axis $y=0$, but the loading is not. We therefore decompose the problem into the two sub-problems illustrated in Figure 5.4(*a,b*).

The problem in Figure 5.4(*a*) is *antisymmetric* in y and hence requires a stress function with only *odd* powers of y, whereas that of Figure 5.4(*b*) is *symmetric* and requires only *even* powers. In fact, the problem of Figure 5.4(*b*) clearly has the trivial solution corresponding to uniform uniaxial compression, $\sigma_{yy}=-p/2$, the appropriate stress function being $\phi=-px^2/4$.

For the problem of Figure 5.4(*a*), the most general fifth degree polynomial which is even in x and odd in y can be written

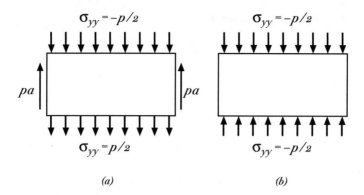

Fig. 5.4 Decomposition of the problem into (*a*) antisymmetric and (*b*) symmetric parts.

$$\phi = C_5 x^2 y + C_7 y^3 + C_{14} x^4 y + C_{16} x^2 y^3 + C_{18} y^5 , \tag{5.57}$$

which has just five degrees of freedom. We have used the same notation for the remaining constants as in (5.32) to aid in comparing the two solutions. The appropriate boundary conditions for this sub-problem are

$$\sigma_{xy} = 0 \ ; \quad y = \pm b \tag{5.58}$$

$$\sigma_{yy} = \mp \frac{p}{2} \ ; \quad y = \pm b \tag{5.59}$$

$$\int_{-b}^{b} \sigma_{xx} dy = 0 \ ; \quad x = \pm a \tag{5.60}$$

$$\int_{-b}^{b} \sigma_{xx} y dy = 0 \ ; \quad x = \pm a . \tag{5.61}$$

Notice that, in view of the symmetry, it is only necessary to satisfy these conditions on one of each pair of edges (e.g. on $y=b$, $x=a$). For the same reason, we do not have to impose a condition on the vertical force at $x=\pm a$, since the symmetry demands that the forces be equal at the two ends and the *total* force must be $2pa$ to preserve global equilibrium, this being guaranteed by the use of the Airy stress function, as in the problem of §5.2.1.

It is usually better strategy to start a manual solution with the strong boundary conditions (equations [5.58, 5.59]), and in particular with those conditions that are homogeneous [in this case equation (5.58)], since these will often require that one or more of the constants be zero, reducing the complexity of subsequent steps. Substituting (5.57) into (5.9)$_{2,3}$, we find

$$\sigma_{xy} = -2C_5 x - 4C_{14} x^3 - 6C_{16} x y^2 \tag{5.62}$$

$$\sigma_{yy} = 2C_5 y + 12C_{14} x^2 y + 2C_{16} y^3 . \tag{5.63}$$

Thus, condition (5.58) requires that

$$4C_{14}x^3 + (2C_5 + 6C_{16}b^2)x = 0 \; ; \quad \text{for all } x$$

and this condition is satisfied if and only if

$$C_{14} = 0 \quad \text{and} \quad 2C_5 + 6C_{16}b^2 = 0 . \tag{5.64}$$

A similar procedure with equation (5.63) and boundary condition (5.59) gives the additional equation

$$2C_5 b + 2C_{16}b^3 = -\frac{p}{2} . \tag{5.65}$$

Equations (5.64, 5.65) have the solution

$$C_5 = -\frac{3p}{8b} \; ; \quad C_{16} = \frac{p}{8b^3} . \tag{5.66}$$

We next determine C_{18} from the condition that the function ϕ is biharmonic, obtaining

$$(24C_{14} + 24C_{16} + 120C_{18})y = 0 \tag{5.67}$$

and hence

$$C_{18} = -\frac{p}{40b^3} , \tag{5.68}$$

from (5.64, 5.66, 5.67).

It remains to satisfy the two weak boundary conditions (5.60, 5.61) on the ends $x = \pm a$. The first of these is satisfied identically in view of the antisymmetry of the stress field and the second gives the equation

$$4C_7 b^3 + 4C_{16}a^2 b^3 + 8C_{18}b^5 = 0 ,$$

which, with equations (5.66, 5.68), serves to determine the remaining constant,

$$C_7 = \frac{p(2b^2 - 5a^2)}{40b^3} .$$

The final solution of the complete problem (the sum of that for Figures 5.4(a) and (b)) is therefore obtained from the stress function

$$\phi = \frac{p}{40b^3}(5x^2 y^3 - y^5 - 15b^2 x^2 y - 5a^2 y^3 + 2b^2 y^3 - 10b^3 x^2) ,$$

as in the 'computer solution' (5.53), and the stresses are therefore given by (5.54–5.56) as before.

5.3 Fourier Series and Transform Solutions

Polynomial solutions can, in principle, be extended to more general loading of the beam edges, as long as the tractions are capable of a power series expansion. However, the practical use of this method is limited by the algebraic complexity encountered for higher order polynomials and by the fact that many important traction distributions do not have convergent power series representations.

A more useful method in such cases is to build up a general solution by components of Fourier form. For example, if we write

$$\phi = f(y)\cos(\lambda x) \quad \text{or} \quad \phi = f(y)\sin(\lambda x) , \tag{5.69}$$

substitution in the biharmonic equation (5.1) shows that $f(y)$ must have the general form

$$f(y) = (A + By)e^{\lambda y} + (C + Dy)e^{-\lambda y} , \tag{5.70}$$

where A, B, C, D are arbitrary constants. Alternatively, by defining new arbitrary constants A', B', C', D' through the relations $A = (A'+C')/2$, $B = (B'+D')/2$, $C = (A'-C')/2$, $D = (B'-D')/2$, we can group the exponentials into hyperbolic functions, obtaining the equivalent form

$$f(y) = (A' + B'y)\cosh(\lambda y) + (C' + D'y)\sinh(\lambda y) . \tag{5.71}$$

The hyperbolic form enables us to take advantage of any symmetry about the line $y = 0$, since $\cosh(\lambda y)$, $y\sinh(\lambda y)$ are even functions of y and $\sinh(\lambda y)$, $y\cosh(\lambda y)$ are odd functions.

More general biharmonic stress functions can be constructed by superposition of terms like (5.70, 5.71), leading to Fourier series expansions for the tractions on the surfaces $y = \pm b$. The theory of Fourier series can then be used to determine the coefficients in the series, using strong boundary conditions on $y = \pm b$. Quite general traction distributions can be expanded in this way, so Fourier series solutions provide a methodology applicable to any problem for the rectangular bar.

5.3.1 Choice of form

The stresses due to the stress function $\phi = f(y)\cos(\lambda x)$ are

$$\sigma_{xx} = f''(y)\cos(\lambda x) ; \quad \sigma_{xy} = \lambda f'(y)\sin(\lambda x) ; \quad \sigma_{yy} = -\lambda^2 f(y)\cos(\lambda x)$$

and the tractions on the edge $x = a$ are

$$\sigma_{xx}(a, y) = f''(y)\cos(\lambda a) ; \quad \sigma_{xy}(a, y) = \lambda f'(y)\sin(\lambda a) . \tag{5.72}$$

It follows that we can satisfy homogeneous boundary conditions on one (but not both) of these tractions in the strong sense, by restricting the Fourier series to specific values of λ. In equation (5.72), the choice $\lambda = n\pi/a$ will give $\sigma_{xy} = 0$ on $x = \pm a$, whilst $\lambda = (2n-1)\pi/2a$ will give $\sigma_{xx} = 0$ on $x = \pm a$, where n is any integer.

Example

We illustrate this technique by considering the rectangular beam $-a < x < a$, $-b < y < b$, simply supported at $x = \pm a$ and loaded by compressive normal tractions $p_1(x)$ on the upper edge $y = b$ and $p_2(x)$ on $y = -b$ — i.e.

$$\sigma_{xy} = 0 \qquad ; \quad y = \pm b \tag{5.73}$$
$$\sigma_{yy} = -p_1(x) \; ; \quad y = b \tag{5.74}$$
$$= -p_2(x) \; ; \quad y = -b \tag{5.75}$$
$$\sigma_{xx} = 0 \qquad ; \quad x = \pm a \, . \tag{5.76}$$

Notice that we have replaced the weak conditions (5.60, 5.61) by the strong condition (5.76). As in §5.2.2, it is not necessary to enforce the remaining weak conditions (those involving the vertical forces on $x = \pm a$), since these will be identically satisfied by virtue of the equilibrium condition.

The algebraic complexity of the problem will be reduced if we use the geometric symmetry of the beam to decompose the problem into four sub-problems. For this purpose, we define

$$f_1(x) = f_1(-x) \equiv \frac{1}{4}\left[p_1(x) + p_1(-x) + p_2(x) + p_2(-x) \right] \tag{5.77}$$

$$f_2(x) = -f_2(-x) \equiv \frac{1}{4}\left[p_1(x) - p_1(-x) + p_2(x) - p_2(-x) \right] \tag{5.78}$$

$$f_3(x) = f_3(-x) \equiv \frac{1}{4}\left[p_1(x) + p_1(-x) - p_2(x) - p_2(-x) \right] \tag{5.79}$$

$$f_4(x) = -f_4(-x) \equiv \frac{1}{4}\left[p_1(x) - p_1(-x) - p_2(x) + p_2(-x) \right] \tag{5.80}$$

and hence

$$p_1(x) = f_1(x) + f_2(x) + f_3(x) + f_4(x) \; ; \quad p_2(x) = f_1(x) + f_2(x) - f_3(x) - f_4(x) \, .$$

The boundary conditions now take the form

$$\sigma_{xy} = 0 \; ; \quad y = \pm b$$
$$\sigma_{yy} = -f_1(x) - f_2(x) - f_3(x) - f_4(x) \; ; \quad y = b$$
$$= -f_1(x) - f_2(x) + f_3(x) + f_4(x) \; ; \quad y = -b$$
$$\sigma_{xx} = 0 \; ; \quad x = \pm a$$

and each of the functions f_1, f_2, f_3, f_4 defines a separate problem with either symmetry or antisymmetry about the x- and y-axes. We shall here restrict attention to the loading defined by the function $f_3(x)$, which is symmetric in x and antisymmetric in y. The boundary conditions of this sub-problem are

$$\sigma_{xy} = 0 \qquad ; \quad y = \pm b \tag{5.81}$$
$$\sigma_{yy} = \mp f_3(x) \quad ; \quad y = \pm b \tag{5.82}$$
$$\sigma_{xx} = 0 \qquad ; \quad x = \pm a \ . \tag{5.83}$$

The problem of equations (5.81–5.83) is even in x and odd in y, so we use a cosine series in x with only the odd terms from the hyperbolic form (5.71) — i.e.

$$\phi = \sum_{n=1}^{\infty} \left[A_n y \cosh(\lambda_n y) + B_n \sinh(\lambda_n y) \right] \cos(\lambda_n x) \ , \tag{5.84}$$

where A_n, B_n are arbitrary constants. The strong condition (5.83) on $x = \pm a$ can then be satisfied in every term by choosing

$$\lambda_n = \frac{(2n-1)\pi}{2a} \ . \tag{5.85}$$

The corresponding stresses are

$$\sigma_{xx} = \sum_{n=1}^{\infty} \left[2A_n \lambda_n \sinh(\lambda_n y) + A_n \lambda_n^2 y \cosh(\lambda_n y) + B_n \lambda_n^2 \sinh(\lambda_n y) \right] \cos(\lambda_n x)$$

$$\sigma_{xy} = \sum_{n=1}^{\infty} \left[A_n \lambda_n \cosh(\lambda_n y) + A_n \lambda_n^2 y \sinh(\lambda_n y) + B_n \lambda_n^2 \cosh(\lambda_n y) \right] \sin(\lambda_n x)$$

$$\sigma_{yy} = -\sum_{n=1}^{\infty} \left[A_n \lambda_n^2 y \cosh(\lambda_n y) + B_n \lambda_n^2 \sinh(\lambda_n y) \right] \cos(\lambda_n x) \tag{5.86}$$

and hence the boundary conditions (5.81, 5.82) on $y = \pm b$ require that

$$\sum_{n=1}^{\infty} \left[A_n \lambda_n \cosh(\lambda_n b) + A_n \lambda_n^2 b \sinh(\lambda_n b) + B_n \lambda_n^2 \cosh(\lambda_n b) \right] \sin(\lambda_n x) = 0$$

$$\tag{5.87}$$

$$\sum_{n=1}^{\infty} \left[A_n \lambda_n^2 b \cosh(\lambda_n b) + B_n \lambda_n^2 \sinh(\lambda_n b) \right] \cos(\lambda_n x) = f_3(x) \ . \tag{5.88}$$

To invert the series, we multiply (5.88) by $\cos(\lambda_m x)$ and integrate from $-a$ to a, obtaining

$$\sum_{n=1}^{\infty} \int_{-a}^{a} \left[A_n \lambda_n^2 b \cosh(\lambda_n b) + B_n \lambda_n^2 \sinh(\lambda_n b) \right] \cos(\lambda_n x) \cos(\lambda_m x) dx$$

$$= \int_{-a}^{a} f_3(x) \cos(\lambda_m x) dx \ . \tag{5.89}$$

The integrals on the left-hand side are all zero except for the case $m=n$ and hence, evaluating the integrals, we find

$$\left[A_m \lambda_m^2 b \cosh(\lambda_m b) + B_m \lambda_m^2 \sinh(\lambda_m b) \right] a = \int_{-a}^{a} f_3(x) \cos(\lambda_m x) dx \ . \tag{5.90}$$

The homogeneous equation (5.87) is clearly satisfied if

$$A_m \lambda_m \cosh(\lambda_m b) + A_m \lambda_m^2 b \sinh(\lambda_m b) + B_m \lambda_m^2 \cosh(\lambda_m b) = 0 \ . \tag{5.91}$$

Solving (5.90, 5.91) for A_m, B_m, we have

$$A_m = \frac{\cosh(\lambda_m b)}{\lambda_m a [\lambda_m b - \sinh(\lambda_m b) \cosh(\lambda_m b)]} \int_{-a}^{a} f_3(x) \cos(\lambda_m x) dx$$

$$B_m = -\frac{(\cosh(\lambda_m b) + \lambda_m b \sinh(\lambda_m b))}{\lambda_m^2 a [\lambda_m b - \sinh(\lambda_m b) \cosh(\lambda_m b)]} \int_{-a}^{a} f_3(x) \cos(\lambda_m x) dx \ , \tag{5.92}$$

where λ_m is given by (5.85). The stresses are then recovered by substitution into equations (5.86).

The corresponding solutions for the functions f_1, f_2, f_4 are obtained in a similar way, but using a sine series for the odd functions f_2, f_4 and the even terms $y \sinh(\lambda y)$, $\cosh(\lambda y)$ in ϕ for f_1, f_2. The complete solution is then obtained by superposing the solutions of the four sub-problems.

The Fourier series method is particularly useful in problems where the traction distributions on the long edges have no power series expansion, typically because of discontinuities in the loading. For example, suppose the beam is loaded only by a concentrated compressive force F on the upper edge at $x=0$, corresponding to the loading $p_1(x)=F\delta(x)$, $p_2(x)=0$ in equations (5.78, 5.79). For the symmetric/antisymmetric sub-problem considered above, we then have

$$f_3(x) = \frac{F\delta(x)}{2}$$

from (5.79), and the integral in equations (5.92) is therefore

$$\int_{-a}^{a} f_3(x) \cos(\lambda_m x) dx = \frac{F}{2} \ ,$$

for all m.

This solution satisfies the end condition on σ_{xx} in the strong sense, but the condition on σ_{xy} only in the weak sense. In other words, the tractions σ_{xy} on the ends add up to the forces required to maintain equilibrium, but we have no control over the exact distribution of these tractions. This represents an improvement over the polynomial solution of §5.2.3, where weak conditions were used for *both* end tractions, so we might be tempted to use a Fourier series even for problems with continuous polynomial loading. However, this improvement is made at the cost of an infinite series solution. If the series were truncated at a finite value of n, errors would be obtained particularly near the ends or any discontinuities in the loading.

5.3.2 Fourier transforms

If the beam is infinite or semi-infinite $(a \to \infty)$, the series (5.84) must be replaced by the integral representation

$$\phi(x, y) = \int_0^\infty f(\lambda, y) \cos(\lambda x) d\lambda , \tag{5.93}$$

where

$$f(\lambda, y) = A(\lambda) y \cosh(\lambda y) + B(\lambda) \sinh(\lambda y) .$$

Equation (5.93) is introduced here as a generalization of (5.69) by superposition, but $\phi(x, y)$ is in fact the Fourier cosine transform of $f(\lambda, y)$, the corresponding inversion being

$$f(\lambda, y) = \frac{2}{\pi} \int_0^\infty \phi(x, y) \cos(\lambda x) dx .$$

The boundary conditions on $y = \pm b$ will also lead to Fourier integrals, which can be inverted in the same way to determine the functions $A(\lambda)$, $B(\lambda)$. For a definitive treatment of the Fourier transform method, the reader is referred to the treatise by Sneddon[3]. Extensive tables of Fourier transforms and their inversions are given by Erdelyi[4]. The cosine transform (5.93) will lead to a symmetric solution. For more general loading, the complex exponential transform can be used.

It is worth remarking on the way in which the series and transform solutions are natural generalizations of the elementary solution (5.69). One of the most powerful techniques in Elasticity — and indeed in any physical theory characterized by linear partial differential equations — is to seek a simple form of solution (often in

---

[3] I. N. Sneddon, *Fourier Transforms*, McGraw-Hill, New York, 1951.

[4] A. Erdelyi, **ed.**, *Tables of Integral Transforms*, Bateman Manuscript Project, California Institute of Technology, Vol. 1, McGraw-Hill, New York, 1954.

separated-variable form) containing a parameter which can take a range of values. A more general solution can then be developed by superposing arbitrary multiples of the solution with different values of the parameter.

For example, if a particular solution can be written symbolically as $\phi = f(x, y, \lambda)$, where λ is a parameter, we can develop a general series form

$$\phi(x, y) = \sum_{i=0}^{\infty} A_i f(x, y, \lambda_i) , \qquad (5.94)$$

or an integral form

$$\phi(x, y) = \int_a^b A(\lambda) f(x, y, \lambda) d\lambda .$$

The series form will naturally arise if there is a discrete set of *eigenvalues*, λ_i for which $f(x, y, \lambda_i)$ satisfies some of the boundary conditions of the problem. Additional examples of this kind will be found in §§6.2, 11.2. In this case, the series (5.94) is most properly seen as an eigenfunction expansion. Integral forms arise most commonly (but not exclusively) in problems involving infinite or semi-infinite domains (see, for example, §§11.3, 32.2.2).

Any particular solution containing a parameter can be used in this way and, since transforms are commonly named after their originators, the reader desirous of instant immortality might like to explore some of those which have not so far been used. Of course, the usefulness of the resulting solution depends upon its *completeness* — i.e. its capacity to represent all stress fields of a given class — and upon the ease with which the transform can be inverted.

Problems

5.1. The beam $-b < y < b$, $0 < x < L$, is built-in at the end $x = 0$ and loaded by a uniform shear traction $\sigma_{xy} = S$ on the upper edge, $y = b$, the remaining edges, $x = L$, $y = -b$ being traction free. Find a suitable stress function and the corresponding stress components for this problem, using the weak boundary conditions on $x = L$.

5.2. The beam $-b < y < b$, $-L < x < L$ is simply supported at the ends $x = \pm L$ and loaded by a shear traction $\sigma_{xy} = Sx/L$ on the lower edge, $y = -b$, the upper edge being traction free. Find a suitable stress function and the corresponding stress components for this problem, using the weak boundary conditions on $x = \pm L$.

5.3. The beam $-b < y < b$, $0 < x < L$, is built-in at the end $x = L$ and loaded by a linearly-varying compressive normal traction $p(x) = Sx/L$ on the upper edge, $y = b$, the remaining edges, $x=0$, $y=-b$ being traction free. Find a suitable stress function and the corresponding stress components for this problem, using the weak boundary conditions on $x=0$.

5.4. The beam $-b<y<b$, $-L<x<L$ is simply supported at the ends $x=\pm L$ and loaded by a compressive normal traction

$$p(x) = S\cos\left(\frac{\pi x}{2L}\right)$$

on the upper edge, $y=b$, the lower edge being traction free. Find a suitable stress function and the corresponding stress components for this problem.

5.5. The beam $-b<y<b$, $0<x<L$, is built-in at the end $x=L$ and loaded by a compressive normal traction

$$p(x) = S\sin\left(\frac{\pi x}{2L}\right)$$

on the upper edge, $y=b$, the remaining edges, $x=0$, $y=-b$ being traction free. Use a combination of the stress function (5.69) and an appropriate polynomial to find the stress components for this problem, using the weak boundary conditions on $x=0$.

5.6. The beam $-a<x<a$, $-b<y<b$ is loaded by a uniform compressive traction p in the central region $-a/2<x<a/2$ of both of the edges $y=\pm b$, as shown in Figure 5.5. The remaining edges are traction free. Use a Fourier series with the appropriate symmetries to obtain a solution for the stress field, using the weak condition on σ_{xy} on the edges $x=\pm a$ and the strong form of all the remaining boundary conditions.

Fig. 5.5

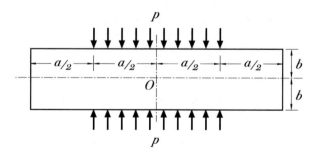

5.7. Use a Fourier series to solve the problem of Figure 5.4(a) in §5.2.3. Choose the terms in the series so as to satisfy the condition $\sigma_{xx}(\pm a, y)=0$ in the strong sense.

If you are solving this problem in Maple or Mathematica, compare the solution with that of §5.2.3 by making a contour plot of the difference between the truncated Fourier series stress function and the polynomial stress function

$$\phi = \frac{p}{40b^3} \left(5x^2y^3 - y^5 - 15b^2x^2y - 5a^2y^3 + 2b^2y^3\right) .$$

Examine the effect of taking different numbers of terms in the series.

5.8. The large plate $y > 0$ is loaded at its remote boundaries so as to produce a state of uniform tensile stress

$$\sigma_{xx} = S \; ; \; \sigma_{xy} = \sigma_{yy} = 0 ,$$

the boundary $y=0$ being traction free. We now wish to determine the perturbation in this simple state of stress that will be produced if the traction-free boundary had a slight waviness, defined by the line

$$y = \epsilon \cos(\lambda x) ,$$

where $\lambda\epsilon \ll 1$. To solve this problem

(i) Start with the stress function

$$\phi = \frac{Sy^2}{2} + f(y)\cos(\lambda x)$$

and determine $f(y)$ if the function ϕ is to be biharmonic.
(ii) The perturbation will be localized near $y=0$, so select only those terms in $f(y)$ that decay as $y \to \infty$.
(iii) Find the stress components and use the stress-transformation equations to determine the tractions on the wavy boundary. Notice that the inclination of the wavy surface to the plane $y=0$ will be everywhere small if $\lambda\epsilon \ll 1$ and hence the trigonometric functions involving this angle can be approximated using $\sin(z) \approx z, \; \cos(z) \approx 1, z \ll 1$.
(iv) Choose the free constants in $f(y)$ to satisfy the traction-free boundary condition on the wavy surface.
(v) Determine the maximum tensile stress and hence the stress-concentration factor as a function of $\lambda\epsilon$.

5.9. A large plate defined by $y > 0$ is subjected to a sinusoidally varying load

$$\sigma_{yy} = S \sin \lambda x \; ; \; \sigma_{xy} = 0$$

at its plane edge $y = 0$.
Find the complete stress field in the plate and hence estimate the depth y at which the amplitude of the variation in σ_{yy} has fallen to 10% of S.
Hint: You might find it easier initially to consider the case of the layer $0 < y < h$, with $y=h$ traction free, and then let $h \to \infty$.

Chapter 6
End Effects

The solution of §5.2.2 must be deemed approximate insofar as the boundary conditions on the ends $x = \pm a$ of the rectangular beam are satisfied only in the weak sense of force resultants, through equations (5.29–5.31). In general, if a rectangular beam is loaded by tractions of finite polynomial form, a finite polynomial solution can be obtained which satisfies the boundary conditions in the strong (i.e. pointwise) sense on two edges and in the weak sense on the other two edges.

The error involved in such an approximation corresponds to the solution of a corrective problem in which the beam is loaded by the difference between the actual tractions applied and those implied by the approximation. These tractions will of course be confined to the edges on which the weak boundary conditions were applied and will be self-equilibrated, since the weak conditions imply that the tractions in the approximate solution have the same force and moment resultants as the actual tractions.

For the particular problem of §5.2.2, we note that the stress field of equations (5.54–5.56) satisfies the boundary conditions on the edges $y = \pm b$, but that there is a self-equilibrated normal traction

$$\sigma_{xx} = \frac{p}{10b^3}(3b^2 y - 5y^3) \tag{6.1}$$

on the ends $x = \pm a$, which must be removed by superposing a corrective solution if we wish to satisfy the boundary conditions of Figure 5.3 in the strong sense.

6.1 Decaying Solutions

In view of Saint-Venant's principle, we anticipate that the stresses in the corrective solution will decay as we move away from the edges where the self-equilibrated tractions are applied. The decay rate is likely to be related to the width of the loaded

© The Author(s), under exclusive license to Springer Nature Switzerland AG 2022
J. R. Barber, *Elasticity*, Solid Mechanics and Its Applications 172,
https://doi.org/10.1007/978-3-031-15214-6_6

region and hence we anticipate that the stresses in the corrective solution will be significant only in two regions near the ends, of linear dimensions comparable to the width of the beam. These regions are, shown shaded in Figure 6.1. It follows that the solution of §5.2.2 will be a good approximation in the *unshaded* region in Figure 6.1.

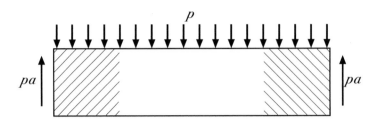

Fig. 6.1 Regions of the beam influenced by end effects.

It also follows that the corrective solutions for the two ends are *uncoupled*, since the corrective field for the end $x = -a$ has decayed to negligible proportions before we reach the end $x = +a$ and vice versa. This implies that, so far as the left end is concerned, the corrective solution is essentially identical to that which would be required in the *semi-infinite* beam, $x > -a$. We can therefore simplify the statement of the problem by considering the corrective solution for the left end only and shifting the origin to $-a$, so that the semi-infinite beam under consideration is defined by $x > 0$, $-b < y < b$.

It is also now clear why we chose to satisfy the strong boundary conditions on the *long* edges, $y = \pm b$. If instead we had imposed strong conditions on $x = \pm a$ and weak conditions on $y = \pm b$, the shaded regions would have overlapped and there would be no region in which the finite polynomial solution was a good approximation to the stresses in the beam. It also follows that the approximation will only be useful when the beam has an aspect ratio significantly different from unity — i.e. $a/b \gg 1$.

6.2 The Corrective Solution

We recall that the stress function ϕ for the corrective solution must (i) satisfy the biharmonic equation (5.1), (ii) have zero tractions on the boundaries $y = \pm b$ — i.e.

$$\sigma_{yy} = \sigma_{yx} = 0 \; ; \quad y = \pm b \tag{6.2}$$

and (iii) have prescribed non-zero tractions [such as those defined by equation (6.1)] on the end(s), which we can write in the form

$$\sigma_{xx}(0, y) = f_1(y) \; ; \quad \sigma_{xy}(0, y) = f_2(y) \,, \tag{6.3}$$

where

$$\int_{-b}^{b} f_1(y)dy = \int_{-b}^{b} f_1(y)y\,dy = \int_{-b}^{b} f_2(y)dy = 0 \, ,$$

since the corrective tractions are required to be self-equilibrated.

We cannot generally expect to find a solution to satisfy all these conditions in closed form and hence we seek a series or transform (integral) solution as suggested in §5.3. However, the solution will be simpler if we can find a *class* of solutions, all of which satisfy some of the conditions. We can then write down a more general solution as a superposition of such solutions and choose the coefficients so as to satisfy the remaining condition(s).

Since the traction boundary condition on the end will vary from problem to problem, it is convenient to seek solutions which satisfy (i) and (ii) — i.e. biharmonic functions for which

$$\frac{\partial^2 \phi}{\partial x^2} = \frac{\partial^2 \phi}{\partial x \partial y} = 0 \; ; \;\; y = \pm b \, . \tag{6.4}$$

6.2.1 Separated-variable solutions

One way to obtain functions satisfying conditions (6.4) is to write them in the separated-variable form

$$\phi = f(x)g(y) \, , \tag{6.5}$$

in which case, (6.4) will be satisfied for all x, provided that

$$g(y) = g'(y) = 0 \; ; \;\; y = \pm b \, . \tag{6.6}$$

Notice that the final corrective solution cannot be expected to be of separated-variable form, but we shall see that it can be represented as the sum of a series of such terms.

If the functions (6.5) are to be biharmonic, we must have

$$g\frac{d^4 f}{dx^4} + 2\frac{d^2 f}{dx^2}\frac{d^2 g}{dy^2} + f\frac{d^4 g}{dy^4} = 0 \, , \tag{6.7}$$

and this equation must be satisfied for all values of x, y. Now, if we consider the subset of points (x, c), where c is a constant, it is clear that $f(x)$ must satisfy an equation of the form

$$A\frac{d^4 f}{dx^4} + B\frac{d^2 f}{dx^2} + Cf = 0 \, ,$$

where A, B, C, are constants, and hence $f(x)$ must consist of exponential terms such as $f(x) = \exp(\lambda x)$. Similar considerations apply to the function $g(y)$. Notice incidentally that λ might be complex or imaginary, giving sinusoidal functions, and there are also degenerate cases where C and/or $B = 0$ in which case $f(x)$ could also be a polynomial of degree 3 or below.

Since we are seeking to represent a field which decays with distance x from the loaded end, we select terms of the form

$$\phi = g(y)e^{-\lambda x} ,$$

in which case, (6.7) reduces to

$$\frac{d^4 g}{dy^4} + 2\lambda^2 \frac{d^2 g}{dy^2} + \lambda^4 g = 0 ,$$

which is a fourth order ordinary differential equation for $g(y)$ with general solution

$$g(y) = (A_1 + A_2 y) \cos(\lambda y) + (A_3 + A_4 y) \sin(\lambda y) . \tag{6.8}$$

6.2.2 The eigenvalue problem

The arbitrary constants A_1, A_2, A_3, A_4 are determined from the boundary conditions (6.2), which in view of (6.6) lead to the four simultaneous equations

$$(A_1 + A_2 b) \cos(\lambda b) + (A_3 + A_4 b) \sin(\lambda b) = 0 \tag{6.9}$$

$$(A_1 - A_2 b) \cos(\lambda b) - (A_3 - A_4 b) \sin(\lambda b) = 0 \tag{6.10}$$

$$(A_2 + A_3 \lambda + A_4 \lambda b) \cos(\lambda b) - (A_1 \lambda + A_2 \lambda b - A_4) \sin(\lambda b) = 0 \tag{6.11}$$

$$(A_2 + A_3 \lambda - A_4 \lambda b) \cos(\lambda b) + (A_1 \lambda - A_2 \lambda b - A_4) \sin(\lambda b) = 0 . \tag{6.12}$$

This set of equations is homogeneous and will generally have only the trivial solution $A_1 = A_2 = A_3 = A_4 = 0$. However, there are some eigenvalues of the exponential decay rate λ, for which the determinant of coefficients is singular and the solution is non-trivial.

A more convenient form of the equations can be obtained by taking sums and differences in pairs — i.e. by constructing the equations (6.9 + 6.10), (6.9 – 6.10), (6.11 + 6.12), (6.11 – 6.12), which after rearrangement and cancellation of non-zero factors yields the set

$$A_1 \cos(\lambda b) + A_4 b \sin(\lambda b) = 0 \tag{6.13}$$

$$A_1 \lambda \sin(\lambda b) - A_4 \{\sin(\lambda b) + \lambda b (\cos(\lambda b)\} = 0 \tag{6.14}$$

$$A_2 b \cos(\lambda b) + A_3 \sin(\lambda b) = 0 \tag{6.15}$$

$$A_2 \{\cos(\lambda b) - \lambda b \sin(\lambda b)\} + A_3 \lambda \cos(\lambda b) = 0 . \tag{6.16}$$

What we have done here is to use the symmetry of the system to partition the matrix of coefficients. The terms $A_1 \cos(\lambda y)$, $A_4 y \sin(\lambda y)$ are symmetric, whereas

$A_2 y \cos(\lambda y)$, $A_3 \sin(\lambda y)$ are antisymmetric. The boundary conditions are also symmetric and hence the symmetric and antisymmetric terms must *separately* satisfy them.

We conclude that the set of equations (6.13–6.16) has two sets of eigenvalues, for one of which the resulting eigenfunctions are symmetric and the other antisymmetric. The symmetric eigenvalues $\lambda^{(S)}$ are obtained by eliminating A_1, A_4 from (6.13, 6.14) with the result

$$\sin(2\lambda^{(S)}b) + 2\lambda^{(S)}b = 0 , \tag{6.17}$$

whilst the antisymmetric eigenvalues $\lambda^{(A)}$ are obtained in the same way from (6.15, 6.16) with the result

$$\sin(2\lambda^{(A)}b) - 2\lambda^{(A)}b = 0 . \tag{6.18}$$

Figure 6.2 demonstrates graphically that the only *real* solution of equations (6.17, 6.18) is the trivial case $\lambda = 0$ (which in fact corresponds to the non-decaying solutions in which an axial force or moment resultant is applied at the end and transmitted along the beam).

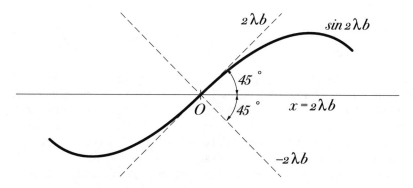

Fig. 6.2 Graphical solution of equations (6.17, 6.18).

However, there is a denumerably infinite set of non-trivial *complex* solutions, corresponding to stress fields which oscillate whilst decaying along the beam. These solutions are fairly easy to find by writing $\lambda b = c + \imath d$, separating real and imaginary parts in the complex equation, and solving the resulting two simultaneous equations for the real numbers, c, d, using a suitable numerical algorithm.

Once the eigenvalues have been determined, the corresponding eigenfunctions $g^{(S)}$, $g^{(A)}$ are readily recovered using (6.8, 6.17, 6.18). We obtain

$$g^{(S)} = C \left[y \sin(\lambda^{(S)}y) \cos(\lambda^{(S)}b) - b \sin(\lambda^{(S)}b) \cos(\lambda^{(S)}y) \right]$$
$$g^{(A)} = D \left[y \sin(\lambda^{(S)}b) \cos(\lambda^{(S)}y) - b \sin(\lambda^{(S)}y) \cos(\lambda^{(S)}b) \right] ,$$

where C, D are new arbitrary constants related to A_1, A_4 and A_2, A_3 respectively. We can then establish a more general solution of the form

$$\phi = \sum_{i=1}^{\infty} C_i g_i^{(S)}(y) e^{-\lambda_i^{(S)}x} + \sum_{i=1}^{\infty} D_i g_i^{(A)}(y) e^{-\lambda_i^{(A)}x} , \qquad (6.19)$$

where $\lambda_i^{(S)}$, $\lambda_i^{(A)}$ represent the eigenvalues of equations (6.17, 6.18) respectively.

The final step is to choose the constants C_i, D_i so as to satisfy the prescribed boundary conditions (6.3) on the end $x=0$. An improved but still approximate solution of this problem can be obtained by truncating the infinite series at some finite value $i = N$, defining a scalar measure of the error $\mathcal{E}$ in the boundary conditions and then imposing the $2N$ conditions

$$\frac{\partial \mathcal{E}}{\partial C_i} = 0 ; \quad \frac{\partial \mathcal{E}}{\partial D_i} = 0 \quad \text{for} \quad i \in [1, N]$$

to determine the unknown constants. An appropriate non-negative quadratic error measure is

$$\mathcal{E} = \int_{-b}^{b} \left\{ \left[f_1(y) - \sigma_{xx}(0, y) \right]^2 + \left[f_2(y) - \sigma_{xy}(0, y) \right]^2 \right\} dy ,$$

where σ_{xx}, σ_{xy} are the stress components defined by the truncated series. Since biharmonic boundary-value problems arise in many areas of mechanics, techniques of this kind have received a lot of attention. Convergence of the truncated series is greatly improved if the boundary conditions are continuous in the corners[1]. It can be shown that the use of the weak boundary conditions automatically defines continuous values of ϕ and its first derivatives in the corners $(0, -b)$, $(0, b)$ in the corrective solution. However, convergence problems are still likely to occur if the shear tractions on the two orthogonal edges in the corner are different, for example if

$$\lim_{x \to 0} \sigma_{yx}(x, b) \neq \lim_{y \to b} \sigma_{xy}(0, y) . \qquad (6.20)$$

Since the tractions on the boundaries are independent, this is a perfectly legitimate physical possibility, but it involves an infinite stress gradient in the corner and leads to a modified Gibbs phenomenon in the series solution, where increase of N leads to greater accuracy over most of the boundary, but to an oscillation of finite amplitude and decreasing wavelength in the immediate vicinity of the corner. The problem can be avoided at the cost of a more complex fundamental problem by extracting a

---

[1] M. I. G. Bloor and M. J. Wilson (2006), An approximate solution method for the biharmonic problem, *Proceedings of the Royal Society of London*, Vol. A462, pp. 1107–1121.

closed-form solution respresenting the discontinuous tractions. We shall discuss this special solution in §11.1.2 below.

Completeness

It is clear that the accuracy of the solution (however defined) can always be improved by taking more terms in the series (6.19) and in particular cases this is easily established numerically. However, it is more challenging to prove that the eigenfunction expansion is complete in the sense that any prescribed self-equilibrated traction on $x=0$ can be described to within an arbitrarily small error by taking a sufficient number of terms in the series, though experience with other eigenfunction expansions (e.g. with expansion of elastodynamic states of a structure in terms of normal modes) suggests that this will always be true. In fact, although the analysis described in this section has been known since the investigations by Papkovich and Fadle in the early 1940s, the formal proof of completeness was only completed by Gregory[2] in 1980.

It is worth noting that, as in many related problems, the eigenfunctions oscillate in y with increasing frequency as λ_i increases and in fact every time we increase i by 1, an extra zero appears in the function $g_i(y)$ in the range $0 < y < b$. Thus, there is a certain similarity to the process of approximating functions by Fourier series and in particular, the residual error in case of truncation will always cross zero once more than the last eigenfunction included.

This is also helpful in that it enables us to estimate the decay rate of the first excluded term. We see from equation (6.8) that the distance between zeros in y in any of the separate terms would be (π/λ_R), where λ_R is the real part of λ. It follows that over a corresponding distance in the x-direction, the field would decay by the factor $\exp(-\pi)=0.0432$. This suggests that we might estimate the decay rate of the end field by noting the distance between zeros in the corresponding tractions[3]. For example, the traction of equation (6.1) has zeros at $y=0, 3b/5$, corresponding to $\lambda_R = (5\pi/3b) = 5.23/b$.

An alternative way of estimating the decay rate is to note that the decay rate for the various terms in (6.19) increases with i and hence as x increases, the leading term will tend to predominate. The tractions of (6.1) are antisymmetric and hence

---

[2] R. D. Gregory (1979), Green's functions, bi-linear forms and completeness of the eigenfunctions for the elastostatic strip and wedge, *J. Elasticity,* Vol. 9, pp. 283–309; R. D. Gregory (1980), The semi-infinite strip $x \geq 0$, $-1 \leq y \leq 1$; completeness of the Papkovich-Fadle eigenfunctions when $\phi_{xx}(0, y)$, $\phi_{yy}(0, y)$ are prescribed, *J. Elasticity,* Vol. 10, pp. 57–80; R. D. Gregory (1980), The traction boundary-value problem for the elastostatic semi-infinite strip; existence of solution and completeness of the Papkovich-Fadle eigenfunctions, *J. Elasticity,* Vol. 10, pp. 295–327. These papers also include extensive references to earlier investigations of the problem.

[3] This assumes that the wavelength of the tractions is the same as that of ϕ, which of course is an approximation, since neither function is purely sinusoidal.

the leading term corresponds to the real part of the first eigenvalue of equation (6.18), which is found numerically to be $\lambda_R b = 3.7$.

Either way, we can conclude that the error associated with the end tractions in the approximate solution of §5.2.2 has decayed to around e^{-4}, i.e. to about 2% of the values at the end, within a distance b of the end. Thus the region affected by the end condition — the shaded region in Figure 6.1 — is quite small.

For problems which are *symmetric* in y, the leading self-equilibrated term is likely to have the form of Figure 6.3(a), which has a longer wavelength than the corresponding antisymmetric form, 6.3(b). The end effects in symmetric problems therefore decay more slowly and this is confirmed by the fact that the real part of the first eigenvalue of the symmetric equation (6.17) is only $\lambda_R b = 2.1$.

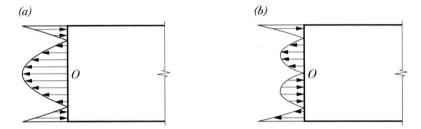

Fig. 6.3 Leading term in self-equilibrated tractions for (a) symmetric loading and (b) antisymmetric loading.

6.3 Other Saint-Venant Problems

The general strategy used in §6.2 can be applied to other curvilinear coördinate systems to correct the errors incurred by imposing the weak boundary conditions on appropriate edges. The essential steps are:-

(i) Define a coördinate system (ξ, η) such that the boundaries on which the strong conditions are applied are of the form, $\eta = $ constant.

(ii) Find a class of separated-variable biharmonic functions containing a parameter (λ in the above case).

(iii) Set up a system of four homogeneous equations for the coefficients of each function, based on the four traction-free boundary conditions for the corrective solution on the edges $\eta = $ constant.

(iv) Find the eigenvalues of the parameter for which the system has a non-trivial solution and the corresponding eigenfunctions, which are then used as the terms in an eigenfunction expansion to define a general form for the corrective field.

(v) Determine the coefficients in the eigenfunction expansion from the prescribed inhomogeneous boundary conditions on the end $\xi = $ constant.

6.4 Mathieu's Solution

The method described in §6.2 is particularly suitable for rectangular bodies of relatively large aspect ratio, since the weak solution is then quite accurate over a substantial part of the domain and the corrections at the two ends are essentially independent of each other. In other cases, and particularly for the square $b = a$, we lose these advantages and the inconvenience of solving the eigenvalue equations (6.17, 6.18) tilts the balance in favour of an alternative method due originally to Mathieu, who represented the solution of the entire problem as the sum of two orthogonal Fourier series of the form (5.84). The general problem can be decomposed into four sub-problems which are respectively either symmetric or antisymmetric with respect to the x- and y-axes. Following Meleshko and Gomilko[4], we restrict attention to the symmetric/symmetric case, defined by the boundary conditions

$$\sigma_{xx}(a, y) = \sigma_{xx}(-a, y) = f(y) ; \quad \sigma_{xy}(a, y) = -\sigma_{xy}(-a, y) = g(y) \quad (6.21)$$
$$\sigma_{yx}(x, b) = -\sigma_{yx}(x, -b) = h(x) ; \quad \sigma_{yy}(x, b) = \sigma_{yy}(x, -b) = \ell(x) , \quad (6.22)$$

where f, ℓ are even and g, h odd functions of their respective variables. We shall also assume that there is no discontinuity in shear traction of the form (6.20) at the corners and hence that $g(b) = h(a)$.

We define a biharmonic stress function with the appropriate symmetries as

$$\phi = A_0 y^2 + \sum_{m=1}^{\infty} (A_m y \sinh(\alpha_m y) + C_m \cosh(\alpha_m y)) \cos(\alpha_m x)$$

$$+ B_0 x^2 + \sum_{n=1}^{\infty} (B_n x \sinh(\beta_n x) + D_n \cosh(\beta_n x)) \cos(\beta_n y) . \quad (6.23)$$

As in §5.3.1, the α_m, β_n can be chosen so as to ensure that each term gives either zero shear tractions or zero normal tractions on two opposite edges, but not both. For example, with

$$\alpha_m = \frac{m\pi}{a} ; \quad \beta_n = \frac{n\pi}{b} , \quad (6.24)$$

---

[4] V. V. Meleshko and A. M. Gomilko (1997), Infinite systems for a biharmonic problem in a rectangle, *Proceedings of the Royal Society of London,* Vol. A453, pp. 2139–2160. The reader should be warned that there are several typographical errors in this paper, some of which are corrected in V. V. Meleshko and A. M. Gomilko (2004), Infinite systems for a biharmonic problem in a rectangle: further discussion, *Proceeedings of the Royal Society of London,* Vol. A460, pp. 807–819.

the second series makes no contribution to the shear traction on $y = b$, whilst the first series (that involving the coefficients A_m, C_m) makes no contribution to the shear traction on $x = a$. The shear traction on $x = a$ is then given by

$$\sigma_{xy}(a, y) = \sum_{n=1}^{\infty} \beta_n^2 \left(B_n \left[a \coth(\beta_n a) + \frac{1}{\beta_n} \right] + D_n \right) \sinh(\beta_n a) \sin(\beta_n y) \, ,$$

using (4.6), and substituting into $(6.21)_2$ and applying the Fourier inversion theorem[5], we have

$$B_n \left[a \coth(\beta_n a) + \frac{1}{\beta_n} \right] + D_n = \frac{1}{\beta_n^2 b \sinh(\beta_n a)} \int_{-b}^{b} g(y) \sin(\beta_n y) dy \, ,$$

which defines a linear relation between each individual pair of constants B_n, D_n. A similar relation between A_m, C_m can be obtained from $(6.22)_1$ and these conditions can then be used to eliminate C_m, D_n in (6.23), giving

$$\phi = A_0 y^2 - \sum_{m=1}^{\infty} A_m B(y, \alpha_m, b) \sinh(\alpha_m b) \cos(\alpha_m x)$$

$$+ B_0 x^2 - \sum_{n=1}^{\infty} B_n B(x, \beta_n, a) \sinh(\beta_n a) \cos(\beta_n y)$$

$$+ \sum_{n=1}^{\infty} \frac{g_n \cosh(\beta_n x) \cos(\beta_n y)}{\beta_n^2 b \sinh(\beta_n a)} + \sum_{m=1}^{\infty} \frac{h_m \cosh(\alpha_m y) \cos(\alpha_m x)}{\alpha_m^2 a \sinh(\alpha_m b)} \, ,$$

where

$$g_n = \int_{-b}^{b} g(y) \sin(\beta_n y) dy \, ; \quad h_m = \int_{-a}^{a} h(x) \sin(\alpha_m x) dx \, ,$$

and

$$B(z, \lambda, h) = \left[h \coth(\lambda h) + \frac{1}{\lambda} \right] \frac{\cosh(\lambda z)}{\sinh(\lambda h)} - z \frac{\sinh(\lambda z)}{\sinh(\lambda h)} \, .$$

The normal stress components are then obtained as

$$\sigma_{xx} = 2A_0 - \sum_{m=1}^{\infty} A_m \alpha_m^2 A(y, \alpha_m, b) \sinh(\alpha_m b) \cos(\alpha_m x)$$

$$+ \sum_{n=1}^{\infty} B_n \beta_n^2 B(x, \beta_n, a) \sinh(\beta_n a) \cos(\beta_n y)$$

$$- \sum_{n=1}^{\infty} \frac{g_n \cosh(\beta_n x) \cos(\beta_n y)}{b \sinh(\beta_n a)} + \sum_{m=1}^{\infty} \frac{h_m \cosh(\alpha_m y) \cos(\alpha_m x)}{a \sinh(\alpha_m b)} \, ,$$

---

[5] See for example equation (5.89).

$$\sigma_{yy} = 2B_0 + \sum_{m=1}^{\infty} A_m \alpha_m^2 B(y, \alpha_m, b) \sinh(\alpha_m b) \cos(\alpha_m x)$$

$$- \sum_{n=1}^{\infty} B_n \beta_n^2 A(x, \beta_n, a) \sinh(\beta_n a) \cos(\beta_n y)$$

$$+ \sum_{n=1}^{\infty} \frac{g_n \cosh(\beta_n x) \cos(\beta_n y)}{b \sinh(\beta_n a)} - \sum_{m=1}^{\infty} \frac{h_m \cosh(\alpha_m y) \cos(\alpha_m x)}{a \sinh(\alpha_m b)} ,$$

where

$$A(z, \lambda, h) = \left[h \coth(\lambda h) - \frac{1}{\lambda} \right] \frac{\cosh(\lambda z)}{\sinh(\lambda h)} - z \frac{\sinh(\lambda z)}{\sinh(\lambda h)} .$$

Imposing the remaining two boundary conditions (6.21)$_1$, (6.22)$_2$ and applying the Fourier inversion theorem to the resulting equations, we then obtain the infinite set of simultaneous equations

$$X_m b B(b, \alpha_m, b) = \sum_{n=1}^{\infty} \frac{4Y_n \alpha_m^2}{(\alpha_m^2 + \beta_n^2)^2} + H_m ,$$

$$Y_n a B(a, \beta_n, a) = \sum_{m=1}^{\infty} \frac{4X_m \beta_n^2}{(\alpha_m^2 + \beta_n^2)^2} + K_n ,$$

for $m, n \geq 1$, where

$$X_m = \frac{(-1)^{m+1} A_m \alpha_m^2 \sinh(\alpha_m b)}{b} ; \quad Y_n = \frac{(-1)^n B_n \beta_n^2 \sinh(\beta_n a)}{a}$$

$$H_m = \frac{(-1)^{m+1} [\ell_m + \coth(\alpha_m b) h_m]}{a} + \sum_{n=1}^{\infty} \frac{2(-1)^n \beta_n g_n}{(\alpha_m^2 + \beta_n^2) ab}$$

$$K_n = \frac{(-1)^n [f_n + \coth(\beta_n a) g_n]}{b} - \sum_{m=1}^{\infty} \frac{2(-1)^m \alpha_m h_m}{(\alpha_m^2 + \beta_n^2) ba}$$

and

$$f_n = \int_{-b}^{b} f(y) \cos(\beta_n y) dy ; \quad \ell_m = \int_{-a}^{a} \ell(x) \cos(\alpha_m x) dx .$$

Also, the zeroth-order Fourier inversion yields the constants A_0, B_0 as

$$A_0 = \frac{f_0}{4b} - \sum_{m=1}^{\infty} \frac{(-1)^m h_m}{2\alpha_m ba} ; \quad B_0 = \frac{\ell_0}{4a} - \sum_{n=1}^{\infty} \frac{(-1)^n g_n}{2\beta_n ab} .$$

With this solution, the coefficients X_m, Y_n generally tend to a common constant value G at large m, n and the series solution exhibits a non-vanishing error near the corners of the rectangle[6]. A more convergent solution can be obtained by defining new constants through the relations

$$X_m = \tilde{X}_m + G \; ; \quad Y_n = \tilde{Y}_n + G \; .$$

so that $\tilde{X}_m$, $\tilde{Y}_n$ decay with increasing m, n. The terms involving G can then be summed explicitly yielding the additional polynomial term

$$\phi_0 = \frac{G}{24} \left[(y^2 - b^2)^2 - (x^2 - a^2)^2 \right] \tag{6.25}$$

in the stress function ϕ. Unfortunately, the required value of G cannot be obtained in closed form except in certain special cases. However, an approximation can be obtained using the Rayleigh-Ritz method[7], leading to the result

$$G = \frac{45}{4(a^4 + b^4)} \left[\frac{1}{b} \int_{-b}^{b} f(y) \left(y^2 - \frac{b^2}{3} \right) dy - \frac{1}{a} \int_{-a}^{a} \ell(x) \left(x^2 - \frac{a^2}{3} \right) dx \right] . \tag{6.26}$$

This agrees with the exact value cited by Meleshko and Gomilko for the special case of the square $b = a$ with $\ell(x)$ a quadratic function of x.

Alternative series solutions

In the technique described above, the parameters α_m, β_n were chosen so as to simplify the satisfaction of the shear traction boundary conditions, after which the normal traction conditions led to an infinite set of algebraic equations. An alternative approach is to reverse this procedure by choosing

$$\alpha_m = \frac{(2m - 1)\pi}{2a} \; ; \quad \beta_n = \frac{(2n - 1)\pi}{2b} \tag{6.27}$$

and omitting the terms with A_0, B_0. Equation (6.23) then defines a stress field in which the first series makes no contribution to the *normal* tractions on $x = a$, and the second series makes no contribution to the normal tractions on $y = b$. These boundary conditions imply a one-to-one relation between A_m, C_m, after which the *shear* traction boundary conditions lead to an infinite set of algebraic equations. One advantage of this version is that it leads to a convergent solution even in the case where $g(b) \neq h(a)$ and hence the shear tractions are discontinuous in the corners[8], though convergence is slow in such cases. An alternative method of treating this

---

[6] Meleshko and Gomilko *loc. cit.*
[7] See §37.5 and Problem 37.5 below.
[8] Meleshko and Gomilko *loc. cit.*

discontinuity is to extract it explicitly, as discussed in §11.1.2 below, and then use the 'even' series defined by equations (6.24) for the corrective problem.

Problems

6.1. Show that if $\zeta = x + \imath y$ and $\sin(\zeta) - \zeta = 0$, where x, y are real variables, then

$$f(x) \equiv \cos x \sqrt{x^2 - \sin^2 x} + \sin x \ln(\sin x) - \sin x \ln\left(x + \sqrt{x^2 - \sin^2 x}\right) = 0 \,.$$

Using Maple or Mathematica to plot the function $f(x)$, find the first six roots of this equation and hence determine the first six values of $\lambda_R b$ for the antisymmetric mode.

6.2. Devise a method similar to that outlined in Problem 6.1 to determine the first six values of $\lambda_R b$ for the symmetric mode.

6.3. A displacement function representation for plane strain problems can be developed[9] in terms of two harmonic functions ϕ, ω in the form

$$2\mu u_x = \frac{\partial \phi}{\partial x} + y\frac{\partial \omega}{\partial x} \;;\;\; 2\mu u_y = \frac{\partial \phi}{\partial y} + y\frac{\partial \omega}{\partial y} - (3 - 4\nu)\omega$$

$$\sigma_{xx} = \frac{\partial^2 \phi}{\partial x^2} + y\frac{\partial^2 \omega}{\partial x^2} - 2\nu\frac{\partial \omega}{\partial y} \;;\;\; \sigma_{xy} = \frac{\partial^2 \phi}{\partial x \partial y} + y\frac{\partial^2 \omega}{\partial x \partial y} - (1 - 2\nu)\frac{\partial \omega}{\partial x}$$

$$\sigma_{yy} = \frac{\partial^2 \phi}{\partial y^2} + y\frac{\partial^2 \omega}{\partial y^2} - 2(1 - \nu)\frac{\partial \omega}{\partial y} \,.$$

Use this representation to formulate the eigenvalue problem of the long strip $x > 0$, $-b < y < b$ whose edges $y = \pm b$ are both bonded to a rigid body. Find the eigenvalue equation for symmetric and antisymmetric modes and comment on the expected decay rates for loading of the strip on the end $x = 0$.

6.4. Use the displacement function representation of Problem 6.3 to formulate the eigenvalue problem for the long strip $x > 0$, $-b < y < b$ whose edges $y = \pm b$ are in frictionless contact with a rigid body (so that the normal displacement is zero, but the frictional (tangential) traction is zero). Find the eigenvalue equation for symmetric and antisymmetric modes and comment on the expected decay rates for loading of the strip on the end $x = 0$.

6.5. Use the displacement function representation of Problem 6.3 to formulate the eigenvalue problem for the long strip $x > 0$, $-b < y < b$ which is bonded to a rigid surface at $y = -b$, the other long edge $y = b$ being traction free. Notice that this

---

[9] See §22.5.3.

problem is not symmetrical, so the problem will not partition into symmetric and antisymmetric modes.

6.6. Use Mathieu's method to approximate the stresses in the square $-a < x < a$, $-a < y < a$ subjected to the tractions

$$\sigma_{xx}(a, y) = \sigma_{xx}(-a, y) = \frac{Sy^2}{a^2} \, ,$$

all the remaining tractions being zero. Start by using (6.25, 6.26) to define a polynomial first approximation to the solution. Then use the series (6.23) to define a *corrective solution* — i.e. the stress function which when added to ϕ_0 defines the complete solution. The constants X_m, Y_n in the corrective solution will then decay with increasing m, n and a good approximation can be found by truncating the series at $m = n = 2$.

Chapter 7
Body Forces

A *body force* is defined as one which acts directly on the interior particles of the body, rather than on the boundary. Since the interior of the body is not accessible, it follows necessarily that body forces can only be produced by some kind of physical process which acts 'at a distance'. The commonest examples are forces due to gravity and magnetic or electrostatic attraction. In addition, we can formulate quasi-static elasticity problems for accelerating bodies in terms of body forces, using d'Alembert's principle (see §7.2.2 below).

7.1 Stress Function Formulation

We noted in §4.2.2 that the Airy stress function formulation satisfies the equilibrium equations if and only if the body forces are identically zero, but the method can be extended to the case of non-zero body forces provided the latter can be expressed as the gradient of a scalar potential, V.

We adopt the new definitions

$$\sigma_{xx} = \frac{\partial^2 \phi}{\partial y^2} + V \tag{7.1}$$

$$\sigma_{yy} = \frac{\partial^2 \phi}{\partial x^2} + V \tag{7.2}$$

$$\sigma_{xy} = = -\frac{\partial^2 \phi}{\partial x \partial y} \, , \tag{7.3}$$

Supplementary Information The online version contains supplementary material available at https://doi.org/10.1007/978-3-031-15214-6_7.

J. R. Barber, *Elasticity*, Solid Mechanics and Its Applications 172,
https://doi.org/10.1007/978-3-031-15214-6_7

in which case the two-dimensional equilibrium equations will be satisfied if and only if

$$p_x = -\frac{\partial V}{\partial x} \quad ; \quad p_y = -\frac{\partial V}{\partial y} , \tag{7.4}$$

i.e.

$$\boldsymbol{p} = -\nabla V . \tag{7.5}$$

Notice that the body force potential V appears in the definitions of the two normal stress components σ_{xx}, σ_{yy} (7.1, 7.2), but not in the shear stress σ_{xy} (7.3). Thus, the modification in these equations is equivalent to the addition of a biaxial hydrostatic tension of magnitude V, which is of course invariant under coördinate transformation (see §1.1.4).

7.1.1 Conservative vector fields

If a force field is capable of being represented as the gradient of a scalar potential, as in equation (7.5), it is referred to as *conservative*. This terminology arises from gravitational theory, since if we move a particle around in a gravitational field and if the force on the particle varies with position, the principle of conservation of energy demands that the work done in moving the particle from point A to point B should be path-independent — or equivalently, the work done in moving it around a closed path should be zero. If this were not the case, we could choose a direction for the particle to move around the path which would release energy and hence have an inexhaustible source of energy.

If all such integrals *are* path-independent, we can use the work done in bringing a particle from infinity to a given point as the definition of a unique local potential. Then, by equating the work done in an infinitesimal motion to the corresponding change in potential energy, we can show that the local force is proportional to the gradient of the potential, thus demonstrating that a conservative force field must be capable of a representation like (7.5). Conversely, if a given force field can be represented in this form, we can show by integration that the work done in moving a particle from A to B is proportional to $V(A) - V(B)$ and is therefore path-independent.

Not all body force fields are conservative and hence the formulation of §7.1 is not sufficiently general for all problems. However, we shall show below that most of the important problems involving body forces can be so treated.

We can develop a condition for a vector field to be conservative in the same way as we developed the compatibility conditions for strains. We argue that the two independent body force components p_x, p_y are defined in terms of a single scalar potential V and hence we can obtain a constraint equation on p_x, p_y by eliminating V between equations (7.4) with the result

$$\frac{\partial p_y}{\partial x} - \frac{\partial p_x}{\partial y} = 0 . \tag{7.6}$$

In three dimensions there are three equations like (7.6), from which we conclude that a vector field p is conservative if and only if curl $p = 0$. Another name for such fields is *irrotational*, since we note that if we replace p by the displacement vector u, the conditions like (7.6) are equivalent to the statement that the rotation ω is identically zero [*cf* equation (1.33)].

If the body force field satisfies equation (7.6), the corresponding potential can be recovered by partial integration. We shall illustrate this procedure in §7.2 below.

7.1.2 The compatibility condition

We demonstrated in §4.3.1 that, in the absence of body forces, the compatibility condition reduces to the requirement that the Airy stress function ϕ be biharmonic. This condition is modified when body forces are present.

We follow the same procedure as in §4.3.1, but use equations (7.1–7.3) in place of (4.6). Substituting into the compatibility equation in terms of stresses (4.7), we obtain

$$\frac{\partial^4 \phi}{\partial y^4} + \frac{\partial^2 V}{\partial y^2} - \nu \frac{\partial^4 \phi}{\partial x^2 \partial y^2} - \nu \frac{\partial^2 V}{\partial y^2} + 2(1+\nu)\frac{\partial^4 \phi}{\partial x^2 \partial y^2}$$
$$+ \frac{\partial^4 \phi}{\partial x^4} + \frac{\partial^2 V}{\partial x^2} - \nu \frac{\partial^4 \phi}{\partial x^2 \partial y^2} - \nu \frac{\partial^2 V}{\partial x^2} = 0 ,$$

i.e.

$$\nabla^4 \phi = -(1-\nu)\nabla^2 V . \tag{7.7}$$

Methods of obtaining suitable functions which satisfy this equation will be discussed in §7.3 below.

7.2 Particular Cases

It is worth noting that the vast majority of mechanical engineering components are loaded principally by boundary tractions rather than body forces. Of course, most components are subject to gravity loading, but the boundary loads are generally so much larger than sefl-weight that gravity can be neglected. This is less true for civil engineering structures such as buildings, where the self-weight of the structure may be much larger than the weight of the contents or wind loads, but even in this case it is important to distinguish between the gravity loading on the individual component and that transmitted to the component from other parts of the structure by way of boundary tractions.

It might be instructive at this point for the reader to draw free-body diagrams for a few common engineering components and identify the sources and relative magnitudes of the forces acting upon them. There are really comparatively few ways of applying a load to a body. By far the commonest is to push against it with another body — in other words to apply the load by *contact*. This is why contact problems occupy a central place in elasticity theory[1]. Significant loads may also be applied by fluid pressure as in the case of turbine blades or aircraft wings. Notice that it is fairly easy to apply a compressive normal traction to a boundary, but much harder to apply tension or shear.

This preamble might be taken as a justification for not studying the subject of body forces at all, but there are a few applications in which they are of critical importance, most notably those dynamic problems in which large accelerations occur. We shall develop expressions for the body force potential for some important cases in the following sections.

7.2.1 Gravitational loading

The simplest type of body force loading is that due to a gravitational field. If the problem is to remain two-dimensional, the direction of the gravitational force must lie in the xy-plane and we can choose it to be in the negative y-direction without loss of generality. The magnitude of the force will be ρg per unit volume, where ρ is the density of the material, so in the notation of §2.1 we have

$$p_x = 0 \; ; \; p_y = -\rho g \; .$$

This force field clearly satisfies condition (7.6) and is therefore conservative, and by inspection we note that it can be derived from the body force potential

$$V = \rho g y \; .$$

It also follows that $\nabla^2 V = 0$ and hence the stress function, ϕ for problems involving gravitational loading is biharmonic, from (7.6).

7.2.2 Inertia forces

It might be argued that Jean d'Alembert achieved immortality simply by moving a term from one side of an equation to the other, since *d'Alembert's principle* consists merely of writing Newton's second law of motion in the form $\boldsymbol{F} - m\boldsymbol{a} = 0$ and treating the term $-m\boldsymbol{a}$ as a fictitious force in order to reduce the dynamic problem to one in statics.

This simple process enables us to formulate elasticity problems for accelerating bodies as elastostatic body force problems, the corresponding body forces being

---

[1] See Chapters 12, 31, 34.

$$p_x = -\rho a_x \quad ; \quad p_y = -\rho a_y \, , \tag{7.8}$$

where a_x, a_y are the local components of acceleration, which may of course vary with position through the body.

7.2.3 Quasi-static problems

If the body were rigid, the accelerations of equation (7.8) would be restricted to those associated with rigid-body translation and rotation, but in a deformable body, the distance between two points can change, giving rise to additional, stress-dependent terms in the accelerations.

These two effects give qualitatively distinct behaviour, both mathematically and physically. In the former case, the accelerations will generally be defined *a priori* from the kinematics of the problem, which therefore reduces to an elasticity problem with body forces. We shall refer to such problems as *quasi-static*.

By contrast, when the accelerations associated with deformations are important, they are not known *a priori*, since the stresses producing the deformations are themselves part of the solution. In this case the kinematic and elastic problems are coupled and must be solved together. The resulting equations are those governing the propagation of elastic waves through a solid body and their study is known as *Elastodynamics*.

In this chapter, we shall restrict attention to quasi-static problems. As a practical point, we note that the characteristic time scale of elastodynamic problems is very short. For example it generally takes only a very short time for an elastic wave to traverse a solid. If the applied loads are applied gradually in comparison with this time scale, the quasi-static assumption generally gives good results. A measure of the success of this approximation is that it works quite well even for the case of elastic impact between bodies, which may have a duration of the order of a few milliseconds[2].

7.2.4 Rigid-body kinematics

The most general acceleration for a rigid body in the plane involves arbitrary translation and rotation. We choose a coördinate system fixed in the body and suppose that,

---

[2] For more information about elastodynamic problems, the reader is referred to the classical texts of J. D. Achenbach, *Wave Propagation in Elastic Solids*, North Holland, Amsterdam, 1973 and A. C. Eringen and E. S. Şuhubi, *Elastodynamics*, Academic Press, New York, 1975. For a more detailed discussion of the impact of elastic bodies, see J. R. Barber, *Contact Mechanics*, Springer, Cham, 2018, Chapter 20, K. L. Johnson, *Contact Mechanics*, Cambridge University Press, Cambridge, 1985, §11.4, W. Goldsmith, *Impact*, Arnold, London, 1960.

at some instant, the origin has velocity v_0 and acceleration a_0 and the body is rotating in the clockwise sense with absolute angular velocity Ω and angular acceleration $\dot\Omega$. The instantaneous acceleration of the point (r, θ) *relative to the origin* can then be written

$$a_r = -\Omega^2 r \; ; \quad a_\theta = -\dot\Omega r \, .$$

Transforming these results into the x, y-coördinate system and adding the acceleration of the origin, we obtain the components of acceleration of the point (x, y) as

$$a_x = a_{0x} - \Omega^2 x + \dot\Omega y \tag{7.9}$$
$$a_y = a_{0y} - \Omega^2 y - \dot\Omega x \tag{7.10}$$

and hence the corresponding body force field is

$$p_x = -\rho(a_{0x} - \Omega^2 x + \dot\Omega y) \tag{7.11}$$
$$p_y = -\rho(a_{0y} - \Omega^2 y - \dot\Omega x) \, . \tag{7.12}$$

The astute reader will notice that the case of gravitational loading can be recovered as a special case of these results by writing $a_{0y} = g$ and setting all the other terms to zero. In fact, a reasonable interpretation of the gravitational force is as a d'Alembert force consequent on resisting the gravitational acceleration. Notice that if a body is in free fall — i.e. if it is accelerating freely in a gravitational field and not rotating — there is no body force and hence no internal stress unless the boundaries are loaded.

Substitution of (7.11, 7.12) into (7.6) shows that the inertia forces due to rigid-body accelerations are conservative if and only if $\dot\Omega = 0$ — i.e. if the angular velocity is constant. We shall determine the body force potential for this special case. Methods of treating the problem with non-zero angular acceleration are discussed in §7.4 below.

From equations (7.4, 7.11, 7.12) with $\dot\Omega = 0$ we have

$$\frac{\partial V}{\partial x} = \rho(a_{0x} - \Omega^2 x) \tag{7.13}$$
$$\frac{\partial V}{\partial y} = \rho(a_{0y} - \Omega^2 y) \, , \tag{7.14}$$

and hence, on partial integration of (7.13)

$$V = \rho\left(a_{0x} x - \frac{1}{2}\Omega^2 x^2\right) + h(y) \, ,$$

where $h(y)$ is an arbitrary function of y only. Substituting this result into (7.14) we obtain the ordinary differential equation

$$\frac{dh}{dy} = \rho(a_{0y} - \Omega^2 y) \tag{7.15}$$

for $h(y)$, which has the general solution

$$h(y) = \rho \left(a_{0y} y - \frac{\Omega^2 y^2}{2} \right) + C ,$$

where C is an arbitrary constant which can be taken to be zero without loss of generality, since we are only seeking a *particular* potential function V.

The final expression for V is therefore

$$V = \rho \left(a_{0x} x + a_{0y} y - \frac{1}{2} \Omega^2 (x^2 + y^2) \right) . \tag{7.16}$$

The reader might like to try this procedure on a set of body forces which *do not* satisfy the condition (7.6). It will be found that the right-hand side of the ordinary differential equation like (7.15) then contains terms which depend on x and hence this equation cannot be solved for $h(y)$.

7.3 Solution for the Stress Function

Once the body force potential V has been determined, the next step is to find a suitable function ϕ, which satisfies the compatibility condition (7.7) and which defines stresses through equations (7.1–7.3) satisfying the boundary conditions of the problem. There are broadly speaking two ways of doing this. One is to choose some suitable form (such as a polynomial) without regard to equation (7.7) and then satisfy the constraint conditions resulting from (7.7) in the same step as those arising from the boundary conditions. The other is to seek a general solution of the inhomogeneous equation (7.7) and then determine the resulting arbitrary constants from the boundary conditions.

7.3.1 The rotating rectangular bar

As an illustration of the first method, we consider the problem of the rectangular bar $-a < x < a$, $-b < y < b$, rotating about the origin at constant angular velocity Ω, all the boundaries being traction free (see Figure 7.1).

The body force potential for this problem is obtained from equation (7.16) as

$$V = -\frac{1}{2} \rho \Omega^2 (x^2 + y^2) , \tag{7.17}$$

and hence the stress function must satisfy the equation

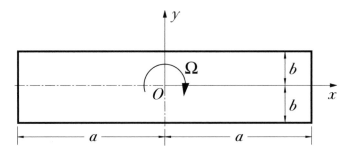

Fig. 7.1 The rotating rectangular bar.

$$\nabla^4\phi = 2\rho(1-\nu)\Omega^2 , \tag{7.18}$$

from (7.7, 7.17).

The geometry suggests a formulation in Cartesian coördinates and equation (7.18) leads us to expect a polynomial of degree 4 in x, y. We also note that V is even in both x and y and that the boundary conditions are homogeneous, so we propose the candidate stress function

$$\phi = A_1x^4 + A_2x^2y^2 + A_3y^4 + A_4x^2 + A_5y^2 , \tag{7.19}$$

which contains all the terms of degree 4 and below with the required symmetry.

The constants $A_1, \ldots, A_5$ will be determined from equation (7.18) and from the boundary conditions

$$\sigma_{yx} = 0 \; ; \quad y = \pm b \tag{7.20}$$

$$\sigma_{yy} = 0 \; ; \quad y = \pm b \tag{7.21}$$

$$\int_{-b}^{b} \sigma_{xx}dy = 0 \; ; \quad x = \pm a , \tag{7.22}$$

where we have applied the weak boundary conditions only on $x = \pm a$. Note also that the other two weak boundary conditions — that there should be no moment and no shear force on the ends — are satisfied identically in view of the symmetry of the problem about $y=0$.

As in the problem of §5.2.3, it is algebraically simpler to start by satisfying the strong boundary conditions (7.20, 7.21). Substituting (7.19) into (7.1–7.3), we obtain

$$\sigma_{xx} = 2A_2x^2 + 12A_3y^2 + 2A_5 - \frac{1}{2}\rho\Omega^2(x^2 + y^2) \tag{7.23}$$

$$\sigma_{yy} = 12A_1x^2 + 2A_2y^2 + 2A_4 - \frac{1}{2}\rho\Omega^2(x^2 + y^2) \tag{7.24}$$

$$\sigma_{yx} = -4A_2xy . \tag{7.25}$$

It follows that conditions (7.20, 7.21) will be satisfied for all x if and only if

$$A_1 = \rho \Omega^2 / 24 \; ; \quad A_2 = 0 \; ; \quad A_4 = \rho \Omega^2 b^2 / 4 \, . \qquad (7.26)$$

The constant A_3 can now be determined by substituting (7.19) into (7.18), with the result

$$24(A_1 + A_3) = 2\rho \Omega^2 (1 - \nu) \, ,$$

which is the inhomogeneous equivalent of the constraint equations (see §5.1), and hence

$$A_3 = \rho \Omega^2 (1 - 2\nu) / 24 \, ,$$

using (7.26).

Finally, we determine the remaining constant A_5 by substituting (7.23) into the weak boundary condition (7.22) and evaluating the integral, with the result

$$A_5 = \rho \Omega^2 \left(\frac{\nu b^2}{6} + \frac{a^2}{4} \right) \, .$$

The final stress field is therefore

$$\sigma_{xx} = \rho \Omega^2 \left[\frac{(a^2 - x^2)}{2} + \frac{\nu(b^2 - 3y^2)}{3} \right] \; ; \quad \sigma_{yy} = \frac{\rho \Omega^2}{2}(b^2 - y^2)$$

$$\sigma_{yx} = 0 \, ,$$

from equations (7.23–7.25).

Notice that the boundary conditions on the ends $x = \pm a$ agree with those of the physical problem except for the second term in σ_{xx}, which represents a symmetric self-equilibrated traction. From §6.2.2, we anticipate that the error due to this disagreement will be confined to regions near the ends of length comparable with the half-length, b.

7.3.2 *Solution of the governing equation*

Equation (7.18) is an inhomogeneous partial differential equation — i.e. it has a known function of x, y on the right-hand side — and it can be solved in the same way as an inhomogeneous *ordinary* differential equation, by first finding a particular solution of the equation and then superposing the general solution of the corresponding homogeneous equation.

In this context, a particular solution is *any* function ϕ that satisfies (7.18). It contains no arbitrary constants. The generality in the general solution comes from arbitrary constants in the *homogeneous* solution. Furthermore, the homogeneous solution is the solution of equation (7.18) modified to make the right-hand side zero — i.e.

$$\nabla^4 \phi = 0$$

Thus, the homogeneous solution is a general biharmonic function and we can summarize the solution method as containing the three steps:-

(i)   finding *any* function ϕ which satisfies (7.18);
(ii) superposing a sufficiently general biharmonic function (which therefore contains several arbitrary constants);
(iii) choosing the arbitrary constants to satisfy the boundary conditions.

In the problem of the preceding section, the particular solution would be any fourth degree polynomial satisfying (7.18) and the homogeneous solution, the most general fourth degree *biharmonic* function with the appropriate symmetry.

Notice that the particular solution (i) is itself a solution of a different physical problem — one in which the body forces are correctly represented, but the correct boundary conditions are not (usually) satisfied. Thus, the function of the homogeneous solution is to introduce additional degrees of freedom to enable us to satisfy the boundary conditions.

Physical superposition

It is often helpful to think of this superposition process in a physical rather than a mathematical sense. In other words, we devise a related problem in which the body forces are the same as in the real problem, but where the boundary conditions are simpler. For example, if a beam is subjected to gravitational loading, a simple physical 'particular' solution would correspond to the problem in which the beam is resting on a rigid foundation and hence the stress field is one-dimensional. To complete the real problem, we would then have to superpose the solution of a corrective problem with tractions equal to the difference between the required tractions and those implied by the particular solution, but with no body force, since this has already been taken into account in the particular solution.

One advantage of this way of thinking is that it is not restricted to problems in which the body force can be represented by a potential. We can therefore use it to solve the problem of the rotationally accelerating beam in the next section.

7.4   Rotational Acceleration

We saw in §7.2.4 that the body force potential cannot be used in problems where the angular acceleration is non-zero. In this section, we shall generate a particular solution for this problem and then generalize it, using the results without body force from Chapter 5.

7.4.1 The circular disk

Consider the rotationally symmetric problem of Figure 7.2. A solid circular disk, radius a, is initially at rest ($\Omega=0$) and at time $t=0$ it is caused to accelerate in a clockwise direction with angular acceleration $\dot{\Omega}$ by shear tractions S uniformly distributed around the edge $r=a$. Note that we could determine the magnitude of these tractions by writing the equation of motion for the disk, but it will not be necessary to do this — the result will emerge from the analysis of the stress field.

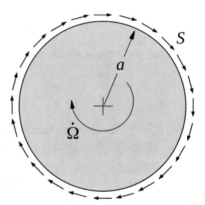

Fig. 7.2 The disk with rotational acceleration.

At any given time, the body forces and hence the resulting stress field will be the sum of two parts, one due to the instantaneous angular velocity and the other to the angular acceleration. The former can be obtained using the body force potential and we therefore concentrate here on the contribution due to the angular acceleration. This is the *only* body force at the beginning of the process, so the following solution can also be regarded as the solution for the instant, $t=0$.

The problem is clearly axisymmetric, so the stresses and displacements must depend on the radius r only, but it is also antisymmetric, since if the sign of the tractions were changed, the problem would become a mirror image of that illustrated. Now suppose some point in the disk had a non-zero outward radial displacement, u_r. Changing the sign of the tractions would change the sign of this displacement — i.e. make it be directed inwards — but this is impossible if the new problem is to be a mirror image of the old. We can therefore conclude from symmetry that $u_r=0$ throughout the disk. In the same way, we can conclude that the stress components

$\sigma_{rr}, \sigma_{\theta\theta}$ are zero everywhere. Incidentally, we also note that if the problem is conceived as being one of plane strain, symmetry demands that σ_{zz} be everywhere zero. It follows that this is one of those lucky problems in which plane stress and plane strain are the same and hence the plane stress assumption involves no approximation.

We conclude that there is only one non-zero displacement component, u_θ, and one non-zero strain component, $e_{r\theta}$. Thus, the number of strain and displacement components is equal and no non-trivial compatibility conditions can be obtained by eliminating the displacement components. (An alternative statement is that all the compatibility equations are satisfied identically, as can be verified by substitution — the compatibility equations in cylindrical polar coördinates are given by Saada[3]).

It follows that the only non-zero stress component, $\sigma_{r\theta}$ can be determined from equilibrium considerations alone. Considering the equilibrium of a small element due to forces in the θ-direction and dropping terms which are zero due to the symmetry of the system, we obtain

$$\frac{d\sigma_{r\theta}}{dr} + \frac{2\sigma_{r\theta}}{r} = -\rho r \dot{\Omega} , \qquad (7.27)$$

which has the general solution

$$\sigma_{r\theta} = -\frac{\rho\dot{\Omega}r^2}{4} + \frac{A}{r^2} ,$$

where A is an arbitrary constant which must be set to zero to retain continuity at the origin. Thus, the stress field in the disk of Figure 7.2 is

$$\sigma_{r\theta} = -\frac{\rho\dot{\Omega}r^2}{4} . \qquad (7.28)$$

In particular, the traction at the surface $r = a$ is

$$S = -\sigma_{r\theta}(a) = \frac{\rho\dot{\Omega}a^2}{4}$$

and the applied moment about the axis of rotation is therefore

$$M = 2\pi a^2 S = \frac{\pi\rho\dot{\Omega}a^4}{2} , \qquad (7.29)$$

since the traction acts over a length $2\pi a$ and the moment arm is a. We know from elementary dynamics that $M = I\dot{\Omega}$, where I is the moment of inertia and we can therefore deduce from (7.29) that

$$I = \frac{\pi\rho a^4}{2} ,$$

---

[3] A. S. Saada, *Elasticity*, Pergamon Press, New York, 1973, §6.9.

which is of course the correct expression for the moment of inertia of a solid disk of uniform density ρ, radius a and unit thickness. Notice however that we were able to deduce the relation between M and $\dot{\Omega}$ without using the equations of rigid-body dynamics. Equation (7.27) ensures that every particle of the body obeys Newton's second law, and this is sufficient to ensure that the complete body satisfies the equations of rigid-body dynamics.

7.4.2 The rectangular bar

We can use the stress field (7.28) as a particular solution for determining the stresses in a body of any shape due to angular acceleration. One way to think of this is to imagine cutting out the real body from an imaginary disk of sufficiently large radius. The stresses in the cut-out body will be the same as those in the disk, provided we arrange to apply tractions to the boundaries of the body that are equal to the stress components on those surfaces before the cut was made. These will not generally be the correct boundary tractions for the real problem, but we can correct the boundary tractions by superposing a homogeneous solution — i.e. a corrective solution for the actual body with prescribed boundary tractions (equal to the difference between those actually applied and those implied by the disk solution) but no body forces (since these have already been taken care of in the disk solution).

As an illustration, we consider the rectangular bar, $-a < x < a$, $-b < y < b$, accelerated by two equal shear forces, F, applied at the ends $x = \pm a$ as shown in Figure 7.3, the other boundaries, $y = \pm b$ being traction free.

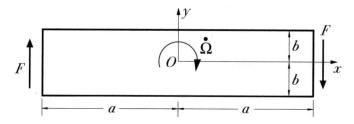

Fig. 7.3 The rectangular bar with rotational acceleration.

We first transform the disk solution into rectangular coördinates, using the transformation relations (1.9–1.11), obtaining

$$\sigma_{xx} = -2\sigma_{r\theta}\sin\theta\cos\theta = \frac{\rho\dot{\Omega}r^2\sin\theta\cos\theta}{2} = \frac{\rho\dot{\Omega}xy}{2} \tag{7.30}$$

$$\sigma_{yy} = 2\sigma_{r\theta}\sin\theta\cos\theta = -\frac{\rho\dot{\Omega}r^2\sin\theta\cos\theta}{2} = -\frac{\rho\dot{\Omega}xy}{2} \tag{7.31}$$

$$\sigma_{xy} = \sigma_{r\theta}(\cos^2\theta - \sin^2\theta) = -\frac{\rho\dot{\Omega}r^2(\cos^2\theta - \sin^2\theta)}{2} = -\frac{\rho\dot{\Omega}(x^2-y^2)}{4} . \tag{7.32}$$

This stress field is clearly odd in both x and y and involves normal tractions on $y=\pm b$ which vary linearly with x. The bending moment will therefore vary with x^3 suggesting a stress function ϕ with a leading term x^5y — i.e. a sixth degree polynomial.

The most general polynomial with the appropriate symmetry is

$$\phi = A_1x^5y + A_2x^3y^3 + A_3xy^5 + A_4x^3y + A_5xy^3 + A_6xy . \tag{7.33}$$

The stress components are the sum of those obtained from the homogeneous stress function (7.33) *using the definitions* (4.6) — remember the homogeneous solution here is one without body force — and those from the disk problem given by equations (7.30–7.32). We find

$$\sigma_{xx} = \frac{1}{2}\rho\dot{\Omega}xy + 6A_2x^3y + 20A_3xy^3 + 6A_5xy \tag{7.34}$$

$$\sigma_{yy} = -\frac{1}{2}\rho\dot{\Omega}xy + 20A_1x^3y + 6A_2xy^3 + 6A_4xy \tag{7.35}$$

$$\sigma_{xy} = -\frac{1}{4}\rho\dot{\Omega}(x^2-y^2) - 5A_1x^4 - 9A_2x^2y^2$$
$$-5A_3y^4 - 3A_4x^2 - 3A_5y^2 - A_6 . \tag{7.36}$$

The boundary conditions are

$$\sigma_{yx} = 0 \ ; \quad y = \pm b \tag{7.37}$$

$$\sigma_{yy} = 0 \ ; \quad y = \pm b \tag{7.38}$$

$$\int_{-b}^{b} y\sigma_{xx}dy = 0 \ ; \quad x = \pm a , \tag{7.39}$$

where we note that weak boundary conditions are imposed on the ends $x=\pm a$. Since the solution is odd in y, only the moment and shear force conditions are non-trivial and the latter need not be explicitly imposed, since they will be satisfied by global equilibrium (as in the example in §5.2.2).

Conditions (7.37, 7.38) have to be satisfied for all x and hence the corresponding coefficients of all powers of x must be zero. It follows immediately that $A_1=0$ and the remaining conditions can be written

$$-\frac{1}{2}\rho\dot{\Omega} + 6A_2b^2 + 6A_4 = 0$$

$$-\frac{1}{4}\rho\dot{\Omega} - 9A_2b^2 - 3A_4 = 0$$

$$\frac{1}{4}\rho\dot{\Omega}b^2 - 5A_3b^4 - 3A_5b^2 - A_6 = 0 \ .$$

We get one additional condition from the requirement that ϕ (equation (7.33)) be biharmonic

$$72A_2 + 120A_3 = 0$$

and another from the boundary condition (7.39), which with (7.34) yields

$$\frac{1}{6}\rho\dot{\Omega} + 2A_2a^2 + 4A_3b^2 + 2A_5 = 0 \ .$$

The solution of these equations is routine, giving the stress function

$$\phi = -\frac{\rho\dot{\Omega}}{60b^2}(5x^3y^3 - 3xy^5 - 10b^2x^3y + 11b^2xy^3 - 5a^2xy^3 + 15a^2b^2xy - 33b^4xy) \ .$$

The complete stress field, including the particular solution terms, is

$$\sigma_{xx} = \rho\dot{\Omega}xy\left[\frac{y^2}{b^2} - \frac{3}{5} + \frac{(a^2 - x^2)}{2b^2}\right] \ ; \quad \sigma_{yy} = \frac{\rho\dot{\Omega}xy}{2}\left(1 - \frac{y^2}{b^2}\right)$$

$$\sigma_{xy} = \rho\dot{\Omega}(b^2 - y^2)\left(\frac{y^2 + a^2 - 3x^2}{4b^2} - \frac{11}{20}\right) \ ,$$

from (7.34–7.36).

Finally, we can determine the forces F on the ends by integrating the shear traction, σ_{xy} over either end as

$$F = -\int_{-b}^{b}\sigma_{xy}(a)dy = \frac{2}{3}\rho\dot{\Omega}b(a^2 + b^2) \ . \tag{7.40}$$

Maple and Mathematica solutions of this problem are given in the files 'S742'. As in §7.4.1, the relation between the applied loading and the angular acceleration has been obtained without recourse to the equations of rigid-body dynamics. However, we note that the moment of inertia of the rectangular bar for rotation about the origin is $I = 4\rho ab(a^2 + b^2)/3$ and the applied moment is $M = 2Fa$, so application of the equation $M = I\dot{\Omega}$ leads to the same expression for F as that found in (7.40).

7.4.3   *Weak boundary conditions and the equation of motion*

We saw in §5.2.1 that weak boundary conditions need only be applied at one end of a stationary rectangular bar, since the stress field defined in terms of the Airy stress function must involve tractions that maintain the body in equilibrium. Similarly, in dynamic problems involving prescribed accelerations, an appropriate set of weak boundary conditions can be omitted. For example, we did not have to specify the value of F in the problem of Figure 7.3. Instead, its value was calculated after the stress field had been determined and necessarily proved to be consistent with the equations of rigid-body dynamics.

If a body is prevented from moving by a statically determinate support, it is natural to treat the support reactions as the 'neglected' weak boundary conditions. Thus, if the body is attached to a rigid support at one boundary, we apply no weak conditions at that boundary, as in §5.2.1. If the body is simply supported at a boundary, we impose the weak boundary condition that there be zero moment applied at the support, leading for example to the conditions stated in footnote 2 on Page 69.

Similar considerations apply in a dynamic problem if the body is attached to a support which moves in such a way as to prescribe the acceleration of the body. However, we may also wish to solve problems in which specified non-equilibrated loads are applied to an unsupported body. For example, we may be asked to determine the stresses in the body of Figure 7.3 due to prescribed end loads F. One way to do this would be first to solve a rigid-body dynamics problem to determine the accelerations and then proceed as in §7.4.2. However, in view of the present discussion, a more natural approach would be to include the angular acceleration $\dot{\Omega}$ as an unknown and use equation (7.40) to determine it in terms of F at the end of the solution procedure.

More generally, if we have a two-dimensional body subjected to prescribed tractions on all edges, we could assume the most general accelerations (7.9, 7.10) and solve the body force problem, treating $a_{0x}, a_{0y}, \dot{\Omega}$ as if they were known. If strong boundary conditions are imposed on two opposite edges and weak boundary conditions on *both* the remaining edges, it will then be found that there are three extra conditions which serve to determine the unknown accelerations.

Problems

7.1. Every particle of an elastic body of density ρ experiences a force

$$F = \frac{C\delta m}{r^2}$$

directed towards the origin, where C is a constant, r is the distance from the origin and δm is the mass of the particle. Find a body force potential V that satisfies these conditions.

7.2. Verify that the body force distribution

$$p_x = Cy \; ; \; p_y = -Cx$$

is non-conservative, by substituting into equation (7.6). Use the technique of §7.2.4 to attempt to construct a body force potential V for this case. Identify the step at which the procedure breaks down.

7.3. To construct a particular solution for the stress components in plane strain due to a non-conservative body force distribution, it is proposed to start by representing the displacement components in the form

$$u_x = \frac{1}{2\mu}\frac{\partial \psi}{\partial y} \; ; \; u_y = -\frac{1}{2\mu}\frac{\partial \psi}{\partial x} \; ; \; u_z = 0 \;.$$

Use the strain-displacement equations (1.37) and Hooke's law (1.54) to find expressions for the stress components in terms of ψ. Substitute these results into the equilibrium equations (2.1, 2.2) to find the governing equations for the stress function ψ.

What is the condition that must be satisfied by the body force distribution p if these equations are to have a solution? Show that this condition is satisfied if the body force distribution can be written in terms of a potential function W as

$$p_x = \frac{\partial W}{\partial y} \; ; \; p_y = -\frac{\partial W}{\partial x} \;.$$

For the special case

$$p_x = Cy \; ; \; p_y = -Cx \;,$$

find a particular solution for ψ in the axisymmetric polynomial form

$$\psi = A(x^2 + y^2)^n \;,$$

where A is a constant and n is an appropriate integer power. Show that this solution can be used to obtain the stress components (7.30–7.32). Suggest ways in which this method might be adapted to give a particular solution for more general non-conservative body force distributions.

7.4. If the elastic displacement u varies in time, there will generally be accelerations $a = \ddot{u}$ and hence body forces $p = -\rho\ddot{u}$, from equation (7.8). Use this result and equation (2.16) to develop the general equation of linear Elastodynamics.

Show that this equation is satisfied by a displacement field of the form

$$u_x = f(x - c_1 t) \; ; \; u_y = g(x - c_2 t) \; ; \; u_z = 0 \;,$$

where f, g are any functions and c_1, c_2 are two constants that depend on the material properties. Find the values of c_1, c_2 and comment on the physical significance of this solution.

7.5. The beam $-b < y < b$, $0 < x < L$ is built-in at the edge $x = L$ and loaded only by its own weight, ρg per unit volume.

Find a solution for the stress field, using weak conditions on the end $x = 0$.

7.6. One wall of a multistory building of height H is approximated as a thin plate $-b < x < b$, $0 < y < H$. During an earthquake, ground motion causes the building to experience a uniform acceleration a in the x-direction. Find the resulting stresses in the wall if the material has density ρ and the edges $x = \pm b$, $y = H$ can be regarded as traction free.

7.7. A tall thin rectangular plate $-a < x < a$, $-b < y < b$ $(b \gg a)$ is supported on the vertical edges $x = \pm a$ and loaded only by its own weight (density ρ). Find the stresses in the plate using weak boundary conditions on the horizontal edges $y = \pm b$ and assuming that the support tractions consist only of uniform shear.

7.8. Figure 7.4(a) shows a triangular cantilever, defined by the boundaries $y = 0$, $y = x \tan \alpha$ and built-in at $x = a$. It is loaded by its own weight, ρg per unit volume. Find a solution for the complete stress field and compare the maximum tensile bending stress with that predicted by the Mechanics of Materials theory.

Would the maximum tensile stress be lower if the alternative configuration of Figure 7.4(b) were used?

7.9. The thin rectangular plate $-a < x < a$, $-b < y < b$ with $a \gg b$ rotates about the y-axis at constant angular velocity Ω. All surfaces of the plate are traction free. Find a solution for the stress field, using strong boundary conditions on the long edges $y = \pm b$ and weak boundary conditions on the ends $x = \pm a$.

7.10. Solve Problem 7.9 for the case where $a \ll b$. In this case you should use strong boundary conditions on $x = \pm a$ and weak boundary conditions on $y = \pm b$.

7.11. A thin triangular plate bounded by the lines $y = \pm x \tan \alpha$, $x = a$ rotates about the axis $x = a$ at constant angular velocity Ω. The inclined edges of the plate $y =$

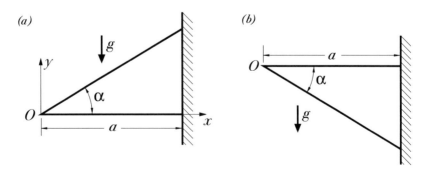

Fig. 7.4 A triangular cantilever

$\pm x \tan \alpha$ are traction free. Find a solution for the stress field, using strong boundary conditions on the inclined edges (weak boundary conditions will then be implied on $x=a$).

7.12. A thin triangular plate bounded by the lines $y= \pm x \tan \alpha$, $x=a$ is initially at rest, but is subjected to tractions at the edge $x=a$ causing an angular acceleration $\dot{\Omega}$ about the perpendicular axis through the point $(a, 0)$. The inclined surfaces $y = \pm x \tan \alpha$ are traction free. Find a solution for the instantaneous stress field, using the technique of §7.4.2.

7.13. The thin square plate $-a < x < a$, $-a < y < a$ rotates about the z-axis at constant angular velocity Ω. Use Mathieu's method (§6.4 and Problem 6.6) to develop a series solution that satisfies all the boundary conditions in the strong sense. Make a contour plot of the maximum principal stress based on truncating the series at $m=n=2$, and identify the location and magnitude of the maximum tensile stress. By what percentage does this value differ from the approximation obtained from §7.3.1?

Chapter 8
Problems in Polar Coördinates

Polar coördinates (r, θ) are particularly suited to problems in which the boundaries can be expressed in terms of equations like $r = a$, $\theta = \alpha$. This includes the stresses in a circular disk or around a circular hole, the curved beam with circular boundaries and the wedge, all of which will be discussed in this and subsequent chapters.

8.1 Expressions for Stress Components

We first have to transform the biharmonic equation (4.9) and the expressions for stress components (4.6) into polar coördinates, using the relations

$$x = r\cos\theta \; ; \quad y = r\sin\theta$$
$$r = \sqrt{x^2 + y^2} \; ; \quad \theta = \arctan\left(\frac{y}{x}\right) . \tag{8.1}$$

The derivation is tedious, but routine. We first note by differentiation that

$$\frac{\partial}{\partial x} = \frac{\partial r}{\partial x}\frac{\partial}{\partial r} + \frac{\partial\theta}{\partial x}\frac{\partial}{\partial\theta} = \cos\theta\frac{\partial}{\partial r} - \frac{\sin\theta}{r}\frac{\partial}{\partial\theta} \tag{8.2}$$

$$\frac{\partial}{\partial y} = \frac{\partial r}{\partial y}\frac{\partial}{\partial r} + \frac{\partial\theta}{\partial y}\frac{\partial}{\partial\theta} = \sin\theta\frac{\partial}{\partial r} + \frac{\cos\theta}{r}\frac{\partial}{\partial\theta} . \tag{8.3}$$

It follows that

Supplementary Information The online version contains supplementary material available at https://doi.org/10.1007/978-3-031-15214-6_8.

J. R. Barber, *Elasticity*, Solid Mechanics and Its Applications 172,
https://doi.org/10.1007/978-3-031-15214-6_8

$$\frac{\partial^2}{\partial x^2} = \left(\cos\theta \frac{\partial}{\partial r} - \frac{\sin\theta}{r} \frac{\partial}{\partial \theta} \right) \left(\cos\theta \frac{\partial}{\partial r} - \frac{\sin\theta}{r} \frac{\partial}{\partial \theta} \right)$$

$$= \cos^2\theta \frac{\partial^2}{\partial r^2} + \sin^2\theta \left(\frac{1}{r} \frac{\partial}{\partial r} + \frac{1}{r^2} \frac{\partial^2}{\partial \theta^2} \right)$$

$$+ 2\sin\theta\cos\theta \left(\frac{1}{r^2} \frac{\partial}{\partial \theta} - \frac{1}{r} \frac{\partial}{\partial r \partial \theta} \right) \tag{8.4}$$

and by a similar process, we find that

$$\frac{\partial^2}{\partial y^2} = \sin^2\theta \frac{\partial^2}{\partial r^2} + \cos^2\theta \left(\frac{1}{r} \frac{\partial}{\partial r} + \frac{1}{r^2} \frac{\partial^2}{\partial \theta^2} \right)$$

$$- 2\sin\theta\cos\theta \left(\frac{1}{r^2} \frac{\partial}{\partial \theta} - \frac{1}{r} \frac{\partial}{\partial r \partial \theta} \right) \tag{8.5}$$

$$\frac{\partial^2}{\partial x \partial y} = \sin\theta\cos\theta \left(\frac{\partial^2}{\partial r^2} - \frac{1}{r} \frac{\partial}{\partial r} - \frac{1}{r^2} \frac{\partial^2}{\partial \theta^2} \right)$$

$$- (\cos^2\theta - \sin^2\theta) \left(\frac{1}{r^2} \frac{\partial}{\partial \theta} - \frac{1}{r} \frac{\partial}{\partial r \partial \theta} \right) . \tag{8.6}$$

Finally, we can determine the expressions for stress components, noting for example that

$$\sigma_{rr} = \sigma_{xx} \cos^2\theta + \sigma_{yy} \sin^2\theta + 2\sigma_{xy} \sin\theta\cos\theta \tag{8.7}$$

$$= \cos^2\theta \frac{\partial^2\phi}{\partial y^2} + \sin^2\theta \frac{\partial^2\phi}{\partial x^2} - 2\sin\theta\cos\theta \frac{\partial^2\phi}{\partial x \partial y}$$

$$= \frac{1}{r} \frac{\partial\phi}{\partial r} + \frac{1}{r^2} \frac{\partial^2\phi}{\partial \theta^2} ,$$

after substituting for the partial derivatives from (8.4–8.6) and simplifying.

The remaining stress components, $\sigma_{\theta\theta}$, $\sigma_{r\theta}$ can be obtained by a similar procedure. We find

$$\sigma_{rr} = \frac{1}{r} \frac{\partial\phi}{\partial r} + \frac{1}{r^2} \frac{\partial^2\phi}{\partial \theta^2} \ ; \quad \sigma_{\theta\theta} = \frac{\partial^2\phi}{\partial r^2} \tag{8.8}$$

$$\sigma_{r\theta} = \frac{1}{r^2} \frac{\partial\phi}{\partial \theta} - \frac{1}{r} \frac{\partial^2\phi}{\partial r \partial \theta} = -\frac{\partial}{\partial r} \left(\frac{1}{r} \frac{\partial\phi}{\partial \theta} \right) . \tag{8.9}$$

If there is a conservative body force p described by a potential V, the stress components are modified to

$$\sigma_{rr} = \frac{1}{r} \frac{\partial\phi}{\partial r} + \frac{1}{r^2} \frac{\partial^2\phi}{\partial \theta^2} + V \ ; \quad \sigma_{\theta\theta} = \frac{\partial^2\phi}{\partial r^2} + V$$

$$\sigma_{r\theta} = \frac{1}{r^2} \frac{\partial\phi}{\partial \theta} - \frac{1}{r} \frac{\partial^2\phi}{\partial r \partial \theta} = -\frac{\partial}{\partial r} \left(\frac{1}{r} \frac{\partial\phi}{\partial \theta} \right) ,$$

where

$$p_r = -\frac{\partial V}{\partial r} \quad ; \quad p_\theta = -\frac{1}{r}\frac{\partial V}{\partial \theta} \ .$$

We also note that the Laplacian operator

$$\nabla^2 \equiv \frac{\partial^2}{\partial x^2} + \frac{\partial^2}{\partial y^2} = \frac{\partial^2}{\partial r^2} + \frac{1}{r}\frac{\partial}{\partial r} + \frac{1}{r^2}\frac{\partial^2}{\partial \theta^2} \ , \tag{8.10}$$

from equations (8.4, 8.5) and hence

$$\nabla^4 \phi \equiv \left(\frac{\partial^2}{\partial r^2} + \frac{1}{r}\frac{\partial}{\partial r} + \frac{1}{r^2}\frac{\partial^2}{\partial \theta^2} \right) \left(\frac{\partial^2 \phi}{\partial r^2} + \frac{1}{r}\frac{\partial \phi}{\partial r} + \frac{1}{r^2}\frac{\partial^2 \phi}{\partial \theta^2} \right) \ . \tag{8.11}$$

Notice that in applying the second Laplacian operator in (8.11) it is necessary to differentiate by parts. The two differential operators cannot simply be multiplied together. In Mathematica and Maple, the easiest technique is to define a new function

$$f \equiv \nabla^2 \phi = \frac{\partial^2 \phi}{\partial r^2} + \frac{1}{r}\frac{\partial \phi}{\partial r} + \frac{1}{r^2}\frac{\partial^2 \phi}{\partial \theta^2}$$

and then obtain $\nabla^4 \phi$ from

$$\nabla^4 \phi = \nabla^2 f = \frac{\partial^2 f}{\partial r^2} + \frac{1}{r}\frac{\partial f}{\partial r} + \frac{1}{r^2}\frac{\partial^2 f}{\partial \theta^2} \ .$$

8.2 Strain Components

A similar technique can be used to obtain the strain-displacement relations in polar coördinates. Writing

$$u_x = u_r \cos\theta - u_\theta \sin\theta$$
$$u_y = u_r \sin\theta + u_\theta \cos\theta$$

and substituting in (8.2), we find

$$e_{xx} = \frac{\partial u_x}{\partial x} = \frac{\partial u_r}{\partial r}\cos^2\theta + \left(\frac{u_\theta}{r} - \frac{\partial u_\theta}{\partial r} - \frac{1}{r}\frac{\partial u_r}{\partial \theta} \right) \sin\theta \cos\theta$$
$$+ \left(\frac{u_r}{r} + \frac{1}{r}\frac{\partial u_\theta}{\partial \theta} \right) \sin^2\theta \ .$$

Using the same method to obtain expressions for e_{xy}, e_{yy} and substituting the results in the strain-transformation relations analogous to (8.7) etc., we obtain the polar coördinate strain-displacement relations

$$e_{rr} = \frac{\partial u_r}{\partial r} \;\; ; \;\; e_{r\theta} = \frac{1}{2}\left(\frac{1}{r}\frac{\partial u_r}{\partial \theta} + \frac{\partial u_\theta}{\partial r} - \frac{u_\theta}{r}\right) \;\; ; \;\; e_{\theta\theta} = \frac{1}{r}\frac{\partial u_\theta}{\partial \theta} + \frac{u_r}{r} \; . \quad (8.12)$$

8.3   Fourier Series Expansion

The simplest problems in polar coördinates are those in which there are no θ-boundaries, the most general case being the disk with a central hole illustrated in Figure 8.1 and defined by $a < r < b$.

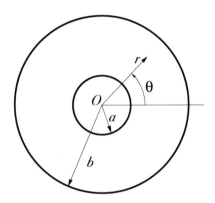

Fig. 8.1 The disk with a central hole.

The stresses and displacements must be single-valued and continuous and hence they must be periodic functions of θ, since $(r, \theta+2m\pi)$ defines the same point as (r, θ), when m is any integer. It is therefore natural to seek a general solution of the problem of Figure 8.1 in the form

$$\phi = \sum_{n=0}^{\infty} f_n(r)\cos(n\theta) + \sum_{n=1}^{\infty} g_n(r)\sin(n\theta) \; . \quad (8.13)$$

Substituting this expression into the biharmonic equation, using (8.11), we find that the functions f_n, g_n must satisfy the ordinary differential equation

$$\left(\frac{d^2}{dr^2} + \frac{1}{r}\frac{d}{dr} - \frac{n^2}{r^2}\right)\left(\frac{d^2 f_n}{dr^2} + \frac{1}{r}\frac{df_n}{dr} - \frac{n^2 f_n}{r^2}\right) = 0 \; , \quad (8.14)$$

which, for $n \neq 0, 1$ has the general solution

$$f_n(r) = A_{n1}r^{n+2} + A_{n2}r^{-n+2} + A_{n3}r^n + A_{n4}r^{-n} \; , \quad (8.15)$$

where $A_{n1}, \dots A_{n4}$ are four arbitrary constants.

When $n=0$, 1, the solution (8.15) develops repeated roots and equation (8.14) has a different form of solution given by

$$f_0(r) = A_{01}r^2 + A_{02}r^2 \ln(r) + A_{03} \ln(r) + A_{04} \qquad (8.16)$$
$$f_1(r) = A_{11}r^3 + A_{12}r \ln(r) + A_{13}r + A_{14}r^{-1} . \qquad (8.17)$$

In all the above equations, we note that the stress functions associated with the constants A_{n3}, A_{n4} are *harmonic* and hence biharmonic *a fortiori*, whereas those associated with A_{n1}, A_{n2} are biharmonic but not harmonic.

8.3.1 Satisfaction of boundary conditions

The boundary conditions on the surfaces $r=a, b$ will generally take the form

$$\sigma_{rr} = F_1(\theta) \; ; \;\; r = a \qquad (8.18)$$
$$= F_2(\theta) \; ; \;\; r = b \qquad (8.19)$$
$$\sigma_{r\theta} = F_3(\theta) \; ; \;\; r = a \qquad (8.20)$$
$$= F_4(\theta) \; ; \;\; r = b , \qquad (8.21)$$

each of which has to be satisfied for all values of θ. This can conveniently be done by expanding the functions $F_1, \ldots F_4$ as Fourier series in θ. i.e.

$$F_j(\theta) = \sum_{n=0}^{\infty} C_{nj} \cos(n\theta) + \sum_{n=1}^{\infty} D_{nj} \sin(n\theta) \; ; \;\; j = 1, \ldots, 4 . \qquad (8.22)$$

In combination with the stress function of equation (8.13), (8.18–8.21) will then give four independent equations for each trigonometric term, $\cos(n\theta)$, $\sin(n\theta)$ in the series and hence serve to determine the four constants A_{n1}, A_{n2}, A_{n3}, A_{n4}. The problem of Figure 8.1 is therefore susceptible of a general solution.

The general solution has some anomolous features for the special values $n = 0$, 1, but before discussing these we shall illustrate the method by solving a simple example.

8.3.2 Circular hole in a shear field

Figure 8.2 shows a large plate in a state of pure shear $\sigma_{xy}=S$, perturbed by a hole of radius a.

A formal statement of the boundary conditions for this problem is

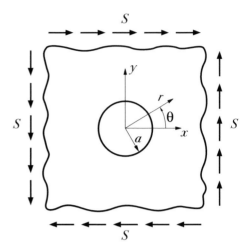

Fig. 8.2 Circular hole in a shear field.

$$\sigma_{rr} = 0 \ ; \quad r = a \tag{8.23}$$

$$\sigma_{r\theta} = 0 \ ; \quad r = a \tag{8.24}$$

$$\sigma_{xx}, \sigma_{yy} \to 0 \ ; \quad r \to \infty \tag{8.25}$$

$$\sigma_{xy} \to S \ ; \quad r \to \infty \ . \tag{8.26}$$

Notice that we describe a body as 'large' when we mean to interpret the boundary condition at the outer boundary as being applied at $r \to \infty$. Notice also that we have stated these 'infinite' boundary conditions in (x, y) coördinates, since this is the most natural way to describe a state of uniform stress. We shall see that the stress function approach makes it possible to use both rectangular and polar coördinates in the same problem without any special difficulty.

This is a typical *perturbation* problem in which a simple state of stress is perturbed by a local geometric feature (in this case a hole). It is reasonable in such cases to anticipate that the stresses distant from the hole will be unperturbed and that the effect of the hole will only be felt at moderate values of r/a. Perturbation problems are most naturally approached by first solving the simpler problem in which the perturbation is absent (in this case the plate *without* a hole) and then seeking a corrective solution which will describe the influence of the hole on the stress field. With such a formulation, we anticipate that the corrective solution will decay with increasing r.

The unperturbed field is clearly a state of uniform shear, $\sigma_{xy} = S$, and this in turn is conveniently described by the stress function

$$\phi = -Sxy = -Sr^2 \sin\theta \cos\theta = -\frac{Sr^2 \sin(2\theta)}{2} \ , \tag{8.27}$$

from equation (4.6). Notice that although the stress function is originally determined in rectangular coördinates, it is easily transformed into polar coördinates using (8.1) and then into the Fourier form of equation (8.13).

The unperturbed solution satisfies the 'infinite' boundary conditions (8.25, 8.26), but will violate the conditions at the hole surface (8.23, 8.24). However, we can correct the stress field by superposing those terms from the series (8.13) which (i) have the same Fourier dependence as (8.27) and (ii) lead to stresses which decay as r increases. There will always be two such terms for any given Fourier component, permitting the two traction boundary conditions to be satisfied. In the present instance, the required terms are those derived from the constants A_{22}, A_{24} in equation (8.15), giving the stress function[1]

$$\phi = -\frac{Sr^2 \sin(2\theta)}{2} + A \sin(2\theta) + \frac{B \sin(2\theta)}{r^2} .$$

The corresponding stress components are obtained by substituting in (8.8, 8.9) with the result

$$\sigma_{rr} = \left(S - \frac{4A}{r^2} - \frac{6B}{r^4} \right) \sin(2\theta) ; \quad \sigma_{r\theta} = \left(S + \frac{2A}{r^2} + \frac{6B}{r^4} \right) \cos(2\theta)$$

$$\sigma_{\theta\theta} = \left(-S + \frac{6B}{r^4} \right) \sin(2\theta) .$$

The two boundary conditions at the hole surface (8.23, 8.24) then yield the two equations

$$\frac{4A}{a^2} + \frac{6B}{a^4} = S ; \quad \frac{2A}{a^2} + \frac{6B}{a^4} = -S$$

for the constants A, B, with solution

$$A = Sa^2 ; \quad B = -\frac{Sa^4}{2} ,$$

and the final stress field is

$$\sigma_{rr} = S \left(1 - 4\frac{a^2}{r^2} + 3\frac{a^4}{r^4} \right) \sin(2\theta) ; \quad \sigma_{r\theta} = S \left(1 + 2\frac{a^2}{r^2} - 3\frac{a^4}{r^4} \right) \cos(2\theta)$$

$$\sigma_{\theta\theta} = S \left(-1 - 3\frac{a^4}{r^4} \right) \sin(2\theta) . \tag{8.28}$$

Notice incidentally that the maximum stress is the *hoop* stress, $\sigma_{\theta\theta} = 4S$ at the point $(a, 3\pi/4)$. At this point there is a state of uniaxial tension, so the maximum *shear* stress is $2S$. Since the unperturbed shear stress has a magnitude S, we say that the hole produces a *stress concentration* of 2. However, for a brittle material,

---

[1] It might be thought that the term involving A_{22} is inappropriate because it does not decay with r. However, it leads to *stresses* which decay with r, which is of course what we require.

we might be inclined to define the stress-concentration factor as the ratio of the maximum tensile stresses in the perturbed and unperturbed solutions, which in the present problem is 4. In general, a stress-concentration factor implies a measure of the severity of the stress field — usually a failure theory — which serves as a standard of comparison and the magnitude of the stress-concentration factor will depend upon the measure used.

The determination of the stress concentration due to holes, notches and changes of section under various loading conditions is clearly a question of considerable practical importance. An extensive discussion of problems of this kind is given by Savin[2] and stress-concentration factors for a wide range of geometries are tabulated by Peterson[3] in a form suitable for use in engineering design.

8.3.3 Degenerate cases

We have already remarked in §8.3 above that the solution (8.15) degenerates for $n = 0, 1$ and must be supplemented by some additional terms. In fact, even the modified stress function of (8.16, 8.17) is degenerate because the stress function

$$\phi = A + Bx + Cy = A + Br \cos \theta + Cr \sin \theta \qquad (8.29)$$

defines a trivial null state of stress (see §5.1) and hence the constants A_{04}, A_{13} in equations (8.16, 8.17) correspond to null stress fields.

This question of degeneracy arises elsewhere in Elasticity and indeed in mathematics generally, so we shall take this opportunity to develop a general technique for resolving it. As we saw in §8.3, the degeneracy often arises from the occurrence of repeated roots to an equation. Thus, in equation (8.15), the terms $A_{n2}r^{-n+2}$, $A_{n3}r^n$ degenerate to the same form when $n = 1$.

Suppose for the moment that we relax the restriction that n be an integer. Clearly the degeneracy in equation (8.15) only arises exactly at the values $n = 0, 1$. There is no degeneracy for $n = 1 + \epsilon$ for any non-zero ϵ, however small. We shall show therefore that we can recover the extra solution at the degenerate point by allowing the solution to tend smoothly to the limit $n \to 1$, rather than setting $n = 1$ *ab initio*.

For $n = 1 + \epsilon$, the two offending terms in (8.15) can be written

$$f(r) = Ar^{1-\epsilon} + Br^{1+\epsilon} ,$$

where A, B, are two arbitrary constants.

Clearly the two terms tend to the same form as $\epsilon \to 0$. However, suppose we construct a new function from the sum and difference of these functions in the form

---

[2] G. N. Savin, *Stress Concentration around Holes*, Pergamon Press, Oxford, 1961.

[3] R. E. Peterson, *Stress Concentration Design Factors*, John Wiley, New York, 1974.

$$f(r) = C(r^{1+\epsilon} + r^{1-\epsilon}) + D(r^{1+\epsilon} - r^{1-\epsilon}) .$$

In this form, the first function tends to $2Cr$ as $\epsilon \to 0$, whilst the second tends to zero. However, we can prevent the second term degenerating to zero, since the constant D is arbitrary. We can therefore choose $D = E\epsilon^{-1}$ in which case the second term will tend to the limit

$$\lim_{\epsilon \to 0} E \left(\frac{r^{1+\epsilon} - r^{1-\epsilon}}{\epsilon} \right) = 2Er \ln(r) , \tag{8.30}$$

where we have used l'Hôpital's rule to evaluate the limit. Notice that this result agrees with the special term included in equation (8.17) above to resolve the degeneracy in ϕ for $n = 1$.

This procedure can be generalized as follows: Whenever a degeneracy occurs at a denumerable set of values of a parameter (such as n in equation (8.15)), we can always make up the defect using additional terms obtained by differentiating the original form with respect to the parameter *before* allowing it to take the special value.

The pairs r^{n+2}, r^{-n+2} and r^n, r^{-n} both degenerate to the same form for $n = 0$, so the new functions are of the form

$$\lim_{n \to 0} \frac{d(r^{n+2})}{dn} = r^2 \ln(r) ; \qquad \lim_{n \to 0} \frac{d(r^n)}{dn} = \ln(r) ,$$

again agreeing with the special terms introduced in (8.16).

Sometimes the degeneracy is of a higher order, corresponding to more than two identical roots, in which case l'Hôpital's rule has to be applied more than once — i.e. we have to differentiate more than once with respect to the parameter. It is always a straightforward matter to check the resulting solutions to make sure they are of the required form.

We now turn our attention to the degeneracy implied by the triviality of the stress function of equation (8.29). Here, the stress function *itself* is not degenerate — i.e. it is a legitimate function of the required Fourier form and it is linearly independent of the other functions of the same form. The trouble is that it gives a null stress field. We must therefore look for stress functions that are *not* of the standard Fourier form, but which give Fourier type stress components.

Once again, the method is to approach the solution as a limit using l'Hôpital's rule, but this time we have to operate on the complete stress function — not just on the part that varies with r. Suppose we consider the axisymmetric degenerate term, $\phi = A$, which can be regarded as the limit of a stress function of the form

$$\phi = Ar^{\epsilon} \cos(\epsilon\theta) + Br^{\epsilon} \sin(\epsilon\theta) ,$$

as $\epsilon \to 0$.

Differentiating this function with respect to ϵ and then letting ϵ tend to zero,we obtain the new function

$$\phi = A \ln(r) + B\theta .$$

The first term is the same one that we found by operating on the function $f_n(r)$ and is of the correct Fourier form, but the second term is not appropriate to a Fourier series. However, when we substitute the second term into equations (8.8, 8.9), we obtain the stress components

$$\sigma_{rr} = \sigma_{\theta\theta} = 0 \quad ; \quad \sigma_{r\theta} = \frac{B}{r^2} ,$$

which *are* of the required Fourier form (for $n=0$), since the stress components do not vary with θ.

In the same way, we can develop special terms to make up the deficit due to the two null terms with $n=1$ [equation (8.29)], obtaining the new stress functions

$$\phi = Br\theta \sin\theta + Cr\theta \cos\theta ,$$

which generate the Fourier-type stress components

$$\sigma_{r\theta} = \sigma_{\theta\theta} = 0 \quad ; \quad \sigma_{rr} = \frac{2B \cos\theta}{r} - \frac{2C \sin\theta}{r} .$$

8.4   The Michell Solution

The preceding results now permit us to write down a general solution of the elasticity problem in polar coördinates, such that the stress components form a Fourier series in θ. We have

$$\begin{aligned}
\phi = {} & A_{01}r^2 + A_{02}r^2 \ln(r) + A_{03} \ln(r) + A_{04}\theta \\
& + \left(A_{11}r^3 + A_{12}r \ln(r) + A_{14}r^{-1} \right) \cos\theta + A_{13}r\theta \sin\theta \\
& + \left(B_{11}r^3 + B_{12}r \ln(r) + B_{14}r^{-1} \right) \sin\theta + B_{13}r\theta \cos\theta \\
& + \sum_{n=2}^{\infty} \left(A_{n1}r^{n+2} + A_{n2}r^{-n+2} + A_{n3}r^n + A_{n4}r^{-n} \right) \cos(n\theta) \\
& + \sum_{n=2}^{\infty} \left(B_{n1}r^{n+2} + B_{n2}r^{-n+2} + B_{n3}r^n + B_{n4}r^{-n} \right) \sin(n\theta) .
\end{aligned}$$

$$(8.31)$$

This solution is due to Michell[4]. The corresponding stress components are easily obtained by substituting into equations (8.8, 8.9). For convenience, we give them in tabular form in Table 8.1, since we shall often wish to select a few components from the general solution in the solution of specific problems. The terms in the Michell solution are also given in the Maple and Mathematica files 'Michell', from which appropriate terms can be cut and pasted as required.

Table 8.1 The Michell solution — stress components

ϕ	σ_{rr}	$\sigma_{r\theta}$	$\sigma_{\theta\theta}$
r^2	2	0	2
$r^2 \ln(r)$	$2\ln(r)+1$	0	$2\ln(r)+3$
$\ln(r)$	$1/r^2$	0	$-1/r^2$
θ	0	$1/r^2$	0
$r^3 \cos\theta$	$2r\cos\theta$	$2r\sin\theta$	$6r\cos\theta$
$r\theta\sin\theta$	$2\cos\theta/r$	0	0
$r\ln(r)\cos\theta$	$\cos\theta/r$	$\sin\theta/r$	$\cos\theta/r$
$\cos\theta/r$	$-2\cos\theta/r^3$	$-2\sin\theta/r^3$	$2\cos\theta/r^3$
$r^3 \sin\theta$	$2r\sin\theta$	$-2r\cos\theta$	$6r\sin\theta$
$r\theta\cos\theta$	$-2\sin\theta/r$	0	0
$r\ln(r)\sin\theta$	$\sin\theta/r$	$-\cos\theta/r$	$\sin\theta/r$
$\sin\theta/r$	$-2\sin\theta/r^3$	$2\cos\theta/r^3$	$2\sin\theta/r^3$
$r^{n+2}\cos n\theta$	$-(n+1)(n-2)r^n\cos n\theta$	$n(n+1)r^n\sin n\theta$	$(n+1)(n+2)r^n\cos n\theta$
$r^{-n+2}\cos n\theta$	$-(n+2)(n-1)r^{-n}\cos n\theta$	$-n(n-1)r^{-n}\sin n\theta$	$(n-1)(n-2)r^{-n}\cos n\theta$
$r^n\cos n\theta$	$-n(n-1)r^{n-2}\cos n\theta$	$n(n-1)r^{n-2}\sin n\theta$	$n(n-1)r^{n-2}\cos n\theta$
$r^{-n}\cos n\theta$	$-n(n+1)r^{-n-2}\cos n\theta$	$-n(n+1)r^{-n-2}\sin n\theta$	$n(n+1)r^{-n-2}\cos n\theta$
$r^{n+2}\sin n\theta$	$-(n+1)(n-2)r^n\sin n\theta$	$-n(n+1)r^n\cos n\theta$	$(n+1)(n+2)r^n\sin n\theta$
$r^{-n+2}\sin n\theta$	$-(n+2)(n-1)r^{-n}\sin n\theta$	$n(n-1)r^{-n}\cos n\theta$	$(n-1)(n-2)r^{-n}\sin n\theta$
$r^n\sin n\theta$	$-n(n-1)r^{n-2}\sin n\theta$	$-n(n-1)r^{n-2}\cos n\theta$	$n(n-1)r^{n-2}\sin n\theta$
$r^{-n}\sin n\theta$	$-n(n+1)r^{-n-2}\sin n\theta$	$n(n+1)r^{-n-2}\cos n\theta$	$n(n+1)r^{-n-2}\sin n\theta$

Notice that there are four independent stress functions for each term in the Fourier series, as required by the argument of §8.3.1. If the disk is solid, there is no inner boundary and we must exclude those components in Table 8.1 that give stresses which go to infinity as $r \to 0$.

In addition to the special stress functions necessitated by the degeneracy discussed in §8.3.3, the cases $n=0$, 1 exhibit other anomolies. In particular, some of the stress functions for $n=0$, 1 correspond to multiple-valued displacement fields and cannot be used for the complete annulus of Figure 8.1. Also, equilibrium requirements place restrictions on the permissible boundary conditions for the terms $n=0$, 1 in the Fourier series (8.22).

---

[4] J. H. Michell (1899), On the direct determination of stress in an elastic solid, with application to the theory of plates, *Proceedings of the London Mathematical Society,* Vol. 31, pp. 100–124.

Detailed discussion of these difficulties will be postponed to §9.3.1, after we have introduced methods of determining the displacements associated with a given stress field. In the present chapter, we shall restrict attention to cases where these difficulties do not arise.

8.4.1   Hole in a tensile field

To illustrate the use of Table 8.1, we consider the case where the the body of Figure 8.2 is subjected to uniform tension at infinity instead of shear, so that the boundary conditions become

$$\sigma_{rr} = 0 \ ; \ \ r = a \tag{8.32}$$

$$\sigma_{r\theta} = 0 \ ; \ \ r = a \tag{8.33}$$

$$\sigma_{xy}, \sigma_{yy} \to 0 \ ; \ \ r \to \infty \tag{8.34}$$

$$\sigma_{xx} \to S \ ; \ \ r \to \infty . \tag{8.35}$$

The unperturbed problem in this case can clearly be described by the stress function

$$\phi = \frac{Sy^2}{2} = \frac{Sr^2 \sin^2 \theta}{2} = \frac{Sr^2}{4} - \frac{Sr^2 \cos(2\theta)}{4} .$$

This function contains both an axisymmetric term and a $\cos(2\theta)$ term, so to complete the solution, we supplement it with those terms from Table 8.1 which have the same form and for which the stresses decay as $r \to \infty$.

The resulting stress function is

$$\phi = \frac{Sr^2}{4} - \frac{Sr^2 \cos(2\theta)}{4} + A \ln(r) + B\theta + C \cos(2\theta) + \frac{D \cos(2\theta)}{r^2}$$

and the corresponding stress components are

$$\sigma_{rr} = \frac{S}{2} + \frac{S \cos(2\theta)}{2} + \frac{A}{r^2} - \frac{4C \cos(2\theta)}{r^2} - \frac{6D \cos(2\theta)}{r^4}$$

$$\sigma_{r\theta} = -\frac{S \sin(2\theta)}{2} + \frac{B}{r^2} - \frac{2C \sin(2\theta)}{r^2} - \frac{6D \sin(2\theta)}{r^4}$$

$$\sigma_{\theta\theta} = \frac{S}{2} - \frac{S \cos(2\theta)}{2} - \frac{A}{r^2} + \frac{6D \cos(2\theta)}{r^4} .$$

The boundary conditions (8.32, 8.33) will be satisfied if and only if the coefficients of both Fourier terms are zero on $r = a$ and hence we obtain the equations

$$\frac{S}{2} + \frac{A}{a^2} = 0$$

$$\frac{B}{a^2} = 0$$

$$\frac{S}{2} - \frac{4C}{a^2} - \frac{6D}{a^4} = 0$$

$$-\frac{S}{2} - \frac{2C}{a^2} - \frac{6D}{a^4} = 0 \,,$$

which have the solution

$$A = -\frac{Sa^2}{2} \;;\; B = 0 \;;\; C = \frac{Sa^2}{2} \;;\; D = -\frac{Sa^4}{4} \,.$$

The final stress field is then obtained as

$$\sigma_{rr} = \frac{S}{2}\left(1 - \frac{a^2}{r^2}\right) + \frac{S\cos(2\theta)}{2}\left(\frac{3a^4}{r^4} - \frac{4a^2}{r^2} + 1\right)$$

$$\sigma_{r\theta} = \frac{S\sin(2\theta)}{2}\left(\frac{3a^4}{r^4} - \frac{2a^2}{r^2} - 1\right)$$

$$\sigma_{\theta\theta} = \frac{S}{2}\left(1 + \frac{a^2}{r^2}\right) - \frac{S\cos(2\theta)}{2}\left(\frac{3a^4}{r^4} + 1\right)\,.$$

The maximum tensile stress is $\sigma_{\theta\theta} = 3S$ at $(a, \pi/2)$ and hence the stress-concentration factor for tensile loading is 3. The maximum stress point and the unperturbed region at infinity are both in uniaxial tension and hence the stress-concentration factor will be the same whatever criterion is used to measure the severity of the stress state. This contrasts with the problem of §8.3.2, where the maximum stress state is uniaxial tension, but the unperturbed field is pure shear.

Problems

8.1. Use equations (8.2, 8.3) to find an expression for the rotation ω of equation (1.32) in polar coördinates (r, θ).

8.2. A large plate with a small central hole of radius a is subjected to in-plane hydrostatic compression $\sigma_{xx} = \sigma_{yy} = -S$, $\sigma_{xy} = 0$ at the remote boundaries. Find the stress field in the plate if the surface of the hole is traction free.

8.3. A large rectangular plate is loaded in such a way as to generate the unperturbed stress field

$$\sigma_{xx} = Cy^2 \;;\; \sigma_{yy} = -Cx^2 \;;\; \sigma_{xy} = 0 \,.$$

The plate contains a small traction-free circular hole of radius a centred on the origin. Find the perturbation in the stress field due to the hole.

8.4. A thin disc of radius b has a small central hole of radius a ($\ll b$). If the disc rotates about its axis at uniform speed Ω, show that the maximum tensile stress is twice as large as that in a similar disc without a hole rotating at the same speed.

8.5. Figure 8.3 shows a thin uniform circular disk, which rotates at constant speed Ω about the diametral axis $y=0$, all the surfaces being traction free. Determine the complete stress field in the disk.

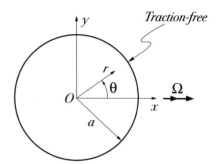

Fig. 8.3 Thin disk rotating about a diametral axis.

8.6. A series of experiments is conducted in which a thin plate is subjected to biaxial tension/compression, σ_1, σ_2, the plane surface of the plate being traction free (i.e. $\sigma_3 = 0$).

Unbeknown to the experimenter, the material contains microscopic defects which can be idealized as a sparse distribution of small holes through the thickness of the plate. Show graphically the relation which will hold at yield between the tractions σ_1, σ_2 applied to the defective plate, if the Tresca (maximum shear stress) criterion applies for the undamaged material.

8.7. The circular disk $0 \leq r < a$ is subjected to uniform compressive tractions $\sigma_{rr} = -S$ in the two arcs $-\pi/4 < \theta < \pi/4$ and $3\pi/4 < \theta < 5\pi/4$, the remainder of the surface $r=a$ being traction free. Expand these tractions as a Fourier series in θ and hence develop a series solution for the stress field. Use Maple or Mathematica to produce a contour plot of the von Mises stress σ_E, using a series truncated at 10 terms.

8.8. The stresses $\sigma_{ij}^{(0)}$ in a large simply-connected plate can be described in terms of a stress function ϕ containing only those terms from Table 8.1 that correspond to bounded stresses at $r = 0$. If this stress field is now perturbed by a small traction-free hole of radius a, show that the resulting hoop stress at the edge of the hole is given by

$$\sigma_{\theta\theta}(a, \theta) = 2\left(\sigma_{\theta\theta}^{(0)} - \sigma_{rr}^{(0)}\right)_{r=a} + \left(\sigma_{rr}^{(0)} + \sigma_{\theta\theta}^{(0)}\right)_{r=0} .$$

8.9. A hole of radius a in a large elastic plate is loaded only by a self-equilibrated distribution of normal pressure $p(\theta)$ that varies around the circumference of the hole. By expanding $p(\theta)$ as a Fourier series in θ and using Table 8.1, show that the hoop stress $\sigma_{\theta\theta}$ at the edge of the hole is given by

$$\sigma_{\theta\theta}(a, \theta) = 2\overline{p} - p(\theta) \, ,$$

where

$$\overline{p} = \frac{1}{2\pi} \int_0^{2\pi} p(\theta)d\theta$$

is the mean value of $p(\theta)$.

8.10. A circular disk of radius a is loaded only by a self-equilibrated distribution of normal pressure $p(\theta)$ that varies around the circumference. By expanding $p(\theta)$ as a Fourier series in θ and using Table 8.1, show that the hoop stress $\sigma_{\theta\theta}$ at $r = a$ is everywhere equal to $p(\theta)$.

8.11. The rotating rectangular bar of §7.3.1 contains a circular hole at the origin of radius $c \ll b$. Find the location and magnitude of the maximum tensile stress in the bar.

8.12. The thin annular disc $a < r < b$ is accelerated in the anticlockwise direction by a uniform shear traction S applied at the inner radius $r = a$, the other boundaries being traction free. Find the complete stress field in the disc.

8.13. A transmission clutch disc $a < r < b$ of thickness $t \ll a$ is compressed between rigid planes which exert a uniform compressive traction p_0 on the plane faces. The disc is now caused to rotate about its axis in the anticlockwise direction by shear tractions applied at the inner radius $r = a$, the outer radius $r = b$ being traction free. If the coefficient of friction between the plane faces is f and the angular acceleration is negligible, find the stresses in the disc. Assume the rotational speed is sufficiently low for inertia forces to be neglected.

    Hint: Since the disc is thin, the frictional tractions can be treated as body forces, but they will be non-conservative.

Chapter 9
Calculation of Displacements

So far, we have restricted attention to the calculation of stresses and to problems in which the boundary conditions are stated in terms of tractions or force resultants, but there are many problems in which displacements are also of interest. For example, we may wish to find the deflection of the rectangular beams considered in Chapter 5, or calculate the stress-concentration factor due to a rigid circular inclusion in an elastic matrix, for which a displacement boundary condition is implied at the bonded interface.

If the stress components are known, the strains can be written down from the stress-strain relations (1.60) and these in turn can be expressed in terms of displacement gradients through (1.37). The problem is therefore reduced to the integration of these gradients to recover the displacement components.

The method is most easily demonstrated by examples, of which we shall give two — one in rectangular and one in polar coördinates.

9.1 The Cantilever with an End Load

We first consider the cantilever beam loaded by a transverse force, F, at the free end (Figure 5.2), for which the stress components were calculated in §5.2.1, being

$$\sigma_{xx} = \frac{3Fxy}{2b^3} \; ; \; \sigma_{xy} = \frac{3F(b^2 - y^2)}{4b^3} \; ; \; \sigma_{yy} = 0 \,,$$

from (5.22).

The corresponding strain components are therefore

Supplementary Information The online version contains supplementary material available at https://doi.org/10.1007/978-3-031-15214-6_9.

$$e_{xx} = \frac{\sigma_{xx}}{E} - \frac{\nu\sigma_{yy}}{E} = \frac{3Fxy}{2Eb^3} \tag{9.1}$$

$$e_{xy} = \frac{\sigma_{xy}(1+\nu)}{E} = \frac{3F(1+\nu)(b^2-y^2)}{4Eb^3} \tag{9.2}$$

$$e_{yy} = \frac{\sigma_{yy}}{E} - \frac{\nu\sigma_{xx}}{E} = -\frac{3F\nu xy}{2Eb^3} \ , \tag{9.3}$$

for plane stress.

We next make use of the strain-displacement relation to write

$$e_{xx} = \frac{\partial u_x}{\partial x} = \frac{3Fxy}{2Eb^3} \ , \tag{9.4}$$

which can be integrated with respect to x to give

$$u_x = \frac{3Fx^2y}{4Eb^3} + f(y) \ , \tag{9.5}$$

where we have introduced an arbitrary function $f(y)$ of y, since any such function would make no contribution to the partial derivative in (9.4). A similar operation on (9.3) yields the result

$$u_y = -\frac{3F\nu xy^2}{4Eb^3} + g(x) \ , \tag{9.6}$$

where $g(x)$ is an arbitrary function of x.

To determine the two functions $f(y)$, $g(x)$, we use the definition of shear strain (1.35) and (9.2) to write

$$e_{xy} = \frac{1}{2}\left(\frac{\partial u_y}{\partial x} + \frac{\partial u_x}{\partial y}\right) = \frac{3F(1+\nu)(b^2-y^2)}{4Eb^3} \ .$$

Substituting for u_x, u_y, from (9.5, 9.6) and rearranging the terms, we obtain

$$\frac{3Fx^2}{8Eb^3} + \frac{1}{2}\frac{dg}{dx} = \frac{3F\nu y^2}{8Eb^3} - \frac{1}{2}\frac{df}{dy} + \frac{3F(1+\nu)(b^2-y^2)}{4Eb^3} \ . \tag{9.7}$$

Now, the left-hand side of this equation is independent of y and the right-hand side is independent of x. Thus, the equation can only be satisfied for all x, y, if *both* sides are independent of both x and y — i.e. if they are equal to a constant, which we shall denote by $\frac{1}{2}C$.

Equation (9.7) can then be partitioned into the two ordinary differential equations

$$\frac{dg}{dx} = -\frac{3Fx^2}{4Eb^3} + C$$

$$\frac{df}{dy} = \frac{3F\nu y^2}{4Eb^3} + \frac{3F(1+\nu)(b^2-y^2)}{2Eb^3} - C \ ,$$

which have the solution

$$g(x) = -\frac{Fx^3}{4Eb^3} + Cx + B$$

$$f(y) = \frac{F\nu y^3}{4Eb^3} + \frac{F(1+\nu)(3b^2 y - y^3)}{2Eb^3} - Cy + A \,,$$

where A, B are two new arbitrary constants.

The final expressions for the displacements are therefore

$$u_x = \frac{3Fx^2 y}{4Eb^3} + \frac{3F(1+\nu)y}{2Eb} - \frac{F(2+\nu)y^3}{4Eb^3} + A - Cy \qquad (9.8)$$

$$u_y = -\frac{3F\nu xy^2}{4Eb^3} - \frac{Fx^3}{4Eb^3} + B + Cx \,. \qquad (9.9)$$

This partial integration process can be performed for any biharmonic function ϕ in Cartesian coördinates, using the Maple and Mathematica files 'uxy'.

9.1.1 Rigid-body displacements and end conditions

The three constants A, B, C define the three degrees of freedom of the cantilever as a rigid body, A, B corresponding to translations in the x-, y-directions respectively and C to a small anticlockwise rotation about the origin.

These rigid-body terms always arise in the integration of strains to determine displacements, and they reflect the fact that a complete knowledge of the stresses and hence the strains throughout the body is sufficient to determine its deformed shape, but not its location in space.

As in Mechanics of Materials, the rigid-body displacements can be determined from appropriate information about the way in which the structure is supported. Since the cantilever is built-in at $x = a$, we would ideally like to specify

$$u_x = u_y = 0 \; ; \quad x = a \,, \quad -b < y < b \,, \qquad (9.10)$$

but the three constants in (9.8, 9.9) do not give us sufficient freedom to satisfy such a strong boundary condition. We should anticipate this deficiency, since the complete solution procedure only permitted us to satisfy weak boundary conditions on the ends of the beam and (9.10), in addition to locating the beam in space, implies a pointwise traction distribution sufficient to keep the end plane and unstretched.

Many authors adapt the Mechanics of Materials support conditions for a cantilever by demanding that the mid-point $(a, 0)$ of the end have zero displacement and the axis of the beam $(y=0)$ have zero slope at that point. This leads to the three conditions

$$u_x = u_y = 0 \; ; \quad \frac{\partial u_y}{\partial x} = 0 \; ; \quad x = a \, , \; y = 0 \tag{9.11}$$

and yields the values

$$A = 0 \; ; \quad B = -\frac{Fa^3}{2Eb^3} \; ; \quad C = \frac{3Fa^2}{4Eb^3} \, ,$$

when (9.8, 9.9) are substituted in (9.11).

However, the displacement u_x at $x = a$ corresponds to the deformed shape of Figure 9.1(a). This might be an appropriate end condition if the cantilever is supported in a horizontal groove, but other built-in support conditions might approximate more closely to the configuration of Figure 9.1(b), where the third of conditions (9.11) is replaced by

$$\frac{\partial u_x}{\partial y} = 0 \; ; \quad x = a \, , \; y = 0 \, .$$

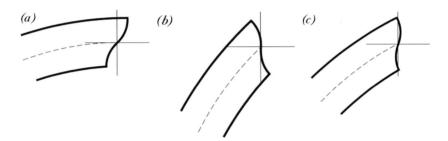

Fig. 9.1 End conditions for the cantilever.

This modified boundary condition leads to the values

$$A = 0 \; ; \quad B = -\frac{Fa^3}{2Eb^3}\left[1 + 3(1+\nu)\frac{b^2}{a^2}\right] \; ; \quad C = \frac{3Fa^2}{4Eb^3}\left[1 + 2(1+\nu)\frac{b^2}{a^2}\right]$$

for the rigid-body displacement coefficients.

The configurations 9.1(a, b) are clearly extreme cases and we might anticipate that the best approximation to (9.10) would be obtained by an intermediate case such as Figure 9.1(c). In fact, the displacement equivalent of the weak boundary conditions of Chapter 5 would be to prescribe

$$\int_{-b}^{b} u_x dy = 0 \; ; \quad \int_{-b}^{b} u_y dy = 0 \; ; \quad \int_{-b}^{b} y u_x dy = 0 \; ; \quad x = a \, . \tag{9.12}$$

The substitution of (9.8, 9.9) into (9.12) gives a new set of rigid-body displacement coefficients, which are

$$A = 0 \; ; \quad B = -\frac{Fa^3}{2Eb^3}\left[1 + \frac{(12 + 11\nu)}{5}\frac{b^2}{a^2}\right] \; ; \quad C = \frac{3Fa^2}{4Eb^3}\left[1 + \frac{(8 + 9\nu)}{5}\frac{b^2}{a^2}\right] \; .$$

9.1.2 Deflection of the free end

The different end conditions corresponding to (a, b, c) of Figure 9.1 will clearly lead to different estimates for the deflection of the cantilever. We can investigate this effect by considering the displacement of the mid-point of the free end, $x=0$, $y=0$, for which

$$u_y(0, 0) = B \; ,$$

from equation (9.9).

Thus, the end deflection predicted with the boundary conditions of (a, b, c) respectively of Figure 9.1 are

$$
\begin{aligned}
u_y(0, 0) &= -\frac{Fa^3}{2Eb^3} & (a) \\
&= -\frac{Fa^3}{2Eb^3}\left[1 + 3(1 + \nu)\frac{b^2}{a^2}\right] & (b) \\
&= -\frac{Fa^3}{2Eb^3}\left[1 + \frac{(12 + 11\nu)}{5}\frac{b^2}{a^2}\right] & (c) \; .
\end{aligned}
\qquad (9.13)
$$

The first of these results (a) is also that predicted by the elementary Mechanics of Materials solution and the second (b) is that obtained when a correction is made for the 'shear deflection', based on the shear stress at the beam axis. Case (c), which is intermediate between (a) and (b), but closer to (b), is probably the closest approximation to the true built-in boundary condition (9.10).

All three expressions have the same leading term and the corrective term in (b, c) will be small as long as $b \ll a$ — i.e. as long as the cantilever can reasonably be considered as a slender beam.

9.2 The Circular Hole

The procedure of §9.1 has some small but critical differences in polar coördinates, which we shall illustrate using the problem of §8.3.2, in which a traction-free circular hole perturbs a uniform shear field. The stresses are given by equations (8.28) and, in plane stress, the strains are therefore

$$e_{rr} = \frac{\sigma_{rr}}{E} - \frac{\nu\sigma_{\theta\theta}}{E} = \frac{S}{E}\left[1 + \nu - \frac{4a^2}{r^2} + \frac{3(1+\nu)a^4}{r^4}\right]\sin(2\theta) \qquad (9.14)$$

$$e_{r\theta} = \frac{(1+\nu)\sigma_{r\theta}}{E} = \frac{S}{E}\left[1 + \nu + \frac{2(1+\nu)a^2}{r^2} - \frac{3(1+\nu)a^4}{r^4}\right]\cos(2\theta) \quad (9.15)$$

$$e_{\theta\theta} = \frac{\sigma_{\theta\theta}}{E} - \frac{\nu\sigma_{rr}}{E} = \frac{S}{E}\left[-1 - \nu + \frac{4\nu a^2}{r^2} - \frac{3(1+\nu)a^4}{r^4}\right]\sin(2\theta) . \quad (9.16)$$

Two of the three strain-displacement relations (8.12) contain both displacement components, so we start with the simpler relation

$$e_{rr} = \frac{\partial u_r}{\partial r} ,$$

substitute for e_{rr} from (9.14) and integrate, obtaining

$$u_r = \frac{S}{E}\left[(1+\nu)r + \frac{4a^2}{r} - \frac{(1+\nu)a^4}{r^3}\right]\sin(2\theta) + f(\theta) , \qquad (9.17)$$

where $f(\theta)$ is an arbitrary function of θ.

Writing the expression (8.12) for $e_{\theta\theta}$ in the form

$$\frac{\partial u_\theta}{\partial\theta} = re_{\theta\theta} - u_r ,$$

we can substitute for $e_{\theta\theta}$, u_r from (9.16, 9.17) respectively and integrate with respect to θ, obtaining

$$u_\theta = \frac{S}{E}\left[(1+\nu)r + \frac{2(1-\nu)a^2}{r} + \frac{(1+\nu)a^4}{r^3}\right]\cos(2\theta) - F(\theta) + g(r) , \quad (9.18)$$

where $g(r)$ is an arbitrary function of r and we have written $F(\theta)$ for $\int f(\theta)d\theta$.

Finally, we substitute for $u_r, u_\theta, e_{r\theta}$ from (9.17, 9.18, 9.15) into the second of equations (8.12) to obtain an equation for the arbitrary functions $F(\theta), g(r)$, which reduces to

$$F(\theta) + F''(\theta) = g(r) - rg'(r) .$$

As in §9.1, this equation can only be satisfied for all r, θ if both sides are equal to a constant[1]. Solving the two resulting ordinary differential equations for $F(\theta), g(r)$ and substituting into equations (9.17, 9.18), remembering that $f(\theta) = F'(\theta)$ we obtain

---

[1] It turns out in polar coördinate problems that the value of this constant does not affect the final expressions for the displacement components.

$$u_r = \frac{S}{E}\left[(1+\nu)r + \frac{4a^2}{r} - \frac{(1+\nu)a^4}{r^3}\right]\sin(2\theta) + A\cos\theta + B\sin\theta$$

$$u_\theta = \frac{S}{E}\left[(1+\nu)r + \frac{2(1-\nu)a^2}{r} + \frac{(1+\nu)a^4}{r^3}\right]\cos(2\theta)$$
$$-A\sin\theta + B\cos\theta + Cr ,$$

for the displacement components.

This partial integration process can be performed for any biharmonic function ϕ in polar coördinates, using the Maple and Mathematica files 'urt'. As before, the three arbitrary constants A, B, C correspond to rigid-body displacements, A, B being translations in the x-, y-directions respectively, whilst C describes a small anticlockwise rotation about the origin.

In the present problem, we might reasonably set A, B, C to zero to preserve symmetry. At the hole surface, the radial displacement is then

$$u_r = \frac{4Sa\sin(2\theta)}{E} ,$$

which implies that the hole distorts into the ellipse shown in Figure 9.2.

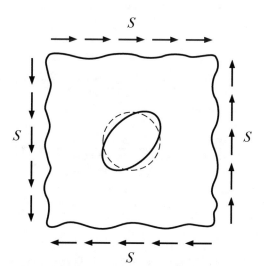

Fig. 9.2 Distortion of a circular hole in a shear field.

The reader should note that, in both of the preceding examples, the third strain-displacement relation gives an equation for two arbitrary functions of one variable which can be partitioned into two ordinary differential equations. The success of the calculation depends upon this partition, which in turn is possible if and only if the strains satisfy the compatibility conditions. In the present examples, this is of course ensured through the derivation of the stress field from a biharmonic stress function.

9.3 Displacements for the Michell Solution

Displacements for all the stress functions of Table 8.1 can be obtained[2] by the procedure of §9.2 and the results[3] are tabulated in Table 9.1.

Table 9.1 The Michell solution — displacement components

ϕ	$2\mu u_r$	$2\mu u_\theta$
r^2	$(\kappa - 1)r$	0
$r^2 \ln(r)$	$(\kappa - 1)r \ln(r) - r$	$(\kappa + 1)r\theta$
$\ln(r)$	$-1/r$	0
θ	0	$-1/r$
$r^3 \cos\theta$	$(\kappa - 2)r^2 \cos\theta$	$(\kappa + 2)r^2 \sin\theta$
$r\theta \sin\theta$	$\frac{1}{2}\{(\kappa - 1)\theta \sin\theta - \cos\theta$	$\frac{1}{2}\{(\kappa - 1)\theta \cos\theta - \sin\theta$
	$+(\kappa + 1)\ln(r)\cos\theta\}$	$-(\kappa + 1)\ln(r)\sin\theta\}$
$r \ln(r)\cos\theta$	$\frac{1}{2}\{(\kappa + 1)\theta \sin\theta - \cos\theta$	$\frac{1}{2}\{(\kappa + 1)\theta \cos\theta - \sin\theta$
	$+(\kappa - 1)\ln(r)\cos\theta\}$	$-(\kappa - 1)\ln(r)\sin\theta\}$
$\cos\theta/r$	$\cos\theta/r^2$	$\sin\theta/r^2$
$r^3 \sin\theta$	$(\kappa - 2)r^2 \sin\theta$	$-(\kappa + 2)r^2 \cos\theta$
$r\theta \cos\theta$	$\frac{1}{2}\{(\kappa - 1)\theta \cos\theta + \sin\theta$	$\frac{1}{2}\{-(\kappa - 1)\theta \sin\theta - \cos\theta$
	$-(\kappa + 1)\ln(r)\sin\theta\}$	$-(\kappa + 1)\ln(r)\cos\theta\}$
$r \ln(r)\sin\theta$	$\frac{1}{2}\{-(\kappa + 1)\theta \cos\theta - \sin\theta$	$\frac{1}{2}\{(\kappa + 1)\theta \sin\theta + \cos\theta$
	$+(\kappa - 1)\ln(r)\sin\theta\}$	$+(\kappa - 1)\ln(r)\cos\theta\}$
$\sin\theta/r$	$\sin\theta/r^2$	$-\cos\theta/r^2$
$r^{n+2} \cos n\theta$	$(\kappa - n - 1)r^{n+1} \cos n\theta$	$(\kappa + n + 1)r^{n+1} \sin n\theta$
$r^{-n+2} \cos n\theta$	$(\kappa + n - 1)r^{-n+1} \cos n\theta$	$-(\kappa - n + 1)r^{-n+1} \sin n\theta$
$r^n \cos n\theta$	$-nr^{n-1} \cos n\theta$	$nr^{n-1} \sin n\theta$
$r^{-n} \cos n\theta$	$nr^{-n-1} \cos n\theta$	$nr^{-n-1} \sin n\theta$
$r^{n+2} \sin n\theta$	$(\kappa - n - 1)r^{n+1} \sin n\theta$	$-(\kappa + n + 1)r^{n+1} \cos n\theta$
$r^{-n+2} \sin n\theta$	$(\kappa + n - 1)r^{-n+1} \sin n\theta$	$(\kappa - n + 1)r^{-n+1} \cos n\theta$
$r^n \sin n\theta$	$-nr^{n-1} \sin n\theta$	$-nr^{n-1} \cos n\theta$
$r^{-n} \sin n\theta$	$nr^{-n-1} \sin n\theta$	$-nr^{-n-1} \cos n\theta$

In this Table, results for plane strain can be recovered by setting $\kappa = (3-4\nu)$, whereas for plane stress, $\kappa = (3-\nu)/(1+\nu)$ [see equation (3.11)]. Note that it is

---

[2] The reader might like to confirm this in a few cases by using the Maple and Mathematica files 'urt'. Notice however that the results for the '$n=0$, 1' functions may differ by a rigid-body displacement from those given in the table.

[3] These results were first compiled by Professor J. Dundurs of Northwestern University and his research students, and are here reprinted with his permission.

always possible to superpose a rigid-body displacement defined by $u_r = A \cos \theta + B \sin \theta$, $u_\theta = -A \sin \theta + B \cos \theta + Cr$, where A, B, C are arbitrary constants.

9.3.1 Equilibrium considerations

In the geometry of Figure 8.1, the boundary conditions (8.18–8.21) on the surfaces $r = a, b$ are not completely independent, since they must satisfy the condition that the body be in equilibrium. This requires that

$$\int_0^{2\pi} \left[F_1(\theta) \cos \theta - F_3(\theta) \sin \theta \right] a d\theta$$

$$- \int_0^{2\pi} \left[F_2(\theta) \cos \theta - F_4(\theta) \sin \theta \right] b d\theta = 0 \tag{9.19}$$

$$\int_0^{2\pi} \left[F_1(\theta) \sin \theta + F_3(\theta) \cos \theta \right] a d\theta$$

$$- \int_0^{2\pi} \left[F_2(\theta) \sin \theta + F_4(\theta) \cos \theta \right] b d\theta = 0 \tag{9.20}$$

$$\int_0^{2\pi} F_3(\theta) a^2 d\theta - \int_0^{2\pi} F_4(\theta) b^2 d\theta = 0 . \tag{9.21}$$

The orthogonality of the terms of the Fourier series ensures that these relations only concern the Fourier terms for $n=0, 1$, but for these cases, there are then only three independent algebraic equations to determine each of the corresponding sets of four constants, $A_{01}, \ldots A_{04}, A_{11}, \ldots A_{14}, B_{11}, \ldots B_{14}$.

The key to this paradox is to be seen in the displacements of Table 9.1. In the terms for $n=0, 1$, some of the displacements include a θ-multiplier. For example, we find $2\mu u_\theta = (\kappa+1)r\theta$ for the stress function $\phi = r^2 \ln(r)$. Now the function θ is multi-valued. We can make it single-valued by defining a *principal value* — e.g., by restricting θ to the range $0 < \theta < 2\pi$, but we would then have a discontinuity at the line $\theta = 0, 2\pi$ which is unacceptable for the continuous body of Figure 8.1. We must therefore restrict our choice of stress functions to a set which defines a single-valued continuous displacement and it turns out that this imposes precisely one additional condition on each of the sets of four constants for $n=0, 1$.

Notice incidentally that if the annulus were incomplete, or if it were cut along the line $\theta = 0$, the principal value of θ would be continuous and single-valued throughout the body and the above restrictions would be removed. However, we would then also lose the equilibrium restrictions (9.19–9.21), since the body would have two new edges (e.g. $\theta = 0, 2\pi$, for the annulus with a cut on $\theta = 0$) on which there may be non-zero tractions.

We see then that the complete annulus has some complications which the incomplete annulus lacks. This arises of course because the complete annulus is *multiply connected*. The results of this section are a direct consequence of the discussion of compatibility in multiply-connected bodies in §2.2.1. These questions and problems in which they are important will be discussed further in Chapter 13.

9.3.2   The cylindrical pressure vessel

Consider the plane strain problem of a long cylindrical pressure vessel of inner radius a and outer radius b, subjected to an internal pressure p_0. The boundary conditions for this problem are

$$\sigma_{rr} = -p_0 \; ; \; \sigma_{r\theta} = 0 \; ; \; r = a \tag{9.22}$$

$$\sigma_{rr} = \sigma_{r\theta} = 0 \; ; \; r = b \; . \tag{9.23}$$

The loading is clearly independent of θ, so we seek a suitable stress function in the $n=0$ row of Tables 8.1, 9.1. However, we must exclude the term $r^2 \ln(r)$, since it gives a multivalued expression for u_θ. We therefore start with

$$\phi = Ar^2 + B \ln(r) + C\theta \; ,$$

for which the stress components are

$$\sigma_{rr} = 2A + \frac{B}{r^2} \; ; \; \sigma_{r\theta} = \frac{C}{r^2} \; ; \; \sigma_{\theta\theta} = 2A - \frac{B}{r^2} \; . \tag{9.24}$$

The boundary conditions (9.22, 9.23) then give the equations

$$2A + \frac{B}{a^2} = -p_0 \tag{9.25}$$

$$\frac{C}{a^2} = 0 \tag{9.26}$$

$$2A + \frac{B}{b^2} = 0 \tag{9.27}$$

$$\frac{C}{b^2} = 0 \; , \tag{9.28}$$

with solution

$$A = \frac{p_0 a^2}{2(b^2 - a^2)} \; ; \; B = -\frac{p_0 a^2 b^2}{2(b^2 - a^2)} \; ; \; C = 0 \; . \tag{9.29}$$

The final stress field is therefore

$$\sigma_{rr} = \frac{p_0 a^2}{(b^2 - a^2)} \left(1 - \frac{b^2}{r^2}\right) ; \quad \sigma_{r\theta} = 0 ; \quad \sigma_{\theta\theta} = \frac{p_0 a^2}{(b^2 - a^2)} \left(1 + \frac{b^2}{r^2}\right) ,$$

from (9.24, 9.29). Notice that the equations (9.25–9.28) are not linearly independent. This occurs because the loading must satisfy the global equilibrium condition (9.21), which here reduces to

$$a^2 \sigma_{r\theta}(a) = b^2 \sigma_{r\theta}(b) .$$

If this condition had not been satisfied — for example if there had been a uniform shear traction on only one of the two surfaces $r = a, b$, the above boundary-value problem would have no solution. Of course, with this loading, the cylinder would experience rotational acceleration about the axis and this would need to be taken into account in formulating the problem, as in the problem of §7.4.1.

Problems

9.1. Find the displacement field corresponding to the stress field of equations (5.54–5.56). The beam is simply-supported at $x = \pm a$. You will need to impose this in the form of appropriate weak conditions on the displacements.

Find the vertical displacement at the points $(0, 0)$, $(0, -b)$, $(0, b)$ and compare them with the predictions of the elementary bending theory.

9.2. The rectangular plate $-a < x < a$, $-b < y < b$ is bonded to rigid supports at $x = \pm a$, the edges $y = \pm b$ being traction free. The plate is loaded only by its own weight in the negative y-direction.

(i) Find a solution for the stress field assuming uniform shear tractions on $x = \pm a$. Use strong boundary conditions on $x = \pm a$ and weak conditions on $y = \pm b$.
(ii) Find the displacements corresponding to this stress field. Do the resulting expressions permit the built-in boundary condition at $x = \pm a$ to be satisfied in the strong sense?
(iii) What are the restrictions on the ratio a/b for this solution to be a reasonable approximation to the physical problem?

9.3. Find the displacements u_x, u_y as functions of x, y for the rotating rectangular plate of Figure 7.1, assuming plane stress conditions.

9.4. The elastic half-space $y > 0$ is deformed by surface loading such that

$$u_x(x, 0) = 0 ; \quad u_y(x, 0) = u_0 \cos(\lambda x) .$$

Find expressions for the displacements throughout the half-space. Also, find the surface tractions needed to achieve this deformed state. Assume plane strain conditions.

9.5. A state of pure shear, $\sigma_{xy} = S$ in a large plate is perturbed by the presence of a rigid circular inclusion in the region $r < a$. The inclusion is perfectly bonded to the plate and is prevented from moving, so that $u_r = u_\theta = 0$ at $r = a$.

Find the complete stress field in the plate and hence determine the stress-concentration factor due to the inclusion based on (i) the maximum shear stress criterion or (ii) the maximum tensile stress criterion.

Is it necessary to apply a force or a moment to the inclusion to prevent it from moving?

9.6. A thin annular disk, inner radius b and outer radius a rotates at constant speed Ω about the axis of symmetry, all the surfaces being traction free. Determine the stress field in the disk.

9.7. Figure 9.3(a) shows a complete circular ring of inner radius a and outer radius $2a$ that is cut along the radius AB, thus making the body simply-connected. The two surfaces are now forced apart and a thin wedge of material of subtended angle $\alpha \ll 1$ is inserted and welded into place as shown in Figure 9.3(b).

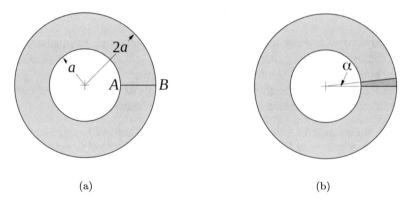

(a)   (b)

Fig. 9.3

Find the stresses in the ring if the material has elastic properties μ, ν and plane stress conditions are assumed.

9.8. A large rectangular plate is subjected to simple bending, such that the stress field is given by

$$\sigma_{xx} = Cy \; ; \quad \sigma_{xy} = \sigma_{yy} = 0 .$$

The plate contains a small traction-free circular hole of radius a centred on the origin. Find the perturbation in the stress field due to the hole.

9.9. A heavy disk of density ρ and radius a is bonded to a rigid support at $r=a$. The gravitational force acts in the direction $\theta=-\pi/2$. Find the stresses and displacements in the disk.

Hint: The easiest method is to find the stresses and displacements for a simple particular solution for the body force and then superpose 'homogeneous' terms using Tables 8.1 and 9.1.

9.10. A heavy disk of density ρ, elastic properties μ, ν and radius $a + \epsilon$ is pressed into a frictionless hole of radius a in a rigid body.

What is the minimum value of ϵ if the disk is to remain in contact with the hole at all points when the gravitational force acts in the direction $\theta=-\pi/2$.

9.11. A rubber bushing comprises a hollow cylinder of rubber $a<r<b$, bonded to concentric thin-walled steel cylinders at $r=a$ and $r=b$, as shown in Figure 9.4. The steel cylinders may be considered as rigid compared with the rubber.

The outer cylinder is held fixed, whilst a force F per unit axial length is applied to the inner cylinder in the x-direction. Find the stiffness of the bushing under this loading — i.e. the relationship between F and the resulting displacement δ of the inner cylinder in the x-direction.

In particular, find the stiffness for a long cylinder (plane strain conditions) and a short cylinder (plane stress), assuming that rubber has a shear modulus μ and is incompressible ($\nu=0.5$).

Plot a graph of the dimensionless stiffness $F/\mu\delta$ as a function of the ratio a/b in the range $F/\mu\delta < 100$. Can you provide a simple physical explanation of the difference between the plane stress and plane strain results?

Hint: The hollow cylinder is a multiply-connected body.

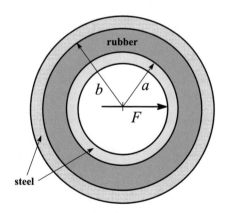

Fig. 9.4 A rubber bushing

9.12. A rigid circular inclusion of radius a in a large elastic plate is subjected to a force F in the x-direction.

Find the stress field in the plate if the inclusion is perfectly bonded to the plate at $r=a$ and the stresses tend to zero as $r \to \infty$.

9.13. It is proposed to cut a circular cylindrical tunnel of radius c at a depth b below the traction-free horizontal surface of a soil mass of density ρ. Find the location and magnitude of the maximum shear stress if the soil mass was originally everywhere in a state of triaxial hydrostatic compression. Assume $c \ll b$.

Chapter 10
Curved Beam Problems

If we cut the circular annulus of Figure 8.1 along two radial lines, $\theta = \alpha, \beta$, we generate a curved beam. The analysis of such beams follows that of Chapter 8, except for a few important differences — notably that (i) the ends of the beam constitute two new boundaries on which boundary conditions (usually weak boundary conditions) are to be applied and (ii) it is no longer necessary to enforce continuity of displacements (see §9.3.1), since a suitable principal value of θ can be defined which is both continuous and single-valued.

10.1 Loading at the Ends

We first consider the case in which the curved surfaces of the beam are traction free and only the ends are loaded. As in §5.2.1, we only need to impose boundary conditions on one end — the Airy stress function formulation will ensure that the tractions on the other end have the correct force resultants to guarantee global equilibrium.

10.1.1 Pure bending

The simplest case is that illustrated in Figure 10.1, in which the beam $a < r < b$, $0 < \theta < \alpha$ is loaded by a bending moment M_0, but no forces. Equilibrium considerations demand that the bending moment M_0 and zero axial force and shear force will be transmitted across all θ-surfaces and hence that the stress field will be independent of θ. We therefore seek the solution in the axisymmetric terms in Table 8.1, using the stress function

$$\phi = Ar^2 + Br^2 \ln(r) + C \ln(r) + D\theta ,$$

© The Author(s), under exclusive license to Springer Nature Switzerland AG 2022
J. R. Barber, *Elasticity*, Solid Mechanics and Its Applications 172,
https://doi.org/10.1007/978-3-031-15214-6_10

the corresponding stress components being

$$\sigma_{rr} = 2A + B(2\ln(r) + 1) + \frac{C}{r^2} \tag{10.1}$$

$$\sigma_{r\theta} = \frac{D}{r^2} \tag{10.2}$$

$$\sigma_{\theta\theta} = 2A + B(2\ln(r) + 3) - \frac{C}{r^2} \ . \tag{10.3}$$

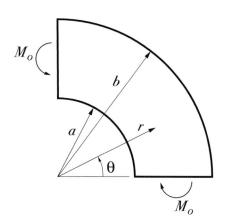

Fig. 10.1 Curved beam in pure bending.

The boundary conditions for the problem of Figure 10.1 are

$$\sigma_{rr} = 0 \ ; \quad r = a, b \tag{10.4}$$

$$\sigma_{r\theta} = 0 \ ; \quad r = a, b \tag{10.5}$$

$$\int_a^b \sigma_{\theta\theta} dr = 0 \ ; \quad \theta = 0 \tag{10.6}$$

$$\int_a^b \sigma_{\theta r} dr = 0 \ ; \quad \theta = 0 \tag{10.7}$$

$$\int_a^b \sigma_{\theta\theta} r \, dr = M_0 \ ; \quad \theta = 0 \ . \tag{10.8}$$

Notice that in practice it is not necessary to impose the zero force conditions (10.6, 10.7), since if there were a non-zero force on the end $\theta=0$, it would have to be transmitted around the beam and this would result in a non-axisymmetric stress field. Thus, our assumption of axisymmetry automatically rules out there being any force resultant.

We first impose the strong boundary conditions (10.4, 10.5), which with (10.1, 10.2) yield the four algebraic equations

$$2A + B\left[2\ln(a) + 1\right] + \frac{C}{a^2} = 0 \tag{10.9}$$

$$2A + B\left[2\ln(b) + 1\right] + \frac{C}{b^2} = 0 \tag{10.10}$$

$$\frac{D}{a^2} = 0 \tag{10.11}$$

$$\frac{D}{b^2} = 0 \,, \tag{10.12}$$

and an additional equation is obtained by substituting (10.3) into (10.8), giving

$$(A + B)(b^2 - a^2) + B\left[b^2 \ln(b) - a^2 \ln(a)\right] - C\ln\left(\frac{b}{a}\right) = M_0 \,.$$

These equations have the solution

$$A = -\frac{M_0}{N_0}\left[b^2 - a^2 + 2b^2 \ln(b) - 2a^2 \ln(a)\right]$$

$$B = \frac{2M_0}{N_0}(b^2 - a^2) \;\; ; \;\; C = \frac{4M_0}{N_0}a^2 b^2 \ln\left(\frac{b}{a}\right) \;\; ; \;\; D = 0 \,,$$

where

$$N_0 \equiv (b^2 - a^2)^2 - 4a^2 b^2 \ln^2\left(\frac{b}{a}\right)$$

The corresponding stress field is readily obtained by substituting these expressions back into equations (10.1–10.3).

The principal practical interest in this solution lies in the extent to which the classical Mechanics of Materials bending theory underestimates the bending stress $\sigma_{\theta\theta}$ at the inner edge, $r = a$ when a/b is small. Timoshenko and Goodier give some numerical values and show that the elementary theory starts to deviate significantly from the correct result for $a/b < 0.5$. The theory of curved beams — based on the application of the principle that plane sections remain plane, but allowing for the fact that the beam elements increase in length as r increases due to the curvature — gives a much better agreement. However, the exact result is quite easy to compute.

An important special case is that of a 'rectangular' bend in a beam, with a small inner fillet radius (see Figure 10.2). This is reasonably modelled as the curved beam shown dotted, since there will be very little stress in the outer corner of the bend.

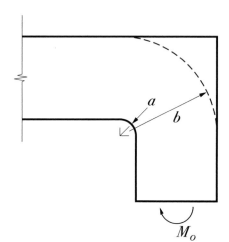

Fig. 10.2 Bend in a rectangular beam with a small fillet radius.

10.1.2   Force transmission

A similar type of solution can be generated for the problem in which a force is applied at the end of the beam. We consider the case illustrated in Figure 10.3, where a shear force is applied at $\theta=0$, resulting in the boundary conditions

$$\int_a^b \sigma_{\theta\theta}dr = 0 \ ; \ \ \theta = 0 \tag{10.13}$$

$$\int_a^b \sigma_{\theta r}dr = F \ ; \ \ \theta = 0 \tag{10.14}$$

$$\int_a^b \sigma_{\theta\theta}rdr = 0 \ ; \ \ \theta = 0 \ , \tag{10.15}$$

which replace (10.6–10.8) in the problem of §10.1.1.

Equilibrium considerations show that the shear force on any section $\theta=\alpha$ will vary as $F\cos\alpha$ and hence we seek the solution in those terms in Table 8.1 in which the shear stress $\sigma_{r\theta}$ depends on $\cos\theta$ — i.e.

$$\phi = (Ar^3 + Br^{-1} + Cr\ln(r))\sin\theta + Dr\theta\cos\theta \ , \tag{10.16}$$

for which the stress components are

$$\sigma_{rr} = \left(2Ar - 2Br^{-3} + Cr^{-1} - 2Dr^{-1}\right)\sin\theta \tag{10.17}$$

$$\sigma_{r\theta} = \left(-2Ar + 2Br^{-3} - Cr^{-1}\right)\cos\theta \tag{10.18}$$

$$\sigma_{\theta\theta} = \left(6Ar + 2Br^{-3} + Cr^{-1}\right)\sin\theta \ . \tag{10.19}$$

The hoop stress $\sigma_{\theta\theta}$ is identically zero on $\theta=0$ and hence the end conditions (10.13, 10.15) are satisfied identically in the strong sense, whilst the strong boundary

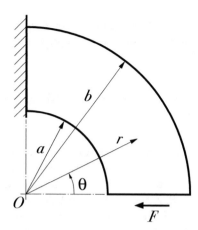

Fig. 10.3 Curved beam with an end load.

conditions on the curved edges (10.4, 10.5) lead to the set of equations

$$2Aa - \frac{2B}{a^3} + \frac{C}{a} - \frac{2D}{a} = 0 \tag{10.20}$$

$$2Ab - \frac{2B}{b^3} + \frac{C}{b} - \frac{2D}{b} = 0 \tag{10.21}$$

$$2Aa - \frac{2B}{a^3} + \frac{C}{a} = 0 \tag{10.22}$$

$$2Ab - \frac{2B}{b^3} + \frac{C}{b} = 0 . \tag{10.23}$$

The final solution that also satisfies the inhomogeneous condition (10.14) is obtained as

$$A = \frac{F}{2N_1} \quad ; \quad B = -\frac{Fa^2b^2}{2N_1} \quad ; \quad C = -\frac{F(a^2 + b^2)}{N_1} \quad ; \quad D = 0 ,$$

where

$$N_1 = (a^2 - b^2) + (a^2 + b^2)\ln\left(\frac{b}{a}\right) . \tag{10.24}$$

As before, the final stress field is easily obtained by back substitution into equations (10.17–10.19).

The corresponding solution for an axial force at the end $\theta = 0$ is obtained in the same way except that we interchange sine and cosine in the stress function (10.16). Alternatively, we can use the present solution and measure the angle from the end $\theta = \pi/2$ instead of from $\theta = 0$. Notice however that the axial force at $\theta = \pi/2$ must have the same line of action as the shear force at $\theta = 0$ in Figure 10.3. In other words, it must act through the origin of coördinates. If we wish to solve a problem in which an axial force is applied with some other, parallel, line of action — e.g. if the force acts through the mid-point of the beam cross section, $r = (a + b)/2$ — we can do so by superposing an appropriate multiple of the bending solution of §10.1.1.

10.2   Eigenvalues and Eigenfunctions

A remarkable feature of the preceding solutions is that in each case we had a set of four simultaneous homogeneous algebraic equations for four unknown constants — equations (10.9–10.12) in §10.1.1 and (10.20–10.23) in §10.1.2 — but we were able to obtain a non-trivial solution, because the equations were not linearly independent[1].

Suppose we were to define an *inhomogeneous* problem for the curved beam in which the curved edges $r = a, b$ were loaded by arbitrary tractions $\sigma_{rr}, \sigma_{r\theta}$. We could then decompose these tractions into appropriate Fourier series — as indicated in §8.3.1 — and use the general solution of equation (8.31) and Table 8.1. There would then be four linearly independent stress function terms for each Fourier component and hence the four boundary conditions would lead to a well-conditioned set of four algebraic equations for four arbitrary constants for each separate Fourier component.

In the special case where there are no tractions of the appropriate Fourier form on the curved surfaces, we should get four homogeneous algebraic equations and we anticipate only the trivial solution in which the corresponding four constants — and hence the stress components — are zero.

This is exactly what happens for all values of n *other than* 0, 1, but, as we have seen above, for these two special cases, there is a non-trivial solution to the homogeneous problem. In a sense, 0, 1 are *eigenvalues* of the general Fourier problem and the corresponding stress fields are *eigenfunctions*.

10.3   The Inhomogeneous Problem

Of course, eigenvalues can be viewed from two different perspectives. They are the values of a parameter at which the homogeneous problem has a non-trivial solution — e.g. the frequency at which a dynamic system will vibrate without excitation — but they are also the values at which the *inhomogeneous* problem has an unbounded solution — excitation at the natural frequency predicts an unbounded (steady-state) amplitude.

In the present instance, we must therefore anticipate difficulties in the inhomogeneous problem if the boundary tractions contain Fourier terms with $n = 0$ or 1.

---

[1] In particular, (10.11, 10.12) can both be satisfied by setting $D = 0$ and (10.20–10.23) reduce to only two independent equations if $D = 0$.

10.3.1 Beam with sinusoidal loading

As an example, we consider the problem of Figure 10.4, in which the curved beam is subjected to a radial normal traction

$$\sigma_{rr} = S \sin \theta \; ; \quad r = b , \tag{10.25}$$

the other boundary tractions being zero.

A 'naïve' approach to this problem would be to use a stress function composed of those terms in Table 8.1 for which the stress component σ_{rr} varies with $\sin \theta$. This of course would lead to the formulation of equations (10.16–10.19) and the strong boundary conditions on the curved edges would give the four equations

$$2Aa - \frac{2B}{a^3} + \frac{C}{a} - \frac{2D}{a} = 0 \tag{10.26}$$

$$2Ab - \frac{2B}{b^3} + \frac{C}{b} - \frac{2D}{b} = S \tag{10.27}$$

$$2Aa - \frac{2B}{a^3} + \frac{C}{a} = 0 \tag{10.28}$$

$$2Ab - \frac{2B}{b^3} + \frac{C}{b} = 0 . \tag{10.29}$$

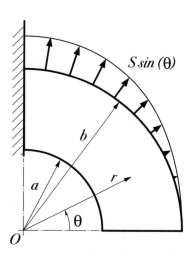

$S \sin (\theta)$

Fig. 10.4 Curved beam with sinusoidal loading.

This looks like a well-behaved set of four equations for four unknown constants, but we know that the coefficient matrix has a zero determinant, since we successfully obtained a non-trivial solution to the corresponding homogeneous problem in §10.1.2 above.

In such cases, we must seek an additional special solution in which the dependence of the stress field on θ differs from that in the boundary conditions. Such solutions are known for a variety of geometries, but are usually presented as a *fait accompli* by the author, so it is difficult to see why they work, or how we might have been expected to determine them without guidance. In this section, we shall present the appropriate function and explain why it works, but we shall then develop a more rational approach to determining the class of solution required.

The special solution can be generated by expressing the original function (10.16) in complex-variable form and then multiplying it by $\ln \zeta$, where $\zeta = x + \iota y$ is the complex variable and the conjugate $x - \iota y$ is denoted by $\bar{\zeta}$. Thus, (10.16) consists of a linear combination of the terms $\Im(\bar{\zeta}\zeta^2, \zeta^{-1}, \zeta \ln \zeta, \bar{\zeta} \ln \zeta)$, so we can generate a new biharmonic function from the terms $\Im(\bar{\zeta}\zeta^2 \ln \zeta, \zeta^{-1} \ln \zeta, \zeta \ln^2 \zeta, \bar{\zeta} \ln^2 \zeta)$, which can be written in the form

$$\phi = A'r^3 \left[\ln(r) \sin \theta + \theta \cos \theta \right] + B'r^{-1} \left[\theta \cos \theta - \ln(r) \sin \theta \right]$$
$$+ C'r \ln(r)\theta \cos \theta + D'r \left[\ln^2(r) \sin \theta - \theta^2 \sin \theta \right] . \qquad (10.30)$$

The corresponding stress components are

$$\sigma_{rr} = \left(2A'r - 2B'r^{-3} + C'r^{-1} - 4D'r^{-1}\right) \theta \cos \theta$$
$$+ \left[2A'r \ln(r) - A'r + 2B'r^{-3} \ln(r) - 3B'r^{-3} \right.$$
$$\left. - 2C'r^{-1} \ln(r) + 2D'r^{-1} \ln(r) - 2D'r^{-1} \right] \sin \theta \qquad (10.31)$$
$$\sigma_{r\theta} = \left(2A'r - 2B'r^{-3} + C'r^{-1}\right) \theta \sin \theta$$
$$+ \left[- 2A'r \ln(r) - 3A'r - 2B'r^{-3} \ln(r) + 3B'r^{-3} \right.$$
$$\left. - C'r^{-1} - 2D'r^{-1} \ln(r) \right] \cos \theta \qquad (10.32)$$
$$\sigma_{\theta\theta} = \left(6A'r + 2B'r^{-3} + C'r^{-1}\right) \theta \cos \theta$$
$$+ \left[6A'r \ln(r) + 5A'r - 2B'r^{-3} \ln(r) + 3B'r^{-3} \right.$$
$$\left. + 2D'r^{-1} \ln(r) + 2D'r^{-1} \right] \sin \theta . \qquad (10.33)$$

Notice in particular that these expressions contain some terms of the required form for the problem of Figure 10.4, but they also contain terms with multipliers of the form $\theta \sin \theta$, $\theta \cos \theta$, which are inappropriate. We therefore get four homogeneous algebraic equations for the coefficients A', B', C', D' from the requirement that these inappropriate terms should vanish in the components $\sigma_{rr}, \sigma_{r\theta}$ on the boundaries $r = a, b$, i.e.

$$2A'a - \frac{2B'}{a^3} + \frac{C'}{a} - \frac{4D'}{a} = 0 \tag{10.34}$$

$$2A'b - \frac{2B'}{b^3} + \frac{C'}{b} - \frac{4D'}{b} = 0 \tag{10.35}$$

$$2A'a - \frac{2B'}{a^3} + \frac{C'}{a} = 0 \tag{10.36}$$

$$2A'b - \frac{2B'}{b^3} + \frac{C'}{b} = 0 . \tag{10.37}$$

Clearly these equations are identical[2] with (10.20–10.23) and are not linearly independent. There is therefore a non-trivial solution to (10.34–10.37), which leaves us with a stress function that can supplement that of equation (10.16) to make the problem well-posed.

From here on, the solution is algebraically tedious, but routine. Adding the two stress functions (10.16, 10.30), we obtain a function with 8 unknown constants which is required to satisfy 9 boundary conditions comprising (i) equations (10.34–10.37), (ii) equations (10.26–10.29) modified to include the $\sin\theta$, $\cos\theta$ terms from equations (10.31, 10.32) respectively and (iii) the weak traction-free condition

$$\int_a^b \sigma_{\theta r} = 0 \; ; \;\; \theta = 0 \tag{10.38}$$

on the end of the beam. Since two of equations (10.34–10.37) are not independent, the system reduces to a set of 8 equations for 8 constants whose solution is

$$A' = \frac{Sb}{4N_1} \; ; \;\; B' = -\frac{Sa^2b^3}{4N_1} \; ; \;\; C' = -\frac{Sb(a^2+b^2)}{2N_1} \; ; \;\; D' = 0$$

$$A = \frac{Sb}{8N_1^2}\left[2(b^2-a^2) - (3a^2+b^2)\ln(b) + (3b^2+a^2)\ln(a)\right.$$

$$\left. -2(a^2\ln(a)+b^2\ln(b))\ln\left(\frac{b}{a}\right)\right] ;$$

$$B = \frac{Sa^2b^3}{8N_1^2}\left[-2(b^2-a^2) + (3b^2+a^2)\ln(b) - (3a^2+b^2)\ln(a)\right.$$

$$\left. -2(a^2\ln(b)+b^2\ln(a))\ln\left(\frac{b}{a}\right)\right]$$

$$C = \frac{Sb}{4N_1^2}\left[2(b^4+a^4)\ln\left(\frac{b}{a}\right) - (b^4-a^4)\right] \tag{10.39}$$

$$D = \frac{Sb}{4N_1}\left[(b^2-a^2) + 2(b^2+a^2)\ln(a)\right] ,$$

where N_1 is given by equation (10.24).

---

[2] except that $4D'$ replaces $2D$.

Of course, it is not fortuitous that the stress function (10.30) leads to a set of equations (10.34–10.37) identical with (10.20–10.23). When we differentiate the function $f(\zeta, \bar{\zeta}) \ln(\zeta)$ by parts to determine the stresses, the extra multiplier $\ln(\zeta)$ is only preserved in those terms where it is not differentiated and hence in which all the differential operations are performed on $f(\zeta, \bar{\zeta})$. But these operations on $f(\zeta, \bar{\zeta})$ are precisely those leading to the stresses in the original solution and hence to equations (10.20–10.23). The reader will notice a parallel here with the procedure for determining the general solution of a differential equation with repeated differential multipliers and with that for dealing with degeneracy of solutions discussed in §8.3.3.

10.3.2   The near-singular problem

Suppose we next consider a more general version of the problem of Figure 10.4 in which the inhomogeneous boundary condition (10.25) is replaced by

$$\sigma_{rr} = S\sin(\lambda\theta) \; ; \quad r = b \; , \tag{10.40}$$

where λ is a constant. In the special case where $\lambda = 1$, this problem reduces to that of §10.3.1. For all other values (excluding $\lambda = 0$), we can use the stress function

$$\phi = \left(Ar^{\lambda+2} + Br^{\lambda} + Cr^{-\lambda} + Dr^{-\lambda+2}\right)\sin(\lambda\theta) \; , \tag{10.41}$$

with stress components

$$\begin{aligned}
\sigma_{rr} = -&\Big[A(\lambda - 2)(\lambda + 1)r^{\lambda} + B\lambda(\lambda - 1)r^{\lambda-2} \\
&+ C\lambda(\lambda + 1)r^{-\lambda-2} + D(\lambda + 2)(\lambda - 1)r^{-\lambda}\Big]\sin(\lambda\theta) \\
\sigma_{r\theta} = &\Big[-A\lambda(\lambda + 1)r^{\lambda} - B\lambda(\lambda - 1)r^{\lambda-2} \\
&+ C\lambda(\lambda + 1)r^{-\lambda-2} + D\lambda(\lambda - 1)r^{-\lambda}\Big]\cos(\lambda\theta) \\
\sigma_{\theta\theta} = &\Big[A(\lambda + 1)(\lambda + 2)r^{\lambda} + B\lambda(\lambda - 1)r^{\lambda-2} \\
&+ C\lambda(\lambda + 1)r^{-\lambda-2} + D(\lambda - 1)(\lambda - 2)r^{-\lambda}\Big]\sin(\lambda\theta) \; .
\end{aligned}$$

The boundary conditions on the curved edges lead to the four equations

$$A(\lambda-2)(\lambda+1)a^{\lambda}+B\lambda(\lambda-1)a^{\lambda-2}+C\lambda(\lambda+1)a^{-\lambda-2}+D(\lambda+2)(\lambda-1)a^{-\lambda} = 0$$
$$A(\lambda-2)(\lambda+1)b^{\lambda}+B\lambda(\lambda-1)b^{\lambda-2}+C\lambda(\lambda+1)b^{-\lambda-2}+D(\lambda+2)(\lambda-1)b^{-\lambda} = -S$$
$$-A\lambda(\lambda+1)a^{\lambda}-B\lambda(\lambda-1)a^{\lambda-2}+C\lambda(\lambda+1)a^{-\lambda-2}+D\lambda(\lambda-1)a^{-\lambda} = 0$$
$$-A\lambda(\lambda+1)b^{\lambda}-B\lambda(\lambda-1)b^{\lambda-2}+C\lambda(\lambda+1)b^{-\lambda-2}+D\lambda(\lambda-1)b^{-\lambda} = 0 \; ,$$

which have the solution

$$A = \frac{Sb^\lambda}{2(\lambda+1)N(\lambda)}\left[(\lambda+1)f(\lambda) - \lambda b^2 f(\lambda-1)\right]$$

$$B = \frac{-Sb^\lambda}{2(\lambda-1)N(\lambda)}\left[\lambda f(\lambda+1) - (\lambda-1)b^2 f(\lambda)\right]$$

$$C = \frac{-Sa^{2\lambda}b^{\lambda+2}}{2(\lambda+1)N(\lambda)}\left[\lambda b^{2\lambda-2}f(1) + f(\lambda)\right] \tag{10.42}$$

$$D = \frac{Sa^{2\lambda-2}b^\lambda}{2(\lambda-1)N(\lambda)}\left[\lambda b^{2\lambda}f(1) + a^2 f(\lambda)\right],$$

where

$$N(\lambda) = \lambda^2 f(\lambda-1)f(\lambda+1) - (\lambda^2-1)f(\lambda)^2$$
$$f(p) = b^{2p} - a^{2p}.$$

On casual inspection, it seems that this problem behaves rather remarkably as λ passes through unity. The stress field is sinusoidal for all $\lambda \neq 1$ and increases without limit as λ approaches unity from either side, since $N(\lambda)=0$ when $\lambda=1$. (Notice that an additional singularity is introduced through the factor $(\lambda-1)$ in the denominator of the coefficients B, D.) However, when λ is exactly equal to unity, the problem has the bounded solution derived in §10.3.1, in which the stress field is not sinusoidal.

However, we shall demonstrate that the solution is not as discontinuous as it looks. The stress field obtained by substituting (10.42) into (10.41) is not a complete solution to the problem since the end condition (10.38) is not satisfied. The solution from (10.41) will generally involve a non-zero force on the end, given by

$$F(\lambda) = \int_a^b \sigma_{r\theta}dr$$
$$= -\lambda\left[Af\left(\frac{\lambda+1}{2}\right) + Bf\left(\frac{\lambda-1}{2}\right) + Cf\left(\frac{-\lambda-1}{2}\right) + Df\left(\frac{-\lambda+1}{2}\right)\right].$$
$$\tag{10.43}$$

To restore the traction-free condition on the end (in the weak sense of zero force resultant) we must subtract the solution of the homogeneous problem of §10.1.2, with F given by (10.43). The final solution therefore involves the superposition of the stress functions (10.16) and (10.41). As λ approaches unity, both components of the solution increase without limit, since (10.43) contains the singular terms (10.42), but they also tend to assume the same form since, for example, $r^{2+\lambda}\sin(\lambda\theta)$ approaches $r^3\sin\theta$. In the limit, the corresponding term takes the form

$$\lim_{\lambda \to 1} \frac{A_1(\lambda) r^{2+\lambda} \sin(\lambda \theta) - A_2(\lambda) r^3 \sin \theta}{N(\lambda)} , \tag{10.44}$$

where

$$A_1(\lambda) = \frac{Sb^\lambda}{2(\lambda + 1)} \left[(\lambda + 1) f(\lambda) - \lambda b^2 f(\lambda - 1) \right] \tag{10.45}$$

$$A_2(\lambda) = \frac{F(\lambda) N(\lambda)}{2 N_1} . \tag{10.46}$$

Setting $\lambda = 1$ in equations (10.45, 10.46), we find

$$A_1(1) = A_2(1) = \frac{Sb(b^2 - a^2)}{2} .$$

The zero in $N(\lambda)$ is therefore cancelled and (10.44) has a bounded limit which can be recovered by using l'Hôpital's rule. We obtain

$$\frac{[A_1'(1) - A_2'(1)] r^3 \sin \theta}{N'(1)} + \frac{A_1(1) g(r, \theta, 1)}{N'(1)} ,$$

where

$$g(r, \theta, \lambda) = \frac{\partial}{\partial \lambda} r^{2+\lambda} \sin(\lambda \theta) = r^{2+\lambda} \left[\ln(r) \sin(\lambda \theta) + \theta \cos(\lambda \theta) \right] ,$$

which in the limit $\lambda = 1$ reduces to $r^3 [\ln(r) \sin \theta + \theta \cos \theta]$. The coefficient of this term will be

$$A' = \frac{A_1(1)}{N'(1)} = \frac{Sb}{4 N_1} ,$$

agreeing with the corresponding coefficient, A', in the special solution of §10.3.1 (see equations (10.39)).

A similar limiting process yields the form and coefficients of the remaining 7 stress functions in the solution of §10.3.1. In particular, we note that the coefficients B, D in equations (10.42) have a second singular term in the denominator and hence l'Hôpital's rule has to be applied twice, leading to stress functions of the form of the last two terms in equation (10.30).

Thus, we find that the solution to the more general problem of equation (10.40) includes that of §10.3.1 as a limiting case and there are no values of λ for which the stress field is singular. The limiting process also shows us a more general way of obtaining the special stress functions required for the problem of §10.3.1. We simply take the stress function (10.41), which degenerates when $\lambda = 1$, and differentiate it with respect to the parameter, λ, before setting $\lambda = 1$. Since the last two terms of (10.16) are already of this differentiated form, a second differentiation is required to generate the last two terms of (10.30).

10.4 Some General Considerations

Similar examples could be found for other geometries in the two-dimensional theory of elasticity. To fix ideas, suppose we consider the body $a < \xi < b$, $c < \eta < d$ in the general system of curvilinear coördinates, ξ, η. In general we might anticipate a class of separated-variable solutions of the form

$$\phi = f(\lambda, \xi)g(\lambda, \eta) , \tag{10.47}$$

where λ is a parameter. An important physical problem is that in which the boundaries $\eta = a, b$ are traction free and the tractions on the remaining boundaries have a non-zero force or moment resultant. In such cases, we can generally use dimensional and/or equilibrium arguments to determine the form of the stress variation in the ξ-direction and hence the appropriate function $f(\lambda_0, \xi)$. The biharmonic equation will then reduce to a fourth order ordinary differential equation for $g(\lambda_0, \eta)$, with four linearly independent solutions. Enforcement of the traction-free boundary conditions will give a set of four homogeneous equations for the unknown multipliers of these four solutions.

If the solution is to be possible, the equations must have a non-trivial solution, implying that the matrix of coefficients is singular. It therefore follows that the corresponding *inhomogeneous* problem cannot be solved by the stress function of equation (10.47) if the tractions on $\eta = a, b$ vary with $f(\lambda_0, \xi)$ — i.e. in the same way as the stresses in the homogeneous (end loaded) problem. However, special stress functions appropriate to this limiting case can be obtained by differentiating (10.47) with respect to the parameter λ and then setting $\lambda = \lambda_0$.

The special solution for $\lambda = \lambda_0$ is not qualitatively different from that at more general values of λ, but appears as a regular limit *once appropriate boundary conditions are imposed on the edges* $\xi = c, d$.

Other simple examples in the two-dimensional theory of elasticity include the curved beam in bending (§10.1.1) and the wedge with traction-free faces (§11.2). Similar arguments can also be applied to three-dimensional axisymmetric problems — e.g. to problems of the cone or cylinder with traction-free curved surfaces.

10.4.1 Conclusions

Whenever we find a classical solution in which there are two traction-free boundaries, we can formulate a related inhomogeneous problem for which the equation system resulting from these boundary conditions will be singular. Special solutions for these cases can be obtained by parametric differentiation of more general solutions for the same geometry. Alternatively, the special case can be recovered as a limit as the parameter tends to its eigenvalue, provided that the remaining boundary conditions are satisfied in an appropriate sense (e.g. in the weak sense of force resultants) before the limit is taken.

Problems

10.1. A curved beam of inner radius a and outer radius $5a$ is subjected to a force F at its end as shown in Figure 10.5. The line of action of the force passes through the mid-point of the beam.

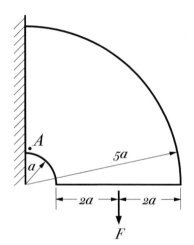

Fig. 10.5 Curved beam
loaded by a central force.

By superposing the solutions for the curved beam subjected to an end force and to pure bending respectively, find the hoop stress $\sigma_{\theta\theta}$ at the point $A(a, \pi/2)$ and compare it with the value predicted by the elementary Mechanics of Materials bending theory.

10.2. The curved beam $a < r < b$, $0 < \theta < \pi/2$ is built in at $\theta = \pi/2$ and loaded only by its own weight, which acts in the direction $\theta = -\pi/2$. Find the stress field in the beam.

10.3. The curved beam $a < r < b$, $0 < \theta < \pi/2$ is built in at $\theta = \pi/2$ and loaded by a uniform normal pressure $\sigma_{rr} = -S$ at $r = b$, the other edges being traction free. Find the stress field in the beam.

10.4. The curved beam $a < r < b$, $0 < \theta < \pi/2$ is built in at $\theta = \pi/2$ and loaded by a uniform shear traction $\sigma_{r\theta} = S$ at $r = b$, the other edges being traction free. Find the stress field in the beam.

10.5. The curved beam $a < r < b, 0 < \theta < \pi/2$ rotates about the axis $\theta = \pi/2$ at speed Ω, the edges $r = a, b$ and $\theta = 0$ being traction free. Find the stresses in the beam.

Hint: The easiest method is probably to solve first the problem of the complete annular ring $a < r < b$ rotating about $\theta = \pi/2$ and then correct the boundary condition at $\theta = 0$ (in the weak sense) by superposing a suitable homogeneous solution (i.e. a solution without body force).

10.6. Figure 10.6 shows a crane hook of thickness t that is loaded by a force F acting through the centre of the curved section. Find the stress field in the curved portion of the hook and compare the maximum tensile stress with that predicted by the elementary bending theory. Neglect the self-weight of the hook.

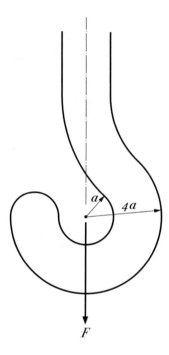

Fig. 10.6 The crane hook.

10.7. The curved beam $a < r < 2a$, $0 < \theta < \pi$ is built in at $\theta = \pi$ and loaded by a compressive normal traction $p_0\theta$ on the inner surface $r = a$, where p_0 is a constant. All the other traction components on the curved surfaces are zero and the end $\theta = 0$ is traction free. Find a solution for the stress field, using weak boundary conditions on $\theta = 0$. Hint: You will need to find some terms in the stress function that are not in Table 8.1. Be guided by the boundary conditions.

10.8. Figure 10.7 shows a composite object comprising four curved beam segments of inner radius a and outer radius b, each of which is connected to a central cross at one end only, the other end being free. The object rotates at uniform speed Ω about the central pivot. Find the stresses in one of the beam segments, using polar coördinates with θ measured from the free end of the segment. The density of the material is ρ.

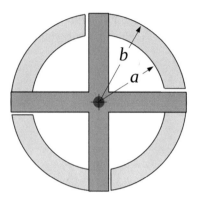

Fig. 10.7

10.9. Find the stress field for the beam segment of Problem 10.8 for the case where the object is instantaneously at rest $[\Omega = 0]$, but is loaded by a central moment such as to produce a clockwise angular acceleration $\dot{\Omega}$.

Chapter 11
Wedge Problems

In this chapter, we shall consider a class of problems for the semi-infinite wedge defined by the lines $\alpha < \theta < \beta$, illustrated in Figure 11.1.

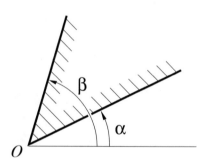

Fig. 11.1 The semi-infinite wedge.

11.1 Power-law Tractions

We first consider the case in which the tractions on the boundaries vary with r^n, in which case equations (8.8, 8.9) suggest that the required stress function will be of the form

$$\phi = r^{n+2} f(\theta) .\tag{11.1}$$

The function $f(\theta)$ can be found by substituting (11.1) into the biharmonic equation (8.11), giving the ordinary differential equation

$$\left(\frac{d^2}{d\theta^2} + (n+2)^2 \right) \left(\frac{d^2}{d\theta^2} + n^2 \right) f = 0 .\tag{11.2}$$

© The Author(s), under exclusive license to Springer Nature Switzerland AG 2022
J. R. Barber, *Elasticity*, Solid Mechanics and Its Applications 172,
https://doi.org/10.1007/978-3-031-15214-6_11

For $n \neq 0, -2$, the four solutions of this equation define the stress function

$$\phi = r^{n+2}\left[A_1 \cos(n+2)\theta + A_2 \cos(n\theta) + A_3 \sin(n+2)\theta + A_4 \sin(n\theta)\right] \quad (11.3)$$

and the corresponding stress and displacement components can be taken from Tables 8.1, 9.1 respectively, with appropriate values of n.

11.1.1 Uniform tractions

Timoshenko and Goodier, in a footnote in §45, state that the uniform terms in the four boundary tractions are not independent and that only three of them may be prescribed. This is based on the argument that in a more general non-singular series solution, the uniform terms represent the stress in the corner $r=0$, and the stress at a point only involves three independent components in two-dimensions.

However, the differential equation (11.2) has four independent non-trivial solutions for $n=0$, corresponding to the stress function[1]

$$\phi = r^2\left[A_1 \cos(2\theta) + A_2 + A_3 \sin(2\theta) + A_4\theta\right] . \quad (11.4)$$

The corresponding stress components are

$$\sigma_{rr} = -2A_1 \cos(2\theta) + 2A_2 - 2A_3 \sin(2\theta) + 2A_4\theta \quad (11.5)$$
$$\sigma_{r\theta} = 2A_1 \sin(2\theta) - 2A_3 \cos(2\theta) - A_4 \quad (11.6)$$
$$\sigma_{\theta\theta} = 2A_1 \cos(2\theta) + 2A_2 + 2A_3 \sin(2\theta) + 2A_4\theta , \quad (11.7)$$

and, in general, these permit us to solve the problem of the wedge with any combination of four independent uniform traction components on the faces[2]. Timoshenko's assertion is therefore incorrect.

Uniform shear on a right-angle wedge

To explore this paradox further, it is convenient to consider the problem of Figure 11.2, in which the right-angle corner $x > 0$, $y > 0$ is subjected to uniform shear tractions on one face defined by

[1] Another way of generating this special solution is to note that the term in (11.2) which degenerates when $n=0$ is $A_4 \sin(n\theta)$. Following §8.3.3, we can find the special solution by differentiating with respect to n, before proceeding to the limit $n \to 0$, giving the result $A_4\theta$.

[2] This problem was investigated by E. Reissner (1944), Note on the theorem of the symmetry of the stress tensor, *Journal of Mathematics and Physics* Vol. 23, pp. 192–194. See also D. B. Bogy and E. Sternberg (1968), The effect of couple-stresses on the corner singularity due to an asymmetric shear loading, *International Journal of Solids and Structures*, Vol. 4, pp. 159–174.

$$\sigma_{xy} = S \; ; \quad \sigma_{xx} = 0 \; ; \quad x = 0$$
$$\sigma_{yy} = 0 \; ; \quad \sigma_{yx} = 0 \; ; \quad y = 0 .$$

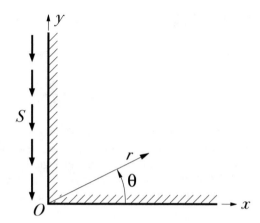

Fig. 11.2 Uniform shear on a right-angle wedge.

Timoshenko's footnote would seem to argue that this is not a well-posed problem, since it implies that $\sigma_{xy} \neq \sigma_{yx}$ at $x = y = 0$.

Casting the problem in polar coördinates, (r, θ), the boundary conditions become

$$\sigma_{\theta r} = -S \; ; \quad \sigma_{\theta\theta} = 0 \; ; \quad \theta = \frac{\pi}{2}$$
$$\sigma_{\theta r} = 0 \; ; \quad \sigma_{\theta\theta} = 0 \; ; \quad \theta = 0 .$$

Using the stress field of equations (11.5–11.7), this reduces to the system of four algebraic equations

$$2A_3 - A_4 = -S$$
$$-2A_1 + 2A_2 + A_4\pi = 0$$
$$-2A_3 - A_4 = 0$$
$$2A_1 + 2A_2 = 0 ,$$

with solution

$$A_1 = \frac{\pi S}{8} \; ; \quad A_2 = -\frac{\pi S}{8} \; ; \quad A_3 = -\frac{S}{4} \; ; \quad A_4 = \frac{S}{2} ,$$

giving the stress function

$$\phi = S \left(\frac{\pi r^2 \cos(2\theta)}{8} - \frac{\pi r^2}{8} - \frac{r^2 \sin(2\theta)}{4} + \frac{r^2 \theta}{2} \right) . \qquad (11.8)$$

Timoshenko's 'paradox' is associated with the apparent inconsistency in the stress components σ_{xy}, σ_{yx} at $x = y = 0$, so it is convenient to recast the stress function in Cartesian coördinates as

$$\phi = S\left(\frac{\pi}{8}(x^2 - y^2) - \frac{\pi(x^2 + y^2)}{8} - \frac{xy}{2} + \frac{(x^2 + y^2)}{2}\arctan\frac{y}{x}\right). \qquad (11.9)$$

We then determine the shear stress σ_{xy} as

$$\sigma_{xy} = -\frac{\partial^2\phi}{\partial x\partial y} = \frac{Sy^2}{x^2 + y^2}.$$

It is easily verified that this expression tends to zero for $y = 0$ and to S for $x = 0$, except of course that it is indeterminate at the point $x = y = 0$. The first three terms in the stress function (11.9) are second degree polynomials in x, y and therefore define a uniform state of stress throughout the wedge, but the fourth term, resulting from $A_4 r^2 \theta$ in (11.4), defines stresses which are uniform along any line $\theta = $ constant, but which vary with θ. Thus, any stress component σ is a bounded function of θ only. It follows that the corner of the wedge is not a singular point, but the stress *gradients* in the θ-direction $(1/r)\partial\sigma/\partial\theta$ increase with r^{-1} as $r \to 0$. This arises because lines of constant θ meet at $r = 0$.

11.1.2   The rectangular body revisited

We noted in §6.4 that convergence problems are encountered in series solutions for the rectangular body when the shear tractions are discontinuous in the corners. One way to circumvent this difficulty is to extract the discontinuity explicitly, using appropriate multiples of the stress function (11.4) centred on each corner. In fact, it is clear from the Cartesian coördinate expression (11.9) that only the term

$$\frac{(x^2 + y^2)}{2}\arctan\frac{y}{x}$$

contributes to the discontinuity. For example, we can construct a particular solution for the symmetric/symmetric problem of equations (6.21, 6.22) in the form

$$\phi_P = C\left[f(a, b) + f(-a, b) + f(a, -b) + f(-a, -b)\right], \qquad (11.10)$$

where

$$f(c, d) = \left(\frac{(x - c)^2 + (y - d)^2}{2}\right)\arctan\left(\frac{y - d}{x - c}\right)$$

and C is a constant. The corresponding shear tractions in the corner (a, b) then have a discontinuity defined by

$$\lim_{y \to b} \sigma_{xy}(a, y) - \lim_{x \to a} \sigma_{yx}(x, b) = C .$$

Thus, if we choose $C = g(b) - h(a)$, the corrective boundary-value problem defined by the stress function $\phi_C = \phi - \phi_P$ will involve continuous shear tractions at all four corners (because of symmetry) and the series solution of §6.4 will converge.

It is interesting to note that the form of the corner stress field defined by equation (11.10) was obtained independently by Meleshko and Gomilko[3] using the series (6.23) with parameters (6.27), by examining the limiting form of the coefficients at large m, n.

11.1.3 *More general uniform loading*

More generally, there is no inconsistency in specifying four independent uniform traction components on the faces of a wedge, though values which are inconsistent with the conditions for stress at a point will give infinite stress gradients in the corner.

For the general case, it is convenient to choose a coördinate system symmetric with respect to the two faces of the wedge, which is then bounded by the lines $\theta = \pm\alpha$.

Writing the boundary conditions

$$\sigma_{\theta r} = T_1 \; ; \;\; \theta = \alpha$$
$$\sigma_{\theta r} = T_2 \; ; \;\; \theta = -\alpha$$
$$\sigma_{\theta\theta} = N_1 \; ; \;\; \theta = \alpha$$
$$\sigma_{\theta\theta} = N_2 \; ; \;\; \theta = -\alpha ,$$

and using the solution (11.5–11.7), we obtain the four algebraic equations

$$2A_1 \sin(2\alpha) - 2A_3 \cos(2\alpha) - A_4 = T_1$$
$$-2A_1 \sin(2\alpha) - 2A_3 \cos(2\alpha) - A_4 = T_2$$
$$2A_1 \cos(2\alpha) + 2A_2 + 2A_3 \sin(2\alpha) + 2A_4\alpha = N_1$$
$$2A_1 \cos(2\alpha) + 2A_2 - 2A_3 \sin(2\alpha) - 2A_4\alpha = N_2 .$$

As in §6.2.2, we can exploit the symmetry of the problem to partition the coefficient matrix by taking sums and differences of these equations in pairs, obtaining the simpler set

[3] V. V. Meleshko and A. M. Gomilko (1997), Infinite systems for a biharmonic problem in a rectangle, *Proceedings of the Royal Society of London*, Vol. A453, pp. 2139–2160.

$$- 4A_3 \cos(2\alpha) - 2A_4 = T_1 + T_2 \qquad (11.11)$$

$$4A_1 \sin(2\alpha) = T_1 - T_2 \qquad (11.12)$$

$$4A_1 \cos(2\alpha) + 4A_2 = N_1 + N_2 \qquad (11.13)$$

$$4A_3 \sin(2\alpha) + 4A_4\alpha = N_1 - N_2 , \qquad (11.14)$$

where we note that the terms involving A_1, A_2 correspond to a symmetric stress field and those involving A_3, A_4 to an antisymmetric field.

11.1.4   Eigenvalues for the wedge angle

The solution of these equations is routine, but we note that there are two eigenvalues for the wedge angle 2α at which the matrix of coefficients is singular. In particular, the solution for the symmetric terms equations (11.12, 11.13) is singular if

$$\sin(2\alpha) = 0 , \qquad (11.15)$$

i.e. at $2\alpha = 180^o$ or 360°, whilst the antisymmetric terms are singular when

$$\tan(2\alpha) = 2\alpha , \qquad (11.16)$$

which occurs at $2\alpha = 257.4^o$. The 180° wedge is the half-plane $x > 0$, whilst the 257.4^o wedge is a reëntrant corner.

As in the problem of §10.3, special solutions are needed for the inhomogeneous problem if the wedge has one of these two special angles. As before, they can be obtained by differentiating the general solution (11.3) with respect to n before setting $n=0$, leading to the new terms, $r^2[\ln(r) \cos(2\theta) - \theta \sin(2\theta)]$ and $r^2 \ln(r)$. A particular problem for which these solutions are required is that of the half-plane subjected to a uniform shear traction over half of its boundary (see Problem 11.1, below).

The homogeneous solution

For wedges of 180° and 257.4^o, the *homogeneous* problem, where $N_1 = N_2 = T_1 = T_2 = 0$ has a non-trivial solution in which the stresses are independent of r. For the 180° wedge (i.e. the half-plane $x > 0$), this solution can be seen by inspection to be one of uniaxial tension, $\sigma_{yy} = S$, which is non-trivial, but involves no tractions on the free surface $x = 0$. The same state of stress is also a non-trivial solution for the 360° wedge, which corresponds to a semi-infinite crack (see Figure 11.6 and §11.2.3 below).

11.2 Williams' Asymptotic Method

Figure 11.3 shows a body with a notch, loaded by tractions on the remote boundaries. Intuitively we anticipate a stress concentration at the notch and we shall show in this section that the stress field there is generally singular — i.e. that the stress components tend to infinity as we approach the sharp corner of the notch.

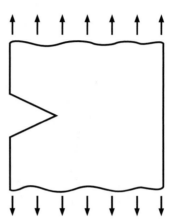

Fig. 11.3 The notched bar in tension.

Williams[4] developed a method of exploring the nature of the stress field near this singularity by defining a set of polar coördinates centred on the corner and expanding the stress field as an asymptotic series in powers of r.

11.2.1 Acceptable singularities

Before developing the asymptotic solution in detail, we must first address the question as to whether singular stress fields are ever acceptable in elasticity problems.

Engineering criteria

This question can be approached from various points of view. The engineer is usually inclined to argue that no real materials are capable of sustaining

---

[4] M. L. Williams (1952), Stress singularities resulting from various boundary conditions in angular corners of plates in extension, *ASME Journal of Applied Mechanics,* Vol. 19, pp. 526–528.

an infinite stress and hence any situation in which such a stress is predicted by an elastic analysis will in practice lead to yielding or some other kind of failure.

We also note that stress singularities are always associated with discontinuities in the geometry or the boundary conditions — for example, a sharp corner, as in the present instance, or a concentrated (delta function) load. In practice, there are no sharp corners and loads can never be perfectly concentrated. If we are to give any meaning to these solutions, they must be considered as limits of more practical situations, such as a corner with a very small fillet radius or a load applied over a very small region of the boundary.

Yet another practical limitation is that of the continuum theory. Real materials are not continua — they have atomic structure and often a larger scale granular structure as well. Thus, it doesn't make much practical sense to talk about the value of quantities nearer to a corner than (say) one atomic distance, since the theory is going to break down there anyway.

Of course, we find this kind of argument whenever we try to idealize a physical system, which means essentially whenever we try to describe its behaviour in mathematical terms. Because there are so many idealizations or approximations involved in the modelling process, it is never certain which one is responsible when mathematical difficulties are encountered in a problem and often there are several ways of reformulating the problem to introduce more physical reality and avoid the difficulty.

Mathematical criteria

A totally different approach to the question — more favoured by mathematicians — is to make choices based on questions of uniqueness, convergence and existence of solutions, rather than on the physical grounds discussed above. In terms of the functional analysis, we choose to locate the theory of elasticity in a function space which guarantees that problems are mathematically well-posed. A function space here denotes a set of functions in which the solution is to be sought, and generally this involves some statement about the strength of acceptable singularities in the solution.

In this context, the engineer's objections outlined above can be addressed by limiting arguments. It is not logically impossible that a material should have a yield strength of any arbitrarily large (but finite) magnitude, that the radius of a corner should be arbitrarily small, but finite, etc. Solutions of problems involving singularities are then to be conceived as the result of allowing these quantities to tend to their limits. Generally the difference between the real problem and the limiting case will be only localized.

In regions of a body or its boundary where no concentrated traction etc. is applied, we adopt the criterion that the only acceptable singularities are those for which the total strain energy in a small region surrounding the singular point vanishes as the size of that region tends to zero.

If the stresses and hence the strains vary with r^a as we approach the point $r = 0$, the strain energy in a two-dimensional problem will be an integral of the form

$$U = \frac{1}{2} \int_0^{2\pi} \int_0^r \sigma_{ij} e_{ij} r \, dr \, d\theta = C \int_0^r r^{2a+1} dr , \qquad (11.17)$$

where C is a constant which depends on the elastic constants and the nature of the stress variation with θ. This integral is bounded if $a > -1$ and otherwise unbounded. Thus, singular stress fields are acceptable if and only if the exponent on the stress components exceeds -1. This question is discussed further in §37.8.1.

Concentrated forces and dislocations[5] involve stress singularities with exponent $a = -1$ and are therefore excluded by this criterion, since the integral in (11.17) would lead to a logarithm, which is unbounded at $r = 0$. However, these solutions can be permitted on the understanding that they are really idealizations of explicitly specified distributions. In this case, it is important to recognize that the resulting solutions are not meaningful at or near the singular point. We also use these and other singular solutions as Green's functions — i.e. as the kernels of convolution integrals to define the solution of a non-singular problem. This technique will be illustrated in Chapters 12 and 13 below.

11.2.2 Eigenfunction expansion

We are concerned only with the stress components in the notch at very small values of r and hence we imagine looking at the corner through a strong microscope, so that we see the wedge of Figure 11.4. The magnification is so large that the other surfaces of the body, including the loaded boundaries, appear far enough away for us to treat the wedge as infinite, with 'loading at infinity'.

The stress field at the notch is of course a complicated function of r, θ, but as in Chapter 6, we seek to expand it as a series of separated-variable terms, each of which satisfies the traction-free boundary conditions on the wedge faces. The appropriate separated-variable form is of course that given by equation (11.3), except that we increase the generality of the solution by relaxing the requirement that the exponent n be an integer — indeed, we shall find that most of the required solutions have *complex* exponents.

Following Williams' notation, we replace n by $(\lambda - 1)$ in equation (11.3), obtaining the stress function

$$\phi = r^{\lambda+1} \left[A_1 \cos(\lambda+1)\theta + A_2 \cos(\lambda-1)\theta + A_3 \sin(\lambda+1)\theta + A_4 \sin(\lambda-1)\theta \right] ,$$
$$(11.18)$$

---

[5] See Chapter 13 below.

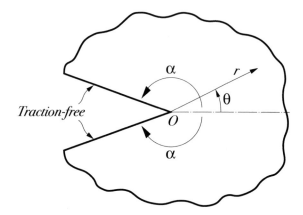

Fig. 11.4 The semi-infinite
notch.

with stress components

$$
\sigma_{rr} = r^{\lambda-1}\Big[- A_1\lambda(\lambda+1)\cos(\lambda+1)\,\theta - A_2\lambda(\lambda-3)\cos(\lambda-1)\,\theta
$$
$$
- A_3\lambda(\lambda+1)\sin(\lambda+1)\,\theta - A_4\lambda(\lambda-3)\sin(\lambda-1)\,\theta\Big] \quad (11.19)
$$
$$
\sigma_{r\theta} = r^{\lambda-1}\Big[A_1\lambda(\lambda+1)\sin(\lambda+1)\,\theta + A_2\lambda(\lambda-1)\sin(\lambda-1)\,\theta
$$
$$
- A_3\lambda(\lambda+1)\cos(\lambda+1)\theta - A_4\lambda(\lambda-1)\cos(\lambda-1)\,\theta\Big] \quad (11.20)
$$
$$
\sigma_{\theta\theta} = r^{\lambda-1}\Big[A_1\lambda(\lambda+1)\cos(\lambda+1)\,\theta + A_2\lambda(\lambda+1)\cos(\lambda-1)\theta
$$
$$
+ A_3\lambda(\lambda+1)\sin(\lambda+1)\,\theta + A_4\lambda(\lambda+1)\sin(\lambda-1)\,\theta\Big] . \quad (11.21)
$$

For this solution to satisfy the traction-free boundary conditions

$$
\sigma_{\theta r} = \sigma_{\theta\theta} = 0; \quad \theta = \pm\alpha \quad (11.22)
$$

we require

$$
\lambda\Big[A_1(\lambda+1)\sin(\lambda+1)\alpha + A_2(\lambda-1)\sin(\lambda-1)\alpha
$$
$$
- A_3(\lambda+1)\cos(\lambda+1)\alpha - A_4(\lambda-1)\cos(\lambda-1)\alpha\Big] = 0
$$
$$
\lambda\Big[- A_1(\lambda+1)\sin(\lambda+1)\alpha - A_2(\lambda-1)\sin(\lambda-1)\alpha
$$
$$
- A_3(\lambda+1)\cos(\lambda+1)\alpha - A_4(\lambda-1)\cos(\lambda-1)\alpha\Big] = 0
$$
$$
\lambda\Big[A_1(\lambda+1)\cos(\lambda+1)\alpha + A_2(\lambda+1)\cos(\lambda-1)\alpha
$$
$$
+ A_3(\lambda+1)\sin(\lambda+1)\alpha + A_4(\lambda+1)\sin(\lambda-1)\alpha\Big] = 0
$$

$$\lambda \Big[A_1(\lambda + 1)\cos(\lambda + 1)\alpha + A_2(\lambda + 1)\cos(\lambda - 1)\alpha$$
$$- A_3(\lambda + 1)\sin(\lambda + 1)\alpha - A_4(\lambda + 1)\sin(\lambda - 1)\alpha \Big] = 0 \; .$$

This is a set of four homogeneous equations for the four constants A_1, A_2, A_3, A_4 and will have a non-trivial solution only for certain eigenvalues of the exponent λ. Since all the equations have a λ multiplier, $\lambda = 0$ must be an eigenvalue for all wedge angles. We can simplify the equations by cancelling this factor and taking sums and differences in pairs to expose the symmetry of the system, with the result

$$A_1(\lambda + 1)\sin(\lambda + 1)\alpha + A_2(\lambda - 1)\sin(\lambda - 1)\alpha = 0 \qquad (11.23)$$
$$A_1(\lambda + 1)\cos(\lambda + 1)\alpha + A_2(\lambda + 1)\cos(\lambda - 1)\alpha = 0 \qquad (11.24)$$
$$A_3(\lambda + 1)\cos(\lambda + 1)\alpha + A_4(\lambda - 1)\cos(\lambda - 1)\alpha = 0 \qquad (11.25)$$
$$A_3(\lambda + 1)\sin(\lambda + 1)\alpha + A_4(\lambda + 1)\sin(\lambda - 1)\alpha = 0 \; . \qquad (11.26)$$

This procedure partitions the coefficient matrix, to yield the two independent matrix equations

$$\begin{bmatrix} (\lambda + 1)\sin(\lambda + 1)\alpha, & (\lambda - 1)\sin(\lambda - 1)\alpha \\ (\lambda + 1)\cos(\lambda + 1)\alpha, & (\lambda + 1)\cos(\lambda - 1)\alpha \end{bmatrix} \begin{bmatrix} A_1 \\ A_2 \end{bmatrix} = \begin{bmatrix} 0 \\ 0 \end{bmatrix} \qquad (11.27)$$

and

$$\begin{bmatrix} (\lambda + 1)\cos(\lambda + 1)\alpha, & (\lambda - 1)\cos(\lambda - 1)\alpha \\ (\lambda + 1)\sin(\lambda + 1)\alpha, & (\lambda + 1)\sin(\lambda - 1)\alpha \end{bmatrix} \begin{bmatrix} A_3 \\ A_4 \end{bmatrix} = \begin{bmatrix} 0 \\ 0 \end{bmatrix} \; . \qquad (11.28)$$

The symmetric terms A_1, A_2 have a non-trivial solution if and only if the determinant of the coefficient matrix in (11.27) is zero, leading to the characteristic equation

$$\lambda \sin 2\alpha + \sin(2\lambda\alpha) = 0 \; , \qquad (11.29)$$

whilst the antisymmetric terms A_3, A_4 have a non-trivial solution if and only if

$$\lambda \sin 2\alpha - \sin(2\lambda\alpha) = 0 \; , \qquad (11.30)$$

from (11.28).

11.2.3 Nature of the eigenvalues

We first note from equations (11.19–11.21) that the stress components are proportional to $r^{\lambda-1}$ and hence λ is restricted to positive values by the energy criterion of §11.2.1. Furthermore, we shall generally be most interested in the lowest possible eigenvalue[6], since this will define the strongest singularity that can occur at the notch and will therefore dominate the stress field at sufficiently small r.

Both equations (11.29, 11.30) are satisfied for all α if $\lambda=0$ and (11.30) is also satisfied for all α if $\lambda=1$.

The case $\lambda=0$ is strictly excluded by the energy criterion, but we shall find in the next chapter that it corresponds to the important case in which a concentrated force is applied at the origin. It can legitimately be regarded as describing the stress field sufficiently distant from a load distributed on the wedge faces near the origin, but as such it is clearly not appropriate to the unloaded problem of Figure 11.3.

The solution $\lambda=1$ of equation (11.30) is a spurious eigenvalue, since if we compute the corresponding eigenfunction, it turns out to have the null form $\phi=A_4 \sin(0)$. The correct limiting form for $\lambda=1$ requires the modified stress function (11.4) and leads to the condition (11.16) for antisymmetric fields, which is satisfied only for $2\alpha=257.4^o$.

Some insight into the nature of the eigenvalues for more general wedge angles can be gained from the graphical representation of Figure 11.5, where we plot the two terms in equations (11.29, 11.30) against $x=2\lambda\alpha$. The sine wave represents the term $\sin(2\lambda\alpha)$ $(=\sin x)$ and the straight lines represent various possible positions for the terms $\pm\lambda\sin(2\alpha)=\pm x(\sin(2\alpha)/2\alpha)$.

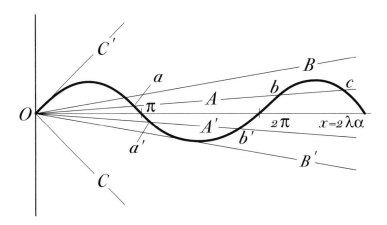

Fig. 11.5 Graphical solution of equations (11.29, 11.30).

The simplest case is that in which $2\alpha=2\pi$ and the wedge becomes the crack illustrated in Figure 11.6. We then have $\sin(2\alpha)=0$ and the solutions of both equations (11.29, 11.30) correspond to the points where the sine wave crosses the horizontal

---

[6] More rigorously, a general stress field near the corner can be expressed as an eigenfunction expansion, but the term with the smallest eigenvalue will be arbitrarily larger than all the other terms in the expansion at sufficiently small values of r.

axis — i.e.

$$\lambda = \frac{1}{2}, \ 1, \ \frac{3}{2}, \ \ldots \tag{11.31}$$

This is an important special case because of its application to fracture mechanics. In particular, we note that the lowest positive eigenvalue of λ is $\frac{1}{2}$ and hence the crack-tip field is square-root singular for both symmetric and antisymmetric loading.

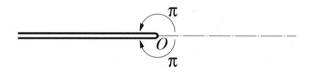

Fig. 11.6 Tip of a crack considered as a 360° wedge.

Suppose we now consider a wedge of somewhat less than 2π, in which case equations (11.29, 11.30) are represented by the intersection between the sine wave and the lines A, A' respectively in Figure 11.5.

As the wedge angle is reduced, the slope $(\sin(2\alpha)/2\alpha)$ of the lines first becomes increasingly negative until a maximum is reached at $2\alpha = 257.4^o$, corresponding to the tangential line B' and its mirror image B. Further reduction causes the slope to increase monotonically, passing through zero again at $2\alpha = \pi$ (corresponding to the case of the half-plane) and reaching the limit C, C' at $2\alpha = 0$.

Clearly, if the lines A, A' are even slightly inclined to the horizontal, they will only intersect a finite number of waves, corresponding to a finite set of real roots, the number of real roots increasing as the slope approaches zero. Increasing the slope causes initially equal-spaced real roots to become closer in pairs such as b, c in Figure 11.5. For any given pair, there will be a critical slope at which the roots coalesce and for further increase in slope, a pair of complex conjugate roots will be developed.

In the antisymmetric case for $360^o > 2\alpha > 257.4^o$, the intersection b' in Figure 11.5 corresponds to the spurious eigenvalue of (11.30), but a' is a meaningful eigenvalue corresponding to a stress singularity that weakens from square-root at $2\alpha = 360^o$ to zero at $2\alpha = 257.4^o$. For smaller wedge angles, a' is the spurious eigenvlaue and b' corresponds to a real but non-singular root.

In the symmetric case, the root a gives a real eigenvalue corresponding to a singular stress field whose strength falls monotonically from square-root to zero as the wedge angle is reduced from 2π to π. The strength of the singularity $(\lambda - 1)$ for both symmetric and antisymmetric terms is plotted against total wedge angle[7] in Figure 11.7. The singularity associated with the symmetric field is always stronger

---

[7] Figure 11.7 is reproduced by courtesy of Professor D. A. Hills of Oxford University.

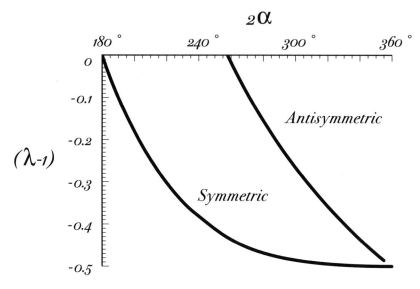

Fig. 11.7 Strength of the singularity in a reëntrant corner.

than that for the antisymmetric field except in the limit $2\alpha = 2\pi$, where they are equal, and there is a range of angles ($257.4^{o} > 2\alpha > 180^{o}$) where the symmetric field is singular, but the antisymmetric field is not.

For non-reëntrant wedges ($2\alpha < \pi$), both fields are bounded, the dominant eigenvalue for the symmetric field becoming complex for $2\alpha < 146^{o}$. Since bounded solutions correspond to eigenfunctions with r raised to a power with positive real part, the stress field always tends to zero in a non-reëntrant corner.

11.2.4 The singular stress fields

If $\lambda = \lambda_S$ is an eigenvalue of (11.29), the two equations (11.23, 11.24) will not be linearly independent and we can satisfy both of them by choosing A_1, A_2 such that

$$A_1 = A(\lambda_S - 1)\sin(\lambda_S - 1)\alpha$$
$$A_2 = -A(\lambda_S + 1)\sin(\lambda_S + 1)\alpha ,$$

where A is a new arbitrary constant. The corresponding (symmetric) singular stress field is then defined through the stress function

$$\phi_S = Ar^{\lambda_S+1}\Big[(\lambda_S - 1)\sin(\lambda_S - 1)\alpha\cos(\lambda_S + 1)\theta$$
$$- (\lambda_S + 1)\sin(\lambda_S + 1)\alpha\cos(\lambda_S - 1)\theta\Big] . \tag{11.32}$$

A similar procedure with equations (11.25, 11.26) defines the antisymmetric singular field through the stress function

$$\phi_A = Br^{\lambda_A+1}\Big[(\lambda_A + 1)\sin(\lambda_A - 1)\alpha\sin(\lambda_A + 1)\theta$$
$$- (\lambda_A + 1)\sin(\lambda_A + 1)\alpha\sin(\lambda_A - 1)\theta\Big] , \tag{11.33}$$

where λ_A is an eigenvalue of (11.30) and B is an arbitrary constant.

For the crack tip of Figure 11.6, the eigenvalues of both equations are given by equation (11.31) and the dominant singular fields correspond to the values $\lambda_S = \lambda_A = \frac{1}{2}$. Substituting this value into (11.32) and the resulting expression into (8.8, 8.9), we obtain the symmetric singular field as

$$\sigma_{rr} = \frac{K_I}{\sqrt{2\pi r}}\left[\frac{5}{4}\cos\left(\frac{\theta}{2}\right) - \frac{1}{4}\cos\left(\frac{3\theta}{2}\right)\right]$$

$$\sigma_{\theta\theta} = \frac{K_I}{\sqrt{2\pi r}}\left[\frac{3}{4}\cos\left(\frac{\theta}{2}\right) + \frac{1}{4}\cos\left(\frac{3\theta}{2}\right)\right]$$

$$\sigma_{r\theta} = \frac{K_I}{\sqrt{2\pi r}}\left[\frac{1}{4}\sin\left(\frac{\theta}{2}\right) + \frac{1}{4}\sin\left(\frac{3\theta}{2}\right)\right] ,$$

where we have introduced the new constant

$$K_I = 3A\sqrt{\frac{\pi}{2}} ,$$

known as the *mode I stress-intensity factor*[8].

The corresponding antisymmetric singular crack-tip field is obtained from (11.33) as

$$\sigma_{rr} = \frac{K_{II}}{\sqrt{2\pi r}}\left[-\frac{5}{4}\sin\left(\frac{\theta}{2}\right) + \frac{3}{4}\sin\left(\frac{3\theta}{2}\right)\right]$$

$$\sigma_{\theta\theta} = \frac{K_{II}}{\sqrt{2\pi r}}\left[-\frac{3}{4}\sin\left(\frac{\theta}{2}\right) - \frac{3}{4}\sin\left(\frac{3\theta}{2}\right)\right]$$

$$\sigma_{r\theta} = \frac{K_{II}}{\sqrt{2\pi r}}\left[\frac{1}{4}\cos\left(\frac{\theta}{2}\right) + \frac{3}{4}\cos\left(\frac{3\theta}{2}\right)\right] ,$$

where K_{II} is the mode II stress-intensity factor.

---

[8] Brittle materials typically fail when the stress-intensity factor at a crack tip reaches a critical value. This will be discussed in more detail in §13.3.

11.2.5   Other geometries

Williams' method can also be applied to other discontinuities in elasticity and indeed in other fields in mechanics. For example, we can extend it to determine the strength of the singularity in a composite wedge of two different materials[9]. A special case is illustrated in Figure 11.8, where two dissimilar materials are bonded[10], leaving a composite wedge of angle $3\pi/2$ at the points A, B. Another composite wedge is the tip of a crack (or debonded zone) at the interface between two dissimilar materials (see Chapter 35).

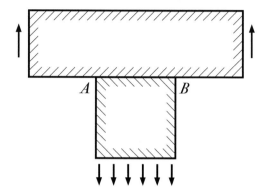

Fig. 11.8 Composite T-bar, with composite notches at the reëntrant corners A, B.

    The method has also been applied to determine the asymptotic fields in contact problems, with and without friction[11] (see Chapter 12 below).

    These problems generally involve both traction and displacement boundary conditions. For example, if the dissimilar wedges $\alpha_1 < \theta < \alpha_0$ and $\alpha_0 < \theta < \alpha_2$ are bonded together at the common plane $\theta = \alpha_0$ and the other faces $\theta = \alpha_1, \alpha_2$ are traction free, the boundary conditions are

---

[9] D. B. Bogy (1971), Two edge-bonded elastic wedges of different materials and wedge angles under surface tractions, *ASME Journal of Applied Mechanics,* Vol. 38, pp. 377–386.

[10] A problem of this kind was solved by G. G. Adams (1980), A semi-infinite elastic strip bonded to an infinite strip, *ASME Journal of Applied Mechanics,* Vol. 47, pp. 789–794.

[11] J. Dundurs and M. S. Lee (1972), Stress concentration at a sharp edge in contact problems, *Journal of Elasticity*, Vol. 2, pp. 109–112, M. Comninou (1976), Stress singularities at a sharp edge in contact problems with friction, *Zeitschrift für angewandte Mathematik und Physik,* Vol. 27, pp. 493–499.

$$\sigma^1_{\theta\theta}(r, \alpha_1) = 0 \tag{11.34}$$

$$\sigma^1_{\theta r}(r, \alpha_1) = 0 \tag{11.35}$$

$$\sigma^1_{\theta\theta}(r, \alpha_0) - \sigma^2_{\theta\theta}(r, \alpha_0) = 0 \tag{11.36}$$

$$\sigma^1_{\theta r}(r, \alpha_0) - \sigma^2_{\theta r}(r, \alpha_0) = 0 \tag{11.37}$$

$$u^1_r(r, \alpha_0) - u^2_r(r, \alpha_0) = 0 \tag{11.38}$$

$$u^1_\theta(r, \alpha_0) - u^2_\theta(r, \alpha_0) = 0 \tag{11.39}$$

$$\sigma^2_{\theta\theta}(r, \alpha_2) = 0 \tag{11.40}$$

$$\sigma^2_{\theta r}(r, \alpha_2) = 0 \tag{11.41}$$

where superscripts 1, 2 refer to the two wedges, respectively. Equations (11.36, 11.37) can be seen as statements of Newton's third law for the tractions transmitted between the two wedges, whilst (11.38, 11.39) express the fact that if two bodies are bonded together, adjacent points on opposite sides of the bond must have the same displacement.

To formulate asymptotic problems of this type, the stress equations (11.19–11.21) must be supplemented by the corresponding expressions for the displacements u_r, u_θ. These can be written down from Table 9.1, since in that table the parameter n does not necessarily have to take integer values. Alternatively, they can be obtained directly by substituting the expression (11.18) into the Maple or Mathematica file 'urt'. Omitting the rigid-body displacements, we obtain

$$2\mu u_r = r^\lambda \left[-A_1(\lambda + 1) \cos(\lambda + 1)\theta + A_2(\kappa - \lambda) \cos(\lambda - 1)\theta \right.$$
$$\left. - A_3(\lambda + 1) \sin(\lambda + 1)\theta + A_4(\kappa - \lambda) \sin(\lambda - 1)\theta \right] \tag{11.42}$$

$$2\mu u_\theta = r^\lambda \left[A_1(\lambda + 1) \sin(\lambda + 1)\theta + A_2(\kappa + \lambda) \sin(\lambda - 1)\theta \right.$$
$$\left. - A_3(\lambda + 1) \cos(\lambda + 1)\theta - A_4(\kappa + \lambda) \cos(\lambda - 1)\theta \right]. \tag{11.43}$$

The results of asymptotic analysis are very useful in numerical methods, since they enable us to predict the nature of the local stress field and hence devise appropriate special elements or meshes. Failure to do this in problems with singular fields will always lead to numerical inefficiency and sometimes to lack of convergence, mesh sensitivity or instability.

To look ahead briefly to three-dimensional problems, we note that when we concentrate our attention on a very small region at a corner, the resulting magnification makes all other dimensions (including radii of curvature etc.) look very large. Thus, a notch in a three-dimensional body will generally have a two-dimensional asymptotic field at small r. The above results and the underlying method are therefore of very general application. Another consequence is that the out-of-plane dimension becomes magnified indefinitely, so that plane strain (rather than plane stress) conditions are appropriate in all three-dimensional asymptotic problems.

11.3   General Loading of the Faces

If the faces of the wedge $\theta=\pm\alpha$ are subjected to tractions that can be expanded as power series in r, a solution can be obtained using a series of terms like (11.1). However, as with the rectangular beam, power series are of limited use in representing a general traction distribution, particularly when it is relatively localized. Instead, we can adapt the Fourier transform representation (5.93) to the wedge, by noting that the stress function (11.18) will oscillate along lines of constant θ if we choose complex values for λ. This leads to a representation as a *Mellin transform* defined as

$$\phi(r,\theta) = \frac{1}{2\pi i} \int_{c-\iota\infty}^{c+\iota\infty} f(s,\theta) r^{-s} ds \ , \tag{11.44}$$

for which the inversion is[12]

$$f(s,\theta) = \int_0^\infty \phi(r,\theta) r^{s-1} dr \ .$$

The integrand of (11.44) is the function (11.18) with $\lambda=-s-1$. The integral is path-independent in any strip of the complex plane in which the integrand is a holomorphic[13] function of s. If we evaluate it along the straight line $s=c+\iota\omega$, it can be written in the form

$$r^c \phi(r,\theta) = \frac{1}{2\pi} \int_{-\infty}^\infty f(c+\iota\omega,\theta) r^{-\iota\omega} d\omega$$
$$= \frac{1}{2\pi} \int_{-\infty}^\infty f(c+\iota\omega,\theta) e^{-\iota\omega \ln(r)} d\omega \ ,$$

which is readily converted to a Fourier integral[14] by the change of variable $t=\ln(r)$. The constant c has to be chosen to ensure regularity of the integrand. If there are no unacceptable singularities in the tractions and the latter are bounded at infinity, it can be shown that a suitable choice is $c=-1$.

The Mellin transform and power series methods can be applied to a wedge of any angle, but it should be remarked that, because of the singular asymptotic fields obtained in §11.2, meaningful results for the reëntrant wedge ($2\alpha>\pi$) can only be obtained if the conditions at infinity are also prescribed and satisfied.

---

[12] I. N. Sneddon, *Fourier Transforms*, McGraw-Hill, New York, 1951.

[13] See Chapter 19 and in particular, §19.5.

[14] The basic theory of the Mellin transform and its relation to the Fourier transform is explained by I. N. Sneddon, *loc. cit.*. For applications to elasticity problems for the wedge, see E. Sternberg and W. T. Koiter (1958), The wedge under a concentrated couple: A paradox in the two-dimensional theory of elasticity, *ASME Journal of Applied Mechanics,* Vol. 25, pp. 575–581; W. J. Harrington and T. W. Ting (1971), Stress boundary-value problems for infinite wedges, *Journal of Elasticity,* Vol. 1, pp. 65–81.

More precisely, we could formulate the problem for the large but finite sector $-\alpha < \theta < \alpha$, $0 < r < b$, with precribed tractions on all edges[15]. The procedure would be first to develop a particular solution for the loading on the faces $\theta = \pm\alpha$ as if the wedge were infinite and then to correct the boundary conditions at $r = b$ by superposing an infinite sequence of eigenfunctions from §11.2 with appropriate multipliers[16]. For sufficiently large b, the stress field near the apex of the wedge would be adequately described by the particular solution and the singular terms from the eigenfunction expansion.

Problems

11.1. Figure 11.9 shows a half-plane, $y < 0$, subjected to a uniform shear traction, $\sigma_{xy} = S$ on the half-line, $x > 0$, $y = 0$, the remaining tractions on $y = 0$ being zero.

Find the complete stress field in the half-plane. **Note**: This is a problem requiring a special stress function with a logarithmic multiplier (see §11.1.4 above).

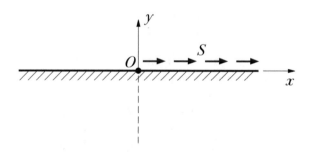

Fig. 11.9 The half-plane with shear tractions.

11.2. The half-plane $y < 0$ is subjected to a uniform normal pressure $\sigma_{yy} = -S$ on the half-line, $x > 0$, $y = 0$, the remaining tractions on $y = 0$ being zero. Find the complete stress field in the half-plane.

11.3. Find the displacements u_r, u_θ due to the stress function (11.8) and hence show that the strains are everywhere bounded, but the rotation ω is logarithmically

---

[15] A problem of this kind was considered by G. Tsamasphyros and P. S. Theocaris (1979), On the solution of the sector problem, *Journal of Elasticity*, Vol. 9, pp. 271–281.

[16] This requires that the eigenfunction series is complete for this problem. The proof is given by R. D. Gregory (1979), Green's functions, bi-linear forms and completeness of the eigenfunctions for the elastostatic strip and wedge, *Journal of Elasticity*, Vol. 9, pp. 283–309.

unbounded at the corner. You can assume without proof that the expression for ω in polar coordinates is

$$\omega = \frac{1}{2}\left[\left(\frac{\partial u_r}{\partial r} - \frac{1}{r}\frac{\partial u_\theta}{\partial \theta}\right)\cos(2\theta) - \left(\frac{\partial u_\theta}{\partial r} + \frac{1}{r}\frac{\partial u_r}{\partial \theta}\right)\sin(2\theta)\right] .$$

11.4. The wedge $-\alpha < \theta < \alpha$ is loaded by a concentrated moment M_0 at the apex, the plane edges being traction free. Use dimensional arguments to show that the stress components must all have the separated-variable form

$$\sigma = \frac{f(\theta)}{r^2} .$$

Use this result to choose a suitable stress function and hence find the complete stress field in the wedge.

11.5. Show that $\phi = Ar^2\theta$ can be used as a stress function and determine the tractions which it implies on the boundaries of the region $-\pi/2 < \theta < \pi/2$. Hence show that the stress function appropriate to the loading of Figure 11.10 is

$$\phi = -\frac{F}{4\pi a}(r_1^2\theta_1 + r_2^2\theta_2)$$

where $r_1, r_2, \theta_1, \theta_2$ are defined in the Figure.
    Determine the principal stresses at the point B.

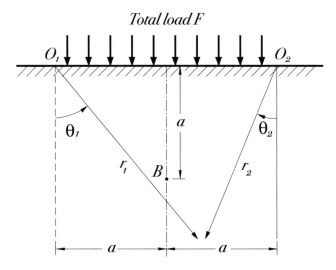

Fig. 11.10 Uniform loading over a discrete region.

11.6. A wedge-shaped concrete dam is subjected to a hydrostatic pressure ρgh varying with depth h on the vertical face $\theta=0$ as shown in Figure 11.11, the other face $\theta = \alpha$ being traction free. The dam is also loaded by self-weight, the density of concrete being $\rho_c = 2.3\rho$.

Find the minimum wedge angle α if there is to be no tensile stress in the dam.

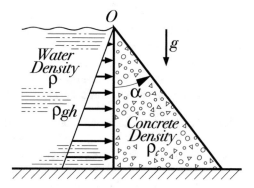

Fig. 11.11 The wedge-shaped dam.

11.7. Figure 11.12 shows a 45° triangular plate ABC built in at BC and loaded by a uniform pressure p_0 on the upper edge AB. The inclined edge AC is traction free.

Find the stresses in the plate, using weak boundary conditions on the edge BC (which do not need to be explicitly enforced). Hence compare the maximum tensile stress with the prediction of the elementary bending theory.

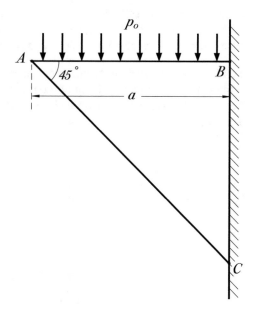

Fig. 11.12 A triangular plate

11.8. The triangular body $-\alpha < \theta < \alpha$, $x = r\cos\theta < a$ is supported at $x = a$ and loaded by uniform antisymmetric shear tractions $\sigma_{\theta r}(r, \pm\alpha) = S$. Find the Cartesian stress components σ_{xx}, σ_{yy}, σ_{xy} in the body using weak boundary conditions on $x = a$ and strong conditions on the inclined surfaces.

What force and/or moment resultants are transmitted across the surface $x = a$?

11.9. The wedge $-\alpha < \theta < \alpha$ is bonded to a rigid body on both edges $\theta = \pm\alpha$. Use the eigenfunction expansion of §11.2.2 to determine the characteristic equations that must be satisfied by the exponent λ in the stress function (11.18) for symmetric and antisymmetric stress fields. Show that these equations reduce to (11.29, 11.30) if $\nu = 0.5$ and plane strain conditions are assumed.

11.10. Find the equation that must be satisfied by λ if the stress function (11.18) is to define a non-trivial solution of the problem of the half-plane $0 < \theta < \pi$, traction free on $\theta = 0$ and in frictionless contact with a rigid plane surface at $\theta = \pi$. Find the lowest value of λ that satisfies this equation and obtain explicit expressions for the form of the corresponding singular stress field near the corner. This solution is of importance in connection with the frictionless indentation of a smooth elastic body by a rigid body with a sharp corner.

11.11. Two large bodies of similar materials with smooth, continuously differentiable curved surfaces make frictionless contact. Use the asymptotic method of §11.2 to examine the asymptotic stress fields near the edge of the resulting contact area. The solution of this problem must also satisfy two *inequalities*: that the contact tractions in the contact region must be compressive and that the normal displacements in the non-contact region must cause a non-negative gap. Show that one or other of these conditions is violated, whatever sign is taken for the multiplier on the first eigenfunction. What conclusion do you draw from this result?

11.12. Find the equation that must be satisfied by λ if the stress function (11.18) is to define a non-trivial solution of the problem of the wedge $0 < \theta < \pi/2$, traction free on $\theta = 0$ and bonded to a rigid plane surface at $\theta = \pi/2$. Do not attempt to solve the equation.

Chapter 12
Plane Contact Problems

In the previous chapter, we considered problems in which the infinite wedge was loaded on its faces or solely by tractions on the infinite boundary. A related problem of considerable practical importance concerns the wedge with traction-free faces, loaded by a concentrated force F at the vertex, as shown in Figure 12.1.[1]

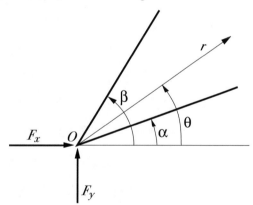

Fig. 12.1 The wedge loaded by a force at the vertex.

12.1 Self-Similarity

An important characteristic of this problem is that there is no inherent length scale. An enlarged photograph of the problem would look the same as the original. The

---

[1] For a more detailed discussion of elastic contact problems, see K. L. Johnson, *Contact Mechanics*, Cambridge University Press, 1985, G. M. L. Gladwell, *Contact Problems in the Classical Theory of Elasticity*, Sijthoff and Noordhoff, Alphen aan den Rijn, 1980, and J. R. Barber *Contact Mechanics*, Springer, Cham, 2018.

© The Author(s), under exclusive license to Springer Nature Switzerland AG 2022
J. R. Barber, *Elasticity*, Solid Mechanics and Its Applications 172,
https://doi.org/10.1007/978-3-031-15214-6_12

solution must therefore share this characteristic and hence, for example, contours of the stress function ϕ must have the same geometric shape at all distances from the vertex. Problems of this type — in which the solution can be mapped into itself after a change of length scale — are described as *self-similar*.

An immediate consequence of the self-similarity is that all the stress components must be capable of expression in the separated-variable form

$$\sigma = f(r)g(\theta) .$$

Furthermore, since the tractions on the line $r = a$ must balance a constant force $\boldsymbol{F}$ for all a, we can deduce that $f(r) = r^{-1}$, because the area available for transmitting the force increases linearly with radius[2].

12.2   The Flamant Solution

Choosing those terms in Table 8.1 that give stresses proportional to r^{-1}, we obtain

$$\phi = C_1 r\theta \sin\theta + C_2 r\theta \cos\theta + C_3 r \ln(r) \cos\theta + C_4 r \ln(r) \sin\theta , \qquad (12.1)$$

for which the stress components are

$$\sigma_{rr} = r^{-1}(2C_1 \cos\theta - 2C_2 \sin\theta + C_3 \cos\theta + C_4 \sin\theta)$$
$$\sigma_{r\theta} = r^{-1}(C_3 \sin\theta - C_4 \cos\theta)$$
$$\sigma_{\theta\theta} = r^{-1}(C_3 \cos\theta + C_4 \sin\theta) .$$

For the wedge faces to be traction free, we require

$$\sigma_{\theta r} = \sigma_{\theta\theta} = 0 \; ; \quad \theta = \alpha, \beta ,$$

which can be satisfied by taking $C_3, C_4 = 0$.

To determine the remaining constants C_1, C_2, we consider the equilibrium of the region $0 < r < a$, obtaining

$$F_x + 2 \int_\alpha^\beta \left(\frac{C_1 \cos\theta - C_2 \sin\theta}{a} \right) a \cos\theta d\theta = 0 \qquad (12.2)$$

$$F_y + 2 \int_\alpha^\beta \left(\frac{C_1 \cos\theta - C_2 \sin\theta}{a} \right) a \sin\theta d\theta = 0 . \qquad (12.3)$$

---

[2] In the same way, we can deduce that the stresses in a cone subjected to a force at the vertex must vary with r^{-2} (see §24.2).

These two equations permit us to choose the constants C_1, C_2 to give the applied force F any desired magnitude and direction. Notice that the radius a cancels in equations (12.2, 12.3) and hence, if they are satisfied for any radius, they are satisfied for all.

This is known as the *Flamant solution*, or sometimes as the *simple radial distribution*, since the act of setting $C_3 = C_4 = 0$ clears all θ-surfaces of tractions, leaving

$$\sigma_{rr} = \frac{2C_1 \cos \theta}{r} - \frac{2C_2 \sin \theta}{r}$$

as the only non-zero stress component.

12.3   The Half-Plane

If we take $\alpha = -\pi$ and $\beta = 0$ in the above solution, we obtain the special case of a force acting at a point on the surface of the half-plane $y < 0$ as shown in Figure 12.2.

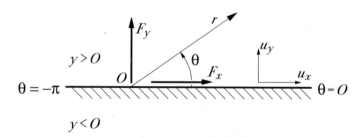

Fig. 12.2 Force acting on the surface of a half-plane.

For this case, performing the integrals in equations (12.2, 12.3), we find

$$F_x + 2 \int_{-\pi}^{0} (C_1 \cos^2 \theta - C_2 \sin \theta \cos \theta) d\theta = F_x + \pi C_1 = 0$$

$$F_y + 2 \int_{-\pi}^{0} (C_1 \cos \theta \sin \theta - C_2 \sin^2 \theta) d\theta = F_y - \pi C_2 = 0$$

and hence

$$C_1 = -\frac{F_x}{\pi} \; ; \; C_2 = \frac{F_y}{\pi} \; .$$

It is convenient to consider these terms separately.

12.3.1   The normal force F_y

The normal (tensile) force F_y produces the stress field

$$\sigma_{rr} = -\frac{2F_y \sin\theta}{\pi r} , \tag{12.4}$$

which reaches a maximum tensile value on the negative y-axis ($\theta=-\pi/2$) — i.e. directly beneath the load — and falls to zero as we approach the surface along any line other than one which passes through the point of application of the force.

The displacement field corresponding to this stress can conveniently be taken from Table 9.1 and is

$$2\mu u_r = \frac{F_y}{2\pi}\left[(\kappa-1)\theta\cos\theta - (\kappa+1)\ln(r)\sin\theta + \sin\theta\right] \tag{12.5}$$

$$2\mu u_\theta = \frac{F_y}{2\pi}\left[-(\kappa-1)\theta\sin\theta - (\kappa+1)\ln(r)\cos\theta - \cos\theta\right]. \tag{12.6}$$

This solution will be used as a Green's function for problems in which a half-plane is subjected to various surface loads, and hence the surface displacements are of particular interest.

On $\theta=0$, $(y=0, \; x>0)$, we have

$$2\mu u_r = 2\mu u_x = 0 \tag{12.7}$$

$$2\mu u_\theta = 2\mu u_y = \frac{F_y}{2\pi}\left[-(\kappa+1)\ln(r)-1\right], \tag{12.8}$$

whilst on $\theta=-\pi$, $(y=0, \; x<0)$, we have

$$2\mu u_r = -2\mu u_x = \frac{F_y}{2}(\kappa-1) \tag{12.9}$$

$$2\mu u_\theta = -2\mu u_y = \frac{F_y}{2\pi}\left[(\kappa+1)\ln(r)+1\right]. \tag{12.10}$$

It is convenient to impose symmetry on the solution by superposing a rigid-body displacement

$$2\mu u_x = \frac{F_y(\kappa-1)}{4} \; ; \; 2\mu u_y = \frac{F_y}{2\pi} ,$$

after which equations (12.7–12.10) can be summarised in the form

$$u_x(x,0) = \frac{F_y(\kappa-1)\mathrm{sgn}(x)}{8\mu} \tag{12.11}$$

$$u_y(x,0) = -\frac{F_y(\kappa+1)\ln|x|}{4\pi\mu} , \tag{12.12}$$

where the signum function $\text{sgn}(x)$ is defined to be $+1$ for $x > 0$ and -1 for $x < 0$.
Note incidentally that $r = |x|$ on $y = 0$.

12.3.2 The tangential force F_x

The tangential force F_x produces the stress field

$$\sigma_{rr} = -\frac{2F_x \cos\theta}{\pi r} \,,$$

which is compressive ahead of the force ($\theta = 0$) and tensile behind it ($\theta = -\pi$) as we
might expect.

The corresponding displacement field is found from Table 9.1 as before and after
superposing an appropriate rigid-body displacement as in §12.3.1, the surface displacements can be written

$$u_x(x, 0) = -\frac{F_x(\kappa + 1) \ln |x|}{4\pi\mu} \quad ; \quad u_y(x, 0) = -\frac{F_x(\kappa - 1)\text{sgn}(x)}{8\mu} \,.$$

12.3.3 Summary

We summarize the results of §§12.3.1, 12.3.2 in the equations

$$u_x(x, 0) = -\frac{F_x(\kappa + 1) \ln |x|}{4\pi\mu} + \frac{F_y(\kappa - 1)\text{sgn}(x)}{8\mu} \tag{12.13}$$

$$u_y(x, 0) = -\frac{F_x(\kappa - 1)\text{sgn}(x)}{8\mu} - \frac{F_y(\kappa + 1) \ln |x|}{4\pi\mu} \,, \tag{12.14}$$

see Figure 12.3.

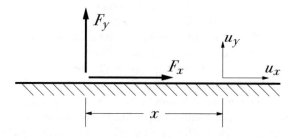

Fig. 12.3 Surface displacements due to a surface force.

12.4   Distributed Normal Tractions

Now suppose that the surface of the half-plane is subjected to a distributed normal load $p(\xi)$ per unit length as shown in Figure 12.4. The stress and displacement field can be found by superposition, using the Flamant solution as a Green's function — i.e. treating the distributed load as the limit of a set of concentrated loads of magnitude $p(\xi)\delta\xi$.

Of particular interest is the distortion of the surface, defined by the normal displacement u_y. At the point $P(x, 0)$, this is given by

$$u_y(x, 0) = -\frac{(\kappa + 1)}{4\pi\mu} \int_{\mathcal{A}} p(\xi) \ln |x - \xi| d\xi , \qquad (12.15)$$

from equation (12.12), where $\mathcal{A}$ is the region over which the load acts.

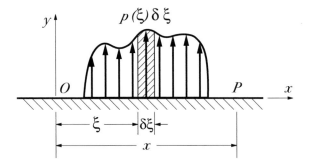

Fig. 12.4 Half-plane subjected to a distributed traction.

The displacement defined by equation (12.15) is logarithmically unbounded as x tends to infinity. In general, if a finite region of the surface of the half-plane is subjected to a non-self-equilibrated system of loads, the stress and displacement fields at a distance $r \gg \mathcal{A}$ will approximate those due to the force resultants applied as concentrated forces — i.e. to the expressions of equations (12.4–12.6) for normal loading. In other words, the stresses will decay with $1/r$ and the displacements (being integrals of the strains) will vary logarithmically.

This means that the half-plane solution can be used for the stresses in a finite body loaded on a region of the surface which is small compared with the linear dimensions of the body, since $1/r$ is arbitrarily small for sufficiently large r. However, the logarithm is not bounded at infinity and hence the rigid-body displacement of the loaded region with respect to distant parts of the body cannot be found without a more exact treatment taking into account the finite dimensions of the body. For this reason, contact problems for the half-plane are often formulated in terms of the displacement *gradient*

$$\frac{du_y}{dx}(x, 0) = -\frac{(\kappa + 1)}{4\pi\mu} \int_A \frac{p(\xi)d\xi}{(x - \xi)} ,$$ (12.16)

thereby avoiding questions of rigid-body translation.

When the point $x = \xi$ lies within A, the integral in (12.16) is interpreted as a Cauchy principal value — for example

$$\int_a^b \frac{p(\xi)d\xi}{(x - \xi)} = \lim_{\epsilon \to 0} \left(\int_a^{x-\epsilon} + \int_{x+\epsilon}^b \right) \frac{p(\xi)d\xi}{(x - \xi)} \quad ; \quad a < x < b .$$

12.5 Frictionless Contact Problems

The results of the last section permit us to develop a solution to the problem of a half-plane indented by a frictionless rigid punch of known profile (see Figure 12.5).

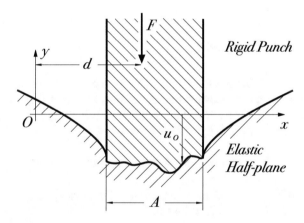

Fig. 12.5 The elastic half-plane indented by a frictionless rigid punch.

We suppose that the load F is sufficient to establish contact over the whole of the contact area A, in which case, the displacement of the half-plane must satisfy the equation

$$u_y(x, 0) = -u_0(x) + C_1 x + C_0 ; \quad x \in A ,$$ (12.17)

where u_0 is a known function of x which describes the profile of the punch.

The constants C_0, C_1 define an unknown rigid-body translation and rotation of the punch respectively, which can generally be assigned in such a way as to give a pressure distribution $p(\xi)$ which is statically equivalent to the force F — i.e.

$$\int_{\mathcal{A}} p(\xi)d\xi = -F \tag{12.18}$$

$$\int_{\mathcal{A}} p(\xi)\xi d\xi = -Fd , \tag{12.19}$$

where d defines the line of action of F as shown in Figure 12.5 and the negative signs are a consequence of the tensile positive convention for $p(\xi)$. These two equations are sufficient to determine the constants C_0, C_1.

If the contact region $\mathcal{A}$ is connected, we can define a coördinate system with origin at the mid-point and denote the half-length of the contact by a. Equations (12.16, 12.17) then give

$$-\frac{du_0}{dx} + C_1 = -\frac{(\kappa+1)}{4\pi\mu} \int_{-a}^{a} \frac{p(\xi)d\xi}{(x-\xi)} \;\; ; \;\; -a < x < a , \tag{12.20}$$

which is a Cauchy singular integral equation for the unknown pressure distribution $p(\xi)$.

12.5.1 Method of solution

Integral equations of this kind arise naturally in the application of the complex-variable method to two-dimensional potential problems and they are extensively discussed in a classical text by Muskhelishvili[3]. However, for our purposes, a simpler solution will suffice, based on the change of variable

$$x = a\cos\phi \;\; ; \;\; \xi = a\cos\theta \tag{12.21}$$

and expansion of the two sides of the equation as Fourier series.

Substituting (12.21) into (12.20), we find

$$\frac{1}{a\sin\phi}\frac{du_0}{d\phi} + C_1 = -\frac{(\kappa+1)}{4\pi\mu} \int_{0}^{\pi} \frac{p(\theta)\sin\theta d\theta}{(\cos\phi - \cos\theta)} \;\; ; \;\; 0 < \phi < \pi ,$$

which can be simplified using the result[4]

$$\int_{0}^{\pi} \frac{\cos(n\theta)d\theta}{(\cos\phi - \cos\theta)} = -\frac{\pi\sin(n\phi)}{\sin\phi} . \tag{12.22}$$

---

[3] N. I. Muskhelishvili, *Singular Integral Equations*, (English translation by J. R. M. Radok, Noordhoff, Groningen, 1953). For applications to elasticity problems, including the above contact problem, see N. I. Muskhelishvili, *Some Basic Problems of the Mathematical Theory of Elasticity*, (English translation by J. R. M. Radok, Noordhoff, Groningen, 1953). For a simpler discussion of the solution of contact problems involving singular integral equations, see K. L. Johnson, *loc. cit*, §2.7.

[4] For a method of proving this result, see Problem 19.7.

Writing

$$p(\theta) = \sum_{n=0}^{\infty} \frac{p_n \cos(n\theta)}{\sin \theta} \tag{12.23}$$

$$\frac{du_0}{d\phi} = \sum_{n=1}^{\infty} u_n \sin(n\phi) , \tag{12.24}$$

we find

$$\sum_{n=1}^{\infty} u_n \sin(n\phi) + C_1 a \sin \phi = \frac{(\kappa + 1)a \sin \phi}{4\pi\mu} \sum_{n=1}^{\infty} \frac{\pi p_n \sin(n\phi)}{\sin \phi}$$

$$= \frac{(\kappa + 1)a}{4\mu} \sum_{n=1}^{\infty} p_n \sin(n\phi)$$

and hence, equating Fourier coefficients,

$$p_n = \frac{4\mu u_n}{(\kappa + 1)a} \quad ; \quad n > 1 \tag{12.25}$$

$$p_1 = \frac{4\mu(C_1 a + u_1)}{(\kappa + 1)a} . \tag{12.26}$$

Thus, if the shape of the punch u_0 is expanded as in equation (12.24), the corresponding coefficients in the pressure series (12.23) can be written down using (12.25), except for $n = 0, 1$.

The two coefficients p_0, p_1 are found by substituting (12.23) into the equilibrium conditions (12.18, 12.19). From (12.18), we have

$$-F = \int_{-a}^{a} p(x')dx' = \int_{0}^{\pi} p(\theta)a \sin \theta d\theta$$

$$= a \sum_{n=0}^{\infty} p_n \int_{0}^{\pi} \cos(n\theta)d\theta = \pi a p_0 ,$$

and from (12.19),

$$-Fd = \int_{-a}^{a} p(x')x'dx' = \int_{0}^{\pi} p(\theta)a^2 \sin \theta \cos \theta d\theta$$

$$= a^2 \sum_{n=0}^{\infty} p_n \int_{0}^{\pi} \cos \theta \cos(n\theta)d\theta = \frac{\pi a^2 p_1}{2} .$$

Hence,

$$p_0 = -\frac{F}{\pi a} \quad ; \quad p_1 = -\frac{2Fd}{\pi a^2} . \tag{12.27}$$

The rigid-body rotation of the punch C_1 can then be found by eliminating p_1 between equations (12.26, 12.27), with the result

$$C_1 = -\frac{Fd(\kappa + 1)}{2\pi\mu a^2} - \frac{u_1}{a} .$$ (12.28)

Alternatively, if the punch is constrained to move vertically without rotation, C_1 will be zero and we can solve for the applied moment Fd as

$$Fd = -\frac{2\pi\mu u_1 a}{(\kappa + 1)} ,$$

from (12.28).

12.5.2   The flat punch

We first consider the case of a flat punch with a symmetric load as shown in Figure 12.6. The profile of the punch is described by the equation

$$u_0 = C \; ; \quad \frac{du_0}{dx} = 0$$ (12.29)

and hence $p_n = 0$ for $n > 1$ from equation (12.25). Furthermore, since the load is symmetric ($d = 0$), p_1 is also zero (equation (12.27)) and p_0 is given by (12.27).

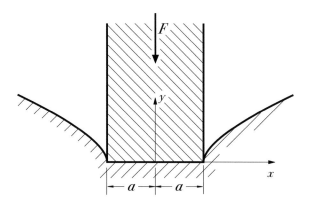

Fig. 12.6 Half-plane indented by a flat rigid punch.

The traction distribution under the punch is therefore

$$p(x) = \frac{p_0}{\sin\phi} = -\frac{F}{\pi a\sqrt{1 - x^2/a^2}}$$

$$= -\frac{F}{\pi\sqrt{a^2 - x^2}} \cdot \tag{12.30}$$

We note that the sharp corners of the punch cause a square root singularity in traction at the edges, $x = \pm a$. This is typical of indentation problems where the punch has a sharp corner and could have been predicted by performing an asymptotic analysis of the local stress field, using the methods developed in §11.2 (see Problem 11.10). The traction distribution is illustrated in Figure 12.7.

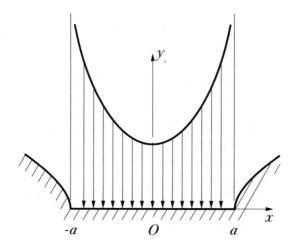

Fig. 12.7 Contact pressure distribution under the flat rigid punch.

12.5.3   The cylindrical punch (Hertz problem)

We next consider the indentation of a half-plane by a frictionless rigid cylinder (Figure 12.8), which is a special case of the contact problem associated with the name of Hertz[5].

This problem differs from the previous example in that the contact area semi-width a depends on the load F and cannot be determined *a priori*. Strictly, a must be determined from the pair of unilateral inequalities

---

[5] For the more general three-dimensional case, see Chapter 34.

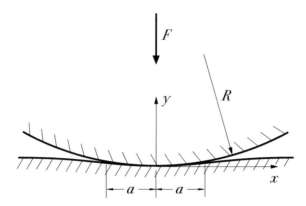

Fig. 12.8 The Hertzian contact problem.

$$p(x) \leq 0 \; ; \quad -a < x < a \tag{12.31}$$

$$-u_y(x) \geq u_0(x) \; ; \quad |x| > a \; , \tag{12.32}$$

which express the physical requirements that the contact traction should not be tensile (12.31), and that there should be no interference between the bodies outside the contact area (12.32). However, it can be shown that these conditions are in most cases formally equivalent to the requirement that the contact traction should not be singular at $x = \pm a$. This question will be discussed more rigorously in connection with three-dimensional contact problems in Chapter 31. Here it is sufficient to remark that a traction singularity such as that obtained at the sharp corner of the flat punch (equation (12.30)) causes a discontinuity in the displacement gradient du_y/dx, which will cause a violation of (12.32) if the punch is smooth.

If the radius of the cylinder is R, we have

$$\frac{d^2 u_0}{dx^2} = -\frac{1}{R} \tag{12.33}$$

and hence

$$u_0 = C_0 - \frac{x^2}{2R} = C_0 - \frac{a^2 \cos(2\phi)}{4R} - \frac{a^2}{4R} \; .$$

Thus

$$\frac{du_0}{d\phi} = \frac{a^2 \sin 2\phi}{2R}$$

and the only non-zero coefficient in the series (12.24) is

$$u_2 = \frac{a^2}{2R} \; .$$

It follows that

$$p_2 = \frac{2\mu a}{R(\kappa + 1)} \,,$$

from (12.25).

For the cylinder, the load must be symmetrical to retain equilibrium and hence $p_1 = 0$. As before, p_0 is given by (12.27) and hence

$$p(\theta) = \left(-\frac{F}{\pi a} + \frac{2\mu a}{R(\kappa + 1)} \cos(2\theta) \right) \Big/ \sin\theta \,.$$

Now this expression will be singular at $\theta = 0, \pi$ ($x = \pm a$) unless we choose a such that

$$\frac{F}{\pi a} = \frac{2\mu a}{R(\kappa + 1)} \,,$$

i.e.

$$a = \sqrt{\frac{F(\kappa + 1)R}{2\pi\mu}} \,. \tag{12.34}$$

Note that for plane strain, $\kappa = (3 - 4\nu)$ and

$$\frac{2\mu}{(\kappa + 1)} = \frac{E}{4(1 - \nu^2)} \,,$$

whilst for plane stress, $\kappa = (3 - \nu)/(1 + \nu)$ and

$$\frac{2\mu}{(\kappa + 1)} = \frac{E}{4} \,.$$

With the value of a from (12.34), we find

$$p(\theta) = -\frac{F\left[1 - \cos(2\theta)\right]}{\pi a \sin\theta} = -\frac{2F\sin\theta}{\pi a}$$

— i.e.

$$p(x) = -\frac{2F\sqrt{a^2 - x^2}}{\pi a^2} \,. \tag{12.35}$$

This pressure distribution is illustrated in Figure 12.9.

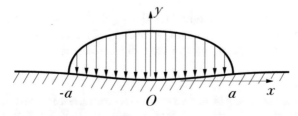

Fig. 12.9 The Hertzian pressure distribution.

12.6   Problems with Two Deformable Bodies

The same method can be used to treat problems involving the contact of two
deformable bodies, provided that they have sufficiently large radii in the vicinity
of the contact area to be approximated by half-planes — i.e. $R_1, R_2 \gg a$, where
R_1, R_2 are the local radii[6] (see Figure 12.10). We shall also take the opportunity to
generalize the formulation to the case where, in addition to the normal tractions p_y,
there are tangential tractions p_x at the interface due to friction.

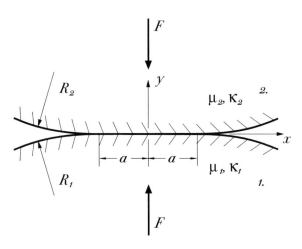

Fig. 12.10   Contact of two curved deformable bodies.

We first note that, by Newton's third law, the tractions must be equal and opposite
on the two surfaces, so the appropriate Green's function corresponds to the force
pairs F_x, F_y of Figure 12.11.

Suppose that the bodies are placed lightly in contact, as shown in Figure 12.12,
and that the initial gap between their surfaces is a known function $g_0(x)$. We now give
the upper body a vertical rigid-body translation C_0 and a small clockwise rotation
C_1 such that, in the absence of deformation, the gap would become

$$g(x) = g_0(x) - C_0 - C_1 x \ .$$

In addition, we assume that the contact tractions will cause some elastic deformation,
represented by the displacements u_1, u_2 in bodies 1,2 respectively.

As a result of these operations, the gap will be modified to

---

[6] In fact the same condition must be satisfied even for the case where the curved body is rigid, since
otherwise the small strain assumption of linear elasticity ($e_{ij} \ll 1$) will be violated near the contact
area.

Fig. 12.11 Green's function
for contact problems.

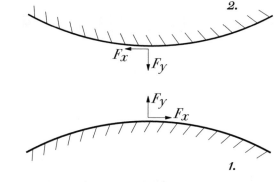

Fig. 12.12 Initial gap
between two unloaded
contacting bodies.

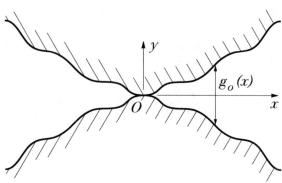

$$g(x) = g_0(x) - u_y^{(1)}(x,0) + u_y^{(2)}(x,0) - C_0 - C_1 x$$

and it follows that the contact condition analogous to (12.17) can be written

$$u_y^{(1)}(x,0) - u_y^{(2)}(x,0) = g_0(x) - C_0 - C_1 x \; ; \quad x \in \mathcal{A} \,, \tag{12.36}$$

since, in a contact region, the gap is by definition zero.

The surface displacement $u_y^{(1)}(x,0)$ of body 1 is simply (12.15) generalized to include the effect of the tangential traction p_x (see equation (12.14)) — i.e.

$$u_y^{(1)}(x,0) = -\frac{(\kappa+1)}{4\pi\mu} \int_{-a}^{a} p_y(\xi) \ln|x-\xi| d\xi$$

$$-\frac{(\kappa-1)}{8\mu} \int_{-a}^{a} p_x(\xi) \mathrm{sgn}(x-\xi) d\xi \tag{12.37}$$

and hence[7]

$$\frac{du_y^{(1)}}{dx} = -\frac{(\kappa+1)}{4\pi\mu}\int_{-a}^{a}\frac{p_y(\xi)d\xi}{(x-\xi)} - \frac{(\kappa-1)}{4\mu}p_x(x)\,. \tag{12.38}$$

We also record the corresponding expression for the tangential displacement $u_x^{(1)}$, which is

$$\frac{du_x^{(1)}}{dx} = -\frac{(\kappa+1)}{4\pi\mu}\int_{-a}^{a}\frac{p_x(\xi)d\xi}{(x-\xi)} + \frac{(\kappa-1)}{4\mu}p_y(x)\,, \tag{12.39}$$

from (12.13).

Comparing Figure 12.11 with Figure 12.3, we see that equations (12.38, 12.39) can be used for the displacements u_x, u_y of body 1, but for the corresponding displacements of body 2 we have to take account of the fact that the tractions p_x, p_y are reversed and that the y-axis is now directed *into* the body. It is easily verified that this can be achieved by changing the signs in the expressions involving u_y, p_x, whilst leaving those expressions with u_x, p_y unchanged. It then follows that

$$\frac{d}{dx}\left(u_y^{(1)}-u_y^{(2)}\right) = -\frac{A}{4\pi}\int_{-a}^{a}\frac{p_y(\xi)d\xi}{(x-\xi)} - \frac{B}{4}p_x(x) \tag{12.40}$$

$$\frac{d}{dx}\left(u_x^{(1)}-u_x^{(2)}\right) = -\frac{A}{4\pi}\int_{-a}^{a}\frac{p_x(\xi)d\xi}{(x-\xi)} + \frac{B}{4}p_y(x)\,, \tag{12.41}$$

where

$$A = \frac{(\kappa_1+1)}{\mu_1} + \frac{(\kappa_2+1)}{\mu_2} \tag{12.42}$$

$$B = \frac{(\kappa_1-1)}{\mu_1} - \frac{(\kappa_2-1)}{\mu_2}\,. \tag{12.43}$$

---

[7] Notice that an alternative representation of $\mathrm{sgn}(x)$ is $2H(x)-1$, where $H(x)$ is the Heaviside step function. It follows that the derivative of $\mathrm{sgn}(x)$ is $2\delta(x)$ and that the derivative of the integral in the second term of (12.37) is simply twice the value of $p_x(\xi)$ at the point $\xi=x$.

12.7 Uncoupled Problems

We can now substitute (12.40) into the derivative of the contact condition (12.36), obtaining

$$g_0'(x) - C_1 = -\frac{A}{4\pi} \int_{-a}^{a} \frac{p_y(\xi)d\xi}{(x - \xi)} - \frac{B}{4} p_x(x) \; ; \quad -a < x < a . \tag{12.44}$$

This equation is similar in form to (12.20), except for the presence of the term involving p_x. It can therefore be solved in the same way if for any reason this term is identically zero. Four important cases where this condition is satisfied are:-

  (i) The contact is frictionless, so $p_x = 0$.
 (ii) The materials are similar ($\kappa_1 = \kappa_2$, $\mu_1 = \mu_2$) and hence $B = 0$.
(iii) Both materials are incompressible ($\nu_1 = \nu_2 = 0.5$, $\kappa_1 = \kappa_2 = 1$, $\mu_1 \neq \mu_2$).
 (iv) One body is rigid ($\mu_1 = \infty$) and the other incompressible ($\kappa_2 = 1$).

Of course, no real materials are even approximately rigid, but the coupling terms in equations (12.40, 12.41) — i.e. those connecting u_y, p_x and u_x, p_y — can reasonably be neglected provided that the ratio

$$\beta = \frac{B}{A} = \left(\frac{(\kappa_1 - 1)}{\mu_1} - \frac{(\kappa_2 - 1)}{\mu_2} \right) \Bigg/ \left(\frac{(\kappa_1 + 1)}{\mu_1} + \frac{(\kappa_2 + 1)}{\mu_2} \right) \ll 1 . \tag{12.45}$$

This will be true if $\mu_1 \gg \mu_2$ and $(\kappa_2 - 1) \ll 1$ and it is a reasonable approximation to the practically important case of rubber in contact with steel. The dimensionless parameter β is one of Dundurs' bimaterial parameters (see §4.3.3).

In the rest of this chapter, we shall restrict attention to problems in which the coupling terms can be neglected for one of the reasons given above.

12.7.1 Contact of cylinders

If the two bodies are cylinders with radii R_1, R_2 (see Figure 12.10), we have

$$g_0''(x) = \frac{1}{R_1} + \frac{1}{R_2} = \frac{R_1 + R_2}{R_1 R_2} \equiv \frac{1}{R} . \tag{12.46}$$

Hence, comparing equations (12.46, 12.33) and (12.44, 12.20), we find that the solution can be written down from equations (12.34, 12.35) by replacing $(\kappa+1)/\mu$ by A of equation (12.42) and R by $R_1 R_2/(R_1 + R_2)$ — i.e.

$$a = \sqrt{\frac{F R_1 R_2}{2\pi (R_1 + R_2)} \left(\frac{(\kappa_1 + 1)}{\mu_1} + \frac{(\kappa_2 + 1)}{\mu_2} \right)} \qquad (12.47)$$

$$p_y(x) = -\frac{2F\sqrt{a^2 - x^2}}{\pi a^2} . \qquad (12.48)$$

12.8   Combined Normal and Tangential Loading

Tangential tractions can be transmitted between contacting bodies only by means of friction and the complete specification of the problem then requires an assumption about the friction law relating the tangential traction to the relative tangential motion at the interface.

As in the normal contact analysis, tangential relative displacement or *shift* $h(x)$ can result from rigid-body motion C and/or elastic deformation, being given by

$$h(x) = u_x^{(2)}(x, 0) - u_x^{(1)}(x, 0) + C ,$$

where a positive shift corresponds to displacement of the upper body to the right relative to the lower body.

We shall define a state of *stick* as one in which the time derivative of the shift, $\dot{h}(x) = 0$. A state with $\dot{h}(x) \neq 0$ will be referred to as *positive* or *negative slip*, depending on the sign of $\dot{h}(x)$.

The simplest frictional assumption is that usually referred to as Coulomb's law, which, in terms of the above notation, can be defined as

$$\dot{h}(x) = \dot{u}_x^{(2)} - \dot{u}_x^{(1)} + \dot{C} = 0 ; \quad f p_y(x) < p_x(x) < -f p_y(x) \qquad (12.49)$$

in stick regions and

$$p_x(x) = -f p_y(x) \operatorname{sgn}(\dot{h}(x)) \qquad (12.50)$$

in slip regions, where f is a constant known as the *coefficient of friction*[8]. We make no distinction between dynamic and static friction coefficients. In interpreting these equations, the reader should recall that the normal traction $p_y(x)$ is always compressive and hence negative in contact problems.

The function $\operatorname{sgn}(\dot{h}(x))$ in (12.50) ensures that the frictional traction opposes the relative motion and hence dissipates energy. This condition and the inequality in (12.49) serve to determine the division of the contact region into positive slip, negative slip and stick zones in much the same way as the inequalities (12.31, 12.32) determine the contact area in the normal contact problem.

---

[8] The more usual symbol μ for the coefficient of friction would lead to confusion with Lamé's constant.

Notice that both of equations (12.49, 12.50) involve time derivatives. Thus, frictional contact problems are *incremental* in nature. It is not generally sufficient to know the final loading condition — we also need to know how that condition was reached. In other words, frictional contact problems are *history-dependent*. They share this and other properties with problems involving another well-known dissipative mechanism — plastic deformation.

12.8.1 Cattaneo and Mindlin's problem

Consider the problem of two elastic cylinders which are first pressed together by a normal compressive force F and then subjected to a monotonically increasing tangential force T as shown in Figure 12.13. We restrict attention to the uncoupled case, $\beta = 0$.

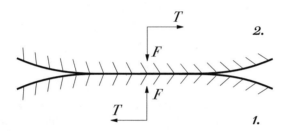

Fig. 12.13 Loading for Cattaneo and Mindlin's problem.

The first phase of the loading is described by the analysis of §12.7.1, the contact semi-width a and the normal tractions $p_y(x)$ being defined by equations (12.47, 12.48) respectively. The condition $\beta = 0$ ensures that these normal tractions produce no tendency for slip (see equations (12.41, 12.45)) and it follows that no tangential tractions are induced and that the whole contact area remains in a state of stick as the normal force F is applied.

The absence of coupling also ensures that the contact area and the normal tractions remain constant during the tangential loading phase. Suppose we first assume that stick prevails everywhere during this phase as well.

Differentiating (12.49) with respect to x and substituting for the displacement derivatives from (12.41) with $B = 0$, we obtain

$$-\frac{A}{4\pi} \int_{-a}^{a} \frac{\dot{p}_x(\xi)d\xi}{(x - \xi)} = 0 \; ; \quad -a < x < a \, .$$

This equation is identical in form to (12.20) and is solved in the same way. The solution is easily shown to be

$$\dot{p}_x(x) = \frac{\dot{T}}{\pi\sqrt{a^2 - x^2}} \, ,$$

by analogy with (12.30) and hence

$$p_x(x) = \frac{T}{\pi\sqrt{a^2 - x^2}} \, ,$$

since a is independent of time during the tangential loading phase.

This result shows that the assumption of stick throughout the contact area $-a < x < a$ leads to a singularity in p_x at the edges $x = \pm a$ and hence the frictional inequality (12.49) must be violated there for any f, since p_y is bounded. We deduce that some slip will occur near the edges of the contact region for any non-zero T, however small.

The problem with slip zones was first solved apparently independently by Cattaneo[9] and Mindlin[10]. In the slip zones, the tractions satisfy the condition $p_x = -f p_y$. We therefore consider the solution for the tangential tractions as the sum of two parts:-

(i) a shear traction

$$p_x(x) = -f p_y(x) = \frac{2 f F \sqrt{a^2 - x^2}}{\pi a^2} \, , \qquad (12.51)$$

from equation (12.48) *throughout* the contact area $-a < x < a$.

(ii) a *corrective* shear traction p_x^*, which must be zero in the slip zones and which is sufficient to restore the condition (12.49) in the stick zone.

As a first step towards finding the corrective traction p_x^*, we find the shift due to the traction distribution (12.51), which is defined by

$$h'(x) = -\frac{d}{dx}\left(u_x^{(1)} - u_x^{(2)}\right) = \frac{A}{4\pi}\int_{-a}^{a}\frac{2 f F \sqrt{a^2 - \xi^2}\,d\xi}{\pi a^2(x - \xi)} = \frac{f F A}{2\pi a^2}\int_0^{\pi}\frac{\sin^2\theta\,d\theta}{(\cos\phi - \cos\theta)}$$

$$= \frac{f F A}{2\pi a}\cos\phi = \frac{f F A x}{2\pi a^2} \, ; \qquad -a < x < a \, . \qquad (12.52)$$

Now, in the *stick* zone, we require the shift to be independent of x and hence we seek corrective shear tractions in the stick zone that will cancel the right-hand side of equation (12.52). By analogy with equations (12.51, 12.52), it is clear that this cancellation can be achieved by the distribution[11]

---

[9] C. Cattaneo (1938), Sul contatto di corpi elastici, *Accademia dei Lincei, Rendiconti,* Ser 6, Vol. 27, pp. 342–348, 433–436, 474–478.

[10] R. D. Mindlin (1949), Compliance of elastic bodies in contact, *ASME Journal of Applied Mechanics,* Vol. 17, pp. 259–268.

[11] Notice that the assumption is that all points in $-c < x < c$ are in a state of stick *throughout the loading process.* It is therefore possible to integrate (12.49) and write the boundary condition in terms of $h(x)$ instead of $\dot{h}(x)$. In general, this is possible as long as no point passes from a state of

$$p_x^*(x) = -\frac{2fF\sqrt{c^2 - x^2}}{\pi a^2} \quad ; \quad -c < x < c \, ,$$

— i.e. a traction similar in form to equation (12.51), but distributed over a smaller centrally located slip zone of width $2c$.

Thus, the complete shear traction distribution is

$$p_x(x) = \frac{2fF}{\pi a^2}\left[\sqrt{a^2 - x^2} - H(c^2 - x^2)\sqrt{c^2 - x^2}\right] \quad ; \quad -a < x < a \, , \quad (12.53)$$

which is illustrated in Figure 12.14.

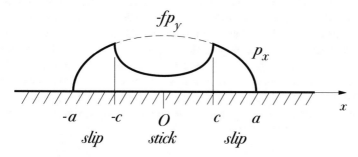

Fig. 12.14 Shear traction distribution for Cattaneo and Mindlin's problem.

The stick zone semi-width c can be determined by requiring

$$T = \int_{-a}^{a} p_x(x)dx = fF - fF\left(\frac{c}{a}\right)^2 \, ,$$

from (12.53) and hence

$$c = a\sqrt{1 - \frac{T}{fF}} \, .$$

As we would expect, the stick zone shrinks to zero as the applied tangential force T approaches fF, after which gross slip (i.e. large scale rigid-body motion) occurs.

---

slip to one of stick during the loading. In particular, the stick zone must not *advance* into the slip zone during loading. For an exhaustive study of the effect of loading history in frictional contact problems, see J. Dundurs and M. Comninou, An educational elasticity problem with friction: Part 1 (1981), Loading and unloading paths for weak friction, *ASME Journal of Applied Mechanics*, Vol. 48, pp. 841–845; Part 2 (1982): Unloading for strong friction and reloading, *ibid.*, Vol. 49, pp. 47–51; Part 3 (1983): General load paths, *ibid*, Vol. 50, pp. 77–84.

Finally, we note that it can be shown that the signs of the displacement derivatives satisfy the condition (12.50), provided that T increases monotonically in time (i.e. $\dot{T} > 0$ for all t)[12].

The related solution for two spheres in contact (also due to Mindlin) has been verified experimentally by observing the damaged regions produced by cyclic microslip between tangentially loaded spheres[13].

A significant generalization of these results has recently been discovered independently by Jäger[14] and Ciavarella[15], who showed that the frictional traction distribution satisfying both equality and inequality conditions for *any* plane frictional contact problem (not necessarily Hertzian) will consist of a superposition of the limiting friction distribution and an opposing distribution equal to the coefficient of friction multiplied by the normal contact pressure distribution at some smaller value of the normal load. Thus, as the tangential force is increased at constant normal force, the stick zone shrinks, passing monotonically through the same sequence of areas as the normal contact area passed through during the normal loading process. These results can be used to predict the size of the slip zone in conditions of fretting fatigue[16]. One consequence of this result is that wear in the sliding regions due to an oscillating tangential load will not change the extent of the adhesive region, so that in the limit the contact is pure adhesive and a singularity develops in the normal traction at the edge of this region[17].

12.8.2   *Steady rolling: Carter's solution*

A final example of considerable practical importance is that in which two cylinders roll over each other whilst transmitting a constant tangential force, T. If we assume that the rolling velocity is V, the solution will tend to a steady-state which is invariant

---

[12] The effect of non-monotonic loading in the related problem of two contacting spheres was considered by R. D. Mindlin and H. Deresiewicz (1953), Elastic spheres in contact under varying oblique forces, *ASME Journal of Applied Mechanics,* Vol. 21, pp. 327–344. The history-dependence of the friction law leads to quite complex arrangements of slip and stick zones and consequent variation in the load-compliance relation. These results also find application in the analysis of oblique impact, where neither normal nor tangential loading is monotonic (see N. Maw, J. R. Barber and J. N. Fawcett (1976), The oblique impact of elastic spheres, *Wear,* Vol. 38, pp. 101–114).

[13] K. L. Johnson (1961), Energy dissipation at spherical surfaces in contact transmitting oscillating forces, *Journal of Mechanical Engineering Science,* Vol. 3, pp. 362–368.

[14] J. Jäger (1997), Half-planes without coupling under contact loading. *Archive of Applied Mechanics* Vol. 67 , pp. 247–259.

[15] M. Ciavarella (1998), The generalized Cattaneo partial slip plane contact problem. I-Theory, II-Examples. *International Journal of Solids and Structures.* Vol. 35, pp. 2349–2378.

[16] D. A. Hills and D. Nowell, *Mechanics of Fretting Fatigue.* Kluwer, Dordrecht, 1994, M. P. Szolwinski and T. N. Farris (1996), Mechanics of fretting fatigue crack formation. *Wear* Vol. 198, pp. 93–107.

[17] M. Ciavarella and D. A. Hills (1999), Some observations on the oscillating tangential forces and wear in general plane contacts, *European Journal of Mechanics A–Solids.* Vol. 18, pp. 491–497.

with respect to the moving coördinate system

$$\xi = x - Vt \ .$$

In this system, we can write the 'stick' condition (12.49) as

$$\dot{h}(x,t) = \frac{d}{dt}\left[u_x^{(2)}(x-Vt) - u_x^{(1)}(x-Vt) + C\right] = V\frac{d}{d\xi}(u_x^{(1)} - u_x^{(2)}) + \dot{C} = 0 \ ,$$
(12.54)

where $\dot{C}$ is an arbitrary but constant rigid-body slip (or creep) velocity. Similarly, the slip condition (12.50) becomes

$$p_x(\xi) = -fp_y(\xi)\,\mathrm{sgn}\left[V\frac{d}{d\xi}\left(u_x^{(1)} - u_x^{(2)}\right)\right] + \dot{C} \ .$$
(12.55)

At first sight, we might think that the Cattaneo traction distribution (12.53) satisfies this condition, since it gives

$$\frac{d}{d\xi}(u_x^{(1)} - u_x^{(2)}) = 0$$

in the central stick zone and $\dot{C}$ can be chosen arbitrarily. However, if we substitute the resulting displacements into the slip condition (12.55), we find a sign error in the leading slip zone. This can be explained as follows: In the Cattaneo problem, as T is increased, positive slip (i.e. $\dot{h}(x) = \dot{u}_{x2} - \dot{u}_{x1} + \dot{C} > 0$) occurs in both slip zones and the magnitude of $h(x)$ increases from zero at the stick-slip boundary to a maximum at $x = \pm a$. It follows that $\frac{d}{d\xi}(u_x^{(1)} - u_x^{(2)})$ is negative in the right slip zone and positive in the left slip zone. Thus, if this solution is used for the steady rolling problem, a violation of (12.55) will occur in the right zone if V is positive, and in the left zone if V is negative. In each case there is a violation in the *leading* slip zone — i.e. in that zone next to the edge where contact is being established.

Now in frictional problems, when we make an assumption that a given region slips and then find that it leads to a sign violation, it is usually an indication that we made the wrong assumption and that the region in question should be in a state of stick. Thus, in the rolling problem, there is no leading slip zone[18]. Carter[19] has shown that the same kind of superposition can be used for the rolling problem as for Cattaneo's problem, except that the corrective traction is displaced to a zone adjoining the leading edge. A corrective traction

---

[18] Note that this applies to the uncoupled problem ($\beta = 0$) only. With dissimilar materials, there is generally a leading slip zone and there can also be an additional slip zone contained within the stick zone. This problem is treated by R. H. Bentall and K. L. Johnson (1967), Slip in the rolling contact of dissimilar rollers, *International Journal of Mechanical Sciences,* Vol. 9, pp. 389–404.

[19] F. W. Carter (1926), On the action of a locomotive driving wheel, *Proceedings of the Royal Society of London,* Vol. A112, pp. 151–157.

$$p_x^*(\xi) = \frac{2fF}{\pi a^2} \sqrt{\left(\frac{a-c}{2}\right)^2 - \left(\xi - \frac{a+c}{2}\right)^2} = \frac{2fF}{\pi a^2} \sqrt{(a-\xi)(\xi-c)}$$

produces a displacement distribution

$$\frac{d}{d\xi}\left(u_x^{(1)} - u_x^{(2)}\right) = \frac{fFA}{2\pi a^2}\left(\xi - \frac{a+c}{2}\right) \quad ; \quad c < \xi < a , \tag{12.56}$$

which cancels the x-varying term in (12.52), leaving only an admissable constant $\dot{C}$ (see equation (12.54)). Hence, the traction distribution for positive V is

$$p_x(\xi) = \frac{2fF}{\pi a^2}\left[\sqrt{a^2 - \xi^2} - H(\xi - c)\sqrt{(a-\xi)(\xi-c)}\right] \quad ; \quad -a < \xi < a ,$$

which is illustrated in Figure 12.15.

Stick occurs in the leading zone $c < \xi < a$ and positive slip in the trailing zone $-a < \xi < c$.

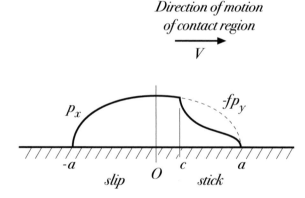

Fig. 12.15 Shear traction distribution for Carter's problem.

The corresponding total tangential load is

$$T = fF\left[1 - \left(\frac{a-c}{2a}\right)^2\right]$$

and hence the stick-slip boundary $\xi = c$ is given by

$$c = a\left(1 - 2\sqrt{1 - \frac{T}{fF}}\right) .$$

An interesting feature of this solution is that the creep velocity $\dot{C}$ is not zero — i.e. there is a small steady-state relative tangential velocity between the two bodies. This has the effect of making the driven roller rotate slightly more slowly than a rigid-body kinematic analysis would lead us to expect. This in turn means that more energy is provided to the driving roller than is recovered from the driven roller, the balance of course being dissipated in the microslip regions in the form of heat. The creep velocity can be calculated by substituting the superposition of (12.52) and (12.56) into (12.54), with the result

$$\dot{C} = -\frac{fVa}{R}\left(1 - \sqrt{1 - \frac{T}{fF}}\right),$$

where R is given by (12.46).

It is interesting to note that Carter's solution was published 20 years before Mindlin's, but there is no evidence that either Mindlin or Cattaneo was aware of it, despite the similarity of the techniques used[20].

Problems

12.1. Figure 12.16 shows a disk of radius a subjected to two equal and opposite forces F at the points O_1, O_2, the rest of the boundary $r = a$ being traction free.

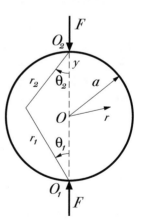

Fig. 12.16 Disk loaded by concentrated forces.

The stress function

$$\phi = -\frac{F}{\pi}(r_1\theta_1 \sin\theta_1 + r_2\theta_2 \sin\theta_2)$$

---

[20] For a more extensive discussion of frictional problems of this type, see K.L.Johnson, *loc. cit.*, Chapters 5,7,8.

is proposed to account for the localized effect of the forces. Find the stress field due to this function and, in particular, find the tractions implied upon the boundary $r = a$. Then complete the solution by superposing appropriate stress functions from Table 8.1, so as to satisfy the traction-free boundary condition.

12.2. The disk $0 \leq r < a$ is accelerated from rest by two concentrated forces F acting in the positive θ-direction at the points $(a, \pm \pi/2)$, as shown in Figure 12.17. Use the Flamant solution in appropriate local coördinates to describe the concentrated forces and the solution of §7.4.1 to describe the inertia forces. Complete the solution by superposing appropriate stress functions from Table 8.1, so as to satisfy the traction-free boundary condition.

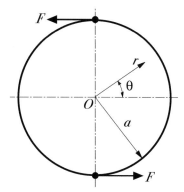

Fig. 12.17 Disk accelerated by concentrated tangential forces.

12.3. Figure 12.18 shows a heavy disk of radius a and density ρ supported by a concentrated force $\pi a^2 \rho g$. Find a solution for the stress field in the disk by combining the Flamant solution with appropriate terms from Table 8.1 in a coördinate system centred on the disk.

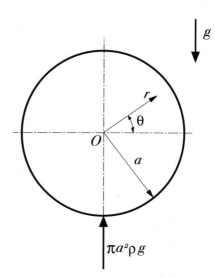

Fig. 12.18 Heavy disk supported by a concentrated force.

$\pi a^2 \rho g$

12.4. A rigid punch in the form of a half cylinder of radius R is pressed into an elastic half-plane such that the plane side of the punch remains vertical, as shown in Figure 12.19. Find the relationship between the indenting force P and the width a of the contact area, and the contact pressure distribution under the punch.

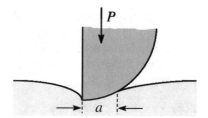

Fig. 12.19 Indentation by a half cylinder.

12.5. A rigid flat punch of width $2a$ is pressed into an elastic half-plane by a force F whose line of action is displaced a distance b from the centreline, as shown in Figure 12.20.

(i) Assuming that the punch makes contact over the entire face $-a < x < a$, find the pressure distribution $p(x)$ and the angle of tilt of the punch. Assume the half-plane is prevented from rotating at infinity.
(ii) Hence find the maximum value of b for which there is contact throughout $-a < x < a$.
(iii) Re-solve the problem, assuming that b is larger than the critical value found in (ii). Contact will now occur in the range $c < x < a$ and the unknown left hand end of the contact region ($x = c$) must be found from a smoothness condition on $p(x)$. Express c and $p(x)$ as functions of F, x, a and b.

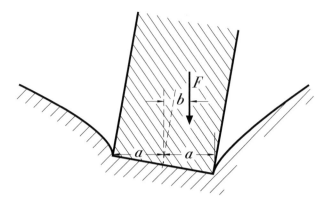

Fig. 12.20 Punch with an eccentric load.

12.6. Two half-planes $y > 0$ and $y < 0$ of the same material are welded together along the section $-a < x < a$ of their common interface $y = 0$. Equal and opposite forces F are now applied at infinity tending to load the weld in tension. Use a symmetry argument to deduce conditions that must be satisfied on the symmetry line $y = 0$ and hence determine the tensile stresses transmitted by the weld as a function of x. Find also the *stress-intensity factor* K_I, defined as

$$K_I \equiv \lim_{x \to a^-} \sigma_{yy}(x, 0)\sqrt{2\pi(a - x)} \; .$$

12.7. A flat rigid punch is pressed into the surface $y = 0$ of the elastic half-plane $y > 0$ by a force F. A tangential force T is then applied to the punch. If Coulomb friction conditions apply at the interface with coefficient f and Dundurs constant $\beta = 0$, show that no microslip will occur until T reaches the value fF at which point gross slip starts.

12.8. Express the stress function

$$\phi = \frac{F_y r \theta \cos \theta}{\pi}$$

in Cartesian coördinates and hence find the stress components $\sigma_{xx}, \sigma_{xy}, \sigma_{yy}$ due to the point force F_y in Figure 12.2.

  Use this result and the integration procedure of §12.4 (Figure 12.4) to determine the stress field in the half-plane $y < 0$ due to the Hertzian traction distribution of equation (12.35). Make a contour plot of the von Mises stress σ_E of equation (1.23) and identify the maximum value and its location.

12.9. From §12.5.2 and equation (12.30), it follows that the contact traction distribution

$$p(x, a) = \frac{1}{\sqrt{a^2 - x^2}} \; ; \quad |x| < a$$
$$= 0 \qquad ; \quad |x| > a$$

produces zero surface slope du_y/dx in $|x| < a$. Use equation (12.16) to determine the corresponding value of slope *outside* the contact area ($|x| > a$) and hence construct the discontinuous function

$$u(x, a) \equiv \frac{du_y}{dx} \; ,$$

such that $p(x, a)$ produces $u(x, a)$.

Linear superposition then shows that the more general traction distribution

$$p(x) = \int_0^b g(a) p(x, a) da$$

produces the surface slope

$$\frac{du_y}{dx} = \int_0^b g(a) u(x, a) da \; ,$$

where $g(a)$ is any function of a. In effect this is a superposition of a range of 'flat punch' traction distributions over different width strips up to a maximum semi-width of b, so the traction will still be zero for $|x| > b$.

Use this representation to solve the problem of the indentation by a wedge of semi-angle $\pi/2 - \alpha$ ($\alpha \ll 1$), for which

$$\frac{du_0}{dx} = -|\alpha| \; ; \quad |x| < b \; ,$$

where b is the semi-width of the contact area. In particular, find the contact traction distribution and the relation between b and the applied force F.

Hint: You will find that the boundary condition leads to an Abel integral equation, whose solution is given in Table 32.2 in Chapter 32.

12.10. Modify the method outlined in Problem 12.9 to solve the problem of the indentation of a half-plane by an unsymmetrical wedge, for which

$$\frac{du_0}{dx} = -\alpha \; ; \quad 0 < x < c$$
$$= \beta \; ; \quad -d < x < 0 \; ,$$

where the unsymmetrical contact area extends from $-d$ to c. In particular, find the contact traction distribution and both c and d as functions of the applied force F.

The modification involves moving the origin to destroy the symmetry once the function $u(x, a)$ has been determined. In other words, instead of superposing a set of symmetrically disposed 'flat punch' distributions, superpose a similar set arranged so that the common point is displaced from the origin.

Chapter 13
Forces, Dislocations and Cracks

In this chapter, we shall discuss the applications of two solutions which are singular at an *interior* point of a body, and which can be combined to give the stress field due to a concentrated force (the *Kelvin solution*) and a dislocation. Both solutions involve a singularity in stress with exponent -1 and are therefore inadmissable according to the criterion of §11.2.1. However, like the Flamant solution considered in Chapter 12, they can be used as Green's functions to describe distributions, resulting in convolution integrals in which the singularity is integrated out. The Kelvin solution is also useful for describing the far field (i.e. the field a long way away from the loaded region) due to a force distributed over a small region.

13.1 The Kelvin Solution

We consider the problem in which a concentrated force F acts in the x-direction at the origin in an infinite body. This is not a perturbation problem like the stress field due to a hole in an otherwise uniform stress field, since the force has a non-zero resultant. Thus, no matter how far distant we make the boundary of the body, there will have to be some traction to oppose the force. In fact, self-similarity arguments like those used in §§12.1, 12.2 show that the stress field must decay with r^{-1}.

The Flamant solution has this behaviour and it corresponds to a concentrated force (see §12.2), but it cannot be used at an interior point in the body, since the

---

Supplementary Information The online version contains supplementary material available at https://doi.org/10.1007/978-3-031-15214-6_13.

J. R. Barber, *Elasticity*, Solid Mechanics and Its Applications 172,
https://doi.org/10.1007/978-3-031-15214-6_13

corresponding displacements [equations (12.5, 12.6)] are multivalued[1]. However, we can construct a solution with the same character and with single-valued displacements from the more general stress function (12.1) by choosing the coefficients in such a way that the multivalued terms cancel.

In view of the symmetry of the problem about $\theta = 0$, we restrict attention to the symmetric terms

$$\phi = C_1 r\theta \sin\theta + C_3 r \ln(r) \cos\theta \qquad (13.1)$$

of (12.1), for which the stress components are

$$\sigma_{rr} = \frac{2C_1 \cos\theta}{r} + \frac{C_3 \cos\theta}{r} \; ; \quad \sigma_{r\theta} = \frac{C_3 \sin\theta}{r} \; ; \quad \sigma_{\theta\theta} = \frac{C_3 \cos\theta}{r} \; , \qquad (13.2)$$

from Table 8.1, and the displacement components are

$$2\mu u_r = \frac{C_1}{2}\left[(\kappa - 1)\theta \sin\theta - \cos\theta + (\kappa + 1)\ln(r)\cos\theta\right]$$

$$+ \frac{C_3}{2}\left[(\kappa + 1)\theta \sin\theta - \cos\theta + (\kappa - 1)\ln(r)\cos\theta\right] \qquad (13.3)$$

$$2\mu u_\theta = \frac{C_1}{2}\left[(\kappa - 1)\theta \cos\theta - \sin\theta - (\kappa + 1)\ln(r)\sin\theta\right]$$

$$+ \frac{C_3}{2}\left[(\kappa + 1)\theta \cos\theta - \sin\theta - (\kappa - 1)\ln(r)\sin\theta\right] \; , \qquad (13.4)$$

from Table 9.1.

Suppose we make an imaginary cut in the plane at $\theta = 0, 2\pi$ and define a principal value of θ such that $0 \le \theta < 2\pi$. This makes θ discontinuous at $\theta = 2\pi$, but the trigonometric functions of course remain continuous. The only potential difficulty is associated with the expressions $\theta \sin\theta, \theta \cos\theta$.

Now $\theta \sin\theta = 0$ at $\theta = 2\pi$ and hence this expression has the same value at the two sides of the cut and is continuous. We can therefore make the whole displacement field continuous by choosing C_1, C_3 such that the terms $\theta \cos\theta$ in u_θ cancel — i.e. by setting[2]

$$C_1(\kappa - 1) + C_3(\kappa + 1) = 0 \; , \qquad (13.5)$$

which for plane stress is equivalent to

$$(1 - \nu)C_1 + 2C_3 = 0 \; . \qquad (13.6)$$

---

[1] This was not a problem for the surface loading problem, since the wedge of Figure 12.1 only occupies a part of the θ-domain and hence a suitable principal value of θ can be chosen to be both single-valued and continuous.

[2] Notice that this choice also has the effect of cancelling the $\theta \sin\theta$ terms in (13.3), so that the complete displacement field, like the stress field, depends on θ only through sine and cosine terms.

This leaves us with one degree of freedom (one free constant) to satisfy the condition that the force at the origin is equal to F. Considering the equilibrium of a small circle of radius r surrounding the origin, we have

$$F + \int_0^{2\pi} (\sigma_{rr} \cos\theta - \sigma_{r\theta} \sin\theta)\, r d\theta = 0 .$$

We now substitute for the stress components from equations (13.2), obtaining

$$C_1 = -\frac{F}{2\pi} , \tag{13.7}$$

after which we recover the constant C_3 from equation (13.6) as

$$C_3 = \frac{(1-\nu)F}{4\pi} .$$

Finally, we substitute these constants back into (13.2) to obtain

$$\sigma_{rr} = -\frac{(3+\nu)F \cos\theta}{4\pi r} ; \quad \sigma_{r\theta} = \frac{(1-\nu)F \sin\theta}{4\pi r} ; \quad \sigma_{\theta\theta} = \frac{(1-\nu)F \cos\theta}{4\pi r} .$$

These are the stress components for Kelvin's problem, where the force acts in the x-direction. The corresponding results for a force in the y-direction can be obtained in the same way, using the antisymmetric terms in (12.1). Alternatively, we can simply rotate the axis system in the above solution by redefining $\theta \to (\theta - \pi/2)$.

13.1.1 Body force problems

Kelvin's problem is a special case of a body force problem — that in which the body force is a delta function at the origin. The solution can also be used to solve more general body force problems by convolution. We consider the body force $p_x \delta x \delta y$ acting on the element $\delta x \delta y$ as a concentrated point force and use the above solution to determine its effect on the stress components at an arbitrary point. Treating the component p_y in the same way and summing over all the elements of the body then gives a double integral representation of the stress field.

This method is not restricted to the infinite body, since we only seek a particular solution of the body force problem. We can therefore use the convolution method to develop a solution for the stresses in an infinite body with the prescribed body force distribution, after which we 'cut out' the shape of the real body and correct the boundary conditions as required, using an appropriate homogeneous solution (i.e. a solution without body forces).

Any body force distribution can be treated this way — the method is not restricted to conservative vector fields. It is generally more algebraically tedious than the methods developed in Chapter 7, but it lends itself naturally to numerical implementation.

For example, it can be used to extend the boundary integral method to body force problems.

13.2 Dislocations

The term *dislocation* has related, but slightly different meanings in Elasticity and in Materials Science. In Elasticity, the material is assumed to be a continuum — i.e. to be infinitely divisible. Suppose we take an infinite continuous body and make a cut along the half-plane $x > 0$, $y = 0$. We next apply equal and opposite tractions to the two surfaces of the cut such as to open up a gap of constant thickness, as illustrated in Figure 13.1. We then slip a thin slice of the same material into the space to keep the surfaces apart and weld the system up, leaving a new continuous body which will now be in a state of residual stress.

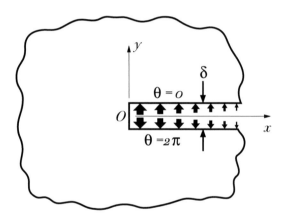

Fig. 13.1 The climb dislocation solution.

The resulting stress field is referred to as the *climb dislocation* solution. It can be obtained from the stress function of equation (13.1) by requiring that there be no net force at the origin, and hence that $C_1 = 0$ [see equation (13.7)]. We therefore have

$$\phi = C_3 r \ln(r) \cos \theta . \tag{13.8}$$

The strength of the dislocation can be defined in terms of the discontinuity in the displacement u_θ on $\theta = 0, 2\pi$, which is also the thickness of the slice of extra material which must be inserted to restore continuity. This thickness is

$$\delta = u_\theta(0) - u_\theta(2\pi) = -\frac{\pi(\kappa + 1)C_3}{2\mu} .$$

Thus, we can define a climb dislocation of strength B_y as one which opens a gap $\delta = B_y$, corresponding to

$$C_3 = -\frac{2\mu B_y}{\pi(\kappa + 1)} \quad \text{or} \quad \phi = -\frac{2\mu B_y r \ln(r) \cos \theta}{\pi(\kappa + 1)} \,, \tag{13.9}$$

where

$$\frac{2\mu}{(\kappa + 1)} = \frac{\mu(1 + \nu)}{2} = \frac{E}{4} \qquad \text{(plane stress)}$$

$$= \frac{\mu}{2(1 - \nu)} = \frac{E}{4(1 - \nu^2)} \qquad \text{(plane strain)}, \tag{13.10}$$

using equation (3.11).

The stress field due to the climb dislocation is

$$\sigma_{rr} = \sigma_{\theta\theta} = -\frac{2\mu B_y \cos \theta}{\pi(\kappa + 1)} \;;\quad \sigma_{r\theta} = -\frac{2\mu B_y \sin \theta}{\pi(\kappa + 1)r} \,. \tag{13.11}$$

We also record the stress components at $y = 0$, — i.e. $\theta = 0, \pi$ — in rectangular coördinates, which are

$$\sigma_{xx} = \sigma_{yy} = -\frac{2\mu B_y}{\pi(\kappa + 1)x} \;;\quad \sigma_{yx} = 0 \,. \tag{13.12}$$

This solution is called a climb dislocation, because it opens a gap on the cut at $\theta = 0, 2\pi$. A corresponding solution can be obtained from the stress function

$$\phi = \frac{2\mu B_x r \ln(r) \sin \theta}{\pi(\kappa + 1)} \tag{13.13}$$

which is discontinuous in the displacement component u_r — i.e. for which the two surfaces of the cut experience a relative tangential displacement but do not separate. This is called a *glide dislocation*. The stress field due to a glide dislocation is given by

$$\sigma_{rr} = \sigma_{\theta\theta} = \frac{2\mu B_x \sin \theta}{\pi(\kappa + 1)r} \;;\quad \sigma_{r\theta} = -\frac{2\mu B_x \cos \theta}{\pi(\kappa + 1)r} \,, \tag{13.14}$$

and on the surface $y = 0$, these reduce to

$$\sigma_{xx} = \sigma_{yy} = 0 \;;\quad \sigma_{yx} = -\frac{2\mu B_x}{\pi(\kappa + 1)x} \,,$$

where $B_x = u_r(0) - u_r(2\pi)$ is the strength of the dislocation.

The solutions (13.11) and (13.14) actually differ only in orientation. The climb dislocation solution defines a glide dislocation if we choose to make the cut along the line $\theta = -3\pi/2,\ \pi/2$ (i.e. along the y-axis instead of the x-axis[3]). The dislocation strengths B_x, B_y can be regarded as the components of a vector $\boldsymbol{B} = \{B_x,\ B_y\}$, known as the *Burgers vector*.

13.2.1   Dislocations in Materials Science

Real materials have a discrete atomic or molecular structure. However, we can follow a similar procedure by imagining cleaving the solid between two sheets of molecules up to the line $x = y = 0$ and inserting *one extra layer* of molecules. When the system is released, there will be some motion of the molecules, mostly concentrated at the end of the added layer, resulting in an imperfection in the regular molecular array. This is what is meant by a dislocation in Materials Science.

Considerable insight into the role of dislocations in material behaviour was gained by the work of Bragg[4] with bubble models. Bragg devised a method of generating a two-dimensional collection of identical size bubbles. Attractive forces between the bubbles ensured that they adopted a regular array wherever possible, but dislocations are identifiable as can be seen in Figure 13.2[5].

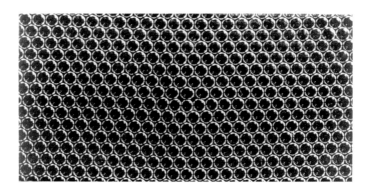

Fig. 13.2   The bubble model — a dislocation. The extra layer is most easily seen by looking along a line inclined at $\pm 30°$ to the vertical.

When forces are applied to the edges of the bubble assembly, the dislocations are found to move in such a way as to permit the boundaries to move. This dislocation motion is believed to be the principal mechanism of plastic deformation in ductile materials. Larger bubble assemblies exhibit discrete regions in which the arrays are

---

[3] In fact, the cut can be made along any (not necessarily straight) line from the origin to infinity. The solution of equation (13.11) will then exhibit a discontinuity in u_y of magnitude B_y at all points along the cut, as long as the principal value of θ is appropriately defined.

[4] L. Bragg and J. F. Nye (1947), A dynamical model of a crystal structure, *Proceedings of the Royal Society of London,* Vol. A190, pp. 474–481.

[5] Figures 13.2 and 13.3 are reproduced by kind permission of the Royal Society of London.

differently aligned as in Figure 13.3. These are analogous with grains in multigranular materials. When the structure is deformed, the dislocations typically move until
they reach a grain boundary, but the misalignment prohibits further motion and the
stiffness of the assembly increases. This pile-up of dislocations at grain boundaries
is responsible for work-hardening in ductile materials. Also, the accumulated dislocations coalesce into larger disturbances in the crystal structure such as voids or
cracks, which function as initiation points for failure by fatigue or fracture.

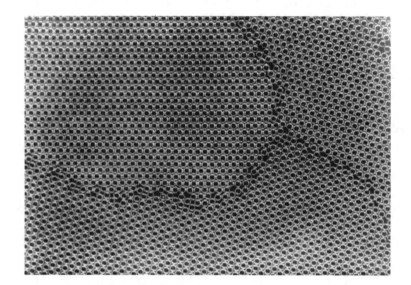

Fig. 13.3 The bubble model — grain boundaries.

13.2.2 Similarities and differences

It is tempting to deduce that the elastic solution of §13.2 describes the stresses due to
the molecular structure dislocation, if we multiply by a constant defining the thickness
of a single layer of molecules. However, the concept of stress is rather vague over
dimensions comparable with interatomic distances. In fact, this is preëminently a
case where the apparent singularity of the mathematical solution is moderated in
reality by the discrete structure of the material.

Notice also that the continuum dislocation of §13.2 can have any strength, corresponding to the fact that B_x, B_y are arbitrary real constants. By contrast, the thickness
of the inserted layer is restricted to one layer of molecules in the discrete theory. This
thickness is sufficiently small to ensure that the 'stresses' due to a material dislocation are very small in comparison with typical engineering magnitudes, except in
the immediate vicinity of the defect. Furthermore, although a stress-relieved metal-

lic component will generally contain numerous dislocations, they will be randomly oriented, leaving the components essentially stress-free on the macroscopic scale.

However, if the component is plastically deformed, the dislocation motion will lead to a more systematic structure, which can result in residual stress. Indeed, one way to represent the stress field due to plastic deformation is as a distribution of mathematical dislocations. For example, the motion of a single dislocation can be represented by introducing a dislocation pair comprising a negative dislocation at the original location and an equal positive dislocation at the final location. This implies a *closure condition* as in §§13.2.3, 13.2.4 below.

13.2.3 *Dislocations as Green's functions*

In the Theory of Elasticity, the principal use of the dislocation solution is as a Green's function to represent localized processes. In this context, it is more natural to think of a *dislocation density* $B(x, y)$ per unit area[6], so that the elemental area $dxdy$ contains a dislocation of Burger's vector $B(x, y)dxdy$.

Suppose we place a distribution of dislocations in some interior domain Ω of the body, which is completely surrounded by elastic material. The resulting stress field obtained by integration will satisfy the conditions of equilibrium everywhere (since the dislocation solution involves no force) and will satisfy the compatibility condition everywhere *except* in Ω. There is of course the possibility that the displacement may be multiple-valued outside Ω, but we can prevent this by enforcing the two *closure conditions*

$$\iint_\Omega B_x(x, y)dxdy = 0 \ ; \quad \iint_\Omega B_y(x, y)dxdy = 0 \ , \qquad (13.15)$$

which state that the total strength of the dislocations in Ω is zero[7].

13.2.4 *Stress concentrations*

A special case of some importance is that in which the enclosed domain Ω represents a hole which perturbs the stress field in an elastic body.

It may seem strange to place dislocations in a region which is strictly not a part of the body. However, we might start with an infinite body with no hole, place

---

[6] P. M. Blomerus, D. A. Hills and P. A. Kelly (1999), The distributed dislocation method applied to the analysis of elastoplastic strain concentrations, *Journal of the Mathematics and Physics of Solids,* Vol.47, pp. 1007–1026.

[7] This does *not* mean that the stress field is null, since the various self-cancelling dislocations have different locations.

dislocations in Ω generating a stress field and then make a cut along the boundary of Ω producing the body with a hole. The stress field will be unchanged by the cut provided we place tractions on its boundary equal to those which were transmitted across the same surface in the original continuous body. In particular, if we choose the dislocation distribution so as to make the boundaries of Ω traction free, we can cut out the hole without changing the stress distribution, which is therefore the solution of the original problem for the body with a hole.

The general idea of developing perturbation solutions by placing singularities in a region where the governing equations (here the compatibility condition) are not required to be enforced is well-known in many branches of Applied Mechanics. For example, the solution for the flow of a fluid around a rigid body can be developed in many cases by placing an appropriate distribution of sources and sinks in the region occupied by the body, the distribution being chosen so as to make the velocity component normal to the body surface be everywhere zero.

The closure conditions (13.15) ensure that acceptable dislocation distributions can be represented in terms of dislocation pairs — i.e. as matched pairs of dislocations of equal magnitude and opposite direction. It follows that acceptable distributions can be represented as distributions of *dislocation derivatives*, since, for example, a dislocation at P and an equal negative dislocation at Q is equivalent to a uniform distribution of derivatives on the straight line joining P and Q[8].

Now the stress components in the dislocation solution [equation (13.11)] decay with r^{-1} and hence those in the *dislocation derivative* solution will decay with r^{-2}. It follows that the dominant term in the perturbation (or corrective) solution due to a hole will decay at large r with r^{-2}. We see this in the particular case of the circular hole in a uniform stress field (§§8.3.2, 8.4.1). The same conclusion applies for the perturbation in the stress field due to an inclusion — i.e. a localized region whose properties differ from those of the bulk material (see Chapter 28).

13.3   Crack Problems

Crack problems are particularly important in Elasticity because of their relevance to the subject of Fracture Mechanics, which broadly speaking is the study of the stress conditions under which cracks grow. For our purposes, a crack will be defined as the limiting case of a hole whose volume (at least in the unloaded case) has shrunk to zero, so that opposite faces touch. It might also be thought of as an interior surface in the body which is incapable of transmitting tension.

---

[8] To see this, think of the derivative as the limit of a pair of equal and opposite dislocations separated by a distance δS whose magnitude is proportional to $1/\delta S$.

In practice, cracks will generally have some finite thickness, but if this is sufficiently small, the crack will behave unilaterally with respect to tension and compression — i.e. it will open if we try to transmit tension, but close in compression, transmitting the tractions by means of contact. For this reason, a cracked body will appear stiffer in compression than it does in tension. Also, the crack acts as a stress concentration in tension, but not in compression, so cracks do not generally propagate in compressive stress fields.

13.3.1 Linear Elastic Fracture Mechanics

We saw in §11.2.3 that the asymptotic stress field at the tip of a crack has a square-root singularity and we shall find this exemplified in the particular solutions that follow. The simplest and most prevalent theory of brittle fracture — that due to Griffith — states in essence that crack propagation will occur when the scalar multiplier on this singular stress field exceeds a certain critical value. More precisely, Griffith proposed the thesis that a crack would propagate when propagation caused a reduction in the total energy of the system. Crack propagation causes a reduction in strain energy in the body, but also generates new surfaces which have *surface energy*. Surface energy is related to the force known as surface tension in fluids and follows from the fact that to cleave a solid body along a plane involves doing work against the interatomic forces across the plane. When this criterion is applied to the stress field in particular cases, it turns out that for a small change in crack length, propagation is predicted when the multiplier on the singular term, known as the *stress-intensity factor*, exceeds a certain critical value, which is a constant for the material known as the *fracture toughness*.

It may seem paradoxical to found a theory of real material behaviour on properties of a singular elastic field, which clearly cannot accurately represent conditions in the precise region where the failure is actually to occur. However, if the material is brittle, non-linear effects will be concentrated in a relatively small *process zone* surrounding the crack tip. Furthermore, the certainly very complicated conditions in this process zone can only be influenced by the surrounding elastic material and hence the conditions for failure must be expressible in terms of the characteristics of the much simpler surrounding elastic field. As long as the process zone is small compared with the other linear dimensions of the body[9] (notably the crack length), it will have only a very localized effect on the surrounding elastic field, which will therefore be adequately characterized by the dominant singular term in the linear elastic solution, whose multiplier (the stress-intensity factor) then determines the conditions for crack propagation.

---

[9] This is often referred to as the *small-scale yielding condition*.

It is notable that this argument requires no assumption about or knowledge of the actual mechanism of failure in the process zone and, by the same token, the success of Linear Elastic Fracture Mechanics (LEFM) as a predictor of the strength of brittle components provides no evidence for or against any particular failure theory[10].

13.3.2 Plane crack in a tensile field

Two-dimensional crack problems are very conveniently formulated using the methods outlined in §§13.2.3, 13.2.4. Thus, we seek a distribution of dislocations on the plane of the crack (which appears as a line in two-dimensions), which, when superposed on the unperturbed stress field, will make the surfaces of the crack traction free. We shall illustrate the method for the simple case of a plane crack in a tensile stress field.

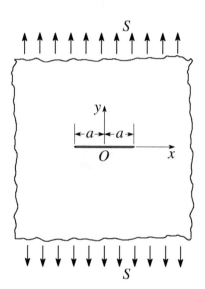

Fig. 13.4 Plane crack in a tensile field.

Figure 13.4 shows a plane crack of width $2a$ occupying the region $-a < x < a$, $y = 0$ in a two-dimensional body subjected to uniform tension $\sigma_{yy} = S$ at its remote boundaries.

---

[10] For more details of the extensive development of the field of Fracture Mechanics, the reader is referred to the many excellent texts on the subject, such as M.F.Kanninen and C.H.Popelar, *Advanced Fracture Mechanics*, Clarendon Press, Oxford, 1985, H.Leibowitz, **ed.**, *Fracture, An Advanced Treatise*, 7 Vols., Academic Press, New York, 1971. Stress-intensity factors for a wide range of geometries are tabulated by G.C.Sih, *Handbook of Stress Intensity Factors*, Institute of Fracture and Solid Mechanics, Lehigh University, Bethlehem, PA, 1973.

Assuming that the crack opens, the boundary conditions for this problem can be stated in the form

$$\sigma_{yx} = \sigma_{yy} = 0 \; ; \quad -a < x < a, \; y = 0$$

$$\sigma_{yy} \to S \; ; \quad \sigma_{xy}, \sigma_{xx} \to 0 \; ; \quad r \to \infty \, .$$

Following the procedure of §8.3.2, we represent the solution as the sum of the stress field in the corresponding body without a crack — here a uniform uniaxial tension $\sigma_{yy} = S$ — and a corrective solution, for which the boundary conditions are therefore

$$\sigma_{yx} = 0 \; ; \quad \sigma_{yy} = -S \; ; \quad -a < x < a, \; y = 0 \qquad (13.16)$$

$$\sigma_{xx}, \sigma_{xy}, \sigma_{yy} \to 0 \; ; \quad r \to \infty \, . \qquad (13.17)$$

Notice that the corrective solution corresponds to the problem of a crack in an otherwise stress-free body opened by uniform compressive normal tractions of magnitude S.

The most general stress field due to the crack would involve both climb and glide dislocations, but in view of the symmetry of the problem about the plane $y=0$, we conclude that there is no relative tangential motion between the crack faces and hence that the solution can be constructed with a distribution of climb dislocations alone. More precisely, we respresent the solution in terms of a distribution $B_y(x)$ of dislocations per unit length in the range $-a < x < a, \; y=0$.

We consider first the traction σ_{yy} at the point $(x, 0)$ due to those dislocations between ξ and $\xi + \delta\xi$ on the line $y=0$. If $\delta\xi$ is small, they can be considered as a concentrated dislocation of strength $B_y(\xi)\delta\xi$ and hence they produce a traction

$$\sigma_{yy} = -\frac{2\mu B_y(\xi)\delta\xi}{\pi(\kappa + 1)(x - \xi)} \, ,$$

from equation (13.12), since the distance from $(\xi, 0)$ to $(x, 0)$ is $(x - \xi)$.

The traction due to the whole distribution of dislocations can therefore be written as the integral

$$\sigma_{yy} = -\frac{2\mu}{\pi(\kappa + 1)} \int_{-a}^{a} \frac{B_y(\xi)d\xi}{(x - \xi)} \, ,$$

and the boundary condition (13.16) leads to the following Cauchy singular integral equation for $B_y(\xi)$

$$\int_{-a}^{a} \frac{B_y(\xi)d\xi}{(x - \xi)} = \frac{\pi(\kappa + 1)S}{2\mu} \; ; \quad -a < x < a \, .$$

This is of exactly the same form as equation (12.20) and can be solved in the same way. Writing

$$x = a \cos \phi \; ; \; \xi = a \cos \theta \, ,$$

we have

$$\int_0^\pi \frac{B_y(\theta) \sin \theta d\theta}{(\cos \phi - \cos \theta)} = \frac{\pi(\kappa + 1)S}{2\mu} \; ; \; 0 < \phi < \pi \, .$$

Now (12.22) with $n = 1$ gives

$$\int_0^\pi \frac{\cos \theta d\theta}{(\cos \phi - \cos \theta)} = -\pi \; ; \; 0 < \phi < \pi$$

and hence

$$B_y(\theta) = -\frac{(\kappa + 1)S \cos \theta}{2\mu \sin \theta} + \frac{A}{\sin \theta}$$

— i.e.

$$B_y(\xi) = -\frac{(\kappa + 1)S\xi}{2\mu\sqrt{a^2 - \xi^2}} + \frac{Aa}{\sqrt{a^2 - \xi^2}} \, . \tag{13.18}$$

The arbitrary constant A is determined from the closure condition (13.15) which here takes the form

$$\int_{-a}^a B_y(\xi) d\xi = 0 \tag{13.19}$$

and leads to the result $A = 0$.

From a Fracture Mechanics perspective, we are particularly interested in the stress field surrounding the crack tip. For example, the stress component σ_{yy} in $|x| > a$, $y = 0$ is given by

$$\sigma_{yy} = -\frac{2\mu}{\pi(\kappa + 1)} \int_{-a}^a \frac{B_y(\xi) d\xi}{(x - \xi)} = \frac{S}{\pi} \int_{-a}^a \frac{\xi d\xi}{(x - \xi)\sqrt{a^2 - \xi^2}}$$

$$= S\left(-1 + \frac{|x|}{\sqrt{x^2 - a^2}}\right) \; ; \; |x| > a, \; y = 0 \tag{13.20}$$

using (13.18) and 3.228.2 of Gradshteyn and Ryzhik[11].

Remembering that this is the *corrective* solution, we add the uniform stress field $\sigma_{yy} = S$ to obtain the complete stress field, which on the line $y = 0$ gives

$$\sigma_{yy} = \frac{S|x|}{\sqrt{x^2 - a^2}} \; ; \; |x| > a, \; y = 0 \, .$$

---

[11] I.S.Gradshteyn and I.M.Ryzhik, *Tables of Integrals, Series and Products*, Academic Press, New York, 1980. See also Problem 19.8 for a method of evaluating this integral.

This tends to the uniform field as it should as $x \to \infty$ and is singular as $x \to a^+$. We define[12] the *mode I stress-intensity factor*, K_I as

$$K_I \equiv \lim_{x \to a^+} \sigma_{yy}(x)\sqrt{2\pi(x-a)} \tag{13.21}$$

$$= \lim_{x \to a^+} \frac{Sx\sqrt{2\pi(x-a)}}{\sqrt{x^2 - a^2}}$$

$$= S\sqrt{\pi a} . \tag{13.22}$$

We can also calculate the crack opening displacement

$$\delta(x) \equiv u_y(x, 0^+) - u_y(x, 0^-) = \int_{-a}^{x} B_y(\xi)d\xi = \frac{(\kappa+1)S}{2\mu}\sqrt{a^2 - x^2} . \tag{13.23}$$

Thus, the crack is opened to the shape of a long narrow ellipse as a result of the tensile field.

13.3.3 Energy release rate

If a crack extends as a result of the local stress field, there will generally be a reduction in the total strain energy U of the body. We shall show that the *strain-energy release rate*, defined as

$$G = -\frac{\partial U}{\partial S}$$

where S is the extent of the crack, is a unique function of the stress-intensity factor. We first note that in the immediate vicinity of the crack tip $x = a$, the stress component σ_{yy} can be approximated by the dominant singular term as

$$\sigma_{yy}(s) = \frac{K_I}{\sqrt{2\pi s}} ; \quad s > 0 , \tag{13.24}$$

from equation (13.21) with $s = (x-a)$. A similar approximation for the crack opening displacement $\delta(s)$ is

$$\delta(s) = \frac{S(\kappa+1)}{2\mu}\sqrt{2a(-s)} = \frac{K_I(\kappa+1)}{2\mu}\sqrt{\frac{2(-s)}{\pi}} ,$$

using (13.22, 13.23).

---

[12] Notice that the factor 2π in this definition is conventional, but essentially arbitrary. In applying fracture mechanics arguments, it is important to make sure that the results used for the stress-intensity factor (a theoretical or numerical calculation) and the fracture toughness (from experimental data) are based on the same definition of K.

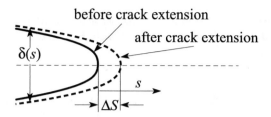

Fig. 13.5 Geometry of crack extension.

Figure 13.5 shows the configuration of the open crack before and after extension of the crack tip by an infinitesimal distance ΔS. The tractions on the crack plane before crack extension are given by (13.24), where s is measured from the initial position of the crack tip. The crack opening displacement in ΔS *after* extension is given by

$$\delta(s) = \frac{K_{\mathrm{I}}(\kappa + 1)}{2\mu} \sqrt{\frac{2(\Delta S - s)}{\pi}} \, , \tag{13.25}$$

since the point $s < \Delta S$ is now a distance $\Delta S - s$ to the left of the new crack tip. The reduction in strain energy during crack extension can be found by following a scenario in which the tractions (13.24) are gradually released, allowing work $W = -\Delta U$ to be done on them in moving through the displacements (13.25). We obtain

$$W = \frac{1}{2} \int_0^{\Delta S} \sigma_{yy}(s)\delta(s)ds = \frac{K_{\mathrm{I}}^2(\kappa + 1)}{4\pi\mu} \int_0^{\Delta S} \sqrt{\frac{(\Delta S - s)}{s}} ds = \frac{K_{\mathrm{I}}^2 \Delta S(\kappa + 1)}{8\mu}$$

and hence

$$G = -\frac{\partial U}{\partial S} = \frac{\partial W}{\partial S} = \frac{K_{\mathrm{I}}^2(\kappa + 1)}{8\mu} \, . \tag{13.26}$$

A similar calculation can be performed for a crack loaded in shear, causing a mode II stress-intensity factor K_{II}. The two deformation modes are orthogonal to each other, so that the energy release rate for a crack loaded in both modes I and II is given by

$$G = \frac{(K_{\mathrm{I}}^2 + K_{\mathrm{II}}^2)(\kappa + 1)}{8\mu} \, .$$

This expression can be written in terms of E, ν, using (13.10). We obtain

$$G = \frac{(K_{\mathrm{I}}^2 + K_{\mathrm{II}}^2)}{E} \qquad \text{(plane stress)} \tag{13.27}$$

$$= \frac{(K_{\mathrm{I}}^2 + K_{\mathrm{II}}^2)(1 - \nu^2)}{E} \qquad \text{(plane strain)}. \tag{13.28}$$

As long as the process zone is small in the sense defined in §13.3.1, a component containing a crack loaded in tension (mode I) will fracture when

$$K_\text{I} = K_\text{Ic} \quad \text{or} \quad G = G_c \ ,$$

where K_Ic, G_c are interrelated material properties, the former being the fracture toughness. The critical energy release rate G_c seems to imply a single failure criterion under combined mode I and mode II loading, but this is illusory, since experiments show that the value of G_c varies with the *mode-mixity ratio* K_II/K_I.

13.4 Disclinations

There is clearly a strong connection between the concept of a dislocation and the compatibility integrals (2.10, 2.13) introduced in §2.2.1. In fact if the integral (2.10) is evaluated clockwise along a closed path S_0 enclosing a dislocation with Burgers vector $\boldsymbol{B}$, we shall obtain

$$\oint_{S_0} \frac{\partial \boldsymbol{u}}{\partial s} ds = \boldsymbol{B}$$

and indeed the term 'dislocation' was coined by Volterra in his pioneering studies of these integral conditions.

We recall that the dislocation solution can be obtained by making a cut so as to render the body simply connected, and then imposing a rigid-body relative displacement between the two faces of the cut. A related solution can be obtained by imposing a relative rigid-body *rotation* between the two faces, so that the integral (2.13) evaluates to a non-zero vector. In two dimensions, rotation ω is a scalar and the corresponding solution is defined by the condition

$$u_\theta(r, 0) - u_\theta(r, 2\pi) = Cr \ . \tag{13.29}$$

By analogy with Figure 13.1, we can also interpret the solution as that in which a thin wedge of subtended angle $C \ll 1$ is welded between the cut faces.

From Table 9.1 we notice that the only stress function supporting a displacement discontinuity of the form (13.29) is $r^2 \ln(r)$, and for the prescribed angle C,

$$\phi = -\frac{\mu C r^2 \ln(r)}{\pi(\kappa + 1)} \ . \tag{13.30}$$

However, there are two objections to using this function in isolation: There is no obvious length scale to render the argument of the logarithm dimensionless, and Table 8.1 shows that the corresponding stresses are unbounded as $r \to \infty$.

Both these difficulties can be avoided by considering a traction-free cylinder or disk of finite radius a. The traction-free condition can be satisfied by superposing a state of uniform two-dimensional hydrostatic stress through the stress function Br^2, and the final stress field is then defined by the non-zero stress components

$$\sigma_{rr} = -\frac{2\mu C}{\pi(\kappa + 1)} \ln\left(\frac{r}{a}\right) \; ; \quad \sigma_{\theta\theta} = -\frac{2\mu C}{\pi(\kappa + 1)} \left[\ln\left(\frac{r}{a}\right) + 1\right] .$$

This stress state is known as a *disclination*.

For $C > 0$, the stress state near the origin is one of logarithmically unbounded hydrostatic tension, whereas near $r = a$ it approaches uniaxial (hoop) compression. The case $C < 0$ corresponds to the removal of a sector of the circle before the faces are rejoined, and involves compression near the origin and uniaxial tension near $r = a$. In the plane stress case, the stresses for $C < 0$ can be completely relaxed by allowing the disk to deform into a thin-walled cone, as is easily confirmed experimentally using a paper or cardboard disc[13].

In two dimensions, dislocations and disclinations are *point defects*, meaning that the differential compatibility equation (2.5) is satisfied everywhere except at a singular point. However, in three dimensions they are *line defects*, since the cut now comprises a surface whose interior boundary is a line along which the stress field will be singular. Also, in three dimensions, the imposed relative rotation at the cut becomes a vector with three components, known as the *Frank vector*, by analogy with the Burgers vector.

13.4.1 Disclinations in a crystal structure

Like dislocations, disclinations can be used to describe defects in an otherwise regular crystal structure, in which case the angular discontinuity C will be dictated by the defect-free structure. For example, in a hexagonal array, each molecule has six nearest neighbours, so to reproduce the cut and paste procedure described above, we would need to add or remove a 60° sector. Clearly, the regular structure could not be preserved distant from the disclination, and this implies that disclinations can arise in molecular arrays only as equal and opposite disclination pairs, also known as *disclination dipoles*.

Figure 13.6 shows an otherwise regular structure where one molecule (A) has only five nearest neighbours, corresponding to a disclination with $C = -\pi/3$, whilst B has seven implying $C = +\pi/3$. However, this same structure can also be seen as a dislocation, with an extra row of molecules entering from the right and terminating at A.

---

[13] You might also like to try the corresponding experiment for the opposite case $C > 0$.

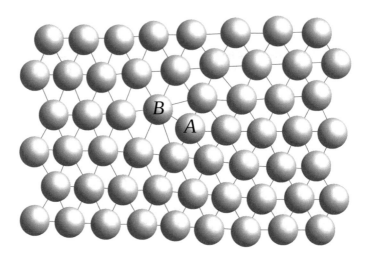

Fig. 13.6 A dislocation: Notice that A has only five nearest neighbours, whereas B has seven. Thus, the dislocation can be viewed as a disclination dipole.

This relationship can also be seen in the Airy stress function representation, since a dipole comprises equal and opposite disclinations separated by a small distance Δ. In the limit $\Delta \to 0$, if the strength of the disclinations increases so as to maintain $C\Delta$ constant, we recover the spatial derivative of equation (13.30) — i.e.

$$\frac{\partial \phi}{\partial x} = -\frac{2\mu C r \ln(r) \cos \theta}{\pi(\kappa + 1)} \quad \text{or} \quad \frac{\partial \phi}{\partial y} = -\frac{2\mu C r \ln(r) \sin \theta}{\pi(\kappa + 1)} , \qquad (13.31)$$

where we have dropped null terms of the form $r \cos \theta, r \sin \theta$. But the expressions (13.31) are precisely the stress functions (13.9, 13.13) defining dislocations with Burgers vector B_y and $-B_x$ respectively. In other words, if the stress function for a unit disclination [$C = 1$ in equation (13.30)] is denoted by ϕ_0, then that for a dislocation of Burgers vector B is given by

$$\phi = \epsilon_{ij} B_j \frac{\partial \phi_0}{\partial x_i} ,$$

where ϵ_{ij} is the two-dimensional alternating tensor, $\epsilon_{11} = \epsilon_{22} = 0$, $\epsilon_{12} = 1$, $\epsilon_{21} = -1$.

13.5 Method of Images

The method described in §13.3.2 applies strictly to the case of a crack in an infinite body, but it will provide a reasonable approximation if the length of the crack is small in comparison with the shortest distance to the boundary of a finite body or to some other geometric feature such as another crack or an interface to a different material. However, the same methodology could be applied to other problems if we could obtain the solution for a dislocation located at an arbitrary point in the uncracked body.

Closed-form solutions exist for bodies of a variety of shapes, including the traction-free half-plane, the infinite body with a traction-free circular hole and the infinite body containing a circular inclusion of a different elastic material[14]. These solutions all depend on the location of image singularities at appropriate points outside the body. The nature and strength of these singularities can be chosen so as to satisfy the required boundary conditions on the boundary of the original body. Earlier treatments of the subject proceeded in an essentially *ad hoc* way until the required boundary conditions were satisfied, but a unified treatment for the case of a plane interface was developed by Aderogba[15].

We consider the case in which the half-plane $y > 0$ with elastic constants μ_1, κ_1 is bonded to the half-plane $y < 0$ with constants μ_2, κ_2. Suppose that a singularity (typically a dislocation or a concentrated force) exists somewhere in $y > 0$ and that the same singularity in an infinite plane would correspond to the Airy stress function $\phi_0(x, y)$ defined throughout the infinite plane. Aderogba showed that the resulting stress field in the bonded bi-material plane is defined by the stress functions

$$
\begin{aligned}
\phi_1(x, y) &= \phi_0(x, y) + \mathcal{L}\{\phi_0(x, -y)\} & y &> 0 \\
\phi_2(x, y) &= \mathcal{J}\{\phi_0(x, y)\} & y &< 0 ,
\end{aligned} \tag{13.32}
$$

where the linear operators $\mathcal{L}, \mathcal{J}$ are defined by

$$
\mathcal{L}\{\cdot\} = A\left[1 - 2y\frac{\partial}{\partial y} + y^2\nabla^2\right] + \frac{(A - B)}{4} \iint \nabla^2\{\cdot\}dydy
$$

$$
\mathcal{J}\{\cdot\} = 1 + A + \frac{(A - B)}{4}\left[\iint \nabla^2\{\cdot\}dydy - 2y\int \nabla^2\{\cdot\}dy\right] ,
$$

with

$$
A = \frac{(\Gamma - 1)}{(\Gamma\kappa_1 + 1)} ; \quad B = \frac{(\Gamma\kappa_1 - \kappa_2)}{(\Gamma + \kappa_2)} ; \quad \Gamma = \frac{\mu_2}{\mu_1} .
$$

These bimaterial constants are related to Dundurs' constants (4.10, 4.11) through the equations

---

[14] See for example J. Dundurs (1963), Concentrated force in an elastically embedded disk, *ASME Journal of Applied Mechanics,* Vol. 30, pp. 568–570; J. Dundurs and G. P. Sendeckyj (1965), Edge dislocation inside a circular inclusion, *Journal of the Mechanics and Physics of Solids,* Vol.13, pp. 141–147.

[15] K. Aderogba (2003), An image treatment of elastostatic transmission from an interface layer, *Journal of the Mechanics and Physics of Solids,* Vol. 51, pp. 267–279.

$$A = \frac{\alpha - \beta}{1 + \beta} \; ; \quad B = \frac{\alpha + \beta}{1 - \beta} \; .$$

Notice that the arbitrary functions of x implied in the indefinite integrals with respect to y in (13.32) must be chosen so as to ensure that the resulting stress functions ϕ_1, ϕ_2 are biharmonic.

The special case of the traction-free half-plane can be recovered in the limit where $\Gamma \to 0$ and hence $A, B \to -1$, giving

$$\mathcal{L}\{\cdot\} = -1 + 2y\frac{\partial}{\partial y} - y^2 \nabla^2 \; . \tag{13.33}$$

Example

As an example, we consider the case where a glide dislocation is situated at the point $(0, a)$ in the traction-free half-plane $y > 0$. The corresponding infinite plane solution is given by equation (13.13) as

$$\phi_0 = \frac{2\mu B_x r \ln(r) \sin\theta}{\pi(\kappa + 1)} \; ,$$

where r, θ are measured from the dislocation. Converting to Cartesian coördinates and moving the origin to the surface of the half-plane, we have

$$\phi_0(x, y) = \frac{\mu B_x (y - a)}{\pi(\kappa + 1)} \ln\left(x^2 + (y - a)^2\right) \; .$$

Substituting into equation (13.33), we have

$$\mathcal{L}\{\phi_0(x, -y)\} = -\frac{\mu B_x}{\pi(\kappa + 1)} \left[(y - a)\ln\left(x^2 + (y + a)^2\right) + \frac{4ay(y + a)}{x^2 + (y + a)^2} \right]$$

and hence the complete stress function is

$$\phi = \frac{\mu B_x}{\pi(\kappa + 1)} \left[(y - a)\ln\left(\frac{x^2 + (y - a)^2}{x^2 + (y + a)^2}\right) - \frac{4ay(y + a)}{x^2 + (y + a)^2} \right] \; .$$

The circular hole

A similar procedure can be used to find the perturbation in a stress field due to the presence of the traction-free circular hole $r = a$. If the unperturbed field is defined by the function $\phi_0(r, \theta)$, the perturbed field can be obtained as

$$\phi = \phi_0(r, \theta) - \mathcal{L}\left[\frac{r^2}{a^2}\phi_0\left(\frac{a^2}{r}, \theta\right)\right] + \frac{C(r^2 - 2a^2 \ln(r))}{4} \; , \tag{13.34}$$

where

$$\mathcal{L}\{\cdot\} = \frac{r^2}{a^2} + \left(1 - \frac{r^2}{a^2}\right) r \frac{\partial}{\partial r} + \frac{a^2}{4} \left(1 - \frac{r^2}{a^2}\right)^2 \nabla^2$$

and the constant C is defined as

$$C = \lim_{r \to 0} \nabla^2 \phi_0(r, \theta) .$$

These mathematical operations are performed by the Maple and Mathematica files 'holeperturbation' and the reader can verify (for example) that they generate the solutions of §8.3.2 and §8.4.1, starting from the stress functions defining the corresponding unperturbed uniform stress fields.

Problems

13.1. Figure 13.7 shows a large body with a circular hole of radius a, subjected to a concentrated force F, tangential to the hole. The remainder of the hole surface is traction free and the stress field is assumed to tend to zero as $r \to \infty$.

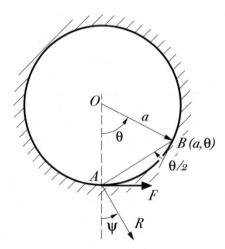

Fig. 13.7

(i) Find the stress components at the point $B(a, \theta)$ on the surface of the hole, due to the candidate stress function

$$\phi = -\frac{F}{\pi} R\psi \cos \psi$$

in polar coördinates R, ψ centred on the point A as shown.
(ii) Transform these stress components into the polar coördinate system r, θ with origin at the centre of the hole, O.
   Note: *For points on the surface of the hole* $(r=a)$, *we have* $R=2a\sin(\theta/2)$, $\sin\psi=\cos(\theta/2)$, $\cos\psi=-\sin(\theta/2)$.
(iii) Complete the solution by superposing stress functions (in r, θ) from Tables 8.1, 9.1 with appropriate Fourier components in the tractions and determining the multiplying constants from the conditions that (a) the surface of the hole be traction free (except at A) and (b) the displacements be everywhere single valued.

13.2. Use a method similar to that outlined in Problem 13.1 to find the stress field if the force F acts in the direction *normal* to the surface of the otherwise traction-free hole $r=a$.

13.3. A large plate is in a state of pure bending such that $\sigma_{xx}=\sigma_{xy}=0$, $\sigma_{yy}=Cx$, where C is a constant.
   We now introduce a crack in the range $-a<x<a$, $y=0$.

(i) Find a suitable corrective solution which, when superposed on the simple bending field, will make the surfaces of the crack free of tractions. [**Hint:** represent the corrective solution by a distribution of climb dislocations in $-a<x<a$. Don't forget the closure condition (13.19)].
(ii) Find the corresponding crack opening displacement as a function of x and show that it has an unacceptable negative value in $-a<x<0$.
(iii) Re-solve the problem assuming that there is frictionless contact in some range $-a<x<b$, where b is a constant to be determined. The dislocations will now have to be distributed only in $b<x<a$ and b is found from a continuity condition at $x=b$. (Move the origin to the mid-point of the range $b<x<a$.)
(iv) For the case with contact, find expressions for (a) the crack opening displacement, (b) the stress-intensity factor at $x=a$, (c) the dimension b and (d) the contact traction in $-a<x<b$, and hence verify that the contact inequalities are satisfied.

13.4. A state of uniform shear $\sigma_{xy}=S$, $\sigma_{xx}=\sigma_{yy}=0$ in a large block of material is perturbed by the presence of the plane crack $-a<x<a$, $y=0$. Representing the perturbation due to the crack by a distribution of glide dislocations along the crack line, find the shear stress distribution on the line $x>a$, $y=0$, the relative motion between the crack faces in $-a<x<a$, $y=0$ and the mode II stress-intensity factor K_{II} defined as

$$K_{II} \equiv \lim_{x\to a^+}\sigma_{yx}(x,0)\sqrt{2\pi(x-a)}.$$

13.5. A state of uniform general bi-axial stress $\sigma_{xx}=S_{xx}$, $\sigma_{yy}=S_{yy}$, $\sigma_{xy}=S_{xy}$ in a large block of material is perturbed by the presence of the plane crack $-a<x<a$,

$y=0$. Note that the crack will close completely if $S_{yy} < 0$ and will be fully open if $S_{yy} > 0$. Use the method proposed in Problem 13.4 to find the mode II stress-intensity factor K_{II} for both cases, assuming that Coulomb friction conditions hold in the closed crack with coefficient f.

If the block contains a large number of widely separated similar cracks of all possible orientations and if the block fails when at any one crack

$$\sqrt{K_I^2 + K_{II}^2} = K_{Ic} ,$$

where the fracture toughness K_{Ic} is a material constant, sketch the biaxial failure surface for the material — i.e. the locus of all failure points in principal biaxial stress space (σ_1, σ_2).

13.6. By treating equation (13.30) as a Green's function or otherwise, find the Airy stress function ϕ defining the stresses due to a disclination of total strength C uniformly distributed over the circle $0 \leq r < c$ at the centre of a traction-free circular disk of radius a [$> c$].

13.7. Find the stress field due to a climb dislocation of unit strength located at the point $(0, a)$ in the traction-free half-plane $y > 0$, using the following method:-

(i) Write the stress function (13.8) in Cartesian coördinates.
(ii) Find the solution for a dislocation at $(0, a)$ in the full plane $-\infty < y < \infty$ by making a change of origin.
(iii) Superpose additional singularities centred on the image point $(0, -a)$ to make the surface $y = 0$ traction free. Notice that these singularities are outside the actual body and are therefore admissible. Appropriate functions [in polar coördinates centred on $(0, -a)$] are

$$C_1 r \ln(r) \cos \theta \; ; \quad \frac{C_2 \cos \theta}{r} \; ; \quad C_3 \sin(2\theta) .$$

13.8. Solve Problem 13.7 using Aderogba's result (13.33, 13.32) for step (iii).

13.9. Find the stress field due to a concentrated force F applied in the x-direction at the point $(0, a)$ in the half-plane $y > 0$, if the surface of the half-plane is traction free. Verify that the results reduce to those of §12.3 in the limit as $a \to 0$.

13.10. Two dissimilar elastic half-planes are bonded on the interface $y=0$ and have material properties as defined in §13.5. Use Aderogba's formula (13.32) to determine the stress field due to a climb dislocation at the point $(0, a)$ in the half-plane $y > 0$. In particular, find the location and magnitude of the maximum shear traction at the interface.

13.11. Use equation (13.34) to solve Problem 8.3.

Chapter 14
Thermoelasticity

Most materials tend to expand if their temperature rises and, to a first approximation, the expansion is proportional to the temperature change. If the expansion is unrestrained and the material is isotropic, all dimensions will expand equally — i.e. there will be a uniform dilatation described by

$$e_{xx} = e_{yy} = e_{zz} = \alpha T \tag{14.1}$$
$$e_{xy} = e_{yz} = e_{zx} = 0 , \tag{14.2}$$

where α is the *coefficient of linear thermal expansion* and T is the temperature change. Notice that no shear strains are induced in unrestrained thermal expansion, so that a body which is heated to a uniformly higher temperature will get larger, but will retain the same shape.

Thermal strains are additive to the elastic strains due to local stresses, so Hooke's law is modified to the form

$$e_{xx} = \frac{\sigma_{xx}}{E} - \frac{\nu \sigma_{yy}}{E} - \frac{\nu \sigma_{zz}}{E} + \alpha T \tag{14.3}$$
$$e_{xy} = \frac{\sigma_{xy}(1+\nu)}{E} . \tag{14.4}$$

14.1 The Governing Equation

The Airy stress function can be used for two-dimensional thermoelasticity, but the governing equation will generally include additional terms associated with the temperature field. Repeating the derivation of §4.3.1, but using (14.3) in place of (1.44), we find that the compatibility condition demands that

$$\frac{\partial^2 \sigma_{xx}}{\partial y^2} - \nu \frac{\partial^2 \sigma_{yy}}{\partial y^2} + E\alpha \frac{\partial^2 T}{\partial y^2} - 2(1+\nu)\frac{\partial^2 \sigma_{xy}}{\partial x \partial y} + \frac{\partial^2 \sigma_{yy}}{\partial x^2} - \nu \frac{\partial^2 \sigma_{xx}}{\partial x^2} + E\alpha \frac{\partial^2 T}{\partial x^2} = 0$$

© The Author(s), under exclusive license to Springer Nature Switzerland AG 2022
J. R. Barber, *Elasticity*, Solid Mechanics and Its Applications 172,
https://doi.org/10.1007/978-3-031-15214-6_14

and after substituting for the stress components from (4.6) and rearranging, we obtain

$$\nabla^4 \phi = -E\alpha \nabla^2 T \ , \tag{14.5}$$

for plane stress.

Plane strain

The corresponding plane strain equations can be obtained by a similar procedure, noting that the restraint of the transverse strain e_{zz} (14.1) will induce a stress

$$\sigma_{zz} = -E\alpha T \tag{14.6}$$

and hence additional in-plane strains $\nu\alpha T$. Equation (14.5) is therefore modified to

$$\nabla^4 \phi = -\frac{E\alpha}{(1-\nu)} \nabla^2 T \ , \tag{14.7}$$

for plane strain and we can supplement the plane stress to plane strain conversions (3.10) with the relation

$$\alpha = \alpha'(1+\nu') \ . \tag{14.8}$$

Equations (14.5, 14.7) are similar in form to that obtained in the presence of body forces (7.7) and can be treated in the same way. Thus, we can seek any particular solution of (14.5) and then satisfy the boundary conditions of the problem by superposing a more general biharmonic function, since the biharmonic equation is the complementary or homogeneous equation corresponding to (14.5, 14.7).

Example

As an example, we consider the case of the thin circular disk, $r < a$, with traction-free edges, raised to the temperature

$$T = T_0 y^2 = T_0 r^2 \sin^2 \theta \ , \tag{14.9}$$

where T_0 is a constant.

Substituting this temperature distribution into equation (14.5), we obtain

$$\nabla^4 \phi = -2E\alpha T_0$$

and a simple particular solution is

$$\phi_0 = -\frac{E\alpha T_0 r^4}{32} .$$

The stresses corresponding to ϕ_0 are

$$\sigma_{rr} = -\frac{E\alpha T_0 r^2}{8} \;\; ; \;\; \sigma_{\theta\theta} = -\frac{3E\alpha T_0 r^2}{8} \;\; ; \;\; \sigma_{r\theta} = 0 ,$$

and the boundary $r=a$ can be made traction free by superposing a uniform hydrostatic tension $E\alpha T_0 a^2/8$, resulting in the final stress field[1]

$$\sigma_{rr} = \frac{E\alpha T_0(a^2 - r^2)}{8} \;\; ; \;\; \sigma_{\theta\theta} = \frac{E\alpha T_0(a^2 - 3r^2)}{8} \;\; ; \;\; \sigma_{r\theta} = 0 .$$

14.2 Heat Conduction

The temperature field might be a given quantity — for example, it might be measured using thermocouples or radiation methods — but more often it has to be calculated from thermal boundary conditions as a separate boundary-value problem. Most materials approximately satisfy the Fourier heat conduction law, according to which the heat flux per unit area q is linearly proportional to the local temperature gradient. i.e.

$$q = -K\nabla T , \tag{14.10}$$

where K is the thermal conductivity of the material. The conductivity is usually assumed to be constant, though for real materials it depends upon temperature. However, the resulting non-linearity is only important when the range of temperatures under consideration is large.

We next apply the principle of conservation of energy to a small cube of material. Equation (14.10) governs the flow of heat across each face of the cube and there may also be heat generated, Q per unit volume, within the cube due to some mechanism such as electrical resistive heating or nuclear reaction etc. If the sum of the heat flowing into the cube and that generated within it is positive, the temperature will rise at a rate which depends upon the *thermal capacity* of the material. Combining these arguments we find that the temperature T must satisfy the equation[2]

$$\rho c \frac{\partial T}{\partial t} = K\nabla^2 T + Q , \tag{14.11}$$

---

[1] It is interesting to note that the stress field in this case is axisymmetric, even though the temperature field (14.9) is not.

[2] More detail about the derivation of this equation and other information about the linear theory of heat conduction can be found in the classical text H. S. Carslaw and J. C. Jaeger, *Conduction of Heat in Solids*, 2nd.ed., Clarendon Press, Oxford, 1959.

where ρ, c are respectively the density and specific heat of the material, so that the product ρc is the amount of heat needed to increase the temperature of a unit volume of material by one degree.

In equation (14.11), the first term on the right-hand side is the net heat flow into the element per unit volume and the second term, Q is the rate of heat generated per unit volume. The algebraic sum of these terms gives the heat available for raising the temperature of the cube.

It is convenient to divide both sides of the equation by K, giving the more usual form of the heat conduction equation

$$\nabla^2 T = \frac{1}{\kappa} \frac{\partial T}{\partial t} - \frac{Q}{K} , \tag{14.12}$$

where

$$\kappa = \frac{K}{\rho c}$$

is the *thermal diffusivity* of the material. Thermal diffusivity has the dimensions area/time and its magnitude gives some indication of the rate at which a thermal disturbance will propagate through the body.

We can substitute (14.12) into (14.5) obtaining

$$\nabla^4 \phi = -E\alpha \left(\frac{1}{\kappa} \frac{\partial T}{\partial t} - \frac{Q}{K} \right) \tag{14.13}$$

for plane stress.

14.3 Steady-state Problems

Equation (14.13) shows that if the temperature is independent of time and there is no internal source of heat in the body ($Q=0$), ϕ will be biharmonic. It therefore follows that in the steady-state problem without heat generation, the in-plane stress field is unaffected by the temperature distribution. In particular, if the boundaries of the body are traction free, a steady-state (and hence harmonic) temperature field will not induce any thermal stresses under plane stress conditions[3].

Furthermore, if there are internal heat sources — i.e. if $Q \neq 0$ — the stress field can be determined directly from equation (14.13), using Q, without the necessity of first solving a boundary-value problem for the temperature T. This also implies that the thermal boundary conditions in two-dimensional steady-state problems have no effect on the thermoelastic stress field.

All of these deductions rest on the assumption that the elastic problem is defined in terms of *tractions*. If some of the boundary conditions are stated in terms of

---

[3] But we recall that the plane stress solution is an approximation. A more exact treatment of the steady-state thermoelastic problem for moderately thick plates is given in §16.4.1.

displacements, constraint of the resulting thermal distortion will induce boundary tractions and the stress field will be affected, though it will still be the same as that which would have been produced by the same tractions if they had been applied under isothermal conditions.

Recalling the arguments of §2.2.1, we conclude that similar considerations apply to the multiply-connected body, for which there exists an implied displacement boundary condition. In other words, multiply-connected bodies will generally develop non-zero thermal stresses even under steady-state conditions with no boundary tractions. However, the resulting stress field is essentially that associated with the presence of a dislocation in the hole and hence can be characterized by relatively few parameters[4].

Appropriate conditions for multiply-connected bodies can be explicitly imposed, but in general it is simpler to obviate the need for such a condition by reverting to a displacement function representation. We shall therefore postpone discussion of thermoelastic problems for multiply-connected bodies until Chapter 23 where such a formulation is introduced.

14.3.1 Dundurs' Theorem

If the conditions discussed in the last section are satisfied and the temperature field therefore induces no thermal stress, the strains will be given by equations (14.1, 14.2). It then follows that

$$\frac{\partial^2 u_y}{\partial x^2} = -\frac{\partial^2 u_x}{\partial x \partial y} \, ,$$

because of (14.2)

$$= -\frac{\partial e_{xx}}{\partial y} = -\alpha \frac{\partial T}{\partial y} = \frac{\alpha q_y}{K} \, , \tag{14.14}$$

from (14.1, 14.10). In view of (14.8), the corresponding result for plane strain can be written

$$\frac{\partial^2 u_y}{\partial x^2} = \frac{\alpha(1+\nu)q_y}{K} \, . \tag{14.15}$$

In this equation, the constant of proportionality $\alpha(1+\nu)/K$ is known as the *thermal distortivity* of the material and is denoted by the symbol δ.

Equations (14.14, 14.15) state that the curvature of an initially straight line segment in the x-direction ($\partial^2 u_y/\partial x^2$) is proportional to the local heat flux across that line segment. This result was first proved by Dundurs[5] and is referred to as *Dundurs'*

---

[4] See for example J. Dundurs (1974), Distortion of a body caused by free thermal expansion, *Mechanics Research Communications*, Vol. 1, pp. 121–124.

[5] J. Dundurs *loc. cit.*

Theorem. It is very useful as a guide to determining the effect of thermal distortion on a structure. Figure 14.1 shows some simple bodies with various thermal boundary conditions and the resulting steady-state thermal distortion.

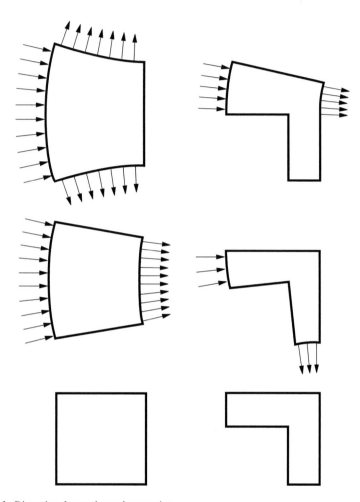

Fig. 14.1 Distortion due to thermal expansion.

Notice that straight boundaries that are unheated remain straight, those that are heated become convex outwards, whilst those that are cooled become concave. The angles between the edges are unaffected by the distortion, because there is no shear strain. Since the thermal field is in the steady-state and there are no heat sources, the algebraic sum of the heat input around the boundary must be zero. Thus, although the boundary is locally rotated by the cumulative heat input from an appropriate starting point, this does not lead to incompatibility at the end of the circuit.

Many three-dimensional structures such as box-sections, tanks, rectangular hoppers etc., are fabricated from plate elements. If the unrestrained thermal distortions of these elements are considered separately, the incompatibilities of displacement developed at the junctions between elements permits the thermal stress problem to be described in terms of dislocations, the physical effects of which are more readily visualized.

Dundurs' Theorem can also be used to obtain some useful simplifications in two-dimensional contact and crack problems involving thermal distortion[6].

Problems

14.1. A direct electric current I flows along a conductor of rectangular cross section $-4a < x < 4a$, $-a < y < a$, all the surfaces of which are traction free. The conductor is made of copper of electrical resistivity ρ, thermal conductivity K, Young's modulus E, Poisson's ratio ν and coefficient of thermal expansion α. Assuming the current density to be uniform and neglecting electromagnetic effects, estimate the thermal stresses in the conductor when the temperature has reached a steady state.

14.2. A fuel element in a nuclear reactor can be regarded as a solid cylinder of radius a. During operation, heat is generated at a rate $Q_0(1 + Ar^2/a^2)$ per unit volume, where r is the distance from the axis of the cylinder and A is a constant.

Assuming that the element is immersed in a fluid at pressure p and that *axial* expansion is prevented, find the radial and circumferential thermal stresses produced in the steady state.

14.3. The instantaneous temperature distribution in the thin plate $-a < x < a$, $-b < y < b$ is defined by

$$T(x, y) = T_0 \left(\frac{x^2}{a^2} - 1 \right) ,$$

where T_0 is a positive constant. Find the magnitude and location of (i) the maximum tensile stress and (ii) the maximum shear stress in the plate if the edges $x = \pm a$, $y = \pm b$ are traction free and $a \gg b$.

14.4. The half-plane $y > 0$ is subject to periodic heating at the surface $y = 0$, such that the surface temperature is

$$T(0, t) = T_0 \cos(\omega t) .$$

Show that the temperature field

$$T(y, t) = T_0 e^{-\lambda y} \cos(\omega t - \lambda y)$$

satisfies the heat conduction equation (14.12) with no internal heat generation, provided that

---

[6] For more details, see J. R. Barber (1980), Some implications of Dundurs' Theorem for thermoelastic contact and crack problems, *Journal of Strain Analysis*, Vol. 22, pp. 229–232.

$$\lambda = \sqrt{\frac{\omega}{2k}} .$$

Find the corresponding thermal stress field as a function of y, t if the surface of the half-plane is traction free.

Using appropriate material properties, estimate the maximum tensile stress generated in a large rock due to diurnal temperature variation, with a maximum daytime temperature of $30^{\circ}C$ and minimum nighttime temperature of $10^{\circ}C$.

14.5. The layer $0 < y < h$ rests on a frictionless rigid foundation at $y = 0$ and the surface $y = h$ is traction free. The foundation is a thermal insulator and the free surface is subjected to the steady state heat input

$$q_y = q_0 \cos(mx) .$$

Use Dundurs' theorem to show that the layer will not separate from the foundation and find the amplitude of the sinusoidal perturbation in the free surface due to thermal distortion.

14.6. Figure 14.2 shows a hollow cylinder of internal radius a and external radius b that has been slit along a plane surface AB at $\theta = 0$. The inner surface $r = a$ is subjected to a heat flux $q_r = q_0(\theta)$, whilst the outer radius $r = b$ is maintained at zero temperature. All the surfaces, including the slit, are traction free. Using Dundurs' theorem, show that in the steady state, the two surfaces at AB will rotate relative to each other so as to open an angle $\psi = Q\delta$, where

$$Q = \int_0^{2\pi} q_0(\theta) a d\theta$$

is the total heat flux transmitted between the inner and outer surfaces and δ is the thermal distortivity defined in equation (14.15).

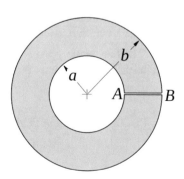

Fig. 14.2

Hence, using the results of §13.4 or otherwise, find the steady-state thermal stresses in a traction-free solid cylinder of radius a due to a line heat source Q at the origin $r = 0$.

Chapter 15
Antiplane Shear

In Chapters 3–14, we considered two-dimensional states of stress involving the in-plane displacements u_x, u_y and stress components $\sigma_{xx}, \sigma_{xy}, \sigma_{yy}$. Another class of two-dimensional stress states that satisfy the elasticity equations exactly is that in which the in-plane displacements u_x, u_y are everywhere zero, whilst the out-of-plane displacement u_z is independent of z — i.e.

$$u_x = u_y = 0 \quad ; \quad u_z = f(x, y) . \tag{15.1}$$

Substituting these results into the strain-displacement relations (1.37) yields

$$e_{xx} = e_{yy} = e_{zz} = 0$$

and

$$e_{xy} = 0 \quad ; \quad e_{yz} = \frac{1}{2}\frac{\partial f}{\partial y} \quad ; \quad e_{zx} = \frac{1}{2}\frac{\partial f}{\partial x} .$$

It then follows from Hooke's law (1.54) that

$$\sigma_{xx} = \sigma_{yy} = \sigma_{zz} = 0 \tag{15.2}$$

and

$$\sigma_{xy} = 0 \quad ; \quad \sigma_{yz} = \mu\frac{\partial f}{\partial y} \quad ; \quad \sigma_{zx} = \mu\frac{\partial f}{\partial x} . \tag{15.3}$$

In other words, the only non-zero stress components are the two shear stresses σ_{zx}, σ_{zy} and these are functions of x, y only. Such a stress state is known as *antiplane shear* or *antiplane strain*.

The in-plane equilibrium equations (2.1, 2.2) are identically satisfied by the stress components (15.2, 15.3) if and only if the in-plane body forces p_x, p_y are zero. However, substituting (15.3) into the out-of-plane equilibrium equation (2.3) yields

$$\mu\nabla^2 f + p_z = 0 . \tag{15.4}$$

© The Author(s), under exclusive license to Springer Nature Switzerland AG 2022
J. R. Barber, *Elasticity*, Solid Mechanics and Its Applications 172,
https://doi.org/10.1007/978-3-031-15214-6_15

Thus, antiplane problems can involve body forces in the axial direction, provided these are independent of z.

In the absence of body forces, equation (15.4) reduces to the Laplace equation

$$\nabla^2 f = 0 \,. \tag{15.5}$$

15.1 Transformation of Coördinates

It is often convenient to regard the two stress components on the z-plane as components of a shear stress vector

$$\boldsymbol{\tau} = \boldsymbol{i}\sigma_{zx} + \boldsymbol{j}\sigma_{zy} = \mu\nabla f \tag{15.6}$$

from (15.3). The vector transformation equations (1.8) then show that the stress components in the rotated coördinate system x', y' of Figure 1.3 are

$$\sigma'_{zx} = \sigma_{zx}\cos\theta + \sigma_{zy}\sin\theta \;\; ; \;\; \sigma'_{zy} = \sigma_{zy}\cos\theta - \sigma_{zx}\sin\theta \,.$$

Equation (15.6) can also be used to define the expressions for the stress components in polar coördinates as

$$\sigma_{zr} = \mu\frac{\partial f}{\partial r} \;\; ; \;\; \sigma_{z\theta} = \frac{\mu}{r}\frac{\partial f}{\partial\theta} \,. \tag{15.7}$$

The maximum shear stress at any given point is the magnitude of the vector $\boldsymbol{\tau}$ and is given by

$$|\boldsymbol{\tau}| = \mu\sqrt{\left(\frac{\partial f}{\partial x}\right)^2 + \left(\frac{\partial f}{\partial y}\right)^2} \,.$$

15.2 Boundary Conditions

Only one boundary condition can be imposed at each point on the boundary. For example, we might prescribe the axial displacement $u_z = f$, or alternatively the traction

$$\sigma_{nz} = \mu\frac{\partial f}{\partial n} \,,$$

where n is the local outward normal to the cross section. Thus, antiplane problems reduce to the solution of the Poisson equation (15.4) [or the Laplace equation (15.5) when there is no body force] with prescribed values of f or $\partial f/\partial n$ on the boundary. This is a classical boundary-value problem that is somewhat simpler than that involved in the solution for the Airy stress function, but essentially similar methods can be used for both. Simple examples can be defined for all of the geometries considered in Chapters 5–13. In the present chapter, we shall consider a few special cases, but a wider range of examples can be found in the problems at the end of this chapter.

15.3 The Rectangular Bar

Figure 15.1 shows the cross section of a long bar of rectangular cross section $2a \times b$, where $a \gg b$. The short edges $x = \pm a$ are built in to rigid supports, the edge $y = b$ is loaded by a uniform shear traction $\sigma_{yz} = S$ and the remaining edge $y = 0$ is traction free.

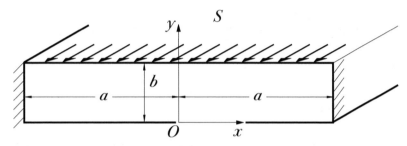

Fig. 15.1 Rectangular bar with shear loading on one edge.

Using equations (15.2, 15.3), the boundary conditions can be written in the mathematical form

$$f = 0 \ ; \quad x = \pm a \tag{15.8}$$

$$\frac{\partial f}{\partial y} = \frac{S}{\mu} \ ; \quad y = b \tag{15.9}$$

$$\frac{\partial f}{\partial y} = 0 \ ; \quad y = 0 \ . \tag{15.10}$$

However, as in Chapter 5, a finite polynomial solution can be found only if the boundary conditions (15.8) on the shorter edges are replaced by the weak form

$$\int_0^b f(x, y)dy = 0 \ ; \quad x = \pm a \ . \tag{15.11}$$

This problem is even in x and a solution can be obtained using the trial function

$$f = C_1 x^2 + C_2 y^2 + C_3 y + C_4 \ .$$

Substitution into equations (15.5, 15.9–15.11) yields the conditions

$$2C_1 + 2C_2 = 0$$

$$2C_2 b + C_3 = \frac{S}{\mu}$$

$$C_3 = 0$$

$$C_1 a^2 b + \frac{C_2 b^3}{3} + \frac{C_3 b^2}{2} + C_4 = 0$$

with solution

$$C_1 = -\frac{S}{2\mu b} \quad ; \quad C_2 = \frac{S}{2\mu b} \quad ; \quad C_3 = 0 \quad ; \quad C_4 = \frac{S(3a^2 - b^2)}{6\mu b} . \qquad (15.12)$$

The final solution for the stresses and displacements is therefore

$$u_z = \frac{S}{2\mu b} \left(y^2 - x^2 + a^2 - \frac{b^2}{3} \right) \quad ; \quad \sigma_{zx} = -\frac{Sx}{b} \quad ; \quad \sigma_{zy} = \frac{Sy}{b} ,$$

from equations (15.12, 15.1, 15.3).

15.4   The Concentrated Line Force

Consider an infinite block of material loaded by a force F per unit length acting along the z-axis. This problem is axisymmetric, in contrast to that solved in §13.1, since the force this time is directed along the axis rather than perpendicular to it. It follows that the resulting displacement function f must be axisymmetric and equation (15.5) reduces to the ordinary differential equation

$$\frac{d^2 f}{dr^2} + \frac{1}{r} \frac{df}{dr} = 0 , \qquad (15.13)$$

since there is no body force except at the origin. Equation (15.13) has the general solution

$$f = C_1 \ln(r) + C_2 ,$$

leading to the stress field

$$\sigma_{rz} = \frac{\mu C_1}{r} \quad ; \quad \sigma_{\theta z} = 0 , \qquad (15.14)$$

from (15.7). The constant C_1 can be determined by considering the equilibrium of a cylinder of material of radius a and unit length. We obtain

$$F + \int_0^{2\pi} \sigma_{rz}(a, \theta) a \, d\theta = 0$$

and hence

$$C_1 = -\frac{F}{2\pi\mu} \ .$$

The corresponding stresses are then obtained from (15.14) as

$$\sigma_{rz} = -\frac{F}{2\pi r} \ ; \quad \sigma_{\theta z} = 0 \ .$$

Of course, the same result could have been obtained without recourse to elasticity arguments, appealing simply to axisymmetry and equilibrium and hence the same stresses would be obtained for a fairly general non-linear or inelastic material, as long as it is isotropic.

The elastic displacement is

$$u_z = f = -\frac{F\ln(r)}{2\pi\mu} + C_2 \ ,$$

where the constant C_2 represents an arbitrary rigid-body displacement.

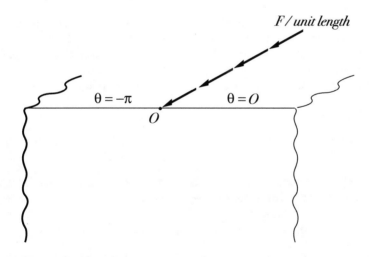

Fig. 15.2 Half-space loaded by an out-of-plane line force.

All θ-surfaces are traction free and hence the same solution can be used for a wedge of any angle loaded by a uniform force per unit length acting at and parallel to the apex, though the factor of 2π in C_1 will then be replaced by the subtended wedge angle. In particular, Figure 15.2 shows the half-space $-\pi < \theta < 0$ loaded by a uniformly distributed tangential force along the z-axis, for which it is easily shown that

$$\sigma_{rz} = -\frac{F}{\pi r} \ ; \quad u_z = -\frac{F\ln(r)}{\pi\mu} + C_2 \ . \tag{15.15}$$

This result can be used to solve antiplane frictional contact problems analogous to those considered in Chapter 12 (see Problems 15.11, 15.12).

15.5   The Screw Dislocation

The dislocation solution in antiplane shear corresponds to the situation in which the infinite body is cut on the half-plane $x > 0$, $y = 0$ and the two faces of the cut experience a relative displacement $\delta = B_z$ in the z-direction. Describing the cut space as the wedge $0 < \theta < 2\pi$, the boundary conditions for this problem can be written

$$u_z(r, 0) - u_z(r, 2\pi) = B_z ,$$

and a suitable solution satisfying equation (15.5) is

$$u_z = -\frac{B_z \theta}{2\pi} \; ; \; \sigma_{zr} = 0 \; ; \; \sigma_{z\theta} = \frac{\mu B_z}{2\pi r} .$$

The corresponding material defect is known as a *screw dislocation*. Notice that the stress components due to a screw dislocation decay with r^{-1} with distance from the origin, as do those for the in-plane climb and glide dislocations discussed in §13.2.

A distribution of screw dislocations can be used to solve problems involving cracks in an antiplane shear field, using the technique introduced in §13.3.

Problems

15.1. The long rectangular bar, $0 < x < a$, $0 < y < b$, $a \gg b$, is built in to a rigid support at $x = a$ and loaded by a uniform shear traction $\sigma_{yz} = S$ at $y = b$. The remaining surfaces are free of traction. Find a solution for the displacement and stress fields, using strong boundary conditions on the edges $y = 0, b$.

15.2. Figure 15.3 shows the cross section of a long bar of equilateral triangular cross section of side a. The three faces of the bar are built in to rigid supports and the axis of the bar is vertical, resulting in gravitational loading $p_z = -\rho g$.

Show that an exact solution for the stress and displacement fields can be obtained using a stress function of the form

$$f = C(y - \sqrt{3}x)(y + \sqrt{3}x)\left(y - \frac{\sqrt{3}a}{2}\right) .$$

Find the value of the constant C and the location and magnitude of the maximum shear stress.

15.3. A state of uniform shear $\sigma_{xz}=S$, $\sigma_{yz}=0$ in a large block of material is perturbed by the presence of a small traction-free hole whose boundary is defined by the equation $r=a$ in cylindrical polar coördinates r, θ, z. Find the complete stress field in the block and hence determine the appropriate stress-concentration factor.

15.4. A state of uniform shear $\sigma_{xz}=S$, $\sigma_{yz}=0$ in a large block of material is perturbed by the presence of a rigid circular inclusion whose boundary is defined by the equation $r=a$ in cylindrical polar coördinates r, θ, z. The inclusion is perfectly bonded to the elastic material and is prevented from moving, so that $u_z=0$ at $r=a$. Find the complete stress field in the block and hence determine the appropriate stress-concentration factor.

15.5. The body defined by $a<r<b$, $0<\theta<\pi/2$, $-\infty<z<\infty$ is built in at $\theta=\pi/2$ and loaded by a shear force F per unit length in the z-direction on the surface $\theta=0$. The curved surfaces $r=a$, b are traction free. Find the stress field in the body, using the strong boundary conditions on the curved surfaces and weak conditions elsewhere.

15.6. The body defined by $a<r<b$, $0<\theta<\pi/2$, $-\infty<z<\infty$ is built in at $\theta=\pi/2$, such that the z-axis is vertical. It is loaded only by its own weight with density ρ, the curved surfaces $r=a$, b being traction free. Find the stress field in the body, using the strong boundary conditions on the curved surfaces and weak conditions elsewhere.

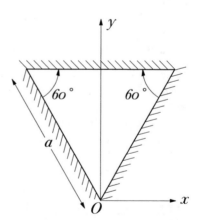

Fig. 15.3 cross section of the triangular bar.

15.7. The body defined by $a<r<b$, $0<\theta<\pi/2$, $-\infty<z<\infty$ is built in at $\theta=\pi/2$ and loaded by the uniform shear traction $\sigma_{rz}=S$ on $r=a$, the other surfaces being traction free. Find the stress field in the body, using the strong boundary conditions on the curved surfaces and weak conditions elsewhere.

Hint: This is a degenerate problem in the sense of §10.3.1 and requires a special stress function. Equilibrium arguments suggest that the stress component $\sigma_{\theta z}$ must increase with θ, implying a function f varying with θ^2. You can construct a suitable function by starting with the harmonic function $r^n \cos(n\theta)$, differentiating twice with respect to n and then setting $n=0$.

15.8. The body $-\alpha < \theta < \alpha$, $0 \le r < a$ is supported at $r=a$ and loaded only by a uniform antiplane shear traction $\sigma_{\theta z} = S$ on the surface $\theta = \alpha$, the other surface being traction free. Find the complete stress field in the body, using strong boundary conditions on $\theta = \pm\alpha$ and weak conditions on $r = a$.

15.9. The half-space $-\pi < \theta < 0$ is bonded to a rigid body in the region $\theta = 0$, the remaining surface $\theta = -\pi$ being traction free. If the rigid body is loaded in the z-direction (out-of-plane), use an asymptotic argument analogous to that in §11.2 to determine the exponent of the most singular term in the stress field near the origin.

15.10. The wedge $-\alpha < \theta < \alpha$ has traction-free surfaces $\theta = \pm\alpha$ and is loaded at large r in antiplane shear. Develop an asymptotic (eigenfunction) expansion for the most general stress field near the apex $r=0$ and plot the exponent of the dominant term for (i) symmetric and (ii) antisymmetric stress fields as functions of the complete wedge angle 2α.

15.11. Two large parallel cylinders each of radius R are pressed together by a normal force P per unit length. An axial tangential force Q per unit length is then applied, tending to make the cylinders slide relative to each other along the axis as shown in Figure 15.4. If the coefficient of friction is f and $Q < fP$, find the extent of the contact region, the extent of the slip and stick regions and the distribution of normal and tangential traction in the contact area[1].

15.12. Solve Problem 15.11 for the case where the cylinders are rolling against each other, as in Carter's problem (§12.8.2). If the rolling speed is Ω, find the relative (axial) creep velocity.

Now suppose that the axes of the cylinders are very slightly misaligned through some angle ϵ which is insufficient to make the contact conditions vary significantly along the axis. If the cylinders run in bearings which prevent relative axial motion, find the axial bearing force per unit length and the maximum misalignment which can occur without there being gross slip.

15.13. A state of uniform shear $\sigma_{xz} = 0$, $\sigma_{yz} = S$ in a large block of material is perturbed by the presence of the plane crack $-a < x < a$, $y = 0$. Representing the perturbation due to the crack by a distribution of screw dislocations along the crack line, find the shear stress distribution on the line $x > a$, $y = 0$, the relative motion between the crack faces in $-a < x < a$, $y = 0$ and the mode III stress-intensity factor K_{III} defined as

$$K_{\mathrm{III}} \equiv \lim_{x \to a^+} \sigma_{yz}(x, 0)\sqrt{2\pi(x - a)} \ .$$

---

[1] This problem is analogous to Cattaneo's problem discussed in §12.8.1.

15.14. The semi-infinite plate $x > 0$, $-c < y < c$ is loaded by self-equilibrated shear tractions $\sigma_{xz} = s(y)$ on $x = 0$, the surfaces $y = \pm c$ being traction free. Use the method of §6.2 to estimate the exponential decay rate of the stress field away from $x = 0$ for symmetric functions $s(y)$.

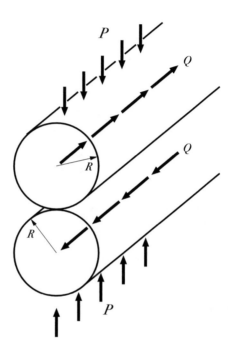

Fig. 15.4 Contacting
cylinders loaded out of plane.

Chapter 16
Moderately Thick Plates

In §3.2, we showed that the plane stress assumption

$$\sigma_{zx} = \sigma_{zy} = \sigma_{zz} = 0$$

leads only to an approximate solution, except in the special case where $\nu = 0$. In particular, the resulting strains do not satisfy the full set of six compatibility equations. In this chapter, we shall show that an exact plane stress solution can be developed if we relax the assumption that the stresses do not vary through the thickness[1]. In other words, we allow the stress state to be three-dimensional.

As in Chapter 4, we can satisfy the two non-trivial equilibrium equations (4.1) by defining a stress function ϕ such that the non-zero stress components

$$\sigma_{xx} = \frac{\partial^2 \phi}{\partial y^2} \; ; \quad \sigma_{yy} = \frac{\partial^2 \phi}{\partial x^2} \; ; \quad \sigma_{xy} = -\frac{\partial^2 \phi}{\partial x \partial y} \; , \tag{16.1}$$

where we note that ϕ will now generally be a function of all three coördinates x, y, z.

The non-zero strain components are then given by

$$e_{xx} = \frac{1}{E}\left(\frac{\partial^2 \phi}{\partial y^2} - \nu \frac{\partial^2 \phi}{\partial x^2}\right) \; ; \quad e_{yy} = \frac{1}{E}\left(\frac{\partial^2 \phi}{\partial x^2} - \nu \frac{\partial^2 \phi}{\partial y^2}\right)$$

$$e_{zz} = -\frac{\nu}{E}\left(\frac{\partial^2 \phi}{\partial x^2} + \frac{\partial^2 \phi}{\partial y^2}\right) \; ; \quad e_{xy} = -\frac{(1+\nu)}{E}\frac{\partial^2 \phi}{\partial x \partial y}$$

and substitution into the six compatibility equations (2.7) yields

$$\frac{\partial^4 \phi}{\partial x^4} + 2\frac{\partial^4 \phi}{\partial x^2 \partial y^2} + \frac{\partial^4 \phi}{\partial y^4} = 0$$

---

[1] See A.E.H.Love, *A Treatise on the Mathematical Theory of Elasticity*, 4th edn., Dover, 1944, §299 *et seq.*.

© The Author(s), under exclusive license to Springer Nature Switzerland AG 2022
J. R. Barber, *Elasticity*, Solid Mechanics and Its Applications 172,
https://doi.org/10.1007/978-3-031-15214-6_16

$$\frac{\partial^2}{\partial z^2}\left(\frac{\partial^2 \phi}{\partial x^2} - \nu \frac{\partial^2 \phi}{\partial y^2}\right) = \nu \frac{\partial^2}{\partial y^2}\left(\frac{\partial^2 \phi}{\partial x^2} + \frac{\partial^2 \phi}{\partial y^2}\right)$$

$$\frac{\partial^2}{\partial z^2}\left(\frac{\partial^2 \phi}{\partial y^2} - \nu \frac{\partial^2 \phi}{\partial x^2}\right) = \nu \frac{\partial^2}{\partial x^2}\left(\frac{\partial^2 \phi}{\partial x^2} + \frac{\partial^2 \phi}{\partial y^2}\right) \qquad (16.2)$$

$$(1 + \nu)\frac{\partial^4 \phi}{\partial x \partial y \partial z^2} = -\nu \frac{\partial^2}{\partial x \partial y}\left(\frac{\partial^2 \phi}{\partial x^2} + \frac{\partial^2 \phi}{\partial y^2}\right)$$

$$\frac{\partial^2}{\partial z \partial y}\left(\frac{\partial^2 \phi}{\partial x^2} + \frac{\partial^2 \phi}{\partial y^2}\right) = \frac{\partial^2}{\partial z \partial x}\left(\frac{\partial^2 \phi}{\partial x^2} + \frac{\partial^2 \phi}{\partial y^2}\right) = 0 .$$

The problem is symmetrical with respect to the mid-plane of the plate so we expect ϕ to be an even function of z. Equations (16.2) therefore suggest a solution of the form

$$\phi(x, y, z) = \phi_1(x, y) + z^2 \phi_2(x, y) . \qquad (16.3)$$

Substituting this expression into (16.2) and equating coefficients of powers of z, we find that all the compatibility equations are satisfied by the choice

$$\phi_2 = -\frac{\nu}{2(1 + \nu)}\nabla^2 \phi_1 \quad \text{and} \quad \nabla^2 \phi_2 = 0 ,$$

implying also that $\nabla^4 \phi_1 = 0$. The non-zero stress components are then obtained as

$$\sigma_{xx} = \frac{\partial^2}{\partial y^2}\left(\phi_1 - \frac{\nu z^2}{2(1 + \nu)}\nabla^2 \phi_1\right)$$

$$\sigma_{yy} = \frac{\partial^2}{\partial x^2}\left(\phi_1 - \frac{\nu z^2}{2(1 + \nu)}\nabla^2 \phi_1\right) \qquad (16.4)$$

$$\sigma_{xy} = -\frac{\partial^2}{\partial x \partial y}\left(\phi_1 - \frac{\nu z^2}{2(1 + \nu)}\nabla^2 \phi_1\right) ,$$

where $\nabla^4 \phi_1 = 0$. This representation reduces to the classical Airy function formulation when $\nu = 0$.

16.1   Boundary Conditions

The two-dimensionally biharmonic function ϕ_1 is sufficient to define two independent quantities at each point s on the boundary, typically the normal and in-plane shear components of the local traction $t(s)$, but the stress components (16.4) imply that t will have the form

$$t(s) = t_1(s) + z^2 t_2(s) .$$

We can therefore satisfy the boundary conditions only in the weak sense

$$\int_{-c}^{c} t(s, z)dz = \int_{-c}^{c} t_0(s, z)dz \,, \tag{16.5}$$

where z is measured from the mid-plane of a plate of thickness $2c$, $t_0(s, z)$ is the traction imposed in the physical problem, and $t(s, z)$ is the corresponding value in the weak solution.

The resulting solution will therefore generally differ from the exact stress state in a region near the boundaries and the error will decay from these edges with a length scale related to the semi-thickness c. The astute reader might reasonably ask at this point whether the approximation involved is any better than that from the simple (z-independent) plane stress solution. We shall discuss this issue in §16.2 below, but first we illustrate the solution process for a simple example.

Example

We consider the problem of §5.2.1 and Figure 5.2 in which the rectangular cantilever $0 < x < a$, $-b < y < b$, $-c < z < c$ is loaded by a traverse force F on the end $x = 0$. The boundary conditions, modified from (5.11–5.14) are

$$\sigma_{xy} = 0 \; ; \quad y = \pm b \tag{16.6}$$
$$\sigma_{yy} = 0 \; ; \quad y = \pm b \tag{16.7}$$
$$\sigma_{xx} = 0 \; ; \quad x = 0 \tag{16.8}$$
$$\int_{-c}^{c} \int_{-b}^{b} \sigma_{xy} dy dz = F \; ; \quad x = 0 \,. \tag{16.9}$$

Since this is a 'correction' on the two-dimensional solution, we start by considering the stress function

$$\phi_1 = C_1 x y^3 + C_2 x y$$

from equation (5.19). However, note that we do not generally expect the constants C_1, C_2 to take the same values as in the two-dimensional solution, because of the boundary condition (16.5). It follows that $\nabla^2 \phi_1 = 6C_1 x y$ and hence the stress components are

$$\sigma_{xx} = 6C_1 x y \; ; \quad \sigma_{yy} = 0 \; ; \quad \sigma_{xy} = -3C_1 y^2 - C_2 + \frac{3C_1 \nu z^2}{(1 + \nu)} \,,$$

from (16.4). These expressions satisfy the boundary conditions (16.7, 16.8), but (16.6) can be satisfied only in the weak sense

$$\int_{-c}^{c} \sigma_{xy}(x, \pm b, z)dz = 0 ,$$

from which

$$C_2 = -3C_1 b^2 + \frac{C_1 \nu c^2}{(1 + \nu)} .$$

The end condition (16.9) then gives

$$C_1 = \frac{F}{8cb^3} ,$$

and the non-zero stress components are obtained as

$$\sigma_{xx} = \frac{3Fxy}{4cb^3} ; \quad \sigma_{xy} = \frac{F}{8cb^3}\left[3(b^2 - y^2) - \frac{\nu(c^2 - 3z^2)}{(1 + \nu)}\right] .$$

We notice that the shear stress distribution is now predicted to vary with z over the cross section, and in particular, the maximum shear stress occurs at the points $(x, 0, \pm c)$ and is

$$\tau_{max} = \frac{3F}{8cb} + \frac{F\nu c}{4b^3(1 + \nu)} . \tag{16.10}$$

In Chapter 18 we shall obtain a solution to this problem using a Fourier series representation to satisfy the strong condition (16.6). In particular, we shall show that the first term in this series is identical to the second 'corrective' term in equation (16.10). For the extreme case of a bar of square cross section ($c = b$) and $\nu = 0.5$, it reduces the percentage error in the elementary two-dimensional solution from 18% to 1%.

16.2   Edge Effects

The error associated with the weak form of the traction boundary condition (16.5) corresponds to the stress field due to locally self-equilibrated tractions of the form

$$t(s, z) = C(s)(c^2 - 3z^2) ; \quad -c < z < c ,$$

and based on Saint Venant's principle, we anticipate that the error will decay exponentially with distance from the boundary. For the normal stress component, the exponential decay rate λ can be estimated using the solution from §6.2.2. In particular, since the problem is symmetric about the mid-plane, $\lambda \approx 2.1/c$. For the shear traction, the corrective problem is one of antiplane strain (see Chapter 15 and particularly Problem 15.14) and the corresponding eigenvalue is $\lambda = \pi/c$.

For the example problem above, the maximum shear stress occurs far from the surface on which the weak boundary condition was imposed, so the solution given here represents a significant improvement on the two-dimensional solution. By contrast, for problems where the maximum stress is expected adjacent to a traction-free surface (notably stress-concentration problems such as §8.4.1), the extra complication of the three-dimensional solution is not justified.

16.3 Body Force Problems

Suppose a thin plate is subjected to a conservative in-plane body-force field $p = -\nabla V$, where the potential V is a function of (x, y) only. Thus the body force is uniform through the thickness. As in Chapter 7, the equilibrium equations can then be satisfied by defining the stress components

$$\sigma_{xx} = \frac{\partial^2 \phi}{\partial y^2} + V ; \quad \sigma_{yy} = \frac{\partial^2 \phi}{\partial x^2} + V ; \quad \sigma_{xy} = -\frac{\partial^2 \phi}{\partial x \partial y} .$$

If these equations are used in place of (16.1) in the derivation of equations (16.2), we obtain

$$\left(\frac{\partial^2}{\partial x^2} + \frac{\partial^2}{\partial y^2} \right)^2 \phi = -(1 - \nu)\nabla^2 V$$

$$\frac{\partial^2}{\partial z^2} \left(\frac{\partial^2 \phi}{\partial x^2} - \nu \frac{\partial^2 \phi}{\partial y^2} \right) = \nu \frac{\partial^2}{\partial y^2} \left(\frac{\partial^2 \phi}{\partial x^2} + \frac{\partial^2 \phi}{\partial y^2} + 2V \right)$$

$$\frac{\partial^2}{\partial z^2} \left(\frac{\partial^2 \phi}{\partial y^2} - \nu \frac{\partial^2 \phi}{\partial x^2} \right) = \nu \frac{\partial^2}{\partial x^2} \left(\frac{\partial^2 \phi}{\partial x^2} + \frac{\partial^2 \phi}{\partial y^2} + 2V \right) \qquad (16.11)$$

$$(1 + \nu) \frac{\partial^4 \phi}{\partial x \partial y \partial z^2} = -\nu \frac{\partial^2}{\partial x \partial y} \left(\frac{\partial^2 \phi}{\partial x^2} + \frac{\partial^2 \phi}{\partial y^2} + 2V \right)$$

$$\frac{\partial^2}{\partial z \partial y} \left(\frac{\partial^2 \phi}{\partial x^2} + \frac{\partial^2 \phi}{\partial y^2} \right) = \frac{\partial^2}{\partial z \partial x} \left(\frac{\partial^2 \phi}{\partial x^2} + \frac{\partial^2 \phi}{\partial y^2} \right) = 0 ,$$

and it can be shown that these six equations have no solution for ϕ, except for the case where $\nabla^2 V$ is independent of x, y. However, this restricted class of body force problems includes the important cases of gravitational loading and uniform rotation within the xy-plane.

Gravitational loading

As in §7.2.1, gravitational loading can be described by the potential $V = \rho g y$, where the y-axis defines the vertically upwards direction. Since V is a linear function of the coördinates, the terms involving V in (16.11) vanish, so the stress components are still given by (16.4) except for the addition of a term $+\rho g y$ in the expressions for σ_{xx} and σ_{yy}.

Uniform rotation

If the plate rotates about the z-axis at constant speed Ω, the body-force field is described by the potential

$$V = -\frac{1}{2}\rho\Omega^2(x^2 + y^2) = -\frac{\rho\Omega^2 r^2}{2} ,$$

from equation (7.17). In this case

$$\nabla^2 V = -2\rho\Omega^2 ,$$

and it is easily verified that equations (16.11) are all satisfied by the particular solution

$$\begin{aligned}
\phi_P &= \frac{\rho\Omega^2(1-\nu)}{32}(x^2 + y^2)^2 - \frac{\rho\Omega^2\nu(1+\nu)}{4(1-\nu)}(x^2 + y^2)z^2 \\
&= \frac{\rho\Omega^2(1-\nu)}{32}r^4 - \frac{\rho\Omega^2\nu(1+\nu)}{4(1-\nu)}r^2 z^2 .
\end{aligned} \qquad (16.12)$$

The general solution is then obtained by superposing a homogeneous solution (i.e. a solution without body force) using equations (16.4).

Example

We illustrate the process for the problem of the circular disk $0 \le r < a$, $-c < z < c$, rotating at constant speed Ω about the z-axis, with traction-free boundary conditions. The particular solution is given by equation (16.12) and for the homogeneous solution we use the axisymmetric biharmonic function that is bounded at $r = 0$ — i.e. $\phi_1 = Ar^2$. We then have

$$\phi_2 = -\frac{\nu}{2(1+\nu)}\nabla^2\phi_1 = -\frac{2\nu A}{(1+\nu)}$$

and hence

$$\phi = \phi_P + \phi_1 + z^2\phi_2 = \frac{\rho\Omega^2(1-\nu)}{32}r^4 - \frac{\rho\Omega^2\nu(1+\nu)}{4(1-\nu)}r^2 z^2 + Ar^2 - \frac{2\nu Az^2}{(1+\nu)} .$$

The boundary condition in this problem is $\sigma_{rr} = 0$ at $r = a$, so we calculate

$$\sigma_{rr} = \frac{1}{r}\frac{\partial \phi}{\partial r} + \frac{1}{r^2}\frac{\partial^2 \phi}{\partial \theta^2} + V = \frac{\rho\Omega^2(1-\nu)}{8}r^2 - \frac{\rho\Omega^2\nu(1+\nu)}{2(1-\nu)}z^2 + 2A - \frac{\rho\Omega^2 r^2}{2} .$$

The traction-free boundary condition can be satisfied only in the weak sense

$$\int_{-c}^{c} \sigma_{rr}(a, z)dz = 0 ,$$

from which we obtain

$$A = \frac{\rho\Omega^2}{4}\left[\frac{(3+\nu)a^2}{4} + \frac{\nu(1+\nu)c^2}{3(1-\nu)}\right] .$$

The final stress field is then given by

$$\sigma_{rr} = \rho\Omega^2\left[\frac{(3+\nu)(a^2-r^2)}{8} + \frac{\nu(1+\nu)(c^2-3z^2)}{6(1-\nu)}\right]$$

$$\sigma_{\theta\theta} = \rho\Omega^2\left[\frac{\{(3+\nu)a^2-(1+3\nu)r^2\}}{8} + \frac{\nu(1+\nu)(c^2-3z^2)}{6(1-\nu)}\right] .$$

The maximum tensile stress (and also the maximum von Mises stress) occurs at the origin and is

$$\sigma_{max} = \rho\Omega^2\left[\frac{(3+\nu)a^2}{8} + \frac{\nu(1+\nu)c^2}{6(1-\nu)}\right] .$$

This location is not close to the boundary where the weak boundary condition was applied, so the three-dimensional solution will give a good approximation to the exact result.

This is a particularly simple example, but it is clear that the same method can be easily applied to other geometries, including the rotating rectangular beam of §7.3.1.

16.4 Normal Loading of the Faces

Suppose that we solve the same two-dimensional boundary-value problem (i) under the plane strain assumptions, and (ii) using the plane stress assumptions, but including the three-dimensional correction described in this chapter. If we now construct the *difference* between these two solutions, the boundary conditions will generally involve normal loading σ_{zz} on the faces of the plate, but the other boundaries will be traction free within the limitations of the weak form of equation (16.5). This approach can be used to generate solutions of the problem where both sides of a thin plate are loaded by equal and opposite normal tractions σ_{zz}, subject to the restriction that these be two-dimensionally harmonic[2].

Suppose that a given two-dimensional plane strain solution is defined by the Airy function $\phi_1(x, y)$, so

$$\sigma_{zz}(x, y) = \nu(\sigma_{xx} + \sigma_{yy}) = \nu\nabla^2\phi_1 , \tag{16.13}$$

from (4.6). Since ϕ_1 is biharmonic, σ_{zz} must be harmonic.

If we now use the same function $\phi_1(x, y)$ to define a three-dimensional plane stress solution using equations (16.3, 16.1), the difference between the two solutions will then include the out-of-plane stresses (16.13), and in-plane stresses defined by the function

$$\phi = -z^2\phi_2(x, y) = \frac{\nu z^2}{2(1+\nu)}\nabla^2\phi_1 = \frac{z^2\sigma_{zz}}{2(1+\nu)} .$$

The in-plane stresses will not generally satisfy weak traction-free boundary conditions, but we can correct this by adding the extra term[3] $A\sigma_{zz}(x, y)$, where A is a constant chosen so as to satisfy the condition

$$\int_{-c}^{c} \phi\, dz = 0 .$$

We obtain

$$\phi = -\frac{(c^2 - 3z^2)\sigma_{zz}}{6(1+\nu)} , \tag{16.14}$$

after which the stresses are recovered from equations (16.1). The resulting stress field will clearly satisfy the condition that all surfaces with normals in the (x, y)-plane will have zero tractions in the weak sense.

16.4.1 Steady-state thermoelasticity

We showed in Chapter 14 that for a two dimensional body with steady-state heat conduction $[\nabla^2 T(x, y) = 0]$, the plane strain solution involves no in-plane stresses, and the out-of-plane stress component is

$$\sigma_{zz} = -E\alpha T(x, y) ,$$

from equation (14.6). It follows immediately from equation (16.14) that the non-zero stress components for a traction-free plate of finite thickness $2c$ can be defined

---

[2] X-F, Li and Z-L. Hu (2020), Generalization of plane stress and plane strain states to elastic plates of finite thickness, *Journal of Elasticity*. These authors also considered problems where the normal surface displacements u_z are prescribed on the faces.

[3] Recall that σ_{zz} is harmonic, so $A\sigma_{zz}(x, y)$ is a legitimate stress function.

through the stress function

$$\phi = \frac{(c^2 - 3z^2)E\alpha T}{6(1 + \nu)}$$

and are

$$\sigma_{xx} = \frac{(c^2 - 3z^2)E\alpha}{6(1 + \nu)} \frac{\partial^2 T}{\partial y^2} \;\; ; \;\; \sigma_{yy} = \frac{(c^2 - 3z^2)E\alpha}{6(1 + \nu)} \frac{\partial^2 T}{\partial x^2}$$

$$\sigma_{xy} = -\frac{(c^2 - 3z^2)E\alpha}{6(1 + \nu)} \frac{\partial^2 T}{\partial x \partial y} . \tag{16.15}$$

Problems

16.1. A curved beam $a < r < b$, $0 < \theta < \pi$, $-c < z < c$ is built in at $\theta = \pi$ and loaded by a tensile force F normal to the surface at $\theta = 0$, whose line of action passes through the origin. Find the distribution of shear stress $\sigma_{\theta r}$ on the cross section at $\theta = \pi/2$. What is the percentage difference in the maximum shear stress at this location, relative to the two-dimensional solution, if $b = 2a$ and $c = a/4$?

16.2. The beam $-b < y < b$, $0 < x < L$, $-c < z < c$ is built-in at the end $x = 0$ and loaded by a uniform shear traction $\sigma_{xy} = S$ on the upper edge, $y = b$, the remaining edges, $x = L$, $y = -b$ being traction free. Find a suitable stress function and the corresponding stress components for this problem, using the weak boundary conditions on $x = L$. (This is the three-dimensional counterpart of Problem 5.1).

16.3. Show that equations (16.11) have solutions if and only if $\nabla^2 V$ is independent of x and y.

16.4. The beam $-b < y < b$, $0 < x < L$, $-c < z < c$ is built-in at the end $x = L$ and is subject only to gravitational loading $p_y = -\rho g$, all the remaining surfaces being traction free. Find the complete stress field in the beam.

16.5. The rectangular block $-a < x < a$, $-b < y < b$, $-c < z < c$ rotates at uniform angular velocity Ω about the z-axis. Estimate the maximum tensile stress in the block if $a \gg b \gg c$.

16.6. An infinite elastic plate of thickness $2c$ contains a non-conducting circular inclusion of radius a whose elastic properties are the same as those in the rest of the plate. Find the magnitude and location of the maximum tensile stress in the plate if the inclusion perturbs a uniform heat flux $q_x = q_0$ and the extremities of the plate are traction free.

Comment on the relation between your solution and that for a plate with a circular hole.

16.7. The solution of §16.4.1 is applied to a plate with a thermally-insulated traction-free boundary at $x = 0$, and the maximum value of the lateral stress σ_{yy} is predicted to be adjacent to this boundary. Show that the correction required to make this boundary exactly traction free is not negligible, and obtain a more accurate expression for the local maximum stress in terms of in-plane derivatives of the temperature field.

16.8. The rectangular block $-a < x < a$, $-b < y < b$, $-c < z < c$ is initially at rest when equal and opposite forces F are applied on the ends $x = \pm a$, as in Figure 7.3. Find the stresses in the block just after the forces are applied, assuming that $a \gg b \gg c$.

Part III
End Loading of the Prismatic Bar

One of the most important problems in the technical application of Elasticity concerns a long[1] bar of uniform cross section loaded only at the ends. Recalling Saint Venant's principle[2], we anticipate a region near the ends where the stress field is influenced by the exact local traction distribution, but these end effects should decay with distance from the ends, leaving the stress field in a central region depending only upon the force resultants transmitted along the bar. In this section we shall consider the problem of determining this 'preferred' form of stress distribution associated with various force resultant end loadings of the bar.

The most general end loading of the bar will comprise an axial force F, shear forces V_x, V_y in the x- and y-directions respectively, bending moments M_x, M_y about the x- and y-axes and a torque T. The axial force and bending moments cause normal stresses σ_{zz} on the cross section, whilst the torque and shear forces cause shear stresses σ_{zx}, σ_{zy}. Notice however that the existence of a shear force implies a bending moment varying linearly with z. These problems are strictly three-dimensional in that the displacements vary with z, but this variation is at most linear or quadratic and the mathematical formulation of the problem therefore reduces to the solution of a boundary-value problem on the two-dimensional cross-sectional domain.

Elementary Mechanics of Materials solutions of these problems are based on *ad hoc* assumptions, most notably the assumption that plane sections remain plane during bending. In Elasticity it is legitimate to use the same approach provided that the final stress field is checked to ensure that it satisfies the equilibrium equations (2.4) and that the corresponding strains satisfy the six compatibility equations (2.7). However, a more satisfactory starting point is to note that in the central region the stress state is independent[3] of z and hence each 'slice' of bar defined by initially parallel cross-sectional planes must deform in exactly the same way. Thus, if initially plane sections deform out of plane, this deformation is independent of z and the distance between corresponding points on two such sections after deformation will be defined by a general rigid-body translation and rotation. This argument is sufficient to establish that the strain e_{zz} is a linear function of x, y, as assumed in the classical

---

[1] Here 'long' implies that the length of the bar is at least several times larger than the largest dimension in the cross section.

[2] See §3.1.2 and Chapter 6

[3] Except for the linear increase of bending stresses due to a shear force.

theory of pure bending. Notice incidentally that the constitutive law is not required for this argument and hence the conclusion holds for a general non-linear or inelastic material[4].

Using these arguments, it is easily shown that the elementary Mechanics of Materials results are exact for the case where the bar is loaded by an axial force and/or a bending moment. It is convenient to locate the origin at the centroid of the cross section, in which case a tensile force F acting along the z-axis will generate the uniform uniaxial stress field $\sigma_{zz} = \dfrac{F}{A}$; $\sigma_{yy} = \sigma_{yy} = \sigma_{xy} = \sigma_{yz} = \sigma_{zx} = 0$,

where A is the area of the cross section. For the bending problem, some simplification results if we choose the coördinate axes to coincide with the principal axes of bending of the bar[5]. The bending stress field is then given by Bending

$$\sigma_{zz} = \frac{M_1 y}{I_1} - \frac{M_2 x}{I_2} \;\; ; \;\; \sigma_{yy} = \sigma_{yy} = \sigma_{xy} = \sigma_{yz} = \sigma_{zx} = 0 \, ,$$

where I_1, I_2 are the principal second moments of area of the cross section and M_1, M_2 are the bending moments about the corresponding axes, chosen to coincide with the x, y-directions respectively.

The problem of the bar loaded in torsion or shear cannot be treated exactly using elementary Mechanics of Materials arguments[6] and will be discussed in the next two chapters.

---

[4] See J. R. Barber, *Intermediate Mechanics of Materials*, Springer, Dordrecht, 2nd. edn. (2011), Chapter 5.

[5] See for example J. R. Barber, *Intermediate Mechanics of Materials*, Springer, Dordrecht, 2nd. edn. (2011), Chapter 4.

[6] Except for the special case of a circular bar loaded only in torsion.

Chapter 17
Torsion of a Prismatic Bar

If a bar is loaded by equal and opposite torques T on its ends, we anticipate that the relative rigid-body displacement of initially plane sections will consist of rotation, leading to a twist per unit length β. These sections may also deform out of plane, but this deformation must be the same for all values of z. These kinematic considerations lead to the candidate displacement field

$$u_x = -\beta z y \; ; \quad u_y = \beta z x \; ; \quad u_z = \beta f(x, y) \; , \tag{17.1}$$

where f is an unknown function of x, y describing the out-of-plane deformation, also known as the *warping function*. Notice that it is convenient to extract the factor β explicitly in u_z, since whatever the exact nature of f, it is clear that the complete deformation field must be linearly proportional to the applied torque and hence to the twist per unit length.

Substituting these results into the strain-displacement relations (1.37) yields

$$e_{xx} = e_{yy} = e_{zz} = e_{xy} = 0 \; ; \quad e_{zx} = \frac{\beta}{2}\left(\frac{\partial f}{\partial x} - y\right) \; ; \quad e_{zy} = \frac{\beta}{2}\left(\frac{\partial f}{\partial y} + x\right)$$

and it follows from Hooke's law (1.54) that

$$\sigma_{xx} = \sigma_{yy} = \sigma_{zz} = 0$$

and

$$\sigma_{xy} = 0 \; ; \quad \sigma_{zx} = \mu\beta\left(\frac{\partial f}{\partial x} - y\right) \; ; \quad \sigma_{zy} = \mu\beta\left(\frac{\partial f}{\partial y} + x\right) \; . \tag{17.2}$$

There are no body forces, so substitution into the equilibrium equations (2.4) yields

$$\nabla^2 f = 0 \; .$$

© The Author(s), under exclusive license to Springer Nature Switzerland AG 2022
J. R. Barber, *Elasticity*, Solid Mechanics and Its Applications 172,
https://doi.org/10.1007/978-3-031-15214-6_17

The torsion problem is therefore reduced to the determination of a harmonic function f such that the stresses (17.2) satisfy the traction-free condition condition on the curved surfaces of the bar. The twist per unit length β can then be determined by evaluating the torque on the cross section Ω

$$T = \iint_{\Omega} (x\sigma_{zy} - y\sigma_{zx})dxdy \ . \tag{17.3}$$

17.1 Prandtl's Stress Function

The imposition of the boundary condition is considerably simplified by introducing Prandtl's stress function φ defined such that

$$\boldsymbol{\tau} \equiv \boldsymbol{i}\sigma_{zx} + \boldsymbol{j}\sigma_{zy} = \text{curl } \boldsymbol{k}\varphi$$

or

$$\sigma_{zx} = \frac{\partial\varphi}{\partial y} \ ; \quad \sigma_{zy} = -\frac{\partial\varphi}{\partial x} \ . \tag{17.4}$$

With this representation, the traction-free boundary condition can be written

$$\boldsymbol{\tau} \cdot \boldsymbol{n} = \sigma_{zn} = \frac{\partial\varphi}{\partial t} = 0 \ , \tag{17.5}$$

where $\boldsymbol{n}$ is the local normal to the boundary of Ω and n, t are a corresponding set of local orthogonal coördinates respectively normal and tangential to the boundary. Thus φ must be constant around the boundary and for simply-connected bodies this constant can be taken as zero without loss of generality giving the simple condition

$$\varphi = 0 \ . \tag{17.6}$$

on the boundary.

The expressions (17.4) satisfy the equilibrium equations (2.4) identically. To obtain the governing equation for φ, we eliminate σ_{zx}, σ_{zy} between equations (17.2, 17.4), obtaining

$$\frac{\partial\varphi}{\partial y} = \mu\beta\left(\frac{\partial f}{\partial x} - y\right) \ ; \quad \frac{\partial\varphi}{\partial x} = -\mu\beta\left(\frac{\partial f}{\partial y} + x\right) \tag{17.7}$$

and then eliminate f between these equations, obtaining

$$\nabla^2\varphi = -2\mu\beta \ . \tag{17.8}$$

Thus, in Prandtl's formulation, the torsion problem reduces to the determination of a function φ satisfying the Poisson equation (17.8), such that $\varphi=0$ on the boundary of the cross section. The final stage is to determine the constant β from (17.3), which here takes the form

$$T = -\iint_\Omega \left(x\frac{\partial\varphi}{\partial x} + y\frac{\partial\varphi}{\partial y} \right) dxdy . \tag{17.9}$$

Integrating by parts and using the fact that $\varphi=0$ on the boundary of Ω, we obtain the simple expression

$$T = 2\iint_\Omega \varphi dxdy . \tag{17.10}$$

17.1.1 Solution of the governing equation

The Poisson equation (17.8) is an inhomogeneous partial differential equation and its general solution can be written

$$\varphi = \varphi_P + \varphi_H ,$$

where φ_P is any particular solution of (17.8) and φ_H is the general solution of the corresponding homogeneous equation

$$\nabla^2\varphi_H = 0 \tag{17.11}$$

— i.e. Laplace's equation. Since (17.8) is a second order equation, it will yield a constant when substituted with any quadratic polynomial function and hence suitable particular solutions in Cartesian or polar coördinates are

$$\varphi_P = -\mu\beta x^2 ; \quad \varphi_P = -\mu\beta y^2 ; \quad \varphi_P = -\frac{\mu\beta r^2}{2} . \tag{17.12}$$

The homogeneous equation (17.11) is satisfied by both the real and imaginary parts of any analytic function of the complex variable

$$\zeta = x + \imath y = re^{\imath\theta}$$

— i.e.

$$\varphi_H = \Re(f(\zeta)) \quad \text{or} \quad \varphi_H = \Im(f(\zeta)) , \tag{17.13}$$

where f is any function. Simple functions obtained by taking $f=\zeta^n$ are

$$C_1 r^n \cos(n\theta) , \quad C_2 r^n \sin(n\theta) , \quad C_3 r^{-n} \cos(n\theta) , \quad C_4 r^{-n} \sin(n\theta) ,$$

where $C_1, ...C_4$ are arbitrary constants. The first two of these correspond to polynomials in Cartesian coördinates, such as

$$x, \quad y, \quad x^2 - y^2, \quad xy, \quad x^3 - 3y^2x, \quad y^3 - 3x^2y$$

etc.

Approximate solutions of the torsion problem for particular cross sections can be obtained by combining functions such as (17.12, 17.13) and adjusting the multiplying constants so as to achieve a contour approximating the required shape. A relatively small number of simple shapes permit exact closed-form solutions, including the circle (Problem 17.1), the equilateral triangle Problem (17.3), a circle with a semicircular groove (Problem 17.6) and the ellipse, which we solve here as an example.

Example — the elliptical cross section

We consider the bar of elliptical cross section defined by the boundary

$$\frac{x^2}{a^2} + \frac{y^2}{b^2} - 1 = 0 ,$$

loaded by a torque T. The quadratic function

$$\varphi = C \left(\frac{x^2}{a^2} + \frac{y^2}{b^2} - 1 \right)$$

clearly satisfies the boundary condition (17.6). Substituting into (17.8), we obtain

$$\nabla^2 \varphi = C \left(\frac{2}{a^2} + \frac{2}{b^2} \right) = -2\mu\beta ,$$

which will be satisfied for all x, y if

$$C = -\frac{\mu\beta a^2 b^2}{(a^2 + b^2)} .$$

The torque T is obtained from (17.10) as

$$T = 2C \int_{-b}^{b} \int_{-a(1-y^2/b^2)}^{a(1-y^2/b^2)} \left(\frac{x^2}{a^2} + \frac{y^2}{b^2} - 1 \right) dx dy = -\pi abC$$

and hence

$$C = -\frac{T}{\pi ab} ; \quad \beta = \frac{T(a^2 + b^2)}{\pi \mu a^3 b^3} .$$

The stresses are then obtained from (17.4) as

$$\sigma_{zx} = \frac{\partial \varphi}{\partial y} = \frac{2Cy}{b^2} = -\frac{2Ty}{\pi ab^3} ; \quad \sigma_{zy} = -\frac{\partial \varphi}{\partial x} = -\frac{2Cx}{a^2} = \frac{2Tx}{\pi a^3 b} .$$

It is also convenient to define the torsional rigidity of the section K such that

$$T = \mu K \beta , \quad \text{so} \quad K = \frac{T}{\mu \beta} .$$ (17.14)

For the elliptical section, we obtain

$$K = \frac{\pi a^3 b^3}{(a^2 + b^2)} ,$$

which reduces to the polar second moment of area $\pi a^4/2$ for the circular section $b = a$.

17.1.2 The warping function

Once the stress function φ is known, the warping function $f(x, y)$ can be found by solving equations (17.7) to obtain

$$\frac{\partial f}{\partial x} = y + \frac{1}{\mu \beta} \frac{\partial \varphi}{\partial y} ; \quad \frac{\partial f}{\partial y} = -x - \frac{1}{\mu \beta} \frac{\partial \varphi}{\partial x} ,$$

followed by integration.

For example, for the elliptical cross section, we obtain

$$\frac{\partial f}{\partial x} = y - \frac{2a^2 y}{a^2 + b^2} = \left(\frac{b^2 - a^2}{a^2 + b^2} \right) y ; \quad \frac{\partial f}{\partial y} = -x + \frac{2b^2 x}{a^2 + b^2} = \left(\frac{b^2 - a^2}{a^2 + b^2} \right) x ,$$

with solution (apart from an arbitrary additive constant)

$$f(x, y) = \left(\frac{b^2 - a^2}{a^2 + b^2} \right) xy .$$

17.2 The Membrane Analogy

Equation (17.8) with the boundary condition (17.6) is similar in form to that governing the displacement of an elastic membrane stretched between the boundaries of the cross-sectional curve and loaded by a uniform normal pressure. This analogy is useful as a means of estimating the effect of the cross section on the maximum shear stress and the torsional rigidity of the bar. You can perform a simple experiment by making a plane wire frame in the shape of the cross section, dipping it into a soap solution to develop a soap film and then blowing gently against one side to see the shape of the deformed film. The stress function is then proportional to the displacement of the film from the plane of the frame.

The transmitted torque is proportional to the integral of the stress function over the cross section (see equation (17.10)) and hence the stiffest cross sections are those which permit the maximum volume to be developed between the deformed film and the plane of the frame for a given pressure.

The shear stress is proportional to the local slope of the film. For sections such as squares, rectangles, circles, ellipses etc, the maximum displacement will occur near the centre of the section and the maximum slope (and hence the maximum shear stress) will be at the point on the boundary nearest the centre. For thin-walled sections, the maximum displacement will occur where the section is thickest and the maximum stress will be at the corresponding point on the boundary. However, if there are reëntrant corners in the section, these will cause locally increased stresses which are unbounded if the corner is sharp.

Notice that it is not really necessary to perform the soap film experiment to come up with these conclusions. A 'thought experiment' is generally sufficient.

17.3   Thin-walled Open Sections

Many structural components consist of thin-walled sections — for example I-beams, channel sections and turbine blades. Figure 17.1 shows a generic thin-walled section of length b, whose thickness t varies with distance ξ along the length. As long as the thickness $t(\xi)$ of the section does not vary too rapidly with ξ, it is clear from the membrance analogy that we can neglect the curvature of the membrane (and hence of the stress function) in the ξ-direction, reducing the governing equation to

$$\frac{d^2\varphi}{d\eta^2} = -2\mu\beta ,$$

where η is a coördinate orthogonal to ξ measured from the local mid-point of the cross section. The solution of this equation satisfying the condition

$$\varphi = 0 \quad ; \quad \eta = \pm\frac{t(\xi)}{2}$$

is

$$\varphi = \mu\beta\left(\frac{t(\xi)^2}{4} - \eta^2\right) \tag{17.15}$$

and the stress field is

$$\sigma_{z\xi} = \frac{\partial\varphi}{\partial\eta} = -2\mu\beta\eta \quad ; \quad \sigma_{z\eta} = 0 .$$

Thus, the shear stress is everywhere parallel to the centreline of the section and varies linearly from that line reaching a maximum

$$\tau_{max}(\xi) = \mu\beta t(\xi)$$

at the edges of the section $\xi = \pm t(\xi)/2$. It follows that the overall maximum shear stress must occur at the point where the thickness is a maximum and is

$$\tau_{max} = \mu\beta t_{max} = \frac{T t_{max}}{K} . \tag{17.16}$$

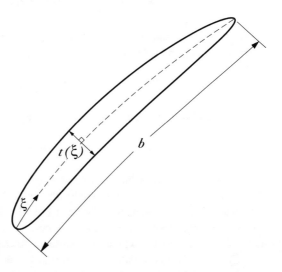

Fig. 17.1 Thin-walled open section with variable thickness.

The torque transmitted by the section is obtained from equations (17.10, 17.15) as

$$T = 2\int_0^b \int_{-t(\xi)/2}^{t(\xi)/2} \varphi d\eta d\xi = \frac{\mu\beta}{3}\int_0^b t(\xi)^3 d\xi ,$$

and hence the torsional rigidity of the section (17.14) is

$$K = \frac{1}{3}\int_0^b t(\xi)^3 d\xi . \tag{17.17}$$

In the special case where the thickness is constant, we have

$$K = \frac{bt^3}{3} \quad \text{and} \quad \tau_{max} = \mu\beta t = \frac{3T}{bt^2} .$$

17.4   The Rectangular Bar

The results of the previous section can be used to obtain an approximate solution for the torsion of the rectangular bar $-a < x < a$, $-b < y < b$, when $b \gg a$. Taking the origin at the centre of the cross section, we obtain

$$\varphi = \mu\beta(a^2 - x^2) \tag{17.18}$$

from (17.15). The corresponding torsional rigidity and maximum shear stress are

$$K = \frac{16a^3 b}{3} \; ; \quad \tau_{\max} = 2\mu\beta a = \frac{3T}{8a^2 b} \; . \tag{17.19}$$

The stress function (17.18) satisfies the governing equation (17.8) and the boundary condition on the long edges

$$\varphi = 0 \; ; \quad x = \pm a \; , \tag{17.20}$$

but it does not satisfy the corresponding boundary condition on the short edges

$$\varphi = 0 \; ; \quad y = \pm b \; . \tag{17.21}$$

To satisfy this condition and hence to obtain an exact solution for all values of the ratio a/b, we need to supplement the stress function (17.18) with appropriate solutions of the homogeneous equation (17.11). These additional functions must preserve the condition (17.20) and we therefore seek separated-variable solutions, as in §6.2.1. Indeed the whole procedure of 'correcting' the approximate solution (17.18) is exactly analogous to that discussed in §6.2. For harmonic functions, separation of variables demands that the functions be trigonometric in one coördinate and exponential in the other and the symmetry of the boundary-value problem about both axes suggests harmonic functions of the form $\cos(\lambda x)\cosh(\lambda y)$. Selecting those values of λ that satisfy (17.20), we obtain the stress function

$$\varphi = \mu\beta(a^2 - x^2) + \sum_{n=1}^{\infty} C_n \cos\left(\frac{(2n-1)\pi x}{2a}\right)\cosh\left(\frac{(2n-1)\pi y}{2a}\right) \; , \tag{17.22}$$

where C_n is a set of arbitrary constants. The function (17.22) will satisfy (17.21) if

$$\sum_{n=1}^{\infty} C_n \cos\left(\frac{(2n-1)\pi x}{2a}\right)\cosh\left(\frac{(2n-1)\pi b}{2a}\right) = -\mu\beta(a^2 - x^2) \; . \tag{17.23}$$

To evaluate the constants C_n, we multiply both sides of equation (17.23) by $\cos((2m-1)\pi x/2a)$ and integrate in $-a < x < a$, obtaining

$$\sum_{n=1}^{\infty} C_n \int_{-a}^{a} \cos\left(\frac{(2n-1)\pi x}{2a}\right) \cosh\left(\frac{(2n-1)\pi b}{2a}\right) \cos\left(\frac{(2m-1)\pi x}{2a}\right) dx$$

$$= -\mu\beta \int_{-a}^{a} (a^2 - x^2) \cos\left(\frac{(2m-1)\pi x}{2a}\right) dx$$

and hence

$$C_m a \cosh\left(\frac{(2m-1)\pi b}{2a}\right) = \frac{32\mu\beta(-1)^m a^3}{\pi^3(2m-1)^3}$$

or

$$C_m = \frac{32\mu\beta(-1)^m a^2}{\pi^3(2m-1)^3 \cosh((2m-1)\pi b/2a)}, \qquad (17.24)$$

where we have used the orthogonality condition

$$\int_{-a}^{a} \cos\left(\frac{(2n-1)\pi x}{2a}\right) \cos\left(\frac{(2m-1)\pi x}{2a}\right) dx = a\delta_{mn}.$$

The transmitted torque is obtained from equations (17.10, 17.22) as

$$T = \frac{16\mu\beta a^3 b}{3} - \sum_{n=1}^{\infty} \frac{32(-1)^n C_n a^2 \sinh((2n-1)\pi b/2a)}{\pi^2(2n-1)^2}$$

$$= \frac{16\mu\beta a^3 b}{3}\left[1 - \sum_{n=1}^{\infty} \frac{192a}{\pi^5(2n-1)^5 b} \tanh\left(\frac{(2n-1)\pi b}{2a}\right)\right], \quad (17.25)$$

using (17.24) to substitute for the constants C_n. The maximum shear stress occurs at the points $(\pm a, 0)$ and is

$$\tau_{max} = \left|\frac{\partial\varphi}{\partial x}(\pm a, 0)\right| = 2\mu\beta a - \sum_{n=1}^{\infty} \frac{16\mu\beta a}{\pi^2(2n-1)^2 \cosh((2n-1)\pi b/2a)}. \quad (17.26)$$

We can define the coördinate system such that $0 < a/b < 1$ without loss of generality and in this range the series (17.25, 17.26) converge extremely rapidly. In fact, taking just one term of each series gives better than 0.5% accuracy. Figure 17.2 shows the effect of the aspect ratio on the torsional rigidity and the maximum shear stress, normalized with respect to the thin-walled expressions (17.19).

17.5 Multiply-connected (Closed) Sections

The traction-free boundary condition (17.5) implies that the stress function remains constant as we move around the boundary Γ and we chose to take this constant as zero in equation (17.6). However, if the bar contains one or more enclosed holes, the

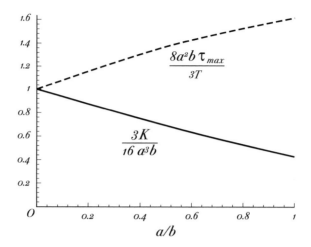

Fig. 17.2 Effect of aspect ratio a/b on the torsional rigidity and maximum shear stress in a rectangular bar.

boundary will be defined by more than one closed curve Γ_0, Γ_1, Γ_2 etc. The condition (17.5) implies that the value of φ on each such curve will be constant — e.g.

$$\varphi = \varphi_0 \text{ on } \Gamma_0 \ ; \quad \varphi = \varphi_1 \text{ on } \Gamma_1 \ ,$$

but we cannot conclude that $\varphi_1 = \varphi_0$, since the curves Γ_1, Γ_0 do not connect.

Since the expressions for the stresses involve derivatives of φ, one of the values $\varphi_0, \varphi_1, \ldots$ can be selected arbitrarily and it is convenient to define $\varphi_0 = 0$, where the corresponding curve Γ_0 defines the external boundary of the bar. If there is a single hole with boundary Γ_1, the constant φ_1 is an additional unknown which must be determined by imposing a Cesaro integral condition, as discussed in §2.2.1. In the present case, this condition is easy to define, since we started from a displacement formulation in equations (17.1).

We require that the displacements be single-valued functions of x, y and hence that the integral

$$\oint_S \frac{\partial f}{\partial s} ds = \oint_S \left(\frac{\partial f}{\partial x} dx + \frac{\partial f}{\partial y} dy \right) = 0 \tag{17.27}$$

for all closed curves S. The stress function representation guarantees this for infinitesimal curves, but as in §2.2.1, we must explicitly enforce the condition around a representative curve S that encircles the hole.

Eliminating f from (17.27) using (17.2), we obtain

$$\frac{1}{\mu} \oint_S (\sigma_{zx}dx + \sigma_{zy}dy) - \beta \oint_S (xdy - ydx) = 0 \ . \tag{17.28}$$

The second integral in (17.28) is equal to twice the area A enclosed by S, so we obtain the condition

$$\oint_S \sigma_{zs}ds = 2\mu\beta A \ , \tag{17.29}$$

which serves to determine the free constant φ_1.

If the section contains several holes, there will be a corresponding number of additional unknowns, but for each hole we can write an integral of the form (17.29) where the corresponding path S is taken to encircle that particular hole.

The torque transmitted by a multiply-connected section can still be calculated by (17.9), provided that we recognize that there will be no contribution from those parts of the total domain Ω within the holes. This can be achieved by adopting the convention that $\varphi = \varphi_1$ throughout the area enclosed by Γ_1 as well as along the boundary. The derivatives in the integrand in (17.9) in this area will then be zero as required. With this convention we can still use the simple expression (17.10) for the torque T, with the result

$$T = 2 \iint_{\bar{\Omega}} \varphi dxdy + 2 \sum A_i \varphi_i \ , \tag{17.30}$$

where A_i is the area of the hole enclosed by the boundary Γ_i on which $\varphi = \varphi_i$, and $\bar{\Omega} = \Omega - \sum A_i$ — i.e. the 'solid' area of the bar excluding the holes.

17.5.1 Thin-walled closed sections

Most practical applications of multiply-connected sections have thin walls, relative to the other dimensions of the section, as shown for example in Figure 17.3. In this case, the stress function φ can reasonably be approximated as a linear function between φ_1 and zero on the two adjacent boundaries, implying a local shear stress

$$\sigma_{zs} = \frac{\varphi_1}{t} \ ,$$

where t is the local wall thickness. The value of φ_1 can then be determined from (17.29) as

$$\varphi_1 = 2\mu\beta A \left/ \oint_S \frac{ds}{t} \right.$$

and the torque is approximately

$$T = 2A\varphi_1$$

from (17.30), where A is the area enclosed by the mean line equidistant between the inner and outer boundary. It follows that the torsional rigidity of a thin-walled closed section is approximately

$$K = 4A^2 \left/ \oint_S \frac{ds}{t} \right. \quad .$$

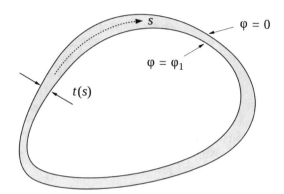

Fig. 17.3 A thin-walled closed section.

These results can also be derived without reference to a stress function, using Mechanics of Materials arguments. For more details, including practical examples, the reader is referred to the many texts on Advanced Mechanics of Materials[1].

A simple experiment

A striking demonstration of the effectiveness of thin-walled closed sections in transmitting torsion can be obtained by taking two identical cylindrical cardboard tubes and slitting one of them along the length to create an open section, as shown in Figure 17.4(a). When the slit tube is loaded in torsion, warping of the cross section causes a relative axial displacement at the cut, as shown in Figure 17.4(b), and significant twist occurs. In the original uncut tube, this relative displacement is prevented by shear tractions $\sigma_{z\theta}$ on the corresponding surfaces, there is no warping, and tube feels almost rigid in torsion.

If the slit tube has mean radius a and wall thickness t, the thin-walled approximation (17.15) can be written

---

[1] See for example J. R. Barber, *Intermediate Mechanics of Materials*, Springer, Dordrecht, 2nd. edn. (2011), Chapter 6, A. P. Boresi, R. J. Schmidt and O. M. Sidebottom, *Advanced Mechanics of Materials,* John Wiley, New York, 5th edn., 1993, Chapter 6, W. B. Bickford, *Advanced Mechanics of Materials,*, Addison Wesley, Menlo Park, 1998, Chapter 3.

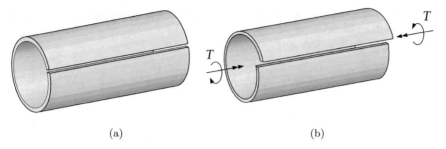

(a)   (b)

Fig. 17.4 (a) A slit cylindrical tube, (b) relative axial displacement at the slit due to loading by a torque T.

$$\varphi = \mu\beta \left[\frac{t^2}{4} - (r - a)^2 \right] .$$

The polar coördinate equivalents of equations (17.4, 17.7) are

$$\sigma_{zr} = \frac{1}{r} \frac{\partial \varphi}{\partial \theta} = \mu\beta \frac{\partial f}{\partial r} \; ; \quad \sigma_{z\theta} = -\frac{\partial \varphi}{\partial r} = \mu\beta \left(\frac{1}{r} \frac{\partial f}{\partial \theta} + r \right) ,$$

so

$$\frac{\partial f}{\partial r} = 0 \quad \text{and} \quad \frac{\partial f}{\partial \theta} = -r^2 - \frac{r}{\mu\beta} \frac{\partial \varphi}{\partial r} = r^2 - 2ar \approx -a^2 ,$$

since $t \ll a$. We therefore obtain $f(r, \theta) = -a^2\theta$ and there will be a discontinuity of $2\pi a^2$ at the cut, corresponding to the relative axial displacement seen in Figure 17.4(b).

Problems

17.1. Find the solution for torsion of a solid circular shaft of radius a by setting $b=a$ in the solution for the elliptical bar. Express the stress function and hence the stress components in polar coördinates and verify that these results and the relation between T and β agree with those given by the elementary Mechanics of Materials theory of torsion. Also, verify that there is no warping of the cross section $f=0$ in this case.

17.2. A bar transmitting a torque T has the cross section of a thin sector $0 < r < a$, $-\alpha < \theta < \alpha$, where $\alpha \ll 1$. Develop an approximate solution for the shear stress distribution and the twist per unit length, using strong boundary conditions on the edges $\theta = \pm\alpha$ only.

17.3. A bar transmitting a torque T has the cross section of an equilateral triangle of side a. Find the equations of the three sides of the triangle in Cartesian coördinates, taking the origin at one corner and the x-axis to bisect the opposite side. Express these equations in the form $f_i(x, y)=0, i=1, 2, 3$.

Show that the function

$$\varphi = Cf_1(x, y)f_2(x, y)f_3(x, y)$$

satisfies the boundary condition (17.6) and can be made to satisfy (17.8) with a suitable choice of the constant C. Hence find the stress field in the bar and make a contour plot of the maximum shear stress. Why can this method not be used for a more general triangular cross section?

17.4. Find the warping function $f(x, y)$ for the equilateral triangular cross section of Problem 17.3.

17.5. A bar transmitting a torque T has the cross section of a sector $0<r<a, -\alpha< \theta<\alpha$, where α is not small compared with unity.

 (i) Find a combination of the functions r^2 and $r^2 \cos(2\theta)$ that satisfies the governing equation (17.8) and the boundary condition (17.6) *on the straight edges $\theta=\pm\alpha$ only.*
 (ii) Superpose a series of harmonic functions of the form $r^\gamma \cos(\gamma\theta)$ where γ is chosen to satisfy (17.6) on $\theta=\pm\alpha$.
(iii) Choose the coefficients on these terms so as to satisfy the boundary condition on the curved edge $r=a$ and hence obtain expressions for the torsional rigidity K and the maximum shear stress.
(iv) Evaluate your results for the special case $\alpha=\pi/6$ and compare them with those obtained using the thin-walled approximation of §17.3.

17.6. Figure 17.5 shows the cross section of a circular shaft of radius a with a circular groove of radius b whose centre lies on the surface of the ungrooved shaft. Construct a suitable stress function for this problem using the particular solution

$$\varphi_P = C(x^2 + y^2 - b^2)$$

and a homogeneous solution φ_H comprising appropriate multipliers of the harmonic functions $r \cos\theta$ and $r^{-1} \cos\theta$ centred on the point B, which in Cartesian coördinates can be written

$$x \quad\text{and}\quad \frac{x}{x^2 + y^2} .$$

Remember that the resulting function must be zero at all points on both the circles $x^2+y^2=b^2$ and $(x-a)^2+y^2=a^2$. Hence find the stress field in the bar and make a contour plot of the maximum shear stress for the case where $b=0.1a$.

17.7. Figure 17.6 shows part of a thin-walled open section that is loaded in torsion. Find the stress function φ in the curved section, assuming that it depends upon r

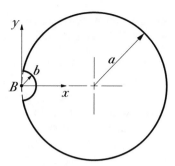

Fig. 17.5 Circular shaft with
a semi-circular groove.

only and hence estimate[2] the maximum shear stress in the corner if the twist per unit
length is β and the shear modulus is μ.

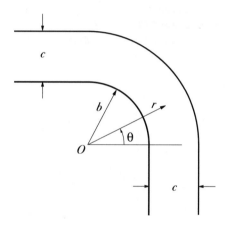

Fig. 17.6 Stress
concentration
due to a bend.

Find the corresponding maximum shear stress in the straight segment, distant
from the corner and hence find the stress-concentration factor due to the corner. Plot
the stress-concentration factor as a function of the ratio of the inner radius to the
section thickness (b/c).

17.8. Approximations to the stresses in a square bar in torsion can be obtained by
adding harmonic functions of the form $r^n \cos(n\theta)$ to the axisymmetric particular
solution (17.12).

Express the function

$$\varphi = r^2 + r^4 \cos(4\theta)$$

in Cartesian coördinates and make a contour plot. Select the contour which is the
best approximation to a square with rounded corners and scale the resulting function
to obtain an approximate solution for the maximum shear stress in a square bar of

---

[2] This solution is approximate because the stress function is not independent of θ in the curved
region. Otherwise there would be a discontinuity at the curved/straight transition.

side a when the twist per unit length is β. Compare the result with that obtained from Figure 17.2 when $b/a = 1$.

17.9. We know from Mechanics of Materials that a bar cross section posesses a *shear centre* which is the point through which a shear force must act if there is to be no twist of the section. Maxwell's reciprocal theorem[3] implies that the shear centre must also be the only point in the section that has no in-plane displacement when the bar is loaded in torsion and hence that equations (17.1) are correct if and only if the shear centre is taken as origin.

Rederive the torsion theory using a more general origin and hence show that the formulation in terms of Prandtl's stress function remains true for all choices of origin.

17.10. A titanium alloy turbine blade is idealized by the section of Figure 17.7. The leading edge is a semicircle of radius 2 mm and the inner and outer edges of the trailing section are cylindrical surfaces of radius 43 mm and 35 mm respectively, with centres of curvature on the line Oy. Find the twist per unit length β and the maximum shear stress when the section is loaded by a torque of 5 Nm. The shear modulus for the alloy is 43 GPa.

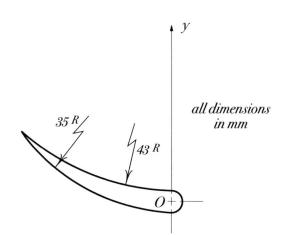

Fig. 17.7 Turbine blade cross section.

17.11. The W200×22 I-beam of Figure 17.8 is made of steel ($G = 80$ GPa) and is 15 m long. One end is built in and the other is loaded by two forces F constituting a torque, as shown. Find the torsional rigidity K for the beam section and hence determine the value of F needed to cause the end to twist through an angle of 20°.

---

[3] See §38.1.

Does the result surprise you? How large an angle of twist[4] could a person of average strength produce in such a beam using his or her hands only?

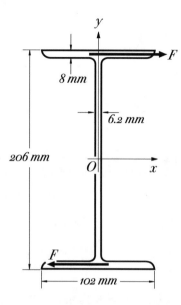

Fig. 17.8 I-beam twisted by two forces.

17.12. Show that the torsional rigidity of a closed thin-walled steel tube of uniform thickness is reduced by the factor

$$\frac{12A^2}{A_s^2}$$

when a longitudinal slit is made down the whole length of the tube to form an open section. In this expression, A is the area enclosed by the mean line of the closed section and A_s is the cross-sectional area of the steel.

Find also by what ratio the maximum shear stress will increase for a given applied torque.

17.13. The outer and inner circular walls of a closed thin-walled circular tube have slightly different centres, as shown in Figure 17.9, so that the thickness of the section varies according to

$$t = t_0(1 - \epsilon \cos \theta) ,$$

where $\epsilon < 1$. The mean radius of the tube is a. Find the torsional rigidity K and the maximum shear stress when the bar is loaded by a torque T. Plot your results as functions of ϵ in the range $0 \leq \epsilon < 1$.

---

[4] If you want to impress your unsuspecting colleagues with your engineering expertise (or your immense strength), ask them to guess the answer to this question and then perform the experiment.

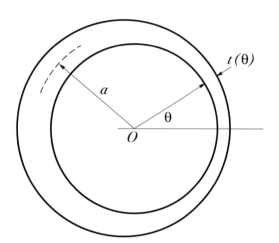

Fig. 17.9 Cylindrical tube
with variable wall thickness.

17.14. Show that the stress function

$$\varphi = -\frac{\mu\beta}{2}\left(x^2 + y^2 - 2a^2 - 2ax + \frac{4a^3x}{(x^2 + y^2)}\right)$$

defines the exact solution for the cross section defined by the intersection of the
circles

$$x^2 + y^2 = 2a^2 \quad \text{and} \quad (x - a)^2 + y^2 = a^2 .$$

Find the corresponding torque T and the maximum shear stress $\tau_{\max}$ and express the
latter as a function of T, a only.

Now find an approximate solution to this problem by assuming that it is a 'thin-
walled section' — i.e. using equations (17.16, 17.17). What is the percentage error
in the approximate solution (i) in the stiffness K and (ii) in $\tau_{\max}$? Are these quantities
overestimated or underestimated by the approximate solution?

17.15. Figure 17.10 shows the cross section of a solid circular bar of radius a which
is slit to the centre along the line $\theta = \pm\pi$. The bar is subjected to a prescribed twist
β per unit length.

(i) Show that the particular solution

$$\varphi_P = -\mu\beta y^2 = -\mu\beta r^2 \sin^2(\theta)$$

satisfies the governing equation and also satisfies the condition that $\varphi = 0$ on the
slit $\theta = \pm\pi$.

(ii) To complete the solution, we need to superpose a series of stress functions each
of which satisfies the homogeneous equation

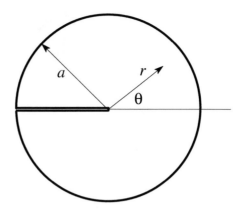

Fig. 17.10

$$\nabla^2 \varphi_H = 0$$

and $\varphi = 0$ on $\theta = \pm\pi$. Show that the functions $r^\gamma \cos(\gamma\theta)$ meet these conditions with a suitable choice of γ.

(iii) Use these solutions to construct a series solution of the problem and use the boundary condition on the curved edge $r = a$ to define equations that the coefficients of the series must satisfy.

(iv) Solve the equations and hence determine the mode III stress-intensity factor K_{III} at $r = 0$, defined as

$$K_{\text{III}} = \lim_{r \to 0} \sigma_{z\theta}(r, 0)\sqrt{2\pi r} \ .$$

Chapter 18
Shear of a Prismatic Bar

In this chapter, we shall consider the problem in which a prismatic bar occupying the region $z > 0$ is loaded by transverse forces F_x, F_y in the negative x- and y-directions respectively on the end $z=0$, the lateral surfaces of the bar being traction free[1]. Equilibrium considerations then show that there will be shear forces

$$V_x \equiv \iint_\Omega \sigma_{zx} dx dy = F_x \; ; \quad V_y \equiv \iint_\Omega \sigma_{zy} dx dy = F_y$$

and bending moments

$$M_x \equiv \iint_\Omega \sigma_{zz} y \, dx dy = z F_y \; ; \quad M_y \equiv - \iint_\Omega \sigma_{zz} x \, dx dy = -z F_x \; , \qquad (18.1)$$

at any given cross section Ω of the bar. In other words, the bar transmits constant shear forces, but the bending moments increase linearly with distance from the loaded end.

18.1 The Semi-inverse Method

To solve this problem, we shall use a semi-inverse approach — i.e. we shall make certain assumptions about the stress distribution and later verify that these assumptions are correct by showing that they are sufficiently general to define a solution that satisfies all the equilibrium and compatibility equations. The assumptions we shall make are

---

[1] In many texts, this topic is referred to as 'bending' or 'flexure' of prismatic bars. However, it should not be confused with 'pure bending' due to the application of equal and opposite bending moments at the two ends of the bar. The present topic is the Elasticity counterpart of that referred to in Mechanics of Materials as 'shear stress distribution in beams'.

© The Author(s), under exclusive license to Springer Nature Switzerland AG 2022
J. R. Barber, *Elasticity*, Solid Mechanics and Its Applications 172,
https://doi.org/10.1007/978-3-031-15214-6_18

(i) The stress component σ_{zz} varies linearly with x, y, as in the elementary bending theory.
(ii) The shear stresses σ_{zx}, σ_{zy} due to the shear forces are indpendent of z.
(iii) The in-plane stresses $\sigma_{xx} = \sigma_{xy} = \sigma_{yy} = 0$.

A more direct justification for these assumptions can be established using results from Chapter 30, where we discuss the more general problem of the prismatic bar loaded on its lateral surfaces (see §30.5).

The bending moments vary linearly with z, so assumption (i) implies that

$$\sigma_{zz} = (Ax + By)z , \tag{18.2}$$

where A, B are two unknown constants and we have chosen to locate the origin at the centroid of Ω. To determine A, B, we substitute (18.2) into (18.1) obtaining

$$zF_y = (AI_{xy} + BI_x)z \;\; ; \;\; -zF_x = -(AI_y + BI_{xy})z , \tag{18.3}$$

where

$$I_x = \iint_\Omega y^2 dxdy \;\; ; \;\; I_y = \iint_\Omega x^2 dxdy \;\; ; \;\; I_{xy} = \iint_\Omega xydxdy$$

are the second moments of area of the cross section. Solving (18.3), we have

$$A = \frac{F_x I_x - F_y I_{xy}}{I_x I_y - I_{xy}^2} \;\; ; \;\; B = \frac{F_y I_y - F_x I_{xy}}{I_x I_y - I_{xy}^2} . \tag{18.4}$$

18.2   Stress Function Formulation

Assumptions (ii,iii) imply that the first two equilibrium equations (2.1, 2.2) are satisfied identically, whilst (2.3) requires that

$$\frac{\partial \sigma_{zx}}{\partial x} + \frac{\partial \sigma_{zy}}{\partial y} + Ax + By = 0 .$$

This equation will be satisfied identically if the shear stresses are defined in terms of a stress function φ through the relations

$$\sigma_{zx} = \frac{\partial \varphi}{\partial y} - \frac{Ax^2}{2} \;\; ; \;\; \sigma_{zy} = -\frac{\partial \varphi}{\partial x} - \frac{By^2}{2} . \tag{18.5}$$

This is essentially a generalization of Prandtl's stress function of equations (17.4, 17.5). The governing equation for φ is then obtained by using Hooke's law to determine the strains and substituting the resulting expressions into the six compatibility equations (2.7). The procedure is algebraically lengthy but routine and will be omitted here. It is found that the three equations of the form (2.5) and one of the three like (2.6) are satisfied identically by the assumed stress field. The two remaining equations yield the conditions

$$\frac{\partial}{\partial x}\nabla^2\varphi = -\frac{B\nu}{(1+\nu)} \quad ; \quad \frac{\partial}{\partial y}\nabla^2\varphi = \frac{A\nu}{(1+\nu)}$$

which imply that

$$\nabla^2\varphi = \frac{\nu}{(1+\nu)}(Ay - Bx) + C , \tag{18.6}$$

where C is an arbitrary constant of integration.

18.3   The Boundary Condition

We require the boundaries of Ω to be traction free and hence

$$\sigma_{zn} = \sigma_{zx}\frac{dy}{dt} - \sigma_{zy}\frac{dx}{dt} = 0 ,$$

where n, t is a Cartesian coördinate system locally normal and tangential to the boundary. Substituting for the stresses from equation (18.5), we obtain

$$\frac{\partial\varphi}{\partial y}\frac{dy}{dt} + \frac{\partial\varphi}{\partial x}\frac{dx}{dt} = \frac{Ax^2}{2}\frac{dy}{dt} - \frac{By^2}{2}\frac{dx}{dt} , \tag{18.7}$$

or

$$\frac{\partial\varphi}{\partial t} = \frac{Ax^2}{2}\frac{dy}{dt} - \frac{By^2}{2}\frac{dx}{dt} . \tag{18.8}$$

18.3.1   Integrability

In order to define a well-posed boundary-value problem for φ, we need to integrate (18.8) with respect to t to determine the value of φ at all points on the boundary. This process will yield a single-valued expression except for an arbitrary constant if and only if the integral

$$\oint_\Gamma \left(\frac{Ax^2}{2}\frac{dy}{dt} - \frac{By^2}{2}\frac{dx}{dt} \right) dt = 0 ,$$

where Γ is the boundary of the cross-sectional domain Ω, which we here assume to be simply connected. Using Green's theorem in the form

$$\iint_\Omega \left(\frac{\partial f}{\partial x} + \frac{\partial g}{\partial y} \right) dxdy = \int_\Gamma \left(f \frac{dy}{dt} - g \frac{dx}{dt} \right) dt ,$$

we obtain

$$\oint_\Gamma \left(\frac{Ax^2}{2} \frac{dy}{dt} - \frac{By^2}{2} \frac{dx}{dt} \right) dt = \iint_\Omega (Ax + By)dxdy . \qquad (18.9)$$

The integral on the right-hand side of (18.9) is a linear combination of the two first moments of area for Ω and it is zero because we chose the origin to be at the centroid.

We conclude that (18.8) can always be integrated and the problem is therefore reduced to the solution of the Poisson equation (18.6) with prescribed boundary values of φ. This problem has a unique solution for all simply-connected[2] domains, thus justifying the assumptions made in §18.1.

18.3.2   Relation to the torsion problem

In the special case where $F_x = F_y = 0$, we have $A = B = 0$ from (18.4) and the problem is defined by the equations

$$\nabla^2 \varphi = C \quad \text{in } \Omega$$
$$\frac{\partial \varphi}{\partial t} = 0 \quad \text{on } \Gamma$$
$$\sigma_{zx} = \frac{\partial \varphi}{\partial y} \; ; \; \sigma_{zy} = -\frac{\partial \varphi}{\partial x} ,$$

which are identical with (17.8, 17.5, 17.4) of the previous chapter if we write $C = -2\mu\beta$. Thus, the present formulation includes the solution of the torsion problem as a special case and the constant C is proportional to the twist of the section and hence to the applied torque.

When the resultant shear force $F \neq 0$, superposing a torque on the loading is equivalent to changing the line of action of F. The constant C in equation (18.6) provides the degree of freedom needed for this generalization. It is generally easier to decompose the problem into two parts — a shear loading problem in which C is set arbitrarily to zero and a torsion problem (in which A and B are zero). These can be combined at the end to ensure that the resultant force has the required line of action. In the special case where the section is symmetric about the axes, setting

---

[2] If the cross section of the bar has one or more enclosed holes, additional unknowns will be introduced corresponding to the constants of integration of (18.8) on the additional boundaries. A corresponding set of additional algebraic conditions is obtained from the Cesaro integrals around each hole, as in §17.5.

$C=0$ will generate a stress field with the same symmetry, corresponding to loading by a shear force acting through the origin.

18.4 Methods of Solution

The problem defined by equations (18.6, 18.8) can be solved by methods similar to those used in the torsion problem and discussed in §17.1.1. A particular solution to (18.6) is readily found as a third degree polynomial and more general solutions can then be obtained by superposing appropriate harmonic functions. Closed-form solutions can be obtained only for a limited number of cross sections, including the circle, the ellipse and the equilateral triangle. Other cases can be treated by series methods.

18.4.1 The circular bar

We consider the case of the circular cylindrical bar bounded by the surface

$$x^2 + y^2 = a^2 \tag{18.10}$$

and loaded by a shear force $F_y = F$ acting through the origin[3]. For the circular cross section, we have

$$I_x = I_y = \frac{\pi a^4}{4} \;\; ; \;\; I_{xy} = 0$$

and hence

$$A = 0 \;\; ; \;\; B = \frac{4F}{\pi a^4} \; ,$$

from (18.4). Substituting into (18.8) and integrating, we obtain

$$\varphi = -\int \frac{2Fy^2 dx}{\pi a^4} \; ,$$

where the integration must be performed on the line (18.10). Substituting for y from (18.10) and integrating, we obtain

$$\varphi = -\int \frac{2F(a^2 - x^2)dx}{\pi a^4} = -\frac{2F(3a^2 x - x^3)}{3\pi a^4} \;\; \text{on } x^2 + y^2 = a^2 \; , \tag{18.11}$$

where we have set the constant of integration to zero to preserve antisymmetry about $x=0$.

---

[3] This problem is solved by a different method in §26.3.1.

Equations (18.6, 18.11) both suggest a third order polynomial for φ with odd powers of x and even powers of y, so we use the trial function

$$\varphi = C_1 x^3 + C_2 xy^2 + C_3 x \ .$$

Substitution into (18.6, 18.11) yields the conditions

$$6C_1 x + 2C_2 x = -\frac{4F\nu x}{\pi a^4(1+\nu)}$$

$$C_1 x^3 + C_2 x(a^2 - x^2) + C_3 x = -\frac{2F(3a^2 x - x^3)}{3\pi a^4} \ ,$$

which will be satisfied for all x if

$$6C_1 + 2C_2 = -\frac{4F\nu}{\pi a^4(1+\nu)} \ ; \quad C_1 - C_2 = \frac{2F}{3\pi a^4} \ ; \quad C_2 a^2 + C_3 = -\frac{2F}{\pi a^2} \ .$$

Solving for C_1, C_2, C_3, we obtain

$$C_1 = \frac{F(1-2\nu)}{6\pi a^4(1+\nu)} \ ; \quad C_2 = -\frac{F(1+2\nu)}{2\pi a^4(1+\nu)} \ ; \quad C_3 = -\frac{F(3+2\nu)}{2\pi a^2(1+\nu)} \ ,$$

and the stress components are then recovered from (18.5) as

$$\sigma_{zx} = -\frac{F(1+2\nu)xy}{\pi a^4(1+\nu)} \ ; \quad \sigma_{zy} = \frac{F\left[(3+2\nu)(a^2-y^2)-(1-2\nu)x^2\right]}{2\pi a^4(1+\nu)} \ .$$

The maximum shear stress occurs at the centre and is

$$\tau_{max} = \frac{F(3+2\nu)}{2\pi a^2(1+\nu)} \ .$$

The elementary Mechanics of Materials solution gives

$$\tau^*_{max} = \frac{4F}{3\pi a^2} \ ,$$

which is exact for $\nu=0.5$ and 12% lower than the exact value for $\nu=0$.

18.4.2   The rectangular bar

As a second example, we consider the rectangular bar whose cross section is bounded by the lines $x=\pm a$, $y=\pm b$, loaded by a force $F_y=F$ acting through the origin in the negative y-direction. As in the corresponding torsion problem of §17.4, we cannot obtain a closed-form solution to this problem. The strategy is to seek a simple solution that satisfies the boundary conditions on the two longer edges in the strong sense and then superpose a series of harmonic functions that leaves this condition unchanged, whilst providing extra degrees of freedom to satisfy the condition on the remaining edges.

For the rectangular cross section, we have

$$I_x = \frac{4ab^3}{3} \; ; \; I_y = \frac{4a^3b}{3} \; ; \; I_{xy} = 0$$

and hence

$$A = 0 \; ; \; B = \frac{3F}{4ab^3} \; , \tag{18.12}$$

from (18.4). The boundary condition (18.8) then requires that

$$\frac{\partial\varphi}{\partial x}(x, \pm b) = -\frac{Bb^2}{2} = -\frac{3F}{8ab} \; ; \; \frac{\partial\varphi}{\partial y}(\pm a, y) = 0 \;. \tag{18.13}$$

The governing equation (18.6) is satisfied by a third degree polynomial and the most general such function with the required symmetry is

$$\varphi = C_1 x^3 + C_2 xy^2 + C_3 x \;. \tag{18.14}$$

Substitution in (18.6) shows that

$$6C_1 + 2C_2 = -\frac{3F\nu}{4ab^3(1+\nu)} \tag{18.15}$$

and on the boundaries of the rectangle we have

$$\frac{\partial\varphi}{\partial x}(x, \pm b) = 3C_1 x^2 + C_2 b^2 + C_3 \; ; \; \frac{\partial\varphi}{\partial y}(\pm a, y) = \pm C_2 ay \;.$$

This solution permits us to satisfy the conditions (18.13) in the strong sense on any two opposite edges, but not on all four edges. In the interests of convergence of the final series solution, it is preferable to use the strong boundary conditions on the long edges at this stage. For example, if $b > a$, we satisfy the condition on $x = \pm a$ by requiring

$$C_2 = 0$$

and hence

$$C_1 = -\frac{F\nu}{8ab^3(1+\nu)} \;,$$

from (18.15). The remaining boundary conditions are satisfied in the weak sense

$$\int_{-a}^{a} \left(3C_1 x^2 + C_2 b^2 + C_3 + \frac{3F}{8ab} \right) dx = 0 \;,$$

from which

$$C_3 = \frac{F\nu a}{8b^3(1+\nu)} - \frac{3F}{8ab}\,.$$

To complete the solution, we superpose a series of harmonic functions φ_n chosen to satisfy the condition $\varphi_n=0$ on $x=\pm a$. In this way, the governing equation and the boundary conditions on $x=\pm a$ are unaffected and the remaining constants can be chosen to satisfy the boundary conditions on $y=\pm b$ in the strong sense. This is exactly the same procedure as was used for the torsion problem in §17.4. Here we need odd functions of x satisfying $\varphi_n(\pm a, y)=0$, leading to the solution

$$\varphi = -\frac{F\nu x^3}{8ab^3(1+\nu)} + \frac{F\nu ax}{8b^3(1+\nu)} - \frac{3Fx}{8ab} + \sum_{n=1}^{\infty} D_n \sin\left(\frac{n\pi x}{a}\right) \cosh\left(\frac{n\pi y}{a}\right),$$

(18.16)

where D_n is a set of arbitrary constants. This function satisfies the first of (18.13) if

$$\sum_{n=1}^{\infty} \frac{n\pi D_n}{a} \cos\left(\frac{n\pi x}{a}\right) \cosh\left(\frac{n\pi b}{a}\right) = \frac{F\nu(3x^2 - a^2)}{8ab^3(1+\nu)} \; ; \quad -a < x < a\,.$$

Multiplying both sides of this equation by $\cos(m\pi x/a)$ and integrating over the range $-a < x < a$, we then obtain

$$m\pi D_m \cosh\left(\frac{m\pi b}{a}\right) = \frac{F\nu}{8ab^3(1+\nu)} \int_{-a}^{a} (3x^2 - a^2) \cos\left(\frac{m\pi x}{a}\right) dx$$

$$= \frac{3(-1)^m F\nu a^2}{2\pi^2 m^2 b^3(1+\nu)}$$

and hence

$$D_m = \frac{3(-1)^m F\nu a^2}{2\pi^3 m^3 b^3(1+\nu)} \Bigg/ \cosh\left(\frac{m\pi b}{a}\right)\,.$$

(18.17)

The complete stress field is therefore

$$\sigma_{zx} = \frac{3F\nu a}{2\pi^2 b^3(1+\nu)} \sum_{n=1}^{\infty} \frac{(-1)^n}{n^2} \sin\left(\frac{n\pi x}{a}\right) \sinh\left(\frac{n\pi y}{a}\right) \Bigg/ \cosh\left(\frac{n\pi b}{a}\right)$$

(18.18)

$$\sigma_{zy} = \frac{3F(b^2 - y^2)}{8ab^3} + \frac{F\nu(3x^2 - a^2)}{8ab^3(1+\nu)}$$

$$- \frac{3F\nu a}{2\pi^2 b^3(1+\nu)} \sum_{n=1}^{\infty} \frac{(-1)^n}{n^2} \cos\left(\frac{n\pi x}{a}\right) \cosh\left(\frac{n\pi y}{a}\right) \Bigg/ \cosh\left(\frac{n\pi b}{a}\right), \quad (18.19)$$

from (18.5, 18.12, 18.16, 18.17)

The maximum shear stress occurs at the points $(\pm a, 0)$ and is

$$\tau_{max} = \frac{3F}{8ab} + \frac{F\nu a}{4b^3(1+\nu)}\left[1 - \frac{6}{\pi^2}\sum_{n=1}^{\infty}1\bigg/n^2\cosh\left(\frac{n\pi b}{a}\right)\right].$$

The first term in this expression is that given by the elementary Mechanics of Materials theory. The error of the elementary theory (in the range $0 < a \le b$) is greatest when $\nu = 0.5$ and $a = b$, for which the approximation underestimates the exact value by 18%. However, the greater part of this correction is associated with the second (closed-form) term in (18.19). If this term is included, but the series is omitted, the error in τ_{max} is only about 1%.

This solution can be used for all values of the ratio a/b, but the series terms would make a more significant contribution in the range $b > a$. For this case, it is better to use the constants in (18.14) to satisfy strong conditions on $y = \pm b$ and weak conditions on $x = \pm a$ (see Problem 18.5).

Problems

18.1. Find an expression for the *local* twist per unit length of the bar, defined as

$$\beta(x, y) = \frac{\partial \omega_z}{\partial z},$$

where ω_z is the rotation defined in equation (1.33). Hence show that the *average* twist per unit length for a general unsymmetrical section will be zero if $C = 0$ in equation (18.6).

18.2. A bar of elliptical cross section defined by the boundary

$$\frac{x^2}{a^2} + \frac{y^2}{b^2} - 1 = 0$$

is loaded by a shear force $F_x = F$, acting along the negative x-axis. Find expressions for the shear stresses σ_{zx}, σ_{zy} on the cross section.

18.3. A bar of equilateral triangular cross section defined by the boundaries

$$y = \sqrt{3}x \ ; \ \ y = -\sqrt{3}x \ ; \ \ y = \frac{\sqrt{3}a}{2}$$

is loaded by a shear force $F_y = F$ acting along the negative y-axis. Find expressions for the shear stresses σ_{zx}, σ_{zy} on the cross section for the special case where $\nu = 0.5$. **Note:** You will need to move the origin to the centroid of the section $(0, a/\sqrt{3})$, since equation (18.2) is based on this choice of origin.

18.4. The bar of problem 18.3 is loaded by a shear force $F_x = F$ acting along the negative x-axis. Find expressions for the shear stresses σ_{zx}, σ_{zy} for the special case where $\nu = 0.5$, assuming the constant $C = 0$ in equation (18.6). Show that the resultant force F acts through the centroid of the triangle.

18.5. Solve the problem of the rectangular bar of §18.4.2 for the case where $a > b$. You will need to satisfy the boundary conditions in the strong sense on $y = \pm b$ and in the weak sense on $x = \pm a$. Compare the solution you get at this stage with the predictions of the elementary Mechanics of Materials theory. Then superpose an appropriate series of harmonic functions (sinusoidal in y and hyperbolic in x) to obtain an exact solution. In what range of the ratio a/b is it necessary to include the series to obtain 1% accuracy in the maximum shear stress?

18.6. Find the distribution of shear stress in a *hollow* cylindrical bar of inner radius b and outer radius a, loaded by a shear force $F_y = F$ acting through the origin. You will need to supplement the polynomial solution by the singular harmonic functions $x/(x^2 + y^2)$ and $(x^3 - 3xy^2)/(x^2 + y^2)^3$ with appropriate multipliers[4].

18.7. A bar has the cross section of a sector $0 < r < a$, $-\alpha < \theta < \alpha$, where α is not small compared with unity. The bar is loaded by a shear force F in the direction $\theta = 0$.

(i) Find a closed-form solution for the shear stresses in the bar, using the strong boundary conditions on the straight edges $\theta = \pm \alpha$ and a weak boundary condition on the curved edge $r = a$.
(ii) Superpose a series of harmonic functions φ_n of the form $r^\gamma \sin(\gamma \theta)$ where γ is chosen to satisfy $\varphi_n = 0$ on $\theta = \pm \alpha$.
(iii) Choose the coefficients on these terms so as to satisfy the boundary condition on the curved edge $r = a$ and hence obtain an expression for the maximum shear stress.

18.8. Solve Problem 18.7 for the case where the force F acts in the direction $\theta = \pi/2$. Assume that the line of action of F is such as to make the constant C in (18.6) be zero.

---

[4] These functions are Cartesian versions of the functions $r^{-1}\cos\theta$ and $r^{-3}\cos3\theta$ respectively.

Part IV
Complex-Variable Formulation

In Parts II and III we have shown how to reduce the in-plane and antiplane problems to boundary-value problems for an appropriate function of the coördinates (x, y). An alternative and arguably more elegant formulation can be obtained by combining x, y in the form of the complex variable $\zeta = x + \iota y$ and expressing the stresses and displacements as functions of ζ and its complex conjugate $\bar{\zeta}$. A significant advantage of this method is that it enables a connection to be made between harmonic boundary-value problems and the theory of Cauchy contour integrals. In certain cases (notably domains bounded by a circle or a straight line), this permits the stresses in the interior of the body to be written down in terms of integrals of the boundary tractions or displacements, thus removing the 'inspired guesswork' that is sometimes needed in the real stress function approach. Furthermore, the scope of this powerful method can be extended to other geometries using the technique of conformal mapping.

In view of these obvious advantages, the reader might reasonably ask why one should not dispense with the whole real stress function approach of Chapters 4 to 15 and instead rely solely on the complex-variable formulation for two-dimensional problems in elasticity. An obvious practical reason is that the latter requires some familiarity with the niceties of complex algebra and the notation tends to reduce the 'transparency' of the physical meaning of the resulting expressions. Also, although we might be able to write down the solution as a definite integral, which is certainly a very satisfactory mathematical outcome, evaluation of the resulting integral can be challenging. Indeed, the cases where the integral is easy to evaluate are usually precisely those where the real stress function approach also permits a very straightforward solution. One final point to bear in mind in this connection is that although Mathematica and Maple have the capacity to handle complex algebra, they do not generally perform well, at least at the present time of writing.

On the positive side, the complex-variable approach extends the scope of exact solutions to a broader class of geometries and boundary conditions and it offers significant advantages of elegance and compactness, which as we shall see later can also be used to simplify some three-dimensional problems. It can be said with some confidence that the development of this theory was the single most significant contribution to the theory of linear elasticity during the twentieth century and no-one with serious pretensions as an elastician can afford to be ignorant of it.

In the next two chapters we shall give merely a brief introduction to the complex-variable formulation. In Chapter 19, we introduce the notation and establish some essential mathematical results, notably concerning the integrals of functions of the complex variable around closed contours and the technique known as conformal mapping. We then apply these results to both in-plane and antiplane elasticity problems in Chapter 20. We shall endeavour to demonstrate connections to the corresponding real stress function results wherever possible, since the two methods can often be used in combination. Readers who wish to obtain a deeper understanding of the subject are referred to the several classical works on the subject[1]. More approachable treatments for the engineering reader are given by A. H. England, *Complex Variable Methods in Elasticity*, John Wiley, London, 1971, S. P. Timoshenko and J. N. Goodier, *loc. cit.* Chapter 6 and D. S. Dugdale and C. Ruiz, *Elasticity for Engineers*, McGraw-Hill, London, 1971, Chapter 2.

---

[1] N. I. Muskhelishvili, *Some Basic Problems of the Mathematical Theory of Elasticity,* P.Noordhoff, Groningen, 1963, A. C. Stevenson (1943), Some boundary problems of two-dimensional elasticity, *Philosophical Magazine*, Vol. 34, pp. 766–793, A. C. Stevenson (1945), Complex potentials in two-dimensional elasticity, *Proceedings of the Royal Society of London*, Vol. A184, pp. 129–179, A. E. Green and W. Zerna, *Theoretical Elasticity,* Clarendon Press, Oxford, 1954, I.S.Sokolnikoff, *Mathematical Theory of Elasticity,* 2nd.ed., McGraw-Hill, New York, 1956. L. M. Milne-Thomson, *Antiplane Elastic Systems*, Springer, Berlin, 1962, L. M. Milne-Thomson, *Plane Elastic Systems*, 2nd edn, Springer, Berlin, 1968.

Chapter 19
Prelinary Mathematical Results

The position of a point in the plane is defined by the two independent coördinates (x, y) which we here combine to form the complex variable ζ and its conjugate $\bar{\zeta}$, defined as

$$\zeta = x + \iota y ; \quad \bar{\zeta} = x - \iota y . \tag{19.1}$$

We can recover the Cartesian coördinates by the relations $x = \Re(\zeta)$, $y = \Im(\zeta)$, but a more convenient algebraic relationship between the real and complex formulations is obtained by solving equations (19.1) to give

$$x = \frac{1}{2}(\zeta + \bar{\zeta}) ; \quad y = -\frac{\iota}{2}(\zeta - \bar{\zeta}) . \tag{19.2}$$

At first sight, this seems a little paradoxical, since if we know ζ, we already know its real and imaginary parts x and y and hence $\bar{\zeta}$. However, for the purpose of the complex analysis, we regard ζ as the indissoluble combination of $x + \iota y$, and hence ζ and $\bar{\zeta}$ act as two independent variables defining position. In this chapter, we shall always make this explicit by writing $f(\zeta, \bar{\zeta})$ for a function that has fairly general dependence on position in the plane.

In polar coördinates (r, θ), we have the alternative expressions

$$\zeta = r \exp(\iota \theta) ; \quad \bar{\zeta} = r \exp(-\iota \theta) \tag{19.3}$$

for the complex variable, with solution

$$r = \left(\zeta \bar{\zeta}\right)^{1/2} ; \quad \theta = \frac{\iota}{2} \ln\left(\frac{\bar{\zeta}}{\zeta}\right) . \tag{19.4}$$

Any real or complex function of x, y can be expressed as a function of $\zeta, \bar{\zeta}$ using (19.2). For example

$$x^2 + y^2 = \frac{1}{4}(\zeta + \bar{\zeta})^2 - \frac{1}{4}(\zeta - \bar{\zeta})^2 = \zeta \bar{\zeta} ; \quad xy = -\frac{\iota}{4}(\zeta + \bar{\zeta})(\zeta - \bar{\zeta}) = \frac{\iota}{4}(\bar{\zeta}^2 - \zeta^2) .$$

© The Author(s), under exclusive license to Springer Nature Switzerland AG 2022
J. R. Barber, *Elasticity*, Solid Mechanics and Its Applications 172,
https://doi.org/10.1007/978-3-031-15214-6_19

The derivative of a function $f(\zeta, \bar{\zeta})$ with respect to ζ can be related to its derivatives with respect to x and y by

$$\frac{\partial f}{\partial \zeta} = \frac{\partial f}{\partial x}\frac{\partial x}{\partial \zeta} + \frac{\partial f}{\partial y}\frac{\partial y}{\partial \zeta} = \frac{1}{2}\left(\frac{\partial f}{\partial x} - \iota\frac{\partial f}{\partial y}\right) \tag{19.5}$$

and by a similar argument

$$\frac{\partial f}{\partial \bar{\zeta}} = \frac{1}{2}\left(\frac{\partial f}{\partial x} + \iota\frac{\partial f}{\partial y}\right). \tag{19.6}$$

Substituting (19.1) into (19.5, 19.6), we find that

$$\frac{\partial \bar{\zeta}}{\partial \zeta} = \frac{\partial \zeta}{\partial \bar{\zeta}} = 0,$$

showing that $(\zeta, \bar{\zeta})$ represent an orthogonal coördinate set.

19.1   Holomorphic Functions

If a function f depends *only* on ζ, so that $\partial f/\partial \bar{\zeta} = 0$, and if it is also infinitely differentiable at all points in some domain Ω, it is referred to as a *holomorphic function* of ζ in Ω. The complex conjugate $\bar{f}$ of a holomorphic function f can be obtained by replacing ζ by $\bar{\zeta}$ and replacing any complex constants by their conjugates. Thus, $\bar{f}$ is a differentiable function of $\bar{\zeta}$ only and will also be described as holomorphic. We shall see in the next section that holomorphic functions always satisfy the Laplace equation and hence are harmonic.

If $f(\zeta) = f_x + \iota f_y$ is a holomorphic function of ζ whose real and imaginary parts are f_x, f_y respectively, we can write

$$\frac{\partial f}{\partial \bar{\zeta}} = \frac{1}{2}\left(\frac{\partial f_x}{\partial x} + \iota\frac{\partial f_x}{\partial y} + \iota\frac{\partial f_y}{\partial x} - \frac{\partial f_y}{\partial y}\right) = 0,$$

using (19.6). For this to be satisfied, both real and imaginary parts must be separately zero, giving the *Cauchy-Riemann relations*

$$\frac{\partial f_x}{\partial x} = \frac{\partial f_y}{\partial y} \; ; \quad \frac{\partial f_x}{\partial y} = -\frac{\partial f_y}{\partial x}. \tag{19.7}$$

Since a holomorphic function $f(\zeta)$ is infinitely differentiable in Ω, it can be replaced by its Taylor series

$$f(\zeta) = \sum_{n=0}^{\infty} C_n (\zeta - \zeta_0)^n \tag{19.8}$$

in the neighbourhood of any point ζ_0 in Ω, where C_n is a set of complex constants.

19.2 Harmonic Functions

It follows from (19.5, 19.6) that

$$4\frac{\partial^2 f}{\partial\zeta\partial\bar{\zeta}} = \frac{\partial^2 f}{\partial x^2} + \frac{\partial^2 f}{\partial y^2} = \nabla^2 f \tag{19.9}$$

and hence the Laplace equation ($\nabla^2\phi=0$) in two dimensions can be written

$$\frac{\partial^2\phi}{\partial\zeta\partial\bar{\zeta}} = 0 . \tag{19.10}$$

The most general solution of (19.10) can be written

$$\phi = f_1 + \bar{f_2} , \tag{19.11}$$

where f_1, $\bar{f_2}$ are arbitrary holomorphic functions of ζ and $\bar{\zeta}$, respectively. Notice that instead of making the argument explicit as in $\bar{f_2}(\bar{\zeta})$, we can identify a holomorphic function of $\bar{\zeta}$ by the condensed notation $\bar{f_2}$.

The function defined by equation (19.11) will generally be complex, but its real and imaginary parts must separately satisfy (19.9) and hence a general *real* harmonic function can be written

$$\phi = \Re\left(f_1 + \bar{f_2}\right) .$$

As in equation (19.2), it is convenient to avoid the necessity of extracting real and imaginary parts by recognizing that the sum of a function and its complex conjugate will always be real. Thus the function

$$\phi = g + \bar{g}$$

represents the most general real solution of the Laplace equation, where g is an arbitrary holomorphic function of ζ only and $\bar{g}$ is its complex conjugate.

19.3 Biharmonic Functions

It follows from (19.9) that the biharmonic equation $\nabla^4\phi=0$ reduces to

$$\frac{\partial^4 \phi}{\partial \zeta^2 \partial \bar{\zeta}^2} = 0$$

and has the general complex solution

$$\phi = f_1 + \bar{f_2} + \bar{\zeta} f_3 + \zeta \bar{f_4} ,$$

where f_1, f_3 are arbitrary holomorphic functions of ζ and $\bar{f_2}$, $\bar{f_4}$ are arbitrary holomorphic functions of $\bar{\zeta}$.

The most general *real* solution of the biharmonic equation can be expressed in the form

$$\phi = g_1 + \bar{g}_1 + \bar{\zeta} g_2 + \zeta \bar{g}_2 , \qquad (19.12)$$

where g_1, g_2 are arbitrary holomorphic functions of ζ and $\bar{g}_1, \bar{g}_2$ are their complex conjugates.

Example

As an example, we can generate a real biharmonic polynomial function of degree n by taking $g_1(\zeta) = (A + \imath B)\zeta^n$, $g_2(\zeta) = (C + \imath D)\zeta^{n-1}$, where A, B, C, D are real arbitrary constants. Equation (19.12) then gives

$$\phi = (A + \imath B)\zeta^n + (A - \imath B)\bar{\zeta}^n + (C + \imath D)\bar{\zeta}\zeta^{n-1} + (C - \imath D)\zeta\bar{\zeta}^{n-1} . \quad (19.13)$$

This expression can be used in place of the procedure of §5.1 to develop the most general biharmonic polynomial in x, y of degree n. For example, for the special case $n = 5$ we have

$$\phi = (A + \imath B)(x + \imath y)^5 + (A - \imath B)(x - \imath y)^5 + (C + \imath D)(x - \imath y)(x + \imath y)^4$$
$$+ (C - \imath D)(x + \imath y)(x - \imath y)^4 .$$

Expanding and simplifying the resulting expressions, we obtain

$$\phi = 2A \left(x^5 - 10x^3 y^2 + 5xy^4\right) - 2B \left(5x^4 y - 10x^2 y^3 + y^5\right)$$
$$+ 2C \left(x^5 - 2x^3 y^2 - 3xy^4\right) - 2D \left(3x^4 y + 2x^2 y^3 - y^5\right) ,$$

which can be converted to the form (5.8) by choosing

$$A = -\frac{(2A_0 + A_2)}{16} \; ; \quad B = \frac{(3A_3 - 2A_1)}{80} \; ; \quad C = \frac{(10A_0 + A_2)}{16}$$

$$D = -\frac{(2A_1 + A_3)}{16} .$$

Alternatively, using the polar coördinate expressions (19.3) in (19.13) we have

$$\phi = (A+\iota B)r^n \exp(\iota n\theta) + (A-\iota B)r^n \exp(-\iota n\theta) + (C+\iota D)r^n \exp(\iota(n-2)\theta)$$
$$+(C-\iota D)r^n \exp(-\iota(n-2)\theta),$$

which expands to

$$\phi = 2Ar^n \cos(n\theta) - 2Br^n \sin(n\theta) + 2Cr^n \cos((n-2)\theta) - 2Dr^n \sin((n-2)\theta).$$

This is of course identical to the general form of the Michell solution for a given leading power r^n, as shown for example by comparison with equation (11.3) with n replaced by $(n-2)$.

19.4 Expressing Real Harmonic and Biharmonic Functions in Complex Form

Suppose we are given a real harmonic function $f(x, y)$ and wish to find the holomorphic function g such that

$$g + \bar{g} = f(x, y).$$

All we need to do is to use (19.2) to substitute for x, y in $f(x, y)$. If f is harmonic and real, the resulting function of $\zeta, \bar{\zeta}$ must reduce to the sum of a holomorphic function of ζ and its conjugate and these functions can then often be identified by inspection. Alternatively, starting from the form

$$g + \bar{g} = f(\zeta, \bar{\zeta}),$$

we differentiate with respect to ζ, obtaining

$$g' = \frac{\partial f}{\partial \zeta}, \tag{19.14}$$

since $\partial \bar{g}/\partial \zeta = 0$. The right-hand side of (19.14) must be a function of ζ only and hence g can be recovered by integration. This operation is performed by the Mathematica and Maple files 'rtoch'.

Example

Consider the function

$$f(x, y) = \frac{x \ln\left(x^2 + y^2\right)}{2} - y \arctan\left(\frac{y}{x}\right),$$

which is easily shown to be harmonic by substitution into the Laplace equation. Using (19.2, 19.4) we have

$$x^2 + y^2 = r^2 = \zeta\bar{\zeta} \; ; \quad \arctan\left(\frac{y}{x}\right) = \theta = \frac{\imath}{2}\ln\left(\frac{\bar{\zeta}}{\zeta}\right)$$

and hence

$$g + \bar{g} = \frac{(\bar{\zeta} + \zeta)\ln(\zeta\bar{\zeta})}{4} + \frac{(\bar{\zeta} - \zeta)}{4}\ln\left(\frac{\bar{\zeta}}{\zeta}\right) = \frac{1}{2}\left(\zeta\ln(\zeta) + \bar{\zeta}\ln(\bar{\zeta})\right)$$

We conclude by inspection that

$$g = \frac{\zeta\ln(\zeta)}{2} \; .$$

19.4.1 Biharmonic functions

The same procedure can also be used to express real biharmonic functions in the form (19.12). We first use (19.2) to express the given real function as a function of $\zeta, \bar{\zeta}$. Equation (19.12) can then be written

$$g_1 + \overline{g_1} + \bar{\zeta}g_2 + \zeta\overline{g_2} = f(\zeta, \bar{\zeta}) \; , \tag{19.15}$$

where $f(\zeta, \bar{\zeta})$ is a known function. Differentiating with respect to ζ and $\bar{\zeta}$, we obtain

$$g_2' + \overline{g_2'} = \frac{\partial^2 f}{\partial\zeta\partial\bar{\zeta}}$$

and the function g_2' and hence g_2 can then be obtained using the above procedure for harmonic functions. Once g_2 is known, it can be substituted into (19.15) and the resulting equation solved for the harmonic function g_1 using the same procedure. The Maple and Mathematica files 'rtocb' perform this operation for any biharmonic function specified in Cartesian coördinates.

19.5 Line Integrals

Suppose that S is a closed contour and that $f(\zeta)$ is holomorphic on S and at all points in the region enclosed by S. It can then be shown[1] that the contour integral

$$\oint_S f(\zeta)d\zeta = 0 \; . \tag{19.16}$$

---

[1] See for example E. T. Copson, *An Introduction to the Theory of Functions of a Complex Variable*, Clarendon Press, Oxford, 1935, §4.21, G. F. Carrier, M. Krook and C. E. Pearson, *Functions of a Complex Variable*, McGraw-Hill, New York, 1966, §2.2.

To prove this result we write $f(\zeta) = f_x + \imath f_y$, $d\zeta = dx + \imath dy$, separate real and imaginary parts and use Green's theorem to convert the resulting real contour integrals into integrals over the area enclosed by S. The integrand of each of these area integrals will then be found to be identically zero because of the Cauchy-Riemann relations (19.7).

Fig. 19.1 Path of the integral
in equation (19.17).

As in §2.2.1, we can use (19.16) to show that the line integral

$$\int_A^B f(\zeta)d\zeta = \int_{\zeta_A}^{\zeta_B} f(\zeta)d\zeta \qquad (19.17)$$

along the line S in Figure 19.1 is not changed by making any infinitesimal change in the path S as long as this does not pass outside the region within which $f(\zeta)$ is holomorphic. It follows that (19.17) is path-independent if S lies wholly within a simply connected region Ω in which $f(\zeta)$ is holomorphic.

19.5.1 The residue theorem

Next suppose that the function $f(\zeta)$ is holomorphic everywhere in a simply-connected domain Ω *except* at a point P, where it is singular. As before, we can use (19.16) to show that the value of the contour integral is not changed by making any infinitesimal deviation in the path S. It follows that all contour integrals whose paths pass around P once and only once must have the same value.

Suppose that P is defined by $\zeta = \zeta_0$ and that the singularity has the form

$$f(\zeta) = \frac{f_1(\zeta)}{(\zeta - \zeta_0)^n} + f_2(\zeta) , \qquad (19.18)$$

where $f_1(\zeta)$, $f_2(\zeta)$ are holomorphic throughout the domain enclosed by S (including at P) and n is an integer. The function $f(\zeta)$ is then said to have a *pole of order n* at $\zeta = \zeta_0$. Since the contour integral is the same for all contours enclosing P, we can evaluate it around an infinitesimal circle of radius ϵ centred on $\zeta = \zeta_0$. For the special case $n = 1$, $f(\zeta)$ has a pole of order unity (also known as a *simple pole*) at $\zeta = \zeta_0$ and if ϵ is sufficiently small, we can replace the continuous function $f_1(\zeta)$ by $f_1(\zeta_0)$ giving

$$\oint_S f(\zeta)d\zeta = f_1(\zeta_0) \oint_S \frac{d\zeta}{(\zeta - \zeta_0)} = f_1(\zeta_0) \ln(\zeta - \zeta_0)\Big|_{\zeta_0 + \epsilon \exp(\iota\theta_1)}^{\zeta_0 + \epsilon \exp(\iota(\theta_1 + 2\pi))}$$

$$= 2\pi\iota f_1(\zeta_0) , \qquad\qquad\qquad\qquad (19.19)$$

where θ_1 is defined such that $\zeta = \zeta_0 + \epsilon \exp(\iota\theta_1)$ is any point on the path S and the integral is evaluated in the anticlockwise direction (increasing θ). Notice that the second term in (19.18) makes no contribution to the integral, because of (19.16). The quantity

$$f_1(\zeta_0) = \lim_{\zeta \to \zeta_0} f(\zeta)(\zeta - \zeta_0) \equiv \mathrm{Res}(\iota_0) \qquad\qquad (19.20)$$

is known as the *residue* at the simple pole P. It then follows from (19.19) that

$$\oint_S f(\zeta)d\zeta = 2\pi\iota \, \mathrm{Res}(\iota_0) .$$

For higher order poles $n > 1$, the limit (19.20) does not exist, but we can extract the residue by successive partial integrations. We obtain

$$\oint \frac{f_1(\zeta)d\zeta}{(\zeta - \zeta_0)^n} = -\frac{f_1(\zeta)}{(n-1)(\zeta - \zeta_0)^{(n-1)}}\Big|_{\epsilon \exp(\iota\theta_1)}^{\epsilon \exp(\iota(\theta_1 + 2\pi))} + \oint \frac{f_1'(\zeta)d\zeta}{(n-1)(\zeta - \zeta_0)^{(n-1)}} ,$$

where we have used the same circular path of radius ϵ. The first term is zero for $n \neq 1$, so

$$\oint \frac{f_1(\zeta)d\zeta}{(\zeta - \zeta_0)^n} = \oint \frac{f_1'(\zeta)d\zeta}{(n - 1)(\zeta - \zeta_0)^{(n-1)}} .$$

Repeating this procedure until the final integral has only a simple pole, we obtain

$$\oint \frac{f_1(\zeta)d\zeta}{(\zeta - \zeta_0)^n} = \oint \frac{f_1^{[n-1]}(\zeta)d\zeta}{(n - 1)!(\zeta - \zeta_0)} ,$$

where

$$f_1^{[n-1]}(\zeta) \equiv \frac{d^{(n-1)}f_1(\zeta)}{d\zeta^{(n-1)}} . \qquad\qquad (19.21)$$

The residue at a pole of order n is therefore

$$\mathrm{Res}(\zeta_0) = \frac{f_1^{[n-1]}(\zeta_0)}{(n - 1)!} , \qquad\qquad (19.22)$$

where $f_1^{[n-1]}$ is defined through equations (19.21, 19.18).

Suppose now that the function $f(\zeta)$ posesses several poles and we wish to determine the value of the contour integral along a path enclosing those at the points $\zeta_1, \zeta_2, ..., \zeta_k$. One such path enclosing three poles at P_1, P_2, P_3 is shown in Figure 19.2. It is clear that the contributions from the straight line segments will be self-

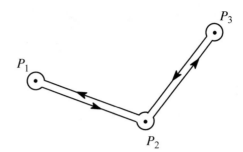

Fig. 19.2 Integral path around three poles at P_1, P_2, P_3.

cancelling, so the integral is simply the sum of those for the circular paths around each separate pole. Thus, the resulting integral will have the value

$$\oint f(\zeta)d\zeta = 2\pi\iota \sum_{j=1}^{k} \text{Res}(\zeta_j) \ . \tag{19.23}$$

This result is extremely useful as a technique for determining the values of complex line integrals. Similar considerations apply to the evaluation of contour integrals around one or more holes in a multiply-connected body.

Example

As a special case, we shall determine the value of the contour integral

$$\oint_S \frac{d\zeta}{\zeta^n(\zeta - a)} \ ,$$

where the integration path S encloses the points $\zeta = 0, a$. The integrand has a simple pole at $\zeta = a$ and a pole of order n at $\zeta = 0$. For the simple pole, we have

$$\text{Res}(a) = \frac{1}{a^n} \ ,$$

from (19.20). For the pole at $\zeta = 0$, we have

$$f_1(\zeta) = \frac{1}{(\zeta - a)} \quad \text{and} \quad f_1^{[n-1]}(\zeta) = \frac{(-1)^{(n-1)}(n - 1)!}{(\zeta - a)^n} \ ,$$

so

$$f_1^{[n-1]}(0) = \frac{(-1)^{(n-1)}(n - 1)!}{(-a)^n} \quad \text{and} \quad \text{Res}(0) = \frac{(-1)^{(n-1)}}{(-a)^n} = -\frac{1}{a^n} \ ,$$

from (19.22).

Using (19.23), we then have

$$\oint_S \frac{d\zeta}{\zeta^n(\zeta - a)} = 2\pi\iota \left(\frac{1}{a^n} - \frac{1}{a^n} \right) = 0 ; \qquad n \geq 1 \qquad (19.24)$$

$$= 2\pi\iota ; \qquad\qquad\qquad n = 0 , \qquad (19.25)$$

since if $n=0$ we have only the simple pole at $\zeta=a$.

19.5.2 The Cauchy integral theorem

Changing the symbols in equation (19.19), we have

$$f(\zeta_0) = \frac{1}{2\pi\iota} \oint_S \frac{f(s)ds}{(s - \zeta_0)} , \qquad (19.26)$$

where the path of the integral, S, encloses the point $s=\zeta_0$. If S is chosen to coincide with the boundary of a simply-connected region Ω within which f is holomorphic, this provides an expression for f at a general point in Ω in terms of its boundary values. We shall refer to this as the *interior problem*, since the region Ω is interior to the contour. Notice that equation (19.19) and hence (19.26) is based on the assumption that the contour S is traversed in the anticlockwise direction. Changing the direction would lead to a sign change in the result. An equivalent statement of this requirement is that the contour is traversed in a direction such that the enclosed simply-connected region Ω always lies on the left of the path S.

A related result can be obtained for the *exterior problem* in which f is holomorphic in the multiply-connected region $\overline{\Omega}$ exterior to the closed contour S and extending to and including the point at infinity. In this case, a general holomorphic function f can be expressed as a *Laurent series*

$$f = C_0 + \sum_{k=1}^{\infty} \frac{C_k}{\zeta^k} , \qquad (19.27)$$

where the origin lies within the hole excluded by S. For the exterior problem to be well posed, we must specify the value of f at all points on S and also at infinity $\zeta \to \infty$. We shall restrict attention to the case where $f(\infty)=0$ and hence $C_0=0$, since the more general case is most easily handled by first solving a problem for the entire plane (with no hole) and then defining a corrective problem for the conditions at the inner boundary S.

To invoke the Cauchy integral theorem, we need to construct a contour that encloses a simply-connected region including the general point $\zeta=\zeta_0$. A suitable contour is illustrated in Figure 19.3, where the path S corresponds to the boundary of the hole and S_1 is a circle of sufficiently large radius to include the point $\zeta=\zeta_0$. Notice that the theorem requires that the contour be traversed in an anticlockwise direction as shown.

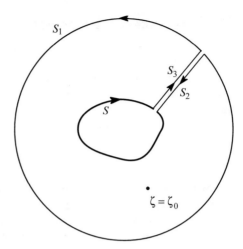

Fig. 19.3 Contour for the region exterior to a hole S.

Equation (19.26) now gives

$$f(\zeta_0) = \frac{1}{2\pi i} \oint_{S+S_1+S_2+S_3} \frac{f(s)ds}{(s-\zeta_0)} . \tag{19.28}$$

In this integral, the contributions from the line segments S_2, S_3 clearly cancel, leaving

$$f(\zeta_0) = -\frac{1}{2\pi i} \oint_S \frac{f(s)ds}{(s-\zeta_0)} + \frac{1}{2\pi i} \oint_{S_1} \frac{f(s)ds}{(s-\zeta_0)} . \tag{19.29}$$

The remaining terms are two separate contour integrals around S and S_1 respectively in Figure 19.3. Notice that we have changed the sign in the integral around S, since S is now a closed contour and the convention demands that it be evaluated in the anticlockwise direction, which is opposite to the direction implied in (19.28) and Figure 19.3.

Using (19.27) and recalling that $C_0 = 0$, we have

$$\oint_{S_1} \frac{f(s)ds}{(s-\zeta_0)} = \sum_{k=1}^{\infty} \oint_{S_1} \frac{C_k ds}{s^k(s-\zeta_0)} = 0 ,$$

from (19.24). Thus, the second term in (19.29) is zero and we have

$$f(\zeta_0) = -\frac{1}{2\pi i} \oint_S \frac{f(s)ds}{(s-\zeta_0)} , \tag{19.30}$$

which determines the value of the function f at a point in the region exterior to the contour S in terms of the values on this contour.

19.6 Solution of Harmonic Boundary-value Problems

At first sight, equations (19.26, 19.30) appear to provide a direct method for finding the value of a harmonic function from given boundary data. However, if the required harmonic function is real, we meet a difficulty, since we know only the real part of the boundary data.

We could write a general real harmonic function as

$$\phi = f + \bar{f} \,, \tag{19.31}$$

in which case we know the boundary values of the function ϕ, but not of the separate functions f, $\bar{f}$. However, we shall show in the next section that a direct solution *can* be developed for both the interior and exterior problems in the special case where the boundary S is circular. Furthermore, this solution can be extended to more general geometries using the technique of conformal mapping.

19.6.1 Direct method for the interior problem for a circle

We consider the simply-connected region interior to the circle $r = a$. The boundary of this circle S is defined by

$$s = a \exp(\iota\theta) \;; \quad \bar{s} = a \exp(-\iota\theta) \tag{19.32}$$

and hence

$$s\bar{s} = a^2 \;; \quad \bar{s} = \frac{a^2}{s} \;; \quad s \in S \,. \tag{19.33}$$

Consider the problem of determining a real harmonic function $\phi(x, y)$ from real boundary data on $r = a$. We choose to write ϕ as the sum of a holomorphic function f and its conjugate as in (19.31). We then expand f inside the circle as a Taylor series

$$f(\zeta) = C_0 + \sum_{k=1}^{\infty} C_k \zeta^k \tag{19.34}$$

in which case

$$\overline{f(s)} = \overline{C}_0 + \sum_{k=1}^{\infty} \frac{\overline{C}_k a^{2k}}{s^k} \;; \quad s \in S \,,$$

using (19.33).

We now apply the Cauchy operator to the boundary data, obtaining

$$\frac{1}{2\pi\iota}\oint\frac{\phi(s)ds}{(s-\zeta)} = \frac{1}{2\pi\iota}\oint\frac{f(s)ds}{(s-\zeta)} + \frac{1}{2\pi\iota}\oint\frac{\overline{f(s)}ds}{(s-\zeta)}$$

$$= \frac{1}{2\pi\iota}\oint\frac{f(s)ds}{(s-\zeta)} + \frac{\overline{C_0}}{2\pi\iota}\oint\frac{ds}{(s-\zeta)} + \frac{1}{2\pi\iota}\sum_{k=1}^{\infty}\overline{C_k}\oint\frac{a^{2k}ds}{s^k(s-\zeta)}$$

$$= f(\zeta) + \overline{C_0},$$

using (19.26) for the first two integrals and noting that every integral in the third term summation is zero in view of (19.24). Thus, we have

$$f(\zeta) = -\overline{C_0} + \frac{1}{2\pi\iota}\oint\frac{\phi(s)ds}{(s-\zeta)}, \qquad (19.35)$$

after which the real harmonic function ϕ can be recovered from (19.31) except for the unknown complex constant C_0. To determine this constant, we equate (19.35) to the Taylor series (19.34) and evaluate it at $\zeta=0$, obtaining

$$C_0 = -\overline{C_0} + \frac{1}{2\pi\iota}\oint\frac{\phi(s)ds}{s},$$

or

$$C_0 + \overline{C_0} = \frac{1}{2\pi\iota}\oint\frac{\phi(s)ds}{s}.$$

Alternatively, writing

$$s = ae^{\iota\theta}; \quad ds = a\iota e^{\iota\theta}d\theta$$

from (19.32), we have

$$C_0 + \overline{C_0} = \frac{1}{2\pi}\int_0^{2\pi}\phi(a,\theta)d\theta. \qquad (19.36)$$

Thus, the real part of the constant C_0 can be determined, but not its imaginary part and this is reasonable since if we add an arbitrary imaginary constant into f it will make no contribution to ϕ in (19.31).

Example

Suppose that $\phi(a,\theta)=\phi_2\cos(2\theta)$, where ϕ_2 is a real constant and we wish to find the corresponding harmonic function $\phi(r,\theta)$ inside the circle of radius a. Using (19.32), we can write

$$\phi_2\cos(2\theta) = \frac{\phi_2(e^{2\iota\theta}+e^{-2\iota\theta})}{2} = \frac{\phi_2}{2}\left(\frac{s^2}{a^2}+\frac{a^2}{s^2}\right)$$

and it follows that

$$\frac{1}{2\pi\iota} \oint \frac{\phi(s)ds}{(s-\zeta)} = \frac{\phi_2}{4\pi a^2\iota} \oint \frac{s^2 ds}{(s-\zeta)} + \frac{\phi_2 a^2}{4\pi\iota} \oint \frac{ds}{s^2(s-\zeta)} = \frac{\phi_2\zeta^2}{2a^2} ,$$

where we have used (19.26) for the first integral and the second integral is zero in view of (19.24).

Using (19.35), we then have

$$f = \frac{\phi_2\zeta^2}{2a^2} - \overline{C}_0 ; \quad \bar{f} = \frac{\phi_2\bar{\zeta}^2}{2a^2} - C_0$$

and

$$\phi = f + \bar{f} = \frac{\phi_2(\zeta^2 + \bar{\zeta}^2)}{2a^2} - C_0 - \overline{C}_0 .$$

Evaluating the integral in (19.36), we obtain

$$C_0 + \overline{C}_0 = \frac{\phi_2}{2\pi} \int_0^{2\pi} \cos(2\theta)d\theta = 0 ,$$

and hence the required harmonic function is

$$\phi = \frac{\phi_2(\zeta^2 + \bar{\zeta}^2)}{2a^2} = \frac{\phi_2 r^2 \cos(2\theta)}{a^2} ,$$

using (19.3). Of course, this result could also have been obtained by writing a general harmonic function as the Fourier series

$$\phi = \sum_{n=0}^{\infty} A_n r^n \cos(n\theta) + \sum_{n=1}^{\infty} B_n r^n \sin(n\theta)$$

and equating coefficients on the boundary $r=a$, but the present method, though longer, is considerably more direct.

19.6.2   Direct method for the exterior problem for a circle

For the exterior problem, we again define ϕ as in (19.31) and apply the Cauchy integral operator to the boundary data, obtaining

$$\frac{1}{2\pi\iota} \oint \frac{\phi(s)ds}{(s-\zeta)} = \frac{1}{2\pi\iota} \oint \frac{f(s)ds}{(s-\zeta)} + \frac{1}{2\pi\iota} \oint \frac{\overline{f(s)}ds}{(s-\zeta)} , \qquad (19.37)$$

where points on the contour are defined by (19.32). We assume that the function ϕ is required to satisfy the condition $\phi \to 0$, $r \to \infty$, in which case the function f, which is holomorphic *exterior* to the circle, can be written as the

Laurent expansion

$$f(\zeta) = \sum_{k=1}^{\infty} \frac{C_k}{\zeta^k} \quad \text{and hence} \quad \overline{f(s)} = \sum_{k=1}^{\infty} \frac{\overline{C}_k s^k}{a^{2k}} ,$$

using (19.33). It follows that the second integral on the right-hand side of (19.37) is

$$\frac{1}{2\pi\iota} \oint \frac{\overline{f(s)}ds}{(s - \zeta)} = \frac{1}{2\pi\iota} \sum_{k=1}^{\infty} \frac{\overline{C}_k}{a^{2k}} \oint \frac{s^k ds}{(s - \zeta)}$$

and this is zero since ζ lies *outside* the circle and hence the integrand defines a holomorphic function in the region *interior* to the circle. Using this result and (19.30) in (19.37), we obtain

$$f(\zeta) = -\frac{1}{2\pi\iota} \oint \frac{\phi(s)ds}{(s - \zeta)} , \tag{19.38}$$

after which ϕ is recovered from (19.31).

The reader might reasonably ask why the method of §§19.6.1, 19.6.2 works for circular contours, but not for more general problems. The answer lies in the relation (19.33) which enables us to define the conjugate $\overline{s}$ in terms of s and the constant radius a. For a more general contour, this relation would also involve the coördinate r in $s = r\exp(\iota\theta)$ and this varies around the contour if the latter is non-circular.

19.6.3   The half-plane

If the radius of the circle is allowed to grow without limit and we move the origin to a point on the boundary, we recover the problem of the half-plane, which can conveniently be defined as the region $y > 0$ whose boundary is the infinite line $-\infty < x < \infty$, $y = 0$. The contour for the integral theorem (19.35) then comprises a circle of infinite radius which lies entirely at infinity except for the line segment $-\infty < x < \infty$, $y = 0$ and we obtain

$$f(\zeta) = -\overline{C}_0 + \frac{1}{2\pi\iota} \int_{-\infty}^{\infty} \frac{\phi(x)dx}{(x - \zeta)} . \tag{19.39}$$

In some cases, an even more direct method can be used to solve the boundary-value problem. We first note that the holomorphic (and hence harmonic) function $f(\zeta)$ reduces to $f(x)$ on $y = 0$. Thus, if we simply replace x by ζ in the boundary function $\phi(x)$, we shall obtain a function of ζ that reduces to $\phi(x)$ on the boundary $y = 0$.

Unfortunately, the resulting function $\phi(\zeta)$ will often not be holomorphic, either because the boundary function $\phi(x)$ is not continuously differentiable, or because

$\phi(\zeta)$ contains one or more poles in the half-plane $y > 0$. In these cases, the integral theorem (19.39) will always yield the correct solution. To illustrate this procedure, we shall give two examples.

Example 1

Consider the boundary-value problem

$$\phi(x, 0) = 1 ; \quad -a < x < a$$
$$= 0 ; \quad |x| > a .$$

Since the function is discontinuous, we must use the integral theorem (19.39), which here yields

$$f(\zeta) = -\overline{C}_0 + \frac{1}{2\pi\iota} \int_{-a}^{a} \frac{dx}{(x - \zeta)} = -\overline{C}_0 + \frac{1}{2\pi\iota} \ln\left(\frac{\zeta - a}{\zeta + a}\right) .$$

The corresponding real function ϕ is then obtained from (19.31) and after using (19.1) and simplifying, we obtain

$$\phi(x, y) = \frac{1}{\pi}\left[\arctan\left(\frac{y}{x - a}\right) - \arctan\left(\frac{y}{x + a}\right)\right] ,$$

where the real constant $C_0 + \overline{C}_0$ can be determined to be zero by equating $\phi(x, y)$ to its value at any convenient point on the boundary.

Example 2

Consider now the case where

$$\phi(x, 0) = \frac{1}{(1 + x^2)} .$$

This function is continuous and differentiable, so a first guess would be to propose the function

$$\phi = \Re\left(f_1(\zeta)\right) \quad \text{where} \quad f_1(\zeta) = \frac{1}{(1 + \zeta^2)} ,$$

which certainly tends to the correct value on $y = 0$. However, f_1 has poles at $\zeta = \pm\iota$, i.e. at the points $(0, 1)$, $(0, -1)$, the former of which lies in the half-plane $y > 0$. Thus $f_1(\zeta)$ is not holomorphic throughout the half-plane and its real part is therefore not harmonic at $(0, 1)$.

As in the previous example, we can resolve the difficulty by using the integral theorem. However, the fact that the boundary function is continuous permits us to evaluate the integral using the residue theorem. We therefore write

$$f(\zeta) = -\overline{C}_0 + \frac{1}{2\pi\iota} \oint_S \frac{ds}{(1+s^2)(s-\zeta)} = -\overline{C}_0 + \frac{1}{2\pi\iota} \oint_S \frac{ds}{(s+\iota)(s-\iota)(s-\zeta)} \, ,$$

where the contour S is the circle of infinite radius which completely encloses the half-plane $y > 0$. The integrand has simple poles at $s = \iota, -\iota, \zeta$, but only the first and third of these lie within the contour. We therefore obtain

$$\frac{1}{2\pi\iota} \oint_S \frac{ds}{(s+\iota)(s-\iota)(s-\zeta)} = \frac{1}{(\zeta+\iota)(\zeta-\iota)} + \frac{1}{2\iota(\iota-\zeta)} = \frac{\iota}{2(\zeta+\iota)}$$

and the real harmonic function ϕ is then recovered as

$$\phi(\zeta, \bar{\zeta}) = f + \bar{f} - C_0 - \overline{C}_0 = \frac{\iota}{2(\zeta+\iota)} - \frac{\iota}{2(\bar{\zeta}-\iota)} - C_0 - \overline{C}_0 \, .$$

The constants can be determined by evaluating this expression at the origin $\zeta = 0$ and equating it to $\phi(0, 0)$, giving

$$\frac{1}{2} + \frac{1}{2} - C_0 - \overline{C}_0 = 1 \quad \text{or} \quad C_0 + \overline{C}_0 = 0 \, .$$

We then use (19.1) to express $\phi(\zeta, \bar{\zeta})$ in terms of (x, y), obtaining

$$\phi(x, y) = \frac{(1+y)}{x^2 + (1+y)^2} \, .$$

This function can be verified to be harmonic and it clearly tends to the required function $\phi(x, 0)$ on $y = 0$.

19.7 Conformal Mapping

In §19.6, we showed that the value of a real harmonic function in the region interior or exterior to a circle or in the half-plane can be written down as an explicit integral of the boundary data, using the Cauchy integral theorem. The usefulness of this result is greatly enhanced by the technique of conformal mapping, which permits us to 'map' a holomorphic function in the circular domain or the half-plane to one in a more general domain, usually but not necessarily with the same connectivity.

Suppose that $w(\zeta)$ is a holomorphic function of ζ in a domain Ω, whose boundary S is the unit circle $|\zeta| = 1$. Each point (x, y) in Ω corresponds to a unique value

$$w = \xi + \iota\eta,$$

whose real and imaginary parts (ξ, η) can be used as Cartesian coördinates to define a point in a new domain Ω^* with boundary S^*. Thus the function $w(\zeta)$ can be used to map the points in Ω into the more general domain Ω^* and points on the boundary S into S^*. Since $\zeta \in \Omega$ and $w \in \Omega^*$, we shall refer to Ω, Ω^* as the ζ-plane and the w-plane respectively.

Next suppose that $f(w)$ is a holomorphic function of w in Ω^*, implying that both the real and imaginary parts of f are harmonic functions of (ξ, η), from (19.10). Since w is a holomorphic function of ζ, it follows that

$$f_\zeta \equiv f(w(\zeta))$$

is a holomorphic function of ζ in Ω and hence that its real and imaginary parts are also harmonic functons of (x, y). Thus, the mapping function $w(\zeta)$ maps holomorphic functions in Ω^* into holomorphic functions in Ω. It follows that a harmonic boundary-value problem for the domain S^* can be mapped into an equivalent boundary-value problem for the unit circle, which in turn can be solved using the direct method of §19.6.1 or §19.6.2.

Example: The elliptical hole

A simple illustration is provided by the mapping function

$$w \equiv \xi + \iota\eta = c\left(\zeta + \frac{m}{\zeta}\right) , \tag{19.40}$$

where c, m are real constants and $m < 1$. Substituting $\zeta = r\exp(\iota\theta)$ into (19.40) and separating real and imaginary parts, we find

$$\xi = c\left(r + \frac{m}{r}\right)\cos\theta ; \quad \eta = c\left(r - \frac{m}{r}\right)\sin\theta .$$

Eliminating θ between these equations, we then obtain

$$\frac{\xi^2}{\alpha^2} + \frac{\eta^2}{\beta^2} = 1 , \tag{19.41}$$

where

$$\alpha = c\left(r + \frac{m}{r}\right) ; \quad \beta = c\left(r - \frac{m}{r}\right) . \tag{19.42}$$

Equation (19.41) with (19.42) shows that the set of concentric circles (lines of constant r) in the ζ-plane maps into a set of confocal ellipses in the w-plane. In particular, the unit circle $r = 1$ maps into the ellipse with semi-axes

$$a = c(1 + m) ; \quad b = c(1 - m) .$$

If $f(w)$ is a holomorphic function in the region *interior* to the ellipse, the mapped function f_ζ will generally be unbounded both at $\zeta=0$ and at $\zeta\to\infty$. This mapping cannot therefore be used for the interior problem for the ellipse. It can however be used for the exterior problem, since in this case, $f(w)$ must posess a Laurent series, every term of which maps into a bounded term in f_ζ.

Equation (19.40) can be solved to give ζ as a function of w. We obtain the two solutions

$$\zeta = \frac{w \pm \sqrt{w^2 - 4mc^2}}{2c} \, . \tag{19.43}$$

The positive square root maps points outside the unit circle into points outside the ellipse, whereas the negative root maps points *inside* the unit circle into points outside the ellipse. Either mapping can be used when combined with the appropriate solution of the boundary-value problem. Thus we can use §19.6.1 and the negative root in (19.43), or §19.6.2 and the positive root.

The present author favours the exterior-to-exterior mapping implied by the positive root, since this preserves the topology of the original problem and hence tends to a more physically intuitive procedure. We then have

$$\zeta = \frac{w + \sqrt{w^2 - 4mc^2}}{2c} \, , \tag{19.44}$$

and it is clear that circles of radius $r>1$ in the ζ-plane map to increasingly circular contours in the w-plane as $r\to\infty$. To solve a given problem, we map the boundary conditions to the unit circle in the ζ-plane using (19.40), solve for the required holomorphic function of ζ using §19.6.2 and then map this function back into the w-plane using (19.44).

Problems

19.1. By writing

$$\zeta^n = e^{n \ln(\zeta)} \, ,$$

show that

$$\frac{\partial^m}{\partial n^m} (\zeta^n) = \zeta^n \ln^m(\zeta) \, .$$

Using equations (19.3), find the real and imaginary parts of the function $\zeta^n \ln^2(\zeta)$ in polar coördinates r, θ and verify that they are both harmonic.

19.2. Verify that the real stress function $\phi=r^2\theta$ satisfies the biharmonic equation using (8.11). Then express it in the complex form (19.15).

19.3. Express the real biharmonic function $\phi=r\theta\sin\theta$ (from equation (8.31) or Table 8.1) in the complex form (19.15).

19.4. Find the imaginary part of the holomorphic function whose real part is the harmonic function $\ln(r)$.

19.5. Use the residue theorem to evaluate the contour integral

$$\oint_S \frac{\zeta^n d\zeta}{(\zeta^2 - a^2)} ,$$

where n is a positive integer and the contour S is the circle $|\zeta| = 2a$.

19.6. Find the real and imaginary parts of the function $\sin(\zeta)$ and hence find the points in the x, y-plane at which $\sin(\zeta) = 0$.
  Use the residue theorem to evaluate the integral

$$\oint_S \frac{d\zeta}{\sin(\zeta)}$$

for a contour S that encloses only the pole at $\zeta = 0$. By choosing S as a rectangle two of whose sides are the lines $x = \pm\pi/2$, $-\infty < y < \infty$, use your result to evaluate the real integral

$$\int_{-\infty}^{\infty} \frac{dy}{\cosh y} .$$

19.7. Show that on the unit circle,

$$\frac{1}{2}\left(\zeta + \frac{1}{\zeta}\right) = \cos\theta ; \quad \frac{1}{2}\left(\zeta^n + \frac{1}{\zeta^n}\right) = \cos(n\theta) .$$

Use this result and the notation $\omega = e^{i\phi}$ to express the real integral (12.22) in terms of an integral around the contour shown in Figure 19.4. Use the residue theorem to evaluate this integral and hence prove equation (12.22). Explain in particular how you determine the contribution from the small semi-circular paths around the poles $\zeta = \omega$, $\zeta = 1/\omega$.

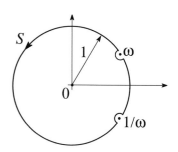

Fig. 19.4

19.8. Use the change of variable $\xi = a\cos\theta$ and $\zeta = e^{i\theta}$ to show that

$$\int_{-1}^{1} \frac{\xi d\xi}{(x - \xi)\sqrt{1 - \xi^2}} = \frac{\iota}{2} \oint_0^{2\pi} \frac{(1 + \zeta^2)d\zeta}{\zeta(\zeta^2 - 2x\zeta + 1)} ,$$

where the contour integral is evaluated around the unit circle. Find the location of the poles of the integrand assuming x is real and greater than unity. Hence, use the residue theorem to evaluate the integral and confirm the expression (13.20).

19.9. Find the real harmonic function $\phi(r, \theta)$ defined inside the circle $0 \leq r < a$ whose boundary values are

$$\phi(a, \theta) = 1 ; \quad -\alpha < \theta < \alpha$$
$$= 0 ; \quad \alpha < |\theta| < \pi .$$

19.10. Find the real harmonic function $\phi(r, \theta)$ defined in the *external* region $r > a$ satisfying the boundary conditions

$$\phi(a, \theta) = \phi_3 \sin(3\theta) ; \quad \phi(r, \theta) \to 0 ; \quad r \to \infty ,$$

where ϕ_3 is a real constant.

19.11. Find the real harmonic function $\phi(x, y)$ in the half-plane $y \geq 0$ satisfying the boundary conditions

$$\phi(x, 0) = \frac{1}{(1 + x^2)^2} ; \quad \phi(x, y) \to 0 ; \quad \sqrt{x^2 + y^2} \to \infty .$$

19.12. Use the mapping function (19.40) to find the holomorphic function $f(w)$ such that the real harmonic function

$$\phi = f + \bar{f}$$

satisfies the boundary condition $\phi = A\xi$ on the ellipse

$$\frac{\xi^2}{a^2} + \frac{\eta^2}{b^2} = 1$$

and $f \to 0$ as $|w| \to \infty$, where A is a real constant. Do not attempt to express ϕ as an explicit real function of ξ, η.

19.13. Show that the function

$$\omega(\zeta) = \zeta^{1/2}$$

maps the quarter plane $\xi \geq 0$, $\eta \geq 0$ into the half-plane $y > 0$. Use this result and appropriate results from the example in §19.6.3 to find the real harmonic function $\phi(\xi, \eta)$ in the quarter plane with boundary values

$$\phi(\xi, 0) = \frac{1}{1 + \xi^4} ; \quad \xi > 0 , \qquad \phi(0, \eta) = \frac{1}{1 + \eta^4} ; \quad \eta > 0 .$$

Chapter 20
Application to Elasticity Problems

20.1 Representation of Vectors

From a mathematical perspective, both two-dimensional vectors and complex numbers can be characterized as ordered pairs of real numbers. It is therefore a natural step to represent the two components of a vector function V by the real and imaginary parts of a complex function. In other words,

$$V \equiv iV_x + jV_y$$

is represented by the complex function

$$V = V_x + \imath V_y \ . \tag{20.1}$$

In the same way, the vector operator

$$\nabla \equiv i\frac{\partial}{\partial x} + j\frac{\partial}{\partial y} = \frac{\partial}{\partial x} + \imath\frac{\partial}{\partial y} = 2\frac{\partial}{\partial\bar{\zeta}} \ , \tag{20.2}$$

from equation (19.6).

If two vectors f, g are represented in the form

$$f = f_x + \imath f_y \ ; \quad g = g_x + \imath g_y$$

the product

$$\bar{f}g + f\bar{g} = (f_x - \imath f_y)(g_x + \imath g_y) + (f_x + \imath f_y)(g_x - \imath g_y) = 2\left(f_x g_x + f_y g_y\right) \ .$$

It follows that the scalar (dot) product

Supplementary Information The online version contains supplementary material available at https://doi.org/10.1007/978-3-031-15214-6_20.

$$f \cdot g = \frac{1}{2}\left(\bar{f}g + f\bar{g}\right) . \tag{20.3}$$

This result can be used to evaluate the divergence of a vector field V as

$$\text{div } V \equiv \nabla \cdot V = \frac{\partial V}{\partial \zeta} + \frac{\partial \overline{V}}{\partial \bar{\zeta}} , \tag{20.4}$$

using (20.2, 20.3).

We also record the complex-variable expression for the vector (cross) product of two in-plane vectors f, g, which is

$$f \times g = (f_x g_y - f_y g_x)k = \frac{\imath}{2}\left(f\bar{g} - \bar{f}g\right)k , \tag{20.5}$$

where k is the unit vector in the out-of-plane direction z. It then follows that

$$\text{curl } V \equiv \nabla \times V = \imath \left(\frac{\partial \overline{V}}{\partial \bar{\zeta}} - \frac{\partial V}{\partial \zeta}\right)k , \tag{20.6}$$

from (20.2, 20.5). Also, the moment of a force $F = F_x + \imath F_y$ about the origin is

$$M = r \times F = \frac{\imath}{2}\left(\bar{\zeta}F - \zeta\overline{F}\right)k , \tag{20.7}$$

where $\zeta = x + \imath y$ represents a position vector defining any point on the line of action of F.

20.1.1   Transformation of coördinates

If a vector $V = V_x + \imath V_y$ is transformed into a new coördinate system (x', y') rotated anticlockwise through an angle α with respect to x, y, the transformed components can be combined into the complex quantity

$$V_\alpha = V_x' + \imath V_y' , \tag{20.8}$$

where
$$V_x' = V_x \cos\alpha + V_y \sin\alpha ; \quad V_y' = V_y \cos\alpha - V_x \sin\alpha ,$$

from (1.8). Substituting these expressions into (20.8), we obtain

$$V_\alpha = (\cos\alpha - \imath\sin\alpha)V_x + (\sin\alpha + \imath\cos\alpha)V_y = e^{-\imath\alpha}V . \tag{20.9}$$

In describing the tractions on a surface defined by a contour S, it is sometimes helpful to define the unit vector

$$\hat{t} = \frac{\Delta s}{|\Delta s|}$$

in the direction of the local tangent, where s is a position vector defining a point on S. If the contour is traversed in the conventional anticlockwise direction, the outward normal from the enclosed area $\hat{n}$ is then defined by a unit vector rotated through $\pi/2$ clockwise from $\hat{t}$ and hence

$$\hat{n} = \hat{t} \times k \quad \text{or} \quad \hat{n} = \hat{t}e^{-\imath\pi/2} = -\imath\hat{t} ,$$

where $\hat{n}$, $\hat{t}$ are complex numbers of unit magnitude representing the unit vectors $\hat{n}$, $\hat{t}$ respectively. For the special case where s represents the circle $s = a \exp(\imath\theta)$, we have

$$\hat{t} = \frac{a\imath \exp(\imath\theta)\Delta\theta}{a\,\Delta\theta} = \imath \exp(\imath\theta) \quad \text{and} \quad \hat{n} = \exp(\imath\theta) .$$

20.2 The Antiplane Problem

To introduce the application of complex-variable methods to elasticity, we first consider the simpler antiplane problem of Chapter 15 and restrict attention to the case where there is no body force ($p_z = 0$). The only non-zero displacement component is u_z, which is constant in direction and hence a scalar function in the xy-plane. From equations (15.1, 15.5), we also deduce that u_z is a real harmonic function, and we can satisfy this condition by writing

$$2\mu u_z = h + \overline{h} , \tag{20.10}$$

where h is a holomorphic function of ζ and $\overline{h}$ is its conjugate. The non-zero stress components σ_{zx}, σ_{zy} can be combined into a vector which we denote as

$$\Psi \equiv \sigma_{zx} + \imath\sigma_{zy} = \mu\left(\frac{\partial u_z}{\partial x} + \imath\frac{\partial u_z}{\partial y}\right) = 2\mu\frac{\partial u_z}{\partial\overline{\zeta}} , \tag{20.11}$$

from (15.3, 19.6). Substituting (20.10) into (20.11) and noting that h is independent of $\overline{\zeta}$, we then obtain

$$\Psi = \overline{h'} . \tag{20.12}$$

We can write the complex function

$$h = h_x + \imath h_y ,$$

where h_x, h_y are real harmonic functions representing the real and imaginary parts of h. Equation (20.12) can then be expanded as

$$\Psi = \frac{1}{2}\left(\frac{\partial}{\partial x} + \imath\frac{\partial}{\partial y}\right)(h_x - \imath h_y) = \frac{1}{2}\left(\frac{\partial h_x}{\partial x} + \frac{\partial h_y}{\partial y} + \imath\frac{\partial h_x}{\partial y} - \imath\frac{\partial h_y}{\partial x}\right)$$

and hence

$$\sigma_{zx} = \frac{1}{2}\left(\frac{\partial h_x}{\partial x} + \frac{\partial h_y}{\partial y}\right) ; \quad \sigma_{zy} = \frac{1}{2}\left(\frac{\partial h_x}{\partial y} - \frac{\partial h_y}{\partial x}\right) .$$

However, the Cauchy-Riemann relations (19.7) show that

$$\frac{\partial h_x}{\partial x} = \frac{\partial h_y}{\partial y} ; \quad \frac{\partial h_x}{\partial y} = -\frac{\partial h_y}{\partial x}$$

so these expressions can be simplified as

$$\sigma_{zx} = \frac{\partial h_x}{\partial x} = \frac{\partial h_y}{\partial y} ; \quad \sigma_{zy} = \frac{\partial h_x}{\partial y} = -\frac{\partial h_y}{\partial x} .$$

Comparison with equations (17.4) then shows that h_y is identical with the real Prandtl stress function φ — i.e.

$$h_y \equiv \Im(h) = \frac{h - \bar{h}}{2\imath} = \varphi .$$

It follows as in equation (17.5) that the boundary traction

$$\sigma_{zn} = \frac{\partial h_y}{\partial t} , \tag{20.13}$$

where t is a real coördinate measuring distance around the boundary in the anticlockwise direction. Also, from (20.10), we have

$$h_x = \frac{h + \bar{h}}{2} = \mu u_z = \mu f(x, y) ,$$

where $f(x, y)$ is the function introduced in equation (15.1). Thus,

$$h = h_x + \imath h_y = \mu f + \imath \varphi ,$$

showing that the real functions μf and φ from Chapters 15, 17 can be combined as the real and imaginary parts of the same holomorphic function, for antiplane problems without body force.

20.2.1 Solution of antiplane boundary-value problems

If the displacement u_z is prescribed throughout the boundary, equation (20.10) reduces the problem to the search for a holomorphic function h such that $h + \bar{h}$ takes specified values on the boundary. If instead the traction σ_{zn} is specified everywhere on the boundary, we must first integrate equation (20.13) to obtain the boundary values of h_y and the problem then reduces to the search for a holomorphic function

h such that $h - \overline{h}$ takes specified values on the boundary. This is essentially the same boundary-value problem, since we can define a new holomorphic function g through the equations

$$h_y = g + \overline{g} \quad \text{with} \quad g = -\frac{\imath h}{2} . \tag{20.14}$$

Example

As an example, we consider Problem 15.3 in which a uniform antiplane stress field $\sigma_{zx} = S$ is perturbed by the presence of a traction-free hole of radius a. As in §8.3.2 and §8.4.1, we start with the unperturbed solution in which $\sigma_{zx} = S$ everywhere. From (20.11, 20.12),

$$\Psi = \overline{h}' = S ,$$

and hence a particular solution (omitting a constant of integration which corresponds merely to a rigid-body displacement) is

$$\overline{h} = S\overline{\zeta} ; \quad h = S\zeta ,$$

where we note that S is a real constant, so $\overline{S} = S$. In particular,

$$h_y = \Im(h) = Sy = Sr \sin \theta .$$

To make the boundary of the hole traction free, we require h_y to be constant around the hole and this constant can be taken to be zero without loss of generality. Thus, the corrective solution must satisfy

$$h_y(a, \theta) = -Sa \sin \theta .$$

It is possible to 'guess' the form of the solution, following arguments analogous to those used for the in-plane problem in §8.3.2, but here we shall illustrate the direct method by using equation (19.38). On the boundary $r = a$, we have

$$s = a \exp(\imath \theta) ; \quad \sin \theta = \frac{(e^{\imath \theta} - e^{-\imath \theta})}{2\imath} = \frac{1}{2\imath} \left(\frac{s}{a} - \frac{a}{s} \right) . \tag{20.15}$$

Since this is an exterior problem, we use (19.38) with (20.14, 20.15) to obtain

$$-\frac{\imath h}{2} = \frac{1}{2\pi\imath} \frac{Sa}{2\imath} \oint \left(\frac{s}{a} - \frac{a}{s} \right) \frac{ds}{(s - \zeta)} ,$$

or

$$h = \frac{S}{2\pi\imath} \oint \left(s - \frac{a^2}{s} \right) \frac{ds}{(s - \zeta)} .$$

We can use the residue theorem (19.23) to compute the contour integral. Since ζ is outside the contour, only the pole a^2/s at $s = 0$ makes a contribution and we obtain

$$\mathrm{Res}(0) = \frac{Sa^2}{2\pi\iota\zeta} \quad \text{giving} \quad h = \frac{Sa^2}{\zeta} \, .$$

The complete stress function (unperturbed + corrective solution) is then

$$h = S\zeta + \frac{Sa^2}{\zeta}$$

and the stress field is obtained from (20.12) as

$$\Psi = \overline{h'} = S - \frac{Sa^2}{\bar{\zeta}^2} \, .$$

As a check on this result, we note that on $\zeta = s$,

$$h = Ss + \frac{Sa^2}{s} = S(s + \bar{s}) \, ,$$

using $a^2 = s\bar{s}$ from (19.33). Thus, on the boundary, h is the sum of a complex quantity and its conjugate which is real, confirming that the imaginary part $h_y = 0$ as required.

20.3   In-plane Deformations

We next turn our attention to the problem of in-plane deformation, which was treated using the Airy stress function in Chapters 4–13. Our strategy will be to express the in-plane displacement vector

$$u = u_x + \iota u_y$$

in terms of complex *displacement functions*, much as the stresses in Chapter 4 were expressed in terms of the real Airy function ϕ. Also, we assume plane strain conditions, so that the out of plane displacement $u_z = 0$.

By choosing to represent the displacements rather than the stresses, we automatically satisfy the compatibility conditions, as explained in §2.2, but the stresses must then satisfy the equilibrium equations and this will place restrictions on our choice of displacement functions.

In vector notation, the equilibrium equations in terms of displacements require that

$$\nabla \mathrm{div}\, u + (1 - 2\nu)\nabla^2 u + \frac{(1 - 2\nu)p}{\mu} = 0 \, ,$$

from (2.16). Using (20.3, 20.2, 20.4, 19.9), we can express the two in-plane components of this condition in the complex-variable form

$$2\frac{\partial}{\partial\bar\zeta}\left(\frac{\partial u}{\partial\zeta}+\frac{\partial\bar u}{\partial\bar\zeta}\right)+4(1-2\nu)\frac{\partial^2 u}{\partial\zeta\partial\bar\zeta}+\frac{(1-2\nu)p}{\mu}=0\,,\qquad(20.16)$$

where the in-plane body force

$$p=p_x+\iota p_y\,.\qquad(20.17)$$

Integrating (20.16) with respect to $\bar\zeta$, we obtain

$$(3-4\nu)\frac{\partial u}{\partial\zeta}+\frac{\partial\bar u}{\partial\bar\zeta}=f-\frac{(1-2\nu)}{2\mu}\int p\,d\bar\zeta\,,\qquad(20.18)$$

where f is an arbitrary holomorphic function of ζ. This complex equation really comprises two separate equations, one for the real part and one for the imaginary part, both of which must be satisfied by u. It follows that the conjugate equation

$$(3-4\nu)\frac{\partial\bar u}{\partial\bar\zeta}+\frac{\partial u}{\partial\zeta}=\bar f-\frac{(1-2\nu)}{2\mu}\int\bar p\,d\zeta\qquad(20.19)$$

must also be satisfied. We can then eliminate $\partial\bar u/\partial\bar\zeta$ between (20.18, 20.19) to obtain

$$8(1-2\nu)(1-\nu)\frac{\partial u}{\partial\zeta}=(3-4\nu)f-\bar f+\frac{(1-2\nu)}{2\mu}\left[\int\bar p\,d\zeta-(3-4\nu)\int p\,d\bar\zeta\right]\,,$$

with solution

$$8(1-2\nu)(1-\nu)u=(3-4\nu)\int f\,d\zeta-\zeta\,\bar f+\bar g$$

$$+\frac{(1-2\nu)}{2\mu}\int\left[\int\bar p\,d\zeta-(3-4\nu)\int p\,d\bar\zeta\right]d\zeta\,,\ (20.20)$$

where $\bar g$ is an arbitrary function of $\bar\zeta$.

The functions f,g are arbitrary and, in particular, $\int f\,d\zeta$ is simply a new arbitrary function of ζ. We can therefore write a more compact form of (20.20) by defining different arbitrary functions through the relations

$$\int f\,d\zeta=\frac{4(1-\nu)(1-2\nu)\chi}{\mu}\ ;\quad\bar g=-\frac{4(1-\nu)(1-2\nu)\bar\theta}{\mu}\,,$$

where $\chi,\bar\theta$ are holomorphic functions of ζ and $\bar\zeta$ only respectively. We then have

$$\bar f=\frac{4(1-\nu)(1-2\nu)\overline{\chi'}}{\mu}\,,\qquad(20.21)$$

where $\overline{\chi'}$ represents the derivative of $\overline\chi$ (which is a function of $\bar\zeta$ only) with respect to $\bar\zeta$.

With this notation, equation (20.20) takes the form

$$2\mu u = (3 - 4\nu)\chi - \zeta\overline{\chi'} - \overline{\theta} + \frac{1}{8(1-\nu)}\int\left[\int \overline{p}d\zeta - (3 - 4\nu)\int pd\overline{\zeta}\right]d\zeta \ .$$

From this point on, we shall restrict attention to the simpler case where the body forces are zero ($p = \overline{p} = 0$), in which case[1]

$$2\mu u = (3 - 4\nu)\chi - \zeta\overline{\chi'} - \overline{\theta} \ . \tag{20.22}$$

It is a simple matter to reintroduce body force terms in the subsequent derivations if required.

20.3.1 Expressions for stresses

The tractions on the x and y-planes are

$$T_x = \sigma_{xx} + \iota\sigma_{xy} \ ; \quad T_y = \sigma_{yx} + \iota\sigma_{yy} \tag{20.23}$$

and these can be combined to form the functions

$$\Phi \equiv T_x + \iota T_y = \sigma_{xx} + 2\iota\sigma_{xy} - \sigma_{yy} \tag{20.24}$$
$$\Theta \equiv T_x - \iota T_y = \sigma_{xx} + \sigma_{yy} \ . \tag{20.25}$$

We note that Θ is a real function and is invariant with respect to coördinate transformation, whilst Φ transforms according to the rule

$$\Phi_\alpha \equiv \sigma'_{xx} + 2\iota\sigma'_{xy} - \sigma'_{yy} = e^{-2\iota\alpha}\Phi \ , \tag{20.26}$$

where Φ_α and the stress components σ' are defined in a coördinate system x', y' rotated anticlockwise through an angle α with respect to x, y. This result is easily established using the stress-transformation equations (1.9–1.11).

Substituting (20.22) into (20.4) and recalling that the state is one of plane strain so $u_z = 0$, we find

$$2\mu \, \mathrm{div} \, \boldsymbol{u} = 2(1 - 2\nu)\left(\chi' + \overline{\chi'}\right)$$

and hence, using the stress-strain relations (1.54)

---

[1] This equation can also be written in the form

$$2\mu u = \kappa\chi - \zeta\overline{\chi'} - \overline{\theta} \ ,$$

where κ is Kolosov's constant defined in equation (3.11). This has the advantage of unifying the plane strain and plane stress formulations.

$$\Theta = 2(\lambda + \mu) \, \text{div} \, \boldsymbol{u}$$
$$= 2(\chi' + \overline{\chi'}) \tag{20.27}$$

$$\Phi = 2\mu \left[\frac{\partial u_x}{\partial x} - \frac{\partial u_y}{\partial y} + \iota \left(\frac{\partial u_x}{\partial y} + \frac{\partial u_y}{\partial x} \right) \right]$$

$$= 4\mu \frac{\partial u}{\partial \bar{\zeta}} = -2(\zeta \overline{\chi''} + \overline{\theta'}) \,. \tag{20.28}$$

These expressions can also be written in terms of the single complex function

$$\psi(\zeta, \bar{\zeta}) = \chi + \zeta \overline{\chi'} + \bar{\theta} \,. \tag{20.29}$$

We then have

$$\frac{\partial \psi}{\partial \zeta} = \chi' + \overline{\chi'} \,; \quad \frac{\partial \psi}{\partial \bar{\zeta}} = \zeta \overline{\chi''} + \overline{\theta'}$$

and hence

$$\Theta = 2 \frac{\partial \psi}{\partial \zeta} \,; \quad \Phi = -2 \frac{\partial \psi}{\partial \bar{\zeta}} \,. \tag{20.30}$$

Notice however that ψ depends on both ζ and $\bar{\zeta}$ and is therefore not holomorphic.

20.3.2 Rigid-body displacement

Equations (20.30) show that all the stress components will be zero everywhere if and only if $\psi(\zeta, \bar{\zeta})$ is a constant. Expanding the holomorphic functions χ, θ as Taylor series

$$\chi = A_0 + A_1\zeta + A_2\zeta^2 + \dots \,; \quad \theta = B_0 + B_1\zeta + B_2\zeta^2 + \dots$$

and substituting into (20.29), we obtain

$$\psi = A_0 + \bar{B}_0 + (A_1 + \bar{A}_1)\zeta + B_1\bar{\zeta} + A_2\zeta^2 + 2\bar{A}_2\bar{\zeta}\zeta + \bar{B}_2\bar{\zeta}^2 + \dots \,.$$

This will be constant if and only if $(A_1 + \bar{A}_1) = 0$ and all the other coefficients are zero except $A_0, \bar{B}_0$. The condition $(A_1 + \bar{A}_1) = 0$ implies that A_1 is pure imaginary, which can be enforced by writing $A_1 = \iota C_1$, where C_1 is a real constant. Substituting these values into (20.22), we then obtain

$$2\mu u = (3 - 4\nu)A_0 - \bar{B}_0 + 4(1 - \nu)\iota C_1\zeta \,. \tag{20.31}$$

The first two terms correspond to an arbitrary rigid-body translation and the third to a rigid-body rotation. In fact, either of the complex constants $A_0, \bar{B}_0$ has sufficient degrees of freedom to define an arbitrary rigid-body translation, so one of them can be set to zero without loss of generality. In the following derivations, we shall therefore generally set $B_0 = \bar{B}_0 = 0$, implying that $\theta(0) = 0$.

20.4   Relation between the Airy Stress Function and the Complex Potentials

Before considering methods for solving boundary-value problems in complex-variable notation, it is of interest to establish some relationships with the Airy stress function formulation of Chapters 4–13. In particular, we shall show that it is always possible to find complex potentials χ, θ corresponding to a given Airy function ϕ. This has the incidental advantage that the resulting complex potentials can then be used to determine the displacement components, so this procedure provides a method for generating the displacements due to a given Airy stress function, which, as we saw in Chapter 9, is not always straightforward in the real stress function formulation.

We start by using the procedure of §19.4.1 to express the real biharmonic function ϕ in terms of two holomorphic functions g_1, g_2, as in equation (19.15).

The stress components are given in terms of ϕ as

$$\sigma_{xx} = \frac{\partial^2 \phi}{\partial y^2} \; ; \quad \sigma_{xy} = -\frac{\partial^2 \phi}{\partial x \partial y} \; ; \quad \sigma_{yy} = \frac{\partial^2 \phi}{\partial x^2} \, ,$$

from equations (4.6). It follows that

$$\Theta = \nabla^2 \phi = 4 \frac{\partial^2 \phi}{\partial \zeta \partial \bar{\zeta}} \tag{20.32}$$

and

$$\Phi = \frac{\partial^2 \phi}{\partial y^2} - 2\iota \frac{\partial^2 \phi}{\partial x \partial y} - \frac{\partial^2 \phi}{\partial x^2} = -\left(\frac{\partial}{\partial x} + \iota \frac{\partial}{\partial y} \right)^2 = -4 \frac{\partial^2 \phi}{\partial \bar{\zeta}^2} \, , \tag{20.33}$$

using (19.9, 19.6).

Substituting for ϕ from (19.15) into (20.32), we then have

$$\Theta = \nabla^2 \phi = 4 \left(g_2' + \overline{g_2'} \right)$$

and comparison with (20.27) shows that

$$\chi = 2g_2 \, . \tag{20.34}$$

Similarly, from (19.15, 20.33) we have

$$\Phi = -4 \left(\overline{g_1''} + \zeta \overline{g_2''} \right) = -2 \left(\zeta \overline{\chi''} + 2\overline{g_1''} \right) \, ,$$

using (20.34). Comparison with (20.28) then shows that

$$\theta = 2g_1' . \tag{20.35}$$

Example

As an example, we consider the problem of §5.2.1 in which the rectangular beam $x > 0$, $-b < y < b$ is loaded by a shear force F at the end $x = 0$. The real stress function for this case is given by (5.21) as

$$\phi = \frac{F(xy^3 - 3b^2xy)}{4b^3} .$$

Using (19.2), we have

$$xy = -\frac{\imath}{4} \left(\zeta^2 - \bar{\zeta}^2\right) \quad \text{and} \quad xy^3 = \frac{\imath}{16} \left(\zeta^2 - \bar{\zeta}^2\right)\left(\zeta - \bar{\zeta}\right)^2 ,$$

from which

$$\phi = \frac{\imath F}{64b^3} \left(\zeta^4 - 2\zeta^3\bar{\zeta} + 2\zeta\bar{\zeta}^3 - \bar{\zeta}^4 + 12b^2\zeta^2 - 12b^2\bar{\zeta}^2\right)$$

Comparison with (19.15) shows that

$$g_1 = \frac{\imath F}{64b^3} \left(\zeta^4 + 12b^2\zeta^2\right) ; \quad g_2 = -\frac{\imath F\zeta^3}{32b^3} .$$

We then have

$$\chi = -\frac{\imath F\zeta^3}{16b^3} ; \quad \theta = \frac{\imath F}{8b^3} \left(\zeta^3 + 6b^2\zeta\right) , \tag{20.36}$$

from (20.34, 20.35), and hence

$$2\mu u = 2\mu(u_x + \imath u_y) = (3 - 4\nu)\chi - \zeta\overline{\chi'} - \overline{\theta}$$

$$= \frac{\imath F}{16b^3} \left(-(3 - 4\nu)\zeta^3 - 3\zeta\bar{\zeta}^2 + 2\bar{\zeta}^3 + 12b^2\bar{\zeta}\right)$$

$$= \frac{F}{4b^3} \left(\left[3(1-\nu)x^2y - (2-\nu)y^3 + 3b^2y\right] + \imath \left[3b^2x - 3\nu xy^2 - (1-\nu)x^3\right]\right) ,$$

agreeing with the plane-strain equivalents of (9.8, 9.9) apart from an arbitrary rigid-body displacement.

20.5   Boundary Tractions

We denote the boundary traction (i.e. the force per unit length along the boundary
S) as $T(s)$, where s is a real coördinate defining position around S. The traction is a
vector which can be expressed in the usual complex form as

$$T(s) \equiv i T_x + j T_y = T_x + \imath T_y \equiv T .$$

Following the conventions of Chapter 19, the coördinate s increases as we traverse
the boundary in the anticlockwise direction as shown in Figure 20.1(a), which also
shows the traction components T_x, T_y. Figure 20.1(b) shows the forces acting on

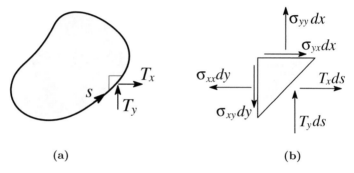

(a)   (b)

Fig. 20.1   (a) The boundary tractions T_x, T_y and (b) equilibrium of a small element at the boundary.

a small triangular element chosen such that x, y, s all increase as we move up the
inclined edge[2]. We recall the convention that the material lies always on the left of
the boundary as s increases.

Equilibrium of this element then requires that

$$T_x ds = \sigma_{xx} dy - \sigma_{yx} dx ; \quad T_y ds = \sigma_{xy} dy - \sigma_{yy} dx$$

and hence

$$T ds = (\sigma_{xx} + \imath \sigma_{xy}) dy - (\sigma_{yx} + \imath \sigma_{yy}) dx = \tau_x dy - \tau_y dx ,$$

using the notation of equation (20.23).

We can solve (20.24, 20.25) for τ_x, τ_y, obtaining

$$\tau_x = \frac{1}{2}(\Theta + \Phi) ; \quad \tau_y = \frac{\imath}{2}(\Theta - \Phi)$$

---

[2] If any other orientation were chosen, one or more of the increments dx, dy, ds would be negative
resulting in a change in direction of the corresponding forces, but the same final result would be
obtained.

and hence

$$T ds = \frac{1}{2}\left[\Theta(dy - \iota dx) + \Phi(dy + \iota dx)\right] = \frac{\iota}{2}\left(\Phi d\bar{\zeta} - \Theta d\zeta\right) . \qquad (20.37)$$

Using (20.30), we then have

$$T ds = \iota\left(-\frac{\partial\psi}{\partial\bar{\zeta}}d\bar{\zeta} - \frac{\partial\psi}{\partial\zeta}d\zeta\right) = -\iota d\psi . \qquad (20.38)$$

In other words,

$$T = -\iota\frac{d\psi}{ds} = -\iota\frac{d}{ds}\left(\chi + \zeta\overline{\chi'} + \bar{\theta}\right) \quad \text{or} \quad \frac{d\psi}{ds} = \iota T . \qquad (20.39)$$

Thus, if the tractions are known functions of s, equation (20.39) can be integrated to yield the values of ψ at all points on the boundary. We also note that ψ must be constant in any part of the boundary that is traction free.

Example

To illustrate these results, we revisit the example in §20.4 in which the rectangular bar $-b < y < b$, $x > 0$ is loaded by a transverse force F on the end $x = 0$, the edges $y = \pm b$ being traction free. The complex potentials are given in (20.36). Substituting them into (20.29), we find

$$\psi = -\frac{\iota F}{16b^3}\left\{(\zeta - \bar{\zeta})^3 + 3\bar{\zeta}\left[(\zeta - \bar{\zeta})^2 + 4b^2\right]\right\} . \qquad (20.40)$$

In complex coördinates, the boundaries $y = \pm b$ correspond to the lines

$$\zeta - \bar{\zeta} = \pm 2\iota b ,$$

from (19.2) and it follows that the second term in the braces in (20.40) is zero on each boundary, whilst the first term takes the value

$$\mp 8\iota b^3 , \quad \text{corresponding to} \quad \psi = \mp\frac{F}{2} .$$

This confirms that ψ is constant along each of the two traction-free boundaries. We also note that as we go down the left end of the bar (thereby preserving the anticlockwise direction for s), ψ increases by F, so

$$F = \int_{\zeta=\iota b}^{\zeta=-\iota b}\frac{d\psi}{ds}ds = \int_{\zeta=\iota b}^{\zeta=-\iota b}\iota T ds ,$$

using (20.39). It follows that the resultant force on the end is

$$\int_{\zeta=\iota b}^{\zeta=-\iota b} T ds = -\iota F ,$$

which corresponds to a force F in the negative y-direction, which of course agrees with Figure 5.2 and equation (5.14).

20.5.1 Equilibrium considerations

The in-plane problem is well posed if and only if the applied loads are self-equilibrated and since we are assuming that there are no body forces, this implies that the resultant force and moment of the boundary tractions must be zero.

The force resultant can be found by integrating the complex traction T around the boundary S, using (20.39). We obtain

$$F_x + \iota F_y = \oint_S T ds = -\iota \oint_S \frac{d\psi}{ds} ds = 0 , \qquad (20.41)$$

which implies that ψ must be single-valued. This condition applies regardless of the shape of the boundary, assuming the body is simply connected.

The contribution to the resultant moment from the traction $T ds$ on an element of boundary ds can be written

$$dM = \frac{\iota}{2} \left(s \overline{T ds} - \overline{s} T ds \right) = -\frac{1}{2} \left(s d\overline{\psi} + \overline{s} d\psi \right) = -\Re \left(\overline{s} d\psi \right) ,$$

using (20.7) and (20.38). The resultant moment can then be written

$$M = \oint_S dM = -\Re \oint_S \overline{s} \frac{\partial \psi}{\partial s} d\overline{s} \qquad (20.42)$$

and integrating by parts, we have

$$\oint_S \overline{s} \frac{\partial \psi}{\partial s} d\overline{s} = \overline{s} \psi |_s - \oint_S \psi d\overline{s} = - \oint_S \psi d\overline{s} ,$$

since $\overline{s} \psi$ is single-valued in view of (20.41). Thus, the condition that the resultant moment be zero reduces to

$$M = \Re \left(\oint_S \psi d\overline{s} \right) = 0 . \qquad (20.43)$$

20.6   Boundary-value Problems

Two quantities must be specified at every point on the boundary S. In the *displacement boundary-value problem*, both components of displacement are prescribed and hence we know the value of the complex function

$$2\mu u(s) = \kappa\chi(s) - s\,\overline{\chi'(s)} - \overline{\theta(s)} \; ; \quad s \in S \, ,$$

where we recall that $\kappa = (3 - 4\nu)$. In the *traction boundary-value problem*, both components of traction $T(s)$ are prescribed and we can integrate equation (20.39) to obtain

$$\psi(s) = \iota \int T(s)ds + C \, , \tag{20.44}$$

where C is an arbitrary constant of integration. The complex potentials must then be chosen to satisfy the boundary condition

$$\chi(s) + s\,\overline{\chi'(s)} + \overline{\theta(s)} = \psi(s) \; ; \quad s \in S \, .$$

Notice that the variable of integration s in (20.44) is a real curvilinear cöordinate measuring the distance traversed around the boundary S, but since we shall later want to apply the Cauchy integral theorem, it is necessary to express the resulting function ψ as a function of the complex cöordinate s defining a general point on S.

   The displacement and traction problems both have the same form and can be combined as

$$\gamma\chi(s) + s\,\overline{\chi'(s)} + \overline{\theta(s)} = f(s) \; ; \quad s \in S \, , \tag{20.45}$$

where $\gamma = -\kappa$, $f(s) = -2\mu u(s)$ for the displacement problem and $\gamma = 1$, $f(s) = \psi(s)$ for the traction problem.

   One significant difference between the displacement and traction problems is that the former is completely defined for all continuous values of the boundary data, whereas the solution of traction problem is indeterminate to within an arbitrary rigid-body displacement as defined in §20.3.2. Also, the permissible traction distributions are restricted by the equilibrium conditions (20.41, 20.42). Notice incidentally that the constant C in (20.44) can be wrapped into the arbitrary rigid-body translation A_0 in (20.31), since a particular solution of the problem $\psi(\zeta, \bar{\zeta}) = C$ on S is $\psi = C$ (constant) throughout Ω.

20.6.1   Solution of the interior problem for the circle

As in §19.6.1, we can use the Cauchy integral theorem to obtain the solution of
the in-plane problem for the region Ω bounded by the circle $r=a$ with prescribed
boundary tractions or displacements.

Applying the Cauchy integral operator to both sides of equation (20.45), we obtain

$$\frac{\gamma}{2\pi\iota}\oint\frac{\chi(s)ds}{(s-\zeta)}+\frac{1}{2\pi\iota}\oint\frac{s\,\overline{\chi'(s)}ds}{(s-\zeta)}+\frac{1}{2\pi\iota}\oint\frac{\overline{\theta(s)}ds}{(s-\zeta)}=\frac{1}{2\pi\iota}\oint\frac{f(s)ds}{(s-\zeta)}.$$

$$(20.46)$$

Since χ is a holomorphic function of ζ in Ω, equation (19.26) gives

$$\frac{1}{2\pi\iota}\oint\frac{\chi(s)ds}{(s-\zeta)}=\chi(\zeta)\,.$$

Also, writing

$$\chi=\sum_{n=0}^{\infty}A_n\zeta^n\quad;\quad\overline{\chi'}=\sum_{n=1}^{\infty}n\bar{A}_n\bar{\zeta}^{n-1}\,,$$

$$(20.47)$$

and noting that $\bar{s}=a^2/s$ on the boundary S, we have

$$s\,\overline{\chi'(s)}=\sum_{n=1}^{\infty}\frac{n\bar{A}_n a^{2n-2}}{s^{n-2}}\,.$$

Using the residue theorem, the second term in (20.46) can then be evaluated as

$$\frac{1}{2\pi\iota}\oint\frac{s\,\overline{\chi'(s)}ds}{(s-\zeta)}=\sum_{n=1}^{\infty}\frac{n\bar{A}_n a^{2n-2}}{2\pi\iota}\oint\frac{ds}{s^{n-2}(s-\zeta)}=\bar{A}_1\zeta+2\bar{A}_2 a^2\,,$$

since all the remaining terms in the series integrate to zero as in (19.24).

A similar argument can be applied to the third integral in (20.46). Expanding

$$\theta(\zeta)=\sum_{k=0}^{\infty}B_k\zeta^k\,,$$

we obtain

$$\frac{1}{2\pi\iota}\oint\frac{\overline{\theta(s)}ds}{(s-\zeta)}=\bar{B}_0\,,$$

and we note from §20.3.2 that this can be set to zero without loss of generality.
Assembling these results, we conclude that

$$\gamma\chi(\zeta)=\frac{1}{2\pi\iota}\oint\frac{f(s)ds}{(s-\zeta)}-\bar{A}_1\zeta-2\bar{A}_2 a^2\,.$$

$$(20.48)$$

To determine the constants $\bar{A}_1$, $\bar{A}_2$, we first replace χ in (20.48) by its Taylor series (20.47) giving

$$\frac{1}{2\pi\iota} \oint \frac{f(s)ds}{(s-\zeta)} = \bar{A}_1\zeta + 2\bar{A}_2 a^2 + \gamma A_0 + \gamma A_1\zeta + \gamma A_2\zeta^2 + \dots .$$

Differentiating twice with respect to ζ, we obtain

$$-\frac{1}{2\pi\iota} \oint \frac{f(s)ds}{(s-\zeta)^2} = \bar{A}_1 + \gamma A_1 + 2\gamma A_2\zeta + \dots$$

$$\frac{1}{\pi\iota} \oint \frac{f(s)ds}{(s-\zeta)^3} = 2\gamma A_2 + 6\gamma A_3\zeta + \dots$$

and evaluating these last two expressions at $\zeta = 0$,

$$\gamma A_2 = \frac{1}{2\pi\iota} \oint \frac{f(s)ds}{s^3} \; ; \quad \bar{A}_1 + \gamma A_1 = -\frac{1}{2\pi\iota} \oint \frac{f(s)ds}{s^2} \; . \tag{20.49}$$

Equations (20.49) are sufficient to determine the required constants for the displacement problem ($\gamma = \kappa$), but a degeneracy occurs for the traction problem where $\gamma = 1$ and the second equation reduces to

$$\bar{A}_1 + A_1 = -\frac{1}{2\pi\iota} \oint \frac{\psi(s)ds}{s^2} \; . \tag{20.50}$$

This implies that the right-hand-side must evaluate to a real constant, which (i) places a restriction on the permissible values of $\psi(s)$ and (ii) leaves the imaginary part of A_1 indeterminate. We saw already in §20.3.2 that the imaginary part of A_1 corresponds to an arbitrary rigid-body rotation. The condition that the right-hand side of (20.50) be real is exactly equivalent to the condition (20.43) enforcing moment equilibrium. To establish the connection, we note that for the circular boundary

$$\bar{s} = \frac{a^2}{s} \quad \text{so that} \quad d\bar{s} = -\frac{a^2 ds}{s^2} \; .$$

Using this result in (20.43), we obtain

$$\Re\left(\oint_S \psi d\bar{s}\right) = -\Re\left(a^2 \oint_S \frac{\psi(s)ds}{s^2}\right) = 0 \; ,$$

showing that the integral is pure imaginary and hence that the right-hand side of (20.50) is real.

Once χ is determined, we can apply a similar method to determine θ. We first write the conjugate of equation (20.45) as

$$\theta(s) = \overline{f(s)} - \gamma\overline{\chi(s)} - \bar{s}\chi'(s) \; ; \quad s \in S \; , \tag{20.51}$$

and apply the Cauchy integral theorem, obtaining

$$\theta(\zeta) = \frac{1}{2\pi\iota} \oint \frac{\theta(s)ds}{(s-\zeta)} = \frac{1}{2\pi\iota} \oint \frac{\overline{f(s)}ds}{(s-\zeta)} - \frac{\gamma}{2\pi\iota} \oint \frac{\overline{\chi(s)}ds}{(s-\zeta)} - \frac{1}{2\pi\iota} \oint \frac{\overline{s}\chi'(s)ds}{(s-\zeta)} .$$

The last two integrals on the right-hand side can be evaluated using (20.47) and the residue theorem leading to

$$\theta(\zeta) = \frac{1}{2\pi\iota} \oint \frac{\overline{f(s)}ds}{(s-\zeta)} - \gamma\bar{A}_0 - \frac{a^2 \left(\chi'(\zeta) - \chi'(0)\right)}{\zeta} .$$

Once this expression has been evaluated, the remaining constant $\bar{A}_0$ can be chosen to satisfy the condition $\theta(0)=0$ as discussed in §20.31. However, there is no essential need to do this, since this constant corresponds merely to a rigid-body displacement and makes no contribution to the stress components.

20.6.2 Solution of the exterior problem for the circle

For the exterior problem, we again apply the Cauchy integral operator, leading to (20.46), which we repeat here for clarity

$$\frac{\gamma}{2\pi\iota} \oint \frac{\chi(s)ds}{(s-\zeta)} + \frac{1}{2\pi\iota} \oint \frac{s\,\overline{\chi'(s)}ds}{(s-\zeta)} + \frac{1}{2\pi\iota} \oint \frac{\overline{\theta(s)}ds}{(s-\zeta)} = \frac{1}{2\pi\iota} \oint \frac{f(s)ds}{(s-\zeta)} .$$
$$(20.52)$$

In this equation, the first term

$$\frac{\gamma}{2\pi\iota} \oint \frac{\chi(s)ds}{(s-\zeta)} = -\gamma\chi(\zeta) ,$$

from (19.30). For the third term, we note that $\theta(\zeta)$ is holomorphic in the exterior region and hence permits a Laurent expansion

$$\theta(\zeta) = \frac{B_1}{\zeta} + \frac{B_2}{\zeta^2} + \dots$$

so

$$\overline{\theta(s)} = \frac{\bar{B}_1}{\bar{s}} + \frac{\bar{B}_2}{\bar{s}^2} + \dots = \frac{\bar{B}_1 s}{a^2} + \frac{\bar{B}_2 s^2}{a^4} + \dots .$$

This can be regarded as the surface value of a function that is holomorphic in the interior region and hence, since ζ is outside the contour S,

$$\frac{1}{2\pi\iota} \oint \frac{\overline{\theta(s)}ds}{(s-\zeta)} = 0 .$$
$$(20.53)$$

For the second term in (20.52), we write

$$\chi(\zeta) = \frac{A_1}{\zeta} + \frac{A_2}{\zeta^2} + \frac{A_3}{\zeta^3} + \dots \tag{20.54}$$

so

$$\overline{\chi'} = -\frac{\bar{A}_1}{\bar{\zeta}^2} - \frac{2\bar{A}_2}{\bar{\zeta}^3} - \frac{3\bar{A}_3}{\bar{\zeta}^4} + \dots ; \quad s\,\overline{\chi'(s)} = -\frac{\bar{A}_1 s^3}{a^4} - \frac{2\bar{A}_2 s^4}{a^6} - \frac{3\bar{A}_3 s^5}{a^8} + \dots$$

and as before, we conclude that

$$\frac{1}{2\pi\iota} \oint \frac{s\overline{\chi'(s)}ds}{(s-\zeta)} = 0 ,$$

so

$$\gamma\chi(\zeta) = -\frac{1}{2\pi\iota} \oint \frac{f(s)ds}{(s-\zeta)} . \tag{20.55}$$

To determine θ, the conjugate equation (20.51) and (19.30) lead to

$$\theta(\zeta)=-\frac{1}{2\pi\iota} \oint \frac{\theta(s)ds}{(s-\zeta)}=-\frac{1}{2\pi\iota} \oint \frac{\overline{f(s)}ds}{(s-\zeta)}+\frac{\gamma}{2\pi\iota} \oint \frac{\overline{\chi(s)}ds}{(s-\zeta)}+\frac{1}{2\pi\iota} \oint \frac{s\overline{\chi'(s)}ds}{(s-\zeta)} ,$$

and the second integral on the right-hand side is clearly of the same form as (20.53) and is therefore zero. For the third integral, we have

$$\frac{1}{2\pi\iota} \oint \frac{s\overline{\chi'(s)}ds}{(s-\zeta)} = \frac{1}{2\pi\iota} \oint \frac{a^2\chi'(s)ds}{s(s-\zeta)} = -\frac{a^2\chi'(\zeta)}{\zeta} ,$$

using (19.30). We conclude that

$$\theta(\zeta) = -\frac{1}{2\pi\iota} \oint \frac{\overline{f(s)}ds}{(s-\zeta)} - \frac{a^2\chi'(\zeta)}{\zeta} . \tag{20.56}$$

Example

To illustrate the procedure, we consider the problem shown in Figure 20.2 in which a circular hole of radius a is loaded by equal and opposite compressive forces F on the diameter.

In this case, we have

$$T = \frac{F\delta(\theta)}{a} - \frac{F\delta(\theta-\pi)}{a}$$

and a suitable integral of (20.39) yields

$$\psi(s) = -\frac{\iota F}{2}\,\mathrm{sgn}(\theta) \quad \text{so} \quad \overline{\psi(s)} = \frac{\iota F}{2}\,\mathrm{sgn}(\theta) , \tag{20.57}$$

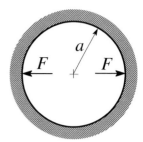

Fig. 20.2 Hole in a large body loaded by equal and opposite forces F.

where $\text{sgn}(\theta) = +1$ in $0 < \theta < \pi$ and -1 in $-\pi < \theta < 0$. Notice that in performing this integral, it is necessary to traverse the contour in the direction that keeps the material of the body on the left, as explained in §20.5. Substituting in equation (20.55) with $\gamma = 1$, $f(s) = \psi(s)$, we obtain

$$\chi(\zeta) = -\frac{1}{2\pi\iota} \oint \frac{\psi(s)ds}{(s-\zeta)} = -\frac{F}{4\pi} \left(\ln(s-\zeta)\Big|_{s=-a}^{s=a} - \ln(s-\zeta)\Big|_{s=a}^{s=-a} \right)$$

$$= \frac{F}{2\pi} \ln\left(\frac{\zeta+a}{\zeta-a}\right) ,$$

from which

$$\chi'(\zeta) = -\frac{Fa}{\pi(\zeta^2 - a^2)} .$$

The integral term in (20.56) is evaluated in the same way as

$$\frac{1}{2\pi\iota} \oint \frac{\overline{\psi(s)}ds}{(s-\zeta)} = \frac{F}{2\pi} \ln\left(\frac{\zeta+a}{\zeta-a}\right)$$

and it follows that

$$\theta(\zeta) = -\frac{F}{2\pi} \ln\left(\frac{\zeta+a}{\zeta-a}\right) + \frac{Fa^3}{\pi\zeta(\zeta^2-a^2)} .$$

Finally, the complex stresses can be recovered by substitution into (20.27, 20.28), giving

$$\Theta = 2\left(\chi' + \overline{\chi'}\right) = -\frac{Fa}{\pi}\left(\frac{1}{(\zeta^2-a^2)} + \frac{1}{(\bar{\zeta}^2-a^2)}\right)$$

$$\Phi = -2\left(\zeta \overline{\chi''} + \overline{\theta'}\right) = \frac{2Fa}{\pi}\left(\frac{2(a^2-\zeta\bar{\zeta})}{(\bar{\zeta}^2-a^2)^2} - \frac{1}{\bar{\zeta}^2}\right) .$$

This problem could have been solved using the real stress function approach of Chapter 13 (for example by superposing the solution of Problem 13.2 for two forces at $\theta = 0$, π), but the present method is considerably more direct.

20.7   Conformal Mapping for In-plane Problems

The technique of conformal mapping introduced in §19.7 can be applied to in-plane problems, but requires some modification because of the presence of the derivative term χ' in the boundary condition (20.45). Suppose that the problem is defined in the domain Ω^* and that we can identify a mapping function $\omega(\zeta)$ that maps each point $\omega \in \Omega^*$ into a point $\zeta \in \Omega$, where Ω represents the domain either interior or exterior to the unit circle $|\zeta| = 1$.

The boundary condition (20.45) requires that

$$\gamma \chi(\omega) + \omega \,\overline{\chi'(\omega)} + \overline{\theta(\omega)} = f(\omega) ; \quad \omega \in S^* , \tag{20.58}$$

where S^* is the boundary of Ω^*. If we substitute $\omega = \omega(\zeta)$, the functions χ, θ will become functions of ζ, but to avoid confusion, we shall adopt the notation

$$\chi_\zeta(\zeta) = \chi(\omega(\zeta)) ; \quad \theta_\zeta(\zeta) = \theta(\omega(\zeta)) ; \quad f_\zeta(\zeta) = f(\omega(\zeta))$$

for the corresponding functions in the ζ-plane. We then have

$$\chi'(\omega) \equiv \frac{d\chi}{d\omega} = \frac{d\chi_\zeta}{d\zeta}\frac{d\zeta}{d\omega} = \frac{\chi'_\zeta}{\omega'} .$$

Using the conjugate of this result in (20.58), we obtain

$$\gamma \chi_\zeta(s) + \frac{\omega(s)}{\overline{\omega'(s)}}\overline{\chi'_\zeta(s)} + \overline{\theta_\zeta(s)} = f_\zeta(s) ; \quad s \in S , \tag{20.59}$$

where S is the unit circle. This defines the boundary-value problem in the ζ-plane.

The mathematical arguments used in §§20.6.1, 20.6.2 can be applied to this equation and will yield identical results except in regard to the second term. To fix ideas, suppose Ω^* is the region exterior to a non-circular hole and $\omega(\zeta)$ maps into the region exterior to the unit circle S. Then, the second term in equation (20.52) must be replaced by

$$\frac{1}{2\pi i} \oint_S \frac{\omega(s)\overline{\chi'_\zeta(s)}ds}{\overline{\omega'(s)}(s - \zeta)} \tag{20.60}$$

and after evaluating the remaining terms as in §20.6.2, we obtain

$$\gamma \chi_\zeta(\zeta) = -\frac{1}{2\pi i} \oint_S \frac{f_\zeta(s)ds}{(s - \zeta)} + \frac{1}{2\pi i} \oint_S \frac{\omega(s)\overline{\chi'_\zeta(s)}ds}{\overline{\omega'(s)}(s - \zeta)} . \tag{20.61}$$

The integral (20.60) will be zero if the integrand has no poles inside the contour and can be evaluated using the residue theorem if it has a finite number of poles. As in §20.6.2, we can represent χ_ζ by its Laurent series (20.54) and hence write

$$\overline{\chi_\zeta'(s)} = -\bar{A}_1 s^2 - 2\bar{A}_2 s^3 - 3\bar{A}_3 s^4 + ..., \tag{20.62}$$

since on the unit circle $s\bar{s}=1$. Equation (20.62) can be regarded as the surface values of a holomorphic function that has no poles within the circle, but the mapping function ω must be holomorphic in the region Ω *exterior* to the circle S, and this generally implies that it will have poles in the interior region.

Since both Ω, Ω^* extend to infinity, it is reasonable to choose a mapping function $\omega(\zeta)$ such that concentric circles in the ζ-plane map into contours that become more and more circular as we go further away from the hole. This is achieved by the function

$$\omega(\zeta) = c\zeta + \frac{B_1}{\zeta} + \frac{B_2}{\zeta^2} + \frac{B_3}{\zeta^3} + ... \, , \tag{20.63}$$

since at large ζ, only the first term remains, which represents simply a linear scaling. The series in (20.63) may be infinite or finite, but in either case $\omega(\zeta)$ clearly posesses poles at the origin. If the series is finite, the integrand in (20.60) will have at most a finite number of poles and the integral can be evaluated using the residue theorem. It follows from (19.22) that the resulting expression will contain unknowns representing the derivatives of $\overline{\chi_\zeta}$ evaluated at the origin, which in view of (20.62) comprises a finite number of the unknown coefficients $\bar{A}_1$, $\bar{A}_2$, ... However, a sufficient number of linear equations for these coefficients can be obtained by constructing the conjugate $\overline{\chi_\zeta}$ from (20.61) and equating it or its derivatives to the corresponding values at the origin. This is of course precisely the same procedure that was used to determine the constants A_1, A_2 in §20.6.1.

This procedure can be generalized to allow ω to be a rational function

$$\omega(\zeta) = \frac{C_1\zeta + C_2\zeta^2 + ... + C_n\zeta^n}{D_1\zeta + D_2\zeta^2 + ... + D_m\zeta^m} \, ,$$

which will posess simple poles at each of the m zeros of the finite polynomial representing the denominator. In this case, the unknowns appearing in the integral (20.60) will be the values of $\overline{\chi_\zeta}$ at each of these poles and a set of equations for determining them is obtained by evaluating (20.61) at each such pole[3].

---

[3] For more details of this procedure, the reader is referred to A. H. England, *loc.cit*, §5.3, I. S. Sokolnikoff, *loc.cit*, §84.

Once χ_ζ has been determined, θ_ζ can be obtained exactly as in §20.6.2. We take the conjugate of (20.59) and apply the Cauchy integral operator obtaining

$$\frac{\gamma}{2\pi\iota}\oint\frac{\overline{\chi_\zeta(s)}ds}{(s-\zeta)}+\frac{1}{2\pi\iota}\oint\frac{\overline{w(s)}\chi'_\zeta(s)ds}{w'(s)(s-\zeta)}+\frac{1}{2\pi\iota}\oint\frac{\theta_\zeta(s)ds}{(s-\zeta)}=\frac{1}{2\pi\iota}\oint\frac{\overline{f_\zeta(s)}ds}{(s-\zeta)}$$

The first integral is zero, as in §20.6.2 and applying (19.30) to the third term, we obtain

$$\theta_\zeta(\zeta)=-\frac{1}{2\pi\iota}\oint\frac{\overline{f_\zeta(s)}ds}{(s-\zeta)}+\frac{1}{2\pi\iota}\oint\frac{\overline{w(s)}\chi'_\zeta(s)ds}{w'(s)(s-\zeta)}. \tag{20.64}$$

At this stage, all the quantities on the right-hand side of (20.64) are known and the integrals can generally be evaluated using the residue theorem.

20.7.1 The elliptical hole

A particularly simple case concerns the elliptical hole for which

$$w=c\left(\zeta+\frac{m}{\zeta}\right), \tag{20.65}$$

from (19.40), where c, m are real constants. We then have

$$\overline{w'(s)}=c\left(1-\frac{m}{s^2}\right)=c\left(1-ms^2\right)$$

and

$$\frac{\overline{w(s)}}{w'(s)}=\frac{1}{s}\left(\frac{s^2+m}{1-ms^2}\right).$$

This expression has a simple pole at the origin, but when introduced into (20.60) this is cancelled by the factor s^2 in (20.62). The simple poles at $s=\pm\sqrt{1/m}$ lie outside S, since $m<1$. It follows that the integral (20.60) is zero and

$$\gamma\chi_\zeta(\zeta)=-\frac{1}{2\pi\iota}\oint_S\frac{f_\zeta(s)ds}{(s-\zeta)}, \tag{20.66}$$

from (20.61). Using (20.65), in (20.64), we obtain

$$\theta_\zeta(\zeta)=-\frac{1}{2\pi\iota}\oint\frac{\overline{f_\zeta(s)}ds}{(s-\zeta)}+\frac{1}{2\pi\iota}\oint\frac{s(1+ms^2)\chi'_\zeta(s)ds}{(s^2-m)(s-\zeta)}$$

and since the integrand in the second term has no poles in the region exterior to the unit circle (except for the Cauchy term $(s - \zeta)$), we can evaluate it using the Cauchy integral theorem (19.30), obtaining

$$\theta_\zeta(\zeta) = -\frac{1}{2\pi\iota} \oint \frac{\overline{f_\zeta(s)}ds}{(s - \zeta)} - \frac{\zeta(1 + m\zeta^2)\chi'_\zeta(\zeta)}{(\zeta^2 - m)} . \tag{20.67}$$

Example: The plane crack opened in tension

The ellipse corresponding to the mapping function (20.65) has semi-axes

$$a = c(1 + m) ; \quad b = c(1 - m) ,$$

from equation (19.42), so with $m = 1$, we have $a = 2c$, $b = 0$. This defines an ellipse with zero minor axis, which therefore degenerates to a plane crack occupying the line $-a < \xi < a$, $\eta = 0$. Equation (20.65) then takes the special form

$$\omega = \frac{a}{2}\left(\zeta + \frac{1}{\zeta}\right) ; \quad \overline{\omega} = \frac{a}{2}\left(\bar{\zeta} + \frac{1}{\bar{\zeta}}\right) . \tag{20.68}$$

We shall use this and the results of §20.7.1 to determine the complete stress field in a cracked body opened by a far-field uniaxial tensile stress $\sigma_{\eta\eta} = S$. This problem was solved using a distribution of dislocations in §13.3.2.

As usual, we start by considering the unperturbed solution, which is a state of uniform uniaxial tension $\sigma_{\xi\xi} = \sigma_{\xi\eta} = 0$, $\sigma_{\eta\eta} = S$ and hence

$$\Theta = S ; \quad \Phi = -S ,$$

from (20.25, 20.24) respectively. It is clear from equations (20.27, 20.28) that a particular solution can be obtained in which the unperturbed potentials χ_0, θ_0 are linear functions of ω. Elementary operations show that the required potentials are

$$\chi_0 = \frac{S\omega}{4} ; \quad \theta_0 = \frac{S\omega}{2} \quad \text{and hence} \quad \psi_0 = \frac{S\omega}{2} + \frac{S\overline{\omega}}{2} ,$$

from (20.29).

For the complete (perturbed) solution, we want the crack S^* to be traction free and hence $\psi(\omega) \equiv \psi_0(\omega) + \psi_1(\omega) = 0$ for $\omega \in S^*$, where ψ_1 is the perturbation due to the crack. Using (20.68) to map the perturbation into the ζ-plane and then setting $\zeta = s$, $\bar{\zeta} = 1/s$, corresponding to the unit circle $|\zeta| = 1$, we have

$$\psi_\zeta(s) = -\frac{Sa}{2}\left(s + \frac{1}{s}\right) .$$

Substituting this result into (20.66) with $\gamma = 1$, $f_\zeta = \psi_\zeta$, we obtain

$$\chi_\zeta(\zeta) = \frac{1}{2\pi\iota} \oint \frac{Sa}{2} \left(s + \frac{1}{s}\right) \frac{ds}{(s - \zeta)} . \tag{20.69}$$

Since this is an exterior problem, ζ is outside the unit circle and the only pole in the integrand is that at $s = 0$. The residue theorem therefore gives

$$\chi_\zeta(\zeta) = \frac{Sa}{2} \frac{1}{(-\zeta)} = -\frac{Sa}{2\zeta} . \tag{20.70}$$

Substituting into (20.67) and noting that

$$\overline{\psi_\zeta(s)} = -\frac{Sa}{2} \left(\frac{1}{s} + s\right) ; \quad \chi_\zeta' = \frac{Sa}{2\zeta^2} ,$$

we then obtain

$$\theta_\zeta(\zeta) = \frac{1}{2\pi\iota} \oint \frac{Sa}{2} \left(\frac{1}{s} + s\right) \frac{ds}{(s - \zeta)} - \frac{Sa(\zeta^2 + 1)}{2\zeta(\zeta^2 - 1)} .$$

The contour integral is identical with that in (20.69) and after routine calculations we find

$$\theta_\zeta(\zeta) = -\frac{Sa\zeta}{(\zeta^2 - 1)} . \tag{20.71}$$

To move back to the w-plane, we use (19.44), which with $m = 1$, $c = a/2$ takes the form

$$\zeta = \frac{w + \sqrt{w^2 - a^2}}{a} .$$

Substituting this result into (20.70, 20.71), adding in the unperturbed solution χ_0, θ_0 and using the identity

$$\frac{1}{w + \sqrt{w^2 - a^2}} = \frac{w - \sqrt{w^2 - a^2}}{w^2 - (w^2 - a^2)} = \frac{w - \sqrt{w^2 - a^2}}{a^2} ,$$

we obtain the complete solution to the crack problem as

$$\chi(w) = -\frac{Sw}{4} + \frac{S\sqrt{w^2 - a^2}}{2} ; \quad \theta(w) = \frac{Sw}{2} - \frac{Sa^2}{2\sqrt{w^2 - a^2}} .$$

Finally, the complex stresses are recovered from (20.25, 20.24) as

$$\Theta = 2\left(\chi' + \overline{\chi'}\right) = -S + \frac{Sw}{\sqrt{w^2 - a^2}} + \frac{S\overline{w}}{\sqrt{\overline{w}^2 - a^2}}$$

$$\Phi = -2\left(w\overline{\chi''} + \overline{\theta'}\right) = -S + \frac{Sa^2(w - \overline{w})}{2(\overline{w}^2 - a^2)^{3/2}} .$$

Notice that these expressions define the stress field throughout the cracked body; not just the tractions on the plane $y = 0$ as in the solution in §13.3.2.

Problems

20.1. A state of uniform antiplane shear $\sigma_{xz} = S$, $\sigma_{yz} = 0$ in a large block of material is perturbed by the presence of a rigid circular inclusion whose boundary is defined by the equation $r = a$ in cylindrical polar coördinates r, θ, z. The inclusion is perfectly bonded to the elastic material and is prevented from moving, so that $u_z = 0$ at $r = a$. Use the complex-variable formulation to find the complete stress field in the block and hence determine the appropriate stress-concentration factor.

20.2. Show that the function (19.40) maps the surfaces of the crack $-a < \xi < a$, $\eta = 0$ onto the unit circle if $m = 1$, $c = a/2$. Use this result and the complex-variable formulation to solve the antiplane problem of a uniform stress field $\sigma_{\eta z} = S$ perturbed by a traction-free crack. In particular, find the mode III stress-intensity factor K_{III}.

20.3. Show that the function $\omega(\zeta) = \zeta^\lambda$ maps the wedge-shaped region $0 < \vartheta < \alpha$ in polar coördinates (ρ, ϑ) into the half-plane $y > 0$ if $\lambda = \alpha/\pi$.

Use this mapping to solve the antiplane problem of the wedge of subtended angle α loaded only by a concentrated out-of-plane force F at the point $\rho = a$, $\vartheta = 0$.

20.4. Use the stress-transformation equations (1.9–1.11) to prove the relation (20.26).

20.5. Express the plane strain equilibrium equations [(2.1, 2.2) with $\sigma_{xz} = \sigma_{yz} = 0$] in terms of Θ, Φ, p and ζ, $\bar\zeta$.

20.6. By analogy with equations (20.24, 20.25), we can define the complex strains e, ε as

$$ e = e_{xx} + e_{yy} ; \quad \varepsilon = e_{xx} + 2\imath e_{xy} - e_{yy} , $$

where we note that $e = \text{div } \boldsymbol{u}$ is the dilatation in plane strain. Express the elastic constitutive law (1.54) as a relation between e, ε and Θ, Φ.

20.7. Express the compatibility equation (2.5) in terms of the complex strains e, ε defined in Problem 20.6 and ζ, $\bar\zeta$.

20.8. Use (20.6) and (1.34) to obtain a complex-variable expression for the in-plane rotation ω_z. Then apply your result to equation (20.31) to verify that the term involving C_1 is consistent with a rigid-body rotation.

20.9. Use the method of §20.4 to determine the displacements due to the Airy stress function $\phi = r^2\theta$. Express the results in terms of the components u_r, u_θ, using the transformation equation (20.9).

20.10. Substitute the Laurent expansion (20.54) into the contour integral

$$\oint_S \frac{\bar{s}\chi'(s)ds}{(s-\zeta)} ,$$

where ζ is a point outside the circular contour S of radius a. Use the residue theorem to evaluate each term in the resulting series and hence verify that

$$\frac{1}{2\pi\imath} \oint \frac{\bar{s}\chi'(s)ds}{(s-\zeta)} = -\frac{a^2\chi'(\zeta)}{\zeta} .$$

20.11. Use the method of §20.6.1 to find the complex stresses Θ, Φ for Problem 12.1, where a disk of radius a is compressed by equal and opposite forces F acting at the points $\zeta = \pm a$.

20.12. A state of uniform uniaxial stress, $\sigma_{xx} = S$ is perturbed by the presence of a traction-free circular hole of radius a. Use the method of §20.6.2 to find the complex stresses Θ, Φ. (Note: the Airy stress function solution of this problem is given in §8.4.1.)

20.13. A state of uniform shear stress, $\sigma_{xy} = S$ is perturbed by the presence of a bonded rigid circular inclusion of radius a. Use the method of §20.6.2 to find the complex stresses Θ, Φ.

20.14. A state of uniform uniaxial stress, $\sigma_{yy} = S$ is perturbed by the presence of a traction-free elliptical hole of semi-axes a, b. Find the complex potentials χ, θ as functions of ω and hence determine the stress-concentration factor as a function of the ratio b/a. Verify that it tends to the value 3 when $b=a$.

20.15. Use the method of §20.7.1 to find the complete stress field for Problem 13.4 in which a state of uniform shear stress $\sigma_{xy} = S$ is perturbed by the presence of a crack occupying the region $-a < x < a$, $y=0$.

Part V
Three-Dimensional Problems

Chapter 21
Displacement Function Solutions

In Part II, we chose a representation for stress which satisfied the equilibrium equations identically, in which case the compatibility condition leads to a governing equation for the potential function. In three-dimensional elasticity, it is more usual to use the opposite approach — i.e. to define a potential function representation for displacement (which therefore identically satisfies the compatibility condition) and allow the equilibrium condition to define the governing equation.

A major reason for this change of method is simplicity. Stress function formulations of three-dimensional problems are generally more cumbersome than their displacement function counterparts, because of the greater complexity of the three-dimensional compatibility conditions. It is also worth noting that displacement formulations have a natural advantage in both two and three-dimensional problems when displacement boundary conditions are involved — particularly those associated with multiply-connected bodies (see §2.2.1).

21.1 The Strain Potential

The simplest displacement function representation is that in which the displacement u is set equal to the gradient of a scalar function — i.e.

$$2\mu u = \nabla\phi .\tag{21.1}$$

The displacement must satisfy the equilibrium equation (2.16), which in the absence of body force reduces to

$$\nabla \operatorname{div} u + (1 - 2\nu)\nabla^2 u = 0 .\tag{21.2}$$

We therefore obtain

$$\nabla\nabla^2\phi = 0 ,$$

© The Author(s), under exclusive license to Springer Nature Switzerland AG 2022
J. R. Barber, *Elasticity*, Solid Mechanics and Its Applications 172,
https://doi.org/10.1007/978-3-031-15214-6_21

and hence

$$\nabla^2 \phi = C \ , \tag{21.3}$$

where C is an arbitrary constant. The displacement function ϕ is generally referred to as the strain or displacement potential.

The general solution of equation (21.3) is conveniently considered as the sum of a (harmonic) complementary function and a particular integral, and for the latter, we can take an arbitrary second-order polynomial corresponding to a state of uniform stress. Thus, the representation (21.1) can be decomposed into a uniform state of stress superposed on a solution for which ϕ is harmonic.

It is easy to see that this representation is not sufficiently general to represent all possible displacement fields in an elastic body, since, for example, the rotation

$$\omega_z = \left(\frac{\partial u_y}{\partial x} - \frac{\partial u_x}{\partial y} \right) = \frac{\partial^2 \phi}{\partial x \partial y} - \frac{\partial^2 \phi}{\partial y \partial x} \equiv 0 \ .$$

In other words, the strain potential can only be used to describe irrotational deformation fields[1]. It is associated with the name of Lamé (whose name also attaches to the Lamé constants λ, μ), who used it to solve the problems of an elastic cylinder and sphere under axisymmetric loading. In the following work it will chiefly be useful as one term in a more general representation.

21.2 The Galerkin Vector

The representation of the previous section is insufficiently general to describe all possible states of deformation of an elastic body, and in seeking a more general form it is natural to examine representations in which the displacement is built up from second rather than first derivatives of a potential function. Using the index notation of §1.1.2, the second derivative term

$$\frac{\partial^2 F}{\partial x_i \partial x_j} \ ,$$

where F is a scalar, will represent a Cartesian tensor if the indices i, j are distinct. Alternatively if we make i and j the same, we obtain

$$\frac{\partial^2 F}{\partial x_i \partial x_i} \equiv \frac{\partial^2 F}{\partial x_1 \partial x_1} + \frac{\partial^2 F}{\partial x_2 \partial x_2} + \frac{\partial^2 F}{\partial x_3 \partial x_3} = \nabla^2 F \ ,$$

---

[1] We encountered the same lack of generality with the body force potential of Chapter 7 (see §7.1.1).

which defines the Laplace operator and returns a scalar. What we want to obtain is a vector, which has one free index, and this can only be done if F also has an index (and hence represents a vector). Then we have a choice of pairing the index on F with one of the derivatives or leaving it free and pairing the two derivatives. The most general form is a linear combination of the two — i.e.

$$2\mu u_i = C \frac{\partial^2 F_i}{\partial x_j \partial x_j} - \frac{\partial^2 F_j}{\partial x_i \partial x_j} \,, \tag{21.4}$$

where C is an arbitrary constant which we shall later assign so as to simplify the representation. The function F_i must be chosen to satisfy the equilibrium equation (21.2), which in index notation takes the form

$$\frac{\partial^2 u_k}{\partial x_i \partial x_k} + (1 - 2\nu) \frac{\partial^2 u_i}{\partial x_k \partial x_k} = 0 \,, \tag{21.5}$$

from (2.15).

Substituting (21.4) into (21.5), and cancelling non-zero factors, we obtain

$$C \frac{\partial^4 F_k}{\partial x_j \partial x_j \partial x_i \partial x_k} - \frac{\partial^4 F_j}{\partial x_k \partial x_j \partial x_i \partial x_k} + C(1 - 2\nu) \frac{\partial^4 F_i}{\partial x_j \partial x_j \partial x_k \partial x_k}$$
$$- (1 - 2\nu) \frac{\partial^4 F_j}{\partial x_i \partial x_j \partial x_k \partial x_k} = 0 \,.$$

Each term in this equation contains one free index i, all the other indices being paired off in an implied sum. In the first, second and fourth term, the index on F is paired with one of the derivatives and the free index appears in one of the derivatives. After expanding the implied summations, these terms are therefore all identical and the representation can be simplified by making them sum to zero. This is achieved by choosing

$$C - 1 - (1 - 2\nu) = 0 \quad \text{or} \quad C = 2(1 - \nu) \,,$$

after which the representation (21.4) takes the form

$$2\mu u_i = 2(1 - \nu) \frac{\partial^2 F_i}{\partial x_j \partial x_j} - \frac{\partial^2 F_j}{\partial x_i \partial x_j} \,, \tag{21.6}$$

or in vector notation

$$2\mu \boldsymbol{u} = 2(1 - \nu) \nabla^2 \boldsymbol{F} - \nabla \operatorname{div} \boldsymbol{F} \,. \tag{21.7}$$

With this choice, the equilibrium equation reduces to

$$\frac{\partial^4 F_i}{\partial x_j \partial x_j \partial x_k \partial x_k} = 0 \quad \text{or} \quad \nabla^4 \boldsymbol{F} = 0 \tag{21.8}$$

and the vector function $\boldsymbol{F}$ is biharmonic.

This solution is due to Galerkin and the function F is known as the Galerkin vector. A more detailed derivation is given by Westergaard[2], who gives expressions for the stress components in terms of F and who uses this representation to solve a number of classical three-dimensional problems, notably those involving concentrated forces in the infinite or semi-infinite body. We also note that the Galerkin solution was to some extent foreshadowed by Love[3], who introduced a displacement function appropriate for an axisymmetric state of stress in a body of revolution. Love's displacement function is in fact one component of the Galerkin vector F.

21.3 The Papkovich-Neuber Solution

A closely related solution is that developed independently by Papkovich and Neuber in terms of harmonic functions. In general, solutions in terms of harmonic functions are easier to use than those involving biharmonic functions, because the properties of harmonic functions have been extensively studied in the context of other physical theories such as electrostatics, gravitation, heat conduction and fluid mechanics.

We define the vector function

$$\psi = -\frac{1}{2}\nabla^2 F \tag{21.9}$$

and since the function F is biharmonic, ψ must be harmonic.

It can then be verified by differentiation that

$$\nabla^2(\mathbf{r}\cdot\psi) \equiv \nabla^2(x\psi_x + y\psi_y + z\psi_z)$$
$$= 2\frac{\partial\psi_x}{\partial x} + 2\frac{\partial\psi_y}{\partial y} + 2\frac{\partial\psi_z}{\partial z}$$
$$= 2\,\mathrm{div}\,\psi\ . \tag{21.10}$$

Substituting for ψ from equation (21.9) into the right-hand side of (21.10), we obtain

$$\nabla^2(\mathbf{r}\cdot\psi) = -\mathrm{div}\,\nabla^2 F = -\nabla^2\mathrm{div}\,F$$

and hence

$$-\,\mathrm{div}\,F = \mathbf{r}\cdot\psi + \phi\ , \tag{21.11}$$

where ϕ is an arbitrary harmonic function.

---

[2] H. M. Westergaard, *Theory of Elasticity and Plasticity*, Dover, New York, 1964, §66.

[3] A. E. H.Love, *A Treatise on the Mathematical Theory of Elasticity*, 4th.edn., Dover, New York, 1944, §188.

Finally, we substitute (21.9, 21.11) into the Galerkin vector representation (21.7) obtaining

$$2\mu u = -4(1 - \nu)\psi + \nabla(r \cdot \psi + \phi) \,, \tag{21.12}$$

where ψ and ϕ are both harmonic functions — i.e.

$$\nabla^2 \psi = 0 \; ; \quad \nabla^2 \phi = 0 \,,$$

but notice that ψ is a vector function, whilst ϕ is a scalar. This is known as the *Papkovich-Neuber solution* and it is widely used in modern treatments of three-dimensional elasticity problems. We note that the scalar function ϕ is the harmonic strain potential introduced in §21.1 above.

21.3.1 Change of coördinate system

One disadvantage of the Papkovich-Neuber solution is that the term $r \cdot \psi$ introduces a dependence on the origin of coördinates, so that the potentials corresponding to a given displacement field will generally change if the origin is changed. Suppose that a given displacement field is characterized by the potentials ψ_1, ϕ_1, so that

$$2\mu u = -4(1 - \nu)\psi_1 + \nabla(r_1 \cdot \psi_1 + \phi_1) \,, \tag{21.13}$$

where r_1 is a position vector defining the distance from an origin O_1. We wish to determine the appropriate potentials in a new coördinate system with origin O_2. If the coördinates of O_1 relative to O_2 are defined by a vector s, the position vector r_2 in the new coördinate system is

$$r_2 = s + r_1 \,. \tag{21.14}$$

Solving (21.14) for r_1 and substituting in (21.13), we have

$$2\mu u = -4(1 - \nu)\psi_1 + \nabla[(r_2 - s) \cdot \psi_1 + \phi_1] \,,$$

or

$$2\mu u = -4(1 - \nu)\psi_2 + \nabla(r_2 \cdot \psi_2 + \phi_2) \,,$$

where

$$\psi_2 = \psi_1 \; ; \quad \phi_2 = \phi_1 - s \cdot \psi_1 \,. \tag{21.15}$$

Thus, the vector potential ψ is unaffected by the change of origin, but there is a change in the scalar potential ϕ. By contrast, the Galerkin vector F is independent of the choice of origin.

21.4 Completeness and Uniqueness

It is a fairly straightforward matter to prove that the representations (21.6, 21.12) are complete — i.e. that they are capable of describing all possible elastic displacement fields in a three dimensional body[4]. The related problem of uniqueness presents greater difficulties. Both representations are redundant, in the sense that there are infinitely many combinations of functions which correspond to a given displacement field. In a non-rigorous sense, this is clear from the fact that a typical elasticity problem will be reduced to a boundary-value problem with three conditions at each point of the boundary (three tractions or three displacements or some combination of the two). We should normally expect such conditions to be sufficient to determine three harmonic potential functions in the interior, but (21.12) essentially contains four independent functions (three components of ψ and ϕ).

Neuber gave a 'proof' that one of these four functions can always be set equal to zero, but it has since been found that his argument only applies under certain restrictions on the shape of the body. A related 'proof' for the Galerkin representation is given by Westergaard in his §70. We shall give a summary of one such argument and show why it breaks down if the shape of the body does not meet certain conditions.

We suppose that a solution to a certain problem is known and that it involves all the three components ψ_x, ψ_y, ψ_z of ψ and ϕ. We wish to find another representation *of the same field* in which one of the components is everywhere zero. We could achieve this if we could construct a 'null' field — i.e. one which corresponds to zero displacement and stress at all points — for which the appropriate component is the same as in the original solution. The difference between the two solutions would then give the same displacements as the original solution (since we have subtracted a null field only) and the appropriate component will have been reduced to zero.

Equation (21.12) shows that ψ, ϕ will define a null field if

$$\nabla(r \cdot \psi + \phi) - 4(1 - \nu)\psi = 0 \ . \tag{21.16}$$

Taking the curl of this equation, we obtain

$$\mathrm{curl}\ \nabla(r \cdot \psi + \phi) - 4(1 - \nu)\mathrm{curl}\ \psi = 0$$

and hence

$$\mathrm{curl}\ \psi = 0 \ , \tag{21.17}$$

since curl ∇ of any function is identically zero.

We also note that by applying the operator div to equation (21.16), we obtain

---

[4] See for example, H. M. Westergaard, *loc. cit.* §69 and R. D. Mindlin (1936), Note on the Galerkin and Papcovich stress functions, *Bulletin of the American Mathematical Society,* Vol. 42, pp. 373–376. See also Problem 21.10.

$$\nabla^2(\boldsymbol{r} \cdot \boldsymbol{\psi} + \phi) - 4(1 - \nu)\operatorname{div} \boldsymbol{\psi} = 0$$

and hence, substituting from equation (21.10) for the first term,

$$2\operatorname{div} \boldsymbol{\psi} - 4(1 - \nu)\operatorname{div} \boldsymbol{\psi} = -2(1 - 2\nu)\operatorname{div} \boldsymbol{\psi} = 0 . \tag{21.18}$$

Now equation (21.17) is precisely the condition which must be satisfied for the function $\boldsymbol{\psi}$ to be expressible as the gradient of a scalar function[5]. Hence, we conclude that there exists a function H such that

$$\boldsymbol{\psi} = \boldsymbol{\nabla} H . \tag{21.19}$$

Furthermore, substituting (21.19) into (21.18), we see that H must be a harmonic function. In order to dispense with one of the components (say ψ_z) of $\boldsymbol{\psi}$ in the given solution, we therefore need to be able to construct a harmonic function H to satisfy the equation

$$\frac{\partial H}{\partial z} = \psi_z . \tag{21.20}$$

If we can do this, it is easily verified that the appropriate null field can be found, since on substituting into equation (21.16) we get

$$\boldsymbol{\nabla}(\boldsymbol{r} \cdot \boldsymbol{\nabla} H + \phi - 4(1 - \nu)H) = 0 ,$$

which can be satisfied by choosing ϕ so that

$$\phi = 4(1 - \nu)H - \boldsymbol{r} \cdot \boldsymbol{\nabla} H .$$

Noting that

$$\nabla^2(\boldsymbol{r} \cdot \boldsymbol{\nabla} H) = \nabla^2 \left(x\frac{\partial H}{\partial x} + y\frac{\partial H}{\partial y} + z\frac{\partial H}{\partial z} \right)$$
$$= 2\nabla^2 H ,$$

we see that ϕ will then be harmonic, as required, since H is harmonic.

21.4.1 Methods of partial integration

We therefore consider the problem of integrating equation (21.20) subject to the constraint that the integral be a harmonic function. We shall examine two methods of doing this which give some insight into the problem.

---

[5] See §7.1.1.

The given solution and hence the function ψ_z is only defined in the finite region occupied by the elastic body which we denote by Ω. We also denote the boundary of Ω (not necessarily connected) by Γ.

One candidate for the required function H is obtained from the integral

$$H_1 = \int_{z_\Gamma}^{z} \psi_z dz ,$$
(21.21)

where the integral is performed from a point on Γ to a point in the body along a line in the z-direction. This function is readily constructed and it clearly satisfies equation (21.20), but it will not in general be harmonic. However,

$$\frac{\partial}{\partial z} \nabla^2 H_1 = \nabla^2 \frac{\partial H_1}{\partial z} = 0$$
(21.22)

and hence

$$\nabla^2 H_1 = f(x, y) .$$

Now a harmonic function H satisfying equation (21.20) can be constructed as

$$H = H_1 + H_2 ,$$

where H_2 is any function of x, y only that satisfies the equation

$$\nabla^2 H_2 = -f(x, y) .$$
(21.23)

A suitable choice for H_2 is the logarithmic potential due to the source distribution $f(x, y)$, which can be explicitly constructed as a convolution integral on the logarithmic (two-dimensional) point source solution. Notice that H_2 is not uniquely determined by (21.23), but we are looking for a particular integral only, not a general solution. Indeed, this freedom of choice of solution could prove useful in extending the range of conditions under which the integration of (21.20) can be performed.

At first sight, the procedure described above seems to give a general method of constructing the required function H, but a difficulty is encountered with the definition of the lower limit z_Γ in the integral (21.21). The path of integration is along a straight line in the z-direction from Γ to the general point (x, y, z). There may exist points for which such a line cuts the surface Γ at more than two points — i.e. for which z_Γ is multivalued. Such a case is illustrated as the point $P(x, y, z)$ in Figure 21.1. The integrand is undefined outside Ω and hence we can only give a meaning to the definition (21.21) if we choose the point A as the lower limit. However, by the same token, we must choose the point C as the lower limit in the integral for the point $Q(x, y, z')$.

It follows that the function $f(x, y)$ is not truly a function of x, y only, since it has different values at the points P, Q, which have the same coördinates (x, y). The $f(x, y)$ appropriate to the region above the line TT' will generally have a

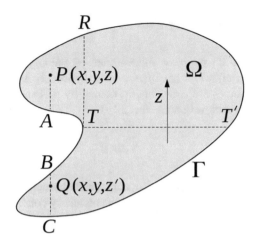

Fig. 21.1 Path of integration for equation (21.21).

discontinuity across the line TR, where T is the point of tangency of Γ to lines in the z-direction (see Figure 21.1). Notice, however that the freedom of choice of the integral of equation (21.23) may in certain circumstances permit us to eliminate this discontinuity.

Another way of developing the function H_1 is as follows:- We first find that distribution of sources which when placed on the surface Γ in an infinite space would give the required potential ψ_z in Ω. This distribution is unique and gives a form of continuation of ψ_z into the rest of the infinite space which is everywhere harmonic *except* on the surface Γ. We then write $-\infty$ for the lower limit in equation (21.21).

The function defined by equation (21.21) will now be harmonic except when the path of integration passes through Γ, in which case $\nabla^2 H_1$ will contain a term proportional to the strength of the singularity at the appropriate point(s) of intersection. As these points will be the same for any given value of (x, y), the resulting function will satisfy equation (21.22), but we note that again there will be different values for the function $f(x, y)$ if the appropriate path of integration cuts Γ in more than two points. Furthermore, this method of construction shows that the difference between the two values is proportional to the strength of the singularities on the extra two intersections with Γ (i.e. at points B, C in Figure 21.1). One of the consequences is that $f(x, y)$ in the upper region will then show a square root singular discontinuity to the left of the line TR, assuming that Γ has no sharp corners. This question is still a matter of active interest in elasticity[6].

---

[6] For further discussion of the question, see R. A. Eubanks and E. Sternberg (1956), On the completeness of the Boussinesq-Papcovich stress functions, *Journal of Rational Mechanics and Analysis,* Vol. 5, pp. 735–746; P. M. Naghdi and C. S. Hsu (1961), On a representation of displacements in linear elasticity in terms of three stress functions, *Journal of Mathematics and Mechanics,* Vol. 10, pp. 233–246; T. Tran Cong and G. P. Steven (1979), On the representation of elastic displacement fields in terms of three harmonic functions, *Journal of Elasticity,* Vol. 9, pp. 325–333.

21.5 Body Forces

So far we have concentrated on the development of solutions to the equation of
equilibrium (21.2) without body forces. As in Chapter 7, problems involving body
forces are conveniently treated by first seeking a particular solution satisfying the
equilibrium equation (2.16)

$$\nabla \operatorname{div} \boldsymbol{u} + (1 - 2\nu)\nabla^2 \boldsymbol{u} + \frac{(1 - 2\nu)\boldsymbol{p}}{\mu} = 0 \tag{21.24}$$

and then superposing appropriate homogeneous solutions — i.e. solutions without
body force — to satisfy the boundary conditions.

21.5.1 Conservative body force fields

If the body force $\boldsymbol{p}$ is conservative, we can write

$$\boldsymbol{p} = -\nabla V \tag{21.25}$$

and a particular solution can always be found using the strain potential (21.1). Substituting (21.1, 21.25) into (21.24) we obtain

$$(1 - \nu)\nabla \nabla^2 \phi - (1 - 2\nu)\nabla V = 0 ,$$

which is satisfied if

$$\nabla^2 \phi = \frac{(1 - 2\nu)V}{(1 - \nu)} . \tag{21.26}$$

Solutions of this equation can be found for all functions V. The corresponding
stress components can then be obtained from the stress-strain and strain-displacement
relations. For example, we have

$$\sigma_{xx} = \lambda \operatorname{div} \boldsymbol{u} + 2\mu \frac{\partial u_x}{\partial x} = \frac{\lambda}{2\mu}\nabla^2 \phi + \frac{\partial^2 \phi}{\partial x^2} = \frac{\nu V}{(1 - \nu)} + \frac{\partial^2 \phi}{\partial x^2} , \tag{21.27}$$

using (1.51–1.53, 21.26). For the shear stresses, we have for example

$$\sigma_{xy} = 2\mu e_{xy} = \mu \left(\frac{\partial u_y}{\partial x} + \frac{\partial u_x}{\partial y} \right) = \frac{\partial^2 \phi}{\partial x \partial y} . \tag{21.28}$$

21.5.2 Non-conservative body force fields

If the body force p is non-conservative, a particular solution can always be obtained using the Galerkin vector representation of equation (21.7). Substituting this into (2.15), we obtain

$$\frac{\partial^4 F_i}{\partial x_j \partial x_j \partial x_k \partial x_k} = -\frac{p_i}{(1-\nu)} ,$$

or

$$\nabla^4 F = -\frac{p}{(1-\nu)} . \tag{21.29}$$

This constitutes three uncoupled equations relating each component of the Galerkin vector to the corresponding component of the body force. Particular solutions can be found for all functions p.

Problems

21.1. Show that the function

$$\psi = y\frac{\partial \phi}{\partial x} - x\frac{\partial \phi}{\partial y}$$

will be harmonic ($\nabla^2 \psi = 0$) if ϕ is harmonic. Show also that ψ will be biharmonic if ϕ is biharmonic.

21.2. Find the most general solution of equation (21.3) that is spherically symmetric — i.e. depends only on the distance $R = \sqrt{x^2 + y^2 + z^2}$ from the origin. Use your solution to find the stress and displacement field in a hollow spherical container of inner radius b and outer radius a, loaded only by internal pressure p_0 at $R = b$. **Note:** Strain-displacement relations and expressions for the gradient and Laplacian operator in spherical polar coördinates are given in §22.4.2 below.

21.3. By expressing the vector operators in index notation, show that

$$\nabla \operatorname{div} \nabla \operatorname{div} V \equiv \nabla^2 \nabla \operatorname{div} V \equiv \nabla \operatorname{div} \nabla^2 V ,$$

where V is any vector field.

21.4. Starting from equation (21.6), use (1.37, 1.54, 1.52) to develop general expressions for the stress components σ_{ij} in the Galerkin formulation, in terms of F_i and the elastic constants μ, ν.

21.5. Verify that the function $\psi = 1/\sqrt{x^2 + y^2 + z^2}$ is harmonic everywhere except at the origin, where it is singular. This function can therefore be used as a Papkovich-

Neuber displacement function for a body that does not contain the origin, such as the infinite body with a spherical hole centred on the origin. Show by partial integration that it is impossible to find a function H such that

$$\frac{\partial H}{\partial z} = \psi$$

and $\nabla^2 H = 0$, except possibly at the origin.

21.6. Repeat the derivation of §21.3 using equation (21.29) in place of (21.8) in order to generalize the Papkovich-Neuber solution to include a body force field p. In particular, find the governing equations for the functions ψ, ϕ.

21.7. Find a particular solution for the Galerkin vector F for the non-conservative body force field

$$p_x = -\rho\dot{\Omega}y \ ; \quad p_y = \rho\dot{\Omega}x \ ; \quad p_z = 0 .$$

21.8. Prove the result (used in deriving (21.17)) that

$$\text{curl } \nabla\phi = 0 ,$$

for any scalar function ϕ.

21.9. (i) Verify that the Galerkin vector

$$F = Cx^3yk$$

satisfies the governing equation (21.8) and determine the corresponding displacement vector u.
   (ii) Show that
$$F_0 = \left((1 - 2\nu)\nabla^2 + \nabla\text{div}\right) f$$

defines a null Galerkin vector (i.e. the corresponding displacements are everywhere zero) if $\nabla^4 f = 0$.
   (iii) Using this result or otherwise, find an alternative Galerkin vector for the displacement field of part (i) that is confined to the xy-plane — i.e. such that $F_z = 0$.

21.10. By writing the equilibrium equation (21.2) as

$$\nabla^2 u = -\frac{1}{(1 - 2\nu)}\nabla\text{div } u ,$$

show that the displacement u can be expressed in the form $\mu u = \nabla\chi + \Psi$, where Ψ is a harmonic vector potential and χ is a scalar potential. Find the equation that must be satisfied by χ and show that the resulting representation of displacement is equivalent to the Papkovich-Neuber solution (21.12). Hence or otherwise, show that this representation is complete.

Chapter 22
The Boussinesq Potentials

The Galerkin and Papkovich-Neuber solutions have the advantage of presenting a general solution to the problem of elasticity in a suitably compact notation, but they are not always the most convenient starting point for the solution of particular three-dimensional problems. If the problem has a plane of symmetry or particularly simple boundary conditions, it is often possible to develop a special solution of sufficient generality in one or two harmonic functions, which may or may not be components or linear combinations of components of the Papkovich-Neuber solution. For this reason, it is convenient to record detailed expressions for the displacement and stress components arising from the several terms separately and from certain related displacement potentials.

The first catalogue of solutions of this kind was compiled by Boussinesq[1] and is reproduced by Green and Zerna[2], whose terminology we use here. Boussinesq identified three categories of harmonic potential, one being the strain potential of §21.1, already introduced by Lamé, and another comprising a set of three scalar functions equivalent to the three components of the Papkovich-Neuber vector, ψ. The third category comprises three solutions particularly suited to torsional deformations about the three axes respectively. In view of the completeness of the Papkovich-Neuber solution, it is clear that these latter functions must be derivable from equation (21.12) and we shall show how this can be done in §22.3 below[3].

Supplementary Information The online version contains supplementary material available at https://doi.org/10.1007/978-3-031-15214-6_22.

---

[1] J. Boussinesq, *Application des potentiels à l'étude de l'équilibre et du mouvement des solides élastiques*, Gauthier-Villars, Paris, 1885.

[2] A. E. Green and W. Zerna, *Theoretical Elasticity*, 2nd.edn., Clarendon Press, Oxford, 1968, pp. 165–176.

[3] Other relations between the Boussinesq potentials are demonstrated by J. P. Bentham (1979), Note on the Boussinesq-Papkovich stress functions, *J.Elasticity,* Vol. 9, pp. 201–206.

365
J. R. Barber, *Elasticity*, Solid Mechanics and Its Applications 172,
https://doi.org/10.1007/978-3-031-15214-6_22

22.1   Solution A: The Strain Potential

Green and Zerna's Solution A is identical with the strain potential defined in §21.1 above (with the restriction that $\nabla^2\phi=0$) and with the scalar potential ϕ of the Papkovich-Neuber solution [equation (21.12)].

We define

$$2\mu u_x = \frac{\partial\phi}{\partial x} \quad ; \quad 2\mu u_y = \frac{\partial\phi}{\partial y} \quad ; \quad 2\mu u_z = \frac{\partial\phi}{\partial z} \ ,$$

and the dilatation

$$2\mu e = 2\mu \operatorname{div} \boldsymbol{u} = \nabla^2\phi = 0 \ .$$

Substituting in the stress-strain relations, we have

$$\sigma_{xx} = \lambda e + 2\mu e_{xx} = \frac{\partial^2\phi}{\partial x^2}$$

and similarly

$$\sigma_{yy} = \frac{\partial^2\phi}{\partial y^2} \quad ; \quad \sigma_{zz} = \frac{\partial^2\phi}{\partial z^2} \ .$$

We also have

$$\sigma_{xy} = 2\mu e_{xy} = \frac{\partial^2\phi}{\partial x\partial y} \quad ; \quad \sigma_{yz} = \frac{\partial^2\phi}{\partial y\partial z} \quad ; \quad \sigma_{zx} = \frac{\partial^2\phi}{\partial z\partial x} \ .$$

22.2   Solution B

Green and Zerna's Solution B is obtained by setting to zero all the functions in the Papkovich-Neuber solution except the z-component of ψ, which we shall denote by ω. In other words, in equation (21.12), we take

$$\psi = \boldsymbol{k}\omega \quad ; \quad \phi = 0 \ .$$

It follows that

$$2\mu u_x = z\frac{\partial\omega}{\partial x} \quad ; \quad 2\mu u_y = z\frac{\partial\omega}{\partial y} \quad ; \quad 2\mu u_z = z\frac{\partial\omega}{\partial z} - (3-4\nu)\omega$$

$$2\mu e = z\nabla^2\omega + \frac{\partial\omega}{\partial z} - (3-4\nu)\frac{\partial\omega}{\partial z} = -2(1-2\nu)\frac{\partial\omega}{\partial z} \ ,$$

since $\nabla^2\omega=0$.

Substituting these expressions into the stress-strain relations and noting that

$$\lambda = \frac{2\mu\nu}{(1 - 2\nu)} \, ,$$

we find

$$\sigma_{xx} = \lambda e + 2\mu e_{xx} = z\frac{\partial^2 \omega}{\partial x^2} - 2\nu\frac{\partial \omega}{\partial z} \; ; \quad \sigma_{yy} = z\frac{\partial^2 \omega}{\partial y^2} - 2\nu\frac{\partial \omega}{\partial z}$$

$$\sigma_{zz} = z\frac{\partial^2 \omega}{\partial z^2} - 2(1 - \nu)\frac{\partial \omega}{\partial z} \; ; \quad \sigma_{xy} = z\frac{\partial^2 \omega}{\partial x \partial y}$$

$$\sigma_{yz} = \frac{1}{2}\left(\frac{\partial \omega}{\partial y} + z\frac{\partial^2 \omega}{\partial y \partial z} + z\frac{\partial^2 \omega}{\partial y \partial z} - (3 - 4\nu)\frac{\partial \omega}{\partial y}\right)$$

$$= z\frac{\partial^2 \omega}{\partial y \partial z} - (1 - 2\nu)\frac{\partial \omega}{\partial y}$$

$$\sigma_{zx} = z\frac{\partial^2 \omega}{\partial z \partial x} - (1 - 2\nu)\frac{\partial \omega}{\partial x} \, .$$

This solution — if combined with solution A — gives a general solution for the torsionless axisymmetric deformation of a body of revolution. The earliest solution of this problem is due to Love[4] and leads to a single biharmonic function which is actually the component F_z of the Galerkin vector (§21.2). Solution B is also particularly suitable for problems in which the plane surface $z=0$ is a boundary, since the expressions for stresses and displacements take simple forms on this surface.

It is clearly possible to write down similar solutions corresponding to Papkovich-Neuber vectors in the x- and y-directions by permuting suffices. Green and Zerna refer to these as Solutions C and D.

22.3   Solution E: Rotational Deformation

Green and Zerna's Solution E can be derived (albeit somewhat indirectly) from the Papkovich-Neuber solution by taking the vector function ψ to be the curl of a vector field oriented in the z-direction — i.e.

$$\psi = \text{curl } \boldsymbol{k}\chi \, .$$

It follows that

$$\boldsymbol{r} \cdot \psi = x\frac{\partial \chi}{\partial y} - y\frac{\partial \chi}{\partial x}$$

and hence

$$\nabla^2(\boldsymbol{r} \cdot \psi) = 2\frac{\partial^2 \chi}{\partial x \partial y} - 2\frac{\partial^2 \chi}{\partial y \partial x} = 0 \, ,$$

— i.e. $\boldsymbol{r} \cdot \psi$ is harmonic. We can therefore choose ϕ such that

---

[4] A. E. H.Love, *loc. cit.*

$$\boldsymbol{r} \cdot \boldsymbol{\psi} + \phi \equiv 0$$

and hence

$$2\mu\boldsymbol{u} = -4(1 - \nu)\mathrm{curl}\,\boldsymbol{k}\chi \ .$$

To avoid unnecessary multiplying constants, it is convenient to write

$$\Psi = -2(1 - \nu)\chi \ ,$$

in which case

$$2\mu u_x = 2\frac{\partial\Psi}{\partial y} \ ; \ \ 2\mu u_y = -2\frac{\partial\Psi}{\partial x} \ ; \ \ 2\mu u_z = 0 \ ,$$

giving

$$2\mu e = 0 \ .$$

Substituting in the stress-strain relations, we find

$$\sigma_{xx} = 2\frac{\partial^2\Psi}{\partial x\partial y} \ ; \ \ \sigma_{yy} = -2\frac{\partial^2\Psi}{\partial x\partial y} \ ; \ \ \sigma_{zz} = 0$$

$$\sigma_{xy} = \left(\frac{\partial^2\Psi}{\partial y^2} - \frac{\partial^2\Psi}{\partial x^2}\right) \ ; \ \ \sigma_{yz} = -\frac{\partial^2\Psi}{\partial x\partial z} \ ; \ \ \sigma_{zx} = \frac{\partial^2\Psi}{\partial y\partial z} \ .$$

As with Solution B, two additional solutions of this type can be constructed by permuting suffices.

22.4   Other Coördinate Systems

The above results have been developed in the Cartesian coördinates x, y, z, but similar expressions are easily obtained in other coördinate systems.

22.4.1   Cylindrical polar coördinates

The cylindrical polar coördinate system r, θ, z is related to the Cartesian system through the equations

$$x = r\cos\theta \ ; \ \ y = r\sin\theta$$

as in the two-dimensional transformation (8.1). The gradient operator takes the form

$$\nabla \equiv \boldsymbol{e}_r\frac{\partial}{\partial r} + \frac{\boldsymbol{e}_\theta}{r}\frac{\partial}{\partial\theta} + \boldsymbol{k}\frac{\partial}{\partial z} \ ,$$

where e_r, e_θ are unit vectors in the r and θ directions respectively, and the two-dimensional Laplacian operator (8.10) aquires a z-derivative term, becoming

$$\nabla^2 \equiv \frac{\partial^2}{\partial r^2} + \frac{1}{r}\frac{\partial}{\partial r} + \frac{1}{r^2}\frac{\partial^2}{\partial \theta^2} + \frac{\partial^2}{\partial z^2}\,. \tag{22.1}$$

The in-plane strain-displacement relations (8.12) need to be supplemented by

$$e_{rz} = \frac{1}{2}\left(\frac{\partial u_r}{\partial z} + \frac{\partial u_z}{\partial r}\right) \quad ; \quad e_{\theta z} = \frac{1}{2}\left(\frac{\partial u_\theta}{\partial z} + \frac{1}{r}\frac{\partial u_z}{\partial \theta}\right) \quad ; \quad e_{zz} = \frac{\partial u_z}{\partial z}\,.$$

Calculation of the stress and displacement components then proceeds as in §§22.1–22.3. The corresponding expressions for both Cartesian and cylindrical polar coördinates are tabulated in Table 22.1. Also, these expressions are listed in the Maple and Mathematica files 'ABExyz' and 'ABErtz'.

22.4.2 Spherical polar coördinates

The spherical polar coördinate system R, θ, β is related to the cylindrical system by the further change of variables

$$r = R\sin\beta \quad ; \quad z = R\cos\beta\,,$$

where

$$R = \sqrt{r^2 + z^2} = \sqrt{x^2 + y^2 + z^2}$$

represents the distance from the origin and the two spherical angles θ, β are equivalent to measures of longitude and lattitude respectively, defining position on a sphere of radius R. Notice however that it is conventional to measure the angle of lattitude from the pole (the positive z-axis) rather than from the equator. Thus, the two poles are defined by $\beta=0, \pi$ and the equator by $\beta=\pi/2$.

The gradient operator takes the form

$$\nabla \equiv e_R \frac{\partial}{\partial R} + \frac{e_\theta}{R\sin\beta}\frac{\partial}{\partial \theta} + \frac{e_\beta}{R}\frac{\partial}{\partial \beta}\,,$$

where e_r, e_θ, e_β are unit vectors in the R, θ and β directions respectively. Transformation of the Laplacian operator, the strain-displacement relations and the expressions for displacement and stress components in Solutions A,B and E is algebraically tedious but mathematically routine and only the principal results are summarized here.

Table 22.1 Green and Zerna's Solutions A, B and E

	Solution A	**Solution B**	**Solution E**
$2\mu u_x$	$\dfrac{\partial\phi}{\partial x}$	$z\dfrac{\partial\omega}{\partial x}$	$2\dfrac{\partial\Psi}{\partial y}$
$2\mu u_y$	$\dfrac{\partial\phi}{\partial y}$	$z\dfrac{\partial\omega}{\partial y}$	$-2\dfrac{\partial\Psi}{\partial x}$
$2\mu u_z$	$\dfrac{\partial\phi}{\partial z}$	$z\dfrac{\partial\omega}{\partial z}-(3-4\nu)\omega$	0
σ_{xx}	$\dfrac{\partial^2\phi}{\partial x^2}$	$z\dfrac{\partial^2\omega}{\partial x^2}-2\nu\dfrac{\partial\omega}{\partial z}$	$2\dfrac{\partial^2\Psi}{\partial x\partial y}$
σ_{xy}	$\dfrac{\partial^2\phi}{\partial x\partial y}$	$z\dfrac{\partial^2\omega}{\partial x\partial y}$	$\dfrac{\partial^2\Psi}{\partial y^2}-\dfrac{\partial^2\Psi}{\partial x^2}$
σ_{yy}	$\dfrac{\partial^2\phi}{\partial y^2}$	$z\dfrac{\partial^2\omega}{\partial y^2}-2\nu\dfrac{\partial\omega}{\partial z}$	$-2\dfrac{\partial^2\Psi}{\partial x\partial y}$
σ_{xz}	$\dfrac{\partial^2\phi}{\partial x\partial z}$	$z\dfrac{\partial^2\omega}{\partial x\partial z}-(1-2\nu)\dfrac{\partial\omega}{\partial x}$	$\dfrac{\partial^2\Psi}{\partial y\partial z}$
σ_{yz}	$\dfrac{\partial^2\phi}{\partial y\partial z}$	$z\dfrac{\partial^2\omega}{\partial y\partial z}-(1-2\nu)\dfrac{\partial\omega}{\partial y}$	$-\dfrac{\partial^2\Psi}{\partial x\partial z}$
σ_{zz}	$\dfrac{\partial^2\phi}{\partial z^2}$	$z\dfrac{\partial^2\omega}{\partial z^2}-2(1-\nu)\dfrac{\partial\omega}{\partial z}$	0
$2\mu u_r$	$\dfrac{\partial\phi}{\partial r}$	$z\dfrac{\partial\omega}{\partial r}$	$\dfrac{2}{r}\dfrac{\partial\Psi}{\partial\theta}$
$2\mu u_\theta$	$\dfrac{1}{r}\dfrac{\partial\phi}{\partial\theta}$	$\dfrac{z}{r}\dfrac{\partial\omega}{\partial\theta}$	$-2\dfrac{\partial\Psi}{\partial r}$
σ_{rr}	$\dfrac{\partial^2\phi}{\partial r^2}$	$z\dfrac{\partial^2\omega}{\partial r^2}-2\nu\dfrac{\partial\omega}{\partial z}$	$\dfrac{2}{r}\dfrac{\partial^2\Psi}{\partial r\partial\theta}-\dfrac{2}{r^2}\dfrac{\partial\Psi}{\partial\theta}$
$\sigma_{r\theta}$	$\dfrac{1}{r}\dfrac{\partial^2\phi}{\partial r\partial\theta}-\dfrac{1}{r^2}\dfrac{\partial\phi}{\partial\theta}$	$\dfrac{z}{r}\dfrac{\partial^2\omega}{\partial r\partial\theta}-\dfrac{z}{r^2}\dfrac{\partial\omega}{\partial\theta}$	$\dfrac{1}{r}\dfrac{\partial\Psi}{\partial r}-\dfrac{\partial^2\Psi}{\partial r^2}+\dfrac{1}{r^2}\dfrac{\partial^2\Psi}{\partial\theta^2}$
$\sigma_{\theta\theta}$	$\dfrac{1}{r}\dfrac{\partial\phi}{\partial r}+\dfrac{1}{r^2}\dfrac{\partial^2\phi}{\partial\theta^2}$	$\dfrac{z}{r}\dfrac{\partial\omega}{\partial r}+\dfrac{z}{r^2}\dfrac{\partial^2\omega}{\partial\theta^2}-2\nu\dfrac{\partial\omega}{\partial z}$	$-\dfrac{2}{r}\dfrac{\partial^2\Psi}{\partial r\partial\theta}+\dfrac{2}{r^2}\dfrac{\partial\Psi}{\partial\theta}$
σ_{rz}	$\dfrac{\partial^2\phi}{\partial r\partial z}$	$z\dfrac{\partial^2\omega}{\partial r\partial z}-(1-2\nu)\dfrac{\partial\omega}{\partial r}$	$\dfrac{1}{r}\dfrac{\partial^2\Psi}{\partial\theta\partial z}$
$\sigma_{\theta z}$	$\dfrac{1}{r}\dfrac{\partial^2\phi}{\partial\theta\partial z}$	$\dfrac{z}{r}\dfrac{\partial^2\omega}{\partial\theta\partial z}-\dfrac{1-2\nu}{r}\dfrac{\partial\omega}{\partial\theta}$	$-\dfrac{\partial^2\Psi}{\partial r\partial z}$

The Laplacian operator is

$$\nabla^2 \equiv \frac{\partial^2}{\partial R^2}+\frac{2}{R}\frac{\partial}{\partial R}+\frac{1}{R^2\sin^2\beta}\frac{\partial^2}{\partial\theta^2}+\frac{1}{R^2}\frac{\partial^2}{\partial\beta^2}+\frac{\cot\beta}{R^2}\frac{\partial}{\partial\beta} \tag{22.2}$$

and the strain-displacement relations[5] are

$$e_{RR} = \frac{\partial u_R}{\partial R} \quad ; \quad e_{\theta\theta} = \frac{u_R}{R} + \frac{u_\beta \cot \beta}{R} + \frac{1}{R \sin \beta} \frac{\partial u_\theta}{\partial \theta}$$

$$e_{\beta\beta} = \frac{u_R}{R} + \frac{1}{R} \frac{\partial u_\beta}{\partial \beta} \quad ; \quad e_{R\theta} = \frac{1}{2} \left(\frac{1}{R \sin \beta} \frac{\partial u_R}{\partial \theta} + \frac{\partial u_\theta}{\partial R} - \frac{u_\theta}{R} \right)$$

$$e_{\theta\beta} = \frac{1}{2} \left(\frac{1}{R} \frac{\partial u_\theta}{\partial \beta} - \frac{u_\theta \cot \beta}{R} + \frac{1}{R \sin \beta} \frac{\partial u_\beta}{\partial \theta} \right)$$

$$e_{\beta R} = \frac{1}{2} \left(\frac{1}{R} \frac{\partial u_R}{\partial \beta} + \frac{\partial u_\beta}{\partial R} - \frac{u_\beta}{R} \right) .$$

The corresponding displacement and stress components in Solutions A,B and E are given in Table 22.2 and are also listed in the Maple and Mathematica files 'ABErtb'.

22.5 Solutions Obtained by Superposition

The solutions obtained in the above sections can be superposed as required to provide a solution appropriate to the particular problem at hand. A case of particular interest is that in which the plane $z=0$ is one of the boundaries of the body — for example the semi-infinite solid or *half-space* $z>0$. We consider here two special cases.

22.5.1 Solution F: Frictionless isothermal contact problems

If the half-space is indented by a frictionless punch, so that the surface $z=0$ is subjected to normal tractions only, a simple formulation can be obtained by combining solutions A and B and defining a relationship between ϕ and ω in order to satisfy identically the condition $\sigma_{zx} = \sigma_{zy} = 0$ on $z=0$.

We write

$$\phi = (1 - 2\nu)\varphi \quad ; \quad \omega = \frac{\partial \varphi}{\partial z} \tag{22.3}$$

in solutions A, B respectively, obtaining

$$\sigma_{zx} = z \frac{\partial^3 \varphi}{\partial x \partial z^2} \quad ; \quad \sigma_{zy} = z \frac{\partial^3 \varphi}{\partial y \partial z^2} ,$$

---

[5] See for example A. S. Saada, *Elasticity*, Pergamon Press, New York, 1973, §6.7.

Table 22.2   Solutions A, B and E in spherical polar coördinates

	Solution A	**Solution B**	**Solution E**
$2\mu u_R$	$\dfrac{\partial\phi}{\partial R}$	$R\cos\beta\dfrac{\partial\omega}{\partial R} - (3-4\nu)\omega\cos\beta$	$\dfrac{2}{R}\dfrac{\partial\Psi}{\partial\theta}$
$2\mu u_\theta$	$\dfrac{1}{R\sin\beta}\dfrac{\partial\phi}{\partial\theta}$	$\cot\beta\dfrac{\partial\omega}{\partial\theta}$	$-2\dfrac{\partial\Psi}{\partial R}\sin\beta - \dfrac{2\cos\beta}{R}\dfrac{\partial\Psi}{\partial\beta}$
$2\mu u_\beta$	$\dfrac{1}{R}\dfrac{\partial\phi}{\partial\beta}$	$\cos\beta\dfrac{\partial\omega}{\partial\beta} + (3-4\nu)\omega\sin\beta$	$\dfrac{2\cot\beta}{R}\dfrac{\partial\Psi}{\partial\theta}$
σ_{RR}	$\dfrac{\partial^2\phi}{\partial R^2}$	$R\cos\beta\dfrac{\partial^2\omega}{\partial R^2} - 2(1-\nu)\dfrac{\partial\omega}{\partial R}\cos\beta$ $+\dfrac{2\nu}{R}\dfrac{\partial\omega}{\partial\beta}\sin\beta$	$\dfrac{2}{R}\dfrac{\partial^2\Psi}{\partial R\partial\theta} - \dfrac{2}{R^2}\dfrac{\partial\Psi}{\partial\theta}$
$\sigma_{\theta\theta}$	$\dfrac{1}{R}\dfrac{\partial\phi}{\partial R} + \dfrac{\cot\beta}{R^2}\dfrac{\partial\phi}{\partial\beta}$ $+\dfrac{1}{R^2\sin^2\beta}\dfrac{\partial^2\phi}{\partial\theta^2}$	$(1-2\nu)\dfrac{\partial\omega}{\partial R}\cos\beta + \dfrac{2\nu}{R}\dfrac{\partial\omega}{\partial\beta}\sin\beta$ $+\dfrac{\cos^2\beta}{R\sin\beta}\dfrac{\partial\omega}{\partial\beta} + \dfrac{\cot\beta}{R\sin\beta}\dfrac{\partial^2\omega}{\partial\theta^2}$	$\dfrac{2}{R^2\sin^2\beta}\dfrac{\partial\Psi}{\partial\theta} - \dfrac{2}{R}\dfrac{\partial^2\Psi}{\partial R\partial\theta}$ $-\dfrac{2\cot\beta}{R^2}\dfrac{\partial^2\Psi}{\partial\beta\partial\theta}$
$\sigma_{\beta\beta}$	$\dfrac{1}{R}\dfrac{\partial\phi}{\partial R} + \dfrac{1}{R^2}\dfrac{\partial^2\phi}{\partial\beta^2}$	$\dfrac{\cos\beta}{R}\dfrac{\partial^2\omega}{\partial\beta^2} + (1-2\nu)\dfrac{\partial\omega}{\partial R}\cos\beta$ $+\dfrac{2(1-\nu)\sin\beta}{R}\dfrac{\partial\omega}{\partial\beta}$	$\dfrac{2\cot\beta}{R^2}\dfrac{\partial^2\Psi}{\partial\theta\partial\beta} - \dfrac{2\cot^2\beta}{R^2}\dfrac{\partial\Psi}{\partial\theta}$
$\sigma_{\theta\beta}$	$\dfrac{1}{R^2\sin\beta}\dfrac{\partial^2\phi}{\partial\theta\partial\beta}$ $-\dfrac{\cot\beta}{R^2\sin\beta}\dfrac{\partial\phi}{\partial\theta}$	$\dfrac{\cot\beta}{R}\dfrac{\partial^2\omega}{\partial\theta\partial\beta} + \dfrac{2(1-\nu)}{R}\dfrac{\partial\omega}{\partial\theta}$ $-\dfrac{1}{R\sin^2\beta}\dfrac{\partial\omega}{\partial\theta}$	$\dfrac{\cot\beta}{R^2\sin\beta}\dfrac{\partial^2\Psi}{\partial\theta^2} - \dfrac{\sin\beta}{R}\dfrac{\partial^2\Psi}{\partial R\partial\beta}$ $-\dfrac{\cos\beta}{R^2}\dfrac{\partial^2\Psi}{\partial\beta^2} + \dfrac{1}{R^2\sin\beta}\dfrac{\partial\Psi}{\partial\beta}$
$\sigma_{\beta R}$	$\dfrac{1}{R}\dfrac{\partial^2\phi}{\partial\beta\partial R} - \dfrac{1}{R^2}\dfrac{\partial\phi}{\partial\beta}$	$(1-2\nu)\dfrac{\partial\omega}{\partial R}\sin\beta + \cos\beta\dfrac{\partial^2\omega}{\partial\beta\partial R}$ $-\dfrac{2(1-\nu)\cos\beta}{R}\dfrac{\partial\omega}{\partial\beta}$	$\dfrac{1}{R^2}\dfrac{\partial^2\Psi}{\partial\theta\partial\beta} + \dfrac{\cot\beta}{R}\dfrac{\partial^2\Psi}{\partial\theta\partial R}$ $-\dfrac{2\cot\beta}{R^2}\dfrac{\partial\Psi}{\partial\theta}$
$\sigma_{R\theta}$	$\dfrac{1}{R\sin\beta}\dfrac{\partial^2\phi}{\partial R\partial\theta}$ $-\dfrac{1}{R^2\sin\beta}\dfrac{\partial\phi}{\partial\theta}$	$\cot\beta\dfrac{\partial^2\omega}{\partial R\partial\theta}$ $-\dfrac{2(1-\nu)\cot\beta}{R}\dfrac{\partial\omega}{\partial\theta}$	$\dfrac{1}{R^2\sin\beta}\dfrac{\partial^2\Psi}{\partial\theta^2} - \sin\beta\dfrac{\partial^2\Psi}{\partial R^2}$ $-\dfrac{\cos\beta}{R}\dfrac{\partial^2\Psi}{\partial R\partial\beta} + \dfrac{2\cos\beta}{R^2}\dfrac{\partial\Psi}{\partial\beta}$ $+\dfrac{\sin\beta}{R}\dfrac{\partial\Psi}{\partial R}$

which vanish as required on the plane $z=0$. The remaining stress and displacement components are listed in Table 22.3 and at this stage we merely note that the normal traction and normal displacement at the surface reduce to the simple expressions

Table 22.3 Solutions F and G

	Solution F	Solution G
	Frictionless isothermal	
$2\mu u_x$	$z\dfrac{\partial^2 \varphi}{\partial x \partial z} + (1-2\nu)\dfrac{\partial \varphi}{\partial x}$	$2(1-\nu)\dfrac{\partial \chi}{\partial x} + z\dfrac{\partial^2 \chi}{\partial x \partial z}$
$2\mu u_y$	$z\dfrac{\partial^2 \varphi}{\partial y \partial z} + (1-2\nu)\dfrac{\partial \varphi}{\partial y}$	$2(1-\nu)\dfrac{\partial \chi}{\partial y} + z\dfrac{\partial^2 \chi}{\partial y \partial z}$
$2\mu u_z$	$z\dfrac{\partial^2 \varphi}{\partial z^2} - 2(1-\nu)\dfrac{\partial \varphi}{\partial z}$	$-(1-2\nu)\dfrac{\partial \chi}{\partial z} + z\dfrac{\partial^2 \chi}{\partial z^2}$
σ_{xx}	$z\dfrac{\partial^3 \varphi}{\partial x^2 \partial z} + \dfrac{\partial^2 \varphi}{\partial x^2} + 2\nu\dfrac{\partial^2 \varphi}{\partial y^2}$	$2(1-\nu)\dfrac{\partial^2 \chi}{\partial x^2} + z\dfrac{\partial^3 \chi}{\partial x^2 \partial z} - 2\nu\dfrac{\partial^2 \chi}{\partial z^2}$
σ_{xy}	$z\dfrac{\partial^3 \varphi}{\partial x \partial y \partial z} + (1-2\nu)\dfrac{\partial^2 \varphi}{\partial x \partial y}$	$2(1-\nu)\dfrac{\partial^2 \chi}{\partial x \partial y} + z\dfrac{\partial^3 \chi}{\partial x \partial y \partial z}$
σ_{yy}	$z\dfrac{\partial^3 \varphi}{\partial y^2 \partial z} + \dfrac{\partial^2 \varphi}{\partial y^2} + 2\nu\dfrac{\partial^2 \varphi}{\partial x^2}$	$2(1-\nu)\dfrac{\partial^2 \chi}{\partial y^2} + z\dfrac{\partial^3 \chi}{\partial y^2 \partial z} - 2\nu\dfrac{\partial^2 \chi}{\partial z^2}$
σ_{xz}	$z\dfrac{\partial^3 \varphi}{\partial x \partial z^2}$	$\dfrac{\partial^2 \chi}{\partial x \partial z} + z\dfrac{\partial^3 \chi}{\partial x \partial z^2}$
σ_{yz}	$z\dfrac{\partial^3 \varphi}{\partial y \partial z^2}$	$\dfrac{\partial^2 \chi}{\partial y \partial z} + z\dfrac{\partial^3 \chi}{\partial y \partial z^2}$
σ_{zz}	$z\dfrac{\partial^3 \varphi}{\partial z^3} - \dfrac{\partial^2 \varphi}{\partial z^2}$	$z\dfrac{\partial^3 \chi}{\partial z^3}$
$2\mu u_r$	$z\dfrac{\partial^2 \varphi}{\partial r \partial z} + (1-2\nu)\dfrac{\partial \varphi}{\partial r}$	$2(1-\nu)\dfrac{\partial \chi}{\partial r} + z\dfrac{\partial^2 \chi}{\partial r \partial z}$
$2\mu u_\theta$	$\dfrac{z}{r}\dfrac{\partial^2 \varphi}{\partial \theta \partial z} + \dfrac{(1-2\nu)}{r}\dfrac{\partial \varphi}{\partial \theta}$	$\dfrac{2(1-\nu)}{r}\dfrac{\partial \chi}{\partial \theta} + \dfrac{z}{r}\dfrac{\partial^2 \chi}{\partial \theta \partial z}$
σ_{rr}	$z\dfrac{\partial^3 \varphi}{\partial r^2 \partial z} + \dfrac{\partial^2 \varphi}{\partial r^2} - 2\nu\left(\dfrac{\partial^2 \varphi}{\partial r^2} + \dfrac{\partial^2 \varphi}{\partial z^2}\right)$	$2(1-\nu)\dfrac{\partial^2 \chi}{\partial r^2} + z\dfrac{\partial^3 \chi}{\partial r^2 \partial z} - 2\nu\dfrac{\partial^2 \chi}{\partial z^2}$
$\sigma_{r\theta}$	$\dfrac{z}{r}\dfrac{\partial^3 \varphi}{\partial r \partial \theta \partial z} - \dfrac{z}{r^2}\dfrac{\partial^2 \varphi}{\partial \theta \partial z}$ $+\dfrac{(1-2\nu)}{r}\left(\dfrac{\partial^2 \varphi}{\partial r \partial \theta} - \dfrac{1}{r}\dfrac{\partial \varphi}{\partial \theta}\right)$	$\dfrac{2(1-\nu)}{r}\left(\dfrac{\partial^2 \chi}{\partial r \partial \theta} - \dfrac{1}{r}\dfrac{\partial \chi}{\partial \theta}\right)$ $+\dfrac{z}{r}\dfrac{\partial^3 \chi}{\partial z \partial r \partial \theta} - \dfrac{z}{r^2}\dfrac{\partial^2 \chi}{\partial z \partial \theta}$
$\sigma_{\theta\theta}$	$-(1-2\nu)\dfrac{\partial^2 \varphi}{\partial r^2} - \dfrac{\partial^2 \varphi}{\partial z^2}$ $-z\dfrac{\partial^3 \varphi}{\partial r^2 \partial z} - z\dfrac{\partial^3 \varphi}{\partial z^3}$	$-2(1-\nu)\dfrac{\partial^2 \chi}{\partial r^2} - 2\dfrac{\partial^2 \chi}{\partial z^2}$ $+\dfrac{z}{r}\dfrac{\partial^2 \chi}{\partial z \partial r} + \dfrac{z}{r^2}\dfrac{\partial^3 \chi}{\partial z \partial \theta^2}$
σ_{rz}	$z\dfrac{\partial^3 \varphi}{\partial r \partial z^2}$	$\dfrac{\partial^2 \chi}{\partial r \partial z} + z\dfrac{\partial^3 \chi}{\partial r \partial z^2}$
$\sigma_{\theta z}$	$\dfrac{z}{r}\dfrac{\partial^3 \varphi}{\partial \theta \partial z^2}$	$\dfrac{1}{r}\dfrac{\partial^2 \chi}{\partial \theta \partial z} + \dfrac{z}{r}\dfrac{\partial^3 \chi}{\partial \theta \partial z^2}$

$$\sigma_{zz} = -\frac{\partial^2 \varphi}{\partial z^2} \; ; \;\; 2\mu u_z = -2(1-\nu)\frac{\partial \varphi}{\partial z} \; .$$

It follows that frictionless indentation problems for the half-space can be reduced to classical boundary-value problems in the harmonic potential function φ. Various examples are considered by Green and Zerna[6], who also show that symmetric problems of the plane crack in an isotropic body can be treated in the same way. These problems are considered in Chapters 31–33 below.

22.5.2   Solution G: The surface free of normal traction

The counterpart of the preceding solution is that obtained by combining Solutions A,B and requiring that the *normal* tractions vanish on the surface $z = 0$. We shall find this solution useful in problems involving a plane interface between two dissimilar materials (Chapter 35) and also in certain axisymmetric crack and contact problems.
   The normal traction on the plane $z = 0$ is

$$\sigma_{zz} = \frac{\partial^2 \phi}{\partial z^2} - 2(1-\nu)\frac{\partial \omega}{\partial z} \; ,$$

from Table 22.1 and hence it can be made to vanish by writing

$$\phi = 2(1-\nu)\chi \; ; \;\; \omega = \frac{\partial \chi}{\partial z} \; ,$$

where χ is a new harmonic potential function.
   The resulting stress and displacement components are given as Solution G in Table 22.3.

22.5.3   A plane strain solution

If we combine Green and Zerna's Solutions A and D[7] and require that the corresponding harmonic functions be independent of z, we obtain the displacement and stress components

---

[6] A. E. Green and W. Zerna, *loc. cit.* §§5.8-5.10.

[7] Solution D can be obtained from Solution B of Table 22.1 by permuting suffices — i.e. making the substitutions $z \to y$, $y \to x$, $x \to z$ in the expressions in Cartesian coördinates.

$$2\mu u_x = \frac{\partial \phi}{\partial x} + y\frac{\partial \omega}{\partial x} \; ; \quad 2\mu u_y = \frac{\partial \phi}{\partial y} + y\frac{\partial \omega}{\partial y} - (3-4\nu)\omega \; ; \quad 2\mu u_z = 0$$

$$\sigma_{xx} = \frac{\partial^2 \phi}{\partial x^2} + y\frac{\partial^2 \omega}{\partial x^2} - 2\nu\frac{\partial \omega}{\partial y} \; ; \quad \sigma_{yy} = \frac{\partial^2 \phi}{\partial y^2} + y\frac{\partial^2 \omega}{\partial y^2} - 2(1-\nu)\frac{\partial \omega}{\partial y} \quad (22.4)$$

$$\sigma_{xy} = \frac{\partial^2 \phi}{\partial x \partial y} + y\frac{\partial^2 \omega}{\partial x \partial y} - (1-2\nu)\frac{\partial \omega}{\partial x} \; ; \quad \sigma_{zx} = \sigma_{zy} = 0 \, ,$$

which describe a state of plane strain, as defined in Chapter 3. This formulation is useful for plane strain problems with displacement boundary conditions[8].

22.6 A Three-dimensional Complex-Variable Solution

We saw in Part IV that the complex-variable notation provides an elegant and powerful way to represent two-dimensional stress fields, and it is natural to ask whether any similar techniques could be extended to three-dimensional problems. Green[9] used a solution of this kind in the development of higher order solutions for elastic plates and we shall see in Chapter 30 that similar techniques can be used for the prismatic bar with general loading on the lateral surfaces.

We combine the displacement and stress components in the same way as in Chapter 20 by defining

$$u = u_x + \iota u_y \; ; \quad \Theta = \sigma_{xx} + \sigma_{yy}$$
$$\Phi = \sigma_{xx} + 2\iota\sigma_{xy} - \sigma_{yy} \; ; \quad \Psi = \sigma_{zx} + \iota\sigma_{zy} \, . \quad (22.5)$$

Following the convention of §20.1 and §22.2, we write the components of the Papkovich-Neuber vector potential as

$$\psi = \psi_x + \iota\psi_y \; ; \quad \omega = \psi_z$$

and hence

$$\boldsymbol{r} \cdot \boldsymbol{\psi} = \zeta\overline{\psi} + \overline{\zeta}\psi + z\omega \, ,$$

from (20.3). Using these results and (20.2) in (21.12), we obtain the in-plane displacements as

$$2\mu u = 2\frac{\partial \phi}{\partial \overline{\zeta}} - (3-4\nu)\psi + \zeta\frac{\partial \overline{\psi}}{\partial \overline{\zeta}} + \overline{\zeta}\frac{\partial \psi}{\partial \overline{\zeta}} + 2z\frac{\partial \omega}{\partial \overline{\zeta}} \, , \quad (22.6)$$

---

[8] see for example, Problems 6.3–6.5.
[9] A. E. Green (1949), The elastic equilibrium of isotropic plates and cylinders, *Proceedings of the Royal Society of London*, Vol. A195, pp. 533–552.

and the antiplane displacement as

$$2\mu u_z = \frac{\partial \phi}{\partial z} + \frac{1}{2}\left(\bar{\zeta}\frac{\partial \psi}{\partial z} + \zeta\frac{\partial \overline{\psi}}{\partial z}\right) + z\frac{\partial \omega}{\partial z} - (3-4\nu)\omega \ .$$

Once the displacements are defined, the stress components can be found from the constitutive law as in §§22.1, 22.2. Combining these as in (22.5), we obtain

$$\Theta = -\frac{\partial^2 \phi}{\partial z^2} - 2\left(\frac{\partial \psi}{\partial \zeta} + \frac{\partial \overline{\psi}}{\partial \overline{\zeta}}\right) - \frac{1}{2}\left(\bar{\zeta}\frac{\partial^2 \overline{\psi}}{\partial z^2} + \zeta\frac{\partial^2 \psi}{\partial z^2}\right)$$

$$-z\frac{\partial^2 \omega}{\partial z^2} - 4\nu\frac{\partial \omega}{\partial z} \tag{22.7}$$

$$\Phi = 4\frac{\partial^2 \phi}{\partial \bar{\zeta}^2} - 4(1-2\nu)\frac{\partial \psi}{\partial \bar{\zeta}} + 2\zeta\frac{\partial^2 \overline{\psi}}{\partial \bar{\zeta}^2} + 2\bar{\zeta}\frac{\partial^2 \psi}{\partial \bar{\zeta}^2} + 4z\frac{\partial^2 \omega}{\partial \bar{\zeta}^2} \tag{22.8}$$

$$\sigma_{zz} = \frac{\partial^2 \phi}{\partial z^2} - 2\nu\left(\frac{\partial \psi}{\partial \zeta} + \frac{\partial \overline{\psi}}{\partial \overline{\zeta}}\right) + \frac{1}{2}\left(\bar{\zeta}\frac{\partial^2 \psi}{\partial z^2} + \zeta\frac{\partial^2 \overline{\psi}}{\partial z^2}\right)$$

$$+z\frac{\partial^2 \omega}{\partial z^2} - 2(1-\nu)\frac{\partial \omega}{\partial z} \tag{22.9}$$

$$\Psi = 2\frac{\partial^2 \phi}{\partial \bar{\zeta}\partial z} - (1-2\nu)\frac{\partial \psi}{\partial z} + \zeta\frac{\partial^2 \overline{\psi}}{\partial \bar{\zeta}\partial z} + \bar{\zeta}\frac{\partial^2 \psi}{\partial \bar{\zeta}\partial z}$$

$$+2z\frac{\partial^2 \omega}{\partial z\partial \bar{\zeta}} - 2(1-2\nu)\frac{\partial \omega}{\partial \bar{\zeta}} \ , \tag{22.10}$$

where $\phi(\zeta, \bar{\zeta}, z), \omega(\zeta, \bar{\zeta}, z)$ are real three-dimensional harmonic functions and $\psi(\zeta, \bar{\zeta}, z)$ is a complex three-dimensional harmonic function. Notice that we can write the three-dimensional Laplace equation in the form

$$\frac{\partial^2 \phi}{\partial z^2} + 4\frac{\partial^2 \phi}{\partial \zeta\partial \bar{\zeta}} = 0 \ ; \quad \frac{\partial^2 \omega}{\partial z^2} + 4\frac{\partial^2 \psi}{\partial \zeta\partial \bar{\zeta}} = 0 \ ; \quad \frac{\partial^2 \psi}{\partial z^2} + 4\frac{\partial^2 \psi}{\partial \zeta\partial \bar{\zeta}} = 0 \ , \tag{22.11}$$

using (19.9). Thus, if the functions ϕ, ψ depend on z, they will not generally be holomorphic with regard to ζ or $\bar{\zeta}$. We shall explore categories of complex-variable solution of equation (22.11) in §25.8 below.

Problems

22.1 Use solution E to solve the elementary torsion problem for a cylindrical bar of radius b transmitting a torque T, the surface $r = b$ being traction free. You will need to find a suitable harmonic potential function — it will be an axisymmetric polynomial.

22.2 Find a way of taking a combination of solutions A,B,E so as to satisfy identically the global conditions

$$\sigma_{zy} = \sigma_{zz} = 0 \; ; \quad \text{all } x, y, \; z = 0$$

Thus, σ_{zx} should be the only non-zero traction component on the surface.

22.3 An important class of frictional contact problems involves the steady sliding of an indenter in the x-direction across the surface of the half-space $z > 0$. Assuming Coulomb's friction law to apply with coefficient of friction f, the appropriate boundary conditions are

$$\sigma_{xz} = -f\sigma_{zz} \; ; \quad \sigma_{zy} = 0 , \tag{22.12}$$

in the contact area, the rest of the surface $z=0$ being traction free.

Show that (22.12) are actually *global* conditions for this problem and find a way of combining Solutions A,B,E so as to satisfy them identically.

22.4 A large number of thin fibres are bonded to the surface $z=0$ of the elastic half-space $z > 0$, with orientation in the x-direction. The fibres are sufficiently rigid to prevent axial displacement u_x, but sufficiently thin to offer no restraint to lateral motion u_y, u_z.

This composite block is now indented by a frictionless rigid punch. Find a linear combination of Solutions A, B and E that satisfies the global conditions of the problem and in particular find expressions for the surface values of the normal displacement u_z and the normal traction σ_{zz} in terms of the one remaining harmonic potential function.

22.5 Show that if ϕ, ω are taken to be independent of x, a linear combination of Solutions A and B defines a state of plane strain in the yz-plane (i.e. $u_x=0$, $\sigma_{xy}=\sigma_{xz}=0$ and all the remaining stress and displacement components are independent of x).

An infinite layer $0 < z < h$ rests on a frictionless rigid foundation at $z=0$, whilst the surface $z=h$ is loaded by the normal traction

$$\sigma_{zz} = -p_0 + p_1 \cos(my) ,$$

where $p_1 < p_0$. Find the distribution of contact pressure at $z=0$.

22.6 A massive asteroid passes close to the surface of the Earth causing gravitational forces that can be described by a known harmonic body force potential $V(x, y, z)$. Assuming that the affected region of the Earth can be approximated as the traction-free half-space $z > 0$, find a representation of the resulting stress and displacement fields in terms of a single harmonic potential function. An appropriate strategy is:-

(i) Using the particular solution of §21.5.1, show that the potential ϕ must be biharmonic.
(ii) Show that this condition can be satisfied by writing $\phi = z\chi$, where χ is a *harmonic* function, and find the relation between χ and V.
(iii) Find the tractions on the surface of the half-space in terms of derivatives of χ.
(iv) Superpose appropriate combinations of Solutions A and/or B so as to render this surface traction free for all functions χ.

Find the resulting stress field for the special case of a spherical asteroid of radius a, for which

$$V = -\frac{4\pi\gamma\rho\rho_0 a^3}{3\sqrt{x^2 + y^2 + (z+h)^2}} \, ,$$

where γ is the universal gravitational constant, ρ, ρ_0 are the densities of the asteroid and the Earth respectively, and the centre of the sphere is at the point $(0, 0, -h)$, a height $h > a$ above the Earth's surface.

If the Earth and the asteroid have the same density ($\rho_0 = \rho$) and the self-weight of the Earth is assumed to cause hydrostatic compression of magnitude $\rho g z$, where g is the acceleration due to gravity, is it possible for the passage of the asteroid to cause tensile stresses anywhere in the Earth's crust. If so, where?

22.7 An elastic half-space is defined by the inequality $z > h(x, y)$, where $h(x, y)$ defines a perturbation from the plane $z = 0$ that is small in the sense that the magnitude of the surface slope $|\nabla h| \ll 1$. The surface $z = h$ is traction free, but the extremities of the half-space are subjected to a uniform biaxial tensile stress $\sigma_{xx} = \sigma_{yy} = \sigma_0$.

Show that Solution G can be used to define an approximation to the perturbation in the stress field due to the surface profile and define the boundary conditions for the harmonic potential function χ in terms of h.

Solve the resulting boundary-value problem for the case where the profile is defined by the two-dimensional depression

$$h(x, y) = \frac{h_0 a^2}{x^2 + a^2} \, ,$$

and hence find the location and magnitude of the maximum tensile stress.

22.8 A general two-dimensional biharmonic function can be written in the form

$$\phi = \psi + y\frac{\partial\chi}{\partial y} \, ,$$

where ψ, χ are harmonic functions of x, y. Use this form to describe the Airy stress function of Chapter 4, and develop expressions for the stress components $\sigma_{xx}, \sigma_{yy}, \sigma_{xy}$ as functions of ψ, χ. Show that the resulting solution can be written in terms of a linear combination of the harmonic functions ϕ, ω of equations (22.4).

Suggest ways of using this relationship to determine the displacements associated with a given Airy stress function ϕ and illustrate the procedure with a simple example.

Chapter 23
Thermoelastic Displacement Potentials

As in the two-dimensional case (Chapter 14), three-dimensional problems of thermoelasticity are conveniently treated by finding a *particular solution* — i.e. a solution which satisfies the field equations without regard to boundary conditions — and completing the general solution by superposition of an appropriate representation for the general isothermal problem, such as the Papkovich-Neuber solution.

In this section, we shall show that a particular solution can always be obtained in the form of a strain potential — i.e. by writing

$$2\mu \boldsymbol{u} = \nabla \phi \,. \tag{23.1}$$

The thermoelastic stress-strain relations (14.3, 14.4) can be solved to give

$$\sigma_{xx} = \lambda e + 2\mu e_{xx} - (3\lambda + 2\mu)\alpha T \tag{23.2}$$

$$\sigma_{xy} = 2\mu e_{xy} \,, \tag{23.3}$$

etc.

Using the strain-displacement relations and substituting for $\boldsymbol{u}$ from (23.1), we then find

$$2\mu e = \nabla^2 \phi \tag{23.4}$$

$$\sigma_{xx} = \frac{\lambda}{2\mu}\nabla^2 \phi + \frac{\partial^2 \phi}{\partial x^2} - (3\lambda + 2\mu)\alpha T \tag{23.5}$$

$$\sigma_{xy} = \frac{\partial^2 \phi}{\partial x \partial y} \,, \tag{23.6}$$

etc., and hence the equilibrium equation (2.1) requires that

Supplementary Information The online version contains supplementary material available at https://doi.org/10.1007/978-3-031-15214-6_23.

J. R. Barber, *Elasticity*, Solid Mechanics and Its Applications 172,
https://doi.org/10.1007/978-3-031-15214-6_23

$$\frac{\partial}{\partial x}\left(\frac{\lambda}{2\mu}\nabla^2\phi + \nabla^2\phi - (3\lambda + 2\mu)\alpha T\right) = 0 .$$

Two similar equations are obtained from (2.2, 2.3) and hence the most general solution of this form must satisfy the condition

$$\nabla^2\phi = \frac{2\mu(3\lambda + 2\mu)\alpha T}{(\lambda + 2\mu)} + C ,$$

where C is an arbitrary constant[1].

Since we are only looking for a particular solution, we choose to set $C = 0$, giving

$$\nabla^2\phi = \frac{2\mu(3\lambda + 2\mu)\alpha T}{(\lambda + 2\mu)} = \frac{2\mu(1 + \nu)\alpha T}{(1 - \nu)} . \tag{23.7}$$

Equation (23.7) defines the potential ϕ due to a source distribution proportional to the given temperature, T. Such a function can be found for all T in any geometric domain[2] and hence a particular solution of the thermoelastic problem in the form (23.1) can always be found.

Substituting for T from equation (23.7) back into the stress equation (23.5), we obtain

$$\sigma_{xx} = \left(\frac{\lambda}{2\mu} - \frac{(\lambda + 2\mu)}{2\mu}\right)\nabla^2\phi + \frac{\partial^2\phi}{\partial x^2} = \frac{\partial^2\phi}{\partial x^2} - \nabla^2\phi$$

$$= -\frac{\partial^2\phi}{\partial y^2} - \frac{\partial^2\phi}{\partial z^2} .$$

The particular solution can therefore be summarized in the equations

$$\nabla^2\phi = \frac{2\mu(1 + \nu)\alpha T}{(1 - \nu)} \tag{23.8}$$

$$2\mu\boldsymbol{u} = \boldsymbol{\nabla}\phi \tag{23.9}$$

$$\sigma_{xx} = -\frac{\partial^2\phi}{\partial y^2} - \frac{\partial^2\phi}{\partial z^2} \ ; \ \sigma_{yy} = -\frac{\partial^2\phi}{\partial z^2} - \frac{\partial^2\phi}{\partial x^2} \ ; \ \sigma_{zz} = -\frac{\partial^2\phi}{\partial x^2} - \frac{\partial^2\phi}{\partial y^2}$$

$$\sigma_{xy} = \frac{\partial^2\phi}{\partial x\partial y} \ ; \ \sigma_{yz} = \frac{\partial^2\phi}{\partial y\partial z} \ ; \ \sigma_{zx} = \frac{\partial^2\phi}{\partial z\partial x} . \tag{23.10}$$

The equivalent expressions in cylindrical polar coördinates are

$$2\mu u_r = \frac{\partial\phi}{\partial r} \ ; \ 2\mu u_\theta = \frac{1}{r}\frac{\partial\phi}{\partial\theta} \ ; \ 2\mu u_z = \frac{\partial\phi}{\partial z}$$

---

[1] We encountered this constant already in equation (21.3) and noted that it corresponds to a state of uniform stress.

[2] For example, as a convolution integral on the point source solution $1/R$.

$$\sigma_{rr} = \frac{\partial^2 \phi}{\partial r^2} - \nabla^2 \phi \;\; ; \;\; \sigma_{\theta\theta} = \frac{1}{r}\frac{\partial \phi}{\partial r} + \frac{1}{r^2}\frac{\partial^2 \phi}{\partial \theta^2} - \nabla^2 \phi \;\; ; \;\; \sigma_{zz} = \frac{\partial^2 \phi}{\partial z^2} - \nabla^2 \phi$$

$$\sigma_{r\theta} = \frac{\partial}{\partial r}\left(\frac{1}{r}\frac{\partial \phi}{\partial \theta}\right) \;\; ; \;\; \sigma_{\theta z} = \frac{1}{r}\frac{\partial^2 \phi}{\partial \theta \partial z} \;\; ; \;\; \sigma_{zr} = \frac{\partial^2 \phi}{\partial z \partial r} \;, \tag{23.11}$$

where $\nabla^2 \phi$ is given by equation (22.1).

In spherical polar coördinates, we have

$$2\mu u_R = \frac{\partial \phi}{\partial R} \;\; ; \;\; 2\mu u_\theta = \frac{1}{R \sin \beta}\frac{\partial \phi}{\partial \theta} \;\; ; \;\; 2\mu u_\beta = \frac{1}{R}\frac{\partial \phi}{\partial \beta}$$

$$\sigma_{RR} = \frac{\partial^2 \phi}{\partial R^2} - \nabla^2 \phi \;\; ; \;\; \sigma_{\theta\theta} = \frac{1}{R}\frac{\partial \phi}{\partial R} + \frac{\cot \beta}{R^2}\frac{\partial \phi}{\partial \beta} - \nabla^2 \phi$$

$$\sigma_{\beta\beta} = \frac{1}{R}\frac{\partial \phi}{\partial R} + \frac{1}{R^2}\frac{\partial^2 \phi}{\partial \beta^2} - \nabla^2 \phi \;\; ; \;\; \sigma_{R\theta} = \frac{1}{R \sin \beta}\frac{\partial^2 \phi}{\partial R \partial \theta}$$

$$\sigma_{\theta\beta} = \frac{1}{R^2 \sin \beta}\frac{\partial^2 \phi}{\partial \theta \partial \beta} \;\; ; \;\; \sigma_{\beta R} = \frac{1}{R}\frac{\partial^2 \phi}{\partial \beta \partial R} - \frac{1}{R^2}\frac{\partial \phi}{\partial \beta} \;, \tag{23.12}$$

where $\nabla^2 \phi$ is given by equation (22.2).

23.1 The Method of Strain Suppression

An alternative approach for obtaining a particular solution for the thermoelastic stress field is to reduce it to a body force problem using the *method of strain suppression*. Suppose we were to constrain every particle of the body so as to prevent it from moving — in other words, we constrain the displacement to be everywhere zero. The strains will then be identically zero everywhere and the constitutive equations (23.2, 23.3) reduce to

$$\sigma_{xx} = \sigma_{yy} = \sigma_{zz} = -(3\lambda + 2\mu)\alpha T = -\frac{2\mu(1+\nu)\alpha T}{(1-2\nu)} \;\; ; \;\; \sigma_{xy} = \sigma_{yz} = \sigma_{zx} = 0 \;. \tag{23.13}$$

Of course, these stresses will not generally satisfy the equilibrium equations, but we can always find a body force distribution that will restore equilibrium — all we need to do is to substitute the non-equilibrated stresses (23.13) into the equilibrium equation with body forces (2.4) and use the latter as a definition of p, obtaining

$$p = \frac{2\mu(1+\nu)}{(1-2\nu)}\alpha \nabla T \;. \tag{23.14}$$

It follows that (23.13) will define the stresses in the body for the specified temperature distribution if (i) body forces (23.14) are applied and (ii) purely normal tractions are applied at the boundaries given by

$$t = -\frac{2\mu(1+\nu)}{(1-2\nu)}\alpha T n \ , \tag{23.15}$$

where T is the local temperature at the boundary and n is the local outward normal unit vector.

To complete the solution, we must now remove the unwanted body forces by superposing the solution of a problem with no thermal strains, but with equal and opposite body forces and with surface tractions that combined with (23.15) satisfy the traction boundary conditions of the original problem.

The resulting body force problem might be solved using the body force potential of §21.5.1, since (23.14) defines a conservative vector field. However, the boundary-value problem would then be found to be identical with that obtained by substituting the original temperature distribution into (23.7), so this does not really constitute a new method. An alternative strategy might be to represent the body force as a distribution of point forces as suggested in §13.1.1, but using the three-dimensional Kelvin solution which we shall develop in the next chapter.

23.2 Boundary-value Problems

If the function ϕ satisfies equation (23.8), the stresses and displacements defined by equations (23.9–23.12) will satisfy the governing equation for all x, y, z, but they will not generally satisfy the displacement and/or traction boundary conditions. To solve a boundary-value problem, we therefore must superpose the general homogeneous solution — i.e. a general state of deformation that can exist in a body without thermal strains.

23.2.1 Spherically-symmetric Stresses

A particularly simple solution can be obtained if the problem is spherically symmetric — i.e. if the temperature depends only on the radius R in spherical polar coördinates (R, θ, β). In this case, equation (23.8) reduces to the ordinary differential equation

$$\nabla^2 \phi = \frac{d^2\phi}{dR^2} + \frac{2}{R}\frac{d\phi}{dR} = \frac{1}{R^2}\frac{d}{dR}\left(R^2\frac{d\phi}{dR}\right) = \frac{2\mu(1+\nu)\alpha T}{(1-\nu)} \ ,$$

which can be integrated to yield

$$\frac{d\phi}{dR} = \frac{2\mu(1+\nu)\alpha}{(1-\nu)R^2} \int TR^2 dR + \frac{A}{R^2} , \tag{23.16}$$

where A is an arbitrary constant. The only non-zero stress and displacement components are

$$2\mu u_R = \frac{d\phi}{dR} ; \quad \sigma_{RR} = -\frac{2}{R}\frac{d\phi}{dR} ; \quad \sigma_{\theta\theta} = \sigma_{\beta\beta} = -\frac{d^2\phi}{dR^2} - \frac{1}{R}\frac{d\phi}{dR} . \tag{23.17}$$

In the most general spherically-symmetric problem, we must satisfy one boundary condition (e.g. normal traction or radial displacement) at each of two boundaries for a hollow sphere, but equation (23.16) contains only one arbitrary constant. An additional degree of freedom can be introduced by superposing an arbitrary state of uniform hydrostatic stress defined by

$$\sigma_{RR} = \sigma_{\theta\theta} = \sigma_{\beta\beta} = \sigma_0 ; \quad 2\mu u_R = \frac{\sigma_0(1-2\nu)R}{(1+\nu)} , \tag{23.18}$$

all other stress and displacement components being zero. Substituting (23.16) into (23.17) and adding (23.18), we obtain

$$2\mu u_R = \frac{2\mu(1+\nu)\alpha}{(1-\nu)R^2} \int TR^2 dR + \frac{A}{R^2} + \frac{\sigma_0(1-2\nu)R}{(1+\nu)}$$

$$\sigma_{RR} = -\frac{4\mu(1+\nu)\alpha}{(1-\nu)R^3} \int TR^2 dR - \frac{2A}{R^3} + \sigma_0$$

$$\sigma_{\theta\theta} = \sigma_{\beta\beta} = \frac{2\mu(1+\nu)\alpha}{(1-\nu)} \left(\frac{1}{R^3} \int TR^2 dR - T \right) + \frac{A}{R^3} + \sigma_0 .$$

For a hollow sphere $b < R < a$, the lower limit in these integrals can be taken as the inner radius b and the boundary conditions at a, b will provide two equations for the unknowns A, σ_0. For a solid sphere, the constant A must be zero to avoid a singularity at the origin.

23.2.2 *More general geometries*

For more general three-dimensional problems, the stresses and displacements defined by equations (23.9–23.12) can be superposed on those defined by the Galerkin vector in §21.2, or on the Boussinesq potentials of Chapter 22. In this connection, we notice that if we set $T = 0$ in (23.8) so that ϕ is harmonic, then the expressions for all the stress and displacement components reduce to those for Solution A of Tables 22.1, 22.2. In other words, Solution A can be construed as the homogeneous solution of (23.8). We therefore have two ways of proceeding, analogous to the two approaches to conservative body-force problems described in §7.3. We can either

1. Find any particular solution ϕ_P of (23.8), find the corresponding stresses and displacements from equations (23.9–23.12) and then use the boundary values to define a corrective solution that can be formulated using Solutions A, B and E.
2. Start with a superposition of equations (23.8–23.12) and Solutions B, E, and satisfy the governing equation (23.8) and the boundary conditions at the same time. In this formulation, notice that ϕ is now interpreted as the general solution of (23.8), including an arbitrary harmonic function whose degrees of freedom are needed in satisfying the boundary conditions.

A similar approach can also be used for conservative body-force problems, using the solution of equations (21.26–21.28) in place of Solution A.

23.3 Plane Problems

If ϕ and T are independent of z, the non-zero stress components in (23.10) reduce to

$$\sigma_{xx} = -\frac{\partial^2 \phi}{\partial y^2} \; ; \; \sigma_{xy} = \frac{\partial^2 \phi}{\partial x \partial y} \; ; \; \sigma_{yy} = -\frac{\partial^2 \phi}{\partial x^2} \tag{23.19}$$

$$\sigma_{zz} = -\frac{\partial^2 \phi}{\partial x^2} - \frac{\partial^2 \phi}{\partial y^2} \tag{23.20}$$

and the solution corresponds to a state of plane strain. In fact, the relations (23.19) are identical to the Airy function definitions, apart from a difference of sign, and similarly, (23.8) reduces to a particular integral of (14.7). However, an important difference is that the present solution also gives explicit relations for the displacements and can therefore be used for problems with displacement boundary conditions, including those arising for multiply-connected bodies.

23.3.1 Axisymmetric problems for the cylinder

We consider the cylinder $b < r < a$ in a state of plane strain with a prescribed temperature distribution. The example in §14.1 and the treatment of isothermal problems in Chapters 8, 9 show that this problem could be solved for a fairly general temperature distribution, but a case of particular importance is that in which the temperature is an axisymmetric function $T(r)$.

If ϕ is also taken to depend upon r only, we can write (23.8) in the form

$$\frac{1}{r}\frac{d}{dr}r\frac{d\phi}{dr} = \frac{2\mu(1+\nu)\alpha T(r)}{(1-\nu)} ,$$

which has the particular integral

$$\frac{d\phi}{dr} = \frac{2\mu(1+\nu)\alpha}{(1-\nu)r} \int_b^r rT(r)dr \;,$$

where we have selected the inner radius b as the lower limit of integration, since the required generality will be introduced later through the superposed isothermal solution.

The corresponding displacement and stress components are then obtained as

$$u_r = \frac{1}{2\mu}\frac{d\phi}{dr} = \frac{(1+\nu)\alpha}{(1-\nu)r} \int_b^r rT(r)dr$$

$$\sigma_{rr} = -\frac{1}{r}\frac{d\phi}{dr} = -\frac{2\mu(1+\nu)\alpha}{(1-\nu)r^2} \int_b^r rT(r)dr$$

$$\sigma_{\theta\theta} = -\frac{d^2\phi}{dr^2} = \frac{2\mu(1+\nu)\alpha}{(1-\nu)} \left(\frac{1}{r^2} \int_b^r rT(r)dr - T(r)\right)$$

$$\sigma_{r\theta} = 0 \;;\quad u_\theta = 0 \;.$$

The general solution is now obtained by superposing those axisymmetric terms[3] from Tables 8.1, 9.1 that give single-valued displacements, with the result

$$u_r = \frac{(1+\nu)\alpha}{(1-\nu)r} \int_b^r rT(r)dr + A(1-2\nu)r - \frac{B}{r}$$

$$\sigma_{rr} = -\frac{2\mu(1+\nu)\alpha}{(1-\nu)r^2} \int_b^r rT(r)dr + 2\mu A + \frac{2\mu B}{r^2}$$

$$\sigma_{\theta\theta} = \frac{2\mu(1+\nu)\alpha}{(1-\nu)} \left(\frac{1}{r^2} \int_b^r rT(r)dr - T(r)\right) + 2\mu A - \frac{2\mu B}{r^2}$$

$$\sigma_{r\theta} = 0 \;;\quad u_\theta = 0 \;.$$

The constants A, B permit arbitrary axisymmetric boundary conditions to be satisfied at $r=a, b$. The same equations can also be used for the solid cylinder or disk $(b=0)$, provided the constant B associated with the singular term is set to zero.

23.3.2 Steady-state plane problems

The definitions of the in-plane stress components in equation (23.19) are identical to those of the Airy stress function except for a difference in sign. Furthermore, if the temperature T is in a steady state, it must be harmonic and hence ϕ must be biharmonic, from (23.8). It follows that if we superpose a homogeneous stress

---

[3] Notice that since this is the plane strain solution, we have used $\kappa = (3-4\nu)$ in Table 9.1.

field derived from an equal Airy stress function $\varphi = \phi$, the resulting in-plane stress components will be everywhere zero. This provides an alternative proof of the result in §14.3 that for steady-state plane problems in simply-connected bodies, there will be no in-plane thermal stress if the boundary tractions are zero.

The displacements of course will not be zero, as we know from §14.3.1. Suppose we are given a temperature distribution T satisfying $\nabla^2 T = 0$ and are asked to determine the corresponding displacement field. If the body is simply connected there are no in-plane stresses, so

$$e_{xx} = \frac{\partial u_x}{\partial x} = \alpha(1+\nu)T \ ; \quad e_{yy} = \frac{\partial u_y}{\partial y} = \alpha(1+\nu)T$$

$$e_{xy} = \frac{1}{2}\left(\frac{\partial u_y}{\partial x} + \frac{\partial u_x}{\partial y}\right) = 0 \ , \tag{23.21}$$

where we have used (14.1, 14.2) modified for plane strain by (14.8). It is convenient to formulate this problem in the complex-variable formalism of Chapter 20. We then have

$$\frac{\partial u}{\partial \bar{\zeta}} = \frac{1}{2}\left(\frac{\partial}{\partial x} + \imath \frac{\partial}{\partial y}\right)(u_x + \imath u_y) = \frac{1}{2}\left(\frac{\partial u_x}{\partial x} - \frac{\partial u_y}{\partial y}\right) + \frac{\imath}{2}\left(\frac{\partial u_y}{\partial x} + \frac{\partial u_x}{\partial y}\right) \ ,$$

from (19.6, 20.3). Comparing with (23.21), we see that both real and imaginary parts of the right-hand side are identically zero, so

$$\frac{\partial u}{\partial \bar{\zeta}} = 0 \ ,$$

implying that u is a holomorphic function of ζ only. We also have

$$\frac{\partial u}{\partial \zeta} = \frac{1}{2}\left(\frac{\partial}{\partial x} - \imath \frac{\partial}{\partial y}\right)(u_x + \imath u_y) = \frac{1}{2}\left(\frac{\partial u_x}{\partial x} + \frac{\partial u_y}{\partial y}\right) + \frac{\imath}{2}\left(\frac{\partial u_y}{\partial x} - \frac{\partial u_x}{\partial y}\right)$$

$$= \frac{e}{2} + \imath \omega_z \tag{23.22}$$

and we know the real part of this holomorphic function since

$$\frac{e}{2} = \alpha(1+\nu)T = \alpha(1+\nu)\left(g(\zeta) + \bar{g}(\bar{\zeta})\right) \ , \tag{23.23}$$

where the (harmonic) temperature is defined as

$$T = g(\zeta) + \bar{g}(\bar{\zeta}) \ , \tag{23.24}$$

as in equation (19.11). The most general holomorphic function of ζ satisfying (23.22, 23.23) is

$$\frac{\partial u}{\partial \zeta} = 2\alpha(1+\nu)g(\zeta) + \imath \omega_0 \ , \tag{23.25}$$

where ω_0 is a real constant representing an arbitrary rigid-body rotation $\omega_z = \omega_0$. Thus, if we know the temperature and hence the function $g(\zeta)$ in equation (23.24), we can determine the displacement field simply by integrating equation (23.25) with respect to ζ. The translational rigid body terms will arise as an arbitrary complex constant during this integration.

23.3.3 Heat flow perturbed by a circular hole

If the body is multiply connected, the function $\varphi = \phi$ may correspond to multiple-valued displacements and hence not represent a valid homogeneous solution. In this case, additional biharmonic terms terms must be added as in §13.1 and the resulting stress field will generally be non-zero.

Consider the case in which a uniform flow of heat $q_x = q_0$ in a large body is perturbed by the presence of an insulated circular hole of radius a. We can write the temperature field in the form

$$T = T_0 + T_1 \; ,$$

where T_0 is the temperature that would be obtained if there were no hole and T_1 is the perturbation in temperature due to the hole. Following the strategy of §8.3.2, we first consider the unperturbed problem of uniform heat flow, in order to determine the appropriate Fourier dependence of the terms describing the perturbation.

The unperturbed problem

In the absence of the hole, the body would be simply connected and we know from §23.3.2 that the in-plane stresses would be everywhere zero. However, as in §8.3.2, an examination of the unperturbed problem enables us to determine the appropriate Fourier dependence of the terms describing the perturbation. The heat flux will then be the same everywhere and we have

$$q_x = -K\frac{\partial T_0}{\partial x} = q_0 \; ; \quad T_0 = -\frac{q_0 x}{K} = -\frac{q_0 r \cos\theta}{K} \; ,$$

where we have chosen the zero of temperature to be that at the origin. This expression varies with $\cos\theta$ and we conclude that the perturbation in both T and ϕ will exhibit similar dependence on θ.

The temperature perturbation

The temperature perturbation T_1 must be harmonic and decay as $r \to \infty$ and the only function with the required Fourier form satisfying these conditions is $\cos\theta/r$. We therefore write

$$T = T_0 + T_1 = -\frac{q_0 r \cos\theta}{K} + \frac{C_1 \cos\theta}{r} , \qquad (23.26)$$

where the constant C_1 must be determined from the condition that the radial heat flux $q_r = 0$ at $r = a$. In other words,

$$-K\frac{\partial T}{\partial r}(a, \theta) = 0 .$$

Substituting for T from (23.26) and solving for C_1, we obtain

$$C_1 = -\frac{q_0 a^2}{K} \quad \text{and hence} \quad T_1 = -\frac{q_0 a^2 \cos\theta}{Kr} .$$

The thermoelastic particular solution

We can now obtain a particular thermoelastic solution ϕ_1 for the perturbation problem by substituting T_1 into (23.8), writing $\phi_1 = f(r)\cos\theta$ and solving for $f(r)$. We obtain

$$\phi_1 = -\frac{\mu(1+\nu)\alpha q_0 a^2 r \ln(r) \cos\theta}{K(1-\nu)}$$

and the corresponding in-plane stress components are

$$\sigma_{rr} = \frac{\mu(1+\nu)\alpha q_0 a^2 \cos\theta}{K(1-\nu)r} \quad ; \quad \sigma_{r\theta} = \frac{\mu(1+\nu)\alpha q_0 a^2 \sin\theta}{K(1-\nu)r}$$

$$\sigma_{\theta\theta} = \frac{\mu(1+\nu)\alpha q_0 a^2 \cos\theta}{K(1-\nu)r} , \qquad (23.27)$$

from (23.11).

The homogeneous solution

The surface of the hole must be traction free — i.e.

$$\sigma_{rr} = \sigma_{r\theta} = 0 \quad ; \quad r = a \qquad (23.28)$$

and these conditions are clearly not satisfied by equations (23.27). We must therefore superpose a homogeneous solution which we construct from the Airy stress function terms

$$\varphi = C_2 r\theta \sin\theta + C_3 r \ln(r)\cos\theta + \frac{C_4 \cos\theta}{r} \, , \qquad (23.29)$$

from Table 8.1. The choice of these terms is dictated by the Fourier dependence of the required stress components and the condition that the perturbation must decay as $r \to \infty$. The corresponding in-plane stresses (including the particular solution) are

$$\sigma_{rr} = \frac{\mu(1+\nu)\alpha q_0 a^2 \cos\theta}{K(1-\nu)r} + \frac{2C_2 \cos\theta}{r} + \frac{C_3 \cos\theta}{r} - \frac{2C_4 \cos\theta}{r^3}$$

$$\sigma_{r\theta} = \frac{\mu(1+\nu)\alpha q_0 a^2 \sin\theta}{K(1-\nu)r} + \frac{C_3 \sin\theta}{r} - \frac{2C_4 \sin\theta}{r^3}$$

$$\sigma_{\theta\theta} = \frac{\mu(1+\nu)\alpha q_0 a^2 \cos\theta}{K(1-\nu)r} + \frac{C_3 \cos\theta}{r} + \frac{2C_4 \cos\theta}{r^3}$$

and the boundary conditions (23.28) then require that

$$\frac{2C_2}{a} + \frac{C_3}{a} - \frac{2C_4}{a^3} = -\frac{\mu(1+\nu)\alpha q_0 a}{K(1-\nu)} \qquad (23.30)$$

$$\frac{C_3}{a} - \frac{2C_4}{a^3} = -\frac{\mu(1+\nu)\alpha q_0 a}{K(1-\nu)} \, . \qquad (23.31)$$

A third equation for the three constants C_2, C_3, C_4 is obtained from the requirement that the displacements be single-valued, as in §13.1. This condition is automatically satisfied by the particular solution, since it is derived from a single-valued displacement function ϕ_1, so we simply record the potentially multiple-valued terms arising from φ. The first two terms in (23.29) are identical to (13.1) except that C_1 is replaced by C_2 and hence the required condition is

$$C_2(\kappa - 1) + C_3(\kappa + 1) = 0 \, ,$$

by comparison with (13.5). The use of the thermoelastic potential independent of z implies plane strain conditions, so we set $\kappa = 3 - 4\nu$ obtaining

$$2(1 - 2\nu)C_2 + 4(1 - \nu)C_3 = 0 \, . \qquad (23.32)$$

We can then solve (23.30, 23.31, 23.32) for C_2, C_3, C_4, obtaining

$$C_2 = C_3 = 0 \; ; \quad C_4 = \frac{\mu\alpha(1+\nu)q_0 a^4}{2K(1-\nu)} \, .$$

The final stress field is then obtained as

$$\sigma_{rr} = \frac{\mu\alpha(1+\nu)q_0a^2(r^2-a^2)\cos\theta}{K(1-\nu)r^3}$$

$$\sigma_{r\theta} = \frac{\mu\alpha(1+\nu)q_0a^2(r^2-a^2)\sin\theta}{K(1-\nu)r^3}$$

$$\sigma_{\theta\theta} = \frac{\mu\alpha(1+\nu)q_0a^2(r^2+a^2)\cos\theta}{K(1-\nu)r^3} \; .$$

In particular, the maximum tensile stress is

$$\sigma_{\max} = \sigma_{\theta\theta}(a,0) = \frac{2\mu\alpha(1+\nu)q_0a}{K(1-\nu)} \; .$$

23.3.4 Plane stress

All the results in §23.3 have been obtained from the three dimensional representation (23.1) under the restriction that ϕ be independent of z and hence $u_z = 0$. They therefore define the plane strain solution. Corresponding plane stress results can be obtained by making the substitutions

$$\mu = \mu' \; ; \quad \nu = \frac{\nu'}{(1+\nu')} \; ; \quad \alpha = \frac{\alpha'(1+\nu')}{(1+2\nu')}$$

and then removing the primes.

23.4 Steady-state Temperature: Solution T

If the temperature distribution is in a steady-state and there are no internal heat sources, the temperature T is harmonic, from equation (14.12). It follows from equation (23.8) that the potential function ϕ is biharmonic and it can conveniently be expressed in terms of a harmonic function χ through the relation

$$\phi = z\chi \; . \tag{23.33}$$

We then have

$$\nabla^2\phi = 2\frac{\partial\chi}{\partial z}$$

and hence

$$T = \frac{(1-\nu)}{\mu\alpha(1+\nu)}\frac{\partial\chi}{\partial z} \; ,$$

from equation (23.8). Expressions for the displacements and stresses are easily obtained by substituting (23.33) into equations (23.9–23.10) and they are tabulated as Solution T in Table 23.1 (for Cartesian and cylindrical polar coördinates) and Table 23.2 (for spherical polar coördinates).

Table 23.1 Solutions T and P

	Solution T Williams' solution	Solution P Thermoelastic Plane Stress
$2\mu u_x$	$z\dfrac{\partial \chi}{\partial x}$	$\dfrac{\partial \psi}{\partial x}$
$2\mu u_y$	$z\dfrac{\partial \chi}{\partial y}$	$\dfrac{\partial \psi}{\partial y}$
$2\mu u_z$	$\chi + z\dfrac{\partial \chi}{\partial z}$	$-\dfrac{\partial \psi}{\partial z}$
σ_{xx}	$z\dfrac{\partial^2 \chi}{\partial x^2} - 2\dfrac{\partial \chi}{\partial z}$	$-\dfrac{\partial^2 \psi}{\partial y^2}$
σ_{xy}	$z\dfrac{\partial^2 \chi}{\partial x \partial y}$	$\dfrac{\partial^2 \psi}{\partial x \partial y}$
σ_{yy}	$z\dfrac{\partial^2 \chi}{\partial y^2} - 2\dfrac{\partial \chi}{\partial z}$	$-\dfrac{\partial^2 \psi}{\partial x^2}$
σ_{xz}	$\dfrac{\partial \chi}{\partial x} + z\dfrac{\partial^2 \chi}{\partial x \partial z}$	0
σ_{yz}	$\dfrac{\partial \chi}{\partial y} + z\dfrac{\partial^2 \chi}{\partial y \partial z}$	0
σ_{zz}	$z\dfrac{\partial^2 \chi}{\partial z^2}$	0
$2\mu u_r$	$z\dfrac{\partial \chi}{\partial r}$	$\dfrac{\partial \psi}{\partial r}$
$2\mu u_\theta$	$\dfrac{z}{r}\dfrac{\partial \chi}{\partial \theta}$	$\dfrac{1}{r}\dfrac{\partial \psi}{\partial \theta}$
σ_{rr}	$z\dfrac{\partial^2 \chi}{\partial r^2} - 2\dfrac{\partial \chi}{\partial z}$	$\dfrac{1}{r}\dfrac{\partial \psi}{\partial r} + \dfrac{1}{r^2}\dfrac{\partial^2 \psi}{\partial \theta^2}$
$\sigma_{r\theta}$	$\dfrac{z}{r}\dfrac{\partial^2 \chi}{\partial r \partial \theta} - \dfrac{z}{r^2}\dfrac{\partial \chi}{\partial \theta}$	$\dfrac{1}{r}\dfrac{\partial^2 \psi}{\partial r \partial \theta} - \dfrac{1}{r^2}\dfrac{\partial \psi}{\partial \theta}$
$\sigma_{\theta\theta}$	$\dfrac{z}{r}\dfrac{\partial \chi}{\partial r} + \dfrac{z}{r^2}\dfrac{\partial^2 \chi}{\partial \theta^2} - 2\dfrac{\partial \chi}{\partial z}$	$-\dfrac{\partial^2 \psi}{\partial r^2}$
σ_{rz}	$\dfrac{\partial \chi}{\partial r} + z\dfrac{\partial^2 \chi}{\partial r \partial z}$	0
$\sigma_{\theta z}$	$\dfrac{z}{r}\dfrac{\partial^2 \chi}{\partial \theta \partial z} + \dfrac{1}{r}\dfrac{\partial \chi}{\partial \theta}$	0
T	$\dfrac{(1-\nu)}{\mu\alpha(1+\nu)}\dfrac{\partial \chi}{\partial z}$	$-\dfrac{1}{2\mu\alpha(1+\nu)}\dfrac{\partial^2 \psi}{\partial z^2}$
q_z	$-\dfrac{(1-\nu)K}{\mu\alpha(1+\nu)}\dfrac{\partial^2 \chi}{\partial z^2}$	$\dfrac{K}{2\mu\alpha(1+\nu)}\dfrac{\partial^3 \psi}{\partial z^3}$
q_r	$-\dfrac{(1-\nu)K}{\mu\alpha(1+\nu)}\dfrac{\partial^2 \chi}{\partial z \partial r}$	$\dfrac{K}{2\mu\alpha(1+\nu)}\dfrac{\partial^3 \psi}{\partial z^2 \partial r}$

Table 23.2 Solution T in spherical polar coördinates

	Solution T
$2\mu u_R$	$R\cos\beta\dfrac{\partial\chi}{\partial R} + \chi\cos\beta$
$2\mu u_\theta$	$\cot\beta\dfrac{\partial\chi}{\partial\theta}$
$2\mu u_\beta$	$\cos\beta\dfrac{\partial\chi}{\partial\beta} - \chi\sin\beta$
σ_{RR}	$R\cos\beta\dfrac{\partial^2\chi}{\partial R^2} + \dfrac{2\sin\beta}{R}\dfrac{\partial\chi}{\partial\beta}$
$\sigma_{\theta\theta}$	$\dfrac{\cot\beta}{R\sin\beta}\dfrac{\partial^2\chi}{\partial\theta^2} - \cos\beta\dfrac{\partial\chi}{\partial R} + \dfrac{(1+\sin^2\beta)}{R\sin\beta}\dfrac{\partial\chi}{\partial\beta}$
$\sigma_{\beta\beta}$	$\dfrac{\cos\beta}{R}\dfrac{\partial^2\chi}{\partial\beta^2} - \dfrac{\partial\chi}{\partial R}\cos\beta$
$\sigma_{\theta\beta}$	$\dfrac{\cot\beta}{R}\dfrac{\partial^2\chi}{\partial\theta\partial\beta} - \dfrac{1}{R\sin^2\beta}\dfrac{\partial\chi}{\partial\theta}$
$\sigma_{\beta R}$	$\cos\beta\dfrac{\partial^2\chi}{\partial R\partial\beta} - \sin\beta\dfrac{\partial\chi}{\partial R}$
$\sigma_{R\theta}$	$\cot\beta\dfrac{\partial^2\chi}{\partial R\partial\theta}$
T	$\dfrac{1-\nu}{\mu\alpha(1+\nu)}\left(\dfrac{\partial\chi}{\partial R}\cos\beta - \dfrac{\sin\beta}{R}\dfrac{\partial\chi}{\partial\beta}\right)$
q_R	$\dfrac{(1-\nu)K}{\mu\alpha(1+\nu)}\left(\dfrac{\sin\beta}{R}\dfrac{\partial^2\chi}{\partial R\partial\beta} - \dfrac{\partial^2\chi}{\partial R^2}\cos\beta - \dfrac{\sin\beta}{R^2}\dfrac{\partial\chi}{\partial\beta}\right)$
q_θ	$\dfrac{(1-\nu)K}{\mu\alpha(1+\nu)}\left(\dfrac{1}{R^2}\dfrac{\partial^2\chi}{\partial\beta\partial\theta} - \dfrac{\cot\beta}{R}\dfrac{\partial^2\chi}{\partial R\partial\theta}\right)$
q_β	$\dfrac{(1-\nu)K}{\mu\alpha(1+\nu)}\left(\dfrac{\sin\beta}{R}\dfrac{\partial\chi}{\partial R} - \dfrac{\cos\beta}{R}\dfrac{\partial^2\chi}{\partial R\partial\beta} + \dfrac{\sin\beta}{R^2}\dfrac{\partial^2\chi}{\partial\beta^2} + \dfrac{\cos\beta}{R^2}\dfrac{\partial\chi}{\partial\beta}\right)$

We also note that the heat flux $\mathbf{q} = -K\nabla T$, and hence (for example)

$$q_z = -K\frac{\partial T}{\partial z} = -\frac{(1-\nu)K}{\mu\alpha(1+\nu)}\frac{\partial^2\chi}{\partial z^2} \; ,$$

where K is the thermal conductivity of the material. These expressions are listed in the Maple and Mathematica files 'Txyz', 'Trtz' 'Trtb' for Cartesian, cylindrical polar and spherical polar coördinates respectively.

The solution here described was first obtained by Williams[4] and was used by him for the solution of some thermoelastic crack problems.

---

[4] W. E. Williams (1961), A solution of the steady-state thermoelastic equations, *Zeitschrift für angewandte Mathematik und Physik,* Vol. 12, pp. 452–455.

23.4.1 Thermoelastic plane stress

An important thermoelastic solution is that appropriate to the steady-state thermo-elastic deformation of a traction-free half-space, which can be obtained by super-posing solutions A,B,T and taking

$$\phi = \psi \; ; \;\; \omega = -\chi = \frac{1}{2(1-\nu)} \frac{\partial \psi}{\partial z} . \tag{23.34}$$

The resulting expressions are given in Table 23.1 as Solution P. A striking result is that the three stress components $\sigma_{zx}, \sigma_{zy}, \sigma_{zz}$ are zero, not merely on the plane $z=0$, where we have forced them to zero by the relations (23.34), but throughout the half-space. For this reason, the solution is also appropriate to the problem of a thick plate $a < z < b$ with traction-free faces. Because of this feature, the solution is henceforth referred to as *thermoelastic plane stress*.

Notice also the similarity of the expressions for the non-zero stresses $\sigma_{xx}, \sigma_{xy}, \sigma_{yy}$ to the corresponding expressions obtained from the Airy stress function. However, the present solution is not two-dimensional — i.e. the temperature and hence all the non-zero stress and displacement components can be functions of all three coördinates — and it is exact, whereas the two-dimensional plane stress solution is generally approximate.

Another result of interest is

$$\frac{\partial^2 u_z}{\partial x^2} + \frac{\partial^2 u_z}{\partial y^2} = -\frac{1}{2\mu} \left(\frac{\partial^3 \psi}{\partial x^2 \partial z} + \frac{\partial^3 \psi}{\partial y^2 \partial z} \right) = \frac{1}{2\mu} \frac{\partial^3 \psi}{\partial z^3}$$

(because ψ and hence $\partial \psi / \partial z$ is harmonic)

$$= \frac{\alpha(1+\nu)q_z}{K} = \delta q_z , \tag{23.35}$$

where δ is the *thermal distortivity* (see §14.3.1).

In other words, the sum of the principal curvatures of any distorted z-plane is pro-portional to the local heat flux across that plane[5]. The corresponding two-dimensional result was proved in §14.3.1.

---

[5] Further consequences of this result are discussed in J.R.Barber (1980), Some implications of Dundurs' theorem for thermoelastic contact and crack problems, *Journal of Strain Analysis,* Vol. 22, pp. 229–232.

Problems

23.1. The temperature field in the elastic half-space $z > 0$ is defined by

$$T(x, y, z) = f(z)\cos(mx) ,$$

where $f(z)$ is a known function of z only. The surface $z=0$ of the half-space is traction free. Show that the normal surface displacement $u_z(x, y, 0)$ can be written

$$u_z(x, y, 0) = u_0\cos(mx) ,$$

where

$$u_0 = 2\alpha(1 + \nu) \int_0^\infty e^{-ms} f(s)ds .$$

Important Note: This problem *cannot* be solved using Solutions T or P, since the temperature field will only be harmonic for certain special functions $f(z)$. To solve the problem, assume a strain potential ϕ of the form

$$\phi = g(z)\cos(mx) ,$$

substitute into (23.8) and solve the resulting ordinary differential equation for $g(z)$. You can then superpose appropriate potentials from solutions A and B of Table 22.1 to satisfy the traction-free boundary condition.

23.2. Due to the decay of radioactive elements, heat is generated in a spherical planet of radius a at a uniform rate of Q_0 per unit volume per unit time. Find the resulting thermal stresses in the planet if the surface $R = a$ is traction free and the temperature field has reached a steady state. You will need the heat conduction equation (14.11).

23.3. If a ductile material is loaded in hydrostatic tension $\sigma_{ij} = \sigma_0\delta_{ij}$, it cannot yield and is therefore expected to fail in a brittle manner at sufficiently high σ_0. Experimentally it is difficult to load a specimen in triaxial tension, but one approach is to heat a solid sphere rapidly at its outer surface, which leads to the generation of a state of triaxial tension at the centre. Show that the instantaneous value of this stress is given by

$$\frac{4\mu(1 + \nu)\alpha}{(1 - \nu)a^3} \int_0^a R^2 \big[T(R) - T(0) \big] dR ,$$

where a is the radius of the sphere, $T(R)$ is the instantaneous temperature at radius R, and the surface $R = a$ is traction free. Comment on possible limitations of this approach.

23.4. During a test on an automotive disk brake, it was found that the temperature T of the disk could be approximated by the expression

$$T = \frac{5}{4}(1 - e^{-\tau}) - \frac{r^2}{2a^2}(1 - e^{-\tau}) + \frac{r^4}{8a^4}\tau e^{-\tau} ,$$

where $\tau = t/t_0$ is a dimensionless time, a is the radius of the disk, r is the distance from the axis, t is time and t_0 is a constant. The disk is continuous at $r = 0$ and all its surfaces can be assumed to be traction free.

Derive expressions for the radial and circumferential thermal stresses as functions of radius and time, and hence show that the maximum value of the stress difference $(\sigma_{\theta\theta} - \sigma_{rr})$ at any given time always occurs at the outer edge of the disk.

23.5. A heat exchanger tube consists of a long hollow cylinder of inner radius a and outer radius b, maintained at temperatures T_a, T_b respectively. Find the steady-state stress distribution in the tube if the curved surfaces are traction free and plane strain conditions can be assumed.

23.6. The steady-state temperature $T(\xi, \eta)$ in the quarter plane $\xi \geq 0$, $\eta \geq 0$ is determined by the boundary conditions

$$T(\xi, 0) = 0 \ ; \quad \xi > 0$$
$$T(0, \eta) = T_0 \ ; \quad 0 < \eta < a$$
$$= 0 \ ; \quad \eta > a \ .$$

Use the function

$$\omega(\zeta) = \zeta^{1/2}$$

to map the problem into the half-plane $y > 0$ and solve for T using the method of §19.6.3.

Then determine the displacement throughout the quarter plane, using equation (23.25).

23.7. Use dimensional arguments and continuity of heat flux to show that the two-dimensional steady-state temperature field in the half-plane $y > 0$ due to a line heat source Q per unit length at the origin is

$$T(x, y) = \frac{Q}{K} \ln(r) + C \ ,$$

where $r = \sqrt{x^2 + y^2}$ is the distance from the origin, K is the thermal conductivity and C is a constant.

Use this result and equation (23.25) to determine the displacement throughout the half-plane. Assume that $C = 0$ and neglect rigid-body displacements. In particular, determine the displacements at the surface $y = 0$ and verify that they satisfy Dundurs' theorem (14.15).

23.8. A uniform heat flux $q_x = q_0$ in a large body in a state of plane strain is perturbed by the presence of a rigid non-conducting circular inclusion of radius a. The inclusion is bonded to the matrix at $r = a$. Find the complete stress field in the matrix and identify the location and magnitude of the maximum tensile stress. Assume plane strain conditions.

23.9. A uniform heat flux $q_x = q_0$ in a large body in a state of plane strain with elastic and thermal properties $\mu_1, \nu_1, \alpha_1, K_1$ is perturbed by the presence of a circular inclusion of radius a with properties $\mu_2, \nu_2, \alpha_2, K_2$. The inclusion is bonded to the matrix at $r = a$. Find the complete stress field in the matrix and the inclusion and identify the location and magnitude of the maximum tensile stress in each. Assume plane strain conditions. **Note:** You will find the algebra for this problem formidable if you do not use Maple or Mathematica.

23.10. A rigid non-conducting body slides at velocity V in the positive x-direction over the surface of the thermoelastic half-space $z > 0$, causing frictional tractions σ_{zx} and frictional heating q_z given by

$$\sigma_{zx} = -f\sigma_{zz} \ ; \quad \sigma_{zy} = 0 \ ; \quad q_z = V\sigma_{zx} \ ,$$

where f is the coefficient of friction. Find a solution in terms of one potential function that satisfies these global boundary conditions in the steady state.

If the half-space remains in contact with the sliding body throughout the plane $z = 0$, we have the additional global boundary condition $u_z = 0$. Show that a non-trivial steady-state solution to this problem can be obtained using the harmonic function $Ce^{-mz}\cos(my)$, where C is an arbitrary constant and m is a parameter which must be chosen to satisfy the displacement boundary condition. Comment on the physical significance of this solution.

23.11. The half-space $z > 0$ is bonded to a rigid body at the surface $z = 0$. Show that in the steady state, the surface temperature $T(x, y, 0)$ and the normal stress $\sigma_{zz}(x, y, 0)$ at the bond are related by the equation

$$\sigma_{zz}(x, y, 0) = -\frac{2\mu\alpha(1 + \nu)T(x, y, 0)}{(3 - 4\nu)} \ .$$

23.12. The traction-free surface of the half-space $z > 0$ is uniformly heated in the annular region $a < r < b$ and unheated elsewhere, so that

$$\begin{aligned} q_z(r, \theta, 0) &= \frac{Q}{\pi(b^2 - a^2)} \ ; \quad a < r < b \\ &= 0 \qquad\qquad ; \quad 0 \le r < a \ \text{and} \ r > b \ . \end{aligned}$$

Use equation (23.35) to determine the slope of the surface

$$\frac{\partial u_z}{\partial r}(r, \theta, 0)$$

in the steady state. By taking the limit of this solution as $a \to b$ or otherwise, find the surface displacement due to a heat input Q uniformly distributed along a circle of radius b. Comment on the implications for steady-state thermoelastic contact problems.

Chapter 24
Singular Solutions

Singular solutions have a special place in the historical development of potential theory in general and of Elasticity in particular. Many of the important early solutions were obtained by appropriate superposition of singular potentials, noting that any form of singularity is permitted provided that the singular point is not a point of the body[1].

The generalization of this technique to allow continuous distributions of singularities in space (either at the boundary of the body or in a region of space not occupied by it) is still one of the most widely used methods of treating three-dimensional problems.

In this chapter, we shall consider some elementary forms of singular solution and examine the problems that they solve when used in the various displacement function representations of Chapter 22. The solutions will be developed in a rather *ad hoc* way, starting with those that are mathematically most straightforward and progressing to more complex forms. However, once appropriate forms have been introduced, a systematic way of developing them from first principles will be discussed in the next chapter.

24.1 The Source Solution

A convenient starting point is the most elementary singular harmonic function

$$\phi = \frac{1}{R} = \frac{1}{\sqrt{x^2 + y^2 + z^2}} \ , \tag{24.1}$$

which is easily demonstrated to be harmonic (except at the origin) by substitution into Laplace's equation. We shall refer to this solution as the 'source' solution, since it describes the steady-state temperature in an infinite body with a point heat source

---

[1] If it *is* a point of the body, certain restrictions must be imposed on the strength of the singularity, as discussed in §11.2.1 and §37.8.1.

© The Author(s), under exclusive license to Springer Nature Switzerland AG 2022
J. R. Barber, *Elasticity*, Solid Mechanics and Its Applications 172,
https://doi.org/10.1007/978-3-031-15214-6_24

at the origin, or alternatively the potential in an infinite space with a point electric charge at the origin. It is clearly spherically symmetric — i.e. the potential depends only on the distance R from the origin — and it can be developed from first principles, starting with Laplace's equation in spherical polar coördinates (22.2) and imposing the condition that ϕ be a function of R only. The governing equation then reduces to

$$\frac{d^2\phi}{dR^2} + \frac{2}{R}\frac{d\phi}{dR} = 0$$

of which (24.1) is clearly the only singular solution.

We shall examine the stress fields corresponding to the use of the source solution in Solutions A and B of Tables 22.1, 22.2.

24.1.1 The centre of dilatation

The strain potential solution (Solution A) is *isotropic* — i.e., the stress and displacement definitions preserve the same form in any Cartesian coördinate transformation — and hence the spherically symmetric function $1/R$ must correspond to a spherically symmetric state of stress and displacement.

Substituting $\phi = 1/R$ into Table 22.2, we obtain

$$2\mu u_R = -\frac{1}{R^2} \quad ; \quad 2\mu u_\theta = 2\mu u_\beta = 0 \tag{24.2}$$

$$\sigma_{RR} = \frac{2}{R^3} \quad ; \quad \sigma_{\theta\theta} = \sigma_{\beta\beta} = -\frac{1}{R^3} \quad ; \quad \sigma_{\theta\beta} = \sigma_{\beta R} = \sigma_{R\theta} = 0 \,.$$

Consider points on the spherical surface $R = a$, whose area is $4\pi a^2$. These points move radially outwards through a distance $u_R(a)$ and hence the material inside the imaginary sphere increases in volume by

$$\Delta V = 4\pi a^2 u_R(a) = -\frac{2\pi}{\mu} \,,$$

using (24.2). Notice that ΔV is independent of a, implying that the volume of the imaginary *hollow* sphere $b < R < a$ remains constant. In other words, the volume change is all concentrated at the origin, and we can construct a solution for a prescribed volume change ΔV through the potential

$$\phi = -\frac{\mu \Delta V}{2\pi R} \,.$$

This defines the singularity known as the *centre of dilatation* or centre of compression[2]. Combined with a state of uniform hydrostatic stress, it can be used to describe

---

[2] S. P. Timoshenko and J. N. Goodier, *loc. cit.*, §136.

the stresses and displacements in a thick-walled spherical pressure vessel loaded by arbitrary uniform pressure at the inner and outer surfaces. This solution was first obtained by Lamé.

The centre of dilatation is not admissable at an interior point of a continuous body — it would correspond to an infinitesimal hole containing a fluid at infinite pressure. However, a *distribution* of centres of dilatation can be used to describe dilatational effects such as those arising from thermal expansion. In fact, the thermoelastic potential of equations (23.10–23.13) could be regarded as being derived from Solution A by distributing sources of strength proportional to αT throughout the space occupied by the body, to take account of the thermal dilatation. A similar technique can be used for dilatation arising from other sources, such as phase transformation.

24.1.2 The Kelvin solution

In Solution B, the stress components contain terms that are first derivatives of the potential ω or second derivatives multiplied by z. Hence, if we substitute $\omega = 1/R$, we anticipate that the stresses will vary inversely with the square of the distance from the origin and hence that the force resultant over a spherical surface of radius a, with centre at the origin, will be independent of a. In fact, this potential corresponds to the three-dimensional Kelvin problem, in which a concentrated force in the z-direction is applied at the origin in an infinite elastic body.

We shall demonstrate this by evaluating the stress components σ_{zz}, σ_{rz} and considering the equilibrium of the region $-h < z < h$, which includes the origin.

Noting that

$$\frac{\partial}{\partial z}\left(\frac{1}{R}\right) = -\frac{z}{R^3} \quad ; \quad \frac{\partial}{\partial r}\left(\frac{1}{R}\right) = -\frac{r}{R^3} ,$$

we obtain

$$\sigma_{zz} = z\frac{\partial^2 \omega}{\partial z^2} - 2(1-\nu)\frac{\partial \omega}{\partial z} = z\left(\frac{3z^2}{R^5} - \frac{1}{R^3}\right) + 2(1-\nu)\frac{z}{R^3}$$

$$= (1-2\nu)\frac{z}{R^3} + \frac{3z^3}{R^5}$$

$$\sigma_{rz} = z\frac{\partial^2 \omega}{\partial r \partial z} - (1-2\nu)\frac{\partial \omega}{\partial r}$$

$$= (1-2\nu)\frac{r}{R^3} + \frac{3rz^2}{R^5} .$$

We now consider the equilibrium of the cylinder $r < a$, $-h < z < h$, which is shown in Figure 24.1.

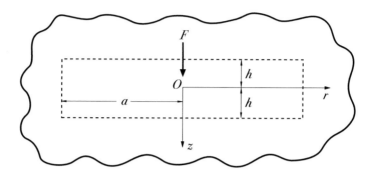

Fig. 24.1 The Kelvin problem

The curved surfaces $r = a$ of this cylinder experience a force in the z-direction because of the stress σ_{rz}, but this force decays to zero as $a \to \infty$, since, although the area increases with a (being equal to $2\pi a h$), the stress component decreases with a^2. We can therefore restrict attention to the surfaces $z = \pm h$ if a is sufficiently large.

The stress component σ_{zz} is odd in z and hence the forces transmitted across the two surfaces are equal and have the total value

$$
\begin{aligned}
F_z &= 2\int_0^\infty 2\pi r \left(\frac{(1-2\nu)h}{R_h^3} + \frac{3h^3}{R_h^5} \right) dr \\
&= 4\pi \left[-\frac{(1-2\nu)h}{R_h} - \frac{h^3}{R_h^3} \right]_{r=0}^{r=\infty} = 8\pi(1-\nu) ,
\end{aligned}
$$

where $R_h = \sqrt{r^2 + h^2}$ and we note that $R_h = h$ when $r = 0$.

It follows that there must be an equal and opposite force at the origin. Hence, the function

$$
\omega = -\frac{F}{8\pi(1-\nu)R} \tag{24.3}
$$

in Solution B corresponds to the problem of a force F acting in the z-direction at the origin in the infinite elastic body. The complete stress field is easily obtained by substituting (24.3) into the appropriate expressions from Table 22.1 or Table 22.2.

24.2 Dimensional Considerations

In the last section, we developed the Kelvin solution 'accidentally' by examining the stress field due to the source potential in Solution B. However, a more deductive development of the solution can be made from equilibrium and dimensional considerations.

We first note that the Kelvin problem is self-similar (see §12.1), since there is no inherent length scale. It follows that the form of the variation of the solution with θ, β must be independent of the distance R from the origin and hence that the solution can be expressed in the separated-variable form

$$f(R, \theta, \beta) = g(R)h(\beta, \theta) ,$$

where f is any field quantity such as a stress or displacement component.

Given this result, we can deduce from equilibrium considerations that the function $g(R)$ appropriate to the stress components must be R^{-2}, since the total resultant force across any one of a class of self-similar surfaces (i.e. surfaces of the same shape but different size) must be the same, and the surface area of such surfaces will be proportional to the square of their linear dimensions. Thus, any stress component must decay with distance from the origin according to R^{-2}.

We now examine solutions A and B to see what type of singular function would satisfy this condition. On dimensional grounds, we find that the potential would have to be of order R^0 in Solution A and of order R^{-1} in Solution B. We shall see in the next section that there *are* singular potentials of order R^0, but that they involve singularities not merely at the origin, but along at least half of the z-axis. Thus, they cannot be used for a problem involving the whole infinite body. We are therefore left with the function $1/R$ — which we know to be harmonic — in Solution B, which can then be developed as in §24.1.2.

24.2.1 The Boussinesq solution

We shall now apply similar arguments to solve the Boussinesq problem, in which a point force F in the z-direction is applied at the origin $R=0$ on the surface $z=0$ of the half-space $z>0$, as shown in Figure 24.2. In particular, it is clear that this problem is also self-similar and equilibrium arguments again demand that the stress components decay with distance from the origin with R^{-2}.

This problem is treated by Timoshenko and Goodier[3] by superposing further singular solutions on the Kelvin solution so as to make the surface $z=0$ free of

---

[3] *loc. cit.*, §138.

tractions except at the origin. A more direct (and modern) approach is to seek a special potential function solution which identically satisfies two of the three boundary conditions at the surface.

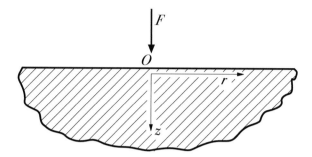

Fig. 24.2 The Boussinesq problem

We note that the force is normal to the surface, so that there is no *tangential* traction at any point on the surface[4] — i.e.

$$\sigma_{zx} = \sigma_{zy} = 0 \; ; \quad \text{all } x, y, \; z = 0 \, .$$

Since this condition applies at all points of the surface $z=0$, we shall refer to it as a *global boundary condition* and it is appropriate to satisfy it by constructing a potential function solution which satisfies it identically, this being Solution F of Table 22.3. It then remains to find a suitable potential function which when used in Solution F will yield stress components which decay with R^{-2}.

Examination of Solution F shows that $1/R$ will not be suitable in the present instance, since the stresses are obtained by two differentiations of the stress function and hence we require φ to vary with R^0. We therefore seek a suitable partial integral of $1/R$ — to be dimensionless in R and singular at the origin, but otherwise to be continuous and harmonic in $z > 0$.

It is easily verified that the function

$$\varphi = \int_{-\infty}^{0} \frac{d\zeta}{\sqrt{x^2 + y^2 + (z - \zeta)^2}} = \ln(R + z)$$

satisfies these requirements. Notice that this function is singular on the negative z-axis, where $R = -z$, but not on the positive z-axis, where $R = z$. In fact, it can be regarded as the potential due to a uniform distribution of sources on the negative z-axis ($r = 0, z < 0$).

---

[4] The normal traction is zero everywhere except at the origin, where there is a delta function loading, i.e. $\sigma_{zz} = -F\delta(x)\delta(y)$.

Substituting into Solution F, we find

$$\sigma_{zz} = z\frac{\partial^3\varphi}{\partial z^3} - \frac{\partial^2\varphi}{\partial z^2} = \frac{3z^3}{R^5}$$

and hence the surface $z=0$ is free of traction except at the origin as required[5].
The force applied at the origin is

$$F = -2\pi\int_0^\infty r\sigma_{zz}(r,h)dr$$

$$= -6\pi h^3\int_0^\infty \frac{rdr}{(r^2+h^2)^{5/2}} = -2\pi$$

and hence the stress field due to a force F in the z-direction applied at the origin is
obtained from the potential

$$\varphi = -\frac{F}{2\pi}\ln(R+z) . \tag{24.4}$$

The displacements at the surface $z=0$ are of particular interest, in view of the
application to contact problems[6]. We have

$$2\mu u_r(r,0) = (1-2\nu)\frac{\partial\varphi}{\partial r} = -\frac{F(1-2\nu)}{2\pi r} \tag{24.5}$$

$$2\mu u_z(r,0) = -2(1-\nu)\frac{\partial\varphi}{\partial z} = \frac{F(1-\nu)}{\pi r} . \tag{24.6}$$

Thus, both displacements vary inversely with distance from the point of application of the force F, as indeed we could have deduced from dimensional considerations. The singularity in displacement at the origin is not serious, since in any
practical application, the force will be distributed over a finite area. If the solution
for such a distributed force is found by superposition using the above result, the
singularity will be integrated out.

We note that the displacements are bounded at infinity, in contrast to the two-
dimensional case (see Chapter 12). It is therefore possible to regard the point at
infinity as a reference for rigid-body displacements, if appropriate.

The radial surface displacement u_r is negative, indicating that a force F directed
into the half-space — i.e. a compressive force — causes the surrounding surface to
move towards the origin[7]. We can define a *surface dilatation* e_s on the surface $z=0$,
such that

---

[5] The condition $\sigma_{zr}=0$ on the surface is guaranteed by the use of Solution F.

[6] cf. Chapter 12.

[7] Most engineers would say that this is what they would intuitively expect, but the validity of
this intuition is suspect, since the corresponding result for a tensile force — for which the radial
displacement is directed away from the origin — is counter-intuitive. Our expectation here is
probably determined by the thought that forces of either direction will stretch the surface and hence

$$e_s \equiv \frac{\partial u_x}{\partial x} + \frac{\partial u_y}{\partial y} = \frac{\partial u_r}{\partial r} + \frac{u_r}{r} + \frac{1}{r}\frac{\partial u_\theta}{\partial \theta} \ .$$

Substituting for the displacement from equation (24.5), we obtain

$$e_s = -\frac{F(1-2\nu)}{4\pi\mu} \left(-\frac{1}{r^2} + \frac{1}{r^2} \right) = 0 \ .$$

In other words, the normal force F does not cause any dilatation of the surface $z=0$, except at the origin. However, if we draw a circle of radius a, centre the origin on the surface, it is clear that the radius of this circle gets smaller by the amount of the inward radial displacement and hence the area of the circle A increases by

$$\Delta A = 2\pi a u_r = -\frac{F(1-2\nu)}{2\mu} \ . \qquad (24.7)$$

This result is independent of a — i.e. all circles reduce in area by the same amount[8] and all of this reduction must therefore be concentrated at the origin, where in a sense a small amount of the surface is lost. We can generalize this result by superposition to state that, if there is an arbitrary distribution of purely normal (compressive) traction on the surface $z=0$ of the half-space, the area A enclosed by any closed curve will change by ΔA of equation (24.7), where F is now to be interpreted as the resultant of the tractions acting *within* the area A.

24.3 Other Singular Solutions

We have already shown how the singular solution $\ln(R+z)$ can be obtained from $1/R$ by partial integration, which of course is a form of superposition. A whole sequence of axially symmetric solutions can be obtained in the same way. Defining

$$\phi_0 = \frac{1}{R} \ , \qquad (24.8)$$

we obtain

$$\phi_{-1} = \ln(R+z) \ ; \quad \phi_{-2} = z\ln(R+z) - R$$
$$\phi_{-3} = \frac{1}{4}\left[(2z^2 - r^2)\ln(R+z) - 3Rz + r^2 \right] \qquad (24.9)$$
$$\phi_{-4} = \frac{1}{36}\left[3(2z^3 - 3zr^2)\ln(R+z) + 9zr^2 - 11z^2R + 4r^2R \right] \ ; \quad \ldots$$

---

draw points towards the origin, but this is a second order (non-linear) effect and cannot be admitted in the linear theory.

[8] as could be argued from the fact that the surface dilatation is zero.

by the operation[9]

$$\phi_{n-1}(r, z) = \int_{-\infty}^{0} \phi_n(r, (z - \zeta))d\zeta . \tag{24.10}$$

All of these functions are harmonic except at the origin and on the negative z-axis, where $R + z \to 0$. A corresponding set of harmonic potentials singular only on the *positive* z-axis can be obtained by setting $z \to -z$ in (24.9), which is equivalent to reflecting the potentials about the plane $z = 0$. It is convenient to define the signs of these functions such that

$$\phi_{-1} = -\ln(R - z) \; ; \quad \phi_{-2} = -z \ln(R - z) - R$$

$$\phi_{-3} = -\frac{1}{4}\left[(2z^2 - r^2)\ln(R - z) + 3Rz + r^2\right] \tag{24.11}$$

$$\phi_{-4} = -\frac{1}{36}\left[3(2z^3 - 3zr^2)\ln(R - z) + 9zr^2 + 11z^2R - 4r^2R\right] \; ; \quad \dots ,$$

since with this convention, the recurrence relation between both sets of potentials is defined by

$$\phi_n = \frac{\partial \phi_{n-1}}{\partial z} . \tag{24.12}$$

The sequence can also be extended to functions with stronger singularities by applying (24.12) to (24.8), obtaining

$$\phi_1 = -\frac{z}{R^3} \; ; \quad \phi_2 = \frac{3z^2}{R^5} - \frac{1}{R^3} \; ; \quad \phi_3 = -\frac{15z^3}{R^7} + \frac{9z}{R^5} \; ; \quad \dots \tag{24.13}$$

These functions are harmonic and bounded everywhere except at the origin.

We can also generate non-axisymmetric potential functions by differentiation with respect to x or y. For example, the function

$$\frac{\partial}{\partial x}\left(-\frac{z}{R^3}\right) = \frac{3xz}{R^5} = \frac{3\sin(2\beta)\cos\theta}{2R^3}$$

in spherical coördinates, is a non-axisymmetric harmonic function. However, these solutions can be obtained more systematically in terms of Legendre polynomials, as will be shown in the next chapter.

The integrated solutions defined by equation (24.10) can be generalized to permit an arbitrary distribution of sources along a line or surface. For example, a fairly general axisymmetric harmonic potential can be written in the form

$$\phi(r, z) = \int_a^b \frac{g(\zeta)dt}{\sqrt{r^2 + (z - \zeta)^2}} , \tag{24.14}$$

---

[9] Some care needs to be taken with the behaviour of the integral at the upper limit to obtain bounded partial integrals by this method. A more systematic method of developing this sequence will be introduced in §25.6 below.

which defines the potential due to a distribution of sources of strength $g(z)$ along the z-axis in the range $a < z < b$. Such a solution defines bounded stresses and displacements in any body which does not include this line segment. When the boundary conditions of the problem are expressed in terms of a representation such as (24.14), the problem is essentially reduced to the solution of an integral equation or a set of integral equations. This technique has a long history in the solution of axisymmetric flow problems around smooth bodies and it can be used for the related elasticity problem in which a uniform tensile stress is perturbed by an axisymmetric cavity. We shall also use an adaptation of the same method for solving crack and contact problems for the half-space in Chapters 32, 33.

24.4 Image Methods

As in the two dimensional case, singular solutions for the half-space can be obtained by placing appropriate image singularities outside the body. We consider the case in which the half-space $z > 0$ with elastic constants μ_1, κ_1 is bonded to the half-space $z < 0$ with constants μ_2, κ_2. Suppose that a singularity such as a concentrated force exists somewhere in the half-space $z > 0$ and that the same singularity in an infinite body can be defined in terms of the complex Papkovich-Neuber solution of §22.6 through the potentials $\phi^{(0)}(\zeta, \bar{\zeta}, z), \psi^{(0)}(\zeta, \bar{\zeta}, z), \omega^{(0)}(\zeta, \bar{\zeta}, z)$. In other words these are the potentials that would apply if the entire space had the same elastic constants μ_1, κ_1.

Aerogba[10] shows that the stress field in the composite body can then be obtained from the potentials

$$
\phi^{(1)} = \phi^{(0)} - A\kappa_1 \left[\phi^{(0)}\right]_{z \to -z} + \frac{(A\kappa_1^2 - B)}{2} \left[\int \omega^{(0)} dz\right]_{z \to -z}
$$
$$
+ (H - A\kappa_1) \left[\mathcal{L}\left\{\int \psi^{(0)} dz\right\}\right]_{z \to -z} + C\left[\mathcal{J}\left\{\iint \psi^{(0)} dz dz\right\}\right]_{z \to -z}
$$
$$
\psi^{(1)} = \psi^{(0)} - H\left[\psi^{(0)}\right]_{z \to -z}
$$
$$
\omega^{(1)} = \omega^{(0)} - A\kappa_1 \left[\omega^{(0)}\right]_{z \to -z} + 2A\left[\frac{\partial \phi^{(0)}}{\partial z}\right]_{z \to -z}
$$
$$
- (H - A\kappa_1) \left[\mathcal{J}\left\{\int \psi^{(0)} dz\right\}\right]_{z \to -z} + 2A\left[\mathcal{L}\left\{\psi^{(0)}\right\}\right]_{z \to -z} \qquad (24.15)
$$

for $z > 0$ and

[10] K. Aderogba (1977), On eigenstresses in dissimilar media, *Philosophical Magazine*, Vol. 35, pp. 281–292.

$$\phi^{(2)} = (A+1)\phi^{(0)} + \frac{(A\kappa_1^2 - B)\Gamma}{2} \int \omega^{(0)} dz + (1 + D - A\kappa_1)\Gamma\mathcal{L}\left\{\int \psi^{(0)} dz\right\}$$

$$+ C\Gamma\mathcal{J}\left\{\iint \psi^{(0)} dz dz\right\}$$

$$\psi^{(2)} = -D\Gamma\psi^{(0)}$$

$$\omega^{(2)} = -(B+1)\phi^{(0)} - (D\Gamma + B + 1)\mathcal{J}\left\{\int \psi^{(0)} dz\right\} \qquad (24.16)$$

for $z < 0$, where

$$\mathcal{J}\{\psi\} = \frac{\partial\psi}{\partial\zeta} + \frac{\partial\bar{\psi}}{\partial\bar{\zeta}} \quad ; \quad \mathcal{L}\{\psi\} = \frac{1}{2}\frac{\partial}{\partial z}\left(\zeta\bar{\psi} + \bar{\zeta}\psi\right) - z\mathcal{J}\{\psi\} \, ,$$

and the composite material properties are defined as

$$\Gamma = \frac{\mu_2}{\mu_1} \quad ; \quad H = \frac{(\Gamma - 1)}{(\Gamma + 1)} \quad ; \quad A = \frac{(\Gamma - 1)}{(\Gamma\kappa_1 + 1)} \quad ; \quad B = \frac{(\Gamma\kappa_1 - \kappa_2)}{(\Gamma + \kappa_2)}$$

$$C = \frac{1}{2}[2H(\kappa_1 + 1) - A\kappa_1(\kappa_1 + 2) - B] \quad ; \quad D = (H - 1)(\kappa_1 + 1)(\kappa_2 + 1) \, .$$

The arbitrary functions implied by the partial integrals in equations (24.15, 24.16) must be chosen so as to ensure that the resulting terms are harmonic and non-singular in the corresponding half-space.

24.4.1   The traction-free half-space

The special case of the traction-free half-space is recovered by setting $\Gamma = 0$. We then have

$$A = B = H = -1 \; ; \quad C = \frac{(\kappa^2 - 1)}{2}$$

and

$$\phi = \phi^{(0)} + \kappa\left[\phi^{(0)}\right]_{z \to -z} - \frac{(\kappa^2 - 1)}{2}\left[\int \omega^{(0)} dz\right]_{z \to -z}$$

$$+ (\kappa - 1)\left[\mathcal{L}\left\{\int \psi^{(0)} dz\right\}\right]_{z \to -z} + \frac{(\kappa^2 - 1)}{2}\left[\mathcal{J}\left\{\iint \psi^{(0)} dz dz\right\}\right]_{z \to -z}$$

$$\psi = \psi^{(0)} + \left[\psi^{(0)}\right]_{z \to -z}$$

$$\omega = \omega^{(0)} + \kappa\left[\omega^{(0)}\right]_{z \to -z} - 2\left[\frac{\partial\phi^{(0)}}{\partial z}\right]_{z \to -z}$$

$$- (\kappa - 1)\left[\mathcal{J}\left\{\int \psi^{(0)} dz\right\}\right]_{z \to -z} - 2\left[\mathcal{L}\{\psi^{(0)}\}\right]_{z \to -z} \qquad (24.17)$$

Example

As an example, we consider the case of the traction-free half-space with a concentrated force F acting in the z-direction at the point $(0, 0, a)$, a distance a below the surface[11]. The unperturbed field for a concentrated force at the origin is given in §24.1.2 by the potential

$$\omega = -\frac{F}{8\pi(1-\nu)R} = -\frac{F}{8\pi(1-\nu)\sqrt{r^2+z^2}} \quad ; \quad \phi = \psi = 0 .$$

For an equal force at the point $(0, 0, a)$ in the infinite body, we need to make use of the results of §21.3.1 with $s = ka$, $\Psi = k\omega$ and $z \to (z - a)$. We obtain

$$\Psi = a\omega$$

and hence, using equation (21.15),

$$\omega^{(0)} = -\frac{F}{8\pi(1-\nu)\sqrt{r^2+(z-a)^2}} \quad ; \quad \phi^{(0)} = \frac{Fa}{8\pi(1-\nu)\sqrt{r^2+(z-a)^2}}$$

$$(24.18)$$

and $\psi^{(0)} = 0$. In constructing the partial integral of $\omega^{(0)}$ for (24.17), we can ensure that the resulting term is harmonic by using the results of §24.3, but we must also ensure that the resulting potential has no singularities in the half-space $z > 0$. We therefore modify the first of equations (24.11) by a change of origin and write

$$\int \omega^{(0)} dz = \frac{F}{8\pi(1-\nu)} \ln \left(\sqrt{r^2+(z-a)^2} - (z-a) \right) ,$$

which after the substitution $z \to -z$ yields

$$\frac{F}{8\pi(1-\nu)} \ln \left(\sqrt{r^2+(z+a)^2} + z + a \right) .$$

This function is singular on the half line $r = 0$, $z < -a$, but this lies entirely outside the half-space $z > 0$ as required. Using this result and (24.18) in (24.17) and substituting $\kappa = (3 - 4\nu)$, we obtain

---

[11] This problem was first solved by R. D. Mindlin (1936), Force at a point in the interior of a semi-infinite solid, *Physics*, Vol. 7, pp. 195–202. A more compendious list of solutions for various singularities in the interior of the half-space was given by R. D. Mindlin and D. H. Cheng (1950), Nuclei of strain in the semi-infinite solid, *Journal of Applied Physics*, Vol. 21, pp. 926–930.

$$\phi = \frac{Fa}{8\pi(1-\nu)\sqrt{r^2+(z-a)^2}} + \frac{F(3-4\nu)a}{8\pi(1-\nu)\sqrt{r^2+(z+a)^2}}$$

$$-\frac{F(1-2\nu)}{2\pi}\ln\left(\sqrt{r^2+(z+a)^2}+z+a\right)$$

$$\omega = -\frac{F}{8\pi(1-\nu)\sqrt{r^2+(z-a)^2}} - \frac{F(3-4\nu)}{8\pi(1-\nu)\sqrt{r^2+(z+a)^2}}$$

$$-\frac{Fa(z+a)}{4\pi(1-\nu)\left(r^2+(z+a)^2\right)^{3/2}} \; .$$

The full stress field is then obtained by substitution into equations (22.6–22.10) and it is easily verified in Maple or Mathematica that the traction components σ_{zr}, $\sigma_{z\theta}$, σ_{zz} go to zero on $z=0$.

In the limit $a \to 0$, we recover the Boussinesq solution of §24.2.1 in which the force F is applied at the surface of the otherwise traction-free half-space. The potentials then reduce to

$$\phi = -\frac{F(1-2\nu)}{2\pi}\ln(R+z) \; ; \quad \omega = -\frac{F}{2\pi R} \; ,$$

agreeing with (24.4) and (22.3).

Problems

24.1. Starting from the potential function solution of Problem 22.2, use dimensional arguments to determine a suitable *axisymmetric* harmonic singular potential to solve the problem of a concentrated *tangential* force F in the x-direction, applied to the surface of the half-space at the origin.

24.2. Find the stresses and displacements[12] corresponding to the use of $\Psi = 1/R$ in Solution E. Show that the surfaces of the cone, $r = z\tan\beta$ are free of traction and hence use the solution to determine the stresses in a conical shaft of semi-angle β_0, transmitting a torque, T.

Find the maximum shear stress at the section $z=c$, and compare it with the maximum shear stress in a cylindrical bar of radius $a = c\tan\beta_0$ transmitting the same torque. Note that (i) the *maximum* shear stress will not generally be either $\sigma_{z\theta}$ or $\sigma_{r\theta}$, but must be found by appropriate coördinate transformation, and (ii) it may not *necessarily* occur at the outer radius of the cone, $r = c\tan\beta_0$.

24.3. An otherwise uniform cylindrical bar of radius b has a small spherical hole of radius a on the axis.

---

[12] This solution is known as the *centre of rotation*.

Combine the function $\Psi = Az/R^3$ in solution E with the elementary torsion solution for the bar without a hole (see, for example, Problem 22.1) and show that, with a suitable choice of the arbitrary constant A, the surface of the hole, $R=a$ can be made traction free. Hence deduce the stress field near the hole for this problem.

Assume that $b \gg a$, so that the perturbation due to the hole has a negligible influence on the stresses near the outer surface, $r=b$.

24.4. The area $\mathcal{A}$ on the surface of an elastic half-space is subjected to a purely normal pressure $p(x, y)$ corresponding to the total force, F. The loaded region is now expanded in a self-similar manner by multiplying all its linear dimensions by the same ratio, λ. Each point in the new contact area is subjected to a self-similar loading such that the pressure at the point $(\lambda x, \lambda y)$ is $Cp(x, y)$, where the constant C is chosen to ensure that the total force F is independent of λ.

Show that the deflection at corresponding points $(\lambda x, \lambda y, \lambda z)$ will then be proportional to F/λ.

24.5. The surface of the half-space $z>0$ is traction free and a concentrated heat source Q is applied at the origin. Determine the stress and displacement field in the half-space in the steady state.

The traction-free condition can be satisfied by using solution P of Table 23.1. You can then use dimensional arguments to determine the dependence of the heat flux on R and hence to choose the appropriate singular potential.

24.6. The otherwise traction-free surface of the elastic half-space $z>0$ is loaded by a uniform tangential force F per unit length along a closed line S directed everywhere along the local outward normal to S. Show that the average normal displacement $\bar{u}_z$ inside the area A enclosed by S is

$$\bar{u}_z = -\frac{F(1-2\nu)}{2\mu} .$$

24.7. Use Aderogba's formula (24.17) to determine the potentials ϕ, ψ, ω defining the stress and displacement fields in the traction-free half-space $z>0$, subjected to a concentrated force in the x-direction acting at the point $(0, 0, a)$.

24.8. Determine the potentials ϕ, ψ, ω defining the stress and displacement field in the traction-free half-space $z>0$ due to a centre of dilatation of strength ΔV located at the point $(0, 0, a)$.

Chapter 25
Spherical Harmonics

The singular solutions introduced in the last chapter are particular cases of a class of functions known as spherical harmonics. In this chapter, we shall develop these functions and some related harmonic potential functions in a more formal way. In particular, we shall identify:-

(i) Finite polynomial potentials expressible in the alternate forms $P_n(x, y, z)$, $P_n(r, z) \cos(m\theta)$ or $R^n P_n(\cos \beta) \cos(m\theta)$, where P_n represents a polynomial of degree n.
(ii) Potentials that are singular only at the origin.
(iii) Potentials including the factor $\ln(R+z)$ that are singular on the negative z-axis $(z < 0, r = 0)$.
(iv) Potentials including the factor $\ln\{(R+z)/(R-z)\}$ that are singular everywhere on the z-axis.
(v) Potentials including the factor $\ln(r)$ and/or negative powers of r that are singular everywhere on the z-axis.

All of these potentials can be obtained in axisymmetric and non-axisymmetric forms.

When used in solutions A,B and E of Tables 22.1, 22.2, the bounded potentials (i) provide a complete set of functions for the solid sphere, cylinder or cone with prescribed surface tractions or displacements on the curved surfaces. These problems are three-dimensional counterparts of those considered in Chapters 5, 8 and 11.

Problems for the hollow cylinder and cone can be solved by supplementing the bounded potentials with potentials (v) and (iv) respectively. *Axisymmetric* problems for the hollow sphere or the infinite body with a spherical hole can be solved using

Supplementary Information The online version contains supplementary material available at https://doi.org/10.1007/978-3-031-15214-6_25.

potentials (i,ii), but these potentials do not provide a complete solution to the non-axisymmetric problem[1].

Potentials (ii) and (iii) are useful for problems of the half-space, including crack and contact problems.

25.1   Fourier Series Solution

The Laplace equation in spherical polar coördinates takes the form

$$\nabla^2\phi \equiv \frac{\partial^2\phi}{\partial R^2} + \frac{2}{R}\frac{\partial\phi}{\partial R} + \frac{1}{R^2\sin^2\beta}\frac{\partial^2\phi}{\partial\theta^2} + \frac{1}{R^2}\frac{\partial^2\phi}{\partial\beta^2} + \frac{\cot\beta}{R^2}\frac{\partial\phi}{\partial\beta} = 0 . \qquad (25.1)$$

We first perform a Fourier decomposition with respect to the variable θ, giving a series of terms involving $\sin(m\theta), \cos(m\theta), \quad m=0, 1, 2, \ldots, \infty$. For the sake of brevity, we restrict attention to the cosine terms, since the sine terms will be of the same form and can be reintroduced at the end of the analysis. We therefore assume that a potential function ϕ can be written

$$\phi = \sum_{m=0}^{\infty} f_m(R, \beta)\cos(m\theta) . \qquad (25.2)$$

Substituting this series into (25.1), we find that the functions f_m must satisfy the equation

$$\frac{\partial^2 f}{\partial R^2} + \frac{2}{R}\frac{\partial f}{\partial R} - \frac{m^2 f}{R^2\sin^2\beta} + \frac{1}{R^2}\frac{\partial^2 f}{\partial\beta^2} + \frac{\cot\beta}{R^2}\frac{\partial f}{\partial\beta} = 0 . \qquad (25.3)$$

25.2   Reduction to Legendre's Equation

We now cast equation (25.3) in terms of the new variable

$$x = \cos\beta ,$$

for which we need the relations

$$\frac{\partial}{\partial\beta} = \frac{\partial}{\partial x}\frac{\partial x}{\partial\beta} = -\sin\beta\frac{\partial}{\partial x} = -\sqrt{1-x^2}\frac{\partial}{\partial x}$$

$$\frac{\partial^2}{\partial\beta^2} = -\sqrt{1-x^2}\frac{\partial}{\partial x}\left(-\sqrt{1-x^2}\frac{\partial}{\partial x}\right) = (1-x^2)\frac{\partial^2}{\partial x^2} - x\frac{\partial}{\partial x} .$$

Substituting into (25.3), we obtain

---

[1] See §27.1.

$$\frac{\partial^2 f}{\partial R^2} + \frac{2}{R}\frac{\partial f}{\partial R} - \frac{m^2 f}{R^2(1-x^2)} - \frac{2x}{R^2}\frac{\partial f}{\partial x} + \frac{(1-x^2)}{R^2}\frac{\partial^2 f}{\partial x^2} = 0 . \qquad (25.4)$$

Finally, noting that (25.4) is homogeneous in powers of R, we expand $f_m(R, x)$ as a power series — i.e.

$$f_m(R, x) = \sum_{n=-\infty}^{\infty} R^n g_{mn}(x) , \qquad (25.5)$$

which on substitution into equation (25.4) requires that the functions $g_{mn}(x)$ satisfy the ordinary differential equation

$$(1-x^2)\frac{d^2 g}{dx^2} - 2x\frac{dg}{dx} + \left(n(n+1) - \frac{m^2}{(1-x^2)} \right) g = 0 . \qquad (25.6)$$

This is the standard form of *Legendre's equation*[2].

25.3   Axisymmetric Potentials and Legendre Polynomials

In the special case where $m=0$, the functions (25.2) will be independent of θ and hence define axisymmetric potentials. Equation (25.6) then reduces to

$$(1-x^2)\frac{d^2 g}{dx^2} - 2x\frac{dg}{dx} + n(n+1)g = 0 , \qquad (25.7)$$

one of whose solutions is a polynomial of degree n, known as a *Legendre polynomial* $P_n(x)$. It can readily be determined by writing g as a general polynomial of degree $n > 0$, substituting into equation (25.7) and equating coefficients. The first few Legendre polynomials are

$P_0(x) = 1$

$P_1(x) = x = \cos\beta$

$P_2(x) = \frac{1}{2}(3x^2 - 1) = \frac{1}{4}\left[3\cos(2\beta) + 1\right]$

$P_3(x) = \frac{1}{2}(5x^3 - 3x) = \frac{1}{8}\left[5\cos(3\beta) + 3\cos\beta\right]$ $\qquad (25.8)$

$P_4(x) = \frac{1}{8}(35x^4 - 30x^2 + 3) = \frac{1}{64}\left[35\cos(4\beta) + 20\cos(2\beta) + 9\right]$

$P_5(x) = \frac{1}{8}(63x^5 - 70x^3 + 15x) = \frac{1}{128}\left[63\cos(5\beta) + 35\cos(3\beta) + 30\cos\beta\right] ,$

---

[2] See for example I. S. Gradshteyn and I. M. Ryzhik, *Tables of Integrals, Series and Products*, Academic Press, New York, 1980, §8.70.

where we have also used the relation $x = \cos \beta$ to express the results in terms of the polar angle β.

The sequence can be extended to higher values of n by using the recurrence relation[3]

$$(n + 1)P_{n+1}(x) - (2n + 1)x P_n(x) + n P_{n-1}(x) = 0 \qquad (25.9)$$

or the differential definition

$$P_n(x) = \frac{1}{2^n n!} \frac{d^n}{dx^n} (x^2 - 1)^n . \qquad (25.10)$$

In view of the above derivations, it follows that the axisymmetric functions

$$R^n P_n(\cos \beta) \qquad (25.11)$$

are harmonic for all n. These functions are known as *spherical harmonics*.

25.3.1   Singular spherical harmonics

Legendre polynomials are defined only for $n \geq 0$, so the corresponding spherical harmonics (25.11) will be bounded at the origin, $R = 0$. However, if we set

$$n = -p - 1 ,$$

where $p \geq 0$, we find that

$$n(n + 1) = (-p - 1)(-p) = p(p + 1)$$

and hence equation (25.6) becomes

$$(1 - x^2)\frac{d^2 g}{dx^2} - 2x\frac{dg}{dx} + \left(p(p + 1) - \frac{m^2}{(1 - x^2)} \right) g = 0 .$$

In other words, the equation is of the same form, with p replacing n. It follows that

$$R^{-n-1} P_n(\cos \beta) \qquad (25.12)$$

are harmonic functions, which are singular at $R = 0$ for $n \geq 0$. If n is allowed to take all non-negative values, the two expressions (25.11, 25.12) define axisymmetric harmonics for all integer powers of R, as indicated formally by the series (25.5). These functions are generated by the Maple and Mathematica files 'sp0', using the recurrence relation (25.9). In combination with solutions A,B and E of Chapter 22, they provide a general solution to the problem of a solid or hollow sphere loaded by arbitrary axisymmetric tractions.

---

[3] See I. S. Gradshteyn and I. M. Ryzhik, *loc. cit.* §§8.81, 8.82, 8.91 for this and more results concerning Legendre functions and polynomials.

25.3.2 Special cases

If we set $n=0$ in (25.12), we obtain

$$\phi = P_0(z/R)R^{-1} = \frac{1}{R} ,$$

which we recognize as the source solution of the previous chapter.
    Furthermore, if we next set $n=1$ in the same expression, we obtain

$$\phi = P_1(z/R)R^{-2} = \frac{z}{R^3} ,$$

which is proportional to the function ϕ_1 of (24.13). Similar relations exist between
the sequence $\phi_2, \phi_3, \ldots$, and the higher order singular spherical harmonics.

25.4 Non-axisymmetric Harmonics

When $m \neq 0$, equation (25.2) will define harmonic functions that are not axisymmetric. Corresponding solutions of Legendre's equation (25.6) can be developed from
the axisymmetric forms by the relations

$$P_n^m(x) = (-1)^m(1-x^2)^{m/2}\frac{d^m}{dx^m}P_n(x) ; \qquad m \leq n \qquad (25.13)$$

$$= (1-x^2)^{-m/2}\int_x^1 \ldots \int_x^1 P_n(x)(dx)^m ; \quad m > n . \qquad (25.14)$$

The functions $P_n^m(x)$ are known as *Legendre functions*. For problems of the sphere,
only the harmonics with $m \leq n$ are useful, since the others involve singularities on
either the positive or negative z-axis.
    The first few non-axisymmetric Legendre functions are

$$P_1^1(x) = -(1-x^2)^{\frac{1}{2}} = -\sin \beta$$

$$P_2^1(x) = -3x(1-x^2)^{\frac{1}{2}} = -\frac{3}{2}\sin(2\beta)$$

$$P_2^2(x) = 3(1-x^2) = \frac{3}{2}\left[1-\cos(2\beta)\right]$$

$$P_3^1(x) = -\frac{3}{2}(5x^2-1)(1-x^2)^{\frac{1}{2}} = -\frac{3}{8}\left[\sin \beta + 5\sin(3\beta)\right]$$

$$P_3^2(x) = 15x(1-x^2) = \frac{15}{4}\left[\cos \beta - \cos(3\beta)\right]$$

$$P_3^3(x) = -15(1-x^2)^{\frac{3}{2}} = -\frac{15}{4}\left[3\sin \beta - \sin(3\beta)\right] .$$

The bounded Legendre functions (25.13) satisfy the orthogonality condition

$$\int_{-1}^{1} P_n^m(x) P_p^q(x)dx = \frac{2}{(2n+1)} \frac{(n+m)!}{(n-m)!} \delta_{np} \delta_{mq} . \tag{25.15}$$

Thus if a general harmonic function for the solid or hollow sphere is written as a double series comprising the spherical harmonics

$$R^n P_n^m(\cos\beta)\cos(m\theta) \quad ; \quad R^n P_n^m(\cos\beta)\sin(m\theta) \tag{25.16}$$

$$R^{-n-1} P_n^m(\cos\beta)\cos(m\theta) \quad ; \quad R^{-n-1} P_n^m(\cos\beta)\sin(m\theta) \tag{25.17}$$

$m \leq n$, the coefficients in the series can be found explicitly as integrals of the boundary values[4].

In combination with solutions A,B and E of Chapter 22 these functions can be used for problems of a solid or hollow sphere loaded by non-axisymmetric tractions, but the solution is complete only for the solid sphere. The functions (25.16, 25.17) are generated by the Maple and Mathematica files 'spn'.

25.5   Cartesian and Cylindrical Polar Coördinates

The bounded Legendre polynomial solutions (25.11, 25.16) also correspond to finite polynomial functions in cylindrical polar coördinates and hence can be used for the problem of the solid circular cylinder loaded by polynomial tractions on its curved boundaries. For example, the axisymmetric functions (25.11) take the form

$$R^n P_n(z/R) \tag{25.18}$$

and the first few can be expanded in cylindrical polar coördinates as

$$R^0 P_0(z/R) = 1$$

$$R^1 P_1(z/R) = z$$

$$R^2 P_2(z/R) = \frac{1}{2}\left(3z^2 - R^2\right) = \frac{1}{2}\left(2z^2 - r^2\right)$$

$$R^3 P_3(z/R) = \frac{1}{2}\left(5z^3 - 3zR^2\right) = \frac{1}{2}\left(2z^3 - 3zr^2\right) \tag{25.19}$$

$$R^4 P_4(z/R) = \frac{1}{8}\left(35z^4 - 30z^2R^2 + 3R^4\right) = \frac{1}{8}\left(8z^4 - 24z^2r^2 + 3r^4\right)$$

$$R^5 P_5(z/R) = \frac{1}{8}\left(63z^5 - 70z^3R^2 + 15zR^4\right) = \frac{1}{8}\left(8z^5 - 40z^3r^2 + 15zr^4\right) ,$$

where we have used the relation $R^2 = r^2 + z^2$. This sequence of functions is generated in the Maple and Mathematica files 'cyl0'.

---

[4] See Problem 25.2.

These functions can also be expanded as finite polynomials in Cartesian co-ordinates, using the relations $x = r \cos \theta$, $y = r \sin \theta$. For example, the function $R^3 P_3^1(z/R) \sin \theta$ expands as

$$R^3 P_3^1(z/R) \sin \theta = -\frac{3}{2}(5z^2 - R^2)\sqrt{R^2 - z^2} \sin \theta = -\frac{3}{2}(4z^2 - r^2)r \sin \theta$$

$$= -\frac{3}{2}(4z^2 y - x^2 y - y^3) .$$

Of course, these polynomial solutions can also be obtained by assuming a general polynomial form of the appropriate order and substituting into the Laplace equation to obtain constraint equations, by analogy with the procedure for biharmonic polynomials developed in §5.1.

25.6 Harmonic Potentials with Logarithmic Terms

Equation (25.7), is a second order ordinary differential equation and as such must have two linearly independent solutions for each value of n, one of which is the Legendre polynomial $P_n(x)$. To find the other solution, we define a new function $h(x)$ through the relation

$$g(x) = P_n(x)h(x) \tag{25.20}$$

and substitute into (25.7), obtaining

$$(1 - x^2)P_n(x)h''(x) + 2[(1 - x^2)P_n'(x) - x P_n(x)]h'(x)$$
$$+ [(1 - x^2)P_n''(x) - 2x P_n'(x) + n(n + 1)P_n(x)]h(x) = 0 .$$

The last term in this equation must be zero, since $P_n(x)$ satisfies (25.7). We therefore have

$$(1 - x^2)P_n(x)h''(x) + 2[(1 - x^2)P_n'(x) - x P_n(x)]h'(x) = 0 . \tag{25.21}$$

This is a homogeneous first order ordinary differential equation for h', which can be solved by separation of variables.

If the non-constant solution of (25.21) is substituted into (25.20), we obtain the logarithmic Legendre function $Q_n(x)$, the first few such functions being

$$Q_0(x) = \frac{1}{2} \ln\left(\frac{1 + x}{1 - x}\right) ; \qquad Q_1(x) = \frac{x}{2} \ln\left(\frac{1 + x}{1 - x}\right) - 1$$

$$Q_2(x) = \frac{1}{4}(3x^2 - 1) \ln\left(\frac{1 + x}{1 - x}\right)$$

$$Q_3(x) = \frac{1}{4}(5x^3 - 3x) \ln\left(\frac{1 + x}{1 - x}\right) - \frac{5x^2}{2} + \frac{2}{3} \tag{25.22}$$

$$Q_4(x) = \frac{1}{16}(35x^4 - 30x^2 + 3) \ln\left(\frac{1+x}{1-x}\right) - \frac{35x^3}{8} + \frac{55x}{24}$$

$$Q_5(x) = \frac{1}{16}(63x^5 - 70x^3 + 15x) \ln\left(\frac{1+x}{1-x}\right) - \frac{63x^4}{8} + \frac{49x^2}{8} - \frac{8}{15}.$$

The recurrence relation

$$(n+1)Q_{n+1}(x) - (2n+1)xQ_n(x) + nQ_{n-1}(x) = 0 \qquad (25.23)$$

can be used to extend this sequence to higher values of n.

The corresponding harmonic functions are logarithmically singular at all points on the z-axis. For example

$$R^0 Q_0(z/R) = \frac{1}{2}\ln\left(\frac{R+z}{R-z}\right)$$

$$R^1 Q_1(z/R) = \frac{z}{2}\ln\left(\frac{R+z}{R-z}\right) - R \qquad (25.24)$$

$$R^2 Q_2(z/R) = \frac{1}{4}(3z^2 - R^2)\ln\left(\frac{R+z}{R-z}\right) - \frac{3Rz}{2}$$

and the factor $(R-z) \to 0$ on the positive z-axis ($r=0$, $z>0$), where $R = \sqrt{r^2+z^2} \to z$, whilst $(R+z) \to 0$ on the negative z-axis ($r=0$, $z<0$), where $R = \sqrt{r^2+z^2} \to -z$. These potentials in combination with the bounded potentials (25.11), permit a general solution of the problem of a hollow cone loaded by prescribed axisymmetric polynomial tractions on the curved surfaces. They are generated in the Maple and Mathematica files 'Qseries', using the recurrence relation (25.23).

The attentive reader will recognize a similarity between the logarithmic terms in this series and the functions $\phi_{-1}, \phi_{-2}, \ldots$ of equation (24.9), obtained from the source solution by successive partial integrations. However, the two sets of functions are not identical. In effect, the functions in (24.9) correspond to distributions of sources along the negative z-axis, whereas those in (25.24) involve distributions of sources along both the positive and negative z-axes. The functions (24.9) are actually more useful, since they are harmonic throughout the region $z>0$ and hence can be applied to problems of the half-space with concentrated loading, as we discovered in §24.2.1.

The functions $Q_n(x)$ can all be written in the form

$$Q_n(x) = \frac{1}{2}P_n(x)\ln\left(\frac{1+x}{1-x}\right) - W_{n-1}(x),$$

where

$$W_{n-1}(x) = \sum_{k=1}^{n}\frac{1}{k}P_{k-1}(x)P_{n-k}(x)$$

is a finite polynomial of degree of degree $(n-1)$. In other words, the multiplier on the logarithmic term is the Legendre polynomial $P_n(x)$ of the same order. This is also true for the functions of equation (24.9), as can be seen by comparing them with equations (25.24). A convenient way to extend the sequence of functions (24.9) is therefore to assume a solution of the form

$$\phi_{-n-1} = R^n P_n(z/R) \ln(R+z) + R_n(R, z) , \qquad (25.25)$$

where R_n is a general polynomial of degree n in R, z, substitute into the Laplace equation

$$\nabla^2 \phi = \frac{\partial^2 \phi}{\partial r^2} + \frac{1}{r}\frac{\partial \phi}{\partial r} + \frac{1}{r^2}\frac{\partial^2 \phi}{\partial \theta^2} + \frac{\partial^2 \phi}{\partial z^2} = 0 , \qquad (25.26)$$

and use the resulting equations to determine the coefficients in R_n. This method is used in the Maple and Mathematica files 'sing'.

25.6.1 Logarithmic functions for cylinder problems

The function $\ln(R+z)$ corresponds to a uniform distribution of sources along the negative z-axis. A similar distribution along the *positive* z-axis would lead to the related harmonic function $\ln(R-z)$. Alternatively, we could add these two functions, obtaining the solution for a uniform distribution of sources along the entire z-axis $(-\infty < z < \infty)$ with the result

$$\ln(R+z) + \ln(R-z) = \ln(R^2 - z^2) = 2\ln(r) .$$

Not surprisingly, this function is independent of z. It is actually the source solution for the two-dimensional Laplace equation

$$\frac{\partial^2 \phi}{\partial r^2} + \frac{1}{r}\frac{\partial \phi}{\partial r} + \frac{1}{r^2}\frac{\partial^2 \phi}{\partial \theta^2} = 0$$

and could have been obtained directly by seeking a singular solution of this equation that was independent of θ.

More importantly, a similar superposition can be applied to the functions (25.25) to obtain a new class of logarithmically singular harmonic functions of the form

$$\phi = R^n P_n(z/R) \ln(r) + S_n(r, z) , \qquad (25.27)$$

where S_n is a polynomial of degree n in r, z. These functions, in combination with the bounded harmonics of equation (25.19) and solutions A,B and E, provide a general solution to the problem of a hollow circular cylinder loaded by axisymmetric polynomial tractions on the curved surfaces. As before, the sequence (25.27) is most conveniently obtained by assuming a solution of the given form with S_n a general polynomial of r, z, substituting into the Laplace equation (25.26), and using the resulting equations to determine the coefficients in S_n. This method is used in the Maple and Mathematica files 'hol0'. The first few functions in the sequence are

$$
\begin{aligned}
\varphi_0 &= \ln(r) \\
\varphi_1 &= z \ln(r) \\
\varphi_2 &= \frac{1}{2}[(2z^2 - r^2)\ln(r) + r^2] \\
\varphi_3 &= \frac{1}{2}[(2z^3 - 3zr^2)\ln(r) + 3zr^2] \\
\varphi_4 &= \frac{1}{8}\left[(8z^4 - 24z^2r^2 + 3r^4)\ln(r) - \frac{9r^4}{2} + 24z^2r^2\right] \\
\varphi_5 &= \frac{1}{8}\left[(8z^5 - 40z^3r^2 + 15zr^4)\ln(r) - \frac{45zr^4}{2} + 40z^3r^2\right] .
\end{aligned}
$$

$$(25.28)$$

Notice that only even powers of r can occur in these functions.

25.7 Non-axisymmetric Cylindrical Potentials

As in §25.4, we can develop non-axisymmetric harmonic potentials in cylindrical polar coördinates in the form

$$
\phi_m = f_m(r, z) \begin{cases} \sin(m\theta) \\ \cos(m\theta) \end{cases} ,
$$

$$(25.29)$$

with $m \neq 0$. Substitution in (25.26) shows that the function f_m must satisfy the equation

$$
\frac{\partial^2 f_m}{\partial r^2} + \frac{1}{r}\frac{\partial f_m}{\partial r} - \frac{m^2 f_m}{r^2} + \frac{\partial^2 f_m}{\partial z^2} = 0 .
$$

$$(25.30)$$

A convenient way to obtain such functions is to note that if ϕ_m is harmonic, the function

$$
\phi_{m+1} = f_{m+1}(r, z) \cos\left((m+1)\theta\right)
$$

$$(25.31)$$

will also be harmonic if

$$
f_{m+1} = \frac{\partial f_m}{\partial r} - \frac{m f_m}{r} \equiv r^m \frac{\partial}{\partial r}\left(\frac{f_m}{r^m}\right) ,
$$

$$(25.32)$$

which can also be written

$$
f_m = r^m \mathcal{L}^m f_0 \quad \text{where} \quad \mathcal{L}\{\cdot\} = \frac{1}{r}\frac{\partial}{\partial r} .
$$

To prove this result, we first note that

$$\frac{\partial^2 f_m}{\partial z^2} = -\frac{\partial^2 f_m}{\partial r^2} - \frac{1}{r}\frac{\partial f_m}{\partial r} + \frac{m^2 f_m}{r^2} , \tag{25.33}$$

from (25.30). We then substitute (25.32) into (25.31) and the resulting expression into (25.26), using (25.33) to eliminate the derivatives with respect to z. The coefficients of all the remaining derivatives will then be found to be identically zero, confirming that (25.31) is harmonic.

The relation (25.32) can be used recursively to generate a sequence of non-axisymmetric harmonic potentials, starting from $m=0$ with any of the axisymmetric functions developed above. For example, starting with the axisymmetric function

$$\phi_0 = f_0 = R^4 P_4(z/R) = \frac{1}{8}\left(8z^4 - 24z^2 r^2 + 3r^4\right) ,$$

from equation (25.19), we can construct

$$f_1 = \frac{\partial f_0}{\partial r} = \frac{3}{2}\left(-4z^2 r + r^3\right) \; ; \; f_2 = \frac{\partial f_1}{\partial r} - \frac{f_1}{r} = 3r^2 ,$$

from which we obtain the harmonic potentials

$$\phi_1 = \frac{3}{2}\left(-4z^2 r + r^3\right)\cos\theta \; ; \; \phi_2 = 3r^2 \cos(2\theta) .$$

Similarly, from the logarithmically singular function

$$\phi_0 = f_0 = \varphi_2 = \frac{1}{2}\left[(2z^2 - r^2)\ln(r) + r^2\right]$$

of equation (25.28), we obtain

$$f_1 = \frac{\partial f_0}{\partial r} = -r\ln(r) + \frac{z^2}{r} + \frac{r}{2} \; ; \; f_2 = \frac{\partial f_1}{\partial r} - \frac{f_1}{r} = -1 - \frac{2z^2}{r^2} ,$$

defining the potentials

$$\phi_1 = \left(-r\ln(r) + \frac{z^2}{r} + \frac{r}{2}\right)\cos\theta \; ; \; \phi_2 = \left(-1 - \frac{2z^2}{r^2}\right)\cos(2\theta) .$$

This process is formalized in the Maple and Mathematica files 'cyln' and 'holn', which generate a sequence of functions F_{mn}, S_{mn} such that the potentials

$$F_{mn}\cos(m\theta) \; ; \quad F_{mn}\sin(m\theta) \; ; \quad S_{mn}\cos(m\theta) \; ; \quad S_{mn}\sin(m\theta) \tag{25.34}$$

are harmonic. These files are easily extended to larger values of m or n if required.

For convenience, we list here the first few functions of this sequence for the case $m = 1$.

$$F_{11} = r ; \quad F_{12} = 3zr ; \quad F_{13} = \frac{3}{2}\left(4z^2r - r^3\right)$$

$$F_{14} = \frac{5}{2}\left(4z^3r - 3zr^3\right) ; \quad F_{15} = \frac{15}{8}\left(8z^4r - 12z^2r^3 + r^5\right) . \tag{25.35}$$

$$S_{10} = \frac{1}{r} ; \quad S_{11} = \frac{z}{r} ; \quad S_{12} = -r\ln(r) + \frac{z^2}{r} + \frac{r}{2}$$

$$S_{13} = -3zr\ln(r) + \frac{z^3}{r} + \frac{3zr}{2} \tag{25.36}$$

$$S_{14} = \frac{3}{2}\left(r^3 - 4z^2r\right)\ln(r) + \frac{z^4}{r} + 3z^2r - \frac{15r^3}{8}$$

$$S_{15} = \frac{5}{2}\left(3zr^3 - 4z^3r\right)\ln(r) + \frac{z^5}{r} + 5z^3r - \frac{75zr^3}{8} .$$

25.8   Spherical Harmonics in Complex-variable Notation

In §22.6 we developed a version of the Papkovich-Neuber solution in which the Cartesian coördinates (x, y, z) were replaced by $(\zeta, \bar{\zeta}, z)$, where $\zeta = x + \imath y$, $\bar{\zeta} = x - \imath y$. In this system, the Laplace equation takes the form

$$\nabla^2\psi \equiv 4\frac{\partial^2\psi}{\partial\zeta\partial\bar{\zeta}} + \frac{\partial^2\psi}{\partial z^2} = 0 . \tag{25.37}$$

We consider solutions of the series form

$$\psi_{2k}(\zeta, \bar{\zeta}, z) = \sum_{j=0}^{k} \frac{z^{2j} f_{k-j}(\zeta, \bar{\zeta})}{(2j)!} . \tag{25.38}$$

Substitution into (25.37) then shows that

$$\sum_{j=0}^{k} \frac{4z^{2j}}{(2j)!}\frac{\partial^2 f_{k-j}}{\partial\zeta\partial\bar{\zeta}} + \sum_{j=1}^{k} \frac{z^{2j-2} f_{k-j}}{(2j-2)!} = 0$$

and equating coefficients of $z^{2k-2i-2}$, we have

$$\frac{\partial^2 f_{i+1}}{\partial\zeta\partial\bar{\zeta}} = -\frac{f_i}{4} ; \quad i = (0, k-1) , \tag{25.39}$$

$$\frac{\partial^2 f_0}{\partial\zeta\partial\bar{\zeta}} = 0 , \tag{25.40}$$

which defines a recurrence relation for the functions f_i. Furthermore, equation (25.40) is the two-dimensional Laplace equation whose general solution can be written

$$f_0 = g_1(\zeta) + \overline{g}_2(\overline{\zeta}) , \qquad (25.41)$$

where $g_1, \overline{g}_2$ are general holomorphic functions of the complex variable ζ and its conjugate $\overline{\zeta}$, respectively. Thus, the most general solution of the form (25.38) can be constructed by successive integrations of (25.39), starting from (25.41). This defines an even function of z, but it is easily verified that the odd function

$$\psi_{2k+1}(\zeta, \overline{\zeta}, z) = \sum_{j=0}^{k} \frac{z^{2j+1} f_{k-j}(\zeta, \overline{\zeta})}{(2j+1)!} \qquad (25.42)$$

also satisfies (25.37), where the functions f_i again satisfy the recurrence relations (25.39, 25.40).

25.8.1 Bounded cylindrical harmonics

The bounded harmonics of equations (25.18, 25.19, 25.35) can be obtained from equations (25.38, 25.42) by choosing

$$f_0(\zeta, \overline{\zeta}) = g_1(\zeta) = \zeta^m .$$

Successive integrations of equation (25.39) then yield

$$f_1(\zeta, \overline{\zeta}) = \frac{\zeta^{m+1}\overline{\zeta}}{(-4)(m+1)} \quad ; \quad f_2(\zeta, \overline{\zeta}) = \frac{\zeta^{m+2}\overline{\zeta}^2}{(-4)^2(m+1)(m+2)(2)} ,$$

or in general

$$f_i(\zeta, \overline{\zeta}) = \frac{m!(\zeta\overline{\zeta})^i \zeta^m}{(-4)^i(m+i)!i!} ,$$

where we have omitted the arbitrary functions of integration. Substituting f_i into (25.38, 25.42), we can then construct the harmonic functions

$$\chi_{2k}^m = \sum_{j=0}^{k} \frac{m! z^{2j} (\zeta\overline{\zeta})^{k-j} \zeta^m}{(-4)^{k-j}(m+k-j)!(k-j)!(2j)!}$$

$$\chi_{2k+1}^m = \sum_{j=0}^{k} \frac{m! z^{2j+1} (\zeta\overline{\zeta})^{k-j} \zeta^m}{(-4)^{k-j}(m+k-j)!(k-j)!(2j+1)!} , \qquad (25.43)$$

the first few functions in the sequence being

$$\chi_0^m(\zeta, \bar{\zeta}, z) = \zeta^m$$
$$\chi_1^m(\zeta, \bar{\zeta}, z) = z\zeta^m$$
$$\chi_2^m(\zeta, \bar{\zeta}, z) = \frac{z^2\zeta^m}{2!} + \frac{(\zeta\bar{\zeta})\zeta^m}{(-4)(m+1)}$$
$$\chi_3^m(\zeta, \bar{\zeta}, z) = \frac{z^3\zeta^m}{3!} + \frac{z(\zeta\bar{\zeta})\zeta^m}{(-4)(m+1)}$$
$$\chi_4^m(\zeta, \bar{\zeta}, z) = \frac{z^4\zeta^m}{4!} + \frac{z^2(\zeta\bar{\zeta})\zeta^m}{(-4)(m+1)(2!)} + \frac{(\zeta\bar{\zeta})^2\zeta^m}{(-4)^2(m+1)(m+2)(2)} \; .$$

(25.44)

The functions (25.43, 25.44) are three-dimensionally harmonic and satisfy the recurrence relations

$$\frac{\partial \chi_n^m}{\partial z} = \chi_{n-1}^m \; ; \quad \frac{\partial \chi_n^m}{\partial \zeta} = m\chi_n^{m-1} \; ; \quad \frac{\partial \chi_n^m}{\partial \bar{\zeta}} = -\frac{\chi_{n-2}^{m+1}}{4(m+1)} \; .$$

(25.45)

or

$$\int \chi_n^m dz = \chi_{n+1}^m \; ; \quad \int \chi_n^m d\zeta = \frac{\chi_n^{m+1}}{(m+1)} \; ; \quad \int \chi_n^m d\bar{\zeta} = -4m\chi_{n+2}^{m-1} \; .$$

The factor

$$(\zeta\bar{\zeta}) = re^{i\theta}re^{-i\theta} = r^2 \quad \text{and} \quad \zeta^m = r^m(\cos(m\theta) + i\sin(m\theta))$$

and hence the functions χ_n^m can be expanded as functions of r, z with Fourier multipliers as in (25.34, 25.35). In fact they are related to the spherical harmonics (25.16) by the expressions

$$\chi_n^m(\zeta, \bar{\zeta}, z) = \frac{(-2)^m m!}{(n+2m)!} R^{m+n} P_{m+n}^m(z/R) \exp(im\theta) \; ,$$

where

$$R = \sqrt{z^2 + \zeta\bar{\zeta}} = \sqrt{x^2 + y^2 + z^2} \; .$$

A few low order potentials that we shall need later are

$$\chi_0^0 = 1 \; ; \quad \chi_1^0 = z \; ; \quad \chi_2^0 = \frac{z^2}{2} - \frac{\zeta\bar{\zeta}}{4} \; ; \quad \chi_3^0 = \frac{z^3}{6} - \frac{z\zeta\bar{\zeta}}{4} \; ,$$

$$\chi_0^1 = \zeta \; ; \quad \chi_1^1 = z\zeta \; ; \quad \chi_2^1 = \frac{z^2\zeta}{2} - \frac{\zeta^2\bar{\zeta}}{8} \; ; \quad \chi_3^1 = \frac{z^3\zeta}{6} - \frac{z\zeta^2\bar{\zeta}}{8} \; ,$$

(25.46)

$$\chi_0^2 = \zeta^2 \; ; \quad \chi_1^2 = z\zeta^2 \; ; \quad \chi_2^2 = \frac{z^2\zeta^2}{2} - \frac{\zeta^3\bar{\zeta}}{12} \; ; \quad \chi_3^2 = \frac{z^3\zeta^2}{6} - \frac{z\zeta^3\bar{\zeta}}{12} \; .$$

25.8.2 Singular cylindrical harmonics

A similar procedure can be used to develop complex versions of the singular harmonics of equations (25.34, 25.36), starting with the function

$$f_0(\zeta, \bar{\zeta}) = \zeta^{-m} \quad ; \quad m \geq 1$$
$$= \ln(\zeta\bar{\zeta}) ; \quad m = 0 .$$

Notice that in the complex-variable formulation of two-dimensional problems, $\ln(\zeta)$ is generally excluded as being multivalued and hence non-holomorphic. However,

$$\ln(\zeta\bar{\zeta}) = \ln(\zeta) + \ln(\bar{\zeta}) = 2 \ln(r)$$

is clearly a real single-valued harmonic function (being of the form (25.41). Singular potentials with $m > 0$ will eventually integrate up to include logarithmic terms when following the procedure (25.39). We see this for example in the real stress function versions (25.36) for $m = 1$.

Problems

25.1. Use the representation (25.10) and integration by parts to show that

$$\int_{-1}^{1} P_n(x) P_m(x) dx = \frac{2\delta_{mn}}{2n + 1} ,$$

and hence that the Legendre polynomials are orthogonal on the domain $(-1, 1)$.

25.2. A general harmonic function $\phi(R, \theta, \beta)$ in the spherical domain $0 < R < a$ can be written as a double series of bounded potentials in the form

$$\phi(R, \theta, \beta) = \sum_{n=0}^{\infty} \sum_{m=0}^{\infty} C_{nm} R^n P_n^m(\cos \beta) \cos(m\theta)$$

$$+ \sum_{n=0}^{\infty} \sum_{m=1}^{\infty} D_{nm} R^n P_n^m(\cos \beta) \sin(m\theta) .$$

Use the orthogonality conditon (25.15), to find the coefficients C_{mn}, D_{mn} in terms of the boundary value $\phi(a, \theta, \beta) = f(\theta, \beta)$.

25.3. Use equation (25.14) to evaluate the function $P_1^2(x)$ and verify that it satisfies Legendre's equation (25.6) with $m = 2$ and $n = 1$. Construct the appropriate harmonic potential function from equation (25.16) and verify that it is singular on the z-axis.

25.4. If $U(x, y, z)$ is a harmonic function of Cartesian coordinates x, y, z, then we can construct additional harmonic functions by differentiation, as in §24.3. This is a form of linear superposition. Expand the bounded axisymmetric potential $\phi = R^5 P_5(z/R)$ of equation (25.19) as a function of x, y, z and then construct a new harmonic potential

$$\psi = \frac{\partial^2 \phi}{\partial x^2} .$$

Show that ψ can be expressed as the sum of an axisymmetric harmonic potential and a non-axisymmetric potential of the form $f(r, z) \cos(2\theta)$.

25.5. The functions (24.9) are singular only on the negative z-axis ($r=0$, $z<0$). Similar functions singular only on the positive z-axis are given in equations (24.11). Develop the first three of (25.28) by superposition, noting that $\ln(R+z)+\ln(R-z)= 2 \ln(r)$.

25.6. Use the methodology of §5.1 to construct the most general harmonic polynomial function of degree 3 in Cartesian coördinates x, y, z. Decompose the resulting polynomial into a set of spherical harmonics.

25.7. Solve equation (25.21) for $h(x)$ with $n = 1$ and hence show that $Q_1(x)$ satisfies Legendre's equation for this case.

25.8. Express the recurrence relation of equations (25.29–25.32) in spherical polar coördinates R, θ, β. Check your result by deriving the first three functions starting from the source solution $\phi_0 = 1/R$ and compare the results with the expressions derived in the Maple or Mathematica file 'spn'.

Chapter 26
Cylinders and Circular Plates

The bounded harmonic potentials of equations (25.19) in combination with solutions A, B and E provide a complete solution to the problem of a solid circular cylinder loaded by axisymmetric polynomial tractions on its curved surfaces. The corresponding problem for the hollow cylinder can be solved by including also the singular potentials of equation (25.28). The method can be extended to non-axisymmetric problems using the results of §25.7. If strong boundary conditions are imposed on the curved surfaces and weak conditions on the ends, the solutions are most appropriate to problems of 'long' cylinders in which $L \gg a$, where L is the length of the cylinder and a is its outer radius. At the other extreme, where $L \ll a$, the same harmonic functions can be used to obtain three-dimensional solutions for in-plane loading and bending of circular plates, by imposing strong boundary conditions on the plane surfaces and weak conditions on the curved surfaces. As in Chapter 5, some indication of the order of polynomial required can be obtained from elementary Mechanics of Materials arguments.

26.1 Axisymmetric Problems for Cylinders

If axisymmetric potentials are substituted into solutions A and B, we find that the circumferential displacement u_θ is zero everywhere, as are the stress components $\sigma_{\theta r}, \sigma_{\theta z}$. There is therefore no torque transmitted across any cross section of the cylinder and no twist. The only force resultant on the cross section is a force F in the z-direction given by

$$F = \int_a^b \int_0^{2\pi} \sigma_{zz} r \, d\theta \, dr = 2\pi \int_a^b r \sigma_{zz} dr , \qquad (26.1)$$

where a, b are the inner and outer radii of the cylinder respectively.

© The Author(s), under exclusive license to Springer Nature Switzerland AG 2022
J. R. Barber, *Elasticity*, Solid Mechanics and Its Applications 172,
https://doi.org/10.1007/978-3-031-15214-6_26

By contrast, substitution of an axisymmetric potential into solution E yields a stress and displacement state where the *only* non-zero displacement and stress components are $u_\theta, \sigma_{\theta r}, \sigma_{\theta z}$. In this case, the axial force $F=0$, but in general there will be a transmitted torque

$$T = \int_a^b \int_0^{2\pi} \sigma_{\theta z} r^2 d\theta dr = 2\pi \int_a^b r^2 \sigma_{\theta z} dr \ . \tag{26.2}$$

Notice that the additional power of r in equation (26.2) compared with (26.1) is the moment arm about the z-axis for the elemental force $\sigma_{\theta z} r dr d\theta$.

Axisymmetric problems are therefore conveniently partitioned into *irrotational* (torsionless) problems requiring solutions A and B only and *rotational* (torsional) problems requiring only solution E. There is a close parallel here with the partition of two-dimensional problems into *in-plane problems* (Chapters 5–14) and *antiplane problems* (Chapter 15). Notice that the rotational problem, like the antiplane problem, involves only one non-zero displacement (u_θ) and requires only a single harmonic potential function (in solution E). By contrast, the irrotational problem involves two non-zero displacements (u_r, u_z) and requires two independent harmonic potentials, one each in solutions A and B. In the same way, the in-plane two-dimensional solution involves two displacements (e.g. u_x, u_y) and requires a single biharmonic potential, which we know to be equivalent to two independent harmonic potentials.

26.1.1 The solid cylinder

Problems for the solid cylinder are solved by a technique very similar to that used in Chapter 5. We shall find that we can always find finite polynomials satisfying polynomial boundary conditions on the curved boundaries in the strong (pointwise) sense, but that the boundary conditions on the ends can then only be satisfied in the weak sense. Corrective solutions for these end effects can be developed from an eigenvalue problem, as in Chapter 6.

A torsional problem

As an example, we consider the solid cylinder ($0 \leq r < a$, $0 < z < L$), built in at the end $z = L$ and subjected to the traction distribution

$$\sigma_{r\theta} = \frac{Sz}{L} \ ; \quad \sigma_{rr} = \sigma_{rz} = 0 \ ; \quad r = a \tag{26.3}$$

$$\sigma_{zr} = \sigma_{z\theta} = \sigma_{zz} = 0 \ ; \quad z = 0 \ . \tag{26.4}$$

In other words, the end $z=0$ is traction free and the curved surfaces are loaded by a linearly varying torsional traction. This is clearly a torsional problem and elementary equilibrium arguments show that the torque must increase with z^2 from the free end. This suggests a leading term proportional to z^2r in the stress component $\sigma_{z\theta}$ and we therefore use solution E with a 5th degree polynomial and below.

We use the trial function[1]

$$\Psi = \frac{C_5}{8}\left(8z^5 - 40z^3r^2 + 15zr^4\right) + \frac{C_4}{8}\left(8z^4 - 24z^2r^2 + 3r^4\right)$$
$$+ \frac{C_3}{2}\left(2z^3 - 3zr^2\right) + \frac{C_2}{2}\left(2z^2 - r^2\right) \quad (26.5)$$

from equation (25.19) and substitute into solution E (Table 22.1), obtaining the non-zero stress and displacement components

$$2\mu u_\theta = 5C_5(4z^3r - 3zr^3) + 3C_4(4z^2r - r^3) + 6C_3zr + 2C_2r \quad (26.6)$$
$$\sigma_{r\theta} = -15C_5zr^2 - 3C_4r^2 \quad (26.7)$$
$$\sigma_{z\theta} = 15C_5\left(2z^2r - \frac{r^3}{2}\right) + 12C_4zr + 3C_3r \ . \quad (26.8)$$

The boundary condition (26.3) must be satisfied for all z, giving

$$C_5 = -\frac{S}{15a^2L} \ ; \quad C_4 = 0 \ . \quad (26.9)$$

The traction-free condition (26.4) can only be satisfied in the weak sense. Substituting (26.8) into (26.2), and evaluating the integral at $z=0$, we obtain

$$T = \frac{\pi a^4(3C_3 - 5C_5a^2)}{2}$$

and hence the weak condition $T = 0$ gives

$$C_3 = \frac{5C_5a^2}{3} = -\frac{S}{9L} \ , \quad (26.10)$$

using (26.9). The complete stress field is therefore

$$\sigma_{r\theta} = \frac{Szr^2}{a^2L} \ ; \quad \sigma_{z\theta} = \frac{Sr}{L}\left(\frac{r^2}{2a^2} - \frac{2z^2}{a^2} - \frac{1}{3}\right) \ .$$

Notice that the constant C_2 is not determined in this solution, since it describes merely a rigid-body displacement and does not contribute to the stress field. In order to determine it, we need to generate a weak form for the built-in condition at $z=L$.

---

[1] Notice that the first two functions from equation (25.19) give no stresses or displacements when substituted into solution E and hence are omitted here.

An appropriate condition is[2]

$$\int_0^a \int_0^{2\pi} u_\theta r^2 d\theta dr = 0 \ ; \quad z = L \ .$$

Substituting for u_θ from (26.6) and using (26.9, 26.10) we obtain

$$C_2 = \frac{2SL^2}{3a^2}$$

and hence

$$2\mu u_\theta = S \left(\frac{zr^3}{a^2 L} - \frac{4z^3 r}{3a^2 L} - \frac{2zr}{3L} + \frac{4L^2 r}{3a^2} \right) \ .$$

A thermoelastic problem

As a second example, we consider the traction-free solid cylinder $0 \le r < a$, $-L < z < L$, subjected to the steady-state thermal boundary conditions

$$q_r(a, z) = -q_0 \ ; \quad T(r, \pm L) = 0 \ . \tag{26.11}$$

In other words, the curved surfaces $r = a$ are uniformly heated, whilst the ends $z = \pm L$ are maintained at zero temperature.

The problem is clearly symmetrical about $z = 0$ and the temperature must therefore be described by even Legendre polynomial functions. Using solution T (Table 23.1) to describe the thermoelastic field, we note that temperature is proportional to $\partial \chi / \partial z$ and hence χ must be comprised of odd functions from equation (25.19) — in particular

$$\chi = \frac{C_3}{2}(2z^3 - 3zr^2) + C_1 z \ . \tag{26.12}$$

Substituting (26.12) into Table 23.1, yields

$$T = \frac{(1 - \nu)}{\mu\alpha(1 + \nu)} \left(\frac{C_3}{2}(6z^2 - 3r^2) + C_1 \right) \tag{26.13}$$

$$q_r = -K\frac{\partial T}{\partial r} = \frac{3(1 - \nu)KC_3 r}{\mu\alpha(1 + \nu)} \tag{26.14}$$

and hence

$$C_3 = -\frac{\mu\alpha(1 + \nu)q_0}{3Ka(1 - \nu)} \ , \tag{26.15}$$

from (26.11, 26.14).

---

[2] See §9.1.1.

The constant C_1 can be determined[3] from the weak form of $(26.11)_2$ — i.e.

$$2\pi \int_0^a T(r, \pm L) r \, dr = \frac{2\pi(1 - \nu)}{\mu a(1 + \nu)} \left[\frac{C_3}{2} \left(3L^2 a^2 - \frac{3a^4}{4} \right) + \frac{C_1 a^2}{2} \right] = 0$$

and hence

$$C_1 = -\frac{3C_3(4L^2 - a^2)}{4} = \frac{\mu a(1 + \nu) q_0 (4L^2 - a^2)}{4Ka(1 - \nu)}. \tag{26.16}$$

It follows that the complete temperature field is

$$T(r, z) = \frac{q_0(4(L^2 - z^2) + 2r^2 - a^2)}{4Ka},$$

from (26.13, 26.15, 26.16).

A particular solution for the thermal stress field is then obtained by substituting (26.15, 26.16, 26.12) into Table 23.1. To complete the solution, we must superpose the homogeneous solution which is here given by solutions A and B, since axisymmetric temperature fields give dilatational but irrotational stress and displacement fields. To determine the appropriate order of polynomials to include in ϕ and ω, we compare Tables 22.1 and 23.1 and notice that whilst solution B involves the same order of differentials as solution T, the corresponding components in solution A involve one further differentiation, indicating the need for a higher order polynomial function. We therefore try the functions

$$\phi = \frac{A_4}{8}(8z^4 - 24z^2 r^2 + 3r^4) + \frac{A_2}{2}(2z^2 - r^2) \; ; \; \omega = \frac{B_3}{2}(2z^3 - 3zr^2) + B_1 z .$$
$$\tag{26.17}$$

Substitution into Tables 22.1, 23.1 then yields the stress components

$$\sigma_{rr} = A_4 \left(\frac{9r^2}{2} - 6z^2 \right) - A_2 - 3B_3(z^2 + 2\nu z^2 - \nu r^2) - 2\nu B_1$$
$$\qquad -3C_3(3z^2 - r^2) - 2C_1$$
$$\sigma_{rz} = -6(2A_4 + \nu B_3 + C_3)zr$$
$$\sigma_{zz} = 6A_4(2z^2 - r^2) + 2A_2 + 3B_3(2^2 - \nu r^2 + 2\nu z^2) + 2\nu B_1 + 6C_3 z^2 .$$

The strong traction-free boundary conditions

---

[3] The constant C_1 serves merely to set the base level for temperature and since no stresses are generated in a traction-free body subject to a uniform temperature rise, we could set $C_1 = 0$ at this stage without affecting the final solution for the stresses. Notice however, that this choice would lead to different values for the constants A_4, A_2, B_3, B_1 in the following analysis. Also, a uniform temperature rise *does* cause dilatation and hence non-zero displacements, so it is essential to solve for C_1 if the displacement field is required.

$$\sigma_{rr}(a, z) = 0 \; ; \; \sigma_{rz}(a, z) = 0$$

yield the three equations

$$-6A_4 - 3B_3(1 + 2\nu) = 9C_3$$

$$\frac{9A_4a^2}{2} - A_2 + 3B_3\nu a^2 - 2\nu B_1 = -3C_3 a^2 + 2C_1 \qquad (26.18)$$

$$2A_4 + \nu B_3 = -C_3 \; ,$$

and another equation is obtained from the weak condition $F = 0$ on the ends $z = \pm L$. Substituting for σ_{zz} into (26.1) and evaluating the integral, we obtain

$$F = \pi a^2 \left[3A_4(4L^2 - a^2) + 3B_3 \left(\frac{(1 - \nu)a^2}{2} + 2\nu L^2 \right) + 2A_2 \right.$$

$$\left. - B_1(1 - \nu) + 6C_3 L^2 \right] = 0 \; . \qquad (26.19)$$

Finally, solving (26.18, 26.19) for A_4, A_2, B_3, B_1 and substituting the resulting expressions into (26.17) and Tables 22.1, 23.1 yields the complete stress field

$$\sigma_{rr} = \frac{q_0 \mu \alpha (a^2 - r^2)}{4Ka} \; ; \; \sigma_{\theta\theta} = \frac{q_0 \mu \alpha (a^2 - 3r^2)}{4Ka} \; ; \; \sigma_{zz} = \frac{q_0 \mu \alpha (2r^2 - a^2)}{2Ka}$$

$$\sigma_{rz} = \sigma_{r\theta} = \sigma_{\theta z} = 0 \; .$$

26.1.2   The hollow cylinder

Problems for the hollow cylinder $a < r < b$ can be solved by supplementing the bounded potentials (25.19) with the logarithmically singular potentials of equation (25.28). The bounded potentials required are the same as those needed for a solid cylinder with similar boundary conditions, but the logarithmic potentials are generally several orders lower, being typically no higher than the order of the most rapidly varying traction on the curved surfaces[4].

   To illustrate the procedure, we consider the hollow cylinder $a < r < b$ subjected to the torsional tractions

$$\sigma_{r\theta} = \frac{Sz}{L} \; ; \; \sigma_{rr} = \sigma_{rz} = 0 \; ; \; r = b \qquad (26.20)$$

---

[4] This arises because when we differentiate the potentials by parts, some of the resulting expressions involve differentiation only of the logarithmic multiplier, leaving higher order polynomial expressions for the stress and displacement components on a given radial surface.

$$\sigma_{r\theta} = \sigma_{rr} = \sigma_{rz} = 0 \; ; \quad r = a \qquad (26.21)$$

$$\sigma_{zr} = \sigma_{z\theta} = \sigma_{zz} = 0 \; ; \quad z = 0 .$$

This is the hollow cylinder equivalent of the torsional example considered in §26.1.1 above and hence the required bounded potentials are given by equation (26.5), though of course the values of the multiplying constants will be different. In the interests of brevity, we shall solve only for the stress field and hence the potential $C_2(2z^2 - r^2)/2$ can be omitted, since it defines only a rigid-body displacement.

The traction varies with z^1, so for the hollow cylinder, we need to add in the logarithmic potentials φ_1, φ_0, giving the potential function

$$\Psi = \frac{C_5}{8}(8z^5 - 40z^3 r^2 + 15zr^4) + \frac{C_4}{8}(8z^4 - 24z^2 r^2 + 3r^4) + \frac{C_3}{2}(2z^3 - 3zr^2)$$
$$+ A_1 z \ln(r) + A_0 \ln(r) .$$

Substituting into solution E (Table 22.1) we obtain

$$\sigma_{r\theta} = -15C_5 zr^2 - 3C_4 r^2 + \frac{2A_1 z}{r^2} + \frac{2A_0}{r^2} \qquad (26.22)$$

$$\sigma_{z\theta} = 15C_5\left(2z^2 r - \frac{r^3}{2}\right) + 12C_4 zr + 3C_3 r - \frac{A_1}{r} . \qquad (26.23)$$

The boundary conditions (26.20, 26.21) must be satisfied for all z, giving the four equations

$$-15C_5 a^2 + \frac{2A_1}{a^2} = 0$$

$$-3C_4 a^2 + \frac{2A_0}{a^2} = 0$$

$$-15C_5 b^2 + \frac{2A_1}{b^2} = \frac{S}{L}$$

$$-3C_4 b^2 + \frac{2A_0}{b^2} = 0 .$$

A fifth equation is obtained by substituting (26.23) into (26.2) and enforcing the weak condition $T{=}0$. Solving these equations for the five constants C_5, C_4, C_3, A_1, A_0 and back substitution into (26.22, 26.23) yields the final stress field

$$\sigma_{r\theta} = \frac{Sb^2 z(r^4 - a^4)}{Lr^2(b^4 - a^4)}$$

$$\sigma_{z\theta} = \frac{Sb^2(3(b^2 + a^2)(a^4 - 4z^2 r^2 + r^4) - 2(4a^4 + a^2 b^2 + b^4)r^2)}{6Lr(b^2 + a^2)(b^4 - a^4)} .$$

26.2 Axisymmetric Circular Plates

Essentially similar methods can be applied to obtain three-dimensional solutions to axisymmetric problems of the circular plate $0 \leq r < a$, $-h/2 < z < h/2$ loaded by polynomial tractions on the surfaces $z = \pm h/2$ (or with a polynomial temperature distribution), where $h \ll a$. In this case, we impose the strong boundary conditions on the plane surfaces $z = \pm h/2$ and weak conditions on $r = a$. As in the corresponding two-dimensional problems of Chapter 5, we generally need to start with a polynomial up to 3 orders higher than that suggested by the applied loading.

Weak boundary conditions at $r = a$ can be stated in terms of force and moment resultants per unit circumference or of averaged displacements or rotations. The force resultants are a membrane tensile force N and shear force V defined by

$$
N = \int_{-h/2}^{h/2} \sigma_{rr} dz \;\; ; \;\; V = \int_{-h/2}^{h/2} \sigma_{rz} dz \; ,
$$

whilst the moment resultant is

$$
M = \int_{-h/2}^{h/2} z \sigma_{rr} dz \; .
$$

Weak displacement boundary conditions can be defined by analogy with equations (9.12).

26.2.1 Uniformly loaded plate on a simple support

As an example, we consider the plate $0 \leq r < a$, $-h/2 < z < h/2$ loaded by a uniform compressive traction p_0 on the surface $z = h/2$, simply-supported at the edge $r = a$ and otherwise traction free. Thus, the traction boundary conditions are

$$
\sigma_{zz} = -p_0 \;\; ; \;\; \sigma_{zr} = 0 \;\; ; \;\; z = h/2 \tag{26.24}
$$

$$
\sigma_{zz} = \sigma_{zr} = 0 \;\; ; \;\; z = -h/2 \tag{26.25}
$$

$$
\int_{-h/2}^{h/2} \sigma_{rr} dz = 0 \;\; ; \;\; \int_{-h/2}^{h/2} z \sigma_{rr} dz = 0 \;\; ; \;\; r = a \; . \tag{26.26}
$$

The loading is uniform suggesting potentials of degree 2 in ϕ and degree 1 in ω, but as in the plane stress problem of Figure 5.3, the loading will generate bending moments and bending stresses, requiring potentials three orders higher than this. We therefore start with the potential functions

$$\phi = \frac{A_5}{8}(8z^5 - 40z^3r^2 + 15zr^4) + \frac{A_4}{8}(8z^4 - 24z^2r^2 + 3r^4) + \frac{A_3}{2}(2z^3 - 3zr^2)$$
$$+\frac{A_2}{2}(2z^2 - r^2)$$
$$\omega = \frac{B_4}{8}(8z^4 - 24z^2r^2 + 3r^4) + \frac{B_3}{2}(2z^3 - 3zr^2) + \frac{B_2}{2}(2z^2 - r^2) + B_1z ,$$

where we have omitted those terms leading only to rigid-body displacements. Substituting into Table 22.1, the stresses are obtained as

$$\sigma_{rr} = \frac{5A_5}{2}(9r^2z - 4z^3) + \frac{3A_4}{2}(3r^2 - 4z^2) - 3A_3z - A_2 - 2B_4(3 + 4\nu)z^3$$
$$+\frac{3B_4}{2}(3 + 8\nu)r^2z - 3B_3(1 + 2\nu)z^2 + 3\nu B_3r^2 - B_2(1 + 4\nu)z - 2B_1\nu$$
$$\sigma_{\theta\theta} = \frac{5A_5}{2}(3r^2z - 4z^3) + \frac{3A_4}{2}(r^2 - 4z^2) - 3A_3z - A_2 - 2B_4(3 + 4\nu)z^3$$
$$+\frac{3B_4}{2}(1 + 8\nu)r^2z - 3B_3(1 + 2\nu)z^2 + 3\nu B_3r^2 - B_2(1 - 4\nu)z - 2B_1\nu$$
$$\sigma_{rz} = \frac{15A_5}{2}(r^3 - 4rz^2) - 12A_4rz - 3A_3r - 6B_4(1 + 2\nu)z^2r$$
$$-\frac{3B_4}{2}(1 - 2\nu)r^3 - 6B_3\nu zr + B_2(1 - 2\nu)r \qquad (26.27)$$
$$\sigma_{zz} = 10A_5(2z^3 - 3r^2z) + 6A_4(2z^2 - r^2) + 6A_3z + 2A_2$$
$$+4B_4(1 + 2\nu)z^3 + 6B_4(1 - 2\nu)r^2z + 6B_3\nu z^2 + 3B_3(1 - \nu)r^2$$
$$-2B_2(1 - 2\nu)z - 2B_1(1 - \nu) .$$

Notice that σ_{zz} is even in r and σ_{rz} is odd in r. The strong boundary conditions (26.24, 26.25) therefore require us to equate coefficients of r^2 and r^0 in σ_{zz} and of r^3 and r^1 in σ_{rz} on each of the boundaries $z = \pm h/2$, leading to the equations

$$\frac{15A_5}{2} - \frac{3B_4}{2}(1 - 2\nu) = 0$$
$$-\frac{15A_5h^2}{2} - 6A_4h - 3A_3 - \frac{3B_4h^2}{2}(1 + 2\nu) - 3B_3\nu h + B_2(1 - 2\nu) = 0$$
$$-\frac{15A_5h^2}{2} + 6A_4h - 3A_3 - \frac{3B_4h^2}{2}(1 + 2\nu) + 3B_3\nu h + B_2(1 - 2\nu) = 0$$
$$-15A_5h - 6A_4 + 3B_4(1 - 2\nu)h + 3B_3(1 - \nu) = 0$$
$$15A_5h - 6A_4 + 3B_4(1 - 2\nu)h - 3B_3(1 - \nu) = 0$$

$$\frac{5A_5h^3}{2} + 3A_4h^2 + 3A_3h + 2A_2 + \frac{B_4(1+2\nu)h^3}{2}$$

$$+\frac{3B_3\nu h^2}{2} - B_2(1-2\nu)h - 2B_1(1-\nu) = -p_0$$

$$-\frac{5A_5h^3}{2} + 3A_4h^2 - 3A_3h + 2A_2 - \frac{B_4(1+2\nu)h^3}{2}$$

$$+\frac{3B_3\nu h^2}{2} + B_2(1-2\nu)h - 2B_1(1-\nu) = 0 \ .$$

Two additional equations are obtained by substituting (26.27) into the weak conditions (26.26) and evaluating the integrals, with the result

$$\frac{A_4(9a^2h - h^3)}{2} - A_2h - \frac{B_3(1+2\nu)h^3}{4} + 3B_3\nu a^2h - 2B_1\nu h = 0$$

$$\frac{A_5(15a^2h^3 - h^5)}{8} - \frac{A_3h^3}{4} - \frac{B_4(3+4\nu)h^5}{40}$$

$$+\frac{B_4(3+8\nu)a^2h^3}{8} - \frac{B_2(1+4\nu)h^3}{12} = 0 \ .$$

Solving these equations for the constants A_5, A_4, A_3, A_2, B_4, B_3, B_2, B_1 and substituting the resulting expressions into (26.27), we obtain the final stress field as

$$\sigma_{rr} = \frac{p_0}{h^3}\left[\frac{3(3+\nu)(r^2-a^2)z}{4} + (2+\nu)\left(\frac{h^2z}{20} - z^3\right)\right]$$

$$\sigma_{\theta\theta} = \frac{p_0}{h^3}\left[\frac{3[(1+3\nu)r^2 - (3+\nu)a^2]z}{4} + (2+\nu)\left(\frac{h^2z}{20} - z^3\right)\right]$$

$$\sigma_{rz} = \frac{3p_0(h^2 - 4z^2)r}{4h^3}$$

$$\sigma_{zz} = \frac{p_0(4z^3 - 3h^2z - h^3)}{2h^3} \ .$$

The maximum tensile stress occurs at the point $0, -h/2$ and is

$$\sigma_{\max} = \frac{3(3+\nu)p_0a^2}{8h^2} + \frac{(2+\nu)p_0}{10} \ .$$

The elementary plate theory predicts the stress field[5]

---

[5] S. Timoshenko and S. Woinowsky-Krieger, *Theory of Plates and Shells*, McGraw-Hill, New York, 2nd edn., 1959, §16.

$$\sigma_{rr} = \frac{3(3+\nu)p_0(r^2 - a^2)z}{4h^3}$$

$$\sigma_{\theta\theta} = \frac{3p_0[(1+3\nu)r^2 - (3+\nu)a^2]z}{4h^3}$$

$$\sigma_{rz} = \sigma_{zz} = 0 \,,$$

with a maximum tensile stress

$$\sigma_{max} = \frac{3(3+\nu)p_0a^2}{8h^2} \,.$$

Thus, the elementary theory provides a good approximation to the more exact solution as long as $h \ll a$. For example, it is in error by only about 0.2% if $h = a/10$.

26.3 Non-axisymmetric Problems

Non-axisymmetric problems can be solved by expanding the loading as a Fourier series in θ and using the potentials of §25.7 in solutions A,B and E. All three solutions are generally required, since there is no equivalent of the partition into irrotational and torsional problems encountered for the axisymmetric case. The method is in other respects similar to that used for axisymmetric problems, but solutions tend to become algebraically complicated, simply because of the number of functions involved. For this reason we shall consider only a very simple example here. More complex examples are readily treated using Maple or Mathematica.

26.3.1 Cylindrical cantilever with an end load

We consider the case of the solid cylinder $0 < r < a$, $0 < z < L$, built in at $z = L$ and loaded only by a shear force F in the x-direction at $z = 0$. Thus, the curved surfaces are traction free

$$\sigma_{rr} = \sigma_{r\theta} = \sigma_{rz} = 0 \;\; ; \;\; r = a \,, \tag{26.28}$$

and on the end $z = 0$ we have $\sigma_{zz} = 0$ and

$$\int_0^a \int_0^{2\pi} (\sigma_{zr} \cos\theta - \sigma_{z\theta} \sin\theta) r \, d\theta \, dr = -F \,, \tag{26.29}$$

from equilibrium considerations.

As in §13.1, loading in the x-direction implies normal stress components varying with $\cos\theta$ and these will be obtained from potentials ϕ, ω with a $\cos\theta$ multiplier and from Ψ with a $\sin\theta$ multiplier[6].

From elementary bending theory, we anticipate bending stresses σ_{zz} varying with $zr\cos\theta$, suggesting fourth order polynomials in ϕ, Ψ and third order in ω. We therefore try

$$\phi = \frac{5A_4}{2}(4z^3r - 3zr^3)\cos\theta + 3A_2zr\cos\theta$$

$$\omega = \frac{3B_3}{2}(4z^2r - r^3)\cos\theta \qquad\qquad (26.30)$$

$$\Psi = \frac{5E_4}{2}(4z^3r - 3zr^3)\sin\theta ,$$

from equations (25.35). Notice that the problem is antisymmetric in z, so only odd powers of z are included in ϕ, Ψ and even powers in ω. Also, we have omitted the second order term in Ψ and the first order term in ω, since these correspond only to rigid-body displacements.

Substituting (26.30) into Table 22.1, we obtain

$$\sigma_{rr} = -3(15A_4 + (3 + 8\nu)B_3 + 10E_4)rz\cos\theta$$
$$\sigma_{r\theta} = 3(5A_4 + B_3 + 10E_4)rz\sin\theta$$
$$\sigma_{\theta\theta} = -3(5A_4 + (1 + 8\nu)B_3 - 10E_4)rz\cos\theta \qquad\qquad (26.31)$$
$$\sigma_{zr} = \frac{3}{2}\Big[5A_4(4z^2 - 3r^2) + 2A_2 + B_3(4(1 + 2\nu)z^2 + 3(1 - 2\nu)r^2)$$
$$+ 5E_4(4z^2 - r^2)\Big]\cos\theta$$
$$\sigma_{z\theta} = -\frac{3}{2}\Big[5A_4(4z^2 - r^2) + 2A_2 + B_3(4(1 + 2\nu)z^2 + (1 - 2\nu)r^2)$$
$$+ 5E_4(4z^2 - 3r^2)\Big]\sin\theta$$
$$\sigma_{zz} = 12(5A_4 - (1 - 2\nu)B_3)rz\cos\theta .$$

Substituting these expressions into (26.29) and evaluating the integral, we obtain the condition

$$\frac{3\pi a^2}{2}(-5A_4a^2 + 2A_2 + B_3(1 - 2\nu)a^2 - 5E_4a^2) = -F \qquad\qquad (26.32)$$

and the strong boundary conditions (26.28) provide the four additional equations

---

[6] Notice that the expressions for normal stresses in solution E involve a derivative with respect to θ, so generally if cosine terms are required in A and B, sine terms will be required in E and *vice versa*.

$$15A_4 + (3 + 8\nu)B_3 + 10E_4 = 0$$
$$5A_4 + B_3 + 10E_4 = 0 \qquad (26.33)$$
$$20A_4 + 4(1 + 2\nu)B_3 + 20E_4 = 0$$
$$-15A_4 a^2 + 2A_2 + 3(1 - 2\nu)B_3 a^2 - 5E_4 a^2 = 0 .$$

The five equations (26.32, 26.33) are not linearly independent, so they can be solved for the four constants A_4, A_2, B_3, E_4. Substituting the resulting expressions into (26.31), we obtain the final stress field

$$\sigma_{rr} = \sigma_{r\theta} = \sigma_{\theta\theta} = 0$$

$$\sigma_{zr} = -\frac{F(3 + 2\nu)\left(a^2 - r^2\right)\cos\theta}{2\pi a^4(1 + \nu)}$$

$$\sigma_{z\theta} = \frac{F\left[(3 + 2\nu)a^2 - (1 - 2\nu)r^2\right]\sin\theta}{2\pi a^4(1 + \nu)}$$

$$\sigma_{zz} = -\frac{4Frz\cos\theta}{\pi a^4} .$$

Problems

26.1. The solid cylinder $0 \le r < a$, $0 < z < L$ is supported at the end $z = L$ and subjected to a uniform shear traction $\sigma_{rz} = S$ on the curved boundary $r = a$. The end $z = 0$ is traction free.

Find the complete stress field in the cylinder using strong boundary conditions on the curved boundary and weak conditions at the ends.

26.2. The hollow cylinder $b < r < a$, $0 < z < L$ is supported at the end $z = L$ and subjected to a uniform shear traction $\sigma_{rz} = S$ on the outer boundary $r = a$, the inner boundary $r = b$ and the end $z = 0$ being traction free.

Find the complete stress field in the cylinder using strong boundary conditions on the curved boundaries and weak conditions at the ends.

26.3. The solid cylinder $(0 \le r < a, -L < z < L)$ is subjected to the traction distribution

$$\sigma_{rr} = -p_0\left(L^2 - z^2\right) ; \quad \sigma_{zr} = 0 ; \quad r = a ;$$
$$\sigma_{zr} = \sigma_{zz} = 0 ; \quad z = \pm L .$$

Find the complete stress field in the cylinder using strong boundary conditions on the curved boundary and weak conditions at the ends.

26.4. The hollow cylinder $b < r < a$, $0 < z < L$ is subjected to the traction distribution

$$\sigma_{rr} = \sigma_{r\theta} = \sigma_{rz} = 0 \; ; \quad r = a$$
$$\sigma_{rz} = S \; ; \quad \sigma_{rr} = \sigma_{r\theta} = 0 \; ; \quad r = b$$
$$\sigma_{zr} = \sigma_{zz} = 0 \; ; \quad z = 0 \; .$$

Find the complete stress field in the cylinder using strong boundary conditions on the curved boundaries and weak conditions at the ends.

26.5. The solid cylinder $0 < z < L$, $0 \leq r < a$ is uniformly heated at a rate q_0 per unit area over the curved surface $r = a$, the end $z = 0$ being insulated. All the surfaces are traction free. Find the complete stress field in the cylinder, using strong boundary conditions on the curved boundary and weak conditions on the ends.

26.6. A heat exchanger tube $b < r < a$, $-L < z < L$ is subjected to the steady-state temperature distribution

$$T = T_0 \left(1 + \frac{z}{10a} \right) \; ; \quad r = b$$
$$= -\frac{T_0 z}{10a} \quad \quad ; \quad r = a \; .$$

Find the complete stress field in the tube if all the surfaces are traction free. Find the location and a general expression for the magnitude of (i) the maximum tensile stress and (ii) the maximum von Mises equivalent stress for the case where $b = 0.8a$ and $L = 20a$. Find the magnitude of these quantities when $T_0 = 50^{\circ}$C and the material is copper for which $E = 121$ GPa, $\nu = 0.33$ and $\alpha = 17 \times 10^{-6}$ per $^{\circ}$C.

26.7. The semi-infinite cylinder $0 \leq r < a$, $z > 0$ is loaded by self-equilibrated axisymmetric torsional tractions $\sigma_{z\theta}(r, \theta, 0) = f(r)$ on the end face $z = 0$, the curved surface $r = a$ being traction free. Use a technique similar to that in §6.2 to determine the rate at which the stress field will decay with distance z from the end.

26.8. The circular plate $0 \leq r < a$, $-h/2 < z < h/2$ is simply supported at $r = a$ and loaded by a shear traction $\sigma_{zr} = Sr$ on the top surface $z = h/2$. All the other tractions on the plane surfaces $z = \pm h/2$ are zero. Find the stresses in the plate, using strong boundary conditions on $z = \pm h/2$ and weak boundary conditions on $r = a$.

26.9. The circular plate $0 \leq r < a$, $-h/2 < z < h/2$ rotates at constant speed Ω about the z-axis, all the surfaces being traction free. Find the stresses in the plate, using strong boundary conditions on $z = \pm h/2$ and weak boundary conditions on $r = a$.

26.10. A circular plate of radius a and thickness h is built in to a rigid support at $r = a$ and is stress free when the temperature is uniform and equal to T_0. The traction-free surface $z = h/2$ is now heated with a uniform heat flux q_0, whilst the opposite surface $z = -h/2$ is insulated. In the steady state, heat flows from the plate into the support which is maintained at the original temperature T_0. Find the complete stress field in the plate.

26.11. Figure 26.1 shows a cylindrical heat exchanger tube of radius a fitted with annular cooling fins of inside radius a, outside radius $2a$ and thickness h. If the fins are cooled with a uniform heat flux q_0 on all exposed surfaces, find the stress field in the fins. Assume that the tube is rigid, and that the fins are rigidly attached to it at $r = a$. Make a suitably normalized contour plot of the von Mises stress σ_E of equation (1.23) for the case where $\nu = 0.3$ and $h = 0.1a$.

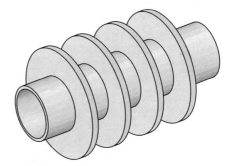

Fig. 26.1 A heat exchanger tube

26.12. The solid cylinder $0 \leq r < a$, $0 < z < L$ with its axis horizontal is built in at $z = L$ and loaded only by its own weight (density ρ). Find the complete stress field in the cylinder.

26.13. The hollow cylinder $b < r < a$, $0 < z < L$ with its axis horizontal is built in at $z = L$ and loaded only by its own weight (density ρ). Find the complete stress field in the cylinder.

26.14. The hollow cylinder $b \leq r < a$, $-L < z < L$ rotates at constant speed Ω about the x-axis, all the surfaces being traction free. Find the complete stress field in the cylinder.

26.15. The solid cylinder $0 \leq r < a$, $-L < z < L$ is accelerated from rest by two equal and opposite forces F applied in the positive and negative y-directions respectively at the ends $z = \pm L$. All the other surface tractions are zero. Find the complete stress field in the cylinder at the instant when the forces are applied[7] (i.e. when the angular velocity is zero).

---

[7] This is a three-dimensional version of the problem considered in §7.4.2.

26.16. The circular plate $0 \leq r < a$, $-h/2 < z < h/2$ rotates at constant speed Ω about the x-axis, all the surfaces being traction free. Find the stresses in the plate, using strong boundary conditions on $z = \pm h/2$ and weak boundary conditions on $r = a$.

Note: This problem is not axisymmetric. It is the three-dimensional version of Problem 8.5.

Chapter 27
Problems in Spherical Coördinates

The spherical harmonics of Chapter 25 can be used in combination with Tables 22.2, 23.2 to treat problems involving bodies whose boundaries are surfaces of the spherical coördinate system, for example the spherical surface $R = a$ or the conical surface $\beta = \beta_0$, where a, β_0 are constants. Examples of interest include the perturbation of an otherwise uniform stress field by a spherical hole or inclusion, the stresses in a solid sphere due to rotation about an axis and problems for a conical beam or shaft.

27.1 Solid and Hollow Spheres

The general solution for a solid sphere with prescribed surface tractions can be obtained using the spherical harmonics of equation (25.16) in solutions A,B and E of Table 22.2. The addition of the singular harmonics (25.17) permits a general solution to the axisymmetric problem of the hollow sphere, but the corresponding non-axisymmetric solution cannot be obtained from equations (25.16, 25.17). To understand this, we recall from §21.4 that the elimination of one of the components of the Papkovich-Neuber vector function ψ is only generally possible when all straight lines drawn in a given direction cut the boundary of the body at only two points. This is clearly not the case for a body containing a spherical hole, since lines in any direction can always be chosen that cut the surface of the hole in two points and the external boundaries of the body at two additional points.

A similar difficulty arises for a body containing a plane crack, such as the penny-shaped crack problem of Chapter 33. There are various ways of getting around it. For example,

Supplementary Information The online version contains supplementary material available at https://doi.org/10.1007/978-3-031-15214-6_27.

(i) We could decompose the loading into symmetric and antisymmetric components with respect to the plane $z = 0$. Each problem would then be articulated for the hollow hemisphere with symmetric or antisymmetric boundary conditions on the cut surface. Since the hollow hemisphere satisfies the conditions of §21.4, a general solution can be obtained using solutions A,B and E, but generally we shall need to supplement the potentials (25.16, 25.17) with those of equations (25.25) that are singular on that half of the z-axis that does not pass through the hemisphere being analyzed.

(ii) We could include all four terms of the Papkovich-Neuber solution[1] or the six terms of the Galerkin vector of section §21.2. The former option is equivalent to the use of solutions A and B in combination with two additional solutions obtained by permuting suffices x, y, z in solution B.

(iii) In certain cases, we may be able to obtain particular non-axisymmetric solutions by superposition of axisymmetric ones. For example, the spherical hole in an otherwise uniform shear field can be solved by superposing the stress fields for the same body subject to equal and opposite tension and compression respectively about orthogonal axes.

27.1.1 The solid sphere in torsion

As a simple example, we consider the solid sphere $0 \leq R < a$ subjected to the torsional tractions

$$\sigma_{R\theta} = S_0 \sin(2\beta) \tag{27.1}$$

on the surface $R = a$, the remaining tractions $\sigma_{RR}, \sigma_{R\beta}$ being zero.

This is clearly an axisymmetric torsion problem, for which we require only solution E and the axisymmetric bounded potentials (25.11). Substitution of the first few potentials into Table 22.2 shows that the problem can be solved using the potential

$$\Psi = E_3 R^3 P_3(\cos \beta) = \frac{E_3}{2}(5 \cos^3 \beta + 3 \cos \beta) \, ,$$

where E_3 is an unknown constant. The corresponding non-zero stress components in spherical polar coördinates are

$$\sigma_{R\theta} = 3E_3 R \sin \beta \cos \beta \; ; \; \sigma_{\beta\theta} = -3E_3 R \sin^2 \beta \, ,$$

so by inspection we see that the boundary condition (27.1) is satisfied by choosing $E_3 = 3S_0/2a$.

---

[1] See for example, A. I. Lur'e, *Three-Dimensional Problems of the Theory of Elasticity*, Interscience, New York, 1964, Chapter 8.

27.1.2 Spherical hole in a tensile field

A more interesting problem is that in which a state of uniaxial tension $\sigma_{zz} = S_0$ is perturbed by the presence of a small, traction-free, spherical hole of radius a — i.e.

$$\sigma_{zz} \to S \; ; \quad R \to \infty \tag{27.2}$$

$$\sigma_{rr}, \sigma_{\theta\theta}, \sigma_{r\theta}, \sigma_{\theta z}, \sigma_{zr} \to 0 \; ; \quad R \to \infty \tag{27.3}$$

$$\sigma_{RR} = \sigma_{R\theta} = \sigma_{R\beta} = 0 \; ; \quad R = a \;. \tag{27.4}$$

This problem is the axisymmetric equivalent of the two-dimensional problem of §8.4.1 and as in that case it is convenient to start by determining a stress function description of the *unperturbed* uniaxial stress field. This is clearly axisymmetric and involves no torsion, so we use solutions A and B with the polynomials (25.19) in cylindrical polar coördinates. The stress components involve two differentiations in solution A and one in solution B, so we need the functions $R^2 P_2(z/R)$ in ϕ and $R^1 P_1(z/R)$ in ω to get a uniform state of stress. Writing

$$\phi = \frac{A_1}{2}(2z^2 - r^2) \; ; \quad \omega = B_1 z$$

and substituting into Table 22.1, we obtain

$$\sigma_{rr} = \sigma_{\theta\theta} = -(A_1 + 2\nu B_1) \; ; \quad \sigma_{zz} = 2A_1 - 2B_1(1 - \nu) \; ; \quad \sigma_{rz} = \sigma_{z\theta} = \sigma_{\theta r} = 0 \;.$$

The conditions (27.2, 27.3) are therefore satisfied by the choice

$$A_1 = \frac{S\nu}{(1+\nu)} \; ; \quad B_1 = -\frac{S}{2(1+\nu)} \;. \tag{27.5}$$

To satisfy the traction-free condition at the hole surface (27.4), we now change to spherical polar coördinates and superpose singular potentials from (25.12) with the same Legendre polynomial form as those in the unperturbed solution — i.e.

$$\phi = A_1 R^2 P_2(\cos \beta) + A_2 R^{-3} P_2(\cos \beta) + A_3 R^{-1} P_0(\cos \beta) \tag{27.6}$$

$$\omega = B_1 R^1 P_1(\cos \beta) + B_2 R^{-2} P_1(\cos \beta) \;. \tag{27.7}$$

Notice that in contrast to the two-dimensional problem of §8.4.1, we need to include also any lower order Legendre polynomial terms that have the same symmetry (in this case even orders in ϕ and odd orders in ω).

Substituting (27.6, 27.7) into Table 22.2, we obtain

$$\sigma_{RR} = \left(3A_1 + \frac{18A_2}{R^5} - 2(1-2\nu)B_1 + \frac{2(5-\nu)B_2}{R^3}\right)\cos^2\beta$$

$$-A_1 - \frac{6A_2}{R^5} + \frac{2A_3}{R^3} - 2\nu B_1 - \frac{2\nu B_2}{R^3}$$

$$\sigma_{R\beta} = \left(-3A_1 + \frac{12A_2}{R^5} - 2(1-2\nu)B_1 + \frac{2(1+\nu)B_2}{R^3}\right)\sin\beta\cos\beta$$

and hence the boundary conditions (27.4) require that

$$3A_1 + \frac{18A_2}{a^5} - 2(1-2\nu)B_1 + \frac{2(5-\nu)B_2}{a^3} = 0$$

$$-A_1 - \frac{6A_2}{a^5} + \frac{2A_3}{a^3} - 2\nu B_1 - \frac{2\nu B_2}{a^3} = 0$$

$$-3A_1 + \frac{12A_2}{a^5} - 2(1-2\nu)B_1 + \frac{2(1+\nu)B_2}{a^3} = 0 .$$

Solving these equations and using (27.5), we obtain

$$A_2 = \frac{Sa^5}{(7-5\nu)} \;\; ; \;\; A_3 = \frac{Sa^3(6-5\nu)}{2(7-5\nu)} \;\; ; \;\; B_2 = \frac{-5Sa^3}{2(7-5\nu)} \; ,$$

and the final stress field is

$$\sigma_{RR} = S\cos^2\beta + \frac{S}{(7-5\nu)}\left(\frac{a^3}{R^3}\left[6-5(5-\nu)\cos^2\beta\right] + \frac{6a^5}{R^5}[3\cos^2\beta - 1]\right)$$

$$\sigma_{\theta\theta} = \frac{3S}{2(7-5\nu)}\left(\frac{a^3}{R^3}\left[5\nu - 2 + 5(1-2\nu)\cos^2\beta\right] + \frac{a^5}{R^5}[1 - 5\cos^2\beta]\right)$$

$$\sigma_{\beta\beta} = S\sin^2\beta$$

$$+ \frac{S}{2(7-5\nu)}\left(\frac{a^3}{R^3}\left[4 - 5\nu + 5(1-2\nu)\cos^2\beta\right] + \frac{3a^5}{R^5}[3 - 7\cos^2\beta]\right)$$

$$\sigma_{R\beta} = S\left(-1 + \frac{1}{(7-5\nu)}\left[-\frac{5a^3(1+\nu)}{R^3} + \frac{12a^5}{R^5}\right]\right)\sin\beta\cos\beta .$$

The maximum tensile stress is $\sigma_{\beta\beta}$ at $\beta = \pi/2$, $R=a$ and is

$$\sigma_{\max} = \frac{3S(9-5\nu)}{2(7-5\nu)} .$$

The stress-concentration factor is $13/6 = 2.17$ for materials with $\nu = 0.5$ and decreases only slightly for other values in the practical range.

27.2 Conical Bars

The equation $\beta = \beta_0$, where β_0 is a constant, defines a conical surface in the spherical polar coördinate system and the potentials developed in Chapter 25 can therefore be used to solve various problems for the solid or hollow conical bar.

Substitution of a potential of the form $R^n f(\theta, \beta)$ into Table 22.2 will yield stress components varying with R^{n-2} in solutions A and E or with R^{n-1} in solution B. The bounded potentials of equations (25.11, 25.16) therefore provide a general solution of the problem of a solid conical bar with polynomial tractions on the curved surfaces.

If the curved surfaces are traction free, but the bar transmits a force or moment resultant, the stresses will increase without limit as the vertex (the origin) is approached, so in these cases we should expect to need singular potentials even for the solid bar.

For the hollow conical bar, we need additional potentials with the same power law variation with R and these are provided by the Q-series of equations (25.24). Notice that the singularity of these solutions on the z-axis is acceptable for the hollow bar, since the axis is not then contained within the body.

27.2.1 Conical bar transmitting an axial force

We consider the solid conical bar $0 \leq \beta < \beta_0$ transmitting a tensile axial force F, the curved surfaces $\beta = \beta_0$ being traction free. We shall impose the strong (homogeneous) conditions

$$\sigma_{\beta R} = \sigma_{\beta \theta} = \sigma_{\beta \beta} = 0 \quad ; \quad \beta = \beta_0 \tag{27.8}$$

on the curved surfaces and weak conditions on the ends. If the bar has plane ends and occupies the region $a < z < b$, we have

$$\int_0^{2\pi} \int_0^{b \tan \beta_0} \sigma_{zz}(r, \theta, b) r \, dr \, d\theta = F$$

and a similar condition[2] at $z = a$. Alternatively, we could consider the equivalent problem of a bar with spherical ends $R_a < R < R_b$, for which the weak boundary

---

[2] As in earlier problems, the weak condition need only be imposed at one end, since the condition at the other end will be guaranteed by the fact that the stress field satisfies the equilibrium condition everywhere.

condition takes the form[3]

$$\int_0^{2\pi}\int_0^{\beta_0}\left[\sigma_{RR}(R_b,\theta,\beta)\cos\beta-\sigma_{R\beta}(R_b,\theta,\beta)\sin\beta\right]R_b^2\sin\beta d\beta d\theta=F\ .\quad(27.9)$$

The cross-sectional area of the ends of the conical bar increase with R^2 (the radius increases with R), so the stresses must decay with R^{-2} as in the point force solutions of §24.1.2 and §24.2.1. This suggests the use of the source solution $1/R$ in solution B and the potential $\ln(R+z)$ in solution A. Notice that this problem is axisymmetric and there is no torsion, so solution E will not be required. Also, notice that the bar occupies only the region $z>0$ in cylindrical polar coördinates, so the singularities on the negative z-axis implied by $\ln(R+z)$ are acceptable.

Writing these potentials in spherical polar coördinates and introducing arbitrary multipliers A, B, we have

$$\phi=A\{\ln(R)+\ln(1+\cos(\beta))\}\ ;\quad\omega=\frac{B}{R}\qquad(27.10)$$

and substitution into Table 22.2 gives the non-zero stress components

$$\sigma_{RR}=-\frac{A}{R^2}+\frac{2(2-\nu)B\cos\beta}{R^2}\qquad(27.11)$$

$$\sigma_{\theta\theta}=\frac{A}{R^2(1+\cos\beta)}-\frac{(1-2\nu)B\cos\beta}{R^2}\qquad(27.12)$$

$$\sigma_{\beta\beta}=\frac{A\cos\beta}{R^2(1+\cos\beta)}-\frac{(1-2\nu)B\cos\beta}{R^2}\qquad(27.13)$$

$$\sigma_{\beta R}=\frac{A\sin\beta}{R^2(1+\cos\beta)}-\frac{(1-2\nu)B\sin\beta}{R^2}\ .\qquad(27.14)$$

The strong conditions (27.8) will therefore be satisfied as long as

$$\frac{A}{(1+\cos\beta_0)}=(1-2\nu)B\qquad(27.15)$$

and the weak condition (27.9), after substitution for the stress components and evaluation of the integral, gives

$$2\pi(1-\cos\beta_0)\left[A+\left(\cos^2\beta_0+\cos\beta_0+2-2\nu\right)B\right]=F\ .\qquad(27.16)$$

Solving equations (27.15, 27.16) for A, B, we obtain

---

[3] To derive this result, note that the element of area on a spherical surface is $(Rd\beta)(rd\theta)=R^2\sin\beta d\beta d\theta$ and the forces on this element due to the stress components $\sigma_{RR},\sigma_{R\beta}$ must be resolved into the axial direction.

$$A = \frac{F(1 - 2\nu)(1 + \cos \beta_0)}{2\pi(1 - \cos \beta_0)\left(1 + 2\nu \cos \beta_0 + \cos^2 \beta_0\right)}$$

$$B = \frac{F}{2\pi(1 - \cos \beta_0)\left(1 + 2\nu \cos \beta_0 + \cos^2 \beta_0\right)} \tag{27.17}$$

and the stress components are then recovered by substituting for A, B into equations (27.11–27.14).

Elementary Mechanics of Materials arguments predict that the axial stress σ_{zz} would be uniform and equal to $F/\pi b^2$ at $z = b$. We can assess the approximation involved in this prediction by recasting the present solution in cylindrical coördinates — i.e. writing

$$\phi = A \ln\left(\sqrt{r^2 + z^2} + z\right) \quad ; \quad \omega = \frac{B}{\sqrt{r^2 + z^2}}$$

with A, B given by (27.17) and substituting into the cylindrical polar form of solutions A and B (Table 22.1). This is done at the end of the Maple file 'conetension'. The normalized stresses on a cross-sectional plane are shown in Figure 27.1 for various cone angles β_0. The axial stress is largest at the axis $r = 0$, but remains fairly uniform across the cross section for $\beta_0 < 10^o$. Significant non-uniformity occurs for larger cone angles. For example, even for a 30° cone, the maximum stress exceeds the elementary prediction by 46%.

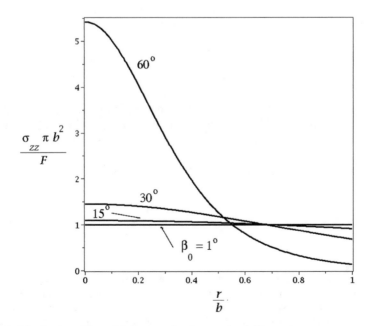

Fig. 27.1 Distribution of the axial stress σ_{zz} for the cone loaded in tension.

In the special case where $\beta_0 = \pi/2$, the cone becomes a half-space and we recover the Boussinesq solution of §24.2.1, albeit expressed in spherical polar coördinates. For values of β in the range $\pi/2 < \beta_0 < \pi$, the solution defines the problem of a large body with a conical notch loaded by a concentrated force at the vertex of the notch.

The related problem of the *hollow* conical bar $\beta_1 < \beta < \beta_2$ can be treated in the same way by supplementing ϕ in solution A by the term $A_2\{\ln(R) + \ln(1 - \cos(\beta))\}$. This is similar to the term in equation (27.10), except that the singularities are now on the positive z-axis, which of course is acceptable for the hollow bar, since this axis does not now lie inside the material of the bar[4].

27.2.2 Inhomogeneous problems

If polynomial tractions are applied to the curved surfaces of the cone, but there are no concentrated loads at the vertex, the stresses will vary with the same power of R as the tractions and the order of the corresponding potential is easily determined[5] by examining the derivatives in Table 22.2.

Example

As a simple example, we consider the hollow cone $\beta_1 < \beta < \beta_2$ loaded by quadratic torsional shear tractions on the outer surface, the other tractions being zero — i.e.

$$\sigma_{\beta\theta} = SR^2 \ ; \quad \sigma_{\beta R} = \sigma_{\beta\beta} = 0 \ ; \quad \beta = \beta_2$$
$$\sigma_{\beta\theta} = \sigma_{\beta R} = \sigma_{\beta\beta} = 0 \ ; \quad \beta = \beta_1 \ . \tag{27.18}$$

This is clearly an axisymmetric torsion problem which requires solution E only. The tractions vary with R^2 and hence Ψ must vary with R^4, from Table 22.2. Since the cone is hollow, we require both P and Q series Legendre functions of this order, giving

$$\Psi = C_1 R^4 P_4(\cos\beta) + C_2 R^4 Q_4(\cos\beta) \ ,$$

---

[4] See Problem 27.13.

[5] As in Chapter 8, some of the lower order bounded potentials correspond merely to rigid-body displacements and involve no stresses. In such cases, we need additional special potentials which are obtained from the logarithmic series of equations (25.25, 24.9), see for example Problem 27.15. Similar considerations apply to the cone loaded by low-order non-axisymmetric polynomial tractions. These special solutions are related to the two-dimensional corner fields described in §§11.1.2, 11.1.3 and are needed because a state of uniform traction on the conical surface is not consistent with a locally homogeneous state of stress near the apex of the cone.

where P, Q are given by equations (25.8, 25.22) respectively. Substituting into Table 22.2 yields the non-zero stress components

$$\sigma_{\theta\beta} = -\frac{R^2}{2}\left[30C_1\sin^2\beta\cos\beta + 15C_2\sin^2\beta\cos\beta\ln\left(\frac{1-\cos\beta}{1+\cos\beta}\right)\right.$$
$$\left. -\frac{C_2\left(30\cos^4\beta - 50\cos^2\beta + 16\right)}{\sin^2\beta}\right] \tag{27.19}$$

$$\sigma_{R\theta} = \frac{R^2\sin\beta}{2}\left[6C_1\left(5\cos^2\beta - 1\right) + 3C_2\left(5\cos^2\beta - 1\right)\ln\left(\frac{1-\cos\beta}{1+\cos\beta}\right)\right.$$
$$\left. +\frac{C_2\left(30\cos^3\beta - 26\cos\beta\right)}{\sin^2\beta}\right] \tag{27.20}$$

and the boundary conditions (27.18) will therefore be satisfied if

$$6C_1\left(5\cos^2\beta_1 - 1\right) + 3C_2\left(5\cos^2\beta_1 - 1\right)\ln\left(\frac{1-\cos\beta_1}{1+\cos\beta_1}\right)$$
$$+\frac{C_2\left(30\cos^3\beta_1 - 26\cos\beta_1\right)}{\sin^2\beta_1} = 0$$

$$6C_1\left(5\cos^2\beta_2 - 1\right) + 3C_2\left(5\cos^2\beta_2 - 1\right)\ln\left(\frac{1-\cos\beta_2}{1+\cos\beta_2}\right)$$
$$+\frac{C_2\left(30\cos^3\beta_2 - 26\cos\beta_2\right)}{\sin^2\beta_2} = \frac{2S}{\sin\beta_2} \quad.$$

Solution of these equations for C_1, C_2 and substitution of the result into (27.19, 27.20) completes the solution of the problem.

27.2.3   Non-axisymmetric problems

Non-axisymmetric problems for the solid or hollow cone can be solved using the potentials[6]

$$R^n P_n^m(\cos\beta)\begin{Bmatrix}\cos(m\theta)\\\sin(m\theta)\end{Bmatrix} \quad ; \quad R^n Q_n^m(\cos\beta)\begin{Bmatrix}\cos(m\theta)\\\sin(m\theta)\end{Bmatrix}$$

$$R^{-n-1}P_n^m(\cos\beta)\begin{Bmatrix}\cos(m\theta)\\\sin(m\theta)\end{Bmatrix} \quad ; \quad R^{-n-1}Q_n^m(\cos\beta)\begin{Bmatrix}\cos(m\theta)\\\sin(m\theta)\end{Bmatrix} \quad .$$

As a simple example, we consider the solid cone $0 \le \beta < \beta_0$ loaded only by a concentrated moment M about the x-axis at the vertex, the curved surfaces $\beta = \beta_0$

---

[6] but see footnote 5 above.

being traction free. There is no length scale, so the problem is self-similar and both the stresses and the potential functions must be of separated-variable form.

The applied moment leads to the transmission of a bending moment M along the cone and we anticipate stress components varying with y and hence with $\sin\theta$. The stress components on a given spherical surface will have moment arms proportional to the radius R and the area of this surface is proportional to R^2, so if the same moment is to be transmitted across all such surfaces, the stresses must decay with R^{-3}. This implies the use of a potential proportional to R^{-2} in solution B and to R^{-1} in solutions A and E. We therefore choose

$$\phi = AR^{-1}P_0^1(\cos\beta)\sin\theta = \frac{A\sin\beta\sin\theta}{R(1+\cos\beta)}$$

$$\omega = BR^{-2}P_1^1(\cos\beta)\sin\theta = \frac{B\sin\beta\sin\theta}{R^2}$$

$$\Psi = CR^{-1}P_0^1(\cos\beta)\cos\theta = \frac{C\sin\beta\cos\theta}{R(1+\cos\beta)}\ ,$$

where we have used (25.14) to evaluate P_0^1.

Substituting these expressions into Table 22.2, we obtain the stress components

$$\sigma_{RR} = \frac{2[A+2C+(5-\nu)B\cos\beta(1+\cos\beta)]\sin\beta\sin\theta}{R^3(1+\cos\beta)}$$

$$\sigma_{\theta\theta} = -\frac{\left[(A+2C)(2+\cos\beta)+3(1-2\nu)B\cos\beta(1+\cos\beta)^2\right]\sin\beta\sin\theta}{R^3(1+\cos\beta)^2}$$

$$\sigma_{\beta\beta} = -\frac{\left[A+2C+(1-2\nu)B(1+\cos\beta)^2\right]\cos\beta\sin\beta\sin\theta}{R^3(1+\cos\beta)^2} \qquad (27.21)$$

$$\sigma_{\theta\beta} = \frac{\left[A+2C+(1-2\nu)B(1+\cos\beta)^2\right]\sin\beta\cos\theta}{R^3(1+\cos\beta)^2}$$

$$\sigma_{\beta R} = -\frac{\left[2A+C(1-3\cos\beta)+2B(1+\cos\beta)\left(1-2\nu+(1+\nu)\cos^2\beta\right)\right]\sin\theta}{R^3(1+\cos\beta)}$$

$$\sigma_{R\theta} = -\frac{\left[2A+C\left(4-3\cos\beta-2\cos^2\beta\right)+2(2-\nu)B\cos\beta(1+\cos\beta)\right]\cos\theta}{R^3(1+\cos\beta)}$$

and hence the traction-free boundary conditions $\sigma_{\beta R}=\sigma_{\beta\theta}=\sigma_{\beta\beta}=0$ on $\beta=\beta_0$ will be satisfied provided

$$2A+C(1-3\cos\beta_0)+2B(1+\cos\beta_0)\left(1-2\nu+(1+\nu)\cos^2\beta_0\right)=0$$

$$A+2C+(1-2\nu)B(1+\cos\beta_0)^2=0$$

$$A+2C+(1-2\nu)B(1+\cos\beta_0)^2=0\ .$$

$$(27.22)$$

These three homogeneous equations have a non-trivial solution, since the last two equations are identical. The resulting free constant is determined from the equilibrium condition

$$\int_0^{\beta_0} \int_0^{2\pi} (\sigma_{R\theta} \cos \theta \cos \beta + \sigma_{R\beta} \sin \theta) R^3 \sin \beta \, d\theta \, d\beta = M ,$$

which states that the resultant of the tractions across any spherical surface of radius R is a moment M about the x-axis. Substituting for the stress components and evaluating the double integral, we obtain

$$\pi \left\{ -2A(1 - \cos \beta_0) + C \cos \beta_0 \sin^2 \beta_0 \right.$$
$$\left. + 2B \left[\cos^3 \beta_0 + (1 - 2\nu) \cos \beta_0 - 2(1 - \nu) \right] \right\} = M . \qquad (27.23)$$

The final stress field is obtained by solving (27.22, 27.23) for A, B, C and substituting into (27.21). These operations are performed in the Maple and Mathematica files 'conebending'.

Problems

27.1. A traction-free solid sphere of radius a and density ρ rotates at constant angular velocity Ω about the z-axis[7]. Find the complete stress field in the sphere. Find the location and magnitude of (i) the maximum tensile stress and (ii) the maximum von Mises stress and also (iii) the increase in the equatorial diameter and (iv) the decrease in the polar diameter due to the rotation.

27.2. A traction-free hollow sphere of inner radius b, outer radius a and density ρ rotates at constant angular velocity Ω about the z-axis. Find the complete stress field in the sphere. Find also the increase in the outer equatorial diameter and the decrease in the outer polar diameter due to the rotation.

27.3. A hollow sphere of inner radius b and outer radius a is filled with a gas at pressure p. Find the complete stress field in the sphere and the increase in the contained volume due to the internal pressure..

27.4. A state of uniaxial tension, $\sigma_{zz} = S$ in a large body is perturbed by the presence of a rigid spherical inclusion in the region $0 \le R < a$. The inclusion is perfectly bonded to the surrounding material, so that the boundary condition at the interface in a suitable frame of reference is one of zero displacement ($u_R = u_\theta = u_\beta = 0$; $R = a$).

Develop a solution for the stress field and find the stress-concentration factor.

---

[7] This problem is of some interest in connection with the stresses and displacements in the earth's crust due to diurnal rotation.

27.5. A rigid spherical inclusion of radius a in a large elastic body is subjected to a force F in the z-direction.

Find the stress field if the inclusion is perfectly bonded to the elastic body at $R=a$ and the stresses tend to zero as $R \to \infty$.

27.6. Investigate the steady-state thermal stress field due to the disturbance of an otherwise uniform heat flux, $q_z = Q$, by an insulated spherical hole of radius a.

You must first determine the perturbed temperature field, noting (i) that the temperature T is harmonic and hence can be represented in terms of spherical harmonics and (ii) that the insulated boundary condition implies that $\partial T/\partial R = 0$ at $R=a$.

The thermal stress field can now be obtained using Solution T supplemented by appropriate functions in Solutions A, B, chosen so as to satisfy the traction-free condition at the hole.

27.7. A state of uniaxial tension, $\sigma_{zz} = S$ in a large body (the matrix) of elastic properties μ_1, ν_1 is perturbed by the presence of a spherical inclusion in the region $0 \le R < a$ with elastic properties μ_2, ν_2. The inclusion is perfectly bonded to the matrix.

Develop a solution for the stress field and find the two stress-concentration factors

$$K_1 = \frac{\sigma_1^{\max}}{S} \quad ; \quad K_2 = \frac{\sigma_2^{\max}}{S} \, ,$$

where $\sigma_1^{\max}$, $\sigma_2^{\max}$ are respectively the maximum tensile stresses in the matrix and in the inclusion.

Plot figures showing how K_1, K_2 vary with the modulus ratio μ_1/μ_2 and with the two Poisson's ratios.

27.8. A sphere of radius a and density ρ rests on a rigid plane horizontal surface. Assuming that the support reaction consists of a concentrated vertical force equal to the weight of the sphere, find the complete stress field in the sphere.

Hint: Start with the solution for a point force acting on the surface of a half-space (the Boussinesq solution of §24.2.1) and determine the tractions on an imaginary spherical surface passing through the point of application of the force. Comparison with the corresponding cylindrical problems 12.1, 12.2, 12.3, would lead one to expect that these tractions could be removed by a finite series of spherical harmonics, but this is not the case. In fact, the tractions are still weakly (logarithmically) singular, though the dominant $(1/R)$ singularity associated with the point force has been removed. A truncated series of spherical harmonics will give an approximate solution, but the series will be rather slowly convergent. A more rapidly convergent solution can be obtained by first removing the logarithmic singularity.

27.9. A sphere of radius a is loaded by two equal and opposite point forces F applied at the poles $\beta = 0, 2\pi$. Find the stress field in the sphere. (Read the 'hint' for Problem 27.8).

27.10. A cylindrical bar of radius b transmits a torque T. It also contains a small spherical hole of radius a ($\ll b$) on the axis. Find the perturbation in the elementary torsional stress field due to the hole.

27.11. A vertical conical tower, $0 < \beta < \beta_0$ of height h and density ρ is loaded only by its own weight. Find the stress field in the tower.

27.12. A vertical hollow conical tower, $\beta_1 < \beta < \beta_2$ of height h and density ρ is loaded only by its own weight. Find the stress field in the tower. Are there any conditions (values of β_1, β_2) for which tensile stresses are developed in the tower? If so, when and where?

27.13. The hollow conical bar $\beta_1 < \beta < \beta_2$ is loaded by a tensile axial force F, the curved surfaces $\beta = \beta_1, \beta_2$ being traction free. Use the stress functions

$$\phi = A_1\{\ln(R) + \ln[1 + \cos(\beta)]\} + A_2\{\ln(R) + \ln[1 - \cos(\beta)]\} \; ; \quad \omega = \frac{B}{R}$$

to obtain a solution for the stress field.

27.14. Find the stress field in the solid conical shaft $0 < \beta < \beta_0$, loaded only by a torque T at the vertex.

27.15. The solid conical bar $0 < \beta < \beta_0$ is loaded by uniform torsional shear tractions $\sigma_{\beta\theta} = S$ at $\beta = \beta_0$, all other traction components being zero. Find the stress field in the bar. **Note:** The 'obvious' stress function for this problem is $R^2 P_2(\cos\beta)$, but it corresponds only to rigid-body rotation and gives zero stress components. As in Chapter 8, we need a special stress function for this case, which is in fact ϕ_{-3} from equation (24.9). You will need to write $z = R\cos\beta$ to express this function in spherical polar coördinates. It will contain the term $R^2\ln(R)$, but as in §10.3, the same degeneracy that causes the elementary potential to give zero stresses will cause there to be no logarithmic terms in the stress components.

27.16. Use spherical polar coördinates to find the stresses in the half-space $z > 0$ due to a concentrated tangential force F in the x-direction, applied at the origin. **Note:** The half-space is equivalent to the cone $0 \le \beta < \pi/2$. This problem is the spherical polar equivalent of Problem 24.1.

27.17. A solid sphere of radius a is loaded only by self-gravitation — i.e. each particle of the sphere is attracted to each other particle by a force $\gamma m_1 m_2 / R^2$, where m_1, m_2 are the masses of the two particles, R is the distance between them and γ is the universal gravitational constant. Find the stress field in the sphere and in particular the location and magnitude of the maximum compressive stress and the maximum shear stress.

Assuming that the solution is to be applied to the earth, express the results in terms of the gravitational acceleration g at the surface.

27.18. Figure 27.2 shows the cross section of a concrete dam which may be approximated as a sector of a truncated hollow cone. The dam is loaded by hydrostatic pressure (water density ρ) and by self weight (concrete density $\rho_c = 2.3\rho$). Assuming the stress field is axisymmetric — i.e. that the tractions at the side support are the same as those that would occur on the equivalent plane in a complete truncated hollow cone loaded all around the circumference — determine the stress field in the dam using strong conditions on the curved surfaces and weak (traction-free) conditions on the truncated end.

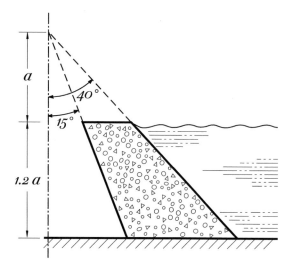

Fig. 27.2 Truncated hollow conical dam.

27.19. The solid cone $0 < \beta < \beta_0$ is loaded by a force in the x-direction applied at the vertex. The surface of the cone is otherwise traction free. Find the stress field in the cone. Notice that Problem 27.16 is a special case of this solution for $\beta_0 = \pi/2$.

27.20. A *horizontal* solid conical bar, $0 < \beta < \beta_0$ of length L and density ρ is built in at $z = L$ and loaded only by its own weight. Find the stress field in the bar and the vertical displacement of the vertex.

27.21. The surface of a solid sphere of radius a is traction free, but is subjected to the steady-state heat flux

$$q_R = q_0 \left[3\cos(2\beta) + 1 \right] .$$

In other words, it is heated near the equator and cooled near the poles.

Find the complete stress and displacement field in the sphere. Also, find the difference between the largest and smallest external diameter of the thermally distorted sphere.

Chapter 28
Eigenstrains and Inclusions

Hooke's law defines the strains in an elastic body due to an imposed stress, but additional inelastic strains can arise from other mechanisms, a simple example being thermal strains which we discussed in Chapters 14 and 23. Other mechanisms include transformation from one crystal structure to another, strains due to plastic deformation, and biological growth strains. Mura[1] coined the term *eigenstrains* to describe strains due to such mechanisms, and this terminology is now widely used.

If the eigenstrains satisfy the compatibility equations (2.7) (including the integral conditions discussed in §2.2.1), the body will be stress free when any external loads have been removed. In all other cases, the unloaded body will remain in a state of *residual stress*. In this chapter, we shall discuss methods of determining the residual stress due to a given distribution of eigenstrains. We shall be mainly concerned with the case where the eigenstrains are non-zero only in some finite region Ω (the *inclusion*) of an infinite body. We shall also consider some cases of the more challenging problem where the inclusion has different elastic properties from those of the surrounding material (the *matrix*).

A special case of some interest is that where the eigenstrain is uniform in the inclusion. Eshelby[2] solved problems of this class by (i) assuming that the inclusion Ω is free to deform, in which case it will be stress-free, since a uniform strain field satisfies all the compatibility equations, (ii) imposing tractions on the boundary Γ of the inclusion sufficient to return it to its original shape, (iii) inserting the deformed inclusion into the matrix, and then (iv) removing the boundary tractions by imposing equal and opposite tractions on Γ in the infinite elastic solid.

Supplementary Information The online version contains supplementary material available at https://doi.org/10.1007/978-3-031-15214-6_28.

---

[1] T. Mura, *Micromechanics of Defects in Solids*, Kluwer, Dordrecht, 2nd edn. 1987.

[2] J. D. Eshelby (1957), The determination of the elastic field of an ellipsoidal inclusion, and related problems, *Proceedings of the Royal Society of London,* Vol. 241, No. 1226,, pp. 376–396.

J. R. Barber, *Elasticity*, Solid Mechanics and Its Applications 172, https://doi.org/10.1007/978-3-031-15214-6_28

For more general distributions of eigenstrain, we need to reconsider the governing kinematic and equilibrium conditions.

28.1 Governing Equations

Suppose that an inelastic process leads to a distribution of eigenstrains $\epsilon_{ij}^*(\boldsymbol{r})$, where the position vector $\boldsymbol{r} = \{x_1, x_2, x_3\}$. The total strain can then be written as the sum

$$e_{ij} = \epsilon_{ij}^* + e_{ij}^e = \frac{1}{2}\left(\frac{\partial u_i}{\partial x_j} + \frac{\partial u_j}{\partial x_i}\right), \tag{28.1}$$

where e_{ij}^e is the elastic strain that is related to the stress σ_{ij} through Hooke's law (1.40) as

$$\sigma_{ij} = c_{ijkl}e_{kl}^e . \tag{28.2}$$

Eliminating e_{ij}^e between equations (28.1, 28.2) we obtain

$$\sigma_{ij} = c_{ijkl}\left(e_{kl} - \epsilon_{kl}^*\right) = \frac{c_{ijkl}}{2}\left(\frac{\partial u_k}{\partial x_l} + \frac{\partial u_l}{\partial x_k}\right) - c_{ijkl}\epsilon_{kl}^* . \tag{28.3}$$

These stress components must satisfy the equilibrium equation (2.4), from which we obtain

$$c_{ijkl}\frac{\partial^2 u_k}{\partial x_l \partial x_j} - c_{ijkl}\frac{\partial \epsilon_{kl}^*}{\partial x_j} + p_i = 0 , \tag{28.4}$$

where we recall that p_i is the body force per unit volume in direction x_i. This shows that the displacement in an otherwise unloaded infinite body due to an eigenstrain distribution ϵ_{ij}^* is the same as that due to a body force distribution

$$p_i = -c_{ijkl}\frac{\partial \epsilon_{kl}^*}{\partial x_j} . \tag{28.5}$$

For an isotropic material, we have

$$c_{ijkl} = \frac{2\mu\nu\delta_{ij}\delta_{kl}}{(1 - 2\nu)} + \mu(\delta_{ik}\delta_{jl} + \delta_{jk}\delta_{il}) , \tag{28.6}$$

from equations (1.55, 1.52), and hence equations (28.3, 28.5) take the form

$$\sigma_{ij} = 2\mu\left(\frac{\nu(e_{kk} - \epsilon_{kk}^*)\delta_{ij}}{(1 - 2\nu)} + (e_{ij} - \epsilon_{ij}^*)\right) \tag{28.7}$$

$$p_i = -2\mu\left(\frac{\nu}{(1 - 2\nu)}\frac{\partial \epsilon_{kk}^*}{\partial x_i} + \frac{\partial \epsilon_{ij}^*}{\partial x_j}\right) . \tag{28.8}$$

Equation (28.8) can be used to reduce the eigenstrain problem to an equivalent body force problem. In the special case where ϵ_{ij}^* is uniform in Ω and zero elsewhere, body forces will be invoked only as boundary tractions on Γ, and these are of course identical to the tractions required at step (iv) of Eshelby's superposition procedure.

This approach is analogous to the method of strain suppression introduced in §23.1 for problems in thermoelasticity. However, a more direct approach is to represent the displacement in equation (28.4) using an appropriate potential function formulation.

28.2 Galerkin Vector Formulation

Using equation (28.6) in (28.4) without body force ($p_i = 0$), we obtain

$$\frac{\partial^2 u_j}{\partial x_j \partial x_i} + (1 - 2\nu)\frac{\partial^2 u_i}{\partial x_k \partial x_k} = 2\nu \frac{\partial \epsilon_{kk}^*}{\partial x_i} + 2(1 - 2\nu)\frac{\partial \epsilon_{ij}^*}{\partial x_j}, \tag{28.9}$$

after cancelling a factor of μ.

We choose to represent the displacement using the Galerkin vector potential F_i of equation (21.6) rather than the Papkovich-Neuber potentials, since this avoids the dependence on the choice of origin of coördinates discussed in §21.3.1. In other words, we substitute

$$2\mu u_i = 2(1 - \nu)\frac{\partial^2 F_i}{\partial x_k \partial x_k} - \frac{\partial^2 F_k}{\partial x_i \partial x_k} \tag{28.10}$$

in (28.9), obtaining

$$\nabla^4 F_i \equiv \frac{\partial^4 F_i}{\partial x_j \partial x_j \partial x_k \partial x_k} = \frac{2\mu}{(1 - \nu)}\left(\frac{\nu}{(1 - 2\nu)}\frac{\partial \epsilon_{kk}^*}{\partial x_i} + \frac{\partial \epsilon_{ij}^*}{\partial x_j}\right). \tag{28.11}$$

If the eigenstrain ϵ_{ij}^* is a differentiable function of $\boldsymbol{r} = \{x_1, x_2, x_3\}$ that is non-zero only in some enclosed region Ω, equation (28.11) defines a well-posed problem for each component F_i of the Galerkin vector, of the general form

$$\nabla^4 \psi = f(\boldsymbol{r}); \quad \boldsymbol{r} \in \Omega$$
$$= 0 \quad ; \quad \boldsymbol{r} \notin \Omega, \tag{28.12}$$

where $f(\boldsymbol{r})$ is a known function.

28.2.1   Non-differentiable eigenstrains

If the eigenstrain ϵ_{ij}^* is non-differentiable, the right-hand side of equation (28.11) cannot be evaluated except in terms of generalized functions. An important special case is that where the eigenstrain is uniform and non-zero in Ω and zero elsewhere, so its derivatives involve delta functions at the boundary.

To solve problems of this class, we first note that if

$$\nabla^4 \chi = f(\mathbf{r}) , \tag{28.13}$$

then

$$\nabla^4 \frac{\partial \chi}{\partial x_i} = \frac{\partial}{\partial x_i} \nabla^4 \chi = \frac{\partial f(\mathbf{r})}{\partial x_i} .$$

Thus, the solution of the equation

$$\nabla^4 \psi = \frac{\partial f(\mathbf{r})}{\partial x_i} \quad \text{is} \quad \psi = \frac{\partial \chi}{\partial x_i} ,$$

where χ is the solution of (28.13).

We conclude that the solution of (28.11) is

$$F_i = \frac{2\mu}{(1-\nu)} \left(\frac{\nu}{(1-2\nu)} \frac{\partial \chi_{mm}}{\partial x_i} + \frac{\partial \chi_{ij}}{\partial x_j} \right) , \tag{28.14}$$

where χ_{ij} is the solution of the equation

$$\nabla^4 \chi_{ij} = \epsilon_{ij}^*(\mathbf{r}) ; \quad \mathbf{r} \in \Omega$$
$$= 0 \quad ; \quad \mathbf{r} \notin \Omega . \tag{28.15}$$

This formulation does not require $\epsilon_{ij}^*(\mathbf{r})$ to be continuous. However, if separate solutions of equations (28.15) are obtained for $\mathbf{r} \in \Omega$ and $\mathbf{r} \notin \Omega$, the degrees of freedom in the homogeneous equation (the biharmonic equation) must be chosen so as to ensure continuity of displacement at the boundary Γ.

The displacements can be found by substituting (28.14) in equation (28.10), and are

$$u_i = \frac{\nu}{(1-\nu)} \frac{\partial^3 \chi_{mm}}{\partial x_i \partial x_k \partial x_k} + 2 \frac{\partial^3 \chi_{ij}}{\partial x_j \partial x_k \partial x_k} - \frac{1}{(1-\nu)} \frac{\partial^3 \chi_{jk}}{\partial x_i \partial x_j \partial x_k} , \tag{28.16}$$

so the solution of (28.15) must be continuous up to and including the third derivatives in χ_{ij}. Such a solution always exists and can be defined formally as a convolution integral on the Green's function of §28.4 below.

28.2.2 *The stress field*

The stresses corresponding to the displacements (28.16) can be obtained by substitution into equations (28.1, 28.7), giving

$$
\frac{\sigma_{ij}}{2\mu} = \frac{\partial^2}{\partial x_k \partial x_k}\left(\frac{\partial^2 \chi_{il}}{\partial x_l \partial x_j} + \frac{\partial^2 \chi_{jl}}{\partial x_l \partial x_i} - \frac{\partial^2 \chi_{ij}}{\partial x_m \partial x_m}\right) - \frac{1}{(1-\nu)}\frac{\partial^4 \chi_{kl}}{\partial x_i \partial x_j \partial x_k \partial x_l}
$$
$$
+ \frac{\nu}{(1-\nu)}\frac{\partial^2}{\partial x_n \partial x_n}\left(\delta_{ij}\frac{\partial^2 \chi_{kl}}{\partial x_k \partial x_l} - \delta_{ij}\frac{\partial^2 \chi_{mm}}{\partial x_k \partial x_k} + \frac{\partial^2 \chi_{mm}}{\partial x_i \partial x_j}\right) , \qquad (28.17)
$$

where we have used equation (28.15) to express ϵ_{ij}^* in terms of derivatives of χ_{ij}.

The stress will be discontinuous at any location where the eigenstrain ϵ_{ij}^* is discontinuous. In particular, if the eigenstrain is uniform inside Ω and zero outside, there will be a stress discontinuity at the boundary Γ. We define this discontinuity as

$$
\Delta\sigma_{ij}(r) = \sigma_{ij}^+(r) - \sigma_{ij}^-(r) \qquad (28.18)
$$

where r denotes a point on the boundary Γ of the inclusion and σ_{ij}^+, σ_{ij}^- denote the stresses at points just outside and just inside Γ respectively. Equilibrium considerations require that there be no discontinuity in those stress components corresponding to tractions on the boundary and hence that

$$
\Delta\sigma_{ij}\ell_i = 0 ,
$$

where $\ell = \{\ell_1, \ell_2, \ell_3\}$ is a unit vector defining the local outward normal to Γ.

If we define a rotated Cartesian coördinate system (n, s, t) such that the n-axis is normal to Γ and s, t are in the tangent plane, continuity of the third derivatives (28.16) implies that fourth derivatives with at least one differentiation in the tangent plane such as

$$
\frac{\partial^4 \chi_{ij}}{\partial s \partial t \partial n^2} \; ; \quad \frac{\partial^4 \chi_{ij}}{\partial s^2 \partial n^2} \; ; \quad \frac{\partial^4 \chi_{ij}}{\partial s \partial n^3}
$$

will be continuous across Γ. The only *discontinuous* derivative is $\partial^4 \chi_{ij}/\partial n^4$ and in view of (28.15), the magnitude of the discontinuity in this derivative is $-\epsilon_{ij}^*$. It follows that the discontinuity

$$
\left(\frac{\partial^4 \chi_{mn}}{\partial x_i \partial x_j \partial x_k \partial x_l}\right)^+ - \left(\frac{\partial^4 \chi_{mn}}{\partial x_i \partial x_j \partial x_k \partial x_l}\right)^- = -\ell_i \ell_j \ell_k \ell_l \, \epsilon_{mn}^* .
$$

Using this result in equations (28.17, 28.18), we obtain

$$\frac{\Delta\sigma_{ij}}{\mu} = \left[\frac{2\nu}{(1-\nu)}\left(\delta_{kl}\delta_{ij} - \delta_{ij}\ell_k\ell_l - \delta_{kl}\ell_i\ell_j\right) + \frac{2}{(1-\nu)}\ell_i\ell_j\ell_k\ell_l\right.$$

$$\left. + \delta_{ik}\delta_{jl} + \delta_{il}\delta_{jk} - \delta_{ik}\ell_j\ell_l - \delta_{il}\ell_j\ell_k - \delta_{jl}\ell_i\ell_k - \delta_{jk}\ell_i\ell_l\right]\epsilon_{kl}^* .$$

$$(28.19)$$

Notice that the stress discontinuity depends only on the *local* value of ϵ_{ij}^* at the boundary. This result therefore applies to inclusions of any shape and to any distribution of eigenstrain in the inclusion.

28.3 Uniform Eigenstrains in a Spherical Inclusion

To illustrate the solution procedure, we first consider the case of a spherical inclusion $0 \le R < a$ inside which the eigenstrains ϵ_{ij}^* are uniform. It then follows from (28.15) that the functions χ_{ij} can be written

$$\chi_{ij} = \epsilon_{ij}^*\psi(R) , \qquad (28.20)$$

where $\psi(R)$ is the solution of the spherically symmetric boundary-value problem

$$\nabla^4\psi(R) = 1 ; \quad 0 \le R < a$$
$$= 0 ; \quad R > a, \qquad (28.21)$$

and $R = |\mathbf{r}| = \sqrt{x_k x_k}$. We recall that χ_{ij} and hence ψ must be continuous up to the third spatial derivative, and it is easily verified that this condition and equation (28.21) are satisfied by the function

$$\psi(R) = \frac{R^4}{120} - \frac{a^2 R^2}{12} \quad ; \quad 0 \le R < a \qquad (28.22)$$

$$= -\frac{a^3 R}{6} - \frac{a^5}{30R} ; \quad R > a . \qquad (28.23)$$

The displacement inside the inclusion is obtained by substituting (28.20, 28.22) into equation (28.16), giving

$$u_i = \frac{1}{15(1-\nu)}\left[(4 - 5\nu)(\epsilon_{ik}^* + \epsilon_{ki}^*)x_k + (5\nu - 1)\epsilon_{kk}^*x_i\right] ; \quad \mathbf{r} \in \Omega .$$

The corresponding strain is

$$e_{ij} = \frac{1}{2}\left(\frac{\partial u_i}{\partial x_j} + \frac{\partial u_j}{\partial x_i}\right) = S_{ijkl}\,\epsilon_{kl}^* ; \quad \mathbf{r} \in \Omega , \qquad (28.24)$$

where

$$S_{ijkl} = \frac{1}{15(1-\nu)}\left[(4-5\nu)\left(\delta_{ik}\delta_{jl}+\delta_{jk}\delta_{il}\right)+(5\nu-1)\delta_{kl}\delta_{ij}\right] \qquad (28.25)$$

is the *Eshelby tensor* for the sphere. Notice that the displacement is a linear function of the coördinates, so the strain is uniform inside the inclusion. We shall show in §28.5 that this remains true for a more general ellipsoidal inclusion with uniform eigenstrains.

The stresses inside the inclusion are therefore also uniform and can be found from (28.17) as

$$\sigma_{ij} = t_{ijkl}\epsilon_{kl}^* , \qquad (28.26)$$

where

$$t_{ijkl} = -\frac{\mu}{15(1-\nu)}\left[(7-5\nu)\left(\delta_{ik}\delta_{jl}+\delta_{il}\delta_{jk}\right)+2(1+5\nu)\delta_{kl}\delta_{ij}\right] .$$

28.3.1 Stresses outside the inclusion

A similar procedure can be used to determine the displacements, strains and stresses outside the inclusion, using equation (28.23). However, these fields all decay to zero as $R \to \infty$, so interest focusses mainly on the maximum stress, which occurs immediately outside the inclusion at $R = a^+$. This can be found by adding the stress discontinuity (28.19) to the internal stresses defined by equation (28.26), noting that for the spherical inclusion,

$$\ell_i = \frac{x_i}{R} .$$

28.4 Green's Function Solutions

For more general problems of the class (28.12), an alternative approach is to express the solution as a convolution integral on an appropriate Green's function. We start by defining the function

$$\phi(R) = \frac{\psi(R)}{V} ; \quad 0 \le R < a , \qquad \text{where} \quad V = \frac{4\pi a^3}{3}$$

is the volume of the sphere and $\psi(R)$ is the function defined by equations (28.22, 28.23). The expression $\nabla^4\phi$ is then given by $1/V$ inside, and zero outside the sphere, so

$$\iiint_{\Omega_0} \nabla^4\phi\, d\Omega = 1 ,$$

where Ω_0 is any volume containing the sphere — i.e. $\Omega \in \Omega_0$. If we then take the limit as $a \to 0$, we obtain the function

$$\phi(R) = \lim_{a \to 0} \frac{3}{4\pi a^3} \left(-\frac{a^3 R}{6} - \frac{a^5}{30 R} \right) = -\frac{R}{8\pi} , \tag{28.27}$$

which defines the appropriate Green's function.

The general solution of equations of the form (28.12) can then be written

$$\psi(r) = \psi_P(r) + \psi_H(r) ,$$

where the homogeneous solution $\psi_H(r)$ is any function which is biharmonic throughout the space, and the particular solution

$$\psi_P(r) = -\frac{1}{8\pi} \iiint_\Omega f(r')|r - r'| d\Omega ,$$

where r' defines a general point in Ω.

For the inclusion problem, we shall generally require that the stresses tend to zero far from the inclusion, and if $\psi(r)$ is used as a component of the Galerkin vector using equation (28.10), this condition is satisfied by the choice $\psi_H(r) = 0$ or $\psi(r) = \psi_P(r)$. Problems where a non-zero stress field is imposed at infinity will be handled by superposition (see §28.6 below).

28.4.1   Nuclei of strain

If the Green's function (28.27) is used to define the function

$$\chi_{jk}(r) = \epsilon^*_{jk} \phi(R) = -\frac{\epsilon^*_{jk}|r|}{8\pi} ,$$

equation (28.14) will define the Galerkin vector due to a *nucleus of strain* ϵ^*_{jk} at the origin. It is convenient to write this solution in the form

$$F_i = \epsilon^*_{jk} H^\epsilon_{ijk}(r) , \tag{28.28}$$

where

$$H^\epsilon_{ijk}(r) = -\frac{\mu}{4\pi(1-\nu)} \left[\frac{\nu}{(1-2\nu)} \delta_{jk} \frac{\partial}{\partial x_i} + \delta_{ik} \frac{\partial}{\partial x_j} \right] |r| . \tag{28.29}$$

In the special case where $\epsilon^*_{jk} = \delta_{jk}$, equation (28.28) will reduce to the Galerkin potential for a centre of dilatation, introduced in §24.1.1.

Using (28.29), we can then write the Galerkin vector due to an arbitrary distribution of eigenstrain $\epsilon^*_{jk}(r')$ in Ω as

$$F_i(r) = \iiint_\Omega \epsilon_{jk}^*(r') H_{ijk}^\epsilon(r - r') d\Omega .$$

The displacement field due to a nucleus of strain is obtained by substituting (28.28) into (28.10) and is

$$u_i(r) = \frac{1}{8\pi(1-\nu)|r|^2} \left[(1-2\nu)\{\ell_j(\epsilon_{ij}^* + \epsilon_{ji}^*) - \ell_i \epsilon_{mm}^*\} + 3\ell_i \ell_j \ell_k \epsilon_{jk}^*\right] ,$$

where ℓ is the unit vector

$$\ell = \frac{r}{|r|} \quad \text{or} \quad \ell_i = \frac{x_i}{\sqrt{x_m x_m}} .$$

An equivalent statement is

$$u_i(r) = \frac{\epsilon_{jk}^* g_{ijk}(\ell)}{8\pi(1-\nu)|r|^2} ,$$

where

$$g_{ijk}(\ell) = (1-2\nu)(\delta_{ij}\ell_k + \delta_{ik}\ell_j - \delta_{jk}\ell_i) + 3\ell_i \ell_j \ell_k . \tag{28.30}$$

With this notation, the displacement field due to a distribution of eigenstrain $\epsilon_{jk}^*(r)$ can then be written

$$u_i(r) = \frac{1}{8\pi(1-\nu)} \iiint_\Omega \frac{\epsilon_{jk}^*(r') g_{ijk}(\ell) d\Omega}{|r - r'|^2} , \tag{28.31}$$

where

$$\ell = \frac{(r - r')}{|r - r'|} .$$

28.5 The Ellipsoidal Inclusion

We found in §28.3 that if the eigenstrain ϵ_{jk}^* is uniform inside a spherical inclusion Ω, the displacement $u_i(r)$ will be a linear function of the coördinates for $r \in \Omega$, and hence the corresponding displacement derivatives, strains and stresses will be uniform. We shall prove that this result also holds for the ellipsoidal inclusion defined by

$$\frac{x_1^2}{a_1^2} + \frac{x_2^2}{a_2^2} + \frac{x_3^2}{a_3^2} \equiv \frac{x_i^2}{a_i^2} \le 1 .$$

We first express the integral (28.31) in spherical polar coördinates centred on a field point $P(x_1, x_2, x_3) \in \Omega$. A point a distance R from P along a straight line in direction ℓ has coördinates $x_i + R\ell_i$, and hence the distance from P to the boundary Γ of Ω along such a line is the positive root R_1 of the quadratic equation

$$\sum_{i=1}^{3} \frac{(x_i + R\ell_i)^2}{a_i^2} = 1 \ . \tag{28.32}$$

This equation also has a negative root R_2 such that $-R_2$ is the distance from P to Γ in the direction $-\ell$, and it follows from (28.32) that the algebraic sum of the roots is given by

$$R_1 + R_2 = -2 \sum_{i=1}^{3} \frac{\ell_i x_i}{a_i^4} \bigg/ \sum_{i=1}^{3} \frac{\ell_i^2}{a_i^4}. \tag{28.33}$$

Suppose we now consider the contribution to the volume integral (28.31) from an infinitesimal cone of solid angle $d\omega$ centred on the above line and with apex at P. We obtain

$$\frac{\epsilon_{jk}^*}{8\pi(1-\nu)} \int_0^{R_1} \frac{g_{ijk}(\ell)(R^2 d\omega)dR}{R^2} = \frac{\epsilon_{jk}^* g_{ijk}(\ell)R_1 d\omega}{8\pi(1-\nu)} \ .$$

If we now add the contribution from the corresponding 'negative' cone of height $-R_2$, we have

$$\frac{\epsilon_{jk}^* d\omega}{8\pi(1-\nu)} \Big[g_{ijk}(\ell)R_1 + g_{ijk}(-\ell)(-R_2) \Big] = \frac{\epsilon_{jk}^* g_{ijk}(\ell)(R_1 + R_2)d\omega}{8\pi(1-\nu)} \ , \tag{28.34}$$

since g_{ijk} is odd in ℓ — i.e. $g_{ijk}(-\ell) = -g_{ijk}(\ell)$, from (28.30). Equation (28.33) shows that this expression is linear in x_i, and this property is preserved in the integration with respect to the solid angle ω.

It follows that the strains inside the inclusion can be expressed in a form similar to equation (28.24) — i.e.

$$e_{ij} = S_{ijkl}\epsilon_{kl}^* \ , \tag{28.35}$$

where the Eshelby tensor S_{ijkl} now depends on the ratios between the semi-axes a_1, a_2, a_3 as well as ν. The corresponding integrals are evaluated by Eshelby and Mura and are[3]

---

[3] *loc. cit.* Here we have adapted the notation of X. Jin, D. Lyu, X. Zhang, Q. Zhou, Q. Wang and L. M. Keer (2016), Explicit analytical solutions for a complete set of the Eshelby tensors of an ellipsoidal inclusion, *ASME Journal of Applied Mechanics*, Vol. 83, Art. 121010, writing $\mathcal{I}_k$ for their $\mathcal{J}_k(0)$ etc.

$$S_{1111} = \frac{3a_1^2}{2(1-\nu)}\mathcal{I}_{11} + \frac{(1-2\nu)}{2(1-\nu)}\mathcal{I}_1 \; ; \quad S_{1122} = \frac{a_2^2}{2(1-\nu)}\mathcal{I}_{12} - \frac{(1-2\nu)}{2(1-\nu)}\mathcal{I}_1$$

$$S_{1212} = \frac{(a_1^2 + a_2^2)}{4(1-\nu)}\mathcal{I}_{12} + \frac{(1-2\nu)}{4(1-\nu)}(\mathcal{I}_1 + \mathcal{I}_2) \, , \ldots \tag{28.36}$$

where

$$\mathcal{I}_j = \frac{a_1 a_2 a_3}{2}\int_0^\infty \frac{dt}{(a_j^2 + t)\sqrt{(a_1^2 + t)(a_2^2 + t)(a_3^2 + t)}}$$

$$\mathcal{I}_{jk} = \frac{a_1 a_2 a_3}{2}\int_0^\infty \frac{dt}{(a_j^2 + t)(a_k^2 + t)\sqrt{(a_1^2 + t)(a_2^2 + t)(a_3^2 + t)}} \, .$$

Other components of the Eshelby tensor can be obtained from (28.36) by permuting suffices. All components not of this form are zero. Note that these expressions satisfy the minor symmetries such as $S_{2121} = S_{1212}$, but do not generally satisfy the major symmetries, so (e.g.) $S_{1122} \neq S_{2211}$. The Maple and Mathematica files 'Eshelby' evaluate all the non-zero components of the Eshelby tensor for given semi-axes a_i.

The integrals $\mathcal{I}_j, \mathcal{I}_{ij}$ satisfy the identities

$$\mathcal{I}_1 + \mathcal{I}_2 + \mathcal{I}_3 = 1 \tag{28.37}$$

$$3\mathcal{I}_{11} + \mathcal{I}_{12} + \mathcal{I}_{13} = \frac{1}{a_1^2} \tag{28.38}$$

$$3a_1^2\mathcal{I}_{11} + a_2^2\mathcal{I}_{12} + a_3^2\mathcal{I}_{13} = 3\mathcal{I}_1 \tag{28.39}$$

and similar identities obtained by permuting suffices. Also, $\mathcal{I}_j$ can be expressed in terms of Carlson elliptic integrals[4]. For example

$$\mathcal{I}_1 = \frac{a_1 a_2 a_3}{3}R_D(a_2^2, a_3^2; a_1^2) \, , \tag{28.40}$$

where

$$R_D(x, y; z) \equiv \frac{3}{2}\int_0^\infty \frac{dt}{(t+z)\sqrt{(t+x)(t+y)(t+z)}} \, . \tag{28.41}$$

The advantage of this representation relative to the better known Legendre elliptic integrals is that Carlson integrals can be evaluated numerically without first arranging the three parameters a_1, a_2, a_3 in order of magnitude.

The integral $\mathcal{I}_{ij}, i \neq j$ can be evaluated by partial fractions as

$$\mathcal{I}_{ij} = \frac{\mathcal{I}_j - \mathcal{I}_i}{(a_i^2 - a_j^2)} \, , \tag{28.42}$$

---

[4] B. C. Carlson (1979), Computing elliptic integrals by duplication, *Numerische Mathematik*, Vol. 33, pp. 1–16. These integrals can be evaluated using the Maple and Mathematica files 'Carlson'.

after which $\mathcal{I}_{11}$ (for example) can be evaluated from the identity (28.38).

If $a_1 = a_2$ (for example), the definition (28.42) for $\mathcal{I}_{12}$ is ill-defined, but in this case $\mathcal{I}_{11} = \mathcal{I}_{12}$ and both can then be expressed in terms of $\mathcal{I}_{13}$ using (28.38). If $a_1 = a_2 = a_3$, the inclusion is spherical and we can use equations (28.25).

28.5.1   The stress field

Since the strains inside the inclusions are uniform, the corresponding stresses are also uniform and we can calculate them by substituting e_{ij} from equation (28.35) into the constitutive law (28.7).

At the boundary of the inclusion, there is a discontinuity in those stress components that act on planes orthogonal to the boundary, which can be determined from equation (28.19), where we recall that ℓ defines the local unit outward normal to the inclusion. For the ellipsoid, this is given by

$$\ell_i = \frac{x_i}{a_i^2} \Big/ \sqrt{\frac{x_1^2}{a_1^4} + \frac{x_2^2}{a_2^4} + \frac{x_3^2}{a_3^4}}. \quad \text{no sum.} \tag{28.43}$$

The stresses then decay with increasing distance from the inclusion, expressions being given by Jin et al. [5].

28.5.2   Anisotropic material

We have developed these results for an isotropic material, because in this case the stress and displacement fields can be expressed in relatively simple algebraic terms, but the underlying result — that a uniform distribution of eigenstrains in an ellipsoidal inclusion produces a uniform state of stress and strain within the inclusion — remains true for a linear elastic material with general anisotropy.

To prove this, we recall that the displacements due to an arbitrary eigenstrain distribution ϵ_{kl}^* are the same as those due to a body force distribution given by

$$p_i = -c_{ijkl} \frac{\partial \epsilon_{kl}^*}{\partial x_j}.$$

An argument similar to that in §28.2.1 then shows that these displacements can be written

---

[5] loc. cit.

$$u = \sum_{j=1}^{3} \frac{\partial u^{(j)}}{\partial x_j}$$

where $u^{(j)}$ is the displacement due to the body force $p_i^{(j)} = -c_{ijkl}\epsilon_{kl}^*$. In particular, if the distribution corresponds to a nucleus of strain at the origin, the displacements will comprise spatial derivatives of those due to a concentrated force. Now self-similarity and equilibrium considerations show that the stresses and hence the strains due to a concentrated force in an infinite body must decay along any given radial line with $|r|^{-2}$, and hence the corresponding displacements must decay with $|r|^{-1}$. It follows that the displacements due to a nucleus of strain must vary with $|r|^{-2}$. In other words, the Green's function for displacements will have the same form as (28.31), except that the dimensionless function $g_{ijk}(\ell)$ will be modified by the anisotropy, but the form of this function does not affect the argument leading to equation (28.34).

28.6 The Ellipsoidal Inhomogeneity

We have shown that if a uniform eigenstrain ϵ_{kl}^* is imposed within an ellipsoidal region Ω of an otherwise unstressed infinite elastic body, the resulting displacements within Ω correspond to strains that are also uniform and given by $e_{ij} = S_{ijkl}\,\epsilon_{kl}^*$. We can use this result to solve the problem of an infinite elastic body that contains an ellipsoidal inclusion with different elastic properties, and that is loaded at infinity.

We start by considering a homogeneous infinite space with elasticity tensor c_{ijkl} in a state of uniform strain e_{ij}^{∞}, and we superpose a fictitious uniform eigenstrain $\epsilon_{ij}^{\ddagger}$ inside the elliptical region Ω. The strain in the inclusion will then be

$$e_{ij} = S_{ijkl}\epsilon_{kl}^{\ddagger} + e_{ij}^{\infty} \; ; \quad r \in \Omega$$

and the corresponding stress is

$$\sigma_{ij} = c_{ijkl}\left(e_{kl} - \epsilon_{kl}^{\ddagger}\right) = c_{ijkl}\left(S_{klmn}\epsilon_{mn}^{\ddagger} - \epsilon_{kl}^{\ddagger} + e_{kl}^{\infty}\right) \; ; \quad r \in \Omega \; . \qquad (28.44)$$

Now compare this with the actual problem where the inclusion has a different elasticity tensor c_{ijkl}^{Ω} and a real eigenstrain ϵ_{ij}^*. The complete strain field is taken to be identical with that in the fictitious homogeneous problem, and hence the stresses in the inclusion are given by

$$\sigma_{ij} = c_{ijkl}^{\Omega}\left(e_{kl} - \epsilon_{kl}^*\right) = c_{ijkl}^{\Omega}\left(S_{klmn}\epsilon_{mn}^{\ddagger} - \epsilon_{kl}^* + e_{kl}^{\infty}\right) \; .$$

This will be identical to (28.44) if we choose $\epsilon_{kl}^{\ddagger}$ so as to satisfy the equation

$$\left(c^{\Omega}_{ijkl} - c_{ijkl}\right)\left(S_{klmn}\epsilon^{\ddagger}_{mn} + e^{\infty}_{kl}\right) = c^{\Omega}_{ijkl}\epsilon^{*}_{kl} - c_{ijkl}\epsilon^{\ddagger}_{kl} . \tag{28.45}$$

This defines six independent linear equations[6], for the six components of the fictitious eigenstrain $\epsilon^{\ddagger}_{kl}$, which can therefore be determined in terms of the eigenstrain ϵ^{*}_{ij} and the imposed strain at infinity e^{∞}_{kl}.

28.6.1 Equal Poisson's ratios

The solution is significantly simplified if the matrix and the inclusion are both isotropic and have the same Poisson's ratio ($\nu^{\Omega} = \nu$). For an isotropic material, the elasticity tensor (28.6) can be written $c_{ijkl} = \mu\tilde{c}_{ijkl}$, where

$$\tilde{c}_{ijkl} = \frac{2\nu\delta_{ij}\delta_{kl}}{(1 - 2\nu)} + \delta_{ik}\delta_{jl} + \delta_{jk}\delta_{il}$$

is a dimensionless tensor depending only on ν. If $\nu^{\Omega} = \nu$, then $\tilde{c}^{\Omega}_{ijkl} = \tilde{c}_{ijkl}$ and we obtain

$$\left(\frac{\mu^{\Omega}}{\mu} - 1\right)\left(S_{klmn}\epsilon^{\ddagger}_{mn} + e^{\infty}_{kl}\right) = \frac{\mu^{\Omega}\epsilon^{*}_{kl}}{\mu} - \epsilon^{\ddagger}_{kl} . \tag{28.46}$$

28.6.2 The ellipsoidal hole

The same method can be used to obtain the solution due to an ellipsoidal hole in a homogeneous body loaded at infinity. In effect, this corresponds to the case $c^{\Omega}_{ijkl} = 0$ and hence

$$\epsilon^{\ddagger}_{kl} = S_{klmn}\epsilon^{\ddagger}_{mn} + e^{\infty}_{kl} , \tag{28.47}$$

from (28.45). In this case, clearly the stress in the 'inclusion' (i.e. the hole) is zero, and one way to explain this approach is to find the fictitious eigenstrain distribution needed to make the stress in the inclusion zero, so that the inclusion can be removed without changing the conditions in the rest of the body.

The result of most interest is the magnitude of the maximum of an appropriate stress measure. This always occurs immediately adjacent to the boundary of the hole and hence it can be found from equations (28.19, 28.43), since the stress just inside the hole, σ^{-}_{ij} is zero.

---

[6] For a proof that the resulting 6×6 matrix can always be inverted, see D. M. Barnett and W. Cai (2018), Properties of the Eshelby tensor and existence of the equivalent ellipsoidal inclusion solution, *Journal of the Mathematics and Physics of Solids*, Vol. 121, pp. 71–80.

In many cases, the maximum will occur at the end of one of the semi-axes of the ellipsoid. For example, at the point $(a_1, 0, 0)$, the normal is $\ell_i = \delta_{i1}$ and we obtain $\sigma_{11} = \sigma_{12} = \sigma_{13} = 0$ and

$$\sigma_{22} = \frac{2\mu(\epsilon_{22}^\ddagger + \nu\epsilon_{33}^\ddagger)}{(1-\nu)} \; ; \quad \sigma_{33} = \frac{2\mu(\epsilon_{33}^\ddagger + \nu\epsilon_{22}^\ddagger)}{(1-\nu)} \; ; \quad \sigma_{23} = 2\mu\epsilon_{23}^\ddagger \; ,$$

where $\epsilon_{ij}^\ddagger$ is the solution of (28.47).

28.7 Energetic Considerations

One of the principal applications of Eshelby's results is to situations where an inclusion experiences a phase transformation. In particular, we would like to determine the conditions under which it is energetically favourable for such an inclusion to grow, or to adopt a particular shape defined in terms of the ratios of the semi-axes. Of course in many cases, the transformed region will not be exactly ellipsoidal, but the ellipsoidal shape provides sufficient degrees of freedom to explore hypotheses about growth processes. In this section, we therefore develop expressions for the total strain energy in the inclusion and the matrix for an inclusion experiencing a uniform eigenstrain.

Suppose, following Eshelby, we first allow the inclusion to deform freely, so that points on the boundary Γ experience displacements $\boldsymbol{u}^* = \epsilon_{ij}^* x_j$. We then apply equal and opposite tractions $\boldsymbol{t}$ and $-\boldsymbol{t}$ to the inclusion and the matrix respectively until continuity is reestablished at a position defined by the displacement $\boldsymbol{u}$. The strain energy U_M in the matrix is then equal to the work done by the tractions $-\boldsymbol{t}$ — i.e.

$$U_M = \frac{1}{2} \iint_\Gamma (-\boldsymbol{t}) \cdot \boldsymbol{u} \, d\Gamma \; , \tag{28.48}$$

and by a similar argument, the strain energy in the inclusion is

$$U_\Omega = \frac{1}{2} \iint_\Gamma \boldsymbol{t} \cdot (\boldsymbol{u} - \boldsymbol{u}^*) \, d\Gamma \; , \tag{28.49}$$

since the traction on the inclusion is $+\boldsymbol{t}$ and points on the boundary have to move from the unrestrained position $\boldsymbol{u}^*$ to the final position $\boldsymbol{u}$. It follows that the total strain energy is

$$U = U_M + U_\Omega = -\frac{1}{2} \iint_\Gamma \boldsymbol{t} \cdot \boldsymbol{u}^* \, d\Gamma \; . \tag{28.50}$$

This result is independent of the final position of the interface defined by the displacement $\boldsymbol{u}$, because the work done by two equal and opposite forces depends only on their *relative* displacement. Notice that if the eigenstrain is tensile, the traction t will be compressive (negative) and *vice versa*, so the strain energy is always positive.

28.7.1   Evaluating the integral

To evaluate the integral in equation (28.50), we note that $t_i = n_j \sigma_{ji}$, where n_j is the outward normal to Ω. Using the divergence theorem, we then have

$$\frac{1}{2} \iint_\Gamma t_i u_i^* \, d\Gamma = \frac{1}{2} \iint_\Gamma n_j \sigma_{ji} u_i^* \, d\Gamma = \frac{1}{2} \iiint_\Omega \frac{\partial}{\partial x_j} \left(\sigma_{ji} u_i^* \right) d\Omega$$

$$= \frac{1}{2} \iiint_\Omega \frac{\partial \sigma_{ji}}{\partial x_j} u_i^* d\Omega + \frac{1}{2} \iiint_\Omega \sigma_{ji} \frac{\partial u_i^*}{\partial x_j} d\Omega . \quad (28.51)$$

If there are no body forces, the equilibrium equation (2.4) shows that the first term in (28.51) is zero, and hence we obtain

$$U = -\frac{1}{2} \iiint_\Omega \sigma_{ji} \epsilon_{ij}^* d\Omega . \quad (28.52)$$

28.7.2   Strain energy in the inclusion

The above expressions define the total strain energy in both the inclusion and the matrix. To find their separate contributions, we can calculate the strain energy in the inclusion alone as

$$U_\Omega = \frac{1}{2} \iiint_\Omega \sigma_{ji} \left(e_{ij} - \epsilon_{ij}^* \right) d\Omega . \quad (28.53)$$

The strain energy in the matrix is then

$$U_M = U - U_\Omega = -\frac{1}{2} \iiint_\Omega \sigma_{ji} e_{ij} d\Omega , \quad (28.54)$$

where (paradoxically), σ_{ij} and e_{ij} are the values in the inclusion.

In these derivations, the expressions (28.48, 28.49) for the work done by the tractions depend on the material obeying Hooke's law, but no other assumptions are made about the constitutive law which could therefore be generally anisotropic and spatially inhomogeneous. The argument is also independent of the shape of the inclusion, and applies equally to a finite body of any shape, provided the external surfaces are traction free.

For the special case of a uniform distribution of eigenstrain in an ellipsoidal inclusion in an infinite body, the stress is uniform in the inclusion and equations

(28.52–28.54) can be replaced by simple products

$$U = -\frac{\sigma_{ji}\epsilon_{ij}^* V}{2} \; ; \quad U_\Omega = \frac{\sigma_{ji}\left(e_{ij} - \epsilon_{ij}^*\right) V}{2} \; ; \quad U_M = -\frac{\sigma_{ji}e_{ij} V}{2}$$

where $V = 4\pi a_1 a_2 a_3/3$ is the volume of the inclusion.

Problems

28.1. Find the Galerkin potential due to a dilatational nucleus of strain $\epsilon_{jk}^* = \delta_{jk}$ and hence find the corresponding displacements. Find the total change in volume of a finite region enclosing the nucleus of strain and comment on the result.

28.2. Equilibrium considerations demand that the traction $t_j = \sigma_{ij}\ell_i$ be continuous across the boundary of the inclusion. Verify that this condition is satisfied by equation (28.19) for all ϵ_{kl}^*.

28.3. An ellipsoidal region with semi-axes a_1, a_2, a_3 in an infinite homogeneous elastic body is raised to a temperature T, the rest of the body being unheated. Find the state of uniform stress in the heated region. Also, find the location and magnitude of the maximum tensile stress outside the heated region (it will be at the end of one of the semi-axes).

28.4. Show that the Carlson integral R_D of equation (28.41) satisfies the relation

$$R_D(\lambda p, \lambda q; \lambda r) = \frac{1}{\lambda^{3/2}} R_D(p, q; r) \ .$$

Hence show that the integral $\mathcal{I}_1$ of equation (28.40) depends only on the ratio of the semi-axes a_1, a_2, a_3, but not on their magnitudes.

28.5. Use the method of §28.6.2 and equation (28.25) to determine the stresses at the pole and the equator for a homogeneous body with a spherical hole loaded by uniaxial tension $\sigma_{xx} = \sigma_0$ at infinity. Verify that the results agree with those in §27.1.2.

28.6. An infinite elastic body containing an ellipsoidal hole of semi-axes $a_1 = a$, $a_2 = 2a$, $a_3 = \lambda a$ is loading at infinity in uniaxial tension $\sigma_{11}^\infty = \sigma_0$, the other stress components being zero. Find the tensile stress $\sigma_{11}(0, 2a^+, 0) \equiv \sigma_1$ just outside the hole and plot the ratio σ_1/σ_0 as a function of λ in the range $0 < \lambda < 3$. Comment on the results in the limits $\lambda \to 0$ and $\lambda \to \infty$.

28.7. Show that if Poisson's ratio for an ellipsoidal inhomogeneity differs from that of the surrounding matrix $[\nu^\Omega \neq \nu]$ and there is no eigenstrain ($\epsilon_{kl}^* = 0$), equation (28.46) takes the more general form

$$\epsilon_{ij}^{\ddagger} = \left[\left(1 - \frac{\mu^{\Omega}}{\mu} \right) \frac{(\delta_{ik}\delta_{jl} + \delta_{il}\delta_{jk})}{2} - \frac{\mu^{\Omega}(\nu^{\Omega} - \nu)\delta_{kl}\delta_{ij}}{\mu(1 - 2\nu^{\Omega})(1 + \nu)} \right] \left(S_{klmn}\epsilon_{mn}^{\ddagger} + e_{kl}^{\infty} \right) .$$

Comment on the interpretation of this equation if the inhomgeneity is incompressible ($\nu^{\Omega} = 0.5$).

28.8. An elastic body is loaded by a uniaxial tensile stress $\sigma_{11} = \sigma_0$. An inclusion Ω within the body of volume V_{Ω} then experiences a uniform uniaxial eigenstrain $\epsilon_{11}^* = \epsilon_0$ as a result of a phase transformation. Find an approximation for the work done by the external tractions during the transformation, under the assumption that the inclusion is much smaller than the body and not close to any boundary.

28.9. A composite material comprises a distribution of ellipsoids of semi-axes a, $3a$, $10a$ and shear modulus 3μ embedded relatively sparsely in a matrix of shear modulus μ, both materials having Poisson's ratio 0.2. The ellipsoids are all oriented such that the long axes are parallel to x_3, but the other axes are randomly oriented in the x_1, x_2-plane.
The composite is now loaded at infinity by a biaxial tensile strain $e_{11}^{\infty} = e_{22}^{\infty} = \epsilon^{\infty}$ and a shear strain $e_{12}^{\infty} = 2\epsilon^{\infty}$, the other strain components being zero. Plot the maximum shear and tensile stress (normalized by $\mu\epsilon^{\infty}$) in an ellipsoid as a function of an appropriate angle defining its orientation.

Chapter 29
Axisymmetric Torsion

We have already remarked in §26.1 that the use of axisymmetric harmonic potential functions in Solution E provides a general solution of the problem of an axisymmetric body loaded in torsion. In this case, the only non-zero stress and displacement components are

$$\mu u_\theta = -\frac{\partial \Psi}{\partial r} \; ; \quad \sigma_{\theta r} = \frac{1}{r}\frac{\partial \Psi}{\partial r} - \frac{\partial^2 \Psi}{\partial r^2} \; ; \quad \sigma_{\theta z} = -\frac{\partial^2 \Psi}{\partial r \partial z} \; . \tag{29.1}$$

A more convenient formulation of this problem can be obtained by considering the relationship between the stress components and the torque transmitted through the body. We first note that the two non-zero stress components (29.1) both act on the cross-sectional θ-plane, on which they constitute a vector field σ_θ of magnitude

$$|\sigma_\theta| = \sqrt{\sigma_{\theta r}^2 + \sigma_{\theta z}^2} \; .$$

This vector field is illustrated in Figure 29.1. It follows that the stress component $\sigma_{\theta n}$ normal to any of the 'flow lines' in this figure is zero and hence the complementary shear stress $\sigma_{n\theta} = 0$. Thus, the flow lines define traction-free surfaces in r, θ, z space.

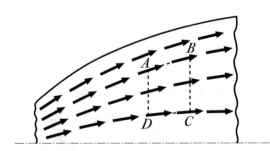

Fig. 29.1 Vector representation of the shear stresses on the cross-sectional plane.

Consider now the equilibrium of the axisymmetric body defined by $ABCD$ in Figure 29.1. The surfaces AB and CD are flow lines and are therefore traction free, so we must conclude that the resultant torque transmitted across the circular surface AD is equal to that transmitted across BC. Thus the flow lines define surfaces enclosing axisymmetric bodies that transmit constant torque along the axis. Conversely, if we use the torque $T(r, z)$ transmitted across the disk of radius r as our fundamental variable, a traction-free surface will then correspond to a line in the cross section along which T is constant. This has many advantages, not the least of which is that we can make contour plots of candidate functions $T(r, z)$ and then adjust parameters in order to approximate the shape of the required body by one of the contours.

29.1 The Transmitted Torque

The torque transmitted across a disk of radius r is

$$T(r, z) = 2\pi \int_0^r r^2 \sigma_{z\theta} dr \ . \tag{29.2}$$

To avoid the factor of 2π, we shall define a stress function ϕ such that

$$T(r, z) = 2\pi\phi(r, z) \ . \tag{29.3}$$

Differentiation of (29.2) with respect to r then yields

$$\sigma_{z\theta} = \frac{1}{2\pi r^2} \frac{\partial T}{\partial r} = \frac{1}{r^2} \frac{\partial \phi}{\partial r} \ . \tag{29.4}$$

The torque transmitted into a disk of thickness δz and radius r across the curved surface is

$$\delta T = -2\pi r^2 \sigma_{r\theta} \delta z$$

and hence

$$\sigma_{r\theta} = -\frac{1}{2\pi r^2} \frac{\partial T}{\partial z} = -\frac{1}{r^2} \frac{\partial \phi}{\partial z} \ . \tag{29.5}$$

29.2 The Governing Equation

Equations (29.4, 29.5) and (29.1) are alternative representations of the stress field and hence

$$-\frac{1}{r^2} \frac{\partial \phi}{\partial z} = \frac{1}{r} \frac{\partial \Psi}{\partial r} - \frac{\partial^2 \Psi}{\partial r^2} \quad ; \quad \frac{1}{r^2} \frac{\partial \phi}{\partial r} = -\frac{\partial^2 \Psi}{\partial r \partial z} \ . \tag{29.6}$$

Multiplying through by r^2 and differentiating these expressions, we obtain

$$\frac{\partial^2 \phi}{\partial z^2} = -r\frac{\partial^2 \Psi}{\partial r \partial z} + r^2\frac{\partial^3 \Psi}{\partial r^2 \partial z} \ ; \ \ \frac{\partial^2 \phi}{\partial r^2} = -2r\frac{\partial^2 \Psi}{\partial r \partial z} - r^2\frac{\partial^3 \Psi}{\partial r^2 \partial z}$$

and hence

$$\frac{\partial^2 \phi}{\partial r^2} + \frac{\partial^2 \phi}{\partial z^2} = -3r\frac{\partial^2 \Psi}{\partial r \partial z} = \frac{3}{r}\frac{\partial \phi}{\partial r} \ ,$$

from (29.6). Thus, the stress function ϕ must satisfy the equation

$$\frac{\partial^2 \phi}{\partial r^2} - \frac{3}{r}\frac{\partial \phi}{\partial r} + \frac{\partial^2 \phi}{\partial z^2} = 0 \ . \tag{29.7}$$

The axisymmetric torsion problem is therefore reduced to the determination of a function ϕ satisfying (29.7), such that ϕ is constant along any traction-free boundary. The stress field is then given by

$$\sigma_{r\theta} = -\frac{1}{r^2}\frac{\partial \phi}{\partial z} \ ; \ \ \sigma_{z\theta} = \frac{1}{r^2}\frac{\partial \phi}{\partial r} \ . \tag{29.8}$$

Recall also that the transmitted torque is related to ϕ through equation (29.3).

29.3 Solution of the Governing Equation

Equation (29.7) is similar in form to the Laplace equation (25.26) and solutions can be obtained in terms of Legendre functions, using a technique similar to that used in Chapter 25. A relationship with harmonic functions can be established by defining a new function f through the equation

$$\phi(r, z) = r^2 f(r, z) \ . \tag{29.9}$$

Substitution into (29.7) then shows that f must satisfy the equation

$$\frac{\partial^2 f}{\partial r^2} + \frac{1}{r}\frac{\partial f}{\partial r} - \frac{4f}{r^2} + \frac{\partial^2 f}{\partial z^2} = 0 \ , \tag{29.10}$$

which is (25.30) with $m=2$. It follows that (29.9) will define a function satisfying (29.7) if the function

$$f(r, z)\cos(2\theta)$$

is harmonic. In particular, we conclude that the functions

$$r^2 R^n P_n^2(\cos\beta) \ ; \ \ r^2 R^n Q_n^2(\cos\beta) \ ; \ \ r^2 R^{-n-1} P_n^2(\cos\beta) \ ; \ \ r^2 R^{-n-1} Q_n^2(\cos\beta)$$

satisfy (29.7). More generally, we can use equation (25.32) recursively to show that
the function

$$f(r, z) = \frac{\partial^2 g}{\partial r^2} - \frac{1}{r}\frac{\partial g}{\partial r} = r\frac{\partial}{\partial r}\left(\frac{1}{r}\frac{\partial g}{\partial r}\right)$$

will satisfy (29.10) if $g(r, z)$ is an axisymmetric harmonic function and hence the
function

$$\phi = r^2\frac{\partial^2 g}{\partial r^2} - r\frac{\partial g}{\partial r} = r^3\frac{\partial}{\partial r}\left(\frac{1}{r}\frac{\partial g}{\partial r}\right) \tag{29.11}$$

satisfies (29.7). This can also be verified directly by substitution (see Problem 29.1).
Any axisymmetric harmonic function g can be used to generate solutions in this
way, including for example those defined by equations (25.11, 25.12, 25.19, 25.24,
25.28).

The functions

$$R^n P_{n-2}^2(\cos\beta) \ ; \ R^n Q_{n-2}^2(\cos\beta) \ ; \ R^{3-n} P_{n-2}^2(\cos\beta) \ ; \ R^{3-n} Q_{n-2}^2(\cos\beta) \tag{29.12}$$

define solutions to equation (29.7) for $n \geq 4$. The P-series functions define the
bounded polynomials

$$\begin{aligned}
\phi_4 &= R^4 \sin^4\beta = r^4 \\
\phi_5 &= R^5 \sin^4\beta\cos\beta = r^4 z \\
\phi_6 &= R^6\left(7\cos^2\beta - 1\right)\sin^4\beta = r^4\left(6z^2 - r^2\right) \\
\phi_7 &= R^7\left(3\cos^2\beta - 1\right)\sin^4\beta\cos\beta = r^4 z\left(2z^2 - r^2\right)
\end{aligned} \tag{29.13}$$

and a corresponding set of functions

$$\begin{aligned}
\phi_{-1} &= R^{-1}\sin^4\beta = \frac{r^4}{R^5} \\
\phi_{-2} &= R^{-2}\sin^4\beta\cos\beta = \frac{r^4 z}{R^7} \\
\phi_{-3} &= R^{-3}\frac{r^4 z}{R^7}\left(7\cos^2\beta - 1\right)\sin^4\beta = \frac{r^4\left(6z^2 - r^2\right)}{R^9} \\
\phi_{-4} &= R^{-4}\left(3\cos^2\beta - 1\right)\sin^4\beta\cos\beta = \frac{r^4 z\left(2z^2 - r^2\right)}{R^{11}} \ .
\end{aligned} \tag{29.14}$$

that are singular only at the origin. The Q-series functions include logarithmic terms
and are singular on the entire z-axis.

The two solutions of (29.7) of degree 0,1,2,3 in R are

$$R^n P_{n-2}^{-2}(\cos\beta) \ ; \ R^n Q_{n-2}^2(\cos\beta) \ , \tag{29.15}$$

which can be expanded as

$$\phi_0 = R^0(1 - \cos\beta)^2(2 + \cos\beta) = \frac{(R-z)^2(2R+z)}{R^3}$$

$$\phi_1 = R^1(1 - \cos\beta)^2 = \frac{(R-z)^2}{R}$$

$$\phi_2 = R^2(1 - \cos\beta)^2 = (R-z)^2 \qquad\qquad (29.16)$$

$$\phi_3 = R^3(1 - \cos\beta)^2(2 + \cos\beta) = (R-z)^2(2R+z)$$

and

$$\varphi_0 = 1$$

$$\varphi_1 = R\cos\beta = z$$

$$\varphi_2 = R^2\cos\beta = Rz \qquad\qquad (29.17)$$

$$\varphi_3 = R^3$$

respectively.

29.4 The Displacement Field

Eliminating Ψ between equations (29.1), we have

$$\sigma_{\theta r} = \mu\left(\frac{\partial u_\theta}{\partial r} - \frac{u_\theta}{r}\right) \;;\; \sigma_{\theta z} = \mu\frac{\partial u_\theta}{\partial z} . \qquad (29.18)$$

Once the stress field is known, these equations can be solved for the only non-zero displacement component u_θ.

An alternative approach is to define a displacement function ψ through the equation

$$u_\theta = r\psi , \qquad\qquad (29.19)$$

in which case

$$\sigma_{\theta r} = \mu r\frac{\partial \psi}{\partial r} \;;\; \sigma_{\theta z} = \mu r\frac{\partial \psi}{\partial z} , \qquad (29.20)$$

from (29.18, 29.19) and more generally the vector traction on the θ-plane is

$$\boldsymbol{\sigma}_\theta = \mu r\boldsymbol{\nabla}\psi . \qquad\qquad (29.21)$$

It follows from (29.21) that on a traction-free surface

$$\frac{\partial \psi}{\partial n} = 0 ,$$

where n is the local normal to the surface. Thus lines of constant ϕ and lines of constant ψ are everywhere orthogonal to each other.

Equations (29.8) and (29.20) are alternative representations of the stress components and hence

$$\mu r \frac{\partial \psi}{\partial r} = -\frac{1}{r^2}\frac{\partial \phi}{\partial z} \; ; \; \; \mu r \frac{\partial \psi}{\partial z} = \frac{1}{r^2}\frac{\partial \phi}{\partial r} \; .$$

Eliminating ϕ between these equations, we obtain

$$\frac{\partial^2 \psi}{\partial r^2} + \frac{3}{r}\frac{\partial \psi}{\partial r} + \frac{\partial^2 \psi}{\partial z^2} = 0 \; ,$$

which is the governing equation that must be satisfied by the displacement function ψ.

29.5 Cylindrical and Conical Bars

The formulation introduced in this chapter is particularly effective for determining the stresses in an axisymmetric bar loaded only by torques on the ends. In this case, the curved surfaces of the bar must be defined by lines of constant ϕ and the problem is reduced to the determination of a solution of (29.7) that is constant along a specified surface. It may be possible to determine this by inspection, using a suitable combination of functions such as (29.12–29.17). In other cases, an approximate solution may be obtained by making a contour plot of a candidate function and adjusting multiplying constants or other parameters until one of the contours approximates the required shape of the bar.

A simple case is the solid cylindrical bar of radius a, transmitting a torque T. Clearly the function

$$\phi = C\phi_4 = Cr^4$$

is constant on $r=a$, so the traction-free condition is satisfied along this boundary. To determine the multiplying constant C, we use equation (29.3) to write

$$T = 2\pi\phi(a, z) - 2\pi\phi(0, z) = 2\pi C a^4$$

and hence

$$C = \frac{T}{2\pi a^4} \; .$$

The stress components are then recovered from (29.8) as

$$\sigma_{z\theta} = 4Cr = \frac{2Tr}{\pi a^4} \; ; \; \; \sigma_{r\theta} = 0 \; ,$$

agreeing of course with the elementary torsion theory.

A more interesting example concerns the conical bar $0 < \beta < \beta_0$ transmitting a torque T. The traction-free condition on the boundary $\beta = \beta_0$ will be satisfied if ϕ is independent of R and hence a function of β only in spherical polar coördinates. This condition is clearly satisfied by the functions ϕ_0, φ_0 of equations (29.16, 29.17), but the latter leads only to a null state of stress, so we take

$$\phi = C\phi_0 = C(1 - \cos \beta)^2(2 + \cos \beta) = \frac{C(R - z)^2(2R + z)}{R^3} \ . \tag{29.22}$$

For the constant C, we have

$$T = 2\pi\phi(R, \beta_0) - 2\pi\phi(R, 0) = 2\pi C(1 - \cos \beta_0)^2(2 + \cos \beta_0) \ ,$$

from (29.3) and hence

$$C = \frac{T}{2\pi(1 - \cos \beta_0)^2(2 + \cos \beta_0)} \ . \tag{29.23}$$

The stress components are then obtained from (29.8) as

$$\sigma_{z\theta} = \frac{3Trz}{2\pi(1 - \cos \beta_0)^2(2 + \cos \beta_0)R^5} \ ; \ \ \sigma_{r\theta} = \frac{3Tr^2}{2\pi(1 - \cos \beta_0)^2(2 + \cos \beta_0)R^5} \ .$$

Simpler expressions for the stresses are obtained in spherical polar coördinates. Vector transformation of the tractions (29.8) on the θ-plane leads to the general expressions

$$\sigma_{\theta R} = \frac{1}{r^2}\frac{1}{R}\frac{\partial \phi}{\partial \beta} = \frac{1}{R^3 \sin^2 \beta}\frac{\partial \phi}{\partial \beta} \ ; \ \ \sigma_{\theta\beta} = -\frac{1}{r^2}\frac{\partial \phi}{\partial R} = -\frac{1}{R^2 \sin^2 \beta}\frac{\partial \phi}{\partial R} \ . \tag{29.24}$$

Substitution for ϕ from (29.22, 29.23) then yields

$$\sigma_{\theta R} = \frac{3T \sin \beta}{2\pi(1 - \cos \beta_0)^2(2 + \cos \beta_0)R^3} \ ; \ \ \sigma_{\theta\beta} = 0 \ , \tag{29.25}$$

showing that the traction on the θ-surface is purely radial. Of course, this follows immediately from the fact that ϕ is a function of β only and hence is constant on any line $\beta = $ constant.

The displacements for the conical bar can be obtained by equating (29.25) and (29.21), which in spherical polar coördinates takes the form

$$\sigma_{\theta R} = \mu R \sin \beta \frac{\partial \psi}{\partial R} \ ; \ \ \sigma_{\theta\beta} = \mu \sin \beta \frac{\partial \psi}{\partial \beta} \ .$$

Using (29.25), we therefore have

$$\frac{\partial \psi}{\partial R} = \frac{3T}{2\pi\mu(1 - \cos \beta_0)^2(2 + \cos \beta_0)R^4} \ ; \ \ \frac{\partial \psi}{\partial \beta} = 0 \ . \tag{29.26}$$

Thus ψ must be a function of R only and integrating the first of (29.26) we have

$$\psi = -\frac{T}{2\pi\mu(1 - \cos\beta_0)^2(2 + \cos\beta_0)R^3} \; .$$

The displacements are then recovered from (29.19) as

$$u_\theta = -\frac{Tr}{2\pi\mu(1 - \cos\beta_0)^2(2 + \cos\beta_0)R^3} = -\frac{T\sin\beta}{2\pi\mu(1 - \cos\beta_0)^2(2 + \cos\beta_0)R^2} \; .$$

29.5.1   The centre of rotation

The stress function (29.22) can also be used to determine the stresses in an infinite body loaded only by a concentrated torque T about the z-axis acting at the origin. Using (29.24), we have

$$\sigma_{\theta R} = \frac{3C\sin\beta}{R^3} \; ; \quad \sigma_{\theta\beta} = 0$$

and the torque transmitted across the spherical surface $R=a$ is

$$T = 2\pi\int_0^\pi \sigma_{R\theta}(a, \beta)a^2\sin^2\beta d\beta = 8\pi C \; .$$

It follows that $C = T/8\pi$ and the stress field is

$$\sigma_{\theta R} = \frac{3T\sin\beta}{8\pi R^3} \; ; \quad \sigma_{\theta\beta} = 0 \; .$$

The same result can be obtained from (29.25) by setting $\beta_0 = \pi$, since a cone of angle π is really an infinite body with an infinitesimal hole along the negative z-axis[1]

29.6   The Saint Venant Problem

As in Chapter 24, bounded polynomial solutions of (29.7) can be used to solve problems of the cylindrical bar with prescribed polynomial tractions on the curved surfaces, but the boundary conditions on the ends can generally only be satisfied only in the weak sense. The corrective solution needed to re-establish strong boundary conditions on the ends is a Saint Venant problem in the sense of Chapter 6 and can be treated by similar methods.

---

[1] This hole does not alter the stress field, since the stresses are zero on the axis.

We first postulate the existence of solutions of (29.7) of the separated-variable form

$$\phi = e^{-\lambda z} f(r) .$$

Substitution in (29.7) then yields the ordinary differential equation

$$\frac{d^2 f}{dr^2} - \frac{3}{r} \frac{df}{dr} + \lambda^2 f = 0$$

for the function $f(r)$. This equation has the general solution

$$f(r) = C_1 r^2 J_2(\lambda r) + C_2 r^2 Y_2(\lambda r) ,$$

where J_2, Y_2 are Bessel functions of the first and second kind respectively of order 2 and C_1, C_2 are arbitrary constants. The boundaries $r = a$, b of the hollow cylinder $a < r < b$ will be traction free if $f(a) = f(b) = 0$ and hence

$$C_1 J_2(\lambda a) + C_2 Y_2(\lambda a) = 0 ; \quad C_1 J_2(\lambda b) + C_2 Y_2(\lambda b) = 0 .$$

These equations have a non-trivial solution for C_1, C_2 if and only if

$$J_2(\lambda a) Y_2(\lambda b) - Y_2(\lambda a) J_2(\lambda b) = 0 , \tag{29.27}$$

which constitutes an eigenvalue equation for the exponential decay rate λ. The function Y_2 is unbounded at $r \to 0$ and hence must be excluded for the solid cylinder $0 \leq r < a$. In this case, the eigenvalue equation reduces to

$$J_2(\lambda a) = 0 . \tag{29.28}$$

As in Chapter 6, a more general solution can then be constructed for the corrective stress field in the form of an eigenfunction expansion.

In contrast to the plane problem, the eigenvalues of equations (29.27) are real. The first eigenvalue of (29.28) is $\lambda a = 5.136$, showing that self-equilibrated corrections to the boundary conditions on the ends decay quite rapidly with distance from the ends.

Problems

29.1. Verify that equation (29.11) defines a function satisfying equation (29.7) by direct substitution, using the result

$$\frac{\partial^2 g}{\partial z^2} = -\frac{\partial^2 g}{\partial r^2} - \frac{1}{r} \frac{\partial g}{\partial r}$$

(which is a restatement of the condition $\nabla^2 g = 0$) to eliminate the derivatives with respect to z.

29.2. The solid cylindrical bar $0 \leq r < a$, $0 < z < L$ is loaded by a uniform torsional shear traction $\sigma_{r\theta} = S$ on the surface $r = a$, the end $z = 0$ being traction free. Find the stress field in the bar, using bounded polynomial solutions to equation (29.7).

29.3. Show that the function

$$\varphi_1 = z$$

satisfies equation (29.7) and defines a state of stress that is independent of z. Combine this solution with a bounded polynomial solution of (29.7) to determine the stress field in the hollow cylindrical bar $a < r < b$, $0 < z < L$ built in at $z = L$ and loaded by a uniform torsional shear traction $\sigma_{r\theta} = S$ on the inner surface $r = a$, the outer surface $r = b$ and the end $z = 0$ being traction free.

29.4. A solid cylindrical bar of radius b transmitting a torque T, contains a small spherical hole of radius a ($\ll b$) centred on the axis. Find a bounded polynomial solution to equation (29.7) to describe the stress field in the bar in the absence of the hole and hence determine the perturbation in this field due to the hole by superposing a suitable singular solution.

29.5. Develop a contour plot in the r, z plane of a function comprising

(i) the function ϕ_4 of equation (29.13),
(ii) a centre of rotation, centred on the point $(0, a)$,
(iii) an equal negative centre of rotation, centred on the point $(0, -a)$.

By experimenting with different multiplying constants on these functions, show that the solution can be obtained for a cylindrical bar with an approximately elliptical central hole, transmitting a torque T. If the bar has outer radius b and the hole has radius $0.1b$ at the mid-plane and is of total axial length $0.4b$, estimate the stress at the point $(0.1b, 0)$.

29.6. Show that the lines $R - z = C$ define a set of parabolas, where C is a constant. Combine suitable functions from equations (29.12–29.17) to obtain a solution of (29.7) that is constant on these lines and hence solve the problem of a paraboloidal bar transmitting a torque T. Comment on the nature of the solution in the region $z \leq 0$ and in particular at the origin. Do your results place any restrictions on the applicability of the solution?

29.7. The solid cylindrical bar $0 \leq r < a$, $z > 0$ is bonded to a rigid body at the curved surface $r = a$ and loaded by torsional tractions on the plane end $z = 0$. Using a displacement function

$$\psi = e^{-\lambda z} f(r) ,$$

find suitable non-trivial solutions for the stress field and hence determine the slowest rate at which the stress field will decay along the bar.

Chapter 30
The Prismatic Bar

We have seen in Chapter 26 that three-dimensional solutions can be obtained to the problem of the solid or hollow cylindrical bar loaded on its curved surfaces, using the Papkovich-Neuber solution with spherical harmonics and related potentials. Here we shall show that similar solutions can be obtained for bars of more general cross section, using the complex-variable form of the Papkovich-Neuber solution from §22.6.

We use a coördinate system in which the axis of the bar is aligned with the z-direction, one end being the plane $z=0$. The constant cross section of the bar then comprises a domain Ω in the xy-plane, which may be the interior of a closed curve Γ, or that part of the region interior to a closed curve Γ_0 that is also *exterior* to one or more closed curves Γ_1, Γ_2, etc. The following derivations and examples will be restricted to the former, simply-connected case, but it will be clear from the methods used that the additional complications associated with multiply-connected cross sections arise only in the solution of two-dimensional boundary-value problems, for which classical methods exist. As in Chapter 26, we shall apply weak boundary conditions at the ends of the bar, which implies that the solutions are appropriate only for relatively long bars in regions that are not too near the ends.

30.1 Power-series Solutions

The most general loading of the lateral surfaces Γ comprises the three traction components T_x, T_y, T_z, which we shall combine in the complex-variable form as

Supplementary Information The online version contains supplementary material available at https://doi.org/10.1007/978-3-031-15214-6_30.

$$T = T_x + \iota T_y , \quad T_z .$$

Consider the problem in which T, T_z can be written as power series in z — i.e.

$$T = \sum_{j=1}^{m-1} f_j(s) z^{j-1} ; \quad T_z = \sum_{j=1}^{m} g_j(s) z^{j-1} , \qquad (30.1)$$

where f, g are arbitrary complex and real functions respectively of the curvilinear coördinate s defining position on Γ. We shall denote this system of tractions by the symbol $T^{(m)}$. Notice that the in-plane tractions T are carried only up to the order z^{m-2} whereas the out-of-plane traction T_z includes a term proportional to z^{m-1}. Practical cases where the highest order term in all three tractions is the same (z^n say) can of course be treated by setting $m = n + 2$ and $g_m(s) = 0$. We describe the problem (30.1) as $\mathcal{P}_m$ and denote the corresponding stress and displacement fields in the bar by

$$\sigma^{(m)} = \left\{ \Theta^{(m)}, \ \Phi^{(m)}, \ \Psi^{(m)}, \ \sigma_{zz}^{(m)} \right\} ; \quad u^{(m)} = \left\{ u^{(m)}, \ u_z^{(m)} \right\} , \qquad (30.2)$$

respectively, using the complex-variable notation of §22.6.

The bar will transmit force resultants comprising an axial force F_z, shear forces F_x, F_y, a torque M_z and bending moments M_x, M_y, which are related to the stress components (30.2) by the equilibrium relations

$$F_z(z) = \iint_\Omega \sigma_{zz} d\Omega ; \quad F(z) \equiv F_x + \iota F_y = \iint_\Omega \Psi d\Omega$$

$$M_z(z) = \frac{\iota}{2} \iint_\Omega (\zeta \overline{\Psi} - \bar{\zeta}\Psi) d\Omega \qquad (30.3)$$

$$M(z) \equiv M_x + \iota M_y = -\iota \iint_\Omega \zeta \sigma_{zz} d\Omega .$$

Consideration of the equilibrium of a segment of the bar shows that these resultants will also take the form of finite polynomials in z, with highest power z^m in F_z, M and z^{m-1} in F, M_z.

The complete definition of problem $\mathcal{P}_m$ requires that we also specify the force resultants $F(0)$, $F_z(0)$, $M(0)$, $M_z(0)$ on the end $z = 0$. However, it is sufficient at this stage to obtain a particular solution with any convenient end condition. Correction of the end condition in the weak sense can then be completed using the techniques of Part III.

30.1.1 Superposition by differentiation

Suppose that the solution to a given problem $\mathcal{P}_m$ is known — i.e. that we have found stress and displacement components $\sigma^{(m)}$, $u^{(m)}$ that reduce to a particular set of polynomial tractions (30.1) on Γ, and that satisfy the quasi-static equations of elasticity in the absence of body forces, which we here represent in the symbolic form

$$\mathcal{L}\left(\sigma^{(m)}, u^{(m)}\right) = 0 , \qquad (30.4)$$

where $\mathcal{L}$ is a set of linear differential operators. Differentiating (30.4) with respect to z, we have

$$\frac{\partial}{\partial z} \mathcal{L}\left(\sigma^{(m)}, u^{(m)}\right) = \mathcal{L}\left(\frac{\partial \sigma^{(m)}}{\partial z}, \frac{\partial u^{(m)}}{\partial z}\right) = 0 .$$

It follows that the new set of stresses and displacements defined by differentiation as

$$\sigma^{(m-1)} = \frac{\partial \sigma^{(m)}}{\partial z} \quad ; \quad u^{(m-1)} = \frac{\partial u^{(m)}}{\partial z}$$

will also satisfy the equations of elasticity and will correspond to the tractions

$$T = \sum_{j=2}^{m-1}(j-1)f_j(s)z^{j-2} \quad ; \quad T_z = \sum_{j=2}^{m}(j-1)g_j(s)z^{j-2} . \qquad (30.5)$$

This process of generating a new particular solution by differentiation with respect to a spatial coördinate can be seen as a form of linear superposition of the original solution on itself after an infinitesimal displacement in the z-direction, and was already used in §24.3 to generate singular spherical harmonics. The resulting tractions (30.5) are polynomials in z with highest order terms z^{m-3}, z^{m-2} in T, T_z respectively and hence are of the form $T^{(m-1)}$. Thus the stress field $\partial \sigma^{(m)}/\partial z$ is the solution of a particular problem of the class $\mathcal{P}_{m-1}$.

30.1.2 The problems $\mathcal{P}_0$ and $\mathcal{P}_1$

Repeating this differential operation m times, we find that the stress and displacement fields

$$\sigma^{(0)} = \frac{\partial^m \sigma^{(m)}}{\partial z^m} \quad ; \quad u^{(0)} = \frac{\partial^m u^{(m)}}{\partial z^m} \qquad (30.6)$$

correspond to the physical problem $\mathcal{P}_0$ in which the tractions $T=0$, and the only non-zero force resultants are the axial force F_z and the bending moment M, which are independent of z. This is a classical problem for which the only non-zero stress component is

$$\sigma_{zz}^{(0)} = A_0 \zeta + \overline{A}_0 \bar{\zeta} + B_0 , \qquad (30.7)$$

where A_0 is a complex constant and B_0 a real constant.

The class $\mathcal{P}_1$ includes the problems of Chapters 17, 18, where an otherwise traction-free bar is loaded by a lateral force $F(0)$ and a torque $T_z(0)$ at the end

$z = 0$. However, it also permits z-independent antiplane tractions T_z on the lateral surfaces, which generally lead to the bending moment M and/or the axial force F_z varying linearly with z. This generalization is described by Milne-Thomson[1] as the 'push' problem. In $\mathcal{P}_1$, the in-plane stress components $\Theta^{(1)}$, $\Phi^{(1)}$ remain zero, the antiplane stress components $\Psi^{(1)}$ are independent of z, and $\sigma_{zz}^{(1)}$ has the form

$$\sigma_{zz}^{(1)} = \left(A_0 \zeta + \overline{A}_0 \overline{\zeta} + B_0 \right) z + A_1 \zeta + \overline{A}_1 \overline{\zeta} + B_1 \ . \tag{30.8}$$

30.1.3 Properties of the solution to $\mathcal{P}_m$

From these results, it is clear that the solution of problem $\mathcal{P}_m$ possesses the following properties:-

(i) The stress and displacement components, $\sigma^{(m)}$, $u^{(m)}$ can be expressed as power series in z.

(ii) The highest order terms in z in the in-plane stress components $\Theta^{(m)}$, $\Phi^{(m)}$ are of order z^{m-2}, since after $m-1$ differentiations with respect to z they are zero.

(iii) The highest order terms in z in the antiplane stress components $\Psi^{(m)}$ are of order z^{m-1}, since after $m-1$ differentiations with respect to z they are independent of z.

(iv) The highest order terms in the stress component $\sigma_{zz}^{(m)}$ are of the form

$$\sigma_{zz}^{(m)} = \frac{\left(A_0 \zeta + \overline{A}_0 \overline{\zeta} + B_0 \right) z^m}{m!} + \frac{\left(A_1 \zeta + \overline{A}_1 \overline{\zeta} + B_1 \right) z^{m-1}}{(m-1)!} \ ,$$

from (30.6, 30.8).

Conclusions (ii,iii) explain why we chose to terminate the traction series (30.1) at z^{m-2} in T and at z^{m-1} in T_z.

30.2 Solution of $\mathcal{P}_m$ by Integration

The process of differentiation elaborated in §30.1.1 shows that for every problem $\mathcal{P}_m$, there exists a set of lower order problems $\mathcal{P}_j$, $j = [0, m-1]$ satisfying the recurrence relations

$$T^{(j)} = \frac{\partial T^{(j+1)}}{\partial z} \ ; \quad \sigma^{(j)} = \frac{\partial \sigma^{(j+1)}}{\partial z} \ ; \quad u^{(j)} = \frac{\partial u^{(j+1)}}{\partial z} \ . \tag{30.9}$$

---

[1] L. M. Milne-Thomson, *Antiplane Elastic Systems*, Springer, Berlin, 1962.

The lowest-order problems in this set, $\mathcal{P}_0$, $\mathcal{P}_1$ can be solved by the methods of Part III, so the more general problem $\mathcal{P}_m$ could be solved recursively if we could devise a method to generate the solution of $\mathcal{P}_{j+1}$ from that of $\mathcal{P}_j$. This process clearly involves partial integration of $\boldsymbol{\sigma}^{(j)}$, $\boldsymbol{u}^{(j)}$ with respect to z. In a formal sense, we can write

$$\boldsymbol{\sigma}^{(j+1)} = \int \boldsymbol{\sigma}^{(j)} dz + \boldsymbol{f}(x, y) , \tag{30.10}$$

where $\boldsymbol{f}(x, y)$ is an arbitrary function of x, y only. Since $\boldsymbol{\sigma}^{(j)}$ is in the form of a finite power series in z, it is always possible to find a particular integral for the first term in (30.10). The function $\boldsymbol{f}$ must then be chosen to satisfy two conditions:-

(i) The complete stress field $\boldsymbol{\sigma}^{(j+1)}$ including $\boldsymbol{f}(x, y)$ must satisfy the equations of elasticity (30.4), and
(ii) the stresses must reduce to the known tractions $\boldsymbol{T}^{(j+1)}$ on Γ.

Condition (ii) involves only the coefficient of z^0 in $\boldsymbol{T}^{(j+1)}$, since the coefficients of higher order terms will have been taken care of at an earlier stage in the recursive process. Similarly, the conditions imposed by the equations of elasticity can only arise in the lowest order terms in the stress field. We shall show that the determination of $\boldsymbol{f}$ is a well-posed two-dimensional problem.

The concept of a recursive solution to problems $\mathcal{P}_m$ was first enunciated by Ieşan[2], who started from a state of arbitrary rigid-body displacement, rather than $\mathcal{P}_0$ of §30.1.2. A rigid-body displacement lies in the class $\mathcal{P}_{-1}$ and since it involves no stresses, it is the same for all domains Ω. The present treatment will place emphasise on the analytical solution of specific problems rather than on general features of the solution method and is based on the paper J. R. Barber (2006), Three-dimensional elasticity solutions for the prismatic bar, *Proceedings of the Royal Society of London*, Vol. A462, pp. 1877–1896.

30.3 The Integration Process

The recursive procedure is most easily formulated in the context of the complex-variable Papkovich-Neuber (P-N) solution of §22.6 and equations (22.6–22.10). The prismatic bar satisfies the condition that all lines drawn in the z-direction cut the boundary of the body at only two points (the ends of the bar), which is the criterion established in §21.4.1 for the elimination of the component ψ_z in the P-N solution without loss of generality. We can therefore simplify the solution by omitting all the terms involving ω in equations (22.6–22.10), giving

---

[2] D. Ieşan, On Saint-Venant's Problem (1986), *Archive for Rational Mechanics and Analysis*, Vol. 91, pp. 363–373.

$$2\mu u = 2\frac{\partial \phi}{\partial \bar{\zeta}} - (3 - 4\nu)\psi + \zeta\frac{\partial \overline{\psi}}{\partial \bar{\zeta}} + \bar{\zeta}\frac{\partial \psi}{\partial \bar{\zeta}} \tag{30.11}$$

$$2\mu u_z = \frac{\partial \phi}{\partial z} + \frac{1}{2}\left(\bar{\zeta}\frac{\partial \psi}{\partial z} + \zeta\frac{\partial \overline{\psi}}{\partial z}\right) \tag{30.12}$$

$$\Theta = -\frac{\partial^2 \phi}{\partial z^2} - 2\left(\frac{\partial \psi}{\partial \zeta} + \frac{\partial \overline{\psi}}{\partial \bar{\zeta}}\right) - \frac{1}{2}\left(\zeta\frac{\partial^2 \overline{\psi}}{\partial z^2} + \bar{\zeta}\frac{\partial^2 \psi}{\partial z^2}\right) \tag{30.13}$$

$$\Phi = 4\frac{\partial^2 \phi}{\partial \bar{\zeta}^2} - 4(1 - 2\nu)\frac{\partial \psi}{\partial \bar{\zeta}} + 2\zeta\frac{\partial^2 \overline{\psi}}{\partial \bar{\zeta}^2} + 2\bar{\zeta}\frac{\partial^2 \psi}{\partial \bar{\zeta}^2} \tag{30.14}$$

$$\sigma_{zz} = \frac{\partial^2 \phi}{\partial z^2} - 2\nu\left(\frac{\partial \psi}{\partial \zeta} + \frac{\partial \overline{\psi}}{\partial \bar{\zeta}}\right) + \frac{1}{2}\left(\bar{\zeta}\frac{\partial^2 \psi}{\partial z^2} + \zeta\frac{\partial^2 \overline{\psi}}{\partial z^2}\right) \tag{30.15}$$

$$\Psi = 2\frac{\partial^2 \phi}{\partial \bar{\zeta}\partial z} - (1 - 2\nu)\frac{\partial \psi}{\partial z} + \zeta\frac{\partial^2 \overline{\psi}}{\partial \bar{\zeta}\partial z} + \bar{\zeta}\frac{\partial^2 \psi}{\partial \bar{\zeta}\partial z} , \tag{30.16}$$

where ϕ, ψ are respectively a real and a complex harmonic function of $\zeta, \bar{\zeta}, z$.

None of the expressions (30.11–30.16) include explicit z-multipliers and hence the integration (30.10) can be performed on the P-N potentials ϕ, ψ, rather than directly on the stress and displacement components. The requirement §30.2(i) that the solution satisfy the equations of elasticity (30.4) will then be met by ensuring that the integrated potentials are three-dimensional harmonic functions.

Suppose that the complex potential ψ in (30.11–30.16) for problem $\mathcal{P}_j$ is given by a known function ψ_j. This must be three-dimensionally harmonic, and hence

$$\nabla^2 \psi_j \equiv \frac{\partial^2 \psi_j}{\partial z^2} + 4\frac{\partial^2 \psi_j}{\partial \zeta\partial \bar{\zeta}} = 0 , \tag{30.17}$$

from (22.11). We next write

$$\psi_{j+1} = h_1(\zeta, \bar{\zeta}, z) + h_2(\zeta, \bar{\zeta}) , \tag{30.18}$$

where

$$h_1 = \int \psi_j(\zeta, \bar{\zeta}, z)dz$$

is any partial integral of ψ_j. The recurrence relations (30.9) then imply that h_2 is independent of z. Applying the Laplace operator (30.17) to the function ψ_{j+1} of equation (30.18), we obtain

$$4\frac{\partial^2 h_2}{\partial \zeta\partial \bar{\zeta}} = -\nabla^2 h_1 ,$$

and the function h_2 can then be recovered by integration with respect to ζ and $\bar{\zeta}$. These integrations will introduce arbitrary holomorphic functions that will be used to satisfy the zeroth-order boundary conditions in problem $\mathcal{P}_{j+1}$.

We know from §30.1.3 that at any given stage in the recurrence procedure, ψ can be written as a finite power series in z and one way to write this series is as a set of terms of the form

$$\psi_{2n}(\zeta, \bar{\zeta}, z) = \sum_{i=0}^{n} \frac{z^{2i} f_{n-i}(\zeta, \bar{\zeta})}{(2i)!} \quad ; \quad \psi_{2n+1}(\zeta, \bar{\zeta}, z) = \sum_{i=0}^{n} \frac{z^{2i+1} f_{n-i}(\zeta, \bar{\zeta})}{(2i+1)!}$$

from §25.8, where the functions f_k satisfy the recurrence relations (25.39, 25.40). It is easily verified that suitable z-integrals of these functions that satisfy the three-dimensional Laplace equation (30.17) are

$$\int \psi_{2n}(\zeta, \bar{\zeta}, z)dz = \psi_{2n+1}(\zeta, \bar{\zeta}, z) \quad ; \quad \int \psi_{2n+1}(\zeta, \bar{\zeta}, z)dz = \psi_{2n+2}(\zeta, \bar{\zeta}, z) ,$$

where ψ_{2n+2} is obtained from ψ_{2n} by replacing n by $n+1$.

The first term of this form is the two-dimensionally harmonic function

$$\psi_0 = f_0(\zeta, \bar{\zeta}) = g_1(\zeta) + \bar{g}_2(\bar{\zeta}) ,$$

from (25.41). Thus, if the above integration process is applied sequentially, the arbitrary functions of ζ and $\bar{\zeta}$ that are introduced at stage i can be expressed in the form of the function ψ_0 and this will integrate up to the corresponding function ψ_{j-i} at stage j.

In the special case where the functions f_k are finite polynomials in $\zeta, \bar{\zeta}$, the potentials can be expressed as a series of the bounded cylindrical harmonics χ_n^m defined in equations (25.43, 25.44), which can be integrated using the relation

$$\int \chi_n^m dz = \chi_{n+1}^m , \tag{30.19}$$

from (25.45).

Similar procedures can be used to produce harmonic partial integrals of the real potential, ϕ_j, in which case the arbitrary holomorphic functions in h_2 must be chosen so as to ensure that the resulting potential is real.

30.4 The Two-dimensional Problem $\mathcal{P}_0$

Problem $\mathcal{P}_0$ corresponds to the elementary stress field in which the lateral surfaces of the bar are traction free and the bar transmits a bending moment M and an axial force F_z. The only non-zero stress component is $\sigma_{zz}^{(0)}$ given by equation (30.7), but to start the recursive process, we need to express this in terms of the P-N solution of §30.3.

It is readily verified by substitution in (30.13–30.16) that this elementary stress field is generated by the choice of potentials

$$\phi_0 = \mathcal{G}_0^0 \; ; \quad \psi_0 = \mathcal{H}_0^0 , \tag{30.20}$$

where we introduce the notation

$$\mathcal{G}_m^n \equiv \frac{(1-2\nu)}{2(1-\nu^2)} \left(A_m \chi_{n+2}^1 + \bar{A}_m \overline{\chi}_{n+2}^{-1} \right) + \frac{B_m \chi_{n+2}^0}{(1+\nu)}$$

$$\mathcal{H}_m^n \equiv \frac{\bar{A}_m \chi_{n+2}^0}{(1-\nu^2)} - \frac{(1-2\nu) A_m \chi_n^2}{8(1-\nu^2)} - \frac{B_m \chi_n^1}{4(1+\nu)}$$

and χ_n^m are the bounded cylindrical harmonics defined in §25.8.1 and equations (25.43, 25.44). The recurrence relation (30.19) implies that

$$\int \mathcal{G}_m^n dz = \mathcal{G}_m^{n+1} \; ; \quad \int \mathcal{H}_m^n dz = \mathcal{H}_m^{n+1} \tag{30.21}$$

and we shall find this useful in subsequent developments.

Since the stress components $\Theta^{(0)}$, $\Phi^{(0)}$, $\Psi^{(0)}$ are identically zero, no tractions are implied on any surface Γ, and hence the potentials (30.20) define the solution of problem $\mathcal{P}_0$ for a bar of any cross section Ω.

30.5 Problem $\mathcal{P}_1$

Problem $\mathcal{P}_1$ could be solved using the methods of Part III, but here we shall develop the solution by integration using the P-N solution, since similar procedures will be needed at higher-order stages of the recursive process. The strategy is first to develop a particular solution describing the most general force resultants that can be transmitted along the bar, but without regard to the exact form of the antiplane tractions on Γ. We then superpose a corrective antiplane solution, in the spirit of Chapter 15, but within the P-N formalism, to satisfy these boundary conditions in the strong sense.

Starting with the solution ϕ_0, ψ_0 of equation (30.20), we first generate a trial solution for ϕ, ψ by integration with respect to z. Using the recurrence relations (30.21) to ensure that the resulting functions are harmonic, we obtain

$$\phi = \mathcal{G}_0^1 \; ; \quad \psi = \mathcal{H}_0^1 . \tag{30.22}$$

The constants A_0, B_0 in these functions provide the degrees of freedom to define the lateral force resultants F, but in problem $\mathcal{P}_1$, we also need to allow for the transmission of a torque M_z. Any particular solution of the torsion problem can be used for this purpose (remember we are going to correct the tractions on the lateral surfaces at the next stage) and the simplest solution is described by the potentials

$$\phi^{(t)} = 0 \; ; \quad \psi^{(t)} = -\frac{\imath C_0 \chi_1^1}{2(1-\nu)} , \tag{30.23}$$

where C_0 is a real constant.

Combining these results, we define the particular antiplane solution as

$$\phi_P = \mathcal{G}_0^1 \; ; \quad \psi_P = \mathcal{H}_0^1 - \frac{\imath C_0 \chi_1^1}{2(1-\nu)} , \tag{30.24}$$

for which the corresponding non-zero stress components are

$$\sigma_{zz} = A_0 z \zeta + \bar{A}_0 z \bar{\zeta} + B_0 z \tag{30.25}$$

$$\Psi = -\frac{A_0(1+2\nu)\zeta^2}{4(1+\nu)} - \frac{\bar{A}_0 \zeta \bar{\zeta}}{2(1+\nu)} - \frac{B_0 \zeta}{2} + \imath C_0 \zeta , \tag{30.26}$$

from (30.24, 30.15, 30.16).

Notice that in this solution, the real constant B_0 corresponds to an axial force F_z that varies linearly with z and this will arise only when there are non-self-equilibrating uniform antiplane tractions on the surface Γ. Thus, if this method is used to solve the problems of Chapters 17 or 18, B_0 can be set to zero.

30.5.1   The corrective antiplane solution

Equation (30.26) defines non-zero values of Ψ and hence generally implies non-zero tractions T_z on Γ, which however are independent of z. We may indeed have non-zero tractions T_z, but these will not generally correspond to those of the particular solution. It is therefore necessary to superpose a corrective solution in which the tractions are changed by some known function ΔT_z without affecting the force resultants implied by the constants A_0, B_0, C_0. Thus, the corrective solution satisfies the conditions

$$\Theta = \Phi = \sigma_{zz} = 0 \; ; \quad \frac{\partial \Psi}{\partial z} = 0 , \tag{30.27}$$

with self-equilibrated tractions ΔT_z, and it is an antiplane solution in the sense of Chapter 15 and §20.2.

A convenient representation of the corrective solution in P-N form can be written in the form

$$\phi = z\left(h + \bar{h}\right) - \frac{z\left(\zeta h' + \bar{\zeta}\bar{h}'\right)}{4(1-\nu)} \; ; \quad \psi = \frac{z\bar{h}'}{2(1-\nu)} , \tag{30.28}$$

where h is a holomorphic function of ζ. These potentials satisfy the Laplace equation (30.17) and substitution into (30.11–30.16) show that the homogeneous conditions (30.27) are satisfied identically. The non-zero displacement and stress components are obtained as

$$2\mu u_z = h + \overline{h} \ ; \quad \Psi = \overline{h}' \ ,$$

agreeing with the notation of equations (20.10, 20.12).

It follows from (20.13) that

$$\Delta T_z = \frac{\partial h_y}{\partial s} \ , \tag{30.29}$$

where s is a real coördinate locally tangential to the boundary Γ, and hence h_y is identical to the real Prandtl stress function φ for the antiplane correction. Since the corrective tractions ΔT_z are known everywhere, we can integrate (30.29) around Γ to determine h_y at all points on the boundary. Interior values of the real two-dimensionally harmonic function h_y and hence the holomorphic function of which it is the imaginary part can then be obtained using the real stress function methods of Chapters 15–18, or the complex-variable method of §20.2. It is important that the integration of (30.29) should yield a single-valued function of s. This is equivalent to the requirement that the tractions in the corrective antiplane problem should sum to a zero axial force F_z. If the applied tractions do not meet this condition, F_z will be a linear function of z and hence the constant B_0 in (30.26) will be non-zero. Thus, the single-valued condition on h_y serves to determine the constant B_0.

30.5.2 The circular bar

To illustrate the process and to record some useful results, we consider the antiplane problem $\mathcal{P}_1$ for the circular bar of radius a, with zero tractions T_z. The potentials ϕ_P, ψ_P define the stress field

$$\sigma_{zr} + \imath\sigma_{z\theta} = e^{-\imath\theta}\Psi(r,\theta) = -\frac{A_0 r^2(1+2\nu)e^{\imath\theta}}{4(1+\nu)} - \frac{\overline{A}_0 r^2 e^{-\imath\theta}}{2(1+\nu)} - \frac{B_0 r}{2} + \imath C_0 r \ ,$$

from (30.26), where we have also used the coördinate transformation relation (20.9). On the boundary $r=a$, we have $\zeta=a\exp(\imath\theta)$, $\overline{\zeta}=a\exp(-\imath\theta)$ and hence the antiplane traction is

$$T_z = \sigma_{zr}(a,\theta) = -\frac{(3+2\nu)a^2\left(\Re\{A_0\}\cos\theta - \Im\{A_0\}\sin\theta\right)}{4(1+\nu)} - \frac{B_0 a}{2} + \frac{1}{a}\frac{\partial h_y}{\partial\theta} \ ,$$

where we have added in the antiplane corrective term from (30.29). Equating T_z to zero, we obtain

$$\frac{\partial h_y}{\partial\theta} = \frac{(3+2\nu)a^3\left(\Re\{A_0\}\cos\theta - \Im\{A_0\}\sin\theta\right)}{4(1+\nu)} + \frac{B_0 a^2}{2}$$

and this integrates to a single-valued function of θ if and only if $B_0=0$, giving

$$h_y = \frac{(3 + 2\nu)a^3 \left(\Re\{A_0\} \sin\theta + \Im\{A_0\} \cos\theta\right)}{4(1 + \nu)} .$$

The holomorphic function h whose imaginary part takes this boundary value can be obtained using the direct method of §20.2.1, but it can also be seen by inspection that the appropriate function is

$$h = \frac{(3 + 2\nu)A_0 a^2 \zeta}{4(1 + \nu)} = \frac{(3 + 2\nu)A_0 a^2 \chi_0^1}{4(1 + \nu)} , \tag{30.30}$$

using (25.46).

Substituting this result into (30.28) and adding the results to (30.22, 30.23) with $B_0 = 0$, we obtain

$$\phi_1 = \mathcal{Q}_0^0 ; \quad \psi_1 = \mathcal{V}_0^0 , \tag{30.31}$$

where we introduce the notation

$$\mathcal{Q}_m^n = \frac{(1 - 2\nu)}{2(1 - \nu^2)} \left(A_m \chi_{n+3}^1 + \bar{A}_m \overline{\chi}_{n+3}^1\right)$$

$$+ \frac{(3 - 4\nu)(3 + 2\nu)a^2(A_m \chi_{n+1}^1 + \bar{A}_m \overline{\chi}_{n+1}^1)}{16(1 - \nu^2)}$$

$$\mathcal{V}_m^n = \frac{\bar{A}_m \chi_{n+3}^0}{(1 - \nu^2)} - \frac{(1 - 2\nu)A_m \chi_{n+1}^2}{8(1 - \nu^2)} + \frac{(3 + 2\nu)\bar{A}_m a^2 \chi_{n+1}^0}{8(1 - \nu^2)}$$

$$- \frac{\imath C_m \chi_{n+1}^1}{2(1 - \nu)} . \tag{30.32}$$

Equation (30.31) defines the solution to $\mathcal{P}_1$ for the traction-free circular cylinder in P-N form. Notice that we have chosen to describe the new polynomial terms arising from (30.30) in terms of the functions χ_n^m of equations (25.43), since this and the notation (30.32) facilitates subsequent integrations through the recurrence relation (30.19), which here implies that

$$\int \mathcal{Q}_m^n dz = \mathcal{Q}_m^{n+1} ; \quad \int \mathcal{V}_m^n dz = \mathcal{V}_m^{n+1} . \tag{30.33}$$

The stress functions (30.31), excluding the torsion term, can be used as an alternative solution to the problem of §18.4.1 and §26.3.1.

30.6 The Corrective In-plane Solution

In higher order problems $\mathcal{P}_j$, $j > 1$, we shall also need to make corrections for zeroth-order in-plane tractions T, using an appropriate form of the two-dimensional plane strain equations. If we use the complex form of the P-N solution (30.11–30.16) and define plane harmonic functions through

$$\phi = f + \bar{f} \; ; \quad \psi = g \, , \tag{30.34}$$

where f, g are holomorphic functions of ζ, we obtain

$$2\mu u = -(3 - 4\nu)g + \zeta\, \overline{g'} + 2\, \overline{f'} \; ; \quad 2\mu u_z = 0 \, .$$

This is identical to the classical form of the plane strain complex-variable solution (20.21) under the notation relation

$$g \Rightarrow -\chi \; ; \quad 2f' \Rightarrow -\theta \, . \tag{30.35}$$

We recall from §20.3.2 that we can set $\theta(0)=0$ without loss of generality, so the implied integration in obtaining f from θ does not require a constant of integration. The corresponding stress components are given by

$$\Theta = -2\left(g' + \overline{g'}\right) \; ; \quad \Phi = 2\left(\zeta\, \overline{g''} + 2\, \overline{f''}\right) \; ; \quad \sigma_{zz} = -2\nu\left(g' + \overline{g'}\right) \; ; \quad \Psi = 0 \, . \tag{30.36}$$

30.7   Corrective Solutions using Real Stress Functions

We have seen in §30.3 that the integration process is greatly facilitated by the use of the complex-variable notation. However, for many geometries, the in-plane and antiplane corrections are easier to perform using the real Airy and Prandtl stress functions of Parts II and III. In Chapter 20 we established simple relations between the real and complex formulations for two-dimensional problems. Here we shall use these relations to convert real stress function expressions to complex P-N form.

30.7.1   Airy function

If we define a corrective field by the Airy stress function φ, we can write it in the form

$$\varphi = g_1 + \bar{g}_1 + \bar{\zeta}g_2 + \zeta\, \bar{g}_2 \, ,$$

using the procedure defined in §19.4.1 or the Maple and Mathematica files 'rtocb', where g_1, g_2 are holomorphic functions. The relations (20.34, 20.35) and (30.35) then imply that the appropriate complex P-N potentials are

$$\phi = -g_1 - \bar{g}_1 \; ; \quad \psi = -2g_2 \, .$$

30.7.2 Prandtl function

If a solution of problem $\mathcal{P}_1$ is defined in terms of the Prandtl function φ, the first
stage is to determine the real constants A, B, C by substituting into equation (18.6)

$$\nabla^2 \varphi = \frac{A\nu y}{(1+\nu)} - \frac{B\nu x}{(1+\nu)} + C$$

and equating coefficients. By expressing the antiplane stress Ψ from equation (30.26)
in terms of σ_{zx}, σ_{zy} and comparing the results with (18.5), we can show that the
complex potentials ϕ_P, ψ_P of equation (30.24) are equivalent to the Prandtl function

$$\varphi_P = -\frac{B\left[3(1+2\nu)y^2 x - (1-2\nu)x^3\right]}{24(1+\nu)} + \frac{A\left[3(1+2\nu)x^2 y - (1-2\nu)y^3\right]}{24(1+\nu)}$$
$$+ \frac{C(x^2 + y^2)}{4}, \tag{30.37}$$

if the constants A_0, B_0, C_0 are replaced by

$$A_0 = \frac{A - \imath B}{2} \ ; \quad B_0 = 0 \ ; \quad C_0 = -\frac{C}{2} \ .$$

Next, we construct the real harmonic function

$$\varphi_H(x, y) = \varphi - \varphi_P \ ,$$

representing the difference between the actual Prandtl function and the particular
solution (30.37). This can be converted to complex form as discussed in §20.2.1. We
first find the holomorphic function g such that $\varphi_H = g + \bar{g}$ (for example, using the
Maple or Mathematica procedure 'rtoch') and then define

$$h = 2\imath g \ .$$

It then follows from $(20.14)_2$ that $\varphi_H = \Im(h)$ and the corresponding complex potentials ϕ_H, ψ_H are obtained by substituting h into (30.28). Finally, the complex potentials appropriate to φ are recovered as

$$\phi = \phi_P + \phi_H \ ; \quad \psi = \psi_P + \psi_H \ .$$

This procedure is contained in the Maple and Mathematica files 'PrandtltoPN'.

30.8 Solution Procedure

We are now in a position to summarize the solution procedure for the three-dimensional problem $\mathcal{P}_m$. We first differentiate the tractions T m times with respect to z in order to define the sub-problems $\mathcal{P}_j$, $j \in [1, m]$. We shall be particularly interested in the zeroth-order tractions in each sub-problem — i.e. the terms in $T^{(j)}$ that are independent of z, since these are the only terms that are active in the incremental solution.

Suppose the potentials ϕ_j, ψ_j corresponding to problem $\mathcal{P}_j$ are known. In other words, if we substitute these potentials into equations (30.11–30.16), the resulting stresses satisfy the boundary conditions on the tractions $T^{(j)}$ on the lateral surfaces in problem $\mathcal{P}_j$. This solution will contain two undetermined constants A_{j-1}, C_{j-1}, corresponding to a lateral force and a torque on the end of the bar, but we do not concern ourselves with end conditions at this stage.

To move up to the solution of problem $\mathcal{P}_{j+1}$, we proceed as follows:-

(i) Define new potentials by adding in the pure bending solution (30.20) with new constants A_j, B_j — i.e.

$$\phi = \phi_j + \mathcal{G}_j^0 \; ; \quad \psi = \psi_j + \mathcal{H}_j^0 \,. \tag{30.38}$$

(ii) Integrate ϕ, ψ with respect to z as in §30.3. Notice that any terms in these potentials of the form $\chi_n^m, \mathcal{G}_m^n, \mathcal{H}_m^n, \mathcal{Q}_m^n, \mathcal{V}_m^n$ can be integrated using the recurrence relations (30.19, 30.21, 30.33).

(iii) Add in the torsion solution (30.23) with a new constant C_j. At this stage, Θ, Φ will generally contain terms up to order $j-1$ in z and Ψ will contain terms up to order j. However, all except the zeroth order terms (those independent of z) will satisfy the boundary conditions on the lateral surfaces, since both the tractions and the stresses were obtained by one integration with respect to z from the given solution $\mathcal{P}_j$.

(iv) To satisfy the boundary conditions on the zeroth order terms in the in-plane tractions $T^{(j+1)}$, we add in the in-plane solution from §30.6 and solve an in-plane boundary-value problem, using the methods of §20.3. Alternatively, we could solve the corrective in-plane problem using the real Airy stress function of Chapters 5–13 and then use the relations in §30.7.1 to convert the resulting biharmonic function to P-N form. Solvability of this in-plane problem requires that the corrective tractions be self-equilibrated in the plane (see for example §20.5.1 and §20.6.1) and this provides a condition for determining the constants A_{j-1}, C_{j-1}.

(v) To satisfy the boundary conditions on the zeroth order term in the antiplane tractions $T_z^{(j+1)}$, we add in the antiplane corrective solution from (30.28) and determine the function $h(\zeta)$, using the procedure outlined in §30.5.1. This solution is possible only if the tractions associated with h alone are self-equilibrating and hence the solution at this stage will also determine the constant B_j.

This completes the solution of problem $\mathcal{P}_{j+1}$, except for the constants A_j, C_j. The same procedure can be adapted to give the solution of problem $\mathcal{P}_1$ by setting $j = 0$ and replacing (30.38) by (30.20).

After repeating the procedure m times, a solution will be obtained that satisfies the traction conditions T completely and which contains two free constants A_{m-1}, C_{m-1}. To complete the solution, we once again add in the zeroth order solution, as in equations (30.38), using constants A_m, B_m. Finally, we determine the constants A_{m-1}, C_{m-1}, A_m, B_m from the weak end conditions (three force and three moment resultants) at $z = 0$ in $\mathcal{P}_m$. More specifically, the conditions on axial force and torque determine the real constants B_m, C_{m-1}, respectively, the conditions on lateral forces determine A_{m-1} (which is a complex constant and hence has two degrees of freedom) and the conditions on bending moments determine A_m.

30.9 Example

The solution procedure described involves a sequence of essentially elementary steps, but for a problem of significant order the number of such steps is large and the solution can become extremely complicated. For this reason, in most cases it is only really practical to undertake the solution using Maple or Mathematica. However, to illustrate the procedure we shall give the solution of a suitably low-order problem.

We consider the circular bar of radius a defined by $0 \le r < a$, $0 < z < L$, in which the end $z = 0$ is traction free and the curved surfaces are subjected to the tractions

$$T = 0 \; ; \quad T_z = Sz \cos \theta ,$$

where S is a real constant. It is clear that the loading is symmetrical about the plane $\theta = 0$, π ($y = 0$) and hence that the bar will not be subject to torsion, so the constants C_j in §30.8(iii) will all be zero and can be omitted.

Comparison with (30.1) shows that this is a problem of class $\mathcal{P}_2$. The sub-problem $\mathcal{P}_1$ corresponds to the tractions

$$T = 0 \; ; \quad T_z = S \cos \theta . \tag{30.39}$$

30.9.1 Problem $\mathcal{P}_1$

The solution of problem $\mathcal{P}_0$ is defined by the functions ϕ_0, ψ_0 of equation (30.20). Integrating with respect to z as in §30.5, but omitting the torsion term, we obtain

$$\phi = \mathcal{G}_0^1 \; ; \quad \psi = \mathcal{H}_0^1 \,. \tag{30.40}$$

There are no in-plane stresses or tractions in problem $\mathcal{P}_1$, so no in-plane correction is required.

Antiplane correction

The antiplane stress components Ψ are then obtained as

$$\Psi = -\frac{A_0(1+2\nu)\zeta^2}{4(1+\nu)} - \frac{\bar{A}_0\zeta\bar{\zeta}}{2(1+\nu)} - \frac{B_0\zeta}{2} \,,$$

from (30.16). The antiplane traction T_z is obtained exactly as in §30.5.2 as

$$T_z = \sigma_{zr}(a,\theta) = -\frac{(3+2\nu)a^2\,(\Re\{A_0\}\cos\theta - \Im\{A_0\}\sin\theta)}{4(1+\nu)} - \frac{B_0 a}{2} + \frac{1}{a}\frac{\partial h_y}{\partial\theta}$$

and equating this to the boundary value (30.39), we obtain

$$\frac{\partial h_y}{\partial\theta} = \frac{(3+2\nu)a^3\,(\Re\{A_0\}\cos\theta - \Im\{A_0\}\sin\theta)}{4(1+\nu)} + \frac{B_0 a^2}{2} + Sa\cos\theta \,.$$

This integrates to a single-valued function of θ if and only if $B_0 = 0$, giving

$$h_y = \frac{(3+2\nu)a^3\,(\Re\{A_0\}\sin\theta + \Im\{A_0\}\cos\theta)}{4(1+\nu)} + Sa\sin\theta$$

and the holomorphic function h whose imaginary part takes this boundary value is

$$h = \frac{(3+2\nu)A_0 a^2\zeta}{4(1+\nu)} + S\zeta \,.$$

Substituting this expression into (30.28) and adding the results to (30.40), we obtain

$$\phi_1 = \mathcal{Q}_0^0 + \frac{(3-4\nu)S\left(\chi_1^1 + \bar{\chi}_1^1\right)}{4(1-\nu)} \; ; \quad \psi_1 = \mathcal{V}_0^0 + \frac{S\chi_1^0}{2(1-\nu)} \,,$$

which defines the solution to $\mathcal{P}_1$ in P-N form.

30.9.2   Problem $\mathcal{P}_2$

The next stage is to substitute ϕ_1, ψ_1 into (30.38) with $j = 1$ and perform a further integration with respect to z, using equations (30.19, 30.21, 30.33) to ensure that the potentials ϕ, ψ are three-dimensionally harmonic. We obtain

$$\phi = \mathcal{Q}_0^1 + \frac{(3-4\nu)S\left(\chi_2^1 + \overline{\chi}_2^1\right)}{4(1-\nu)} + \mathcal{G}_1^1 \ ; \quad \psi = \mathcal{V}_0^1 + \frac{S\chi_2^0}{2(1-\nu)} + \mathcal{H}_1^1 \ . \tag{30.41}$$

In-plane correction

There are no z-independent components in the in-plane tractions T for problem $\mathcal{P}_2$ and we anticipate requiring a corrective solution to satisfy this condition. Substituting the partial solution (30.41) into equations (30.13, 30.14), we obtain the in-plane stress components

$$\Theta = \frac{\left[2(1-\nu)\zeta\overline{\zeta} - (3+2\nu)(3-4\nu)a^2\right](A_0\zeta + \overline{A}_0\overline{\zeta})}{16(1-\nu^2)} - \frac{S(3-4\nu)(\zeta + \overline{\zeta})}{4(1-\nu)}$$

$$\Phi = \frac{(1+4\nu)A_0\zeta^3}{24(1+\nu)} + \frac{\left[2(1-\nu)\zeta\overline{\zeta} - (3+2\nu)a^2\right]\overline{A}_0\zeta}{16(1-\nu^2)} - \frac{S\zeta}{4(1-\nu)} \ . \tag{30.42}$$

From equation (20.37), we have

$$T ds = \frac{\iota}{2}\left(\Phi d\overline{\zeta} - \Theta d\zeta\right)$$

and on $r = a$,

$$\zeta = ae^{\iota\theta} \ ; \quad \overline{\zeta} = ae^{-\iota\theta} \ ; \quad d\zeta = \iota ae^{\iota\theta}d\theta \ ; \quad d\overline{\zeta} = -\iota ae^{-\iota\theta}d\theta \ , \tag{30.43}$$

so the integral around a portion of the boundary of the complex traction T is

$$\int T ds = \frac{a}{2}\int \left(\Phi e^{-\iota\theta} + \Theta e^{\iota\theta}\right)d\theta \ . \tag{30.44}$$

Substituting the boundary values (30.43) in (30.42), we obtain

$$\Phi e^{-\iota\theta} + \Theta e^{\iota\theta} = -\frac{(19 - 18\nu - 16\nu^2)A_0a^3e^{2\iota\theta}}{48(1-\nu^2)} - \frac{Sa(3-4\nu)e^{2\iota\theta}}{4(1-\nu)}$$
$$- \frac{\overline{A}_0a^3}{2} - Sa \ .$$

The in-plane corrective problem is solvable if and only if the tractions on the boundary are self-equilibrating, which requires that the integral (30.44) return a single-valued function of θ. This condition requires that

$$-\frac{\overline{A}_0a^3}{2} - Sa = 0 \quad \text{and hence} \quad \overline{A}_0 = -\frac{2S}{a^2} = A_0 \ . \tag{30.45}$$

Notice that since the constant $\bar{A}_0$ is real it is also equal to A_0. Using this result, we then obtain

$$\Phi e^{-\iota\theta} + \Theta e^{\iota\theta} = \frac{Sa(1 - 12\nu + 8\nu^2)e^{2\iota\theta}}{24(1 - \nu^2)} . \qquad (30.46)$$

To satisfy the homogeneous in-plane boundary conditions on Γ in $\mathcal{P}_2$, we superpose holomorphic functions f, g through equations (30.34, 30.36) and choose them so as to reduce (30.46) to zero for all θ. This can be done using the direct method of §20.6.1, but the form of (30.46) makes it clear that low order polynomial terms in f, g will give an expression of the appropriate form. Choosing

$$f = 0 \; ; \quad g = C\zeta^2 = C\chi_0^2 \quad \text{and hence} \quad \phi = 0 \; ; \quad \psi = C\chi_0^2 \qquad (30.47)$$

from (30.34), where C is a real constant, we obtain the additional stress components

$$\Theta = -4C(\zeta + \bar{\zeta}) \; ; \quad \Phi = 4C\zeta \, ,$$

from (30.36) and on the surface $r = a$,

$$\Phi e^{-\iota\theta} + \Theta e^{\iota\theta} = -4Cae^{2\iota\theta} \, .$$

The boundary condition $T = 0$ will therefore be satisfied for all θ if

$$C = \frac{S(1 - 12\nu + 8\nu^2)}{96(1 - \nu^2)} .$$

Antiplane correction

Since there are no zeroth-order antiplane tractions in $\mathcal{P}_2$, the antiplane correction is identical to that in §30.5.2. In particular, we have $B_1 = 0$ and the additional potentials are derived from (30.30) with A_0 replaced by A_1.

Applying this and the in-plane correction (30.47) to (30.41) and using (30.45) to eliminate A_0, we then have

$$\phi_2 = -\frac{S(1 - 2\nu)}{a^2(1 - \nu^2)}\left(\chi_4^1 + \overline{\chi}_4^1\right) - \frac{S(3 - 4\nu)}{8(1 - \nu^2)}\left(\chi_2^1 + \overline{\chi}_2^1\right) + \mathcal{Q}_1^0 \qquad (30.48)$$

$$\psi_2 = -\frac{2S\chi_4^0}{a^2(1 - \nu^2)} + \frac{S(1 - 2\nu)\chi_2^2}{4a^2(1 - \nu^2)} - \frac{S\chi_2^0}{4(1 - \nu^2)}$$

$$+ \frac{S(1 - 12\nu + 8\nu^2)\chi_0^2}{96(1 - \nu^2)} + \mathcal{V}_1^0 \, , \qquad (30.49)$$

which completes the solution of problem $\mathcal{P}_2$.

30.9.3 End conditions

It remains to satisfy the weak traction-free conditions on the end $z=0$. We add the pure bending solution (30.20) with new constants A_2, B_2 into (30.48, 30.49), which is achieved by adding the terms $\mathcal{G}_2^0$, $\mathcal{H}_2^0$ into ϕ, ψ respectively. We then substitute the resulting potentials into (30.15, 30.16) to obtain the complex stress components σ_{zz}, Ψ. In particular, the tractions on the end of the bar $z=0$ are obtained as

$$\sigma_{zz}(0, \zeta, \bar{\zeta}) = \frac{S(2+\nu)\zeta\bar{\zeta}(\zeta+\bar{\zeta})}{4a^2(1+\nu)} - \frac{S(3-2\nu)(\zeta+\bar{\zeta})}{6} + A_2\zeta + \bar{A}_2\bar{\zeta} + B_2$$

$$\Psi(0, \zeta, \bar{\zeta}) = \frac{\bar{A}_1[(3+2\nu)a^2 - 2\zeta\bar{\zeta}] - A_1(1+2\nu)\zeta^2}{4(1+\nu)}.$$

Clearly the shear tractions $\Psi = \sigma_{zx} + \iota\sigma_{zy}$ on $z = 0$ can be set to zero in the strong sense by choosing $A_1 = 0$. For the normal tractions, we use (30.3) to calculate the force and moment resultants

$$F_z = \int_0^a \int_0^{2\pi} \sigma_{zz}(0, r, \theta)r \, d\theta dr = \pi B_2 a^2$$

$$M_x + \iota M_y = \int_0^a \int_0^{2\pi} \sigma_{zz}(0, r, \theta)r^2 e^{\iota\theta} dr d\theta = -\frac{\pi S(1-2\nu^2)a^4}{12(1+\nu)} + \frac{\pi \bar{A}_2 a^4}{2}.$$

Thus, the weak traction-free conditions are satisfied by setting

$$A_2 = \bar{A}_2 = \frac{S(1-2\nu^2)}{6(1+\nu)} \quad ; \quad B_2 = 0.$$

Substituting these constants into the potentials and simplifying, we obtain

$$\phi = -\frac{S(1-2\nu)}{a^2(1-\nu^2)}\left(\chi_4^1 + \bar{\chi}_4^1\right) - \frac{S(7+\nu-8\nu^2-8\nu^3)}{24(1-\nu^2)(1+\nu)}\left(\chi_2^1 + \bar{\chi}_2^1\right)$$

$$\psi = -\frac{2S\chi_4^0}{a^2(1-\nu^2)} + \frac{S(1-2\nu)\chi_2^2}{4a^2(1-\nu^2)} - \frac{S(1+3\nu+4\nu^2)\chi_2^0}{12(1-\nu^2)(1+\nu)} - \frac{S(1+7\nu)\chi_0^2}{96(1-\nu^2)(1+\nu)},$$

which defines the complete solution of the problem. The complex stress components can be recovered by substitution into equations (30.13–30.16) as

$$\Theta = \frac{S(\zeta+\bar{\zeta})}{3} - \frac{S\zeta\bar{\zeta}(\zeta+\bar{\zeta})}{4(1+\nu)a^2}$$

$$\Phi = \frac{S\zeta[2a^2 - 3\zeta\bar{\zeta} - (1+4\nu)\zeta^2 + 2(1-2\nu)a^2]}{12(1+\nu)a^2}$$

$$\sigma_{zz} = \frac{S(2+\nu)(3\zeta\bar{\zeta} - 2a^2)(\zeta + \bar{\zeta})}{12(1+\nu)a^2} - \frac{Sz^2(\zeta + \bar{\zeta})}{a^2}$$

$$\Psi = \frac{Sz\left[(1+2\nu)\zeta^2 + 2\zeta\bar{\zeta} - a^2\right]}{2(1+\nu)a^2}.$$

We have deliberately given a very detailed and hence lengthy solution of this example problem in order to clarify the steps involved, but the astute reader will realize that many steps might reasonably have been omitted as trivial. For example, the original problem is even in z, which immediately allows us to omit the terms involving the constants A_1, $\bar{A}_1$, B_1, causing the in-plane correction to be trivial in problems $\mathcal{P}_j$ where j is odd and the antiplane correction to be trivial when j is even.

Problems

30.1. A circular cylindrical bar of radius a is loaded by the tangential tractions $\sigma_{r\theta} = S\cos\theta$ on $r = a$. The end of the bar $z = 0$ is traction free. Find the complete stress field in the bar.

30.2. A circular cylindrical bar of radius a is loaded by a uniform radially-inward load F_0 per unit length along the line $\theta = 0$. The end of the bar $z = 0$ is traction free. Find the complete stress field in the bar.

30.3. A circular cylindrical bar of radius a is loaded by a uniform tangential load F_0 per unit length along the line $\theta = 0$. The load acts in the positive θ-direction and the end of the bar $z = 0$ is traction free. Find the complete stress field in the bar.

30.4. The lateral surfaces Γ of a bar of equilateral trianglular cross section are defined by the equations $x = \pm\sqrt{3}y$, $x = a$. The bar is twisted by in-plane tractions T on Γ that are independent of z, the end $z = 0$ being traction free. The tractions on the surface $x = a$ are defined by

$$\sigma_{xx} = 0 \; ; \quad \sigma_{xy} = S$$

and those on the other two surfaces are similar under a rotation of $\pm 120°$.

This problem lies in class $\mathcal{P}_2$. Use the strategy suggested in Problem 17.3 to find a Prandtl stress function solution of the corresponding problem $\mathcal{P}_1$, convert it to P-N form as in §30.7.2 and then integrate with respect to z to obtain a particular solution of $\mathcal{P}_2$. Determine the tractions on the surface $x = a$ implied by this solution and discuss possible strategies for completing the in-plane correction, but do not attempt it.

30.5. The rectangular bar $-a < x < a$, $-b < y < b$, $0 < z < L$ is built in at $z = L$ and loaded only by the uniform shear tractions

$$\sigma_{xy}(a, y, z) = S \; ; \quad \sigma_{xy}(-a, y, z) = -S$$

on the surfaces $x = \pm a$. Find the complete stress field in the bar using weak conditions on the end $z = 0$. Assume that $b \gg a$ and use weak boundary conditions on the edges $y = \pm b$.

Hint: The problem is even in x and odd in y, which implies that $\Re(A_0) = C_0 = 0$. The easiest way to impose these symmetries on the rest of the solution is to use an Airy function for the in-plane corrective solution.

30.6. A bar of square cross section $-a < x < a$, $-a < y < a$, $0 < z < L$ is built in at $z = L$ and loaded only by the uniform tractions

$$\sigma_{yy}(x, a, z) = -p_0$$

on the surface $y = a$. Find the complete stress field in the bar using only the closed-form terms from the solution of §18.4.2 for the problem $\mathcal{P}_1$ and an appropriate finite polynomial as an Airy stress function for the in-plane correction in $\mathcal{P}_2$. Plot appropriate stress components to estimate the errors in the elementary bending solution.

Chapter 31
Frictionless Contact

As we noted in §22.5.1, Green and Zerna's Solution F is ideally suited to the solution of frictionless contact problems for the half-space $z > 0$, since it identically satisfies the condition that the shear tractions be zero at the surface $z=0$. In fact the *surface tractions* for this solution take the form

$$\sigma_{zz} = -\frac{\partial^2 \varphi}{\partial z^2} \ ; \ \ \sigma_{zx} = \sigma_{zy} = 0 \ ; \ \ z=0 , \tag{31.1}$$

whilst the surface displacements are

$$u_x = \frac{(1-2\nu)}{2\mu}\frac{\partial \varphi}{\partial x} \ ; \ \ u_y = \frac{(1-2\nu)}{2\mu}\frac{\partial \varphi}{\partial y} \ ; \ \ u_z = -\frac{(1-\nu)}{\mu}\frac{\partial \varphi}{\partial z} \ ; \ \ z=0 , \tag{31.2}$$

from Table 22.3.

31.1 Boundary Conditions

Following §12.5, we can now formulate the general problem of indentation of the half-space by a frictionless rigid punch whose profile is described by a function $u_0(x, y)$ (see Figure 12.5). Notice that since the problem is three-dimensional, the punch profile is now a function of two variables x, y and the contact area will be an extended region of the plane $z=0$, which we denote by $\mathcal{A}$.

Within the contact area, the normal displacement of the half-space must conform to the shape of the punch and hence

$$u_z = u_0(x, y) + C_0 + C_1 x + C_2 y \ ; \ \ (x, y) \in \mathcal{A} , \tag{31.3}$$

where the constants C_0, C_1, C_2 define an arbitrary rigid-body displacement of the punch. Outside the contact area, there must be no normal tractions. i.e.

© The Author(s), under exclusive license to Springer Nature Switzerland AG 2022
J. R. Barber, *Elasticity*, Solid Mechanics and Its Applications 172,
https://doi.org/10.1007/978-3-031-15214-6_31

$$\sigma_{zz} = 0 ; \quad (x, y) \in \bar{\mathcal{A}} ,$$

where we denote the complement of $\mathcal{A}$ — i.e. that part of the surface $z=0$ which is *not* in contact — by $\bar{\mathcal{A}}$.

We now use equations (31.1, 31.2) to write these conditions in terms of the potential function φ with the result

$$\frac{\partial \varphi}{\partial z} = -\frac{\mu}{(1 - \nu)} u_0(x, y) ; \quad (x, y) \in \mathcal{A} \tag{31.4}$$

$$\frac{\partial^2 \varphi}{\partial z^2} = 0 ; \qquad\qquad (x, y) \in \bar{\mathcal{A}} , \tag{31.5}$$

where for brevity we have omitted the rigid-body displacement terms in equation (31.3), since these can easily be reintroduced as required or subsumed under the function $u_0(x, y)$.

31.1.1 Mixed boundary-value problems

Equations (31.4, 31.5) define a *mixed boundary-value problem* for the potential function φ in the region $z > 0$. The word 'mixed' here refers to the fact that different derivatives of the function are specified at different parts of the boundary. More specifically, it is a *two-part* mixed boundary-value problem[1], since the boundary conditions are specified over two complementary regions of the boundary, $\mathcal{A}$, $\bar{\mathcal{A}}$.

Similar mixed boundary-value problems arise in many fields of engineering mechanics, such as heat conduction, electrostatics and fluid mechanics. For example, in heat conduction, a mathematically similar problem arises if the region $\mathcal{A}$ is raised to a prescribed temperature whilst $\bar{\mathcal{A}}$ is insulated. They can be reduced to integral equations using a Green's function formulation. Thus, considering the normal traction $\sigma_{zz}(x, y, 0) \equiv -p(x, y)$ in $\mathcal{A}$ as comprising a set of point forces $F = p(x, y)dxdy$ and using equation (24.6) for the normal surface displacement due to a concentrated normal force at the surface, we can write

$$u_z(x, y, 0) = \frac{(1 - \nu)}{2\pi\mu} \iint_{\mathcal{A}} \frac{p(\xi, \eta)d\xi d\eta}{r(x, y, \xi, \eta)} ,$$

where

$$r(x, y, \xi, \eta) = \sqrt{(x - \xi)^2 + (y - \eta)^2}$$

---

[1] Strictly this terminology is restricted to cases where the two regions $\mathcal{A}$, $\bar{\mathcal{A}}$ are connected. A special case where this condition is not satisfied is the indentation by an annular punch for which $\bar{\mathcal{A}}$ has two unconnected regions — one inside the annulus and one outside. This would be referred to as a three-part problem.

is the distance between the points $(\xi, \eta, 0)$ and $(x, y, 0)$. Substitution in (31.3) then yields a double integral equation for the unknown contact pressure $p(x, y)$. This is of course a generalization of the method used in Chapter 12. However, in three-dimensional problems, the resulting double integral equation is generally of rather intractable form. We shall find that other methods of solution are more efficient (see Chapter 32 below).

The fact that Solution F is described in terms of a single harmonic function enables us to use results from potential theory to prove some interesting results. For example, we note that the derivative $\partial\varphi/\partial z$ must also be harmonic in the half-space $z > 0$. Suppose we wish to locate the point where $\partial\varphi/\partial z$ is a maximum. It cannot be inside the region $z > 0$, since at a such a maximum we would need

$$\nabla^2 \frac{\partial\varphi}{\partial z} < 0$$

and the left-hand side of this expression is zero everywhere. It follows that the maximum must occur somewhere on the boundary $z = 0$ or else at infinity. Furthermore, if it occurs on the boundary, all immediately adjacent points must have lower values and in particular, we must have

$$\frac{\partial^2\varphi}{\partial z^2} < 0$$

at the maximum point.

Recalling equations (31.1, 31.2), we conclude that the maximum value of the surface displacement u_z must occur either at infinity or at a point where the contact traction σ_{zz} is compressive. If we choose a frame of reference such that the displacement at infinity is zero, it follows that the maximum can only occur there if the surface displacement is negative throughout the finite domain, which can only arise if the net force on the half-space is tensile. Thus, when the half-space is loaded by a net compressive force, the maximum surface displacement must occur in a region where the surface traction is compressive, and it must be positive.

31.2 Determining the Contact Area

In many contact problems, the contact area is not known *a priori*, but has to be determined as part of the solution[2]. As in Chapter 12, the contact area has to be chosen so as to satisfy the two *inequalities*

$$\sigma_{zz} \leq 0 ; \qquad\qquad\qquad (x, y) \in \mathcal{A} \qquad\qquad (31.6)$$
$$u_z > u_0(x, y) + C_0 + C_1 x + C_2 y ; \quad (x, y) \in \bar{\mathcal{A}} , \qquad (31.7)$$

---

[2] See for example §12.5.3.

which state respectively that the contact traction should never be tensile and that
the gap between the indenter and the half-space should always be positive. It is
worth noting that physical considerations demand that these conditions be satisfied
in *all* problems — not only those in which the contact area is initially unknown.
However, if the indenter is rigid and has sharp corners, the contact area will usually
be identical with the plan-form of the indenter and conditions (31.6, 31.7) need not be
explicitly imposed. Notice however that if there is also distortion due to thermoelastic
effects or if there is a sufficiently large rigid-body rotation as in Problem 12.5, the
contact area may not be coextensive with the plan-form of the indenter. In general,
conditions (31.6, 31.7) define a *solution space* in which the mathematical solution
will be physically meaningful. If any further restrictions are imposed in the interests
of simplicity, it is always wise to check the solution at the end to ensure that the
inequalities are satisfied.

In treating the Hertzian problem in §12.5.3, we were able to satisfy the inequalities
by choosing the contact area in such a way that the contact traction tended to zero at
the edge. A similar technique can be applied in three-dimensional problems provided
that the profile of the indenter is smooth, but now we have to determine the equation
of a boundary line in the plane $z = 0$, rather than two points defining the ends of a
strip. This is generally a formidable problem and closed-form solutions are known
only for the case where the contact area is either an ellipse or a circle.

Boundedness of the contact traction at the edge of the contact area is a necessary
but not sufficient condition for the inequalities to be satisfied. Suppose for example
the indenter has concave regions as shown in Figure 31.1. For some values of the
load, there will be contact in two discrete regions near to A and B. However, we
could construct a mathematical solution assuming contact near B only, in which the
traction was bounded throughout the edge of the contact area and in which there
would be interpenetration (i.e. violation of inequality (31.7)) near A. In general,
the boundedness condition may be insufficient to determine a unique solution if the
physically correct contact area is not simply connected.

In a numerical solution, this difficulty can be avoided by using the inequalities
directly to determine the contact area by iteration. A simple approach is to guess
the extent of the contact area $\mathcal{A}$, solve the resulting contact problem numerically
and then examine the solution to determine whether the inequalities (31.6, 31.7) are
violated at any point. Any points in $\mathcal{A}$ at which the contact traction is tensile are then
excluded from $\mathcal{A}$ in the next iteration, whereas any points in $\bar{\mathcal{A}}$ at which the gap is
negative are included in $\mathcal{A}$.

This method is found to converge rapidly, but it involves the calculation of tractions
or displacements at all surface nodes. An alternative method is to formulate a function
which is minimized when the solution satisfies the inequalities and then treat the

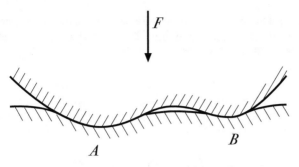

Fig. 31.1 Contact problem which will have two unconnected contact areas in an intermediate load range.

iterative process as an optimization problem. Kalker[3] and Kikuchi and Oden[4] have explored this technique at length and used it to develop finite element solutions to the contact problem.

A simpler method is available for the special problem of the indentation of an elastic half-space by a rigid indenter. Suppose we select a value at random for the contact area $\mathcal{A}$, solve the resulting indentation problem, and then determine the corresponding total indentation force

$$F(\mathcal{A}) = - \iint_{\mathcal{A}} \sigma_{zz}(x, y, 0) dx dy \ ,$$

where σ_{zz} is the contact traction associated with the 'wrong' contact area, $\mathcal{A}$. It can be shown that the maximum value of $F(\mathcal{A})$ occurs when $\mathcal{A}$ is chosen such that the inequalities (31.6, 31.7) are satisfied — i.e. when $\mathcal{A}$ takes its correct value. Thus, we can formulate the iteration process in terms of an optimization problem for $F(\mathcal{A})$. The proof of the theorem is as follows:-

Proof

Consider the effect of increasing $\mathcal{A}$ by a small element $\delta \mathcal{A}$.

If $F(\mathcal{A})$ is thereby increased, the corresponding *differential* contact traction distribution — i.e. the difference between the final and the initial traction — amounts to a net compressive force on the half-space. Now we proved in §31.1.1 that in such cases, the maximum surface displacement must occur in a region where the traction is compressive and be positive in sign. Thus, the maximum differential surface displacement δu_z must occur somewhere in $\mathcal{A}$ or $\delta \mathcal{A}$ and be positive in sign. However,

[3] J. J. Kalker (1977), Variational principles of contact elastostatics, *Journal of the Institute for Mathematics and its Applications,* Vol. 20, pp. 199–221.
[4] N. Kikuchi and J. T. Oden, *Contact Problems in Elasticity: A Study of Variational Inequalities and Finite Element Methods,* SIAM, Philadelphia, 1988.

δu_z is zero throughout $\mathcal{A}$, since the process of including a further region in the contact area does not affect the displacement boundary condition in $\mathcal{A}$. It therefore follows that the differential displacement in $\delta\mathcal{A}$ is positive and that the contact traction there is compressive.

We know that the final value of u_z in $\delta\mathcal{A}$ is $f(x, y) + C_0 + C_1 x + C_2 y$. The effect of the differential contact traction distribution has therefore been shown to be an *increase* of u_z at $\delta\mathcal{A}$ to the value $f(x, y) + C_0 + C_1 x + C_2 y$ and it follows that $\delta\mathcal{A}$ must be an area for which initially

$$u_z < f(x, y) + C_0 + C_1 x + C_2 y \tag{31.8}$$

— i.e. for which the gap is negative.

Hence, an increase in $F(\mathcal{A})$ can only be achieved by increasing $\mathcal{A}$ if there are some areas $\delta\mathcal{A}$ outside $\mathcal{A}$ satisfying (31.8). By a similar argument, we can show that an increase in $F(\mathcal{A})$ can follow from a *reduction* in $\mathcal{A}$ only if there exist areas $\delta\mathcal{A}$ *inside* $\mathcal{A}$ for which

$$\sigma_{zz} > 0 \tag{31.9}$$

— i.e. for which the contact traction is tensile.

It follows that the maximum value of $F(\mathcal{A})$ must occur when $\mathcal{A}$ is chosen so that there are no regions satisfying (31.8, 31.9) — i.e. when the original inequalities (31.6, 31.7) are satisfied everywhere[5].

The above method has the formal advantage that it replaces the intractable inequality conditions by a variational statement, but in order to use it we need to have a way of determining the total load $F(\mathcal{A})$ on the indenter for an arbitrary contact area[6] $\mathcal{A}$. We shall show in Chapter 38 how the reciprocal theorem can be used to simplify this problem[7].

31.3 Contact Problems Involving Adhesive Forces

At very small length scales, the attractive forces between molecules (e.g. van der Waals forces) become significant and must be taken into account in the solution of contact problems. This subject has increased in importance in recent years with the emphasis on micro- and nano-scale systems. One approach is to apply the same arguments as in the Griffith theory of brittle fracture (§13.3.1) and postulate that the

---

[5] For more details of this argument see J. R. Barber (1974), Determining the contact area in elastic contact problems, *Journal of Strain Analysis,* Vol. 9, pp. 230–232.

[6] Even then, the variational problem is far from trivial.

[7] See also R. T. Shield (1967), Load-displacement relations for elastic bodies, *Zeitschrift für angewandte Mathematik und Physik,* Vol. 18, pp. 682–693.

contact area will adopt the value for which the total energy (comprising elastic strain energy, potential energy of external forces and surface energy) is at a minimum. Perturbing about this minimum energy state, we then obtain

$$G = \Delta\gamma \,, \tag{31.10}$$

where G is the energy release rate introduced in §13.3.3 and $\Delta\gamma$ is the *interface energy* of the two contacting materials — i.e. the work that must be done per unit area against interatomic forces at the interface to separate two bodies with atomically plane surfaces[8]. It follows from equation (13.26) that the contact tractions will exhibit a tensile square-root singularity at the edge of the contact area characterized by a stress-intensity factor

$$K_I = \lim_{s \to 0} \sigma_{zz}(s)\sqrt{2\pi s} \,, \tag{31.11}$$

where $\sigma_{zz}(s)$ is the normal contact traction (tensile positive) at a distance s from the edge of the contact area. For the more general case of frictionless contact between dissimilar materials, equation (13.26) is modified to

$$G = \frac{K_I^2}{16}\left(\frac{\kappa_1+1}{\mu_1} + \frac{\kappa_2+1}{\mu_2}\right)$$

and for three-dimensional problems, the local conditions at the edge of the contact are always those of plane strain ($\kappa = 3-4\nu$), so (31.10) gives

$$K_I = \sqrt{2E^*\Delta\gamma} \,, \tag{31.12}$$

where E^* is the composite contact modulus defined such that

$$\frac{1}{E^*} = \frac{1-\nu_1}{2\mu_1} + \frac{1-\nu_2}{2\mu_2} \,. \tag{31.13}$$

This approach to adhesive contact problems was first introduced by Johnson, Kendall and Roberts[9] and has come to be known as the JKR theory. Notice that it predicts that the stress-intensity factor is constant around the edge of the contact region, regardless of its shape. Of course, in the special case where adhesive forces and hence surface energy are neglected, this stress-intensity factor will be zero and the theory reduces to the classical condition that the contact tractions are square-root bounded at the edge of the contact area.

The JKR theory is approximate in that it neglects the attractive forces between the surfaces in the region where the gap is small but non-zero. More exact numerical solutions using realistic force separation laws show that the theory provides a good

---

[8] This will generally be less than the sum of the surface energies $\gamma_1 + \gamma_2$ of the two contacting materials, except in the case of similar materials.

[9] K. L. Johnson, K. Kendall and A. D. Roberts (1971), Surface energy and the contact of elastic solids, *Proceedings of the Royal Society of London*, Vol. A324, pp. 301–313.

limiting solution in the range where the dimensionless parameter

$$\Psi \equiv \left[\frac{R}{\varepsilon^3} \left(\frac{\Delta\gamma}{E^*} \right)^2 \right]^{1/3} \gg 1 \,,$$

where R is a representative radius of the contacting surfaces and ε is the equilibrium intermolecular distance[10].

An alternative criterion, analogous to the small-scale yielding criterion for elastic-plastic fracture (See §13.3.1), is that the width

$$s_0 = \frac{E^* \Delta\gamma}{\pi \sigma_0^2}$$

of the region in which the tensile traction is predicted to exceed the theoretical strength σ_0 should be small compared with the linear dimensions of the contact area.

Problems

31.1. Using the result of Problem 24.4 or otherwise, show that if the punch profile $u_0(x, y)$ has the form

$$u_0 = r^p f(\theta)$$

where $p > 0$ and $f(\theta)$ is any function of θ, the resulting frictionless contact problems at different indentation forces will be self-similar[11].

Show also that if l is a representative dimension of the contact area, the indentation force F and the rigid-body indentation d will vary according to

$$F \sim l^{p+1} \;;\; d \sim l^p \,,$$

and hence that the indentation has a stiffening load-displacement relation

$$F \sim d^{1+\frac{1}{p}} \,.$$

Verify that the Hertzian contact relations (§32.2.5 below) agree with this result.

---

[10] J. A. Greenwood (1997), Adhesion of elastic spheres, *Proceedings of the Royal Society of London,* Vol. A453, pp. 1277–1297.

[11] D. A. Spence (1973), An eigenvalue problem for elastic contact with finite friction, *Proceedings of the Cambridge Philosophical Society,* Vol. 73, pp. 249–268, has shown that this argument also extends to problems with Coulomb friction at the interface, in which case, the zones of stick and slip also remain self-similar with monotonically increasing indentation force.

31.2. A frictionless rigid body is pressed into an elastic body of shear modulus μ and Poisson's ratio ν by a normal force F, establishing a contact area $\mathcal{A}$ and causing the rigid body to move a distance δ. Use the results of §31.1 to define the boundary-value problem for the potential function φ corresponding to the *incremental problem*, in which an infinitesimal additional force increment ΔF produces an incremental rigid-body displacement $\Delta\delta$.

Suppose that the rigid body is a perfect electrical conductor at potential V_0 and the 'potential at infinity' in the elastic body is maintained at zero. Define the boundary-value problem for the potential V in the elastic body, noting that the current density vector i is given by Ohm's law

$$i = -\frac{1}{\rho}\nabla V,$$

where ρ is the electrical resistivity of the material. Include in your statement an expression for the total current I transmitted through the contact interface.

By comparing the two potential problems or otherwise, show that the electrical contact resistance $R = V_0/I$ is related to the incremental contact stiffness $dF/d\delta$ by the equation

$$\frac{1}{R} = \frac{(1-\nu)}{\rho\mu}\frac{dF}{d\delta}.$$

31.3. Use equations (19.8, 19.10) to show that the function

$$f(x, z) = \Im\left\{\arcsin\left(\frac{x + \imath z}{a}\right)\right\}$$

satisfies Laplace's equation. Show that the function

$$\frac{\partial\varphi}{\partial z} = A + Bf(x, z)$$

can be used to solve the plane strain equivalent of the flat punch problem of §12.5.2 with suitable choices for the constants A, B. In particular, obtain expressions for the contact pressure distribution and for the normal surface displacement u_z outside the contact area.

Note: For complex arguments, we can write[12]

$$\arcsin(\zeta) = \frac{1}{\imath}\ln(\imath\zeta + \sqrt{1 - \zeta^2}).$$

31.4. Show that for frictionless contact problems for the half-space $z > 0$, the normal stress components near the surface $z = 0$ satisfy the relation

---

[12] See I. S. Gradshteyn and I. M. Ryzhik, *Tables of Integrals, Series and Products*, Academic Press, New York, 1980, §1.622.1.

$$\sigma_{xx} + \sigma_{yy} = (1 + 2\nu)\sigma_{zz} \,,$$

where σ_{zz} is the applied traction.

31.5. Using results from §24.2.1, find the curvature

$$\frac{\partial^2 u_z}{\partial x^2} + \frac{\partial^2 u_z}{\partial z^2}$$

of the surface of the half-space $z > 0$ due to the concentrated normal force F in Figure 24.2.

Using this result or otherwise, show that if a smooth frictionless indenting body is everywhere convex, meaning

$$\frac{\partial^2 u_0}{\partial x^2} + \frac{\partial^2 u_0}{\partial y^2} < 0 \quad \text{all } (x, y),$$

then the contact area must be simply connected for all applied normal forces.

31.6. Figure 31.2 shows a rigid punch with two parallel plane faces A_1, A_2, pressed into an elastic half-space by a force F sufficient to cause all points in both of these areas to to be in contact. The punch is constrained to move vertically. Show that the work done by F during loading is least when the two areas have the same height [i.e. when they lie in the same plane]. Hence or otherwise show that of all punches of given planform $\mathcal{A}$ [convex or concave], the work done loading to a given force F [sufficient to establish full contact] is least when the punch is flat.

Hint: It might help to consider the electrostatic problem corresponding to the same mixed boundary-value problem.

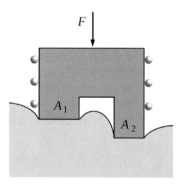

Fig. 31.2 A rigid punch with
two plane faces.

Chapter 32
The Boundary-value Problem

The simplest frictionless contact problem of the class defined by equations (31.3–31.5) is that in which the contact area $\mathcal{A}$ is the circle $0 < r < a$ and the indenter is axisymmetric, in which case we have to determine a harmonic function $\varphi(r, z)$ to satisfy the mixed boundary conditions

$$\frac{\partial \varphi}{\partial z} = -\frac{\mu}{(1-\nu)} u_0(r) \; ; \quad 0 \le r < a \, , \, z = 0 \tag{32.1}$$

$$\frac{\partial^2 \varphi}{\partial z^2} = 0 \; ; \qquad\qquad r > a \, , \, z = 0 \, . \tag{32.2}$$

32.1 Hankel Transform Methods

This a classical problem and many solution methods have been proposed. The most popular is the Hankel transform method developed by Sneddon[1] and discussed in the application to contact problems by Gladwell[2]. The function

$$f(r, z) = \exp(-\xi z) J_0(\xi r) \, ,$$

where ξ is a constant, is harmonic and hence a more general axisymmetric harmonic function can be written in the form

$$f(r, z) = \int_0^\infty A(\xi) \exp(-\xi z) J_0(\xi r) d\xi \, ,$$

where $A(\xi)$ is an unknown function to be determined from the boundary conditions.

---

[1] I. N. Sneddon (1947), Note on a boundary-value problem of Reissner and Sagoci, *Journal of Applied Physics*, Vol. 18, pp. 130–132.

[2] G. M. L. Gladwell, *Contact Problems in the Classical Theory of Elasticity*, Sijthoff and Noordhoff, Alphen aan den Rijn, 1980.

© The Author(s), under exclusive license to Springer Nature Switzerland AG 2022
J. R. Barber, *Elasticity*, Solid Mechanics and Its Applications 172,
https://doi.org/10.1007/978-3-031-15214-6_32

Using this representation for φ and substituting into equations (32.1, 32.2), we find that

$$-\int_0^\infty \xi A(\xi) J_0(\xi r) d\xi = -\frac{\mu}{(1-\nu)} u_0(r) \ ; \ \ 0 \le r < a$$

$$\int_0^\infty \xi^2 A(\xi) J(\xi r) d\xi = 0 \ ; \qquad\qquad r > a \ .$$

These constitute a pair of *dual integral equations* for the function $A(\xi)$. Sneddon used the method of Titchmarsh[3] and Busbridge[4] to reduce equations of this type to a single equation, but a more recent solution by Sneddon[5] and formalized by Gladwell[6] effects this reduction more efficiently for the classes of equation considered here.

32.2   Collins' Method

A related method, which has the advantage of yielding a single integral equation in elementary functions directly, was introduced by Green and Zerna[7] and applied to a wide range of axisymmetric boundary-value problems by Collins[8].

32.2.1   Indentation by a flat punch

To introduce Collins' method, we first examine the simpler problem in which the punch is flat and hence $u_0(r)$ is a constant. This was first solved by Boussinesq in the 1880s.

A particularly elegant solution was developed by Love[9], using a series of complex harmonic potential functions generated from the real Legendre polynomial solutions

---

[3] E. C. Titchmarsh, *Introduction to the Theory of Fourier Integrals,* Clarendon Press, Oxford, 1937.

[4] I. W. Busbridge (1938), Dual integral equations, *Proceedings of the London Mathematical Society,* Ser. 2 Vol. 44, pp. 115–129

[5] I. N. Sneddon (1960), The elementary solution of dual integral equations, *Proceedings of the Glasgow Mathematical Association,* Vol. 4, pp. 108–110.

[6] G. M. L. Gladwell, *loc. cit.*, Chapters 5, 10.

[7] A. E. Green and W. Zerna, *loc. cit.*.

[8] W. D. Collins (1959), On the solution of some axisymmetric boundary-value problems by means of integral equations, II: Further problems for a circular disc and a spherical cap, *Mathematika,* Vol. 6, pp. 120–133.

[9] A. E. H. Love (1939), Boussinesq's problem for a rigid cone, *Quarterly Journal of Mathematics,* Vol. 10, pp. 161–175.

of Chapter 25 by substituting $(z+\iota a)$ for z. This is tantamount to putting the origin at the 'imaginary' point $(0, -\iota a)$. The real and imaginary parts of the resulting functions are separately harmonic and have discontinuities at $r=a$ on the plane $z=0$.

For example, if we start with the harmonic function $\ln(R+z)$ from equation (24.9) and replace z by $z+\iota a$, we can define the new harmonic function

$$\phi = \phi_1 + \iota\phi_2 = \ln(R^* + z + \iota a) ,$$

where

$$R^* = \sqrt{r^2 + (z+\iota a)^2} .$$

We also record the first derivative of ϕ which is

$$\frac{\partial\phi}{\partial z} = \frac{1}{R^*} .$$

On the plane $z=0$, $R^* \to \sqrt{r^2-a^2}$ and hence

$$\phi(r, 0) = \ln\left(\sqrt{r^2 - a^2} + \iota a\right)$$

$$\frac{\partial\phi}{\partial z}(r, 0) = \frac{1}{\sqrt{r^2 - a^2}} .$$

Both the real and imaginary parts of these functions have discontinuities at $r=a$ on the plane $z=0$. For example

$$\phi_2(r, 0) \equiv \Im\{\phi(r, 0)\} = \frac{\pi}{2} ; \qquad 0 \le r < a \qquad (32.3)$$

$$= \sin^{-1}\frac{a}{r} ; \quad r > a , \qquad (32.4)$$

whilst

$$\frac{\partial\phi_2}{\partial z}(r, 0) = -\frac{1}{\sqrt{a^2 - r^2}} ; \quad 0 \le r < a \qquad (32.5)$$

$$= 0 ; \qquad\qquad r > a . \qquad (32.6)$$

Comparing these results with the boundary conditions (32.1, 32.2), we see that the function

$$\frac{\partial\varphi}{\partial z} = -\frac{2\mu u_0}{\pi(1 - \nu)}\Im\left\{\ln\left(R^* + z + \iota a\right)\right\} \qquad (32.7)$$

satisfies both boundary conditions identically for the case where the function $u_0(r)$ is a constant — i.e. if the rigid indenter is flat and is pressed a distance u_0 into the half-space, as shown in Figure 32.1.

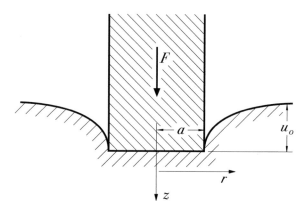

Fig. 32.1 Indentation by a flat-ended cylindrical rigid punch.

The contact pressure distribution is then immediately obtained from equations (31.1, 32.5, 32.7) and is

$$p(r) = -\sigma_{zz}(r, 0) = \frac{2\mu u_0}{\pi(1 - \nu)\sqrt{a^2 - r^2}} \quad ; \quad 0 \le r < a . \qquad (32.8)$$

We also use (32.4) to record the surface displacement of the half-space *outside* the contact area, which is

$$u_z(r, 0) = u_0 \sin^{-1}\frac{a}{r} \quad ; \quad r > a ,$$

and the total indenting force

$$F = 2\pi \int_0^a r p(r) dr = \frac{4\mu u_0}{(1 - \nu)} \int_0^a \frac{r \, dr}{\sqrt{a^2 - r^2}} = \frac{4\mu u_0 a}{(1 - \nu)} .$$

The remaining stress and displacement components can be obtained by substituting (32.7) into the corresponding expressions for Solution F from Table 22.3.

This is Love's solution of the flat punch problem. To gain an appreciation of its elegance, it is really necessary to compare it with the original Hankel transform solution[10] which is extremely complicated.

---

[10] I. N. Sneddon (1946), Boussinesq's problem for a flat-ended cylinder, *Proceedings of the Cambridge Philosophical Society,* Vol. 42, pp. 29–39.

32.2.2 Integral representation

Green and Zerna[11] extended the method of the last section to give a general solution
of the boundary-value problem of equations (32.1, 32.2), using the integral representation

$$\frac{\partial \varphi}{\partial z} = \Re \int_0^a \frac{g(t)dt}{\sqrt{r^2 + (z + \iota t)^2}} = \frac{1}{2} \int_{-a}^a \frac{g(t)dt}{\sqrt{r^2 + (z + \iota t)^2}} , \qquad (32.9)$$

where $g(t)$ is an even function of t. It can be shown that equation (32.9) satisfies
(32.2) identically and the remaining boundary condition (32.1) will then yield an
integral equation for the unknown function $g(t)$.

In effect, equation (32.9) is a superposition of solutions of the form $\Re(r^2 + (z + \iota t)^2)^{-1/2}$ derived from the source solution $1/R$. We could therefore describe φ as the
potential due to an arbitrary distribution $g(t)$ of point sources along the *imaginary*
z-axis between $(0,0)$ and $(0, \iota a)$. It is therefore a logical development of the classical
method of obtaining axisymmetric potential functions by distributing singularities
along the axis of symmetry[12]. A closely related solution is given by Segedin[13],
who develops it as a convolution integral of an arbitrary kernel function with the
Boussinesq solution of §32.2.1. He uses this method to obtain the solution for a
power law punch ($u_0(r) \sim r^n$) and treats more general problems by superposition
after expanding the punch profile as a power series in r. It should be noted that
Segedin's solution is restricted to indentation by a punch of continuous profile[14], in
which case the contact pressure tends to zero at $r = a$. The idea of representing a
general solution by superposition of Boussinesq-type solutions for different values
of a has also been used as a direct numerical method by Maw et al.[15].

Green's method was extensively developed by Collins, who used it to treat many
interesting problems including the indentation problem for an annular punch[16] and
a problem involving 'radiation' boundary conditions[17].

---

[11] A. E. Green and W. Zerna, *loc.cit.*, §§5.8–5.10.

[12] see §24.3 and particularly equation (24.14).

[13] C. M. Segedin (1957), The relation between load and penetration for a spherical punch, *Mathematika*, Vol. 4, pp. 156–161.

[14] see §31.2.

[15] N. Maw, J. R. Barber and J. N. Fawcett (1976), The oblique impact of elastic spheres, *Wear,* Vol. 38, pp. 101–114.

[16] W. D. Collins (1963), On the solution of some axisymmetric boundary-value problems by means of integral equations, VIII: Potential problems for a circular annulus, *Proceedings of the Edinburgh Mathematical Society,* Vol. 13, pp. 235–246.

[17] W. D. Collins (1959), On the solution of some axisymmetric boundary-value problems by means of integral equations, II: Further problems for a circular disc and a spherical cap, *Mathematika*, Vol. 6, pp. 120–133.

32.2.3   Basic forms and surface values

To represent harmonic potential functions we shall use suitable combinations of the four basic forms

$$\phi_1 = \Re \int_0^a g_1(t)\mathcal{F}(r,z,t)dt \; ; \qquad \phi_2 = \Re \int_a^\infty g_2(t)\mathcal{F}(r,z,t)dt$$

$$\phi_3 = \Im \int_0^a g_3(t)\mathcal{F}(r,z,t)dt \; ; \qquad \phi_4 = \Im \int_a^\infty g_4(t)\mathcal{F}(r,z,t)dt \qquad (32.10)$$

where

$$\mathcal{F}(r,z,t) = \ln\left(\sqrt{r^2 + (z+\imath t)^2} + z + \imath t\right) . \qquad (32.11)$$

The square root in (32.11) is interpreted as

$$\sqrt{r^2 + (z+\imath t)^2} = \rho e^{\imath \upsilon/2} ,$$

where

$$\rho = \sqrt[4]{(r^2 + z^2 - t^2)^2 + 4z^2 t^2} \; ; \quad \upsilon = \tan^{-1}\left(\frac{2zt}{r^2 + z^2 - t^2}\right)$$

and $\rho \geq 0, \; 0 \leq \upsilon < \pi$.

Equations (32.10) can be written in two alternative forms which for ϕ_1 are

$$\phi_1 = \frac{1}{2} \int_0^a g_1(t)\{\mathcal{F}(r,z,t) + \mathcal{F}(r,z,-t)\}dt \qquad (32.12)$$

and

$$\phi_1 = \frac{1}{2} \int_{-a}^a g_1(t)\mathcal{F}(r,z,t)dt . \qquad (32.13)$$

We note that (32.12) is exactly equivalent to (32.13) if and only if g_1 is an even function of t. If the boundary values of ϕ, $\partial\phi/\partial z$ etc. specified at $z=0$ are even in r, it will be found that g_1, g_2 are even and g_3, g_4 odd functions of t and forms like (32.13) can be used. The majority of problems fall into this category, but there are important exceptions such as the conical punch (where u_z is proportional to r) and problems with Coulomb friction for which σ_{zr} is proportional to σ_{zz} in some region. In these problems, forms like (32.13) can only be used if $g_i(t)$ is extended into $t < 0$ by a definition with the required symmetry.

Expressions for the important derivatives of the functions ϕ_i at the surface $z=0$ are given in Table 32.1. In certain cases, higher derivatives are required — notably in thermoelastic problems, where heat flux is proportional to $\partial^3\phi/\partial z^3$ (see Solution P, Table 23.1). Higher derivatives are most easily obtained by differentiating *within* the plane $z=0$, making use of the fact that for an axisymmetric harmonic function $f(r,z)$,

$$\frac{\partial^2 f}{\partial z^2} = -\frac{1}{r}\frac{\partial}{\partial r}r\,\frac{\partial f}{\partial r}\;. \qquad (32.14)$$

The reader can verify that this result permits the expressions for $\partial^2\phi_i/\partial z^2$ in Table 32.1 to be obtained from those for $\partial\phi_i/\partial r$.

Table 32.1 Surface values of the derivatives of the functions ϕ_i (Equations (32.10)).

	$0 < r < a$	$r > a$
$\dfrac{\partial\phi_1}{\partial r}$	$\dfrac{1}{r}\displaystyle\int_0^a g_1(t)dt - \dfrac{1}{r}\displaystyle\int_r^a \dfrac{t g_1(t)dt}{\sqrt{t^2-r^2}}$	$\dfrac{1}{r}\displaystyle\int_0^a g_1(t)dt$
$\dfrac{\partial\phi_1}{\partial z}$	$\displaystyle\int_0^r \dfrac{g_1(t)dt}{\sqrt{r^2-t^2}}$	$\displaystyle\int_0^a \dfrac{g_1(t)dt}{\sqrt{r^2-t^2}}$
$\dfrac{\partial^2\phi_1}{\partial z^2}$	$\dfrac{1}{r}\dfrac{d}{dr}\displaystyle\int_r^a \dfrac{t g_1(t)dt}{\sqrt{t^2-r^2}}$	0
$\dfrac{\partial\phi_2}{\partial r}$	$\dfrac{1}{r}\displaystyle\int_a^\infty g_2(t)dt - \dfrac{1}{r}\displaystyle\int_a^\infty \dfrac{t g_2(t)dt}{\sqrt{t^2-r^2}}$	$\dfrac{1}{r}\displaystyle\int_a^\infty g_2(t)dt - \dfrac{1}{r}\displaystyle\int_r^\infty \dfrac{t g_2(t)dt}{\sqrt{t^2-r^2}}$
$\dfrac{\partial\phi_2}{\partial z}$	0	$\displaystyle\int_a^r \dfrac{g_2(t)dt}{\sqrt{r^2-t^2}}$
$\dfrac{\partial^2\phi_2}{\partial z^2}$	$\dfrac{1}{r}\dfrac{d}{dr}\displaystyle\int_a^\infty \dfrac{t g_2(t)dt}{\sqrt{t^2-r^2}}$	$\dfrac{1}{r}\dfrac{d}{dr}\displaystyle\int_r^\infty \dfrac{t g_2(t)dt}{\sqrt{t^2-r^2}}$
$\dfrac{\partial\phi_3}{\partial r}$	$-\dfrac{1}{r}\displaystyle\int_0^r \dfrac{t g_3(t)dt}{\sqrt{r^2-t^2}}$	$-\dfrac{1}{r}\displaystyle\int_0^a \dfrac{t g_3(t)dt}{\sqrt{r^2-t^2}}$
$\dfrac{\partial\phi_3}{\partial z}$	$-\displaystyle\int_r^a \dfrac{g_3(t)dt}{\sqrt{t^2-r^2}}$	0
$\dfrac{\partial^2\phi_3}{\partial z^2}$	$\dfrac{1}{r}\dfrac{d}{dr}\displaystyle\int_0^r \dfrac{t g_3(t)dt}{\sqrt{r^2-t^2}}$	$\dfrac{1}{r}\dfrac{d}{dr}\displaystyle\int_0^a \dfrac{t g_3(t)dt}{\sqrt{r^2-t^2}}$
$\dfrac{\partial\phi_4}{\partial r}$	0	$-\dfrac{1}{r}\displaystyle\int_a^r \dfrac{t g_4(t)dt}{\sqrt{r^2-t^2}}$
$\dfrac{\partial\phi_4}{\partial z}$	$-\displaystyle\int_a^\infty \dfrac{g_4(t)dt}{\sqrt{t^2-r^2}}$	$-\displaystyle\int_r^\infty \dfrac{g_4(t)dt}{\sqrt{t^2-r^2}}$
$\dfrac{\partial^2\phi_4}{\partial z^2}$	0	$\dfrac{1}{r}\dfrac{d}{dr}\displaystyle\int_a^r \dfrac{t g_4(t)dt}{\sqrt{r^2-t^2}}$

32.2.4   Reduction to an Abel equation

Table 32.1 shows that we can satisfy the boundary condition (32.2) identically if we represent φ in the form of ϕ_1 of equations (32.10). Substitution in the remaining boundary condition (32.1) then gives the integral equation

$$\int_0^r \frac{g_1(t)dt}{\sqrt{r^2 - t^2}} = -\frac{\mu}{(1 - \nu)} u_0(r) ; \quad 0 \leq r < a , \tag{32.15}$$

for the unknown function $g_1(t)$. Notice that there is a one-to-one correspondence between the points $0 \leq t < a$ in which $g_1(t)$ is defined and the domain $0 \leq r < a$ in which the integral equation is to be satisfied. This is a necessary condition for the equation to be solvable[18].

Equation (32.15) is an *Abel integral equation*. Abel equations can be inverted explicitly. We operate on both sides of the equation with

$$\int_0^s \frac{r dr}{\sqrt{s^2 - r^2}} ,$$

obtaining

$$\int_0^s \int_0^r \frac{r g_1(t)dt dr}{\sqrt{(s^2 - r^2)(r^2 - t^2)}} = -\frac{\mu}{(1 - \nu)} \int_0^s \frac{r u_0(r)dr}{\sqrt{s^2 - r^2}} ; \quad 0 \leq r < a . \tag{32.16}$$

Since the unknown function, $g_1(t)$ does not depend upon r, it is advantageous to reverse the order of integration on the left-hand side of (32.16) and perform the inner integration. To obtain the limits on the new double integral, we consider r, t as coördinates defining the two-dimensional region of Figure 32.2, in which the domain of integration is shown shaded. It follows from this figure that when we reverse the order of integration the new limits are

$$\int_0^s \int_0^r \frac{r g_1(t)dt dr}{\sqrt{(s^2 - r^2)(r^2 - t^2)}} = \int_0^s g_1(t)dt \int_t^s \frac{r dr}{\sqrt{(s^2 - r^2)(r^2 - t^2)}} . \tag{32.17}$$

The inner integral can now be evaluated as

$$\int_t^s \frac{r dr}{\sqrt{(s^2 - r^2)(r^2 - t^2)}} = \frac{\pi}{2} \tag{32.18}$$

and hence from equations (32.17, 32.18) we have

$$\int_0^s g_1(t)dt = -\frac{2\mu}{\pi(1 - \nu)} \int_0^s \frac{r u_0(r)dr}{\sqrt{s^2 - r^2}} .$$

---

[18] It is akin to the condition that the number of unknowns and the number of equations in a system of linear equations be equal.

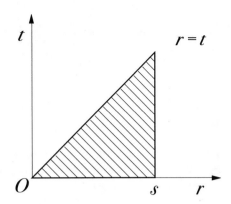

Fig. 32.2 Indentation by a flat-ended cylindrical rigid punch.

Finally, differentiating with respect to s, we have

$$g_1(s) = -\frac{2\mu}{\pi(1-\nu)} \frac{d}{ds} \int_0^s \frac{ru_0(r)dr}{\sqrt{s^2 - r^2}} \,, \tag{32.19}$$

which is the explicit solution of the Abel equation (32.16).

In general, we shall find that the use of the standard forms of equations (32.10) in two-part boundary-value problems leads to the four types of Abel equations whose solutions — obtained by a process similar to that given above — are given in Table 32.2.

Table 32.2 Inversions of Abel integral equations.

If $f(x) =$	$\int_0^x \frac{g(t)dt}{\sqrt{x^2 - t^2}}$	$\int_x^a \frac{g(t)dt}{\sqrt{t^2 - x^2}}$
then $g(x) =$	$\frac{2}{\pi} \frac{d}{dx} \int_0^x \frac{tf(t)dt}{\sqrt{x^2 - t^2}}$	$-\frac{2}{\pi} \frac{d}{dx} \int_x^a \frac{tf(t)dt}{\sqrt{t^2 - x^2}}$
If $f(x) =$	$\int_a^x \frac{g(t)dt}{\sqrt{x^2 - t^2}}$	$\int_x^\infty \frac{g(t)dt}{\sqrt{t^2 - x^2}}$
then $g(x) =$	$\frac{2}{\pi} \frac{d}{dx} \int_a^x \frac{tf(t)dt}{\sqrt{x^2 - t^2}}$	$-\frac{2}{\pi} \frac{d}{dx} \int_x^\infty \frac{tf(t)dt}{\sqrt{t^2 - x^2}}$

Once g_1 is known, the complete stress field can be written down using equation (32.10) and Solution F of Table 22.3. In particular, the contact pressure

$$p(r) = -\sigma_{zz}(r, 0) = \frac{\partial^2 \varphi}{\partial z^2}(r, 0) = \frac{1}{r} \frac{d}{dr} \int_r^a \frac{tg_1(t)dt}{\sqrt{t^2 - r^2}} \,, \tag{32.20}$$

from Table 32.1, and the total indenting force

$$F = 2\pi \int_0^a rp(r)dr = 2\pi \int_0^a \left(\frac{1}{r}\frac{d}{dr}\int_r^a \frac{tg_1(t)dt}{\sqrt{t^2 - r^2}}\right)rdr = 2\pi \left[\int_r^a \frac{tg_1(t)dt}{\sqrt{t^2 - r^2}}\right]_{r=0}^{r=a}$$

$$= -2\pi \int_0^a g_1(t)dt \ . \tag{32.21}$$

Example

As an example, we consider the problem in which a half-space is indented by a rigid cylindrical punch of radius b with a spherical end of radius $R \gg b$. The function u_0 is then defined by

$$u_0 = d - \frac{r^2}{2R} \ , \tag{32.22}$$

where d is the indentation at the centre of the punch. We suppose that the force is sufficient to ensure that the punch makes contact over the entire surface $0 \le r < b$.

Substituting (32.22) in (32.19) and performing the integration and differentiation, we obtain

$$g_1(s) = -\frac{2\mu}{\pi(1 - \nu)}\left(d - \frac{s^2}{R}\right) \ . \tag{32.23}$$

Using this result in (32.20) the contact pressure distribution is obtained as

$$p(r) = \frac{2\mu}{\pi(1 - \nu)R}\left(\frac{Rd + b^2 - 2r^2}{\sqrt{b^2 - r^2}}\right) \ , \tag{32.24}$$

and the total force is

$$F = \frac{4\mu}{(1 - \nu)}\int_0^b \left(d - \frac{t^2}{R}\right)dt = \frac{4\mu b}{(1 - \nu)}\left(d - \frac{b^2}{3R}\right) \ ,$$

from (32.21, 32.23).

Equation (32.15) defines a positive contact pressure for all $r < b$ if and only if $d > b^2/R$ and hence

$$F \ge \frac{8\mu b^3}{3(1 - \nu)R} \ . \tag{32.25}$$

For smaller values of F, contact will occur only in a smaller circle of radius $a < b$. In this case, the material of the punch outside $r = a$ plays no rôle in the problem and the contact pressure can be determined by replacing b by a in (32.24). Furthermore, the smooth contact condition requires that $p(r)$ be bounded at $r = a$ and this is equivalent to enforcing the equality in (32.25) with b replaced by a, and then solving for a giving

$$a = \left(\frac{3(1 - \nu)FR}{8\mu}\right)^{1/3} \ .$$

However, we shall introduce a more systematic approach to problems of this class in the next section.

32.2.5 Smooth contact problems

If the indenter is smooth, the extent of the contact area must be found as part of the solution by enforcing the unilateral inequalities (31.6, 31.7). One way to do this is first to solve the problem assuming a is an independent variable and then determine it from the condition that $p(r)$ be non-singular at the boundary of the contact area $r = a$, as in the two-dimensional Hertz problem of §12.5.3. Alternatively, we recall from Chapter 31 that the appropriate contact area is that which maximizes the indentation load F, so for a circular contact area of radius a,

$$\frac{dF}{da} = 0 \quad \text{and hence} \quad g_1(a) = 0 , \tag{32.26}$$

from (32.21).

However, more direct relationships can be established between the contact radius, the indenting force and the shape of the indenter, defined merely through its *slope* $u_0'(r)$. Starting with the right-hand side of equation (32.20), we integrate by parts, obtaining

$$\int_r^a \frac{tg_1(t)dt}{\sqrt{t^2 - r^2}} = g_1(a)\sqrt{a^2 - r^2} - \int_r^a \sqrt{t^2 - r^2}\, g_1'(t)dt$$

and hence

$$p(r) = -\frac{g_1(a)}{\sqrt{a^2 - r^2}} + \int_r^a \frac{g_1'(t)dt}{\sqrt{t^2 - r^2}} .$$

The second term is bounded at $r \to a$ and hence the tractions will be bounded there if and only if $g_1(a) = 0$, as in (32.26). Using this result, we can then write a simpler expression for the contact pressure in smooth contact problems as

$$p(r) = \int_r^a \frac{g_1'(t)dt}{\sqrt{t^2 - r^2}} . \tag{32.27}$$

Since the contact pressure depends only on the derivative g_1', we can also achieve some simplification by performing a similar integration by parts on the right-hand side of (32.19). We obtain

$$\int_0^s \frac{ru_0(r)dr}{\sqrt{s^2 - r^2}} = u_0(0)s + \int_0^s \sqrt{s^2 - r^2}\, u_0'(r)dr$$

and hence

$$g_1(s) = -\frac{2\mu}{\pi(1-\nu)}\left[d + s\int_0^s \frac{u_0'(r)dr}{\sqrt{s^2 - r^2}}\right], \tag{32.28}$$

where $d = u_0(0)$ represents the rigid-body indentation of the punch. We can determine d by setting $g_1(a) = 0$ in (32.28), giving

$$d = -a\int_0^a \frac{u_0'(r)dr}{\sqrt{a^2 - r^2}} \tag{32.29}$$

and hence

$$g_1(s) = \frac{2\mu}{\pi(1-\nu)}\left[a\int_0^a \frac{u_0'(r)dr}{\sqrt{a^2 - r^2}} - s\int_0^s \frac{u_0'(r)dr}{\sqrt{s^2 - r^2}}\right]. \tag{32.30}$$

Notice that this result depends only on the shape of the punch as defined by the derivative $u_0'(r)$. Also, differentiating (32.30), we obtain

$$g_1'(s) = -\frac{2\mu}{\pi(1-\nu)}\frac{d}{ds}\left[s\int_0^s \frac{u_0'(r)dr}{\sqrt{s^2 - r^2}}\right]. \tag{32.31}$$

Substituting (32.30) into (32.21), we have

$$F = -\frac{4\mu}{(1-\nu)}\left[a^2\int_0^a \frac{u_0'(r)dr}{\sqrt{a^2 - r^2}} - \int_0^a s\int_0^s \frac{u_0'(r)drds}{\sqrt{s^2 - r^2}}\right].$$

Changing the order of integration in the second term and performing the inner integral, we have

$$\int_0^a s\int_0^s \frac{u_0'(r)drds}{\sqrt{s^2 - r^2}} = \int_0^a u_0'(r)\int_r^a \frac{sdsdr}{\sqrt{s^2 - r^2}} = \int_0^a \sqrt{a^2 - r^2}\,u_0'(r)dr$$

and hence, after some algebraic simplification,

$$F = -\frac{4\mu}{(1-\nu)}\int_0^a \frac{r^2 u_0'(r)dr}{\sqrt{a^2 - r^2}}. \tag{32.32}$$

In a typical problem, the shape of the punch $u_0'(r)$ is a known function and either the force F or the indentation d will be prescribed. The contact radius a can then be determined by evaluating the integrals in (32.32) or (32.29) and solving the resulting equation. The function $g_1'(s)$ and the contact pressure can then be obtained from equations (32.31) and (32.27) respectively.

Example – The Hertz problem

As an example, we consider the problem in which a half-space is indented by a rigid sphere of radius R, in which case the slope of the indenter is defined by

$$u_0' = -\frac{r}{R} \cdot$$

Substituting in equation (32.32) and performing the integration, we obtain

$$F = \frac{4\mu}{(1-\nu)R} \int_0^a \frac{r^3(r)dr}{\sqrt{a^2-r^2}} = \frac{8\mu a^3}{3(1-\nu)R} \cdot \tag{32.33}$$

Also, the rigid-body indentation is obtained from (32.29) as

$$d = \frac{a}{R} \int_0^a \frac{rdr}{\sqrt{a^2-r^2}} = \frac{a^2}{R} \cdot \tag{32.34}$$

Alternatively, we can eliminate a between equations (32.33, 32.34), obtaining

$$F = \frac{8\mu R^{1/2} d^{3/2}}{3(1-\nu)} ,$$

which exhibits a stiffening characteristic with increasing indentation.

For the contact pressure distribution, we first note that

$$\int_0^s \frac{u_0'(r)dr}{\sqrt{s^2-r^2}} = -\frac{1}{R} \int_0^s \frac{rdr}{\sqrt{s^2-r^2}} = -\frac{s}{R} \cdot$$

Substituting this result into the right-hand side of (32.31), we then obtain

$$g_1'(s) = \frac{4\mu s}{\pi(1-\nu)R} ,$$

after which (32.27) gives

$$p(r) = \frac{4\mu}{\pi(1-\nu)R} \int_r^a \frac{tdt}{\sqrt{t^2-r^2}} = \frac{4\mu\sqrt{a^2-r^2}}{\pi(1-\nu)R} \cdot \tag{32.35}$$

32.2.6 Choice of form

Each of the functions ϕ_i in Table 32.1 has a zero in $\partial\phi/\partial z$ or $\partial^2\phi/\partial z^2$ either in $0 < r < a$ or in $r > a$. Thus, in the above examples it would have been possible to satisfy (32.2) by choosing $\partial\varphi/\partial z = \phi_3$ instead of $\varphi = \phi_1$.

The choice is best made by examining the requirements of continuity imposed at $r = a$, $z = 0$ by the physical problem. It can be shown that, if $g_i(a)$ is bounded, the expressions in Table 32.1 will define continuous values of $\partial\phi_i/\partial z$, but $\partial^2\phi_i/\partial z^2$ will be discontinuous unless $g_i(a) = 0$. Thus if, as in the present case, the function $\partial\varphi/\partial z$ represents a physical quantity like displacement or temperature which is required to be continuous, it is appropriate to choose $\partial\varphi/\partial z = \partial\phi_1/\partial z$. The alternative choice of $\partial\varphi/\partial z = \phi_3$ imposes too strong a continuity condition at $r = a$, since it precludes discontinuities in second derivatives and hence in the stress components.

Of course, in problems involving contact between smooth surfaces, the normal traction σ_{zz} must also be continuous at the edge of the contact region. The most straightforward treatment is that given in §32.2.5, in which the continuity condition furnishes an extra condition to determine the radius of the contact region, which is not known *a priori*. However, it would also be possible to enforce the required continuity through the formulation by using $\partial\varphi/\partial z = \phi_3$ in which case the contact radius is prescribed and the rigid-body indentation d of the punch must be allowed to float. This is essentially the technique used by Segedin[19].

32.3   Non-axisymmetric Problems

The method described in this chapter can also be applied to problems in which non-axisymmetric boundary conditions are imposed interior to and exterior to the circle $r = a$ — for example, if the punch has a profile $u_0(r, \theta)$ that depends on both polar coördinates. The first step is to perform a Fourier decomposition of the profile, so that the boundary conditions can be expressed as a series of terms of the form $f_m(r)\cos(m\theta)$ or $f_m(r)\sin(m\theta)$, where m is an integer. These terms can then be treated separately using linear superposition.

Copson[20] shows that if a harmonic potential function $\phi(r, \theta, z)$ satisfies the mixed boundary conditions

$$\phi(r, \theta, 0) = f(r)\cos(m\theta) ; \quad 0 \leq r < a$$
$$\frac{\partial\phi}{\partial z}(r, \theta, 0) = 0 ; \qquad\qquad r > a , \tag{32.36}$$

the surface value of the derivative in $0 \leq r < a$ can be written

$$\frac{\partial\phi}{\partial z}(r, \theta, 0) = r^{m-1}\cos(m\theta)\frac{d}{dr}\int_r^a \frac{tg(t)dt}{\sqrt{t^2 - r^2}} , \tag{32.37}$$

where

$$g(t) = \frac{2}{\pi t^{2m}}\frac{d}{dt}\int_0^t \frac{s^{m+1}f(s)ds}{\sqrt{t^2 - s^2}} . \tag{32.38}$$

This result can clearly be used to solve non-axisymmetric problems of the form (32.1, 32.2) by setting

---

[19] C. M. Segedin (1957), The relation between load and penetration for a spherical punch, *Mathematika*, Vol. 4, pp. 156–161.

[20] E. T. Copson (1947), On the problem of the electrified disc, *Proceedings of the Edinburgh Mathematical Society*, Vol. 8, pp. 14–19.

$$\phi = \frac{\partial \varphi}{\partial z} . \tag{32.39}$$

With this notation, equations (32.36, 32.37) reduce to the equivalent expressions for ϕ_1 in Table 32.1 in the axisymmetric case $m = 0$.

Example: The tilted flat punch

We suppose that a cylindrical rigid flat punch of radius a is pressed into the surface of the elastic half-space $z > 0$ by an offset force F, whose line of action passes through the point $(\epsilon, 0, 0)$ causing the punch to tilt through some small angle α. We restrict attention to the case where the offset is sufficiently small to ensure that the complete face of the punch makes contact with the half-space. The boundary condition in $0 \le r < a$ is then

$$u_z(r, \theta, 0) = d + \alpha r \cos \theta \tag{32.40}$$

and since the problem is linear, we can decompose the solution into an axisymmetric and a non-axisymmetric term, the former being identical to that given in §32.2.1. To complete the solution, we therefore seek a harmonic function φ satisfying the boundary conditions

$$\frac{\partial \varphi}{\partial z} = -\frac{\mu \alpha r \cos \theta}{(1 - \nu)} ; \quad 0 \le r < a , z = 0$$

$$\frac{\partial^2 \varphi}{\partial z^2} = 0 ; \qquad r > a , z = 0 , \tag{32.41}$$

from (32.1, 32.2). Using (32.39), these conditions are equivalent to (32.36) with $m = 1$ and

$$f(r) = -\frac{\mu \alpha r}{(1 - \nu)} .$$

Equation (32.38) then gives

$$g(t) = -\frac{2 \mu \alpha}{\pi (1 - \nu) t^2} \frac{d}{dt} \int_0^t \frac{s^3 f(s) ds}{\sqrt{t^2 - s^2}} = -\frac{4 \mu \alpha}{\pi (1 - \nu)}$$

and substituting into (32.37), we obtain

$$\frac{\partial^2 \varphi}{\partial z^2}(r, \theta, 0) = \cos \theta \frac{d}{dr} \int_r^a \frac{t g(t) dt}{\sqrt{t^2 - r^2}} = \frac{4 \mu \alpha r \cos \theta}{\pi (1 - \nu) \sqrt{a^2 - r^2}} .$$

The contact pressure distribution under the punch is then obtained as

$$p(r, \theta) = \frac{\partial^2 \varphi}{\partial z^2} = \frac{2 \mu}{\pi (1 - \nu) \sqrt{a^2 - r^2}} (d + 2 \alpha r \cos \theta) ,$$

where we have superposed the axisymmetric contribution due to the term d in (32.40) using (32.8). This solution of the tilted punch problem was first given by Green[21].

Contact will occur over the complete face of the punch if and only if $p(r, \theta) > 0$ for all (r, θ) in $0 \leq r < a$. This condition will fail at $(a, -\pi)$, implying local separation, if

$$\alpha > \frac{d}{2a} \,. \tag{32.42}$$

The rigid-body motion of the punch defined by the constants d, α can be related to the force F and its offset ϵ by equilibrium considerations. We obtain

$$F = \int_0^a \int_0^{2\pi} p(r, \theta) r \, d\theta \, dr = \frac{4\mu a d}{(1 - \nu)}$$

$$F\epsilon = \int_0^a \int_0^{2\pi} p(r, \theta) r^2 \cos\theta \, d\theta \, dr = \frac{8\mu a^3 \alpha}{3(1 - \nu)}$$

and using (32.42), we conclude that for full-face contact, we require

$$\epsilon \leq \frac{a}{3} \,.$$

In other words, the line of action of the indenting force must lie within a central circle of radius $a/3$.

32.3.1 The full stress field

In his solution of the tilted punch problem, Green used a semi-intuitive method similar to that introduced in §32.2.1 to deduce the form of the harmonic potential φ through-out the half-space from the surface values obtained above. A more formal approach is to use the recurrence relation (25.31, 25.32) to relate the non-axisymmetric prob-lem to an axisymmetric problem that can be solved using the method of §32.2.2 and Table 32.1.

For example, in the case of the tilted punch problem, the non-axisymmetric term in $\varphi(r, \theta, z)$ can be written $f_1(r, z) \cos\theta$, where f_1 satisfies the boundary conditions

$$\frac{\partial f_1}{\partial z} = -\frac{\mu \alpha r}{(1 - \nu)} \quad ; \quad 0 \leq r < a \,, \; z = 0$$

$$\frac{\partial^2 f_1}{\partial z^2} = 0 \quad ; \qquad\qquad r > a \,, \; z = 0 \,, \tag{32.43}$$

[21] A. E. Green (1949), On Boussinesq's problem and penny-shaped cracks, *Mathematical Proceed-ings of the Cambridge Philosophical Society,* Vol. 45, pp. 251–257.

from (32.41). Setting $m=0$ in (25.32), we can define an axisymmetric harmonic function $f_0(r, z)$ such that

$$f_1(r, z) = \frac{\partial f_0}{\partial r} . \tag{32.44}$$

Substituting this relation into (32.43)$_1$ and integrating with respect to r, we conclude that f_0 must satisfy the boundary condition

$$\frac{\partial f_0}{\partial z}(r, 0) = \int \frac{\partial f_1}{\partial z}(r, 0)dr = -\frac{\mu\alpha r^2}{2(1-\nu)} + C \; ; \; 0 \le r < a , z = 0 ,$$

where C is a constant of integration. Since the stress field must decay at large values of r, the corresponding integration of the outer boundary condition (32.43)$_2$ leads simply to

$$\frac{\partial^2 f_0}{\partial z^2} = 0 \; ; \; r > a , z = 0 . \tag{32.45}$$

Equation (32.45) can be satisfied identically by representing the axisymmetric harmonic $f_0(r, z)$ in the form ϕ_1 of equation (32.10). Table 32.1 then gives

$$\int_0^r \frac{g_1(t)dt}{\sqrt{r^2 - t^2}} = -\frac{\mu\alpha r^2}{2(1-\nu)} + C$$

and the Abel equation solution is obtained from Table 32.2 as

$$g_1(x) = \frac{2}{\pi}\frac{d}{dx}\int_0^x \left(-\frac{\mu\alpha t^3}{2(1-\nu)} + Ct\right)\frac{dt}{\sqrt{x^2 - t^2}} = \frac{2}{\pi}\left(-\frac{\mu\alpha x^2}{(1-\nu)} + C\right) .$$

The constant C must be chosen so as to satisfy the condition of continuity of displacement at $r=a$. As explained in §32.2.6, the representations (32.10) ensure continuity of the derivative $\partial\phi_i/\partial z$ at $r=a$ as long as the corresponding function $g_i(t)$ is bounded. In the present case, this means that $\partial f_0/\partial z$ will be continuous for all values of C. However, the differentiation in (32.44) will introduce a square-root singularity in $\partial f_1/\partial z$ unless the constant C is chosen so as to satisfy the stronger condition $g_1(a)=0$, giving

$$g_1(x) = \frac{2\mu\alpha(a^2 - x^2)}{\pi(1-\nu)} .$$

The function $f_0(r, z)$ is then given by (32.10)$_1$ with this value of g_1 after which $f_1(r, z)$ is defined by (32.44).

For higher-order problems [larger values of m in $\varphi = f_m(r, z)\cos(m\theta)$], additional constants of integration will appear in the boundary conditions for $f_0(r, z)$ and stronger continuity conditions must be imposed at $r=a$. In fact $g_1(t)$ and its first $m-1$ derivatives must be set to zero at $t=a$.

Problems

32.1. A harmonic potential function $w(r, z)$ in the region $z > 0$ is defined by the surface values $w(r, 0) = A(a^2 - r^2)$; $0 \le r < a$, $w(r, 0) = 0$; $r > a$. Select a suitable form for the function using Table 32.1 and then use the boundary conditions to determine the corresponding unknown function $g_i(t)$.

32.2. Use equation (32.14) to derive the expressions for $\partial^2 \phi_1 / \partial z^2$ in Table 32.1 from those for $\partial \phi_1 / \partial r$.

32.3. An elastic half-space is indented by a rigid cylindrical punch of radius a with a *concave* spherical end of radius $R \gg a$. Find the contact pressure distribution $p(r)$ and hence determine the minimum force F_0 required to maintain contact over the entire punch surface. Do not attempt to solve the problem for $F < F_0$.

32.4. A rigid flat punch of radius a is bonded to the surface of the elastic half-space $z > 0$. A torque T is then applied to the punch. Use Solution E to describe the stress field in the half-space and Collins' method to solve the resulting boundary-value problem for Ψ. In particular, find the distribution of traction between the punch and the half-space and the rotation of the punch due to the torque.

32.5. A rigid sphere of radius R is pressed into an elastic half-space of shear modulus μ and Poisson's ratio ν. Assuming that the interface energy for this material combination is $\Delta \gamma$, find a relation between the indenting force F and the radius of the contact area a. You should use the conditions (31.11, 31.12) to determine the unknown constant d in (32.22) and then (32.21) to determine F.

Hence determine the contact radius when the sphere is unloaded ($F = 0$) and the tensile force required to pull the sphere out of contact[22].

32.6. (i) An elastic half-space is indented by a conical frictionless rigid punch as shown in Figure 32.3, the penetration of the apex being d. Find the contact pressure distribution, the total load F and the contact radius a. The semi-angle of the cone is $\frac{\pi}{2} - \alpha$, where $\alpha \ll 1$.

(ii) What is the nature of the singularity in pressure at the vertex of the cone (point A).

(iii) Now suppose that the cone is truncated as shown in Figure 32.4. *Without finding the new contact pressure distribution*, find the new relationship between F and a. **Note:** Take the radius of the truncated end of the cone as b and refer the profile to the point A where the vertex would have been before truncation.

(iv) Would you expect a singularity in contact pressure at B and if so of what form. (The process of determining $p(r)$ is algebraically tedious. Try to find a simpler way to answer this question.)

---

[22] This solution was first given by K. L. Johnson, K. Kendall and A. D. Roberts (1971), Surface energy and the contact of elastic solids, *Proceedings of the Royal Society of London*, Vol.A324, pp. 301–313.

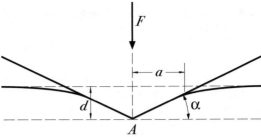

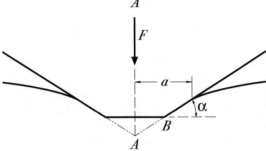

Fig. 32.3 The conical punch.

Fig. 32.4 The truncated
conical punch.

32.7. A rigid flat punch has rounded edges, as shown in Figure 32.5. The punch
is pressed into an elastic half-space by a force F. Assuming that the contact is
frictionless, find the relation between F, the indentation depth u_0 and the radius a
of the contact area.

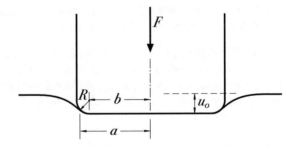

Fig. 32.5 Flat punch with
rounded corners.

32.8. An elastic half-space is indented by a smooth axisymmetric rigid punch with
the power law profile Cr^s, so the displacement in the contact area is

$$u_0(r) = d - Cr^s ,$$

where C, s are constants. Show that the indentation force F, the contact radius a and
the indentation depth d are related by the equation

$$F = \frac{4\mu ad}{(1 - \nu)} \left(\frac{s}{s + 1} \right) .$$

32.9. The profile of a smooth axisymmetric frictionless rigid punch is described by the power law

$$u_0(r) = A_n r^{2n} ,$$

where n is an integer. The punch is pressed into an elastic half-space by a force F. Find the indentation depth d, the radius of the contact area a and the contact pressure distribution $p(r)$. Check your results by comparison with the Hertz problem of §32.2.5 and give simplified expressions for the case of the fourth order punch

$$u_0(r) = A_2 r^4 .$$

32.10. For a smooth axisymmetric contact problem, the contact pressure must be continuous at $r = a$. Usually this condition is imposed explicitly and it serves to determine the radius a of the contact region. However, it can also be guaranteed by 'choice of form' as suggested in §32.2.6. Use this strategy to solve the Hertzian contact problem, for which $u_0'(r) = -r/R$, where R is the radius of the indenter.

You will need to find a suitable representation of the form

$$\frac{\partial \varphi}{\partial z} = \phi_i ,$$

where ϕ_i is one of the four functions in equation (32.10), chosen so as to satisfy the homogeneous boundary condition (zero contact pressure in $r > a$). Then use Tables 32.1, 32.2 to set up and solve an Abel equation for the unknown function $g_i(t)$. In particular, find the contact pressure distribution $p(r)$ and the total load F needed to establish a circular contact area of radius a.

32.11. The flat end of a rigid cylindrical punch of radius a is pressed into the curved surface of an elastic cylinder of radius $R \gg a$ by a centric force F sufficient to ensure contact throughout the punch face. Use Copson's formula (32.36–32.38) to determine the contact pressure distribution and hence find the minimum value of F required for full contact.

32.12. Solve Problem 32.11 by relating the non-axisymmetric component of $\varphi(r, \theta, z)$ to a corresponding axisymmetric harmonic function $f_0(r, z)$, as in §32.3.1. Notice that the required function varies with $\cos(2\theta)$, so you will need to apply equation (25.32) twice recursively to relate $f_2(r, z)$ to $f_0(r, z)$. This will introduce two arbitrary constants in the boundary conditions for f_0 and these must be determined from the continuity conditions $g_1(a) = g_1'(a) = 0$.

Chapter 33
The Penny-shaped Crack

As in the two-dimensional case, we shall find considerable similarities in the formulation and solution of contact and crack problems. In particular, we shall find that problems for the plane crack can be reduced to boundary-value problems which in the case of axisymmetry can be solved using the method of Green and Collins developed in §32.2.

33.1 The Penny-shaped Crack in Tension

The simplest axisymmetric crack problem is that in which a state of uniform tension $\sigma_{zz} = S$ in an infinite isotropic homogeneous solid is perturbed by a plane crack occupying the region $0 \leq r < a$, $z = 0$. Thus, the crack has the shape of a circular disk. This geometry has come to be known as the *penny-shaped* crack[1].

As in Chapter 13, we seek the solution in terms of an unperturbed uniform stress field $\sigma_{zz} = S$ (constant) and a corrective solution which tends to zero at infinity. The boundary conditions on the corrective solution are therefore

$$\sigma_{zz} = -S; \quad \sigma_{zr} = 0 \ ; \quad 0 \leq r < a , \quad z = 0$$
$$\sigma_{zz}, \sigma_{rz}, \sigma_{rr}, \sigma_{\theta\theta} \to 0 \ ; \quad R \to \infty .$$

The corrective solution corresponds to the problem in which the crack is opened by an internal pressure of magnitude S and the body is not loaded at infinity.

The problem is symmetrical about the plane $z = 0$ and it follows that on that plane there can be no shear stress σ_{zx}, σ_{zy} and no normal displacement u_z. We can therefore reduce the problem to a boundary-value problem for the half-space $z > 0$ defined by the boundary conditions

---

[1] I. N. Sneddon (1946), The distribution of stress in the neighbourhood of a crack in an elastic solid, *Proceedings of the Royal Society of London,* Vol. A187, pp. 226–260.

© The Author(s), under exclusive license to Springer Nature Switzerland AG 2022
J. R. Barber, *Elasticity*, Solid Mechanics and Its Applications 172,
https://doi.org/10.1007/978-3-031-15214-6_33

$$\sigma_{zx} = \sigma_{zy} = 0 \; ; \quad \text{all } r, \; z = 0 \tag{33.1}$$

$$\sigma_{zz} = -S \; ; \quad 0 \le r < a, \; z = 0 \tag{33.2}$$

$$u_z = 0 \; ; \quad r > a, \; z = 0 . \tag{33.3}$$

Equation (33.1) is a global condition and can be satisfied identically[2] by using Solution F of Table 22.3. The remaining conditions (33.2, 33.3) then define the mixed boundary-value problem

$$\frac{\partial^2 \varphi}{\partial z^2} = S \; ; \quad 0 \le r < a, \; z = 0 \tag{33.4}$$

$$\frac{\partial \varphi}{\partial z} = 0 \; ; \quad r > a, \; z = 0 . \tag{33.5}$$

This problem can be solved by the method of §32.2. We note from Table 32.1 that (33.5) can be satisfied identically by representing φ in the form of ϕ_3 of (32.10) — i.e.

$$\varphi = \Im \int_0^a g(t) \mathcal{F}(r, z, t) dt$$

and the remaining boundary condition (33.4) then reduces to

$$\frac{1}{r} \frac{d}{dr} \int_0^r \frac{t g(t) dt}{\sqrt{r^2 - t^2}} = S \; ; \quad 0 \le r < a .$$

We now multiply both sides of this equation by r and integrate, obtaining

$$\int_0^r \frac{t g(t) dt}{\sqrt{r^2 - t^2}} = \frac{Sr^2}{2} + C \; ; \quad 0 \le r < a , \tag{33.6}$$

where C is an arbitrary constant.

Equation (33.6) is an Abel integral equation similar to (32.15) and can be inverted in the same way (see Table 32.2), with the result

$$x g(x) = \frac{2}{\pi} \frac{d}{dx} \int_0^x \left(\frac{Sr^2}{2} + C \right) \frac{r dr}{\sqrt{x^2 - r^2}}$$

and hence

$$g(x) = \frac{2Sx}{\pi} + \frac{2C}{\pi x} . \tag{33.7}$$

The second term in this expression is singular at $x \to 0$ and it is readily verified from Table 32.1 that $\partial \varphi / \partial z$ and hence u_z would be unbounded at the origin if $C \ne 0$.

---

[2] More generally, any problem involving a crack of arbitrary cross section $\mathcal{A}$ on the plane $z = 0$ and loaded by a uniform tensile stress σ_{zz} at $z = \pm \infty$ has the same symmetry and can also be formulated using Solution F. In the general case, we obtain a two-part boundary-value problem for the half-space $z > 0$, in which conditions (33.4) and (33.5) are to be satisfied over $\mathcal{A}$ and $\bar{\mathcal{A}}$ respectively, where $\bar{\mathcal{A}}$ is the complement of $\mathcal{A}$ in the plane $z = 0$.

We therefore set $C=0$ to retain continuity of displacement at the origin.

We can then recover the expression for the stress σ_{zz} in the region $r>a$, $z=0$ which is

$$\sigma_{zz} = -\frac{\partial^2\varphi}{\partial z^2} = -\frac{2S}{\pi r}\frac{d}{dr}\int_0^a \frac{x^2 dx}{\sqrt{r^2-x^2}}$$

$$= -\frac{2S}{\pi}\left(\sin^{-1}\frac{a}{r} - \frac{a}{\sqrt{r^2-a^2}}\right),$$

using Table 32.1 and (33.7).

This of course is the corrective solution. To find the corresponding stress in the original crack problem, we must superpose the uniform tensile stress $\sigma_{zz}=S$, with the result

$$\sigma_{zz} = \frac{2S}{\pi}\left(\cos^{-1}\frac{a}{r} + \frac{a}{\sqrt{r^2-a^2}}\right) ; \quad r>a, \ z=0 .$$

The stress-intensity factor is defined as

$$K_{\mathrm{I}} = \lim_{r\to a^+} \sigma_{zz}\sqrt{2\pi(r-a)} = 2S\sqrt{\frac{a}{\pi}} . \tag{33.8}$$

We could have obtained this result directly from $g(x)$, without computing the complete stress distribution. We have

$$\sigma_{zz} = -\frac{1}{r}\frac{d}{dr}\int_0^a \frac{xg(x)dx}{\sqrt{r^2-x^2}} ,$$

and integrating by parts before performing the differentiation as in §32.2.5, we obtain

$$\sigma_{zz} = \frac{g(a)}{\sqrt{r^2-a^2}} - \frac{g(0)}{r} - \int_0^a \frac{g'(x)dx}{\sqrt{r^2-x^2}} . \tag{33.9}$$

Now, unless $g'(x)$ is singular at $x=a$, the only singular term in (33.9) will be the first, and the stress-intensity factor is therefore

$$K_{\mathrm{I}} = \lim_{r\to a^+} \sigma_{zz}\sqrt{2\pi(r-a)} = g(a)\sqrt{\frac{\pi}{a}} .$$

In the present example, $g(a)=2Sa/\pi$, leading to (33.8) as before.

We also compute the displacement u_z at $0\le r<a$, $z=0^+$, which is

$$u_z(r,0^+) = -\frac{(1-\nu)}{\mu}\frac{\partial\varphi}{\partial z} = \frac{(1-\nu)}{\mu}\int_r^a \frac{g(x)dx}{\sqrt{x^2-r^2}}$$

$$= \frac{2(1-\nu)S\sqrt{a^2-r^2}}{\pi\mu} .$$

Since the crack opens symmetrically on each face, it follows that the crack-opening displacement is

$$\delta(r) = u_z(r, 0^+) - u_z(r, 0^-) = \frac{4(1 - \nu)S\sqrt{a^2 - r^2}}{\pi\mu} .$$

33.2   Thermoelastic Problems

In addition to acting as a stress concentration in an otherwise uniform stress field, a crack will obstruct the flow of heat and generate a perturbed temperature field in components of thermal machines. The simplest investigation of this effect assumes that the crack acts as a perfect insulator, so that the heat is forced to flow around it. For the penny-shaped crack[3], the appropriate thermal boundary conditions are then

$$q_z = -K\frac{\partial T}{\partial z} = 0 \; ; \;\; 0 \le r < a, \;\; z = 0 \qquad (33.10)$$

$$\rightarrow q_0 \; ; \;\; R \rightarrow \infty , \qquad (33.11)$$

where q_0 is a constant defining the unperturbed heat flux.

As in isothermal problems, we construct the temperature field as the sum of a uniform heat flux and a corrective solution, the boundary conditions for which become

$$q_z = -K\frac{\partial T}{\partial z} = -q_0 \; ; \;\; 0 \le r < a, \;\; z = 0$$

$$\rightarrow 0 \; ; \;\;\;\; R \rightarrow \infty .$$

This problem is antisymmetric about the plane $z=0$ and hence, with a suitable choice of datum, the temperature must be zero in the region $r > a$, $z=0$. We can therefore convert the heat conduction problem into a boundary-value problem for the half-space $z > 0$ with boundary conditions

$$q_z = -q_0 \; ; \;\; 0 \le r < a \qquad (33.12)$$

$$T = 0 \;\;\; ; \;\; r > a . \qquad (33.13)$$

The temperature field and a particular thermoelastic solution can be constructed using Williams' solution [Solution T of Table 23.1], in terms of which (33.12, 33.13) define the mixed boundary-value problem

---

[3] This problem was first solved by A. L. Florence and J. N. Goodier (1963), The linear thermoelastic problem of uniform heat flow disturbed by a penny-shaped insulated crack, *International Journal of Engineering Science,* Vol. 1, pp. 533–540.

$$\frac{\partial^2 \chi}{\partial z^2} = \frac{\mu\alpha(1+\nu)q_0}{(1-\nu)K} \ ; \quad 0 \le r < a \tag{33.14}$$

$$\frac{\partial \chi}{\partial z} = 0 \ ; \qquad\qquad r > a \ . \tag{33.15}$$

We require that the temperature and hence $\partial\chi/\partial z$ be continuous at $r=a$, $z=0$ and hence in view of the arguments of §32.2.6, we choose to satisfy (33.15) using the function ϕ_3 — i.e.

$$\chi = \Im \int_0^a g_3(t)\mathcal{F}(r,z,t)dt \ .$$

Condition (33.14) then defines the Abel equation

$$\frac{1}{r}\frac{d}{dr}\int_0^r \frac{tg_3(t)dt}{\sqrt{r^2-t^2}} = \frac{\mu\alpha(1+\nu)q_0}{(1-\nu)K} \ ; \quad 0 \le r < a \ ,$$

which has the solution

$$g_3(t) = \frac{2\mu\alpha(1+\nu)q_0 t}{\pi(1-\nu)K} \ . \tag{33.16}$$

This defines a particular thermoelastic solution. We recall from Chapter 14 that the homogeneous solution corresponding to thermoelasticity is the general solution of the isothermal problem. Thus we now must superpose a suitable isothermal stress field to satisfy the mechanical boundary conditions of the problem.

The mechanical boundary conditions are that (i) the surfaces of the crack be traction free and (ii) the stress field should decay to zero at infinity. However, the antisymmetry of the problem once again permits us to define boundary conditions on the half-space $z > 0$ which are

$$\sigma_{zz} = \sigma_{zr} = 0 \ ; \quad 0 \le r < a, \ z = 0 \tag{33.17}$$
$$\sigma_{zz} = 0 \ ; \quad u_r = 0 \ ; \quad r > a, \ z = 0 \ . \tag{33.18}$$

Equations (33.17) state that the crack surfaces are traction free and (33.18) are symmetry conditions.

Notice that taken together, these conditions imply that $\sigma_{zz} = 0$ throughout the plane $z = 0$. This is therefore a global condition and can be satisfied by the choice of form. It is already satisfied by the particular solution (see Table 23.1) and hence it must also be satisfied by the additional isothermal solution, for which we therefore use Solution G of Table 22.3. Notice however that we shall use the symbol ϕ for the potential in Solution G to avoid confusion with χ in Solution T.

The complete solution is therefore obtained by superposing Solutions T and G, the surface values of σ_{zr} and u_r being

$$\sigma_{zr} = \frac{\partial \chi}{\partial r} + \frac{\partial^2 \phi}{\partial z \partial r} \; ; \quad 2\mu u_r = 2(1-\nu)\frac{\partial \phi}{\partial r} \; .$$

The mixed conditions in (33.17, 33.18) then require

$$\frac{\partial^2 \phi}{\partial r \partial z} = -\frac{\partial \chi}{\partial r} = \frac{1}{r}\int_0^r \frac{tg_3(t)dt}{\sqrt{r^2-t^2}}$$

$$= \frac{\mu\alpha(1+\nu)q_0 r}{2(1-\nu)K} \; ; \quad 0 \le r < a, \; z=0 \; , \tag{33.19}$$

and

$$\frac{\partial \phi}{\partial r} = 0 \; ; \quad r > a, \; z=0 \; , \tag{33.20}$$

where we used Table 32.1 and (33.16) to evaluate the right-hand side of (33.19).
   We can satisfy (33.20) by defining ϕ such that

$$\phi = \Re \int_0^a g_1(t)\mathcal{F}(r,z,t)dt \; , \tag{33.21}$$

with the auxiliary condition

$$\int_0^a g_1(t)dt = 0 \; , \tag{33.22}$$

(see Table 32.1).
   The other boundary condition (33.19) can then be integrated to give

$$\frac{\partial \phi}{\partial z} = \frac{\mu\alpha(1+\nu)q_0 r^2}{4(1-\nu)K} + C \; ; \quad 0 \le r < a, \; z=0 \; ,$$

where C is an arbitrary constant of integration, and hence

$$\int_0^r \frac{g_1(t)dt}{\sqrt{r^2-t^2}} = \frac{\mu\alpha(1+\nu)q_0 r^2}{4(1-\nu)K} + C \; ; \quad 0 \le r < a \; , \tag{33.23}$$

using (33.21) and Table 32.1. The solution of this Abel equation is

$$g_1(x) = \frac{2}{\pi}\left(\frac{\mu\alpha(1+\nu)q_0 x^2}{2(1-\nu)K} + C\right) \; , \tag{33.24}$$

using Table 32.2, and the auxiliary condition (33.22) can then be used to determine C giving

$$C = -\frac{\mu\alpha(1+\nu)q_0 a^2}{6(1-\nu)K} \; . \tag{33.25}$$

Finally, we can recover the expression for the stress σ_{zr} on $r > a$, $z=0$, which is

$$\sigma_{zr}(r,0) = \frac{\partial^2 \phi}{\partial r \partial z} + \frac{\partial \chi}{\partial r} = \frac{d}{dr} \int_0^a \frac{g_1(t)dt}{\sqrt{r^2 - t^2}} - \frac{1}{r} \int_0^a \frac{t g_3(t)dt}{\sqrt{r^2 - t^2}}$$

$$= -\frac{2\mu\alpha(1+\nu)q_0 a^3}{3\pi(1-\nu)Kr\sqrt{r^2 - a^2}},$$

from Table 32.1 and equations (33.16, 33.24, 33.25).

Notice that the antisymmetry of the problem ensures that the crack tip is loaded in shear only — we argued earlier that σ_{zz} would be zero throughout the plane $z=0$. It also follows from a similar argument that the crack remains closed — there is a displacement u_z at the crack plane, but it is the same on both sides of the crack.

The stress-intensity factor is of mode II (shear) form and is

$$K_{\mathrm{II}} = \lim_{r \to a^+} \sigma_{zr} \sqrt{2\pi(r-a)} = -\frac{2\mu\alpha(1+\nu)q_0 a^{3/2}}{3\sqrt{\pi}(1-\nu)K}.$$

Problems

33.1. An infinite homogeneous body contains an axisymmetric *external* crack — i.e. the crack extends over the region $r > a$, $z=0$. Alternatively, the external crack can be considered as two half-spaces bonded together over the region $0 \leq r < a$, $z=0$.

The body is loaded in tension, such that the total tensile force transmitted is F. Find an expression for the stress field and in particular determine the mode I stress-intensity factor K_{I}.

33.2. An infinite homogeneous body containing a penny-shaped crack is subjected to torsional loading $\sigma_{z\theta} = Cr$, $R \to \infty$, where C is a constant[4]. Use Solution E (Table 22.1) to formulate the problem and solve the resulting boundary-value problem using the methods of §32.2.

33.3. The axisymmetric external crack of radius a (see Problem 33.1) is unloaded, but is subjected to the steady-state thermal conditions

$$T(r,z) \to T_1 \; ; \quad z \to \infty$$
$$\to T_2 \; ; \quad z \to -\infty$$
$$q_z(r,0) = 0 \; ; \quad r > a.$$

In other words the extremities of the body are maintained at different temperatures T_1, T_2 causing heat to flow through the ligament $z=0$, $0 \leq r < a$ and the crack faces are insulated.

Find the mode II stress-intensity factor K_{II}.

---

[4] This is known as the Reissner-Sagoci problem.

33.4. The antisymmetry of the thermoelastic penny-shaped crack problem of §33.2 implies that there is no crack-opening displacement and hence the assumption of insulation (33.10) is arguably rather unrealistic. A more realistic assumption might be that the heat flux across the crack faces is proportional to the local temperature difference — i.e.

$$q_z(r, 0) = h \left(T(r, 0^-) - T(r, 0^+) \right) ; \quad 0 \le r < a .$$

Use the methods of Chapter 32 to formulate the heat conduction problem. In particular, express the temperature in terms of one of the forms (32.10) and find the Abel integral equation which must be satisfied by the function $g_3(t)$. Do not attempt to solve this equation.

33.5. A long rectangular beam defined by the cross section $-c < x < c$, $-d < y < d$ transmits a bending moment M about the positive x-axis. The beam contains a small penny-shaped crack of radius a in the cross-sectional plane $z = 0$ with its centre at the point $(0, b, 0)$ — i.e. the crack surface is defined by $0 \le \sqrt{x^2 + (y-b)^2} < a$, $z = 0$. Assuming that the crack opens completely and that $c, d, d-b$ are all large compared with a, find the variation of K_{I} around the crack edge and hence determine the minimum value of b for the open-crack assumption to be valid. Find also the crack-opening displacement as a function of position within the crack.

Notice that the corrective problem involves both axisymmetric and non-axisymmetric terms. For the latter, you will need to apply the methods introduced in §32.3.

33.6. The otherwise plane surface of an elastic half-space contains an axisymmetric dimple whose depth below the plane can be described by the function

$$h(r) = h_0 \left(1 - \frac{r^2}{a^2} \right)^2 ; \quad 0 \le r < a ,$$

where h_0 is the depth at $r = 0$. This body is placed in contact with an unblemished elastic half-space and experiences adhesive forces whose effect can be described using the JKR model of §31.3.

The bodies are now pulled apart by a far-field tensile stress $\sigma_{zz} = \sigma_0$. Assuming that the contact area is defined by $r > b$, where $0 < b < a$, find the relation between $b, \sigma_0, \Delta\gamma$ and the elastic properties of the materials. Assume that the contact is frictionless.

Plot an appropriate dimensionless measure of σ_0 as a function of b/a for some representative values of h_0 and comment on the implications of these plots for the stability of the corresponding contact configurations.

Chapter 34
Hertzian Contact

In this chapter, we shall consider the frictionless contact of two elastic bodies with profiles that can be described by general quadratic functions. This problem was first solved by Hertz[1] and has numerous practical applications. Special cases where the bodies are two-dimensional or axisymmetric were already considered in §12.7.1 and §32.2.5 respectively.

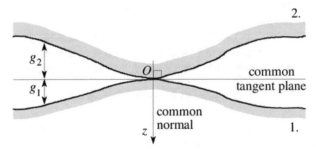

Fig. 34.1 Definition of profiles through gap functions.

Figure 34.1 shows two bodies with fairly general profiles that are placed in contact, but not loaded, so that contact occurs only at a single point O. The two surfaces must share a common tangent plane at O, since any other assumption would imply local interpenetration of material. We therefore define Cartesian coördinates such that x, y lie in this plane and z is the normal pointing into the lower body 1.

Supplementary Information The online version contains supplementary material available at https://doi.org/10.1007/978-3-031-15214-6_34.

---

[1] H. Hertz (1882), Über die Berührung fester elastischer Körper (On the contact of stiff elastic solids), *Journal für die reine und angewandte Mathematik*, Vol 92, pp. 156–171. (in German).

J. R. Barber, *Elasticity*, Solid Mechanics and Its Applications 172,
https://doi.org/10.1007/978-3-031-15214-6_34

We then define the profiles of the two bodies by two gap functions $g_1(x, y)$, $g_2(x, y)$ as shown in Figure 34.1, such that $g_i(x, y)$, $i = 1, 2$ defines the gap if body i is placed in contact with a rigid plane. The total gap as defined in Figure 12.12 is then given by

$$g_0(x, y) = g_1(x, y) + g_2(x, y) ,$$

and if both bodies have quadratic gap functions, g_0 will also be a quadratic function of x and y. Furthermore, since we take the initial contact point as origin, and since the gap functions are assumed to be twice differentiable (i.e. no discontinuous changes of slope), we must have

$$g_0(0, 0) = 0 \ ; \quad \frac{\partial g_0}{\partial x}(0, 0) = 0 \ ; \quad \frac{\partial g_0}{\partial y}(0, 0) = 0 .$$

Thus, the second derivatives of the gap function

$$\left\{ \frac{\partial^2 g_0}{\partial x^2} , \ \frac{\partial^2 g_0}{\partial x \partial y} , \ \frac{\partial^2 g_0}{\partial y \partial x} , \ \frac{\partial^2 g_0}{\partial y^2} \right\} \tag{34.1}$$

can be considered as defining the first non-zero terms in a Taylor series expansion of a general function $g_0(x, y)$, and since the contact region is generally expected to be small, the Hertzian theory provides a good approximation to the contact problem for fairly general smooth convex surfaces — not just those where the contacting surfaces are strictly quadratic.

The second derivatives (34.1) define a second order Cartesian tensor, so there is no loss of generality in taking the principal directions as the coördinate directions, in which case the most general quadratic gap function can be written

$$g_0(x, y) = Ax^2 + By^2 . \tag{34.2}$$

We recover the two-dimensional case of §12.7.1 if $B = 0$, and the axisymmetric case of §32.2.5 if $A = B$. For all other values, we shall show that the contact area $\mathcal{A}$ is an ellipse which we shall define as

$$\frac{x^2}{a^2} + \frac{y^2}{b^2} < 1 .$$

When a normal contact force F is applied, the bodies will approach each other by some distance Δ and there will be elastic displacements $u^{(1)}$, $u^{(2)}$ relative to the respective undeformed configurations of bodies 1 and 2. As in §12.7.1, these effects change the gap to

$$g(x, y) = g_0(x, y) + u_z^{(1)}(x, y, 0) - u_z^{(2)}(x, y, 0) - \Delta$$

and the condition that the gap be zero in the contact area leads to the equation

$$u_z^{(1)}(x, y, 0) - u_z^{(2)}(x, y, 0) = \Delta - g_0(x, y) = \Delta - Ax^2 - By^2 ; \quad (x, y) \in \mathcal{A} . \tag{34.3}$$

To solve this contact problem, we shall use a semi-inverse approach in which we assume provisionally that the contact pressure over the ellipse has the form

$$p(x, y) = p_0 \sqrt{1 - \frac{x^2}{a^2} - \frac{y^2}{b^2}} , \tag{34.4}$$

which is consistent with the equations (12.48) and (32.35) in the two-dimensional and axisymmetric cases respectively. We shall demonstrate that for the more general case, the resulting displacements have precisely the form (34.3) and we can therefore obtain three equations for the three unknowns p_0, a, b by equating coefficients.

34.1 Elastic Deformation

To find the elastic displacements due to the pressure distribution (34.4), we shall use a Green's function method similar to that in §12.7.1, except that we must now use the three-dimensional Boussinesq solution of §24.2.1 in place of the Flamant solution. In particular, a compressive force F at the origin on the surface of the half-space $z > 0$ produces a normal surface displacement

$$u_z(r, 0) = \frac{F(1 - \nu)}{2\pi \mu r} , \tag{34.5}$$

from equation (24.6). It follows that an equal and opposite compressive force pair at the origin will produce

$$u_z^{(1)}(x, y, 0) - u_z^{(2)}(x, y, 0) = \frac{F(1 - \nu_1)}{2\pi \mu_1 \sqrt{x^2 + y^2}} + \frac{F(1 - \nu_2)}{2\pi \mu_2 \sqrt{x^2 + y^2}}$$

$$= \frac{F}{\pi E^* \sqrt{x^2 + y^2}} , \tag{34.6}$$

where the composite contact modulus[2] E^* is defined as

$$\frac{1}{E^*} = \frac{(1 - \nu_1^2)}{E_1} + \frac{(1 - \nu_2^2)}{E_2} = \frac{(1 - \nu_1)}{2\mu_1} + \frac{(1 - \nu_2)}{2\mu_2} .$$

Using (34.6) as a Green's function, we can then write the contact condition (34.3) as an integral equation

$$\frac{1}{\pi E^*} \iint_A \frac{p(\xi, \eta) d\xi d\eta}{\sqrt{(x - \xi)^2 + (y - \eta)^2}} = \Delta - Ax^2 - By^2 ; \quad (x, y) \in A .$$

---

[2] Notice that for plane strain [$\kappa = (3 - 4\nu)$], the constant A of equation (12.42) is equal to $8/E^*$.

34.1.1 Field-point integration

In evaluating this integral, it is advantageous to move the origin to a field point $P(x, y)$ and revert to polar coördinates (r, θ) centred on P, since then the factor 'r' in the elemental area $r\,d\theta\,dr$ cancels with that in the Green's function (34.5). We obtain

$$\frac{1}{\pi E^*} \int_{-\pi/2}^{\pi/2} \int_{S_1}^{S_2} p(r, \theta)\,dr\,d\theta = \Delta - Ax^2 - By^2 , \qquad (34.7)$$

where the points S_1, S_2 are defined in Figure 34.2. Notice that we include contributions from the entire line $S_1 S_2$ in the inner integral, so that r should be interpreted as negative in the segment $S_1 P$.

Fig. 34.2 Elliptical contact area.

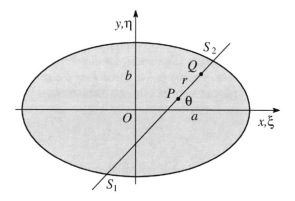

The coördinates of a general point $Q(\xi, \eta)$ are then defined through

$$\xi = x + r\cos\theta \; ; \quad \eta = y + r\sin\theta ,$$

so the pressure distribution (34.4) is

$$p(r, \theta) = p_0 \sqrt{1 - \frac{(x + r\cos\theta)^2}{a^2} - \frac{(y + r\sin\theta)^2}{b^2}}$$

$$= p_0 \sqrt{C_0 - C_1(\theta)r - C_2(\theta)r^2} , \qquad (34.8)$$

where

$$C_0 = 1 - \frac{x^2}{a^2} - \frac{y^2}{b^2} \; ; \quad C_1(\theta) = 2\left(\frac{x\cos\theta}{a^2} + \frac{y\sin\theta}{b^2}\right) \; ; \quad C_2(\theta) = \frac{\cos^2\theta}{a^2} + \frac{\sin^2\theta}{b^2} .$$

$$(34.9)$$

Also, the limits r_1, r_2 for the inner integral correspond to the two roots of the quadratic equation

$$C_0 - C_1(\theta)r - C_2(\theta)r^2 = 0 \, ,$$

since $p(r, \theta) = 0$ at S_1 and S_2, from (34.4).

If (34.8) is substituted into (34.7), the inner integral can be evaluated in terms of elementary functions as

$$\int_{S_1}^{S_2} p(r, \theta)dr = p_0 \int_{r_1}^{r_2} \sqrt{C_0 - C_1(\theta)r - C_2(\theta)r^2}\,dr$$

$$= \frac{\pi p_0}{2\sqrt{C_2}} \left(C_0 + \frac{C_1^2}{4C_2} \right) \, ,$$

and hence

$$\frac{p_0}{2E^*} \int_{-\pi/2}^{\pi/2} \left(C_0 + \frac{C_1^2}{4C_2} \right) \frac{d\theta}{\sqrt{C_2}} = \Delta - Ax^2 - By^2 \, .$$

In view of the definitions (34.9), the left-hand side of this equation will clearly evaluate to an even quadratic function of x and y, so it can be satisfied exactly if a, b and p_0 are chosen appropriately. Equating coefficients of x^2, y^2 and constants respectively, we obtain

$$A = \frac{p_0ab}{E^*} \int_0^{\pi/2} \frac{\sin^2\theta d\theta}{(b^2\cos^2\theta + a^2\sin^2\theta)^{3/2}} \qquad (34.10)$$

$$B = \frac{p_0ab}{E^*} \int_0^{\pi/2} \frac{\cos^2\theta d\theta}{(b^2\cos^2\theta + a^2\sin^2\theta)^{3/2}} \qquad (34.11)$$

$$\Delta = \frac{p_0ab}{E^*} \int_0^{\pi/2} \frac{d\theta}{\sqrt{b^2\cos^2\theta + a^2\sin^2\theta}} \, . \qquad (34.12)$$

Greenwood[3] shows that these results can be expressed in terms of Carlson elliptic integrals using the change of variable $t = \cot^2\theta$. We obtain

$$A = \frac{p_0a}{3E^*b^2} R_D \left(0, 1; \frac{a^2}{b^2} \right) \; ; \qquad B = \frac{p_0b}{3E^*a^2} R_D \left(0, 1; \frac{b^2}{a^2} \right) \; ; \qquad (34.13)$$

$$\Delta = \frac{p_0a}{E^*} R_F \left(0, 1, \frac{a^2}{b^2} \right) = \frac{p_0b}{E^*} R_F \left(0, 1, \frac{b^2}{a^2} \right) \, , \qquad (34.14)$$

where

---

[3] J. A. Greenwood (2018), Hertz theory and Carlson elliptic integrals, *Journal of the Mechanics and Physics of Solids*, Vol. 119, pp. 240–249. Notice that Greenwood uses a modified definition of R_D without the factor of 3, which he then denotes by $R_D^\otimes$.

$$R_F(p, q, r) = \frac{1}{2} \int_0^\infty \frac{dt}{\sqrt{(t + p)(t + q)(t + r)}}$$

$$R_D(p, q; r) = \frac{3}{2} \int_0^\infty \frac{dt}{(t + r)\sqrt{(t + p)(t + q)(t + r)}} \ .$$

The expressions (34.10–34.12) can also be defined in terms of the more familiar Legendre elliptic integrals

$$K(k) = \int_0^{\pi/2} \frac{d\theta}{\sqrt{1 - k^2 \cos^2 \theta}} \quad ; \quad E(k) = \int_0^{\pi/2} \sqrt{1 - k^2 \cos^2 \theta} d\theta$$

as

$$A = \frac{p_0 b}{E^* a^2 k^2} [K(k) - E(k)] \quad ; \quad B = \frac{p_0 b}{E^* a^2 k^2} \left(\frac{E(k)}{(1 - k^2)} - K(k) \right) \qquad (34.15)$$

$$\Delta = \frac{p_0 b K(k)}{E^*} \ , \quad \text{where} \quad k = \sqrt{1 - \frac{b^2}{a^2}}$$

is the eccentricity of the contact ellipse. The definitions of Legendre's integrals require that the axes be oriented such that $b \le a$ and hence $B \ge A$, whereas the original expressions (34.10–34.12) and the Carlson integral expressions (34.13, 34.14) are clearly symmetric with respect to the axis directions, and hence do not require this preliminary step.

34.2   Solution Procedure

In a typical technical problem, the coefficients A and B will be determined from the shape of the contacting bodies, and either the rigid-body displacement Δ, or the applied force

$$F = \iint_A p(x, y) dx dy = \frac{2\pi p_0 ab}{3} \qquad (34.16)$$

will be prescribed. Equations (34.13) and either (34.14) or (34.16) then define three equations for the unknowns a, b, p_0.

From equations (34.15) or (34.13), we obtain

$$\frac{B}{A} = \frac{1}{[K(k) - E(k)]} \left(\frac{E(k)}{(1 - k^2)} - K(k) \right) = \left(\frac{b}{a} \right)^3 \frac{R_D(0, 1; b^2/a^2)}{R_D(0, 1; a^2/b^2)} \ , \qquad (34.17)$$

which shows that k, or equivalently the ratio b/a, is a unique function of B/A. Thus the shape of the ellipse remains unchanged as the load or the rigid-body displacement is increased, and it can be found by solving this non-linear equation. This relationship

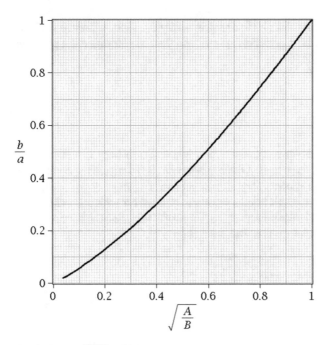

Fig. 34.3 Relation between $\sqrt{A/B}$ and b/a.

is shown in Figure 34.3, where we note that $\sqrt{A/B}$ represents the ratio of the semi-axes of the elliptical contours of the undeformed composite surface $g_0(x, y)$ from equation (34.2). We note that b/a is always less than $\sqrt{A/B}$ and hence the contact ellipse has a larger eccentricity than that of the undeformed contours.

34.2.1 Axisymmetric bodies

In many applications, the contacting bodies will be axisymmetric — typically cylinders, spheres or cones — and the individual gap functions $g_1(x, y)$, $g_2(x, y)$ are then of particularly simple form. For example, for a sphere of radius R we have

$$g(x, y) = \frac{x^2 + y^2}{2R} \, ,$$

and for a cylinder of radius R with its axis aligned with the x-axis,

$$g(x, y) = \frac{y^2}{2R} \, .$$

However, if the gap functions g_1, g_2 thereby determined are in different Cartesian coördinate systems as in Problem 34.1, it will be necessary to use the tensor transformation rules (1.16) or equivalently the Mohr's circle equations (1.9–1.11) to rotate them into a common system and then to determine the principal values A, B in equation (34.2).

Example

Figure 34.4 shows a cross-sectional view of a railway wheel in contact with a rail. The radius of the wheel at the contact point is 0.5 m and the contact surface is a cone of angle 5° as shown. The rail contact surface can be approximated as a cylinder of radius 100 mm. Both bodies are made of steel for which $E = 210$ GPa, $\nu = 0.3$.

If the wheel transmits a vertical force of 10 kN, find the maximum contact pressure p_0 and the dimensions of the contact area.

Fig. 34.4 Wheel-rail contact

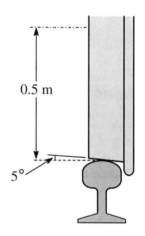

0.5 m

5°

Both materials are of steel, so

$$E^* = \frac{E}{2(1 - \nu^2)} = \frac{210}{2 \times 0.91} = 115 \text{ GPa} .$$

The contact force must be

$$F = 10,000/\cos(5°) = 10,038 \text{ N}$$

in order to have a vertical component of 10 kN.

The local radius of the conical surface is

$$\frac{500}{\cos(5°)} = 502 \text{ mm} ,$$

so if we define the x-direction as parallel with the rail, we have

$$g_1(x, y) = \frac{y^2}{200} ; \quad g_2(x, y) = \frac{x^2}{1004}$$

and

$$g_0(x, y) = g_1(x, y) + g_2(x, y) = Ax^2 + By^2$$

with

$$A = 0.000996 \text{ mm}^{-1} ; \quad B = 0.005 \text{ mm}^{-1} .$$

Noting that $B > A$, we can solve equation (34.17) for k, obtaining

$$k = 0.9386 \quad \text{or} \quad \frac{b}{a} = \sqrt{1 - k^2} = 0.345 .$$

Maple and Mathematica codes for this solution are available in the files 'Hertz'. Alternatively, we could use the above values of A, B and estimate b/a from Figure 34.3.

Eliminating p_0 between equations (34.16) and (34.15)$_1$, we obtain

$$a = \left(\frac{3F[K(k) - E(k)]}{2\pi AE^*k^2} \right)^{1/3} = \left(\frac{3 \times 10,038 \times [K(0.939) - E(0.939)]}{2\pi \times 0.000996 \times 115 \times 10^3 \times 0.939^2} \right)^{1/3}$$
$$= 4.10 \text{ mm}$$

so

$$b = 0.345 \times 4.10 = 1.41 \text{ mm}$$

and

$$p_0 = \frac{3F}{2\pi ab} = \frac{3 \times 10,038}{2\pi \times 4.10 \times 1.41} = 828 \text{ MPa} ,$$

noting that 1 MPa = 1 N/mm^2.

Problems

34.1. Two identical cylinders of radius R are pressed together by a force F. If the axes of the two cylinders are inclined at an angle of 45°, find the ratio of the semi-axes b/a and the maximum contact pressure p_0 in terms of F, R and E^*.

34.2. Find a relation between the force F and the indentation depth Δ that depends only on the shape of the contacting bodies and E^*. Hence show that the incremental stiffness

$$\frac{dF}{d\Delta} = CF^{1/3}$$

where C is a constant.

34.3. An incompressible elastic half-space with modulus $\mu = 400$ MPa has a sinusoidal profile defined by

$$g(x) = -2\cos(0.1x) ,$$

where g and x are in mm. A rigid sphere of radius 40 mm is pressed into the trough of the sinusoid by a force F. Find the eccentricity of the resulting elliptical contact and estimate the value of F beyond which the Hertzian theory can be expected to be significantly in error.

34.4. For a given Hertzian contact problem, the applied force F, the contact pressure $p(x, y)$ and the displacements $u_z(x, y)$ are related by equations (34.16, 34.4, 34.3, 34.13, 34.14). Find the corresponding quantities for a force $F + \Delta F$ applied to the same elastic bodies and construct the difference between these two solutions — e.g. $p(x, y, F + \Delta F) - p(x, y, F)$ and the corresponding incremental displacements. By taking the limit as $\Delta F \to 0$, find the contact pressure distribution and the indentation depth for a flat rigid punch of elliptical cross section with semi-axes a, b.

34.5. Figure 34.5 shows a ball bearing with an inner race of radius 36 mm, an outer race of radius 48 mm and 16 balls of 12 mm diameter. Both inner and outer race surfaces have a (concave) groove of radius 12 mm.

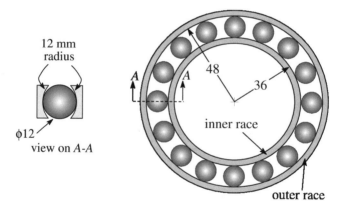

Fig. 34.5 A ball bearing

Find the eccentricity k and the semi-axis ratio a/b of the contact area between the balls and the inner race and also for the contact area at the outer race.

Chapter 35
The Interface Crack

The problem of a crack at the interface between dissimilar elastic media is of considerable contemporary importance in Elasticity, because of its relevance to the problem of debonding of composite materials and structures. Figure 35.1 shows the case where such a crack occurs at the plane interface between two elastic half-spaces. We shall use the suffices 1,2 to distinguish the stress functions and mechanical properties for the lower and upper half-spaces, $z>0$, $z<0$, respectively.

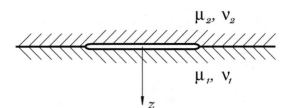

μ_2, ν_2

μ_1, ν_1

Fig. 35.1 The plane interface crack.

z

The difference in material properties destroys the symmetry that we exploited in the previous chapter and we therefore anticipate shear stresses as well as normal stresses at the interface, even when the far field loading is a state of uniform tension. However, we shall show that we can still use the same methods to reduce the problem to a mixed boundary-value problem for the half-space.

35.1 The Uncracked Interface

The presence of the interface causes the problem to be non-trivial, even if there is no crack and the two bodies are perfectly bonded.

Consider the tensile loading of the composite bar of Figure 35.2(a).

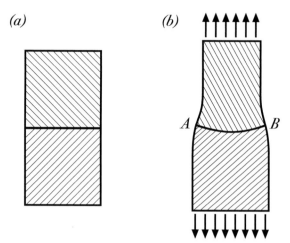

Fig. 35.2 The composite bar in tension.

If we assume a state of uniform uniaxial tension $\sigma_{zz} = S$ exists throughout the bar, the Poisson's ratio strains, $e_{xx} = e_{yy} = -\nu_i S / E_i$ will imply an inadmissible discontinuity in displacements u_x, u_y at the interface, unless $\nu_1 / E_1 = \nu_2 / E_2$. In all other cases, a locally non-uniform stress field will be developed involving shear stresses on the interface and the bar will deform as shown in Figure 35.2(b). Indeed, for many material combinations, there will be a stress singularity at the edges A, B, which can be thought of locally as two bonded orthogonal wedges[1] and analyzed by the method of §11.2.

We shall not pursue the perfectly-bonded problem here — a solution for the corresponding two-dimensional problem is given by Bogy[2] — but we note that as long as the solution for the uncracked interface is known, the corresponding solution when a crack is present can be obtained as in §§13.3.2, 33.1, by superposing a corrective solution in which tractions are imposed on the crack face, equal and opposite to those transmitted in the uncracked state, and the distant boundaries of the body are traction free. Furthermore, if the crack is small compared with the other linear dimensions of the body — notably the distance from the crack to the free boundary — it can be conceived as occurring at the interface between two bonded half-spaces, as in Figure 35.1.

[1] D. B. Bogy (1968), Edge-bonded dissimilar orthogonal elastic wedges under normal and shear loading, *ASME Journal of Applied Mechanics,* Vol. 35, pp. 460–466.

[2] D. B. Bogy (1975), The plane solution for joined dissimilar elastic semistrips under tension, *ASME Journal of Applied Mechanics.,* Vol. 42, pp. 93–98.

35.2 The Corrective Solution

If we assume that the loading conditions are such as to cause the crack to open fully and therefore to be traction free, the corrective solution will be defined by the boundary conditions

$$
\begin{aligned}
\sigma_{xz}^{(1)}(x, y, 0^+) &= \sigma_{xz}^{(2)}(x, y, 0^-) = -S_1(x, y) \\
\sigma_{yz}^{(1)}(x, y, 0^+) &= \sigma_{yz}^{(2)}(x, y, 0^-) = -S_2(x, y) \\
\sigma_{zz}^{(1)}(x, y, 0^+) &= \sigma_{zz}^{(2)}(x, y, 0^-) = -S_3(x, y) ,
\end{aligned} \tag{35.1}
$$

for $(x, y) \in \mathcal{A}$ and

$$
\sigma_{xx}^{(i)}, \sigma_{yy}^{(i)}, \sigma_{zz}^{(i)}, \sigma_{xy}^{(i)}, \sigma_{yz}^{(i)}, \sigma_{zx}^{(i)} \to 0 \; ; \quad R \to \infty , \tag{35.2}
$$

where $\mathcal{A}$ is the region of the interfacial plane $z = 0$ occupied by the crack and S_1, S_2, S_3 are the tractions that would be transmitted across the interface in the absence of the crack.

We shall seek a potential function solution of the problem in the half-space, so we supplement (35.1–35.2) by the continuity and equilibrium conditions

$$
\sigma_{xz}^{(1)}(x, y, 0^+) = \sigma_{xz}^{(2)}(x, y, 0^-) \tag{35.3}
$$

$$
\sigma_{yz}^{(1)}(x, y, 0^+) = \sigma_{yz}^{(2)}(x, y, 0^-) \tag{35.4}
$$

$$
\sigma_{zz}^{(1)}(x, y, 0^+) = \sigma_{zz}^{(2)}(x, y, 0^-) \tag{35.5}
$$

$$
u_x^{(1)}(x, y, 0+) = u_x^{(2)}(x, y, 0-) \tag{35.6}
$$

$$
u_y^{(1)}(x, y, 0+) = u_y^{(2)}(x, y, 0-) \tag{35.7}
$$

$$
u_z^{(1)}(x, y, 0+) = u_z^{(2)}(x, y, 0-) , \tag{35.8}
$$

for $(x, y) \in \bar{\mathcal{A}}$ — i.e. the uncracked part of the interface[3].

It is clear from (35.1) that (35.3–35.5) apply throughout the interface and are therefore global conditions. We can exploit this fact by expressing the fields in the two bonded half-spaces by separate sets of stress functions and then developing simple symmetric relations between the sets.

35.2.1 Global conditions

The global conditions involve the three stress components σ_{zx}, σ_{zy}, σ_{zz} at the surface $z = 0$, so it is convenient to choose a formulation in which these components take

---

[3] Notice that only those stress components that act on the interface (and hence have a z-suffix) are continuous across the interface, because of the equilibrium requirement. The remaining three components will generally be discontinuous because of (35.6–35.8) and the dissimilar elastic properties.

simple forms at the surface and are as far as possible uncoupled. This can be achieved by superposing Solutions E,F,G of Chapter 22, for which

$$\sigma_{zx}^{(i)}(x, y, 0) = \frac{\partial^2 \Psi_i}{\partial y \partial z} + \frac{\partial^2 \chi_i}{\partial x \partial z} \tag{35.9}$$

$$\sigma_{zy}^{(i)}(x, y, 0) = -\frac{\partial^2 \Psi_i}{\partial x \partial z} + \frac{\partial^2 \chi_i}{\partial y \partial z} \tag{35.10}$$

$$\sigma_{zz}^{(i)} = -\frac{\partial^2 \varphi_i}{\partial z^2} , \tag{35.11}$$

from Tables 22.1, 22.3, where the index i takes the value 1,2 for bodies 1,2 respectively.

The global condition (35.5) now reduces to

$$\frac{\partial^2 \varphi_1}{\partial z^2}(x, y, 0^+) = \frac{\partial^2 \varphi_2}{\partial z^2}(x, y, 0^-)$$

and, remembering that φ_1 is defined only in $z > 0$ and φ_2 only in $z < 0$, we can satisfy it by imposing the symmetry relation

$$\varphi_1(x, y, z) = \varphi_2(x, y, -z) \equiv \varphi(x, y, z) \tag{35.12}$$

throughout the half-spaces. In the same way[4], the two remaining global conditions can be satisfied by demanding

$$\Psi_1(x, y, z) = -\Psi_2(x, y, -z) \equiv \Psi(x, y, z) \tag{35.13}$$
$$\chi_1(x, y, z) = -\chi_2(x, y, -z) \equiv \chi(x, y, z) , \tag{35.14}$$

where the negative sign in (35.13, 35.14) arises from the fact that symmetry of the derivatives $\partial \Psi / \partial z$, $\partial \chi / \partial z$ implies *antisymmetry* of Ψ, χ.

35.2.2   Mixed conditions

We can now use the remaining boundary conditions (35.1) and (35.6–35.8) to define a mixed boundary-value problem for the three potentials φ, Ψ, χ in the half-space $z > 0$.

For example, using the expressions from Tables 22.1, 22.3 and the definitions (35.12–35.14), the boundary condition (35.6) reduces to

---

[4] This idea can also be extended to steady-state thermoelastic problems by adding in Solution P of Table 23.1, which leaves the expressions (35.9–35.11) for the interface stresses unchanged and requires a further symmetry relation between ψ_1, ψ_2 to satisfy the global condition of continuity of heat flux, $q_z^{(1)}(x, y, 0^+) = q_z^{(2)}(x, y, 0^-)$. See Problem 35.2.

$$\frac{1}{\mu_1}\frac{\partial\Psi}{\partial y}+\frac{(1-2\nu_1)}{2\mu_1}\frac{\partial\varphi}{\partial x}+\frac{(1-\nu_1)}{\mu_1}\frac{\partial\chi}{\partial x}=-\frac{1}{\mu_2}\frac{\partial\Psi}{\partial y}+\frac{(1-2\nu_2)}{2\mu_2}\frac{\partial\varphi}{\partial x}-\frac{(1-\nu_2)}{\mu_2}\frac{\partial\chi}{\partial x}$$

in $\bar{A}$, $z=0$.

Rearranging the terms and dividing by the non-zero factor $[(1-\nu_1)/\mu_1+(1-\nu_2)/\mu_2]$ (which is $A/4$ in the terminology of equation (12.42) for plane strain), we obtain

$$\gamma\frac{\partial\Psi}{\partial y}+\beta\frac{\partial\varphi}{\partial x}+\frac{\partial\chi}{\partial x}=0 \; ; \; \text{in } \bar{A} , \tag{35.15}$$

where β is Dundurs' constant (see §§4.3.3, 12.7) and

$$\gamma=\left(\frac{1}{\mu_1}+\frac{1}{\mu_2}\right)\Bigg/\left(\frac{(1-\nu_1)}{\mu_1}+\frac{(1-\nu_2)}{\mu_2}\right) .$$

A similar procedure applied to the boundary conditions (35.7, 35.8) yields

$$-\gamma\frac{\partial\Psi}{\partial x}+\beta\frac{\partial\varphi}{\partial y}+\frac{\partial\chi}{\partial y}=0 \tag{35.16}$$

$$\frac{\partial\varphi}{\partial z}+\beta\frac{\partial\chi}{\partial z}=0 , \tag{35.17}$$

in $\bar{A}$.

In addition to (35.15, 35.16, 35.17), three further conditions are obtained from (35.1), using (35.9–35.11) and (35.12–35.14). These are

$$\frac{\partial^2\Psi}{\partial y\partial z}+\frac{\partial^2\chi}{\partial x\partial z}=-S_1(x, y) \tag{35.18}$$

$$-\frac{\partial^2\Psi}{\partial x\partial z}+\frac{\partial^2\chi}{\partial y\partial z}=-S_2(x, y) \tag{35.19}$$

$$-\frac{\partial^2\varphi}{\partial z^2}=-S_3(x, y) , \tag{35.20}$$

in $\mathcal{A}$. The six conditions (35.15, 35.16–35.20) define a well-posed boundary-value problem for the three potentials φ, Ψ, χ, when supplemented by the requirements that (i) displacements should be continuous at the crack boundary (between $\mathcal{A}$ and $\bar{A}$) and (ii) the stress and displacement fields should decay as $R\rightarrow\infty$ in accordance with (35.2).

The two boundary conditions (35.15, 35.16) involve only two independent functions Ψ and $(\beta\varphi+\chi)$ and hence we can eliminate either by differentiation with the result

$$\frac{\partial^2 \Psi}{\partial x^2} + \frac{\partial^2 \Psi}{\partial y^2} = -\frac{\partial^2 \Psi}{\partial z^2} = 0 \qquad (35.21)$$

$$\left(\frac{\partial^2}{\partial x^2} + \frac{\partial^2}{\partial y^2} \right) (\beta\varphi + \chi) = -\beta \frac{\partial^2 \varphi}{\partial z^2} - \frac{\partial^2 \chi}{\partial z^2} = 0 \qquad (35.22)$$

in $\bar{\mathcal{A}}$, where we have used the fact that the potential functions are all harmonic to express the boundary conditions in terms of a single derivative normal to the plane.

In the same way, (35.18, 35.19) yield

$$\frac{\partial^3 \Psi}{\partial z^3} = \frac{\partial S_1}{\partial y} - \frac{\partial S_2}{\partial x} \qquad (35.23)$$

$$\frac{\partial^3 \chi}{\partial z^3} = \frac{\partial S_1}{\partial x} + \frac{\partial S_2}{\partial y} , \qquad (35.24)$$

in $\mathcal{A}$.

Equations (35.21, 35.23) now define a two-part boundary-value problem for Ψ, similar to those solved in Chapter 32 for the axisymmetric geometry. It should be noted however that it is not necessary, nor is it generally possible, to satisfy the requirement of continuity of in-plane displacement u_x, u_y at the crack boundary in the *separate* contributions from the potential functions φ, Ψ, χ, as long as the *superposed* fields satisfy this requirement. Thus, the homogeneous problem for Ψ obtained by setting the right-hand side of (35.23) to zero has non-trivial solutions which are physically unacceptable in isolation, since they correspond to displacement fields which are discontinuous at the crack boundary, but they are an essential component of the solution of more general problems, where they serve to cancel similar discontinuities resulting from the fields due to φ, χ. This problem also arises for the crack in a homogeneous medium with shear loading[5] and in non-axisymmetric thermoelastic problems[6]

The remaining boundary conditions define a coupled two-part problem for φ, χ. As in the two-dimensional contact problems of Chapter 12, the coupling is proportional to the Dundurs' constant β. If the material properties are such that $\beta = 0$ (see §12.7), the problems of normal and shear loading are independent and (for example) the crack in a shear field has no tendency to open or close.

---

[5] See for example J. R. Barber (1975), The penny-shaped crack in shear and related contact problems, *International Journal of Engineering Science*, Vol. 13, pp. 815–832.

[6] L. Rubenfeld (1970), Non-axisymmetric thermoelastic stress distribution in a solid containing an external crack, *International Journal of Engineering Science*, Vol. 8, pp. 499–509, J. R. Barber (1975), Steady-state thermal stresses in an elastic solid containing an insulated penny-shaped crack, *Journal of Strain Analysis*, Vol. 10, pp. 19–24.

35.3 The Penny-shaped Crack in Tension

We now consider the special case of the circular crack $0 \leq r < a$, subjected to uniform tensile loading $S_3 = S$, $S_1 = S_2 = 0$. The boundary-value problem for φ, χ is then defined by the conditions[7]

$$\frac{\partial^3 \chi}{\partial z^3} = 0 \ ; \ z = 0, \ 0 \leq r < a \tag{35.25}$$

$$\frac{\partial^2 \varphi}{\partial z^2} = S \ ; \ z = 0, \ 0 \leq r < a \tag{35.26}$$

$$\frac{\partial \varphi}{\partial z} + \beta \frac{\partial \chi}{\partial z} = 0 \ ; \ z = 0, \ r > a \tag{35.27}$$

$$\beta \frac{\partial^2 \varphi}{\partial z^2} + \frac{\partial^2 \chi}{\partial z^2} = 0 \ ; \ z = 0, \ r > a \ . \tag{35.28}$$

From Table 32.1, we see that we can satisfy (35.27, 35.28) by writing[8]

$$\beta \varphi + \chi = \phi_1 \ , \ \varphi + \beta \chi = \phi_3$$

— i.e.

$$\varphi = \frac{\phi_3 - \beta \phi_1}{(1 - \beta^2)} \ , \ \chi = \frac{\phi_1 - \beta \phi_3}{(1 - \beta^2)} \ , \tag{35.29}$$

where ϕ_1, ϕ_3 are defined by equation (32.10).

The remaining boundary conditions then define two coupled integral equations of Abel-type for the unknown functions $g_1(t)$, $g_3(t)$ in the range $0 \leq t < a$. To develop these equations, we first note that the condition $\nabla^2 \chi = 0$ and the fact that the solution is axisymmetric enable us to write (35.25) in the form

$$\frac{1}{r} \frac{d}{dr} r \frac{d}{dr} \frac{\partial \chi}{\partial z}(r, 0) = 0 \ ; \ z = 0, \ 0 \leq r < a \ .$$

This can be integrated within the plane $z = 0$ to give

$$\frac{\partial \chi}{\partial z} = C \ ; \ z = 0, \ 0 \leq r < a \ , \tag{35.30}$$

---

[7] This problem was first considered by V. I. Mossakovskii and M. T. Rybka (1964), Generalization of the Griffith-Sneddon criterion for the case of a non-homogeneous body, *Journal of Applied Mathematics and Mechanics,* Vol. 28, pp. 1277–1286, and by F. Erdogan (1965), Stress distribution on bonded dissimilar materials containing circular or ring-shaped cavities, *ASME Journal of Applied Mechanics,* Vol. 32, pp. 829–836. A formulation for the penny-shaped crack with more general loading was given by J. R. Willis (1972), The penny-shaped crack on an interface,*Quarterly Journal of Mechanics and Applied Mathematics,* Vol. 25, pp. 367–382.

[8] This method would fail for the special case $\beta = \pm 1$, but materials with positive Poisson's ratio are restricted by energy considerations to values in the range $-\frac{1}{2} \leq \beta \leq \frac{1}{2}$.

where C is a constant which will ultimately be chosen to ensure continuity of displacements at $r = a$, and we have eliminated a logarithmic term to preserve continuity of displacements at the origin.

Table 32.1 and (35.29, 35.30) now enable us to write

$$\int_0^r \frac{g_1(t)dt}{\sqrt{r^2 - t^2}} + \beta \int_r^a \frac{g_3(t)dt}{\sqrt{t^2 - r^2}} = (1 - \beta^2)C \; ; \;\; 0 \le r < a \,, \qquad (35.31)$$

whilst (35.26, 35.29) and Table 32.1 give

$$\frac{1}{r}\frac{d}{dr}\int_0^r \frac{tg_3(t)dt}{\sqrt{r^2 - t^2}} - \frac{\beta}{r}\frac{d}{dr}\int_r^a \frac{tg_1(t)dt}{\sqrt{t^2 - r^2}} = (1 - \beta^2)S \; ; \;\; 0 \le r < a \,, \quad (35.32)$$

which can be integrated to give

$$\int_0^r \frac{tg_3(t)dt}{\sqrt{r^2 - t^2}} - \beta \int_r^a \frac{tg_1(t)dt}{\sqrt{t^2 - r^2}} = \frac{(1 - \beta^2)Sr^2}{2} + B \; ; \;\; 0 \le r < a \,, \quad (35.33)$$

where B is an arbitrary constant.

We now treat (35.31) as an Abel equation for $g_1(t)$, carrying the g_3 integral onto the right-hand side and using the inversion rules (Table 32.2) to obtain

$$g_1(x) = \frac{2\beta}{\pi} \int_0^a \frac{tg_3(t)dt}{(x^2 - t^2)} + \frac{2C(1 - \beta^2)}{\pi} \,, \qquad (35.34)$$

where we have simplified the double integral term in g_3 by changing the order of integration and performing the resulting inner integral.

In the same way, treating (35.33) as an equation for $tg_3(t)$, we obtain

$$xg_3(x) = -\frac{2\beta}{\pi} \int_0^a \frac{t^2 g_1(t)dt}{(x^2 - t^2)} + \frac{2B}{\pi} + \frac{2S(1 - \beta^2)x^2}{\pi} \,. \qquad (35.35)$$

The function $g_3(x)$ must be bounded at $x = 0$ and hence B must be chosen so that

$$B = -\beta \int_0^a g_1(t)dt \,.$$

Thus, (35.35) reduces to

$$g_3(x) = -\frac{2\beta x}{\pi} \int_0^a \frac{g_1(t)dt}{(x^2 - t^2)} + \frac{2S(1 - \beta^2)x}{\pi} \,. \qquad (35.36)$$

35.3.1 Reduction to a single equation

We observe from equations (35.34, 35.36) that g_1 is an even function of x, whereas g_3 is odd. Using this result and expanding the integrands as partial fractions, we arrive at the simpler expressions[9]

$$g_1(x) = \frac{\beta}{\pi} \int_{-a}^{a} \frac{g_3(t)dt}{(x-t)} + \frac{2C(1-\beta^2)}{\pi} \quad ; \quad -a < x < a \qquad (35.37)$$

$$g_3(x) = -\frac{\beta}{\pi} \int_{-a}^{a} \frac{g_1(t)dt}{(x-t)} + \frac{2S(1-\beta^2)x}{\pi} \quad ; \quad -a < x < a . \qquad (35.38)$$

Two methods are available for reducing (35.37, 35.38) to a single equation. The most straightforward approach is to use (35.38) (for example) to substitute for g_3 in (35.37), resulting after some manipulations in the integral equation

$$g_1(x) + \frac{\beta^2}{\pi^2(1-\beta^2)} \int_{-a}^{a} \left(\ln\left|\frac{a-t}{a+t}\right| - \ln\left|\frac{a-x}{a+x}\right| \right) \frac{g_1(t)dt}{(t-x)}$$

$$= -\frac{2\beta S}{\pi^2}\left(2a + x \ln\left|\frac{a-x}{a+x}\right| \right) + \frac{2C}{\pi} \quad ; \quad -a < x < a . \qquad (35.39)$$

The kernel of this equation is bounded at $t = x$, the limiting form being obtainable by l'Hôpital's rule as $2a/(x^2-a^2)$. Equations of this kind with definite integration limits and bounded kernels are known as *Fredholm integral equations*[10].

However, the identity of kernel and range of integration in equations (35.37, 35.38) permits a more direct approach. Multiplying (35.38) by $\imath$ and adding the result to (35.37), we find that both equations are contained in the complex equation

$$g(x) + \frac{\imath\beta}{\pi} \int_{-a}^{a} \frac{g(t)dt}{(x-t)} = \frac{2(1-\beta^2)}{\pi}(C + \imath Sx) ,$$

where $g(x) \equiv g_1(x) + \imath g_3(x)$. Furthermore, this equation can be inverted explicitly[11] in the form of an integral of the right-hand side, leading to a closed-form solution for $g(x)$. The tractions at the interface can then be recovered by separating $g(x)$ into its real and imaginary parts and substituting into the expressions

---

[9] *cf.* equation (32.13) and the associated discussion.

[10] F. G. Tricomi, *Integral Equations*, Interscience, New York, 1957, Chapter 2.

[11] N. I. Muskhelishvili, *Singular Integral Equations*, (English translation by J. R. M. Radok, Noordhoff, Groningen, 1953)

$$\sigma_{zr} = \frac{1}{(1-\beta^2)} \frac{d}{dr} \int_0^a \frac{g_1(t)dt}{\sqrt{r^2-t^2}} \quad ; \quad z = 0, \ r > a \tag{35.40}$$

$$\sigma_{zz} = -\frac{1}{(1-\beta^2)r} \frac{d}{dr} \int_0^a \frac{tg_3(t)dt}{\sqrt{r^2-t^2}} \quad ; \quad z = 0, \ r > a \ . \tag{35.41}$$

35.3.2   Oscillatory singularities

An asymptotic analysis of the integrals (35.40, 35.41) at $r = a + s$, $s \ll a$, shows that they have the form

$$\sigma_{zz} + \iota\sigma_{zr} = SK\left(\frac{s}{2a}\right)^{-\frac{1}{2}+\iota\epsilon} + O(1)$$

$$= SK\left(\frac{s}{2a}\right)^{-\frac{1}{2}}\left[\cos\left(\epsilon\ln\frac{s}{2a}\right) + \iota\sin\left(\epsilon\ln\frac{s}{2a}\right)\right] + O(1) , \tag{35.42}$$

where

$$\epsilon = \frac{1}{2\pi}\ln\left(\frac{1+\beta}{1-\beta}\right)$$

and Kassir and Bregman[12] have found the complex constant K to be

$$K = \frac{1}{\sqrt{\pi}}\frac{\Gamma(2+\iota\epsilon)}{\Gamma(\frac{1}{2}+\iota\epsilon)} \ .$$

Equation (35.42) shows that the stresses are square-root singular at the crack tip, but that they also oscillate with increasing frequency as r approaches a and $\ln(s/2a) \to -\infty$. This behaviour is common to all interface crack problems for which $\beta \neq 0$ and can be predicted from an asymptotic analysis similar to that of §11.2.2, from which a complex leading singular eigenvalue $\lambda = \frac{1}{2}\pm\iota\epsilon$ is obtained.

35.4   The Contact Solution

A more disturbing feature of this behaviour is that the crack opening displacement, $(u_z^{(1)}(r,0) - u_z^{(2)}(r,0))$, also oscillates as $r \to a^-$ and hence there are infinitely many regions of interpenetration of material near the crack tip.

---

[12] M. K. Kassir and A. M. Bregman (1972), The stress-intensity factor for a penny-shaped crack between two dissimilar materials, *ASME Journal of Applied Mechanics,* Vol. 39, pp. 308–310.

This difficulty was first resolved by Comninou[13] for the case of the plane crack, by relaxing the superficially plausible assumption that the crack will be fully open and hence have traction-free faces in a tensile field, using instead the more rigorous requirements of frictionless unilateral contact — i.e. permitting contact to occur, the extent of the contact zones (if any) to be determined by the inequalities (31.6, 31.7). She found that a very small contact zone is established adjacent to each crack tip, the extent of which is of the same order of magnitude as the zone in which interpenetration is predicted for the original 'open crack' solution. At the closed crack tip, only the shear stresses in the bonded region are singular, leading to a 'mode II' (shear) stress-intensity factor, by analogy with equation (13.21). However, the contact stresses show a proportional compressive singularity in $r \to a^-$. Both singularities are of the usual square-root form, without the oscillatory character of equations (35.42).

Solutions have since been found for other interface crack problems with contact zones, including the plane crack in a combined tensile and shear field[14]. These original solutions of the interface crack problem used numerical methods to solve the resulting integral equation, but more recently an analytical solution has been found by Gautesen and Dundurs[15].

The problem of §35.3 — the penny-shaped crack in a tensile field — was solved in a unilateral contact formulation by Keer et al.[16]. They found a small annulus of contact $b < r < a$ adjacent to the crack tip. In fact, the formulation of §35.3 is easily adapted to this case. We note that the crack opening displacement must be zero in the contact zone, so the range of equation (35.27) is extended inwards to $b < r$, whilst (35.26) — setting the total normal tractions at the interface to zero — now only applies in the open region of the crack $0 \le r < b$. This can be accommodated by replacing a by b in the definitions of ϕ_3, so that (35.31, 35.32) are replaced by

$$\int_0^r \frac{g_1(t)dt}{\sqrt{r^2 - t^2}} + \beta \int_r^b \frac{g_3(t)dt}{\sqrt{t^2 - r^2}} = (1 - \beta^2)C \; ; \;\; 0 \le r < a$$

$$\frac{1}{r}\frac{d}{dr}\int_0^r \frac{t g_3(t)dt}{\sqrt{r^2 - t^2}} - \frac{\beta}{r}\frac{d}{dr}\int_r^a \frac{t g_1(t)dt}{\sqrt{t^2 - r^2}} = (1 - \beta^2)S \; ; \;\; 0 \le r < b \; ,$$

and (35.37, 35.38) by

---

[13] M. Comninou (1977), The interface crack, *ASME Journal of Applied Mechanics,* Vol. 44, pp. 631–636.

[14] M. Comninou and D. Schmueser (1979), The interface crack in a combined tension-compression and shear field, *ASME Journal of Applied Mechanics,* Vol. 46, pp. 345–358.

[15] A. K. Gautesen and J. Dundurs (1987), The interface crack in a tension field, *ASME Journal of Applied Mechanics,* Vol. 54, pp. 93–98; A. K. Gautesen and J.Dundurs (1988), The interface crack under combined loading, *ibid.,* Vol. 55, pp. 580–586.

[16] L. M. Keer, S. H. Chen and M. Comninou (1978), The interface penny-shaped crack reconsidered, *International Journal of Engineering Science,* Vol. 16, pp. 765–772.

$$g_1(x) = \frac{\beta}{\pi} \int_{-b}^{b} \frac{g_3(t)dt}{(x-t)} + \frac{2C(1-\beta^2)}{\pi} \; ; \quad -a < x < a \qquad (35.43)$$

$$g_3(x) = -\frac{\beta}{\pi} \int_{-a}^{a} \frac{g_1(t)dt}{(x-t)} + \frac{2S(1-\beta^2)x}{\pi} \; ; \quad -b < x < b \; . \qquad (35.44)$$

The difference in range in the integrals in (35.43, 35.44) now prevents us from combining them in a single complex equation, but it is still possible to develop a Fredholm equation in either function[17] by substitution, as in equation (35.39).

35.5   Implications for Fracture Mechanics

When an interface crack is loaded in tension, the predicted contact zones are very much smaller than the crack dimensions and it must therefore be possible to characterize all features of the local crack-tip fields, including the size of the contact zone, in terms of features of the elastic field further from the crack tip, where the open and contact solutions are essentially indistinguishable. Hills and Barber[18] have used this argument to derive the contact solution from the simpler 'open-crack' solution for a plane crack loaded in tension.

The contact zones may also be smaller than the fracture process zone defined in §13.3.1, in which case the conditions for fracture must also be capable of characterization in terms of the open-crack asymptotic field. If the Griffith fracture criterion applies, fracture will occur when the strain-energy release rate G of §13.3.3 exceeds a critical value for the material. For a homogeneous material under in-plane loading, the energy release rate is given by equation (13.28) and is proportional to $K_I^2 + K_{II}^2$. For the open solution of the interface crack, the oscillatory behaviour prevents the stress-intensity factors being given their usual meaning, but the energy release rate depends on the asymptotic behaviour of $\sigma_{zr}^2 + \sigma_{zz}^2$, which is non-oscillatory. The contact solution gives almost the same energy release rate, as indeed it must do, since the two solutions only differ in a very small region at the tip.

Experiments with homogeneous materials show that the energy release rate required for fracture varies significantly with the ratio K_{II}/K_I, which is referred to as the *mode mixity*[19]. This is probably more indicative of our over-idealization of the crack geometry than of any limitation on Griffith's theory, since real cracks are

---

[17] The best choice here is to eliminate $g_1(x)$, since the opposite choice will lead to an integral equation on the range $-a < x < a$, which might have discontinuities in at the points $x = \pm b$. These could cause problems with convergence in the final numerical solution.

[18] D. A. Hills and J. R. Barber (1993), Interface cracks, *International Journal of Mechanical Sciences*, Vol. 35, pp. 27–37.

[19] See for example, H. A. Richard (1984), Examination of brittle fracture criteria for overlapping mode I and mode II loading applied to cracks, in G. C. Sih, E. Sommer and W. Dahl, **eds.**, *Application of Fracture Mechanics to Structures*, Martinus Nijhoff, The Hague, 1984, pp. 309–316.

seldom plane and the attempt to shear them without significant opening will probably lead to crack face contact and consequent frictional forces.

It is difficult to extend the concept of mode mixity to the open interface crack solution, since, for example, the ratio σ_{zr}/σ_{zz} passes through all values, positive and negative, in each cycle of oscillation. However, the change in the ratio with s is very slow except in the immediate vicinity of the crack tip, leading Rice[20] to suggest that some specific distance from the tip be agreed by convention at which the mode mixity be defined.

An interesting feature of the asymptotics of the open solution is that, in contrast to the case $\beta = 0$ (including the crack in a homogeneous material), all possible fields can be mapped into each other by a change in length scale and in one linear multiplier (e.g. the square root of the energy release rate). Thus, an alternative way of characterizing the conditions at the tip for fracture experiments would be in terms of the energy release rate and a characteristic length, which could be taken as the maximum distance from the tip at which the open solution predicts interpenetration. Conditions approach most closely to the classical mode I state when this distance is small.

It must be emphasised that these arguments are all conditional upon the characteristic length being small compared with all the other dimensions of the system, *including the anticipated process zone dimension*. If this condition is not satisfied, it is essential to use the unilateral contact formulation.

Problems

35.1. Use equations (12.40, 12.41) to formulate the problem of two dissimilar half-planes $y < 0$, $y > 0$, bonded over the region $-a < x < a$ and pulled apart by a force F per unit length. Assume plane strain conditions with elastic properties μ_1, ν_1 and μ_2, ν_2 in $y < 0$, $y > 0$, respectively. Show that the resulting integral equations can be combined as in §35.3.1, but do not attempt to solve the resulting equation.

35.2. The interface penny-shaped crack of Figure 35.1 is unloaded, but the crack acts as an insulated obstruction to an otherwise uniform heat flux $q_z = q_0$. Represent the temperature and displacement fields by appropriate potential functions using functions of the form (32.10), but do not attempt to solve the resulting integral equations.

35.3. The end $z = 0$ of the semi-infinite cylinder $0 \leq r < a$, $z > 0$ is protected by a thermal barrier coating of thickness t. The bodies are stress free at the assembly

---

[20] J. R. Rice (1988), Elastic fracture mechanics concepts for interfacial cracks, *ASME Journal of Applied Mechanics,* Vol. 55, pp. 98–103.

temperature T_0, but we wish to investigate the stress state at a uniform higher temperature T, in particular to assess the risk of delamination. For simplicity, suppose that both materials have the same elastic properties E, ν, but that the thermal expansion coefficients of the cylinder and the coating are α_1, α_2 respectively, with $\alpha_1 > \alpha_2$.

Show that no tractions will be transmitted across the interface if an appropriate normal traction is applied to the curved surfaces of the coating $r = a$, $-t < z < 0$. If this traction is not applied, will the stress state at $r = a$, $z = 0$ involve a singularity, and if so, of what form.

35.4. It is proposed to formulate the two-dimensional interface crack problem using a distribution of dislocations, as in §13.3. Find the Airy stress function for a climb dislocation at the interface — i.e. the solution of the problem

$$u_y(x, 0^+) - u_y(x, 0^-) = B_y H(x) ; \quad u_x(x, 0^+) - u_x(x, 0^-) = 0 ,$$

if the elastic constants are μ_1, κ_1 for $y < 0$ and μ_2, κ_2 for $y > 0$.

35.5. If $\beta \neq 0$, equation (35.42) defines an oscillatory singularity at the tip of an open interface crack. Show that if a frictionless contact region is interposed between the bonded and separation regions, the singularity at the crack tip will be square-root singular in mode II, but not oscillatory. For simplicity, consider only the case where one body is rigid. Show also that there will then be a square-root singular contact pressure in the contact region, except when $\beta = 0$.

Chapter 36
Anisotropic Elasticity

Up to now, we have considered only isotropic materials, except in the development of certain general relations. However, many practical applications involve materials with significant anisotropy. This can arise by design, as in the development of fibre-reinforced composites or metamaterials, or result from a manufacturing process such as rolling which causes grains to become elongated in the rolling direction. Also many naturally occurring materials are anisotropic, notably those resulting from a growth process or from biological adaption to the need for preferential strength or stiffness in a specific direction (for example the grain in wood).

Crystalline materials are necessarily anisotropic at the scale of a single grain, though with random orientation of grains, a multigranular material will appear isotropic at length scales significantly larger than the average grain size. For this reason, it becomes more important to consider anisotropy when describing microme-chanical effects — for example, in approximating the local state of stress due to a dislocation, or at a grain boundary.

36.1 The Constitutive Law

The elasticity and compliance tensors (c_{ijkl}, s_{ijkl}) of equation (1.40) each have 81 components, but these are constrained by the symmetries (e.g. $\sigma_{ij} = \sigma_{ji}$) of the stress and strain tensors, and by Maxwell's reciprocal theorem (see Chapter 38), so there are at most only 21 independent constants. However, this is still a sufficiently large number to impose a serious algebraic cost to the solution of elasticity problems for generally anisotropic materials.

The algebraic relations can be simplified somewhat by using *Voigt notation* in which the six independent stresses are treated as components of a vector. The constitutive law can then be written in the form of the matrix equations

$$\sigma = C e ; \quad e = S \sigma , \tag{36.1}$$

© The Author(s), under exclusive license to Springer Nature Switzerland AG 2022
J. R. Barber, *Elasticity*, Solid Mechanics and Its Applications 172,
https://doi.org/10.1007/978-3-031-15214-6_36

where

$$\boldsymbol{\sigma} = \{\sigma_{xx}, \sigma_{yy}, \sigma_{zz}, \sigma_{yz}, \sigma_{zx}, \sigma_{xy}\}^T \tag{36.2}$$

$$\boldsymbol{e} = \{e_{xx}, e_{yy}, e_{zz}, 2e_{yz}, 2e_{zx}, 2e_{xy}\}^T \ , \tag{36.3}$$

and $\boldsymbol{C}$, $\boldsymbol{S}$ are 6×6 symmetric matrices derived from c_{ijkl}, s_{ijkl} respectively. Notice the factor of 2 which is conventionally included in the shear strains in equation (36.3). Since the strain-energy density of equation (37.3) must be positive for all states of stress or strain, the matrices $\boldsymbol{C}$ and $\boldsymbol{S}$ are positive definite.

Analytical methods of solution for anisotropic materials mostly depend on performing a linear transformation on the coördinates so as to reduce the governing equations to a standard form, usually the Laplace equation, or in two-dimensional problems, to allow solutions to be defined as holomorphic functions of a modified complex variable. We shall describe several illustrations of this technique in this chapter. If the characteristic equation defining these transformations has distinct roots (eigenvalues), a linear combination of the corresponding stress or displacement fields provides a general solution to the problem. In some cases, the equation has repeated roots and special techniques are required. A simple example where this occurs is discussed in §36.3.2 below.

36.2 Two-dimensional Solutions

As in isotropic problems, we can identify states of deformation such that the displacements and hence the strains and stresses are independent of z. However, if all the coefficients in $\boldsymbol{C}$, $\boldsymbol{S}$ are non-zero, the normal-shear coupling (e.g. between σ_{zy} and e_{xx}) implies that we should now expect non-zero antiplane shear stresses σ_{zx}, σ_{zy} and displacements u_z even if the boundaries are loaded only by in-plane tractions. In effect, the plane strain problem of Chapters 4–13 and the antiplane problem of Chapter 15 become coupled.

The most general case in which these problems remain *uncoupled* is that in which the z-plane is a plane of symmetry. It then follows that

$$C_{ij} = 0 \ ; \quad S_{ij} = 0 \qquad \text{if } i \in \{1, 2, 3, 6\} \text{ and } j \in \{4, 5\} \text{ or } \textit{vice versa.} \tag{36.4}$$

Materials satisfying these conditions but with no additional symmetries are known as *monoclinic*.

For the more general case where the z-plane is *not* a plane of symmetry, the in-plane and antiplane problems cannot be decoupled and we must satisfy all three equilibrium conditions (2.4), which in the absence of body forces have the reduced form

$$\frac{\partial \sigma_{xx}}{\partial x} + \frac{\partial \sigma_{xy}}{\partial y} = 0 \ ; \quad \frac{\partial \sigma_{yx}}{\partial x} + \frac{\partial \sigma_{yy}}{\partial y} = 0 \ ; \quad \frac{\partial \sigma_{zx}}{\partial x} + \frac{\partial \sigma_{zy}}{\partial y} = 0 \ . \quad (36.5)$$

Also, since the displacements are independent of z, we have

$$2e_{zx} = \frac{\partial u_z}{\partial x} \ ; \quad 2e_{zy} = \frac{\partial u_z}{\partial y} \ ; \quad e_{zz} = 0 \ , \quad (36.6)$$

and since all z-derivatives in the compatibility equations (2.7) are zero, three of these are satisfied identically, and the remaining three reduce to

$$\frac{\partial^2 e_{xx}}{\partial y^2} - 2\frac{\partial^2 e_{xy}}{\partial x \partial y} + \frac{\partial^2 e_{yy}}{\partial x^2} = 0 \ ; \quad \frac{\partial e_{zx}}{\partial y} - \frac{\partial e_{yz}}{\partial x} = 0 \ . \quad (36.7)$$

Broadly speaking, there are two approaches to the solution of the resulting two-dimensional problem. Lekhnitskii[1] satisfies the equilibrium equations by representing the stress components in terms of the Airy and Prandtl stress functions of equations (4.6, 18.5), and then develops a governing equation by substituting the resulting strains into the compatibility conditions (36.7). Stroh[2] defines the displacement as a holomorphic function of a linearly transformed complex variable and obtains conditions for the transformation parameters from the equilibrium equations (36.5). We shall discuss both these methods in §§36.4, 36.5, below, but to introduce the topic we first consider the simpler uncoupled case of an orthotropic material under in-plane loading.

36.3 Orthotropic Material

An orthotropic material is one with three mutually orthogonal symmetry planes. If we use these to define a Cartesian coördinate system, equation $(36.1)_2$ takes the form

$$\begin{bmatrix} e_{xx} \\ e_{yy} \\ e_{zz} \\ 2e_{yz} \\ 2e_{zx} \\ 2e_{xy} \end{bmatrix} = \begin{bmatrix} S_{11} & S_{12} & S_{13} & 0 & 0 & 0 \\ S_{21} & S_{22} & S_{23} & 0 & 0 & 0 \\ S_{31} & S_{32} & S_{33} & 0 & 0 & 0 \\ 0 & 0 & 0 & S_{44} & 0 & 0 \\ 0 & 0 & 0 & 0 & S_{55} & 0 \\ 0 & 0 & 0 & 0 & 0 & S_{66} \end{bmatrix} \begin{bmatrix} \sigma_{xx} \\ \sigma_{yy} \\ \sigma_{zz} \\ \sigma_{yz} \\ \sigma_{zx} \\ \sigma_{xy} \end{bmatrix} \ , \quad (36.8)$$

and in view of the symmetry of the compliance matrix, there are nine independent elastic constants $S_{11}, S_{12}, S_{13}, S_{22}, S_{23}, S_{33}, S_{44}, S_{55}, S_{66}$.

---

[1] S. G. Lekhnitskii, *Theory of Elasticity of an Anisotropic Elastic Body,* Holden-Day, San Francisco, 1963.

[2] A. N. Stroh (1958), Dislocations and cracks in anisotropic elasticity, *Philosophical Magazine,* Vol. 3, pp. 625–646, A. N. Stroh (1962), Steady-state problems in anisotropic elasticity, *Journal of Mathematics and Physics,* Vol. 41, pp. 77–103.

Notice that this simplification of the constitutive law applies only when the coördinate system is chosen to align with the material symmetries. If we use equation (1.43) to rotate into another coördinate system, the coefficients C'_{ij}, S'_{ij} will generally all be non-zero. This also implies that the directions of principal stresses and principal strains will not be aligned, except when these coincide with the material symmetries.

Equation (36.8) and the condition $e_{zz} = 0$ give

$$\sigma_{zz} = -\frac{S_{13}}{S_{33}}\sigma_{xx} - \frac{S_{23}}{S_{33}}\sigma_{yy} ,$$

and hence the in-plane strains are

$$e_{xx} = \beta_{11}\sigma_{xx} + \beta_{12}\sigma_{yy} \; ; \quad e_{yy} = \beta_{12}\sigma_{xx} + \beta_{22}\sigma_{yy} \; ; \quad 2e_{xy} = \beta_{66}\sigma_{xy} ,$$

where, following Lekhnitskii, we define the 5×5 reduced compliance matrix β with components

$$\beta_{ij} = S_{ij} - \frac{S_{3i}S_{j3}}{S_{33}} \; ; \quad i, j \neq 3 . \tag{36.9}$$

Substituting these expressions into the compatibility condition $(36.7)_1$ and using the Airy function representation

$$\sigma_{xx} = \frac{\partial^2 \phi}{\partial y^2} \; ; \quad \sigma_{yy} = \frac{\partial^2 \phi}{\partial x^2} \; ; \quad \sigma_{xy} = -\frac{\partial^2 \phi}{\partial x \partial y}$$

to satisfy the in-plane equilibrium equations $(36.5)_{1,2}$, we obtain the governing equation

$$\beta_{22}\frac{\partial^4 \phi}{\partial x^4} + (2\beta_{12} + \beta_{66})\frac{\partial^4 \phi}{\partial x^2 \partial y^2} + \beta_{11}\frac{\partial^4 \phi}{\partial y^4} = 0 . \tag{36.10}$$

This equation can be factorized as

$$\beta_{22}\left(\frac{\partial^2}{\partial x^2} + \lambda_1^2\frac{\partial^2}{\partial y^2}\right)\left(\frac{\partial^2}{\partial x^2} + \lambda_2^2\frac{\partial^2}{\partial y^2}\right)\phi = 0 , \tag{36.11}$$

where λ_1^2, λ_2^2 are the two roots of the quadratic equation

$$\beta_{22}X^2 - (2\beta_{12} + \beta_{66})X + \beta_{11} = 0 . \tag{36.12}$$

If the roots are both real and positive, we can then define two linear transformations of $(x, y) \leftrightarrow (\xi, \eta)$ through

$$x = \xi_1 \; ; \quad y = \lambda_1 \eta_1 \quad \text{and} \quad x = \xi_2 \; ; \quad y = \lambda_2 \eta_2 ,$$

so that equation (36.11) becomes

$$\beta_{22} \left(\frac{\partial^2}{\partial \xi_1{}^2} + \frac{\partial^2}{\partial \eta_1{}^2} \right) \left(\frac{\partial^2}{\partial \xi_2{}^2} + \frac{\partial^2}{\partial \eta_2{}^2} \right) \phi = 0 ,$$

and the general solution can be written as the sum of two harmonic functions in the respective transformed coördinates — i.e.

$$\phi(x, y) = \phi_1(\xi_1, \eta_1) + \phi_2(\xi_2, \eta_2) . \tag{36.13}$$

Equivalently, we can write a general solution as

$$\phi = \Re \left\{ f_1(\zeta_1) + f_2(\zeta_2) \right\} \quad \text{where} \quad \zeta_i = \xi_i + \iota \eta_i ,$$

and f_1, f_2 are holomorphic functions.

36.3.1 Normal loading of the half-plane

As an example, consider the problem where the half-plane $y < 0$ is loaded by purely normal tractions on the edge $y = 0$. The shear stress

$$\sigma_{xy} = -\frac{\partial^2 \phi}{\partial x \partial y} = -\frac{1}{\lambda_1} \frac{\partial^2 \phi_1}{\partial \xi_1 \partial \eta_1} - \frac{1}{\lambda_2} \frac{\partial^2 \phi_2}{\partial \xi_2 \partial \eta_2} ,$$

and the line $y = 0$ corresponds to both $\eta_1 = 0$ and $\eta_2 = 0$, so the shear traction on this edge will be zero for all x if

$$\phi_2(\xi, \eta) = -\frac{\lambda_1}{\lambda_2} \phi_1(\xi, \eta) \quad \text{or} \quad \phi = f\left(x, \frac{y}{\lambda_1} \right) - \frac{\lambda_1}{\lambda_2} f\left(x, \frac{y}{\lambda_2} \right) ,$$

where f is any harmonic function.

The particular case of a concentrated force F_y at the origin, as in §12.3 and Figure 12.2, can be solved using the harmonic function

$$f(\xi, \eta) = C \left[\eta \ln \left(\xi^2 + \eta^2 \right) + 2\xi \arctan \left(\frac{\eta}{\xi} \right) \right] ,$$

where C is an arbitrary constant which can be related to F_y by considering the equilibrium of the strip $-h < y < 0$. We obtain

$$C = \frac{\lambda_1 F_y}{2\pi(\lambda_2 - \lambda_1)} , \tag{36.14}$$

and the corresponding stress components are

$$\sigma_{xx} = \frac{F_y(\lambda_1 + \lambda_2)x^2 y}{\pi(y^2 + \lambda_1^2 x^2)(y^2 + \lambda_2^2 x^2)} \quad ; \quad \sigma_{yy} = \frac{F_y(\lambda_1 + \lambda_2)y^3}{\pi(y^2 + \lambda_1^2 x^2)(y^2 + \lambda_2^2 x^2)}$$

$$\sigma_{xy} = \frac{F_y(\lambda_1 + \lambda_2)xy^2}{\pi(y^2 + \lambda_1^2 x^2)(y^2 + \lambda_2^2 x^2)} \, . \tag{36.15}$$

36.3.2 Degenerate cases

This solution technique cannot be used if the two roots of equation (36.12) are equal ($\lambda_1^2 = \lambda_2^2$), since the two functions in equation (36.13) could then be combined to form a single arbitrary harmonic function, and this does not provide enough degrees of freedom (e.g. arbitrary constants) to satisfy the boundary conditions. However in this case, equation (36.11) can be converted to the biharmonic equation by a single transformation $y = \lambda\eta$ and the solution can then be obtained as in Chapters 4–13, a special case being that of an isotropic material, for which $\lambda_1^2 = \lambda_2^2 = 1$.

It is interesting to note that if we set $\lambda_1^2 = \lambda_2^2 = 1$ in the final expressions for the stresses (36.15), we recover the Cartesian equivalent of the isotropic Flamant solution. In other words, if we pretend that the material is orthotropic with $\lambda_1^2 \neq \lambda_2^2$, and proceed to the isotropic limit only at the last step, the correct solution is obtained, even though intermediate steps such as equation (36.14) are ill-defined in this limit.

This situation is clearly similar to that described in §§8.3.3, 10.3.1, where an otherwise general technique breaks down at one or more special values of a parameter. We could formalize the limiting process in the present case by defining (say) $\lambda_2 = \lambda_1 + \epsilon$ and taking the limit as $\epsilon \to 0$. It is significant that the constant C in (36.14) is unbounded when $\lambda_2^2 \to \lambda_1^2$ which compensates for the harmonic functions becoming equal, exactly as in equation (8.30). An alternative approach is to obtain the limit by differentiation, defining a second function by

$$\phi_2 = \frac{\partial}{\partial\lambda} f_2(\xi, \eta)) = -\frac{\eta}{\lambda}\frac{\partial}{\partial\eta} f(\xi, \eta) \, .$$

Since the spatial derivative of a harmonic function is itself harmonic, this is equivalent to replacing (36.13) by

$$\phi(x, y) = \phi_1(\xi, \eta) + \eta\phi_2(\xi, \eta) \, ,$$

which defines a general two-dimensionally biharmonic function in (ξ, η) coördinates if ϕ_1, ϕ_2 are arbitrary harmonic functions.

Related degeneracies occur for more general anisotropic materials in both Lekhnitskii and Stroh solutions when two or more of the transformations required for the solution coincide. They can be resolved by similar techniques to those described above.

36.4 Lekhnitskii's Formalism

We now consider the case of a general anisotropic material, where any or all of the coefficients in C, S may be non-zero. We define the stress components

$$\sigma_{xx} = \frac{\partial^2 \phi}{\partial y^2} \;\; ; \;\; \sigma_{yy} = \frac{\partial^2 \phi}{\partial x^2} \;\; ; \;\; \sigma_{xy} = -\frac{\partial^2 \phi}{\partial x \partial y}$$

$$\sigma_{zx} = \frac{\partial \psi}{\partial y} \;\; ; \;\; \sigma_{zy} = -\frac{\partial \psi}{\partial x} \, , \tag{36.16}$$

where ϕ, ψ are the Airy and Prandtl stress functions of Chapters 4, 17 respectively[3]. If there is no body force, this representation satisfies the equilibrium equations identically, so it remains to satisfy the non-trivial compatibility equations (36.7).

As in §36.3, we use the condition $e_{zz} = 0$ to eliminate the stress component σ_{zz}, so that there are only five independent stress and strain components in equations (36.2, 36.3), which we combine as

$$\varsigma = \{\sigma_{xx}, \sigma_{yy}, \sigma_{yz}, \sigma_{zx}, \sigma_{xy}\}^T \;\; ; \;\; \epsilon = \{e_{xx}, e_{yy}, 2e_{yz}, 2e_{zx}, 2e_{xy}\}^T \, .$$

The constitutive law $(36.1)_2$ then reduces to

$$\epsilon = \beta \varsigma \, , \tag{36.17}$$

where the reduced compliance matrix β is defined by equation (36.9).

Substituting (36.17) into the compatibility equations (36.7) and using the stress function representation (36.16) for the stress components, we then obtain

$$\mathcal{L}_4 \phi + \mathcal{L}_3 \psi = 0 \;\; ; \quad \mathcal{L}_3 \phi + \mathcal{L}_2 \psi = 0 \, , \tag{36.18}$$

where the differential operators

$$\mathcal{L}_2 \equiv \beta_{44} \frac{\partial^2}{\partial x^2} - 2\beta_{45} \frac{\partial^2}{\partial x \partial y} + \beta_{55} \frac{\partial^2}{\partial y^2}$$

$$\mathcal{L}_3 \equiv -\beta_{24} \frac{\partial^3}{\partial x^3} + (\beta_{25} + \beta_{46}) \frac{\partial^3}{\partial x^2 \partial y} - (\beta_{56} + \beta_{14}) \frac{\partial^3}{\partial x \partial y^2} + \beta_{15} \frac{\partial^3}{\partial y^3}$$

$$\mathcal{L}_4 \equiv \beta_{22} \frac{\partial^4}{\partial x^4} - 2\beta_{26} \frac{\partial^4}{\partial x^3 \partial y} + (2\beta_{12} + \beta_{66}) \frac{\partial^4}{\partial x^2 \partial y^2} - 2\beta_{16} \frac{\partial^4}{\partial x \partial y^3} + \beta_{11} \frac{\partial^4}{\partial y^4} \, .$$

If the stiffness matrix satisfies the monoclinic condition (36.4), the differential operator $\mathcal{L}_3 \equiv 0$, and the problem then decouples into an in-plane problem and an antiplane problem governed by $\mathcal{L}_4 \phi = 0$ and $\mathcal{L}_2 \psi = 0$ respectively. In particular, for the orthotropic case, the in-plane equation $\mathcal{L}_4 \phi = 0$ reduces to (36.10)

---

[3] Notice that we are here using ψ for the Prandtl function, to avoid possible confusion between ϕ and φ when writing solutions by hand.

For the coupled case, simultaneous solution of equations (36.18) shows that both ϕ and ψ must satisfy the sixth order partial differential equation

$$\mathcal{L}_6\{\phi, \psi\} = 0 \quad \text{where} \quad \mathcal{L}_6 \equiv \mathcal{L}_4\mathcal{L}_2 - \mathcal{L}_3\mathcal{L}_3 . \tag{36.19}$$

36.4.1 Polynomial solutions

Polynomial solutions to problems in rectangular coördinates can be found using exactly the same techniques as in Chapter 5. We first identify the order of the polynomial in ϕ as in §5.2.2, noting that for general anisotropy we cannot appeal to geometric symmetry to eliminate any of the terms. Equation $(36.18)_2$ then implies that we shall also generally require a polynomial for ψ of one order less than that for ϕ.

Example

As a simple example, we consider the plane strain equivalent of Figure 5.2, where the surfaces $y = \pm b$ are traction free, and the end $x = 0$ is loaded by a shear force F per unit length in the negative y-direction. Since we anticipate non-zero contributions in all the stress components, the weak boundary conditions on $x = 0$ must include both forces and moments, comprising[4]

$$\int_{-b}^{b} \sigma_{xx}(x, 0)\, dy = 0 \; ; \quad \int_{-b}^{b} \sigma_{xy}(x, 0)\, dy = F \; ; \quad \int_{-b}^{b} \sigma_{xz}(x, 0)\, dy = 0 \; ;$$

$$\int_{-b}^{b} \sigma_{xx}(x, 0)\, y\, dy = 0 . \tag{36.20}$$

We anticipate a dominant stress component $\sigma_{xx} \sim xy$, so we start with a fourth degree polynomial for ϕ and hence a third degree polynomial for ψ — i.e.

---

[4] It is tempting here to add the weak condition

$$\int_{-b}^{b} \sigma_{xz}(x, 0)\, y\, dy = 0 ,$$

corresponding to the absence of a distributed torque on the end $x = 0$. However, if such a torque exists, its release would imply twist of the cross section, which violates the plane strain condition $\partial u_y / \partial z = 0$. For more discussion of the torsion problem, see §36.6.2 below.

$$\phi = A_1x^4 + A_2x^3y + A_3x^2y^2 + A_4xy^3 + A_5y^4 + A_6x^3 + A_7x^2y + A_8xy^2$$
$$+ A_9y^3 + A_{10}x^2 + A_{11}xy + A_{12}y^2 ,$$
$$\psi = B_1x^3 + B_2x^2y + B_3xy^2 + B_4y^3 + B_5x^2 + B_6xy + B_7y^2 + B_8x + B_9y .$$

$$(36.21)$$

The stress components are then defined by equations (36.16), and substitution into the boundary conditions

$$\sigma_{yx} = \sigma_{yy} = \sigma_{yz} = 0 ; \quad \text{all } x, \ y = \pm b ,$$

yields six polynomial equations. Two additional equations are obtained by substituting (36.21) into equations (36.18). Equating coefficients of powers of x and y in these equations, and imposing the end conditions (36.20), we obtain a total of 29 linear algebraic equations which can be solved for the 21 constants A_i, B_i. It will be appreciated that even a problem as simple as this is practical only when using software for the algebraic calculations!

The final stress field is given by[5]

$$\sigma_{xx} = \frac{3Fxy}{2b^3} + \frac{F\alpha_1}{b}\left(1 - \frac{3y^2}{b^2}\right) ; \quad \sigma_{xy} = \frac{3F}{4b}\left(1 - \frac{y^2}{b^2}\right) ; \quad \sigma_{yy} = 0$$

$$\sigma_{xz} = \frac{F\alpha_3}{b}\left(1 - \frac{3y^2}{b^2}\right) - \frac{3F\alpha_2xy}{2b^3} ; \quad \sigma_{yz} = -\frac{3F\alpha_2}{4b}\left(1 - \frac{y^2}{b^2}\right) ,$$

where

$$\alpha_1 = \frac{\beta_{55}\beta_{15}\beta_{56} + \beta_{55}\beta_{14}\beta_{15} - \beta_{45}\beta_{15}^2 - \beta_{16}\beta_{55}^2}{2\beta_{55}(\beta_{55}\beta_{11} - \beta_{15}^2)} ; \quad \alpha_2 = \frac{\beta_{15}}{\beta_{55}}$$

$$\alpha_3 = \frac{2\beta_{15}(\beta_{55}\beta_{16} + \beta_{45}\beta_{11}) - (\beta_{56} + \beta_{14})(\beta_{55}\beta_{11} + \beta_{15}^2)}{4\beta_{55}(\beta_{55}\beta_{11} - \beta_{15}^2)}$$

Notice that the anisotropy introduces additional terms into these stress components compared with equations (5.22), and in particular that the resulting solution has no symmetry about $y = 0$.

36.4.2 Solutions in linearly transformed space

It can be verified by substitution that equation (36.19) has a particular solution of the form

$$\phi = f(\zeta) + \overline{f(\zeta)} = 2\Re\{(f(\zeta)\} \quad \text{where} \quad \zeta = x + py ,$$

---

[5] The final stress component σ_{zz} can be found by imposing the condition $e_{zz} = 0$, but this step is omitted here in the interests of brevity.

if p is chosen to satisfy the sixth-order polynomial equation

$$\ell_6(p) \equiv \ell_4(p)\ell_2(p) - \ell_3(p)\ell_3(p) = 0 , \tag{36.22}$$

and

$$\begin{aligned}
\ell_2(p) &= \beta_{44} - 2\beta_{45}p + \beta_{55}p^2 \\
\ell_3(p) &= -\beta_{24} + (\beta_{25} + \beta_{46})p - (\beta_{56} + \beta_{14})p^2 + \beta_{15}p^3 \\
\ell_4(p) &= \beta_{22} - 2\beta_{26}p + (2\beta_{12} + \beta_{66})p^2 - 2\beta_{16}p^3 + \beta_{11}p^4 .
\end{aligned}$$

Lekhnitskii[6] has shown that the polynomial equation (36.22) has no real roots, and since the coefficients are all real, the six solutions must comprise three complex conjugate pairs, each of which defines a linear transformation of the in-plane coördinates. For example, if $p = b + \iota c$ and we write $\zeta = \xi + \iota\eta$, then equation $(36.23)_2$ maps (x, y) into (ξ, η) through

$$\xi = x + by ; \quad \eta = cy .$$

In the following derivations, we shall sometimes omit the real part operator $\Re\{\cdot\}$ in the interests of brevity.

Suppose we choose one of the roots p_α of equation (36.22) and write

$$\phi_\alpha = 2\,\Re\{f(\zeta_\alpha)\} ; \quad \psi_\alpha = 2\,\Re\{g(\zeta_\alpha)\} ,$$

where $\zeta_\alpha = x + p_\alpha y$. Then $(36.18)_2$ requires that

$$\mathcal{L}_2(\psi_\alpha) = \ell_2(p_\alpha)g''(\zeta_\alpha) = -\mathcal{L}_3(\phi_\alpha) = -\ell_3(p_\alpha)f'''(\zeta_\alpha)$$

and a particular solution is

$$g(\zeta_\alpha) = -\lambda_\alpha f'(\zeta_\alpha) \quad \text{where} \quad \lambda_\alpha = \frac{\ell_3(p_\alpha)}{\ell_2(p_\alpha)} = \frac{\ell_4(p_\alpha)}{\ell_3(p_\alpha)} .$$

Using these results, we can then write a general solution as

$$\phi = 2\,\Re\left\{\sum_{\alpha=1}^{3} f_\alpha(\zeta_\alpha)\right\} ; \quad \psi = -2\,\Re\left\{\sum_{\alpha=1}^{3} \lambda_\alpha f_\alpha'(\zeta_\alpha)\right\} .$$

The three independent holomorphic functions $f_\alpha(\zeta_\alpha)$, $\alpha = (1, 2, 3)$ then allow us to satisfy three boundary conditions at each point on the boundary of the two-dimensional domain.

Notice that the equation $\zeta_\alpha = x + p_\alpha y$ defines a different linear transformation of the two-dimensional space for each value of α. In particular, the circle transforms into three different ellipses, and hence requires the use of the conformal mapping of

---

[6] S. G. Lekhnitskii, *loc. cit.*

§§19.7, 20.7.1. However, straight lines remain straight under linear transformation, so (for example) boundary-value problems for wedge-shaped domains can be solved using techniques similar to those in Chapter 11.

36.5   Stroh's Formalism

Stroh's formalism[7] starts by considering the conditions under which the equilibrium conditions can be satisfied by a displacement field of the form[8]

$$\boldsymbol{u} = \boldsymbol{a} f(\zeta) + \overline{\boldsymbol{a} f(\zeta)} = 2\,\Re\,\{\boldsymbol{a} f(\zeta)\} \qquad \text{where} \qquad \zeta = x + py. \tag{36.23}$$

The displacement gradients are then given by

$$\frac{\partial \boldsymbol{u}}{\partial x} = 2\,\Re\,\{\boldsymbol{a} f'(\zeta)\} \;\; ; \quad \frac{\partial \boldsymbol{u}}{\partial y} = 2\,\Re\,\{\boldsymbol{a} p f'(\zeta)\} \;\; ; \quad \frac{\partial \boldsymbol{u}}{\partial z} = \boldsymbol{0}\,,$$

and the strain components are

$$e_{xx} = 2\,\Re\,\{a_1 f'(\zeta)\} \;\; ; \quad e_{yy} = 2\,\Re\,\{a_2 p f'(\zeta)\} \;\; ; \quad 2e_{xy} = 2\,\Re\,\{(a_1 p + a_2) f'(\zeta)\}$$

$$2e_{zx} = 2\,\Re\,\{a_3 f'(\zeta)\} \;\; ; \quad 2e_{yz} = 2\,\Re\,\{a_3 p f'(\zeta)\} \;\; ; \quad e_{zz} = 0\,. \tag{36.24}$$

Substituting these expressions into the constitutive law $(36.1)_1$ and reverting to the index notation for stress components, we obtain

$$\sigma_{i1} = 2\,\Re\,\{(Q_{ik} + p R_{ik}) a_k f'(\zeta)\} \;\; ; \quad \sigma_{i2} = 2\,\Re\,\{(R_{ki} + p T_{ik}) a_k f'(\zeta)\}\,, \tag{36.25}$$

where $i, k = 1, 2, 3$ and

$$\boldsymbol{Q} = \begin{bmatrix} C_{11} & C_{16} & C_{15} \\ C_{61} & C_{66} & C_{65} \\ C_{51} & C_{56} & C_{55} \end{bmatrix} \;\; ; \quad \boldsymbol{R} = \begin{bmatrix} C_{16} & C_{12} & C_{14} \\ C_{66} & C_{62} & C_{64} \\ C_{56} & C_{52} & C_{54} \end{bmatrix} \;\; ; \quad \boldsymbol{T} = \begin{bmatrix} C_{66} & C_{26} & C_{46} \\ C_{62} & C_{22} & C_{42} \\ C_{64} & C_{24} & C_{44} \end{bmatrix}\,,$$

---

[7] This method is based on earlier results due to J. D. Eshelby, W. T. Read and W. Shockley (1953), Anisotropic elasticity with applications to dislocation theory, *Acta Metallurgica*, Vol. 1(3), pp. 251–259, but Stroh's contributions to its development are so significant that it is now generally referred to by his name. For a much more detailed and rigorous development of the method, see T. C. T. Ting, *Anisotropic Elasticity: Theory and Applications*, Oxford University Press, New York, 1996.

[8] Notice that although this representation has some similarity to that used in Chapter 20, we cannot take the further step of identifying the displacement components with the real and imaginary parts of a complex function because there are now three displacement components u_x, u_y, u_z.

or equivalently, $Q_{ik} = c_{i1k1}$, $R_{ik} = c_{i1k2}$, $T_{ik} = c_{i2k2}$. We can also obtain an expression for the stress component σ_{33} from (36.24) and (36.1)$_1$, but this plays no rôle in the solution procedure.

The three equilibrium equations (36.5) can be written concisely in the form

$$\frac{\partial \sigma_{i1}}{\partial x_1} + \frac{\partial \sigma_{i2}}{\partial x_2} = 0 \quad \text{and hence} \quad \frac{\partial \sigma_{i1}}{\partial \zeta} + p \frac{\partial \sigma_{i2}}{\partial \zeta} = 0 , \tag{36.26}$$

and these will be satisfied by the stress components (36.25) if

$$\left[\boldsymbol{Q} + p \left(\boldsymbol{R} + \boldsymbol{R}^T \right) + p^2 \boldsymbol{T} \right] \boldsymbol{a} f''(\zeta) = \boldsymbol{0} . \tag{36.27}$$

Equation (36.27) has non-trivial solutions if and only if the determinant

$$\left| \boldsymbol{Q} + p \left(\boldsymbol{R} + \boldsymbol{R}^T \right) + p^2 \boldsymbol{T} \right| = 0 , \tag{36.28}$$

and this yields a sextic equation for p which for physically realistic material properties always has three pairs of complex conjugate roots, those with positive imaginary part being denoted by p_1, p_2, p_3 respectively. Barnett and Kirchner[9] have shown that the roots of equation (36.28) are identical to those of Lekhnitskii's polynomial equation (36.22).

If the roots (eigenvalues) p_1, p_2, p_3 are distinct, a general solution of the two-dimensional problem can then be constructed by superposition in the form

$$\boldsymbol{u} = 2 \Re \left\{ \sum_{\alpha=1}^{3} \boldsymbol{a}^{(\alpha)} f_\alpha (\zeta_\alpha) \right\} , \tag{36.29}$$

where $\zeta_\alpha = x + p_\alpha y$ and $\boldsymbol{a}^{(\alpha)}$ is the eigenvector of (36.27) corresponding to the eigenvalue p_α. The three independent holomorphic functions f_α provide sufficient generality to satisfy three conditions at each point on the boundary and hence can represent any two-dimensional stress state.

The equilibrium equation (36.26)$_1$ implies that the stress components can be expressed in terms of a vector stress function $\boldsymbol{\phi}$, where

$$\sigma_{i1} = -\frac{\partial \phi_i}{\partial x_2} \quad ; \quad \sigma_{i2} = \frac{\partial \phi_i}{\partial x_1} \quad ; \quad i = 1, 2, 3 . \tag{36.30}$$

Equations (36.25) then imply that

$$\boldsymbol{\phi} = 2 \Re \left\{ \sum_{\alpha=1}^{3} \boldsymbol{b}^{(\alpha)} f_\alpha (\zeta_\alpha) \right\} , \tag{36.31}$$

---

[9] D. M. Barnett and H. O. K. Kirchner (1997), A proof of the equivalence of the Stroh and Lekhnitskii sextic equations for plane anisotropic elastostatics, *Philosophical Magazine A*, Vol. 76 (1), pp. 231–239.

where

$$b^{(\alpha)} = \left(R^T + p_\alpha T\right) a^{(\alpha)} = -\frac{1}{p_\alpha} \left(Q + p_\alpha R\right) a^{(\alpha)} . \tag{36.32}$$

Since $a^{(\alpha)}$, $b^{(\alpha)}$ each comprise sets of three vectors ($\alpha = 1, 2, 3$), it is often convenient to combine them into matrices by defining

$$A = \begin{bmatrix} a_1^{(1)} & a_1^{(2)} & a_1^{(3)} \\ a_2^{(1)} & a_2^{(2)} & a_2^{(3)} \\ a_3^{(1)} & a_3^{(2)} & a_3^{(3)} \end{bmatrix} ; \quad B = \begin{bmatrix} b_1^{(1)} & b_1^{(2)} & b_1^{(3)} \\ b_2^{(1)} & b_2^{(2)} & b_2^{(3)} \\ b_3^{(1)} & b_3^{(2)} & b_3^{(3)} \end{bmatrix} . \tag{36.33}$$

Equations (36.29, 36.31) can then be written concisely as

$$\boldsymbol{u} = 2\,\Re\{A\boldsymbol{f}\} ; \quad \boldsymbol{\phi} = 2\,\Re\{B\boldsymbol{f}\} \quad \text{where} \quad \boldsymbol{f} = \{f_1(\zeta_1), f_2(\zeta_2), f_3(\zeta_3)\}^T . \tag{36.34}$$

In many applications, the form of the three functions $f_\alpha(\zeta_\alpha)$ will be the same, so that

$$\boldsymbol{f} = \langle f(\zeta) \rangle \boldsymbol{q} \quad \text{where} \quad \langle f(\zeta) \rangle = \mathrm{diag}\left[f(\zeta_1), f(\zeta_2), f(\zeta_3) \right] , \tag{36.35}$$

and $\boldsymbol{q}$ is a constant (position-independent) vector.

36.5.1 The eigenvalue problem

Routine matrix operations on equations (36.32) show that the vectors $a^{(\alpha)}$, $b^{(\alpha)}$ satisfy the eigenvalue equations

$$\begin{bmatrix} N_1 & N_2 \\ N_3 & N_4 \end{bmatrix} \begin{bmatrix} a \\ b \end{bmatrix} = p \begin{bmatrix} a \\ b \end{bmatrix} \quad \text{and} \quad [\,b, a\,] \begin{bmatrix} N_1 & N_2 \\ N_3 & N_4 \end{bmatrix} = p\,[\,b, a\,] , \tag{36.36}$$

where

$$N_1 = -T^{-1}R^T ; \quad N_2 = T^{-1} ; \quad N_3 = RT^{-1}R^T - Q ; \quad N_4 = N_1^T .$$

In other words, $[a^{(\alpha)}, b^{(\alpha)}]^T$ and $[b^{(\alpha)}, a^{(\alpha)}]$ are respectively the right and left eigenvectors of the matrix

$$N = \begin{bmatrix} N_1 & N_2 \\ N_3 & N_4 \end{bmatrix} .$$

It follows that the eigenfunctions corresponding to distinct eigenvalues $\alpha = i$ and $\alpha = j$ with $i \neq j$ satisfy the orthogonality conditions

$$\left[\, \boldsymbol{b}^{(i)}, \boldsymbol{a}^{(i)}\, \right]\begin{bmatrix} \boldsymbol{a}^{(j)} \\ \boldsymbol{b}^{(j)} \end{bmatrix} = \left[\, \boldsymbol{b}^{(i)}, \boldsymbol{a}^{(i)}\, \right]\begin{bmatrix} \overline{\boldsymbol{a}}^{(j)} \\ \overline{\boldsymbol{b}}^{(j)} \end{bmatrix} = \left[\, \overline{\boldsymbol{b}}^{(i)}, \overline{\boldsymbol{a}}^{(i)}\, \right]\begin{bmatrix} \boldsymbol{a}^{(j)} \\ \boldsymbol{b}^{(j)} \end{bmatrix} = 0 \,.$$

In other words, the matrix product

$$\begin{bmatrix} \boldsymbol{B}^T & \boldsymbol{A}^T \\ \overline{\boldsymbol{B}}^T & \overline{\boldsymbol{A}}^T \end{bmatrix}\begin{bmatrix} \boldsymbol{A} & \overline{\boldsymbol{A}} \\ \boldsymbol{B} & \overline{\boldsymbol{B}} \end{bmatrix}$$

is diagonal.

Since equations (36.36) are homogeneous, any linear (possibly complex) multiplier of the eigenfunctions will also satisfy them, provided p_α satisfies (36.28), so it is convenient to normalize the matrices $\boldsymbol{A}$, $\boldsymbol{B}$ so as to satisfy the condition[10]

$$\begin{bmatrix} \boldsymbol{B}^T & \boldsymbol{A}^T \\ \overline{\boldsymbol{B}}^T & \overline{\boldsymbol{A}}^T \end{bmatrix}\begin{bmatrix} \boldsymbol{A} & \overline{\boldsymbol{A}} \\ \boldsymbol{B} & \overline{\boldsymbol{B}} \end{bmatrix} = \begin{bmatrix} \boldsymbol{I} & 0 \\ 0 & \boldsymbol{I} \end{bmatrix}\,, \tag{36.37}$$

where $\boldsymbol{I}$ is the 3×3 identity matrix $\delta_{\alpha\beta}$. It then follows that the two 6×6 matrices on the left-hand side of (36.37) are mutual inverses, and also therefore that the matrix product commutes.

36.5.2   Solution of boundary-value problems

The matrices $\boldsymbol{Q}, \boldsymbol{R}, \boldsymbol{T}$, and hence the eigenvalues p_α and vectors $\boldsymbol{a}^{(\alpha)}, \boldsymbol{b}^{(\alpha)}$ depend only on the Voigt stiffness matrix $\boldsymbol{C}$, so they can be calculated once and for all for a given anisotropic material. The problem is therefore reduced to a boundary-value problem similar to those treated in Chapter 20. However, as with the Lekhnitskii formulation of §36.4.2, three different conformal mappings will generally be required[11], one for each eigenvalue p_α.

An important exception comprises problems for the infinite plane or the half-plane, since the boundary $x_2 = 0$ ($y = 0$) corresponds to the x-axis for all values of p in equation $(36.23)_2$. This class includes two-dimensional contact problems (Chapter 12), the Kelvin problem (§13.1), dislocations (§13.2) and plane crack problems (§13.3).

Suppose the tractions $t_i(x)$ are known functions of x on the boundary $y = 0$ of the half-plane $y > 0$. Equations $(36.30)_2$ and (36.31) then show that

$$t_i(x) = -2\,\Re\left\{ \sum_{\alpha=1}^{3} b_i^{(\alpha)} f_\alpha'(x) \right\}\,, \tag{36.38}$$

---

[10] However, notice that with this normalization, $\boldsymbol{A}$ and $\boldsymbol{B}$ will have the dimensions of stress$^{-1/2}$ and stress$^{1/2}$ respectively.

[11] Problems of this kind are discussed by T. C. T. Ting, *loc. cit.*, Chapter 10.

where the negative sign arises because the unit outward normal to the surface is defined by $-\delta_{i2}$. Equation (36.38) comprises three simultaneous equations for the three functions $f'_\alpha(x)$. We can then use equation (19.39) to determine the holomorphic functions $f'_\alpha(\zeta_\alpha)$, after which $f_\alpha(\zeta_\alpha)$ is determined by integration. The complete stress and displacement fields are then given by equations (36.29–36.31).

Example

Suppose the half-plane $y > 0$ is loaded by a uniform traction s_i in $-c < x < c$, so that $t_i(x) = s_i\, g(x)$, where

$$g(x) = 1; \quad -c < x < c$$
$$= 0; \quad |x| > c .$$

Equation (36.38) will then be satisfied if

$$f'_\alpha(\zeta_\alpha) = -q_\alpha h(\zeta_\alpha) \quad \text{where} \quad \boldsymbol{Bq} = \boldsymbol{s} \quad \text{and} \quad h(x) + \bar{h}(x) = g(x) .$$

Using the results from Example 1 of §19.6.3, we then obtain

$$f'_\alpha(\zeta_\alpha) = -\frac{q_\alpha}{2\pi\iota}\ln\left(\frac{\zeta_\alpha - c}{\zeta_\alpha + c}\right)$$

and hence, on integration,

$$f_\alpha(\zeta_\alpha) = \frac{q_\alpha}{2\pi\iota}\left[(\zeta_\alpha + c)\ln(\zeta_\alpha + c) - (\zeta_\alpha - c)\ln(\zeta_\alpha - c)\right] .$$

36.5.3 The line force solution

In the limit where $s = F/2c$ and $c \to 0$ in the above example, we recover the solution for a line force F at the origin, which is

$$f_\alpha(\zeta_\alpha) = \frac{q_\alpha}{2\pi\iota}\ln(\zeta_\alpha) \quad \text{where} \quad \boldsymbol{Bq} = \boldsymbol{F} .$$

The corresponding displacement $\boldsymbol{u}$ and the stress function $\boldsymbol{\phi}$ are then given by equations (36.34, 36.35) as

$$\boldsymbol{u} = \frac{1}{\pi}\Im\left\{\boldsymbol{A}\langle\ln(\zeta)\rangle\boldsymbol{q}\right\}; \quad \boldsymbol{\phi} = \frac{1}{\pi}\Im\left\{\boldsymbol{B}\langle\ln(\zeta)\rangle\boldsymbol{q}\right\} .$$

In particular, the surface displacement is

$$u(x, 0) = \frac{1}{\pi} \Im \{Aq \ln(x)\} = \frac{1}{\pi} \Im \left\{ AB^{-1} F \ln(x) \right\}$$

$$= -\frac{1}{\pi} \Re \{DF \ln(x)\} \ ,$$

where we define $D \equiv M^{-1} = \iota AB^{-1}$ as the inverse of the *impedance tensor*[12] $M = -\iota BA^{-1}$. Since $\ln(x) = \ln(|x|) + \iota \pi H(-x)$ and F is real, we then have

$$u(x, 0) = -\frac{1}{\pi} \Re \{D\} \, F \ln(|x|) - \frac{1}{2} \Im \{D\} F \, \mathrm{sgn}(x) \ , \qquad (36.39)$$

where we have added a rigid-body translation to restore symmetry, as in Chapter 12. Maxwell's reciprocal theorem §38.1 shows that $\Re \{D\}$ must be symmetric and $\Im \{D\}$ must be skew symmetric, so the matrices D and M are Hermitian[13] (i.e. $M^T = \overline{M}$).

36.5.4 Internal forces and dislocations

A similar approach can be used for the Kelvin problem of §13.1. If a concentrated force F per unit length (along the x_3-axis) acts at the origin in the infinite space, we know from equilibrium considerations and self-similarity that the stress components will decay with r^{-1}, which in view of the representation (36.30, 36.31) implies that the functions $f_\alpha(\zeta_\alpha)$ should be multiples of the holomorphic function $\ln(\zeta_\alpha)$. Using (36.34, 36.35), we therefore write

$$u = 2 \Re \{A \langle \ln(\zeta) \rangle q\} \ ; \quad \phi = 2 \Re \{B \langle \ln(\zeta) \rangle q\} \ . \qquad (36.40)$$

Equilibrium of a cylindrical region of radius a then gives the condition

$$F_i + \int_0^{2\pi} \sigma_{ij} n_j a d\theta = 0 \quad \text{where} \quad n = \{\cos \theta, \sin \theta, 0\} \qquad (36.41)$$

is the outward normal. Also, since points on the circle are defined by $x_1 = a \cos \theta$, $x_2 = a \sin \theta$, we have

$$\frac{\partial x_1}{\partial \theta} = -a \sin \theta \ ; \quad \frac{\partial x_2}{\partial \theta} = a \cos \theta \ .$$

Using these results and (36.30), we can write (36.41) in the form

$$F_i + \int_0^{2\pi} \left(-\frac{\partial \phi_i}{\partial x_2} \frac{\partial x_2}{\partial \theta} - \frac{\partial \phi_i}{\partial x_1} \frac{\partial x_1}{\partial \theta} \right) d\theta = 0 \quad \text{or} \quad F_i = \phi_i \big|_0^{2\pi} \ . \qquad (36.42)$$

---

[12] See T. C. T. Ting *loc. cit.*, Chapter 5.

[13] This can be proved directly from the properties of the matrices A, B.

From the properties of $\ln(\zeta)$ and equations (36.40, 36.42), we conclude that

$$F = 2\,\Re\,\{2\pi\iota\,Bq\} = 2\,\Re\{Bh\} \quad \text{where} \quad h = 2\pi\iota q \,. \tag{36.43}$$

We also note that the displacement defined by equation (36.29) will generally exhibit a discontinuity on the cut $\theta = 0, 2\pi$, defining a dislocation with Burgers' vector[14]

$$\mathcal{B} \equiv \{B_x, B_y, B_z\}^T = -u\Big|_0^{2\pi} = -2\,\Re\{2\pi\iota\,Aq\} = -2\,\Re\{Ah\} \,. \tag{36.44}$$

Equations (36.43, 36.44) can be combined in the matrix equation

$$\begin{bmatrix} A & \overline{A} \\ B & \overline{B} \end{bmatrix} \begin{bmatrix} h \\ \overline{h} \end{bmatrix} = \begin{bmatrix} -\mathcal{B} \\ F \end{bmatrix} \,,$$

and if the force F and the Burgers' vector $\mathcal{B}$ are prescribed, this equation can be solved, using the inverse property (36.37), to obtain

$$\begin{bmatrix} h \\ \overline{h} \end{bmatrix} = \begin{bmatrix} B^T & A^T \\ \overline{B}^T & \overline{A}^T \end{bmatrix} \begin{bmatrix} -\mathcal{B} \\ F \end{bmatrix} \,.$$

36.5.5 Planar crack problems

Equations (36.40, 36.43, 36.44) can be used exactly as in §13.3.2 to solve the problem of the plane crack $-a < x < a$, $y=0$ in a body loaded by uniform tractions $\sigma_{i2} \to T$ at the remote boundaries. Alternatively, the corrective solution — i.e. the solution for the same crack loaded only by tractions $-T, T$ on the crack faces $y = 0^-$, $y = 0^+$ respectively — is easily obtained in the form

$$u = 2\,\Re\,\{A\,\langle f(\zeta)\rangle\,q\} \,; \quad \phi = 2\,\Re\,\{B\,\langle f(\zeta)\rangle\,q\} \,,$$

with

$$f(\zeta) = \sqrt{\zeta^2 - a^2} - \zeta \,.$$

The tractions on the plane $y = 0^-$ are then given by

$$t(x, 0^-) = \sigma_{i2}(x, 0) = \frac{\partial\phi}{\partial x}(x, 0) = 2\,\Re\left\{Bq\left(\frac{|x|}{\sqrt{x^2 - a^2}} - 1\right)\right\} \,,$$

---

[14] not to be confused with the matrix B of equation (36.33). Ting *loc. cit.* defines the Burger's vector as a displacement discontinuity on the *negative* x-axis — i.e. as $u|_{-\pi}^{\pi}$ —, leading to the opposite sign on $\mathcal{B}$. Here we follow the same convention as in Chapter 13, where the discontinuity is placed on the positive x-axis.

and this satisfies the boundary condition $t(x, 0^-) = -T$ in $-a < x < a$ if Bq is real and given by

$$-2Bq = -T \quad \text{or} \quad q = \frac{B^{-1}T}{2} .$$

The corrective tractions on the plane $y = 0$ *outside* the crack ($|x| > a$) are then given by

$$t(x, 0) = 2Bq \left(\frac{|x|}{\sqrt{x^2 - a^2}} - 1 \right) = T \left(\frac{|x|}{\sqrt{x^2 - a^2}} - 1 \right) .$$

Notice that this traction and hence also the stress-intensity factors depend only on the far-field loading and are independent of the anisotropic constants.

36.5.6   The Barnett-Lothe tensors

The method of solution described in this section requires the determination of the complex eigenvalues p_α and eigenvectors $a^{(\alpha)}, b^{(\alpha)}$ from equations (36.28, 36.27, 36.32). However, in many problems, important results can be expressed in terms of the real Barnett-Lothe tensors

$$S = \imath \left(2AB^T - I \right) ; \quad H = 2\imath AA^T ; \quad L = -2\imath BB^T ,$$

where A, B are normalized so as to satisfy equation (36.37). Furthermore, S, H and L can be determined from the real matrices Q, R and T without explicit solution of the eigenvalue problem. We obtain[15]

$$S = -\frac{1}{\pi} \int_0^\pi T^{-1}(\theta) R^T(\theta) d\theta ; \quad H = \frac{1}{\pi} \int_0^\pi T^{-1}(\theta) d\theta$$

$$L = -\frac{1}{\pi} \int_0^\pi \left[R(\theta) T^{-1}(\theta) R^T(\theta) - Q(\theta) \right] d\theta ,$$

where

$$Q(\theta) = Q \cos^2 \theta + \left(R + R^T \right) \sin \theta \cos \theta + T \sin^2 \theta$$
$$R(\theta) = R \cos^2 \theta + (T - Q) \sin \theta \cos \theta - R^T \sin^2 \theta$$
$$T(\theta) = T \cos^2 \theta - \left(R + R^T \right) \sin \theta \cos \theta + Q \sin^2 \theta .$$

---

[15] For a proof of this result, see T. C. T. Ting *loc. cit.*, Chapter 7, or A. F. Bower, Applied Mechanics of Solids, CRC Press, Boca Raton, 2010, §5.5.11.

Once S, H and L have been determined, the impedance tensor and its inverse can be obtained as

$$M = -\iota B A^{-1} = H^{-1} + \iota H^{-1} S$$
$$M^{-1} = D = \iota A B^{-1} = L^{-1} - \iota S L^{-1} \; . \tag{36.45}$$

These results enable us to write equation (36.39) (for example) in terms of real matrices as

$$u(x, 0) = -\frac{1}{\pi} L^{-1} F \ln(|x|) + \frac{1}{2} S L^{-1} F \, \mathrm{sgn}(x) \; .$$

36.6 End Loading of the Prismatic Bar

If a long bar of uniform cross section is loaded only at the ends by a torque, a bending moment or an axial force, the stresses distant from the ends will be independent of z, but the displacements will contain terms that vary with z. Thus, these problems do not fall into the category defined in §36.2 and treated in §§36.3–36.5.

36.6.1 Bending and axial force

If we postulate an elementary stress state where the only non-zero stress component is $\sigma_{zz} = Ax + By + C$, the resulting strains will be linear functions of x, y and hence all six compatibility equations will be satisfied identically. Also, the tractions on the lateral surfaces of the bar will be zero, so this elementary stress state defines an exact solution to the problem of a prismatic bar loaded only by equal and opposite bending moments and axial forces at the two ends[16]. In particular, the stress components depend only on the force and moment resultants and the cross section of the bar and are independent of the elastic constants.

However, if the monoclinic condition (36.4) is not satisfied — i.e. if the in-plane and antiplane problems are coupled — the axial displacement u_z will generally be a quadratic function of x, y and hence all initially plane sections will warp into identical quadratic surfaces. Also the bending and torsion problems are then coupled — bending moments cause twist and torques cause the axis of the bar to bend.

---

[16] If the tractions on the ends differ from those implied by this solution but have the same force and moment resultants, we anticipate exponentially decaying modifications to the stress fields near the ends that can be found by the technique introduced in Chapter 6.

36.6.2 Torsion

Suppose the bar transmits a torque T and the axis is constrained to remain straight by the imposition of appropriate bending moments on the ends. We assume that there will be a uniform twist per unit length β, and, as in equation (17.1), this implies additional z-dependent terms $-\beta zy$ and βzx in u_x, u_y respectively. Equation (36.6) is therefore modified to

$$2e_{zx} = \frac{\partial u_z}{\partial x} - \beta y \; ; \quad 2e_{zy} = \frac{\partial u_z}{\partial y} + \beta x \; ,$$

and the second compatibility equation $(36.7)_2$ becomes

$$\frac{\partial e_{zx}}{\partial y} - \frac{\partial e_{yz}}{\partial x} + \beta = 0 \; . \tag{36.46}$$

Substitution of (36.17) into the compatibility equations $(36.7)_1$ and (36.46) then yields the differential equations

$$\mathcal{L}_4\phi + \mathcal{L}_3\psi = 0 \; ; \quad \mathcal{L}_3\phi + \mathcal{L}_2\psi = -2\beta \; .$$

As in Chapter 17, a convenient method of solution is to write $\psi = \psi_H + \psi_P$, where the particular solution ψ_P satisfies the equation

$$\mathcal{L}_2\psi_P = -2\beta \; .$$

Any convenient quadratic function of x, y can be chosen for ψ_P, in which case $\mathcal{L}_3\psi_P = 0$. The functions ϕ and ψ_H will then satisfy (36.18), and general solutions can be obtained as in §36.4.2. It is arguably simpler conceptually to (i) determine the stress field due to ψ_P alone (with $\phi = 0$), (ii) find the tractions implied on the lateral surfaces, and then (iii) use these to define a torsion-free corrective solution that can be treated using the method of §36.4.

If the in-plane and antiplane problems are coupled ($\mathcal{L}_3 \neq 0$), the stress component $\sigma_{zz}(x, y)$ will be non-zero, so non-zero force and moment resultants F, M_x, M_y are generally required on the ends of the bar. If the ends are actually traction free, an approximate solution can be obtained by cancelling these resultants by superposing an elementary stress state, as discussed in §3.1.1.

36.7 Three-dimensional Problems

Three-dimensional problems for anisotropic bodies are extremely challenging and very little progress can be made using purely analytical methods. The potential function approach used in Chapters 21–33 has no anisotropic counterpart except where the anisotropy is axisymmetric, including the special case of transverse isotropy (see §36.8).

Lekhnitskii's formalism can be used for the problem of an end loaded cantilever[17] — i.e. for the anisotropic solution of the problems considered in Chapter 18 — and the recursive technique of Chapter 30 can be applied to Stroh's formalism to provide the general solution of the prismatic bar loaded by polynomial tractions on the lateral surfaces[18].

36.7.1 Concentrated force on a half-space

If a concentrated normal force is applied at the origin on the otherwise traction-free surface of the half-space $z > 0$, the stress state is self-similar and equilibrium considerations show that the stress components will vary with R^{-2} and the displacement components with R^{-1}. In particular, the normal surface displacements must take the form

$$u_z(r, \theta, 0) = \frac{F h(\theta)}{r} \, , \tag{36.47}$$

where $h(\theta)$ depends on the elastic constants c_{ijkl} or equivalently on the stiffness matrix C of equation (36.1). Willis[19] used this argument to show that the contact area in Hertzian contact remains elliptical for general anisotropy, and the contact pressure distribution is still given by equation (34.4). This result is easily verified using the field-point integration method of §34.1.1 since the function $h(\theta)$ appears only in the outer integral in the anisotropic counterpart of equation (34.7).

However, the parameters a, b, p_0 defining the semi-axes of the ellipse and the maximum contact pressure p_0 depend on the function $h(\theta)$ in equation (36.47). Willis solved the concentrated force problem using a double Fourier transform, but an arguable more elegant solution was provided by Sveklo[20], using the Smirnov-Sobolev transform. A remarkable result from this solution is that

$$h\left(\frac{\pi}{2}\right) = \frac{D_{22}}{2\pi} \quad \text{where} \quad D = M^{-1} = L^{-1} - \imath S L^{-1} \tag{36.48}$$

---

[17] S. G. Lekhnitskii, *loc. cit.*, Chapter 5.

[18] J. R. Barber and T. C. T. Ting (2007), Three-dimensional solutions for general anisotropy, *Journal of the Mechanics and Physics of Solids*, Vol. 55, pp. 1993–2006.

[19] J. R. Willis (1966), Hertzian contact of anisotropic bodies, *Journal of the Mechanics and Physics of Solids,* Vol. 14 (3), pp. 163–176.

[20] V. A. Sveklo (1964), Boussinesq type problems for the anisotropic half-space, *Journal of Applied Mathematics and Mechanics,* Vol. 28 (5), pp. 1099–1105.

is the inverse of the impedance matrix[21]. In other words, the displacement at $\theta = \pi/2$ is proportional to the multiplier on the logarithmic term in the corresponding two-dimensional solution of equation (36.39) with normal loading only. Thus an alternative expression for $h(\theta)$ is

$$h(\theta) = \frac{1}{2\pi} D_{22} \left(\theta - \frac{\pi}{2} \right) ,$$

where $D(\alpha)$ is the inverse impedance matrix in a coördinate system rotated through α about the z-axis.

36.8 Transverse Isotropy

For a transversely isotropic material, the most general form of Hooke's law can be written

$$\begin{bmatrix} \sigma_{xx} \\ \sigma_{yy} \\ \sigma_{zz} \\ \sigma_{yz} \\ \sigma_{zx} \\ \sigma_{xy} \end{bmatrix} = \begin{bmatrix} C_{11} & C_{12} & C_{13} & 0 & 0 & 0 \\ C_{21} & C_{11} & C_{13} & 0 & 0 & 0 \\ C_{31} & C_{31} & C_{33} & 0 & 0 & 0 \\ 0 & 0 & 0 & C_{44} & 0 & 0 \\ 0 & 0 & 0 & 0 & C_{44} & 0 \\ 0 & 0 & 0 & 0 & 0 & (C_{11}-C_{12})/2 \end{bmatrix} \begin{bmatrix} e_{xx} \\ e_{yy} \\ e_{zz} \\ 2e_{yz} \\ 2e_{zx} \\ 2e_{xy} \end{bmatrix} . \tag{36.49}$$

This suggests that solutions might be obtained using an appropriate linear scaling of the z-axis, so in the spirit of §21.1, we look for conditions under which a solution can be defined in terms of a displacement function ϕ through the relations

$$u_x = \frac{\partial \phi}{\partial x} \;\; ; \;\; u_y = \frac{\partial \phi}{\partial y} \;\; ; \;\; u_z = \lambda \frac{\partial \phi}{\partial z} ,$$

where λ is a scalar parameter.

Substituting these definitions into the strain-displacement relations and the resulting strains into (36.49), we obtain

$$\sigma_{xx} = C_{11} \frac{\partial^2 \phi}{\partial x^2} + C_{12} \frac{\partial^2 \phi}{\partial y^2} + C_{13} \lambda \frac{\partial^2 \phi}{\partial z^2} \;\; ; \;\; \sigma_{yy} = C_{12} \frac{\partial^2 \phi}{\partial x^2} + C_{11} \frac{\partial^2 \phi}{\partial y^2} + C_{13} \lambda \frac{\partial^2 \phi}{\partial z^2}$$

$$\sigma_{zz} = C_{13} \frac{\partial^2 \phi}{\partial x^2} + C_{13} \frac{\partial^2 \phi}{\partial y^2} + C_{33} \lambda \frac{\partial^2 \phi}{\partial z^2}$$

$$\sigma_{yz} = C_{44}(1+\lambda) \frac{\partial^2 \phi}{\partial y \partial z} \;\; ; \;\; \sigma_{zx} = C_{44}(1+\lambda) \frac{\partial^2 \phi}{\partial z \partial x} \;\; ; \;\; \sigma_{xy} = (C_{11} - C_{12}) \frac{\partial^2 \phi}{\partial x \partial y} .$$

These stress components will satisfy the three equilibrium equations (2.4) with no body force if ϕ satisfies the two equations

---

[21] See §36.5.3 and equation(36.45). Also, note that D is Hermitian, so D_{22} is real.

$$C_{11}\nabla_1^2\phi + \left[C_{13}\lambda + C_{44}(1+\lambda)\right]\frac{\partial^2\phi}{\partial z^2} = 0 \tag{36.50}$$

$$\left[C_{13} + C_{44}(1+\lambda)\right]\nabla_1^2\phi + C_{33}\lambda\frac{\partial^2\phi}{\partial z^2} = 0 , \tag{36.51}$$

where we write

$$\nabla_1^2 \equiv \left(\frac{\partial^2}{\partial x^2} + \frac{\partial^2}{\partial y^2}\right)$$

for the two-dimensional Laplacian operator. The two conditions (36.50, 36.51) become identical if we choose λ such that

$$\left[C_{13} + C_{44}(1+\lambda)\right]\left[C_{13}\lambda + C_{44}(1+\lambda)\right] = C_{11}C_{33}\lambda .$$

This is a quadratic equation for λ with two roots

$$\lambda_1, \lambda_2 = c \pm \sqrt{c^2 - 1} \quad \text{where} \quad c = \frac{C_{11}C_{33} - C_{13}^2}{2C_{44}(C_{13} + C_{44})} - 1 .$$

When $\lambda = \lambda_i$ $(i = 1, 2)$, the two equations (36.50, 36.51) both reduce to

$$\nabla_1^2\phi + p_i\frac{\partial^2\phi}{\partial z^2} = 0 , \tag{36.52}$$

where

$$p_i = \frac{C_{44} + (C_{13} + C_{44})\lambda_i}{C_{11}} .$$

If we then make the further transformation

$$z = Z_i\sqrt{p_i} ,$$

equation (36.52) is reduced to Laplace's equation

$$\frac{\partial^2\phi}{\partial x^2} + \frac{\partial^2\phi}{\partial y^2} + \frac{\partial^2\phi}{\partial Z_i^2} = 0$$

in the coördinate system (x, y, Z_i).

A more general solution can be constructed by superposing the solutions associated with functions ϕ_1, ϕ_2, corresponding to the eigenvalues λ_1, λ_2. These two functions can then be used in the same way as Solutions A and B of Table 22.1 to formulate boundary-value problems for the half-space. They can also be conveniently combined to construct solutions analogous to Solutions F and G of Table 22.3 in which either the shear traction or the normal traction is identically zero throughout the surface $z = 0$. For non-axisymmetric problems, a third potential function is required, analogous to Solution E, of Table 22.1. The development of this solution is left as an exercise for the reader (Problem 36.11).

In the special case of isotropy, $c = 1$, so $\lambda_1 = \lambda_2 = 1$ and $p_1 = p_2 = 1$. However, as in §36.3.2, if we approach the isotropic case as a limit $c \to 1^+$, two independent solutions in terms of harmonic functions can be obtained.

Problems

36.1. Use the Prandtl stress function φ of equations (17.4) to satisfy the antiplane equilibrium equation (36.5)$_3$ and then use the compatibility equation (36.7)$_2$ to develop the governing equation for φ for the orthotropic material of equation (36.8). Use this formulation to find the stress components for the problem of Figure 15.2.

36.2. Use the transformation equations (1.43) and the two dimensional direction cosines (1.14) to find the components C'_{16}, C'_{26} for the orthotropic material of equation (36.8) in a coördinate system rotated through an angle θ about the z-axis. Hence show that these coefficients are zero for all θ if and only if

$$C_{11} = C_{22} \quad \text{and} \quad 2C_{66} = C_{11} - C_{12} .$$

If these condition are *not* satisfied, how many planes (values of θ in $0 \le \theta < \pi$) exist on which $C'_{16} = C'_{26} = 0$?

36.3. Transform the stress components (36.15) into polar coördinates (r, θ) centred on the point of application of the force F_y and hence show that $\sigma_{\theta r} = \sigma_{\theta\theta} = 0$ for all (r, θ) as in the isotropic case. Can you find a proof of this without using the orthotropic solution of §36.3.1?

Use your results to make plots of the function $\sigma_{rr}(r, \theta)/\sigma_{rr}(r, -\pi/2)$ in the range $-\pi < \theta < 0$ (see Figure 12.2) for the cases $(\lambda_1, \lambda_2) = (1, 2), (1, 0.5)$, and the isotropic case $(1, 1)$, and comment on the results.

36.4. Show that the antiplane stress field near a traction-free notch $-\beta < \theta < \beta$ in a monoclinic material is characterized by a singular field with a dominant term proportional to $r^{\lambda-1}$, where

$$\lambda = \frac{\pi}{2\gamma} \quad \text{and} \quad \tan\gamma = \frac{\sqrt{C_{44}C_{55} - C_{45}^2}\, \tan\beta}{(C_{44} + C_{45}\tan\beta)} .$$

36.5. Use the Lekhnitskii formalism to find the stress components in Cartesian coördinates associated with the generalized Flamant solution, where a force $F = \{F_x, F_y, F_z\}^T$ per unit length acts along the line $x = y = 0$ on the otherwise traction-free half-plane $y > 0$. Assume that the parameters p_α, λ_α are known for the material of the half-plane.

36.6. Suppose that for a particular anisotropic material, equation (36.22) has a pair of repeated complex roots, so that $p_1 = p_2 \equiv p \ne p_3$. Show that a general solution of the two dimensional elasticity problem can then be written in the form

$$\phi = 2\,\Re\,\left\{ f_1(x + py) + (x + \overline{p}y) f_2(x + py) + f_3(x + p_3 y) \right\} ,$$

Discuss how you would then find the corresponding function ψ, but do not complete the solution.

36.7. The half-plane $y > 0$ is loaded by a normal traction

$$\sigma_{yy}(x, 0) = S \cos(\lambda x) .$$

Find expressions for the functions $f_\alpha(\zeta_\alpha)$ in equation (36.31) and hence show that the normal surface displacement is

$$u_y(x, 0) = \frac{S D_{22} \cos(\lambda x)}{\lambda} \qquad \text{where} \qquad \boldsymbol{D} = \imath \boldsymbol{A} \boldsymbol{B}^{-1} .$$

36.8. If the material is monoclinic — i.e. the stiffness matrix satisfies the condition (36.4) — show that the matrix equation (36.27) partitions into an in-plane problem for $\boldsymbol{a}^{(1)}$, $\boldsymbol{a}^{(2)}$ and an antiplane problem for $\boldsymbol{a}^{(3)}$. Solve the resulting scalar equation for the antiplane eigenvalues p_3 and show that they form a complex conjugate pair for all physically reasonable materials.

36.9. Use Maxwell's reciprocal theorem (§38.1) and equation (36.39) to show that the matrix $\boldsymbol{D}$ is Hermitian.

36.10. Use an argument similar to that in §13.3 and the solution of §36.5.5 to find the energy-release rate for a plane crack as a function of the vector stress-intensity factor $\boldsymbol{K} = \{K_I, K_{II}, K_{III}\}^T$.

36.11. For a generally anisotropic material with non-uniform temperature T, the Fourier heat conduction law (14.10) and the constitutive law (1.42) take the more general form

$$q_i = -K_{ij} \frac{\partial T}{\partial x_j} \quad ; \quad \sigma_{ij} = c_{ijkl} \frac{\partial u_k}{\partial x_l} + \beta_{ij} T ,$$

where $\boldsymbol{K}$ is the conductivity matrix and $\boldsymbol{\beta}$ is the stress-temperature matrix, both of which are real and symmetric. If the temperature is steady-state and there are no internal heat sources, show that a general two dimensional temperature distribution can be written in the form

$$T(x_1, x_2) = 2\,\Re\{f(\zeta_T)\} \qquad \text{where} \qquad \zeta_T = x_1 + \kappa x_2 ,$$

f is a holomorphic function and κ is a complex parameter. Then show that a particular solution for the thermoelastic displacement $\boldsymbol{u}$ can be written $\boldsymbol{u} = 2\,\Re\{\boldsymbol{c} g(\zeta_T)\}$ and find the governing equations for $\boldsymbol{c}$ and $g(\zeta_T)$.

36.12. Show that the relations (36.32) imply that a, b satisfy the eigenvalue equations (36.36).

36.13. Find the strain components for a generally anisotropic material corresponding to the elementary state of stress where $\sigma_{zz} = By$ is the only non-zero stress component. Then integrate the strain-displacement relations to find the corresponding displacements. In particular, by examining the terms that vary with z, show that this state of stress causes twist about the z-axis as well as bending about the x-axis, but that there is no bending about the y axis. Find the ratio between the twist per unit length and the curvature.

36.14. A solid circular cylinder of radius a with general anisotropy is loaded by a torque T. Find the twist per unit length β and the force and moment resultants on the ends needed to ensure that the axis of the cylinder remains straight.

36.15. Use Maxwell's theorem (§38.1) to prove that the function $h(\theta)$ in equation (36.47) satisfies the condition $h(\theta + \pi) = h(\theta)$. Then prove equation (36.48) by applying Betti's reciprocal theorem (§38.2) to the concentrated force solution, using the two-dimensional sinusoidal solution of Problem 36.6 as auxiliary solution.

36.16. For the transversely isotropic material of equation (36.49), develop a potential function solution analogous to Solution E of Table 22.1. Start by assuming that the displacements can be represented in the form

$$u_x = 2\frac{\partial \psi}{\partial y} \;\; ; \;\; u_y = -2\frac{\partial \psi}{\partial x} \;\; ; \;\; u_z = 0 \, ,$$

determine the corresponding stress components, and hence find the equation that must be satisfied by the potential function ψ. Show that with a suitable scaling of the z-coördinate, this equation can be transformed into Laplace's equation.

36.17. A transversely isotropic half-space $z > 0$ is loaded by a normal concentrated compressive force F applied at the origin. Find the surface displacements $u_r(r, 0), u_z(r, 0)$, where r is distance from the origin.

36.18. A hexagonal close-packed crystal transforms into an identical pattern under a rotation of $\pi/3$ about an axis perpendicular to a basal plane. Show that this implies that the material is transversely isotropic in the basal plane.

Chapter 37
Variational Methods

Energy or variational methods have an important place in Solid Mechanics both as an alternative to the more direct method of solving the governing partial differential equations and as a means of developing convergent approximations to analytically intractable problems. They are particularly useful in situations where only a restricted set of results is required — for example, if we wish to determine the resultant force on a cross section or the displacement of a particular point, but are not interested in the full stress and displacement fields. Indeed, such results can often be obtained in closed form for problems in which a solution for the complete fields would be intractable.

From an engineering perspective, it is natural to think of these methods as a consequence of the principle of conservation of energy or the first law of thermodynamics. However, conservation of energy is in some sense guaranteed by the use of Hooke's law and the equilibrium equations. Once these physical premises are accepted, the energy theorems we shall present here are purely mathematical consequences. Indeed, the finite element method, which is one of the more important developments of this kind, can be developed simply by applying arguments from approximation theory (such as a least-squares fit) to the governing equations introduced in previous chapters. For this reason, these techniques are now more often referred to as *Variational Methods*, meaning that instead of seeking to solve the governing partial differential equations directly, we seek to define a scalar function of the physical parameters which is stationary (generally maximum or minimum) in respect to infinitesimal variations about the solution.

In this chapter, we shall introduce only the principal theorems of this kind and indicate some of their applications[1]. We shall present most of the derivations in the context of general anisotropy, using the index notation of §1.1.2, since this leads to a more compact presentation as well as demonstrating the generality of the results.

---

[1] For a more comprehensive review of variational methods in elasticity, the reader is referred to S. G. Mikhlin, *Variational Methods in Mathematical Physics*, Pergamon, New York, 1964 and S. P. Timoshenko and J. N. Goodier, *loc. cit.*, Chapter 8.

© The Author(s), under exclusive license to Springer Nature Switzerland AG 2022
J. R. Barber, *Elasticity*, Solid Mechanics and Its Applications 172,
https://doi.org/10.1007/978-3-031-15214-6_37

37.1 Strain Energy

When a body is deformed, the external forces move through a distance associated with the deformation and hence do work. The fundamental premise of linear elasticity is that the deformation is linearly proportional to the load, so if the load is now removed, the deformation must pass through the same sequence of states as it did during the loading phase. It follows that the work done during loading is exactly recovered during unloading and must therefore be stored in the deformed body as *strain energy*.

Suppose that the body Ω with boundary Γ is subjected to a body force distribution p_i and boundary tractions $t_i = \sigma_{ij} n_j$ (1.15), where n_j defines a unit vector in the direction of the local outward normal to Γ. If the loads are applied sufficiently slowly for inertia effects to be negligible[2], and if the final elastic displacements are u_i, the stored strain energy U in the body in the loaded state will be equal to the work W done by the external forces during loading, which is

$$U = W = \frac{1}{2} \iiint_{\Omega} p_i u_i \, d\Omega + \frac{1}{2} \iint_{\Gamma} t_i u_i \, d\Gamma \ . \tag{37.1}$$

37.1.1 Strain energy density

The same principle can be applied to determine the strain energy in an infinitesimal rectangular element of material subjected to a uniform state of stress σ_{ij}. We conclude that the *strain-energy density* — i.e. the strain energy stored per unit volume — is given by

$$U_0 = \frac{1}{2} \sigma_{ij} e_{ij} \ . \tag{37.2}$$

Substituting for the stress or strain components from the generalized Hooke's law (1.40, 1.42), we obtain the alternative expressions

$$U_0 = \frac{1}{2} c_{ijkl} e_{ij} e_{kl} = \frac{1}{2} c_{ijkl} \frac{\partial u_i}{\partial x_j} \frac{\partial u_k}{\partial x_l} = \frac{1}{2} s_{ijkl} \sigma_{ij} \sigma_{kl} \ . \tag{37.3}$$

Notice incidentally that U_0 must be positive for all possible states of stress or deformation and this places some inequality restrictions on the elasticity tensors c_{ijkl}, s_{ijkl}.

If the material is isotropic, these expressions reduce to

---

[2] This implies that the stress field is always quasi-static in the sense of Chapter 7 and hence that all the work done by the external loads appears as elastic strain energy in the body, rather than partially as kinetic energy.

$$U_0 = \frac{1}{2E} \left[(\sigma_{xx}^2 + \sigma_{yy}^2 + \sigma_{zz}^2) - 2\nu(\sigma_{yy}\sigma_{zz} + \sigma_{zz}\sigma_{xx} + \sigma_{xx}\sigma_{yy}) \right.$$
$$\left. + 2(1+\nu) \left(\sigma_{yz}^2 + \sigma_{zx}^2 + \sigma_{xy}^2 \right) \right]$$
$$= \mu \left[\frac{\nu e^2}{(1-2\nu)} + e_{xx}^2 + e_{yy}^2 + e_{zz}^2 + 2e_{yz}^2 + 2e_{zx}^2 + 2e_{xy}^2 \right]. \tag{37.4}$$

37.2 Conservation of Energy

The strain energy U stored in the entire body Ω can be obtained by summing that stored in each of its individual particles, giving

$$U = \iiint_{\Omega} U_0 d\Omega . \tag{37.5}$$

This expression and equation (37.1) must clearly yield the same result and hence

$$\frac{1}{2} \iiint_{\Omega} p_i u_i d\Omega + \frac{1}{2} \iint_{\Gamma} t_i u_i d\Gamma = \iiint_{\Omega} U_0 d\Omega . \tag{37.6}$$

This argument appeals to the principle of conservation of energy, but this principle is implicit in Hooke's law, which guarantees that the loading is reversible. Thus, (37.6) can be derived from the governing equations of elasticity without explicitly invoking conservation of energy.

To demonstrate this, we first substitute (1.15) into the second term on the left-hand side of (37.6) and apply the divergence theorem, obtaining

$$\frac{1}{2} \iint_{\Gamma} t_i u_i d\Gamma = \frac{1}{2} \iint_{\Gamma} n_j \sigma_{ji} u_i d\Gamma = \frac{1}{2} \iiint_{\Omega} \frac{\partial}{\partial x_j} (\sigma_{ji} u_i) d\Omega$$
$$= \frac{1}{2} \iiint_{\Omega} \frac{\partial \sigma_{ji}}{\partial x_j} u_i d\Omega + \frac{1}{2} \iiint_{\Omega} \sigma_{ji} \frac{\partial u_i}{\partial x_j} d\Omega . \tag{37.7}$$

Finally, we use the equilibrium equation (2.4) in the first term on the right-hand side of (37.7) and Hooke's law (1.42) in the second term to obtain

$$\frac{1}{2} \iint_{\Gamma} t_i u_i d\Gamma = -\frac{1}{2} \iiint_{\Omega} p_i u_i d\Omega + \frac{1}{2} \iiint_{\Omega} c_{ijkl} \frac{\partial u_k}{\partial x_l} \frac{\partial u_i}{\partial x_j} d\Omega ,$$

from which (37.6) follows after using (37.2) in the last term.

37.3   Potential Energy of the External Forces

We can also construct a *potential energy* of the external forces which we denote by V. The reader is no doubt familiar with the concept of potential energy as applied to gravitational forces, but the definition can be extended to more general force systems as *the work that the forces would do if they were allowed to return to their reference positions*. For a single concentrated force $\boldsymbol{F}$ displaced through $\boldsymbol{u}$ this is defined as

$$V = -\boldsymbol{F} \cdot \boldsymbol{u} = -F_i u_i \ .$$

It follows by superposition that the potential energy of the boundary tractions and body forces is given by

$$V = - \iint_{\Gamma_t} t_i u_i \, d\Gamma - \iiint_{\Omega} p_i u_i \, d\Omega \ ,$$

where Γ_t is that part of the boundary over which the tractions are prescribed. We can then define the *total potential energy* Π as the sum of the stored strain energy and the potential energy of the external forces — i.e.

$$\Pi \equiv U + V = \frac{1}{2} \iiint_{\Omega} c_{ijkl} \frac{\partial u_i}{\partial x_j} \frac{\partial u_k}{\partial x_l} d\Omega - \iint_{\Gamma_t} t_i u_i \, d\Gamma - \iiint_{\Omega} p_i u_i \, d\Omega \ .$$
$$(37.8)$$

37.4   Theorem of Minimum Total Potential Energy

Suppose that the displacement field u_i satisfies the equilibrium equations (2.14) for a given set of boundary conditions and that we then perturb this state by a small variation δu_i. The corresponding perturbation in Π is

$$\delta \Pi = \iiint_{\Omega} c_{ijkl} \frac{\partial \delta u_i}{\partial x_j} \frac{\partial u_k}{\partial x_l} d\Omega - \iint_{\Gamma} t_i \delta u_i \, d\Gamma - \iiint_{\Omega} p_i \delta u_i \, d\Omega \ , \qquad (37.9)$$

from (37.8). Notice that $\delta u_i = 0$ in any region Γ_u of Γ in which the displacement is prescribed and hence the domain of integration Γ_t in the second term on the right-hand side of (37.8) can be replaced by $\Gamma = \Gamma_u + \Gamma_t$.

Substituting for t_i from (1.15) into the second term on the right-hand side of (37.9) and then applying the divergence theorem, we have

$$\iint_{\Gamma} t_i \delta u_i d\Gamma = \iint_{\Gamma} \sigma_{ij} n_j \delta u_i d\Gamma = \iiint_{\Omega} \frac{\partial}{\partial x_j} \left(\sigma_{ij} \delta u_i \right) d\Omega$$

$$= \iiint_{\Omega} \frac{\partial \sigma_{ij}}{\partial x_j} \delta u_i d\Omega + \iiint_{\Omega} \frac{\partial \delta u_i}{\partial x_j} \sigma_{ij} d\Omega \ .$$

Finally, using the equilibrium equation (2.4) in the first term and Hooke's law (1.42) in the second, we obtain

$$\iint_{\Gamma} t_i \delta u_i d\Gamma = - \iiint_{\Omega} p_i \delta u_i d\Omega + \iiint_{\Omega} c_{ijkl} \frac{\partial \delta u_i}{\partial x_j} \frac{\partial u_k}{\partial x_l} d\Omega \ ,$$

and comparing this with (37.9), we see that $\delta \Pi = 0$. In other words, the equilibrium equation requires that the total potential energy must be stationary with regard to any kinematically admissible[3] small variation δu_i in the displacement field u_i, or

$$\frac{\partial \Pi}{\partial u_i} = 0 \ . \tag{37.10}$$

A more detailed second order analysis shows that the total potential energy must in fact be a minimum and this is intuitively reasonable, since if some variation δu_i from the equilibrium state could be found which reduced Π, the surplus energy would take the form of kinetic energy and the system would therefore generally accelerate away from the equilibrium state. However, we emphasise that the above derivation of equation (37.10) makes no appeal to the principle of conservation of energy.

37.5 Approximate Solutions — the Rayleigh-Ritz Method

We remarked in Chapter 2 that the problem of elasticity is to determine a stress field satisfying appropriate boundary conditions such that (i) the stresses satisfy the equilibrium equations (2.4) and (ii) the corresponding elastic strains satisfy the compatibility conditions (2.7). The compatibility conditions can be satisfied by defining the problem in terms of the displacements u_i and we have just shown that the equilibrium equations are equivalent to the principal of minimum total potential energy. Thus, an alternative formulation is to seek a kinematically admissible displacement field u_i such that the total potential energy Π is a minimum.

This approach is particularly useful as a method of generating approximate solutions to problems that might otherwise be analytically intractable. Suppose that we approximate the displacement field in the body by the finite series

$$u_i(x_1, x_2, x_3) = \sum_{k=1}^{n} A_k f_i^{(k)}(x_1, x_2, x_3) \ , \tag{37.11}$$

---

[3] i.e. any perturbation that is a continuous function of position within the body and is consistent with the displacement boundary conditions.

where the $f_i^{(k)}$ are a set of approximating functions and A_k are arbitrary constants constituting the *degrees of freedom* in the approximation. The functions $f_i^{(k)}$ must be continuous and single-valued and must also be chosen so as to ensure that equation (37.11) satisfies any displacement boundary conditions of the problem for all A_k, but they are otherwise unrestricted as to form.

We can then substitute (37.11) into (37.8) and perform the integrals over Ω, obtaining the total potential energy as a quadratic function of the A_k. The theorem (37.10) demands that

$$\frac{\partial \Pi}{\partial A_k} = 0 ; \quad k \in [1, n] , \tag{37.12}$$

which defines n linear equations for the n unknown degrees of freedom A_k. The corresponding stress components can then be found by substituting (37.11) into Hooke's law (1.42).

If the approximating functions $f_i^{(k)}$ are defined over the entire body Ω, this typically leads to series solutions (e.g. power series or Fourier series) and the method is known as the *Rayleigh-Ritz method*. It is particularly useful in structural mechanics applications such as beam problems, where the displacement is a function of a single spatial variable, but it can also be applied (e.g.) to the problem of a rectangular plate, using double Fourier series or power series.

Example

To illustrate the method, we consider the problem of a rectangular plate $-a < x < a$, $-b < y < b$ which makes frictionless contact with fixed rigid planes at $y = \pm b$ and which is loaded by compressive tractions

$$\sigma_{xx}(\pm a, y) = -\frac{S y^2}{b^2}$$

at $x = \pm a$. The problem is symmetric about both axes, so u_x must be odd in x and even in y, whilst u_y is even in x and odd in y. Also, the displacement boundary condition requires that $u_y = 0$ at $y = \pm b$. The most general third order polynomial approximation satisfying these conditions is

$$u_x = A_1 x^3 + A_2 x y^2 + A_3 x ; \quad u_y = A_4 (b^2 - y^2)$$

and the corresponding strains are

$$e_{xx} = \frac{\partial u_x}{\partial x} = 3 A_1 x^2 + A_2 y^2 + A_3 ; \quad e_{yy} = \frac{\partial u_y}{\partial y} = -2 A_4 y$$

$$e_{xy} = \frac{1}{2} \left(\frac{\partial u_y}{\partial x} + \frac{\partial u_x}{\partial y} \right) = A_2 x y .$$

Substituting these results into (37.4) and then evaluating the integral (37.5), we obtain

$$U = \int_{-b}^{b} \int_{-a}^{a} U_0(x, y)dxdy = \frac{4\mu ab(1 - \nu)}{15(1 - 2\nu)} \left[(27A_1^2 a^4 + 3A_2^2 b^4 + 15A_3^2 + 20A_4^2 b^2 \right.$$

$$\left. + 10A_2 A_3 b^2 + 30A_1 A_3 a^2 + 10A_1 A_2 a^2 b^2) \right] + \frac{8\mu A_2^2 a^3 b^3}{9} .$$

The potential energy of the tractions on the boundaries $x = \pm a$ is

$$V = 2 \int_{-b}^{b} \frac{Sy^2(A_1 a^3 + A_2 ay^2 + A_3 a)}{b^2} dy = 4Sab \left(\frac{A_1 a^2}{3} + \frac{A_2 b^2}{5} + \frac{A_3}{3} \right) ,$$

where the factor of 2 results from there being identical expressions for each boundary. Calculating $\Pi = U + V$ and imposing the condition (37.12) for each degree of freedom A_i, $i = 1, 4$, we obtain four linear equations whose solution is

$$A_1 = 0 ; \quad A_2 = -\frac{S(1 - 2\nu)}{\mu[5(1 - 2\nu)a^2 + 2(1 - \nu)b^2]}$$

$$A_3 = -\frac{5Sa^2(1 - 2\nu)^2}{6\mu(1 - \nu)[5(1 - 2\nu)a^2 + 2(1 - \nu)b^2]} ; \quad A_4 = 0 .$$

Rayleigh-Ritz solutions are conceptually straightforward, but they tend to generate lengthy algebraic expressions, as illustrated in this simple example. However, this is no bar to their use provided the algebra is performed in Mathematica or Maple. Also, if Fourier series are used in place of power series, the orthogonality of the corresponding integrals will often lead to significant simplifications.

However, if high accuracy is required it is often more effective to use a set of piecewise continuous functions for the $f_i^{(k)}(x_1, x_2, x_3)$ of equation (37.11). The body is thereby divided into a set of *elements* and the displacement in each element is described by one or more *shape functions* multiplied by degrees of freedom A_k. Typically, the shape functions are defined such that the A_k represent the displacements at specified points or *nodes* within the body and the $f_i^{(k)}$ are zero except in those elements contiguous to node k. They must also satisfy the condition that the displacement be continuous between one element and the next for all A_k. Once the approximation is defined, equation (37.12) once again provides n linear equations

for the n nodal displacements. This is the basis of the *finite element method*[4]. Since
the theorem of minimum total potential energy is itself derivable from Hooke's law
and the equilibrium equation, an alternative derivation of the finite element method
can be obtained by applying approximation theory directly to these equations. To
develop a set of n linear equations for the A_k, we substitute the approximate form
(37.11) into the equilibrium equations, multiply by n *weight functions*, integrate over
the domain Ω and set the resulting n linear functions of the A_k to zero. These equations will be identical to (37.12) if the weight functions are chosen to be identical to
the shape functions.

37.6   Castigliano's Second Theorem

The strain energy U can be written as a function of the stress components, using the
final expression in (37.3). We obtain

$$U = \frac{1}{2} \iiint_\Omega s_{ijkl}\sigma_{ij}\sigma_{kl}d\Omega \ .$$

We next consider the effect of perturbing the stress field by a small variation $\delta\sigma_{ij}$,
chosen so that the perturbed field $\sigma_{ij} + \delta\sigma_{ij}$ still satisfies the equilibrium equation
(2.4) — i.e.

$$\frac{\partial\sigma_{ij}}{\partial x_j} + p_i = 0 \quad \text{and} \quad \frac{\partial}{\partial x_j}(\sigma_{ij} + \delta\sigma_{ij}) + p_i = 0 \ . \tag{37.13}$$

The corresponding perturbation in U will be

$$\delta U = \iiint_\Omega s_{ijkl}\sigma_{kl}\delta\sigma_{ij}d\Omega = \iiint_\Omega \frac{\partial u_i}{\partial x_j}\delta\sigma_{ij}d\Omega \ . \tag{37.14}$$

The divergence theorem gives

$$\iint_\Gamma u_i\delta\sigma_{ij}n_jd\Gamma = \iiint_\Omega \frac{\partial}{\partial x_j}\left(u_i\delta\sigma_{ij}\right)d\Omega$$

$$= \iiint_\Omega \frac{\partial u_i}{\partial x_j}\delta\sigma_{ij}d\Omega + \iiint_\Omega \frac{\partial\delta\sigma_{ij}}{\partial x_j}u_id\Omega \tag{37.15}$$

and by taking the difference between the two equations (37.13), we see that the second
term on the right-hand side must be zero. Using (37.14, 37.15) and $\delta\sigma_{ij}n_j = \delta t_i$, we
then have

---

[4] For more detailed discussion of the finite element method, see O. C. Zienkiewicz, *The Finite
Element Method*, McGraw-Hill, New York, 1977, K. J. Bathe, *Finite Element Procedures in Engi-
neering Analysis*, Prentice-Hall, Englewood Cliffs, NJ, 1982, T. J. R. Hughes, *The Finite Element
Method*, Prentice Hall, Englewood Cliffs, NJ, 1987.

$$\delta U = \iint_{\Gamma_u} u_i \delta\sigma_{ij} n_j d\Gamma = \iint_{\Gamma_u} u_i \delta t_i d\Gamma ,$$

where the integral is taken only over that part of the boundary Γ_u in which displacement u_i is prescribed, since no perturbation in traction is permitted in Γ_t where t_i is prescribed. It follows that the *complementary energy*

$$C \equiv U - \iint_{\Gamma_u} u_i t_i d\Gamma$$

must be stationary with respect to all self-equilibrated variations of stress $\delta\sigma_{ij}$, or

$$\frac{\partial C}{\partial \sigma_{ij}} = 0 . \qquad (37.16)$$

This is *Castigliano's second theorem*. It enables us to state a second alternative formulation of the elasticity problem as the search for a state of stress satisfying the equilibrium equation (2.4), such that the complementary energy C is stationary. It is interesting to note that the theorem of minimum potential energy and Castigliano's second theorem each substitutes for one of the two conditions equilibrium and compatibility identified in Chapter 2.

37.7 Approximations using Castigliano's Second Theorem

We recall that both the Airy stress function of Chapters 5–13 and the Prandtl stress function of Chapters 15, 17, 18 satisfy the equilibrium equations identically. In fact, these representations define the most general states of in-plane and antiplane stress respectively that satisfy the equilibrium equations without body force. Thus, if we describe the solution in terms of such a stress function, Castigliano's second theorem reduces the problem to the search for a function satisfying the boundary conditions and equation (37.16). This leads to a very effective approximate method for both in-plane and antiplane problems in which the stress function ϕ is represented in the form

$$\phi(x, y) = \sum_{k=1}^{n} A_k f_k(x, y) \qquad (37.17)$$

by analogy with (37.11), where each of the functions f_k satisfies the traction boundary conditions. This can generally be fairly easily achieved by making use of the mathematical description of the boundary line Γ. We shall start the discussion by considering the torsion problem of Chapter 17, for which this procedure is particularly straightforward. We shall then adapt it for the in-plane problem in §37.7.2.

37.7.1 The torsion problem

If the torsion problem is described in terms of Prandtl's stress function φ, the traction-free boundary condition for a simply-connected region Ω can be satisfied by requiring that $\varphi=0$ on the boundary Γ. Suppose this boundary is defined by a set of m line segments $f_i(x, y)=0$, $i \in [1, m]$. Then the stress function

$$\varphi = f_1(x, y) f_2(x, y)...f_m(x, y) g(x, y) \tag{37.18}$$

will go to zero at all points on Γ for any function $g(x, y)$. We can therefore choose a function $g(x, y)$ with an appropriate number of arbitrary constants $A_k, k \in [1, n]$ and then use Castigliano's second theorem to develop n equations for the n unknowns.

For example, the equations $x=0$ and $x^2+y^2-a^2=0$ define the boundaries of a semicircular bar of radius a whose centre is at the origin. The stress function

$$\varphi = x(x^2 + y^2 - a^2)(A_1 + A_2 x)$$

therefore satisfies the traction-free condition at all points on the boundary and contains two degrees of freedom, A_1, A_2.

Once a function with an appropriate number of degrees of freedom has been chosen, we calculate the stress components

$$\sigma_{zx} = \frac{\partial \varphi}{\partial y} \; ; \quad \sigma_{zy} = -\frac{\partial \varphi}{\partial x} \, ,$$

from (17.4) and hence the strain-energy density from (37.4), which in this case reduces to

$$U_0 = \frac{\sigma_{zx}^2 + \sigma_{zy}^2}{2\mu} = \frac{1}{2\mu}\left[\left(\frac{\partial \varphi}{\partial x}\right)^2 + \left(\frac{\partial \varphi}{\partial y}\right)^2\right].$$

The torque corresponding to this stress distribution is given by equation (17.10) as

$$T = 2 \iint_{\Omega} \varphi d\Omega \tag{37.19}$$

and hence the complementary energy per unit length of bar is

$$C = \iint_{\Omega} U_0 d\Omega - T\beta$$
$$= \frac{1}{2\mu} \iint_{\Omega} \left[\left(\frac{\partial \varphi}{\partial x}\right)^2 + \left(\frac{\partial \varphi}{\partial y}\right)^2 - 4\mu\beta\varphi\right] d\Omega \, , \tag{37.20}$$

where β is the twist per unit length. Castigliano's second theorem (37.16) then demands that

$$\frac{\partial C}{\partial A_k} = 0 ; \quad k \in [1, n] ,$$

which defines n linear equations for the unknown constants A_k.

Example

To illustrate the method, we consider a bar of square cross section defined by $-a < x < a$, $-a < y < a$. The boundaries are defined by the four lines $x \pm a = 0$, $y \pm a = 0$ and the simplest function of the form (37.18) is

$$\varphi = A(x^2 - a^2)(y^2 - a^2) ,$$

which has just one degree of freedom — the multiplying constant A. Substituting into equation (37.20) and evaluating the integrals, we have

$$C = \frac{1}{2\mu} \left(\frac{256 A^2 a^8}{45} - \frac{64 \mu \beta A a^6}{9} \right) .$$

To determine A, we impose the condition

$$\frac{\partial C}{\partial A} = 0 \quad \text{giving} \quad A = \frac{5\mu\beta}{8a^2} ,$$

after which the torque is recovered from (37.19) as

$$T = \frac{20\mu\beta a^4}{9} .$$

The exact result can be obtained by substituting $b = a$ in equation (17.25) and summing the series. The present one-term approximation underestimates the exact value by only about 1.3%. Closer approximations can be obtained by adding a few extra degrees of freedom, preserving the symmetry of the system with respect to x and y. For example, Timoshenko and Goodier[5] give a solution of the same problem using the two degree of freedom function

$$\varphi = (x^2 - a^2)(y^2 - a^2)[A_1 + A_2(x^2 + y^2)] ,$$

which gives a result for the torque within 0.15% of the exact value.

---

[5] *loc. cit.* Art.111.

37.7.2   *The in-plane problem*

We showed in Chapter 4 that the most general in-plane stress field satisfying the equations of equilibrium (2.4) in the absence of body force can be expressed in terms of the scalar Airy stress function ϕ through the relations

$$\sigma_{xx} = \frac{\partial^2 \phi}{\partial y^2} \ ; \quad \sigma_{xy} = -\frac{\partial^2 \phi}{\partial x \partial y} \ ; \quad \sigma_{yy} = \frac{\partial^2 \phi}{\partial x^2} \ . \tag{37.21}$$

We also showed in §4.3.3 that the traction boundary-value problem — i.e. the problem of a two-dimensional body with prescribed tractions on the boundaries — has a solution in which the stress components are independent of Poisson's ratio. We can therefore simplify the following treatment by considering the special case of plane stress ($\sigma_{zx} = \sigma_{zy} = \sigma_{zz} = 0$) with $\nu = 0$ for which equation (37.4) reduces to

$$\begin{aligned}
U_0 &= \frac{\left(\sigma_{xx}^2 + \sigma_{yy}^2\right)}{4\mu} + \frac{\sigma_{xy}^2}{2\mu} \\
&= \frac{1}{4\mu}\left[\left(\frac{\partial^2 \phi}{\partial x^2}\right)^2 + \left(\frac{\partial^2 \phi}{\partial y^2}\right)^2 + 2\left(\frac{\partial^2 \phi}{\partial x \partial y}\right)^2\right] \ .
\end{aligned}$$

We express the trial stress function ϕ in the form

$$\phi = \phi_P + \phi_H \ ,$$

where ϕ_P is any particular function of x, y that satisfies the traction boundary conditions and ϕ_H is a function satisfying homogeneous (i.e. traction-free) boundary conditions and that contains one or more degrees of freedom A_k. The stress components (37.21) will define a traction-free boundary if

$$\phi = 0 \quad \text{and} \quad \frac{\partial \phi}{\partial n} = 0 \ , \tag{37.22}$$

on the boundary, where n is the local normal. In order to satisfy these conditions, we note that if $f_i(x, y) = 0$ defines a line segment that is part of the boundary Γ, and if f_i is a continuous function in the vicinity of the line segment, then we can perform a Taylor expansion about any point on the boundary. It then follows that the function $[f_i(x, y)]^2$ will satisfy both conditions (37.22), since the first non-zero term in its Taylor expansion at the boundary will be quadratic. Thus, the conditions on ϕ_H can be satisfied by defining

$$\phi_H = \Big[f_1(x, y) f_2(x, y) ... f_m(x, y)\Big]^2 g(x, y) \tag{37.23}$$

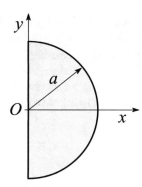

Fig. 37.1 The semi-circular
plate.

where $f_i(x, y) = 0$ $i \in [1, m]$ define line segments comprising the boundary Γ and $g(x, y)$ is any function containing one or more degrees of freedom.

Example

Consider the semicircular plate of Figure 37.1 subjected to the self-equilibrated tractions

$$\sigma_{xx} = S\left(1 - \frac{3y^2}{a^2}\right)$$

on the straight boundary $x = 0$, the curved boundary $r^2 - a^2 = x^2 + y^2 - a^2 = 0$ being traction free.

A particular solution maintaining the traction-free condition on the circle can be written

$$\phi_P = B(r^2 - a^2)^2 = B(x^4 + y^4 + 2x^2y^2 - 2a^2x^2 - 2a^2y^2 + a^4),$$

where B is a constant. The stress components

$$\sigma_{xx} = \frac{\partial^2 \phi_P}{\partial y^2} = B(12y^2 + 4x^2 - 4a^2) \;; \quad \sigma_{xy} = -\frac{\partial^2 \phi_P}{\partial x \partial y} = -8Bxy$$

and on the boundary $x = 0$,

$$\sigma_{xx} = B(3y^2 - a^2) \;; \quad \sigma_{xy} = 0.$$

Thus, the boundary conditions can be satisfied by the choice

$$B = -\frac{S}{4a^2} \;; \quad \phi_P = -\frac{S}{4a^2}(r^2 - a^2)^2.$$

Notice that if the tractions on $x = 0$ had been higher order polynomials in y, we could have found an appropriate particular solution satisfying the traction-free condition on the curved edge using the form $\phi_P = (r^2 - a^2)^2 g(x, y)$, where $g(x, y)$ is a polynomial in x, y of appropriate order.

The stress function ϕ_P provides already a first order approximation to the stress field, but to improve it we can add the homogeneous term

$$\phi_H = A_1 x^2 (r^2 - a^2)^2 = \frac{A_1 r^2 (r^2 - a^2)^2 (1 + \cos 2\theta)}{2} ,$$

which satisfies traction-free conditions on both straight and curved boundaries in view of (37.23). The stress components in polar coördinates are

$$\sigma_{rr} = S \left(1 - \frac{r^2}{a^2} \right) + A_1 (3r^4 - 4a^2 r^2 + a^4) + A_1 (r^4 - a^4) \cos 2\theta$$

$$\sigma_{\theta\theta} = S \left(1 - \frac{3r^2}{a^2} \right) + A_1 (15r^4 - 12a^2 r^2 + a^4)(1 + \cos 2\theta) \qquad (37.24)$$

$$\sigma_{r\theta} = A_1 (5r^4 - 6a^2 r^2 + a^4) \sin 2\theta$$

from (8.8, 8.9) and the total strain energy is then calculated as

$$U = \frac{1}{4\mu} \int_{-\pi/2}^{\pi/2} \int_0^a \left(\sigma_{rr}^2 + \sigma_{\theta\theta}^2 + 2\sigma_{r\theta}^2 \right) r \, dr \, d\theta$$

$$= \frac{\pi a^2}{30\mu} \left(14a^8 A_1^2 - 5a^4 A_1 S + 5S^2 \right)$$

(using Mathematica or Maple). There are no displacement boundary conditions in this problem, so $C = U$ and the degree of freedom A_1 is determined from the condition

$$\frac{\partial C}{\partial A_1} = 0 \quad \text{or} \quad A_1 = \frac{5S}{28a^4} ,$$

after which the stress field is recovered by substitution in (37.24).

37.8   Uniqueness and Existence of Solution

A typical elasticity problem can be expressed as the search for a displacement field u satisfying the equilibrium equations (2.14), such that the stress components defined through equations (1.42) and the displacement components satisfy appropriate boundary conditions. Throughout this book, we have tacitly assumed that this is a well-posed problem — i.e. that if the problem is *physically* well defined in the sense that we could conceive of loading a body of the given geometry in the laboratory, then there exists one and only one solution.

To examine this question from a mathematical point of view, suppose provisionally that the solution is non-unique, so that there exist two distinct stress fields σ_1, σ_2, both satisfying the field equations and the same boundary conditions. We can then construct a new stress field $\Delta\sigma = \sigma_1 - \sigma_2$ by taking the difference between these fields, which is a form of linear superposition. The new field $\Delta\sigma$ clearly involves no external loading, since the same external loads were included in each of the constituent solutions *ex hyp.* We therefore conclude from (37.1) that the corresponding total strain energy U associated with the field $\Delta\sigma$ must be zero. However, U can also be written as a volume integral of the strain-energy density U_0 as in (37.5), and U_0 must be everywhere positive or zero. The only way these two results can be reconciled is if U_0 is zero everywhere, implying that the stress is everywhere zero, from (37.3). Thus, the difference field $\Delta\sigma$ is null, the two solutions must be identical *contra hyp.* and only one solution can exist to a given elasticity problem.

The question of existence of solution is much more challenging and will not be pursued here. A short list of early but seminal contributions to the subject is given by Sokolnikoff[6] who states that "the matter of existence of solutions has been satisfactorily resolved for domains of great generality." More recently, interest in more general continuum theories including non-linear elasticity has led to the development of new methodologies in the context of functional analysis[7].

37.8.1 Singularities

In §11.2.1, we argued that stress singularities are acceptable in the mathematical solution of an elasticity problem if and only if the strain energy in a small region surrounding the singularity is bounded. In the two-dimensional case (line singularity), this leads to equation (11.17) and hence to the conclusion that the stresses can vary with r^a as $r \to 0$ only if $a > -1$. We can now see the reason for this restriction, since if there were any points in the body where the strain energy was not integrable, the total strain energy (37.5) would be ill-defined and the above proof of uniqueness would fail.

If these restrictions on the permissible strength of singularities are not imposed, it is quite easy to generate examples in which a given set of boundary conditions permits multiple solutions. Consider the flat punch problem of §12.5.2 for which the contact pressure distribution is given by equation (12.30). If we differentiate the stress and displacement fields with respect to x, we shall generate a field in which the contact traction is

---

[6] I. S. Sokolnikoff, *Mathematical Theory of Elasticity*, McGraw-Hill, New York, 2nd.ed. 1956, §27.

[7] See for example, J. E. Marsden and T. J. R. Hughes, *Mathematical Foundations of Elasticity*, Prentice-Hall, Englewood Cliffs, 1983, Chapter 6.

$$p(x) = \frac{d}{dx}\left(-\frac{F}{\pi\sqrt{a^2 - x^2}}\right) = -\frac{Fx}{\pi(a^2 - x^2)^{3/2}} \quad ; \quad -a < x < a$$

and the normal surface displacement is

$$u_y(x, 0) = \frac{d}{dx}(C) = 0 \quad ; \quad -a < x < a$$

from (12.29). The traction-free condition is retained in $|x| > a$, $y = 0$. We can now add this differentiated field to the solution of any frictionless contact problem over the contact region $-a < x < a$ without affecting the satisfaction of the displacement boundary condition (12.17), so solutions of such problems become non-unique. However, the superposed field clearly involves stresses varying with $r^{3/2}$ near the singular points $(\pm a, 0)$ and if (non-integrable) singularities of this strength are precluded, uniqueness is restored.

The three-dimensional equivalent of equation (11.17) would involve a stress field that tends to infinity with R^a as we approach the point $R = 0$. In this case, the integral in spherical polar coördinates centred on the singular point would take the form

$$U = \frac{1}{2}\int_0^{2\pi}\int_0^{\pi}\int_0^R \sigma_{ij}e_{ij}R^2\sin\beta\,dR\,d\beta\,d\theta = C\int_0^R R^{2a+2}dR \;,$$

where C is a constant. This integral is bounded if and only if $a > -\frac{3}{2}$.

In both two and three-dimensional cases, if a concentrated force F is applied to the body, either at a point on the boundary or at an interior point, the resulting stress field violates this energy criterion. An 'engineering' argument can be made that the real loading in such cases consists of a distribution of pressure or body force over a small finite region $\mathcal{A}$ that is statically equivalent to F, in which case no singularity is involved. Furthermore, a unique solution is obtained for each member of a regular sequence of such problems in which $\mathcal{A}$ is progressively reduced. Thus, if we conceive of the point force as the limit of this set of distributions, the uniqueness theorem still applies. Saint-Venant's principle §3.1.2 implies that only the stresses close to $\mathcal{A}$ will be changed as $\mathcal{A}$ is reduced, so the concentrated force solution also represents an approximation to the stress field distant from the loaded region when $\mathcal{A}$ is finite. Sternberg and Eubanks[8] proposed an extension of the uniqueness theorem to cover these cases. Similar arguments can be applied to the stronger singularity due to a concentrated moment. However, there is now some ambiguity in the limit depending on the detailed sequence of finite states through which it is approached[9].

---

[8] E. Sternberg and R. A. Eubanks (1955), On the concept of concentrated loads and an extension of the uniqueness theorem in the linear theory of elasticity, *Journal of Rational Mechanics and Analysis*, Vol. 4, pp. 135–168.

[9] E. Sternberg and V. Koiter (1958), The wedge under a concentrated couple: A paradox in the two-dimensional theory of elasticity, *ASME Journal of Applied Mechanics*, Vol.25, pp. 575–581; X. Markenscoff (1994), Some remarks on the wedge paradox and Saint Venant's principle, *ASME Journal of Applied Mechanics*, Vol.61, pp. 519–523.

Problems

37.1. For an isotropic material, find a way to express the strain-energy density U_0 of equation (37.4) in terms of the stress invariants of equations (1.20–1.22).

37.2. An annular ring $a < r < b$ is loaded by uniform (axisymmetric) shear tractions $\sigma_{r\theta}$ at $r = a$ that are statically equivalent to a torque T per unit thickness. By considering the equilibrium of the annular region $a < r < s$ where $s < b$, find the shear stress $\sigma_{r\theta}$ as a function of radius and hence find the total strain energy stored in the annulus. Use this result to determine the rotation of the outer boundary relative to the inner boundary, assuming the material has a shear modulus μ.

37.3. The isotropic semi-infinite strip $-a < y < a$, $0 < x < \infty$ is attached to a rigid support at $y = \pm a$, whilst the end $x = 0$ is subject to the antiplane displacement

$$u_z(0, y) = u_0 \left(1 - \frac{y^2}{a^2} \right) .$$

Since the resulting deformation will decay with x, approximate the displacement in the form

$$u_z(x, y) = u_0 \left(1 - \frac{y^2}{a^2} \right) e^{-\lambda x}$$

and use the minimum total potential energy theorem to determine the optimal value of the decay rate λ.

37.4. The rectangular plate $-a < x < a$, $-b < y < b$ is subject to the tensile tractions

$$\sigma_{xx} = S \left(1 - \frac{y^2}{b^2} \right)$$

on the edges $x = \pm a$, all the other tractions being zero. Using appropriate third-order polynomials for the displacement components u_x, u_y, find an approximation for the displacement field and hence estimate the distribution of the stress component σ_{xx} on the plane $x = 0$. Comment on the way this result is affected by the aspect ratio b/a.

37.5. Use the Maple or Mathematica file 'uxy' or the method of §9.1 to find the displacement field associated with the Airy stress function

$$\phi = \frac{G}{24} \left[(y^2 - b^2)^2 - (x^2 - a^2)^2 \right] .$$

Using this displacement field, find the value of the single degree of freedom G that best approximates the stress field in the rectangular bar $-a < x < a$, $-b < y < b$ subjected only to the normal tractions

$$\sigma_{xx} = f(y) ; \quad x = \pm a .$$

37.6. A simple two-dimensional finite element mesh can be obtained by joining a series of nodes (x_i, y_i), $i \in [1, N]$ by straight lines, forming a set of triangular elements. If the antiplane displacement u_z takes the values u_i at the three nodes $i = 1, 2, 3$, develop an expression for a linear function of x, y that interpolates these nodal values. In other words determine the three constants A, B, C in

$$u_z(x, y) = Ax + By + C$$

such that $u_z(x_i, y_i) = u_i$ for $i = 1, 2, 3$. Hence determine the strain energy U stored in the element as a function of u_i and use the principle of stationary total potential energy to determine the element stiffness matrix k relating the nodal forces F_i at the nodes to the nodal displacements through the equation

$$F_i = k_{ij} u_j .$$

37.7. Use a one degree-of-freedom approximation and Castigliano's second theorem to estimate the twist per unit length β of a bar whose triangular cross section is defined by the three corners $(0, 0)$, $(a, 0)$, $(0, a)$ and which is loaded by a torque T. Also, estimate the location and magnitude of the maximum shear stress.

37.8. Using the single degree of freedom stress function

$$\varphi = Ax(a^2 - x^2 - y^2) ,$$

estimate the torsional stiffness of the semicircular bar of radius a.

37.9. Figure 37.2 shows the cross section of a bar defined by four symmetrically disposed circular arcs of radius a, separated horizontally and vertically by a distance a. Use Castigliano's second theorem to estimate the torsional stiffness and the maximum shear stress when the bar transmits a torque T.

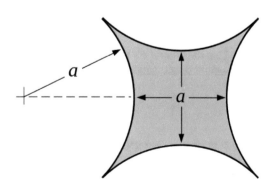

Fig. 37.2

37.10. The rectangular plate $-a < x < a$, $-b < y < b$ is subject to the tractions

$$\sigma_{xx} = \frac{Sy^2}{b^2}$$

on the edges $x = \pm a$, all the other tractions being zero. Find an approximate solution for the stress field using Castigliano's second theorem. An appropriate particular solution is

$$\phi_P = \frac{Sy^4}{12b^2} \, .$$

Use a single degree of freedom approximation of the form (37.23) for the homogeneous solution.

37.11. A plate defined by the region $a < r < 2a$, $-\alpha < \theta < \alpha$ is loaded only by the uniform tensile tractions

$$\sigma_{rr}(a, \theta) = \frac{F}{2a\alpha} \; ; \quad \sigma_{rr}(2a, \theta) = \frac{F}{4a\alpha}$$

on the curved edges $r = a$, $2a$. Find an approximate solution for the stress field using Castigliano's second theorem. Use a single degree of freedom approximation of the form (37.23) for the homogeneous solution ϕ_H.

Chapter 38
The Reciprocal Theorem

In the previous chapter, we exploited the idea that the strain energy U stored in an elastic structure must be equal to the work done by the external loads, if these are applied sufficiently slowly for inertia effects to be negligible. Equation (37.1) is based on the premise that the loads are all increased at the same time and in proportion, but the final state of stress cannot depend on the exact history of loading and hence the work done W must also be history-independent. This conclusion enables us to establish an important result for elastic systems, known as the reciprocal theorem.

38.1 Maxwell's Theorem

The simplest form of the theorem states that if a linear elastic body is kinematically supported and subjected to an external force F_1 at the point P, which produces a displacement $u_1(Q)$ at another point Q, then a force F_2 at Q would produce a displacement $u_2(P)$ at P where

$$F_1 \cdot u_2(P) = F_2 \cdot u_1(Q) .$$

To prove the theorem, we consider two scenarios — one in which the force F_1 is applied slowly and then held constant whilst F_2 is applied slowly, and the other in which the order of application of the forces is reversed. Suppose that the forces F_1, F_2 produce the displacement fields u_1, u_2, respectively. If we apply F_1 first, the work done will be

$$W_{11} = \frac{1}{2} F_1 \cdot u_1(P) .$$

If we now superpose the second force F_2 causing additional displacements u_2, the additional work done will be $W_{12} + W_{22}$, where

© The Author(s), under exclusive license to Springer Nature Switzerland AG 2022   615
J. R. Barber, *Elasticity*, Solid Mechanics and Its Applications 172,
https://doi.org/10.1007/978-3-031-15214-6_38

$$W_{12} = F_1 \cdot u_2(P)$$

is the work done by the loads F_1 'involuntarily' moving through the additional displacement $u_2(P)$ and

$$W_{22} = \frac{1}{2} F_2 \cdot u_2(Q)$$

is the work done by the force F_2. Notice that there is no factor of $\frac{1}{2}$ in W_{12}, since the force F_1 remains at its full value throughout the deformation. By contrast, F_2 only increases gradually to its final value during the loading process so that the work it does W_{22} is the area under a linear load displacement curve, which introduces a factor of $\frac{1}{2}$. The total work done in this scenario is

$$W_1 = W_{11} + W_{12} + W_{22} = \frac{1}{2} F_1 \cdot u_1(P) + F_1 \cdot u_2(P) + \frac{1}{2} F_2 \cdot u_2(Q) \qquad (38.1)$$

Consider now a second scenario in which the force F_2 is applied first, followed by force F_1. Clearly the total work done can be written down from equation (38.1) by interchanging the suffices 1,2 and the locations P, Q — i.e.

$$W_2 = W_{22} + W_{21} + W_{11} = \frac{1}{2} F_2 \cdot u_2(Q) + F_2 \cdot u_1(Q) + \frac{1}{2} F_1 \cdot u_1(P) \ .$$

In each case, since the forces were applied gradually, the work done will be stored as strain energy in the deformed body. But the final state in each case is the same, so we conclude that $W_1 = W_2$. Comparing the two expressions, we see that they will be equal if and only if $W_{12} = W_{21}$ and hence

$$F_1 \cdot u_2(P) = F_2 \cdot u_1(Q) \ ,$$

proving the theorem. Another way of stating Maxwell's theorem is *"The work done by a force F_1 moving through the displacements u_2 due to a second force F_2 is equal to the work done by the force F_2 moving through the displacements u_1 due to F_1."*

38.1.1   Example: Mindlin's problem

We recall from §24.2.1 that if a point force F is applied in the positive z-direction at the origin on the surface of the otherwise traction-free half-space $z > 0$, the resulting stresses and displacements are described by the potential

$$\varphi = -\frac{F}{2\pi} \ln(R + z) \ ,$$

in Solution F of Table 22.3, where $R = \sqrt{x^2 + y^2 + z^2}$. In particular, the displacement component u_x is given by

$$u_x = \frac{1}{2\mu}\left(z\frac{\partial^2\varphi}{\partial x\partial z} + (1-2\nu)\frac{\partial\varphi}{\partial x}\right) = \frac{F}{4\pi\mu}\left(\frac{xz}{R^3} - \frac{(1-2\nu)x}{R(R+z)}\right).$$

Maxwell's theorem then tells us that the same expression defines the normal surface displacement u_z at the origin due to an equal force F applied in the positive x-direction at the point (x, y, z). A simple translation of the coördinate system then shows that if the force is applied in the x-direction at the point $(0, 0, h)$, the normal surface displacement at $(x, y, 0)$ will be

$$u_z(x, y, 0) = \frac{F}{4\pi\mu}\left(-\frac{hx}{R_h^3} + \frac{(1-2\nu)x}{R_h(R_h+h)}\right),$$

where $R_h = \sqrt{x^2 + y^2 + h^2}$. Thus, we obtain a general expression defining the shape of the deformed surface due to a subsurface tangential force, without the necessity of solving that quite challenging problem explicitly.

38.2 Betti's Theorem

Maxwell's theorem is useful in Mechanics of Materials, where point forces are used to represent force resultants (e.g. on the cross section of a beam), but in elasticity, the reciprocal theorem is more useful in the generalized form due to Betti, which relates the displacement fields due to two different distributions of surface traction and/or body force.

Suppose that the surface tractions t_i and body forces p_i produce the displacement fields u_i, where i takes the values 1 and 2 respectively. Once again we consider two scenarios, one in which the loads t_1, p_1 are applied first, followed by t_2, p_2 and the other in which the order of loading is reversed. An argument exactly parallel to that in §38.1 then establishes that *"The work done by the loads t_1, p_1 moving through the displacements u_2 due to t_2, p_2 is equal to the work done by the loads t_2, p_2 moving through the displacements u_1 due to t_1, p_1."*

The mathematical statement of Betti's theorem for a body Ω with boundary Γ is

$$\iint_\Gamma t_1\cdot u_2\,d\Gamma + \iiint_\Omega p_1\cdot u_2\,d\Omega = \iint_\Gamma t_2\cdot u_1\,d\Gamma + \iiint_\Omega p_2\cdot u_1\,d\Omega. \tag{38.2}$$

Use of the theorem

Betti's theorem defines a relationship between two different stress and displacement states for the same body. In most applications, one of these states corresponds to the problem under investigation, whilst the second is an *auxiliary solution*, for which the tractions and surface displacements are known. Often the auxiliary solution will be a simple state of stress for which the required results can be written down by inspection. The art of using the reciprocal theorem lies in choosing the auxiliary solution so that equation (38.2) yields the required result. This is best illustrated by examples.

38.2.1 Change of volume

(a)   *(b)*

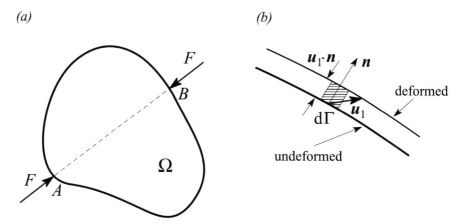

Fig. 38.1 Increase in volume of a body due to two colinear forces.

Suppose we wish to determine the change in volume of a body Ω due to a pair of equal and opposite colinear concentrated forces, as shown in Figure 38.1(*a*). Figure 38.1(*b*) shows an enlarged view of a region of the boundary in the deformed and undeformed states, from which it is clear that the change in volume associated with some small region $d\Gamma$ of the boundary is $dV = \boldsymbol{u}_1 \cdot \boldsymbol{n}\, d\Gamma$, where $\boldsymbol{n}$ is the unit vector defining the outward normal to the surface at $d\Gamma$. Thus, the change in volume of the entire body is

$$\Delta V_1 = \iint_\Gamma \boldsymbol{u}_1 \cdot \boldsymbol{n}\, d\Gamma \ .$$

Now this expression can be made equal to the right-hand side of (38.2) if we choose $\boldsymbol{t}_2 = \boldsymbol{n}$ and $\boldsymbol{p}_2 = 0$. In other words, the auxiliary solution corresponds to the case where the body is subjected to a purely normal (tensile) traction of unit magnitude throughout its surface and there is no body force.

The solution of this auxiliary problem is straightforward, since the tractions are those associated with a state of uniform hydrostatic tension

$$\sigma_{xx} = \sigma_{yy} = \sigma_{zz} = 1 \;\; ; \;\; \sigma_{xy} = \sigma_{yz} = \sigma_{zx} = 0 \tag{38.3}$$

throughout the body, and the corresponding strains are uniform and hence satisfy all the compatibility conditions. The stress-strain relations then give[1]

$$e_{xx} = e_{yy} = e_{zz} = \frac{(1 - 2\nu)}{E} \;\; ; \;\; e_{xy} = e_{yz} = e_{zx} = 0 \tag{38.4}$$

and hence, in state 2, the body will simply deform into a larger, geometrically similar shape, any linear dimension L increasing to $L(1+(1-2\nu)/E)$.

It remains to calculate the *left-hand side* of (38.2), which we recall is the work done by the tractions t_1 in moving through the surface displacements u_2. In this case, t_1 consists of the two concentrated forces, so the appropriate work is $F\delta_2$, where δ_2 is the amount that the length L of the line AB shrinks *in state 2*, permitting the points of application of the forces to approach each other.

From the above argument,

$$\delta_2 = -\frac{(1 - 2\nu)L}{E}$$

and hence we can collect results to obtain

$$\begin{aligned} \Delta V_1 &= \iint_\Gamma u_1 \cdot n \, d\Gamma = \iint_\Gamma u_1 \cdot t_2 \, d\Gamma \\ &= \iint_\Gamma u_2 \cdot t_1 \, d\Gamma = F\delta_2 = -\frac{F(1 - 2\nu)L}{E} \,, \end{aligned}$$

which is the required expression for the change in volume.

Clearly the same method would enable us to determine the change in volume due to any other self-equilibrated traction distribution t_1, either by direct substitution in the integral $\iint_\Gamma u_2 \cdot t_1 d\Gamma$ or by using the force pair of Figure 38.1(a) as a Green's function to define a more general distribution through a convolution integral. It is typical of the reciprocal theorem that it leads to the proof of results of some generality.

---

[1] The equation appears to be dimensionally inconsistent, but we must remember that the loading comprises a traction of unit magnitude. The dimensions could be made explicit by writing $\sigma_{xx} = \sigma_{yy} = \sigma_{zz} = S$, where S is a constant with dimensions of stress which will be found to cancel in the final expression for ΔV_1.

38.2.2   A tilted punch problem

For any given auxiliary solution, Betti's theorem will yield only a single result, which can be the value of the displacement at a specific point, or more often, an integral quantity such as a force resultant or an average displacement. Indeed, when interest in the problem is restricted to such integrals, the reciprocal theorem should always be considered first as a possible method of solution.

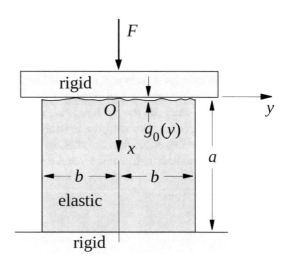

Fig. 38.2 The tilted punch problem.

As a second example, we shall consider the two-dimensional plane stress contact problem of Figure 38.2, in which a rectangular block with a slightly irregular surface is pressed down by a frictionless flat rigid punch against a frictionless rigid plane. We suppose that the force F is large enough to ensure that contact is established throughout the end of the block and that the line of action of the applied force passes through the mid-point of the block. The punch will therefore generally tilt through some small angle, α, and the problem is to determine α.

We suppose that the profile of the block is defined by an initial gap function $g_0(y)$ when the punch is horizontal and hence, if the punch moves downward a distance u_0 and rotates clockwise through α, the contact condition can be stated as

$$g(y) = g_0(y) + u_1(y) - u_0 - \alpha y = 0 \;, \tag{38.5}$$

where $u_1(y)$ is the downward normal displacement of the block surface, as shown in Figure 38.3(a). Since the punch is frictionless, the traction t_1 is also purely normal, as shown.

The key to the solution lies in the fact that the line of action of F, which is also the resultant of t_1, passes through the centre of the block. Taking moments about this point, we therefore have

$$\int_{-b}^{b} y t_1(y) dy = 0 .$$

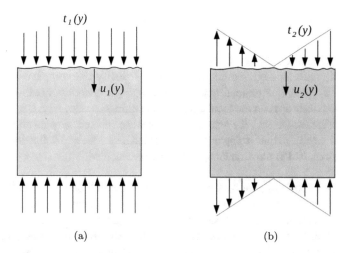

$t_1(y)$

$u_1(y)$

$t_2(y)$

$u_2(y)$

(a)

(b)

Fig. 38.3 Tractions for (a) the original problem, and (b) the auxiliary problem for the tilted punch.

We can make this expression look like the left-hand side of (38.2) by choosing the auxiliary solution so that the normal displacement $u_2(y) = y$, which corresponds to a simple bending distribution with

$$t_2(y) = \frac{Ey}{a} ,$$

as shown in Figure 38.3(b).

Betti's theorem now gives

$$\int_{-b}^{b} \frac{Eyu_1(y)dy}{a} = \int_{-b}^{b} yt_1(y)dy = 0 \qquad (38.6)$$

and hence, treating (38.5) as an equation for $u_1(y)$ and substituting into (38.6), we obtain

$$\int_{-b}^{b} \{u_0 + \alpha y - g_0(y)\} y dy = 0$$

— i.e.

$$\alpha = \frac{3}{b^3} \int_{-b}^{b} y g_0(y) dy ,$$

which is the required result.

This result is relevant to the choice of end conditions in the problem of §9.1, since it shows that the end of the beam can be restored to a vertical plane by a self-equilibrated purely normal traction if and only if

$$\int_{-b}^{b} y g_0(y) dy = 0 ,$$

which is equivalent to the condition $(9.12)_3$.

However, unfortunately we cannot rigorously justify the complete set of conditions (9.12) by this argument, since the full built-in boundary conditions (9.10) also involve constraints on the local extensional strain $e_{yy} = \partial u_y / \partial y$. It is clear from energy considerations that if these constraints were relaxed, any resulting motion would involve the end load doing positive work and hence the end deflection obtained with conditions (9.12) must be larger than that obtained with the strong conditions[2] (9.10).

In the above solution, we have tacitly assumed that the only contribution to the two integrals in (38.2) comes from the tractions and displacements at the top surface of the block. This is justified, since although there are also normal tractions t_1, t_2 at the lower surface, they make no contribution to the work integrals because the corresponding normal *displacements* (u_2, u_1 respectively) are zero. However, it is important to consider all surfaces of the body in the application of the theorem, including a suitably distant boundary in problems involving infinite or semi-infinite regions.

38.2.3   Indentation of a half-space

We demonstrated in §31.2 that the contact area $\mathcal{A}$ between a frictionless rigid punch and an elastic half-space is that value which maximizes the total load $F(\mathcal{A})$ on the punch. Shield[3] has shown that this can be determined without solving the complete contact problem, using Betti's theorem.

The total load on the punch is

$$F(\mathcal{A}) = \iint_{\mathcal{A}} p_1(x, y) dx dy ,$$

---

[2] J. D. Renton (1991), Generalized beam theory applied to shear stiffness, *International Journal of Solids and Structures*, Vol. 27, pp. 1955–1967, argues that the correct shear stiffness for a beam of any cross section should be that which equates the strain energy associated with the shear stresses σ_{xz}, σ_{yz} and the work done by the shear force against the shear deflection. For the rectangular beam of §9.1, this is equivalent to replacing the multiplier on b^2/a^2 in $(9.13)_{2,3}$ by $2.4(1+\nu)$. This estimate differs from $(9.13)_3$ by at most 3%.

[3] R. T. Shield (1967), Load-displacement relations for elastic bodies, *Zeitschrift für angewandte Mathematik und Physik*, Vol. 18, pp. 682–693.

where $p_1(x, y)$ is the contact pressure. We can make the left-hand side of (38.2) take this form by choosing the auxiliary solution such that the normal displacement $u_2(x, y)$ in $\mathcal{A}$ is uniform and of unit magnitude.

The simplest result is obtained by using as auxiliary solution that corresponding to the indentation of the half-space by a frictionless flat rigid punch of plan-form $\mathcal{A}$, defined by the boundary conditions

$$
\begin{aligned}
u_2 &\equiv u_z(x, y) = 1 &&; \quad (x, y) \in \mathcal{A} \\
p_2 &\equiv -\sigma_{zz}(x, y) = 0 &&; \quad (x, y) \in \bar{\mathcal{A}} \\
\sigma_{xz} &= \sigma_{yz} = 0 &&; \quad (x, y) \in \mathcal{A} \cup \bar{\mathcal{A}} .
\end{aligned}
$$

Since the only non-zero tractions in both solutions are the normal pressures in the contact area $\mathcal{A}$, equation (38.2) reduces to

$$
F(\mathcal{A}) \equiv \iint_{\mathcal{A}} p_1(x, y) dx dy = \iint_{\mathcal{A}} p_2(x, y) u_1(x, y) dx dy , \tag{38.7}
$$

where $u_1(x, y)$ is the normal displacement under the punch, which is defined through the contact condition (31.3) in terms of the initial punch profile $u_0(x, y)$ and its rigid-body displacement.

In problems involving infinite domains, it is important to verify that no additional contribution to the work integrals in (38.2) is made on the 'infinite' boundaries. To do this, we consider a hemispherical region of radius b much greater than the linear dimensions of the loaded region $\mathcal{A}$. The stress and displacement fields in both solutions will become self-similar at large R, approximating those of the point force solution (§24.2.1). In particular, the stresses will tend asymptotically to the form $R^{-2} f(\theta, \alpha)$ and the displacements to $R^{-1} g(\theta, \alpha)$.

If we now construct the integral $\iint t_1 \cdot u_2 d\Gamma$ over the hemisphere surface, it will therefore have the form

$$
\int_0^{\pi/2} \int_0^{2\pi} b^{-2} f_1(\theta, \alpha) b^{-1} g_2(\theta, \alpha) b^2 d\theta d\alpha ,
$$

which approaches zero with b^{-1} as $b \to \infty$, indicating that the distant surfaces make no contribution to the integrals in (38.2).

Of course, we can only use (38.7) to find $F(\mathcal{A})$ if we already know the solution $p_2(x, y)$ for the flat punch problem with plan-form $\mathcal{A}$, and exact solutions[4] are only known for a limited number of geometries, such as the circle, the ellipse and the strip.

---

[4] An *approximate* solution to this problem for a punch of fairly general plan-form is given by V. I. Fabrikant (1986), Flat punch of arbitrary shape on an elastic half-space, *International Journal of Engineering Science*, Vol. 24, pp. 1731–1740. Fabrikant's solution and the above reciprocal theorem argument are used as the bases of an approximate solution of the general smooth punch problem by J. R. Barber and D. A. Billings (1990), An approximate solution for the contact area and elastic compliance of a smooth punch of arbitrary shape, *International Journal of Mechanical Sciences*, Vol. 32, pp. 991–997.

However, (38.7) does for example enable us to write down the indenting force for a circular punch of arbitrary profile[5] and hence, using the above theorem, to determine the contact area for the general axisymmetric contact problem.

38.3   Eigenstrain Problems

We saw in §28.1, equation (28.4) that the displacements u_i due to an eigenstrain distribution $\epsilon_{kl}^*(x, y, z)$ in an infinite body can be completely suppressed by the application of a body force distribution

$$p_i = c_{ijkl} \frac{\partial \epsilon_{kl}^*}{\partial x_j} \; , \tag{38.8}$$

in which case the stress field will be defined by

$$\sigma_{ij} = -c_{ijkl}\epsilon_{kl}^* \; ,$$

from (28.3) with $u_i = 0$.

If the body is finite, the same result applies, except that we must now also apply tractions on the boundary given by

$$t_i = \sigma_{ij}n_j = -c_{ijkl}\epsilon_{kl}^* n_j \; , \tag{38.9}$$

where n_j is the local outward normal to the boundary.

38.3.1   Deformation of a traction-free body

These results imply that the displacements in a traction-free body due to an eigenstrain distribution ϵ_{kl}^* are identical to those in an initially stress-free body due to body forces and tractions equal and opposite to those in equations (38.8, 38.9)— i.e.

$$p_i = -c_{ijkl} \frac{\partial \epsilon_{kl}^*}{\partial x_j} \; ; \quad t_i = c_{ijkl}\epsilon_{kl}^* n_j \; . \tag{38.10}$$

Writing equation (38.2) in index notation, we have

$$\iint_\Gamma t_i^{(1)} u_i^{(2)} d\Gamma + \iiint_\Omega p_i^{(1)} u_i^{(2)} d\Omega = \iint_\Gamma t_i^{(2)} u_i^{(1)} d\Gamma + \iiint_\Omega p_i^{(2)} u_i^{(1)} d\Omega$$

---

[5] R. T. Shield, *loc. cit.*.

and using (38.10) for $p_i^{(1)}, t_i^{(1)}$

$$c_{ijkl} \left(\iint_\Gamma \epsilon_{kl}^* n_j u_i^{(2)} d\Gamma - \iiint_\Omega \frac{\partial \epsilon_{kl}^*}{\partial x_j} u_i^{(2)} d\Omega \right)$$

$$= \iint_\Gamma t_i^{(2)} u_i^{(1)} d\Gamma + \iiint_\Omega p_i^{(2)} u_i^{(1)} d\Omega . \qquad (38.11)$$

Notice that the first (traction) term on the left-hand side will be non-zero only if the eigenstrain distribution extends to the boundary. However, there will be additional internal boundary integrals if the eigenstrain distribution has local discontinuities.

Example

Suppose the traction-free half-space Ω is subject to uniform eigenstrains ϵ_{kl}^* inside an inclusion Ω_I which is completely contained within Ω. Equation (38.11) then reduces to

$$c_{ijkl} \epsilon_{kl}^* \iint_{\Gamma_I} n_j u_i^{(2)} d\Gamma = \iint_\Gamma t_i^{(2)} u_i^{(1)} d\Gamma + \iiint_\Omega p_i^{(2)} u_i^{(1)} d\Omega ,$$

where Γ_I is the boundary of the inclusion with outward normal n_j, and Γ, Ω refer to the boundary and volume of the half-space. If we then define the auxiliary solution as the Boussinesq solution of §24.2.1 with a force $F = 1$ applied at the point $(0, x, y)$ on the surface of the half-space $z > 0$, so that

$$p_i^{(2)} = 0 \; ; \quad t_i^{(2)} = \delta_{i3}\delta(x_1 - x)\delta(x_2 - y) ,$$

we obtain an integral expression for the normal surface displacement due to ϵ_{kl}^* — i.e.

$$u_3(x, y, 0) = c_{ijkl}\epsilon_{kl}^* \iint_{\Gamma_I} n_j u_i^{(2)} d\Gamma .$$

38.3.2 Displacement constraints

Equation (38.11) was developed under the assumption that the body containing the eigenstrains is traction free. For the more general case where it is also subjected to prescribed body forces $p_i^{(1)}$ and boundary tractions $t_i^{(1)}$, these must be included explicitly, leading to the equation

$$c_{ijkl}\left(\iint_{\Gamma}\epsilon_{kl}^{*}n_{j}u_{i}^{(2)}d\Gamma - \iiint_{\Omega}\frac{\partial\epsilon_{kl}^{*}}{\partial x_{j}}u_{i}^{(2)}d\Omega\right) + \iint_{\Gamma}t_{i}^{(1)}u_{i}^{(2)}d\Gamma$$

$$+ \iiint_{\Omega}p_{i}^{(1)}u_{i}^{(2)}d\Omega = \iint_{\Gamma}t_{i}^{(2)}u_{i}^{(1)}d\Gamma + \iiint_{\Omega}p_{i}^{(2)}u_{i}^{(1)}d\Omega \ .$$

$$(38.12)$$

This generalization is most useful in problems where the body is mechanically supported — i.e. the exposed surfaces of the body are traction free, but there are parts of the boundary Γ_{u} where homogeneous displacement boundary conditions are imposed. In this case, the third term in equation (38.12) will reduce to

$$\iint_{\Gamma_{u}}t_{i}^{(1)}u_{i}^{(2)}d\Gamma \ ,$$

and an appropriate choice of $u_{i}^{(2)}$ will allow this term to represent the reaction force or moment at the support[6].

38.4   Thermoelastic Problems

The reciprocal theorem can be extended to problems in thermoelasticity by describing the local unrestrained thermal expansion as an eigenstrain distribution. For an isotropic material, we then have

$$\epsilon_{kl}^{*} = \alpha T \delta_{kl} \quad \text{and} \quad c_{ijkl}\epsilon_{kl}^{*} = \frac{2\mu(1+\nu)\alpha T \delta_{ij}}{(1-2\nu)} \ ,$$

from (1.55, 1.52), so equation (38.12) gives

$$\frac{2\mu(1+\nu)\alpha}{(1-2\nu)}\left(\iint_{\Gamma}Tn_{i}u_{i}^{(2)}d\Gamma - \iiint_{\Omega}\frac{\partial T}{\partial x_{i}}u_{i}^{(2)}d\Omega\right) + \iint_{\Gamma}t_{i}^{(1)}u_{i}^{(2)}d\Gamma$$

$$+ \iiint_{\Omega}p_{i}^{(1)}u_{i}^{(2)}d\Omega = \iint_{\Gamma}t_{i}^{(2)}u_{i}^{(1)}d\Gamma + \iiint_{\Omega}p_{i}^{(2)}u_{i}^{(1)}d\Omega \ .$$

$$(38.13)$$

From the divergence theorem, we have

$$\iint_{\Gamma}Tu_{i}^{(2)}n_{i}d\Gamma = \iiint_{\Omega}\frac{\partial}{\partial x_{i}}\left(Tu_{i}^{(2)}\right)d\Omega$$

$$= \iiint_{\Omega}\frac{\partial T}{\partial x_{i}}u_{i}^{(2)}d\Omega + \iiint_{\Omega}T\frac{\partial u_{i}^{(2)}}{\partial x_{i}}d\Omega \ ,$$

so (38.13) becomes

---

[6] See for example, Problem 38.10.

$$\frac{2\mu(1+\nu)\alpha}{(1-2\nu)} \iiint_\Omega T \frac{\partial u_i^{(2)}}{\partial x_i} d\Omega + \iint_\Gamma t_i^{(1)} u_i^{(2)} d\Gamma + \iiint_\Omega p_i^{(1)} u_i^{(2)} d\Omega$$

$$= \iint_\Gamma t_i^{(2)} u_i^{(1)} d\Gamma + \iiint_\Omega p_i^{(2)} u_i^{(1)} d\Omega ,$$

or in vector notation

$$\frac{2\mu(1+\nu)\alpha}{(1-2\nu)} \iiint_\Omega T \operatorname{div} \boldsymbol{u}_2 \, d\Omega + \iint_\Gamma \boldsymbol{t}_1 \cdot \boldsymbol{u}_2 d\Gamma + \iiint_\Omega \boldsymbol{p}_1 \cdot \boldsymbol{u}_2 d\Omega$$

$$= \iint_\Gamma \boldsymbol{t}_2 \cdot \boldsymbol{u}_1 d\Gamma + \iiint_\Omega \boldsymbol{p}_2 \cdot \boldsymbol{u}_1 d\Omega . \tag{38.14}$$

This is known as the *thermoelastic reciprocal theorem*[7].

This theorem can be used in exactly the same way as Betti's theorem to develop general expressions for quantities that are expressible as integrals. For example, the change in the volume of a traction-free body due to an arbitrary temperature distribution can be found using the auxiliary solution (38.3, 38.4), for which $\boldsymbol{t}_2 = \boldsymbol{n}$ and $\boldsymbol{p}_2 = 0$. From (38.4) we have

$$\operatorname{div} \boldsymbol{u}_2 = e_{xx} + e_{yy} + e_{zz} = \frac{3(1-2\nu)}{E} = \frac{3(1-2\nu)}{2\mu(1+\nu)}$$

and hence (38.14) reduces to

$$\Delta V_1 = \iint_\Gamma \boldsymbol{u}_1 \cdot \boldsymbol{n} d\Gamma = \frac{2\mu(1+\nu)\alpha}{(1-2\nu)} \iiint_\Omega T \operatorname{div} \boldsymbol{u}_2 \, d\Omega$$

$$= 3\alpha \iiint_\Omega T d\Omega .$$

In other words, the total change in volume of the body is the sum of that which would be produced by free thermal expansion of all its constituent particles if there were no thermal stresses. For additional applications of the thermoelastic reciprocal theorem, see J. N. Goodier, *loc. cit.* or S. P. Timoshenko and J. N. Goodier, *loc. cit.*, §155.

Problems

38.1. Figure 38.4 shows an elastic cube of side a which just fits between two parallel frictionless rigid walls. The block is now loaded by equal and opposite concentrated forces, F, at the mid-point of two opposite faces, as shown. 'Poisson's ratio' strains

---

[7] J.N.Goodier, *Proceedings of the 3rd U.S. National Congress on Applied Mechanics*, (1958), pp. 343–345.

will cause the block to exert normal forces P on the walls. Find the magnitude of these forces.

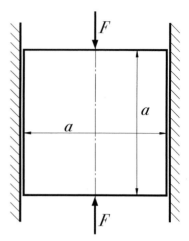

Fig. 38.4 Loading of the constrained cube.

Would your result be changed if (i) the forces F were applied away from the mid-point or (ii) if their common line of action was inclined to the vertical?

38.2. Use the method of §31.2, §38.2.3 to find the relations between the indentation force F, the contact radius a and the maximum indentation depth d, for a rigid frictionless punch with a fourth order profile $u_0 = -Cr^4$ indenting an elastic half-space.

38.3. The penny-shaped crack $0 \le r < a$, $z = 0$ is opened by equal and opposite normal forces F applied to the two crack faces at a radius r_1. Find the volume of the opened crack.

38.4. A long elastic cylinder of constant but arbitrary cross-sectional area and length L rests horizontally on a plane rigid surface. Show that gravitational loading will cause the cylinder to increase in length by an amount

$$\delta = \frac{\nu \rho g L h}{E} \, ,$$

where ρ, E, ν are the density, Young's modulus and Poisson's ratio respectively of the material and h is the height of the centre of gravity of the cylinder above the plane.

38.5. A pressure vessel of arbitrary shape occupies a space of volume V_1 and encloses a hole of volume V_2, so that the volume of solid material is $V_M = V_1 - V_2$. If the vessel is now subjected to an external pressure p_1 and an internal pressure p_2, show that the total change in V_M is

$$\delta V_M = \frac{p_2 V_2 - p_1 V_1}{K_b} ,$$

where K_b is the bulk modulus of equation (1.57).

38.6. The long rectangular bar $0 < x < a$, $-b < y < b$, $a \gg b$ is believed to contain residual stresses σ_{xx} that are functions of y only, except near the ends, the other residual stress components being zero. To estimate the residual stress distribution, it is proposed to clamp the bar at $x = 0$, cut a groove as shown in Figure 38.5, using laser machining, and measure the vertical displacement u_y (if any) of the end $x = a$ due to the resulting stress relief.

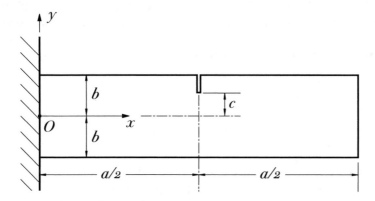

Fig. 38.5 Groove cut in a bar with residual stresses.

What would be the auxiliary problem we would need to solve to establish an integral relation between the residual stress distribution and the displacement u_y, using the reciprocal theorem. (Do not attempt to solve the auxiliary problem!)

38.7. By differentiating the solution of the axisymmetric Hertz problem of §32.2.5 with respect to x, show that the contact pressure distribution

$$p(r, \theta) = \frac{Cr \cos \theta}{\sqrt{a^2 - r^2}} ; \quad 0 \leq r < a$$

produces the normal surface displacement

$$u_z(r, \theta) = \frac{\pi C (1 - \nu) r \cos \theta}{4\mu} ; \quad 0 \leq r < a, \ z = 0 .$$

Use this result and the reciprocal theorem to determine the angle through which the punch in §38.2.3 will tilt about the y-axis if the force F is applied along a line passing through the origin.

38.8. Suppose that the force F in Figure 38.2 is not sufficient to ensure contact throughout the end of the block, so there are some regions where the gap $g(y) \neq 0$. Show that the mean gap

$$\bar{g} = \frac{1}{2b} \int_{-b}^{b} g(y) dy$$

is then given by

$$\bar{g} = \bar{g}_0 + \frac{Fa}{2Eb} - u_0$$

where $\bar{g}_0$ is the mean gap when $F = 0$.

38.9. A traction-free body Ω of initial volume V experiences a differentiable distribution of eigenstrain $\epsilon_{ij}^*(r)$ in an inclusion Ω_I that does not extend to boundary of Ω.

  Show that the change in volume due to the eigenstrain is

$$\Delta V = \iiint_{\Omega_I} \epsilon_{kk}^* d\Omega \ .$$

Discuss the extension of your result to the case where the eigenstrain distribution is not differentiable.

38.10. An elastic cylinder $0 < z < L,\ 0 \leq r < a$ is constrained between two frictionless rigid planes at $z = 0, L$. The cylinder then experiences a material transformation leading to a differentiable eigenstrain distribution ϵ_{ij}^* in an internal inclusion Ω_I. Assuming there is no loss of contact at the ends, show that the axial force F developed in the cylinder is

$$F = -\frac{E}{L} \iiint_{\Omega_I} \epsilon_{33}^* d\Omega \ .$$

38.11. A long hollow cylinder of inner radius a and outer radius b has traction-free surfaces, but it is subjected to a non-axisymmetric temperature increase $T(r, \theta)$. Find the change in the cross-sectional area of the hole due to thermoelastic distortion.

38.12. The long solid cylinder $0 \leq r < a,\ 0 < z < L,\ L \gg a$ is built in at $z = L$ and all the remaining surfaces are traction free. If the cylinder is now heated to an arbitrary non-uniform temperature T, find an integral expression for the displacement u_y at the free end.

38.13. A body Ω is raised to an arbitrary temperature field $T(x, y, z)$, but the boundaries are all free of traction. Show that the average value of the bulk stress $\bar{\sigma}$ of equation (1.56) is zero — i.e.

$$\iiint_{\Omega} \bar{\sigma} d\Omega = 0 \ .$$

Appendix A
Using Maple and Mathematica

The algebraic manipulations involved in solving problems in elasticity are relatively routine, but they can become lengthy and complicated, particularly for three-dimensional problems. There are therefore significant advantages in using a symbolic programming language such as Maple or Mathematica. Before such programs were available, the algebraic complexity of manual calculations often made it quicker to use a finite element formulation, even when it was clear that an analytical solution could in principle be obtained. Apart from the undoubted saving in time, Maple or Mathematica also make it easy to plot stress distributions and perform parametric studies to determine optimum designs for components. From the engineering perspective, this leads to a significantly greater insight into the nature of the subject.

Using these languages efficiently necessitates a somewhat different approach from that used in conventional algebraic solutions. Anyone with more than a passing exposure to computer programs will know that it is almost always quicker to try a wide range of possible options, rather than trying to construct a logical series of steps leading to the required solution. This trial and error process also works well in the computer solution of elasticity problems. For example, if you are not quite sure which stress function to use in the problem, the easiest way to find out is to use your best guess, run the calculation and see what stresses are obtained. Often, the output from this run will make it clear that minor modifications or additional terms are required, but this is easily done. Remember that it is extremely easy and quick to make a small change in the problem formulation and then re-run the program. By contrast, in a conventional algebraic solution, it is essential to be very careful in the early stages because repeating a calculation with even a minor change in the initial formulation is very time consuming.

The best way to learn how to use these programs is also by trial and error. Don't waste time reading a compendious instruction manual or following a tutorial, since these will tell you how to do numerous things that you probably will never need. In the Supplementary Material and at the website http://www-personal.umich.edu/ jbarber/elasticity/book4.html we provide the source code and the resulting output for the

J. R. Barber, *Elasticity*, Solid Mechanics and Its Applications 172,
https://doi.org/10.1007/978-3-031-15214-6

two-dimensional problem treated in §5.2.2. This file has sufficient embedded explanation to show what is being calculated and you can easily copy the files and make minor changes to gain experience in the use of the method. The web site also includes an explanation of the commands you will need most often in the solution of elasticity problems and electronic versions of the Tables in the present Chapters 22 and 23 and of the recurrence relations for generating spherical harmonics.

Index

© The Editor(s) (if applicable) and The Author(s), under exclusive license to Springer Nature Switzerland AG 2022
J. R. Barber, *Elasticity*, Solid Mechanics and Its Applications 172,
https://doi.org/10.1007/978-3-031-15214-6

Printed in the United States
by Baker & Taylor Publisher Services